Biochar for Environmental Remediation
Principles, Applications, and Prospects

Biochar for Environmental Remediation
Principles, Applications, and Prospects

Edited by

Willis Gwenzi
Formerly Alexander von Humboldt Fellow and Guest Full Professor, Leibniz-Institut für Agrartechnik und Bioökonomie e.V. (ATB), Potsdam, Germany; Formerly Alexander von Humboldt Fellow and Guest Full Professor, Grassland Grassland Science and Renewable Plant Resources, Faculty of Organic Agricultural Sciences, Universität Kassel, Witzenhausen, Germany; Biosystems and Environmental Engineering Research Group, Marlborough, Harare, Zimbabwe

ELSEVIER

Elsevier
Radarweg 29, PO Box 211, 1000 AE Amsterdam, Netherlands
125 London Wall, London EC2Y 5AS, United Kingdom
50 Hampshire Street, 5th Floor, Cambridge, MA 02139, United States

Copyright © 2025 Elsevier Inc. All rights are reserved, including those for text and data mining, AI training, and similar technologies.

For accessibility purposes, images in electronic versions of this book are accompanied by alt text descriptions provided by Elsevier. For more information, see https://www.elsevier.com/about/accessibility.

Publisher's note: Elsevier takes a neutral position with respect to territorial disputes or jurisdictional claims in its published content, including in maps and institutional affiliations.

No part of this publication may be reproduced or transmitted in any form or by any means, electronic or mechanical, including photocopying, recording, or any information storage and retrieval system, without permission in writing from the publisher. Details on how to seek permission, further information about the Publisher's permissions policies and our arrangements with organizations such as the Copyright Clearance Center and the Copyright Licensing Agency, can be found at our website: www.elsevier.com/permissions.

This book and the individual contributions contained in it are protected under copyright by the Publisher (other than as may be noted herein).

Notices

Knowledge and best practice in this field are constantly changing. As new research and experience broaden our understanding, changes in research methods, professional practices, or medical treatment may become necessary.

Practitioners and researchers must always rely on their own experience and knowledge in evaluating and using any information, methods, compounds, or experiments described herein. In using such information or methods they should be mindful of their own safety and the safety of others, including parties for whom they have a professional responsibility.

To the fullest extent of the law, neither the Publisher nor the authors, contributors, or editors, assume any liability for any injury and/or damage to persons or property as a matter of products liability, negligence or otherwise, or from any use or operation of any methods, products, instructions, or ideas contained in the material herein.

ISBN: 978-0-323-99889-5

For Information on all Elsevier publications
visit our website at https://www.elsevier.com/books-and-journals

Publisher: Candice Janco
Acquisitions Editor: Jessica Mack
Editorial Project Manager: Sara Valentino
Production Project Manager: Sruthi Satheesh
Cover Designer: Miles Hitchen

Typeset by MPS Limited, Chennai, India

Dedication

To my beloved daughter, Isabella "Isa" Sibongile Gwenzi, and my beloved son, Vusumuzi "Vuso" Nkosilathi Gwenzi:

It always feels great to read a good book, but it feels even better and more fulfilling to write your own books and see them being read and displayed in libraries worldwide.

Please develop the habit of telling your own story by writing books.
This book is dedicated to you!
Dad

—Willis Gwenzi

Contents

List of contributors	xvii
Preface	xxi
Acknowledgments	xxv

Part 1
Biochar technology: fundamental principles

1. Biochar for environmental remediation at a glance: principles, applications, and prospects

Willis Gwenzi

1.1	Introduction	3
1.2	Purpose, motivation, and novelty	4
	1.2.1 The origin and evolution of biochar technology	4
	1.2.2 Recent shifts and expansions in biochar research	5
	1.2.3 The case for biochar in environmental remediation	5
	1.2.4 Gaps in the existing literature and the need for a comprehensive resource	5
	1.2.5 Addressing the need for a comprehensive book	5
	1.2.6 Novel aspects of the book	6
1.3	The book at a glance: layout and content	6
	1.3.1 Thematic sections	6
	1.3.2 Overview of the chapters	7
1.4	Future perspectives and prospects	8
1.5	Summary and concluding remarks	9
Acknowledgments		10
References		10

2. Feedstocks, preparation, and characteristics of pristine biochars

Terrence Wenga, Munyaradzi Mtisi, Irvine Nyaguwa, Kudzanayi Andrew Marondedze, Albert Gumbo and Nhamo Chaukura

2.1	Introduction	13
2.2	Types of biomass feedstocks for biochar preparation	14
	2.2.1 Agricultural, forest, and aquatic biomass	14
	2.2.2 Plastics	15
2.3	Biomass quantification approaches	16
	2.3.1 Agricultural waste	16
	2.3.2 Municipal solid waste	17
	2.3.3 Animal manure	17
	2.3.4 Municipal sewage sludge	17
2.4	Preparation of biochar	17
	2.4.1 Pyrolysis systems for biochar production	17
	2.4.2 Effects of pyrolysis type/temperature on changes in functional groups	19
	2.4.3 Hydrothermal carbonization of biomass	21
	2.4.4 Gasification of biomass	21
	2.4.5 Torrefaction of biomass	22
2.5	Biochar physicochemical properties and characterization techniques	22
	2.5.1 Biochar physicochemical characterization	22
	2.5.2 Biochar physicochemical properties	23
2.6	Functional characterization	24
	2.6.1 Analytical methods	24
	2.6.2 Artificial intelligence	25
2.7	Applications of biochar	26
	2.7.1 Agriculture	26
	2.7.2 Composite development	26
	2.7.3 Environmental remediation	26
2.8	Summary and outlook	27
References		28

3. Development of novel engineered/functionalized biochars

Nhamo Chaukura, Jeremia Shale Sefadi, Nonhlangabezo Mabuba, Soraya Malinga, Abimbola Oluwalana-Sanusi and Wisdom Archford Munzeiwa

3.1	Introduction	35
3.2	Biochar synthesis routes	36

3.3 Activation techniques 37
    3.3.1 Physical activation 38
    3.3.2 Chemical activation 39
3.4 Environmental remediation applications 40
    3.4.1 Removal of organic contaminants 40
    3.4.2 Removal of inorganic contaminants 40
3.5 Novel characteristics of engineered biochars 48
3.6 Contaminant removal mechanisms 48
    3.6.1 Organic contaminants 48
    3.6.2 Inorganic contaminants 50
3.7 Economic feasibility studies of biochar production and application 51
3.8 Future outlook and conclusion 52
    3.8.1 Synthesis and fabrication 53
    3.8.2 Testing and evaluation 53
    3.8.3 Regeneration and disposal 53
References 53

4. Design, characterization, and evaluation of biochar: recent advances, applications, and future research directions

Abimbola Oluwalana-Sanusi, Wisdom Archford Munzeiwa, Silas Verkijika and Nhamo Chaukura

4.1 Introduction 59
4.2 Characterization of biochar and biochar-contaminant systems 60
    4.2.1 Biochar surface properties/phenomena 61
    4.2.2 Internal microstructure 62
    4.2.3 3-D micro-CT analysis 63
    4.2.4 Synchrotron X-ray microtomography and multifractal analysis 63
4.3 Design and evaluation of biochar systems 64
    4.3.1 In-silico-computational modeling or computer-aided design approach 65
    4.3.2 Artificial intelligence and machine learning tools 66
    4.3.3 Current and potential applications 68
4.4 Future perspectives and research directions 69
4.5 Conclusions 70
References 70

5. Harnessing biochar for sustainable catalysis in environmental applications

Lucas Meili, Rodolfo Junqueira Brandão and Thais Logetto Caetité Gomes

5.1 Introduction 75
5.2 Nature of biochar catalysts 76
5.3 Preparation and characterization of biochar catalysts 76
5.4 Mechanisms of biochar catalysis 80
    5.4.1 Fenton system 81
    5.4.2 Persulfate activation system 82
    5.4.3 Photocatalytic system 83
5.5 Biochar catalysts regeneration 84
5.6 Environmental applications of biochar catalysis 85
5.7 Future research directions 86
5.8 Conclusions 86
AI Disclosure 86
References 86

Part 2
Biochar for contaminated land remediation

6. Biochar remediation of inorganic contaminants in soils

Terrence Wenga, Albert Gumbo, Irvine Nyaguwa, Munyaradzi Mtisi and Kudzanayi Andrew Marondedze

6.1 Introduction 91
6.2 Occurrence of inorganic contaminants in soils 92
    6.2.1 Metal and nutrient-contaminated soils 93
    6.2.2 Wastewater and sludge-amended soils 94
    6.2.3 Occurrence of inorganic contaminants in munition fields 95
6.3 Biochar removal of inorganic contaminants 95
6.4 Large-scale remediation of inorganics by biochars 96
6.5 Mechanisms for biochar removal of inorganic contaminants in soils 97
    6.5.1 Adsorption and immobilization mechanisms 97
    6.5.2 Synergistic interactions of biochar with other remediation technologies 99
6.6 Factors affecting capacity of biochar in soil remediation 100
    6.6.1 Physiochemical attributes of polluted soils 100
    6.6.2 Physicochemical characteristics of biochars and removal efficacy 100
    6.6.3 Application rate and particle sizes 101
6.7 Behavior and the fate of contaminants in biochar-amended soils 101

		6.7.1	Properties influencing the behavior of contaminants in biochar-amended soils	102
		6.7.2	Other factors influencing the bioavailability of contaminants in biochar-amended soils	102
	6.8	Conclusion and outlook		102
	References			103

7. Biochars for the remediation and repurposing of postmining landscapes and metalliferous substrates: applications and future perspectives

Willis Gwenzi

7.1	Introduction		109
7.2	The case for biochar-based land remediation		110
	7.2.1	The rationale and context	110
	7.2.2	Biochar feedstocks and production systems	110
7.3	The nature and extent of contaminated lands		111
	7.3.1	Postmining landscapes	111
	7.3.2	Metal-contaminated lands	112
	7.3.3	Serpentinitic geological systems	112
	7.3.4	Sludge and wastewater-amended soils	113
7.4	Biochar-based remediation of contaminated lands		113
	7.4.1	Revegetation of postmining landscapes	113
	7.4.2	Metal-contaminated soils	114
	7.4.3	Toxic geogenic contaminants in serpentines	116
	7.4.4	Sludge and wastewater-amended soils	116
	7.4.5	Mechanisms of biochar remediation of mine wastes and metalliferous substrates	117
	7.4.6	Design of biochar-based remediation systems	118
7.5	Future research and perspectives		119
	7.5.1	Increasing Africa's research footprint on biochar-based remediation systems	119
	7.5.2	Long-term behavior and fate of contaminants	119
	7.5.3	Remediation of organic contaminants	120
	7.5.4	Biochar-based extraction and recovery systems for essential elements	120
	7.5.5	Large-scale pilot field studies	120
	7.5.6	Technical and economic feasibility studies	120
	7.5.7	Repurposing postmining landscapes as biomass sources for a circular bioeconomy	120
	7.5.8	Metal-enriched biomass from metalliferous substrates as a unique biomass feedstock	121
	7.5.9	Building Africa' biochar research capacity	121
	7.5.10	The need for biochar research funding and collaboration	121
7.6	Conclusions and outlook		121
References			122

8. Biochar remediation of conventional and emerging organic contaminants in soils

Terrence Wenga, Miranda Mpeta and Phenias Sadondo

8.1	Introduction		127
8.2	Conventional organic contaminants		128
	8.2.1	Types of conventional organic contaminants and their effects	128
	8.2.2	Physico-chemical characteristics of conventional organic contaminants	129
	8.2.3	Sources and pathways of conventional organic contaminants for contaminating the soil	129
8.3	Emerging organic contaminants		131
	8.3.1	Pharmaceuticals and personal care products	131
	8.3.2	Molecular structure of some pharmaceuticals and personal care products and contamination pathways	132
8.4	Biochar removal of conventional and emerging organic contaminants from contaminated soils		132
8.5	Biochar removal mechanisms of organic contaminants in soils		134
	8.5.1	Adsorption mechanisms	134
	8.5.2	Factors affecting adsorption efficiency	137
8.6	Behavior and fate of organic contaminants in soils		138
	8.6.1	Adsorption and desorption	139
	8.6.2	Leaching	139
	8.6.3	Transformation and degradation	139
	8.6.4	Persistence	140

8.6.5	Volatilization	140
8.6.6	Washed away by runoff and erosion	140
8.6.7	Accumulation and Bioavailability	140

8.7 Fate of biochar after organic contaminants removal — 140
8.8 Biochar and other soil remediation methods — 141
8.9 Challenges and opportunities of biochar in removal of organic pollutants — 141
    8.9.1 Challenges of biochar removal of organic pollutants — 141
    8.9.2 Research opportunities — 141
8.10 Conclusion — 142
References — 142

9. Biochar for remediation of petroleum hydrocarbons in soil, sediments, and sludge

Musa Manga, Herbert Cirrus Kaboggoza, Swaib Semiyaga, Lauren Sprouse, Jiahui Guo, Anais Gentles, Yashraj Banga, Sarah Lebu and Chimdi Muoghalu

9.1 Introduction — 149
9.2 Petroleum hydrocarbons in soil, sediments, and sludges — 151
    9.2.1 Classes and sources of petroleum hydrocarbons — 151
    9.2.2 Ecological and human health impacts of petroleum hydrocarbons in soils, sediments, and sludges — 152
9.3 Biochar for remediation of petroleum hydrocarbons in soils, sediments, and sludges — 152
    9.3.1 Use and suitability of biochar for PHC remediation — 152
    9.3.2 Methods of applying biochar to soil, sediments, and sludges — 153
    9.3.3 Factors that influence the ability of biochar to remediate petroleum hydrocarbon-contaminated solid matrices — 153
    9.3.4 Mechanisms for remediation of petroleum hydrocarbon contaminated solid matrices using biochar — 153
    9.3.5 Scale of application for biochar remediation of petroleum hydrocarbon contaminated solid matrices — 155
    9.3.6 Synergistic interaction of biochar with other remediation technologies — 158
    9.3.7 Indicators of remediation efficiency of biochar for petroleum hydrocarbons in soil, sediments, and sludge — 158
9.4 Ecotoxicity and health risks of biochar-remediated solid matrices — 158
    9.4.1 Ecotoxicity of biochar — 158
    9.4.2 Human health effects of biochar — 159
    9.4.3 Safety measures for biochar application — 159
9.5 Future perspectives and directions — 160
    9.5.1 Key challenges — 160
    9.5.2 Considerations for future research — 160
9.6 Conclusions — 160
References — 160

Part 3
Biochar for water and wastewater treatment

10. Removal of contaminants in drinking water using biochars

Nnanake-Abasi O. Offiong, Odunayo T. Ore, Deborah O. Aderibigbe, Ajibola Abiodun Bayode, Olawale S. Dabo and Olaniran Kolawole Akeremale

10.1 Introduction — 169
10.2 Removal of inorganic contaminants — 170
    10.2.1 Removal of radionuclides from water using biochar — 170
    10.2.2 Removal of nutrients from water using biochar — 171
    10.2.3 Removal of toxic metals from water — 171
    10.2.4 Mechanism of removal of inorganic contaminants — 173
10.3 Removal of organic contaminants — 174
10.4 Removal of microbial contaminants — 177
10.5 Conceptual designs of biochar water filters — 178
    10.5.1 Conceptual designs — 178
    10.5.2 Overview of performance — 179
10.6 Efficient disposal management of spent biochars — 180
10.7 Large scale/field applications of biochar in Africa — 181
10.8 Future perspectives and conclusion — 182
References — 182

11. Biowaste-derived biochars for treatment of wastewater contaminated by dyes

Ebuka Chizitere Emenike, Hussein K. Okoro, Kingsley O. Iwuozor, Abel U. Egbemhenghe, Kingsley Chidiebere Okwu, Adewale George Adeniyi, Sujata Paul, Akshaya K and Rangabhashiyam Selvasembian

11.1	Introduction	191
11.2	Biowaste-derived biochar	192
	11.2.1 Biowaste Feedstock	192
	11.2.2 Biochar production from biowaste	193
	11.2.3 Activation methods for the production of engineered biochar	195
11.3	Applications and performance	197
	11.3.1 Dye removal	197
	11.3.2 Adsorption Modeling	200
	11.3.3 Mechanism of adsorption	204
11.4	Regeneration and reuse	205
11.5	Future directions and conclusion	207
	11.5.1 Future directions	207
	11.5.2 Conclusion	208
Disclosure statements		208
Consent to publish		208
Conflict of interest		208
Funding		208
Compliance with ethical standards		208
Data availability		208
References		208

12. Removal of per- and polyfluoroalkyl substances in environmental matrices by biochars: mechanisms, fate, and research needs

Bashir Adelodun, Oyebankole Agbelusi, Qudus Adeyi, Abdulhamid Yusuf, Fidelis Odedishemi Ajibade, Aminu Abdullahi, Golden Odey, Pankaj Kumar, Temitope Fausat Ajibade, Tarun Pal, Abdulwaheed Mohammed and Timothy Denen Akpenpuun

12.1	Introduction	215
12.2	Occurrence of PFAS and its sources in different environmental matrices	216
	12.2.1 Classification and occurrence of PFAS	216
	12.2.2 Sources, fate, and transport of PFAS in environmental matrices	217
12.3	Remediation and treatment methods and processes of PFAS in the environment	220
	12.3.1 PFAS removal technologies and techniques	220
	12.3.2 Biodegradation and transformation of PFAS in the environment	220
12.4	Biochar production processes and its usage for removal of PFAS in the environmental matrices	226
	12.4.1 General processes of biochar production	226
	12.4.2 Modification of biochar for PFAS removal from environmental matrices	227
	12.4.3 Removal mechanism of PFAS from environmental matrices using biochars	229
	12.4.4 Performance of biochars on the removal of PFAS	231
12.5	Existing regulations on the PFAS in the environmental matrices and use of biochar	231
12.6	Knowledge gaps, future research needs, and concluding remarks	233
References		234

13. Application of biochar for the treatment of urban stormwater: processes and future directions

Ahmed Abdelhafez, Mohamed Abbas, Shawky Metwally, Ahmed Al-Hossainy, Sedky Hassan, Hassan Abbas and Abdel Aziz Tantawy

13.1	Introduction	241
13.2	Urban stormwater challenges	242
13.3	Role of biochar in stormwater management	243
	13.3.1 Biochar characteristics relevant to stormwater treatment	243
	13.3.2 High surface area and porosity	243
	13.3.3 Adsorption properties	244
	13.3.4 Surface functional groups	244
13.4	Versatile sorption of contaminants and contaminant-specific adsorption	244
13.5	Influence of solution chemistry on adsorption	244
	13.5.1 Solution chemistry parameters	245
13.6	Urban stormwater contaminants	245
	13.6.1 Contaminants in urban stormwater	245
	13.6.2 Inorganic contaminants	246
	13.6.3 Conventional organic contaminants	246
13.7	Treating emerging contaminants with biochar	247

	13.7.1 Biochar for microbial contaminant removal	248
13.8	Fate and behavior of contaminants in stormwater	249
13.9	Biochar catalytic reactions	250
13.10	Mechanisms of contaminant removal from stormwater using biochars	251
	13.10.1 Ion exchange processes	251
	13.10.2 Precipitation and coprecipitation	251
13.11	Microbial interactions	252
13.12	Catalytic degradation	252
13.13	Adsorption capacity and selectivity	253
13.14	Pivotal role of biochar in influencing soil structure and water retention	253
13.15	Challenges and considerations of biochar for treating stormwater	254
	13.15.1 Adsorption capacity and selectivity	254
	13.15.2 Impact on soil structure	254
	13.15.3 Interactions with microbial communities	254
	13.15.4 Long-term stability and persistence	255
	13.15.5 Scale-up and implementation challenges	255
13.16	Environmental implications of using biochar for treating stormwater	255
13.17	Future directions	256
	13.17.1 Integrated approaches	256
	13.17.2 Synergies with green infrastructure	256
	13.17.3 Optimization of biochar production methods	256
	13.17.4 Scalability and cost-effectiveness	256
	13.17.5 Collaborative research and policy integration	257
13.18	Conclusion	257
References		258

14. Treatment of industrial and municipal wastewaters using biochar: performance, mechanisms, and research needs

Abudu Ballu Duwiejuah, Abubakari Zarouk Imoro, Noel Bakobie, Abdul-Aziz Bawa, Joseph Payne and Emmanuel Okoampah

14.1	Introduction	265
14.2	Wastewater treatment	266
14.3	Biochar technology in wastewater treatment	266
14.4	Type of wastewater treated by biochar	267
14.5	Industrial wastewater treatment	267
14.6	Municipal wastewater treatment	267
14.7	Adsorption of inorganic pollutants onto biochar in wastewater	268
14.8	Adsorption of organic pollutants onto biochar in wastewater	269
14.9	Adsorption mechanisms of biochar for inorganic pollutants in wastewater	269
14.10	Adsorption mechanisms of biochar for organic pollutants in wastewater	270
14.11	Some applications of biochar for wastewater treatment	271
14.12	Current and emerging constraints and gaps	272
14.13	Future research in biochar application for wastewater treatment	273
14.14	Conclusion	274
References		274

Part 4
Biochar sytems for clean and renewable energy

15. Electrochemical properties of biochar for environmental applications: advances, challenges, and perspectives

Vineet Kumar, Shivali Sharma, Sunny Sharma and Gaurav Sharma

15.1	Introduction	281
15.2	Electrochemical properties of biochar	283
	15.2.1 Conductivity	283
	15.2.2 Redox activity	283
	15.2.3 Surface chemistry	284
15.3	Environmental applications of biochar	284
	15.3.1 Agricultural soil amendment	284
	15.3.2 Remediation of polluted wastewater	286
	15.3.3 Carbon sequestration	289
	15.3.4 Biochar as an electrode material for the development of microbial fuel cells	294
15.4	Direct interspecies electron transfer pathways	295
15.5	Challenges	295
15.6	Future perspectives	298
15.7	Conclusion	298
References		299

16. Biochar in bioelectrochemical systems: applications and future directions

Wilgince Apollon, Tatiana Kuleshova, Willis Gwenzi, Felipe Caballero-Briones and Sathish Kumar Kamaraj

16.1	Introduction	307
16.2	Materials and methods	308
	16.2.1 Bioelectrochemical systems: microbial fuel cells	308
	16.2.2 Principles of microbial fuel cells	308
	16.2.3 Designs and types of microbial fuel cells	308
16.3	Constraints and limitations of bioelectrochemical systems	312
	16.3.1 Constraints and limitations	312
16.4	Biochar properties and their potential applications in microbial fuel cells	312
	16.4.1 Electrochemical properties of biochar	312
	16.4.2 Applications in microbial fuel cells	316
16.5	Performance and Mechanisms of biochar function in microbial fuel cells	318
	16.5.1 Overview of performance	318
	16.5.2 Mechanisms of biochar enhancement of microbial fuel cells	318
16.6	Future directions and perspectives	319
	16.6.1 Knowledge gaps	319
	16.6.2 A roadmap for the advancement of biochar-based microbial fuel cells	320
16.7	Conclusions	321
Acknowledgment		321
References		321

17. Thermochemical treatment of human excreta to energy and biochar: recent advances, applications, and future directions

Flávio Lopes Francisco Bittencourt and Marcio Ferreira Martins

17.1	Introduction	329
17.2	Human excreta as an energy feedstock	330
	17.2.1 Overview of thermochemical properties of human excreta	330
	17.2.2 Rationale for thermochemical conversion of human excreta	332
17.3	Thermochemical conversion	332
	17.3.1 Principles of thermochemical conversion processes	332
	17.3.2 Types of thermochemical conversion processes	333
	17.3.3 Comparison of thermochemical conversion methods	334
17.4	Current state of applications	336
	17.4.1 Case study: a portrait of a Brazilian Amazon community	336
	17.4.2 Other thermochemical sanitation technologies	338
17.5	Future research directions and conclusions	339
	17.5.1 Future directions	339
	17.5.2 Conclusions	340
References		340

Part 5
Biochar systems for air pollution control

18. Clean and efficient biochar cookstoves for mitigation of indoor air pollution

Zia Ur Rahman, Abdul Gani Abdul Jameel and Li Songlin

18.1	Introduction	347
18.2	Understanding indoor air pollution	348
18.3	Sources of indoor air pollution	348
	18.3.1 Traditional cooking methods	348
	18.3.2 Biomass combustion and emissions	348
18.4	Biochar cookstoves: an innovative solution	349
18.5	The evolution of cookstove technology	349
	18.5.1 From traditional to modern designs	349
18.6	Types and designs of biochar cookstoves	349
	18.6.1 Top-lit updraft biochar stoves	349
	18.6.2 Rocket stoves	351
	18.6.3 Top-lit updraft gasifier stoves	351
	18.6.4 Institutional-scale biochar cook stoves	354
	18.6.5 Top-lit updraft rice husk gasifier stoves	354
18.7	Biochar cookstoves: efficiency and emission reduction potential	355
18.8	Biochar cookstoves as a multifaceted solution for low-income settings	355

	18.9	Environmental benefits of biochar cookstoves	356
	18.10	Challenges and solutions	356
	18.11	Case studies and success stories	357
	18.12	Future directions and innovations	358
	18.13	Conclusion	359
		18.13.1 Recap of key points	359
		18.13.2 Call to action for sustainable cooking practices	359
	18.14	AI disclosure	359
	References		359

19. Biochars for the removal of toxic gaseous contaminants: state-of-the-art and future directions

Robinah Kulabako, Swaib Semiyaga, Charles Niwagaba, Chimdi Muoghalu and Musa Manga

19.1	Introduction	361
19.2	Sources of toxic gaseous contaminants	362
19.3	The need for gaseous contaminant removal	363
19.4	Removal of toxic gaseous contaminants from air/flue gases	363
19.5	Biochar properties and mechanisms for toxic gaseous contaminant removal	365
	19.5.1 Biochar properties	365
	19.5.2 Mechanisms of toxic gaseous contaminant removal by biochars	367
	19.5.3 Adsorption kinetics, isotherm, and thermodynamics	370
19.6	Engineered biochars for enhanced gaseous contaminants removal	371
	19.6.1 Physical activation	371
	19.6.2 Chemical activation	371
	19.6.3 Metal impregnation	371
	19.6.4 Heteroatoms doping	371
19.7	Regeneration of spent biochars	372
19.8	Future directions and recommendations	372
19.9	Conclusion	373
References		373

Part 6
Assessment of biochar systems

20. Regeneration, recycling, and disposal of spent biochars

Miranda Mpeta, Terrence Wenga, Kudzanayi Andrew Marondedze and Phenias Sadondo

20.1	Introduction	379

20.2	Regeneration of spent biochars		381
	20.2.1	Magnetic separation	381
	20.2.2	Filtration	382
	20.2.3	Thermal desorption and decomposition	382
	20.2.4	Chemical desorption	383
	20.2.5	Supercritical fluid desorption	383
	20.2.6	Advanced oxidation processes	383
	20.2.7	Microbial-assisted regeneration	384
	20.2.8	Microwave irradiation regeneration	384
20.3	Reuse, recycling, and disposal of spent biochar for a circular economy and environmental sustainability	384
	20.3.1	Reuse as a soil amendment	384
	20.3.2	Composting	385
	20.3.3	Land application	385
	20.3.4	Regeneration through pyrolysis	385
	20.3.5	Contaminant removal and landfill disposal	385
	20.3.6	Incineration	386
	20.3.7	Fillers in novel construction materials	386
	20.3.8	Catalyst precursors for trace metal-rich spent biochars	386
	20.3.9	Incorporation in solid fuels	386
	20.3.10	Codisposal of sludge	387
20.4	Behavior and fate of contaminants in spent biochar	387
20.5	Summary and future research directions	388
References		389

21. Life cycle assessment for biochar systems: a review

Simone Marzeddu, Francesca Lazzari, Annarita Cepollaro, Andrea Cappelli and Maria Rosaria Boni

21.1	Introduction	395
21.2	Life cycle assessment as a sustainability tool	397
21.3	Comparative life cycle assessment of biochar production systems	400
21.4	Goal and scope definition	402
21.5	Life cycle inventory	406
21.6	Life cycle impact assessment	416
21.7	Conclusion	427
References		428

22. Potential environmental and human health risks of biochar systems: a call for comprehensive health risk assessments

Willis Gwenzi

22.1	Introduction	433
22.2	Environmental health risks	434
	22.2.1 Soil pollution	435
	22.2.2 Water pollution	436
	22.2.3 Air pollution	436
	22.2.4 Radiative forcing/radiative forcing by black carbon/biochar	436
	22.2.5 Ecological health risks	437
22.3	Human exposure and health risks	437
22.4	A call for health risk assessments and mitigation	437
	22.4.1 Health risk assessment framework	437
	22.4.2 Principles of health risk assessment	438
	22.4.3 Qualitative risk assessment	438
	22.4.4 Quantitative risk assessment	439
	22.4.5 Quantitative human health risk assessment tools	439
	22.4.6 Ecotoxicological risk assessments	441
	22.4.7 Biochar certification systems	441
	22.4.8 Status of research on risk assessment of biochar systems	441
	22.5 Future research directions	442
22.6	Conclusions and outlook	442
References		443

23. Techno-economic assessment of biochar systems: state-of-the-art and future research directions

Sita Koné, Xavier Galiegue and Willis Gwenzi

23.1	Introduction	447
23.2	Techno-economic assessment of biochar systems: tools and evidence	448
	23.2.1 Techno-economic assessment tools	448
	23.2.2 Overview of the evidence	449
23.3	Key factors influencing techno-economic feasibility of biochar	450
	23.3.1 Biomass feedstock	450
	23.3.2 Thermo-hydro-chemical conversion technology	450
	23.3.3 Market price of biochar	451
	23.3.4 Biochar application	452
23.4	Challenges in techno-economic assessments of biochar systems	452
23.5	Discussion and future perspectives	453
	23.5.1 Discussion	453
	23.5.2 Future research directions	455
23.6	Conclusion	457
References		457

Part 7
Looking ahead: future perspectives and epilogue

24. Biochar for environmental remediation and beyond: a "twin or multiple solutions" heuristic framework for uptake, adoption, and impact

Willis Gwenzi

24.1	Introduction	463
24.2	The biochar "twin or multiple solutions" framework	465
	24.2.1 Overview of the "twin or multiple solutions" concept	465
	24.2.2 A handful of biochar "twin or multiple solutions" targeting the water-energy-food-environment nexus	466
24.3	Discussion and perspectives	469
	24.3.1 On the role of "twin or multiple solutions" in biochar uptake, adoption, and impact	470
	24.3.2 The "twin or multiple solutions" framework as a primer for biochar adoption	470
	24.3.3 Moving from biochar as a "silver bullet" to biochar application domains	472
	24.3.4 Moving from "one-size-fits-all" and "blanket salesmen" to biochar application domains	472
	24.3.5 Confronting biochar critics and skeptics with the "twin or multiple solutions" concept?	472
	24.3.6 Coupling biochar systems to address multiple challenges in light of the energy-water-food-environment nexus	473
	24.3.7 Linking the twin or multiple solutions framework to emerging concepts	473
24.4	Concluding remarks and outlook	473
Acknowledgments		474
Declaration of conflict of interest		474
References		475

25. Moving ahead with biochar for environmental remediation and beyond: future research directions, roadmap, and epilogue

Willis Gwenzi

25.1 Introduction	477
25.2 Biochar for environmental remediation: a synoptic overview of the state-of-the-art	479
25.3 Looking ahead: future research directions and perspectives	480
25.3.1 Twenty knowledge gaps to advance biochar for environmental remediation	481
25.3.2 A handful of "grand questions" on biochar technology	484
25.4 Advancing biochar for environmental remediation: a conceptual roadmap	486
25.4.1 Preparation and characterization of biochars	486
25.4.2 Laboratory-scale prototyping and evaluation	487
25.4.3 Pilot-scale prototyping and evaluation	487
25.4.4 Field-scale application	487
25.4.5 Monitoring, evaluation, and feedback	488
25.4.6 Approval and commercialization	488
25.5 Epilogue and outlook	488
Acknowledgments	488
Declaration of conflict of interest	489
References	489
Index	491

List of contributors

Hassan Abbas Department of Geology, Faculty of Science, New Valley University, New Valley, Egypt

Mohamed Abbas Department of Soils and Water Science, Faculty of Agriculture, Benha University, Benha, Egypt

Ahmed Abdelhafez Department of Soils and Water, Faculty of Agriculture, New Valley University, New Valley, Egypt; National Committee of Soil Science, Academy of Scientific Research and Technology, Egypt, Giza, Egypt

Abdul Gani Abdul Jameel Interdisciplinary Research Center for Refining and Advanced Chemicals, King Fahd University of Petroleum & Minerals, Dhahran, Saudi Arabia; Chemical Engineering Department, King Fahd University of Petroleum & Minerals, Dhahran, Saudi Arabia

Aminu Abdullahi Department of Biotechnology, Modibbo Adama University, Yola, Nigeria

Bashir Adelodun Arusha Climate and Environmental Research Center, Aga Khan University, Arusha, Tanzania; Department of Agricultural Civil Engineering, Kyungpook National University, Daegu, Korea; Department of Agricultural and Biosystems Engineering, University of Ilorin, Ilorin, Nigeria

Adewale George Adeniyi Department of Chemical Engineering, University of Ilorin, Ilorin, Nigeria

Deborah O. Aderibigbe Department of Pure and Applied Chemistry, Ladoke Akintola University of Technology, Ogbomoso, Oyo State, Nigeria

Qudus Adeyi Department of Agricultural Civil Engineering, Kyungpook National University, Daegu, Korea

Oyebankole Agbelusi Safeguards and Compliance Department (SNSC), African Development Bank, Abuja, Nigeria (RDNG), Abuja, Nigeria

Fidelis Odedishemi Ajibade Department of Civil and Environmental Engineering, Federal University of Technology, Akure, Nigeria

Temitope Fausat Ajibade Department of Civil and Environmental Engineering, Federal University of Technology, Akure, Nigeria

Olaniran Kolawole Akeremale Department of Science and Technology Education, Bayero University, Kano, Kano State, Nigeria

Timothy Denen Akpenpuun Department of Agricultural and Biosystems Engineering, University of Ilorin, Ilorin, Nigeria

Ahmed Al-Hossainy Department of Chemistry, Faculty of Science, New Valley University, New Valley, Egypt

Wilgince Apollon Instituto Politécnico Nacional (IPN), Centro de Investigación en Ciencia Aplicada y Tecnología Avanzada (CICATA), Unidad Altamira, Altamira, Tamaulipas, Mexico

Noel Bakobie University for Development Studies, Environment, Water and Waste Engineering, Tamale, Ghana

Yashraj Banga Department of Environmental Sciences and Engineering, Gillings School of Global Public Health, University of North Carolina at Chapel Hill, Chapel Hill, NC, United States

Abdul-Aziz Bawa Spanish Laboratory Complex, University for Development Studies, Tamale, Ghana

Ajibola Abiodun Bayode Department of Chemical Sciences, Redeemer's University, Ede, Osun State, Nigeria; Department of Chemical Engineering, Sichuan University of Science and Engineering, Zigong, Sichuan Province, P.R. China

Flávio Lopes Francisco Bittencourt Federal Institute of Education, Science, and Technology of Espírito Santo, Piúma, Espírito Santo, Brazil

Maria Rosaria Boni Department of Civil, Constructional and Environmental Engineering (DICEA), Faculty of Civil and Industrial Engineering, Sapienza University of Rome, Rome, Italy

Rodolfo Junqueira Brandão Federal University of Alagoas, Flowlab (Fluid Dynamics Laboratory), Center of Technology, Maceió, AL, Brazil; Federal University of Alagoas, Laboratory of Processes, Center of Technology, Maceió, AL, Brazil

Felipe Caballero-Briones Instituto Politécnico Nacional (IPN), Centro de Investigación en Ciencia Aplicada y Tecnología Avanzada (CICATA), Unidad Altamira, Altamira, Tamaulipas, Mexico

Andrea Cappelli Department of Chemical Engineering Materials Environment (DICMA), Faculty of Civil and Industrial Engineering, Sapienza University of Rome, Rome, Italy

Annarita Cepollaro Department of Chemical Engineering Materials Environment (DICMA), Faculty of Civil and Industrial Engineering, Sapienza University of Rome, Rome, Italy

Nhamo Chaukura Department of Physical and Earth Sciences, Sol Plaatje University, Kimberley, South Africa

Olawale S. Dabo Department of Pure and Applied Chemistry, Ladoke Akintola University of Technology, Ogbomoso, Oyo State, Nigeria

Abudu Ballu Duwiejuah Biotechnology and Molecular Biology, University for Development Studies, Tamale, Ghana

Abel U. Egbemhenghe Department of Chemistry and Biochemistry, College of Art and Science, Texas Tech University, Lubbock, TX, United States

Ebuka Chizitere Emenike Department of Pure and Industrial Chemistry, Nnamdi Azikiwe University, Awka, Nigeria

Marcio Ferreira Martins Multiphysics Modeling Laboratories (MM Labs), Department of Mechanical Engineering, Federal University of Espírito Santo (UFES), Vitória, Espírito Santo, Brazil

Xavier Galiegue Faculté de Droit d'Economie et de Gestion, University of Orleans, Orleans, France

Anais Gentles Department of Environmental Sciences and Engineering, Gillings School of Global Public Health, University of North Carolina at Chapel Hill, Chapel Hill, NC, United States

Albert Gumbo Department of Land and Water Resources Management, Faculty of Agriculture, Environment and Natural Resources Management, Midlands State University, Gweru, Zimbabwe

Jiahui Guo Department of Environmental Sciences and Engineering, Gillings School of Global Public Health, University of North Carolina at Chapel Hill, Chapel Hill, NC, United States

Willis Gwenzi Formerly Alexander von Humboldt Fellow and Guest Full Professor, Leibniz-Institut für Agrartechnik und Bioökonomie e.V. (ATB), Potsdam, Germany; Formerly Alexander von Humboldt Fellow and Guest Full Professor, Grassland Grassland Science and Renewable Plant Resources, Faculty of Organic Agricultural Sciences, Universität Kassel, Witzenhausen, Germany; Biosystems and Environmental Engineering Research Group, Marlborough, Harare, Zimbabwe

Sedky Hassan Department of Biology, College of Science, Sultan Qaboos University, Muscat, Oman; Department of Botany and Microbiology, Faculty of Science, New Valley University, New Valley, Egypt

Abubakari Zarouk Imoro University for Development Studies, Environment, Water and Waste Engineering, Tamale, Ghana

Kingsley O. Iwuozor Department of Pure and Industrial Chemistry, Nnamdi Azikiwe University, Awka, Nigeria

Akshaya K Department of Environmental Science and Engineering, School of Engineering and Sciences, SRM University-AP, Amaravati, Andhra Pradesh, India

Herbert Cirrus Kaboggoza Department of Environmental Sciences and Engineering, Gillings School of Global Public Health, University of North Carolina at Chapel Hill, Chapel Hill, NC, United States

Sathish Kumar Kamaraj Instituto Politécnico Nacional (IPN), Centro de Investigación en Ciencia Aplicada y Tecnología Avanzada (CICATA), Unidad Altamira, Altamira, Tamaulipas, Mexico

Sita Koné International Center for Biosaline Agriculture, Dubai, United Arab Emirates

Robinah Kulabako Department of Civil and Environmental Engineering, School of Engineering, College of Engineering, Design, Art and Technology, Makerere University, Kampala, Uganda

Tatiana Kuleshova Agrophysical Research Institute, Department of Plant Lightphysiology and Agroecosystem Bioproductivity, Saint-Petersburg, Grazhdanskiy pr., Russia

Pankaj Kumar Agro ecology and Pollution Research Laboratory, Department of Zoology and Environmental Science, Gurukula Kangri (Deemed to be University), Haridwar, Uttarakhand, India; Research and Development Division, Society for AgroEnvironmental Sustainability, Dehradun, India

Vineet Kumar Department of Microbiology, School of Life Sciences, Central University of Rajasthan, Bandar Sindri, Ajmer, Rajasthan, India

Francesca Lazzari Department of Civil, Constructional and Environmental Engineering (DICEA), Faculty of Civil and Industrial Engineering, Sapienza University of Rome, Rome, Italy

Sarah Lebu Department of Environmental Sciences and Engineering, Gillings School of Global Public Health, University of North Carolina at Chapel Hill, Chapel Hill, NC, United States

Thais Logetto Caetité Gomes Federal University of Alagoas, Flowlab (Fluid Dynamics Laboratory), Center of Technology, Maceió, AL, Brazil; Federal University of Alagoas, Laboratory of Processes, Center of Technology, Maceió, AL, Brazil

Nonhlangabezo Mabuba Department of Chemical Sciences, University of Johannesburg, Johannesburg, South Africa

Soraya Malinga Department of Chemical Sciences, University of Johannesburg, Johannesburg, South Africa

Musa Manga Department of Environmental Sciences and Engineering, Gillings School of Global Public Health, University of North Carolina at Chapel Hill, Chapel Hill, NC, United States; Department of Construction Economics and Management, College of Engineering, Design, Art and Technology (CEDAT), Makerere University, Kampala, Uganda

Kudzanayi Andrew Marondedze Department of Soil Science and Environment, Faculty of Agriculture, Environment and Food Systems, University of Zimbabwe, Mount Pleasant, Harare, Zimbabwe; Department of Geography, School of Agricultural, Earth, and Environmental Sciences, University of KwaZulu Natal, Pietermaritzburg, South Africa

Simone Marzeddu Department of Civil, Constructional and Environmental Engineering (DICEA), Faculty of Civil and Industrial Engineering, Sapienza University of Rome, Rome, Italy

Lucas Meili Federal University of Alagoas, Center of Technology, Maceió, AL, Brazil; Federal University of Alagoas, Laboratory of Processes, Center of Technology, Maceió, AL, Brazil

Shawky Metwally Department of Soils and Water, Faculty of Technology and Development, Zagazig University, Zagazig, Egypt

Abdulwaheed Mohammed Kwara State Ministry of Agriculture and Natural Resources, Ilorin, Nigeria

Miranda Mpeta Department of Environmental Engineering, School of Engineering Sciences and Technology, Chinhoyi University of Technology, Chinhoyi, Zimbabwe

Munyaradzi Mtisi Department of Environmental Protection, Hazardous Substances & Hazardous Waste Unit. Environmental Management Agency, Bluffhill, Harare, Zimbabwe; Department of Soil Science and Environment, Faculty of Agriculture, Environment and Food Systems, University of Zimbabwe, Mount Pleasant, Harare, Zimbabwe

Wisdom Archford Munzeiwa Department of Physical and Earth Sciences, Centre for Applied Data Science, Sol Plaatje University, Kimberley, South Africa

Chimdi Muoghalu Department of Environmental Sciences and Engineering, Gillings School of Global Public Health, University of North Carolina at Chapel Hill, Chapel Hill, NC, United States

Charles Niwagaba Department of Civil and Environmental Engineering, School of Engineering, College of Engineering, Design, Art and Technology, Makerere University, Kampala, Uganda

Irvine Nyaguwa Department of Environmental Protection, Hazardous Substances & Hazardous Waste Unit. Environmental Management Agency, Bluffhill, Harare, Zimbabwe

Golden Odey Department of Agricultural Civil Engineering, Kyungpook National University, Daegu, Korea

Nnanake-Abasi O. Offiong Department of Chemical Sciences, Topfaith University, Mkpatak, Akwa Ibom State, Nigeria

Emmanuel Okoampah Biochemistry, University for Development Studies, Tamale, Ghana

Hussein K. Okoro Department of Industrial Chemistry, University of Ilorin, Ilorin, Nigeria

Kingsley Chidiebere Okwu Department of Chemical/Petrochemical Engineering, Rivers State University, Port Harcourt, Nigeria

Abimbola Oluwalana-Sanusi Department of Physical and Earth Sciences, Centre for Global Change, Sol Plaatje University, Kimberley, South Africa

Odunayo T. Ore Department of Chemical Sciences, Achievers University, Owo, Ondo State, Nigeria; Department of Chemistry and Industrial Chemistry, Kogi State University, Kabba, Kogi State, Nigeria

Tarun Pal School of Bioengineering and Food Technology, Faculty of Applied Sciences and Biotechnology, Shoolini University, Solan, Himachal Pradesh, India

Sujata Paul Department of Environmental Science and Engineering, School of Engineering and Sciences, SRM University-AP, Amaravati, Andhra Pradesh, India

Joseph Payne Biotechnology and Molecular Biology, University for Development Studies, Tamale, Ghana

Zia Ur Rahman Interdisciplinary Research Center for Refining and Advanced Chemicals, King Fahd University of Petroleum & Minerals, Dhahran, Saudi Arabia

Phenias Sadondo Department of Soil Science and Environment, Faculty of Agriculture, Environment and Food Systems, University of Zimbabwe, Mount Pleasant, Harare, Zimbabwe

Jeremia Shale Sefadi Department of Physical and Earth Sciences, Sol Plaatje University, Kimberley, South Africa

Rangabhashiyam Selvasembian Department of Environmental Science and Engineering, School of Engineering and Sciences, SRM University-AP, Amaravati, Andhra Pradesh, India

Swaib Semiyaga Department of Civil and Environmental Engineering, School of Engineering, College of Engineering, Design, Art and Technology, Makerere University, Kampala, Uganda

Gaurav Sharma Rani Lakshmi Bai Central Agricultural University, Jhansi, Uttar Pradesh, India

Shivali Sharma Rani Lakshmi Bai Central Agricultural University, Jhansi, Uttar Pradesh, India

Sunny Sharma Department of Horticulture, School of Agriculture, Lovely Professional University, Phagwara, Punjab, India

Li Songlin School of Environmental Science and Engineering, Hainan University, Haikou, Hainan, P.R. China

Lauren Sprouse Department of Environmental Sciences and Engineering, Gillings School of Global Public Health, University of North Carolina at Chapel Hill, Chapel Hill, NC, United States

Abdel Aziz Tantawy Department of Geology, Faculty of Science, New Valley University, New Valley, Egypt

Silas Verkijika Department of Physical and Earth Sciences, Centre for Applied Data Science, Sol Plaatje University, Kimberley, South Africa

Terrence Wenga Department of Soil Science and Environment, Faculty of Agriculture, Environment and Food Systems, University of Zimbabwe, Mount Pleasant, Harare, Zimbabwe; Key Laboratory of Agro-Forestry Environmental Processes and Ecological Regulation of Hainan Province, School of Environmental Science and Engineering, Hainan University, Haikou, P.R. China

Abdulhamid Yusuf Department of Plant Science and Biotechnology, Federal University Dutsin–Ma, Dutsin–Ma, Nigeria; School of Water and Environment, Chang'an University, Xi'an, P.R. China; Key Laboratory of Subsurface Hydrology and Ecological Effects in Arid Region, Ministry of Education, Chang'an University, Xi'an, P.R. China

Preface

1 Motivation: the case for biochar in environmental remediation

Biochar technology has its origins in agriculture but in recent years, there has been a significant increase in research on biochar applications for environmental remediation, including soil, water, and air pollution control. New applications have emerged in bioenergy, such as anaerobic digestion, gasifier cookstoves for household heating and cooking, and bioelectrochemical systems or microbial fuel cells. Due to their unique electrochemical properties, biochars have also been utilized as novel green biomaterials in electrodes and supercapacitors for energy storage applications. Additionally, biochar catalysts have been developed for various environmental and industrial purposes.

This book posits that biochar for environmental remediation offers greater opportunities for uptake, adoption, and upscaling compared to agricultural applications. Two key reasons support this claim: (1) the quantities required for environmental remediation (e.g., water and wastewater treatment, bioenergy applications, air quality control) are significantly smaller than those needed for agriculture and (2) biochar has multiple environmental application domains, including soil remediation, water and wastewater treatment, air quality control, environmental catalysis, and bioenergy applications. These diverse applications present opportunities for innovation and the development of biochar-based environmental technologies. In contrast, agricultural applications are predominantly focused on biochar soil amendments, which require large quantities to achieve beneficial effects.

Compared to agricultural applications, the use of biochar for environmental remediation is a relatively emerging field. Consequently, existing books on biochar provide limited coverage of its applications in remediating contaminated drinking water, wastewater, industrial air pollution, indoor air pollution from biomass cookstoves, and contaminated lands/soils. Data on these applications remain scattered across individual articles. To the editor's knowledge, no single book comprehensively covers biochar's role in remediating inorganic, organic, microbial, and emerging contaminants in various environmental media, including drinking water, industrial wastewater, urban stormwater, industrial and indoor air pollution, and contaminated lands and soils. Additionally, earlier books have paid limited attention to biochar's applications in bioenergy, such as bioelectrochemical systems, gasifier cookstoves, thermochemical conversion of fecal matter to energy and biochar, and biochar catalysis for environmental applications. Therefore a comprehensive resource synthesizing the evidence on biochar applications in environmental remediation has been missing. This book seeks to fill that gap.

Currently, students, researchers, engineers, practitioners, and decision and policymakers interested in biochar for environmental remediation must search through scattered and disparate articles for information. Additionally, a clear roadmap guiding future research on biochar-based systems for environmental remediation has been lacking. This book directly addresses these gaps in a logical and systematic approach, tracking biochar along the value chain from biomass feedstocks to postapplication regeneration, recycling, and disposal of spent biochars.

2 Novel aspects of the book

The following ten novel aspects distinguish this book from its predecessors:

1. Adopts a value-chain approach covering the entire biochar cycle from biomass feedstocks, preparation, applications, and regeneration, recycling, and final disposal.
2. Covers key decision-support tools, specifically life cycle assessment, techno-economic assessment, health risk assessment, and regeneration, recycling, and disposal technologies.
3. Covers a broad range of inorganic, organic, microbial, and emerging contaminants.
4. Provides a comprehensive discussion of biochar applications across the soil−water−atmosphere−human environmental continuum.

5. Includes diverse environmental media such as drinking water, industrial wastewater, urban stormwater, contaminated soils, and outdoor and indoor air pollution control.
6. Features contributions from over 90 experts with diverse but relevant perspectives and experience in biochar technology.
7. Contains over 50 figures and tables summarizing data from various studies.
8. Is well-structured with clearly laid out sections and chapters that enable readers to easily find information.
9. Covers several topics excluded in earlier books, such as biochar catalysis, biochar application in bioelectrochemical systems, thermochemical conversion of human excreta to energy and biochar, health risk assessment, regeneration, recycling and disposal technologies, biochar/gasifier cookstoves, and air pollution control.
10. Presents a conceptual framework for the uptake, adoption, and upscaling of biochar, as well as a comprehensive list of knowledge gaps and future prospects.

3 The book at a glance: layout and content

To provide a comprehensive understanding of biochar technology, this book is logically structured into seven sequential parts, each with two to five chapters. An overview of the focus of the 26 chapters under the seven sections is presented below.

- **Part 1: Biochar Technology: Fundamental Principles**
- This section contains five chapters on the fundamental principles of biochar technology, including an editorial introduction, biomass feedstocks, biochar production, conventional and advanced characterization techniques, and properties of pristine biochars and biochar catalysts.
- **Part 2: Biochar for Contaminated Land Remediation**
- This section has four chapters discussing biochar application in land remediation, focusing on inorganic contaminants in soils, rehabilitation/reclamation of postmining landscapes and metal-enriched serpentine soils, and remediation of conventional and emerging organic contaminants and petroleum hydrocarbons in soils.
- **Part 3: Biochar for Water and Wastewater Treatment**
- This section has five chapters on applications of biochars for the removal of contaminants in drinking water and wastewaters, including dyes, per-and polyfluoroalkyl substances (PFAS) in environmental matrices, treatment of urban stormwater, and industrial and municipal wastewaters.
- **Part 4: Biochar Systems for Clean and Renewable Energy**
- This section consists of three chapters discussing the electrochemical properties and potential biochar applications in bioenergy, bioelectrochemical systems or microbial fuel cells, and thermochemical conversion of human excreta to generate bioenergy and biochar.
- **Part 5: Biochar Systems for Air Pollution Control**
- This section comprises three chapters on biochar applications in air quality/pollution control, with emphasis on gasifier or biochar cookstoves as a clean technology for household energy provision and mitigation of indoor air pollution, removal of toxic gaseous contaminants using biochars as air filter media, and an overview of biochar for mitigation of greenhouse gas emissions.
- **Part 6: Assessment of Biochar Systems**
- This section has four chapters dedicated to decision-support tools for assessing biochar systems, focusing on regeneration, recycling and disposal technologies, life cycle assessment, health risks and risk assessment, and techno-economic assessment of biochar systems.
- **Part 7: Looking Ahead: Future Perspectives and Epilogue**
- Two chapters discuss a heuristic framework for biochar promotion, uptake, and adoption, future research directions, a roadmap for further advancements of biochar, and an epilogue.

Overall, this book provides comprehensive and systematic coverage of various aspects of biochar applications in environmental remediation.

4 Target audience

The primary audience of this book includes undergraduate and postgraduate students, research students, academics, and researchers with interest in biochars, and their application in environmental remediation. Public health engineers, water and sanitation practitioners, renewable energy engineers, water and wastewater treatment engineers, process engineers, policymakers, environmental practitioners, chemists, chemical engineers, material scientists, and civil engineers will also find this book valuable.

Specifically, this book will be an ideal resource for undergraduate and postgraduate students in the following programs:

1. Environmental sciences
2. Renewable energy systems
3. Environmental engineering
4. Remediation technology/engineering
5. Chemistry/chemical engineering
6. Process engineering
7. Civil and public health engineering
8. Materials sciences
9. Environmental and occupational health
10. Green chemistry/materials

I hope you find this book a valuable resource and enjoy reading it. Finally, I welcome any suggestions for improvement.

Willis Gwenzi

Acknowledgments

I am indebted to several individuals and institutions who provided inspiration, support, and immense contributions to the success of this book. Special thanks to the following:

- The Alexander von Humboldt Foundation generously funded my research stay in Germany as a Georg Forster Research Fellow. This book is a culmination of that productive research stay, for which I will forever be grateful to the Alexander von Humboldt Foundation.
- Thanks to the Grassland Science and Renewable Plant Resources Group in the Faculty of Organic Agricultural Sciences at the University of Kassel, Witzenhausen and the Technology Assessment Department at Leibniz Institute for Agricultural Engineering and Bioeconomy e.V. (ATB), Potsdam, Germany for co-hosting my Alexander von Humboldt Fellowship. I will forever cherish the excellent support, facilities, and interactions I had during my stay in Germany, which opened new opportunities, including some of the topics covered in this book.
- The International Foundation for Science (IFS), Sweden, provided me with two research grants (grant numbers C/5266-1 and C/5266-2) that supported my earliest works on biochar and motivated subsequent research. This work served as the launch pad for my biochar research, culminating in this book.
- Several individuals from the German research community, including PhD students, postdoctoral fellows, researchers, and academics working on biochar and allied topics, interacted with me during my stay. They are too many to mention by name. Their insights shaped some of the ideas covered in this book.
- The more than 100 contributing authors who spent sleepless nights brainstorming, reviewing the literature, and writing the chapters under tight deadlines. I greatly value your commitment and effort throughout the book-writing process—you wrote the book, and it was great working with you all.
- The team at Elsevier, particularly Sara Valentino (senior editorial project manager), for the effective management of the book project and for motivating me to complete the book, especially when the temptation to quit was high. Thanks very much Sara for successfully taking me through my second book with Elsevier!
- Special thanks to the following individuals for committing time and effort to review some of the book chapters:

  −Dr. Jiahui Hu, University of Kassel, Germany
  −Dr. Bernard Owuraku Kanin, Kwame Nkrumah University of Science & Technology, Ghana

- The anonymous reviewers of the book proposal who provided detailed comments that greatly improved the structure and content of the book.
- For the second time, my beloved 13-year-old daughter, Isabella "Isa" Sibongile Gwenzi, was the only family member who was always at home when I wrote and finalized the book. I thank her for all the support, for always checking my "pulse," and for praying for me during the writing process.

Several other people contributed directly and indirectly to the completion of this book, and I might have inadvertently omitted some names. However, your support is greatly acknowledged and appreciated.

I thank you all!

Willis Gwenzi

Part 1

Biochar technology: fundamental principles

Chapter 1

Biochar for environmental remediation at a glance: principles, applications, and prospects

Willis Gwenzi[1,2,3]

[1]*Formerly Alexander von Humboldt Fellow and Guest Full Professor, Leibniz-Institut für Agrartechnik und Bioökonomie e.V. (ATB), Potsdam, Germany,*
[2]*Formerly Alexander von Humboldt Fellow and Guest Full Professor, Grassland Grassland Science and Renewable Plant Resources, Faculty of Organic Agricultural Sciences, Universität Kassel, Witzenhausen, Germany,* [3]*Biosystems and Environmental Engineering Research Group, Marlborough, Harare, Zimbabwe*

Chapter outline

1.1 Introduction	3	1.2.6 Novel aspects of the book	6
1.2 Purpose, motivation, and novelty	4	1.3 The book at a glance: layout and content	6
1.2.1 The origin and evolution of biochar technology	4	1.3.1 Thematic sections	6
1.2.2 Recent shifts and expansions in biochar research	5	1.3.2 Overview of the chapters	7
1.2.3 The case for biochar in environmental remediation	5	1.4 Future perspectives and prospects	8
1.2.4 Gaps in the existing literature and the need for a comprehensive resource	5	1.5 Summary and concluding remarks	9
		Acknowledgments	10
1.2.5 Addressing the need for a comprehensive book	5	References	10

1.1 Introduction

Rapid industrialization and population growth have led to increased demand for clean energy, food, and water, as well as environmental pollution (Javan et al., 2023; Rasul, 2016). Water, energy, food, and environmental degradation are often interlinked, forming a complex nexus (Javan et al., 2023). This water-energy-food-environment nexus is particularly strong in low and middle-income countries. These nexus problems should be addressed simultaneously rather than individually, and biochar technology has the potential to contribute significantly to addressing this nexus (Gwenzi et al., 2015, 2017). This unique attribute accounts for the novelty of biochar technology.

Biochar is a broad term referring to carbon-rich biomaterials produced through the thermochemical and hydrothermal conversion of biomass under no or limited oxygen conditions (Gwenzi et al., 2015). Thermochemical conversion processes include pyrolysis, gasification, and hydrothermal carbonization, among other related processes. A large body of evidence, including reviews exists on the biochar system, comprising various types of biomass feedstocks, thermochemical conversion processes, and characterization of the biochar and its co-products (Gwenzi et al., 2015, 2017; Wang & Wang, 2019).

Several studies have investigated various aspects of biochar applications in environmental remediation. These studies have focused on the remediation of (1) contaminated land/soils (Sizmur et al., 2016), (2) drinking water and wastewater treatment (Chaukura et al., 2017a, 2017b; Chaukura, Chiworeso, et al., 2020; Chaukura, Masilompane, et al., 2020; Gwenzi et al., 2014; Maya et al., 2020; 2017), (3) clean and renewable bioenergy applications (Hu et al., 2023, 2024; Ma et al., 2020; Pérez-Rodríguez et al., 2021), (4) air pollution control (Bamdad, 2019; Bamdad et al., 2018; Gwenzi et al., 2021; Mohamed et al., 2022), and (5) greenhouse gas mitigation (Zhang et al., 2019). Other studies have also investigated the assessment of biochar systems using various decision-support tools, including techno-economic assessment, environmental and social impact assessment, and life cycle assessment (Azzi et al., 2021; Bergman et al., 2022; Sahoo et al., 2021). However, the results of these studies largely remain scattered across individual articles and reviews.

Earlier books on biochar have mainly focused on its application in agroecosystems, addressing biochar's effects on soil fertility, physical, chemical, and microbiological properties, crop productivity, and greenhouse gas mitigation (Anderson et al., 2010; Bates, 2010; Lehmann & Joseph, 2015; Ok et al., 2015; Ralebitso-Senior & Orr, 2016). These aspects are beyond the scope of this book, except for greenhouse gas mitigation, which is briefly discussed under air quality control. However, compared to agroecosystems, existing books have paid little attention to biochar applications in environmental remediation, particularly land remediation, water and wastewater treatment, air pollution control, and the use of biochar technology in bioenergy applications.

Therefore, a comprehensive book synthesizing the evidence on biochar applications in environmental remediation has been lacking. Specifically, a one-stop book discussing biochar applications in soil remediation, water and wastewater treatment, air pollution control, clean bioenergy provision, and decision-support tools for assessing biochar systems is needed. To bridge this gap, this book offers a one-stop resource on biochar for environmental remediation, covering principles, applications, and prospects. This is the first book to adopt such a systematic and broad approach to biochar applications in environmental remediation.

This editorial introduction to the book seeks to provide an overview. The chapter addresses four objectives: (1) to discuss the fundamental principles of biochar technology and its key components, including biomass feedstocks, thermochemical conversion processes, characterization techniques, properties, and biochar catalysis, (2) to present an overview of the various biochar applications in environmental remediation, (3) to discuss decision-support tools for assessing biochar systems, and (4) to present an overview of future research directions and prospects for biochar-based environmental remediation. Finally, the layout and structure of the book and the chapters in each part are highlighted. Fig. 1.1 presents the thematic focal areas of the book.

1.2 Purpose, motivation, and novelty

1.2.1 The origin and evolution of biochar technology

Biochar technology originated in agriculture, inspired by the *terra preta* (or *terra de indios*) soils of the Amazon (Bezerra et al., 2019; Glaser & Birk, 2012). Thus, historically, research and applications of biochar have focused primarily on agriculture, with the earliest literature emphasizing its production and benefits in agroecosystems

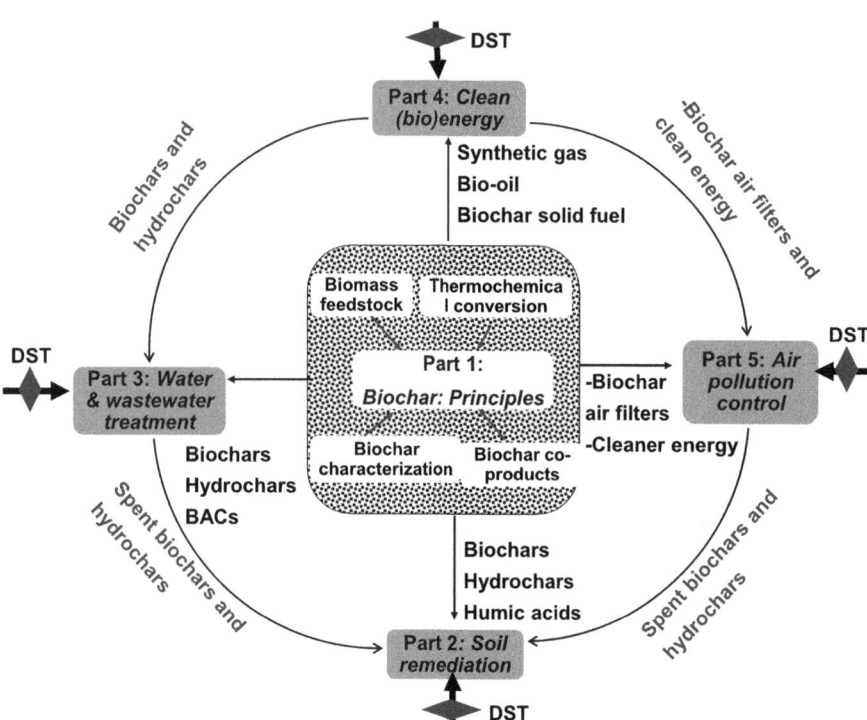

FIGURE 1.1 Depiction of the biochar system, applications of biochar in environmental remediation, and decision-support tools (DST; for instance, life cycle assessment, techno-economic assessment, health risk assessment) for assessing performance of biochar technology. *BACs*: biogenic activated carbons.

(Anderson et al., 2010; Bates, 2010; Lehmann & Joseph, 2015; Ok et al., 2015). These publications primarily explored biochar's effects on soil fertility, physical and chemical properties, microbiological health, and crop productivity. In these earlier works, environmental remediation discussions were often limited to the mitigation of greenhouse gas emissions within agricultural contexts. However, scaling biochar technology for agricultural use remains challenging due to the vast quantities required to cover large agricultural areas effectively (Gwenzi et al., 2015).

1.2.2 Recent shifts and expansions in biochar research

In recent years, there has been a significant increase in research on biochar applications for environmental remediation, including soil, water, and air pollution control (Bamdad, 2019; Bamdad et al., 2018; Gwenzi et al., 2014; Gwenzi et al., 2021; Hu et al., 2024; Ma et al., 2020; Maya et al., 2020; Mohamed et al., 2022; Pérez-Rodríguez et al., 2021; Sizmur et al., 2016; Bamdad et al., 2017). New applications have emerged in bioenergy, such as anaerobic digestion, gasifier cookstoves for household heating and cooking, and bioelectrochemical systems or microbial fuel cells (Hu et al., 2023, 2024; Patwardhan et al., 2022; Pérez-Rodríguez et al., 2021; Sundberg et al., 2020) (a, b). Due to their unique electrochemical properties, biochars have also been utilized as novel green biomaterials in electrodes and supercapacitors for energy storage applications (Rawat et al., 2023). Additionally, biochar catalysts have been developed for various environmental and industrial purposes (Lee et al., 2017; Shan et al., 2020).

1.2.3 The case for biochar in environmental remediation

This book posits that biochar for environmental remediation offers greater opportunities for uptake, adoption, and upscaling compared to agricultural applications. Two key reasons support this notion: (1) the quantities required for environmental remediation (e.g., water and wastewater treatment, bioenergy applications, air quality control) are significantly smaller than those needed for agriculture, and (2) besides agriculture, biochar has multiple environmental application domains, including soil remediation, water and wastewater treatment, air quality control, environmental catalysis, and bioenergy applications. These diverse applications present opportunities for innovation and the development of biochar-based environmental technologies. In contrast, agricultural applications are predominantly focused on biochar soil amendments, which require large quantities to achieve beneficial effects.

1.2.4 Gaps in the existing literature and the need for a comprehensive resource

Compared to agricultural applications, the use of biochar for environmental remediation is a relatively emerging field. Consequently, existing books on biochar provide limited coverage of its applications in remediating contaminated drinking water, wastewater, industrial air pollution, indoor air pollution from biomass cookstoves, and contaminated lands/soils. Data on these applications remain scattered across individual articles. To the editor's knowledge, no single book comprehensively covers biochar's role in remediating inorganic, organic, microbial, and emerging contaminants in various environmental media, including drinking water, industrial wastewater, urban stormwater, industrial and indoor air pollution, and contaminated lands and soils. Additionally, earlier books have paid limited attention to biochar's applications in bioenergy, such as bioelectrochemical systems, gasifier cookstoves, thermochemical conversion of fecal matter to energy and biochar, and biochar catalysis for environmental applications. Therefore, a comprehensive resource synthesizing the evidence on biochar applications in environmental remediation has been missing. This book seeks to fill that gap.

1.2.5 Addressing the need for a comprehensive book

Currently, students, researchers, engineers, practitioners, and decision- and policy-makers interested in biochar for environmental remediation must search through scattered and disparate articles for information. Additionally, a clear roadmap guiding future research on biochar-based systems for environmental remediation has been lacking. This book directly addresses these gaps in a logical and systematic approach, tracking biochar along the value chain from biomass feedstocks to postapplication regeneration, recycling, and disposal of spent biochar. To provide a comprehensive understanding of biochar technology, the book is logically structured into seven sequential parts/sections: (1) fundamental principles of biochar technology (**Section** 1), (2) contaminated land remediation (**Section** 2), (3) water and wastewater treatment (**Section** 3), (4) air quality/pollution control (**Section** 4), (5) clean and renewable bioenergy (**Section** 5), (6) cross-cutting decision-support tools for assessing biochar systems (**Section** 6), and (7) future perspectives and epilogues (**Section** 7). Overall, the book provides a comprehensive and systematic coverage of various aspects of biochar applications in environmental remediation.

1.2.6 Novel aspects of the book

The following 10 novel aspects distinguish this book from its predecessors:

1. Adopts a value-chain approach covering the entire biochar cycle from biomass feedstocks, preparation, applications, regeneration, recycling, and final disposal.
2. Covers key decision-support tools, specifically life cycle assessment, techno-economic assessment, health risk assessment, and regeneration, recycling, and disposal technologies.
3. Covers a broad range of inorganic, organic, microbial, and emerging contaminants.
4. Provides a comprehensive discussion of biochar applications across the soil-water-atmosphere-human environmental continuum.
5. Includes diverse environmental media such as drinking water, industrial wastewater, urban stormwater, contaminated soils, and outdoor and indoor air pollution control.
6. Features contributions from over 90 experts with diverse but relevant perspectives and experience in biochar technology.
7. Contains over 50 figures and tables summarizing data from various studies.
8. Is well-structured with clearly laid out sections and chapters that enable readers to easily find information.
9. Covers several topics excluded in earlier books, such as biochar catalysis, biochar application in bioelectrochemical systems, thermochemical conversion of human excreta to energy and biochar, health risk assessment, regeneration, recycling and disposal technologies, biochar/gasifier cookstoves, and air pollution control.
10. Presents a conceptual framework for the uptake, adoption, and upscaling of biochar, as well as a comprehensive list of knowledge gaps and future prospects.

1.3 The book at a glance: layout and content

The book is systematically structured around the biochar value chain, spanning biomass feedstocks to regeneration, recycling, and disposal of postapplication spent biochars. Therefore, the book covers:

1. Fundamental principles of biochar technology (biomass feedstocks, biochar production, characterization techniques, properties including biochar catalysis)
2. Biochar applications in land, water, and air pollution remediation
3. Biochar applications in clean and renewable bioenergy systems
4. Decision-support systems for assessing biochar systems (techno-economic assessment, life cycle assessment, health risk assessment), and regeneration, recycling, and disposal technologies

An overview of the thematic parts/sections and chapters is presented below.

1.3.1 Thematic sections

In summary, the book has seven thematic parts, each with two to five chapters. The parts are arranged in a logical sequence as follows:

- **Section 1: Biochar Technology—Fundamental Principles**

    This section sets the scene and presents an overview of biochar application for environmental remediation, biomass feedstocks, preparation of pristine and engineered biochars, conventional and emerging characterization methods, biochar properties, and biochar catalysis.
- **Section 2: Biochar for Contaminated Land Remediation**

    This section focuses on biochar application in land remediation, addressing conventional and emerging contaminants, including petroleum hydrocarbons, organic and inorganic pollutants, and the rehabilitation of postmining landscapes and serpentine soils.
- **Section 3: Biochar for Water and Wastewater Treatment**

    This section is dedicated to the application of biochar for removing various contaminants in drinking water and wastewater, including industrial, municipal, and urban stormwater.
- **Section 4: Biochar for Clean and Renewable Bioenergy**

    This section covers biochar applications in bioenergy systems, focusing on the electrochemical properties of biochar and their potential applications in (bio)energy, bioelectrochemical systems, and the thermochemical conversion of human excreta to energy and biochar.

- **Section 5: Biochar for Air Quality Control**
  This section addresses biochar applications in air quality or pollution control, focusing on gasifier/biochar cookstoves for mitigating indoor air pollution, and the removal of toxic gases by biochars.
- **Section 6: Decision-Support Tools for Assessment of Biochar Systems**
  This section focuses on decision-support systems for assessing biochar systems, specifically life cycle assessment, techno-economic assessment, health risk assessments, and biochar regeneration, recycling, and disposal technologies.
- **Section 7: Future Perspectives and Outlook**
  This section covers future perspectives and outlook for biochar application in environmental remediation. It focuses on a heuristic framework for biochar promotion, uptake and adoption, future research directions, and an epilogue.

1.3.2 Overview of the chapters

An overview of the focus of the 26 chapters under the seven sections is presented below.

- **Section 1: Biochar Technology: Fundamental Principles**
  This section contains five chapters on the fundamental principles of biochar technology:
  Chapter 1: An editorial introduction to the book highlighting the purpose motivation, novelty, and an overview of the book
  Chapter 2: Discusses the biomass feedstocks, preparation, conventional characterization techniques, and properties of pristine biochars
  Chapter 3: Covers the development or synthesis of engineered or functionalized biochars with unique properties for various applications.
  Chapter 4: Covers recent or emerging characterization and analytical tools for the design, characterization, and evaluation of biochars
  Chapter 5: Discusses the preparation, characterization, and environmental applications of biochar catalysts.
- **Section 2: Contaminated Land Remediation**
  This section has four chapters discussing biochar application in land remediation:
  Chapter 6: Focuses on the applications of biochars for the remediation of various inorganic contaminants in soils.
  Chapter 7: Discusses the application of biochar for the rehabilitation/reclamation of postmining landscapes and metal-enriched serpentine soils
  Chapter 8: Covers biochar use in the remediation of conventional and emerging organic contaminants in soils
  Chapter 9: Discusses the application of biochars for the remediation of soils, sediments, and sludge contaminated with petroleum hydrocarbons.
- **Section 3: Water and Wastewater Treatment**
  This section has five chapters on applications of biochars for the removal of contaminants in drinking water and wastewaters:
  Chapter 10: Discusses biochar application as a filter media in drinking water treatment systems.
  Chapter 11: Covers the removal of dyes in wastewater using biochars derived from biowaste.
  Chapter 12: Focuses on the removal of per–and polyfluoroalkyl substances (PFAS) in environmental matrices by biochars.
  Chapter 13: Covers biochar application for the treatment of urban stormwater
  Chapter 14: Discusses the use of biochar for the treatment of various industrial and municipal wastewater.
- **Section 4: Clean and Renewable Bioenergy**
  This section consists of three chapters discussing electrochemical properties and potential biochar applications in bioenergy:
  Chapter 15: Discusses the electrochemical properties of biochars and their potential applications as novel green in energy (supercapacitors and electrodes) and environmental applications.
  Chapter 16: Covers recent advances in biochar applications in bioelectrochemical systems focusing on microbial fuel cells.
  Chapter 17: Discusses the thermochemical conversion of human excreta to generate bioenergy and biochar, including field evidence from Brazil.
- **Section 5: Air Quality Control**
  This section comprises three chapters on biochar applications in air quality/pollution control:
  Chapter 18: Discusses gasifier or biochar cookstoves as a clean technology for household energy provision and mitigation of indoor air pollution from biomass cooking systems.

Chapter 19: Covers the removal of industrial gaseous contaminants using biochars as air filter media.
- **Section 6: Decision-Support Tools for Assessment of Biochar Systems**
  This section has four chapters dedicated to decision-support tools for assessing biochar systems:
  Chapter 20: Covers regeneration, recycling, and disposal technologies for post-application spent biochars particularly those used as adsorbents.
  Chapter 21: Discusses life cycle assessment as a tool for assessing the sustainability and environmental footprints of biochar systems.
  Chapter 22: Covers the environmental, ecological, and human health risks of biochar systems, and the associated risk assessment tools.
  Chapter 23: Discusses techno-economic assessment as a decision-support tool for biochar systems
- **Section 7: Future Perspectives and Outlook**
  Two chapters discuss a heuristic framework for biochar promotion, uptake and adoption, future research directions, and an epilogue.
  Chapter 24: Presents a heuristic conceptual framework for the uptake, adoption, and upscaling of biochar based on the dual (twin) or multiple applications of biochar to address nexus problems.
  Chapter 25: Synthesises and presents future research directions and prospects for biochar, including knowledge gaps, a road-map for further advancements, and an epilogue.

1.4 Future perspectives and prospects

Despite recent advances in biochar research, several knowledge gaps still exist on various aspects of biochar systems. In summary, future research directions include the following thematic topics:

(1) *(Re)Designing and Evaluation of Prototypes of Biochar Devices*
Detailed designs for biochar devices such as gasifier cookstoves and water and wastewater filtration systems either have limitations or are still lacking. Thus, there is a need to (re)design and evaluate the next-generation of biochar devices to address these limitations and enhance the uptake and adoption of the technology.

(2) Long-term Fate of Contaminants Following Remediation using Biochars
Most studies on biochar applications for remediating contaminants in soils, water, and air are limited to short-term, small-scale laboratory experiments conducted under controlled conditions. These studies provide little or no information on the long-term fate of contaminants after removal by biochars. For example, it remains unknown whether contaminants are remobilized and released into the environment when biochar ages and undergoes decomposition induced by changes in biogeochemical conditions. Therefore, there is a need for studies investigating the long-term fate of contaminants following removal by biochars under environmentally relevant conditions. Such studies will provide realistic data for assessing the sustainability of biochar as a remediation technology.

(3) *Biochar and the Problem of Temporal and Spatial Scales*
Most data on biochar technology and its applications in environmental remediation and bioenergy are based on small-scale and point-scale measurements often taken over short periods. Many studies assume that these short-term measurements of impacts and benefits scale linearly to large temporal and spatial scales, including ecosystem goods and services such as biogeochemical cycling. However, data on scale issues are still lacking, yet the potential impacts and benefits of technology at scale are critical for decision and policy-making.

(4) *Moving from Biochar "Silver Bullet" to Biochar Application Domains*
Biochars are heterogeneous biomaterials, and some inconsistencies in performance have been reported. This calls for research to (1) define the biophysical limits or boundaries of biochar application domains and (2) develop a database of well-characterized biochars for tailor-made applications. This requires a shift from the current "one-size-fits-all" approach. Research should focus on identifying biochar application domains or niches and developing tailor-made biochar solutions for these domains.

(5) *Life Cycle Assessment of the Environmental and Social Footprints of Biochar Systems*
To date, only a few life cycle assessment studies have investigated the environmental footprints of biochar technology, and these are mostly limited to energy applications. Comprehensive life cycle assessments investigating the environmental and social footprints of biochars applied in multiple contexts are still lacking.

(6) *Biochar or Black Carbon-Induced Radiative Forcing and Associated Climatic Risks*
Similar to black carbon, biochars applied to soil may alter soil thermal properties, including the albedo and temperature regimes, and induce radiative forcing. However, the potential climatic risks associated with biochar

or black carbon-induced radiative forcing in large-scale biochar adoption have received limited attention. Thus, there is a need to investigate the potential impacts of large-scale biochar adoption on climatic forcing and atmospheric feedbacks at regional and continental scales.

(7) *Ecological Health Risks of Biochar Systems*

The ecological health risks of biochar systems, including feedstock collection (e.g., litter) on habitat and biodiversity loss remain unclear. Moreover, the impacts of biochar systems on ecological functions and processes at various levels of biological organization spanning individuals, populations, communities, trophic interactions, and ecosystem goods and services, have received limited research attention.

(8) **Human Exposure and Health Risks**

Biochar systems release particulate and gaseous emissions at various points along the value chain, including feedstock collection and logistics, biochar production, application, and postapplication disposal. However, human exposure and health risks of biochar systems remain understudied. Qualitative and quantitative risk assessment studies are needed to better understand these health risks.

(9) *Techno-Economic Assessment of Biochar Systems*

Techno-economic assessments (TEAs) of biochars are critical for decision and policy-making, including technology uptake and adoption. However, the available TEA studies are limited to biochar production. Comparatively, studies investigating the techno-economic assessment of several industrial applications of biochar in environmental and bioenergy applications are scarce. Further work is required to compare the techno-economic performance of biochar for environmental remediation to competing technologies.

(10) *Technology Readiness Assessment of biochar systems*

Technology readiness assessment (TRA) seeks to evaluate the development level of a technology using a standardized scale with various levels. TRA is critical for identifying technologies ready for uptake, adoption, and promotion, as well as those still in their infancy stages. Biochar applications in environmental remediation and bioenergy are likely at different technology readiness levels. However, studies on TRA for biochar systems are still lacking.

Additional future directions, including knowledge gaps and prospects, are covered in the individual chapters and Chapter 25 on Future research directions and epilogue.

1.5 Summary and concluding remarks

Biochars are novel biomaterials, and research on biochar technology and its industrial applications is rapidly evolving. Until now, a comprehensive book dedicated to biochar applications in environmental remediation has been lacking. To address this gap, this book provides a one-stop comprehensive discussion of the principles, applications, and prospects of biochar applications in environmental remediation. In summary, the book contributes the following:

1. **Section 1 (Biochar technology: fundamental principles):** Five chapters cover an overview of biochar application for environmental remediation, biomass feedstocks, preparation of pristine and engineered biochars, conventional and emerging characterization methods, biochar properties, and biochar catalysis.
2. **Section 2 (Contaminated land remediation):** Four chapters discuss biochar application in land remediation, focusing on conventional and emerging contaminants, including petroleum hydrocarbons, organic and inorganic pollutants, and the rehabilitation of post-mining landscapes and metal-enriched substrates and serpentine soils.
3. **Section 3 (Water and wastewater treatment):** Five chapters are dedicated to the treatment of drinking water and various wastewaters, including industrial, municipal, and urban stormwater, and those contaminated with PFAS and dyes.
4. **Section 4 (Clean and renewable bioenergy):** Three chapters discuss biochar's electrochemical properties and their potential applications in energy, biochar application in bioelectrochemical systems, and the thermochemical conversion of human excreta to energy and biochar.
5. **Section 5 (Air quality control):** Two chapters cover biochar cookstoves for the mitigation of indoor air pollution, and the removal of toxic gases by biochars.
6. **Section 6 (Decision-support tools for assessment of biochar systems):** Four chapters cover life cycle assessment, techno-economic assessment, health risk assessments, and biochar regeneration, recycling, and disposal technologies.
7. **Section 7 (Future perspectives and epilogue):** Two chapters discuss a heuristic framework for biochar promotion, uptake and adoption, future research directions, and an epilogue.

In summary, the book comprises 25 main chapters providing a comprehensive discussion of various applications of biochar in environmental remediation and the associated decision-support tools for assessing the technology. Finally, the book closes with future directions and an epilogue presenting a summary of the book, outstanding gaps, and the next logical steps. The knowledge gaps highlighted in the individual chapters, and in the chapter on future research directions, should stimulate further work on biochar technology in the coming years. This is the first book to provide such a systematic discussion of biochar applications in environmental remediation. I hope that this book will appeal to a broad audience with an interest in environmental remediation and bioenergy using novel biochar materials.

Acknowledgments

I thank the Alexander von Humboldt Stiftung/Foundation, Germany for a Georg Forster Fellowship for Experienced Researchers which funded the project titled, *"Environmental and Climate Footprints of a Biochar and Biogas Technology versus Current Practices in Sub-Saharan Africa: A Life Cycle Analysis."* I also thank the University of Kassel and the Leibniz Institute for Agricultural Engineering and Bioeconomy e.V. (ATB) for co-hosting me as a Geord Forster fellow and for providing facilities and a conducive environment to work on the book.

I am also grateful to the International Foundation for Science — IFS, Sweden for providing seed research grants for biochar research [grant numbers: C-5266−1, and C/5266−2]. The projects formed the basis and motivated the current paper. However, the views expressed in the current paper are solely those of the author. IFS and the Alexander von Humboldt Foundation played no role whatsoever in the research, write-up and decision to publish the work.

References

Anderson, P., Bates, A., & Frogner, K. (2010). *The Biochar revolution: transforming agriculture & environment.* Global.

Azzi, E. S., Karltun, E., & Sundberg, C. (2021). Assessing the diverse environmental effects of biochar systems: An evaluation framework. *Journal of Environmental Management, 286,* 112154. Available from https://doi.org/10.1016/j.jenvman.2021.112154.

Bamdad, H. (2019). A theoretical and experimental study on biochar as an adsorbent for removal of acid gases (CO_2 and H_2S) (doctoral dissertation).

Bamdad, H., Hawboldt, K., & MacQuarrie, S. (2018). A review on common adsorbents for acid gases removal: Focus on biochar. *Renewable and Sustainable Energy Reviews, 81,* 1705−1720. Available from https://doi.org/10.1016/j.rser.2017.05.261, https://www.journals.elsevier.com/renewable-and-sustainable-energy-reviews.

Bates, A. (2010). *The biochar solution: carbon farming and climate change.* New Society Publishers.

Bergman, R., Sahoo, K., Englund, K., & Mousavi-Avval, S. H. (2022). Lifecycle assessment and techno-economic analysis of biochar pellet production from forest residues and field application. MDPI, United States. *Energies, 15*(4). Available from https://doi.org/10.3390/en15041559, https://www.mdpi.com/1996-1073/15/4/1559/pdf.

Bezerra, J., Turnhout, E., Vasquez, I. M., Rittl, T. F., Arts, B., & Kuyper, T. W. (2019). The promises of the Amazonian soil: shifts in discourses of Terra Preta and biochar. *Journal of Environmental Policy and Planning, 21*(5), 623−635. Available from https://doi.org/10.1080/1523908X.2016.1269644, http://www.tandf.co.uk/journals/titles/1523908x.html.

Chaukura, N., Chiworeso, R., Gwenzi, W., Motsa, M. M., Munzeiwa, W., Moyo, W., Chikurunhe, I., & Nkambule, T. T. (2020). A new generation low-cost biochar-clay composite 'biscuit' ceramic filter for point-of-use water treatment. *Applied Clay Science, 185,* 105409.

Chaukura, N., Masilompane, T. M., Gwenzi, W., & Mishra, A. K. (2020). Biochar-Based Adsorbents for the Removal of Organic Pollutants from Aqueous Systems. *Emerging Carbon-Based Nanocomposites for Environmental Applications,* 147−174.

Chaukura, N., Murimba, E. C., & Gwenzi, W. (2017a). Sorptive removal of methylene blue from simulated wastewater using biochars derived from pulp and paper sludge. *Environmental Technology & Innovation, 8,* 132−140.

Chaukura, N., Murimba, E. C., & Gwenzi, W. (2017b). Synthesis, characterisation and methyl orange adsorption capacity of ferric oxide−biochar nano-composites derived from pulp and paper sludge. *Applied Water Science, 7,* 2175−2186.

Glaser, B., & Birk, J. J. (2012). State of the scientific knowledge on properties and genesis of Anthropogenic Dark Earths in Central Amazonia (terra preta de índio). *Geochimica et Cosmochimica Acta, 82,* 39−51. Available from https://doi.org/10.1016/j.gca.2010.11.029.

Gwenzi, W., Chaukura, N., Mukome, F. N. D., Machado, S., & Nyamasoka, B. (2015). Biochar production and applications in sub-Saharan Africa: Opportunities, constraints, risks and uncertainties. *Journal of Environmental Management, 150,* 250−261. Available from https://doi.org/10.1016/j.jenvman.2014.11.027, https://www.sciencedirect.com/journal/journal-of-environmental-management.

Gwenzi, W., Chaukura, N., Noubactep, C., & Mukome, F. N. (2017). Biochar-based water treatment systems as a potential low-cost and sustainable technology for clean water provision. *Journal of Environmental Management, 197,* 732−749.

Gwenzi, W., Chaukura, N., Wenga, T., & Mtisi, M. (2021). Biochars as media for air pollution control systems: Contaminant removal, applications and future research directions. *Science of the Total Environment, 753,* 142249.

Gwenzi, W., Musarurwa, T., Nyamugafata, P., Chaukura, N., Chaparadza, A., & Mbera, S. (2014). Adsorption of $Zn2+$ and $Ni2+$ in a binary aqueous solution by biosorbents derived from sawdust and water hyacinth (Eichhornia crassipes). *Water Science and Technology, 70*(8), 1419−1427. Available from https://doi.org/10.2166/wst.2014.391, http://www.iwaponline.com/wst/07008/1419/070081419.pdf.

Hu, J., Stenchly, K., Gwenzi, W., Wachendorf, M., & Kaetzl, K. (2023). Critical evaluation of biochar effects on methane production and process stability in anaerobic digestion. *Frontiers in Energy Research, 11*, 1205818.

Hu, J., Wachendorf, M., Gwenzi, W., Joseph, B., Stenchly, K., & Kaetzl, K. (2024). Improving acid-stressed anaerobic digestion processes with biochar-towards a combined biomass and carbon management system. *Environmental Research Communications, 6*(3), 035010.

Javan, K., Altaee, A., BaniHashemi, S., Darestani, M., Zhou, J., & Pignatta, G. (2023). A review of interconnected challenges in the water−energy−food nexus: Urban pollution perspective towards sustainable development. *Science of the Total Environment*, 169319.

Lee, J., Kim, K. H., & Kwon, E. E. (2017). Biochar as a catalyst. *Renewable and Sustainable Energy Reviews, 77*, 70−79. Available from https://doi.org/10.1016/j.rser.2017.04.002, https://www.journals.elsevier.com/renewable-and-sustainable-energy-reviews.

Lehmann, J., & Joseph, S. (Eds.), (2015). *Biochar for environmental management: science, technology and implementation*. Routledge.

Ma, Y., Yao, D., Liang, H., Yin, J., Xia, Y., Zuo, K., & Zeng, Y.-P. (2020). Ultra-thick wood biochar monoliths with hierarchically porous structure from cotton rose for electrochemical capacitor electrodes. *Electrochimica Acta, 352*, 136452. Available from https://doi.org/10.1016/j.electacta.2020.136452.

Maya, M., Gwenzi, W., & Chaukura, N. (2020). A biochar-based point-of-use water treatment system for the removal of fluoride, chromium and brilliant blue dye in ternary systems. *Environmental Engineering and Management Journal, 19*(1), 143−156. Available from http://www.eemj.eu/index.php/EEMJ/article/view/4040/3978.

Mohamed, G. O., Saleh, M. E., Shalaby, E. A., & Elsafty, A. S. (2022). Using biochar to control nitric oxide air pollution. *Journal of Physics: Conference Series, 2305*(1), 012029. Available from https://doi.org/10.1088/1742-6596/2305/1/012029.

Ok, Y. S., Uchimiya, S. M., Chang, S. X., & Bolan, N. (Eds.), (2015). *Biochar: Production, characterization, and applications*. CRC press.

Patwardhan, S. B., Pandit, S., Kumar Gupta, P., Kumar Jha, N., Rawat, J., Joshi, H. C., Priya, K., Gupta, M., Lahiri, D., Nag, M., Kumar Thakur, V., & Kumar Kesari, K. (2022). Recent advances in the application of biochar in microbial electrochemical cells. *Fuel, 311*. Available from https://doi.org/10.1016/j.fuel.2021.122501, http://www.journals.elsevier.com/fuel/.

Pérez-Rodríguez, S., Pinto, O., Izquierdo, M. T., Segura, C., Poon, P. S., Celzard, A., Matos, J., & Fierro, V. (2021). Upgrading of pine tannin biochars as electrochemical capacitor electrodes. *Journal of Colloid and Interface Science, 601*, 863−876. Available from https://doi.org/10.1016/j.jcis.2021.05.162, http://www.elsevier.com/inca/publications/store/6/2/2/8/6/1/index.htt.

Ralebitso-Senior, T. K., & Orr, C. H. (2016). *Biochar Application: Essential Soil Microbial Ecology Biochar Application: Essential Soil Microbial Ecology* (pp. 1−323). United Kingdom: Elsevier Inc. Available from http://www.sciencedirect.com/science/book/9780128034330.

Rasul, G. (2016). Managing the food, water, and energy nexus for achieving the Sustainable Development Goals in South Asia. *Environmental Development, 18*, 14−25. Available from https://doi.org/10.1016/j.envdev.2015.12.001.

Rawat, S., Wang, C. T., Lay, C. H., Hotha, S., & Bhaskar, T. (2023). Sustainable biochar for advanced electrochemical/energy storage applications. *Journal of Energy Storage, 63*. Available from https://doi.org/10.1016/j.est.2023.107115, http://www.journals.elsevier.com/journal-of-energy-storage/.

Sahoo, K., Upadhyay, A., Runge, T., Bergman, R., Puettmann, M., & Bilek, E. (2021). Life-cycle assessment and techno-economic analysis of biochar produced from forest residues using portable systems. *The International Journal of Life Cycle Assessment, 26*(1), 189−213. Available from https://doi.org/10.1007/s11367-020-01830-9.

Shan, R., Han, J., Gu, J., Yuan, H., Luo, B., & Chen, Y. (2020). A review of recent developments in catalytic applications of biochar-based materials. *Resources, Conservation and Recycling, 162*, 105036. Available from https://doi.org/10.1016/j.resconrec.2020.105036.

Sizmur, T., Quilliam, R., Puga, A. P., Moreno-Jiménez, E., Beesley, L., & Gomez-Eyles, J. L. (2016). Application of biochar for soil remediation. *Agricultural and environmental applications of biochar: Advances and barriers, 63*, 295−324.

Sundberg, C., Karltun, E., Gitau, J. K., Kätterer, T., Kimutai, G. M., Mahmoud, Y., Njenga, M., Nyberg, G., Roing de Nowina, K., Roobroeck, D., & Sieber, P. (2020). Biochar from cookstoves reduces greenhouse gas emissions from smallholder farms in Africa. *Mitigation and Adaptation Strategies for Global Change, 25*(6), 953−967. Available from https://doi.org/10.1007/s11027-020-09920-7, http://www.wkap.nl/journalhome.htm/1381-2386.

Wang, J., & Wang, S. (2019). Preparation, modification and environmental application of biochar: A review. *Journal of Cleaner Production, 227*, 1002−1022. Available from https://doi.org/10.1016/j.jclepro.2019.04.282, https://www.journals.elsevier.com/journal-of-cleaner-production.

Zhang, C., Zeng, G., Huang, D., Lai, C., Chen, M., Cheng, M., Tang, W., Tang, L., Dong, H., Huang, B., Tan, X., & Wang, R. (2019). Biochar for environmental management: Mitigating greenhouse gas emissions, contaminant treatment, and potential negative impacts. *Chemical Engineering Journal, 373*, 902−922. Available from https://doi.org/10.1016/j.cej.2019.05.139, http://www.elsevier.com/inca/publications/store/6/0/1/2/7/3/index.htt.

Chapter 2

Feedstocks, preparation, and characteristics of pristine biochars

Terrence Wenga[1,2], Munyaradzi Mtisi[1], Irvine Nyaguwa[3], Kudzanayi Andrew Marondedze[1], Albert Gumbo[4] and Nhamo Chaukura[5]

[1]*Department of Soil Science and Environment, Faculty of Agriculture, Environment and Food Systems, University of Zimbabwe, Mount Pleasant, Harare, Zimbabwe,* [2]*Key Laboratory of Agro-Forestry Environmental Processes and Ecological Regulation of Hainan Province, School of Environmental Science and Engineering, Hainan University, Haikou, China,* [3]*Department of Environmental Protection, Hazardous Substances & Hazardous Waste Unit. Environmental Management Agency, Bluffhill, Harare, Zimbabwe,* [4]*Department of Land and Water Resources Management, Faculty of Agriculture, Environment and Natural Resources Management, Midlands State University, Gweru, Zimbabwe,* [5]*Department of Physical and Earth Sciences, Sol Plaatje University, Kimberley, South Africa*

Chapter outline

2.1 Introduction	13
2.2 Types of biomass feedstocks for biochar preparation	14
2.2.1 Agricultural, forest, and aquatic biomass	14
2.2.2 Plastics	15
2.3 Biomass quantification approaches	16
2.3.1 Agricultural waste	16
2.3.2 Municipal solid waste	17
2.3.3 Animal manure	17
2.3.4 Municipal sewage sludge	17
2.4 Preparation of biochar	17
2.4.1 Pyrolysis systems for biochar production	17
2.4.2 Effects of pyrolysis type/temperature on changes in functional groups	19
2.4.3 Hydrothermal carbonization of biomass	21
2.4.4 Gasification of biomass	21
2.4.5 Torrefaction of biomass	22
2.5 Biochar physicochemical properties and characterization techniques	22
2.5.1 Biochar physicochemical characterization	22
2.5.2 Biochar physicochemical properties	23
2.6 Functional characterization	24
2.6.1 Analytical methods	24
2.6.2 Artificial intelligence	25
2.7 Applications of biochar	26
2.7.1 Agriculture	26
2.7.2 Composite development	26
2.7.3 Environmental remediation	26
2.8 Summary and outlook	27
References	28

2.1 Introduction

The fast rate of the growing population worldwide has resulted in various adverse effects on the environment which include soil degradation, water contamination, and air pollution, due to the increase in the utilization of various chemicals (dos Reis et al., 2023; Wen et al., 2023). Various conventional methods are available for the treatment of each of the environmental contaminants. For example, contamination of water bodies by dyes from industrial wastes can be removed by methods such as nanofiltration (David et al., 2020), reverse osmosis (Nataraj et al., 2009), electrocoagulation, and advanced oxidation methods (dos Reis et al., 2023). Nevertheless, such strategies demand very high operating costs, sometimes lead to the formation of toxic by-products, and oftentimes are inefficient.

Biochar, a carbon-rich solid, stable, and highly aromatic-containing material with different physicochemical and adsorptive properties produced from pyrolysis of biomass wastes in limited or oxygen-free conditions, provides a cheap and efficient alternative method of removing environmental contaminants (Sánchez-Sánchez et al., 2015). Besides, its chemical stability, surface functionalities, high porosity, and high specific surface area (SSA) are essential attributes for the efficient removal of contaminants during the adsorption process (Dos Reis et al., 2021). Moreover, the physicochemical characteristics of biochar can be improved for efficient applications by doping the heteroatoms with other elements during the preparation.

Biochar offers several advantages such as recyclability, biocompatibility, and low costs, which makes it an ideal solution for net-zero greenhouse gas emissions and carbon retention (You et al., 2017). Among the multitude of applications of biochar, it is preferred as an adsorbent due to its superior properties such as high specific surface area, high porosity etc. (Wu et al., 2018), making it used in applications such as wastewater treatment, soil decontamination, and air pollutants removal thereby providing an effective way to utilize biomass waste resources (Ji et al., 2022). Generally, the higher the temperature at which organic material is pyrolyzed, the higher the surface area of the resulting biochar, although there is evidence that some biochar pores collapse and decrease surface area at very high pyrolysis temperatures (Lua et al., 2004). Several efforts are available on the preparation of biochar from different biomass as well as their applications, which warrant a comprehensive summarization of the available literature on biochar preparation.

Therefore, this chapter aims to provide a comprehensive summary of (1) types of biomass feedstock suitable for biochar production; (2) various methods (pyrolysis process, hydrothermal carbonization (HTC), and gasification) for producing biochar; (3) physicochemical properties and functional characterization of biochar; (4) application of biochar in various areas; and (5) summary and outlook for further research directions.

2.2 Types of biomass feedstocks for biochar preparation

2.2.1 Agricultural, forest, and aquatic biomass

Several biomasses and biowaste materials are readily available as feedstock for the production of biochar. Principally, the abundant materials, which are generally regarded as waste and are obtained at a low or no cost, are considered for biochar production (Ravindran et al., 2019). These include agricultural wastes such as waste wood, peanut shells, hazelnut, rice hush, and wheat straw (Kumar et al., 2023), forestry residues, animal manure, sewage sludge, organic-food wastes, and MSW (Al-Rumaihi et al., 2022), as shown in Fig. 2.1 to produce biochar via thermal processes within reaction temperature of 300°C–900°C in an oxygen-free environment. These feedstock types have low to no economic value and generally do not compete with the food crops for land requirements; they are good candidates for biochar production.

In addition, purpose-grown biomasses are available, and these include switchgrass and miscanthus, which produce moderately high yield and energy content and usually require very low maintenance relative to other grown crops. These crops (switchgrass and miscanthus) have low moisture content, approximately < 10%, at the time of harvesting,

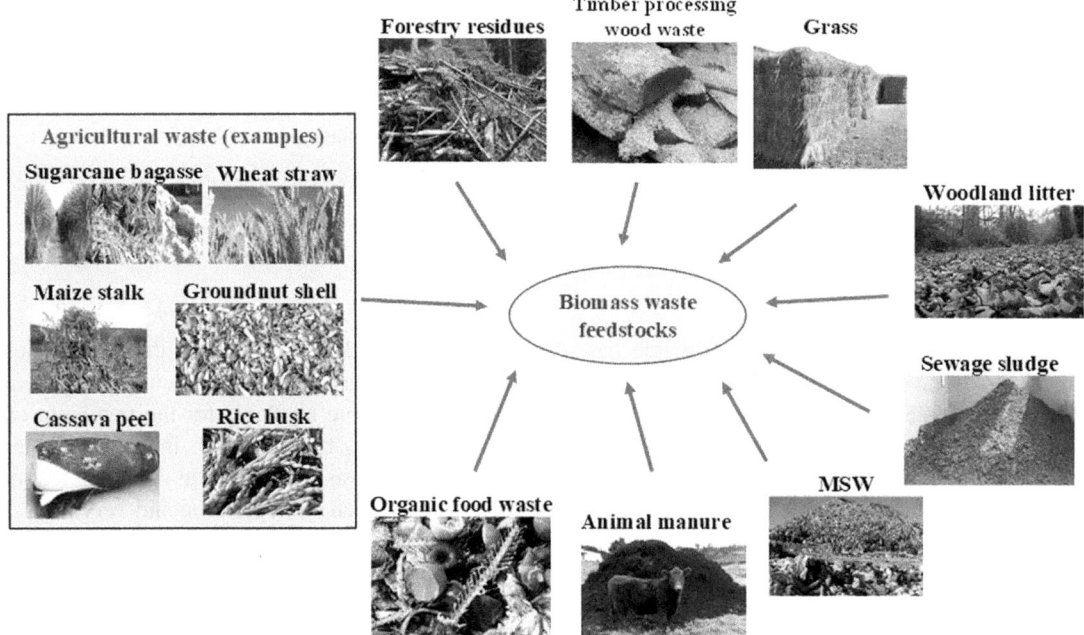

FIGURE 2.1 Different types of biomass waste for biochar production. Other biomass often missed but are suitable for biochar production include green waste, aquacultural waste, invasive plants, processing biowaste, micro/macro-algae, and macrophytes.

which eliminates the drying process of the biomass. Nevertheless, the harvesting time significantly affects the amount of ash that can be produced from biomass, which negatively impacts its combustion behavior; hence, it is much more focused on sustainable energy production. Biomass from plant wastes, as well as their components such as cellulose, hemicellulose, and lignin, are included in agricultural waste biomass that can be differentiated into paper waste, wood residues, and grasses (Mukhopadhyay et al., 2022) and can yield a different amount of biochar, as shown in Table 2.1.

2.2.2 Plastics

Aside from the biogenic waste, polymeric (plastic) materials, including polypropylene and polyester, are also considered as another feedstock that can be used to produce biochar due to the large carbon content contained in the hydrocarbon chain (Al-Rumaihi et al., 2022). Multiple researchers have investigated the potential of polyethylene terephthalate (PET) in pyrolysis. However, the acidic nature of pyrolysis products is unfavorable because of their corrosive nature, which reduces the quality of the fuel produced. Additionally, sublimation of benzoic acid would clog the heat exchanger and the piping, thereby increasing costs during industrial application (Çepelioğullar & Pütün, 2013).

As with PET, high-density polyethylene (HDPE), low-density polyethylene (LDPE), polypropylene (PP), and polystyrene (PS) have also been widely studied in biochar production. The co-pyrolysis of PP with rice bran shows coke or char formation occurred at higher temperatures (Akancha et al., 2019). A 2016 study by Chattopadhyay et al. (2016) examined the pyrolysis of PP at a greater temperature of 380 °C, and the bio-oil was 80% of the total product yield, while the char and gaseous products were 13.3 wt% and 6.6 wt%, respectively. FakhrHoseini & Dastanian (2013) achieved a lower char yield of 8.98 wt% when PP was pyrolyzed at 500 °C. Increasing temperature above 500 °C has been reported to result in a reduced char and liquid oil yield while promoting the gaseous products. The negative impact of temperature increase on char yield was also observed by Demirbaş (2005) during the pyrolysis of PP at 740 °C using a batch reactor, which resulted in 1.6 wt% char, 49.6 wt% gaseous materials and 48.8 wt% of liquid oil. Miandad et al. (2017) examined the effect of reaction time and temperature on the quality and yield amount of char and liquid oil from PP pyrolysis. The maximum yield of

TABLE 2.1 Types of biomass used and the amount of char produced and other properties.

Biomass feedstock	Pyrolysis Temp (°C)	C	N	H	O	biochar Yield (%)	Ash (%)	Gas (%)	References
Algae matter	300–700	50.4	10.6	7.54	30.7	40–90	4.8	–	Chang et al., (2015)
Bamboo	500–900	54.49	0.19	6.15	37.1	–	–	–	Tong et al. (2020)
Corn stover	600	70.5	–	–	–	–	16.6	23.6	Lehmann et al. (2011)
Animal dung	800	27.78	1.67	3.98	20.3	53.1	34	13	Zhou et al. (2020)
Pinewood	400	74.1	0.06	4.95	20.9	35.3	1.5	36.4	Keiluweit et al. (2010)
Wheat straw	500	62.9	–	–	–	29.8	18.0	17.6	Zhao et al. (2014)
Walnut shell	900	55.3	0.47	0.89	1.6	–	40.4	–	Mukome et al. (2013)
Turkey litter	700–800	15.6	0.78	0.83	4.4	–	64	–	Mukome et al. (2013)
Timothy grass	450	67.5	1.9	2.3	28.2	43	3.5	7.5	Mohanty et al. (2013)
Sugarcane bagasse	600	76.5	3.03	2.93	19.8	–	–	–	Kuan et al. (2013)
Rice husk	400	37.2	1.3	1.2	12.4	–	47.9	38.2	Kameyama et al. (2016)

bio-oil, gas, and char was 76 wt%, 8 wt%, and 16 wt%, respectively, at a temperature of 400 °C and a residence time of 75 min. However, when the temperature was raised to 450 °C, the gaseous yield was 13 wt%, while the char yield was reduced to 6.2 wt%, and the liquid yield increased to 80.8 wt%. Recently, Elnour et al. (2019) produced a large amount of char, about 78 wt% when PP was pyrolyzed at 300°C. This implies that PP is a suitable material for char production at relatively low operating temperatures ranging from 380°C to 500°C. Researchers have also examined the use of PS as potential waste material for the pyrolysis process. Unlike other plastic waste, pyrolysis of PVC has been limited because of the potential production of hydrogen chloride (HCl), which is a highly corrosive gas. Table 2.2 summarizes the amount of char produced from various plastic materials and different operating conditions.

2.3 Biomass quantification approaches

Although large quantities of biomass are generated in different countries and regions, the robust data on various feedstock quantities is often not available at national and regional levels, making it difficult to assess biochar. However, some alternative methods and approaches can be implemented to quantify the amount of biomass available for biochar and bioenergy production for a particular country or region, as discussed in this section.

2.3.1 Agricultural waste

Agricultural residues are used for many different purposes, including cattle feed, animal bedding, and cooking fuel, as well as soil conditioning (Wenga et al., 2023). To quantify the residues available for biochar and bioenergy production, firstly, the gross crop residues (GCR) should be determined using Eq. (2.1):

$$GCR = \sum_{i=1}^{n} A_i \times Y_i \times RPR_i \tag{2.1}$$

here, GCR represents the gross crop residue potential from n number of available crops (tons); A_i represents the land area cultivated for a particular ith crop in ha; Y_i designates the total amount of produce of ith crop in tons/ha; while RPR_i signifies the residue production ratio of the ith crop. It should be noted that yield and RPR vary spatially due to variations in weather conditions, soil type, crop variety, and even agricultural practices within the country and should be taken into account when quantifying the GCR. After quantifying the GCR, all the quantities used in other competing uses should be removed such that the surplus amount of crop residues is the amount available for biochar and bioenergy production (Hiloidhari et al., 2014; Wenga et al., 2023). The surplus crop residue (SCR) is then calculated using Eq. (2.2):

$$SCR = \sum_{i=1}^{n} GCR_i \times SF_i \tag{2.2}$$

TABLE 2.2 Pyrolysis of various types of plastics to produce biochar.

Plastic type	Temp (°C)	Char (%wt)	Liquid (% wt)	Gas (% wt)	References
PET	500	8.98	52.13	39.89	FakhrHoseini & Dastanian (2013)
HDPE	300–400	33.05–0.54	80.88	–	Ahmad et al. (2015)
HDPE	600	34.7	18.1	28.9	Zhang et al. (2019)
PP	300	78.8	21.8	7.2	Elnour et al. (2019)
PVC	500–700	–	29.65–0.38	–	Dai et al. (2018)
LDPE	350–600	12.7	74.4	36.8	Dewangan et al. (2016)
PP	380	13.3	80.1	6.6	Chattopadhyay et al. (2016)
PP	740	1.6	48.8	49.6	Demirbaş (2005)
PP	400	16	76	8	Miandad et al. (2017)

here, SCR represents the surplus amount of crop residues produced from the total number (*n*) grown crops (tons); GCR*i* represents the GCR for *i*th cultivated crops (tons); while SF$_i$ represents the surplus residue fraction of *i*th crop. It should be noted that the methods of harvesting and threshing affect the SCRs. For example, the manual method of harvesting leads to different heights of cutting crops and therefore affects the quantities available as surplus (Hiloidhari et al., 2014).

2.3.2 Municipal solid waste

The quantity of municipal solid waste (MSW) generated can be quantified by weighing truck loads collecting and transporting MSW to the dumping site or landfills. Official data can then be collected from the local authorities. In the absence of a proper waste management system where MSW is not properly collected and quantified, the quantities of MSW generated by a particular country can be estimated using Eq. (2.3):

$$\text{MSW} = \text{Per capita waste generation}(kg/d) \times \text{Total population} \times \% \text{ waste collected} \quad (2.3)$$

Like crop residues, MSW has competing uses, and the percentage of waste not collected is considered to have been used in other competing uses such as composting, pit dumping, and even open dumping (Shonhiwa, 2013).

2.3.3 Animal manure

The quantities of dung produced by animals are affected by various factors, including animal body weight, physiological state, and feed type and quality. Moreover, the accessibility of dung plays a crucial role in the overall amount of the dung. If livestock freely range in the bush, the dung is not readily available. Therefore the total quantities of dung generated and available for biochar production can be quantified using Eq. (2.4):

$$\text{Dung production (t/d)} = \text{Per animal dung production (kg/d)} \times \text{Animal population} \times \text{Dung collection efficiency} \quad (2.4)$$

Some animal dung can be utilized for enhancing soil fertility in agricultural lands, whereas others are just left in pastures. As such, the competing uses should be quantified and subtracted from the total dung collected for biochar production (Wenga et al., 2023).

2.3.4 Municipal sewage sludge

Sewage sludge can be quantified by measuring the dry solids removed from wastewater treatment plants. The volume of municipal sewage sludge (MSS) produced from wastewater treatment plants varies due to the sludge treatment process. Primary and secondary treatment processes generate approximately 940 grams of dry solids/3.79 m^3 of MSS that is treated. The dry solids from sewage works can be estimated using Eq. (2.5) (Wenga et al., 2023):

$$\text{MSS}\left(Drysolids \frac{tons}{day}\right) = \frac{0.94 \, kg \, of \, dry \, solids}{3.79 m^3} \times Amount \, of \, waste \, water \, generated \, per day \quad (2.5)$$

2.4 Preparation of biochar

2.4.1 Pyrolysis systems for biochar production

2.4.1.1 The process of pyrolysis

Pyrolysis is an irrevocable thermochemical process that treats complex solid fuel materials at raised temperatures in an atmosphere that is free from oxygen (Adeniyi et al., 2022). In the course of pyrolysis, the molecules are exposed to elevated temperatures which leads to very high molecular pulsations at which the molecules are shaken and stretched until they begin to break down into smaller molecules (Amutio et al., 2012). The thermal putrefaction and dehydrogenation reactions are subsequently followed by secondary isomerization and polymerization reactions of the primary volatiles. The magnitude of the secondary reactions is influenced generally, by the conditions of pyrolysis as well as the reactor type. Secondary reactions are stimulated by high temperatures and high residence times. Practically, a completely oxygen-free environment is difficult to achieve as some amounts of oxygen are present leading to some oxidation reactions. The quantities of the products are a result of a combination of the primary decomposition of the biomass feedstock and the ensuing secondary reactions of the volatiles leading to the formation of either solid char, noncondensable gases, or condensable pyrolytic liquid oils. The products of pyrolysis are affected by operating conditions, including feedstock, pressure, particle size, reactor type, heating rate, residence time, and temperature (Wang et al., 2014). Pyrolysis process generally takes place at temperatures ranging from 300°C to 800°C. As the temperature changes during the process, the distribution of the products is also altered. Lower temperatures produce a larger percentage of the solid and liquid products whereas higher pyrolysis temperatures produce more gaseous products. Moreover, operating at high pressures leads to the production of greater amounts of biochar and gases, whereas lower operating pressures or vacuum yield large amounts of liquid products. Although the size of the feedstock particle does not greatly influence the product distribution, larger particle size increases the liquid product at higher pyrolysis temperatures while smaller particle sizes favor internal heat transfer within the particles (Setter et al., 2020).

It is important to note that pyrolysis systems exist at various scales from small-scale household units such as cookstoves to medium-scale and advanced automatic systems. These systems provide energy, biochar, and other valuable products while promoting sustainable development and reducing poverty in low-income settings (Yao et al., 2018). Small-scale pyrolysis systems are used in small-scale industrial applications or low-income rural development projects, processing a few hundred to a few thousand kilograms of biomass per day. On the other hand, medium-scale pyrolysis systems are used in medium-scale industrial applications, processing tens of thousands to hundreds of thousands of kilograms of biomass per day. The last is large-scale pyrolysis systems that are used in large industrial applications, processing millions to tens of millions of kilograms of biomass per day (Liao et al., 2018). These scales vary in terms of capacity, complexity, and application, and are used in various industries, including energy production, chemical manufacturing, and waste management.

2.4.1.2 Types of pyrolysis
2.4.1.2.1 Slow pyrolysis

Slow pyrolysis is characterized by heating biomass in an oxygen-restricted or oxygen-deprived environment, with ideal heating rates ranging from 1 to 30°C/min (Lua et al., 2004). Slow pyrolysis requires atmospheric pressure conditions, and the process heat is usually supplied from an outside energy source — commonly from the burning of the produced gases or by partial combustion of the biomass feedstock (Laird et al., 2009). These conditions yield biochar usually up to 30 wt.%, on a dry feedstock weight basis. However, the properties and yield of the resultant biochar are determined by several factors, including the residence time, type of feedstock, heating rate, and pyrolysis temperature. Amongst these parameters, it has been recognized that the peak treatment temperature has the utmost overall effect on the ultimate product physiognomies (Lua et al., 2004).

Biochar prepared from numerous organic and non-organic materials, such as agricultural residue, forest residue, algal biomass, scrap tire, and heavy crude oil, have been utilized extensively as the precursor of biochar through slow pyrolysis (Liao et al., 2018). Biomass undergoes decomposition at a relatively moderate temperature (350°C–500°C) and slow heating rate (0.1°C–0.8°C/s) and has a long residence time (5–30 min or more) which provides sufficient residence time for biomass pyrolysis vapor and increases its secondary cracking level. In addition, the quality of biochar is correlated to its pH value, nutrients, carbon content, porosity, and SSA. Among these, biochar quality is more closely related to its carbon content (Yao et al., 2018). Higher quality biochar (high carbon content) is obtained based on these pyrolysis parameters, including relatively longer residence time, high pyrolysis temperature, and lower heating rate. The content of carbon for biochar obtained from the slow pyrolysis of red cedar wood reached up to 88% at 500°C pyrolysis temperature and 6 °C/min heating rate. Meanwhile, a higher heating value of biochar was realized at 32.95 MJ/kg, indicating a high quality of biochar (Yang et al., 2016). In addition, reducing the rate of heating favors ample heat conduction, which is conducive to the carbon deposition reaction and thus to the increase in the production of biochar (Veses et al., 2015).

2.4.1.2.2 Fast pyrolysis

Fast pyrolysis occurs in elevated temperatures (range of 300°C–700°C) at a faster heating rate (10°C–200°C/s), with fine particle size (<1 mm) of feedstock and with a short residence time (0.5–10 s) (Yao et al., 2018). Consequently, the feedstock putrefies to produce mostly aerosols, vapors, and some char. The production of liquid requires a very low vapor residence time of typically 1 s to minimize secondary reactions, even though acceptable yields can be attained at residence times of up to 5 s if the vapor temperature is retained below 400°C (Yang et al., 2016). After the processes of cooling and condensation, a mobile dark brown liquid is formed which has a heating value about half that of conventional oil fuel. Fast pyrolysis is related to the traditional pyrolysis processes for making char, only that it is an advanced process that is judiciously controlled to give higher yields of liquid. Research has shown that maximum liquid yields are achieved with high heating rates, at reaction temperatures around 500°C, and with short vapor residence times to minimize secondary reactions.

2.4.1.2.3 Flash pyrolysis

Flash pyrolysis is different from other pyrolysis types in that slow pyrolysis involves a slow heating rate that favors secondary reactions and yields quality char, whereas fast pyrolysis favors the production of gases and bio-oil ahead of char, flash pyrolysis occurs at a higher heating rate and temperature coupled with shorter residence time than fast pyrolysis. Flash pyrolysis is characterized by short vapor residence time < 1 s, high temperature and heating rates, and rapid cooling of the pyrolysis vapor coupled with rapid removal of char from the system (Amutio et al., 2012). As a result of the reduction in secondary cracking reactions in flash pyrolysis, the liquid product yield is maximized to about 60–75 wt% (Amutio et al., 2012). Flash pyrolysis produces bio-oil that is different from conventional fossil oil. The bio-oil has a heating value that is almost half of the conventional fossil oil, it is very rich in oxygen-containing functional groups which impacts it with poor properties such as viscosity, high acidity and non-volatility, lower energy density, thermal instability, and tendency to polymerize in the presence of air, and this is dependent on the raw material as well as the pyrolysis process conditions (Adeniyi et al., 2022). These factors mean the resultant oil needs upgrading before it's used as a transportation fuel.

2.4.2 Effects of pyrolysis type/temperature on changes in functional groups

Pyrolysis temperature has significant effects on the properties and functional groups of biochar. Several authors have investigated and examined the effects of pyrolysis temperature/type using various feedstock for biochar production (Liu & Han, 2015; Sun et al., 2014; Wang et al., 2015) by subjecting wood and non-woody biomass to pyrolysis with a varying heating rate between 300°C and 600°C. The biochar produced from both wood and non-woody decreased upon the increase in pyrolysis temperature, as shown in Fig. 2.2. However, non-woody biochar produced from corn cob showed the highest biochar yield than woody-based biochar such as pinewood. Liu & Balasubramanian (2014) explained the observation that the biochar produced at a higher heating rate, shorter residence time, and larger particle size is unfavorable to heat transfer during the pyrolysis process than to those with a lower heating rate, longer residence time, and smaller particle size.

Some authors (Cárdenas-Aguiar et al., 2024; Wang et al., 2021) further analyzed the elemental composition of biochar produced at varying temperatures. The amount of total carbon in biochars showed an increasing trend as pyrolysis temperatures increased. As the temperature increased from 300°C to 500°C, the C content elevated from 60.72% to 72.18% and from 52.49% to 54.45% in *P.australis* and *S.alterniflora* biomass, respectively (Wang et al., 2021). However, the amounts of H and O in biochars reduced with an increase in pyrolysis temperature. Relative to the

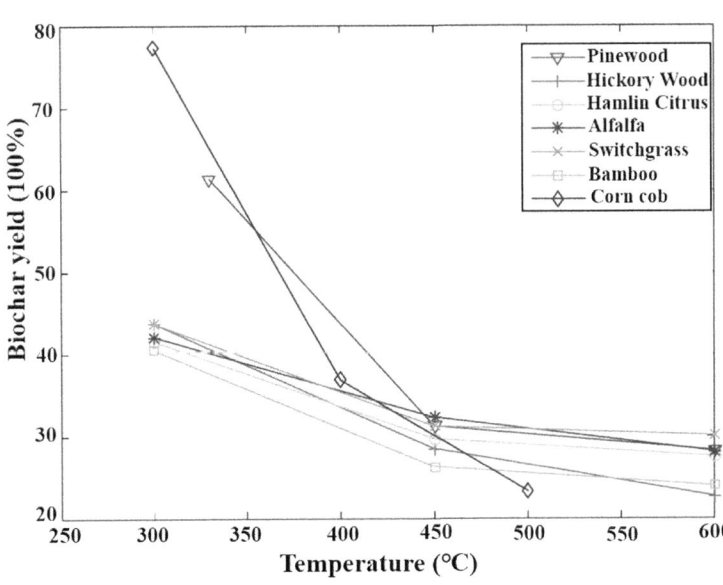

FIGURE 2.2 Effect of pyrolysis temperature on the biomass-derived biochar yields. Pyrolysis temperature affects the amount of biochar produced. *From Jafri, N., Wong, W. Y., Doshi, V., Yoon, L. W. & Cheah, K. H. (2018). A review on production and characterization of biochars for application in direct carbon fuel cells. Process Safety and Environmental Protection, 118, 152–166. https://doi.org/10.1016/j.psep.2018.06.036.*

biochars produced at 300°C, the amount of O in biochars obtained at 500°C reduced by 62.63% in *P. australis* while by 50.17% *S. alterniflora* biomass. The amount of H lowered from 5.03% to 2.95%, and from 4.49% to 2.35% in *P.australis* and *S.alterniflora* biomass, respectively. The decrease in H content as the pyrolysis temperature and residence time increased was attributed to the polymerization and condensation reactions leading to more aromatic structures and a decreased content of aliphatic groups (Cárdenas-Aguiar et al., 2024). N contents in the two biochars increased before decreasing as the temperature rose with the highest N content appearing at 350°C. Another study using different biomasses showed that the N content increased as the temperature and residence time increased (Cárdenas-Aguiar et al., 2024). The amount of C also was found to increase with pyrolysis temperature although it was lower than in other studies (Dumroese et al., 2020). The amount of S in the broom-derived biochars was lower in biochars compared to the amount in the broom feedstock. In contrast, the S content in another feedstock studied, which is gorse-derived biochars, did not show any significant differences between the biochars and the feedstock. This implies that different feedstock has different S content. The (O + N)/C ratio reduced as pyrolysis temperature increased, indicating the loss of nitrogenated and oxygenated functional groups during pyrolysis and the polycondensation reactions that take place during pyrolysis (Cárdenas-Aguiar et al., 2024). A different study confirmed the trend using different feedstock (Wang et al., 2021). The authors observed that as temperature increased from 300°C to 500°C, the (O + N)/C ratio declined from 0.31 to 0.10 and from 0.34 to 0.17 in *P. australis* and *S. alterniflora* biomass, respectively.

O/C and H/C molecular ratios are often analyzed using the van Krevelen diagram (Moiseenko et al., 2021). The H/C molecular ratio represents the aromaticity of biochars, a low H/C ratio indicates that the compounds have a large aromatic structure and high stability. The O/C molecular ratio represents the degree of aging of biochars which is important for estimating the stability of the obtained biochars for environmental remediation (Tu et al., 2022). Fig. 2.3 shows the van Krevelen diagram showing the influence of pyrolysis temperature on the H/C and O/C ratio and biochar stability. From analyzing, it can be observed that the biomass feedstocks have the highest H/C and O/C molecular ratios. Subsequently, due to decarboxylated, demethylation, and dehydration, the O- and H-containing functional groups disappeared gradually, resulting in a decrease in H/C and O/C with the increase in temperature. The O/C molecular ratios for the biochars produced at 300°C resemble those of low-grade sub-bituminous coals and lignite coal while the biochars obtained at 600°C exhibited low H/C and O/C ratios, with H/C values close to those for anthracite (Cárdenas-Aguiar et al., 2024). A lower O/C molecular ratio also reflects the degradation of carbon compounds and has the possibility of carbon sequestration.

According to the literature (Almutairi et al., 2023), biochars with an O/C ratio of <0.2 are regarded as stable, 0.2 and 0.6 are moderately stable, while biochars with an O/C ratio of >0.6 are relatively unstable. Similarly, when biochar has H/C ratios < 0.7, it shows that it has more C aromatic structures and is thermochemically stable relative to biochar having H/C ratios > 0.7 (Spokas, 2010). In addition, if the biochar has a H/C ratio < 0.3, it has been proven that it is effective and has great potential to mitigate N_2O. Biochars produced at pyrolysis temperature above 500 °C are effective in mitigating N_2O emission with a half-life ranging between 100–1000 years after soil application because at that temperature biochars are moderately altered thermochemically (Almutairi et al., 2023).

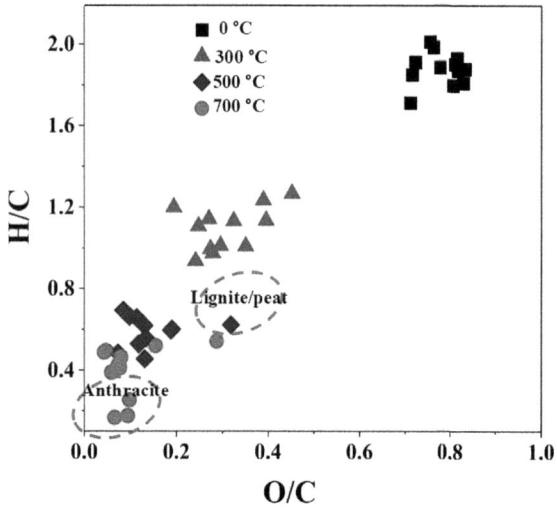

FIGURE 2.3 van Krevelen diagram showing effects of temperature on pyrolysis. Different temperatures result in the pyrolysis products with different functional groups. *From Yan, C., Wang, W., Nie, M., Ding, M., Wang, P., Zhang, H. & Huang, G. (2023). Characterization of copper binding to biochar-derived dissolved organic matter: Effects of pyrolysis temperature and natural wetland plants. Journal of Hazardous Materials, 442, 130076. https://doi.org/10.1016/j.jhazmat.2022.130076.*

2.4.3 Hydrothermal carbonization of biomass

HTC, also referred to as wet pyrolysis, is gaining attention as the most widely used pyrolysis method which is a result of the thermochemical process of lignocellulosic biomass conversion in the presence of water, under moderate temperatures (180°C–260°C) and pressures ranging between 2–10 MPa (Liu & Balasubramanian, 2014). HTC has significant advantages over pyrolysis since it does not require energy-intensive drying procedures or pretreatment, low-temperature operation and it poses high conversion efficiency. Thus, HTC has the capacity to process biomass with high moisture content (e.g., raw sewage sludge contains about 80% moisture) without any pretreatment procedures therefore reducing energy consumption. Of concern, these techniques play a pivotal role in waste management, especially on farms that produce a significant amount of livestock manure. There is a huge environmental risk from leachate emanating from large deposits of organic farm manure as they contain large proportions of phosphorous and nitrogen which could burden nearby surface and underground water, respectively. As such, thermal application processes immobilize nitrogen and phosphorous reducing their impact when applied to soil in a biochar medium for soil remediation. The HTC produces biochar also simply referred to as hydrochars and their properties significantly vary due to feedstock characteristics and conditions (He et al., 2013). Other valuable characteristics affected by the applied feedstock and HTC conditions include carbon content, texture (pore size), and the nutritive quality exhibited in calorific value.

However, both pyrolysis and HTC produce biochar through the conversion of biomass for use as adsorbents, soil growth media, and other functions as construction material (Gwenzi et al., 2017). HTC processes were observed to significantly reduce the leaching of salts from hydrochars compared to leachate observed from raw pig manure, indicating their potential to reduce the solubility of salts and other inorganic elements during thermal treatment (Reza et al., 2013). Studies have highlighted the availability of higher concentrations of trace elements (Cd, Pb, Cr, Ni, Zn, Fe, and Cu) in biochar and hydrochar compared to pig, cattle, and poultry manure. This is attributed to metal concentration in the ashes as the temperature rises and further, trace elements are revealed to migrate to the liquids during the HTC process (Huang & Xing-zhong, 2016). Hydrochars produced at 220°C and 240°C from pig manure showed almost 95% high field capacity, with water potential retained at over 24% higher than pyrolysis processes. These hydro-physical properties, including the macroporosity nature obtained between 220°C and 240°C during the HTC process of hydrochar, explicitly indicate its capacity to support plant growth when applied as a soil growing media.

An increase in HTC temperature was observed to increase carbon content (C-C bonds) and energy density, with oxygen observed to be decreasing at alarming rates (Liu et al., 2013). As such these changes are recommended since they improve the combustion properties of HTC-derived biochar. Another important property of this HTC-derived biochar is the low ratios of O-C and H-C bonds due to temperature increases. Biochar with such properties is desirable due to its little energy loss and minimized smoke production during combustion. Additionally, the HTC can significantly reduce volatile matter, ash, and oxygen in combustible carbon, which is a potential indicator of the reduced release of inorganic vapors during combustion of the biochar compared to coconut and eucalyptus biomass. Due to minimal operational costs and residence time HTC processes can significantly be a panacea on waste-to-energy production, with further advantages of bio-oil production at high temperatures (Nizamuddin et al., 2017), as well soil growth media which are in solid phase at low temperatures.

2.4.4 Gasification of biomass

Gasification is a high-temperature thermochemical process for converting biomass materials into products such as solid residue biochar, syngas, and a liquid product (Iwuozor et al., 2023). The process occurs at high temperatures and comprises partially oxidizing the biomass material using various gasifying agents such as oxygen, air, or steam. This thermochemical process involves other sub-processes such as pyrolysis, cracking, reduction, and combustion. It is influenced by a number of significant parameters, including heating rate, temperature, reaction time, energy yield, and mass yield. These parameters have a great effect on the quality and efficiency of the gasification, which in turn affect the conversion efficiency, composition of the syngas that are generated, and the production of biochar and tars. As a result, careful selection, control, and management of these influencing parameters are fundamental for the optimization of the gasification process for the production of biochar (Alptekin & Celiktas, 2022).

The preparation of biochar through gasification involves a number of steps illustrated in Fig. 2.4 which starts with the pretreatment of biomass. The pretreatment stage involves processes such as drying, sorting, and grinding of the organic material. The purpose is to lower the moisture content, eliminate impurities, and increase the surface area of the biomass feedstock for improved efficiency of the gasification process. After the pretreatment stage, the ensuing step is the treatment stage. This step encompasses the actual process of gasification, in which the biomass materials undergo thermal

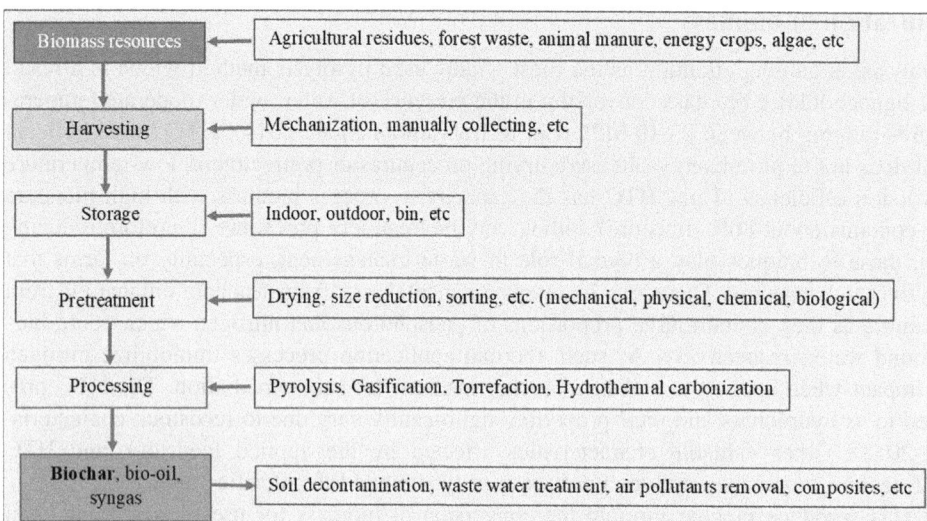

FIGURE 2.4 Steps involved in the preparation of biochar. Biochar preparation methods involve various steps starting from the harvesting to the biochar production.

decomposition in the gasifying reactor producing syngas, biochar, and other by-products. Gasification process can be conducted using various types of gasification reactors, for example, fixed-bed, fluidized-bed, and entrained-flow reactors. Aside from physical parameters that can influence the gasification process, gasification can also be affected by the gasifying agent e.g., steam, air, or CO_2 that are employed and can alter the qualities of the produced biochar. The produced biochar can be utilized in a wide range of applications owing to its enviable characteristics (Iwuozor et al., 2023).

The potential of preparing biochar from various biomass feedstocks, for example, Delonix regia pod, via gasification technique has been studied by Emenike et al. (2022) who evaluated the effectiveness of their biochar produced from gasification in the removal of phenol from water. Biochar was prepared using a retort-heated gasifier at 375 °C and tested for its ability to remove phenol in batch adsorption studies under varying conditions. The results indicated that at a biochar dose of 13 g/L for 4 h, and a solution pH of 7, maximum phenol elimination in equilibrium experiments was obtained.

2.4.5 Torrefaction of biomass

Torrefaction is a low-temperature biomass thermal decomposition process that is also used in the preparation and production of carbon-rich biochars. It is a pretreatment method for upgrading biomass for further use instead of direct use in its raw form. Biomass partly decomposes, during this process as water and volatile compounds escape from the compound, generating both condensable and noncondensable gases (Nakason et al., 2019). After torrefaction, the resulting product is a solid substance rich in carbon, referred to as biochar, "torrefied biomass" or "biocarbon." The torrefaction process has been seen to improve several biochar properties which include lower moisture content, good hydrophobicity, improved grind ability, higher energy density, mass density, and heating value, also reducing oxygen-to-carbon ratio compared to untorrefied biochars. Temperature and retention time are two main parameters that influence torrefaction process efficiency. It is usually conducted at temperatures between 200–300°C and the process temperature is maintained for 15–60 min. There are two main methods of torrefaction which are conventional and microwave torrefaction. Conventional torrefaction is practiced in a tube furnace with a quartz tube powered by electricity. The reaction temperature can be controlled precisely but a longer duration is required. Microwave torrefaction involves heating assisted by electromagnetic irradiation creating an extremely fast rating of the biomass interior, in a matter of seconds, and has a highly tunable heating rate (Hanoğlu et al., 2019). Microwave heating has some advantages over conventional heating in the torrefaction process such as rapid, selective, and non-contact heating, abrupt start and stop automation, and a high level of safety (Pulka et al., 2019).

2.5 Biochar physicochemical properties and characterization techniques

2.5.1 Biochar physicochemical characterization

A study by Xu et al. (2012) indicated that pH in biochar can be measured using a supernatant of aqueous solution with a solid-to-water ratio of 1:5 which will then be assayed using a digital pH meter. Other studies by Brewer et al. (2012)

have also shown that the pH of biochar can be measured using 24-h equilibrated solution of biochar and deionized water with a solid/liquid ratio of 1:20 (w/v), and then assayed using a pH meter. Other methods of pH determination involve the use of twenty grams of biochar mixed with deionized water for approximately 2 h. The resulting solution will then be filtered two times, and then measure pH level using a pH meter (Prens et al., 2023). Electrical conductivity in biochars can be measured using a conductivity meter in a supernatant of solution (solid–water ratio of 1:5) shaken at 200 rpm for 5 min and filtered.

The elemental analyzer can be employed to determine concentrations of metals like heavy metals, rare earth elements, and other trace elements in biochar. The bioavailable DTPA-extractable trace nutrient elements can be measured using a buffered DTPA solution, such as 0.005 mol L^{-1} DTPA, 0.01 mol L^{-1} $CaCl_2$, and 0.1 mol L^{-1} $C_6H_{15}NO_3$, according to Chinese Standard NY/T 890−2004 method (Yuan et al., 2015). Heavy metals like Fe, Cr, Mn, Cu, Ni, and Pb can be measured using ICP-AES or AAS (Zhao et al., 2013). The chemical oxygen demand can be measured using the potassium dichromate method, while the total carbon and total organic carbon (TOC) can be determined using a TOC analyzer. On the other hand, Fourier transform infrared (FTIR) spectrophotometry can be used to confirm the existence of several kinds of functional groups on the surfaces of biochars. In this case, about 1 g of biochar is mixed with a few grams of KBr powder pressed to form a KBr pellet, and put in a transparent disk. The disk will then be heated at 110°C in an electrical oven to evaporate moisture content and then taken through FTIR spectroscopy. The main inorganic elements of biochars can be measured by acid digestion of the samples followed by inductively-coupled plasma atomic emission spectroscopy analysis. Nitrate and ammonium nitrogen concentrations in biochar types are measured using the indophenol blue method where their concentrations are determined in 2.0 mol L^{-1} KCl (1:5) by UV–vis spectrophotometer, according to the indophenol blue method (Yuan et al., 2015).

The SSA and pore size distribution of biochars can be determined using N_2 sorption isotherms run on an automated surface area and pore size analyzer. The SSAs and pore size distribution are then taken from adsorption isotherms using the Brunauer, Emmett, and Teller equation. SSAs can also be obtained from nitrogen adsorption-desorption isotherms measured at 77 K on a Quantachrome Autosorb-1 analyzer, with all samples outgassed at 200°C prior to analysis for a minimum of 8 h (Lee et al., 2013). Other instruments like the Brunauer−Emmett−Teller (BET) porosimetry can also be used to measure the SSA and pore size distribution of biochars (Feng et al., 2023). A BET isotherm model is used to calculate the surface area of the solid material by physically adsorbing inert gas, most commonly utilizing nitrogen on the solid surface. As the gas is used to calculate the surface area of material, the most crucial factor affecting the gas is temperature. Therefore, the whole experiment is performed at a constant temperature. The BET results also give information about pore size and pore volume. Phase structures of biochar composites can be determined using X-ray diffraction (XRD) and the Debye-Scherrer formula can then be used for calculating the average grain size (Feng et al., 2023). The porosity of biochars can be approximately evaluated based on the common definition of porosity of solid materials total volume of the generated voids divided by the volume of its total mass (Wibawa et al., 2023). Moisture content, volatile matter, and fixed carbon can be determined by heating biochar at 100°C, 500°C, and 900°C, respectively to determine mass loss due to the release of moisture content, volatile matter, and fixed carbon, respectively (Prens et al., 2023). The structural properties of biochar are investigated using Raman spectroscopy and to assess functional groups present, graphite structures, and the amorphous character of biochar, researchers use Raman spectroscopy.

The cation exchange capacity (CEC) of biochars can be determined according to a modified barium chloride compulsive exchange 81 method (Yang et al., 2018). Following the initial CEC determination for barium loading at pH 8.5, the suspensions have to be adjusted to lower pH by the addition of 0.010 M H_2SO_4 and the CEC determined again. To assess the crystalline carbon structures, Raman spectroscopy analysis is employed to examine the carbon nanostructures of biochar. Using the ratio of D and G band intensities (ID/IG) of carbon nanostructures in biochar, the in-plane crystallite size distribution (La) can be calculated (Jorio et al., 2012).

2.5.2 Biochar physicochemical properties

Biochar is generally considered to be a pyrogenous, carbon-rich, and highly porous material synthesized by processes such as pyrolysis, HTC, and gasification of biomass feedstocks or related organic wastes. Studies note that the physicochemical and elemental properties of biochar are mainly influenced by feedstock composition and technological parameters such as temperature, heating rate, particle size, and residence time. These variables yield biochars with different characteristics ranging from SSA, pH, pore volume, volatile matter, and carbon to ash content. Due to its organic origins, biochars are mainly composed of carbon, nitrogen, sulfur, hydrogen, oxygen, and ash.

2.5.2.1 Chemical properties

Chemical properties of biochars are chiefly influenced by the temperature and heating rate utilized in their synthesis. Studies report that there is a positive correlation between temperature and pH of the resultant biochar (Pariyar et al., 2020). Yuan et al. (2011) attributed the relationship between temperature and pH to the subsequent separation of organic materials from alkali salts owing to the increased synthesis temperature. Mukome et al. (2013) show that biochar pH correlated best with O content ($R^2 = 0.7$), corroborating previous findings that biochar basicity resulted from oxygen-rich functional groups such as γ-pyrone-type, chromene, diketone, or quinine groups (Montes-Morán et al., 2004). Additionally, studies have established a positive relationship between synthesis temperature and carbon content accompanied by a negative relationship with oxygen and hydrogen content (Tu et al., 2022). Research by Demirbas (2004) postulated that the decrease in oxygen and hydrogen content could be attributed to the cracking of weak molecular bonds. The study highlighted the importance of temperature in the overall chemical properties of biochars from different feedstocks.

Pariyar et al. (2020) having investigated the physicochemical properties of different biochars reported that the findings were consistent with prior studies. CEC decreased with increasing temperature. Increased temperature was established to yield loss in the surface organic functional group responsible for CEC in biochar. Additionally, the relationship between temperature and CEC has been explained by the oxidation of aromatic C occurring at increased temperatures as well as the formation of carboxyl groups (Liang et al., 2006). Biochar chemical, mineralogical, and surface morphology are functions of pyrolysis temperature owing to the liberation of volatile compounds and the formation of transitional melts. Biochars synthesized from feedstocks such as sewage sludge, chicken manure, and crop straws have demonstrated reduced available metal content (Yang et al., 2018). This reduction in metal concentrations has been attributed to mechanisms such as precipitation, ion exchange, electrostatic attraction, and surface binding of functional groups. Additionally, biochars derived from biosolids are noted to exhibit neutral to alkaline properties which are influenced largely by synthesis temperature.

2.5.2.2 Physical properties

The physical properties of biochar are recognized as being very important for their various applications. Studies suggest that these properties are functions of operating parameters such as heating rate, pressure, reaction vessel regime, pretreatment, treatment temperature, flow rates of inputs, and resident time (Jien, 2019). Biochar surface properties such as charge density, surface area, pore structure, and distribution are imperative determinants of microbial activity, nutrients, and retention capacity of water. Parameters such as hydrophilic capacity have also been observed to be largely influenced by the highest operating temperature. Additionally, interplanar distance has been observed to decrease as ordering and molecular organization increase, resulting in a larger surface area/volume. Increased surface area and micropores yielded from increased temperatures are recognized as potentially important for water retention (Jien, 2019). Increased pyrolysis temperature has been noted to affect surface area and porosity alterations attributed to organic matter decomposition leading to micropore formation (Tomczyk et al., 2020). The same study postulated that the combined effect of ester groups and aliphatic alkyl destruction coupled with exposure of lignin core may also contribute to increased surface area. Fig. 2.5 shows an example of the surface structure of biochar produced from rice husk and wood at different temperatures. It can be noted that increasing temperature from 400 °C to 600 °C reduced the pore size of both biochar types with a significant effect observed on wood biomass. The reduction in pore sizes of the biochar increases the SSA which in turn improves the adsorption efficiency.

2.6 Functional characterization

2.6.1 Analytical methods

A number of techniques have been used to determine the characteristics of biochar. These include classical methods, instrumental techniques, and, more recently, artificial intelligence (AI) techniques. On the one hand, classical methods include the determination of surface charge using the pH-drift method, CEC, ash content, and oxygenated groups (Dai et al., 2021; Xiong et al., 2022). The surface charge is influenced by the surface functional groups, and it determines the interactions between the biochar matrix and fugitive molecules. Besides, the surface charge is pH dependent, and can be determined by either the classical pH-drift method (Li et al., 2023), or by using a zeta-sizer instrument (Li et al., 2022). Thus, the pH of the biochar or polluted medium can be tailored to generate a conducive surface charge. Introducing metal oxides such as TiO_2 or CeO_2 into the biochar matrix can form defects that generate regions of low

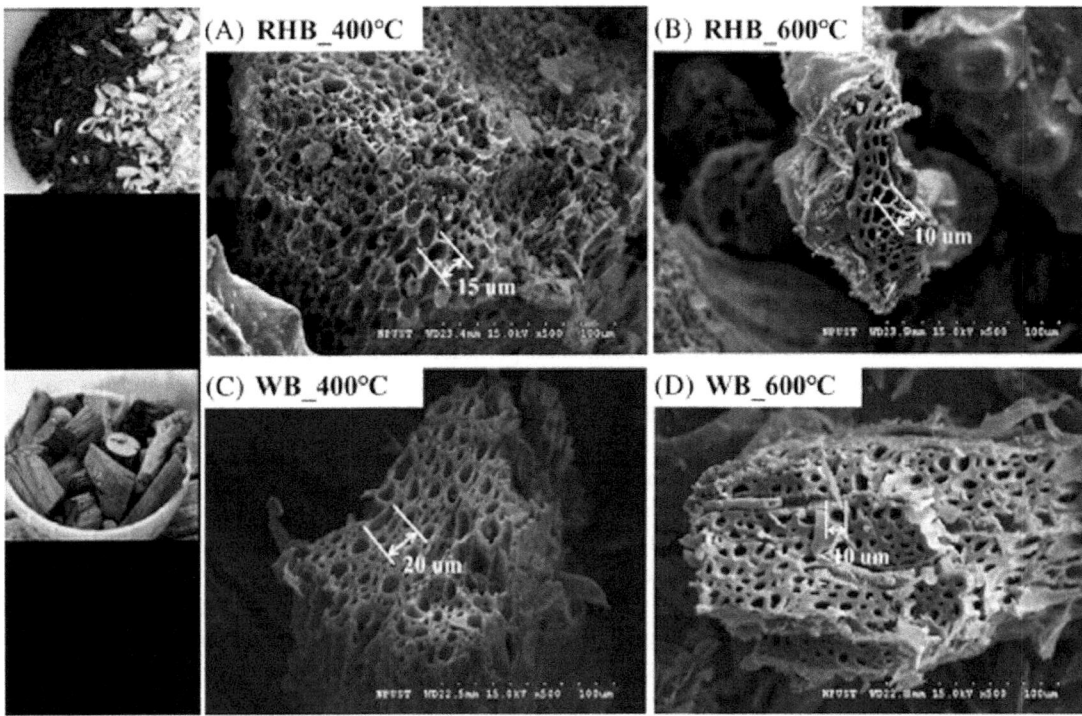

FIGURE 2.5 Characteristics of surface structures of rice husk biochar (RHB) and wood biochar (WB) synthesized at different temperatures. Increasing the temperature from 400 °C to 600 °C reduced the pore size of both biochar. *From Jien, S. H. (2019). Physical characteristics of biochars and their effects on soil physical properties (pp. 21−35). Elsevier BV. https://doi.org/10.1016/b978-0-12-811729-3.00002-9*

and high charge density, thus altering the electron distribution (Guo et al., 2023). Such structural defects can be analyzed using Raman spectroscopy. For instance, Raman spectra data showed that the presence of CeO_2 nanoparticles in biochar derived from Jerusalem artichoke increased the oxidation and disorderliness of biochar graphitic sheets (Wang et al., 2022).

On the other hand, instrumental analytical techniques such as FTIR spectroscopy, XRD, electron microscopy, BET, and X-ray photoelectron spectroscopy have been used to determine the surface functional groups, crystallinity, morphology, textural properties, and the elemental composition of biochar, respectively (Li et al., 2022; Li et al., 2023). While thermal stability can be determined using thermal gravimetric analysis and differential scanning calorimetric techniques, deeper insights into the functional groups can be obtained using the thermogravimetric analysis coupled to FTIR spectroscopy technique (Wang et al., 2022).

Surface functional groups may undergo a transformation as they interact with guest molecules during the application of biochar. These transformations can be determined using characterization techniques, allowing the kinetics and thermodynamics of the performance of the biochar to be followed by either taking measurements before and after application or making real-time measurements during the course of performance. Measurements can also be taken during the preparation stage to determine the kinetics of the pyrolysis process, and hence reactor design parameters. Ultimately, the data can be used to get insights into possible mechanisms involved in pyrolysis, pollutant removal, and catalytic processes. Such knowledge can then be used to design preparation processes tailored for particular applications.

2.6.2 Artificial intelligence

Until recently, AI has not been used in studying the preparation, characteristics, and application of materials, not least for biochar. The technique itself is emerging and largely confined to data analysis in predicting weather patterns, occurrence of diseases, and recognition of images. However, AI is rapidly evolving, and its application in materials design, preparation, characterization, and application is gaining prominence (Chen et al., 2023; Chung et al., 2023). Large amounts of historical data can be analyzed and used to predict possible preparation pathways, biochar characteristics, and application outcomes. This has revolutionized the optimization of pyrolysis and conditions suitable for application. For instance, the characteristics of hydrothermal biochar were predicted using machine learning models (Chen et al., 2023), while biochar yield was predicted from the characteristics of biomass (Ma et al., 2023). In general, the use of AI presents new possibilities for rapid screening of methods and ultimately reduces design time, manual dexterity, and costs associated with biochar preparation and application optimization. In view of this, future studies should invest in the use of AI in the design, preparation, characterization, and application of biochar.

2.7 Applications of biochar

2.7.1 Agriculture

Biochar has been established and confirmed to have significant potential for improving soil health for enhancing agricultural productivity. The application of biochar in farming activities has been due to its ability to increase nutrient retaining, improve soil condition, and decrease environmental pollution (Adeniyi, Abdulkareem, et al., 2023). Amongst the benefits of biochar in agriculture, the improvement of soil fertility and nutrient availability are the most important ones. This is due to the fact that agricultural productivity is completely reliant on soil fertility and nutrient availability. The soil fertility and nutrient availability is aided by the porous structure of biochar which provides a suitable habitat for soil microorganisms that play a fundamental role in nutrient cycling. In addition, the high SSA and the CEC of biochar allow it to hold and gradually/slowly release nutrients such as phosphorus (P), nitrogen (N), and potassium (K) into the soil, henceforth minimizing the need and application of synthetic fertilizers. Pouangam Ngalani et al. (2023) evaluated the influence of biochar produced from cocoa pods on the chemical properties of soil. The authors found that the addition of biochar to the acidic soil increased the soil's pH, electrical conductivity, and available phosphorus.

Biochar can also improve soil water holding capacity and hence reduce irrigation requirements, particularly in semiarid regions. The porous structure of biochar assists in the creation of a more stable soil structure which increases the water-retention capacity; thus, water runoff is reduced, and plant growth is enhanced. Bahrun et al. (2018) studied the effect of biochar prepared from cocoa pods on the growth of cocoa seedlings. The authors observed that the use of biochar reduced the watering frequency while the growth of the cocoa seedlings was enhanced. Another advantage of biomass-based biochar is its ability to sequester carbon and therefore mitigate climate change. The stability of biochar makes it remain in the soil for more than hundreds of years, locking up carbon while reducing greenhouse gas emissions. Nevertheless, biochar can fragment in the soil, enabling higher surface adsorption, and it becomes more recalcitrant (Meng et al., 2014).

2.7.2 Composite development

Biochar has also demonstrated potential application in the composite's development. Adding biochar to polymers such as thermosetting and thermoplastics to produce biochar-polymer composites improves their mechanical and thermal strength (Adeniyi, Emenike, et al., 2023). One of the primary merits of employing biochar as a filler in composites is its low cost compared to other traditional fillers. Apart from that, biochar has a higher SSA and a lower density, which helps to improve the properties of the composites. Moreover, the properties of the biochar-polymer composites can be enhanced by using different quantities and sizes of the biochar particles.

Several researchers have studied the application of biomass-based biochar in the development of composites. Adeniyi et al. (2022) investigate the influence of biochar prepared from Delonix regia pod as a filler in a plastic-based composite. The author found that there was a good interaction between the polystyrene matrix and biochar. The same group of authors evaluated the morphological and mechanical properties of composites produced from flamboyant seed pod-based biochar (Adeniyi, Emenike, et al., 2023). The addition of biochar to polystyrene resin increased the hardiness of the composite. Aside from the improvement of mechanical and thermal properties, the application of biochar in composites provides an environmentally friendly solution for waste management as agricultural waste can be used in a more sustainable way (Adeniyi, Iwuozor, et al., 2023; Pouangam Ngalani et al., 2023).

2.7.3 Environmental remediation

Due to its unique properties such as large surface area, high porosity, and ability to adsorb various contaminants, biochar has been progressively explored in the remediation of various environmental pollution and contaminants. Biochar can be used in the reclamation of contaminated soils by adsorbing organic pollutants, heavy metals, and nutrients. The high SSA of biochar enables a greater adsorption capacity, whereas the porous structure promotes the growth of beneficial microbes that help break down contaminants (Ohale et al., 2023). Research has demonstrated that adding biochar to contaminated soil significantly lowers the bioavailability of heavy metals e.g., cadmium, lead, and arsenic, as well as improves soil fertility. Ogunkunle et al. (2021) evaluated the potential of biochar prepared from cocoa in the immobilization of cadmium in soil. Biochar soil amendment reduced the bioavailability of cadmium in the soil. A similar finding was also observed by Pinzon-Nuñez et al. (2022) where an adsorption capacity of 21.58 mg/g for cadmium from the polluted soil was obtained.

Apart from soil decontamination, biochar can be utilized in the treatment of water contaminated with heavy metals, organic pollutants, and/or pathogens. Due to the adsorption capacities of biochar, it can remove a wide variety of contaminants from water, whereas its pores provide a perfect habitat for harboring useful microbes that can degrade the pollutants. Moreover, the application of biochar in the treatment of water is cost-effective compared to traditional water treatment techniques, such as chemical coagulation and chlorination. Abbey et al. (2023) investigated the removal of mercury, lead, and cadmium from aqueous media using biochar prepared from cacao plant. The removal efficiency of > 99% for all three heavy metals was observed. Mohamed & Mahmoud (2020) applied nanosorbent biochar synthesized from Pisum sativum in the removal of naproxen and chromium (VI) ions from aqueous solution. A maximum adsorption capacity of 309 mg/g and 420 mg/g, respectively, was realized. Córdova et al. (2020) employed biochar made from cocoa for the removal of different types of dyes, such as brilliant green, rhodamine B, and methyl orange, from textile wastewater. The maximum adsorption capacity of 147 mg/g, 3450 mg/g, and 666 mg/g, respectively, were observed, which showed the usefulness of biochar in the decontamination of wastewater. Meanwhile, active research is currently underway on the application of biochar in the removal of air pollutants. The summary of the interrelationship between feedstock, biochar preparation, properties, and applications is illustrated in Fig. 2.6.

2.8 Summary and outlook

This chapter has explored a variety of feedstock that can be used for the fabrication of biochar for a range of applications. These include agro-waste, forestry residues, municipal biosolids, and industrial biowates. The choice of pyrolysis

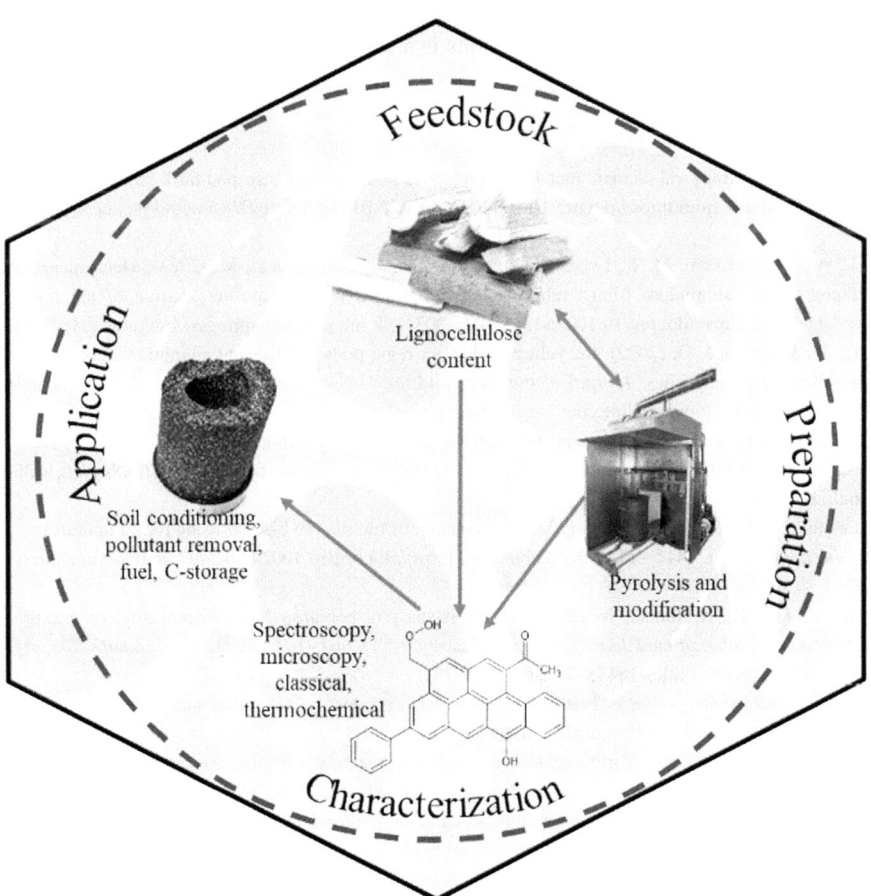

FIGURE 2.6 Interrelationship between feedstock, biochar preparation, properties, and applications. Different biomass and preparation condition leads to biochar with different properties for various applications.

process depends on the feedstock, while the properties of biochar will, to a great extent, be influenced by the preparation method. It is, therefore, important to judiciously select feedstock that produces biochar of the desired properties for a particular application.

Slow pyrolysis in batch reactors is a commonly used preparation method for biochar. This results in a highly porous structure with a wide range of applications. Continuous methods are relatively less researched. This could be because of the more sophisticated engineering design and the cost involved. However, continuous processes produce higher throughputs and are therefore likely to be up-taken for large-scale operation.

Owing to its highly porous structure which presents opportunities for modification with various moieties, biochar is rapidly gaining prominence in pollutant removal from various environmental compartments. Consequently, remediation of contaminated water, wastewater, and soils has been widely researched and applied. However, the removal of air pollutants remains largely unexplored. Apart from pollutant removal applications, biochar can also be potentially used for carbon capture and as catalyst support and activator. This opens a vast avenue for use in reducing the carbon footprint of various industrial processes, photocatalysis, oligomerization of short-chain olefins to useful fuels, and catalytic degradation of engine exhausts. Because it is a store of carbon and energy, biochar can also be used as a solid fuel. Although some of these applications have been explored at laboratory scale, there is scope for up-scaling them to industrial scale. This requires laboratory investigations that use environmentally relevant variables and reactors with parameters that can be up-scaled to pilot scale and eventually field studies.

The properties of biochar depend to a large extent on the feedstock and pyrolysis process. These, in turn, determine its application. Although biochar is endowed with high porosity and intrinsic functionalities, it is often necessary to tailor its properties by introducing specific functional groups. Methods such as sulphonation, amination, and, more recently, the inclusion of nanoparticles, have been used. To determine the properties of biochar, methods such as measurement of pH, CEC, surface charge, SSA, morphology, surface functional groups, and crystallinity have been used. Through these methods, it is possible to precisely determine the characteristics and consequently engineer certain functionalities of the biochar matrix. Interestingly, AI techniques have shown great promise in predicting the properties of biochar. Because of its powerful predictive prowess using empirical and *a priori* information, it removes the unnecessary burden and cost of performing many laboratory experiments while focusing on methods that confer desired properties. For this reason, the use of AI in the design of biochar and its applications is an area that deserves further research.

References

Abbey, C. Y. B., Duwiejuah, A. B., & Quianoo, A. K. (2023). Removal of toxic metals from aqueous phase using cacao pod husk biochar in the era of green chemistry. *Applied Water Science*, 13(2). Available from https://doi.org/10.1007/s13201-022-01863-5, https://www.springer.com/journal/13201.

Adeniyi, A. G., Abdulkareem, S. A., Adeyanju, C. A., Abdulkareem, M. T., Iwuozor, K. O., Emenike, E. C., & Ndagi, M. (2023). Mechanical and morphological analyses of flamboyant seed pod biochar/aluminium filings reinforced hybrid polystyrene composite. *Journal of the Indian Academy of Wood Science*, 20(1), 28–36. Available from https://doi.org/10.1007/s13196-023-00311-4, https://www.springer.com/journal/13196.

Adeniyi, A. G., Abdulkareem, S. A., Adeyanju, C. A., & Ighalo, J. O. (2022). Recycling of delonix regia pods biochar and aluminium filings in the development of thermally conducting hybrid polystyrene composites. *Journal of Polymers and the Environment*, 30(8), 3150–3162. Available from https://doi.org/10.1007/s10924-022-02413-5, http://www.kluweronline.com/issn/1566-2543/.

Adeniyi, A. G., Emenike, E. C., Iwuozor, K. O., & Saliu, O. D. (2023). Solvated polystyrene resin: a perspective on sustainable alternative to epoxy resin in composite development. *Materials Research Innovations*, 27(7), 490–502. Available from https://doi.org/10.1080/14328917.2023.2199597, http://www.tandfonline.com/loi/ymri20#.VwHdSE1f1Qs.

Adeniyi, A. G., Iwuozor, K. O., Muritala, K. B., Emenike, E. C., & Adeleke, J. A. (2023). Conversion of biomass to biochar using top-lit updraft technology: a review. *Biofuels, Bioproducts and Biorefining*, 17(5), 1411–1424. Available from https://doi.org/10.1002/bbb.2497, http://onlinelibrary.wiley.com/journal/10.1002/(ISSN)1932-1031.

Ahmad, I., Khan, M. I., Khan, H., Ishaq, M., Tariq, R., Gul, K., & Ahmad, W. (2015). Pyrolysis study of polypropylene and polyethylene into premium oil products. Bellwether Publishing, Ltd., Pakistan. *International Journal of Green Energy*, 12(7), 663–671. Available from https://doi.org/10.1080/15435075.2014.880146, http://www.tandf.co.uk/journals/titles/15435075.asp.

Akancha., Kumari, N., & Singh, R. K. (2019). Co-pyrolysis of waste polypropylene and rice bran wax- production of biofuel and its characterization. *Journal of the Energy Institute*, 92(4), 933–946. Available from https://doi.org/10.1016/j.joei.2018.07.011.

Almutairi, A. A., Ahmad, M., Rafique, M. I., & Al-Wabel, M. I. (2023). Variations in composition and stability of biochars derived from different feedstock types at varying pyrolysis temperature. *Journal of the Saudi Society of Agricultural Sciences*, 22(1), 25–34. Available from https://doi.org/10.1016/j.jssas.2022.05.005, https://www.journals.elsevier.com/journal-of-the-saudi-society-of-agricultural-sciences/.

Alptekin, F. M., & Celiktas, M. S. (2022). Review on catalytic biomass gasification for hydrogen production as a sustainable energy form and social, technological, economic, environmental, and political analysis of catalysts. *ACS Omega*, 7(29), 24918–24941. Available from https://doi.org/10.1021/acsomega.2c01538, pubs.acs.org/journal/acsodf.

Al-Rumaihi, A., Shahbaz, M., Mckay, G., Mackey, H., & Al-Ansari, T. (2022). A review of pyrolysis technologies and feedstock: A blending approach for plastic and biomass towards optimum biochar yield. *Renewable and Sustainable Energy Reviews, 167*. Available from https://doi.org/10.1016/j.rser.2022.112715, https://www.journals.elsevier.com/renewable-and-sustainable-energy-reviews.

Amutio, M., Lopez, G., Aguado, R., Bilbao, J., & Olazar, M. (2012). Biomass oxidative flash pyrolysis: Autothermal operation, yields and product properties. *Energy and Fuels, 26*(2), 1353−1362. Available from https://doi.org/10.1021/ef201662x.

Bahrun, A., Fahimuddin, M. Y., Rakian, T. C., Safuan, L. O., & Kilowasid, L. O. M. H. (2018). Cocoa Pod Husk Biochar Reduce Watering Frequency and Increase Cocoa Seedlings Growth. *International Journal of Environment, Agriculture and Biotechnology, 3*(5), 1635−1639. Available from https://doi.org/10.22161/ijeab/3.5.9.

Brewer, C. E., Hu, Y. Y., Schmidt-Rohr, K., Loynachan, T. E., Laird, D. A., & Brown, R. C. (2012). Extent of pyrolysis impacts on fast pyrolysis biochar properties. *Journal of Environmental Quality, 41*(4), 1115−1122. Available from https://doi.org/10.2134/jeq2011.0118, https://www.agronomy.org/publications/jeq/pdfs/41/4/1115.

Çepelioğullar, Ö., & Pütün, A. E. (2013). Thermal and kinetic behaviors of biomass and plastic wastes in co-pyrolysis. *Energy Conversion and Management, 75*, 263−270. Available from https://doi.org/10.1016/j.enconman.2013.06.036.

Chang, Y. M., Tsai, W. T., & Li, M. H. (2015). Chemical characterization of char derived from slow pyrolysis of microalgal residue. *Journal of Analytical and Applied Pyrolysis, 111*, 88−93. Available from https://doi.org/10.1016/j.jaap.2014.12.004.

Chattopadhyay, J., Pathak, T. S., Srivastava, R., & Singh, A. C. (2016). Catalytic co-pyrolysis of paper biomass and plastic mixtures (HDPE (high density polyethylene), PP (polypropylene) and PET (polyethylene terephthalate)) and product analysis. *Energy, 103*, 513−521. Available from https://doi.org/10.1016/j.energy.2016.03.015.

Chen, C., Wang, Z., Ge, Y., Liang, R., Hou, D., Tao, J., Yan, B., Zheng, W., Velichkova, R., & Chen, G. (2023). Characteristics prediction of hydrothermal biochar using data enhanced interpretable machine learning. *Bioresource Technology, 377*, 128893. Available from https://doi.org/10.1016/j.biortech.2023.128893.

Chung, B. Y. H., Ang, J. C., Tang, J. Y., Chong, J. W., Tan, R. R., Aviso, K. B., Chemmangattuvalappil, N. G., & Thangalazhy-Gopakumar, S. (2023). Rough set approach to predict biochar stability and pH from pyrolysis conditions and feedstock characteristics. *Chemical Engineering Research and Design, 198*, 221−233. Available from https://doi.org/10.1016/j.cherd.2023.09.003, http://www.elsevier.com/wps/find/journaldescription.cws_home/713871/description#description.

Cárdenas-Aguiar, E., Méndez, A., Gascó, G., Lado, M., & Paz-González, A. (2024). The effects of feedstock, pyrolysis temperature, and residence time on the properties and uses of biochar from broom and gorse wastes. *Appl. Sci., 14*.

Córdova, B. M., Santa Cruz, J. P., Ocampo, T. V. M., Huamani-Palomino, R. G., & Baena-Moncada, A. M. (2020). Simultaneous adsorption of a ternary mixture of brilliant green, rhodamine B and methyl orange as artificial wastewater onto biochar from cocoa pod husk waste. Quantification of dyes using the derivative spectrophotometry method. *New Journal of Chemistry, 44*(20), 8303−8316. Available from https://doi.org/10.1039/d0nj00916d, http://pubs.rsc.org/en/journals/journal/nj.

Dai, L., Lu, Q., Zhou, H., Shen, F., Liu, Z., Zhu, W., & Huang, H. (2021). Tuning oxygenated functional groups on biochar for water pollution control: A critical review. *Journal of Hazardous Materials, 420*, 126547. Available from https://doi.org/10.1016/j.jhazmat.2021.126547.

Dai, M., Xu, H., Yu, Z., Fang, S., Chen, L., Gu, W., & Ma, X. (2018). Microwave-assisted fast co-pyrolysis behaviors and products between microalgae and polyvinyl chloride. *Applied Thermal Engineering, 136*, 9−15. Available from https://doi.org/10.1016/j.applthermaleng.2018.02.102, http://www.journals.elsevier.com/applied-thermal-engineering/.

David, P. S., Karunanithi, A., & Fathima, N. N. (2020). Improved filtration for dye removal using keratin−polyamide blend nanofibrous membranes. *Environmental Science and Pollution Research, 27*(36), 45629−45638. Available from https://doi.org/10.1007/s11356-020-10491-y, https://link.springer.com/journal/11356.

Demirbas, A. (2004). Effects of temperature and particle size on bio-char yield from pyrolysis of agricultural residues. *Journal of Analytical and Applied Pyrolysis, 72*(2), 243−248. Available from https://doi.org/10.1016/j.jaap.2004.07.003.

Demirbaş, A. (2005). Recovery of chemicals and gasoline-range fuels from plastic wastes via pyrolysis. *Energy Sources, 27*(14), 1313−1319. Available from https://doi.org/10.1080/009083190519500.

Dewangan, A., Pradhan, D., & Singh, R. K. (2016). Co-pyrolysis of sugarcane bagasse and low-density polyethylene: Influence of plastic on pyrolysis product yield. *Fuel, 185*, 508−516. Available from https://doi.org/10.1016/j.fuel.2016.08.011, http://www.journals.elsevier.com/fuel/.

Dumroese, R. K., Page-Dumroese, D. S., & Pinto, J. R. (2020). Biochar potential to enhance forest resilience, seedling quality, and nursery efficiency. *Tree Plant, 63*, 61−68.

Elnour, A. Y., Alghyamah, A. A., Shaikh, H. M., Poulose, A. M., Al-Zahrani, S. M., Anis, A., & Al-Wabel, M. I. (2019). Effect of pyrolysis temperature on biochar microstructural evolution, physicochemical characteristics, and its influence on biochar/polypropylene composites. *Applied Sciences (Switzerland), 9*(6). Available from https://doi.org/10.3390/app9061149, https://res.mdpi.com/applsci/applsci-09-01149/article_deploy/applsci-09-01149.pdf.

Emenike, E. C., Ogunniyi, S., Ighalo, J. O., Iwuozor, K. O., Okoro, H. K., & Adeniyi, A. G. (2022). Delonix regia biochar potential in removing phenol from industrial wastewater. *Bioresource Technology Reports, 19*. Available from https://doi.org/10.1016/j.biteb.2022.101195, https://www.journals.elsevier.com/bioresource-technology-reports.

FakhrHoseini, S. M., & Dastanian, M. (2013). Predicting Pyrolysis Products of PE, PP, and PET Using NRTL Activity Coefficient Model. *Journal of Chemistry, 2013*, 1−5. Available from https://doi.org/10.1155/2013/487676.

Feng, C., Huang, M., & Huang, C. P. (2023). Specific chemical adsorption of selected divalent heavy metal ions onto hydrous γ-Fe_2O_3-biochar from dilute aqueous solutions with pH as a master variable. *Chemical Engineering Journal, 451*. Available from https://doi.org/10.1016/j.cej.2022.138921, http://www.elsevier.com/inca/publications/store/6/0/1/2/7/3/index.htt.

Guo, L., Zhao, L., Tang, Y., Zhou, J., & Shi, B. (2023). Chrome shaving-derived biochar as efficient persulfate activator: Ti-induced charge distribution modulation for 1O2 dominated nonradical process. *Science of The Total Environment, 862*, 160838. Available from https://doi.org/10.1016/j.scitotenv.2022.160838.

Gwenzi, W., Chaukura, N., Noubactep, C., & Mukome, F. N. D. (2017). Biochar-based water treatment systems as a potential low-cost and sustainable technology for clean water provision. *Journal of Environmental Management, 197*, 732–749. Available from https://doi.org/10.1016/j.jenvman.2017.03.087, https://www.sciencedirect.com/journal/journal-of-environmental-management.

Hanoğlu, A., Çay, A., & Yanık, J. (2019). Production of biochars from textile fibres through torrefaction and their characterisation. *Energy, 166*, 664–673. Available from https://doi.org/10.1016/j.energy.2018.10.123, http://www.elsevier.com/inca/publications/store/4/8/3/.

He, C., Giannis, A., & Wang, J. Y. (2013). Conversion of sewage sludge to clean solid fuel using hydrothermal carbonization: Hydrochar fuel characteristics and combustion behavior. *Applied Energy, 111*, 257–266. Available from https://doi.org/10.1016/j.apenergy.2013.04.084, http://www.elsevier.com/inca/publications/store/4/0/5/8/9/1/index.htt.

Hiloidhari, M., Das, D., & Baruah, D. C. (2014). Bioenergy potential from crop residue biomass in India. *Renewable and Sustainable Energy Reviews, 32*, 504–512. Available from https://doi.org/10.1016/j.rser.2014.01.025, https://www.journals.elsevier.com/renewable-and-sustainable-energy-reviews.

Huang, H.-jun, & Xing-zhong, Y. (2016). The migration and transformation behaviors of heavy metals during the hydrothermal treatment of sewage sludge. *Bioresource Technology, 200*, 991–998. Available from https://doi.org/10.1016/j.biortech.2015.10.099.

Iwuozor, K. O., Emenike, E. C., Omonayin, E. O., Bamigbola, J. O., Ojo, H. T., Awoyale, A. A., Eletta, O. A. A., & Adeniyi, A. G. (2023). Unlocking the hidden value of pods: A review of thermochemical conversion processes for biochar production. *Bioresource Technology Reports, 22*. Available from https://doi.org/10.1016/j.biteb.2023.101488, https://www.journals.elsevier.com/bioresource-technology-reports.

Ji, M., Wang, X., Usman, M., Liu, F., Dan, Y., Zhou, L., Campanaro, S., Luo, G., & Sang, W. (2022). Effects of different feedstocks-based biochar on soil remediation: A review. *Environmental Pollution, 294*. Available from https://doi.org/10.1016/j.envpol.2021.118655, https://www.journals.elsevier.com/environmental-pollution.

Jien, S.-H. (2019). *Physical characteristics of biochars and their effects on soil physical properties* (pp. 21–35). Elsevier BV. Available from https://doi.org/10.1016/b978-0-12-811729-3.00002-9.

Jorio, A., Ribeiro-Soares, J., Cançado, L. G., Falcão, N. P. S., Dos Santos, H. F., Baptista, D. L., Martins Ferreira, E. H., Archanjo, B. S., & Achete, C. A. (2012). Microscopy and spectroscopy analysis of carbon nanostructures in highly fertile Amazonian anthrosoils. *Soil and Tillage Research, 122*, 61–66. Available from https://doi.org/10.1016/j.still.2012.02.009.

Kameyama, K., Miyamoto, T., Iwata, Y., & Shiono, T. (2016). Influences of feedstock and pyrolysis temperature on the nitrate adsorption of biochar. *Soil Science and Plant Nutrition, 62*(2), 180–184. Available from https://doi.org/10.1080/00380768.2015.1136553, http://www.tandfonline.com/toc/tssp20/current.

Keiluweit, M., Nico, P. S., Johnson, M., & Kleber, M. (2010). Dynamic molecular structure of plant biomass-derived black carbon (biochar). *Environmental Science and Technology, 44*(4), 1247–1253. Available from https://doi.org/10.1021/es9031419.

Kuan, W. H., Huang, Y. F., Chang, C. C., & Lo, S. L. (2013). Catalytic pyrolysis of sugarcane bagasse by using microwave heating. *Bioresource Technology, 146*, 324–329. Available from https://doi.org/10.1016/j.biortech.2013.07.079, http://www.elsevier.com/locate/biortech.

Kumar, J. A., Sathish, S., Prabu, D., Renita, A. A., Saravanan, A., Deivayanai, V. C., Anish, M., Jayaprabakar, J., Baigenzhenov, O., & Hosseini-Bandegharaei, A. (2023). Agricultural waste biomass for sustainable bioenergy production: Feedstock, characterization and pre-treatment methodologies. *Chemosphere, 331*. Available from https://doi.org/10.1016/j.chemosphere.2023.138680, http://www.elsevier.com/locate/chemosphere.

Laird, D. A., Brown, R. C., Amonette, J. E., & Lehmann, J. (2009). Review of the pyrolysis platform for coproducing bio-oil and biochar. *Biofuels, Bioproducts and Biorefining, 3*(5), 547–562. Available from https://doi.org/10.1002/bbb.169, http://www3.interscience.wiley.com/cgi-bin/fulltext/122589511/PDFSTART.

Lee, Y., Park, J., Ryu, C., Gang, K. S., Yang, W., Park, Y. K., Jung, J., & Hyun, S. (2013). Comparison of biochar properties from biomass residues produced by slow pyrolysis at 500°C. *Bioresource Technology, 148*, 196–201. Available from https://doi.org/10.1016/j.biortech.2013.08.135, http://www.elsevier.com/locate/biortech.

Lehmann, J., Rillig, M. C., Thies, J., Masiello, C. A., Hockaday, W. C., & Crowley, D. (2011). Biochar effects on soil biota - A review. *Soil Biology and Biochemistry, 43*(9), 1812–1836. Available from https://doi.org/10.1016/j.soilbio.2011.04.022.

Li, B., Liu, X., Wang, A., Tan, C., Sun, K., Deng, L., Fan, M., Cui, J., Xue, J., Jiang, J., & Yao, D. (2022). Biochar with inherited negative surface charges derived from Enteromorpha prolifera as a promising cathode material for capacitive deionization technology. *Desalination, 539*, 115955. Available from https://doi.org/10.1016/j.desal.2022.115955.

Li, X., Wang, T., Li, Y., Liu, T., Ma, X., Han, X., & Wang, Y. (2023). Aged biochar for simultaneous removal of Pb and Cd from aqueous solutions: Method and mechanism. *Environmental Technology & Innovation, 32*, 103368. Available from https://doi.org/10.1016/j.eti.2023.103368.

Liang, B., Lehmann, J., Solomon, D., Kinyangi, J., Grossman, J., O'Neill, B., Skjemstad, J. O., Thies, J., Luizão, F. J., Petersen, J., & Neves, E. G. (2006). Black carbon increases cation exchange capacity in soils. *Soil Science Society of America Journal, 70*(5), 1719–1730. Available from https://doi.org/10.2136/sssaj2005.0383.

Liao, F., Yang, L., Li, Q., Li, Y. R., Yang, L. T., Anas, M., & Huang, D. L. (2018). Characteristics and inorganic N holding ability of biochar derived from the pyrolysis of agricultural and forestal residues in the southern China. *Journal of Analytical and Applied Pyrolysis, 134*, 544–551. Available from https://doi.org/10.1016/j.jaap.2018.08.001, https://www.journals.elsevier.com/journal-of-analytical-and-applied-pyrolysis.

Liu, Z., & Balasubramanian, R. (2014). Upgrading of waste biomass by hydrothermal carbonization (HTC) and low temperature pyrolysis (LTP): A comparative evaluation. *Applied Energy, 114*, 857–864. Available from https://doi.org/10.1016/j.apenergy.2013.06.027, http://www.elsevier.com/inca/publications/store/4/0/5/8/9/1/index.htt.

Liu, Z., & Han, G. (2015). Production of solid fuel biochar from waste biomass by low temperature pyrolysis. *Fuel, 158*, 159–165. Available from https://doi.org/10.1016/j.fuel.2015.05.032, http://www.journals.elsevier.com/fuel/.

Liu, Z., Quek, A., Kent Hoekman, S., & Balasubramanian, R. (2013). Production of solid biochar fuel from waste biomass by hydrothermal carbonization. *Fuel, 103*, 943–949. Available from https://doi.org/10.1016/j.fuel.2012.07.069.

Lua, A. C., Yang, T., & Guo, J. (2004). Effects of pyrolysis conditions on the properties of activated carbons prepared from pistachio-nut shells. *Journal of Analytical and Applied Pyrolysis, 72*(2), 279–287. Available from https://doi.org/10.1016/j.jaap.2004.08.001.

Ma, J., Zhang, S., Liu, X., & Wang, J. (2023). Machine learning prediction of biochar yield based on biomass characteristics. *Bioresource Technology, 389*, 129820. Available from https://doi.org/10.1016/j.biortech.2023.129820.

Meng, C. P., Hanif, A. H. M., Wahid, S. A., & Abdullah, L. C. (2014). Short-term field decomposition and physico-chemical transformation of jatropha pod biochar in acidic mineral soil. *Open Journal of Soil Science, 04*(07), 226–234. Available from https://doi.org/10.4236/ojss.2014.47025.

Miandad, R., Barakat, M. A., Rehan, M., Aburiazaiza, A. S., Ismail, I. M. I., & Nizami, A. S. (2017). Plastic waste to liquid oil through catalytic pyrolysis using natural and synthetic zeolite catalysts. *Waste Management, 69*, 66–78. Available from https://doi.org/10.1016/j.wasman.2017.08.032.

Mohamed, A. K., & Mahmoud, M. E. (2020). Nanoscale Pisum sativum pods biochar encapsulated starch hydrogel: A novel nanosorbent for efficient chromium (VI) ions and naproxen drug removal. *Bioresource Technology, 308*. Available from https://doi.org/10.1016/j.biortech.2020.123263, http://www.elsevier.com/locate/biortech.

Mohanty, P., Nanda, S., Pant, K. K., Naik, S., Kozinski, J. A., & Dalai, A. K. (2013). Evaluation of the physiochemical development of biochars obtained from pyrolysis of wheat straw, timothy grass and pinewood: Effects of heating rate. *Journal of Analytical and Applied Pyrolysis, 104*, 485–493. Available from https://doi.org/10.1016/j.jaap.2013.05.022.

Moiseenko, K. V., Glazunova, O. A., Savinova, O. S., Vasina, D. V., Zherebker, A. Y., Kulikova, N. A., Nikolaev, E. N., & Fedorova, T. V. (2021). Relation between lignin molecular profile and fungal exo-proteome during kraft lignin modification by Trametes hirsuta LE-BIN 072. *Bioresource Technology, 335*. Available from https://doi.org/10.1016/j.biortech.2021.125229, http://www.elsevier.com/locate/biortech.

Montes-Morán, M. A., Suárez, D., Menéndez, J. A., & Fuente, E. (2004). On the nature of basic sites on carbon surfaces: an overview. *Carbon, 42*(7), 1219–1225. Available from https://doi.org/10.1016/j.carbon.2004.01.023.

Mukhopadhyay, S., Masto, R. E., Sarkar, P., & Bari, S. (2022). Biochar washing to improve the fuel quality of agro-industrial waste biomass. *Journal of the Energy Institute, 102*, 60–69. Available from https://doi.org/10.1016/j.joei.2022.02.011, http://www.journals.elsevier.com/journal-of-the-energy-institute.

Mukome, F. N. D., Zhang, X., Silva, L. C. R., Six, J., & Parikh, S. J. (2013). Use of chemical and physical characteristics to investigate trends in biochar feedstocks. *Journal of Agricultural and Food Chemistry, 61*(9), 2196–2204. Available from https://doi.org/10.1021/jf3049142.

Nakason, K., Pathomrotsakun, J., Kraithong, W., Khemthong, P., & Panyapinyopol, B. (2019). Torrefaction of agricultural wastes: Influence of lignocellulosic types and treatment temperature on fuel properties of biochar. *International Energy Journal, 19*(4), 253–266. Available from http://www.rericjournal.ait.ac.th/index.php/reric/article/view/2163/728.

Nataraj, S. K., Hosamani, K. M., & Aminabhavi, T. M. (2009). Nanofiltration and reverse osmosis thin film composite membrane module for the removal of dye and salts from the simulated mixtures. *Desalination, 249*(1), 12–17. Available from https://doi.org/10.1016/j.desal.2009.06.008.

Nizamuddin, S., Baloch, H. A., Griffin, G. J., Mubarak, N. M., Bhutto, A. W., Abro, R., Mazari, S. A., & Ali, B. S. (2017). An overview of effect of process parameters on hydrothermal carbonization of biomass. *Renewable and Sustainable Energy Reviews, 73*, 1289–1299. Available from https://doi.org/10.1016/j.rser.2016.12.122, https://www.journals.elsevier.com/renewable-and-sustainable-energy-reviews.

Ogunkunle, C. O., Falade, F. O., Oyedeji, B. J., Akande, F. O., Vishwakarma, V., Alagarsamy, K., Ramachandran, D., & Fatoba, P. O. (2021). Short-term aging of pod-derived biochar reduces soil cadmium mobility and ameliorates cadmium toxicity to soil enzymes and tomato. *Environmental Toxicology and Chemistry, 40*(12), 3306–3316. Available from https://doi.org/10.1002/etc.4958, http://onlinelibrary.wiley.com/journal/10.1002/(ISSN)1552-8618.

Ohale, P. E., Igwegbe, C. A., Iwuozor, K. O., Emenike, E. C., Obi, C. C., & Białowiec, A. (2023). A review of the adsorption method for norfloxacin reduction from aqueous media. *MethodsX, 10*. Available from https://doi.org/10.1016/j.mex.2023.102180, http://www.journals.elsevier.com/methodsx/.

Pariyar, P., Kumari, K., Jain, M. K., & Jadhao, P. S. (2020). Evaluation of change in biochar properties derived from different feedstock and pyrolysis temperature for environmental and agricultural application. *Science of the Total Environment, 713*. Available from https://doi.org/10.1016/j.scitotenv.2019.136433, http://www.elsevier.com/locate/scitotenv.

Pinzon-Nuñez, D. A., Adarme-Durán, C. A., Vargas-Fiallo, L. Y., Rodriguez-Lopez, N., & Rios-Reyes, C. A. (2022). Biochar as a waste management strategy for cadmium contaminated cocoa pod husk residues. *International Journal of Recycling of Organic Waste in Agriculture, 11*(1), 101–115. Available from https://doi.org/10.30486/ijrowa.2021.1920124.1192, http://ijrowa.khuisf.ac.ir/article_686239_48c5e2fdd1ea04ec0e297d8c47a10009.pdf.

Pouangam Ngalani, G., Dzemze Kagho, F., Peguy, N. N. C., Prudent, P., Ondo, J. A., & Ngameni, E. (2023). Effects of coffee husk and cocoa pods biochar on the chemical properties of an acid soil from West Cameroon. *Archives of Agronomy and Soil Science, 69*(5), 744–758. Available from https://doi.org/10.1080/03650340.2022.2033733, http://www.tandf.co.uk/journals/titles/03650340.asp.

Prens, J., Kurt, Z., James Rivas, A. M., & Chen, J. (2023). Production and Characterization of Wild Sugarcane (Saccharum spontaneum L.) Biochar for Atrazine Adsorption in Aqueous Media. *Agronomy, 13*(1). Available from https://doi.org/10.3390/agronomy13010027, http://www.mdpi.com/journal/agronomy/.

Pulka, J., Manczarski, P., Koziel, J. A., & Białowiec, A. (2019). Torrefaction of sewage sludge: Kinetics and fuel properties of biochars. *Energies, 12*(3). Available from https://doi.org/10.3390/en12030565, https://www.mdpi.com/1996-1073/12/3.

Ravindran, B., Nguyen, D. D., Chaudhary, D. K., Chang, S. W., Kim, J., Lee, S. R., Shin, J. D., Jeon, B. H., Chung, S. J., & Lee, J. J. (2019). Influence of biochar on physico-chemical and microbial community during swine manure composting process. *Journal of Environmental Management*, *232*, 592–599. Available from https://doi.org/10.1016/j.jenvman.2018.11.119, http://www.elsevier.com/inca/publications/store/6/2/2/8/7/1/index.htt.

dos Reis, G. S., Bergna, D., Grimm, A., Lima, E. C., Hu, T., Naushad, M., & Lassi, U. (2023). Preparation of highly porous nitrogen-doped biochar derived from birch tree wastes with superior dye removal performance. *Colloids and Surfaces A: Physicochemical and Engineering Aspects*, *669*. Available from https://doi.org/10.1016/j.colsurfa.2023.131493, http://www.elsevier.com/locate/colsurfa.

Dos Reis, G. S., Larsson, S. H., Thyrel, M., Pham, T. N., Lima, E. C., de Oliveira, H. P., & Dotto, G. L. (2021). Preparation and application of efficient biobased carbon adsorbents prepared from spruce bark residues for efficient removal of reactive dyes and colors from synthetic effluents. *Coatings*, *11*(7). Available from https://doi.org/10.3390/coatings11070772, https://www.mdpi.com/2079-6412/11/7/772/pdf.

Reza, M. T., Lynam, J. G., Uddin, M. H., & Coronella, C. J. (2013). Hydrothermal carbonization: Fate of inorganics. *Biomass and Bioenergy*, *49*, 86–94. Available from https://doi.org/10.1016/j.biombioe.2012.12.004.

Setter, C., Silva, F. T. M., Assis, M. R., Ataíde, C. H., Trugilho, P. F., & Oliveira, T. J. P. (2020). Slow pyrolysis of coffee husk briquettes: Characterization of the solid and liquid fractions. *Fuel*, *261*, 116420. Available from https://doi.org/10.1016/j.fuel.2019.116420.

Shonhiwa, C. (2013). An assessment of biomass residue sustainably available for thermochemical conversion to energy in Zimbabwe. *Biomass and Bioenergy*, *52*, 131–138. Available from https://doi.org/10.1016/j.biombioe.2013.02.024.

Spokas, K. A. (2010). Review of the stability of biochar in soils: Predictability of O:C molar ratios. *Carbon Management*, *1*(2), 289–303. Available from https://doi.org/10.4155/cmt.10.32.

Sun, Y., Gao, B., Yao, Y., Fang, J., Zhang, M., Zhou, Y., Chen, H., & Yang, L. (2014). Effects of feedstock type, production method, and pyrolysis temperature on biochar and hydrochar properties. *Chemical Engineering Journal*, *240*, 574–578. Available from https://doi.org/10.1016/j.cej.2013.10.081.

Sánchez-Sánchez, Á., Suárez-García, F., Martínez-Alonso, A., & Tascón, J. M. D. (2015). Synthesis, characterization and dye removal capacities of N-doped mesoporous carbons. *Journal of Colloid and Interface Science*, *450*, 91–100. Available from https://doi.org/10.1016/j.jcis.2015.02.073.

Tomczyk, A., Sokołowska, Z., & Boguta, P. (2020). Biochar physicochemical properties: pyrolysis temperature and feedstock kind effects. *Reviews in Environmental Science and Biotechnology*, *19*(1), 191–215. Available from https://doi.org/10.1007/s11157-020-09523-3, http://www.kluweronline.com/issn/1569-1705.

Tong, W., Cai, Z., Liu, Q., Ren, S., & Kong, M. (2020). Effect of pyrolysis temperature on bamboo char combustion: Reactivity, kinetics and thermodynamics. *Energy*, *211*. Available from https://doi.org/10.1016/j.energy.2020.118736, https://www.journals.elsevier.com/energy.

Tu, P., Zhang, G., Wei, G., Li, J., Li, Y., Deng, L., & Yuan, H. (2022). Influence of pyrolysis temperature on the physicochemical properties of biochars obtained from herbaceous and woody plants. *Bioresources and Bioprocessing*, *9*(1). Available from https://doi.org/10.1186/s40643-022-00618-z, https://bioresourcesbioprocessing.springeropen.com.

Veses, A., Aznar, M., López, J. M., Callén, M. S., Murillo, R., & García, T. (2015). Production of upgraded bio-oils by biomass catalytic pyrolysis in an auger reactor using low cost materials. *Fuel*, *141*, 17–22. Available from https://doi.org/10.1016/j.fuel.2014.10.044.

Wang, J., Zhang, M., Chen, M., Min, F., Zhang, S., Ren, Z., & Yan, Y. (2014). Catalytic effects of six inorganic compounds on pyrolysis of three kinds of biomass. *Thermochim Acta*, *444*(2014), 110–114.

Wang, L., Xue, G., Li, T., Ye, T., Ma, X., Ju, X., Ma, P., Liu, J., & Lei, H. (2022). The pyrolysis behavior and biochar characteristics of Jerusalem artichoke straw with cerium nitrate. *Journal of Analytical and Applied Pyrolysis*, *168*, 105768. Available from https://doi.org/10.1016/j.jaap.2022.105768.

Wang, S., Gao, B., Zimmerman, A. R., Li, Y., Ma, L., Harris, W. G., & Migliaccio, K. W. (2015). Physicochemical and sorptive properties of biochars derived from woody and herbaceous biomass. *Chemosphere*, *134*, 257–262. Available from https://doi.org/10.1016/j.chemosphere.2015.04.062, http://www.elsevier.com/locate/chemosphere.

Wang, W., Bai, J., Lu, Q., Zhang, G., Wang, D., Jia, J., Guan, Y., & Yu, L. (2021). Pyrolysis temperature and feedstock alter the functional groups and carbon sequestration potential of Phragmites australis- and Spartina alterniflora-derived biochars. *GCB Bioenergy*, *13*(3), 493–506. Available from https://doi.org/10.1111/gcbb.12795, http://onlinelibrary.wiley.com/journal/10.1111/(ISSN)1757-1707.

Wen, C., Liu, T., Wang, D., Wang, Y., Chen, H., Luo, G., Zhou, Z., Li, C., & Xu, M. (2023). Biochar as the effective adsorbent to combustion gaseous pollutants: Preparation, activation, functionalization and the adsorption mechanisms. *Progress in Energy and Combustion Science*, *99*. Available from https://doi.org/10.1016/j.pecs.2023.101098, https://www.journals.elsevier.com/progress-in-energy-and-combustion-science.

Wenga, T., Chinyama, S. R., Gwenzi, W., & Jamro, I. A. (2023). Quantification of bio-wastes availability for bioenergy production in Zimbabwe. *Scientific African*, *20*. Available from https://doi.org/10.1016/j.sciaf.2023.e01634, https://www.journals.elsevier.com/scientific-african.

Wibawa, P. J., Ningrum, H. A. S., Damayanti, P., Al-Hasan, Z. U. F., Suhartana., & Pardoyo. (2023). Sequentially citric acid-KMnO$_4$-modified surface of activated carbon microparticles to enhance the capability of loading silver nanoparticles as a bacterial sensor material. *Diamond and Related Materials*, *136*, 109900. Available from https://doi.org/10.1016/j.diamond.2023.109900.

Wu, X., Ba, Y., Wang, X., Niu, M., & Fang, K. (2018). Evolved gas analysis and slow pyrolysis mechanism of bamboo by thermogravimetric analysis, Fourier transform infrared spectroscopy and gas chromatography-mass spectrometry. *Bioresource Technology*, *266*, 407–412. Available from https://doi.org/10.1016/j.biortech.2018.07.005, http://www.elsevier.com/locate/biortech.

Xiong, X., Liu, Z., Zhao, L., Huang, M., Dai, L., Tian, D., Zou, J., Zeng, Y., Hu, J., & Shen, F. (2022). Tailoring biochar by PHP towards the oxygenated functional groups (OFGs)-rich surface to improve adsorption performance. *Chinese Chemical Letters*, *33*(6), 3097–3100. Available from https://doi.org/10.1016/j.cclet.2021.09.099, http://www.elsevier.com/wps/find/journaldescription.cws_home/997/description#description.

Xu, R. K., Zhao, A. Z., Yuan, J. H., & Jiang, J. (2012). pH buffering capacity of acid soils from tropical and subtropical regions of China as influenced by incorporation of crop straw biochars. *Journal of Soils and Sediments, 12*(4), 494–502. Available from https://doi.org/10.1007/s11368-012-0483-3.

Yang, Y., Meehan, B., Shah, K., Surapaneni, A., Hughes, J., Fouché, L., & Paz-Ferreiro, J. (2018). Physicochemical properties of biochars produced from biosolids in Victoria, Australia. *International Journal of Environmental Research and Public Health, 15*(7). Available from https://doi.org/10.3390/ijerph15071459, http://www.mdpi.com/1660-4601/15/7/1459/pdf.

Yang, Z., Kumar, A., Huhnke, R. L., Buser, M., & Capareda, S. (2016). Pyrolysis of eastern redcedar: Distribution and characteristics of fast and slow pyrolysis products. *Fuel, 166*, 157–165. Available from https://doi.org/10.1016/j.fuel.2015.10.101, http://www.journals.elsevier.com/fuel/.

Yao, Z., You, S., Ge, T., & Wang, C.-H. (2018). Biomass gasification for syngas and biochar co-production: Energy application and economic evaluation. *Applied Energy, 209*, 43–55. Available from https://doi.org/10.1016/j.apenergy.2017.10.077.

You, S., Ok, Y. S., Chen, S. S., Tsang, D. C. W., Kwon, E. E., Lee, J., & Wang, C. H. (2017). A critical review on sustainable biochar system through gasification: Energy and environmental applications. *Bioresource Technology, 246*, 242–253. Available from https://doi.org/10.1016/j.biortech.2017.06.177, http://www.elsevier.com/locate/biortech.

Yuan, H., Lu, T., Huang, H., Zhao, D., Kobayashi, N., & Chen, Y. (2015). Influence of pyrolysis temperature on physical and chemical properties of biochar made from sewage sludge. *Journal of Analytical and Applied Pyrolysis, 112*, 284–289. Available from https://doi.org/10.1016/j.jaap.2015.01.010.

Yuan, J. H., Xu, R. K., & Zhang, H. (2011). The forms of alkalis in the biochar produced from crop residues at different temperatures. *Bioresource Technology, 102*(3), 3488–3497. Available from https://doi.org/10.1016/j.biortech.2010.11.018.

Zhang, Q., Khan, M. U., Lin, X., Cai, H., & Lei, H. (2019). Temperature varied biochar as a reinforcing filler for high-density polyethylene composites. *Composites Part B: Engineering, 175*. Available from https://doi.org/10.1016/j.compositesb.2019.107151, https://www.journals.elsevier.com/composites-part-b-engineering.

Zhao, L., Cao, X., Mašek, O., & Zimmerman, A. (2013). Heterogeneity of biochar properties as a function of feedstock sources and production temperatures. *Journal of Hazardous Materials, 256-257*, 1–9. Available from https://doi.org/10.1016/j.jhazmat.2013.04.015.

Zhao, X., Wang, W., Liu, H., Ma, C., & Song, Z. (2014). Microwave pyrolysis of wheat straw: Product distribution and generation mechanism. *Bioresource Technology, 158*, 278–285. Available from https://doi.org/10.1016/j.biortech.2014.01.094, http://www.elsevier.com/locate/biortech.

Zhou, Y., Chen, Z., Gong, H., Wang, X., & Yu, H. (2020). A strategy of using recycled char as a co-catalyst in cyclic in-situ catalytic cattle manure pyrolysis for increasing gas production. *Waste Management, 107*, 74–81. Available from https://doi.org/10.1016/j.wasman.2020.04.002, http://www.elsevier.com/locate/wasman.

Chapter 3

Development of novel engineered/functionalized biochars

Nhamo Chaukura[1], Jeremia Shale Sefadi[1], Nonhlangabezo Mabuba[2], Soraya Malinga[2], Abimbola Oluwalana-Sanusi[3] and Wisdom Archford Munzeiwa[4]

[1]*Department of Physical and Earth Sciences, Sol Plaatje University, Kimberley, South Africa,* [2]*Department of Chemical Sciences, University of Johannesburg, Johannesburg, South Africa,* [3]*Department of Physical and Earth Sciences, Centre for Global Change, Sol Plaatje University, Kimberley, South Africa,* [4]*Department of Physical and Earth Sciences, Centre for Applied Data Science, Sol Plaatje University, Kimberley, South Africa*

Chapter outline

3.1 Introduction	35
3.2 Biochar synthesis routes	36
3.3 Activation techniques	37
3.3.1 Physical activation	38
3.3.2 Chemical activation	39
3.4 Environmental remediation applications	40
3.4.1 Removal of organic contaminants	40
3.4.2 Removal of inorganic contaminants	40
3.5 Novel characteristics of engineered biochars	48
3.6 Contaminant removal mechanisms	48
3.6.1 Organic contaminants	48
3.6.2 Inorganic contaminants	50
3.7 Economic feasibility studies of biochar production and application	51
3.8 Future outlook and conclusion	52
3.8.1 Synthesis and fabrication	53
3.8.2 Testing and evaluation	53
3.8.3 Regeneration and disposal	53
References	53

3.1 Introduction

Biochar is a carbon material derived from the pyrolysis of biomass feedstock under an oxygen-starved environment. A collection of suitable biomass includes industrial biowastes, chitin, municipal biosolids, and agro-wastes such as banana peels, macadamia nut shells, peanut shells, rice straw, poultry litter, and forest residues such as pine cones and blue gum wood (Shakya et al., 2022). Besides, there are some energy plants grown specifically for biochar preparation e.g., algae. The advantages of such plants are that they survive in diverse climatic conditions, and they do not compete against nutritious crops for agricultural land. In most developing countries, the climatic conditions are generally suitable for agriculture, thus, there is an abundance of agro-wastes and other biowastes that can potentially be used as biochar feedstock (Tripathi et al., 2016). The key parameter that determines the characteristics of the generated biochar is the lignocellulose component. In this regard, hardwood produces more robust biochar due to its high lignocellulose content (Hassan et al., 2020). Because the feedstock is predominantly biowaste, biochar is a low-cost material that is economically feasible even in developing countries. Besides the cost of pyrolysis, harvesting/collection of the biowaste, haulage, and preparation determine the overall cost associated with biochar technology. Therefore, the cost of each of these components needs to be judiciously controlled (Zhu et al., 2022).

Following harvesting/collection, biomass can be cleaned up, dried, and reduced in size before undergoing pyrolysis. These steps greatly influence pyrolysis; for instance, humidity levels and particle size determine heat distribution in the feedstock throughout pyrolysis, and subsequently, the characteristics of biochar. Pyrolysis reactors can be batch or continuous, with batch processes being predominant, especially in small-scale operations. This is because they are easier to design, fabricate, manage, and control (Osman et al., 2022). However, the throughput from such processes can be lower relative to continuous processes. Thus, continuous reactors can be automated and are more suited for large-scale processes that produce large quantities of biochar. Pyrolysis commonly involves carbonization, which converts lignocellulosic matter to a carbon-rich structure. Either fast pyrolysis or slow pyrolysis can be used, resulting in biochar with varied characteristics. Slow pyrolysis at temperatures exceeding 600°C results in a highly porous carbon framework, while fast pyrolysis limits the formation of the porous framework (Brown et al., 2011). The pyrolysis temperature and heating rate are key parameters that affect the biochar characteristics. Hydrothermal carbonization (HTC) and gasification are other biochar preparation methods that are used, and these will be explored in more detail later. The biochar production process flow and possible areas of application are depicted in Fig. 3.1.

The application of biochar depends on its properties, for instance, highly porous biochar has found applications as catalyst supports in adsorption and filtration processes. Biochar, environmental remediation applications include various inorganic and organic contaminants removal in aquatic systems, and more recently, the remotion of volatile organic compounds from flue gases and air. To enhance performance and target particular fugitive molecules, biochar may require modification to introduce desired functionalities. This includes the activation and introduction of sulfonic groups, amine moieties, and nanoparticles as surface functional groups. While functional groups will interact with pollutants through various mechanisms, nanoparticles are sometimes introduced to enhance the surface area and confer catalytic properties to the biochar. Consequently, nanoparticle-modified biochar can be used in the photocatalytic conversion of pollutants into benign compounds or less harmful metabolites. This chapter gives an analysis of the recent advances in biochar engineering and design by looking at synthesis methods, characterization, and modification toward intended applications.

3.2 Biochar synthesis routes

Various biowastes, including bones, cotton seed hulls, forest plant biomass, manures, municipal biosolids, paper mill sludge, pine chips, peanut hulls, poultry litter, and tires, have been used as feedstock (Gupta et al., 2020). The variety of feedstock and pyrolysis conditions regulates the characteristics of the resulting biochar. Higher lignin content achieves a higher biochar yield, whereas higher hemicellulose or cellulose content leads to simpler modifications (Gupta et al., 2020; Jung et al., 2019). During pyrolysis, biomass is thermally degraded, and this modifies reactive groups and the porous framework, eventually altering the sorption capacities of the biochar.

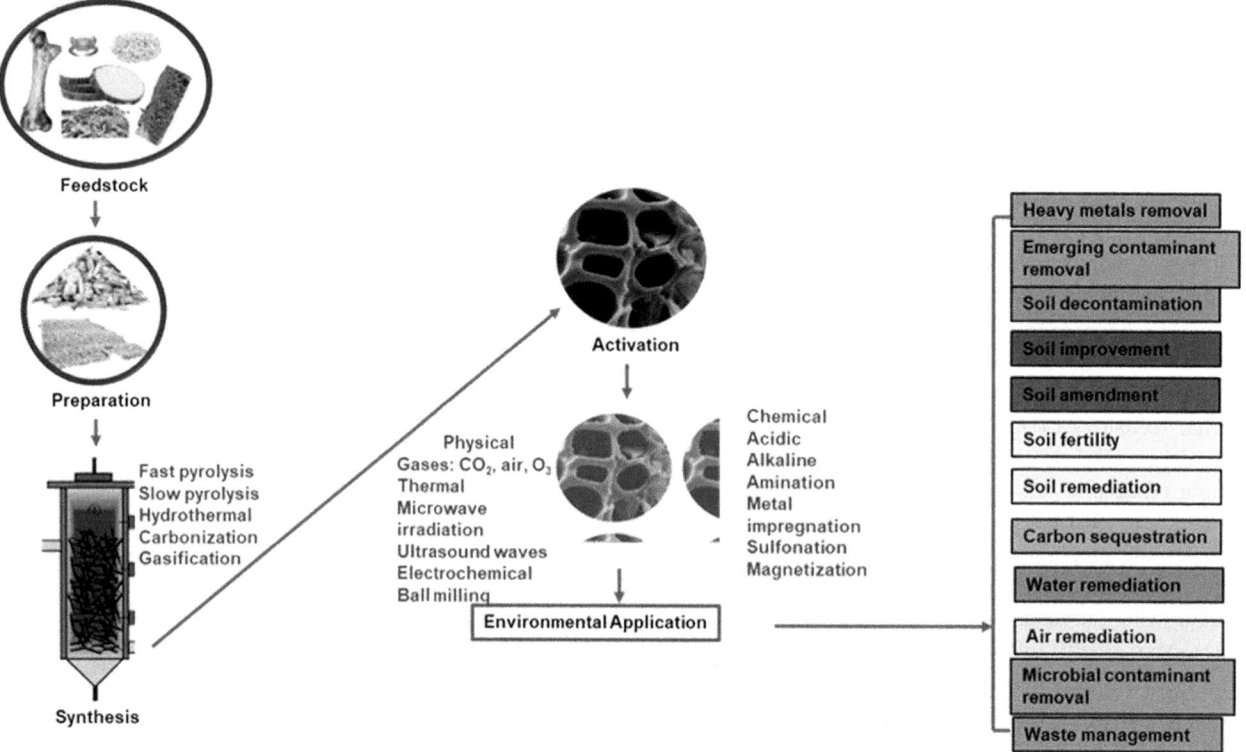

FIGURE 3.1 Processing biomass feedstock into biochar.

Techniques such as gasification, fast pyrolysis, slow pyrolysis, and hydrothermal methods can be used to prepare pristine biochar. The synthesis of biochar is generally performed via slow pyrolysis, which involves pyrolyzing biomass within 500°C–800°C under an inert atmosphere and kept at that temperature for a given duration (Deng et al., 2020). Afterward, the biochar is cooled to room temperature. Slow pyrolysis produces higher biochar yields, particularly when operated using a feedstock of large particle size, high lignin, and ash load. Feedstocks such as wheat straw, poplar wood, spruce wood, vegetable waste, pine cones, and oil palm waste have high lignin and have been used to generate biochar of excellent quality using slow pyrolysis (Rego et al., 2022; Ronsse et al., 2013; Zakaria et al., 2023). The increase in biochar yield is due to a reduction in bio-oil production through cracking. Moreover, slow pyrolysis is a simple, cheap, and robust process suited for small-scale biochar production (Zhang et al., 2019). The major parameter determining the biochar porous structure is the thermal treatment (Bedia et al., 2018). Reactors such as drum kilns, Kon Tiki, and Elsa stoves have been applied for this purpose (Maya et al., 2020).

HTC is conducted at 180°C–250°C in an aqueous medium in a closed reactor under high pressure for 3–24 h (Bamdad et al., 2018; Bedia et al., 2018). The operating conditions and lack of drying requirements significantly reduce the energy demand and cost of the process. In the HTC process, decarboxylation and dehydration reactions occur as exothermic processes (Bedia et al., 2018). This produces hydrochar, bio-oil/water emulsion, and a small proportion of short-chain hydrocarbons and CO_2, the relative abundances of which are determined by the operating conditions and the type of biomass feedstock (Bamdad et al., 2018). HTC could be suitable for feedstock with high moisture content; for instance, food waste, dairy cheese whey, brewery spent grains, sewage residue, and livestock waste (Bardhan et al., 2021; Pecchi et al., 2022).

Gasification produces a mixture of synthetic gases (H_2, CO, CO_2, and CH_4) through the partial combustion of biomass at elevated temperatures (>700 °C) with the aid of an oxidant (O_2 or steam) (Zhang et al., 2019). Because the majority of organic material is converted to gases, the biochar yield in gasifiers is low ($\leq 10\%$) (Bamdad et al., 2018; Zhang et al., 2019). Notably, biochar from the gasification process has a high alkali and alkaline earth metal content and polyaromatic hydrocarbons, which are themselves toxic (Zhang et al., 2019). Torrefaction is mild pyrolysis, which uses lower temperatures (200°C–300°C) with a period of residence in the range of 30 min–2 h to improve biomass characteristics (Gan et al., 2018). The solid residue contains some volatile organic components, hence its characteristics are intermediate between raw biomass and biochar. Feedstock for gasification must be abundant, renewable, produce combustible gases with low residues, have sufficient calorific value, and be environmentally acceptable (Suryawanshi et al., 2023). Examples of such feedstocks are cypress mulch, hardwood pellets, pine bark nuggets, corn stover pellets, and pine pellets (Gutiérrez et al., 2022; Uroić Štefanko and Leszczynska, 2020). The biochar synthetic and activation processes are illustrated in Fig. 3.2.

3.3 Activation techniques

The utilization of raw biomass presents challenges like poor adsorption capacity, poor mechanical properties, and discharge of organic substances; this inevitably raises BOD, COD, and TOC in treated water. Thus, biomass pretreatment and activation are required to overcome these limitations and enhance adsorption performance and selectivity relative to raw biomass. Through pretreatment and activation, ion-exchange capacity, the formation of reactive groups, and the quantity of adsorption points are increased, and toxic metal adsorption is enhanced as a result (Gupta et al., 2020). Heat treatment increases the amount of basic functional groups in biochar. One way of achieving this is to thermally treat biomass to 800–900°C for 1–2 h, and afterward introduce air, argon, or hydrogen to form new surface functional groups. Heat treatment of biochar thus improves the hydrophobic nature of the surface by removing hydrophilic groups, while H_2 stabilizes the surface of biochar through inactivation of certain adsorption points, leading to a highly stable basic carbon framework (Ahmed et al., 2016; Liu et al., 2016).

Although biochar generally possesses an excellent capability for adsorption, the insufficient content of active area limits the capacity of pristine biochar for toxic metal removal (Maya et al., 2020; Peter et al., 2021). For instance, owing to its low reduction capacity and pH_{PZC}, pristine biochar has a low Cr(VI) adsorption capability at pH ≤ 7 (Su et al., 2021). Consequently, a number of biochar modification approaches including chemical modification and impregnation with metal/metal oxides have been investigated (Agrafioti et al., 2014b; Su et al., 2021). In the case of Cr(VI) adsorption, loading pristine biochar with Fe increases the pH_{PZC} and reduction capacity and enhances removal efficiency (Su et al., 2021). Besides the poor ion exchange capacity of most bochars, the adsorption of As is limited by electrostatic repulsion between the As oxyanions and biochar (Gao et al., 2020). Therefore, to achieve appropriate physicochemical attributes like surface morphology, specific surface area, and porosity that provide greater adsorptive

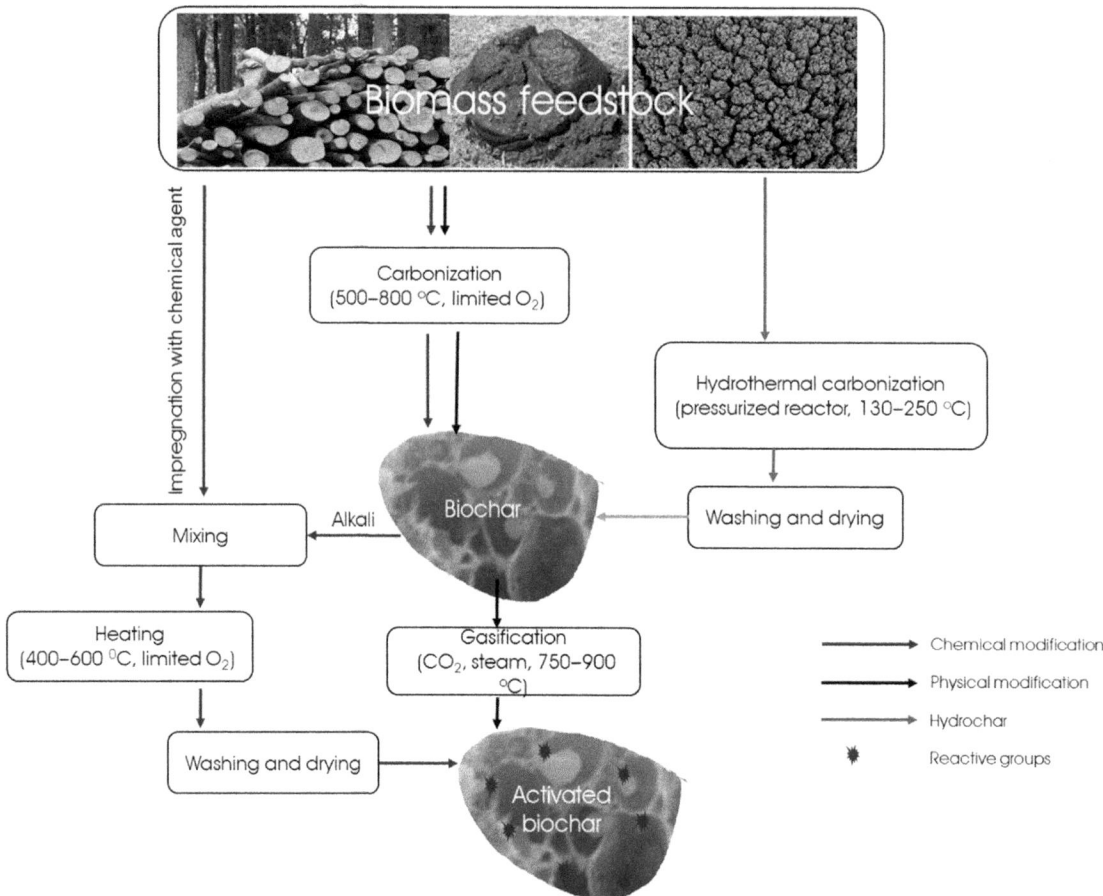

FIGURE 3.2 A schematic diagram of biochar synthesis and modification.

performance, pristine biochar requires modification to produce engineered biochar (Ahmed et al., 2021; Murtaza et al., 2022). Surface functional groups such as hydroxyl, phenols, amines, carboxyl, and phosphonate are introduced to the biochar, providing sorption sites for different pollutants including As and Cr.

3.3.1 Physical activation

Various processes including physical activation using steam have been explored to tune the properties of biochar for targeted applications. Generally, the activation method entails pyrolysis of biomass at 300°C–700°C for 1–2 h in a limited oxygen atmosphere, followed by steam activation at 800°C–900°C for a given time. Steam alters the characteristics of biochar by purging the stuck residual products of partial combustion generated from pyrolysis, and oxidizing the carbon through evolving CO_2, CO, and H_2 (Ahmed et al., 2016). This improves porosity and increases oxygenated moieties on biochar surfaces (Ahmed et al., 2016), resulting in a rise in the hydrophilic character of biochar despite steam being a weaker oxidizing agent. Carbon dioxide activation has been used in the same manner (Wang et al., 2020). In recent studies, ultrasonic pretreatment has been utilized to modify the adsorption nature of softwood-derived biochar (Peter et al., 2021). Physical activation depends on temperature, activation extent, feedstock type, and agent (CO_2, air, stem). Higher temperatures lead to better porosity growth, but increased pore size dispersion. The use of air contains carbon therefore it is important to control the amount of air to avoid combustion as this can lower activated carbon yield (Sakhiya et al., 2020). For physical activation, it is important that the feedstock used is softwood such as coconut shell, wheat straw, corn hull, bur-cucumber, and corn stover (Bardha et al., 2024; Di Stasi et al., 2019). The merit of physical activation is the increase in pore size distribution and unclogging pores. However, the challenge experienced by researchers is the formation of ash, which reduces the biochar yield and the inconsistency in yield of different feedstock despite having the same burn-off energy (Sakhiya et al., 2020). The financial and economic effectiveness of biochar processing operations may be estimated using techno-economic assessment and discounted cash-flow rate frameworks (Patra et al., 2021). Studies have shown that physically activated biochar (1.41 USD kg^{-1}) is cheaper compared to chemically activated biochar (2.57 USD kg^{-1}) (Sakhiya et al., 2021). As a result, the assessment of biochar through the conversion process might play a significant impact in regulating how it is marketed, thereby influencing its monetary value and that of the by-products.

3.3.2 Chemical activation

Alkaline or acidic treatment and the utilization of metal salts also increases the biochar surface area. Conversely, acid modification of biochar can be accomplished through different oxidizing agents that intensify the acidity by leaching out mineral components, thus increasing the hydrophilicity of biochar (Liu et al., 2016). This is normally achieved by soaking or suspending biochar in an acidic solution at a biochar-to-acid ratio of up to 1:10 at 25°C–120°C for a certain time, subsequently washing, drying, and further pyrolysis (Ahmed et al., 2016). Pine tree sawdust treated with dilute H_3PO_4 resulted in increased surface pore volume and area (Zhao et al., 2017). When biochar is exposed to acidic media, oxygen-based groups are attached to the surface and the chemical interaction with positively charged pollutants is increased (Qian et al., 2013). The carboxyl functional group can be placed on a biochar surface via an esterification process with weak acids, and different weak acids were investigated (Ahmed et al., 2021; Deng et al., 2020). Although successful at laboratory-scale, large-scale acid modification of biochar is costly and has negative environmental impacts, particularly in the disposal of spent adsorbents. Since most biochars are acidic, they can potentially increase soil acidity for example. Chemical activation of biochar is desirable for the improvement of pore framework and hydrophilic properties as this enhances its capabilities for the adsorption process and elimination of pollutants from water (Usevičiūtė et al., 2021). However, the source of feedstock needs to be highly considered for chemical activation, as hardwoods like Jarah, birch, and eucalyptus stem chips and softwoods like kola nutshell, fox nutshell, acorn shell, and waste wood have been reported to have improved adsorption attributes and better porous structure. (Vijayaraghavan, 2019; Jiang et al., 2017).

Alkaline modification confers positive charges on biochars, which are key in the adsorption of negatively charged pollutants. Basic activation is performed by immersing or floating biochar in an alkaline solution at ambient temperature up to 100°C for a given duration. Following washing and drying, the biochar is pyrolyzed again at 300°C–700°C under oxygen-starved conditions for a certain time (Ahmed et al., 2016). In one study, pyrolyzing pine tree sawdust biochar with KOH solution at 350°C–550°C resulted in the reopening of trapped pores, pore enlargement, and an enhanced surface area (Goswami et al., 2016). Relative to acids, bases or alkalis enhance both the surface area and oxygen-carrying moieties in biochar (Deng et al., 2020). Alternatively, strong oxidizing agents such as $KMnO_4$ and H_2O_2 can also be used to enhance oxygenated surface groups on biochars (Deng et al., 2020). For instance, biochar produced using corn straw was impregnated with $KMnO_4$, resulting in an increased surface area and functional groups (Wang et al., 2023).

Although these chemical modification approaches can augment the properties of biochar and enhance the toxic metal adsorption efficiency, some are only applicable at the laboratory scale because they are expensive and there is a likelihood of secondary pollution. Strong acids and alkalis potentially alter water pH and disrupt ecological balance (Deng et al., 2020). Strong oxidizing agents are costly, and their ecotoxicological impacts are not well known. Accordingly, the use of cheaper, more environmentally friendly, and efficient chemical reagents to modify biochar deserves further research. In this regard, carbon-rich and other nanocomposites hold some promise. Biochar can be doped with metal oxides or nonmetals via the use of metal salts. A previous study synthesized an engineered biochar adsorbent by a one-step pyrolysis of Fe_3O_4-impregnated feedstock and used magnetic biochar for sorption of Pb from aqueous solution (Yang et al., 2019). Highly dispersed Fe_3O_4 particles were deposited on magnetic biochar resulting in a high surface area and improved magnetization. In addition, more oxygenated groups such as the $C=O$ group were introduced to the magnetic biochar. Another study used mixed Cu-Mn oxides anchored on a biochar matrix generated using impregnation (Yi et al., 2018).

Owing to their fascinating physicochemical properties, especially enhanced surface area and porosity, nanomaterials exhibit increased sorption capability for toxic metals and can circumvent many challenges faced by conventional materials. As a result, carbon-rich nanomaterials have been trialed for the scavenging of toxic metals (Zhang & Gao, 2013). Previous studies have reported that biochar decorated with nanoparticles can enhance its adsorptive effectiveness (Luo et al., 2019). Of these, the commonly studied nanoparticle is nano zerovalent iron (nZVI) (Stefaniuk et al., 2016). Besides being cost-effective and eco-friendly, nZVI is capable of oxidizing and degrading pollutants (Patanjali et al., 2019). In addition, compared to granular ZVI, nZVI induces fast kinetics because of its higher specific area of surface, allowing enhanced contact of Fe^0 to the pollutant (Taghizadeh et al., 2013). For instance, an investigation using nZVI-modified biochar for the adsorption of Cr(VI) showed that the C = C and C = O bonds formed while C-O and COOH groups were cleaved upon combining nanoparticles with the biochar (Deng et al., 2020). This signified that the C-O-Fe played the role of an electron acceptor during the reduction.

Overall, because of lower operation temperatures, a higher yield, and fast activation kinetics, chemical activation is preferable over physical activation. However, the mechanism involved in chemical activation is poorly understood, although the chemical activation agent is likely to form micro- and meso-pore clusters that considerably improve the specific surface area and activate oxygenated moieties (Peter et al., 2021). Biomass pretreatment or treatment of biochar to activate it can introduce additional costs. Therefore, further work is required to better understand whether the additional contaminant adsorption capacity gained via activation justifies the cost of such activation processes.

3.4 Environmental remediation applications

Global industrialization and urbanization generate organic and inorganic contaminants such as pesticides (Ilager et al., 2023), synthetic organic dyes and agrochemicals (Gautam et al., 2021), petroleum hydrocarbons (Dike et al., 2022), endocrine-disrupting compounds (Kundu et al., 2024), nanoparticles (Reddy et al., 2020), pharmaceuticals (Guo et al., 2022; Shetti et al., 2019), personal care products, and heavy metals (Qiu et al., 2021), which have resulted in chronic environmental pollution. Consequently, scientists are focusing on the development and practical implementation of affordable detoxification methods (Monga et al., 2022). Several carbon-based materials like carbon nanotubes, activated carbon, and biochar have been explored for remediation applications. Biochar demonstrated superior attributes (comparative sorptive properties, relative abundance, eco-friendliness, and cost-effectiveness) as a viable substitute, offering novel material for abatement of organic and inorganic pollutants (Yang et al., 2021). The organic and inorganic pollutant removal capability of biochar is due to its different physicochemical properties such as surface chemical groups, porosity, pH, pH_{zpc}, surface area, and mineral concentration, which facilitate its several removal mechanisms.

3.4.1 Removal of organic contaminants

The unregulated releases of numerous organic contaminants over the past decades have resulted in environmental deterioration and numerous public health concerns. Adsorption methods are among many approaches that have been designed for decontaminating organics toxins from wastewater and polluted soil, including adsorption (Xu et al., 2016), photocatalytic degradation (Luo, Li, et al., 2022; Luo, Yao, et al., 2022), improved oxidation processes (Miao et al., 2022), and biodegradation (Zhao et al., 2020). Presently, biochar has been explored for the remediation of organic contaminants such as antibiotics (El-Azazy et al., 2023), personal care products (PCPs) (Choudhary & Philip, 2022; Feng et al., 2022), insecticides (Jacob et al., 2020), pesticides (Shi et al., 2022), organic dyes (Al-Mahbashi et al., 2022; Navya et al., 2020), steroid hormones, polycyclic aromatic hydrocarbons, endocrine disrupting compounds (Ćwieląg-Piasecka et al., 2023), and polychlorinated biphenyls (Cimirro et al., 2022). Several factors affect the organic contaminants' adsorption process onto the biochar, and these include the feedstock used (Section 3.2), the activation process (Section 3.3), and the mechanism of removal (Section 3.6.1). Studies have shown that physisorption, electrostatic attraction, and hydrogen bonding are the main dominant adsorptive interactions between organic contaminants and biochar (Table 3.1).

3.4.2 Removal of inorganic contaminants

A number of laboratory experiments have demonstrated the efficacy of biochar in the scavenging of As and Cr from polluted aquatic systems (Table 3.2). Biochar adsorbents fabricated from rice husk and municipal biosolids were used as adsorbents for the removal of As(V), Cr(III), and Cr(VI) from water, where removal efficiencies of more than 95% Cr(III), 89% Cr(VI), and 53% As(V) were reported (Agrafioti et al., 2014a). Using batch studies, KH_2PO_4-modified biochar exhibited a maximum adsorption capacity (q_{max}) of 30.76 mg/g for As(III) (Ahmed et al., 2021). However, an iron oxide/bamboo biochar composite had a much lower q_{max} of 7.5–8.2 mg/g (Alchouron et al., 2021). A separate study used *Tectona grandis* sawdust-derived biochar activated with Fe and Fe/Zr and achieved 98% As removal (Sahu et al., 2021). In another study, the removal of Cr(VI) was 100% using Fe-decorated poplar wood-derived biochar (Su et al., 2021). Overall, removal capacities depend on the type of biochar, its physicochemical properties, and the experimental conditions. Because of the variations in biochar characteristics and experimental conditions used, the comparison of different removal performances reported in various studies is problematic, and the results of such comparison

TABLE 3.1 Removal of organic pollutants using biochar.

Biochar type and fabrication	Pollutant	Properties	Removal efficiency	Kinetic models	Isotherm models	Removal mechanism	References
Biochar derived from mango seeds kernel (MSK) by pyrolysis in a muffle furnace at 500°C (1 h). Its analog was laden with nanoceria (Ce-Py-MSK) by microemulsion synthesis	Antibiotics (rifampicin (RIFM) and tigecycline (TIGC))	MSK is morphous while Ce-Py-MSk is crystalline. BET: 24.72–33.83 m²/g pH_{zpc}: 8.20–9.11	Ce-Py-MSK: 92.36% RIFM 90.13% TIGC MSK: q_e:6.32—RIFM q_e:3.09—TIGC	Pseudo-2nd order Elovich	Temkin D-R	Physisorption, chemisorption, electrostatic interactions, hydrogen-bonding, intraparticle and surface diffusion.	El-Azazy et al. (2023)
Paper mill sludge was pyrolyzed at 700°C the resultant biochar was then modified with $FeSO_4 \cdot 7H_2O$ and $NaBH_4$ as a reducing agent to generate ZVI-MBC	Pentachlorophenol	Pore Volume Biochar: 0.083 cm³/g ZVI-MBC:0.079cm³/g SBET: biochar 67 m²/g ZVI-MBC: 101.23 m²/g Pore diameter Biochar: 31.7 Å ZVI-MBC: 47.83 Å Biochar: fluffy material ZVI-MBC: spherical morphology	Biochar: 70.5% ZVI-MBC: 100%	—	—	Simultaneous adsorption, electrostatic attraction	Devi & Saroha (2014)
Bagasse powder was pyrolyzed at 450°C to get biochar	Insecticides (chlorpyrifos)	Amorphous material Moisture content: 1 ± 0.47% Ash content: 2 ± 0.47% Volatile content: 0 ± 0.47% Carbon content: 97 ± 0.47% pH_{zpc}: 7.8	q_{max}: 3.20 mg/g (85%)	Pseudo-2nd order	Freundlich	Physisorption	Jacob et al. (2020)

(Continued)

TABLE 3.1 (Continued)

Biochar type and fabrication	Pollutant	Properties	Removal efficiency	Kinetic models	Isotherm models	Removal mechanism	References
Cassava stem was pyrolyzed at 400°C in a chemical vapor deposition chamber under argon and activated with 1% (w/v) of oxalic acid to obtain cassava stem biochar	Mixed reactive dyes (reactive red and drimarene turquoise)	Porous morphology	q_{max}: 49.75 mg/g	Pseudo-1st order	Langmuir Dubinin-Radushkevich (D-R) Temkin	Electrostatic, physisorption	Navya et al. (2020)
MAGB was obtained by soaking powdered garlic skin in FeCl$_3$ solution afterwards mixed with ZnCl$_2$ and pyrolyzed at 800°C for 2 h under N$_2$ flow	Pharmaceuticals (oxytetracycline, metronidazole, paraben)	Crystalline structure S_{BET}: 1008.97 m^2/g Pore volume: 0.51 cm^3/g Pore radius: 25–30 nm	Oxytetracycline- q_{max}: 822 mg/g Metronidazole- q_{max}: 287 mg/g Paraben- q_{max}: 415 mg/g	Pseudo-2nd order	Freundlich	Pore adsorption, electrostatic effects, π-π stacking action, hydrogen bonding	Feng et al. (2022)
TMFBC-750A biochar was obtained by the pyrolysis of T. Molitor larvae fruss at 500°C under N$_2$ for 1 h, further activation was done by mixing with KOH in 1:4 mass ratio	Neonicotinoid pesticides (nitenpyram, thiaclopid, dinotefuran)	S_{BET}: 1858.80 m^2/g Pore volume: 1.11 cm^3/g Pore radius: 2 nm pH$_{zpc}$: 8.71	Nitenpyram- q_{max}: 195.86 mg/g Thiaclopid- q_{max}: 155.08 mg/g Dinotefuran- q_{max}: 325.81 mg/g	Pseudo-1st order	Langmuir	Micropore filling, π-π electron donor-acceptor interactions, hydrophobic interaction, hydrogen bonding, covalent bonding	Shi et al. (2022)
Deashed wheat straw biochar Pyrolysis conditions: 550°C for 30 s under limited oxygen and further activated chemically	Endocrine-disrupting compounds (carbaryl, carbofuran, metolachlor, 2,4-dichlorophenoxyacetic (2,4-D), 4-chloro-2-methylphenoxyacetic acid (MCPA))	Randomly shaped structures with sharp-edged particles. S_{BET}: 250 m^2/g Pore volume: 0.24 cm^3/g Ash content: 4.30% Carbon content: 78.18%	2,4-D (94.4%); q_{max}: 2.32 mg/g MCPA (97.5%); q_{max}: 2.03 mg/g Metolachlor (96.4%); q_{max}: 2.02 mg/g Carbaryl (96.3%); q_{max}: 3.30 mg/g Carbofuran (92.7%); q_{max}: 1.84 mg/g	—	—	Pore filling, physisorption, chemisorption, hydrophobic interaction, hydrogen bonding	Ćwielag-Piasecka et al. (2023)

Preparation	Pollutant	Characteristics	Performance	Kinetic model	Isotherm model	Mechanism	Reference
Pinus elliottii sawdust was chemically activated with ZnCl$_2$ then pyrolyzed at 600 °C (1.5 h) under N$_2$	Diphenols (hydroquinone (HYD), catechol (CAT), resorcinol (RES))	Flower-like morphology with some rugosity and small tortuous pores. S_{BET}: 1473 m^2/g Pore radius: 1.72–2.19 nm Pore volume: 0.707 cm^3/g Carbon content: 79.48% Ash content: 2.61% pH$_{zpc}$: 6.43	CAT- q_{max}: 419.8 mg/g RES- q_{max}: 263.8 mg/g HYD- q_{max}: 500.9 mg/g 95.08%–95.97%	Avrami-fractional order	Liu	Hydrogen bonds, pores filling, π-π interactions, physisorption	Cimirro et al. (2022)
Sewage sludge was carbonized in a muffle furnace (700°C, 1 h) to obtain the biochar	Batik industrial effluent (containing: naphthol, indigo soluble, remazol, and reactive dyes)	S_{BET}: 117.7 m^2/g Pore radius: 6.01 nm Pore volume: 0.007 cm^3/g	q_{max}: 42.30 mg/g at 16 mL/h flowrate and 12 cm bed height	Thomas	Langmuir	Electrostatic interactions, ion exchange, π-π interaction	Al-Mahbashi et al. (2022)
Cedrella fissilis sawdust was pyrolyzed in a quart tube under N$_2$ at 800°C for 1 h to obtain biochar	Herbicides (atrazine)	S_{BET}: 27.96 m^2/g Pore radius: 1.13 nm Pore volume: 0.018 cm^3/g Carbon content: 88.40%	q_{max}: 4.68 mg/g 76.58% Removal efficiency decreases after 3rd cycle	Linear driving force	Langmuir	π-π interaction, n-π interaction, hydrogen bonding	Hernandes et al. (2022)
Using water hyacinth, slow pyrolysis under N$_2$ at 300 °C produced citric acid (CA)-modified biochar (CAWB).	Methylene blue	Smooth homogeneous surface morphology. S_{BET}: 57.08 m^2/g Pore volume: 0.0759 cm^3/g Pore size: 5.319 nm Introduction of COOH group on the modified biochar from the citric acid	99.5% q_{max}: 395 mg/g 85% efficiency at 5th cycle	Pseudo-2nd order	Langmuir	Electrostatic attraction, chemisorption, ion exchange, chelation.	Xu et al. (2016)

(Continued)

TABLE 3.1 (Continued)

Biochar type and fabrication	Pollutant	Properties	Removal efficiency	Kinetic models	Isotherm models	Removal mechanism	References
Red mud coconut shell biochar (RM-BC) was synthesized by hydrothermal treatment (200°C, 8 h) followed by pyrolysis in tube furnace (800°C, 2 h)	Arbidol, chloroquine phosphate, hydroxychloroquine phosphate, acyclovir	Honey-comb morphology S_{BET}: 246.47 m²/g Pore volume: 0.3056 cm³/g Crystalline structure with O-H stretching bond of free hydroxy group	Arbidol: 100% Chloroquine phosphate: 84.8% Hydroxychloroquine phosphate: 87.2% Acyclovir: 90.8% Stable in reusability in 4th run	Pseudo-1st order	–	Hydrogen-bonding, π-π electron donor and acceptor.	Guo et al. (2022)
Jamun seed was calcinated at 750°C in a carbolite tube furnace under N_2 to obtain the biochar	Antiviral (lamivudine)	S_{BET}: 220.8 m²/g Pore radius: 19.44 Å Pore volume: 0.215 cc/g Carbon content: 76.21%	84.9%	–	Freundlich	Surface charge	Ripanda et al. (2023)
B-CuFe-CS was produced with the use of date kernel-derived biochar pyrolyzed in a muffle furnace (700°C, 4 h) followed by ultrasonication with 2:1 mol ratio of Cu^{2+}:Fe^{2+} and chitosan. The resulting biochar was then co-precipitated.	Eriochrome black T	S_{BET}: 286.32 m²/g Pore radius: 3.05 nm Pore volume: 0.21 cm³/g pH_{zpc}: 6.36	q_{max}: 806.4 mg/g	Pseudo-2nd order	Langmuir	Electrostatic interactions, hydrogen-bonding, and metal complexation	Zubair et al. (2022)
Empty palm bunch was slowly pyrolyzed at 450°C and further treated with H_2SO_4 to obtain PEBC450-A	Pharmaceuticals and PCPs (ibuprofen (IBP), triclosan (TCS), carbamazepine (CZP), methylparaben (MPB))	Percentage yield: 38.1 ± 0.41% Volatile matter: 22.2 ± 0.39% Carbon content: 75.76% pH_{zpc}: 7.5 ± 0.7	IBP- q_{max}: 38.8 mg/g (70.2%) TCS- q_{max}: 35.4 mg/g (74.3%) CZP- q_{max}: 51.7 mg/g (79.9%) MPB- q_{max}: 60.2 mg/g (80.93%) Five cycles via thermal regeneration was recommended for reusability.	Pseudo-2nd order	Langmuir	π-π and n-π forces, hydrophobic, channel diffusion, hydrogen bonding	Choudhary & Philip (2022)

Preparation	Adsorbate	Properties	Performance	Kinetics	Isotherm	Mechanism	Reference
CHB-Fe$_3$O$_4$ was obtained by burning coffee husk in a muffle furnace for 2 h at 600°C under N$_2$ followed by mixing with FeSO$_4 \cdot$7H$_2$O and FeCl$_3$ at 60°C for 1 h. The resulting biochar was then filtered and dried at 60°C for 2 h	Herbicide (Glyphosate)	S$_{BET}$: 142.32 m^2/g; pH$_{zpc}$: 2.00	(99.64%) q$_{max}$: 22.44 mg/g	Pseudo-2nd order	Freundlich	Pore diffusion, electrostatic interactions, hydrogen bonding	Lita et al. (2023)
Hydrothermal carbonization at 180°C (12 h) was used to prepare litchi peel biochar and activated in the muffle furnace (850°C, 1 h)	Congo red Malachite green	Porous morphology, amorphous structure S$_{BET}$: 1006 m^2/g Pore volume: 0.588 cm^3/g pH$_{zpc}$: 2.00	Congo red (100%) q$_{max}$: 404 mg/g Malachite green (84%) q$_{max}$: 2468 mg/g	Elovich	Freundlich	Hydrogen bonding, pore occupation, electrostatic effect, π-π forces	Wu et al. (2020)
Ground coffee residue mixed without/with NaOH were pyrolyzed in a tubular furnace under N$_2$ for 2 h at 800°C to obtain GCRB and GBRB-N respectively.	Herbicides (alachlor, diuron, simazine)	GCRB S$_{BET}$: 3.83 m^2/g Pore volume: 0.014 cm^3/g GCRB-N S$_{BET}$: 405.33 m^2/g Pore volume: 0.293 cm^3/g	GCRB Alachlor: (6%); q$_{max}$:11.74 μmol/g Diuron: (4.7%); q$_{max}$:9.95 μmol/g Simazine: (3.1%); q$_{max}$:6.53 μmol/g GCRB-N Alachlor: (56.6%); q$_{max}$:122.71 μmol/g Diuron: (80.6%); q$_{max}$: 166.42 μmol/g Simazine: (47.4%); q$_{max}$:99.16 μmol/g	Pseudo-2nd order	Freundlich	Chemisorption, hydrophobic, and electrostatic interaction	Lee et al. (2021)

TABLE 3.2 Removal of Cr and As using biochars.

Biochar type and fabrication	Properties	Removal efficiency	Kinetic models	Isotherm models	Thermodynamics	Removal mechanism	References
Ca and Fe modified biochar. Rice husk was impregnated with CaO, while both rice husk and municipal biosolids were impregnated with Fe^0 and Fe^{3+}. Pyrolysis was conducted at 300°C in a muffle furnace under N_2 atmosphere.	–	>95% As(V) and 58% for municipal biosolids and rice husk impregnated with Fe^0, respectively.	–	Freundlich	–	Co-precipitation, electrostatic forces	Agrafioti et al. (2014a)
Biochars are derived from rice husk, organic solid wastes and sewage sludge. Pyrolysis in a muffle furnace at 300°C under N_2.	–	>95% Cr(III), 89% of Cr(VI) and 53% of As(V)	Pseudo-2nd order	Freundlich	–	Electrostatic interactions, ion exchange.	Agrafioti et al. (2014a)
Phosphorus-modified biochar	Crystalline and smooth structure which became amorphous following pyrolysis, uniform layered structure, O–H stretches of carboxylic groups, (Ca, Mg)CO_3	q_{max}: As(III) 30.76 mg/g	Pseudo-2nd order	Langmuir	–	Physisorption, surface precipitation, electrostatic attraction, ion exchange.	Ahmed et al. (2021)
Iron oxide/bamboo biochar composite produced by slow pyrolysis at 700°C for 1 h, and its analog with Fe_3O_4 nanoparticles.	S_{BET}: 6.7 m^2/g	As(V) q_{max}: 7.5–8.2 mg/g	–	–	–	Surface chemisorption, Electrostatic attractions, hydrogen bonding, precipitation	Alchouron et al. (2021)
Raw corncobs pyrolyzed at 550°C. The modified biochar was synthesized by sol–gel method after ultrasonic treatment.	Increase in S_{BET} from 3.38 to 62.75 m^2/g, pore volume from 0.008 to 0.038 cm^3/g, O-H, Ti-O, C-H functional groups.	As q_{max}: 118.06 mg/g	Pseudo-2nd order	Langmuir	$\Delta G^o < 0$	Ion exchange and chelation	Luo et al. (2019)

Material	Properties	Removal/Capacity	Kinetics	Isotherm	Thermodynamics	Mechanism	Reference
Tectona grandis sawdust-derived biochar activated with Fe and Fe/Zr.	NH_2, C-O, -OH, Fe-O, Zr-O functional groups, amorphous, acidic, micrometer-sized agglomerated particles, 20.195–50.468 m^2/g	98.8% As removal	Pseudo-2nd order	Langmuir	$\Delta G^0 < 0$	Ionic attractions	Sahu et al. (2021)
Poplar wood-derived biochar pretreated with Fe and pyrolyzed at 900 °C for 2 h to produce Fe-modified biochar	—	100% Cr(VI) removal	Pseudo-2nd order	Langmuir: aerobic conditions; Freundlich: anaerobic conditions	—	Electrostatic adsorption, chemical reduction, complex precipitation	Su et al. (2021)
magnetically modified biochar prepared by pyrolysis of $FeCl_3$-impregnated bamboo at 600 °C for 1 h	243.4–317.8 m^2/g, rough surface.	Cr(VI) q_{max}: 48 mg/g	Pseudo-2nd order	Temkin	—	Electrostatic complexation	Wang et al. (2017)
KOH-modified biochar derived from sewage sludge	Alkaline, electrical conductivity: 4.0–6.1 µS/cm, pH_{PZC}: 2.7–3.4, S_{BET}: 5.7–7.9 m^2/g	76%–92% As (III)	—	—	—	Chemisorption	Wongrod et al. (2019)
Sugarcane bagasse impregnated with steel pickling waste liquor and pyrolyzed at 600 °C for 2 h to produce magnetic biochar	C = O groups, S_{BET}: 29.918–36.812 m^2/g, magnetization: 31.31–37.26	Cr(VI) q_{max}: 43.122 mg/g	Pseudo-2nd order	Freundlich	—	Electrostatic interaction, complexation	Yi et al. (2019)
Biochar/AlOOH nanocomposite prepared by impregnating biomass with $AlCl_3$ and pyrolyzing at 600 °C for 1 h	Uneven and porous surface, randomly oriented nanostructures, wrinkled AlOOH nano-flakes	As q_{max}: 17410 mg/kg	Elovich	Freundlich	—	Chemisorption	Zhang & Gao, 2013
Bismuth-modified biochar was prepared from Bi_2O_3-loaded wheat straw and pyrolyzed at 400°C, 500°C, and 600°C for 1 h.	Bi-O, O-P-O, -OH groups	Cr(VI) q_{max}: 12.23 mg/g As (III) q_{max}: 6.21 mg/g	Pseudo-2nd order	Langmuir: As; Freundlich: Cr	$\Delta G^0 < 0$	Ligand exchange, electrostatic incorporation, electrostatic interactions	Zhu et al. (2016)

should be interpreted with caution. Ideally, to draw valid comparative results, the comparison of biochar adsorbents should be done under similar operation conditions (e.g., initial concentrations and agitation times).

While laboratory experiments are useful in generating data that indicate the toxic metal performance of biochar, there is a need to upscale to industrial scale for technology utilization to benefit the affected population. Yet pilot-scale studies of biochar-based remediation systems are still lacking. In order to achieve this, the following should be considered: (1) the use of environmentally relevant conditions such as toxic metal concentrations, (2) experimentation with multicomponent aqueous solutions that closely mimic real contaminated media, and (3) the use of large columns that generate data that can be easily upscaled.

3.5 Novel characteristics of engineered biochars

The adsorbent nature and physicochemical properties largely influence the adsorption dynamics. Chemical groups, porosity, surface area, and surface charge influence the adsorbent/adsorbate interactions, and consequently the pollutants removal capacity (Gayathri et al., 2021; Sahu et al., 2021). Biomass feedstock and composition, synthesis protocols and the modifications employed, and operation parameters of the synthesis process (i.e., material size, pH, contact time, and temperature) also influence the physicochemical characteristics and the ultimate removal efficiency of biochar (Gupta et al., 2020; Yi et al., 2019). The biochar surface area is usually increased at high thermal treatment temperatures, and the degree of magnetization is also a function of pyrolysis temperature, which influences the sorption capacity (Gupta et al., 2020). Thus, biochars synthesized from different feedstock possess different characteristics of ash content, electrical conductivity, porosity, surface area, and functionalities (Tables 3.1 and 3.2).

Compared to raw biomass precursor materials, biochar possesses desirable properties such as high biological and chemical stability, a well-developed porous structure, a large specific surface area, and multiple surface functional groups (Arabi et al., 2021). The chemical constituents of biochar include carbon, hydrogen, oxygen, nitrogen, sulfur, and ash (Panwar et al., 2019). The carbon matrix of biochar is usually composed of honey-comb geometries or slit-like pores, with pore sizes varying from subnanometer to micrometer dimensions (Madzaki & KarimGhani, 2016). Chemical and physical activation techniques may be used to control porosity and enhancement of the surface area (Jung et al., 2019). For example, previous studies have used elevated temperature CO_2-NH_3 gaseous-enhanced biochar production approach and a substantial rise in the surface area up to approximately 627 m^2/g was observed (Zhang et al., 2014). The ammonia component functionalized the biochar surface with N-containing groups, while the CO_2 improved the micropore structure. Oxygenated functional groups have a strong affinity for toxic metals, which are removed via chelation, electrostatic forces, and ion exchange (Deng et al., 2020). In addition, the presence of $S_2O_8^{2-}$ or H_2O_2 biochar can enhance persistent free radicals production, which can accelerate the disintegration of pollutants (Deng et al., 2020). This could be useful in degrading organic forms of As, for instance. Metal-loaded biochar such as magnetic biochar incorporates metal-based nanoparticles such as Fe, Ni, Cr, Ti, Zr, Zn, Cu, salts, and oxides, which improve the surface area, porosity, thermal resistance, crystallinity, and surface chemical functional moieties, resulting in increased adsorption capacity and pollutant cleanup (Gupta et al., 2020). Metal/metal oxide biochars act as hybrid or dual adsorbents, where contaminants accumulate on both the carbonaceous and oxide phases (Benis et al., 2020; Wongrod et al., 2019). This hybrid nature explains the relatively high adsorption capacity of metal/metal oxide-biochar composites relative to pristine biochars.

3.6 Contaminant removal mechanisms

3.6.1 Organic contaminants

The extraction of organic compounds is guided by the major surface functional groups that act as adsorption active sites, and these include lactonic, hydroxyl, amide, amine, and carboxylic groups among others (Chen et al., 2022). Other surface physical properties such as noncarbonized sites (crystalline or amorphous carbon), carbonized crystalline, and graphene-like active sites contribute to the overall adsorption efficiency (Ambaye et al., 2021). The adsorption process is complex, and multiple mechanisms are simultaneously involved (Fig. 3.3). The major driving forces in organic pollutants' adsorption mechanism include partition and pore filling, hydrophobic and van der Waals forces, hydrogen bonding, and π-effects (Ambaye et al., 2021; Chen et al., 2022).

Hydrophobic and van der Waals interactions are usually witnessed in biochar sorbents and pollutants systems with limited functional groups. Neutral and hydrophobic pollutants can be removed using hydrophobic interactions (Ambaye et al., 2021). Besides, partitioning and pore-filling mechanisms also contribute to the overall removal of pollutants using

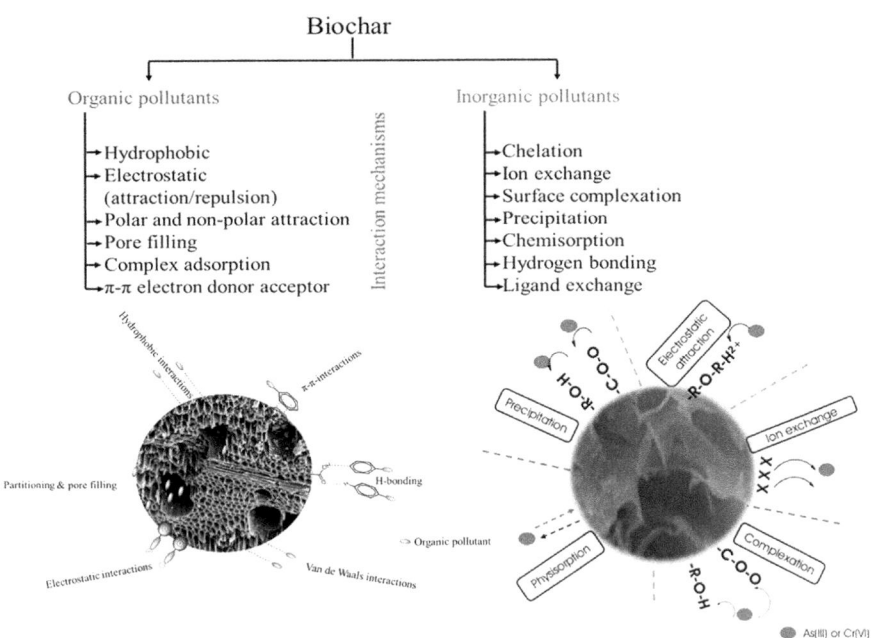

FIGURE 3.3 Adsorption mechanisms for the removal of organic and inorganic pollutants.

biochar. The presence of pores on biochar surface and the inherent specific surface area control the pore-filling mechanism of organic pollutant adsorption. Biochar pyrolyzed at lower temperatures, usually 200°C–350°C, has a high noncarbonized fraction, and this portion can act as a sink for fugitive molecules (Murtaza et al., 2022). Organic contaminants can easily diffuse and partition within the noncarbonized surfaces. Activated biochar can also exhibit micropores ($<$ 2 nm) and mesopores (2–50 nm) where contaminants can diffuse, trap, and fill in the pores, and this depends on the polarity of the pollutants (Goswami et al., 2022). Generally, the smaller organic pollutants are most likely eliminated through the pore-filling mechanism, while those with greater size are excluded (Dong et al., 2023). Pyrolysis conditions can also affect the aromaticity of the biochar surfaces. High temperatures usually present the graphene-like polyaromatic crystalline structure, which can have an electron-rich/deficient π-system. When pyrolysis temperatures of >500 °C are used, the π aromatic framework has electron acceptor tendencies, while those pyrolyzed above 500 °C have electron donor characteristics (Ahmad et al., 2022). The interaction between the π-clouds of the adsorbent and pollutant aromatic system results in π-π interactions, which enable the removal of organic pollutants. The other π-effects such as n-π interactions, which are also referred to as n-π electron donor-acceptor (EDA) interactions due to adsorbent surface carbonyl oxygen heteroatoms acting as electron donors, while the aromatic phenyl rings serve as electron acceptors (Ambaye et al., 2021). Besides, cationic-π and anionic-π interactions are also involved where charged pollutant species interact with the BC aromatic system. A study demonstrated that doping biochar with nitrogen resulted in the creation of a positive hole on the graphitic layer of biochar, which acts as a π-electron acceptor for the pollutant aromatic rings (Wang et al., 2018). Dyes such as acid red and methyl blue were enriched onto the biochar surface through such interactions.

Hydrogen bonding and halogen σ-hole-π-electron bonding are also key driving forces that act between organic contaminants and biochar surfaces. The bonding can arise from interactions between biochar H-atoms on donor atoms and H-accepter functionalities on the contaminants (Chen et al., 2022). Alternatively, H-donor groups on the biochar surface can interact with the aromatic rings on contaminants. The uptake of organic contaminants via electrostatic interactions is premised on the pH of the bulk solution. The different assortment of environmental pollutants, such as dyes, pesticides, phenolics, and pharmaceuticals, exhibit different pKa values. Hence, reliant on the solution pH, the organics can exist as either neutral, anionic, or cationic forms. The net surface charge is zero at pH = pH_{pzc}, but when the pH is less than pH_{pzc}, the surface adsorbs more protons and becomes more positive (Ambaye et al., 2021). This entails that negative pollutant species are removed at the expense of positive species due to repulsion. Likewise, when pH is greater than the pH_{pzc}, the surface assumes a negative charge due to the ionization of the acidic groups and adsorption of the more negative hydroxy ions, which accelerates the sorption of positively charged contaminant species (Dong et al., 2023). For example, antibiotics tend to be decontaminated using electrostatic interactions due to their sensitivity to pH changes. Ionic strength also affects electrostatic interactions, as the ions also vie for the available exchange sites on the biochar surface. Biochar containing metal ions and metal chelation can also be another avenue for organic pollutants containing hetero donor atoms. The pollutants form complexes and are precipitated on the biochar surfaces (Murtaza et al., 2022).

Generally, biochar produced from different feedstock has different sorption capabilities, hence exhibiting different sorption mechanisms. For nonpolar and hydrophobic pollutants, e.g., polyaromatic hydrocarbons and petroleum hydrocarbons, pore filling, partitioning, and hydrophobic interactions dominate, while for polar and ionized organic contaminants, H-bonding, electrostatic forces, and precipitation are the major removal mechanisms.

3.6.2 Inorganic contaminants

Though challenging, an understanding of the mechanisms is vital and can provide information on how to enhance the feasibility of the sorption process through improvements in the design of biochar, and regeneration and recyclability conditions (Gusain et al., 2020). Fig. 3.3 summarizes the removal mechanisms of As and Cr ions in aqueous systems. Most adsorption processes are endothermic, and the mechanisms involved can be generally categorized as chemisorption and physisorption (Gupta et al., 2020; Gusain et al., 2020). On the one hand, physisorption includes reversible weak intermolecular physical interactions between biochar and toxic metals. It involves surface adsorption without disrupting the biochar and toxic metal electronic orbitals. Interactions and processes involved include diffusion, electrostatic, hydrophobic, hydrogen bonding, and van der Waals interactions. On the other hand, chemisorption is an irreversible phenomenon that is characterized by the formation of chemical bonds between the pollutants to the biochar surface functionalities. This entails the participation of orbitals and valence forces between biochar and toxic metals. Thus, the chemisorption mechanism typically involves chelation, complexation, covalent bonding, proton shift, and redox processes (Gusain et al., 2020).

Adsorption mechanisms are explained in terms of biochar properties like functional groups, organic composition, and surface charge (Gupta et al., 2020). Biochar modification methods influence adsorption mechanisms; hence, the choice of suitable modification methods is important to attain favorable stability, adsorption efficiency, regeneration/recyclability, and safe disposal (Agrafioti et al., 2014b; Gupta et al., 2020). Chemically modified biochar removes toxic metals through ligand exchange, or via electrostatic attraction between the negatively ionized biochar surface and metal cations, proton migration between biochar surface and toxic metal cations, protonation of oxygenated functional groups (e.g., hydroxyl, lactonic, and carboxylic groups), inorganic phases (e.g., ash and metal oxides), and basic N-moieties in the biochar matrix (Table 3.1) (Agrafioti et al., 2014b; Ahmed et al., 2021).

Unlike other toxic metals, As and Cr are redox-sensitive; thus, their removal mechanisms are influenced by oxidation-reduction reactions. For instance, in the removal of Cr, Cr(VI) is reduced to Cr(III) and subsequently complexed and precipitated with Cr(III). Previous studies used nZVI-biochar for Cr(VI) removal and it was observed that the C-O-Fe moiety played the role of an electron acceptor in the reduction of Cr(VI) to Cr(III) (Deng et al., 2020). In this system, Fe acted as an anode and supplied electrons, while biochar was the cathode. Apart from reduction, other mechanisms like complexation, electrostatic attraction, ion exchange, and precipitation have also been reported in the removal of Cr(VI) (Fig. 3.2, Deng et al., 2020). In contrast, the reduction of As(V) to As(III) or oxidation of As(III) to As(V) is more complex and poorly understood. However, most studies suggest similar removal mechanisms to Cr(VI). Fig. 3.3 summarizes the key scavenging mechanisms of inorganic pollutants from aqueous systems.

To fully comprehend the adsorption mechanisms of toxic metal ions on biochar, an investigation of kinetics, isotherms, and thermodynamics parameters of the adsorption phenomenon is performed (Table 3.2). A detailed treatment of the specific mathematical equations is provided elsewhere (Tran et al., 2017). The influence of residence time on the adsorption capacity of biochar is used to understand the kinetic behavior of the adsorption process (Peter et al., 2021). Studying the kinetic model provides information on the reaction pathway, rate of adsorption, mechanism of removal, mass transport, physiochemical interactions, and rate-limiting steps (Gayathri et al., 2021). This information, in turn, is useful in the design of industrial toxic metal scavenging processes. Various kinetic models such as 1st order, pseudo 1st order, 2nd order, pseudo 2nd order, and Elovich models have been commonly investigated. The examples cited in Table 3.2 show that most of the adsorption processes fit pseudo-2nd-order kinetics. This implies chemisorption, where the number of toxic metal ions on the biochar surface plays a decisive part, in governing the operational mechanism (Gupta et al., 2020; Luo et al., 2019).

Adsorption isotherms such as the Langmuir, Freundlich, and Temkin models are used to elucidate the nature of interactions between biochar and toxic metals in aqueous solution (Gusain et al., 2020). They shed light on the sorption mechanisms of toxic metals on the biochar surface and help determine the synthesis strategy and optimal adsorption conditions. An analysis of data from previous studies (Tables 3.1 and 3.2) shows a variation in the isotherm models followed. It seems the model depends on factors that include the physicochemical properties of the biochar. Commonly investigated thermodynamic variables such as enthalpy (ΔH^0), entropy (ΔS^0), and Gibb's free energy (ΔG^0) are used to determine the adsorption conditions that are thermodynamically favorable (Gayathri et al., 2021; Peter et al., 2021). If $\Delta G^0 < 0$, this indicates a spontaneous adsorption process, while $\Delta G^0 > 0$ indicates nonspontaneity. For $\Delta G^0 < 0$, the system becomes more spontaneous when there is external energy supplied (Peter et al., 2021). Experimental studies do not frequently investigate thermodynamic parameters. However, the few studies that reported thermodynamic data show the adsorption of As and Cr was spontaneous, indicating the possibility of easily upscaling the process with a lower energy demand. Overall, kinetic, isotherm, and thermodynamic data help improve the deployment of biochar for a particular application.

In order to understand the adsorption mechanisms of toxic metals on biochar, an investigation of the kinetics, isotherms, and thermodynamics of the adsorption phenomenon is performed (Table 3.2). A detailed treatment of the specific mathematical equations is provided elsewhere (Tran et al., 2017). The influence of contact time on the adsorption capacity of biochar is used to understand the kinetic behavior of the adsorption process (Peter et al., 2021). Studying the kinetic model provides information on the reaction pathway, rate of adsorption, mechanism of removal, mass transport, physiochemical interactions, and rate-limiting steps (Gayathri et al., 2021). This information, in turn, is useful in the design of industrial toxic metal scavenging processes. Various kinetic models such as 1st order, pseudo 1st order, 2nd order, pseudo 2nd order, and Elovich models have been commonly investigated. The examples cited in Table 3.2 show the majority of the adsorption processes followed pseudo-2nd order kinetics. This implies that chemisorption, where the concentration of toxic metal on the biochar surface plays an important part, was the governing mechanism (Gupta et al., 2020; Luo et al., 2019).

Adsorption isotherms such as the Langmuir, Freundlich, and Temkin models are used to elucidate the nature of interactions between biochar and toxic metals in aqueous solution (Gusain et al., 2020). They shed light on the sorption mechanisms of toxic metals on the biochar surface and help determine the synthesis strategy and optimal adsorption conditions. An analysis of data from previous studies (and 2) shows a variation in the isotherm models followed. It seems the model depends on factors that include the physicochemical properties of the biochar. Commonly investigated thermodynamic parameters such as change in enthalpy (ΔH^0), change in entropy (ΔS^0), and Gibb's free energy (ΔG^0) are used to determine the adsorption conditions, which are thermodynamically favorable (Gayathri et al., 2021, Peter et al., 2021). If $\Delta G^0 < 0$, this indicates a spontaneous adsorption process, while $\Delta G^0 > 0$ indicates nonspontaneity. For $\Delta G^0 < 0$, the system becomes more spontaneous when there is external energy supplied (Peter et al., 2021). Experimental studies do not frequently investigate thermodynamic parameters. However, the few studies that reported thermodynamic data show the adsorption of As and Cr was spontaneous, indicating the possibility of easily upscaling the process with a lower energy demand. Overall, kinetic, isotherm, and thermodynamic data help improve the deployment of biochar for a particular application.

3.7 Economic feasibility studies of biochar production and application

Activated carbon is the predominantly used adsorbent for pollutant remediation. However, activated carbon is too expensive and thus not practically viable (Praveen et al., 2022). Although biochar holds great promise in the remediation of contaminated aquatic systems, the limited data on its economic feasibility has hampered its large-scale uptake (Zhu et al., 2023). The choice of adsorbent depends not only on a high adsorption capacity but also on factors such as the unit equipment costs, abundance of feedstock, ease of regeneration, and its economic feasibility (Badran et al., 2023) A previous study reported an internal project rate of return of 25.91%, at 1.58 years payback period, for biowaste (0.2 tons/d) producing biochar (60.12 kg/d) and bio-oil (75.64 kg/d) (Manmeen et al., 2023). An efficient and economical regeneration method can decrease the transportation cost of fresh biochar and of the pollutant-laden biochar (Badran et al., 2023). Naturally, high pyrolysis temperatures will be more expensive. Therefore, moderate pyrolysis temperatures have been recommended (Amalina et al., 2022).

To enhance the adsorption efficiency of biochar, the prepared feedstock or pristine biochar is commonly modified with chemical reagents. In addition, the regeneration of spent biochar can use expensive chemicals and potentially discharge pollutants into the environment. The use of additional reagents and procedures will increase production expenditure and potentially harm the environment (Zhao et al., 2024). In addition to the fabrication costs, the adsorption efficiency also affects the profitability of the technology. A techno-economic analysis is therefore required, taking into account the feedstock availability and properties, production costs, adsorption performance, and renewability of biochar. In this regard, a number of studies have performed techno-economic analyses to assess the financial aspect and technical feasibility of the entire biochar technology in the removal of pollutants from the environment (Fawzy et al., 2022; Mishra and Mohanty, 2018; Nematian et al., 2021)

Pyrolysis reactors can be designed to optimize the use of thermal energy. In fact, integrated heat and power units that produce biochar, syngas, and bio-oil can be used. (Mishra and Mohanty, 2018). These byproducts can then be used to fire the reactor and reduce production costs. A previous study recommended the use of portable units to generate market interest for forest residues as biochar feedstock (Mishra and Mohanty, 2018). Waste CO_2 is usually generated during the pyrolysis process, and this can be captured and used for other applications such as fuel synthesis (Kang et al., 2024). Thus, despite fitting well within the circular economy concept, economic feasibility and affordability are important to conduce large-scale production and industrial uptake. Logically, total economic output increases with high throughput and cost levels.

Improved insights into production costs facilitate the development of a competitive advantage and ultimately stimulate demand for biochar technology. Despite being important in generating useful information for scaling up biochar production, techno-economic assessments are seldom reported in the literature. In some cases, conclusions are derived from lab-scale experiments, which might be atypical of large-scale production conditions (Zhu et al., 2022) Thus, industrially relevant trials deserve further research.

3.8 Future outlook and conclusion

Adsorption using biochar could be useful in scavenging As and Cr from contaminated aquatic systems. Lately, there has been a rise in the research on the application of biochar in the removal of As and Cr. This arises from the inherent low production expenses and their eco-friendliness since they are fabricated from abundant and cheap biowastes. The development of suitable synthesis procedures is key to achieving the desired properties in the biochar, resulting in enhanced toxic metal adsorptive efficiency. These fabrication methods also govern the overall process's economic feasibility and environmental impact (Gupta et al., 2020). In order to improve adsorptive efficiency, pristine biochar requires modification, and this can be achieved in multiple ways including chemical activation. However, chemical activation methods are only useful at laboratory scale due to the likelihood of secondary pollution and the associated high financial capital. The use of strong bases and acids, for instance, may disrupt the ecological balance by altering environmental pH. Strong oxidizing agents are costly, and little data exists on their ecotoxicological impacts. Thus, it is important to develop low-cost, environmentally friendly, and high-efficiency chemical reagents for biochar synthesis. Such methods could include introducing nanoparticles into the biochar microstructure (Deng et al., 2020). Further research could also: (1) explore combining appropriate modification strategies to produce biochars that can efficiently and safely remove toxic metals from aquatic systems, and (2) develop economically viable biochar modification processes that improve adsorption performance relative to pristine biochar and raw biomass (Benis et al., 2020).

During adsorption, biochar transfers toxic pollutants from the aquatic phase to the solid phase, which raises utilization worries relating to the spent biochar safe disposition. The disposal of exhausted adsorbent is a challenge since this can potentially cause secondary pollution (Ahmed et al., 2021). It is therefore vital to find an economical strategy to convert spent biochar to a valuable material. Although the regeneration potential of spent biochar using chelating agents, alkalis, and acids has been demonstrated at a laboratory scale, the practicality of the process on an industrial scale is uncertain (Gupta et al., 2020). Further research needs to fill the void between laboratory scale evaluations and industrial technology utilization by the use of large columns that closely mimic real-life situations. Biochar has plenty of biomolecules such as lignin and cellulose. Thus, incineration of spent biochar decreases the mass and volume and helps recover thermal energy and valuable metals. However, inappropriate incineration can generate pollutants in the atmosphere. Landfilling as a disposal method is simple and cheap. To minimize secondary pollution, biochar laden with toxic metals could be desorbed prior to landfilling (Gupta et al., 2020). Because biochar feedstock consists of biowastes, organics in these materials can improve soil quality. Thus, spent biochar can find use as slow-releasing fertilizer and for soil conditioning (Gupta et al., 2020). Alternatively, certain plants can be used for phytoremediation of toxic metals, and such plants can be subsequently used as feedstock for biochar synthesis. Previous studies have suggested using spent biochar for construction materials (Chaukura et al., 2016). This could include some components for road construction and the fabrication of bricks and pavers.

Considering that most previous studies on the scavenging of As and Cr from polluted water were laboratory-based, it is therefore imperative to carry out performance evaluation to investigate the scalability of industrial-scale applications. For instance, comprehensive multicomponent evaluation, adsorption experiments at environmentally relevant concentrations, real water experimentation, and large-scale column studies deserve further study (Alchouron et al., 2021). In a previous review, we provided a detailed evaluation of some problems and opportunities that biochar presents in the remediation of polluted gaseous streams (Maya et al., 2020). Overall, the challenges associated with using biochar for As and Cr and toxic metals remediation in general can be summarized by three key thematic areas (Fig. 3.4).

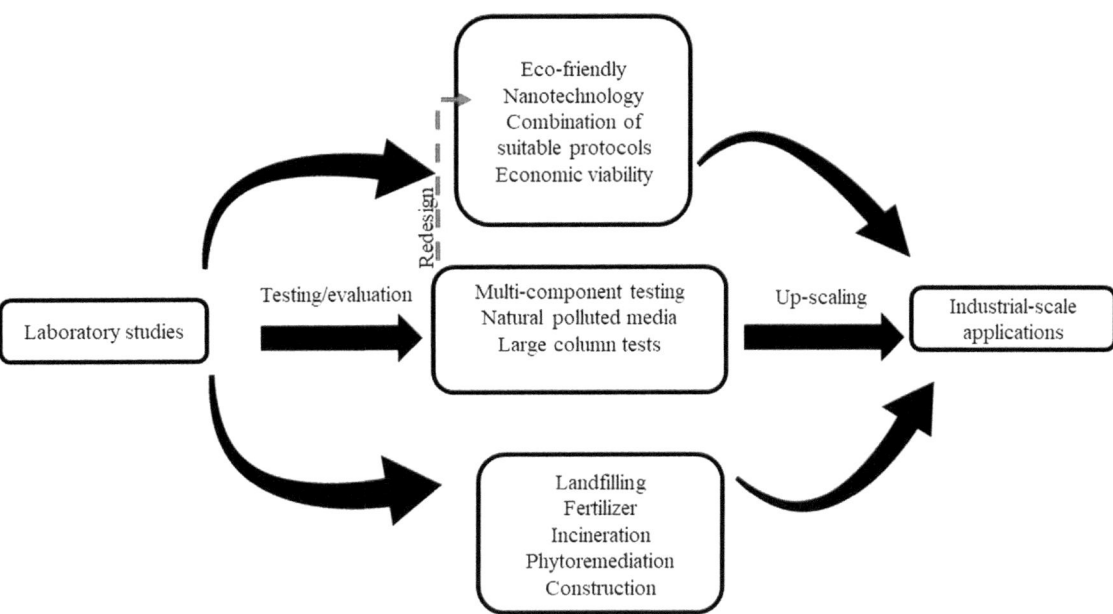

FIGURE 3.4 Challenges and future research on biochar application for remediation of aquatic contaminants.

3.8.1 Synthesis and fabrication

An eco-friendly and cost-effective strategy that could possibly take advantage of the superior adsorption properties of nanomaterials should be sought. In addition, alternative, green pathways could replace chemical activation reagents. Preferably, a combination of a number of methods could be used to achieve the desired adsorption capacity while avoiding environmental pollution.

3.8.2 Testing and evaluation

Biochar adsorbents should be tested under environmentally relevant conditions in order to generate data to scale up the technology to an industrial scale. In this regard, multicomponent solutions, large column tests, and naturally polluted aquatic samples are useful.

3.8.3 Regeneration and disposal

The safe disposal of spent biochar is of primary importance. Several possible ways of safe disposal have been discussed. These include landfilling, incineration, use as fertilizer, and construction applications.

The uptake of biochar for the remediation of As and Cr on an industrial scale depends largely on (1) adsorption performance, (2) recyclability and multiple reuse, and (3) economic feasibility. With an increasing amount of data being generated from laboratory experiments, the large-scale utilization of biochar is promising.

References

Agrafioti, E., Kalderis, D., & Diamadopoulos, E. (2014a). Arsenic and chromium removal from water using biochars derived from rice husk, organic solid wastes and sewage sludge. *Journal of Environmental Management, 133*, 309–314.

Agrafioti, E., Kalderis, D., & Diamadopoulos, E. (2014b). Ca and Fe modified biochars as adsorbents of arsenic and chromium in aqueous solutions. *Journal of Environmental Management, 146*, 444–450.

Ahmed, M. B., Zhou, J. L., Ngo, H. H., & Guo, W. (2016). Insight into biochar properties and its cost analysis. *Biomass and Bioenergy, 84*, 76–86.

Ahmed, W., Mehmood, S., Núñez-Delgado, A., Ali, S., Qaswar, M., Shakoor, A., Maitlo, A. A., & Chen, D. Y. (2021). Adsorption of arsenic (III) from aqueous solution by a novel phosphorus-modified biochar obtained from Taraxacum mongolicum Hand-Mazz: Adsorption behavior and mechanistic analysis. *Journal of Environmental Management, 292*. Available from https://doi.org/10.1016/j.jenvman.2021.112764, https://www.sciencedirect.com/journal/journal-of-environmental-management.

Ahmad, S., Gao, F., Lyu, H., Ma, J., Zhao, B., Xu, S., Ri, C., & Tang, J. (2022). Temperature-dependent carbothermally reduced iron and nitrogen doped biochar composites for removal of hexavalent chromium and nitrobenzene. *Chemical Engineering Journal, 450*, 138006.

Alchouron, J., Navarathna, C., Rodrigo, P. M., Snyder, A., Chludil, H. D., Vega, A. S., Bosi, G., Perez, F., Mohan, D., Pittman, C. U., & Mlsna, T. E. (2021). Household arsenic contaminated water treatment employing iron oxide/bamboo biochar composite: An approach to technology transfer. *Journal of Colloid and Interface Science, 587*, 767–779. Available from https://doi.org/10.1016/j.jcis.2020.11.036, http://www.elsevier.com/inca/publications/store/6/2/2/8/6/1/index.htt.

Al-Mahbashi, N. M. Y., Kutty, S. R. M., Bilad, M. R., Huda, N., Kobun, R., Noor, A., Jagaba, A. H., Al-Nini, A., Ghaleb, A. A. S., & Al-dhawi, B. N. S. (2022). Bench-scale fixed-bed column study for the removal of dye-contaminated effluent using sewage-sludge-based biochar. *Sustainability, 14*, 6484.

Amalina, F., Abd Razak, A. S., Krishnan, S., Sulaiman, H., Zularisam, A. W., & Nasrullah, M. (2022). Biochar production techniques utilizing biomass waste-derived materials and environmental applications–A review. *Journal of Hazardous Materials Advances, 7*, 100134.

Ambaye, T. G., Vaccari, M., van Hullebusch, E. D., Amrane, A., & Rtimi, S. J. I. J. O. E. S. (2021). Mechanisms and adsorption capacities of biochar for the removal of organic and inorganic pollutants from industrial wastewater. *International Journal of Environmental Science and Technology, 18*(10), 3273–3294.

Arabi, Z., Rinklebe, J., El-Naggar, A., Hou, D., Sarmah, A. K., & Moreno-Jiménez, E. (2021). Im) mobilization of arsenic, chromium, and nickel in soils via biochar: a meta-analysis. *Environmental Pollution, 286*, 117199.

Badran, A. M., Utra, U., Yussof, N. S., & Bashir, M. J. (2023). Advancements in Adsorption Techniques for Sustainable Water Purification: A Focus on Lead Removal. *Separations, 10*(11), 565.

Bamdad, H., Hawboldt, K., & MacQuarrie, S. (2018). A review on common adsorbents for acid gases removal: Focus on biochar. *Renewable and Sustainable Energy Reviews, 81*, 1705–1720.

Bardha, A., Kurian, J. K., Gariépy, Y., Prasher, S., Cornejo, R. L., Khirpin, C., Mehlem, J., & Dumont, M. J. (2024). Physicochemical Characterization of Different Steam Activations of Corn Stover Biochar and Their Reinforcement of Styrene-Butadiene Rubber Composites. *Waste and Biomass Valorization, 15*(4), 2285–2298.

Bardhan, M., Novera, T. M., Tabassum, M., Islam, M. A., Islam, M. A., & Hameed, B. H. (2021). Co-hydrothermal carbonization of different feedstocks to hydrochar as potential energy for the future world: A review. *Journal of Cleaner Production, 298*, 126734.

Bedia, J., Peñas-Garzón, M., Gómez-Avilés, A., Rodriguez, J. J., & Belver, C. (2018). A review on the synthesis and characterization of biomass-derived carbons for adsorption of emerging contaminants from water. *C, 4*(4), 63.

Benis, K. Z., Damuchali, A. M., Soltan, J., & McPhedran, K. N. (2020). Treatment of aqueous arsenic–A review of biochar modification methods. *Science of The Total Environment, 739*, 139750.

Brown, T. R., Wright, M. M., & Brown, R. C. (2011). Estimating profitability of two biochar production scenarios: slow pyrolysis vs fast pyrolysis. *Biofuels, Bioproducts and Biorefining, 5*(1), 54–68.

Chaukura, N., Gwenzi, W., Tavengwa, N., & Manyuchi, M. M. (2016). Biosorbents for the removal of synthetic organics and emerging pollutants: opportunities and challenges for developing countries. *Environmental Development, 19*, 84–89.

Chen, H., Chen, S., Zhang, Z., Sheng, L., Zhao, J., Fu, W., Xi, S., Si, R., Wang, L., Fan, M., & Yang, B. (2022). Single-atom-induced adsorption optimization of adjacent sites boosted oxygen evolution reaction. *Acs Catalysis, 12*(21), 13482–13491.

Choudhary, V., & Philip, L. (2022). Sustainability assessment of acid-modified biochar as adsorbent for the removal of pharmaceuticals and personal care products from secondary treated wastewater. *Journal of Environmental Chemical Engineering, 10*, 107592.

Cimirro, N. F., Lima, E. C., Cunha, M. R., Thue, P. S., Grimm, A., dos Reis, G. S., Rabiee, N., Saeb, M. R., Keivanimehr, F., & Habibzadeh, S. (2022). Removal of diphenols using pine biochar. Kinetics, equilibrium, thermodynamics, and mechanism of uptake. *Journal of Molecular Liquids, 364*, 119979.

Ćwieląg-Piasecka, I., Jamroz, E., Medyńska-Juraszek, A., Bednik, M., Kosyk, B., & Polláková, N. (2023). Deashed wheat-straw biochar as a potential superabsorbent for pesticides. *Materials, 16*(6), 2185.

Deng, R., Huang, D., Wan, J., Xue, W., Wen, X., Liu, X., Chen, S., Lei, L., & Zhang, Q. (2020). Recent advances of biochar materials for typical potentially toxic elements management in aquatic environments: A review. *Journal of Cleaner Production, 255*, 119523.

Deng, Y., Huang, Q., Gu, W., & Li, S. (2020). Application of sludge-based biochar generated by pyrolysis: A mini review. *Energy Sources, Part A: Recovery, Utilization, and Environmental Effects*, 1–10.

Devi, P., & Saroha, A. K. (2014). Synthesis of the magnetic biochar composites for use as an adsorbent for the removal of pentachlorophenol from the effluent. *Bioresource Technology, 169*, 525–531.

Di Stasi, C., Alvira, D., Greco, G., González, B., & Manyà, J. J. (2019). Physically activated wheat straw-derived biochar for biomass pyrolysis vapors upgrading with high resistance against coke deactivation. *Fuel, 255*, 115807.

Dike, C. C., Hakeem, I. G., Rani, A., Surapaneni, A., Khudur, L., Shah, K., & Ball, A. S. (2022). The co-application of biochar with bioremediation for the removal of petroleum hydrocarbons from contaminated soil. *Science of The Total Environment, 849*, 157753.

Dong, M., He, L., Jiang, M., Zhu, Y., Wang, J., Gustave, W., Wang, S., Deng, Y., Zhang, X., & Wang, Z. (2023). Biochar for the removal of emerging pollutants from aquatic systems: a review. *International Journal of Environmental Research and Public Health, 20*(3), 1679.

El-Azazy, M., El-Shafie, A. S., Al-Mulla, R., Hassan, S. S., & Nimir, H. I. (2023). Enhanced adsorptive removal of rifampicin and tigecycline from single system using nano-ceria decorated biochar of mango seed kernel. *Heliyon, 9*(5).

Fawzy, S., Osman, A. I., Mehta, N., Moran, D., Ala'a, H., & Rooney, D. W. (2022). Atmospheric carbon removal via industrial biochar systems: A techno-economic-environmental study. *Journal of Cleaner Production, 371*, 133660.

Feng, Z., Zhai, X., & Sun, T. (2022). Sustainable and efficient removal of paraben, oxytetracycline and metronidazole using magnetic porous biochar composite prepared by one step pyrolysis. *Separation and Purification Technology, 293*, 121120.

Gan, Y. Y., Ong, H. C., Show, P. L., Ling, T. C., Chen, W. H., Yu, K. L., & Abdullah, R. (2018). Torrefaction of microalgal biochar as potential coal fuel and application as bio-adsorbent. *Energy Conversion and Management, 165*, 152–162.

Gao, X., Peng, Y., Guo, L., Wang, Q., Guan, C. Y., Yang, F., & Chen, Q. (2020). Arsenic adsorption on layered double hydroxides biochars and their amended red and calcareous soils. *Journal of Environmental Management, 271*, 111045.

Gautam, R. K., Goswami, M., Mishra, R. K., Chaturvedi, P., Awashthi, M. K., Singh, R. S., Giri, B. S., & Pandey, A. (2021). Biochar for remediation of agrochemicals and synthetic organic dyes from environmental samples: A review. *Chemosphere, 272*, 129917.

Goswami, L., Kushwaha, A., Kafle, S. R., & Kim, B. S. (2022). Surface modification of biochar for dye removal from wastewater. *Catalysts, 12*(8), 817.

Goswami, R., Shim, J., Deka, S., Kumari, D., Kataki, R., & Kumar, M. (2016). Characterization of cadmium removal from aqueous solution by biochar produced from Ipomoea fistulosa at different pyrolytic temperatures. *Ecological Engineering, 97*, 444–451.

Gayathri, R., Gopinath, K., & Kumar, P. S. (2021). Adsorptive separation of toxic metals from aquatic environment using agro waste biochar: Application in electroplating industrial wastewater. *Chemosphere, 262*, 128031.

Guo, Z., Zhang, Y., Gan, S., He, H., Cai, N., Xu, J., Guo, P., Chen, B., Pan, X. 2022. Effective degradation of COVID-19 related drugs by biochar-supported red mud catalyst activated persulfate process: Mechanism and pathway. *Journal of Cleaner Production, 340*, 130753.

Gupta, S., Sireesha, S., Sreedhar, I., Patel, C. M., & Anitha, K. L. (2020). Latest trends in heavy metal removal from wastewater by biochar based sorbents. *Journal of Water Process Engineering, 38*, 101561. Available from https://doi.org/10.1016/j.jwpe.2020.101561.

Gusain, R., Kumar, N., & Ray, S. S. (2020). Recent advances in carbon nanomaterial-based adsorbents for water purification. *Coordination Chemistry Reviews, 405*, 213111.

Gutiérrez, J., Rubio-Clemente, A., & Pérez, J. F. (2022). Analysis of biochars produced from the gasification of Pinus patula pellets and chips as soil amendments. *Maderas. Ciencia y Tecnología, 24*, 1–24. Available from https://doi.org/10.4067/s0718-221x2022000100449.

Hassan, M., Liu, Y., Naidu, R., Parikh, S. J., Du, J., Qi, F., & Willett, I. R. (2020). Influences of feedstock sources and pyrolysis temperature on the properties of biochar and functionality as adsorbents: A meta-analysis. *Science of the Total Environment, 744*, 140714.

Hernandes, P. T., Franco, D. S., Georgin, J., Salau, N. P., & Dotto, G. L. (2022). Investigation of biochar from Cedrella fissilis applied to the adsorption of atrazine herbicide from an aqueous medium. *Journal of Environmental Chemical Engineering, 10*, 107408.

Ilager, D., Pai, A. M., Kalanur, S. S., Pandiaraj, S., & Shetti, N. P. (2023). Trace level detection of 2, 4-dichloro phenoxy acetic acid and 4-cholrophenoxy acetic acid pesticides at iron-doped WO3 intercalated carbon matrix modified electrode. *Journal of Environmental Chemical Engineering, 11*(6), 111461.

Jacob, M. M., Ponnuchamy, M., Kapoor, A., & Sivaraman, P. (2020). Bagasse based biochar for the adsorptive removal of chlorpyrifos from contaminated water. *Journal of Environmental Chemical Engineering, 8*(4), 103904.

Jiang, S., Nguyen, T. A., Rudolph, V., Yang, H., Zhang, D., Ok, Y. S., & Huang, L. (2017). Characterization of hard-and softwood biochars pyrolyzed at high temperature. *Environmental geochemistry and Health, 39*, 403–415.

Jung, S., Park, Y.-K., & Kwon, E. E. (2019). Strategic use of biochar for CO2 capture and sequestration. *Journal of CO2 Utilization, 32*, 128–139.

Kang, B. S., Farooq, A., Valizadeh, B., Lee, D., Seo, M. W., Jung, S. C., Hussain, M., Kim, Y. M., Khan, M. A., Jeon, B. H., & Rhee, G. H. (2024). Valorization of sewage sludge via air/steam gasification using activated carbon and biochar as catalysts. *International Journal of Hydrogen Energy, 54*, 284–293.

Kundu, S., Ray, A., Gupta, S. D., Biswas, A., Roy, S., Tiwari, N. K., Kumar, V. S., & Das, B. K. (2024). Environmental bisphenol A disrupts methylation of steroidogenic genes in the ovary of Paradise threadfin Polynemus paradiseus via abnormal DNA methylation: Implications for human exposure and health risk assessment. *Chemosphere, 351*, 141236.

Lita, A. L., Hidayat, E., Mohamad Sarbani, N. M., Harada, H., Yonemura, S., Mitoma, Y., Herviyanti., & Gusmini. (2023). Glyphosate Removal from Water Using Biochar Based Coffee Husk Loaded Fe3O4. *Water, 15*, 2945.

Liu, H.-B., Yang, B., & Xue, N.-D. (2016). Enhanced adsorption of benzene vapor on granular activated carbon under humid conditions due to shifts in hydrophobicity and total micropore volume. *Journal of Hazardous Materials, 318*, 425–432.

Luo, M., Lin, H., He, Y., Li, B., Dong, Y., & Wang, L. (2019). Efficient simultaneous removal of cadmium and arsenic in aqueous solution by titanium-modified ultrasonic biochar. *Bioresource Technology, 284*, 333–339. Available from https://doi.org/10.1016/j.biortech.2019.03.108.

Luo, S., Li, S., Zhang, S., Cheng, Z., Nguyen, T. T., & Guo, M. (2022). Visible-light-driven Z-scheme protonated g-C3N4/wood flour biochar/BiVO4 photocatalyst with biochar as charge-transfer channel for enhanced RhB degradation and Cr (VI) reduction. *Science of The Total Environment, 806*, 150662.

Luo, Z., Yao, B., Yang, X., Wang, L., Xu, Z., Yan, X., Tian, L., Zhou, H., & Zhou, Y. (2022). Novel insights into the adsorption of organic contaminants by biochar: a review. *Chemosphere, 287*, 132113.

Madzaki, H., & KarimGhani, W. A. W. A. (2016). Carbon dioxide adsorption on sawdust biochar. *Procedia engineering, 148*, 718–725.

Manmeen, A., Kongjan, P., Palamanit, A., & Jariyaboon, R. (2023). Biochar and pyrolysis liquid production from durian peel by using slow pyrolysis process: Regression analysis, characterization, and economic assessment. *Industrial Crops and Products, 203*, 117162.

Maya, M., Gwenzi, W., & Chaukura, N. (2020). A biochar-based point-of-use water treatment system for the removal of fluoride, chromium and brilliant blue dye in ternary systems. *Environmental Engineering & Management Journal (EEMJ)*, 19.

Miao, Q., Li, M., Gao, G., Zhang, W., Zhang, J., & Yan, B. (2022). Improved Process for Separating TiO2 from an Oxalic-Acid Hydrothermal Leachate of Vanadium Slag. *Metals, 13*(1), 20.

Mishra, R. K., & Mohanty, K. (2018). An Overview of Techno-economic Analysis and Life-Cycle Assessment of Thermochemical Conversion of Lignocellulosic Biomass. *Recent Advancements in Biofuels and Bioenergy Utilization*, 363–402.

Monga, A., Fulke, A. B., & Dasgupta, D. (2022). Recent developments in essentiality of trivalent chromium and toxicity of hexavalent chromium: implications on human health and remediation strategies. *Journal of Hazardous Materials Advances*, *7*, 100113.

Murtaza, G., Ahmed, Z., Dai, D.-Q., Iqbal, R., Bawazeer, S., Usman, M., Rizwan, M., Iqbal, J., Akram, M. I., & Althubiani, A. S. (2022). A review of mechanism and adsorption capacities of biochar-based engineered composites for removing aquatic pollutants from contaminated water. *Frontiers in Environmental Science*, *10*, 2155.

Navya, A., Nandhini, S., Sivamani, S., Vasu, G., Sivarajasekar, N., & Hosseini-Bandegharaei, A. (2020). Preparation and characterization of cassava stem biochar for mixed reactive dyes removal from simulated effluent. *Desalination and Water Treatment*, *189*, 440–451.

Nematian, M., Keske, C., & Ng'ombe, J. N. (2021). A techno-economic analysis of biochar production and the bioeconomy for orchard biomass. *Waste Management*, *135*, 467–477.

Osman, A. I., Fawzy, S., Farghali, M., El-Azazy, M., Elgarahy, A. M., Fahim, R. A., Maksoud, M. A., Ajlan, A. A., Yousry, M., Saleem, Y., & Rooney, D. W. (2022). Biochar for agronomy, animal farming, anaerobic digestion, composting, water treatment, soil remediation, construction, energy storage, and carbon sequestration: a review. *Environmental Chemistry Letters*, *20*(4), 2385–2485.

Panwar, N., Pawar, A., & Salvi, B. (2019). Comprehensive review on production and utilization of biochar. *SN Applied Sciences*, *1*, 1–19.

Patanjali, P., Singh, R., Kumar, A., & Chaudhary, P. (2019). *Nanotechnology for water treatment: A green approach. Green synthesis, characterization and applications of nanoparticles* (pp. 485–512). Elsevier.

Patra, B. R., Mukherjee, A., Nanda, S., & Dalai, A. K. (2021). Biochar production, activation and adsorptive applications: a review. *Environmental Chemistry Letters*, *19*, 2237–2259.

Pecchi, M., Baratieri, M., Maag, A. R., & Goldfarb, J. L. (2023). Uncovering the transition between hydrothermal carbonization and liquefaction via secondary char extraction: A case study using food waste. *Waste Management*, *168*, 281–289.

Peter, A., Chabot, B., & Loranger, E. (2021). Enhanced activation of ultrasonic pre-treated softwood biochar for efficient heavy metal removal from water. *Journal of Environmental Management*, *290*, 112569.

Praveen, S., Jegan, J., Bhagavathi Pushpa, T., Gokulan, R., & Bulgariu, L. (2022). Biochar for removal of dyes in contaminated water: an overview. *Biochar*, *4*(1), 10.

Qian, L., Chen, B., & Hu, D. (2013). Effective alleviation of aluminum phytotoxicity by manure-derived biochar. *Environmental Science & Technology*, *47*, 2737–2745.

Qiu, X., Zhou, G., Wang, H., & Wu, X. (2021). The behavior of antibiotic-resistance genes and their relationships with the bacterial community and heavy metals during sewage sludge composting. *Ecotoxicology and Environmental Safety*, *216*, 112190.

Reddy, S. J. (2020). The recent advances in the nanotechnology and its applications-A review. *Nanotechnology*, *50*(5).

Rego, F., Xiang, H., Yang, Y., Ordovás, J. L., Chong, K., Wang, J., & Bridgwater, A. (2022). Investigation of the role of feedstock properties and process conditions on the slow pyrolysis of biomass in a continuous auger reactor. *Journal of Analytical and Applied Pyrolysis*, *161*, 105378.

Ripanda, A., Rwiza, M. J., Nyanza, E. C., Bakari, R., Miraji, H., Njau, K. N., Vuai, S. A. H., & Machunda, R. L. (2023). Removal of lamivudine from synthetic solution using jamun seed (Syzygium cumini) biochar adsorbent. *Emerging Contaminants*, *9*, 100232.

Ronsse, F., Van Hecke, S., Dickinson, D., & Prins, W. (2013). Production and characterization of slow pyrolysis biochar: influence of feedstock type and pyrolysis conditions. *Gcb Bioenergy*, *5*(2), 104–115.

Sahu, N., Singh, J., & Koduru, J. R. (2021). Removal of arsenic from aqueous solution by novel iron and iron–zirconium modified activated carbon derived from chemical carbonization of Tectona grandis sawdust: Isotherm, kinetic, thermodynamic and breakthrough curve modelling. *Environmental Research*, *200*, 111431.

Sakhiya, A. K., Anand, A., & Kaushal, P. (2020). Production, activation, and applications of biochar in recent times. *Biochar*, *2*, 253–285.

Sakhiya, A. K., Baghel, P., Anand, A., Vijay, V. K., & Kaushal, P. (2021). A comparative study of physical and chemical activation of rice straw derived biochar to enhance Zn^{+2} adsorption. *Bioresource Technology Reports*, *15*, 100774.

Shakya, A., Vithanage, M., & Agarwal, T. (2022). Influence of pyrolysis temperature on biochar properties and Cr (VI) adsorption from water with groundnut shell biochars: Mechanistic approach. *Environmental Research*, *215*, 114243.

Shetti, N. P., Malode, S. J., Nayak, D. S., Reddy, C. V., & Reddy, K. R. (2019). Novel biosensor for efficient electrochemical detection of methdilazine using carbon nanotubes-modified electrodes. *Materials Research Express*, *6*(11), 116308.

Shi, Y., Wang, S., Xu, M., Yan, X., Huang, J., & Wang, H. W. (2022). Removal of neonicotinoid pesticides by adsorption on modified Tenebrio molitor frass biochar: Kinetics and mechanism. *Separation and Purification Technology*, *297*, 121506.

Stefaniuk, M., Oleszczuk, P., & Ok, Y. S. (2016). Review on nano zerovalent iron (nZVI): from synthesis to environmental applications. *Chemical Engineering Journal*, *287*, 618–632.

Su, C., Wang, S., Zhou, Z., Wang, H., Xie, X., Yang, Y., Feng, Y., Liu, W., & Liu, P. (2021). Chemical processes of Cr (VI) removal by Fe-modified biochar under aerobic and anaerobic conditions and mechanism characterization under aerobic conditions using synchrotron-related techniques. *Science of The Total Environment*, *768*(2021).

Suryawanshi, S. S., Kamble, P. P., Gurav, R., Yang, Y. H., & Jadhav, J. P. (2023). Statistical comparison of various agricultural and non-agricultural waste biomass-derived biochar for methylene blue dye sorption. *Biomass Conversion and Biorefinery*, *13*(6), 5353–5366.

Taghizadeh, M., Kebria, D., Darvishi, G., & Kootenaei, F. (2013). The Use of Nano Zero Valent Iron in Remediation of Contaminated Soil and Groundwater. *International Journal of Scientific Research in Environmental Sciences*, *1*(7), 152–157.

Tran, H. N., You, S. J., Hosseini-Bandegharaei, A., & Chao, H. P. (2017). Mistakes and inconsistencies regarding adsorption of contaminants from aqueous solutions: A critical review. *Water Research*, *120*, 88–116. Available from https://doi.org/10.1016/j.watres.2017.04.014, http://www.elsevier.com/locate/watres.

Tripathi, M., Sahu, J. N., & Ganesan, P. (2016). Effect of process parameters on production of biochar from biomass waste through pyrolysis: A review. *Renewable and sustainable energy reviews, 55*, 467–481.

Uroić Štefanko, A., & Leszczynska, D. (2020). Impact of biomass source and pyrolysis parameters on physicochemical properties of biochar manufactured for innovative applications. *Frontiers in Energy Research, 8*, 138.

Usevičiūtė, L., Baltrėnaitė-Gedienė, E., & Baltrėnas, P. (2021). Hydrophilicity enhancement of low-temperature lignocellulosic biochar modified by physical–chemical techniques. *Journal of Material Cycles and Waste Management, 23*(5), 1838–1854.

Vijayaraghavan, K. (2019). Recent advancements in biochar preparation, feedstocks, modification, characterization and future applications. *Environmental Technology Reviews, 8*(1), 47–64.

Wang, F., Liu, L. Y., Liu, F., Wang, L. G., Ouyang, T., & Chang, C. T. (2017). Facile one-step synthesis of magnetically modified biochar with enhanced removal capacity for hexavalent chromium from aqueous solution. *Journal of the Taiwan Institute of Chemical Engineers, 81*, 414–418. Available from https://doi.org/10.1016/j.jtice.2017.09.035, http://www.elsevier.com/wps/find/journaldescription.cws_home/715607/description#description.

Wang, L., Yan, W., He, C., Wen, H., Cai, Z., Wang, Z., Chen, Z., & Liu, W. (2018). Microwave-assisted preparation of nitrogen-doped biochars by ammonium acetate activation for adsorption of acid red 18. *Applied Surface Science, 433*, 222–231.

Wang, H., Ren, T., Yang, H., Feng, Y., Feng, H., Liu, G., Yin, Q., & Shi, H. (2020). Research and application of biochar in soil CO_2 emission, fertility, and microorganisms: A sustainable solution to solve China's agricultural straw burning problem. *Sustainability, 12*, 1922.

Wang, H., Zang, S., Teng, H., Wang, X., Xu, J., & Sheng, L. (2023). Characteristic of $KMnO_4$-modified corn straw biochar and its application in constructed wetland to treat city tail water. *Environmental Science and Pollution Research, 30*(17), 49948–49962.

Wongrod, S., Simon, S., van Hullebusch, E. D., Lens, P. N. L., & Guibaud, G. (2019). Assessing arsenic redox state evolution in solution and solid phase during As(III) sorption onto chemically-treated sewage sludge digestate biochars. *Bioresource Technology, 275*, 232–238. Available from https://doi.org/10.1016/j.biortech.2018.12.056, http://www.elsevier.com/locate/biortech.

Xu, Y., Liu, Y., Liu, S., Tan, X., Zeng, G., Zeng, W., Ding, Y., Cao, W., & Zheng, B. (2016). Enhanced adsorption of methylene blue by citric acid modification of biochar derived from water hyacinth (Eichornia crassipes). *Environmental Science and Pollution Research, 23*, 23606–23618.

Yang, F., Zhang, S., Sun, Y., Du, Q., Song, J., & Tsang, D. C. (2019). A novel electrochemical modification combined with one-step pyrolysis for preparation of sustainable thorn-like iron-based biochar composites. *Bioresource Technology, 274*, 379–385.

Yang, Y., Ye, S., Zhang, C., Zeng, G., Tan, X., Song, B., Zhang, P., Yang, H., Li, M., & Chen, Q. (2021). Application of biochar for the remediation of polluted sediments. *Journal of Hazardous Materials, 404*, 124052.

Yi, Y., Li, C., Zhao, L., Du, X., Gao, L., Chen, J., Zhai, Y., & Zeng, G. (2018). The synthetic evaluation of CuO-MnO x-modified pinecone biochar for simultaneous removal formaldehyde and elemental mercury from simulated flue gas. *Environmental Science and Pollution Research, 25*, 4761–4775.

Yi, Y., Tu, G., Zhao, D., Tsang, P. E., & Fang, Z. (2019). Biomass waste components significantly influence the removal of Cr(VI) using magnetic biochar derived from four types of feedstocks and steel pickling waste liquor. *Chemical Engineering Journal, 360*, 212–220. Available from https://doi.org/10.1016/j.cej.2018.11.205, http://www.elsevier.com/inca/publications/store/6/0/1/2/7/3/index.htt.

Zakaria, M. R., Farid, M. A. A., Andou, Y., Ramli, I., & Hassan, M. A. (2023). Production of biochar and activated carbon from oil palm biomass: current status, prospects, and challenges. *Industrial Crops and Products, 199*116767.

Zhang, M., & Gao, B. (2013). Removal of arsenic, methylene blue, and phosphate by biochar/AlOOH nanocomposite. *Chemical engineering journal, 226*, 286–292.

Zhang, X., Zhang, S., Yang, H., Feng, Y., Chen, Y., Wang, X., & Chen, H. (2014). Nitrogen enriched biochar modified by high temperature CO_2–ammonia treatment: Characterization and adsorption of CO_2. *Chemical Engineering Journal, 257*, 20–27.

Zhang, Y., Xu, X., Zhang, P., Zhao, L., Qiu, H., & Cao, X. (2019). Pyrolysis-temperature depended quinone and carbonyl groups as the electron accepting sites in barley grass derived biochar. *Chemosphere, 232*, 273–280.

Zhao, N., Zhao, C., Lv, Y., Zhang, W., Du, Y., Hao, Z., & Zhang, J. (2017). Adsorption and coadsorption mechanisms of Cr (VI) and organic contaminants on H_3PO_4 treated biochar. *Chemosphere, 186*, 422–429.

Zhao, L., Xiao, D., Liu, Y., Xu, H., Nan, H., Li, D., Kan, Y., & Cao, X. (2020). Biochar as simultaneous shelter, adsorbent, pH buffer, and substrate of Pseudomonas citronellolis to promote biodegradation of high concentrations of phenol in wastewater. *Water Research, 172*, 115494.

Zhao, W., Zhang, Z., Xin, Y., Xiao, R., Gao, F., Wu, H., Wang, W., Guan, Q., & Lu, K. (2024). Na_2S-modified biochar for Hg (II) removal from wastewater: a techno-economic assessment. *Fuel, 356*, 129641.

Zhu, N., Yan, T., Qiao, J., & Cao, H. (2016). Adsorption of arsenic, phosphorus and chromium by bismuth impregnated biochar: Adsorption mechanism and depleted adsorbent utilization. *Chemosphere, 164*, 32–40. Available from https://doi.org/10.1016/j.chemosphere.2016.08.036.

Zhu, H., An, Q., Nasir, A. S. M., Babin, A., Saucedo, S. L., Vallenas, A., Li, L., Baldwin, S. A., Lau, A., & Bi, X. (2023). Emerging applications of biochar: A review on techno-environmental-economic aspects. *Bioresource Technology*, 129745.

Zubair, M., Aziz, H. A., Ihsanullah, I., Ahmad, M. A., & Al-Harthi, M. A. (2022). Enhanced removal of Eriochrome Black T from water using biochar/layered double hydroxide/chitosan hybrid composite: Performance evaluation and optimization using BBD-RSM approach. *Environmental Research, 209*, 112861.

Chapter 4

Design, characterization, and evaluation of biochar: recent advances, applications, and future research directions

Abimbola Oluwalana-Sanusi[1], Wisdom Archford Munzeiwa[2], Silas Verkijika[2] and Nhamo Chaukura[3]

[1]Department of Physical and Earth Sciences, Centre for Global Change, Sol Plaatje University, Kimberley, South Africa, [2]Department of Physical and Earth Sciences, Centre for Applied Data Science, Sol Plaatje University, Kimberley, South Africa, [3]Department of Physical and Earth Sciences, Sol Plaatje University, Kimberley, South Africa

Chapter outline

4.1 Introduction	59
4.2 Characterization of biochar and biochar-contaminant systems	60
4.2.1 Biochar surface properties/phenomena	61
4.2.2 Internal microstructure	62
4.2.3 3-D micro-CT analysis	63
4.2.4 Synchrotron X-ray microtomography and multifractal analysis	63
4.3 Design and evaluation of biochar systems	64
4.3.1 In-silico-computational modeling or computer-aided design approach	65
4.3.2 Artificial intelligence and machine learning tools	66
4.3.3 Current and potential applications	68
4.4 Future perspectives and research directions	69
4.5 Conclusions	70
References	70

4.1 Introduction

Biochar derives its prominence from ancient societies where its main use was for energy, agricultural practices and as a traditional remedy for abdominal conditions (Osman et al., 2022). Its relevance was rekindled as global trends were navigating toward sustainable farming, climate change mitigation, and food security. This renewed interest over the years has ushered in new innovations in the biochar production process optimization, characterization techniques, and applications in agronomy and animal husbandry, ecological remediation, and renewable energy (Osman et al., 2022). The inherent biochar properties such as large surface area, stable carbon structure, porosity, and conductivity make it a green material of choice in industrial applications. Biochar attributes are majorly controlled through conditions of pyrolysis and the type of feedstock (Shaikh et al., 2023).

Earliest production methods involved burning the feedstock in oxygen-starved molded ground pits. The advancement of biochar production has witnessed the development of different types of kilns and more sophisticated reactors for large-scale production. Recently, the widely used method of biochar production is pyrolysis under fast or slow conditions (Wang et al., 2020). Slow pyrolysis is low temperature (300°C–500°C) conversion, while fast pyrolysis happens at elevated temperatures (500°C–800°C) with production of biochar, bio-oil and other byproducts. Of late, hydrothermal carbonization (HT), microwave pyrolysis, gasification, torrefaction, methods have gained prominence (Jeyasubramanian et al., 2021) Advances in research are now paying attention to optimizing production processes to maximize yields, reduce environmental effects and improve the physicochemical properties of biochar.

Biochar has a heterogeneous structure which comprises of the carbonaceous component, mineral phases, pore structure, cellular residues, and surface functional groups (SFGs). The chemical composition is dominated by carbon and hydrogen, along with other heteroatoms such as oxygen, nitrogen, and sulfur (Hassan et al., 2020). Various characterization techniques have been used to discern the properties and performance of biochar in a bid to optimize the

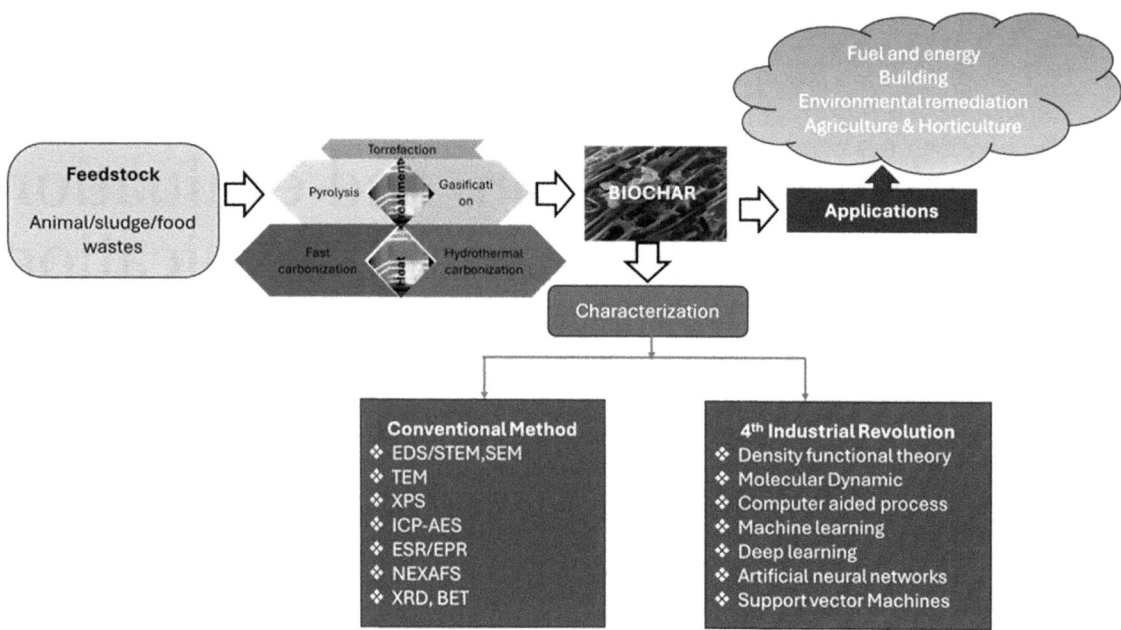

FIGURE 4.1 Biochar fabrication and characterization techniques.

production process and applicability. Spectroscopy methods (such as Fourier transform infrared spectrometry [FTIR], Raman, and X-ray diffraction [XRD]), microscopy, surface area analysis, and elemental analysis have conventionally been used for biochar characterization (Jang et al., 2018). Modern research and development have seen the evolution of computer aided designing and engineering of biochar systems.

Specifically, artificial intelligence (AI) and machine learning (ML) have revolutionized the industrial landscape in material design (Wang et al., 2024). Its adoption in the context of biochar has had a tremendous effect in understanding and designing biochar systems that are suitable for a wide range of applications. Despite enormous developments in biochar exploration, there are still knowledge gaps that need to be addressed. This chapter explores the design, characterization, and evaluation of biochar systems by evaluating current state-of-art techniques that will advance the acceptability and usability of biochar material in a sustainable manner (Fig. 4.1).

4.2 Characterization of biochar and biochar-contaminant systems

Biochar has been used for carbon sequestration and soil amelioration, as well as an adsorbent for inorganic and organic contaminants. Nevertheless, the utilization of biochar in contaminant systems is based on the production parameters as well as the feedstock and reactor type as they affect its physicochemical properties, structure, and chemical attributes. Hence, to determine the appropriateness of biochar for agricultural/environmental usage thorough characterization must be done. Conventional characterization techniques such as FTIR, electron spin resonance (ESR), BET, and XRD among others have been utilized to determine the functional surface properties, surface area, pore size, pore volume and size distribution (Fig. 4.2).

Proximate analysis like pH, cation exchange capacity (CEC), ash content, and Boehm titration are commonly utilized for the determination of functional groups while elemental composition is detected using energy-dispersive X-ray spectroscopy (Jafri et al., 2018). Since biochars are made from biomass, the decrease in the ratio of hydrogen/carbon and oxygen/carbon indicates the level of aromaticity and this is measured with nuclear magnetic resonance (Chen, Gao, et al., 2022; Chen, Zhao, et al., 2022). Upon functionalization, biochar physicochemical properties change based on the modification method. Pristine biochar has lower surface area, pore volume, pH value and cationic exchange concentration compared to functionalized biochar. The waste for wood biochar pyrolyzed at 500°C had a surface area of 5.7 m^2 g^{-1} and upon functionalization with Fe(II), it significantly increased to 37.8 m^2 g^{-1} (Dong et al., 2022). There is a recent emergence of synchrotron-based techniques technique combined with machine learning for data processing which includes micro-X-ray florescence, 3-D micro-computed tomography (μCT), and near edge X-ray absorption fine structure spectroscopy (NEXAFS). These techniques are capable of providing insights into the adsorption mechanisms

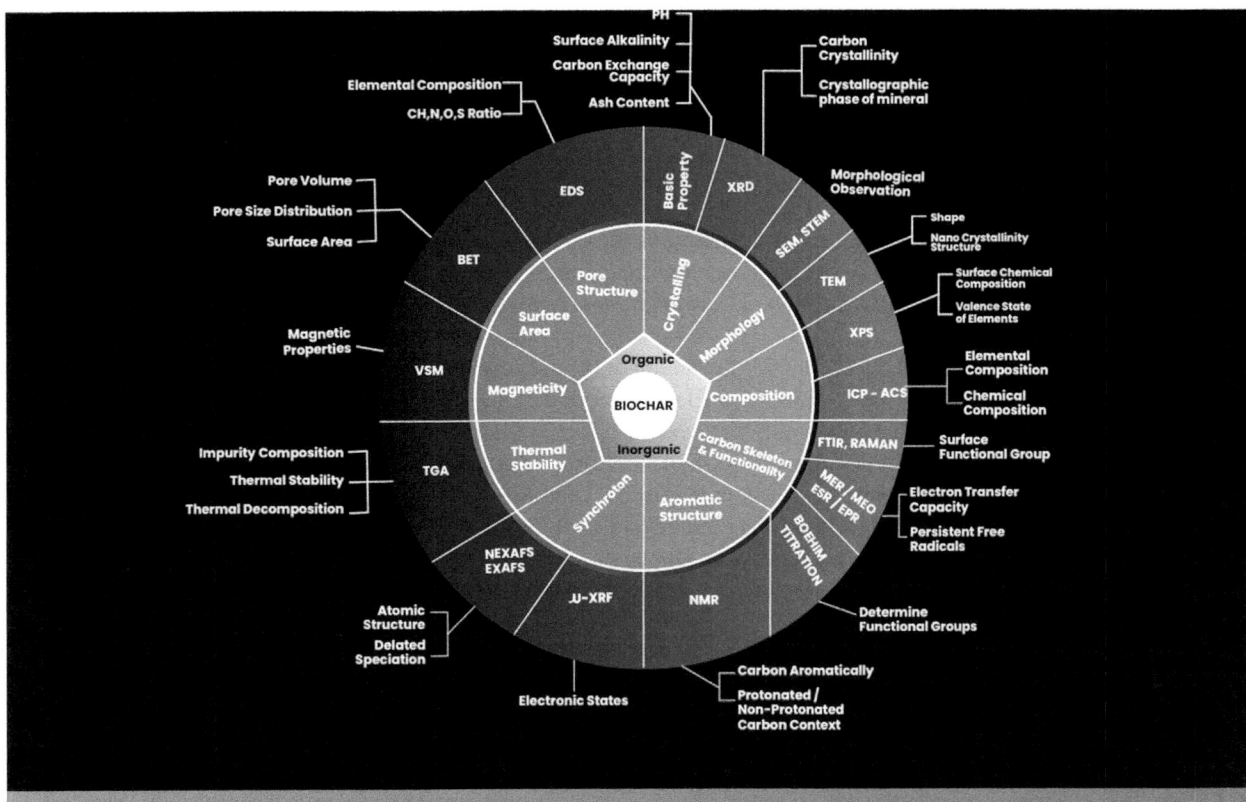

FIGURE 4.2 Characteristics and characterization techniques for biochar

of biochar and contaminants. However, these synchrotron-based techniques have not been fully utilized in biochar characterization and they are going to be explored more in this chapter.

4.2.1 Biochar surface properties/phenomena

Biochars usually exhibit different surface properties which are dependent on biomass chemical composition and pyrolysis conditions, and they are indexed as total surface area (TSA), volume/size of pores, and SFGs. It has been proved that treating biochars and biomass with strong acids (H_2SO_4, H_2PO_4), alkalis (KOH, NaOH), and salts ($ZnCl_2$), can increase porosity and surface area (Berhanu et al., 2018; Zhang, Huang, et al., 2022).

During pyrolysis the volatile compounds from the biomass are released leaving pores and carbon rich residues. The physical and chemical processes that take place leave the carbonaceous material and cause aromatization. The disordered aromatic rings form an amorphous phase and ordered polyaromatic rings form the crystalline phase. The inherent biomass cellular structure of the starting materials also contributes to the overall surface and pore structure (Tan et al., 2021).

Biochar surfaces can also exhibit hydrophilic and hydrophobic characteristics which are controlled by the biochar functionality and pyrolysis temperature. Increase in pyrolysis temperature tends to reduce the surface density polar functional groups (Pak et al., 2023). High crystalline biochar materials tend to be water-repellent because of the highly organized carbon structure which is void of free space and surface area for water penetration. On the other hand amorphous portion because of its disordered structure, existence of polar chemical moieties and large surface area tends to be more hydrophilic (Clurman et al., 2020).

Biochar surfaces have different SFGs which can be categorized based on the heteroatom present (i.e. oxygen, nitrogen, or sulfur), which can be further subdivided as acidic, neutral, and basic (Kumar et al., 2023). The oxygen-containing functional groups are the most dominant because of the native O-containing components (cellulose, hemicellulose, and lignin). During pyrolysis different thermochemical reactions generate O-containing chemical groups, (-COOH, alcoholic-OH, or phenolic-OH, and keto-RC(O)R) (Weber & Quicker, 2018).

Nitrogenous abundant biomass, which is enriched with proteins, amino acids, and nucleic acids, usually possesses a high fraction nitrogen nitrogen-containing functional groups. These include amines (NH_2), amides ($COHN_2$), nitriles (CN), heterocyclic pyridinic, and pyrolytic and N-oxides. Functional groups can also be introduced by pre- and post-treatment methods. Methods such as steam activation and oxidation (hammers method) tend to raise the surface density of oxygen-containing chemical moieties. Surface properties are also enhanced by impregnation with different inorganic nanoparticles such as iron oxide and zinc oxide (Zhang, Huang, et al., 2022).

The acidic and basic groups contribute to the overall surface charge which is dependent on the pH of the surroundings. The surface charges can be quantified using Zeta potentials. The magnitude of the values reflects the polarity of the biochar surfaces at a specific pH. At pH values below the point of zero charge (PZC), zeta potential is negative signifying negatively charged surfaces. Conversely, above PZC, zeta potentials are positive, pointing to positively charged surfaces (Weber & Quicker, 2018).

Biochar surface phenomena denote the numerous interactions and processes that happen at its surface with the surroundings. The presence of functional groups facilitates adsorption at biochar surfaces via physio- or chemisorption. Depending on the nature of contaminants within the system different interaction mechanisms exist. Organic pollutants usually interact with biochar *via* van der Waals forces, electrostatic interactions, dipole-dipole interactions, London dispersion forces, π-π stacking, and H-bonding (Tan et al., 2021). Inorganic species usually interact with biochar surfaces via electrostatic interactions, complexation effects, and physio-sorption. Some biochar surfaces potentially promote catalytic processes due to mineral phases and functional groups that can facilitate electron transfer processes. Biochar surfaces can also exhibit electrochemical behavior as an electron acceptor, an important attribute for energy storage applications (Faheem et al., 2020)(.

4.2.2 Internal microstructure

The pyrolysis of feedstock results in porous materials with macropores, mesopores, and micropores, and these various degree pore sizes provide a hierarchical pore architecture (Mrad & Chehab, 2019). These pores depending on process variables can interconnect and form a porous channel network. The temperature of pyrolysis is the most important factor that causes the loss of carbon and directly affects the pore structure and internal microstructure (Xie et al., 2021). The pore sizes affect its application as adsorbents, carbon sequestering, soil amendments, and as internal curing agent in mortar as the greater the pore size, the quicker water can penetrate the particles. The internal microstructure such as pore surface and morphology has been predicted to influence the adsorption property of biochar in the gaseous and aqueous phases. Gas-enhanced pyrolysis often improves the pore morphology and surface area. Carbon dioxide-assisted rice husk pyrolysis at 500°C resulted in pristine biochar with a smooth morphology in comparison to the activated biochar (Arif et al., 2023). The surface area of the activated biochar was higher (95 $m^2 g^{-1}$) compared to the pristine (74 $m^2 g^{-1}$) and this affected the adsorption capacity. The pores in the activated biochar assisted in the surface adsorption of azithromycin *via* pore-filling mechanism.

Biochar modification also contributes to the overall biochar microstructure and surface area. Recently, sulfur-modified biochar (SBC) prepared Na_2S from sawdust exhibited tiny pores compared to the unmodified biochar (UBC) and this was attributed to H_2S gas formation and volatilization. It was also observed that the yields of SBC (15%) were lower than UBC (42%) but this did not affect the Hg(II) removal efficiency as SBC efficiency was 96.4% while UBC was 73%, the high efficiency of SBC was due to additional Na_2S adsorption sites (Zhao et al., 2024). In addition to gases and inorganic compounds can be used for activating/enhancing biochar; for example, hybrid nanocomposite (ZnO/SiO_2) have been used to dope biochar derived from kitchen waste. A rough surface and folded flaky porous structure were obtained, and this increases its surface area and adsorption capacity. The BET analysis shows the biochar has large porous cavities that are mesoporous and macroporous, with a pore volume of 8.17×10^{-3} $cm^3 g^{-1}$ and surface area of 45 $m^2 g^{-1}$. This promoted the efficient removal of tetracycline (76.04 ± 1.086%) (Shaikh et al., 2023). A residue of a common Chinese herb (*Flueggea suffruticosa*) was pyrolyzed at 500°C and further modified with zinc chloride. The internal structure of BC and zinc functionalized (ZnBC) showed a porous and rough surface, with a higher surface area of 1556 $m^2 g^{-1}$ and pore volume of 0.871 $cc\ g^{-1}$ compared to pristine BC and this is consistent with previous studies (Wang, Lou, et al., 2021). Recent research utilized pharmaceutical sludge to prepare biochar, and the influence of pyrolysis temperature and modification were investigated (Wu et al., 2022). BET surface area (S_{Bet}) attributes were enhanced with increased temperature, however, a drop was noticed at 900°C making 800°C the optimal temperature. Further modification with KOH, $ZnCl_2$, and CO_2 showed $ZnCl_2$ modification to be the most effective in removing levofloxacin (LEV) at 159.26 $mg\ g^{-1}$ (99.9%) (Wu et al., 2022). Modification of biochar enhances the adsorption properties through pore enhancement and increased surface area.

4.2.3 3-D micro-CT analysis

To interpret the mechanism behind biochar adsorption mechanism pertaining to soil remediation and removal of contaminants in water, it is important to understand its pore architecture. The arrangement of the pore space plays an important role in interaction between the biochar and its environment in cases such as the flow of water containing contaminants, diffusion of gases, and retention of nutrients in soil (Singh et al., 2021). Techniques such as atomic force microscopy, scanning electron microscopy X-ray photoelectron spectroscopy, and electron dispersive spectrometry have been used and are still in used for visualization analysis of heterogenous adsorption process but are limited in determining the suitability of the model and adsorption procedure of adsorbents like biochar (Zhang, Tian, et al., 2022). Hence, researchers developed a 3D visualization model that can observe the distribution of elements in the sample while observing the micromorphological interaction of the adsorbent and the contaminant system. A novel high-resolution X-ray micro-computed tomography (μ-CT) is used to investigate the unchanged internal micropores of biochar as the technique utilizes nondestructive 3D-imaging technology (Schlüter et al., 2020). Simulations obtained from μCT scan are in 3D dimension facilitating its utilization in porous research and analysis (Zhang et al., 2023). The merit of μCT includes a direct pore permeability view thus enabling proper pore structure examination and quantitative identification of pores thus the use of μCT scan to explore the micropore architecture of macroscopic clusters in biochar (Yu & Lu, 2020). However, only a small number of research has endeavored to evaluate dynamic activities using actual CT data. Simulation tools help in deriving larger, reliable outcomes. While exploring theoretical situations can help us understand specific mechanisms, algorithms, and simulations might not precisely replicate real-world scenarios in the environments. Because of computational constraints, integrating biochar pore structure with adsorption process/remediation takes time and effort.

Contaminants and biochar have different density matrices, and micro-CT can be used to characterize their adsorption. A previous study used 3D μ-CT to observe the adsorption behavior of Pb(II) on biochar, the result shows that Pb(II) was first absorbed on biochar surface before intra-particle diffusion took place proving the existence of a complex adsorption process (Zhang, Tian, et al., 2022). X-ray μ-CT coupled with K-means clustering algorithm was utilized to comprehend the mechanism of adsorption of biochar derived from wheat straw, coconut shell, and rice husk (Zhang, Li, et al., 2022). The authors reported that the different morphologies of the chars affected the adsorption sites for Pb(II) even though the adsorption occurred mainly on the chars surface; they also stipulated the different porous structures of the chars contributed to the different distribution sites. Understanding the mechanism of μCT works by comparing the macrostructure of the biochar before and after applying it to a contaminant system as it is a technique that can extract the different biochar macropore sizes through the image segmentation method. Research on the use of four different char on soil amendments was carried out and the application of μCT and advanced analytical methods to understand the mechanism indicated the soil was improved through the addition of pore space in the soil known as expansion effect and occupying the initial pore space of the soil known as occupying effect (Yu & Lu, 2020). However, despite μ-CT being able to analyze the mechanism between biochar and contaminant systems, it still requires monitored machine-learning algorithms which is still a bottleneck.

4.2.4 Synchrotron X-ray microtomography and multifractal analysis

Techniques that utilize the screening of structural properties and the variation of biochar preparation will assist in the mass production as simple adjustment of pyrolysis temperature or quantity, or pyrolysis time and feedstock cannot predetermine biochar pore structure. Hence, the quantitative micrometer-scale pore-structure of feedstock for biochar before and after pyrolysis is being studied through the utilization of image analysis and X-ray microtomography (μCT) (Suuronen & Jyske, 2019) even though there is limited data available. μCT is like medical tomography which requires images from different directions, and it is a nondestructive visualization method. Synchrotron of μCT radiographs makes it easy to identify phase contrast and postprocess data (Suuronen & Jyske, 2019). Therefore, personalized biochar generation and optimization require a thorough understanding of biomass metamorphosis during pyrolysis, as well as the effect of pre- or post-pyrolysis treatments. This includes implications for biochar internal structure, porosity, and surface area, all of these might affect how it behaves in the surroundings or technological uses like diffusion of gases in biofilters, water storage and transportation in soil.

Incorporating biochar in soil may enhance its characteristics and uses, which include moisture and nutrient retention, while also sequestering carbon from the atmosphere. Furthermore, the application of biochar as an agent for the elimination of pollutants from contaminated waters or as a porous material in biofilters has lately received substantial interest (Shaheen et al., 2019). Utilization of biochar relies on its chemical and physical qualities, and these are defined by biomass characteristics and the parameters of conversion mechanism. Therefore, it is crucial to comprehend the transformation that occurs in biomass during pyrolysis and its effects on specific applications; this is where synchrotron X-ray microtomography is needed due to its qualitative analysis of biochar pore, network, and geometry distribution network and it is a nondestructive technique. μCT was used to study wood waste at pyrolyzed at 400°C and 700°C, it was discovered that pyrolysis temperature did not affect the pores and macrostructure of the biochar and the morphology resembles that of the initial wood which is an interconnected longitudinal pores (Berhanu et al., 2018). In another study, μCT was used to investigate biochars prepared from softwood pellets at 0.87 μm voxel resolution, it was discovered that the biochar had residue properties of the original wood. However, the pore structure was modified by thermochemical decomposition and pelleting (Srocke et al., 2021) as pyrolysis occurs the biomass shrunk which resulted in slit-shaped pores of 30–150 μm. The authors also performed multifractal analysis on the porosity frequency and a monofractal behavior could be explained by the linear form of the Rényi spectra which also implies a homogeneous pore structure with little fluctuation. At 3.7 μm voxel resolution, research has shown that soil porosity and macroaggregates increase with the addition of wood chip biochar based on stimulation done in reconstructed 3D pore structures for water flow, it was discovered that water and nutrients could flow in a definite direction thus facilitating interaction between soil and biochar particles (Yu et al., 2016; Zhou et al., 2018). Willow biochar's pore arrangement, influenced by its vascular tissue architecture, exhibits anisotropy in the micrometer-range. Pyrolysis at different temperatures alters size distribution and pore shape but does not affect porosity (Hyväluoma, Hannula, et al., 2018). Pelagic and Benthic brown algae were evaluated for preparation of biochar and its activation, the internal structure of the activated biochar, biochar, and biomass were analyzed using X-ray μCT and the result shows the formation of new pore space through pyrolysis (Pak et al., 2023). The surface area, porosity, inner pore architecture, pore size configuration, and anisotropy of biochar prepared from a variety of biomass and varying pyrolysis temperatures have been studied using X-ray μCT (Hyväluoma, Kulju, et al., 2018; Watanabe, 2018). However, interpretation of the results, was limited because structural modifications are indirectly measurable since the samples studied are different in chemical composition. The utilization of synchrotron excitation sources, to create a tunable, semi-coherent x-ray beam of exceptional brightness, allows for the acquisition of phase distinction pictures enabling analysis of biochar properties to submicron levels (Fusseis et al., 2014). Excellent quality 3D images captured with synchrotron x-ray XμCT give helpful data regarding the modifications occurring in biomass going through thermochemical breakdown at higher temperatures, as well as the microstructural attributes of biochar derived from diverse feedstocks and pyrolyzed at different temperatures (Fusseis et al., 2014).

Carbonization temperature impact on straw of rice husk, miscanthus, oilseed rape, and wheat was investigated by collecting 3D scans at 50°C intervals from 50°C to 800°C (Edeh et al., 2023). The findings demonstrate that pyrolysis temperature and feedstock biomass had a significant impact on porosity, TSA, total void volume, and pore architecture. The pore linkage was depended on the type of biomass; nevertheless, for all biomass, pore linkage reduced for biochar prepared above 550°C (Edeh et al., 2023). Using 4D image visualization and SXμCT data processing, a deeper knowledge of the parameters that influence biochar pore growth was gained.

4.3 Design and evaluation of biochar systems

The complex engineering and designing processes for biochar systems in informed by biomass composition, availability, and the ultimate biochar application. Depending on different factors different types of biomass contain different lignocellulosic material, moisture, and ash content as well as the elemental composition. The production process is vital as it ensures optimum biochar properties, consistency, and standards (Leng et al., 2021). The initial step in the production line is biomass selection followed by conditioning mainly by drying, particles size reduction and sieving, pretreatment and blending, and palletization. The drying process helps to control the moisture levels which enhances pyrolysis thereby improving on the quality of biochar. Uniform smaller feedstock particle sizes are achieved by grinding and sieving can ensure uniform thermal heat distribution thereby controlling the overall biochar properties. The biomass can be modified for specific biochar applications by pretreatment or by blending. Different types of pretreatments exist which include chemical and biological utilizing acids or bases and enzymes and/or microbes, respectively.

The biomass is usually transformed into the carbonaceous biochar material *via* thermochemical approaches such as hydrothermal, gasification, and pyrolysis among other methods. Pyrolysis has been the method of choice to produce biochar because of its flexibility, proficiency, low carbon footprint, high biochar yields, and quality. To engineer biochar with specific surface and physical properties the pyrolysis conditions (heating rate, pyrolysis temperature, pressure, residence time, gas flow rate) are optimized (Dadi et al., 2023). The proficiency of the production process is evaluated by considering different metrics such as biochar yield, ash content, and physiochemical properties such as stability, surface area, pore dimensions, elemental rations, and SFGs. Application parameters such as adsorption capacity are also correlated to the biochar physiochemical properties (Hassan et al., 2020).

Conventional methods of optimizing biochar production and applications are time-consuming as they require repeated experiments. The Fourth Industrial Revolution (4IR) has brought impetus in transforming industrialization and civilization, by incorporating digital tools, automation, artificial intelligence (AI), and Internet of Things (IoT). Optimization of engineered biochar systems production and application has also been made easy by these computer-aided tools. The use of AI and experimental data has revolutionized the production of biochar systems with predetermined properties for specific applications.

4.3.1 In-silico-computational modeling or computer-aided design approach

Computational methods have been used to transform the engineering and design of biochar systems by optimizing the production phases, biochar properties, and applications. Computational tools such as Computational fluid dynamics (CFD), Density Functional Theory (DFT), Molecular Dynamics (MD) Simulations, have been widely explored in the design of biochar systems. Computer-aided process simulation software's (e.g., Aspen Plus, COMSOL Multiphysics, and ANSYS Fluent) and computational software (Materials Studio, CHARMM, and LAMMPS) commonly used for optimizing the production and application processes.

Computational fluid dynamics (CFD) simulation models are usually applied to the production stages mainly to optimize reactor conditions, predicting biochar properties and applicability. To optimize the biochar yield and quality CFD methods can be used to simulate pyrolysis conditions such as temperature, residence period, and gas flux, which will also inform in the designing of a suitable pyrolysis chamber (Dadi et al., 2023). A model utilizing fluid dynamics (CFD) simulation in co-combustion of biomass with propane agreed with the experimental data and concluded that the rate of heat exchange among thermal oxidizer and pyrolysis unit and minor water content in the biomass were the determining factors in producing good biochar (Khodaei et al., 2021).

Scaling up of the pilot laboratory setups can also be handled by CFD methods quickening the production process. A CFD-DEM model toolsets have been used for scaling-up an autothermal pyrolysis systems for industrial adoption. Optimization on corn stover biomass considering proximate and ultimate compositions resonated with experimental results, and parameters like equivalence ratio, injection point, and size distribution on autothermal pyrolysis were observed to be the determining factors (Oyedeji et al., 2022).

Thermal treatment of biomass results in an array of products which include solid biochar, carbon oxides, and other organic compounds. CFD theoretical calculations can be used to model the emission of volatile compounds and other gases that contribute to the depletion of the environment; hence a sustainable method can be developed (Dadi et al., 2023). Numerical simulations of downdraft gasification imputing pyrolysis conditions showed that an increase in airflow enhanced the emission of volatile components and boost heat exchange, among other factors, leading to an increase in biochar surface area and pore volume, but density diminished by about 10%. These characteristics are attributed to increased emission of volatile compounds as airflow increased, which elevated the thermal temperature. (González & Pérez, 2019). Hydrogen and greenhouse gases are also produced with the solid biochar. CFD simulations can be used to optimize the amount of emitted gases. CFD model on an updraft gasification reactor showed that hydrogen concentration and thermal ramp of producer gas amplified when all inputted variables were increased, but the yield of tar, and green-house gases subsided (Ismail & El-Salam, 2017).

DFT is another computational tool based on quantum mechanical methods which can be used to correlate electronic structure and material properties at the atomic level. DFT modeling tools can be used to characterize and correlate biochar physical properties to the application process. Valuable insights into biochar porosity can be deduced from mathematical modeling using DFT methods (Jeyasubramanian et al., 2021). Pore size, distribution, connectivity, and the ultimate specific surface area can be calculated to get the structural-property relationships. DFT can be used to characterize the chemical properties of biochar systems. Surface functional group, elemental composition, and redox properties can be correlated to the adsorption and reactivity process and pollutant selectivity of the biochar (Jeyasubramanian et al., 2021). DFT has been used as a vital tool to complement experimental data in understanding the adsorption mechanisms of different pollutants which include, heavy metals and organic compounds. For example, the caffeine adsorption process and kinetics were determined to be due to molecule stability and high surface area to volume ratio (de Almeida et al., 2021). Results from quantum chemical calculations for a biochar system for adsorption of Pb and Zn showed that oxygen adsorption sites were responsible for adsorption of ions from the solution (Zhao et al., 2020).

MD simulation can also be used to study the morphological and structural properties of biochar materials. Pollutant adsorption mechanisms and chemical group distribution can be deduced from MD simulations (Sochacki et al., 2024).

A combined simulation approach using ReaxFF MD simulations in conjunction with DFT calculations proved to be a versatile tool for understanding biochar properties (Wang, Li, et al., 2021). An investigation of the applicability of agricultural biomass-derived biochar confirmed that the dominant interactions were π-π stacking the overall mechanism was dependent on ionic strength using experimental and simulation data (Mrozik et al., 2021).

4.3.2 Artificial intelligence and machine learning tools

AI has revolutionized the production and application of biochar engineering and designing, facilitating rapid production optimization, properties prediction, and decision-making. AI is the science of using computer systems to do responsibilities that naturally use human intelligence. The responsibilities include communicating, decoding trends, learning from prior exposure, troubleshooting, and decision-making (Lakshmi et al., 2021).

AI tools are mainly used to optimize biochar physicochemical properties via optimizing the thermochemical production process and feedstock selection process. An array of different concepts of AI and techniques are available and have been used in the context of biochar engineering. The most frequently used tools include ML DL, data mining and pattern recognition, and decision support systems (DSS) (Nguyen et al., 2024b). AI depends on software, library databases, and other applications that are vital for data manipulation, model directing, evaluation, deployment, and monitoring. ML is a subclass of AI that rely on algorithms and statistical models that permit computers to learn, predict and make decisions based on literature data without being programmed (Wang et al., 2024).

The AI tools use different algorithms and different statistical models to process the input variables to generate the output parameters. The algorithms used have varying degrees of suitability and the selected algorithm greatly affects the outcome (Wang et al., 2024). Examples of different algorithms used are shown in Fig. 4.3.

The selection of algorithm usually depends on what one wants to achieve, available database, and type of data. The most used algorithms among others in biochar engineering and design include random forest (RF), support vector machines (SVM), artificial neural networks (ANN), gradient boosting machines (GBM), and k-means (Nguyen et al., 2024b). These algorithms can be used for optimizing the production process, surface characteristics, biochar property prediction, and characterization through image analysis and instrumentation data. Availability of data is a huge setback in terms of obtaining reliable data. For huge data RF is preferred and SVM can handle linear and nonlinear data while GBM for heterogeneous data and ANNs for sequential data. k-mean is a clustering algorithm that can be used to assemble biochar based on their physicochemical properties and determine trends (Leng et al., 2021; Nguyen et al., 2024b).

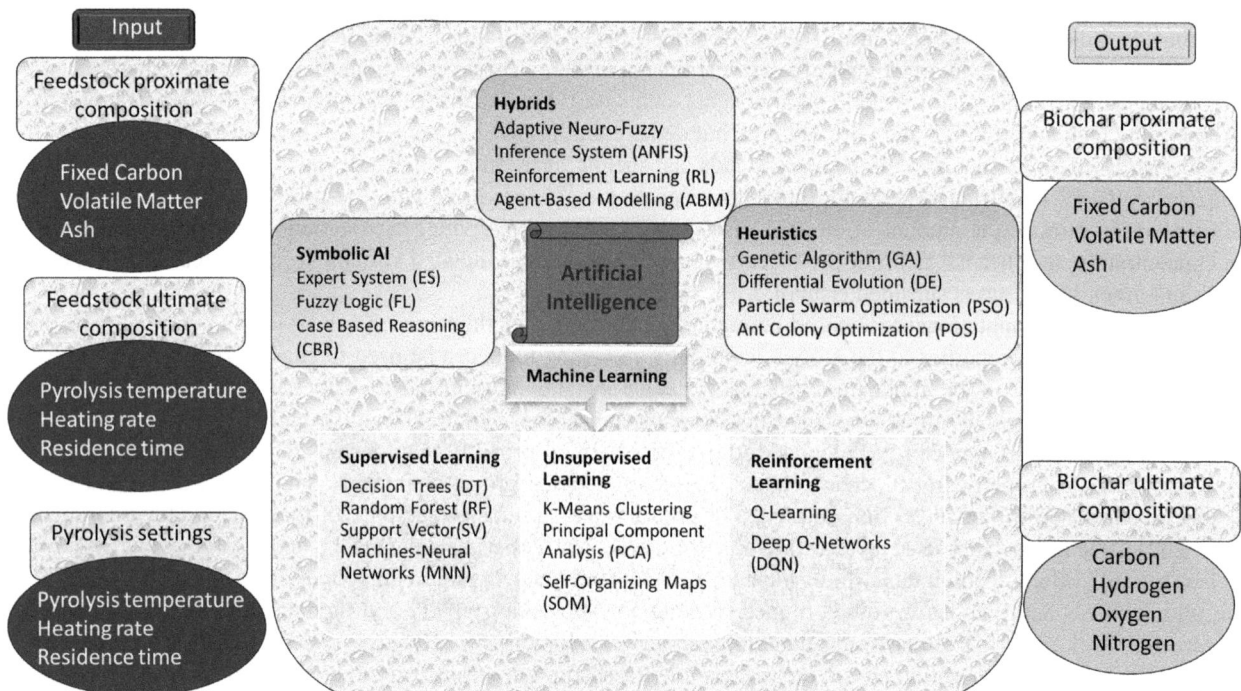

FIGURE 4.3 Schematic diagram showing common AI and ML algorithms and model input and output variables.

AI and ML algorithms are compared for suitable fit using different parameter metrics which are a culmination of specific tasks, such as classification, regression, or clustering, and the objectives. Examples include mean absolute error (MAE), root mean squared error (RMSE), R-squared (R^2), and F1 Score (Nguyen et al., 2024b). A model is said to be highly accurate for R^2 close to a unit and an MAE value close to zero. Biochar yields have been predicted with accuracy using different models with input biomass composition and pyrolysis conditions. A hybrid model of ANNs and metaheuristic algorithms showed that ANN with Rao-2 algorithm (heuristic) was the best ($R^2 \sim 0.93$, RMSE $\sim$ 1.74%) (Khan et al., 2022). Adoption of ML algorithms RF and gradient boosting regression (GBR) to envisage specific surface area (SSA), N content, and yield of biochar using biomass elemental, proximate, and biopolymer compositions and pyrolysis conditions showed that GBR outclassed RF for most predictions ($R^2 = 0.90-0.95$) (Leng et al., 2022). Table 4.1 summarizes some recent applications of AI and ML which were used to accurately predict and optimize biochar properties.

Generally, in-silico methods have presented numerous advantages in material design and engineering. The key merits include time-saving by streamlining the process by focusing on promising materials and eliminating lab testing. There is also expenditure saving and resource preservation, resources can be channeled in optimizing only those materials which have shown potential and the need for very expensive techniques for characterization is also negated (Nguyen et al., 2024a; Zhu et al., 2019). Another advantage is that there in enhanced innovation and risk mitigation, novel materials can be designed and tested to understand the properties and predict failure in advance that might not have been achieved using the conventional physical methods. However, despite these merits, there are still some challenges, in the applicability of these computer-aided tools. Biochar production process is intricate and depends on various variables and parameters, and simulation of these process parameters require advanced models and algorithms. This is also further constrained by the lack of high-quality data representative of all the production parameters (Ambaye et al., 2021; Wei et al., 2014).

In the context of low-income countries, many institutions lack financial resources to purchase some state-of-te-art characterization techniques and put together computational resources. This is also further worsened by the lack of professionals who have the prerequisite skills to operate the few characterization instruments that are available and there is also limited expertise personnel in AI.

TABLE 4.1 Recent studies of AI and ML-based biochar engineering.

Predictions	Input variables	Data type	AI and ML algorithms	Metrics	References
Heavy metal immobilization efficiencies	Biochar properties Exp conditions Soil Properties	Secondary	RF, SVR, NN	$R^2 = 0.98-0.99$ RMSE =	Palansooriya et al. (2022)
Biochar yield and C-content	Biochemical composition (cellulose, lignin, and hemicellulose), ash, C:H:O:N, particle size of feedstock, pyrolysis setting (pyrolysis temperature, heating rate, and residence time)	Secondary	RF	$R^2 = 0.85$ RMSE = 3.40	Zhu, Li and Wang (2019)
Specific surface area, total pore volume, and yield of biochar.	volatile matter ash, fixed carbon, and C:H:N:S:O Pyrolysis conditions (temperature, retention time and heating rate)	Secondary	RF, GBR*	$R^2 = 0.89-0.94$ RMSE =	Li et al. (2023)
Yield of 314 feedstock samples, costs and lifecycle emissions	Mass and energy balances, operation and capital costs, lifecycle economic costs And feedstock composition.	Secondary	ROM	$R^2 = 0.99$ RMSE = 0.0–0.163	Olafasakin et al. (2021)

GBR, Gradient boosting regression; *NN*, neural networks; *RF*, random forest; *RMSE*, root mean square error; *ROM*, reduced order model; *SVR*, support vector regression.

4.3.3 Current and potential applications

Biochar has found use in many areas because of its physicochemical properties which include high surface area, huge porosity, diverse SFGs, and electroconductivity. The major areas include agricultural, environmental remediation, energy, and electrochemistry (Fig. 4.4).

Biochar has offered enormous benefits in agricultural production and environmental sustainability mainly in soil amendment (Nguyen et al., 2024a). The addition of biochar to the soils improves carbon sequestration, nutrient control, pH regulation, microbial ecosystem, water balance, and soil degradation. The presence of biochar in agricultural land modifies the soil porosity structure, thereby inhibiting rapid disintegration of soil organic carbon matter (Osman et al., 2022). Biochar can act as a pH maintenance system that regulate the availability of nutrients and microbial communities. This helps in reducing methane emissions and availability of micronutrients which are vital for plant growth. The moisture retention capability of biochar is because of its pore network and large surface area which trap water molecules and allow capillary suction and storing the water within the reach of plants thereby promoting rapid growth. The efficiency of biochar in boosting different soils has been demonstrated in different crops (Osman et al., 2022).

Biochar can improve soil clusters, which allow water to infiltrate deep reducing runoff thereby taming earth surface degradation. The addition of biochar soils and fertilizers can enhance nutrient preservation and avert nutrient leakage helping in higher crop yields and drought tolerance. In animal production, biochar is generally added as a feed enhancer to encourage digestive process and nutrient uptake. The biochar acts by binding toxins; it reduces methane and ammonia emissions, improves feed efficiency, and supports the immune system (Ukoba & Jen, 2022).

Biochar surfaces contain different surface groups that confer different interactions, and which can potentially adsorb different classes of compounds and find widespread application in wastewater treatment and environmental remediation. In wastewater treatment packed bed or column filtration systems are effectively used to remove pollutants such as heavy metals, pesticides, pharmaceuticals, dyes, and other organic compounds (Nighojkar et al., 2023). '*Opuntia ficus-indica*' (OFI) derived char could adsorb both malachite green dye, Cu(II) and Ni(II) with maximum adsorption capacity of 1341 mg g^{-1}, 49 mg g^{-1} and 44 mg g^{-1}, respectively (Choudhary et al., 2020). Biochar can also function in a duplex way both as an adsorber and photodegradation catalyst. In environmental remediation, biochar can be used to mobilize contaminants via adsorption such that their detrimental effects are reduced. It has been observed that biochar incorporation in soils stabilize cationic ions such as Cd(II), Cu(II), Ni(II), Pb(II), and Zn(II), and on the other hand, it mobilizes anionic toxic elements (e.g., $Cr_2O_2^-$, $Sb(OH)^{7-}$ and CrO_2^{4-}, AsO^{3-} and AsO^{3-}) (Guo et al., 2020). Because of its

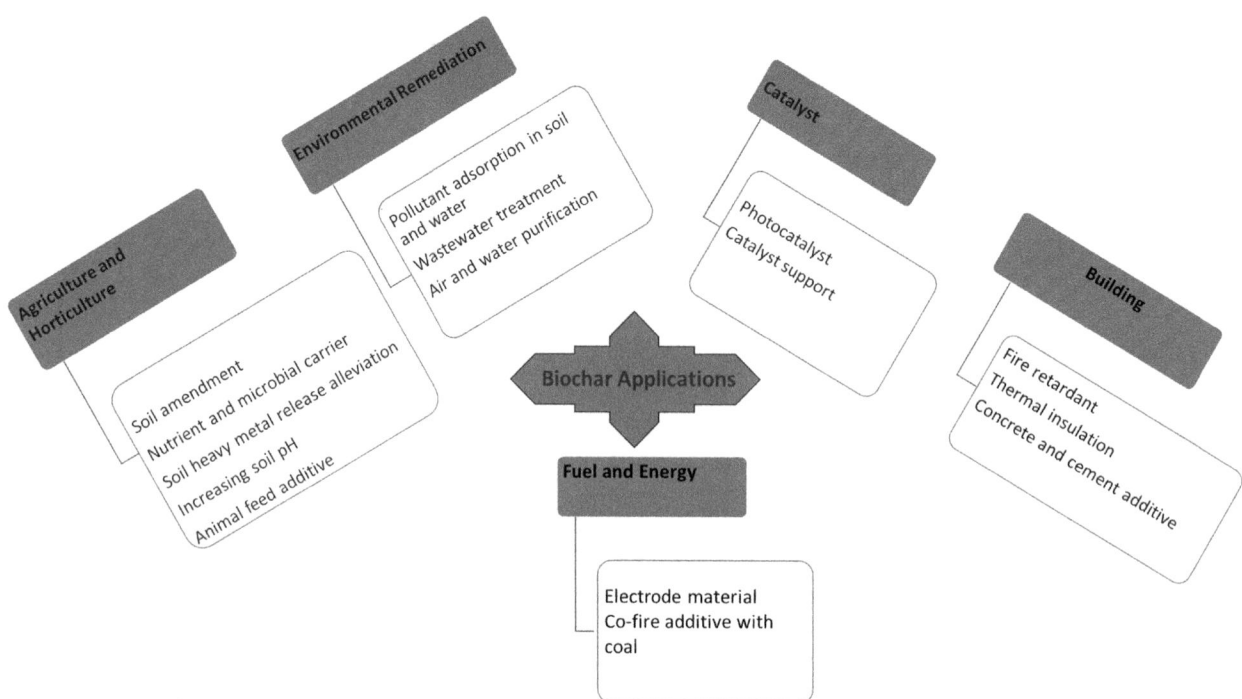

FIGURE 4.4 Illustration of different applications of biochar.

kinetic stability, biochar can also act as a sink for greenhouse gas emissions and help in carbon sequestration thereby mitigating the effects of climate change.

Biochar has found widespread application in the building and construction industry. Biochar has been added to composite building materials to improve thermal and mechanical strength, moisture resistance, and fire retardancy to reduce weight. Biochar has been included in cementitious composites in various percentages by weight. The addition of as low as 5% rice husk biochar resulted in improved compressive and about a 78% rise in tensile strength (Zeidabadi et al., 2018). The type of feedstock also tends to affect the properties of materials. The inclusion of biochar derived from three feedstocks; sugarcane bagasse, water hyacinth and yellow oleander in thermal modifier organic phase change materials (PCM) showed that sugarcane bagasse-derived biochar lowered the thermal conductivity (Bordoloi et al., 2022). The fire retardancy and thermal properties of bio-based building material were also improved by adding biochar (Choi et al., 2023) and a gypsum-biochar composited proved to be a potential electromagnetic radiation shield (Natalio et al., 2020). In road construction biochar-modified asphalt exhibited less road surface degradation due to moisture damage and UV radiation.

Biochar has shown great potential as a heterogeneous catalyst and/or catalyst support material for different chemical reactions. Some of the reactions include aerobic digestion, advanced oxidation, hydrogen production, and biomass conversion. The as-prepared biochar tends to have low catalytic properties, but functionalization tends to enhance the catalytic properties. For instance, sulfonic acid functionalized biochar achieved 95% transesterification of algal oil to biodiesel (Behera et al., 2020). Nickel-loaded biochar could achieve a higher yield of syngas (109.848 mmol g^{-1}) compared to the pristine (Wang et al., 2022). Carefully, engineered biochar can exhibit photocatalytic activity which can enable redox processes generating reactive oxygen species (ROS) which initiate the disintegration of organic pollutants and disinfection microorganisms from polluted water. For example, wood biochar encapsulated with Prussian blue exhibited 95% efficiency degradation of methylene blue (20 mg L^{-1}) compared to 52% of the unmodified (Zhao et al., 2024).

Engineered biochar materials have shown widespread potential application in the energy sector. Different biochar materials have been observed to enhance the efficiency of direct carbon solid oxide fuel cells (DC-SOFCs). Sesame straw biochar electrode material exhibited an output of 228 $mWcm^{-2}$, while that of eggplant straw biochar produced 214 $mWcm^{-2}$ outclassing activated carbon-fueled cell (173 $mW\ cm^{-2}$) (Miao et al., 2024). Biochar has also shown promising potential as a co-fuel in iron and steel production. A meta-analysis study has shown that substituting a fraction of coal-derived fuel in the kiln can diminish greenhouse gas discharges into the atmosphere and also the heat storage effect can be improved in the ore sintering process (Meng et al., 2024).

Biochar has the potential to replace carbon black in many electrochemical applications. It has demonstrated potential as electrocatalyst and photocatalyst in water splitting and anergy storage material for supercapacitance. An S-dopped biochar demonstrated dual function in water splitting and supercapacitance. Overpotentials of 154 (H_2) and 362 mV (O_2) at 10 mA cm^{-2} were obtained and high energy density (34.54 Wh kg^{-1}) and power density of 1600 W kg^{-1} was achieved by the energy storage device 10.1002/cey2.207. Because of its high electrical conductivity, biochar materials have been used as electrode materials, membranes, and catalysts in fuel cells (e.g., microbial fuel cells [MFCs] or bioelectrochemical systems [BESs]). It was observed that the use of clay/char MFC exchange improved the water uptake and the chemical oxygen demand giving an output voltage of 297 mV (Verma et al., 2023). Electrode materials from bamboo-biochar could generate a power density (12.9 W m^{-3}) comparable to activated carbon and it simultaneously degrade xylan in MFC (Kumar et al., 2023).

4.4 Future perspectives and research directions

Research in biochar design and engineering has evolved and expanded over time due to widespread application in energy generation and storage, agronomy, environmental, and engineering. It has been highlighted that the intrinsic properties of biochar majorly depend on thermal production processes. More research should be done to optimize the production process in designing more proficient pyrolysis strategies, such as microwave-assisted pyrolysis, co-pyrolysis, and wet pyrolysis (Wang et al., 2020). Investigations should also navigate toward designing processes that harness the energy, and the byproducts produced during pyrolysis in an eco-friendly and cost-sustainable way. The feedstock biomass also influences the ultimate biochar and proximate composition; hence, the biochar exhibits different properties. In literature, most of the studies have been done on a small number of feedstock sources, but a comprehensive multiple biomass comparison is required to determine the cooperative behavior in different applications (Hassan et al., 2020). Understanding the chemical mechanisms of biochar sorption behavior is important in informing the designing of surface and structural properties of engineered biochar. From previous studies, the sorption characteristics for different classes of compounds which include inorganic, and organic species have been reported. The bulk of studies have been mainly on heavy metals, dyes, and pharmaceuticals. Limited data is available on persistent organic and emerging organic pollutants, endocrine disruptors, microplastics, and other organic compounds. There is also some inconsistency with regards to the interactions of pollutants. The elucidation of the underlying behavior and mechanisms can be further explored using the current robust potential of AI and ML toward the designing of tailor-made biochar for specific applications. There is still a lack of data and standardized experimental parameter settings hence there is a need to generate large database to accelerate unbiased prediction models through AI and ML and more generalized models need to be developed in order to accommodate recent data being generated.

Biochar properties descriptors are pivotal in assessing biochar quality, functionality, and potential applications. Research has focused more on surface properties such as porosity, yield, elemental and proximate composition. Investigation of more descriptors such as SFGs and the ultimate acid-base properties, ion exchange efficiency, and electrical properties are worth exploring. The use of biochar in medical applications such as drug delivery and biosensors is still in its infancy and many aspects are still not clear; hence, this can be a hotspot of research in the near future. The ecosystem is so dynamic and it is also imperative to investigate the biochar interactions in aquatic environments to understand its sedimentation characteristics, pollutant transport, and toxicology. The development of biochar-based functional materials, such as composites with nanoparticles and other carbon nanomaterials, is currently being exploited and the majority of the studies explore the characterization of composite adsorption applications. The retrieval (e.g., via iron modified biochar's) and regeneration and the environment and human health risk should also be the focus as related hazards are not yet fully understood. Proper planning and management are required for large-scale application of biochar as the fine particles can potentially spread adsorbed pollutants and contaminate surface and groundwater, posing health risks to the humans and other organisms. Currently, available studies of engineered biochar are lab-scale experiments and theoretical models. Industrial-scale production of engineered biochar and the application process are in early stages, and this warrants further development. The hazards and control of ambiguity factors need also to be weighed prior to the large-scale application of engineered biochar.

4.5 Conclusions

Biochar descriptors which are classified under physical, chemical, thermal, sorption, and adsorption can sometimes be inconsistent regardless of the same feedstock source if the pyrolysis conditions are different. Hence, there is a need to design a pyrolysis reactor with advanced heating control to reproduce the same properties. AI and ML tools are vital in completing the success story of biochar engineering and the development of models for each phase in the biochar lifecycle must be optimized so as to extract the economic and sustainable benefits for the application of biochar (Wang et al., 2024). The associated environmental risks of biochar to the ecosystems must be considered for full exploitation of biochar potential.

References

Ambaye, T. G., Rene, E. R., Nizami, A. S., Dupont, C., Vaccari, M., & van Hullebusch, E. D. (2021). Beneficial role of biochar addition on the anaerobic digestion of food waste: a systematic and critical review of the operational parameters and mechanisms. *Journal of Environmental Management, 290*, 112537.

Arif, M., Liu, G., Zia ur Rehman, M., Mian, M. M., Ashraf, A., Yousaf, B., Rashid, M. S., Ahmed, R., Imran, M., & Munir, M. A. M. (2023). Impregnation of biochar with montmorillonite and its activation for the removal of azithromycin from aqueous media. *Environmental Science and Pollution Research, 30*(32), 78279–78293.

Behera, B., Dey, B., & Balasubramanian, P. (2020). Algal biodiesel production with engineered biochar as a heterogeneous solid acid catalyst. *Bioresource Technology, 310*, 123392.

Berhanu, S., Hervy, M., Weiss-Hortala, E., Proudhon, H., Berger, M., Chesnaud, A., Faessel, M., King, A., Minh, D. P., Villot, A., Gérente, C., Thorel, A., Le Coq, L., & Nzihou, A. (2018). Advanced characterization unravels the structure and reactivity of wood-based chars. *Journal of Analytical and Applied Pyrolysis, 130*, 79–89. Available from https://doi.org/10.1016/j.jaap.2018.01.024.

Bordoloi, U., Das, D., Kashyap, D., Patwa, D., Bora, P., Muigai, H. H., & Kalita, P. (2022). Synthesis and comparative analysis of biochar based form-stable phase change materials for thermal management of buildings. *Journal of Energy Storage, 55*, 105801.

Chen, H., Gao, Y., Li, J., Fang, Z., Bolan, N., Bhatnagar, A., Gao, B., Hou, D., Wang, S., & Song, H. (2022). Engineered biochar for environmental decontamination in aquatic and soil systems: a review. *Carbon Research, 1*, 4.

Chen, T., Zhao, L., Gao, X., Li, L., & Qin, L. (2022). Modification of carbonation-cured cement mortar using biochar and its environmental evaluation. *Cement and Concrete Composites, 134*, 104764.

Choi, J. Y., Kim, Y. U., Nam, J., Kim, S., & Kim, S. (2023). Enhancing the thermal stability and fire retardancy of bio-based building materials through pre-biochar system. *Construction and Building Materials, 409*, 134099.

Choudhary, M., Kumar, R., & Neogi, S. (2020). Activated biochar derived from Opuntia ficus-indica for the efficient adsorption of malachite green dye, Cu^{+2} and Ni^{+2} from water. *Journal of Hazardous Materials, 392*, 122441.

Clurman, A. M., Rodríguez-Narvaez, O. M., Jayarathne, A., De Silva, G., Ranasinghe, M. I., Goonetilleke, A., & Bandala, E. R. (2020). Influence of surface hydrophobicity/hydrophilicity of biochar on the removal of emerging contaminants. *Chemical Engineering Journal, 402*, 126277.

Dadi, V. S., Veluru, S., Tanneru, H. K., Busigari, R. R., Potnuri, R., Kulkarni, A., Mishra, G., & Basak, T. (2023). Recent advancements of CFD and heat transfer studies in pyrolysis: A review. *Journal of Analytical and Applied Pyrolysis, 175*, 106163.

de Almeida, A. D. S. V., Vieira, W. T., Bispo, M. D., de Melo, S. F., da Silva, T. L., Balliano, T. L., Vieira, M. G. A., & Soletti, J. I. (2021). Caffeine removal using activated biochar from açaí seed (Euterpe oleracea Mart): Experimental study and description of adsorbate properties using Density Functional Theory (DFT). *Journal of Environmental Chemical Engineering, 9*(1), 104891.

Dong, J., Shen, L., Shan, S., Liu, W., Qi, Z., Liu, C., & Gao, X. (2022). Optimizing magnetic functionalization conditions for efficient preparation of magnetic biochar and adsorption of Pb (II) from aqueous solution. *Science of the Total Environment, 806*, 151442. Available from https://doi.org/10.1016/j.scitotenv.2021.151442.

Edeh, I. G., Masek, O., & Fusseis, F. (2023). 4D structural changes and pore network model of biomass during pyrolysis. *Scientific Reports, 13*, 22863.

Faheem, Du, J., Kim, S.H., Hassan, M.A., Irshad, S. and Bao, J., 2020. Application of biochar in advanced oxidation processes: supportive, adsorptive, and catalytic role. Environmental Science and Pollution Research, 27, pp.37286-37312.

Fusseis, F., Xiao, X., Schrank, C., & De Carlo, F. (2014). A brief guide to synchrotron radiation-based microtomography in (structural) geology and rock mechanics. *Journal of Structural Geology, 65*, 1–16.

González, W. A., & Pérez, J. F. (2019). CFD analysis and characterization of biochar produced via fixed-bed gasification of fallen leaf pellets. *Energy, 186*.

Guo, M., Song, W., & Tian, J. (2020). Biochar-Facilitated Soil Remediation: Mechanisms and Efficacy Variations. *Frontiers in Environmental Science, 8*, 521512.

Hassan, M., Liu, Y., Naidu, R., Parikh, S. J., Du, J., Qi, F., & Willett, I. R. (2020). Influences of feedstock sources and pyrolysis temperature on the properties of biochar and functionality as adsorbents: A meta-analysis. *Science of The Total Environment, 744*, 140714. Available from https://doi.org/10.1016/j.scitotenv.2020.140714.

Hyväluoma, J., Hannula, M., Arstila, K., Wang, H., Kulju, S., & Rasa, K. (2018). Effects of pyrolysis temperature on the hydrologically relevant porosity of willow biochar. *Journal of Analytical and Applied Pyrolysis, 134*, 446–453.

Hyväluoma, J., Kulju, S., Hannula, M., Wikberg, H., Källi, A., & Rasa, K. (2018). Quantitative characterization of pore structure of several biochars with 3D imaging. *Environmental Science and Pollution Research, 25*, 25648–25658.

Ismail, T. M., & El-Salam, M. A. (2017). Parametric studies on biomass gasification process on updraft gasifier high temperature air gasification. *Applied Thermal Engineering, 112*, 1460–1473.

Jafri, N., Wong, W. Y., Doshi, V., Yoon, L. W., & Cheah, K. H. (2018). A review on production and characterization of biochars for application in direct carbon fuel cells. *Process Safety and Environmental Protection, 118*, 152–166. Available from https://doi.org/10.1016/j.psep.2018.06.036.

Jang, H. M., Yoo, S., Park, S., & Kan, E. (2018). Engineered biochar from pine wood: Characterization and potential application for removal of sulfamethoxazole in water. *Environmental Engineering Research, 24*. Available from https://doi.org/10.4491/eer.2018.358.

Jeyasubramanian, K., Thangagiri, B., Sakthivel, A., Raja, J. D., Seenivasan, S., Vallinayagam, P., Madhavan, D., Devi, S. M., & Rathika, B. (2021). A complete review on biochar: Production, property, multifaceted applications, interaction mechanism and computational approach. *Fuel, 292*, 120243. Available from https://doi.org/10.1016/j.fuel.2021.120243.

Khan, M., Ullah, Z., Masek, O., Raza Naqvi, S., & Nouman Aslam Khan, M. (2022). Artificial neural networks for the prediction of biochar yield: A comparative study of metaheuristic algorithms. *Bioresour Technol, 355*, 127215.

Khodaei, H., Gonzalez, L., Chapela, S., Porteiro, J., Nikrityuk, P., & Olson, C. (2021). CFD-based coupled multiphase modeling of biochar production using a large-scale pyrolysis plant. *Energy, 217*, 119325.

Kumar, A., Bhattacharya, T., Shaikh, W. A., Roy, A., Chakraborty, S., Vithanage, M., & Biswas, J. K. (2023). Multifaceted applications of biochar in environmental management: a bibliometric profile. *Biochar, 5*(1), 11.

Lakshmi, D., Akhil, D., Kartik, A., Gopinath, K. P., Arun, J., Bhatnagar, A., Rinklebe, J., Kim, W., & Govarthanan, M. (2021). Artificial intelligence (AI) applications in adsorption of heavy metals using modified biochar. *Science of the Total Environment, 801*, 149623.

Leng, L., Xiong, Q., Yang, L., Li, H., Zhou, Y., Zhang, W., Jiang, S., Li, H., & Huang, H. (2021). An overview on engineering the surface area and porosity of biochar. *Sci Total Environ, 763*, 144204.

Leng, L., Yang, L., Lei, X., Zhang, W., Ai, Z., Yang, Z., Zhan, H., Yang, J., Yuan, X., Peng, H., & Li, H. (2022). Machine learning predicting and engineering the yield, N content, and specific surface area of biochar derived from pyrolysis of biomass. *Biochar, 4*(1), 63.

Li, H., Ai, Z., Yang, L., Zhang, W., Yang, Z., Peng, H., & Leng, L. (2023). Machine learning assisted predicting and engineering specific surface area and total pore volume of biochar. *Bioresource Technology, 369*, 128417.

Meng, F., Rong, G., Zhao, R., Chen, B., Xu, X., Qiu, H., Cao, X., & Zhao, L. (2024). Incorporating biochar into fuels system of iron and steel industry: carbon emission reduction potential and economic analysis. *Applied Energy, 356*, 122377.

Miao, K., Han, T., Wu, Y., Yu, L., Xie, Y., Zhang, J., Yu, F., & Yang, N. (2024). Highly efficient utilization of crop straw-derived biochars in direct carbon solid oxide fuel cells for electricity generation. *International Journal of Hydrogen Energy, 61*, 39–46.

Mrad, R., & Chehab, G. (2019). Mechanical and microstructure properties of biochar-based mortar: An internal curing agent for PCC. *Sustainability*, *11*, 2491.

Mrozik, W., Minofar, B., Thongsamer, T., Wiriyaphong, N., Khawkomol, S., Plaimart, J., Vakros, J., Karapanagioti, H., Vinitnantharat, S., & Werner, D. (2021). Valorisation of agricultural waste derived biochars in aquaculture to remove organic micropollutants from water - experimental study and molecular dynamics simulations. *J Environ Manage*, *300*, 113717.

Natalio, F., Corrales, T. P., Feldman, Y., Lew, B., & Graber, E. R. (2020). Sustainable lightweight biochar-based composites with electromagnetic shielding properties. *ACS omega*, *5*, 32490–32497.

Nguyen, V. G., Sharma, P., Ağbulut, Ü., Le, H. S., Tran, V. D., & Cao, D. N. (2024b). Precise prognostics of biochar yield from various biomass sources by Bayesian approach with supervised machine learning and ensemble methods. *International Journal of Green Energy*, *21*(9), 2180–2204.

Nguyen, V. G., Sharma, P., Ağbulut, Ü., Le, H. S., Truong, T. H., Dzida, M., Tran, M. H., Le, H. C., & Tran, V. D. (2024a). Machine learning for the management of biochar yield and properties of biomass sources for sustainable energy. *Biofuels, Bioproducts and Biorefining*, *18*(2), 567–593.

Nighojkar, A., Pandey, S., Naebe, M., Kandasubramanian, B., Soboyejo, W. W., Plappally, A., & Wang, X. (2023). Using machine learning to predict the efficiency of biochar in pesticide remediation. *npj Sustainable Agriculture*, *1*(1), 1.

Olafasakin, O., Chang, Y., Passalacqua, A., Subramaniam, S., Brown, R. C., & Mba Wright, M. (2021). Machine learning reduced order model for cost and emission assessment of a pyrolysis system. *Energy & Fuels*, *35*(12), 9950–9960. Available from https://doi.org/10.1021/acs.energyfuels.1c00490.

Osman, A. I., Fawzy, S., Farghali, M., et al. (2022). Biochar for agronomy, animal farming, anaerobic digestion, composting, water treatment, soil remediation, construction, energy storage, and carbon sequestration: a review. *Environmental Chemistry Letters*, *20*, 2385–2485. Available from https://doi.org/10.1007/s10311-022-01424-x.

Oyedeji, O. A., Brennan Pecha, M., Finney, C. E. A., Peterson, C. A., Smith, R. G., Mills, Z. G., Gao, X., Shahnam, M., Rogers, W. A., Ciesielski, P. N., Brown, R. C., & Parks Ii, J. E. (2022). CFD–DEM modeling of autothermal pyrolysis of corn stover with a coupled particle- and reactor-scale framework. *Chemical Engineering Journal*, *446*, 136920.

Pak, T., Gomari, K. E., Bose, S., Tonon, T., Hughes, D., Gronnow, M., & Macquarrie, D. (2023). Biochar from brown algae: Production, activation, and characterisation. *Bioresource Technology Reports*, *24*, 101688. Available from https://doi.org/10.1016/j.biteb.2023.101688.

Palansooriya, K. N., Li, J., Dissanayake, P. D., Suvarna, M., Li, L., Yuan, X., Sarkar, B., Tsang, D. C. W., Rinklebe, J., Wang, X., & Ok, Y. S. (2022). Prediction of Soil Heavy Metal Immobilization by Biochar Using Machine Learning. *Environmental Science and Technology*, *56*(7), 4187–4198. Available from https://doi.org/10.1021/acs.est.1c08302. Available from: http://pubs.acs.org/journal/esthag.

Schlüter, S., Albrecht, L., Schwärzel, K., & Kreiselmeier, J. (2020). Long-term effects of conventional tillage and no-tillage on saturated and near-saturated hydraulic conductivity–Can their prediction be improved by pore metrics obtained with X-ray CT? *Geoderma*, *361*, 114082.

Shaheen, S. M., Niazi, N. K., Hassan, N. E., Bibi, I., Wang, H., Tsang, D. C., Ok, Y. S., Bolan, N., & Rinklebe, J. (2019). Wood-based biochar for the removal of potentially toxic elements in water and wastewater: a critical review. *International Materials Reviews*, *64*(4), 216–247.

Shaikh, W. A., Chakraborty, S., Kumar, A., Biswas, J. K., Jha, A. K., Bhattacharya, T., Vithanage, M., Ansar, S., & Hossain, N. (2023). Tailor-made biochar-based nanocomposite for enhancing aqueous phase antibiotic removal. *Journal of Water Process Engineering*, *55*, 104215.

Singh, A., Sharma, R., Pant, D., & Malaviya, P. (2021). Engineered algal biochar for contaminant remediation and electrochemical applications. *Science of the Total Environment*, *774*, 145676.

Sochacki, A., Lebrun, M., Minofar, B., Pohorely, M., Vithanage, M., Sarmah, A. K., Boserle Hudcova, B., Buchtelik, S., & Trakal, L. (2024). Adsorption of common greywater pollutants and nutrients by various biochars as potential amendments for nature-based systems: Laboratory tests and molecular dynamics. *Environ Pollut*, *343*, 123203.

Srocke, F., Han, L., Dutilleul, P., Xiao, X., Smith, D. L., & Mašek, O. (2021). Synchrotron X-ray microtomography and multifractal analysis for the characterization of pore structure and distribution in softwood pellet biochar. *Biochar*, *3*, 671–686.

Suuronen, J.-P., & Jyske, T. (2019). Noninvasive investigation of phloem structure by 3D synchrotron X-ray microtomography. *Phloem: Methods and Protocols*, 37–54.

Wang, H., Lou, X., Hu, Q., & Sun, T. (2021). Adsorption of antibiotics from water by using Chinese herbal medicine residues derived biochar: Preparation and properties studies. *Journal of Molecular Liquids*, *325*, 114967.

Wang, Y., Li, Y., Zhang, C., Yang, L., Fan, X., & Chu, L. (2021). A study on co-pyrolysis mechanisms of biomass and polyethylene via ReaxFF molecular dynamic simulation and density functional theory. *Process Safety and Environmental Protection*, *150*, 22–35.

Wang, L., Ok, Y. S., Tsang, D. C. W., Alessi, D. S., Rinklebe, J., Wang, H., Mašek, O., Hou, R., O'Connor, D., Hou, D., & Nicholson, F. (2020). New trends in biochar pyrolysis and modification strategies: feedstock, pyrolysis conditions, sustainability concerns and implications for soil amendment. *Soil Use and Management*, *36*, 358–386.

Tan, X., Zhu, S., Wang, R., Chen, Y., Show, P., Zhang, F., & Ho, S. (2021). Role of biochar surface characteristics in the adsorption of aromatic compounds: Pore structure and functional groups. *Chinese Chemical Letters*, *32*(10), 2939–2946. Available from https://doi.org/10.1016/j.cclet.2021.04.059.

Ukoba, K., & Jen, T. (2022). Biochar and Application of Machine Learning: A Review. In book: Biochar - Productive Technologies, Properties and Application. Intech. https://doi.org/10.5772/intechopen.108024.

Wang, W., Chang, J. S., & Lee, D.-J. (2024). Machine learning applications for biochar studies: A mini-review. *Bioresource Technology*.

Wang, Y., Huang, L., Zhang, T., & Wang, Q. (2022). Hydrogen-rich syngas production from biomass pyrolysis and catalytic reforming using biochar-based catalysts. *Fuel*, *313*, 123006.

Watanabe, H. (2018). X-ray computed tomography visualization of the woody char intraparticle pore structure and its role on anisotropic evolution during char gasification. *Energy & fuels*, *32*, 4248–4254.

Weber, K., & Quicker, P. (2018). Properties of biochar. *Fuel*, *217*, 240–261.

Wei, L., Shutao, W., Jin, Z., & Tong, X. (2014). Biochar influences the microbial community structure during tomato stalk composting with chicken manure. *Bioresource technology*, *154*, 148–154.

Wu, Q., Zhang, Y., Cui, M. H., Liu, H., Liu, H., Zheng, Z., Zheng, W., Zhang, C., & Wen, D. (2022). Pyrolyzing pharmaceutical sludge to biochar as an efficient adsorbent for deep removal of fluoroquinolone antibiotics from pharmaceutical wastewater: Performance and mechanism. *Journal of Hazardous Materials*, *426*, 127798.

Xie, R., Zhu, Y., Zhang, H., Zhang, P., & Han, L. (2021). Effects and mechanism of pyrolysis temperature on physicochemical properties of corn stalk pellet biochar based on combined characterization approach of microcomputed tomography and chemical analysis. *Bioresource Technology*, *329*, 124907.

Yu, X., & Lu, S. (2020). Double effects of biochar in affecting the macropore system of paddy soils identified by high-resolution X-ray tomography. *Science of the total environment*, *720*, 137690.

Yu, X., Wu, C., Fu, Y., Brookes, P., & Lu, S. (2016). Three-dimensional pore structure and carbon distribution of macroaggregates in biochar-amended soil. *European Journal of Soil Science*, *67*, 109–120.

Zeidabadi, Z. A., Bakhtiari, S., Abbaslou, H., & Ghanizadeh, A. R. (2018). Synthesis, characterization and evaluation of biochar from agricultural waste biomass for use in building materials. *Construction and Building Materials*, *181*, 301–308.

Zhang, H., Li, Y., Xie, R., Zhu, Y., Shi, S., Yang, Z., & Han, L. (2022). A particle scale micro-CT approach for 3D in-situ visualizing the Pb (II) adsorption in different crop residue-derived chars. *Bioresource Technology*, *344*126269.

Zhang, H., Tian, S., Zhu, Y., Zhong, W., Qiu, R., & Han, L. (2022). Insight into the adsorption isotherms and kinetics of Pb (II) on pellet biochar via in-situ non-destructive 3D visualization using micro-computed tomography. *Bioresource Technology*, *358*127406.

Zhang, W., Huang, W., Tan, J., Huang, D., Ma, J., Wu, B. 2022. Modeling, optimization and understanding of adsorption process for pollutant removal via machine learning: Recent progress and future perspectives.

Zhang, J., Wen, N., Sun, Q., Horton, R., & Liu, G. (2023). The effect of macropore morphology of actual anecic earthworm burrows on water infiltration: A COMSOL simulation. *Journal of Hydrology*, *618*, 129261.

Zhao, M., Dai, Y., Zhang, M., Feng, C., Qin, B., Zhang, W., Zhao, N., Li, Y., Ni, Z., Xu, Z., Tsang, D. C. W., & Qiu, R. (2020). Mechanisms of Pb and/or Zn adsorption by different biochars: Biochar characteristics, stability, and binding energies. *Science of The Total Environment*, *717*, 136894.

Zhao, W., Zhang, Z., Xin, Y., Xiao, R., Gao, F., Wu, H., Wang, W., Guan, Q., & Lu, K. (2024). Na2S-modified biochar for Hg (II) removal from wastewater: a techno-economic assessment. *Fuel*, *356*129641.

Zhou, H., Yu, X., Chen, C., Zeng, L., Lu, S., & Wu, L. (2018). Evaluating hydraulic properties of biochar-amended soil aggregates by high-performance pore-scale simulations. *Soil Science Society of America Journal*, *82*, 1–9.

Zhu, X., Li, Y., & Wang, X. (2019). Machine learning prediction of biochar yield and carbon contents in biochar based on biomass characteristics and pyrolysis conditions. *Bioresource Technology*, *288*, 121527. Available from https://doi.org/10.1016/j.biortech.2019.121527.

Chapter 5

Harnessing biochar for sustainable catalysis in environmental applications

Lucas Meili[1,3], Rodolfo Junqueira Brandão[2,3] and Thais Logetto Caetité Gomes[2,3]
[1]*Federal University of Alagoas, Center of Technology, Maceió, AL, Brazil,* [2]*Federal University of Alagoas, Flowlab (Fluid Dynamics Laboratory), Center of Technology, Maceió, AL, Brazil,* [3]*Federal University of Alagoas, Laboratory of Processes, Center of Technology, Maceió, AL, Brazil*

Chapter outline

5.1 Introduction	75
5.2 Nature of biochar catalysts	76
5.3 Preparation and characterization of biochar catalysts	76
5.4 Mechanisms of biochar catalysis	80
5.4.1 Fenton system	81
5.4.2 Persulfate activation system	82
5.4.3 Photocatalytic system	83
5.5 Biochar catalysts regeneration	84
5.6 Environmental applications of biochar catalysis	85
5.7 Future research directions	86
5.8 Conclusions	86
AI Disclosure	86
References	86

5.1 Introduction

In the context of sustainable development, governments and societies face significant environmental challenges. These include the need to enhance waste management practices, mitigate greenhouse gas emissions, utilize water resources sparingly, and conserve them. Simultaneously, there is a pressing demand to meet energy needs and produce sufficient, high-quality food to promote human development and quality of life.

Addressing these challenges requires a fundamental shift from the current model of resource exploitation, which assumes natural resources are infinite. Therefore, adopting new technologies is essential to achieve sustainable development across social, environmental, and economic dimensions. In this scenario, biochar composites are at the frontier of materials development for current environmental and engineering challenges in contaminant sorption and degradation, capacitive deionization, and supercapacitor research. However, the parameter space for optimization of such composites is vast, spanning from the choice of feedstocks and synthesis procedure to postprocessing.

Biochar and its composites have been extensively researched for the remediation of organic and inorganic pollutants from groundwater and soil. The primary environmental applications of biochar and its derivatives include the degradation of organic pollutants such as p-nitrophenol, polychlorinated biphenyls (PCBs), chlorobenzene, and 2-chlorobiphenyl, which offer significant risks to human health and the environment. Additionally, biochar and biochar-based catalysts can help reduce the toxicity of inorganic contaminants. These materials are effective, cost-efficient, and environmentally friendly, employing mechanisms such as oxidation, reduction, and photocatalysis to remove pollutants (Lyu et al., 2020).

Biochar can be the primary product of biomass carbonization or a by-product of biomass gasification and fast pyrolysis processes. New methods like microwave pyrolysis and hydrothermal carbonization (HTC) provide additional options for biochar production. Moreover, the physical and chemical properties of biochar can be easily customized for specific applications. Scientists have developed various modification techniques to enhance its functionality. These qualities make biochar highly promising in various fields, including catalysis, energy storage, and environmental remediation. Research highlights biochar's effectiveness as a soil amendment, carbon capture medium, pollutant absorber, electrode material, catalyst, and catalyst support (Cheng & Li, 2018).

According to Lee et al. (2017), catalysts have played a crucial role in developing technologies to convert not only conventional carbonaceous feedstocks (e.g., coal, natural gas, and petroleum) but also renewable feedstocks (e.g., biomass) into value-added products such as fuels and chemicals. Researchers have valued carbon-based materials in various industrial catalytic processes for many years. These materials offer beneficial properties for both catalyst support and acting as catalysts to apply in industrial processes (Rodríguez-Reinoso, 1998). Biochar has gained particular interest due to its inherent properties, such as porosity and carbon content, making it a promising alternative to traditional solid carbon-based catalysts, which often have disadvantages like high cost and environmental impact (Lee et al., 2017).

The growing interest in using biochar as a catalyst and catalyst support has spurred initial investigations into its effectiveness and functionality in various catalytic processes. This chapter will examine biochar's applications in catalysis, exploring different preparation methods, comparing its properties and costs to traditional catalysts, addressing the challenges faced in its use, and identifying potential future research directions.

5.2 Nature of biochar catalysts

The elemental composition of biochar, which typically includes carbon, hydrogen, oxygen, nitrogen, and inorganic elements, varies based on the feedstock and its conditions of production. For example, increasing pyrolysis temperatures typically produce biochar with a larger surface area, a higher pH, and greater porosity. Biochar's pH can range from neutral to alkaline, and it tends to rise with higher production temperatures as acidic components decompose. (Bartoli et al., 2023). Functional organic groups like hydroxyl, aldehyde, and ketone can contribute to high pH (Lee et al., 2010). Furthermore, these groups can affect the biochar's hydrophobicity, hydrophilicity, and adsorptive properties. Additionally, they can reduce its negative charge, thereby increasing its cation exchange capacity (CEC) (Cheng & Li, 2018).

The focus of research has shifted towards developing environmentally friendly catalysts known as green catalysts, derived from biomass. These catalysts are not only cost-effective but also renewable. Biochar-based catalysts and nanocatalysts show significant promise for producing biofuels in a sustainable and socially responsible manner (Chellappan et al., 2018; Dehkhoda et al., 2010). Biochar's surface chemistry provides several advantages. For instance, the presence of inorganic elements such as potassium (K) and iron (Fe) promotes tar cracking during biofuel production. Moreover, functional groups on biochar facilitate the adsorption of metal precursors, thereby enhancing its catalytic abilities (Velusamy et al., 2021). This focus on green catalysts, particularly nano- and biochar-based options, stems from their attractive features: simple synthesis procedures, affordability, ease of disposal, reusability, and the potential for increased biofuel yields (Cheng & Li, 2018).

Compared to traditional catalysts, biochar can exhibit higher thermal stability and resistance to deactivation. The presence of inherent mineral content (nitrogen, phosphorus, sulfur, calcium, magnesium, and potassium), derived from the original biomass, can also contribute to its catalytic properties, providing a more sustainable and environmentally friendly option. Notably, biochar has a high surface area and porosity, enhancing its catalytic efficiency by providing abundant active sites for reactions. Its surface allows for the incorporation of various functional groups, enhancing its versatility in catalysis (Singh, 2021).

Besides that, biochar catalysts are generally more cost-effective; their production utilizes waste biomass, which is often inexpensive and readily available, leading to lower raw material costs. In contrast, conventional catalysts often rely on precious metals or complex manufacturing processes, resulting in higher costs. Biochar's extended lifespan and resistance to deactivation result in reduced replacement frequency, leading to lower operational costs over time. Its combination of low raw material costs, straightforward production methods, and durable performance makes biochar a highly economical alternative to traditional catalysts (Akpasi et al., 2022).

The essential properties such as surface area, pore volume, pore size, and acidity directly affect the biochar's catalytic ability. A catalyst's effectiveness heavily relies on the accessibility of its active sites, which are often distributed within internal pores. Raw biochar often has limitations as a catalyst due to its pore structure and morphology and lack of beneficial functional groups, requiring the use of methods for pretreatment and activation before its application (Zhao et al., 2023).

5.3 Preparation and characterization of biochar catalysts

Biochar can be produced from the thermochemical conversion of biomass in an oxygen-limited environment. Among the production methods can be mentioned torrefaction, pyrolysis, HTC, and gasification. A significant advantage of thermochemical processes lies in their relative simplicity, leading to lower capital costs (Xiu et al., 2017). Below are presented some particularities of each method.

- **Torrefaction:** It is a process carried out at temperatures ranging from 200 to 300°C in an inert atmosphere that removes moisture and volatile organic compounds, resulting in a biochar with improved energy density and hydrophobic properties, higher ash and fixed carbon content, and also greater calorific value and lower moisture content (Dai et al., 2019). In general, it is a thermal pretreatment to improve biomass quality before being used in the pyrolysis or gasification processes (Bartoli et al., 2023).
- **Pyrolysis or Carbonization:** It is the most common method for biochar production and occurs at higher temperatures, typically between 300 and 700°C, also in an oxygen-limited environment. The specific temperature and residence time can significantly affect the yield and properties of the biochar, including its carbon content, surface area, and porosity. Pyrolysis can be further classified into slow and fast pyrolysis, depending on the heating rate and residence time (Arni, 2018; Ramos et al., 2022; Tan et al., 2021). Slow pyrolysis typically operates at lower temperatures and slower heating rates, resulting in higher biochar yields but longer processing times. This method enhances biochar's stability and carbon content, making it suitable for soil amendment and carbon sequestration (Arni, 2018). Fast pyrolysis, on the other hand, involves rapid heating rates and higher temperatures, producing biochar, bio-oil, and syngas in a matter of seconds. This process maximizes the production of bio-oil and syngas, with biochar as a by-product. The biochar produced through fast pyrolysis typically has lower carbon content and stability but higher surface area and porosity, making it effective for applications like filtration and adsorption (Tan et al., 2021).
- **Gasification:** It is conducted at even higher temperatures, above 700°C, in the presence of a controlled amount of oxidizing agent (e.g., oxygen or steam). Primarily, this process produces syngas, a mixture of hydrogen and carbon monoxide, with biochar as a by-product. Gasification results in biochar with high surface area and porosity, which is effective for adsorption applications. However, the biochar yield is generally lower compared to pyrolysis due to the higher degree of biomass conversion (Ramos et al., 2022; Yaashikaa et al., 2020).
- **HTC:** A process that utilizes heat and pressure in the presence of water to transform biomass into a carbonaceous material known as hydrochar. Unlike traditional biomass conversion methods such as pyrolysis and gasification, HTC operates at relatively moderate temperatures (180°C–280°C) and pressures (10–20 bar), making it an energy-efficient and environmentally friendly approach. HTC is particularly advantageous for feedstock with high moisture levels, as it does not require drying before processing. In comparison with biochar, hydrochar contains less carbon, ash, surface area, and a smaller pore volume. However, hydrochar is enriched with functional groups, leading to a higher CEC than biochar (Ramos et al., 2022).

Additionally, several approaches have been developed for synthesizing biochar-based catalysts from various feedstocks. Key technologies include calcination, hydrothermal treatment, sol–gel, and impregnation, as presented in Table 5.1.

Calcination typically occurs under nitrogen or air conditions at 300°C–700°C, akin to pyrolysis. This method is crucial for synthesizing biochar-based catalysts, where multiple solid phases combine for composite multifunctionalization. Compared to pyrolysis, calcination enhances material hardness, prevents breakage, and alters textural and mineralogical characteristics (Peng et al., 2019). Temperature, gas composition, and thermal stability significantly impact the calcination process. For instance, TiO_2/pBC composites prepared via calcination have lower specific surface areas (SBET) due to semiconductor coverage on the biochar surface (Mian et al., 2019). Higher calcination temperatures, such as 500°C, can degrade biochar structure, reducing SBET, whereas calcination at 600°C–800°C can increase surface area in some composites. Therefore, selecting appropriate feedstock and temperature is essential.

In the hydrothermal treatment, compared to pyrolysis chars, hydrochars have more oxygen-containing functional groups (OFGs), making them better catalysts. Hydrothermal treatment for biochar-based catalysts occurs at 90°C–220°C for 2–24 hours, capable of producing materials with nanosized structures. Hydrothermal treatment is cost-efficient compared to calcination due to milder conditions and simpler processes.

In the impregnation method, active metallic or non-metallic species (such as N, S, P, and B) are incorporated into biochar by mixing feedstock with precursors, creating active interfaces and binding sites. This enhances biochar-based catalysts' adsorption capacity, catalytic performance, and magnetism (metallic impregnation). For example, magnetic nitrogen-doped biochar catalysts are prepared by impregnating biochar with $Co(NO_3)_2$ and $FeSO_4$ under oxygen-limited conditions, followed by calcination at 400°C–800°C, resulting in the formation of crystals on the biochar surface and stabilizing the pore structure. Higher temperatures also improve graphitization, aiding charge transfer and electron promotion.

In the nonmetallic impregnation method, due to the different atomic radius, atomic orbitals, and electronegativity of these elements, this method could promote the electron density, reactive-active moieties, defective sites, and reusability by chemically modifying and modulating the inherent carbon configuration (Tan et al., 2017). For instance,

TABLE 5.1 Key technologies for synthesizing biochar-based catalysts.

Biomass types	Preparation methods	Preparation conditions	Catalysts	References
Pine needles	Pyrolysis	300°C; 10°C·min−1; 6 h	Pine needle biochar	Fang et al. (2015)
Cotton straw	Pyrolysis; activation	350°C; 10°C·min−1; 2 h; NH_2Cl treatment	Cotton straw biochar	Wang et al. (2019)
Rice straw	Impregnation; pyrolysis	Salt solution impregnation; 800°C for 4 h; 5°C·min^{-1}	N-biochar @ $CoFe_2O_4$	
Camellia seed husks	Impregnation; pyrolysis; co-precipitation	400°C for 2 h; HNO_3(68%) H_2SO_4 (70%) soaking 1 h +	oxidation biochar- Fe_3O_4	Pi et al. (2022)
Wheat husks, paper sludge	Pyrolysis; impregnation; hydrothermal treatment	BC: pyrolysis at 500°C for 20 min; $Zn(NO_3)_2$ solution; hydrothermal: 90°C for 2 h	ZnO-biochar	Gholami et al. (2019)
Biochar	Oxidation; hydro-thermal method	Oxidation by H_2SO_4, HNO_3; hydrothermal method at 90°C for 24 h	Fe-Cu-LDH / biochar	Gholami et al. (2020)
Hemp stem	Sol–gel; calcine	500°C for 2 h in N_2	TiO_2-CuO/HSC	Peng et al. (2019)
Hog fuel, demolition waste	Pyrolysis; hydrothermal	600°C for 30 min; 180°C for 3 h	Cu_2O-CuO @ biochar	Khataee et al. (2019)
Sewage sludge	Impregnation; thermal decomposition	800°C for 1 h; 10°C·min−1 under N_2	TiO_2/Fe/Fe_3C-biochar	Mian et al. (2019)
Olive pits	Sol–gel; hydrothermal; magnetization	500°C; 180°C for 6 h	TiO_2-OP @Fe_3O_4	Djellabi et al. (2019)

Source: Yang, B., Dai, J., Zhao, Y., Wu, J., Ji, C. & Zhang, Y. (2022). Advances in preparation, application in contaminant removal, and environmental risks of biochar-based catalysts: a review. *Biochar*, 4(1). https://doi.org/10.1007/s42773-022-00169-8; Zhou, X., Zhu, Y., Niu, Q., Zeng, G., Lai, C., Liu, S., Huang, D., Qin, L., Liu, X., Li, B., Yi, H., Fu, Y., Li, L., Zhang, M., Zhou, C., & Liu, J. (2021). New notion of biochar: A review on the mechanism of biochar applications in advanced oxidation processes. *Chemical Engineering Journal*, 416, Article 129027. https://doi.org/10.1016/j.cej.2021.129027. Copyright 2021, with permission of Elsevier.

biochar-based catalysts doping S could improve the persulfate activation capacity of sludge biochar-based catalysts by enhancing the electron shuttling capacity in nonradical pathways (Yun et al., 2018). The doping of P could introduce functionality onto biochars to realize chemical activation based on the low electronegativity and the large atomic radius of P (Zhou et al., 2022).

Biochar properties can be improved through modification methods. By increasing the porosity and surface area, biochar becomes a more effective support for active substances. Additionally, the introduction of desired functional groups onto the biochar's surface can create active sites that drive catalytic reactions. Therefore, selecting the appropriate modification method (such as physical or chemical modification) is essential to optimizing the catalytic performance of biochar-based catalysts (Fig. 5.1).

Physical activation or gas activation is a process that utilizes gases like steam, carbon dioxide, or even ozone at high temperatures (above 700°C) to enhance the biochar's properties. This process typically involves two stages. The first stage removes selectively less structured portions of the biochar and simultaneously opens up smaller pores within the remaining carbon structure. This effectively increases the internal surface area. In the second stage, activation reactions deplete the biochar of crystalline carbon, further expanding these smaller pores into larger ones. The formation of these larger pores is directly linked to carbon depletion during the activation process (Cha et al., 2016). Physical activation characteristics are related to the biochar-activated type, activating gas, and reaction conditions, and its effectiveness in creating pores depends on two key factors: the amount of ash remaining in the biochar and the existing pore structure (Zhao et al., 2023).

Chemical activation modifies biochar by introducing a more porous structure with a larger surface area. This process involves treating the biochar with chemicals at high temperatures in an inert gas flow (Zhao et al., 2023). Common activating agents include strong bases like potassium hydroxide (KOH) or sodium hydroxide (NaOH), including acids like phosphoric

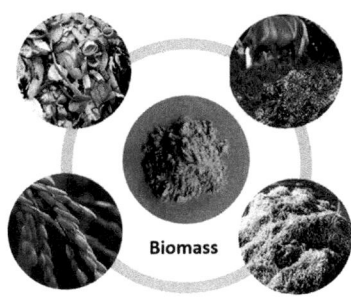
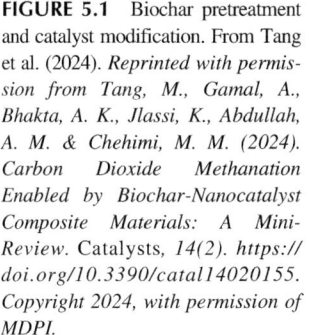
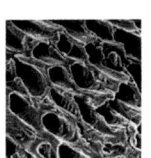

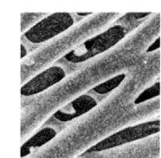

FIGURE 5.1 Biochar pretreatment and catalyst modification. From Tang et al. (2024). *Reprinted with permission from Tang, M., Gamal, A., Bhakta, A. K., Jlassi, K., Abdullah, A. M. & Chehimi, M. M. (2024). Carbon Dioxide Methanation Enabled by Biochar-Nanocatalyst Composite Materials: A Mini-Review. Catalysts, 14(2). https://doi.org/10.3390/catal14020155. Copyright 2024, with permission of MDPI.*

acid (H_3PO_4) or sulfuric acid (H_2SO_4). Salts such as potassium carbonate (K_2CO_3), zinc chloride ($ZnCl_2$), and even ammonia (NH_3) can also be used for this purpose. The specific chemical reactions between these agents and biochar can be complex, but they ultimately lead to a more effective material for various applications (Cha et al., 2016). Eq. (5.1)–(5.4) present the reaction of biochar activation with an alkali metal activating agent (Zhao et al., 2023):

Reaction of biochar activation with an alkali metal activating agent

$$2KOH + CO_2 \rightarrow K_2CO_3 + H_2O \uparrow \qquad (5.1)$$

$$2C + 2KOH \rightarrow 2CO\uparrow + 2K\uparrow + H_2\uparrow \qquad (5.2)$$

$$K_2CO_3 + C \rightarrow K_2O + 2CO\uparrow \qquad (5.3)$$

$$K_2O + C \rightarrow 2K\uparrow + CO\uparrow \qquad (5.4)$$

At high temperatures, biochar releases volatile substances like water and carbon compounds that can react with KOH to form potassium carbonate (represented by Eq. 5.1). This reaction also releases gaseous carbon oxides and hydrogen to the surrounding environment (Eq. 5.2). As these gases escape, they leave behind a porous structure within the biochar. Further increasing the temperature causes the potassium carbonate to react with the remaining carbon, releasing additional gaseous products like carbon monoxide and potassium (Eqs. 5.3 and 5.4). This secondary reaction creates even larger pores in the biochar (Zhao et al., 2023).

Several parameters influence the outcome of biochar's chemical activation, including the type and amount of activating agent and the activation temperature employed. The choice of activating agent significantly impacts the biochar's properties, as different chemicals can induce distinct structural changes. The quantity of activating agents also plays a role, as higher concentrations generally lead to more extensive modifications. Finally, the activation temperature is crucial, as it affects the rate and extent of the chemical reactions that occur (Bartoli et al., 2023).

A study by Tay et al. (2009) investigated how different chemical activation conditions affect the properties of biochar derived from soybean oil cake. The researchers compared the effects of two activating agents, K_2CO_3 and KOH, and different temperatures (ranging from 600°C to 800°C). They found that K_2CO_3 activation resulted in a higher yield of activated biochar with a greater pore volume and lower ash and sulfur content compared to KOH activation. Interestingly, both activating agents increased the specific surface area of the biochar more significantly at higher temperatures. However, the effect of temperature was much more pronounced with K_2CO_3. When K_2CO_3 was used, the total pore volume, microporosity, and mesoporosity all increased with higher activation temperatures. In contrast, KOH activation led to a decrease in microporosity at higher temperatures. At 800°C, the maximum specific surface area was obtained for activated carbon with K_2CO_2, whose performance was similar to commercially available activated carbons.

In general, chemical activation offers a more efficient way to boost biochar's performance than physical activation. However, this method comes with challenges, such as corrosion of equipment caused by the chemicals, difficulty in recovering the activating agents, and their high cost.

The potential application of biochar as a catalyst or based catalyst is closely related to its physicochemical properties, which depend on the raw material, method, and operational conditions used for production. From this perspective, the appropriate biochar characterization becomes essential to comprehend and predict its catalytic performance.

As described by Lyu et al. (2020), many characterization methods have been used to determine the physiochemical properties of biochar catalysts, whose techniques are based on the structure, surface functional groups, and elemental analysis, such as photoluminescence spectroscopy, scanning electron microscope, energy dispersive X-ray analysis (EDAX), energy dispersive spectrometer (EDS), X-ray diffraction (XRD), Fourier-transform infrared spectroscopy (FTIR), BET adsorption method, X-ray photoelectron spectroscopy (XPS), elemental analysis, and ultraviolet-visible diffuse reflectance spectroscopy (UV-Vis Drs), as shown in Fig. 5.2.

5.4 Mechanisms of biochar catalysis

It is widely recognized that the functional structures of biochar, such as oxygen-containing groups, defects, and persistent free radicals (PFRs), serve as active sites in heterogeneous catalytic systems. Furthermore, the incorporation of

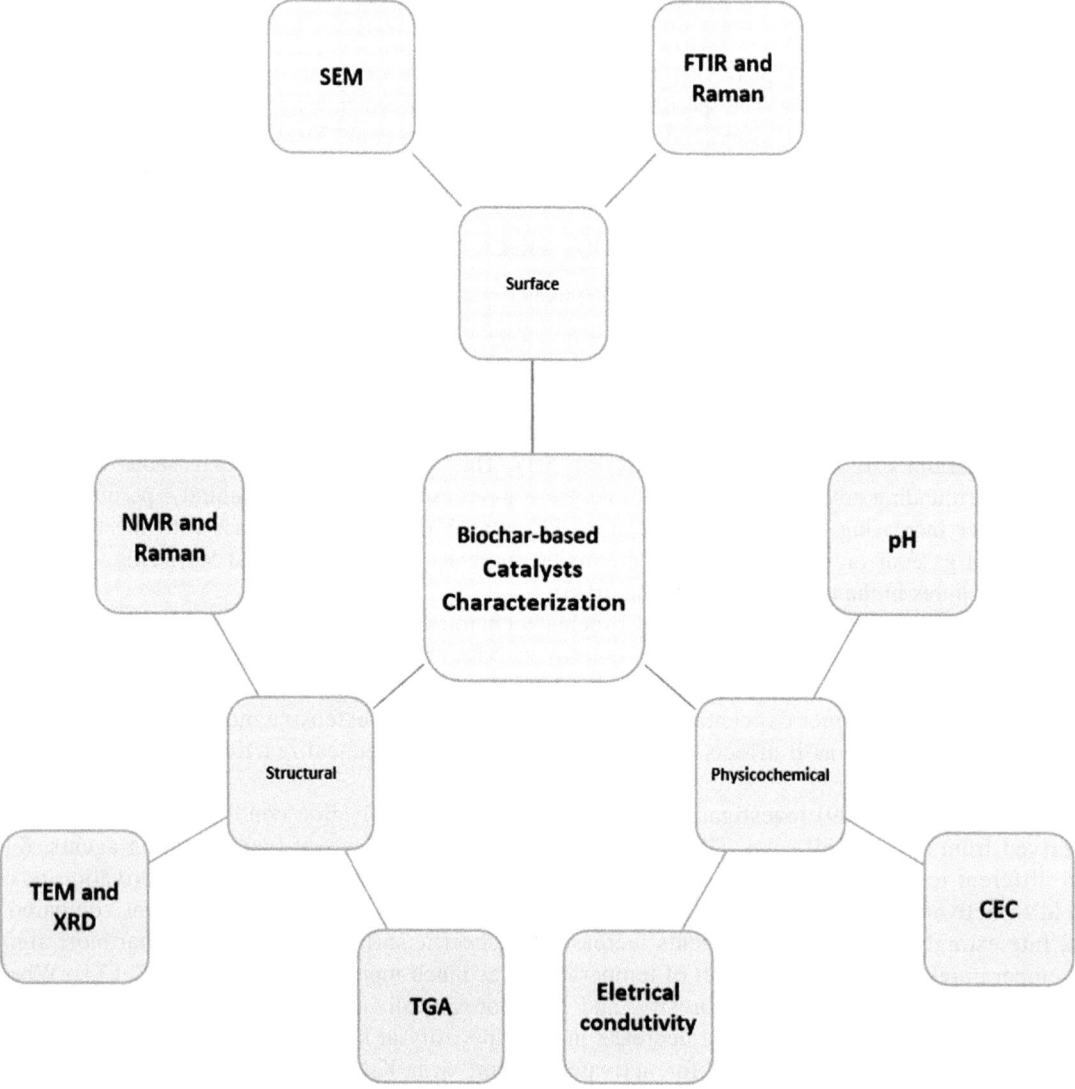

FIGURE 5.2 Biochar-based catalyst's physicochemical, surface, and structural characterization.

metal components and heteroatom doping not only tunes the inherent properties of biochar but also provides additional active sites for catalysis.

$$C - OH + H_2O_2 \rightarrow COOH + H_2O \tag{5.5}$$

$$C - OH + H_2O_2 \rightarrow CO^{\bullet} + {}^{\bullet}OH + H_2O \tag{5.6}$$

Understanding the physics and biochemical behaviors of biochar helps explaining or predicting its reactivity. Controlling the interactions maximizes the number of active sites, enabling targeted processes for specific substrates and influent characteristics. Biochar's performance is intricately influenced by its intrinsic redox potential, electron storage and transfer capabilities, conjugated conductive network, and the formation of heterojunction interfaces. Morphology, surface texture, specific binding sites, and active centers also play crucial roles in shaping its effectiveness. Notably, electron pooling and charge transfer facilitated by biochar's semiconducting structure are critical for most catalytic interactions. Semiconducting biochar can be structured to act as a controllable charge trapper, restraining charge recombination and transferring them to different interfaces for intended redox reactions involving holes (h^+) or electrons (e^-).

The reaction rate is often governed by the diffusion and mobility of substrates and reactive species to and from catalytically active sites. This rate can be influenced by several factors: (1) the disparity between sorption and reaction kinetics; (2) the competition for binding or catalytic sites; or (3) self-activated reactions or those sensitive to external stimuli. Additionally, scattered functionalities on the biochar surface can disrupt catalytic reactions within the pores. Semiconducting biochar can be engineered to function as a controllable charge trap, reducing charge recombination and directing them to various interfaces for specific redox reactions involving h^+ or e^-. Proper design enhances both background and light absorption, enabling biochar to act as a reservoir and sensitizer to capture excitons and prevent their recombination. Furthermore, increased porosity was found to facilitate the sequestration-destruction-desorption process.

Zhou et al. (2021) summarized the steps of the mechanisms of biochar catalysis: (1) oxygen-containing groups and PFRs on original biochar enhance the catalytic process by generating radicals; (2) when metal is loaded on biochar, it becomes the primary active site; (3) doped heteroatoms can function as active sites or induce other Lewis base sites on biochar, promoting catalysis, and are usually associated with electron transfer processes; (4) increased graphitization and specific surface area improve the adsorption capacity of biochar, thereby facilitating the catalytic process.

The following topics elucidate the primary mechanisms discussed in the literature:

5.4.1 Fenton system

The importance of oxygen-containing groups in biochar adsorption has been consistently emphasized. These groups serve as active sites in the Fenton-like systems. Generally, oxygen-containing groups on biochar are categorized into two types: acidic groups (such as −OH, −COOH, and C=O) and basic groups (such as chromene species and ketone pyrone).

Although oxygen-containing groups are commonly encountered, PFR in the Fenton-like system have been reported frequently, generating active sites with a long lifetime from days to months. PFRs are formed by electron transfer from organic components (such as phenols, and quinones) to transition metals under combustion conditions (Fig. 5.3).

It is also reported in the literature that the addition of carbonaceous materials improves the removal efficiency of the metal catalyst in heterogeneous catalytic systems, and the metal plays a major role in most conditions. As one of the

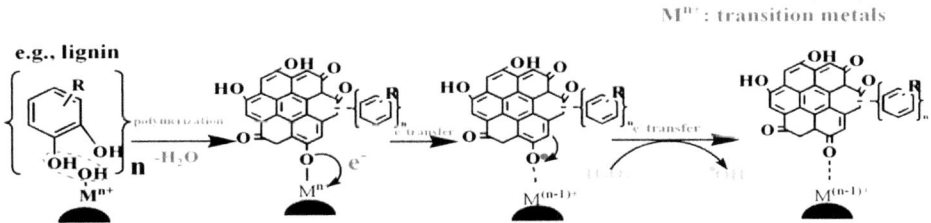

FIGURE 5.3 Persistent free radicals (PFRs) form primarily during the thermal treatment of carbon-based materials. From Fang et al. (2014). *Reprinted with permission from Fang, G., Zhu, C., Dionysiou, D. D., Gao, J., & Zhou, D. (2015). Mechanism of hydroxyl radical generation from biochar suspensions: Implications to diethyl phthalate degradation. Bioresource Technology, 176, 210−217. https://doi.org/10.1016/j.biortech.2014.11.032. Copyright 2014, with permission of American Chemical Society.*

key parts of the traditional Fenton reaction, Fe is a conventional doped element in the Fenton-like system, including Fe(III), Fe(II), and nanoscale zero-valent iron (nZVI). Efficient degradation of organic matter was achieved through the valence reaction of iron (Fe). In these conditions, biochar typically enhances the removal rate by preventing the aggregation of metal components and boosting the adsorption capacity of catalysts. Table 5.2 synthesizes the relationships between functional structures and mechanisms in the Fenton system. Table 5.3 shows the environmental application of biochar catalysts using the Fenton mechanism.

5.4.2 Persulfate activation system

Persulfate (PS) is known to produce sulfate radicals ($SO_4^{\bullet-}$), which possess a redox potential comparable to hydroxyl radicals ($^{\bullet}OH$). PS is advantageous for transportation and storage, offering improvements over traditional Fenton-like systems. As a result, PS activation has quickly become a significant area of study in advanced oxidation processes (AOPs). When biochar, produced through pyrolysis, serves as the catalyst in PS activation systems, $SO_4^{\bullet-}$ and $^{\bullet}OH$ are

TABLE 5.2 Relationship between functional structures and mechanisms in Fenton system. From Yang et al. (2022).

Functional structure	Mechanism
-OH	$^{\bullet}OH$
Quinone moieties	$O_2^{\bullet-}$
Metals	$^{\bullet}OH, O_2^{\bullet-}$
Oxygen-centered PRFs	$^{\bullet}OH$

Source: Yang, B., Dai, J., Zhao, Y., Wu, J., Ji, C. & Zhang, Y. (2022). Advances in preparation, application in contaminant removal, and environmental risks of biochar-based catalysts: a review. *Biochar*, 4(1). https://doi.org/10.1007/s42773-022-00169-8. Copyright 2022, with permission of Springer.

TABLE 5.3 Active sites and mechanisms of different biochar in Fenton like system. From Zhou et al. (2021).

Catalyst	Feedstock	Pollutant	Reported active sites	Mechanism	References
nZVI/BC	rice hull	Trichloroethylene	nZVI, −OH	$^{\bullet}OH$	Yan et al. (2017)
Fe-impregnated BC	sugarcane	Orange G	Fe oxide particle	$^{\bullet}OH$	Park et al. (2018)
Pristine biochar	rice husk	Chlorobenzene	quinones and phenolic−OH	$^{\bullet}OH, O_2^{\bullet-}$	Zhang et al. (2018)
Pristine biochar	pine needles	Diethyl phthalate bisphenol A	PFRs	$^{\bullet}OH$	Fang et al. (2015)
Cu(II) doped or Fe(III) doped BC	pine needles	Bisphenol A	Oxygen-centered PFRs	$^{\bullet}OH, O_2^{\bullet-}$	Ruan et al. (2018)
Fe doped biochar	sugarcane bagasse	Metronidazole	Fe	$^{\bullet}OH$	Yi et al. (2020)
Pristine biochar	Swine manure	Sulfamethazine	Oxygen-centered PFRs	$^{\bullet}OH$	Deng et al. (2020)
MnFe$_2$O$_4$/biochar	pine needles	Tetracycline	Fe-Mn binary oxide	$^{\bullet}OH$	Li et al. (2019)

Source: Zhou, X., Zhu, Y., Niu, Q., Zeng, G., Lai, C., Liu, S., Huang, D., Qin, L., Liu, X., Li, B., Yi, H., Fu, Y., Li, L., Zhang, M., Zhou, C. & Liu, J. (2021). New notion of biochar: A review on the mechanism of biochar applications in advanced oxidation processes. *Chemical Engineering Journal*, 416, 129027. https://doi.org/10.1016/j.cej.2021.129027.

typically the main reactive species. Additionally, the oxygen-containing functional groups on biochar, such as hydroxyl (−OH) and carboxyl (−COOH) groups, act as active sites as shown in Eqs. (5.7) and (5.8), respectively (Zhou et al., 2021).

$$\text{BIOCHAR} - \text{OH} + S_2O_8^{2-} \rightarrow \text{BIOCHAR} - \text{O}\cdot + SO_4^- + HSO_4^- \quad (5.7)$$

$$\text{BIOCHAR} - \text{OH} + S_2O_8^{2-} \rightarrow \text{BIOCHAR} - \text{OO}\cdot + SO_4^- + HSO_4^- \quad (5.8)$$

Table 5.4 summarizes the relationship between functional structures and mechanisms in the persulfate activation system. Table 5.5 details the environmental applications of biochar catalysts utilizing persulfate activation mechanisms.

5.4.3 Photocatalytic system

Photocatalytic systems rely on the light sensitivity of semiconductors. The primary active species in these systems include h^+, •OH, $O_2^{\cdot-}$, and O_2. The processes induced by light irradiation follow specific equations numbered as Eq. (5.9)–(5.16). In these systems, •OH and $O_2^{\cdot-}$ are produced at the valence band (VB) and conduction band (CB), respectively. When the CB potential of the photosensitizer is more negative than that of $O_2/O_2^{\cdot-}$ (-0.33 eV/NHE), electrons at the CB interact with O_2 to form $O_2^{\cdot-}$. Additionally, $O_2^{\cdot-}$ can further generate •OH. Moreover, h + at the VB can react with OH− or H_2O in the aqueous phase to produce •OH, provided the VB is more positive than 2.4 eV/NHE or 2.72 eV/NHE (OH−/•OH, H_2O/•OH). h + can also react with $O_2^{\cdot-}$ to produce O_2. Since h + is directly generated by

TABLE 5.4 Relationship between functional structures and mechanisms in Persulfate activation system. From Zhou et al. (2021).

Functional Structure	Mechanism
C = O	O_2
−OH	•OH, $SO_4^{\cdot-}$
−COOH	•OH, $SO_4^{\cdot-}$
PFRs	•OH, $SO_4^{\cdot-}$, $O_2^{\cdot-}$

TABLE 5.5 Active sites and mechanisms of different biochar in Persulfate activation system. From Zhou et al. (2021).

Catalyst	Feedstock	Pollutant	Activator	Reported active sites	Mechanism	References
Pristine biochar	sawdust	Acid orange 7	$Na_2S_2O_8$	PFRs	•OH, $SO_4^{\cdot-}$	He et al. (2019)
Magnetic N-doped sludge biochar	sewage sludge	Tetracycline hydrochloride	$Na_2S_2O_8$	Fe, C = O, doped N	•OH, $SO_4^{\cdot-}$	Yu et al. (2019)
ZVI/BC	sawdust	Bisphenol A	Peroxy-monosulfate	ZVI, PFRs	Electron transfer process, $SO_4^{\cdot-}$, $O_2^{\cdot-}$	Jiang et al. (2019)
nZVI/BC	rice husk	Nonylphenol	Peroxydisulfate	nZVI, −OH, −COOH	•OH, $SO_4^{\cdot-}$	Hussain et al. (2017)
Pristine biochar	corn stalk	Norfloxacin	$K_2S_2O_8$	−OH, −COOH	•OH, $SO_4^{\cdot-}$	Wang et al. (2019)

light irradiation at the VB of the semiconductor, it is a crucial factor in photocatalytic systems. Generally, the positions of VB and CB are determined by the photosensitizer species.

$$\text{semicondutor} + h\nu \rightarrow \text{semicondutor} + (h^+ + e^-) \tag{5.9}$$

$$h^+ + H_2O \rightarrow {}^\bullet OH + H^+ \tag{5.10}$$

$$h^+ + OH^- \rightarrow {}^\bullet OH \tag{5.11}$$

$$e^- + O_2 \rightarrow O_2^{\bullet -} \tag{5.12}$$

$$h^+ + O_2^{\bullet -} \rightarrow O_2 \tag{5.13}$$

$$O_2^{\bullet -} + H_2O \rightarrow {}^\bullet HO_2 + OH^- \tag{5.14}$$

$$^\bullet HO_2 + H_2O \rightarrow H_2O_2 + {}^\bullet OH \tag{5.15}$$

$$H_2O_2 \rightarrow 2{}^\bullet OH \tag{5.16}$$

In contrast to its role in PS activation and Fenton-like systems, biochar typically plays a secondary role in photocatalytic systems. Its main functions in these systems include:

- Serving as a support structure to prevent the aggregation of photosensitizers.
- Creating a hydrophobic microenvironment around active sites, thereby enhancing the adsorption capacity of catalysts.
- Acting as an electron shuttle to reduce electron-hole recombination.
- Integrating into the structure of photosensitizers to lower the band gap.

Table 5.6 synthesizes the relationship between functional structures and mechanisms in the photocatalytic system (Zhou et al., 2021).

5.5 Biochar catalysts regeneration

Biochar catalysts often undergo deactivation due to factors such as the accumulation of reaction intermediates, coking, and poisoning by various contaminants. This way, its regeneration and reutilization are essential to guaranteeing the greatest economic and environmental benefits. The primary methods for biochar regeneration include thermal treatment, chemical treatment, and biological treatment.

1. **Thermal treatment**: The spent biochar is typically heated in an inert atmosphere, such as nitrogen or argon, to prevent oxidation. The temperature range for effective regeneration usually lies between 400°C and 800°C. This process may be carried out in a furnace or a specialized reactor designed to handle high temperatures and prevent contamination. In general, this mechanism involves the desorption and decomposition of contaminants that have accumulated on the catalyst during its use. By heating the biochar, volatile organic compounds and other impurities are driven off, thereby cleaning its surface and pores. Thus, thermal regeneration can restore the biochar's surface area and pore structure, which are critical for its catalytic performance. A significant challenge of this method is the risk of structural alterations in the biochar when exposed to high temperatures, which can alter its catalytic functionality. Moreover, frequent cycles of thermal regeneration can result in the progressive degradation of the biochar material.
2. **Chemical treatment**: Chemical reagents such as acids or bases have been used to clean the biochar surface. For acid treatment, common acids such as hydrochloric acid (HCl) or H_2SO_4 are used to remove inorganic impurities and heavy metals that may be adsorbed on the biochar surface. The acid dissolves these impurities, which are then

TABLE 5.6 Relationship between functional structures and mechanisms in Photocatalytic system. From Zhou et al. (2021).

Functional Structure	Mechanism
Oxygen-centered	Reducing the band gap, generating radicals
PRFs	Generating radicals

washed away with water. Also, acid treatment can help to increase the biochar's surface area and porosity, enhancing its catalytic efficiency. On the other hand, in the base treatment, an alkali compound (e.g., NaOH or KOH) is used to remove organic contaminants and improve the surface characteristics of biochar. This process can introduce more functional groups (such as carboxyl and hydroxyl groups) on the biochar surface, enhancing its reactivity and adsorption capacity. The base treatment also helps in neutralize acidic groups on the biochar, making it more suitable for catalytic applications. Sometimes, a sequential acid-base treatment is employed to maximize the removal of both organic and inorganic impurities. This combined approach can significantly improve the structural and surface properties of biochar, making it more effective as a catalyst.

3. **Biological treatment**: Microorganisms are utilized to degrade and remove contaminants from biochar. This environmentally friendly method can improve the catalytic properties of biochar by introducing beneficial microorganisms. However, it generally operates at a slower pace compared to thermal and chemical treatments and may not be universally effective against all types of contaminants.

Each method can be optimized depending on the type of contaminants, the original biochar properties, and the intended reuse application.

5.6 Environmental applications of biochar catalysis

Biochar catalysts have been utilized for a variety of applications, including water and soil remediation, as stated in Tables 5.3 - 5.5, and holds promise for various catalytic processes such as biodiesel production, bio-oil upgrading, reforming, and the synthesis of specialty chemicals.

Biochar catalysis is interesting for environmental applications due to its high efficiency in removing pollutants and sustainability. Also, it offers an eco-friendly alternative to traditional catalysts, significantly reducing harmful pollutants such as heavy metals, organic contaminants, and greenhouse gases. The porous structure and high surface area of biochar enhance its ability to adsorb and degrade pollutants effectively, making it an ideal choice for soil and water remediation (Wang et al., 2017).

Economic advantages also highlight its importance, since biochar catalysts are typically produced from agricultural and forestry waste, which are abundant and inexpensive. This cost-effectiveness, combined with biochar's long-term stability and low maintenance requirements, makes it a sustainable solution for large-scale environmental remediation projects.

The following case demonstrates the comparison between biochar and conventional catalysts. The use of traditional catalysts for converting biomass (such as vegetable oils) into biodiesel has been widely studied; these catalysts often require expensive metal precursors. In contrast, sulfonated biochars have been employed due to their cost-effectiveness and versatility. Sulfonated biochar has been reported to achieve a maximum yield of 88% biodiesel from vegetable oil through simultaneous esterification of free fatty acids (FFAs) and transesterification of triglycerides (TGs) at 100°C over a 15-hour period. However, it was noted that after five cycles of reuse, the methyl ester yield decreased from 88% to 80% because of the leaching of $\cdot SO_3H$ groups (Akpasi et al., 2022, Luque & Clark, 2011).

The ester yields (TGs and FFAs) using biochar catalysts are comparable to those obtained from nonbiochar catalysts. Nevertheless, to make biochar catalysts more viable for biodiesel production, their stability must be improved to eliminate the need for posttreatment processes to remove sulfur or calcium residues from the catalyst (Akpasi et al., 2022).

To highlight the potential of biochar as a catalyst in biodiesel production reactions, Table 5.7 compares the different reaction yields when using waste oil as raw material and applying rice husk biochar and non-biochar-based catalysts.

TABLE 5.7 Comparison of biodiesel production efficiency using rice-husk biochar and non-biochar-based catalysts.

Biochar-based Catalyst	Feedstock	Temperature (K)	Biodiesel Yield	References
Rice husk-biochar	Cooking oil waste	383	88%	Shen et al. (2015)
Al(HSO$_4$)$_3$	Vegetable oil waste	493	81%	Shen et al. (2014)
Zeolite beta	Cooking oil waste	353	25%	Mani et al. (2013)
SO_4^{2-}/ZrO_2	Cooking oil waste	353	44%	Wang et al. (2012)

5.7 Future research directions

Biochar is viewed as a sustainable resource for tackling various environmental issues, including the cleanup of contaminants from soil, water, and air. Enhancing biochar through activation is a key area of research aimed at improving its effectiveness in removing specific pollutants. Ongoing studies are needed to discover new activation techniques and to understand the adsorption and desorption processes for different pollutants.

The interactions between soil microbes and biochar require further investigation. Detailed research is necessary to examine how biochar affects the growth and development of microbial communities. When mixed with soil, biochar not only aids in pollutant removal and soil fertility but also supplies essential nutrients. More studies are needed to evaluate microbial activity during the mineralization process and soil remediation. The binding mechanisms between biochar and soil need to be thoroughly understood.

The processes by which biochar removes contaminants from wastewater are not yet fully clear. Recent research has suggested the potential for converting solid carbon materials into electricity using direct carbon fuel cells. However, more work is required to understand the reaction kinetics and oxidation processes at the anode/electrode interface, as well as the interactions between solid carbon and electrolytes.

Despite its many benefits, biochar may contain harmful substances such as dioxins, chlorinated hydrocarbons, and polycyclic aromatic hydrocarbons, depending on the biomass used. The efficiency of biochar as supercapacitors also warrants further research. Comprehensive life-cycle assessments are essential to determining the biochar's economic and environmental impacts. Advances in biochar characterization techniques are crucial for optimizing its properties and activation to achieve maximum efficiency. Standardized characterization procedures should be implemented to understand biochar properties better.

5.8 Conclusions

In conclusion, biochar catalysis presents a promising and sustainable solution for various environmental challenges. The versatility of biochar, derived from its inherent properties such as high surface area, porosity, and functional group availability, makes it an effective catalyst for degrading organic and inorganic pollutants. This chapter reviewed various preparation methods, highlighting the importance of optimizing production processes to enhance catalytic performance. Biochar's cost-effectiveness and reduced environmental impact compared to traditional catalysts highlight its potential for broad adoption. Future research efforts should prioritize overcoming challenges such as catalyst poisoning and regeneration to fully leverage biochar's capabilities in environmental remediation.

AI Disclosure

During the preparation of this work, the author(s) used ChatGPT in order to improve the readability of the chapter. After using this tool/service, the author(s) reviewed and edited the content as needed and take(s) full responsibility for the content of the publication.

References

Akpasi, S. O., Anekwe, I. M. S., Adedeji, J., & Kiambi, S. L. (2022). Biochar development as a catalyst and its application. *In book: Biochar - Productive Technologies, Properties and Application.* IntechOpen.

Arni, S. A. (2018). Comparison of slow and fast pyrolysis for converting biomass into fuel. *Renewable Energy, 124,* 197–201.

Bartoli, M., Giorcelli, M., & Tagliaferro, A. (2023). *Biochar productive technologies, properties and applications.* IntechOpen. Available from 10.5772/intechopen.105439.

Cha, J. S., Park, S. H., Jung, S.-C., Ryu, C., Jeon, J.-K., Shin, M.-C., & Park, Y.-K. (2016). Production and utilization of biochar: A review. *Journal of Industrial and Engineering Chemistry, 40,* 1–15. Available from https://doi.org/10.1016/j.jiec.2016.06.002.

Chellappan, S., Nair, V., Sajith, V., & Aparna, K. (2018). Synthesis, optimization and characterization of biochar based catalyst from sawdust for simultaneous esterifcation and transesterifcation. *Chinese Journal of Chemical Engineering, 26*(12), 2654–2663. Available from https://doi.org/10.1016/j.cjche.2018.02.034.

Cheng, F., & Li, X. (2018). Preparation and application of biochar-based catalysts for biofuel production. *Catalysts, 8*(9), 346. Available from https://doi.org/10.3390/catal8090346.

Dai, L., Dai, L., Wang, Y., Liu, Y., Ruan, R., He, C., Yu, Z., Jiang, L., Zeng, Z., & Tian, X. (2019). Integrated process of lignocellulosic biomass torrefaction and pyrolysis for upgrading bio-oil production: A state-of-the-art review. *Renewable and Sustainable Energy Reviews, 107,* 20–36.

Dehkhoda, A. M., West, A. H., & Ellis, N. (2010). Biochar based solid acid catalyst for biodiesel production. *Applied Catalysis A: General*, *382*, 197–204. Available from https://doi.org/10.1016/j.apcata.2010.04.051.

Deng, R., Luo, H., Huang, D., & Zhang, C. (2020). Biochar-mediated Fenton-like reaction for the degradation of sulfamethazine: Role of environmentally persistent free radicals. *Chemosphere*, *255*, 126975. Available from https://doi.org/10.1016/j.chemosphere.2020.126975.

Djellabi, R., Yang, B., Wang, Y., Cui, X., & Zhao, X. (2019). Carbonaceous biomass-titania composites with TiOC bonding bridge for efficient photocatalytic reduction of Cr(VI) under narrow visible light. *Chemical Engineering Journal*. Available from https://doi.org/10.1016/j.cej.2019.02.035.

Fang, G., Gao, J., Liu, C., Dionysiou, D., Wang, Y., & Zhou, D. (2014). Key role of persistent free radicals in hydrogen peroxide activation by biochar: Implications to organic contaminant degradation. *Environmental Science & Technology*, *48*(3), 1902–1910. Available from https://doi.org/10.1021/es4048126.

Fang, G., Zhu, C., Dionysiou, D. D., Gao, J., & Zhou, D. (2015). Mechanism of hydroxyl radical generation from biochar suspensions: Implications to diethyl phthalate degradation. *Bioresource Technology*, *176*, 210–217. Available from https://doi.org/10.1016/j.biortech.2014.11.032.

Gholami, A., Pourfayaz, F., Hajinezhad, A., & Mohadesi, M. (2019). Biodiesel production from Norouzak (Salvia leriifolia) oil using choline hydroxide catalyst in a microchannel reactor. *Renewable Energy*. Available from https://doi.org/10.1016/j.renene.2019.01.057.

Gholami, A., Pourfayaz, F., & Maleki, A. (2020). Recent Advances of Biodiesel Production Using Ionic Liquids Supported on Nanoporous Materials as Catalysts: A Review. *Frontiers in Energy Research*.

He, J., Xiao, Y., Tang, J., Chen, H., & Sun, H. (2019). Persulfate activation with sawdust biochar in aqueous solution by enhanced electron donor-transfer effect. *Science of the Total Environment*, *690*, 768–777. Available from https://doi.org/10.1016/j.scitotenv.2019.07.043.

Hussain, I., Li, M., Zhang, Y., Li, Y., Huang, S., Du, X., Liu, G., Hayat, W., & Anwar, N. (2017). Insights into the mechanism of persulfate activation with nZVI/BC nanocomposite for the degradation of nonylphenol. *Chemical Engineering Journal*, *311*, 163–172. Available from https://doi.org/10.1016/j.cej.2016.11.085.

Jiang, S. F., Ling, L. L., Chen, W. J., Liu, W. J., Li, D. C., & Jiang, H. (2019). High efficient removal of bisphenol A in a peroxymonosulfate/iron functionalized biochar system: Mechanistic elucidation and quantification of the contributors. *Chemical Engineering Journal*, *359*, 572–583. Available from https://doi.org/10.1016/j.cej.2018.11.124.

Khataee, A., Sajjadi, S., Hasanzadeh, A., & Joo, S. W. (2019). Synthesis of magnetically reusable Fe_3O_4 nanospheres-N, S co-doped graphene quantum dots enclosed CdSe its application as a photocatalyst. *Journal of Industrial and Engineering Chemistry*. Available from https://doi.org/10.1016/j.jiec.2019.01.048.

Lee, J. W., Kidder, M., Evans, B. R., Paik, S., Buchanan, A. C., Garten, C. T., & Brown, R. C. (2010). Characterization of biochars produced from cornstovers for soil amendment. *Environmental Science & Technology*, *44*(20), 7970–7974.

Lee, J., Kim, K.-H., & Kwon, E. E. (2017). Biochar as a Catalyst. *Renewable and Sustainable Energy Reviews*, *77*, 70–79.

Li, L., Lai, C., Huang, F., Cheng, M., Zeng, G., Huang, D., Li, B., Liu, S., Zhang, M., Qin, L., Li, M., He, J., Zhang, Y., & Chen, L. (2019). Degradation of naphthalene with magnetic bio-char activate hydrogen peroxide: Synergism of bio-char and Fe–Mn binary oxides. *Water Research*, *160*, 238–248. Available from https://doi.org/10.1016/j.watres.2019.05.081.

Luque, R., & Clark, J. H. (2011). Biodiesel-like biofuels from simultaneous transesterification/esterification of waste oils with a biomass-derived solid acid catalyst. *ChemCatChem*, *3*(3), 594–597. Available from https://doi.org/10.1002/cctc.201000280.

Lyu, H., Zhang, Q., & Shen, B. (2020). Application of biochar and its composites in catalysis. *Chemosphere*, *240*, 124842. Available from https://doi.org/10.1016/j.chemosphere.2019.124842.

Mani, S., Kastner, J. R., & Juneja, A. (2013). Catalytic decomposition of toluene using a biomass derived catalyst. *Fuel Processing Technology*, *114*, 118–125. Available from https://doi.org/10.1016/j.fuproc.2013.03.015.

Mian, M, Liu, G., Yousaf, B., Fu, B., Ahmed, R., Abbas, Q., Munir, M. A. M., & Ruijia, Liu (2019). One-step synthesis of N-doped metal/biochar composite using NH_3-ambiance pyrolysis for efficient degradation and mineralization of Methylene Blue. *Journal of Environmental Sciences*. Available from https://doi.org/10.1016/j.jes.2018.06.014.

Park, J. H., Wang, J. J., Xiao, R., Tafti, N., DeLaune, R. D., & Seo, D. C. (2018). Degradation of Orange G by Fenton-like reaction with Fe-impregnated biochar catalyst. *Bioresource Technology*, *249*, 368–376. Available from https://doi.org/10.1016/j.biortech.2017.10.030.

Peng, Y., Sun, Y., Sun, R., Zhou, Y., Tsang, D. C. W., & Chen, Q. (2019). Optimizing the synthesis of Fe/Al (Hydr)oxides-Biochars to maximize phosphate removal via response surface model. *Journal of Cleaner Production*. Available from https://doi.org/10.1016/j.jclepro.2019.117770.

Pi, Y., Duan, C., Zhou, Y., Sun, S., Yin, Z., Zhang, H., Liu, C., & Zhao, Ye (2022). The effective removal of Congo Red using a bio-nanocluster: Fe_3O_4 nanoclusters modified bacteria. *Journal of Hazardous Materials*. Available from https://doi.org/10.1016/j.jhazmat.2021.127577.

Ramos, R., Abdelkader-Fernández, V. K., Matos, R., Peixoto, A. F., & Fernandes, D. M. (2022). Metal-supported biochar catalysts for sustainable biorefinery, electrocatalysis, and energy storage applications: A review. *Catalysts*, *12*, 207. Available from https://doi.org/10.3390/catal12020207.

Rodríguez-Reinoso, F. (1998). The role of carbon materials in heterogeneous catalysis. *Carbon*, *36*, 159–175.

Ruan, X., Liu, Y., Wang, G., Frost, R. L., Qian, G., & Tsang, D. C. W. (2018). Transformation of functional groups and environmentally persistent free radicals in hydrothermal carbonisation of lignin. *Bioresource Technology*, *270*, 223–229. Available from https://doi.org/10.1016/j.biortech.2018.09.027.

Shen, Y., Zhao, P., Shao, Q., Ma, D., Takahashi, F., & Yoshikawa, K. (2014). In-situ catalytic conversion of tar using rice husk char-supported nickel-iron catalysts for biomass pyrolysis/gasification. *Applied Catalysis B: Environmental*, *152*, 140–151. Available from https://doi.org/10.1016/j.apcatb.2014.01.032.

Shen, Y., Zhao, P., Shao, Q., Takahashi, F., & Yoshikawa, K. (2015). In situ catalytic conversion of tar using rice husk char/ash supported nickel–iron catalysts for biomass pyrolytic gasification combined with the mixing-simulation in fluidized-bed gasifier. *Applied Energy*, *160*, 808–819. Available from https://doi.org/10.1016/j.apenergy.2014.10.074.

Singh, P. (2021). *Biochar as a Catalytic Material. In book: Catalysis for Clean Energy and Environmental Sustainability* (pp. 767–801). Springer International Publishing. Available from https://doi.org/10.1007/978-3-030-65017-9_24.

Tan, H., Lee, C. T., Ong, P. Y., Wong, K. Y., Bong, C. P. C., Li, C., & Gao, Y. (2021). A review on the comparison between slow pyrolysis and fast pyrolysis on the quality of lignocellulosic and lignin-based biochar. *IOP Conference Series: Materials Science and Engineering, 1051*(1), 012075. Available from https://doi.org/10.1088/1757-899x/1051/1/012075.

Tan, Z., Wang, Y., Zhang, L., & Huang, Q. (2017). Study of the mechanism of remediation of Cd-contaminated soil by novel biochars. *Environmental Science and Pollution Research*. Available from https://doi.org/10.1007/s11356-017-0109-9.

Tang, M., Gamal, A., Bhakta, A. K., Jlassi, K., Abdullah, A. M., & Chehimi, M. M. (2024). Carbon dioxide methanation enabled by biochar nanocatalyst composite materials: A mini-review. *Catalysts, 14,* 155. Available from https://doi.org/10.3390/catal14020155.

Tay, T., Ucar, S., & Karagoz, S. (2009). Preparation and characterization of activated carbon from waste biomass. *Journal of Hazardous Materials, 165,* 481–485. Available from https://doi.org/10.1016/j.jhazmat.2008.10.011.

Velusamy, K., Devanand, J., Senthil Kumar, P., Soundarajan, K., Sivasubramanian, V., Sindhu, J., & Vo, D.-V. N. (2021). A review on nano-catalysts and biochar-based catalysts for biofuel production. *Fuel, 306,* 121632.

Wang, B. O., Li, Y. N., & Wang, L. I. (2019). Metal-free activation of persulfates by corn stalk biochar for the degradation of antibiotic norfloxacin: Activation factors and degradation mechanism. *Chemosphere, 237,* 124454. Available from https://doi.org/10.1016/j.chemosphere.2019.124454.

Wang, H., Feng, L., & Chen, Y. (2012). Advances in biochar production from wastes and its applications. *Chemical Industry and Engineering Progress*.

Wang, B., Gao, B., & Fang, J. (2017). Recent advances in engineered biochar productions and applications. *Critical Reviews in Environmental Science and Technology, 47*(22), 2158–2207. Available from https://doi.org/10.1016/j.jece.2021.106869.

Xiu, S., Shahbazi, A., & Li, R. (2017). Characterization, modification and application of biochar for energy storage and catalysis: A review. *Trends in Renewable Energy, 3*(1), 86–101. Available from https://doi.org/10.17737/tre.2017.3.1.0033.

Yaashikaa, P. R., Kumar, P. S., Varjani, S., & Saravanan, A. (2020). A critical review on the biochar production techniques, characterization, stability and applications for circular bioeconomy. *Biotechnology Reports, 28,* e00570.

Yan, J., Qian, L., Gao, W., Chen, Y., Ouyang, D., & Chen, M. (2017). Enhanced fenton-like degradation of trichloroethylene by hydrogen peroxide activated with nanoscale zero valent iron loaded on biochar. *Science Reports, 7,* 43051. Available from https://doi.org/10.1038/srep43051.

Yang, B., Dai, J., Zhao, Y., Wu, J., Ji, C., & Zhang, Y. (2022). Advances in preparation, application in contaminant removal, and environmental risks of biochar-based catalysts: A review. *Biochar, 4.* Available from https://doi.org/10.1007/s42773-022-00169-8, Springer.

Yi, Y., Tu, G., Tsang, P. E., & Fang, Z. (2020). Insight into the influence of pyrolysis temperature on Fenton-like catalytic performance of magnetic biochar. *Chemical Engineering Journal., 380*122518. Available from https://doi.org/10.1016/j.cej.2019.122518.

Yu, J. F., Tang, L., Pang, Y., Zeng, G. M., Wang, J. J., Deng, Y. C., Liu, Y. N., Feng, H. P., Chen, S., & Ren, X. Y. (2019). Magnetic nitrogen-doped sludge-derived biochar catalysts for persulfate activation: Internal electron transfer mechanism. *Chemical Engineering Journal, 364,* 146–159. Available from https://doi.org/10.1016/j.cej.2019.01.163.

Yun, E.-T., Moon, G.-H., Lee, H., Jeon, T. H., Lee, C., Choi, W., & Lee, J. (2018). Oxidation of organic pollutants by peroxymonosulfate activated with low-temperature-modified nanodiamonds: Understanding the reaction kinetics and mechanismShow affiliations. *Appl. Catal.* Available from https://doi.org/10.1016/j.apcatb.2018.04.067.

Zhang, K., Sun, P., Faye, M. C. A. S., & Zhang, Y. (2018). Characterization of biochar derived from rice husks and its potential in chlorobenzene degradation. *Carbon, 130,* 730–740. Available from https://doi.org/10.1016/j.carbon.2018.01.036.

Zhao, C., Xu, Q., Gu, Y., Nie, X., & Shan, R. (2023). Review of Advances in the Utilization of Biochar-Derived Catalysts for Biodiesel Production. *ACS Omega, 8,* 8190–8200. Available from https://doi.org/10.1021/acsomega.2c07909.

Zhou, X., Jin, H., Ma, Z., Li, N., Li, G., Zhang, T., Lu, P., & Gong, X. (2022). Biochar sacrificial anode assisted water electrolysis for hydrogen production. *International Journal of Hydrogen Energy*. Available from https://doi.org/10.1016/j.ijhydene.2022.08.190.

Zhou, X., Zhu, Y., Niu, Q., Zeng, G., Lai, C., Liu, S., Huang, D., Qin, L., Liu, X., Li, B., Yi, H., Fu, Y., Li, L., Zhang, M., Zhou, C., & Liu, J. (2021). New notion of biochar: A review on the mechanism of biochar applications in advannced oxidation processes. *Chemical Engineering Journal, 416,* 129027. Available from https://doi.org/10.1016/j.cej.2021.129027.

Part 2

Biochar for contaminated land remediation

Chapter 6

Biochar remediation of inorganic contaminants in soils

Terrence Wenga[1,2], Albert Gumbo[3], Irvine Nyaguwa[4], Munyaradzi Mtisi[4] and Kudzanayi Andrew Marondedze[1]

[1]*Department of Soil Science and Environment, Faculty of Agriculture, Environment and Food Systems, University of Zimbabwe, Mount Pleasant, Harare, Zimbabwe,* [2]*Key Laboratory of Agro-Forestry Environmental Processes and Ecological Regulation of Hainan Province, School of Environmental Science and Engineering, Hainan University, Haikou, P.R. China,* [3]*Department of Land and Water Resources Management, Faculty of Agriculture, Environment and Natural Resources Management, Midlands State University, Gweru, Zimbabwe,* [4]*Department of Environmental Protection, Hazardous Substances & Hazardous Waste Unit. Environmental Management Agency, Bluffhill, Harare, Zimbabwe*

Chapter outline

6.1 Introduction	91
6.2 Occurrence of inorganic contaminants in soils	92
6.2.1 Metal and nutrient-contaminated soils	93
6.2.2 Wastewater and sludge-amended soils	94
6.2.3 Occurrence of inorganic contaminants in munition fields	95
6.3 Biochar removal of inorganic contaminants	95
6.4 Large-scale remediation of inorganics by biochars	96
6.5 Mechanisms for biochar removal of inorganic contaminants in soils	97
6.5.1 Adsorption and immobilization mechanisms	97
6.5.2 Synergistic interactions of biochar with other remediation technologies	99
6.6 Factors affecting capacity of biochar in soil remediation	100
6.6.1 Physiochemical attributes of polluted soils	100
6.6.2 Physicochemical characteristics of biochars and removal efficacy	100
6.6.3 Application rate and particle sizes	101
6.7 Behavior and the fate of contaminants in biochar-amended soils	101
6.7.1 Properties influencing the behavior of contaminants in biochar-amended soils	102
6.7.2 Other factors influencing the bioavailability of contaminants in biochar-amended soils	102
6.8 Conclusion and outlook	102
References	103

6.1 Introduction

The contamination of soils by inorganic pollutants such as metals and nutrients released from different types of industries, including mining, manufacturing, and agriculture, is currently one of the environmental problems (Zama et al., 2018). These contaminants pose a severe risk to untargeted organisms, cause soil slumping, change soil structure, and lead to soil salinization, which subsequently affects crop quality and yield, and leads to the deterioration of the overall air and water quality (Yang et al., 2023). Moreover, due to biomagnification and bioaccumulation processes, inorganic contaminants pose significant human health risks as some of them are carcinogenic, genotoxic, mutagenic, and teratogenic, and can adversely affect various living species through the food chain (Nguyen, Sherpa, et al., 2023). A 2011 study reported that soil contamination affects approximately 33% of soils globally (Zaharia, 2011), indicating the urgent need for addressing this problem.

Scientists, researchers, and policymakers have continuously been seeking novel methods to address soil contamination, particularly inorganic chemicals. The soil contaminants remedial techniques such as precipitation, phytoremediation, membrane filtration, ion exchange, and coagulation are among the most applied methods for decontaminating soils (Nguyen, Sherpa, et al., 2023). However, these approaches have some drawbacks, including exorbitant costs and creating secondary soil contamination due to secondary reactions. Therefore, amid these challenges, researchers have been prompted to investigate other cost-effective and environmentally friendly methods of remediating soils contaminated by inorganic pollutants.

In light of this, several researchers have explored the application of biochar to minimize soil contamination (Ahmad et al., 2014; Azeem et al., 2021; Gnanasundar & Akshai Raj, 2021; Wang et al., 2020). Biochar has proved to be an efficient and attractive material for the capture of potentially toxic elements (PTEs), the sorption of suspended particulate matter, as well as the degradation of biological pathogens (Reddy, Xie, & Dastgheibi, 2014). In addition, biochar has been reported to improve soil properties (including physical, chemical, and biological properties) and reduce the hazardous effects of soil acidification. Due to its remarkable properties, such as low cost, stability, abundance, negative charge, high internal surface area, resistance to degradation, and high contaminant removal efficiency, biochar has gained much recognition as an environmental remediation technique (Nguyen, Nguyen, et al., 2023). A considerable amount of negative charge over the biochar surface attracts positively charged metals to the internal biochar surface from the soil solution (Mukherjee, Zimmerman, & Harris, 2011). In doing so, the concentration of metals and inorganic contaminants in the soil solution can be reduced, along with their current availability for uptake by plants and organisms, that is bioavailability, or their potential to become available for uptake by living species, that is bioaccessibility (Houben, Evrard, & Sonnet, 2013).

The surface of biochar produced at lower pyrolysis temperatures between 200°C and 400°C is rich in oxygen-containing functional groups that enable the creation of surface complexes between cations such as Ni^{2+} and Cu^{2+}, and the biochar surface (Beesley & Marmiroli, 2011). The negative charge can also increase soil pH when biochar is added to acidic soils since the negative surfaces of the biochar attract hydrogen ions from the soil solution. A higher soil pH further increases the sorption of metals from the solution because of the deprotonation of pH-dependent cation exchange sites on soil surfaces, especially in acidic soils (Rees, Simonnot, & Morel, 2014).

Some biochars contain a considerable mineral ash component, for example, up to 50% for manure-derived or even 85% for bone meal–derived source materials (Sizmur et al., 2015). Minerals, including phosphates, carbonates, and sulfates, can cause some toxic elements, such as lead, to precipitate in the soil solution. Since the precipitates are insoluble, this mechanism can substantially contribute to the remedial efficacy of biochar. Moreover, biochar degrades very slowly, with predicted carbon half-lives ranging from 102 to 107 years (Zimmerman, 2010). Hence, throughout the time period the soil reclamation projects take place ($\sim$10 to 100 years), biochar is regarded as an inert material. This stability of biochar gives it the potential to sequester soil contaminants for a long period of time. Nevertheless, oxygen-containing functional groups on the surface of the biochar may release cations into the solution over time as they are replaced by hydrogen ions (Kim et al., 2013).

To date, the addition of biochar to soil has largely been focused on its ability to improve soil's physical, chemical, and biological properties rather than its ability to remediate contamination. Several studies have been carried out in the past decades on the biochar remediation of soil contamination and require a comprehensive summarization of the research advances in the decontamination of soils by biochar. Therefore, this chapter aims to provide an overview of the occurrence of inorganic contaminants in soils, the mechanisms for the removal of inorganic pollutants, as well as their behavior and fate. Lastly, a summary and outlook for future research perspectives are presented. The sources of inorganic pollutants, soil contamination, and benefits of biochar decontamination of contaminated soils are depicted in Fig. 6.1.

6.2 Occurrence of inorganic contaminants in soils

Soil contamination by inorganic compounds and elements occurs when their concentrations exceed their background concentrations with detrimental impacts on any part of the ecosystem (FA UNEP, 2021; Patinha et al., 2018). Inorganic compounds and elements are noted to occur in nature; however, anthropogenic activities are the principal and foremost drivers of soil contamination by inorganic pollutants (Núñez, 2023). For instance, undertakings, including agricultural application of pesticides and fertilizers, mining, various industrial activities, waste management practices, and domestic greywater, are major sources of inorganic contamination in soils (Patinha et al., 2018). Inorganic contaminants are largely nonbiodegradable and occur chiefly as trace elements/heavy metals, metalloids, salts, dissolved ions, organo-metallic complexes, radionuclides, sulfides, and oxides (FA UNEP, 2021). The presence of inorganic contaminants in soils poses adverse ecological and human health impacts, and perturbations in soil aspects such as structure, productivity, and usability while also presenting high remediation and restoration costs (Patinha et al., 2018). The contaminants in soils can be categorized into inorganic (metals and nonmetals) and organic pollutants. Inorganic contaminants are discussed in this chapter while organic contaminants are out of the scope of this chapter.

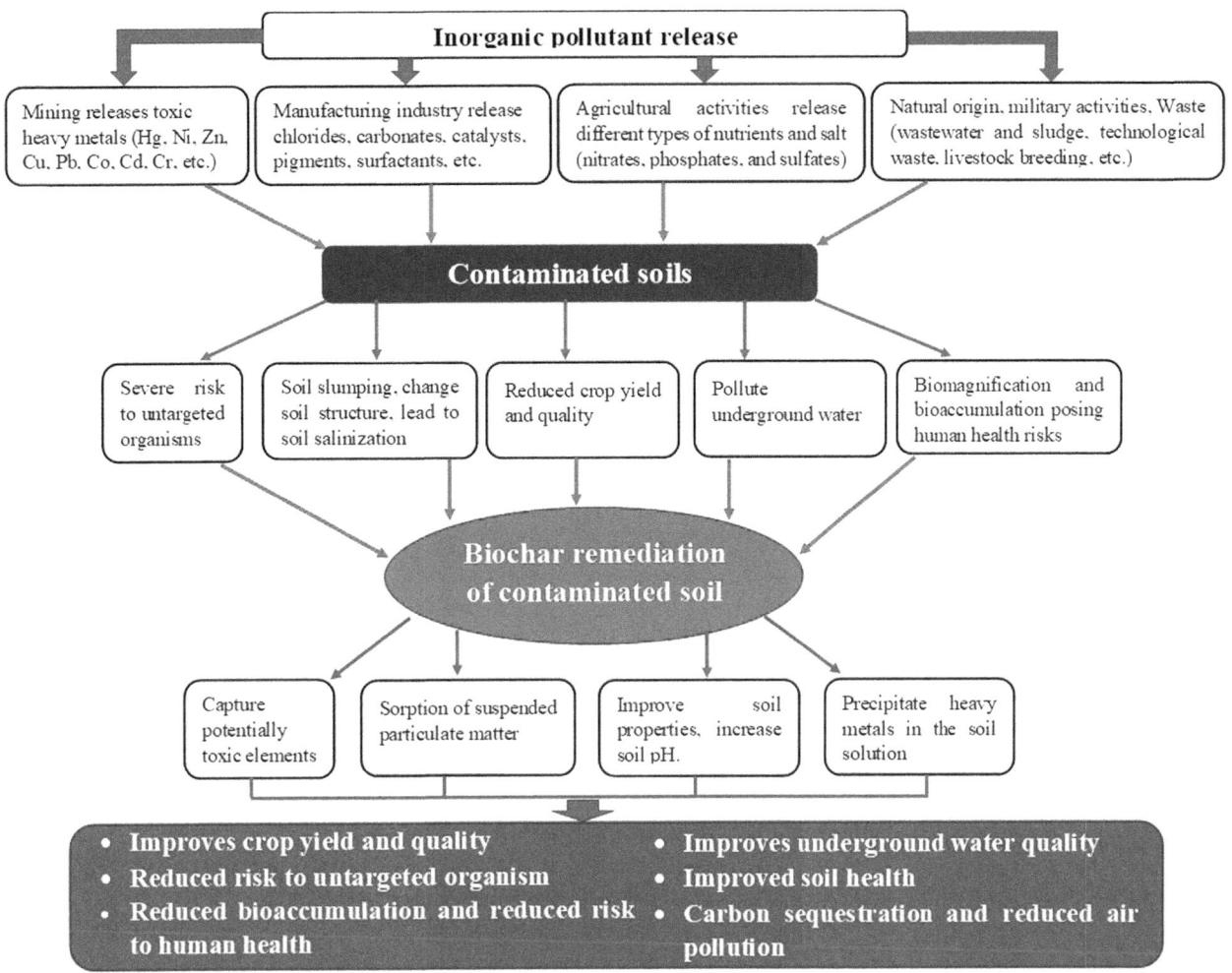

FIGURE 6.1 Sources of inorganic pollutants and contamination. Overview of sources of inorganic pollutants, soil contamination, and benefits of biochar decontamination of contaminated soils. Inorganic pollutants come from different sources polluting the environment.

6.2.1 Metal and nutrient-contaminated soils

Anthropogenic sources of inorganic contaminants, such as toxic heavy metals, radionuclides, asbestos, and explosives, are significant determinants of soil quality (Murtaza et al., 2014). Trace elements and heavy metals which are naturally ubiquitous and essential micronutrients for the soil ecosystem pose adverse impacts when exceed the minimum concentrations. Such trace elements and toxic metals are listed in Table 6.1. These metals also referred to as PTEs are recognized to naturally occur in nature from the natural weathering of crustal rocks and form their geochemical background (Patinha et al., 2018). The geochemical background of these elements has non-contributory effects on toxic hazards (Weldeslassie et al., 2017). Concentrations of these toxic elements have been noted to have sharply increased in soils across the globe after the industrial revolution, demonstrating the anthropogenic effect of soil contamination by inorganic pollutants, as shown in Table 6.1.

The main activities contributing to the increase in metal concentration in soils are municipal waste, mining, combustion of fossil fuels, smelting and foundry operations, incineration, chlor-alkali plants, steel industry, industrial waste disposal, atmospheric deposition, production and application of pesticides and fertilizers, and domestic use of metals and metal-bearing compounds (Morel et al., 2005; Patinha et al., 2018). Metals of high concern due to high toxicity at low concentrations include mercury, lead, zinc, copper, chromium, nickel, cadmium, arsenic, and manganese.

Due to their nonbiodegradability, metals persist in soils, negatively impacting soil properties and plant growth, and posing adverse impacts to microbial communities and the associated ecosystem (Wuana & Okieimen, 2011). The occurrence of these elements in soils is accompanied by potential speciation and bioavailability which influence their chemodynamics. These processes redistribute the elements into various chemical forms whose toxicity, bioavailability, and

TABLE 6.1 Estimated concentrations of toxic elements in soils.

Inorganic element (potentially toxic metal)	Min–Max (mg/kg)	Average range (mg/kg)	Median (mg/kg)
Arsenic	<0.1–93	9.3–21	5
Cadmium	<0.01–41	0.8–1.5	0.3
Cobalt	0.1–469	12	10
Chromium	1–1100	47–83	80
Copper	<1–1508	24–38	25
Mercury	0.005–1.1	0.26–0.28	0.05
Nickel	0.14–487	26–34	20
Lead	0.95–16338	44–51	17
Zinc	0.7–3648	100–117	70

Source: From Patinha, C., Armienta, A., Argyraki, A., & Durães, N. (2018). Inorganic pollutants in soils. In *Soil Pollution* (pp. 127–159). Elsevier BV. https://doi.org/10.1016/b978-0-12-849873-6.00006-6.

mobility vary (Wuana & Okieimen, 2011). The distribution of elements in soils is further influenced by processes such as complexation, plant uptake, ion exchange, biological immobilization and mobilization, mineral precipitation, dissolution, adsorption, and desorption. The presence of these metals impedes organic toxins biodegradation in soils. Additionally, elevated levels of these metals negate plant growth due to bioaccumulation and biomagnification, resulting in phytotoxicity (Wuana & Okieimen, 2011).

6.2.2 Wastewater and sludge-amended soils

Sewage sludge is a composite and heterogeneous mixture of inadequately digested organics such as botany remains, cellulose, oils, excrement, micro-organisms, inorganic materials, micronutrients, and heavy metals. Also, it can contain various fuel types, including hydrogen, syngas, biodiesel, bio-oil, and bioplastics, as well as proteins, bio-pesticides, volatile acids, and bio-fertilizers (Fijalkowski et al., 2017). Sewage sludge is most commonly managed via its application in agricultural fields, land reclamation, and plantations due to its rich content of macro- and micropollutants which are necessary for plant growth. However, both nutrients (C, N, and P) for plant growth and toxic compounds are introduced into the soil as a consequence of sewage sludge application as fertilizer or soil conditioner. The main inorganic toxicants are heavy metals, as listed in Table 6.1. Aside from heavy metals, metalloids such as arsenic, selenium, or nonmetals and light metals such as aluminum have similar toxicity to the environment. The major influencers of heavy metals in sewage sludge are typically surface runoff and waste effluent from industrial activities while the total concentration of these metals varies from 0.5 to 2% of dry sludge (Fijalkowski et al., 2017).

Due to the nonbiodegradability of these inorganic pollutants, they accrue in soils leading to food chain contamination which in turn results in environmental bioaccumulation (Sieciechowicz et al., 2014). A study investigating the occurrence of heavy metals in soil fertilized by sewage sludge (Maliszewska-Kordybach & Smreczak, 2003) suggested that higher concentrations in soils fertilized with sewage sludge over five seasons relative to the concentrations prior to the fertilizer application. Once in the soil, heavy metals occur or exist in different forms. A study by Lasheen & Ammar (2009) showed that Ni, Mn, and Zn are present in soil as the exchangeable, carbonate, with Fe/Mn-oxide as the most mobile compounds. Cu, Cd, Cr, and Fe mainly occur as sulfide and organic, exhibiting some degree of mobility, as well as residual form, i.e., inert phase corresponding to less mobilization. Dąbrowska & Rosińska (2012) did not observe the accumulation of mobile fractions (exchangeable and carbonate) due to the effect of thermophilic digestion of sewage sludge except for nickel while Cu, Zn, Cd, and Cr concentrations increased and were observed in the form of organic-sulfide compounds.

Agronomic application of sewage sludge potentially improves soil organic matter (OM), bulk density, porosity, microbial biomass, mean weight diameter, dehydrogenase activity, basal respiration, enzymatic properties, and nitrogen and phosphorus content (Dhanker et al., 2021). In some instances, inappropriate application may add salts, leading to increased soil salinity. Some studies investigating the impacts of sewage sludge as an amendment have established decreased microbial biomass and enzyme activity. While interventions such as stabilization techniques are used to reduce pathogens, organic compounds, and heavy metals, long-term application of sludge in agronomics still poses adverse impacts (Nunes et al., 2021). With respect to heavy metal contamination, pH plays the most important role in influencing heavy metal bioavailability. The majority of sludge introduced heavy metals are restricted to the topsoil while simultaneously adsorptive properties of the soil are altered (Alloway & Jackson, 1991).

6.2.3 Occurrence of inorganic contaminants in munition fields

Military activities have been seen to significantly affect soil properties, mainly through physical and chemical disturbances during military training and warfare. Most inorganic chemicals, particularly metal elements and their compounds used in military ammunition and explosives have the potential to contaminate soil and surface waters, which may later have long-term adverse effects on the environment and human health. As a result, war zones with intense conflict, military training areas, shooting sports zones, and explosives and ammunition manufacturing or disposal locations are among the major sources of inorganic contamination in terrestrial ecosystems (Jenkins et al., 2001). Previous studies by Broomandi et al. (2020) have reported high-level concentrations of heavy metals in soil samples collected from war-impacted areas and military training grounds. Heavy metal remains are among the longest-remaining war residuals in conflict-impacted zones. The residence time of these metals depends mainly on soil redox properties, pH, and cation exchange capacity (CEC) (Albright, 2011). Bullets contain significant amounts of lead, and therefore munition fields are significantly contaminated by lead, due to chemical processes occurring in the soil, some of the lead precipitates as insoluble minerals and some can be absorbed by plants, resulting in bioaccumulation. Pb partitioned in different soil fractions may be initially inert but later become reactive following changing soil conditions like pH, soil moisture content, and soil OM, or when its quantities in soil exceed the soil-holding capacity. Other toxic elements released into the soil by weapon residues include Sb, C, As, Hg, Ni, Zn, and Cd (Major et al., 1992).

The distribution of the contaminants in ammunition-contaminated soils is affected by firing activity, soil properties, soil exposure time, and climate (Meyers et al., 2007). In a study by Diaz et al. (2007), extremely contaminated sites are mostly sandy, having acidic-neutral pH (from 6.11 to 6.72), low OM content (0.21%–1.01% wt.), and moderate CEC (CEC: from 8.34 cmol/kg to 24.8 cmol/kg). Samples collected from these sites had the most Pb accumulation (60%–70%) in very coarse sand fraction (1.0–2.0 mm), with concentrations being high ranging from 10,068 mg/kg to 70,350 mg/kg. In another study by Broomandi et al. (2020) soil samples collected from another ammunition field had a slightly acidic pH (as low as 5.3), and the locations with no vegetation cover were among the most contaminated, whereas the samples from areas covered by forest vegetation and hummus were the least contaminated with toxic metals. Fayiga (2019) examined the soil type contaminated by inorganic pollutants and the authors observed that As, Zn, Cu, and Pb S were dominantly found in coarse and fine sand fractions, while K, Al, and Si were mainly present in clay loam. As concentrations were extremely high, however, they can also be attributed to geogenic sources. Other authors (Walsh, Walsh, & Hewitt, 2010; Zhang et al., 2022) argued that extremely high concentrations reported were way beyond the ranges in soils containing geogenic As. Furthermore, the presence of As has also been linked to military activities.

The pH value of soil has a great influence on the behavior of heavy metals, determining mainly their solubility and availability. An alkaline pH can, but does not always, positively affect the immobilization of heavy metals, whereas in an acidic environment, metallic cations are more mobile, that is larger quantities can be released into the soil solution and thus potentially induce toxicity to soil biota. Slightly alkaline or neutral environments generally provide the highest heavy metal retention. Soil OM is also another key soil property which may be significantly affected by land use management practices which also directly controls the mobility of toxic metals in the soil (Zhang et al., 2022). The specific characteristics of acidic soils, sandy texture, low CEC, and low OM (separately or in combination) increase the leaching and mobilization processes governing the mobility of PTEs and thus result in higher elemental concentrations in soil solution; whereas alkaline soils, loam/silty loam/clay texture, high CEC, and high OM (separately or in combination) limit heavy metal mobility and thus restrain their concentrations in soil solution and subsequent translocation into plant tissues (Walsh, Walsh, & Hewitt, 2010).

6.3 Biochar removal of inorganic contaminants

Inorganic contaminants threaten soil health and crop productivity, food chain contamination, groundwater pollution, and eutrophication. Biochar immobilizes heavy metals and other inorganics on site and improves soil conditions. Table 6.2 shows the inorganic pollutant removal efficacy of biochar prepared from various feedstocks. A study

TABLE 6.2 Soil inorganic pollutants removal by biochar.

References	Feedstock for biochar preparation	Preparation Temperature	Pollutant removed	Removal effect
Meier et al. (2017)	Chicken dung	500°C	Cu	Reduced Cu uptake by 43% on spiked soil, increasing biomass
O'Connor et al. (2018)	Sulfur-modified rice husk biochar	550°C, at rate of 15°C/min	Hg	Reduced freely accessible Hg by 73%. Enhanced Hg adsorption capacity,
Ibrahim et al. (2022)	Casuarina biochar	500°C	Cd, Co, Cr, Cu, Ni, Pb, Zn	4% Casuarina biochar is most efficient in reducing heavy metal adsorption by roots and shoots. The removal efficiency varied.
Zhang et al. (2019)	Sewage sludge biochar (SSB)	600°C	Hg	Reduced MeHg by 73.4% and THg by 81.9%
Beesley and Dickinson, (2010); Beesley and Marmiroli, (2011); Kumpiene, Lagerkvist, & Maurice (2008)	Hard wood	450	As, Cu, Zn, and Cd	Cadmium was reduced by ten-fold; Cd and Zn concentrations were reduced by 300 and 45-fold, respectively
Beesley & Dickinson (2010)	Hardwood	450	As, Cu, Pb, Zn, and Cd	As and Cu mobility in soil profile enhanced; Pb and Cd little effect
Debela, Thring, and Arocena (2012)	Wood	200 and 400	Cd and Zn	>90% reduction of Cd and Zn

conducted by Meier et al. (2017) showed that amending spiked soils with 5% biochar produced from chicken manure resulted in approximately 43% reduction in Cu absorbed by *Oenothera picensis* plants in Cu-mine-polluted soil. The concentration of Cu absorbed decreased from 66.9 mg/kg to 36.6 mg/kg suggesting biochar reduced the uptake of Cu by plants. Another study by O'Connor et al. (2018) observed that the incorporation of sulfur-modified biochar obtained from rice husk stabilized Hg in contaminated soils, making them easier to alleviate. The application of this S-modified biochar resulted in a 73% increase in biochar's Hg^{2+} adsorptive capacity (q_{max}). Soil contaminated with 12924 mg/kg Hg was collected from a paddy rice field in China and two types of biochars were obtained from rice hull and wheat straw at 480°C–660°C and 350°C–450°C, respectively, were supplied to the contaminated soil (Xing et al., 2019). A significant decrease in Hg was observed in rice, the stalk, bran, and hull after soil treatment with rice hull biochar relative to the biochar produced from wheat straw. The amount of Hg in rice cultivated on soil treated with wheat straw biochar was also below the threshold of 20 ng/g. In addition, three different types of biochar produced from mango, casuarina, and Salix were used to amend the contaminated soils with various heavy metals at two application rates of 2% and 4% by w/w to investigate the effects of biochar on the plant bioaccumulation, availability of heavy metal in soil and the growth of summer squash in highly polluted soil (Ibrahim et al., 2022). The application of 4% Casuarina biochar led to a decrease in the concentration of heavy metals, including Co, Cd, Ni, Cr, Zn Cu, and Pb, in the root by a percentage ranging from 12.1% to 85.0%, compared to soil that was not treated with biochar. The findings from various studies indicated that the incorporation of biochar into contaminated soils had significant efficacy in reducing the uptake of inorganic elements by both the roots and shoots, as also shown in Table 6.2.

6.4 Large-scale remediation of inorganics by biochars

Although the biochar decontamination of soils contaminated by inorganic pollutants has been largely conducted on a laboratory scale, several field experiments were conducted exploring the effectiveness of biochar for soil decontamination. Bian et al. (2014) carried out a three-year field study, that is, from 2010 to 2012, to investigate the decontamination of rice paddy soils contaminated by Cd and Pb in southern China. The bioavailability of Pb and Cd in soil was measured and monitored after the soils were amended with wheat straw biochar at a dosage of 40 t/ha. The authors observed that the addition of wheat straw biochar significantly elevated the soil pH and the total organic carbon and reduced the amount of Cd by about 70.9%, which was extractable by $CaCl_2$ and Pb by about 79.6%. Consequently, the amount of Cd, which was absorbed by rice (O. sativa L.), decreased by about 67.3%, and the amount of Pb taken by rice also reduced by about 69.0%. Meanwhile, the yield increased by approximately 18.3% at the end of the third year. L. Cui, Pan, et al. (2016) conducted an almost similar study. The authors carried out a five-year field experiment to investigate the effectiveness of biochar in transforming Cd and Pb into insoluble form. Biochar converted the exchangeable fractions of Pb and Cd into rather more stable ones. The exchangeable fractions of Pb and Cd were reduced by approximately 14.2%–50.3% and 8.0%–44.6%, respectively, whereas the residual amounts were elevated by approximately 14.9%–39.6% and 4.0%–32.4% for Pb and Cd, respectively, over the 5 years. The attraction of metal contaminants was ascribed to the profuse functional groups on biochar. Although biochar can remove Cd from contaminated soils, secondary reactions with other heavy metals can occur, leading to secondary pollution and toxicity. For example, the simultaneous immobilization of Cd and As is obstructed by the different geochemical behaviors of Cd and As in soils such as rice paddy soils. The amendments of such soils by biochar result in the effective immobilization of Cd while also facilitating the reduction of As(V) into As(III), leading to an increased As toxicity and leaching in contaminated fields (Chen et al., 2016; Gong, Zhao, & Wang, 2018; Vithanage et al., 2017). To further examine the simultaneous immobilization of heavy metals, other researchers Yu et al. (2018) produced the zero-valent iron (ZVI)-biochar composite and investigated the simultaneous immobilization of Cd and As in rice paddy fields over a three-year period, that is, from 2013 to 2015, in Guangdong Province, China. The ZVI showed a high sorption capacity for As while biochar demonstrated a high capacity for Cd. Upon application of about 2250 kg per acre in the field, the amendments decreased the accumulation of Cd and As in rice by 48% and 24%, respectively (Gong, Zhao, & Wang, 2018). (Houben, Evrard, & Sonnet, 2013) further studied the influence of different dosages of biochar, that is, 1%, 5%, and 10%, on the removal of Zn, Cd, and Pb in a contaminated field in Belgium. The authors found that the amount of exchangeable metals by $CaCl_2$ in soil was reduced as the dosage rate of biochar increased after 56 days. Shen et al. (2016) carried out a field decontamination to study the long-term influence of biochar on the sorption of inorganic metals in contaminated fields in Castleford, UK. After adding biochar at a dosage of about 0.5%–2%, the concentrations of Zn and Ni were decreased by 83%–98% after a three-year period. The addition of biochar improved the residual fraction of Zn and Ni in soils from 7% to 27%–35% and from 51% to 61%–66%, respectively. Nevertheless, the field experiment carried out by H. Cui, Fan, et al. (2016) indicated that the effectiveness of the biochar reduced with time, in which constant addition of biochar was required for effective sorption of heavy metals. Meanwhile, the application of biochar is hindered by the limited affinity and adsorption capacity, soil deliverability, and reaction rate and is only suitable for shallow soil contaminated with heavy metals. There is also limited information available on the long-term stability of metals immobilized at the field scale.

6.5 Mechanisms for biochar removal of inorganic contaminants in soils

The application of biochar in soils for metal stabilization involves different mechanisms. The mechanisms of metal removal by biochar in soils and aqueous systems share some similarities but also have differences. It is worth noting that in soils, biochar's interactions with OM, pH, and microbial communities play a larger role, while in aqueous systems, electrostatic and coagulation effects are more significant (Kumar et al., 2021).

6.5.1 Adsorption and immobilization mechanisms

Biochar is characterized by a porous structure, a high pH, active functional groups aromatic components. These attributes play an essential role in the process of reclamation of metals in the soil, such as complexation, precipitation, electrostatic interaction, redox, ion exchange, and physical adsorption.

6.5.1.1 Physical adsorption

The process of physical adsorption is brought about by intermolecular forces that interact on biochar surfaces and the inorganic contaminants. The process is largely reversible. The process is influenced by surface energy, pore volume, and specific surface area (SSA) of the biochar (Alipour et al., 2022). Biochars produced at higher temperatures are generally suitable for the removal of organic pollutants due to their hydrophobicity and microporosity while biochar produced at relatively low pyrolysis temperatures contain greater pore volume and SSA, contributing a considerably large contact area to the heavy metal ions, hence improving the physical biochar adsorption capacity. For example, pine wood-derived and switchgrass biochar at 300 °C and 700 °C can competently immobilize uranium and copper with physical adsorption. Heavy metal ions, including cadmium, zinc, and arsenic, are immobilized on the biochar surface by van der Waals adsorption.

6.5.1.2 Ion exchange

Ion exchange signifies the selective exchange of metal ions that are transferable, such as Na^+, K^+, Mg^{2+}, and Ca^{2+}, on the biochar surface through metal ions. The ion exchange efficacy largely depends on the chemical properties of the biochar surface. The ion exchange capacity between metal cations and biochar particles can be improved by a higher CECCEC. The higher CEC of biochar is observed at 200°C–350°C pyrolysis temperatures because higher temperatures diminish the acidic carbon/oxygen and oxygen-comprising functional groups, which decreases the CEC of biochar (Fedeli et al., 2023). Kumar et al. (2021) studied the reclamation of mercury and zinc through shell-derived biochar prepared at 170°C–185°C. They found that the acidic oxygen-containing functional groups on the surface of biochar, including -OH and -COOH, can exchange with Hg^+ and Zn^{2+} ions to discharge ionizable protons, according to reactions (1) and (2).

$$2 - COOH + Zn^{2+} = -(COO)_2Zn + H^+ \tag{6.1}$$

$$2 - COH + Zn^{2+} = -(CO)_2Zn + 2H^+ \tag{6.2}$$

The ion exchange capacity of biochar is closely associated with the pH of the soil. When the soil solution pH is lower than the biochar's pH at the point of zero charge, more metal ions are attracted to the biochar surface by the ion exchange method (Fedeli et al., 2023). According to Tomczyk, Sokołowska, & Boguta (2020) investigation, the biochar derived by the hydrothermal process has plenty of oxygen-containing functional groups, which helps in the adsorption of copper (Cu^{2+}) ions through physical adsorption and ion exchange. Oxygen-containing functional groups can enhance the adsorption and enrichment of the contaminants near the cathode, thereby improving the degradation efficiency.

6.5.1.3 Complexation

Biochar contains a lot of functional groups that contain oxygen such as -COH, -OH, and -COOH. These functional groups form stable complexes when they react with metal ions. Biochar prepared at a low pyrolysis temperature holds many oxygen-containing functional groups which immobilize the heavy metals more efficiently through metal complexation. The amount of oxygen-containing functional groups in biochar augments with time, which is caused by the carboxyl formation and oxidation of the biochar surface (Hamid et al., 2020; Oni, Oziegbe, & Olawole, 2019; Sun et al., 2020). Complexation can be formed between the C = O ligand of oxygen-comprising functional groups and positively charged metal cations. For example, Pb (II) ions surface complexation with free carboxyl and hydroxyl functional groups and inner-sphere complexation of Pb (II) ions with hydroxyl functional groups of mineral oxides, as shown in reactions (6.3)–(6.5) (Azeem et al., 2021).

$$-COOH + Pb^{2+} + H_2O \rightarrow -COOPb^+ + H_3O^+ \tag{6.3}$$

$$-OH + Pb^{2+} + H_2O \rightarrow -OPb^+ + H_3O^+ \tag{6.4}$$

$$C - COOH + M^n \rightarrow >C - COOM^+ + H_3O \tag{6.5}$$

M = metal

The oxygen-containing functional groups in biochar considerably augment the ligands on the soil surface to arrest various metals by establishing heavy metal-ligand complexes (Das, Ghosh, & Avasthe, 2023; Pandey, Suthar, & Chand, 2022). In another experiment, Hasan, Kubra, et al. (2023) examined the processes of reclamation of Cr through biochar derived from sugar beet tailing. They found that complexation is the primary process accountable for Cr reclamation.

6.5.1.4 Precipitation

Biochars co-precipitate with heavy metal cations to produce insoluble carbonates and phosphates that immobilize heavy metals in soils (Jain et al., 2016; Pande et al., 2022). A higher pyrolysis temperature (more than 400°C) of cellulose and hemicellulose in plant feedstock mostly produces alkaline biochar to facilitate metal precipitation in soil (Kumar et al., 2018). On the other hand, biochar produced from animal manure contains higher ash contents, namely sulfur, phosphorus, silicon, sodium, potassium, calcium, and magnesium, which can form insoluble minerals after reacting with heavy metals (Ambaye et al., 2021; Montagnoli et al., 2021; Rana et al., 2021). For example, cow-manure biochar possesses an abundant amount of phosphates that can immobilize lead in the soil due to pyromorphite formation. Another study presented that biochar derived from dairy manure adsorbs Pb via surface sorption (13 to 16%) and precipitation (84 to 87%) (Allohverdi et al., 2021). The same author compared the effects and mechanisms of various metals, including copper, cadmium, lead, and zinc, adsorption through cow- and rice-bone-derived biochar. The leading adsorption process was found to be precipitation between metal cations and phosphate or carbonate.

6.5.1.5 Reduction-oxidation process

The reduction-oxidation (redox) reactions are indispensable processes through which heavy metals are immobilized by biochar. The functional groups on biochar surfaces can undergo redox reactions with metal ions, which, consequently alter their toxicity. For example, biochar can diminish the elevated toxic Cr (VI) to reasonably less toxic Cr (III) and then immobilize Cr (III) on the surface via a complexation process (Liu et al., 2022). In this process, biochar acts as an electron donor to offer electrons from surface functional groups and graphitic structures to Cr (VI). In consequence, during the remediation of contaminated soils, the biochar electron-giving capacity can reduce metals, including Sb, As, and Tl, which can raise their bioavailability (Stefaniuk et al., 2018). Moreover, nobbling biochar with some metals augments its removal efficacy. For example, some scholars have observed an improved performance for the adsorption of arsenic on biochar modified with Fe and Mn oxides (Taraqqi-A-kamal et al., 2021). The procedure is that manganese oxide on the biochar oxidizes As (II) to form As (V), and then manganese arsenate precipitates and causes them to be adsorbed onto the surface of the biochar. Amin et al. (2023) showed that As (III) was adsorbed with a composite of BC-Mn-Fe, and its adsorption capacity was four times better than the pristine biochar, but the adsorption mechanism was significantly affected by pH because the sorption impact of the composite for arsenic was weakened by an electrostatic repulsion under alkaline environments. A soil contaminated with Zn and Cd was amended with a hardwood-derived biochar, and the concentration of these heavy metals in pore water was significantly reduced (Beesley & Dickinson, 2010). Fig. 6.2 summarizes the mechanism for the removal of inorganic contaminants using biochar.

6.5.2 Synergistic interactions of biochar with other remediation technologies

Biochar and phytoremediation are two heavy metal remediation technologies that work synergistically together. Phytoremediation uses plants to remove pollutants from the soil, air, and water (Ghosh & Maiti, 2021). Biochar can enhance phytoremediation in the following ways:

- Improving soil properties as it increases soil water retention, aeration, and nutrient availability, thereby benefiting plant growth,

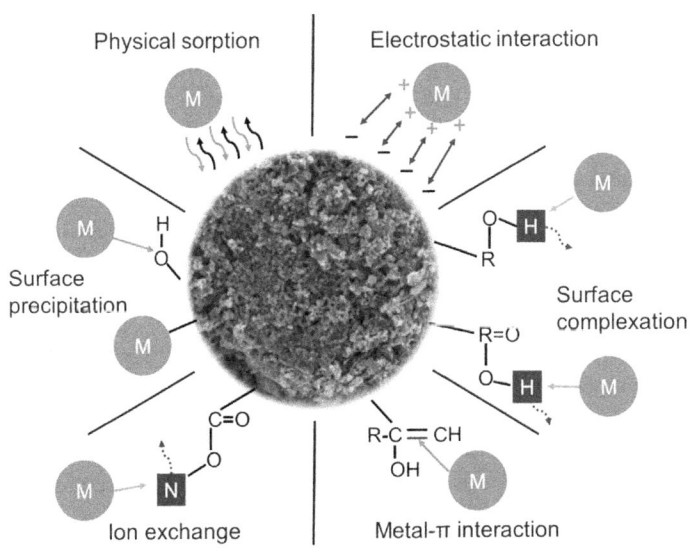

FIGURE 6.2 Inorganic pollutants removal mechanism. Mechanism for the biochar removal of inorganic contaminants. Inorganic pollutants are removed via various mechanisms, including complexation and reduction-oxidation process. *From Nguyen, T. B., Sherpa, K., Bui, X. T., Nguyen, V. T., Vo, T. D. H., Ho, H. T. T., Chen, C. W. & Dong, C. D. (2023). Biochar for soil remediation: A comprehensive review of current research on pollutant removal.* Environmental Pollution, *337. https://doi.org/10.1016/j.envpol.2023.122571.*

- Biochar reduces metal availability as it adsorbs metals reducing their phytoavailability and toxicity to plants hence improving soil health and fertility,
- Biochar supports microbial activity that enhances plant-microbe interactions and pollutant degradation, and
- Biochar can increase plant uptake of pollutants resultantly enhancing phytoremediation efficiency resulting in reduced remediation time (Ghosh & Maiti, 2021; Valizadeh et al., 2021).

Integrating biochar and phytoremediation creates more effective and sustainable remediation strategies. The synergies between biochar and phytoremediation offer a promising approach to enhance environmental remediation. The synergies can be applied to various environmental scenarios. However, further research and field-scale applications are needed to fully realize the potential of biochar-phytoremediation synergies and to scale up this innovative approach for real-world environmental challenges.

6.6 Factors affecting capacity of biochar in soil remediation

Biochar's impact on various inorganic pollutants, such as heavy metals and nutrients in soils, depends on soil attributes, biochar characteristics, particle size, and the amount of biochar added, as well as biochar pyrolysis conditions from different types of biomasses.

6.6.1 Physiochemical attributes of polluted soils

The pH of the soil is the most significant parameter in the pollutant's stabilization process. Under a lower pH environment, a large concentration of hydrogen ions exists in the soil contributing to its electrostatic repulsion with positively charged metal cations, and hydrogen ions compete with these cations for sorption sites. Thus, the mobility of metals in contaminated soils with lower pH is typically higher (Chen et al., 2021). The alkaline, carbonyl, and hydroxyl groups released via biochar in contact with water in soils elevate the soil pH (Carvalho et al., 2022). Under alkaline conditions, heavy metals are liable to undergo sorption reactions with O-comprising functional groups in biochar and generate precipitates with phosphate and carbonate. An increase in soil pH can increase the biochar stabilization capacity for heavy metals. Nonetheless, not all detrimental compounds can be immobilized in a higher soil pH condition (Amin et al., 2023). For instance, high concentrations of OH- in alkaline nature soils would undergo competitive sorption with the negatively charged oxyanion. As creatine is easily desorbed from the soil particle surface under a higher pH environment (Brtnicky et al., 2021), this shows that soil pH is a critical parameter affecting the impacts of biochar.

Various redox conditions can control heavy metal's adsorption via biochar addition. Several studies on alterations in redox potential were conducted in flooded conditions. Many researchers observed significant alterations in pH after the addition of biochar to upper mining-contaminated soil, but no effect when biochar was applied to lower mining-polluted soils, and hypothesized that the hydric regime hydration process might change the biochar impacts on the pH of the soil. Hamid et al. (2019) reported that biochar derived from sewage sludge created an apparent rise in residual constituents of cadmium under flooding environments, and thus, they concluded that hydrophobic environments were the main factor influencing the effective metals immobilization via biochar application. Additionally, continuous drying–wetting cycles of soil can accelerate the biochar aging mechanism, which may enhance the surface O-enriched functional groups, therefore maintaining the metals' immobilization efficiency of biochar (Ai et al., 2019).

6.6.2 Physicochemical characteristics of biochars and removal efficacy

Various factors may affect the chemical attributes of biochars, such as feedstock types, pyrolysis conditions, dissolved organic carbon content, and the SSA of biochar. These aspects can influence the biochar performance of polluted soils. The feedstock can influence the biochar attributes and affect its removal process and outcome on various metals. Different feedstocks lead to changes in biochar ash content, which, in turn, impacts its pH and remediation ability (Rahim, Akbar, & Alatalo, 2022). The pyrolysis temperature is strongly correlated with changes in the structure and physicochemical properties of biochar. A higher pyrolysis temperature resulted in an increase in surface area, carbonized fractions, pH, and volatile matter, and a decrease in CEC and the content of surface functional groups (Jain et al., 2016). It has been found that increasing the pyrolysis temperature causes changes in the biochar surface area and porosity. This is most likely due to the decomposition of OM and the formation of micropores. Moreover, the destruction of aliphatic alkyl and ester groups and the exposure of the aromatic lignin core under higher pyrolysis temperatures result in increased surface area. The heating to temperatures of 350°C–650°C breaks and rearranges the chemical bonds in the biomass, forming new functional groups, including carboxyl, lactone, lactol, quinine, chromene, anhydride, phenol, ether, pyrone, pyridine, pyridine, and pyrrole (Oni, Oziegbe, & Olawole, 2019). On the other hand, biochar produced at lower temperatures (300°C–400°C) displays a more diversified organic character due to the occurrence of aliphatic and cellulose-type structures. As a result, the structure of biochar appears to have more organized C layers (such as graphene structure) and less content of surface functional groups when pyrolysis temperature increases (Cameselle, Gouveia, & Cabo, 2021). Biochar pyrolyzed at a high temperature has high porosity, surface area, and polar functional groups, which can remove the mercury more effectively (Hasan, Salman, et al., 2023). Biochar with a larger surface area has a greater contact surface with the soil solution, leading to more reactions with pollutants (Hamid et al., 2020).

6.6.3 Application rate and particle sizes

Biochar application rates can alter the speciation of heavy metals in polluted soils to various degrees, which can decrease the heavy metal concentration in plant tissues. The addition rate of biochar is negatively interrelated with the amount of contaminating pollutants in polluted soils (Rana et al., 2021). Pandey, Suthar, & Chand (2022) observed that with enhancing biochar application rates of 1%–4% w/w, the Pb and Cd amount of plant shoots reduced from 2.81 mg/kg and 22.6 mg/kg to 2.37 mg/kg and 15.5 mg/kg, respectively, while those of the roots reduced from 15.7 mg/kg and 16.1 mg/kg to 8.42 mg/kg and 11.5 mg/kg in that order. A 5% biochar application rate could enhance the plant shoot biomass by 29.3%, which might be due to the heavy metal reduction and the improvement of nutrients and OM (Cameselle, Gouveia, & Cabo, 2021). It is hypothesized that benzoic acid and ethylene in biochar could accelerate plant growth and seedling development, as well as reduce toxicity (Cara et al., 2022; Carvalho et al., 2022). Therefore, it is essential to find an appropriate biochar application rate for phyto-ecology promotion in a cost-effective way in polluted soils.

Biochar with different particle sizes can also affect soil remediation efficiency, and mostly, small particle sizes have excellent effects on remediation (Brtnicky et al., 2021). Biochar derived from pinewood, with particle sizes of 0.1 and 0.4 mm, was observed to decrease soil pore water lead concentration by 86 and 69%, respectively, in contaminated soil (Rahim, Akbar, & Alatalo, 2022). This was attributed to the finer size biochar having a higher surface area, which also showed that biochar particle size was more efficient in decreasing the organic-bound metals but did not influence metal species residual. Moreover, combining the biochar with other treatment measures could play an effective role in contaminated soil. The mixing of lime with 5% biochar caused a significant enhancement of microbial activity in the soil and decreased the extractable Zn, Cu, and Al amount compared with biochar application alone (Hamid et al., 2020). The mixing of technosol and biochar significantly reduced the mobility of Pb from 17 to 2.1, Ni from 47 to 2.3, and Cu from 18 to 1.6. Taraqqi-A-kamal et al. (2021) reported that a combination of iron sulfate and biochar added to soil accelerated the arsenic release and immobilized the arsenic effectively. In general, the interaction between heavy metals and biochar depends not only on biomass type and pyrolysis temperature but also on physiochemical attributes and the soil pollution intensity. Different application rate also plays a significant role in pollutant mobilization and soil improvement.

6.7 Behavior and the fate of contaminants in biochar-amended soils

The movement of inorganic compounds is governed by their physicochemical properties, environmental conditions, and the properties of the medium to which they are exposed, for instance, the soil and water matrices (Cheng et al., 2013). The sorption properties of biochar are predominantly associated with structural and physical properties, including SSA, pore size and distribution, and particle size. As such the biochar surface area is directly linked with its capacity to bind compounds; therefore, increasing their surface area and carbonization properties directly increases the sorption capacity of trace elements. Studies reveal that reduced particle sizes in biochar result in higher sorption capacity (Cabrera et al., 2011), contributing to the declining rates of bioavailable contaminants and reduced leaching of contaminants into the immediate environment. The introduction of biochar influences the in situ inorganic pollutants stabilization via enhanced regulation of their bioavailability and bioaccessibility through sorption and entrapment processes. Biochar influence can be observed through the rates and extents of sorption, desorption, and biodegradation of contaminants in the biochar-amended soils against the control unamended media (Cabrera et al., 2011).

6.7.1 Properties influencing the behavior of contaminants in biochar-amended soils

The affinity of biochar to metal elements is facilitated by high SSA, microporosity, high CEC, and prolonged residence time in soil (Ahmad et al., 2014; Hilber et al., 2017). In order to improve biochar surfaces for enhanced adsorption process, modifications can be done during pyrolysis or by post-treatments with reagents (Xue et al., 2012). For instance, biochar produced relatively at high temperatures (> 500°C) poses few functional groups (e.g., hydroxyl, carboxyl, carbonyl, or phenolic groups). As such, subsequent reactions with oxidizing additives, microbial metabolism, or enzymatic breakdown could result in the creation of more functional groups for enhanced binding (Chen et al., 2011). Biochar produced at lower temperatures preserves more functional groups due to incomplete pyrolysis, however, they are susceptible to degradation and are likely to increase desorption rates (Mukherjee, Zimmerman, & Harris, 2011). The charged surface functional groups of biochar enhance its capacity to bind polar compounds such as the rhizospheric heavy metals and are immobilized using such unique characteristics, therefore remediating soils and reducing their impacts on crops and the environment (Spokas et al., 2009).

Biochar is predominant on cationic metals, indicating its high affinity and immobilization capacity compared to its behavior on anionic metalloids which are rather mobilized (Xu et al., 2022). In the view of determining the contaminants' fate and behavior in amended soils, it is important to consider factors such as pyrolysis temperature, as it is a determinant of surface area and micro-porosity. The higher the temperature, the greater the surface area and microporosity in biochar and the reverse is true for low-temperature-produced biochar. Unequivocally, biochar produced at high temperatures tends to lose the H- and O- functional groups, reducing their capacity to immobilize polar organic pollutants (Uchimiya, Chang, & Klasson, 2011).

6.7.2 Other factors influencing the bioavailability of contaminants in biochar-amended soils

Biochar nutrient properties vary depending on the biomass waste used and are widely known to enhance plant growth and microbial activity, additionally, they have been shown to enhance the biodegradation of bio-accessible and bioavailable contaminants (Cheng et al., 2013). Cara et al. (2022) reiterate that the biodegradation of pesticides can be enhanced in biochar-amended soils due to mimicked microbial activity resulting from biochar's presence. Additionally, biochar produced from corn stalks provides easily accessible energy for soil microbes, therefore accelerating the fate of nutrients in soil through degradation. Pyrolysis temperature is also a contributing factor in determining the fate of contaminants in soil. For example, Ren et al. (2016) reported that enhanced biodegradation of carbaryl on soil treated with biochar at temperatures of 350 °C compared to biotic removal at 700°C. This is attributed to less resistive or easily degradable substances on biochar produced under low temperatures which therefore favor microbial population. This positively enhances the reduction of groundwater contamination from heavy metals, through leaching (Cabrera et al., 2011). Further, some contaminants are soluble, therefore the introduction of biochar could immensely reduce underground water contamination through reduced leaching processes. Additionally, biochars produced at lower temperatures (< 500°C) are heavily linked with the removal of inorganic contaminants due to increased surface functional groups. Therefore, significantly curbing environmental effects driven by the leaching of inorganic contaminants into the underground and surface waters. Overall, several studies have indicated the efficacy and high affinity of biochar in the removal of contaminants, including heavy metals, due to their high adsorption efficiency (Ahmad et al., 2012); however, there is a need for a further long-term study on the fate of adsorbed-desorbed contaminants. Thus, there is paucity due to a longer retention period of contaminants on biochar before desorption that could result in causal environmental impact and prompt increased control costs.

6.8 Conclusion and outlook

Biochar can be used to remediate soil contaminated with inorganic pollutants by reducing the bioavailability of heavy metals in contaminated soils, as well as by changing the soil properties. The efficiency of biochar in the reclamation of contaminated soils is largely dependent on the operating conditions, e.g., feedstock type, temperature, and residence time, that are employed during the preparation of biochar, which has much influence on the biochar characteristics, including pore size distribution, surface groups, and ion-exchange capacity. Moreover, the efficacy of biochar for soil contaminants removal relies on the application rate of biochar to contaminated soils. Biochar produced at high temperatures, for example, at ~800°C, contained more porosity and large SSA, therefore offering more adsorption capacity for contaminants. The reduction-oxidation and electrostatic adsorption processes contributed more to the adsorption of oxyanions, whereas ion exchange, complexation, and precipitation are largely involved in the adsorption of cations.

Furthermore, soil pH is the central parameter that influences the efficiency of biochar in the remediation of inorganic contaminated soils. Although biochar can remediate contaminated soil, the spent biochar loaded with heavy metals has potential problems, including (1) the presence of metals can impede organic toxins biodegradation; (2) elevated concentrations of these metals in biochar may reduce plant growth due to bioaccumulation and biomagnification, resulting in phytotoxicity. It is not clear whether the spent biochar loaded with heavy metals can actually lower plant growth, which necessitates future work to research that aspect.

References

Ahmad, M., Lee, S. S., Dou, X., Mohan, D., Sung, J. K., Yang, J. E., & Ok, Y. S. (2012). Effects of pyrolysis temperature on soybean stover- and peanut shell-derived biochar properties and TCE adsorption in water. *Bioresource Technology, 118*, 536–544. Available from https://doi.org/10.1016/j.biortech.2012.05.042.

Ahmad, M., Rajapaksha, A. U., Lim, J. E., Zhang, M., Bolan, N., Mohan, D., Vithanage, M., Lee, S. S., & Ok, Y. S. (2014). Biochar as a sorbent for contaminant management in soil and water: A review. *Chemosphere, 99*, 19–33. Available from https://doi.org/10.1016/j.chemosphere.2013.10.071, http://www.elsevier.com/locate/chemosphere.

Ai, S., Yang, Y., Ding, J., Yang, W., Bai, X., Bao, X., Ji, W., & Zhang, Y. (2019). Metal Exposure Risk Assessment for Tree Sparrows at Different Life Stages via Diet from a Polluted Area in Northwestern China. *Environmental Toxicology and Chemistry, 38*(12), 2785–2796. Available from https://doi.org/10.1002/etc.4576, http://onlinelibrary.wiley.com/journal/10.1002/(ISSN)1552-8618.

Albright, R. (2011). *Cleanup of Chemical and Explosive Munitions: Location, Identification and Environmental Remediation Cleanup of Chemical and Explosive Munitions: Location, Identification and Environmental Remediation* (pp. 1–305). United States: Elsevier. Available from https://www.elsevier.com/books/cleanup-of-chemical-and-explosive-munitions/albright/978-1-4377-3477-5, 10.1016/C2010-0-67036-2.

Alipour, M., Asadi, H., Chen, C., & Besalatpour, A. A. (2022). Fate of organic pollutants in sewage sludge during thermal treatments: Elimination of PCBs, PAHs, and PPCPs. *Fuel, 319*.

Allohverdi, T., Mohanty, A. K., Roy, P., & Misra, M. (2021). A Review on Current Status of Biochar Uses in Agriculture. *Molecules (Basel, Switzerland), 26*(18), 5584. Available from https://doi.org/10.3390/molecules26185584.

Alloway, B. J., & Jackson, A. P. (1991). The behaviour of heavy metals in sewage sludge-amended soils. *Science of the Total Environment, The, 100*(C), 151–176. Available from https://doi.org/10.1016/0048-9697(91)90377-Q.

Ambaye, T. G., Vaccari, M., van Hullebusch, E. D., Amrane, A., & Rtimi, S. (2021). Mechanisms and adsorption capacities of biochar for the removal of organic and inorganic pollutants from industrial wastewater. *International Journal of Environmental Science and Technology, 18*(10), 3273–3294. Available from https://doi.org/10.1007/s13762-020-03060-w, http://www.springerlink.com/content/1735-1472.

Amin, M. A., Haider, G., Rizwan, M., Schofield, H. K., Qayyum, M. F., Zia-ur-Rehman, M., & Ali, S. (2023). Different feedstocks of biochar affected the bioavailability and uptake of heavy metals by wheat (Triticum aestivum L.) plants grown in metal contaminated soil. *Environmental Research, 217*. Available from https://doi.org/10.1016/j.envres.2022.114845, https://www.sciencedirect.com/journal/environmental-research.

Azeem, M., Ali, A., Arockiam Jeyasundar, P. G. S., Li, Y., Abdelrahman, H., Latif, A., Li, R., Basta, N., Li, G., Shaheen, S. M., Rinklebe, J., & Zhang, Z. (2021). Bone-derived biochar improved soil quality and reduced Cd and Zn phytoavailability in a multi-metal contaminated mining soil. *Environmental Pollution, 277*. Available from https://doi.org/10.1016/j.envpol.2021.116800, https://www.journals.elsevier.com/environmental-pollution.

Beesley, L., & Dickinson, N. (2010). Carbon and trace element mobility in an urban soil amended with green waste compost. *Journal of Soils and Sediments, 10*(2), 215–222. Available from https://doi.org/10.1007/s11368-009-0112-y.

Beesley, L., & Marmiroli, M. (2011). The immobilisation and retention of soluble arsenic, cadmium and zinc by biochar. *Environmental Pollution, 159*(2), 474–480. Available from https://doi.org/10.1016/j.envpol.2010.10.016.

Bian, R., Joseph, S., Cui, L., Pan, G., Li, L., Liu, X., Zhang, A., Rutlidge, H., Wong, S., Chia, C., Marjo, C., Gong, B., Munroe, P., & Donne, S. (2014). A three-year experiment confirms continuous immobilization of cadmium and lead in contaminated paddy field with biochar amendment. *Journal of Hazardous Materials, 272*, 121–128. Available from https://doi.org/10.1016/j.jhazmat.2014.03.017, http://www.elsevier.com/locate/jhazmat.

Broomandi, P., Guney, M., Kim, J. R., & Karaca, F. (2020). Soil contamination in areas impacted by military activities: A critical review. *Sustainability, 12*(21). Available from https://doi.org/10.3390/su12219002.

Brtnicky, M., Datta, R., Holatko, J., Bielska, L., Gusiatin, Z. M., Kucerik, J., Hammerschmiedt, T., Danish, S., Radziemska, M., Mravcova, L., Fahad, S., Kintl, A., Sudoma, M., Ahmed, N., & Pecina, V. (2021). A critical review of the possible adverse effects of biochar in the soil environment. *Science of The Total Environment, 796*148756. Available from https://doi.org/10.1016/j.scitotenv.2021.148756.

Cabrera, A., Cox, L., Spokas, K. A., Celis, R., Hermosín, M. C., Cornejo, J., & Koskinen, W. C. (2011). Comparative sorption and leaching study of the herbicides fluometuron and 4-chloro-2-methylphenoxyacetic acid (MCPA) in a soil amended with biochars and other sorbents. *Journal of Agricultural and Food Chemistry, 59*(23), 12550–12560. Available from https://doi.org/10.1021/jf202713q.

Cameselle, C., Gouveia, S., & Cabo, A. (2021). Enhanced Electro Kinetic Remediation for the Removal of Heavy Metals from Contaminated Soils. *Applied Sciences, 11*(4). Available from https://doi.org/10.3390/app11041799.

Cara, I. G., Țopa, D., Puiu, I., & Jitareanu, G. (2022). Biochar a promising strategy for pesticide-contaminated soils. *Agriculture, 12*(10). Available from https://doi.org/10.3390/agriculture12101579.

Carvalho, J., Nascimento, L., Soares, M., Valério, N., Ribeiro, A., Faria, L., Silva, A., Pacheco, N., Araújo, J., & Vilarinho, C. (2022). Life cycle assessment (LCA) of biochar production from a circular economy perspective. *Processes, 10*(12). Available from https://doi.org/10.3390/pr10122684, http://www.mdpi.com/journal/processes.

Chen, C. K., Chen, J. J., Nguyen, N. T., Le, T. T., Nguyen, N. C., & Chang, C. T. (2021). Specifically designed magnetic biochar from waste wood for arsenic removal. *Sustainable Environment Research, 31*(1). Available from https://doi.org/10.1186/s42834-021-00100-z, https://link.springer.com/journal/42834.

Chen, J. L., Liu, H. B., Wu, W., & Xie, D. T. (2011). Estimation of monthly solar radiation from measured temperatures using support vector machines - A case study. *Renewable Energy, 36*(1), 413–420. Available from https://doi.org/10.1016/j.renene.2010.06.024.

Chen, Z., Wang, Y., Xia, D., Jiang, X., Fu, D., Shen, L., Wang, H., & Li, Q. B. (2016). Enhanced bioreduction of iron and arsenic in sediment by biochar amendment influencing microbial community composition and dissolved organic matter content and composition. *Journal of Hazardous Materials, 311*, 20–29. Available from https://doi.org/10.1016/j.jhazmat.2016.02.069, http://www.elsevier.com/locate/jhazmat.

Cheng, G., Sun, L., Jiao, L., Peng, L.-X., Lei, Z.-H., Wang, Y.-X., & Lin, J. (2013). Adsorption of methylene blue by residue biochar from copyrolysis of dewatered sewage sludge and pine sawdust. *Desalination and Water Treatment, 51*(37-39), 7081–7087. Available from https://doi.org/10.1080/19443994.2013.773265.

Cui, H., Fan, Y., Fang, G., Zhang, H., Su, B., & Zhou, J. (2016). Leachability, availability and bioaccessibility of Cu and Cd in a contaminated soil treated with apatite, lime and charcoal: A five-year field experiment. *Ecotoxicology and Environmental Safety, 134*, 148–155. Available from https://doi.org/10.1016/j.ecoenv.2016.07.005, http://www.elsevier.com/inca/publications/store/6/2/2/8/1/9/index.htt.

Cui, L., Pan, G., Li, L., Bian, R., Liu, X., Yan, J., Quan, G., Ding, C., Chen, T., Liu, Y., Liu, Y., Yin, C., Wei, C., Yang, Y., & Hussain, Q. (2016). Continuous immobilization of cadmium and lead in biochar amended contaminated paddy soil: A five-year field experiment. *Ecological Engineering, 93*, 1–8. Available from https://doi.org/10.1016/j.ecoleng.2016.05.007, http://www.elsevier.com/inca/publications/store/5/2/2/7/5/1.

Das, S. K., Ghosh, G. K., & Avasthe, R. (2023). Application of biochar in agriculture and environment, and its safety issues. *Biomass Conversion and Biorefinery, 13*(2), 1359–1369. Available from https://doi.org/10.1007/s13399-020-01013-4, https://www.springer.com/journal/13399.

Debela, F., Thring, R. W., & Arocena, J. M. (2012). Immobilization of Heavy Metals by Co-pyrolysis of Contaminated Soil with Woody Biomass. *Water, Air, & Soil Pollution, 223*(3), 1161–1170. Available from https://doi.org/10.1007/s11270-011-0934-2.

Dhanker, R., Chaudhary, S., Goyal, S., & Garg, V. K. (2021). Influence of urban sewage sludge amendment on agricultural soil parameters. *Environmental Technology & Innovation, 23*101642. Available from https://doi.org/10.1016/j.eti.2021.101642.

Diaz, E., Brochu, S., Thiboutot, S., Ampleman, G., Marois, A., & Gagnon, A. (2007). *Energetic Materials and Metals Contamination at CFB/ASU Wainwright, Alberta Phase I (2007)*. DRDC Valcartier TR.

Dąbrowska, L., & Rosińska, A. (2012). Change of PCBs and forms of heavy metals in sewage sludge during thermophilic anaerobic digestion. *Chemosphere, 88*(2), 168–173. Available from https://doi.org/10.1016/j.chemosphere.2012.02.073.

F.A UNEP, Global assessment of soil pollution. (2021), 2021.

Fayiga, A. O. (2019). Remediation of inorganic and organic contaminants in military ranges. *CSIRO, United States Environmental Chemistry, 16*(2), 81–91. Available from https://doi.org/10.1071/EN18196, http://www.publish.csiro.au/nid/188.htm.

Fedeli, R., Alexandrov, D., Celletti, S., Nafikova, E., & Loppi, S. (2023). Biochar improves the performance of Avena sativa L. grown in gasoline-polluted soils. *Environmental Science and Pollution Research.*, 28791–28802. Available from https://doi.org/10.1007/s11356-022-24127-w.

Fijalkowski, K., Rorat, A., Grobelak, A., & Kacprzak, M. J. (2017). The presence of contaminations in sewage sludge – The current situation. *Journal of Environmental Management, 203*, 1126–1136. Available from https://doi.org/10.1016/j.jenvman.2017.05.068, https://www.sciencedirect.com/journal/journal-of-environmental-management.

Ghosh, D., & Maiti, S. K. (2021). Biochar assisted phytoremediation and biomass disposal in heavy metal contaminated mine soils: a review. *International Journal of Phytoremediation, 23*(6), 559–576. Available from https://doi.org/10.1080/15226514.2020.1840510, https://www.tandfonline.com/loi/bijp20.

Gnanasundar, V. M., & Akshai Raj, R. (2021). Remediation of inorganic contaminants in soil using electrokinetics, phytoremediation techniques. *Materials Today: Proceedings, 45*, 950–956. Available from https://doi.org/10.1016/j.matpr.2020.03.038.

Gong, Y., Zhao, D., & Wang, Q. (2018). An overview of field-scale studies on remediation of soil contaminated with heavy metals and metalloids: Technical progress over the last decade. *Water Research, 147*, 440–460. Available from https://doi.org/10.1016/j.watres.2018.10.024, http://www.elsevier.com/locate/watres.

Hamid, Y., Tang, L., Hussain, B., Usman, M., Gurajala, H. K., Rashid, M. S., He, Z., & Yang, X. (2020). Efficiency of lime, biochar, Fe containing biochar and composite amendments for Cd and Pb immobilization in a co-contaminated alluvial soil. *Environmental Pollution, 257*. Available from https://doi.org/10.1016/j.envpol.2019.113609, https://www.journals.elsevier.com/environmental-pollution.

Hamid, Y., Tang, L., Sohail, M. I., Cao, X., Hussain, B., Aziz, M. Z., Usman, M., He, Z. L., & Yang, X. (2019). An explanation of soil amendments to reduce cadmium phytoavailability and transfer to food chain. *Science of the Total Environment, 660*, 80–96. Available from https://doi.org/10.1016/j.scitotenv.2018.12.419, http://www.elsevier.com/locate/scitotenv.

Hasan, M. M., Kubra, K. T., Hasan, M. N., Awual, M. E., Salman, M. S., Sheikh, M. C., & Awual, M. R. (2023). Sustainable ligand-modified based composite material for the selective and effective cadmium (II) capturing from wastewater. *Journal of Molecular Liquids., 371*.

Hasan, M. N., Salman, M. S., Hasan, M. M., Kubra, K. T., Sheikh, M. C., Rehan, A. I., & Awual, M. R. (2023). Assessing sustainable Lutetium (III) ions adsorption and recovery using novel composite hybrid nanomaterials. *Journal of Molecular Structure, 1276*.

Hilber, I., Bastos, A. C., Loureiro, S., Soja, G., Marsz, A., Cornelissen, G., & Bucheli, T. D. (2017). The different faces of biochar: contamination risk versus remediation tool. *Journal of Environmental Engineering and Landscape Management, 25*(2), 86–104. Available from https://doi.org/10.3846/16486897.2016.1254089, http://www.tandfonline.com/loi/teel20.

Houben, D., Evrard, L., & Sonnet, P. (2013). Mobility, bioavailability and pH-dependent leaching of cadmium, zinc and lead in a contaminated soil amended with biochar. *Chemosphere, 92*(11), 1450–1457. Available from https://doi.org/10.1016/j.chemosphere.2013.03.055.

Ibrahim, E. A., El-Sherbini, M. A. A., & Selim E.-M. (2022). Effects of biochar on soil properties, heavy metal availability and uptake, and growth of summer squash grown in metal-contaminated soil. *Scientia Horticulturae* 301, 111097.

Jain, S., Mishra, D., Khare, P., Yadav, V., Deshmukh, Y., & Meena, A. (2016). Impact of biochar amendment on enzymatic resilience properties of mine spoils. *Science of The Total Environment, 544*, 410–421. Available from https://doi.org/10.1016/j.scitotenv.2015.11.011.

Jenkins, T.F., Pennington, J.C., Ranney, T.A., Berry, T.E., Miyares, P.H., Walsh, M.E., Hewitt, A.D., Perron, N.M., Parker, L.V., Hayes, C.A., & Wahlgren, E.G. (2001). Characterization of explosives contamination at military firing ranges. US Army Corps of Engineers, Engineer Research and Development Center, Technical Report ERDC TR-01-5.

Kim, P., Johnson, A. M., Essington, M. E., Radosevich, M., Kwon, W. T., Lee, S. H., Rials, T. G., & Labbé, N. (2013). Effect of pH on surface characteristics of switchgrass-derived biochars produced by fast pyrolysis. *Chemosphere, 90*(10), 2623–2630. Available from https://doi.org/10.1016/j.chemosphere.2012.11.021, http://www.elsevier.com/locate/chemosphere.

Kumar, A., Joseph, S., Tsechansky, L., Privat, K., Schreiter, I. J., Schüth, C., & Graber, E. R. (2018). Biochar aging in contaminated soil promotes Zn immobilization due to changes in biochar surface structural and chemical properties. *Science of the Total Environment, 626*, 953–961. Available from https://doi.org/10.1016/j.scitotenv.2018.01.157, http://www.elsevier.com/locate/scitotenv.

Kumar, R., Verma, A., Shome, A., Sinha, R., Sinha, S., Jha, P. K., & Prasad. (2021). Impacts of plastic pollution on ecosystem services, sustainable development goals, and need to focus on circular economy and policy interventions. *Sustainability, 13*.

Kumpiene, J., Lagerkvist, A., & Maurice, C. (2008). Stabilization of As, Cr, Cu, Pb and Zn in soil using amendments – A review. *Waste Management, 28*(1), 215–225. Available from https://doi.org/10.1016/j.wasman.2006.12.012.

Lasheen, M. R., & Ammar, N. S. (2009). Assessment of metals speciation in sewage sludge and stabilized sludge from different Wastewater Treatment Plants, Greater Cairo, Egypt. *Journal of Hazardous Materials, 164*(2-3), 740–749. Available from https://doi.org/10.1016/j.jhazmat.2008.08.068.

Liu, M., Almatrafi, E., Zhang, Y., Xu, P., Song, B., Zhou, C., Zeng, G., & Zhu, Y. (2022). A critical review of biochar-based materials for the remediation of heavy metal contaminated environment: Applications and practical evaluations. *Science of The Total Environment, 806*150531. Available from https://doi.org/10.1016/j.scitotenv.2021.150531.

Major, M. A., Checkai, R. T., Phillips, C. T., Wentsel, R. S., & Nwanguma, R. O. (1992). Method for screening and analysis of residues common to munition open burning/open detonation (OB/OD) Sites. *International Journal of Environmental Analytical Chemistry, 48*(3-4), 217–227. Available from https://doi.org/10.1080/03067319208027402.

Maliszewska-Kordybach, B., & Smreczak, B. (2003). Habitat function of agricultural soils as affected by heavy metals and polycyclic aromatic hydrocarbons contamination. *Environment International, 28*(8), 719–728. Available from https://doi.org/10.1016/s0160-4120(02)00117-4.

Meier, S., Curaqueo, G., Khan, N., Bolan, N., Cea, M., Eugenia, G. M., Cornejo, P., Ok, Y. S., & Borie, F. (2017). Chicken-manure-derived biochar reduced bioavailability of copper in a contaminated soil. *Journal of Soils and Sediments, 17*(3), 741–750. Available from https://doi.org/10.1007/s11368-015-1256-6, http://www.springerlink.com/content/1439-0108.

Meyers, S. K., Deng, S., Basta, N. T., Clarkson, W. W., & Wilber, G. G. (2007). Long-term explosive contamination in soil: Effects on soil microbial community and bioremediation. *Soil and Sediment Contamination, 16*(1), 61–77. Available from https://doi.org/10.1080/15320380601077859.

Montagnoli, A., Baronti, S., Alberto, D., Chiatante, D., Scippa, G. S., & Terzaghi, M. (2021). Pioneer and fibrous root seasonal dynamics of Vitis vinifera L. are affected by biochar application to a low fertility soil: A rhizobox approach. *Science of The Total Environment, 751*141455. Available from https://doi.org/10.1016/j.scitotenv.2020.141455.

Morel, J. L., Schwartz, C., Florentin, L., & de Kimpe, C. (2005). *Urban Soils* (pp. 202–208). Elsevier BV. Available from 10.1016/b0-12-348530-4/00305-2.

Mukherjee, A., Zimmerman, A. R., & Harris, W. (2011). Surface chemistry variations among a series of laboratory-produced biochars. *Geoderma, 163*(3-4), 247–255. Available from https://doi.org/10.1016/j.geoderma.2011.04.021.

Murtaza, G., Murtaza, B., Niazi, N. K., & Sabir, M. (2014). *Soil Contaminants: Sources, Effects, and Approaches for Remediation* (pp. 171–196). Springer Science and Business Media LLC. Available from 10.1007/978-1-4614-8824-8_7.

Nguyen, T. B., Nguyen, V. T., Hoang, H. G., Cao, N. D. T., Nguyen, T. T., Vo, T. D. H., Nguyen, N. K. Q., Pham, M. D. T., Nghiem, D. L., Vo, T. K. Q., Dong, C. D., & Bui, X. T. (2023). Recent development of algal biochar for contaminant remediation and energy application: A state-of-the art review. *Current Pollution Reports, 9*(1), 73–89. Available from https://doi.org/10.1007/s40726-022-00243-6, https://www.springer.com/journal/40726.

Nguyen, T. B., Sherpa, K., Bui, X. T., Nguyen, V. T., Vo, T. D. H., Ho, H. T. T., Chen, C. W., & Dong, C. D. (2023). Biochar for soil remediation: A comprehensive review of current research on pollutant removal. *Environmental Pollution, 337*. Available from https://doi.org/10.1016/j.envpol.2023.122571, https://www.journals.elsevier.com/environmental-pollution.

Nunes, N., Ragonezi, C., Gouveia, C. S. S., & Pinheiro de Carvalho, M. Â. A. (2021). Review of Sewage Sludge as a Soil Amendment in Relation to Current International Guidelines: A Heavy Metal Perspective. *Sustainability, 13*(4), 2317. Available from https://doi.org/10.3390/su13042317.

Núñez, P. (2023). *Urban Agriculture & Regional Food Systems, 8*, 2023.

O'Connor, D., Peng, T., Li, G., Wang, S., Duan, L., Mulder, J., Cornelissen, G., Cheng, Z., Yang, S., & Hou, D. (2018). Sulfur-modified rice husk biochar: A green method for the remediation of mercury contaminated soil. *Science of the Total Environment, 621*, 819–826. Available from https://doi.org/10.1016/j.scitotenv.2017.11.213, http://www.elsevier.com/locate/scitotenv.

Oni, B. A., Oziegbe, O., & Olawole, O. O. (2019). Significance of biochar application to the environment and economy. *Annals of Agricultural Sciences, 64*(2), 222–236. Available from https://doi.org/10.1016/j.aoas.2019.12.006, http://www.elsevier.com/journals/annals-of-agricultural-sciences/0570-1783/.

Pande, V., Pandey, S. C., Sati, D., Bhatt, P., & Samant, M. (2022). Microbial interventions in bioremediation of heavy metal contaminants in agroecosystem. *Front. Microbiol., 13*.

Pandey, B., Suthar, S., & Chand, N. (2022). Effect of biochar amendment on metal mobility, phytotoxicity, soil enzymes, and metal-uptakes by wheat (Triticum aestivum) in contaminated soils. *Chemosphere, 307*.

Patinha, C., Armienta, A., Argyraki, A., & Durães, N. (2018). *Inorganic Pollutants in Soils* (pp. 127–159). Elsevier BV. Available from 10.1016/b978-0-12-849873-6.00006-6.

Rahim, H. U., Akbar, W. A., & Alatalo, J. M. (2022). A comprehensive literature review on cadmium (Cd) status in the soil environment and its immobilization by biochar-based materials. *Agronomy, 12*.

Rana, A., Sindhu, M., Kumar, A., Dhaka, R. K., Chahar, M., Singh, S., & Nain, L. (2021). Restoration of heavy metal-contaminated soil and water through biosorbents: A review of current understanding and future challenges. *Physiologia Plantarum, 173*(1), 394–417. Available from https://doi.org/10.1111/ppl.13397, http://onlinelibrary.wiley.com/journal/10.1111/(ISSN)1399-3054.

Reddy, K. R., Xie, T., & Dastgheibi, S. (2014). Evaluation of Biochar as a Potential Filter Media for the Removal of Mixed Contaminants from Urban Storm Water Runoff. *Journal of Environmental Engineering, 140*(12). Available from https://doi.org/10.1061/(asce)ee.1943-7870.0000872.

Rees, F., Simonnot, M. O., & Morel, J. L. (2014). Short-term effects of biochar on soil heavy metal mobility are controlled by intra-particle diffusion and soil pH increase. *European Journal of Soil Science, 65*(1), 149–161. Available from https://doi.org/10.1111/ejss.12107.

Ren, X., Zhang, P., Zhao, L., & Sun, H. (2016). Sorption and degradation of carbaryl in soils amended with biochars: influence of biochar type and content. *Environmental Science and Pollution Research., 23*(3), 2724–2734. Available from https://doi.org/10.1007/s11356-015-5518-z.

Shen, Z., Som, A. M., Wang, F., Jin, F., McMillan, O., & Al-Tabbaa, A. (2016). Long-term impact of biochar on the immobilisation of nickel (II) and zinc (II) and the revegetation of a contaminated site. *Science of the Total Environment, 542*, 771–776. Available from https://doi.org/10.1016/j.scitotenv.2015.10.057, http://www.elsevier.com/locate/scitotenv.

Sieciechowicz, A., Sadecka, Z., Myszograj, S., Włodarczyk-Makuła, M., Wiśniowska, E., & Turek, A. (2014). Occurrence of heavy metals and PAHs in soil and plants after application of sewage sludge to soil. *Desalination and Water Treatment, 52*(19-21), 4014–4026. Available from https://doi.org/10.1080/19443994.2014.922292, http://www.tandfonline.com/toc/tdwt20/current.

Sizmur, T., Quilliam, R., Puga, A. P., Moreno-Jiménez, E., Beesley, L., & Gomez-Eyles, J. L. (2015). *Application of biochar for soil remediation Agricultural and Environmental Applications of Biochar: Advances and Barriers* (pp. 295–324). United Kingdom: Wiley. Available from http://onlinelibrary.wiley.com/book/9780891189671, 10.2136/sssaspecpub63.2014.0046.5.

Spokas, K. A., Koskinen, W. C., Baker, J. M., & Reicosky, D. C. (2009). Impacts of woodchip biochar additions on greenhouse gas production and sorption/degradation of two herbicides in a Minnesota soil. *Chemosphere, 77*(4), 574–581. Available from https://doi.org/10.1016/j.chemosphere.2009.06.053.

Stefaniuk, M., Tsang, D. C. W., Ok, Y. S., & Oleszczuk, P. (2018). A field study of bioavailable polycyclic aromatic hydrocarbons (PAHs) in sewage sludge and biochar amended soils. *Journal of Hazardous Materials, 349*, 27–34. Available from https://doi.org/10.1016/j.jhazmat.2018.01.045, http://www.elsevier.com/locate/jhazmat.

Sun, J., Cui, L., Quan, G., Yan, J., Wang, H., & Wu, L. (2020). Effects of biochar on heavy metals migration and fractions changes with different soil types in column experiments. *BioResources, 15*(2), 4388–4406. Available from https://doi.org/10.15376/biores.15.2.4388-4406.

Taraqqi-A-kamal, A., Atkinson, C. J., Khan, A., Zhang, K., Sun, P., Akther, S., & Zhang, Y. (2021). Biochar remediation of soil: Linking biochar production with function in heavy metal contaminated soils. *Plant, Soil and Environment, 67*(4), 183–201. Available from https://doi.org/10.17221/544/2020-PSE, https://www.agriculturejournals.cz/publicFiles/544_2020-PSE.pdf.

Tomczyk, A., Sokołowska, Z., & Boguta, P. (2020). Biochar physicochemical properties: pyrolysis temperature and feedstock kind effects. *Reviews in Environmental Science and Bio/Technology, 19*(1), 191–215. Available from https://doi.org/10.1007/s11157-020-09523-3.

Uchimiya, M., Chang, S. C., & Klasson, K. T. (2011). Screening biochars for heavy metal retention in soil: Role of oxygen functional groups. *Journal of Hazardous Materials, 190*(1-3), 432–441. Available from https://doi.org/10.1016/j.jhazmat.2011.03.063.

Valizadeh, S., Lee, S. S., Baek, K., Choi, Y. J., Jeon, B.-H., Rhee, G. H., Lin, K.-Y. A., & Park, Y.-K. (2021). Bioremediation strategies with biochar for polychlorinated biphenyls (PCBs)-contaminated soils: A review. *Environmental Research, 200*111757. Available from https://doi.org/10.1016/j.envres.2021.111757.

Vithanage, M., Herath, I., Joseph, S., Bundschuh, J., Bolan, N., Ok, Y. S., Kirkham, M. B., & Rinklebe, J. (2017). Interaction of arsenic with biochar in soil and water: A critical review. *Carbon, 113*, 219–230. Available from https://doi.org/10.1016/j.carbon.2016.11.032, http://www.journals.elsevier.com/carbon/.

Walsh, M. R., Walsh, M. E., & Hewitt, A. D. (2010). Energetic residues from field disposal of gun propellants. *Journal of Hazardous Materials, 173*(1-3), 115–122. Available from https://doi.org/10.1016/j.jhazmat.2009.08.056.

Wang, Y., Liu, Y., Zhan, W., Zheng, K., Wang, J., Zhang, C., & Chen, R. (2020). Stabilization of heavy metal-contaminated soils by biochar: Challenges and recommendations. *Science of The Total Environment, 729*.

Weldeslassie, T., Naz, H., Singh, B., & Oves, M. (2017). *Chemical contaminants for soil, air and aquatic ecosystem Modern Age Environmental Problems and their Remediation* (pp. 1–22). Eritrea: Springer International Publishing. Available from http://doi.org/10.1007/978-3-319-64501-8, 10.1007/978-3-319-64501-8_1.

Wuana, R. A., & Okieimen, F. E. (2011). Heavy metals in contaminated soils: A review of sources, chemistry, risks and best available strategies for remediation. *ISRN Ecology, 2011*, 1–20. Available from https://doi.org/10.5402/2011/402647.

Xing, Y., Wang, J., Xia, J., Liu, Z., Zhang, Y., Du, Y., & Wei, W. (2019). A pilot study on using biochars as sustainable amendments to inhibit rice uptake of Hg from a historically polluted soil in a Karst region of China. *Ecotoxicology and Environmental Safety, 170*, 18–24. Available from https://doi.org/10.1016/j.ecoenv.2018.11.111, http://www.elsevier.com/inca/publications/store/6/2/2/8/1/9/index.htt.

Xu, Q., Xu, Q., Zhu, H., Li, H., Yin, W., Feng, K., Wang, S., & Wang, X. (2022). Does biochar application in heavy metal-contaminated soils affect soil micronutrient dynamics? *Chemosphere, 290*.

Xue, Y., Gao, B., Yao, Y., Inyang, M., Zhang, M., Zimmerman, A. R., & Ro, K. S. (2012). Hydrogen peroxide modification enhances the ability of biochar (hydrochar) produced from hydrothermal carbonization of peanut hull to remove aqueous heavy metals: Batch and column tests. *Chemical Engineering Journal, 200-202*, 673–680. Available from https://doi.org/10.1016/j.cej.2012.06.116.

Yang, S., Sun, L., Sun, Y., Song, K., Qin, Q., Zhu, Z., & Xue, Y. (2023). Towards an integrated health risk assessment framework of soil heavy metals pollution: Theoretical basis, conceptual model, and perspectives. *Environmental Pollution, 316*. Available from https://doi.org/10.1016/j.envpol.2022.120596, https://www.journals.elsevier.com/environmental-pollution.

Yu, H. Y., Cui, J. H., Qiao, J. T., Liu, C. P., & Li, F. B. (2018). Principle and technique of arsenic and cadmium pollution control in paddy field. *Journal of Agro-Environment Science, 37*(7), 1418–1426. Available from https://doi.org/10.11654/jaes.2018-0730, http://www.aes.org.cn/nyhjkxxben/ch/reader/create_pdf.aspx?file_no = 20180714&year_id = 2018&quarter_id = 7&falg = 1.

Zaharia, C. (2011). Assessing the impact of some industrial and transport activities on soil by the global pollution index. *Environmental Engineering and Management Journal, 10*(3), 387–391. Available from https://doi.org/10.30638/eemj.2011.056.

Zama, E. F., Reid, B. J., Arp, H. P. H., Sun, G. X., Yuan, H. Y., & Zhu, Y. G. (2018). Advances in research on the use of biochar in soil for remediation: a review. *Journal of Soils and Sediments, 18*(7), 2433–2450. Available from https://doi.org/10.1007/s11368-018-2000-9, http://www.springerlink.com/content/1439-0108.

Zhang, H., Zhu, Y., Wang, S., Zhao, S., Nie, Y., Liao, X., Cao, H., Yin, H., & Liu, X. (2022). Contamination characteristics of energetic compounds in soils of two different types of military demolition range in China. *Environmental Pollution, 295*118654. Available from https://doi.org/10.1016/j.envpol.2021.118654.

Zhang, J., Wu, S., Xu, Z., Wang, M., Man, Y. B., Christie, P., Liang, P., Shan, S., & Wong, M. H. (2019). The role of sewage sludge biochar in methylmercury formation and accumulation in rice. *Chemosphere, 218*, 527–533. Available from https://doi.org/10.1016/j.chemosphere.2018.11.090, http://www.elsevier.com/locate/chemosphere.

Zimmerman, A. R. (2010). Abiotic and microbial oxidation of laboratory-produced black carbon (biochar). *Environmental Science and Technology, 44*(4), 1295–1301. Available from https://doi.org/10.1021/es903140c.

Chapter 7

Biochars for the remediation and repurposing of postmining landscapes and metalliferous substrates: applications and future perspectives

Willis Gwenzi[1,2,3]

[1]*Formerly Alexander von Humboldt Fellow and Guest Full Professor, Leibniz-Institut für Agrartechnik und Bioökonomie e.V. (ATB), Potsdam, Germany,*
[2]*Formerly Alexander von Humboldt Fellow and Guest Full Professor, Grassland Grassland Science and Renewable Plant Resources, Faculty of Organic Agricultural Sciences, Universität Kassel, Witzenhausen, Germany,* [3]*Biosystems and Environmental Engineering Research Group, Marlborough, Harare, Zimbabwe*

Chapter outline

7.1 Introduction	109
7.2 The case for biochar-based land remediation	110
7.2.1 The rationale and context	110
7.2.2 Biochar feedstocks and production systems	110
7.3 The Nature and extent of contaminated lands	111
7.3.1 Postmining landscapes	111
7.3.2 Metal-contaminated lands	112
7.3.3 Serpentinitic geological systems	112
7.3.4 Sludge and wastewater-amended soils	113
7.4 Biochar-based remediation of contaminated lands	113
7.4.1 Revegetation of postmining landscapes	113
7.4.2 Metal-contaminated soils	114
7.4.3 Toxic geogenic contaminants in serpentines	116
7.4.4 Sludge and wastewater-amended soils	116
7.4.5 Mechanisms of biochar remediation of mine wastes and metalliferous substrates	117
7.4.6 Design of biochar-based remediation systems	118
7.5 Future research and perspectives	119
7.5.1 Increasing Africa's research footprint on biochar-based remediation systems	119
7.5.2 Long-term behavior and fate of contaminants	119
7.5.3 Remediation of organic contaminants	120
7.5.4 Biochar-based extraction and recovery systems for essential elements	120
7.5.5 Large-scale pilot field studies	120
7.5.6 Technical and economic feasibility studies	120
7.5.7 Repurposing postmining landscapes as biomass sources for a circular bioeconomy	120
7.5.8 Metal-enriched biomass from metalliferous substrates as a unique biomass feedstock	121
7.5.9 Building Africa' biochar research capacity	121
7.5.10 The need for biochar research funding and collaboration	121
7.6 Conclusions and outlook	121
References	122

7.1 Introduction

Contaminated lands, such as postmining landscapes and metalliferous substrates, pose significant human and ecological health risks. Contaminants include inorganic pollutants (toxic metals, radionuclides, asbestos), organic pollutants (pharmaceuticals, pesticides), and microbiological contaminants (pathogens). These contaminants originate from anthropogenic sources (industrial activities) and geogenic sources (naturally enriched geological systems like serpentines). Due to their environmental health risks, remediation of these lands has garnered significant public and research attention.

Remediation techniques include in-situ and ex-situ methods, classified into bioremediation, physical methods (thermal treatment), and chemical treatment. The choice of method depends on remediation objectives, contaminant nature, and feasibility. Conventional methods can be expensive and energy-intensive, while bioremediation is often slow. Recent interest has focused on sustainable, low-cost, and renewable biochar-based remediation systems.

Biochar, a carbon-rich material produced via biomass pyrolysis, has demonstrated effectiveness in removing inorganic, organic, and microbiological contaminants in aqueous systems. Studies have shown biochar's capability to remove toxic metals, radionuclides, anionic contaminants, pharmaceuticals, pesticides, and pathogens. Motivated by this, biochar-based systems are now used for contaminated lands and soils.

Evidence shows biochar's effectiveness in revegetation of postmining landscapes, stabilization of metals in contaminated soils, and removal of microbial and nutrient contaminants from sludge-amended lands. Despite this, comprehensive reviews of biochar applications for land remediation are lacking. There is a need for perspectives on biochar's potential in diverse contaminated lands, including postmining landscapes, geogenic contaminant-enriched geological systems, and sludge-amended soils. Biochar-based systems offer a sustainable technology for remediating these lands and mitigating associated health risks.

This chapter aims to: (1) provide an overview of contaminated lands, (2) summarize biochar applications for land remediation, and (3) present future perspectives and knowledge gaps. Fig. 7.1 summarizes biochar-based remediation systems, including feedstocks, pyrolysis methods, nature of contaminated lands, remediation mechanisms, and outputs.

7.2 The case for biochar-based land remediation

7.2.1 The rationale and context

The application of biochar-based systems for the remediation of contaminated lands is motivated by several factors. First, a diverse range of contaminated lands exists, including postmining landscapes, metal-contaminated soils, sludge and wastewater-amended soils, and geological systems enriched with toxic geogenic contaminants. Second, current remediation methods, such as physical and chemical techniques, are often expensive and knowledge-intensive. Consequently, implementing such advanced and costly remediation technologies in resource-poor settings poses significant challenges.

Given the low-cost and renewable nature of biochar-based remediation systems, they have the potential to overcome these challenges. Large quantities of cheap and readily available biowastes and biomass feedstocks for biochar production exist in many countries, including those in Africa (Duku et al., 2011; Gwenzi et al., 2015). For example, estimates show that Zimbabwe alone produces approximately 9.9 million tons per year of biochar feedstock, capable of producing 3.5 million tons per year of biochar (Gwenzi et al., 2015). Comparable quantities of biochar feedstocks have also been estimated in other African countries, such as Ghana (Duku et al., 2011).

Currently, such biowastes are improperly disposed of in nonsanitary waste dumps or burnt in open spaces. This improper disposal releases toxic contaminants, posing human and ecological health risks, and greenhouse gases, contributing to climate change. These conditions create an ideal environment for biochar-based remediation systems. However, the application of biochar technology for the remediation of vast tracts of contaminated lands remains under-exploited.

This chapter presents the nature and extent of contaminated lands, principles, and potential applications of biochar-based remediation systems, including biochar production.

7.2.2 Biochar feedstocks and production systems

Biochar is considered a cost-effective and easy-to-produce material for the remediation of contaminated land. Its production hinges on two key factors: (1) the availability of cheap, renewable biomass feedstocks and (2) the use of appropriate pyrolysis systems. Potential biomass sources include manures, sewage sludges, forestry residues (e.g., litter and twigs), invasive plants, crop residues (e.g., maize stover and rice husks), and municipal solid wastes (Fig. 7.1). Detailed discussions on various feedstocks are available in earlier reviews (Duku et al., 2011; Gwenzi et al., 2015).

Pyrolysis systems range from small-scale manual setups to large-scale automated systems (Gwenzi et al., 2015). Low-cost small-scale designs include Kon Tiki kilns, pit kilns, and household pyrolytic or biochar cookstoves (Fig. 7.1). Large-scale systems include screw/auger systems, fluidized-bed reactors, and rotary and drum kilns (Gwenzi et al., 2015; Brown, 2012; Fig. 7.1). Detailed descriptions of these systems, including operating conditions, are available in earlier reviews (Gwenzi et al., 2015; Sohi et al., 2009; Brown, 2012). The choice of feedstock and pyrolysis system depends on cost, technology availability, operational expertise, and the quantity of biochar needed.

Field-scale applications often require large quantities of biochar, which can be bulky due to their low density. To minimize high transport costs, biochar feedstocks and pyrolysis systems should be located near the application site. In regions with limited technological development and financial resources, systems should ideally be easy to design,

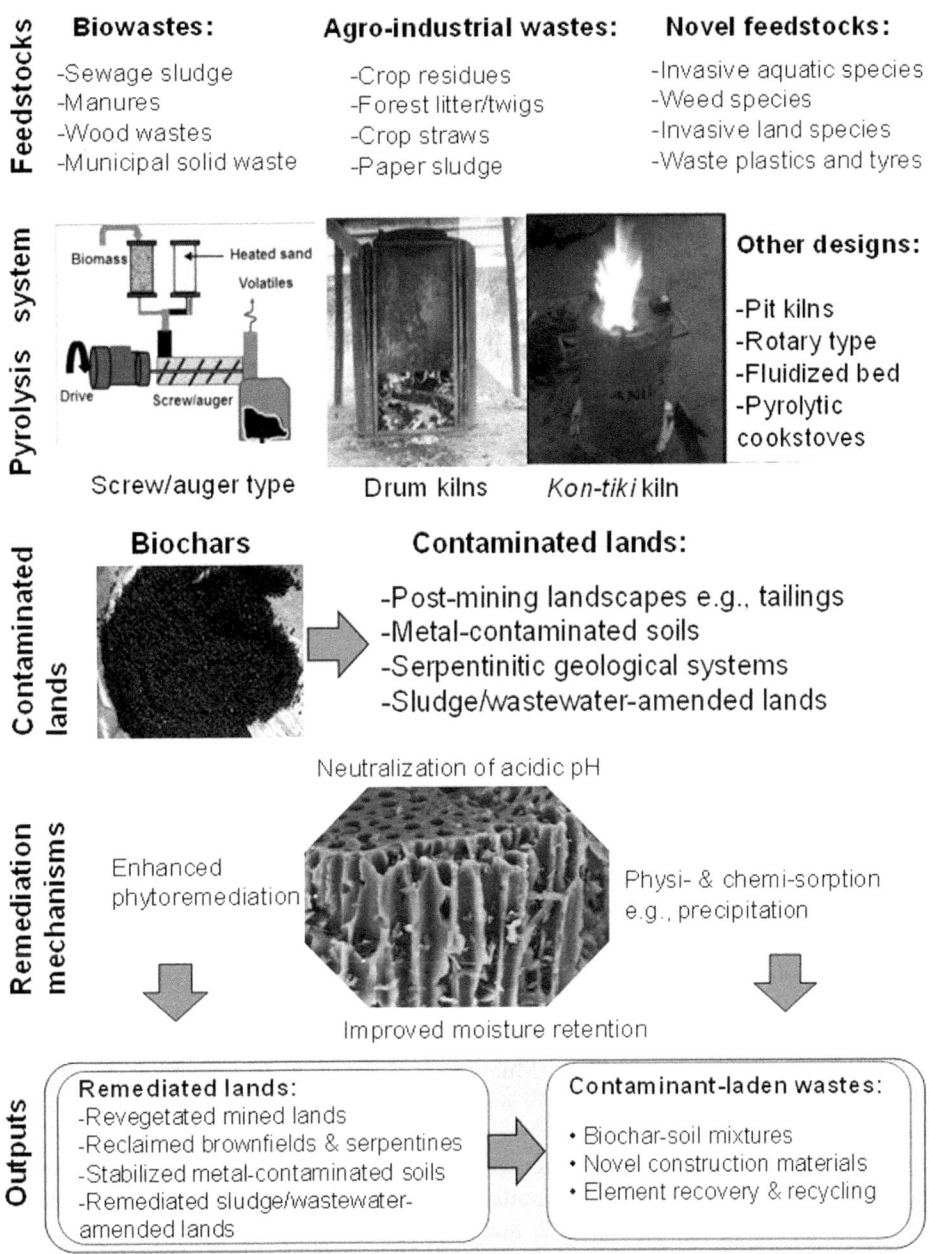

FIGURE 7.1 Biochar-based remediation systems depicting feedstocks, pyrolysis methods, nature of contaminated lands, remediation mechanisms, and outputs (Gwenzi et al., 2015). *Images of pyrolysis systems obtained with permission from Gwenzi, W., Chaukura, N., Mukome, F. M., Machado, S., Nyamasoka, B. (2015). Biochar production and applications in sub-Saharan Africa: Opportunities, constraints, risks and uncertainties.* Journal of Environmental Management, *150, 250–261.*

construct, and operate, with low installation and maintenance costs. They should also ideally function without a reliable electric power supply and produce the required quantities within a reasonable time.

7.3 The nature and extent of contaminated lands

7.3.1 Postmining landscapes

Geological systems rich in minerals such as gold, platinum, coal, diamonds, copper, and nickel are found globally, including in Africa (Festin et al., 2019; Venkateswarlu et al., 2016). These resources attract significant investment, leading to extensive mining and mineral processing operations. However, mining disrupts ecosystems and generates large quantities of waste, including tailings and waste rock. Postmining landscapes are prevalent worldwide and in Africa, including Zimbabwe, South Africa, and Zambia (Festin et al., 2019). For instance, South Africa alone has approximately 6,150 abandoned mines covering about 321 km^2 (Venkateswarlu et al., 2016). In Zambia's Copperbelt Province, mining wastes cover around 30,438 ha (Sikaundi, 2016). These figures, combined with the extensive history of artisanal and large-scale mining, indicate significant areas of severely degraded postmining landscapes.

The extent and severity of human and ecological health risks associated with these degraded landscapes remain poorly understood due to a lack of surveillance data. Risks are particularly high in low-income regions like Africa due to factors such as (Gwenzi, 2020): (1) the use of less advanced mining and processing methods, (2) reliance on small-scale and informal miners using hazardous chemicals (e.g., mercury) that are banned in developed countries, and (3) weakly enforced environmental and occupational health regulations.

Moreover, the mining of highly toxic minerals such as asbestos, which have been banned in developed countries is still prevalent in several low-income countries especially in Africa (e.g., Zimbabwe) (Gwenzi, 2020). Thus, African communities are more likely to be exposed to human health risks associated with severely degraded postmining landscapes than their developed counterparts.

Without proper rehabilitation, postmining landscapes pose significant human and ecological health risks. Yet the rehabilitation and restoration of severely degraded postmining landscapes is currently limited. For example, a recent review of severely degraded postmining landscapes in Africa showed that only about 20 studies existed on the subject in 2018 (Festin et al., 2019). The same study showed that out of nearly 56 countries in Africa, existing studies were drawn from just 8 countries (i.e., the Democratic Republic of Congo, South Africa, Zambia, Kenya, Zimbabwe, Ghana, Rwanda, and Morocco). The lack of remediation could probably be indicative of the lack of appropriate technology, or that current efforts have not been successful. The application of biochars for the remediation of postmining landscapes has been reported in developed countries such as the USA (Ippolito et al., 2017) and Italy (Fellet et al., 2011, 2014). However, studies on the application of biochars for the remediation of postmining landscapes in Africa are still lacking. Here, the potential applications of biochars in revegetation, phytostabilization and immobilization of contaminants on postmining landscapes are discussed.

7.3.2 Metal-contaminated lands

Apart from postmining landscapes, multimetal contamination of soils is also prevalent in several countries. Metal-contaminated soils have been reported in various nations, including Zambia, Zimbabwe, and Nigeria (Mungazi & Gwenzi, 2019; Odekunle et al., 2018; Ogundiran, Mekwunyei & Adejumo, 2018). Trace metals of particular public health concern include lead, mercury, cadmium, zinc, and chromium due to their toxicities and potential for bioaccumulation and biomagnification along trophic levels. Sources of these metals include industrial solid wastes (e.g., mine and battery wastes), effluents, and particulate emissions (Mungazi & Gwenzi, 2019; Ippolito et al., 2017; Odekunle et al., 2018; Ogundiran et al., 2018). Metals are emitted from sources such as smelters, industrial production plants, and combustion systems like furnaces and crematoria.

Once in soils, these metals can spread into surface and groundwater systems, which may serve as drinking water sources. Metals in soils can be taken up and bioaccumulated by plants, including edible crops. Consumption of metal-contaminated foods can transfer these toxins into the human food chain. Human exposure to these metals can occur through ingestion of contaminated food and water, and inhalation of contaminated particulates. The health risks associated with exposure to toxic metals are well-documented (Adriano, 2001). Remediation strategies for metal-contaminated soils include bioremediation, phytoremediation, and stabilization. Biochars have shown potential to enhance these remediation processes. Recent studies have explored the use of biochars for the remediation of metal-contaminated soils in Africa (Odekunle et al., 2018; Ogundiran et al., 2018) (e.g., Nigeria,) and other regions (Ippolito et al., 2017). This section summarizes the literature on biochar use for metal-contaminated soil remediation and the mechanisms involved.

7.3.3 Serpentinitic geological systems

Serpentinitic geological systems, or serpentines, and their precursor ultramafic rocks are geochemically enriched with toxic geogenic contaminants that pose significant human and ecological health risks. These systems form through the hydrothermal alteration of ultramafic rocks, which are the precursors for serpentines (Oze et al., 2008; Sleep et al., 2004). The three main serpentine minerals are lizardite, antigorite, and chrysotile (Gwenzi, 2020). Serpentines are distributed patchily but are found in several parts of Africa, often associated with various minerals and mining operations

(Gwenzi, 2020). They have been reported in regions including (1) Southern Africa (e.g., Great Dyke in Zimbabwe, Zambia, South Africa) (Cooper, 2002; Holwell et al., 2017; Stripp et al., 2006); (2) Central Africa (e.g., Cameroon) (Dzemua & Gleeson, 2012; Dzemua et al., 2011); (3) East Africa (e.g., Mozambique, Ethiopia, Tanzania, Kenya) (Blades et al., 2019; Odhiambo & Howarth, 2002; Prochaska & Pohl, 1983; Tadesse & Allen, 2005); and (4) North Africa (e.g., Egypt, Morocco, Tunisia) (Ater et al., 2000; Azer & Khalil, 2005; Gahlan et al., 2006; Kort et al., 2015). In Zimbabwe, serpentines are prominent in the Great Dyke region (Prendergast & Wilson, 1989; Stribrny et al., 2000; Wilson & Prendergast, 1989; Oberthür et al., 1998; Oberthür et al., 2002). Chrysotile asbestos, once extensively mined in the Shabani-Mashava mine, is also found in the Great Dyke (e.g., Ethel Mine) and in over 60 sites across Masvingo, Insiza, Gwanda, Mberengwa, and Shurugwi (Gwenzi, 2020; Mugumbate et al., 2001; ZGS, 2019).

Serpentines exhibit unique geochemistry with high concentrations of toxic geogenic contaminants, including: (1) redox-active trace metals such as iron (Fe), chromium (Cr), nickel (Ni), zinc (Zn), manganese (Mn), and cobalt (Co); and (2) naturally occurring asbestos in the form of chrysotile. The elevated chromium levels in serpentines derive from chromium-bearing minerals or spinels like magnesiochromite ($MgCr_2O_4$) and chromite ($FeCr_2O_4$) (Oze et al., 2004, 2008).

Redox-active trace metals (Fe, Cr, Ni, Zn, Mn, Co) undergo redox reactions that produce highly reactive oxygen species or radicals (Gwenzi, 2020; Papanikolaou & Pantopoulos, 2005; Puntarulo, 2005). These radicals can oxidize cell biomolecules, such as lipid membranes and DNA, leading to oxidative stress, carcinogenicity, and genotoxicity (Papanikolaou & Pantopoulos, 2005; Puntarulo, 2005). Chrysotile asbestos, a fibrous mineral found in high concentrations in serpentines, is a human carcinogen (WHO, 2014). The human health risks and toxicity mechanisms of chrysotile are well-documented (Oury et al., 2014; Stayner et al., 2013; WHO, 2014). A comprehensive review of toxic geogenic contaminants in serpentines, including their occurrence, behavior, exposure, and health risks, is available (Gwenzi, 2020).

Mining, mineral processing, agriculture, construction, and natural weathering and erosion release these toxic contaminants into the environment. The widespread presence of serpentines and associated mining activities pose potential health risks. To mitigate human exposure, biochars can be used to immobilize toxic trace metals and facilitate the chemical degradation of chrysotile asbestos. Detailed discussions on these mechanisms and case studies are available in the literature (Bandara, Herath, Kumarathilaka, Hseu, et al., 2017, Bandara, Herath, Kumarathilaka, Seneviratne, et al., 2017, Herath et al., 2015, 2017; Kumarathilaka & Vithanage, 2017). This section summarizes the use of biochar for these applications, including chrysotile asbestos degradation.

7.3.4 Sludge and wastewater-amended soils

Industrial processes, including wastewater treatment, often generate large quantities of sludge and wastewater/effluents, which are frequently disposed of via land application. In low-income countries, particularly in urban and peri-urban areas near centralized wastewater treatment plants, the land application of sewage sludge and the reuse of raw or partially treated wastewater in agriculture is common due to limited advanced wastewater treatment systems (Abel, 2014, Gwenzi & Munondo, 2008a, 2008b). Sewage sludges and wastewater contain a mixture of inorganic, organic, and microbial contaminants. Thus, their application can introduce various contaminants into the soil system. Inorganic contaminants may include trace metals and nutrients (Gwenzi & Munondo, 2008a, 2008b). Organic contaminants in these soils may include natural organic matter, conventional pollutants like pesticides, and emerging contaminants such as pharmaceuticals, endocrine-disrupting compounds, and industrial chemicals, which have gained attention in Africa and elsewhere (Gwenzi & Chaukura, 2018; Gwenzi, 2022). Microbial contaminants include indicator organisms and human pathogens from fecal contamination (Gwenzi & Munondo, 2008a). Once in the soil, these contaminants can be mobilized and spread into surface and groundwater systems through processes like infiltration, leaching, groundwater recharge, and surface and subsurface runoff. Contaminants in sludge and wastewater-amended soils can pose significant health risks. Biochars can act as effective barriers to attenuate and immobilize these contaminants. Methods for using biochars to remediate contaminants in sludge and wastewater-amended soils are discussed in subsequent sections.

7.4 Biochar-based remediation of contaminated lands

7.4.1 Revegetation of postmining landscapes

Postmining landscapes, such as tailings and waste rock dumps, present significant challenges for plant establishment and growth (Anawar et al., 2015). Vegetation establishment is often hindered by adverse physico-chemical, hydraulic, and hydrological conditions. Physico-chemical constraints include: (1) extreme substrate pH, such as strong acidity in acidic pyritic substrates (Mungazi & Gwenzi, 2019) and highly alkaline conditions in bauxite residues (Gwenzi, Hinz, et al., 2011; Gwenzi, Veneklaas et al., 2011); (2) phytotoxic elements; and (3) high salinity from dissolved solutes. Physical constraints include poorly weathered substrates and high bulk densities that restrict root development. Hydraulic and hydrological constraints may involve low moisture retention, low plant-available water, and extremely high hydraulic conductivity (Gwenzi, Hinz, et al., 2011). Due to their unique properties, biochars can help overcome several of these constraints. For instance, biochars, with their high organic carbon content, can improve soil structure and increase the water retention capacity of substrates like mine tailings and waste rock.

Biochars can enhance the revegetation of postmining landscapes through the following processes: (1) improving substrate physico-chemical properties, including moisture retention and soil structure (Abel et al., 2013; Sun & Lu, 2014); (2) ameliorating soil acidity in acid-generating mine substrates (Ippolito et al., 2017); (3) improving soil fertility, nutrient retention, and reducing leaching (Gwenzi et al., 2015; Novak et al., 2009); and (4) reducing the bioavailability and phytotoxicity of toxic elements (Gwenzi et al., 2016; Park et al., 2011). In turn, vegetation stabilizes postmining landscapes through phytostabilization, which involves root reinforcement and protection of the soil against water and wind erosion. Vegetation can also take up, translocate, and bioaccumulate elements in aboveground biomass, a process known as phytoextraction. Phytoextraction helps reduce the concentration of bioavailable elements in the substrate. However, studies on agricultural soils show that the effects of biochars on soil physico-chemical properties are inconsistent (Werner et al., 2019). For instance, significant reductions in nutrient leaching have been reported in soils subjected to normal fertilization without sludge or wastewater irrigation (Laird et al., 2010; Major, Rondon, et al., 2012; Major, Steiner, et al., 2012). Conversely, a study in Ghana found that biochar had no effect on nutrient leaching in urban agricultural systems with wastewater irrigation due to high nutrient loading rates (Werner et al., 2019). Thus, the impact of biochars on physico-chemical properties and plant uptake of elements likely depends on both biochar and soil characteristics, as well as sludge and wastewater management practices.

Biochars can be used to ameliorate soil acidity in acid-generating mine tailings and waste rock, thereby promoting plant establishment and growth. Increasing pH also reduces metal bioavailability in postmining landscapes and other metal-contaminated substrates (Park et al., 2011). Biochars are often inherently alkaline due to the presence of carbonates, $-COO-$, and $-O-$ functional groups (Ippolito et al., 2017; Yuan et al., 2011). Additionally, engineered or designer biochars can be specifically designed to have alkaline pH values through manipulation of pyrolysis conditions or chemical activation (Gwenzi et al., 2017). For example, pyrolysis of lodgepole pine and tamarisk at 300°C–500°C produced biochars with alkaline pH values of 9.1 and 10.4, respectively. Application of these biochars to mine tailings with initial pH values ranging from 3.97 to 5.44 significantly increased the substrate pH by more than 1 unit. Given that pH is expressed on a logarithmic scale, this represents a more than 10-fold reduction in acidity. Significant pH increases following biochar application have been reported in several studies focusing on contaminated lands (Gwenzi et al., 2016; Kelly et al., 2014; Zhu et al., 2015). The acid-neutralizing capacity of biochars suggests they could serve as an alternative to acid-neutralizing chemicals like lime. However, the required biochar application rates to achieve desirable pH levels for plant growth and contaminant immobilization must be determined on a case-by-case basis.

7.4.2 Metal-contaminated soils

Human and ecological health risks associated with metal-contaminated soils are often linked to the bioavailable and bioaccessible fractions of metals, rather than their total concentrations. Immobilizing metals by transforming them from bioavailable and bioaccessible forms to insoluble adsorbed or particulate phases can potentially reduce health risks. Despite the widespread prevalence of metal-contaminated lands, literature on biochar application for their remediation is limited compared to other regions. Notable exceptions include: (1) a pot experiment investigating the effects of sewage sludge biochar on metal bioavailability and plant uptake in agricultural soils in Zimbabwe (Gwenzi et al., 2016); and (2) studies on metal immobilization by biochars applied to soils contaminated from industrial sources in Nigeria (Ndor et al., 2016; Odekunle et al., 2018; Ogundiran, Mekwunyei & Adejumo, 2018). For example, in Nigeria, biochar-based methods for remediating metal-contaminated soils have included: (1) biochar alone (Ndor et al., 2016); (2) biochar combined with other organic amendments like compost (Ogundiran et al., 2018) and (3) biochar, compost, and phytoremediation (Ogundiran et al., 2018). These studies demonstrate that biochars can immobilize metals, reducing their bioavailability and potential ecotoxicity. The findings from these African studies align with results from research conducted elsewhere (e.g., USA, Italy) (Fellet et al., 2011; Ippolito et al., 2017; Uchimiya et al., 2010, 2011).

Outside Africa, the remediation of metal-contaminated soils using biochars is relatively well-documented (Fellet et al., 2011; Ippolito et al., 2012, 2017). Several studies report significant reductions in metal bioavailability following biochar application to contaminated soils (Fellet, Marmiroli & Marchiol, 2014; Ippolito et al., 2017). For example, reductions in bioavailable Cd, Cu, Pb, and Zn were observed in mine tailings from Colorado and Idaho after applying 5%, 10%, and 15% biochar (Ippolito et al., 2017; Fig. 7.2). The reduction in metal bioavailability was more pronounced at higher biochar application rates (10%, 15%) compared to 5%. Another study found that pelletized biochars from Douglas fir tree (Abies sp.), orchard pruning residues, and manure applied at various rates (0%, 1.5%, and 3% w/w) to mine tailings significantly reduced the bioavailability of Zn, Cd, and Pb (Fellet, Marmiroli & Marchiol, 2014). Significant reductions in metal bioavailability were also observed with orchard pruning residue biochar at application rates of 0%, 1%, 5%, and 10% w/w (Fellet et al., 2011). On soils contaminated with Cd, Pb, and Zn from a smelting facility, biochars derived from miscanthus (Miscanthus sinensis) straw significantly increased soil pH, attributed to the dissolution of hydroxides, oxides, and carbonates from the biochars (Houben & Sonnet, 2015). The same study observed that biochar reduced Cd and Zn concentrations in the exchangeable pool and increased their concentrations in the carbonate pool, similar to other studies (e.g., Zhu et al., 2015). These findings illustrate the capacity of biochars to sorb metals in soils, reducing their bioavailability and potential ecotoxicity. Three mechanisms could explain the reduction of metal bioavailability in biochar-amended substrates: (1) physicochemical sorption of metals on biochars due to their high surface areas and cation exchange capacity, and (2) pH-dependent adsorption and precipitation caused by the alkaline pH of biochars (Beesley & Marmiroli, 2011; Beesley, Moreno-Jiménez & Gomez-Eyles, 2010; Ippolito et al., 2012; Lu et al., 2012). On vegetated metal-contaminated substrates, phytoextraction by plants—entailing uptake, translocation, and subsequent bioaccumulation—can also reduce bioavailable metal concentrations (Ogundiran et al., 2018).

However, mixed results have been reported in some studies investigating the effects of biochars on metal behavior in contaminated soils. For example, Kelly et al. (2014) applied biochars derived from lodgepole pine killed by beetles at rates of 0%, 10%, 20%, and 30% w/w to two mine soils.

In the same study, biochar significantly reduced the concentrations of Zn, Pb, Cu, and Cd in leachate from one mine soil but increased the concentrations of Zn and Cd in the second mine soil. The increased concentrations of Cd and Zn in the second mine soil were attributed to two factors (Kelly et al., 2014): (1) biochar had no significant effect on soil pH, resulting in no substantial pH-induced metal precipitation and adsorption, and (2) the second mine soil had inherently higher concentrations of Zn and Cd compared to the first mine soil. In another study, the addition of varying amounts of biochar derived from willow (Salix sp.) to smelter-impacted soils immobilized Cu and Pb through sorption (Trakal et al., 2011). However, no changes were observed in the concentrations of Cd and Zn in this study. Thus, the capacity of biochars to immobilize and reduce metal bioavailability appears to depend on: (1) the inherent physicochemical properties of the metal-contaminated soils, particularly pH and initial metal concentrations, and (2) biochar properties, including its acid-neutralization capacity, which depends on the concentrations of soluble acid-neutralizing species, particularly calcium and magnesium oxides, hydroxides, and carbonates.

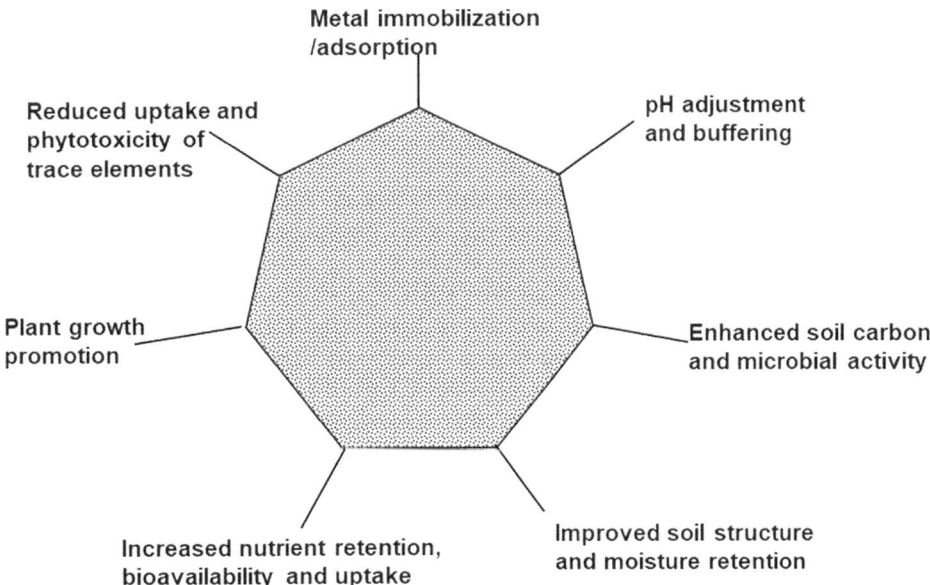

FIGURE 7.2 Overview of mechanisms of biochar enhancement of remediation and rehabilitation of post-mining landscapes and metalliferous substrates.

However, the long-term behavior and fate of metals immobilized by biochars remain poorly understood. Additionally, due to their predominantly alkaline pH, biochars may have limited capacity to remediate metal bioavailability and toxicity in alkaline metal-contaminated substrates such as bauxite residues. Bauxite residues are characterized by highly alkaline pH conditions and salinity (Gwenzi, Hinz, et al., 2011; Gwenzi, Veneklaas et al., 2011). These highly alkaline conditions induce micronutrient deficiencies and restrict root growth in bauxite residues (Gwenzi, 2020). To date, few studies have investigated the capacity of biochars to remediate alkaline metal-contaminated substrates like bauxite residues.

7.4.3 Toxic geogenic contaminants in serpentines

Similar to other organic amendments, biochars can also be used to stabilize and immobilize contaminants such as toxic trace metals in serpentines. The mechanisms involved are similar to those discussed earlier for metal-contaminated lands, such as mine tailings. The use of biochars for the remediation of toxic metals and chrysotile asbestos in serpentine soils and mine wastes like tailings and waste rock dumps is potentially promising, though only a few studies have investigated this aspect. Notable exceptions include studies conducted in Sri Lanka, which showed that biochar effectively immobilizes and reduces the bioavailability of toxic metals in serpentine soils (Bandara, Herath, Kumarathilaka, Hseu, et al., 2017; Bandara, Herath, Kumarathilaka, Seneviratne, et al., 2017; Herath et al., 2015, 2017; Kumarathilaka & Vithanage, 2017). In these studies, significant reductions in metal bioavailability were observed in biochar-amended serpentine soils compared to unamended soils (controls).

Evidence indicates that strong organic acids can induce the chemical degradation of asbestos, including chrysotile (Holmes & Lavkulich, 2014; Holmes et al., 2012). Organic acids cause the loss of magnesium from asbestos fibers, altering the microstructure and surface charge of asbestos from positive to negative (Holmes & Lavkulich, 2014; Holmes et al., 2012; Seshan, 1983). The loss of magnesium under acidic conditions promotes the dissolution and degradation of chrysotile asbestos through loss of crystallinity (Seshan, 1983). Therefore, applying biochars to chrysotile-rich serpentines and the subsequent release of organic acids could promote asbestos degradation. This approach is consistent with literature recommendations (Holmes et al., 2012). For instance, Holmes et al. (2012) recommended that organic amendments such as compost, sawdust, and peat can effectively reclaim serpentines and minimize environmental pollution. Biochars or mixtures of biochars with other organic amendments could enhance the degradation reaction kinetics of chrysotile through organic acid release. Thus, applying biochars to remediate both toxic metals and chrysotile asbestos presents opportunities for improving serpentines. However, further research is needed to assess the large-scale feasibility of such remediation at a pilot scale.

7.4.4 Sludge and wastewater-amended soils

There are several options for leveraging biochar-based remediation systems in the disposal of industrial sludges and wastewaters. The need to dispose of large quantities of sludges and wastewaters has led to a global increase in the land application of these materials (Drechsel et al., 2010). Sludges, such as biowastes from industrial and wastewater treatment processes, can be used as feedstock for biochar production. Converting raw biowaste/sludge into biochar stabilizes and reduces the mobility of contaminants, including toxic metals (Gwenzi et al., 2016). For instance, Gwenzi et al. (2016) compared soil applications of raw sewage sludge with its biochar and found that biochar significantly reduced the bioavailability and plant uptake of toxic metals more effectively than the raw sludge. Thus, land application of biochars potentially reduces contaminant mobility and transport into aquatic systems compared to raw sewage and industrial sludges. This reduced mobility is attributed to the high content of stabilized carbon, surface area, and adsorption sites in biochars (Gwenzi et al., 2017). Besides organic and inorganic contaminants, using sewage sludges and wastewaters in agriculture is also constrained by potential human health risks due to the transfer of pathogenic organisms into the human food chain (Drechsel et al., 2010). Therefore, pyrolysis of sewage sludges and other biowastes, such as manures, is crucial for inactivating pathogenic organisms like protozoa, helminths, viruses, and bacteria often found in these materials (Gwenzi et al., 2016).

Biochars produced from sewage sludges and other biomass feedstocks can be applied to soils to simulate remediation systems similar to soil aquifer treatment. Soil aquifer treatment involves passing contaminated water or wastewater through porous soil media, which removes contaminants via adsorption and filtration (Abel, 2014). In this context, biochars can be incorporated into soils to act as reactive media, enhancing contaminant removal and reducing their leaching into groundwater systems. Research in urban areas in Ghana showed that biochars effectively removed contaminants and reduced microbial load in wastewater (Akoto-Danso et al., 2019). Therefore, this contaminant removal capability can be extended to contaminated lands or soils. Biochars will be incorporated into the soil at a predetermined application rate, either alone or as part of a mixture with other organic amendments such as sludges. The biochar or biochar/sludge mixture will then immobilize and remove contaminants in sludge as well as those applied through wastewater irrigation. To date, the capacity of biochars to reduce the bioavailability and plant uptake of contaminants in sludge-amended soils (Gwenzi et al., 2016) and wastewater intended for irrigation (Werner et al., 2019; Kaetzl et al., 2018) has been demonstrated experimentally in Africa. However, large-scale field applications of this concept are still lacking. The use of biochar as a reactive filter media in soil aquifer systems is intriguing. However, literature on the design and evaluation of biochar-based remediation systems resembling soil aquifer treatment systems is scarce in Africa and elsewhere. Further research is needed to address this knowledge gap.

7.4.5 Mechanisms of biochar remediation of mine wastes and metalliferous substrates

Several mechanisms account for the capacity of biochar to remediate postmining landscapes and metalliferous substrates. A detailed discussion of these mechanisms and processes is presented in the literature on soils and mine substrates (Kelly et al., 2014; Kim et al., 2024; Major, Rondon, et al., 2012; Major, Steiner, et al., 2012; Mohamed et al., 2017; Penido et al., 2019; Schulz, Dunst & Glaser, 2013, Lehmann et al., 2011). These mechanisms account for the application of biochars in:

1 **Mine Tailings Rehabilitation**: Application of biochar to mine tailings can immobilize contaminants, improve soil properties, and promote the establishment of vegetation.
2 **Phytoremediation Enhancement**: Biochar can be used in combination with phytoremediation strategies to enhance the growth of metal-tolerant plants that can extract or stabilize contaminants.
3 **Contaminant Filtration**: Biochar can be used in constructed wetlands or as a filter medium to capture contaminants from leachate and runoff from mine tailings.
4 **Remediation of Metal-Enriched Serpentine Soils**: biochars have been used to remediate metal-enriched serpentine soils as discussed elsewhere in this chapter.

Here, an overview of the mechanisms is presented (Fig. 7.2).

7.4.5.1 Immobilization of trace metals

Biochar immobilizes trace metals through two primary mechanisms:

- **Adsorption**: Biochar's porous structure and large surface area allow it to adsorb heavy metals from mine tailings. Functional groups on biochar surfaces, such as carboxyl, hydroxyl, and phenolic groups, can bind heavy metals, reducing their mobility and bioavailability.
- **Complexation and Precipitation**: Biochar can facilitate the formation of stable complexes or precipitates with heavy metals, thereby immobilizing them and preventing leaching into the environment.

7.4.5.2 pH modification

Most biochars have an alkaline pH and a strong pH buffering capacity due to the presence of ash content, which can neutralize acidic mine tailings and substrates. This pH adjustment can precipitate metals as hydroxides and carbonates, thereby reducing their solubility and mobility. Biochar's capacity to neutralize acidity may also reduce the phytotoxicity of some trace elements, such as aluminum.

7.4.5.3 Soil structure improvement

Biochar is a carbon-rich biomaterial consisting of stable carbon forms that can persist in soils for centuries, helping to sequester carbon. This, in turn, can improve soil structure via enhanced aggregation:

- **Aggregation**: Biochar can improve soil structure by promoting the formation of soil aggregates. This enhanced structure improves water infiltration and retention, reduces erosion, and increases root penetration.
- **Porosity and water-holding capacity**: The high porosity of biochar can increase the water-holding capacity of soils, which is particularly beneficial in arid environments where mine tailings are often located.

7.4.5.4 Nutrient retention, bioavailability, and uptake

Biochar can enhance the cation exchange capacity (CEC) of soils, improving their ability to retain and bioavailability of essential nutrients to plants. This is particularly important in mine substrates, such as tailings, which are often nutrient-deficient. Moreover, nutrient-enriched biochar can act as a slow-release fertilizer, gradually supplying nutrients such as potassium, phosphorus, and nitrogen to plants, thereby enhancing plant establishment and growth on mine tailings and metalliferous substrates.

7.4.5.5 Enhancement of microbial activity

Biochar provides a habitat for beneficial soil microorganisms, which can aid in the degradation of organic pollutants and the transformation of heavy metals into less toxic forms. Biochar can enhance the growth of mycorrhizal fungi and other beneficial microbes that form symbiotic relationships with plants, improving nutrient uptake and plant growth in contaminated soils.

7.4.5.6 Plant growth promotion

By improving soil structure and nutrient availability, biochar can promote healthier root development and overall plant growth. Biochar can help plants tolerate environmental stresses, such as heavy metal toxicity and drought, by improving soil conditions and reducing the bioavailability of toxic metals. Collectively, this may promote the revegetation of mine wastes and metalliferous substrates.

7.4.5.7 Reduction of toxic metal uptake by plants

Biochar may reduce toxic metal uptake and phytotoxicity via two mechanisms:

- **Root barrier formation**: Biochar can form a barrier around plant roots, reducing the uptake of toxic metals and preventing them from entering the food chain.
- **Chelation and complexation**: Biochar can chelate metals in the soil, forming complexes that are less likely to be absorbed by plants.

However, further work is required to understand the dominant process between enhanced bioavailability and uptake versus reduced uptake in biochar-amended substrates.

In summary, biochar offers a multifaceted approach to the rehabilitation and remediation of mine tailings and metalliferous substrates. By immobilizing trace metals, improving soil structure, enhancing microbial activity, and promoting plant growth, biochar can play a crucial role in mitigating the environmental impact of mining activities and restoring degraded landscapes. However, most of these mechanisms are inferred from studies focusing on soils. Therefore, there is a need for studies focusing on the role of various mechanisms on mine wastes and metalliferous substrates. This is because substrates such as mine tailings are artificial and lack the properties of typical soils (Gwenzi, 2021). For example, several studies conducted on mine substrates show that they have unique hydraulic properties and substrate chemistry, leading to equally unique vegetation attributes, including shallow root systems and low water uptake and transpiration (Gwenzi, 2010; Gwenzi, Hinz, et al., 2011; Gwenzi, Veneklaas et al., 2011; Gwenzi et al., 2012; Gwenzi, Hinz, et al., 2014; Gwenzi, Musarurwa, et al., 2014). These unique properties call for dedicated studies to understand how biochar enhances the remediation of mine tailings and metalliferous substrates.

7.4.6 Design of biochar-based remediation systems

The application of biochar-based systems for the remediation of contaminated lands in Africa and elsewhere appears quite promising. As discussed in previous sections, the remediation mechanisms in biochar-based systems depend on several factors. This includes the physico-chemical and micromorphology of the biochars, including pH and acid-neutralizing capacity, surface area, and charge (Gwenzi et al., 2015, 2017). The physico-chemical properties of the contaminated media, including the nature, speciation, and concentrations of the contaminant are also important. Because biochars have predominantly negatively charged surfaces, they have a high capacity to remove positively charged contaminants such as metals. Therefore, a detailed understanding of these factors and the remediation mechanisms is critical in the design and application of biochar-based remediation systems. However, unlike conventional remediation methods based on physico-chemical methods, detailed procedures for the design and implementation of biochar-based systems are still lacking. For example, a comprehensive database on the optimum application rates of biochars for the various remediation applications is still lacking. Moreover, detailed feasibility studies are required to estimate the costs and benefits of biochar-based remediation systems relative to alternative competing systems. The need for detailed feasibility studies, including considerations of transport costs, coupled with the requirement to produce biochars close to the point of application are often overlooked in literature discussing biochar-based remediation systems. Similarly, it remains unclear whether biochars should be applied on a once-off basis or frequently at a pre-determined application interval to achieve its remediation efficacy. This lack of design procedures is not only limited to the remediation of contaminated lands. Rather, this has also been discussed in the case of biochar-based water and wastewater treatment systems. Therefore, the development of detailed designs and the associated design parameters represent the next research frontier in biochar-based remediation systems. Such designs should take into account the remediation objectives, costs, and long-term performance of the system.

To date, the bulk of literature on the application of biochars for the remediation of contaminated lands is limited to studies investigating its capacity and efficacy to immobilize contaminants (Ahmad et al., 2016). The production and transport costs associated with the application of biochars for large-scale remediation of contaminated are rarely addressed in existing literature. The studies that have attempted to directly address this aspect are limited to those that have used low-cost and readily available feedstocks such as biowastes (e.g., orchard prunings; Fellet et al., 2011). Similarly, the use of biomass from plants/trees killed by pests and pathogens as biochar feedstocks also represents another effort to address the cost aspect. Examples include the use of beetle-killed lodgepole pine or willows as biochars feedstock in the USA (Kelly et al., 2014; Trakal et al., 2011), and biomass from invasive plant species such as tamarisk (Ippolito et al., 2017). The use of such novel feedstocks is particularly attractive if they are available in close proximity to the point of remediation application of the biochars. This will reduce the high putative transport costs associated with biochars given their low densities and high application rates in the order of tonnes per hectare (Ippolito et al., 2017). The rationale for using novel feedstocks such as biomass from trees/plants killed by pathogens or pests has been discussed in an earlier study (Ippolito et al., 2017). Similarly, the use of other novel feedstocks such as invasive and weed plant species has been discussed (Gwenzi et al., 2015; Ippolito et al., 2017). In the case of Africa, these novel feedstocks include invasive and noxious plants species and weeds such as *Lantana camara*, and aquatic ones such as water hyacinth (*Eichhornia crassipes*) (Gwenzi et al., 2015). In Africa this has been demonstrated in the case of biochars derived from invasive aquatic species such as water hyacinth (*Eichhornia crassipes*) used for the remediation of metal-contaminated water and wastewater (Gwenzi, Hinz, et al., 2014, Gwenzi, Musarurwa, et al., 2014). As reported in the USA (Ippolito et al., 2017), such invasive species are often characterized by high seed and biomass productivity, which make them ideal candidates for use as biochar feedstocks. However, the development of a comprehensive database of well-characterized feedstocks, biochars, and pyrolysis systems, and their associated cost estimates for large-scale remediation of contaminated lands requires further investigation.

7.5 Future research and perspectives

The application of biochars for the remediation of contaminated lands is an emerging field of research. Hence, further research is required to address several knowledge gaps, including: (1) lack of large-scale pilot studies, (2) the long-term behavior and fate of contaminants in biochar-remediation systems, (3) technical and economic feasibility, and (4) extraction, recovery and recycling of essential elements. The need to address the critical shortage of research capacity and funding in Africa is also discussed.

7.5.1 Increasing Africa's research footprint on biochar-based remediation systems

The bulk of the evidence on the applications of biochars for the remediation of contaminated lands is drawn from other regions such as Asia, Europe, and North America. By comparison, barring a few studies, the contribution of the African research community to the global research on biochar-based remediation of contaminated lands remains low. Therefore, both fundamental and applied research based on laboratory and field experiments are required to validate the research findings reported in other regions.

7.5.2 Long-term behavior and fate of contaminants

Current literature on biochar-based remediation of contaminants lands is dominated by short-term studies. For example, toxic metals are nonbiodegradable, hence biochar-based remediation systems only transfer the metals from one pool (i.e., mobile phase) to another (i.e., immobile or adsorbed phase). Moreover, biochars are highly dynamic, thus, their physico-chemical and microstructure change over time as they undergo aging. Yet the behavior, fate, and ecotoxicology of the contaminants in biochar-amended soils as biochars undergo aging over time remain poorly understood. This calls for further research investigating the long-term behavior, ecotoxicology, and fate of various contaminants in biochar-amended contaminated soils.

7.5.3 Remediation of organic contaminants

Barring data on metals, there is a paucity of studies investigating the remediation of lands contaminated with conventional and emerging organic contaminants. Thus, the potential application of biochars for the remediation organic contaminants in soils is based on inferential evidence drawn from studies conducted on aqueous systems. Therefore, the capacity of biochars to effectively remediate conventional and emerging contaminants in soil systems warrants further research. Such research should also investigate the removal mechanisms as well as the long-term behavior and fate of such contaminants.

7.5.4 Biochar-based extraction and recovery systems for essential elements

Several metals and rare earth elements occurring in contaminated lands are of economic value and, hence can be recovered and recycled. In this regard, biochars can be used to facilitate the (phyto)-extraction of such essential elements from contaminants soils, and their subsequent recovery and recycling. Such novel extraction, recovery, and recycling systems may reduce the need and demand to mine raw or pristine elements. However, the application of biochars in the extraction, recovery, and recycling of essential elements from contaminated media remains largely unexplored. Hence, the potentially novel application of biochars requires further investigation in light of increasing public and research interest in recycling and the circular economy.

7.5.5 Large-scale pilot field studies

Current literature on biochar-based land remediation is limited to laboratory and small-scale pilot studies, which provide limited evidence on the large-scale feasibility of the technology. The establishment and monitoring of large-scale pilot experiments to investigate the best biochar-based remediation technologies identified through laboratory and small-scale pilot studies is urgently required. Such studies will provide critical information on contaminant removal and behavior at environmentally relevant spatial and temporal scales. Such scale effects cannot be investigated in small-scale pilot and laboratory studies currently dominating the literature.

7.5.6 Technical and economic feasibility studies

Available studies on biochar-based remediation of contaminated land often overlook the technical and economic feasibility of the technology. Hence, large-scale pilot field studies investigating contaminant removal and behavior should be coupled with detailed technical and economic feasibility studies. In this regard, comparative investigations of the technical and economic feasibility of the biochar-based remediation systems to alternative remediation technologies are required. Depending on the feedstock and pyrolysis process, biochars may also contain potentially toxic chemicals such as metals and even poly-aromatic hydrocarbons. However, the human and ecological health risks of biochar-derived contaminants remain poorly understood. Hence comparative studies investigating biochar-derived contaminants to those in conventional biosolids such as manures and sludges are required.

7.5.7 Repurposing postmining landscapes as biomass sources for a circular bioeconomy

The application of biochar to remediate or rehabilitate postmining landscapes and other metalliferous substrates unsuitable for plant growth could provide a pathway for repurposing these areas for biomass production. This biomass can then be used as feedstock in various biomass conversion pathways and bioenergy production, contributing to a circular bioeconomy strategy. However, limited studies have explored the use of such postmining landscapes and metalliferous substrates as biomass sources for a circular bioeconomy. This highlights the need for further research on this aspect.

7.5.8 Metal-enriched biomass from metalliferous substrates as a unique biomass feedstock

The production of biomass on biochar-amended postmining landscapes and metalliferous substrates is likely to result in biomass enriched with trace metals. Without proper harvesting and valorization, biochar-based remediation efforts may reintroduce metals back into the system. Therefore, it is essential to develop strategies to harvest and valorize the metal-enriched biomass. Potential applications of such biomass include:

- **Biomass-to-energy systems**: This includes anaerobic digestion, pyrolysis to produce biochar and incineration.
- **Biochar catalysis**: Metal-enriched biomass could serve as an excellent precursor material for the development of biochar catalysts.

However, these applications of metal-enriched biomass from contaminated sites have received limited research attention.

7.5.9 Building Africa' biochar research capacity

The limited contribution of Africa to global research on biochar-based remediation systems could reflect the critical shortage of research capacity in the field of biochar technology. Therefore, there is an urgent need to undertake capacity building in biochar research at various levels, including biochar production, characterization, applications, and evaluation. Such capacity-building should including training of researchers, technical staff, and undergraduate and postgraduate students. Key disciplines to be included in such training include chemical and processing engineering, energy engineering, environmental engineering, water and wastewater engineering, environmental sciences, materials science, agronomy, soil science, and socio-economics, among others.

7.5.10 The need for biochar research funding and collaboration

The lack of comprehensive research funding mechanisms in Africa in the form of grants is touted as one of the key factors hindering research and development in Africa. Hence, similar to other regions, African countries should take an active role in funding biochar research. Such funding is critical for the development of critical research infrastructure including engineering and analytical laboratories. Moreover, research funding is critical for supporting capacity-building including training of postgraduate research students.

Several isolated research groups in African countries such as Nigeria, Zimbabwe, South Africa, Kenya, and Zambia work on biochar, in some cases in collaboration with international partners in developed countries. However, these research groups work independently, with limited collaboration and coordination of the research. Accordingly, Africa's biochar research community fails to harness the collective benefits associated with collaboration, including the sharing of scarce research resources such as scientific expertise, laboratory facilities, and funding. Hence, better coordination and collaboration are required among the African research community in order to attract local, regional, and international funding. Such collaborations should include the development and submission of joint research proposals as well as co-supervision of postgraduate research students.

7.6 Conclusions and outlook

The present chapter discussed the potential application of biochars as amendments for the remediation of various contaminated lands prevalent in several countries. First, the potential methods for the production of biochars via pyrolysis were highlighted. The dominant contaminated lands include; (1) postmining landscapes such as waste rock dumps and mine tailings, (2) serpentinitic geological systems naturally enriched in toxic metals and chrysotile asbestos, and (3) metal-contaminated soils, including those amended with industrial and wastewater sludges. Biochars have several potential applications in the remediation of contaminated lands. This includes the amelioration of physico-chemical constraints posed by contaminated substrates during revegetation/rehabilitation of postmining landscapes. Biochar application to serpentinitic geological systems reduces contaminant mobility and bioavailability through phytostabilization and immobilization of toxic metals. Biochars also phytostabilize and phytoextract metals from metal-contaminated soils. Application of biochars to serpentinitic geological systems may also release organic acids that induce the chemical breakdown and subsequent dissolution of chrysotile asbestos. The potential to use biochar for the phytoextraction and subsequent recovery of metals was also discussed, although limited evidence exists on such applications. In summary, the key mechanisms involved in contaminant removal in biochar-based remediation systems include: (1)

phytostabilization and phytoextraction, (2) adsorption via chemisorption and physisorption, (3) neutralization of acidity and subsequent contaminant removal via precipitation. Several limitations and knowledge gaps pertaining to biochar-based land remediation systems were discussed. These include: (1) limited pilot-scale applications demonstrating large-scale economic and technical feasibility, (2) lack of long-term data on the behavior, fate, and ecotoxicity of contaminants in biochar-remediation systems, (3) the potential release of phytotoxic compounds from biochars into the soil system and their ecological and ecotoxicological impacts, and (4) limited evidence demonstrating the capacity to use biochars for metal phytoextraction and subsequent recovery from metal-contaminated soils. Therefore, further research is required to address these limitations and knowledge gaps before the potential benefits of biochar application in the remediation of contaminated lands are realized.

References

Abel, C. D. T. (2014). Soil Aquifer Treatment Assessment and Applicability of Primary Effluent Reuse in Developing Countries. UNESCO-IHE.

Abel, S., Peters, A., Trinks, S., Schonsky, H., Facklam, M., & Wessolek, G. (2013). Impact of biochar and hydrochar addition on water retention and water repellency of sandy soil. *Geoderma, 202*, 183–191.

Adriano, D. C. (2001). *Trace elements in terrestrial environments: Biogeochemistry, bioavailability and risk of metals.* Springer Verlag.

Ahmad, M., Lee, S. S., Lee, S. E., Al-Wabel, M. I., Tsang, D. C. W., & Ok, Y. S. (2016). Biochar-induced changes in soil properties affected by immobilization/mobilization of metals/metalloids in contaminated soils. *Journal of Soils and Sediments.* Available from https://doi.org/10.1007/s11368-015-1339-4.

Akoto-Danso, E. K., Manka'abusi, D., Steiner, C., Werner, S., Häring, V., Nyarko, G., Marschner, B., Drechsel, P., & Buerkert, A. (2019). Agronomic effects of biochar and wastewater irrigation in urban crop production of Tamale, northern Ghana. *Nutrient Cycling in Agroecosystems, 115*(2), 231–247. Available from https://doi.org/10.1007/s10705-018-9926-6, http://www.wkap.nl/journalhome.htm/1385-1314.

Anawar, H. M., Akter, F., Solaiman, Z. M., & Strezov, V. (2015). Biochar: An emerging panacea for remediation of soil contaminants from mining, industry and sewage wastes. *Pedosphere, 25*(5), 654–665. Available from https://doi.org/10.1016/S1002-0160(15)30046-1, http://pedosphere.issas.ac.cn.

Ater, M., Lefèbvre, C., Gruber, W., & Meerts, P. (2000). A phytogeochemical survey of the flora of ultramafic and adjacent normal soils in North Morocco. *Plant and Soil, 218*(1-2), 127–135. Available from https://doi.org/10.1023/a:1014925007960, http://www.wkap.nl/journalhome.htm/0032-079X.

Azer, M. K., & Khalil, A. E. S. (2005). Petrological and mineralogical studies of Pan-African serpentinites at Bir Al-Edeid area, central Eastern Desert, Egypt. *Journal of African Earth Sciences, 43*(5), 525–536.

Bandara, T., Herath, I., Kumarathilaka, P., Hseu, Z. Y., Ok, Y. S., & Vithanage, M. (2017). Efficacy of woody biomass and biochar for alleviating heavy metal bioavailability in serpentine soil. *Environmental Geochemistry and Health, 39*(2), 391–401.

Bandara, T., Herath, I., Kumarathilaka, P., Seneviratne, M., Seneviratne, G., Rajakaruna, N., Vithanage, M., & Ok, Y. S. (2017). Role of woody biochar and fungal-bacterial co-inoculation on enzyme activity and metal immobilization in serpentine soil. *Journal of Soils and Sediments, 17*(3), 665–673.

Beesley, L., & Marmiroli, M. (2011). The immobilisation and retention of soluble arsenic, cadmium and zinc by biochar. *Environmental Pollution, 159*(2), 474–480. Available from https://doi.org/10.1016/j.envpol.2010.10.016.

Beesley, L., Moreno-Jiménez, E., & Gomez-Eyles, J. L. (2010). Effects of biochar and greenwaste compost amendments on mobility, bioavailability and toxicity of inorganic and organic contaminants in a multi-element polluted soil. *Environmental Pollution, 158*(6), 2282–2287. Available from https://doi.org/10.1016/j.envpol.2010.02.003.

Blades, M. L., Foden, J., Collins, A. S., Alemu, T., & Woldetinsae, G. (2019). The origin of the ultramafic rocks of the Tulu Dimtu Belt, western Ethiopia–do they represent remnants of the Mozambique Ocean? *Geological Magazine, 156*(1), 62–82. Available from https://doi.org/10.1017/S0016756817000802.

Brown, R. (2012). Biochar production technology. In J. Lehmann, & S. Joseph (Eds.), *Biochar for environmental management* (pp. 159–178). London: Routledge.

Cooper, G. R. C. (2002). Oxidation and toxicity of chromium in ultramafic soils in Zimbabwe. *Applied Geochemistry, 17*(8), 981–986. Available from https://doi.org/10.1016/s0883-2927(02)00014-8.

Drechsel, P., Scott, C. A., Raschid-Sally, L., Redwood, M., & Bahri, A. (2010). Wastewater irrigation and health: assessing and mitigating risk in low-income countries. IWMI/IDRC, IWMI/IDRC. Earthscan, London

Duku, M. H., Gu, S., & Hagan, E. B. (2011). Biochar production potential in Ghana – a review. *Renewable and Sustainable Energy Reviews., 15*, 3539–3551.

Dzemua, G. L., & Gleeson, S. A. (2012). Petrography, Mineralogy, and Geochemistry of the Nkamouna Serpentinite: Implications for the formation of the cobalt-manganese laterite deposit, southeast cameroon. *Economic Geology, 107*(1), 25–41. Available from https://doi.org/10.2113/econgeo.107.1.25.

Dzemua, G. L., Mees, F., Stoops, G., & Van Ranst, E. (2011). Micromorphology, mineralogy and geochemistry of lateritic weathering over serpentinite in Southeast Cameroon. *Journal of African Earth Sciences, 60*(1-2), 38–48. Available from https://doi.org/10.1016/j.jafrearsci.2011.01.011.

Fellet, G., Marchiol, L., Delle Vedove, G., & Peressotti, A. (2011). Application of biochar on mine tailings: Effects and perspectives for land reclamation. *Chemosphere, 83*(9), 1262–1267. Available from https://doi.org/10.1016/j.chemosphere.2011.03.053.

Fellet, G., Marmiroli, M., & Marchiol, L. (2014). Elements uptake by metal accumulator species grown on mine tailings amended with three types of biochar. *Science of The Total Environment, 468-469*, 598–608. Available from https://doi.org/10.1016/j.scitotenv.2013.08.072.

Festin, E. S., Tigabu, M., Chileshe, M. N., Syampungani, S., & Odén, P. C. (2019). Progresses in 974 restoration of post-mining landscape in Africa. *Journal of Forestry Research, 30*(2), 381–975, 396.

Gahlan, H. A., Arai, S., Ahmed, A. H., Ishida, Y., Abdel-Aziz, Y. M., & Rahimi, A. (2006). Origin of magnetite veins in serpentinite from the Late Proterozoic Bou-Azzer ophiolite, Anti-Atlas, Morocco: An implication for mobility of iron during serpentinization. *Journal of African Earth Sciences, 46*(4), 318–330. Available from https://doi.org/10.1016/j.jafrearsci.2006.06.003.

Gwenzi, W. (2010). Vegetation and soil controls on water redistribution on recently constructed ecosystems in water-limited environments- PhD Dissertation. University of Western Australia, Perth, Australia.

Gwenzi, W. (2020). Occurrence, behaviour, and human exposure pathways and health risks of toxic geogenic contaminants in serpentinitic ultramafic geological environments (SUGEs): A medical geology perspective. *Science of The Total Environment, 700*, 134622. Available from https://doi.org/10.1016/j.scitotenv.2019.134622.

Gwenzi, W. (2021). Rethinking restoration indicators and end-points for post-mining landscapes in light of novel ecosystems. *Geoderma, 387*, 114944.

Gwenzi, W. (Ed.), (2022). *Emerging Contaminants in the Terrestrial-Aquatic-Atmosphere Continuum: Occurrence, Health Risks and Mitigation*. Elsevier. Available from https://doi.org/10.1016/C2020-0-02112-2.

Gwenzi, W., & Chaukura, N. (2018). Organic contaminants in African aquatic systems: Current knowledge, health risks, and future research directions. *Science of the Total Environment, 619*, 1493–1514.

Gwenzi, W., Chaukura, N., Mukome, F. M., Machado, S., & Nyamasoka, B. (2015). Biochar production and applications in sub-Saharan Africa: opportunities, constraints, risks and uncertainties. *Journal of Environmental Management, 150*, 250–261. Available from https://doi.org/10.1016/j.jenvman.2014.11.027.

Gwenzi, W., Muzava, M., Mapanda, F., & Tauro, T. (2016). Comparative short-term effects of sewage sludge and its biochar on soil properties, maize growth and uptake of nutrients on a tropical clay soil in Zimbabwe. *Journal of Integrative Agriculture, 15*(6), 1395–1406. Available from https://doi.org/10.1016/52095-3119(15)61154-6.

Gwenzi, W., Chaukura, N., Noubactep, C., & Mukome, F. N. D. (2017). Biochar-based water treatment systems as a potential low-cost and sustainable technology for clean water provision. *Journal of Environmental Management, 197*, 732–749. Available from https://doi.org/10.1016/j.jenvman.2017.03.087, https://www.sciencedirect.com/journal/journal-of-environmental-management.

Gwenzi, W., Veneklaas, E. J., Bleby, T. M., Yunusa, I. A. M., & Hinz, C. (2012). Transpiration and plant water relations of evergreen woody vegetation on a recently constructed artificial ecosystem under seasonally dry conditions in Western Australia. *Hydrological Processes, 26*(21), 3281–3292. Available from https://doi.org/10.1002/hyp.8330.

Gwenzi, W., Hinz, C., Holmes, K., Philips, I. R., & Mullins, I. J. (2011). Field-scale spatial variability of saturated hydraulic conductivity on a recently constructed artificial ecosystem. *Geoderma, 166*(1), 43–56. Available from https://doi.org/10.1016/j.geoderma.2011.06.010.

Gwenzi, W., & Munondo, R. (2008a). Long-term impacts of pasture irrigation with treated sewage effluent on nutrient status of a sandy soil. *Nutrient Cycling in Agroecosystems, 82*, 197–207. Available from https://doi.org/10.1007/s10705-008-9181-3.

Gwenzi, W., & Munondo, R. (2008b). Long-term impacts of pasture irrigation with treated sewage effluent on shallow groundwater quality. *Water Science & Technology, 58*(12), 2443–2452. Available from https://doi.org/10.2166/wst.2008.583.

Gwenzi, W., Veneklaas, E., Holmes, K., Bleby, T. M., Phillips, I. R., & Hinz, C. (2011). Spatial analysis of root distribution on a recently constructed ecosystem in a water-limited environment. *Plant and Soil, 344*(1-2), 255–272. Available from https://doi.org/10.1007/s11104-011-0744-8.

Gwenzi, W., Hinz, C., Bleby, T. M., & Veneklaas, E. J. (2014). Transpiration and water relations of evergreen shrub species on an artificial landform for mine waste storage versus an adjacent natural site in semi-arid Western Australia. *Ecohydrology, 7*(3), 965–981.

Gwenzi, W., Musarurwa, T., Nyamugafata, P., Chaukura, N., Chaparadza, A., & Mbera, S. (2014). Adsorption of Zn^{2+} and Ni^{2+} in a binary aqueous solution by biosorbents derived from sawdust and water hyacinth (*Eichhornia crassipes*). *Water Science and Technology*, 1419–1427.

Herath, I., Kumarathilaka, P., Navaratne, A., Rajakaruna, N., & Vithanage, M. (2015). Immobilization and phytotoxicity reduction of heavy metals in serpentine soil using biochar. *Journal of Soils and Sediments, 15*(1), 126–138. Available from https://doi.org/10.1007/s11368-014-0967-4.

Herath, I., Iqbal, M. C. M., Al-Wabel, M. I., Abduljabbar, A., Ahmad, M., Usman, A. R., Ok, Y. S., & Vithanage, M. (2017). Bioenergy-derived waste biochar for reducing mobility, bioavailability, and phytotoxicity of chromium in anthropized tannery soil. *Journal of soils and sediments, 17*(3), 731–740.

Holmes, E. P., & Lavkulich, L. M. (2014). The effects of naturally occurring acids on the surface properties of chrysotile asbestos. *Journal of Environmental Science and Health - Part A Toxic/Hazardous Substances and Environmental Engineering, 49*(12), 1445–1452. Available from https://doi.org/10.1080/10934529.2014.928558, http://www.tandf.co.uk/journals/titles/10934529.asp.

Holmes, E. P., Wilson, J., Schreier, H., & Lavkulich, L. M. (2012). Processes affecting surface and chemical properties of chrysotile: Implications for reclamation of asbestos in the natural environment. *Canada Canadian Journal of Soil Science, 92*(1), 229–242. Available from https://doi.org/10.4141/CJSS2010-014, http://pubs.aic.ca/doi/pdf/10.4141/cjss2010-014.

Holwell, D. A., Mitchell, C. L., Howe, G. A., Evans, D. M., Ward, L. A., & Friedman, R. (2017). The Munali Ni sulfide deposit, southern Zambia: A multi-stage, mafic-ultramafic, magmatic sulfide-magnetite-apatite-carbonate megabreccia. *Ore Geology Reviews, 90*, 553–575. Available from https://doi.org/10.1016/j.oregeorev.2017.02.034, http://www.sciencedirect.com/science/journal/01691368.

Houben, D., & Sonnet, P. (2015). Impact of biochar and root-induced changes on metal dynamics in the rhizosphere of Agrostis capillaris and Lupinus albus. *Chemosphere, 139*, 644–651. Available from https://doi.org/10.1016/j.chemosphere.2014.12.036, http://www.elsevier.com/locate/chemosphere.

Ippolito, J. A., Berry, C. M., Strawn, D. G., Novak, J. M., Levine, J., & Harley, A. (2017). Biochars reduce mine land soil bioavailable metals. *Journal of Environmental Quality, 46*(2), 411–419. Available from https://doi.org/10.2134/jeq2016.10.0388, https://dl.sciencesocieties.org/publications/jeq/pdfs/46/2/411.

Ippolito, J. A., Strawn, D. G., Scheckel, K. G., Novak, J. M., Ahmedna, M., & Niandou, M. A. S. (2012). Macroscopic and molecular investigations of copper sorption by a steam-activated biochar. *Journal of Environmental Quality, 41*(4), 1150–1156. Available from https://doi.org/10.2134/jeq2011.0113.

Kaetzl, K., Lübken, M., Gehring, T., & Wichern, M. (2018). Efficient low-cost anaerobic treatment of wastewater using biochar and woodchip filters. *Water, 10*(7), 818.

Kelly, C. N., Peltz, C. D., Stanton, M., Rutherford, D. W., & Rostad, C. E. (2014). Biochar application to hardrock mine tailings: Soil quality, microbial activity, and toxic element sorption. *Applied Geochemistry, 43*, 35–48. Available from https://doi.org/10.1016/j.apgeochem.2014.02.003.

Kim, H. B., Kim, J. G., Alessi, D. S., & Baek, K. (2024). Temporal changes in the mobility of As, Pb, Zn, and Cu due to differences in biochar stability caused by lignin content. *Chemical Engineering Journal, 493*. Available from https://doi.org/10.1016/j.cej.2024.152567, https://www.sciencedirect.com/science/journal/13858947.

Kort, H. M., El Asmi, A. M., Ouazaa, N. L., Gasquet, D., & Saidi, M. (2015). Hydrothermal history in the eastern margin of Tunisia: inferred magmatic rocks alterations, new paragenesis and associated gas occurrences. *Arabian Journal of Geosciences, 8*(10), 8927–8942. Available from https://doi.org/10.1007/s12517-015-1836-1, http://www.springer.com/geosciences/journal/12517?cm_mmc = AD-_-enews-_-PSE1892-_-0.

Kumarathilaka, P., & Vithanage, M. (2017). Influence of Gliricidia sepium Biochar on Attenuate Perchlorate-Induced Heavy Metal Release in Serpentine Soil. *Journal of Chemistry, 2017*, 1–8. Available from https://doi.org/10.1155/2017/6180636.

Laird, D., Fleming, P., Wang, B., Horton, R., & Karlen, D. (2010). Biochar impact on nutrient leaching from a Midwestern agricultural soil. *Geoderma, 158*(3-4), 436–442. Available from https://doi.org/10.1016/j.geoderma.2010.05.012.

Lu, H., Zhang, W., Yang, Y., Huang, X., Wang, S., & Qiu, R. (2012). Relative distribution of Pb^{2+} sorption mechanisms by sludge-derived biochar. *Water Research, 46*(3), 854–862. Available from https://doi.org/10.1016/j.watres.2011.11.058, https://www.sciencedirect.com/science/journal/00431354.

Major, J., Rondon, M., Molina, D., Riha, S. J., & Lehmann, J. (2012). Nutrient leaching in a colombian savanna oxisol amended with biochar. *Journal of Environmental Quality, 41*, 1076–1086.

Major, J., Steiner, C., Downie, A., & Lehmann, J. (2012). Biochar effects on nutrient leaching. In *Biochar for Environmental Management* (pp. 303–320). Routledge.

Mohamed, B. A., Ellis, N., Kim, C. S., & Bi, X. (2017). The role of tailored biochar in increasing plant growth, and reducing bioavailability, phytotoxicity, and uptake of heavy metals in contaminated soil. *Environmental Pollution, 230*, 329–338. Available from https://doi.org/10.1016/j.envpol.2017.06.075, http://www.elsevier.com/inca/publications/store/4/0/5/8/5/6.

Mugumbate, F., Oesterlen, P. M., Masiyambiri, S., & Dube, W. (2001). Industrial minerals and rock deposits of Zimbabwe. *Mineral Resources Series, 27*.

Mungazi, A. A., & Gwenzi, W. (2019). Cross-layer leaching of coal fly ash and mine tailings to control acid generation from mine wastes. *Mine Water and the Environment, 38*(3), 602–616. Available from https://doi.org/10.1007/s10230-019-00618-0.

Ndor, E., Jayeoba, O., & Ogara, J. (2016). Effect of Biochar Amendment on Heavy Metals Concentration in Dumpsite Soil and their Uptake by Amaranthus (Amaranthus cruentus). *Journal of Applied Life Sciences International, 9*(1), 1–7. Available from https://doi.org/10.9734/jalsi/2016/29948.

Novak, J. M., Busscher, W. J., Laird, D. L., Ahmedna, M., Watts, D. W., & Niandou, M. A. S. (2009). Impact of biochar amendment on fertility of a southeastern coastal plain soil. *Soil Science, 174*(2), 105–112. Available from https://doi.org/10.1097/SS.0b013e3181981d9a.

Oberthür, T., Davis, D. W., Blenkinsop, T. G., & Höhndorf, A. (2002). Precise U-Pb mineral ages, Rb-Sr and Sm-Nd systematics for the Great Dyke, Zimbabwe – constraints on crustal evolution and metallogenesis of the Zimbabwe Craton. *Precambr. Res., 113*, 293–306.

Oberthür, T., Weiser, Th., Müller, P., Lodziak, Je., & Cabri, L. J. (1998). New observations on the distribution of platinum group elements (PGE) and minerals (PGM) in the MSZ at Hartley Mine, Great Dyke, Zimbabwe. In: Eighth International Platinum Symposium. *S. Afr. Instn. Mining and Metallurgy, Johannesburg, Symposium Series S, 18*, 293–296.

Odekunle, D., Olutona, G. O., Oshunsanya, S. O., & Fagbenro, J. A. (2018). Effect of organic-based amendments on soil chemical properties and stabilization of cadmium (Cd) and lead (Pb) in battery-waste contaminated soil. *BiocharTec Journal., 1*, 32–35.

Odhiambo, B. D. O., & Howarth, P. J. (2002). Chromium concentrations in geobotanical samples from West Pokot District, Kenya. *Environmental Geochemistry and Health, 24*(2), 111–122. Available from https://doi.org/10.1023/A:1014295022616.

Ogundiran, M. B., Mekwunyei, N. S., & Adejumo, S. A. (2018). Compost and biochar assisted phytoremediation potentials of Moringa oleifera for remediation of lead contaminated soil. *Journal of Environmental Chemical Engineering, 6*(2), 2206–2213. Available from https://doi.org/10.1016/j.jece.2018.03.025.

Oury, T. D., Sporn, T. A., & Roggli, V. L. (2014). *Pathology of Asbestos-Associated Diseases*. Berlin Heidelberg: Springer. Available from 10.1007/978-3-642-41193-9.

Oze, C., Fendorf, S., Bird, D. K., & Coleman, R. G. (2004). Chromium geochemistry in serpentinized ultramafic rocks and serpentine soils from the Franciscan complex of California. *American Journal of Science, 304*(1), 67–101. Available from https://doi.org/10.2475/ajs.304.1.67, http://www.ajsonline.org/.

Oze, C., Skinner, C., Schroth, A. W., & Coleman, R. G. (2008). Growing up green on serpentine soils: Biogeochemistry of serpentine vegetation in the Central Coast Range of California. *Applied Geochemistry, 23*(12), 3391–3403. Available from https://doi.org/10.1016/j.apgeochem.2008.07.014.

Papanikolaou, G., & Pantopoulos, K. (2005). Iron metabolism and toxicity. *Toxicology and Applied Pharmacology, 202*(2), 199–211. Available from https://doi.org/10.1016/j.taap.2004.06.021.

Park, J. H., Choppala, G. K., Bolan, N. S., Chung, J. W., & Chuasavathi, T. (2011). Biochar reduces the bioavailability and phytotoxicity of heavy metals. *Plant and Soil, 348*(1-2), 439–451. Available from https://doi.org/10.1007/s11104-011-0948-y.

Penido, E. S., Martins, G. C., Mendes, T. B. M., Melo, L. C. A., do Rosário Guimarães, I., & Guilherme, L. R. G. (2019). Combining biochar and sewage sludge for immobilization of heavy metals in mining soils. *Ecotoxicology and Environmental Safety, 172*, 326–333. Available from https://doi.org/10.1016/j.ecoenv.2019.01.110, http://www.elsevier.com/inca/publications/store/6/2/2/8/1/9/index.htt.

Prendergast, M. D., & Wilson, A. H. (1989). The Great Dyke of Zimbabwe - II: mineralization and mineral deposits. *Magmatic sulphides - the Zimbabwe volume, 21*–42.

Prochaska, W., & Pohl, W. (1983). Petrochemistry of some mafic and ultramafic rocks from the Mozambique Belt, northern Tanzania. *Journal of African Earth Sciences, 1*(3-4). Available from https://doi.org/10.1016/s0731-7247(83)80002-0, 183–191.

Puntarulo, S. (2005). Iron, oxidative stress and human health. *Molecular Aspects of Medicine, 26*(4-5), 299–312. Available from https://doi.org/10.1016/j.mam.2005.07.001.

Schulz, H., Dunst, G., & Glaser, B. (2013). Positive effects of composted biochar on plant growth and soil fertility. *Agronomy for Sustainable Development, 33*(4), 817–827. Available from https://doi.org/10.1007/s13593-013-0150-0.

Seshan, K. (1983). How are the physical and chemical properties of chrysotile asbestos altered by a 10-year residence in water and up to 5 days in simulated stomach acid? *Environmental Health Perspectives, 53*, 143–148. Available from https://doi.org/10.1289/ehp.8353143.

Sikaundi, G. (2016). *Copper mining industry in Zambia: environmental challenges* (p. 28) Zambia: Lusaka, Zambia: Environmental Council of.

Sleep, N. H., Meibom, A., Fridriksson, T., Coleman, R. G., & Bird, D. K. (2004). H 2 -rich fluids from serpentinization: Geochemical and biotic implications. *Proceedings of the National Academy of Sciences, 101*(35), 12818–12823. Available from https://doi.org/10.1073/pnas.0405289101.

Sohi, S., Lopez-Capel, E., Krull, E., & Bol, R. (2009). Biochar Climate Change and Soil: A Review to Guide Future Research. *CSIRO Land and Water Science Report Rep.* Australia: CSIRO, Glen Osmond.

Stayner, L., Welch, L. S., & Lemen, R. (2013). The worldwide pandemic of asbestos-related diseases. *Annual Review of Public Health, 34*, 205–216. Available from https://doi.org/10.1146/annurev-publhealth-031811-124704.

Stribrny, B., Wellmer, F.-W., Burgath, K.-P., Oberthür, T., Tarkian, M., & Pfeiffer, T. (2000). Unconventional PGE occurrences and PGE mineralization in the Great Dyke: metallogenic and economic aspects. *Mineralium Deposita, 35*(2-3), 260–280. Available from https://doi.org/10.1007/s001260050019.

Stripp, G. R., Field, M., Schumacher, J. C., Sparks, R. S. J., & Cressey, G. (2006). Post-emplacement serpentinization and related hydrothermal metamorphism in a kimberlite from Venetia, South Africa. *Journal of Metamorphic Geology, 24*(6), 515–534. Available from https://doi.org/10.1111/j.1525-1314.2006.00652.x.

Sun, F., & Lu, S. (2014). Biochars improve aggregate stability, water retention, and pore-space properties of clayey soil. *Journal of Plant Nutrition and Soil Science, 177*(1), 26–33. Available from https://doi.org/10.1002/jpln.201200639.

Tadesse, G., & Allen, A. (2005). Geology and geochemistry of the Neoproterozoic Tuludimtu Ophiolite suite, western Ethiopia. *Journal of African Earth Sciences, 41*(3), 192–211. Available from https://doi.org/10.1016/j.jafrearsci.2005.04.001.

Trakal, L., Komárek, M., Száková, J., Zemanová, V., & Tlustoš, P. (2011). Biochar application to metal-contaminated soil: Evaluating of Cd, Cu, Pb and Zn sorption behavior using single- and multi-element sorption experiment. *Plant, Soil and Environment, 57*(8), 372–380. Available from https://doi.org/10.17221/155/2011-pse.

Uchimiya, M., Lima, I. M., Thomas Klasson, K., Chang, S., Wartelle, L. H., & Rodgers, J. E. (2010). Immobilization of heavy metal ions (CuII, CdII, NiII, and PbII) by broiler litter-derived biochars in water and soil. *Journal of Agricultural and Food Chemistry, 58*(9), 5538–5544. Available from https://doi.org/10.1021/jf9044217.

Uchimiya, M., Klasson, K. T., Wartelle, L. H., & Lima, I. M. (2011). Influence of soil properties on heavy metal sequestration by biochar amendment: 1. Copper sorption isotherms and the release of cations. *Chemosphere, 82*, 1431–1437. Available from https://doi.org/10.1016/j.chemosphere.2010.11.050.

Venkateswarlu, K., Nirola, R., Kuppusamy, S., Thavamani, P., Naidu, R., & Megharaj, M. (2016). Abandoned metalliferous mines: ecological impacts and potential approaches for reclamation. *Reviews in Environmental Science and Biotechnology, 15*(2), 327–354. Available from https://doi.org/10.1007/s11157-016-9398-6, http://www.kluweronline.com/issn/1569-1705.

Werner, S., Akoto-Danso, E. K., Manka'abusi, D., Steiner, C., Haering, V., Nyarko, G., Buerkert, A., & Marschner, B. (2019). Nutrient balances with wastewater irrigation and biochar application in urban agriculture of Northern Ghana. *Nutrient Cycling in Agroecosystems, 115*(2), 249–262. Available from https://doi.org/10.1007/s10705-019-09989-w, http://www.wkap.nl/journalhome.htm/1385-1314.

WHO. (2014). World Health Organization *Chrysotile Asbestos*. Geneva: World Health Organization.

Yuan, I. H., Xu, R. K., & Zhang, H. (2011). The forms of alkalis in the biochar produced from crop residues at different temperatures. *Bioresource Technology, 102*(3), 3488–3497. Available from https://doi.org/10.1016/j.biortech.2010.11.018.

Wilson, A. H., Prendergast, M. D. (1989). The Great Dyke of Zimbabwe: I: tectonic setting, stratigraphy, petrology, structure, emplacement and crystallization. Magmatic sulphides. 3 August 1987, Harare, Zimbabwe. 1–20. IMM, London

ZGS (Zimbabwe Geological Survey), 2019. Minerals of Zimbabwe. Available at: http://www.mines.gov.zw/?q = minerals-zimbabwe. Accessed 23 August 2020.

Zhu, Q., Wu, J., Wang, L., Yang, G., & Zhang, X. (2015). Effect of biochar on heavy metal speciation of paddy soil. *Water, Air, and Soil Pollution, 226*(12). Available from https://doi.org/10.1007/s11270-015-2680-3, http://www.kluweronline.com/issn/0049-6979/.

Chapter 8

Biochar remediation of conventional and emerging organic contaminants in soils

Terrence Wenga[1,2], Miranda Mpeta[3] and Phenias Sadondo[1]

[1]Department of Soil Science and Environment, Faculty of Agriculture, Environment and Food Systems, University of Zimbabwe, Mount Pleasant, Harare, Zimbabwe, [2]Key Laboratory of Agro-Forestry Environmental Processes and Ecological Regulation of Hainan Province, School of Environmental Science and Engineering, Hainan University, Haikou, P.R. China, [3]Department of Environmental Engineering, School of Engineering Sciences and Technology, Chinhoyi University of Technology, Chinhoyi, Zimbabwe

Chapter outline

8.1 Introduction	127
8.2 Conventional organic contaminants	128
8.2.1 Types of conventional organic contaminants and their effects	128
8.2.2 Physico-chemical characteristics of conventional organic contaminants	129
8.2.3 Sources and pathways of conventional organic contaminants for contaminating the soil	129
8.3 Emerging organic contaminants	131
8.3.1 Pharmaceuticals and personal care products	131
8.3.2 Molecular structure of some pharmaceuticals and personal care products and contamination pathways	132
8.4 Biochar removal of conventional and emerging organic contaminants from contaminated soils	132
8.5 Biochar removal mechanisms of organic contaminants in soils	134
8.5.1 Adsorption mechanisms	134
8.5.2 Factors affecting adsorption efficiency	137
8.6 Behavior and fate of organic contaminants in soils	138
8.6.1 Adsorption and desorption	139
8.6.2 Leaching	139
8.6.3 Transformation and degradation	139
8.6.4 Persistence	140
8.6.5 Volatilization	140
8.6.6 Washed away by runoff and erosion	140
8.6.7 Accumulation and Bioavailability	140
8.7 Fate of biochar after organic contaminants removal	140
8.8 Biochar and other soil remediation methods	141
8.9 Challenges and opportunities of Biochar in removal of organic pollutants	141
8.9.1 Challenges of biochar removal of organic pollutants	141
8.9.2 Research opportunities	141
8.10 Conclusion	142
References	142

8.1 Introduction

Due to increased industrialization, manufacturing, farming, pharmaceuticals, and utilization of various chemicals, concerns regarding the release of organic contaminants (OCs) into the environment dramatically increased (Das et al., 2023; Sanderson et al., 2023). As the major sink for chemicals in the environment, the soil suffers much contamination from various pollutants, including organic chemicals (Huang et al., 2012). These organic chemicals are either conventional organic contaminants (COCs), which include pesticides and herbicides, or emerging organic contaminants (EOCs), which comprise a different group of thousands of chemical compounds, such as pharmaceuticals and personal care products (PPCPs), flame retardants, plasticizers, surfactants, metabolites, and industrial additives, etc. (García et al., 2020). The contaminants contain chemical compounds such as benzene, toluene, ethylbenzene, and xylenes, polychlorinated biphenyls (PCBs), polycyclic aromatic hydrocarbons (PAHs), for example, phenanthrene, naphthalene, and benzo[a]pyrene. The presence of COCs and EOCs in the soil results in contamination of groundwater and accumulation of pollutants in the plants and food systems, posing a great threat to numerous nontargeted living species and public health, as the contaminants are toxic, mutagenic, and carcinogenic (Pazos et al., 2010; Semple et al., 2003).

Because of their low degradability and recalcitrant properties, OCs resist chemical, biological, and photolytic breakdown. As a result, these contaminants remain in soils for long periods of time and penetrate various other environmental media. Hence, they migrate and spread over a large area while being active in the soil (Khalid et al., 2020). The nonbiodegradability of OCs results in an increase in the quantities of contaminant residues in the environment; for example, pesticides reach up to 16%−20% greater than the quality index for main agricultural products, rendering them unsafe for human consumption (Huang et al., 2012). The total PAHs in soils have been estimated to range from 0.001 to 300,000 mg/kg (Mukome et al., 2020). Chlorophenols are frequently used as synthetic byproducts in biocides, pesticides, dyes, and polymers, leading to widespread contamination affecting soil microbial populations, which in turn affects some soil biological processes such as nitrogen turnover and hence soil fertility (Shahzad et al., 2018). Moreover, OCs (e.g., PAHs and chlorophenols) harm or destroy the physiology of plant species (Haider et al., 2022). Therefore, the reclamation of contaminated soils to reduce contamination and diminish downstream damage has increasingly attracted worldwide attention and has become a research hotspot in environmental science.

Several approaches and strategies have been employed to decontaminate soil contaminated by organic pollutants, such as soil vapor extraction, soil flushing, soil washing, etc. (Anawar et al., 2015). Nevertheless, these traditional methods, when applied in situ, are generally costly and/or may pose secondary problems, such as soil erosion and fertility loss. Other environmentally friendly and cheap approaches like bioremediation, phytoremediation, and ecological remediation are currently gaining much attention (Li et al., 2019; Pouangam Ngalani et al., 2023; Sanderson et al., 2023). Biochar remediation, one of the most green bioremediation technologies, has been applied for the last two decades for the treatment of soils contaminated by OCs. It has the capability of immobilizing/degrading pesticides and other organic pollutants and rendering them harmless. This capability of biochar depends mainly on its porous structure and the presence of different kinds of functional groups, such as phenolic, hydroxyl, and carboxyl groups. These functional groups enable biochar to interact with organic pollutants primarily via the van der Waals attraction, hydrophobic bonding, hydrogen interactions, intermolecular forces, or ligand exchange (Khalid et al., 2020; Kuśmierz et al., 2016). The electronic associations enable biochar to adsorb both hydrophilic and hydrophobic organic compounds. Biochar adsorption of organic molecules is further influenced by the relative fraction of the carbonaceous content for adsorption and noncarbonized fractions for partition (Lan et al., 2022). The adsorption process competes with other coexisting organic compounds, while the partition is a noncompetitive and linear process (Haider et al., 2022). Moreover, the graphitic and semiquinone structures of biochar allow it to gain or release electrons as well as free radicals, thereby stimulating the breakdown and redox reactions with organic molecules (Lan et al., 2022). Also, the higher the carbonaceous content and high surface area of the biochar, the more effective the biochar is for remediating soil polluted with organic pollutants such as sulfamethazine (Teixidó et al., 2013). Low pyrolysis temperatures, around 400 °C, enhance the surface area of biochar, which improves the adsorption of organic molecules. The surface area of biochar for the adsorption of organic pollutants can further be boosted by coating with oxides such as Al_2O_3, $CaCO_3$, Fe_2O_3, $CaSO_4$, and SiO_2 (Goswami et al., 2022), which improves the efficiency of biochar for remediating contaminated soils. The advantages of biochar soil remediation include: (1) raising the soil organic carbon content and pH; (2) elevating the soil water-holding capacity; (3) reducing a total load of contaminants; (4) leading to increased yields from agricultural crops; (5) obstructing the uptake and accumulation of pollutants; and (6) lowering the cost (Cheng et al., 2020; Qiu et al., 2022).

Several good reviews are available on the influence and interactions of biochar with various types of organic pollutants, for example, pesticides and emerging organic pollutants in the soil as well as their fate (El-Naggar et al., 2018; Khalid et al., 2020; Qiu et al., 2022). Therefore, an overview highlighting the recent advances in the biochar application to soils for the remediation of both conventional and EOCs is a worthy contribution effort, and it is of paramount importance to better understand the mechanism of biochar removal of organic pollutants and their fate in the soils.

Therefore, this chapter aims to provide a comprehensive overview of different types of COCs and EOCs. In addition, the occurrence of COCs and EOCs in soils and their removal mechanism by biochar will be discussed. Moreover, the behavior and fate of organic pollutants in soil will be presented. Finally, the challenges and hotspots for future research directions are highlighted.

8.2 Conventional organic contaminants

8.2.1 Types of conventional organic contaminants and their effects

COCs, also known as synthetic or chemical organics, refer to the use of synthetic pesticides, herbicides, insecticides, fungicides, and nematicides, as well as other similar substances in agriculture and farming practices (Ruomeng et al.,

2023). These COCs include artificial chemical compounds designed to kill or control pests, insects, fungi, weeds, and other agricultural nuisances (Agboola et al., 2022). There are different mechanisms by which pesticides kill pests. However, these mechanisms are out of the scope of this chapter, but if the reader is interested in the mechanism for killing pests, they are referred to the work of (Ruomeng et al., 2023). Pesticides and herbicides, e.g., chlordane, cypermethrin, maneb, azoxystrobin, endosulfan, lindane, aldrin, atrazine, and heptachlor, are excessively used in agricultural farms. Accumulation and leaching of these chemicals in the soil pose a significant threat to human health and the ecosystem. Pesticides enter the underground water and the surface water via leaching, runoff, and accidental leakage, affecting the quality of water and, consequently, the health of people (Haider et al., 2022). Moreover, COCs result in a large range of changes in the molecular, morphological, and physiological structures of plants, adversely reducing the plant's productivity and pest resistance. Aside from agricultural pesticides, several other COCs are derived from various processes such as wood preservation, fire-retardant manufacture, gasoline stations, etc.

After the chemicals are applied, various residual particles of COCs are suspended in the environment. They react with nitrogen oxides, forming ozone and affecting air quality. They can be washed into water bodies, and eventually, they accumulate in the soil, causing environmental pollution, contamination of water sources, soil degradation, harm to beneficial organisms, toxic residues, and pose a potential health risk. A 2010 study by Bahlai et al. (2010) revealed that organic insecticides may have lower efficacy and similar or even greater negative impact than synthetic insecticides on several natural enemy species in laboratory studies and were more detrimental to biological control organisms in field experiments. Although it is claimed that, due to high selectivity, trace residues of pesticides have apparently no significant effects on nontarget species. Toxicological studies from vitro to vivo conducted by (Tan et al., 2023) revealed that pesticides that target various systems, such as the endocrine, nervous system, and reproduction, pose a great threat to nontarget organisms in many aspects after long-term accumulation. Although it is purported that COCs affect nontargeted organisms such as people, there is a lack of solid epidemiological evidence to support the effects on human health (Xiong et al., 2023). Nevertheless, the likelihood cannot be ruled out. Recently, Tan et al. (2023) analyzed the occurrence of 32 current-use pesticides in grown 10 edible plants and their ecological/human health risks. A total of six trace residues of pesticide, namely pyraclostrobin, carbendazim, methyl, acetamiprid, thiophanate phoxim, and imidacloprid, were detected in 72.7% of samples, with concentrations ranging from 0.0021 to 13.5 mg kg^{-1} implying that when consumed by people, definitely, they will cause some health issues.

Sophia A & Lima (2018) noted that most COCs could lead to lethal effects on life even at small concentrations. The pollution from COCs and inorganic contaminants such as pesticides, explosives, plastics, and lubricants may also affect humans and biodiversity in the environment, resulting in toxicity and ailments. In humans, the COCs can result in both acute and chronic health effects. COCs can cause a variety of health problems, including cancer, reproductive problems, neurological disorders, developmental delays, and endocrine disruption (Thakur & Pathania, 2019). In the environment, COCs can pollute air, water, and soil. COCs can bioaccumulate in the food chain, making them more concentrated in organisms at higher trophic levels. COCs can harm wildlife, including fish, birds, and mammals.

8.2.2 Physico-chemical characteristics of conventional organic contaminants

OCs enter the environment via multiple pathways, either as part of an industrial discharge or agriculture or as a component of runoff and exhibit a wide range of physical and chemical properties, as shown in Table 8.1. These properties influence their behavior and the extent to which the contaminant stays in the soil. Once in the soil, some contaminants may undergo natural biodegradation or photolysis breakdown. However, the complete mineralization of COCs is rare; they instead undergo biotransformation whereby there is an internal rearrangement of chemical bonds within COCs, resulting in changes to the chemical composition and molecular structure. Some compounds are bio-transformed into innocuous products that may then enter a particular plant's system. Other compounds can form daughter products that may be more or less toxic than the parent compound, while others may prove to be generally recalcitrant under the prevailing conditions and persist in one or more phases (Byrns, 2001).

8.2.3 Sources and pathways of conventional organic contaminants for contaminating the soil

Sources of COCs that eventually lead to the contamination of soils can be categorized into (1) point sources and (2) diffuse sources of pollution. Point sources originate from discrete locations that release the chemicals into the environment — soil system and can often be defined in a spatially discrete manner. The spatial extent of pollution is consequently more localized. Examples of point sources of pollution include industrial sites (such as gasoline stations, wood

TABLE 8.1 Physico-chemical properties of conventional organic contaminants.

Compound	Molecular weight	Vapor pressure (Pa)	Half-life $T_{1/2}$ (d)	Solubility (g/m^3)	Log K_{ow}
Benzene	78	12,700	5–16	1800	2.13
Toluene	92.15	2933	4–22	515	2.73
Xylene	106.18	880	7–28	178	3.13
Napthalene	128	11.3	1–20	32	3.36
Acenapthylene	152.2	3.866	12–102	3.93	3.7
Acenapthene	154.21	0.20664	102	3.42	4.0
Anthracene	178.24	0.002266	50–460	0.044	4.45
Pyrene	202	0.00033	210–1900	0.132	5.2
B[a]P	252.32	0.000000746	530	0.0038	6.06
Acetone	58.09	35,995	1–7	Inf.	−0.24
Dichloromethane	84.93	46,527	28	20,000	1.30
Dichloroethene	96.94	79,989	180	2250	1.84
Chlorobenzene	112.56	1560	150	500	2.84
1,2-dichlorobenzene	147	133.3	28–180	100	3.36
1,1-dichloroethane	98.96	23,997	154	5500	1.79
DDT	354.48	0.00002533	730–5708	0.0055	6.19
Dieldrin	380.9	0.00002373	175–1080	0.195	3.5
Dibutylphthalate	278.38	0.00133	1–23	13	5.6
2,3,7,8-TCDD	321.96	0.00000000149	590	0.00000791	6.72
Bis(2-ethylhexyl) phthalate	391	0.00086	23	0.34	7.30

Source: From Byrns, Geoff. (2001). The fate of xenobiotic organic compounds in wastewater treatment plants. *Water Research*, 35(10), 2523–2533. https://doi.org/10.1016/s0043-1354(00)00529-7.

preservation sites, manufacturing plants, and food processing plants), resource extraction (mining), waste disposal sites (landfill sites, industrial impoundments, farm waste lagoons), and buried septic tanks (Lapworth et al., 2012).

On the other hand, diffuse pollution originates from poorly defined, diffuse sources that generally occur over broad geographical scales. Examples of diffuse sources of pollution are the application of agricultural pesticides and herbicides, manufacturing sites of chemicals, explosive industry, military facilities, leakage from reticulated urban sewerage systems, and diffuse aerial deposition, etc. (Buerge et al., 2011). These types of pollution sources are characterized by coverage of larger spatial scales, lower environmental loading relative to point sources, and are poorly defined with less obvious/direct links that can be traced back to the polluter (Lapworth et al., 2012), henceforth, they continue posing a challenge to monitor, regulate, and evaluate their impact on the soil system. When released either from point sources or diffuse sources, there are multiple pathways in which COCs find their way into soils, as illustrated in Fig. 8.1.

Application of herbicides and manure in agricultural fields, as well as bio-solids from sewage sludge processing plants, as a way to enhance soil fertility and nutrient levels, which improve agricultural productivity, results in an increase and accumulation of contaminant residues in the soils (Clarke & Smith, 2011). Due to their fairly high concentrations in bio-solids as well as their relatively high solubility, halogenated hydrocarbons such as perfluorochemicals and polychlorinated alkanes further diffuse to contaminate underground water, and the concentrations detected in groundwater can be explained in terms of solubility and K_{ow} shown in Table 8.1.

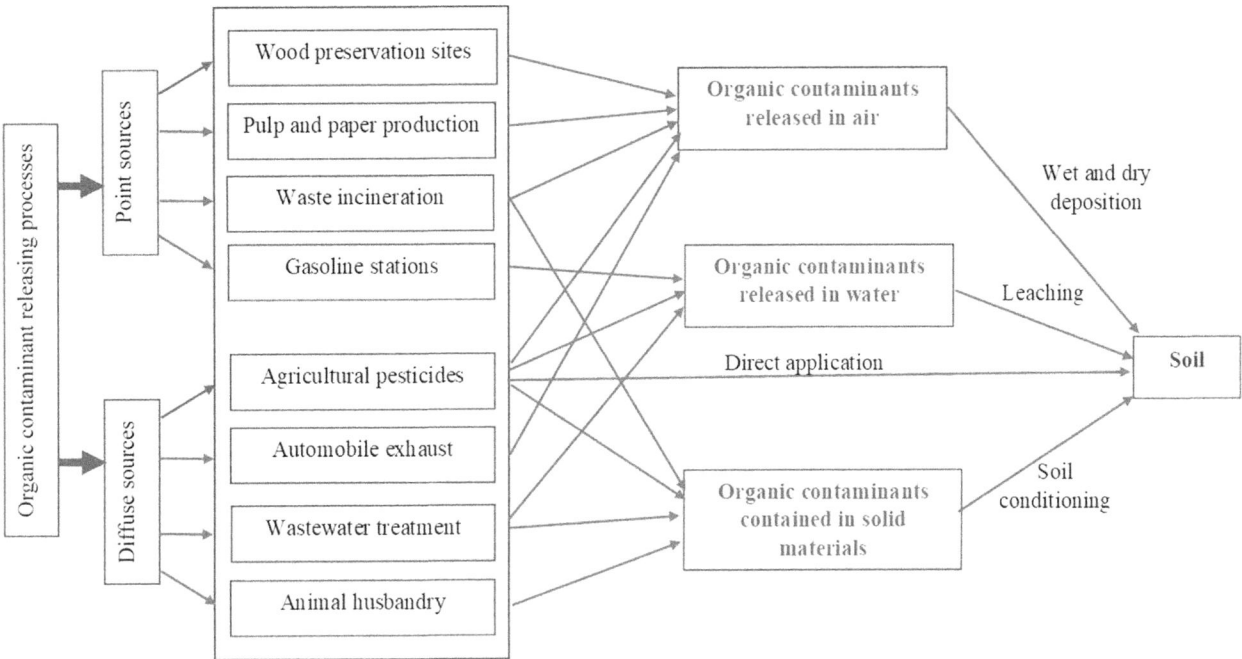

FIGURE 8.1 Sources-pathway-receptor process for contamination of soil by organic contaminants.

8.3 Emerging organic contaminants

Emerging pollutants is an umbrella term that refers to a wide range of compounds that do not have any regulations yet, or require monitoring or reporting (Bell et al., 2011; Wells et al., 2009). Some of these compounds include PPCPs, antibiotic resistance genes (ARGs), and endocrine-disrupting chemicals. In recent years, attention has been given to these products due to their potential effect on human health and environmental threats (Liu et al., 2018). These emerging pollutants have been detected in the groundwater and surface water sources, with most conventional treatment plants having difficulties removing them from water and ending up in the soil as the water is used up in activities such as irrigation. This has raised concerns among different populations as these compounds are used in day-to-day activities across the globe.

8.3.1 Pharmaceuticals and personal care products

PPCPs can be categorized into two major groups based on their application. The first one is pharmaceuticals, which are largely used to diagnose, prevent, or treat diseases in humans and animals to restore their health and improve functions (Patel et al., 2019). The second one is personal care products (PCPs) comprising a group of chemicals to protect humans from potential harm, thereby improving quality of life (Brausch & Rand, 2011). Examples of typical PPCPs are broad-spectrum antibiotics, analgesics [e.g., ibuprofen (Advil), acetaminophen (Tylenol), and aspirin], nonsteroidal antiinflammatory drugs, β-blockers, hormones, lipid regulators, preservatives, mood regulators, disinfectants, insect repellents, cosmetics, fragrances, and other chemical substances used widely in daily life for different purposes. Excessive consumption of PPCPs causes them to enter into the environment either directly or indirectly via different human activities such as livestock breeding, sewage discharge, aquaculture, compost fertilizing, and landfill, eventually ending up in the soil, which is the contaminant sink. Most of the EOCs are persistent, polar, bioaccumulation, highly bioactive toxicity, and disruption of the endocrine system (Chaturvedi et al., 2021; Zhang et al., 2019). The removal efficiencies showed that many technologies were unable to effectively remove most of the PPCPs investigated (Kosma et al., 2014). Paracetamol, caffeine, trimethoprim, sulfamethoxazole, carbamazepine, diclofenac, and salicylic acid were the dominant compounds, while tolfenamic acid, fenofibrate, and simvastatin were the less frequently detected compounds with concentrations in effluents below the limit of quantification (Kosma et al., 2014).

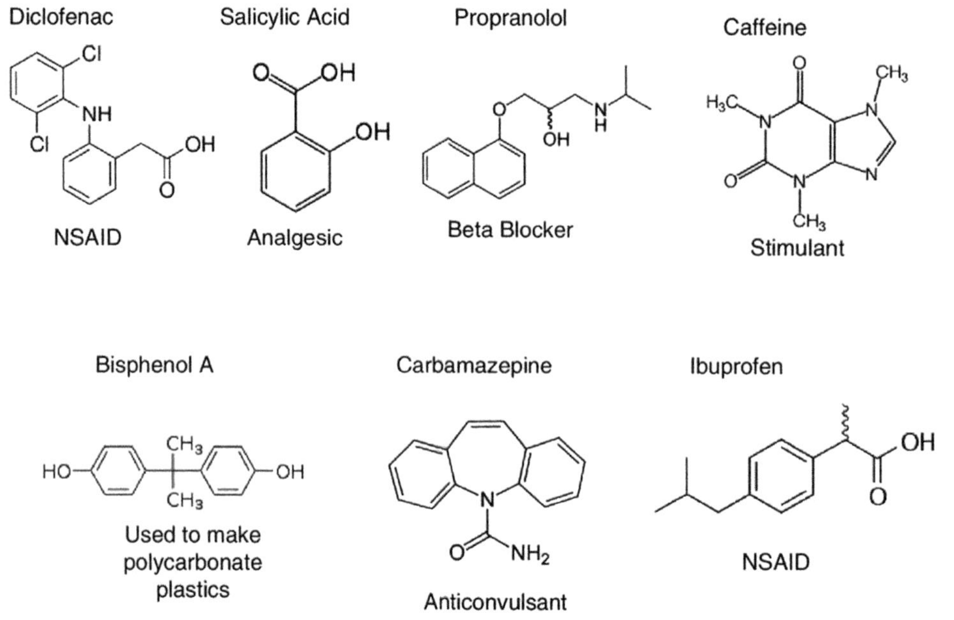

FIGURE 8.2 Molecular structure of some common EOCs. EOCs are generally formed from aromatic compounds. *From Chaturvedi, P., Shukla, P., Giri, B. S., Chowdhary, P., Chandra, R., Gupta, P. & Pandey, A. (2021). Prevalence and hazardous impact of pharmaceutical and personal care products and antibiotics in environment: A review on emerging contaminants. Environmental Research, 194. https://doi.org/10.1016/j.envres.2020.110664.*

8.3.2 Molecular structure of some pharmaceuticals and personal care products and contamination pathways

PPCPs have complex molecules, and most of them appear to be a readily available combination of several active compounds. Each human uses a variety of consumer substances that seem to contribute to PPCP persistence pollutants in soil systems. Shampoos, detergents, household cleaners, sanitizers, biocontrol, and cosmetic products are all examples of PCP pollutants that enter soil habitats directly. The molecular structure of some examples of PPCPs is shown in Fig. 8.2. Improper disposal of unutilized and lapsed medical products in toilet facilities and drainages, unmetabolized medicines, unnecessary use of medicines, and pesticide application in agriculture, livestock rearing, and veterinary purposes all play a significant role as active contributors for emerging organic compounds or pollutants (Wilkinson et al., 2017). Such highly persistent contaminants return back to human beings through the food chain and food web and have been linked to reproductive anomalies, an increase in cancer rates, as well as antibacterial drug resistance (Jelic et al., 2012). The constant emergence of smart complexes in the chemical and pharmaceutical industries and relative ease of connectivity to PPCPs on the market are making a contribution tremendously to the availability and accumulation of even more chemical compounds in water sources, which have very distinguishable forms of metabolic activity that are unexplained clearly.

The pathways for the contamination of soils by PPCPS are shown in Fig. 8.1. Soils are contaminated via (1) effluent discharge from wastewater treatment plants into streams and rivers. Due to persistence, when pharmaceuticals are consumed, they are not fully metabolized. Unmetabolized drugs end up in the sewer system. Wastewater treatment facilities do not completely degrade the organic chemicals in pharmaceutical drugs; rather, they are just transformed from one form to another with different toxicity. Discharging into rivers subsequently leads to the contamination of the soil by the OCs; (2) Leaching and run-off. Disposing of PCPs and pharmaceutical residues in landfills and open dumpsites or burying them leads to the washing of chemicals (expired or contaminated drugs) into the soil, henceforth contaminating the soil via surface run-off, percolation, and leaching; (3) Direct application to soil via irrigation, compost, or soil conditioner.

8.4 Biochar removal of conventional and emerging organic contaminants from contaminated soils

Several studies have demonstrated that biochar's remedial impacts are unique to various pesticides. The efficiency for the alleviation of the negative impacts of pesticides in soil is illustrated in Table 8.2. The breakdown of pesticides and other COCs in soil requires oxidation, biodegradation, hydrolysis, and photolysis (Muter et al., 2014). Biochar plays a fundamental role in the degradation of pesticides in the soil. It stimulates higher microbial activity in the soil, resulting

TABLE 8.2 Influence of biochar on the removal of conventional organic contaminants in the soil.

Feedstock used for biochar preparation	Temp (°C)	Chemical	Pollutants removed	Toxicity conc.	App. Rate (%; w/w)	Remediation potential (%)	References
Wood chip residues	850	Insecticides	Carbofuran and chlorpyrifos	50 mg/kg	1.00	51.00 and 44.00	Yu et al. (2009)
Dairy manure	450	Herbicide	Atrazine	25 mg/L	5.00	66.00	Cao et al. (2011)
Hardwood residues	600	Herbicide	Simazine	5 μg/kg	0.33	85.65	Jones et al. (2011)
Bamboo residues	600	Fungicide	Pentachlorophenol	600 mg/kg	5.00	42.00	Xu et al. (2012)
Wheat straw	300	Herbicide	MCPA (2-methyl-4-chlorophenoxyacetic-acid)	22 mg/L	1.00	35.00	Tatarková et al. (2013)
Hardwood chip	540	Herbicide	Bentazone and aminocyclopyrachlor	3 mg/L	10.00	50.00	Cabrera et al. (2014)
Beech wood residues	550	Herbicide	Imazamox	10 μg/L	1.50	5.00	Dechene et al. (2014)
plant bur cucumber	700	Antibacterial	Sulfamethazine	50 mg/kg	5.00	63.00	Ahmad et al. (2014); Rajapaksha et al. (2014)
Pine wood residues	750	Herbicide	Atrazine	126 mg/L	10.00	52.00	Delwiche et al. (2014)
Wooden box residues	725	Herbicide	MCPA	50 mg/kg	5.30	150.00	Muter et al. (2014)
Wheat straw	600	Herbicide	Atrazine, imidacloprid, and isoproturon	100 mg/L	20.00	76.79	Jin et al. (2016)
Pinewood residues	400	Herbicides and fungicide	Boscalid, bentazone, and pyrimethanil	12.8 mg/kg	5.00	5.84, 19.10, and 1.93	Mukherjee et al. (2016)
Bamboo residues	820	Insecticide	Diethyl-phthalate	10 mg/L	0.50	98.00	Zhang et al. (2016)
Wood chip residue	400		Phenanthrene and 2–4-dichlorophenol	5.3 mg/kg	5% w/w	43.70% and 84.7%	Gu et al. (2016)

in an increased microbial population, which facilitates the biodegradation of pesticides. However, if biochar-amended soils have a higher sorption rate than the desorption rate of pesticides, this lowers the availability of pesticides to microbes, hence reducing the biodegradation of pesticides (Khorram et al., 2016). An experiment was conducted to amend the soil with 1%–2% biochar produced from red-gum-woodchip. The authors observed that there was a decline in carbofuran, chlorpyrifos, and isoproturon herbicides biodegradation due to increased sorption and lower desorption rate. The differences in the biochar remedial efficiency of COCs are due to differences in the biochar preparation process, pyrolysis medium, feedstock type, application rate, as well as the environmental condition where the experiment was performed as shown in Table 8.2. Even though residual pesticide contaminants persist in the soil rhizosphere for quite long periods, the uptake of pesticides by plants reduces with the increase in the biochar application rate in contaminated soil (Tan et al., 2016). The soil amended with 1% biochar substantially decreased the absorption of chlorpyrifos and carbofuran in plants by 10% and 25%, respectively than the soil not treated with the biochar (Yu et al., 2009).

Aside from the alleviation of COCs, studies on the remediation of EOCs are also available. Biochar produced at higher pyrolysis temperatures greater than 500 °C significantly absorbed antibiotics (ceftiofur and florfenicol) more efficiently than produced at low pyrolysis temperatures (Mitchell et al., 2015). The same authors found successful and efficient adsorption of sulfamethoxazole, diclofenac, and carbamazepine to the surface area of activated powdered carbon. The presence of various functional groups on the surfaces of biochar, the net negative charges, higher surface area, diverse surface sites, and microporous structure enable biochar to be an effective adsorbent for various kinds of EOCs.

A single and two-stage stirred adsorber was employed to investigate the adsorption efficiency of tetracycline by biochar produced from chicken bones. A 75% removal efficiency of tetracycline was observed in a single-stage system, and it used around 63.0 g of chicken-bone biochar over 12 h. The magnetically-produced biochar significantly removes various antibiotics and it has been reported to have adsorption potential for hormones ranging between (3.46–169.7 mg/g), sulfamethoxazole (5.19–212.8 mg/g), and tetracyclines (33.1–297.61 mg/g) (Yi et al., 2020). In another investigation, willow biochar activated by H_2O and CO_2 was utilized to remediate PAHs in soil (Kołtowski et al., 2016). The activated biochar substantially decreased the PAHs concentration dissolved in the bitumen plant soil and coal plant soil from 174 to 24 mg/L and 153–22 mg/L, respectively, and the PAHs concentration in soil was reduced by 86% (Yang et al., 2012). In addition to PAH remediation, alleviation of soil contaminated by dyes has been conducted, and the removal efficiency has been impressive. For example, a one- and two-stage stirred adsorber was utilized to examine the removal efficiency of rhodamine B dye by chicken-bone biochar (Oladipo & Ifebajo, 2018). The authors observed that approximately 75% efficiency of rhodamine B dye was removed in a one-stage system, taking 63.0 g of magnetic chicken-bone biochar over a 12-hour period. Whereas in a two-stage system, 33.2 g and 22.2 g of magnetic chicken-bone biochar removed approximately 96% of rhodamine B dye within 180 min (Xu et al., 2011). The adsorption capabilities of several biochars for EOCs are shown in Table 8.3.

8.5 Biochar removal mechanisms of organic contaminants in soils

Biochar has been used effectively to remove a wide range of PPCPs from different environmental media, including soil. Studies have shown that biochar removal efficiencies for PPCPs can be as high as 90%. This makes biochar a promising adsorbent material for the treatment of pharmaceutical residues. This can be attributed to the physicochemical characteristics of biochar described in Chapter 2. The highly porous structure of biochar gives it a large surface area for the process of adsorption, this in turn increases its potential to remove target pollutants and increases removal efficiency. In addition, the oxygen-containing functional groups, such as carboxyl, hydroxyl, and carbonyl, on the biochar surface increase its capacity for adsorption. The hydrogen bonds between unionized surface phenolic groups and carboxylate anions enable biochar to have the maximum adsorption ability. The electrical conductivity of biochar gives it the ability to facilitate electron transfer, required especially for persulfate-based advanced oxidation processes (AOPs). Biochar removes organic pollutants through a number of different mechanisms encompassing physical (adsorption) mechanisms, chemical adsorption, electrostatic attractions, and Hydrogen bonding.

8.5.1 Adsorption mechanisms

This is a process in which the adsorbate (organic pollutant) settles on the surface of the adsorbent (biochar). This takes place using a number of different mechanisms including hydrophobic interaction, electrostatic interaction, pore filling, partitioning, hydrogen bonding and electron donor–acceptor interaction (EDA) or pi-pi interactions, as illustrated in Fig. 8.3.

TABLE 8.3 Influence of biochar on the removal of emerging organic contaminants in the soil.

Feedstock used for biochar preparation	Temp (°C)	Experimental condition	Pollutant removed	Toxicity conc.	App. Rate (%; w/w)	Remediation potential (%)	References
Wood chip residue	850–900		Tylosin		10% w/w	Remedial efficiency was 18%	Jeong et al. (2012)
Wheat straw	300	Incubation experiment	PAHs	100 mg/kg	10% w/w	Degrade the PAHs in soil by 8.46% and improve dehydrogenase activity and polyphenol oxidase activity by 74.81% and 21.71%, respectively.	Cao et al. (2016)
Wood chip residue	700	Pot experiment	Di-chlorophenyl-dichloroethylene (DDE), pyrene & polychlorinated biphenyl	1.6 µg/g	10% w/w	Degrade pyrene and polychlorinated biphenyl by 37% and 41%, respectively.	Bielská et al. (2018)
Sawdust and wheat straw	300–500	Incubation experiment	PAHs	1399.63 mg/kg	5% w/w	Improved the sorption of microbes and PAHs, which degraded the 3-ring PAHs by 69.95% and 4-, 5-, and 6-ring PAHs by 45.96%, 37.92%, and 30.66%, respectively.	Kong et al. (2018)
Maize straw and bamboo residues	500	Field experiment	PAHs	150 mg/kg	16.5 t/ha	PAHs in soil were reduced by 84.31% while soil organic carbon, cation exchange capacity, and soil pH of contaminated soil were improved by 87.59%, 51.54%, and 23.4%, respectively	Rombolà et al. (2019)
Pinewood	550		Acid Blue 264		0.1 g/ 0.1 dm³	The removal efficiency of dye was 19.63%	Tseng et al. (2003)
Wood chip residues	630	Pot experiment	PAHs	0.46 µg/kg	5% w/w	Reduced the PAHs bioavailability in soils by 92%. Enhanced the soil organic carbon and microbial biomass in soil by 22.08% and 125%, respectively.	Ukalska-Jaruga et al. (2020)

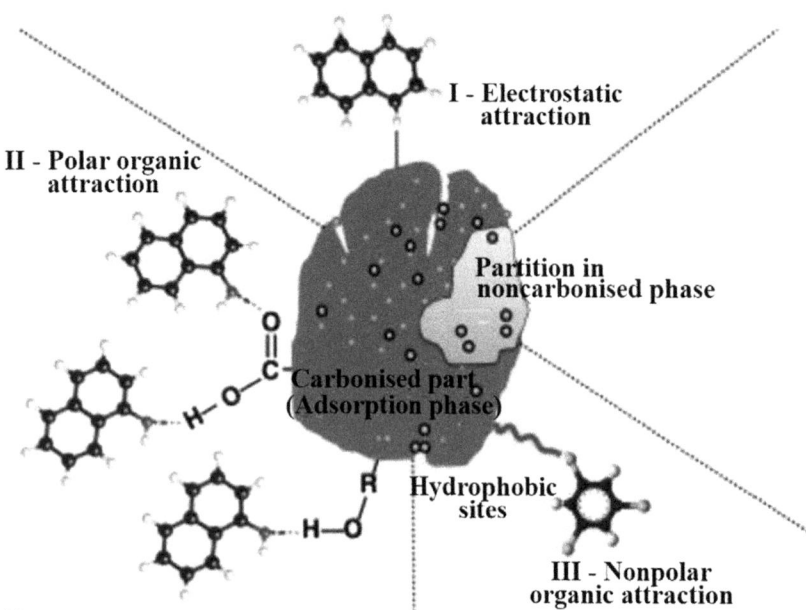

FIGURE 8.3 Removal mechanism. Illustration for biochar removal mechanism of organic pollutants. Biochar removes organic contaminants via different mechanisms including hydrophobic and partition. *From Ahmad, M., Rajapaksha, A. U., Lim, J. E., Zhang, M., Bolan, N., Mohan, D., Vithanage, M., Lee, S. S. & Ok, Y. S. (2014). Biochar as a sorbent for contaminant management in soil and water: A review.* Chemosphere, 99, 19–33. https://doi.org/10.1016/j.chemosphere.2013.10.071.

8.5.1.1 Hydrophobic interactions

This is one of the main adsorption processes or mechanisms of biochar and organic compounds. This has been attributed to the nature of organic compounds, particularly their hydrophobicity (Li et al., 2018). OCs form less energy-consuming bonds with the surface of the biochar. During pyrolysis, the increase in temperature leads to less polar functional groups on the biochar surface, thereby increasing its hydrophobic qualities. Its hydrophobicity causes it to adsorb nonpolar organic molecules. Hydrophobic interactions normally partner with the partitioning mechanism of some organic molecules, as explained in the next section.

8.5.1.2 Partitioning

The process of partitioning involves the adsorbate EOCs material diffusing into the pores of the noncarbonized portion of the adsorbent (biochar). This portion of the biochar forms interactions with the organic adsorbate thereby forming an attachment to it. This mechanism is also dependent on the characteristics of the noncarbonized biochar and of carbonized crystalline (graphene) portions of the biochar then dissolve in the organic matter matrix (Zhang et al., 2013). The partitioning mechanism has been seen to be more efficient when there is high volatile content biochar and correspondingly high concentration of OCs (Keiluweit et al., 2010). When compared to hydrophobic interactions, partitioning interactions require more energy than the hydrophobic interactions, attributed to the competitive attractions of non polar molecules and water molecules that also exist.

8.5.1.3 π-π electron-donor acceptor interactions

The pi-pi EDA interactions are brought about by the changes in the biochar structure during pyrolysis. During the process of pyrolysis, irregular charge sharing between the aromatic rings of biochar leads to variations in the electron density of biochar leading to the formation of a π-electron enriched or π-electron deficient medium (Zheng et al., 2013). At temperatures below 500 °C, biochar is prepared below the π aromatic system and therefore will be an electron acceptor (Abdul et al., 2017). This is because of the abundance of highly polar functional groups. However, at temperatures above 500°C, the biochar acts as an electron donor by binding escaped electron molecules (Abdul et al., 2017; Sun et al., 2011). Crystalline structures in biochar (as exhibited in the graphene structure) expand as pyrolysis temperature increase thereby increasing the π-π interaction (Abdul et al., 2017). Different functional groups have been reported to act in various ways as far as pi-pi interactions are concerned. For instance, (1) nitro groups in some contaminants can reduce the heterocyclic ring electron density and thus enhance the π-electron deficiency which resulted in stronger π-π interactions (Ahmad et al., 2014); (2) π-π interactions are formed by the conjugation of the C − C single bonds of composites with the benzene rings of pollutants; (3) the quantity of organic pollutants taken up by biochar increases with an increase in amount of oxygen-containing functional groups, partly because of π-π electron donor-receptor interactions;

(4) carboxylic acid on the biochar surface, with nitro groups as electron acceptors, and ketones and aromatic molecules also form π-π electron donor—acceptor interactions, thereby improving the adsorption of aromatic molecules; and (5) some hydroxyl and amine groups in biochar can also be considered as π-π electron donor sites.

8.5.1.4 Electrostatic interactions

Electrostatic interactions occur between the ionizable and ionic organic chemicals on biochar. The mechanism and magnitude of the electrostatic forces are affected by pH to a great extent as well as the ionic strength. This is due to the attraction between charged contaminants and oppositely charged sites on the biochar surfaces. The organic compounds or pollutants are attracted towards the opposite charge of the biochar surface. Additionally, the presence of clay minerals with high ion exchange capacity also contributes to the adsorption of pollutants like dyes and PPCPs by ion exchange. Electrostatic attraction between atoms depends on the distance between the two atoms magnitude of the charge of each atom.

8.5.1.5 Pore filling

Pore filling is a process in which the adsorbate EOCs are condensed into the pore of the adsorbent (biochar), Due to its porous nature and large specific surface area, pore filling is a vital mechanism when using biochar as an adsorbent. The high porosity of the biochar provides stronger solid surface adhesion and many active sites.

Small particle size gives an advantage to those organic compounds that are small size, as biochar mainly has micropores (<2 nm) and mesopores (2–50 nm) (Ambaye et al., 2021). Organic molecules which have large pore sizes are normally not easily attracted into the pores due to their sizes. The pore filling mechanism of biochar depends on the size of the organic molecules and the pore size of the adsorbent (Solanki & Boyer, 2019). The pore filling mechanism has a double-pronged effect on microbial degradation of organic pollutants: (1) organic pollutants get into the pores of biochar and cannot be directly used by microorganisms, reducing the degradation rate of organic pollutants by microorganisms; (2) pores are a good habitat for microorganisms including algae, bacteria, and fungi. This enhances microorganisms' activity (Li et al., 2015). For this mechanism, the high porosity and large surface area of micropores affect the rate of adsorption and its capacity for some EOCs. In terms of the rate of the reactions or mechanism, pore filling is fast, hence it helps achieve a higher initial removal rate. However, when biochar space is filled with hydrophobic compounds, active sites begin to clog.

8.5.1.6 Hydrogen bonding

Hydrogen bonding is an interaction that can be intermolecular or intramolecular between hydrogen atoms and other atoms including oxygen, fluorine, and nitrogen. Two types of interactions take place, namely: (1) dipole-dipole attractions between hydrogen atoms of the OH functional group of biochar and atoms of the EOC; (2) Yoshida H-bonding interaction between OH − functional groups on the biochar surface and aromatic ring in EOC molecule (Tran et al., 2020). Biochar has many polar groups, which make biochar and organic compounds contain electronegative elements bond by hydrogen bonds (Ersan et al., 2016). Hydrogen bonding is one of the main mechanisms of polar organic compound adsorption on biochar.

Organic pollutants and their biochar complexes have functional groups such as hydroxyl, amino, and other groups. These functional groups of the biochar composite and the pollutant interact leading to the formation of hydrogen bonds. If there are compounds with functional groups containing oxygen, pollutants are then removed through hydrogen bonding and the formation of complexes between biochar and EOCs. Hydrogen bonding occurs between contaminants and biochar functional groups. For instance, the hydrogen bond is the electrostatic force between the hydrogen nucleus on the strong polar bond (A-H) and the highly electronegative and partially negative atom B (Liang et al., 2019).

8.5.2 Factors affecting adsorption efficiency

The efficiency of the adsorption process in removing organic pollutants is affected by a number of factors, including:

- Adsorbate (Biochar) characteristics: The ability of biochar to adsorb depends on its surface area, porosity, and surface functional groups.
- Adsorbent (target PPCPs): The molecular structure, size, and functional groups of the PPCP influence the ability for them to be adsorbed onto biochar.

- pH of solution: The solution pH affects the surface charge of the biochar and the ionization state of the PPCPs, thereby affecting the adsorption process.
- Coexisting ions of heavy metals and other target pollutants: The presence of other ions in solution may offer competition for adsorption sites on the biochar surface, thereby reducing the efficiency of the target PPCPs. Physicochemical properties of biochar cause it to adsorb a wide variety of pollutants. Adsorption mechanism of organic pollutants is not the same as that of heavy metals, but their coexistence causes competition for adsorption sites between the organic pollutants and heavy metals in solution (Oliveira et al., 2005).
- Adsorbent dosage: At high PPCP concentrations, an increase in the biochar dosage can meaningfully increase the removal of emerging pollutants (Chen et al., 2022). This is due to the increased number of binding sites and an increased larger specific surface area. This will result in more PPCP molecules being bound to the biochar (Chen et al., 2022). Above a certain concentration, a continuous addition of biochar may cause an agglomeration effect, thereby reducing the total specific surface area of the biochar and inhibiting the exposure of binding sites, thereby weakening adsorption capacity (Zhang et al., 2019).
- Adsorption contact time: The contact time between adsorbate and adsorbent governs the sorption kinetics and adsorption process as a whole. The adsorptive removal capacity of the pollutant is directly affected by the contact time. The higher the contact time, the higher the adsorption. This can be attributed to the fact that more time allows for molecules to attach to the binding sites (Singh et al., 2013).
- Adsorption temperature: Adsorption is inversely proportional to temperature. Adsorption theory states that adsorption decreases with an increase in temperature and molecules adsorbed earlier on a surface tend to desorb from the surface at elevated temperatures.

8.6 Behavior and fate of organic contaminants in soils

OCs enter the soil media of the environment mainly through deliberate application and uses, via leakages and spillages, and via atmospheric deposition. Consequently, the soil is a sink for OCs and plays a vital role in the fate of OCs in the environment. The various processes that OCs undergo when they enter the soils are shown in Fig. 8.4. While OCs may be lost from the soil, significant concentrations may be retained within soils. Consequently, the fate and behavior of OCs in soils have been the subject of intensive research, with particular interest directed at the bioavailability of contaminants in the soil.

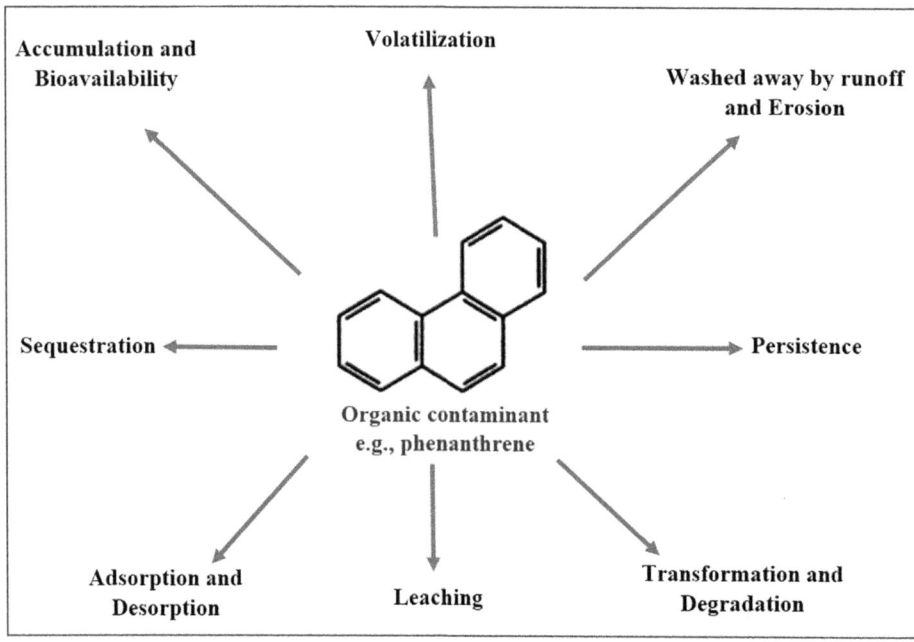

FIGURE 8.4 Fate of organic contaminants in soil. Organic contaminants pass through various processes which determine their fate.

8.6.1 Adsorption and desorption

OCs applied to soils can undergo adsorption, where they bind to soil particles or organic matter (Matter et al., 2020). Pesticide residues, including fungicides present in both organic and conventional agricultural soils and negatively affect microbial soil life (Riedo et al., 2021). The extent of adsorption depends on factors such as chemical properties of the compound and soil characteristics. For example, the adsorption of organophosphate pesticides on soil can be influenced by the addition of organic amendments, however, in literature the exact correlation between pesticide sorption and soil organic carbon content could not be established (Gaonkar et al., 2019). As established by Haider et al. (2022), the soil application of biochar has a significant impact on the biodegradation, leaching, and sorption/desorption of OCs. The sorption/desorption of OCs was noted to be influenced by chemical composition and structure, porosity, surface area, pH, elemental ratios, and surface functional groups of biochar (Haider et al., 2022). Aljumaili and Abdul-Aziz (2023) suggested that, due to its high specific surface area and surface functional groups, biochar can adsorb organic pollutants in soil, aiding in environmental remediation. Biochar amendments enhance the sorption of pesticides by increasing soil organic carbon content, while dissolved organic matter from biochar amendments can facilitate their leaching. For example, a study by Haider et al. (2022) found that biochar increased the stability and dispersibility of $MnFe_2O_4$, leading to synergistic adsorption and oxidation of organic pollutants in the soil. A study by Zhang et al. (2016) concluded that soil organic content and the biochar aging process are important factors in the adsorption capacity of biochar. As suggested by results from (Demisie et al., 2014), biochar application increases total organic carbon, stimulates microbial activities, in turn, increases macro-aggregation, and thus soil quality. Liu et al. (2018) revealed that biochar with high surface area and low dissolved organic carbon content generally increases pesticide sorption in soils as compared to the nonamended soil, and aging process usually causes lower sorption capacity of biochar.

8.6.2 Leaching

Depending on their chemical properties, organic pollutants can be susceptible to leaching through the soil profile. Adsorbed organic pollutants can potentially leach into groundwater if they are not strongly sorbed (Scow & Johnson, 1996). With respect to pesticides, their sorption, decomposition, and movement influence fate in the soils (Sadegh-Zadeh et al., 2017). Pesticides, including herbicides, are commonly detected in agricultural soils, including those managed organically with the number and concentration of pesticide residues are significantly higher in conventional fields compared to organic fields (Katagi, 2013; Matter et al., 2020). Leaching occurs when water carrying these compounds percolates through the soil, potentially reaching groundwater or nearby water bodies. Sorption of pesticides increases by amending soils with organic matter. The mobility of conventional organics can vary, with some being highly mobile and others less so, and this mobility is also influenced by organic matter in the soil (Sadegh-Zadeh et al., 2017). Therefore, it is important to monitor and assess the presence of pesticides and herbicides in soil to prevent their leaching and potential contamination of groundwater.

8.6.3 Transformation and degradation

In soils, organic pollutants can undergo various transformation processes. Microorganisms play a crucial role in the degradation of these compounds, breaking them down into simpler forms (Kumar & Ghosh Sachan, 2020). For example, fungi and their associated enzymes have been found to be effective in the degradation of various agricultural organic pollutants, including pesticides and herbicides (Chaurasia et al., 2022). Factors like temperature, moisture, pH, and the presence of specific microbial populations influence the rate and extent of degradation (Muskus et al., 2022). Microorganisms have the capability to biodegrade and biotransform organic pollutants, converting them into less toxic by-products (Sarangi & Rajkumar, 2022). A study by Ali et al. (2014) found that temperature, moisture, pH, and microbial activity influenced the degradation of pesticides during composting. Siddiqa & Faisal (2021) revealed that microbial degradation of complex organic pollutants, such as hydrocarbons and phenolic compounds, could lead to their complete mineralization. Fungi and their associated enzymes, such as laccases, have shown efficacy in degrading various agricultural organic pollutants, including pesticides and herbicides (Chaurasia et al., 2022). Microbes, such as fungi, bacteria, and cyanobacteria, secrete metabolites and enzymes that aid in the transformation of pesticides and organic pollutants into environmentally safe compounds (Khaire et al., 2022). Mohammadi-Moghadam et al. (2022) found that microbial consortia enriched by vermicompost, containing Proteobacteria phylum bacterial species such as Pseudomonas and Citrobacter, were able to degrade PAHs in contaminated soils. In addition, microbes such as *Pseudomonas aeruginosa, Pseudomonas japonica, Pseudomonas fluorescens, Alcaligenes faecalis, Brevibacterium iodinum*, and *Saccharomyces cerevisiae* have been found in various studies to be effective in the bioremediation of contaminated soils (Ghosh et al., 2021). However, a study by Harju et al. (2021) showed the importance of chemical dissipation in the natural cleansing capacity of the environment, suggesting that microbial degradation may not always be the primary mechanism for pesticide dissipation. Hence, the degradation of organic pollutants requires a combination of both biological and chemical factors for it to be effective.

8.6.4 Persistence

The persistence of conventional organics in soils refers to the length of time they remain active or detectable (Garrido et al., 2015). Some compounds are relatively stable and can persist in soils for extended periods, while others degrade more rapidly. For example, a study by Jablonowski et al. (2012) established that the pesticides ethidimuron, methabenzthiazuron, and anilazine persisted in soils for 9–17 years, indicating long-term persistence. Meanwhile, an investigation by Mukherjee et al. (2016) found the persistence of the herbicides florasulam and halauxifen-methyl in alluvial and saline alluvial soils to have half-lives ranging from 7.13 to 17.60 days. Persistence can vary greatly depending on chemical structure, soil conditions, and management practices (Ghafoor et al., 2011). The persistence of organic pollutants is influenced by their physico-chemical properties, structure, and stability (Pereira, 2014). A study by Matei et al. (2023) established those persistent organic pollutants (POPs) are usually halogenated, lipophilic, organic compounds with a stable chemical structure that are resistant to degradation, leading to their persistence in the environment. Owing to their halogenated structure and physco-chemical properties, polychlorinated diphenyl ethers (PCDEs) organic pollutants were found to be potentially persistent (Huang et al., 2004).

8.6.5 Volatilization

Certain volatile conventional organics have the potential to vaporize from soil surfaces into the atmosphere. For example, some organochlorine pesticides (OCPs) have the potential to volatilize from soil surfaces into the atmosphere (Fang et al., 2017). The extent of volatilization depends on factors such as vapor pressure, temperature, soil moisture, and organic matter content. Volatilization can contribute to air pollution or indirect contamination of other environmental compartments (Guarda et al., 2020). In a study by Ong et al. (1992), volatilization of organic vapors was shown to be favored for contaminants with high Henry's law constants and low aqueous-phase partitioning coefficients. Lee et al. (2004) revealed that the volatilization rates of organic pollutants are closely related to molecular weight (solubility) in the existence of mixing (surfactants).

8.6.6 Washed away by runoff and erosion

When conventional organics are applied to soils, there is a potential risk of their being transported by surface runoff or erosion. Runoff can carry the compounds to nearby water bodies, leading to water pollution. Soil erosion can also transport these chemicals to other locations, potentially affecting nontarget areas (Guimarães et al., 2019).

8.6.7 Accumulation and Bioavailability

Conventional organics that persist in soils can accumulate over time (Mamy & Barriuso, 2007). This accumulation can pose a risk to organisms in the soil ecosystem and affect their health and biodiversity (Sarkar et al., 2020). Additionally, the bioavailability of these compounds, which refers to their potential for uptake and impact on living organisms, depends on their chemical properties and interactions with soil constituents. A study by Cao et al. (2016) found that PCBs biomagnify consistently in both aquatic and terrestrial food webs, while short-chain chlorinated paraffins biomagnify more in terrestrial organisms. (Wu et al., 2022) investigated the biomagnification of POPs from invertebrates to songbirds and found that the biomagnification factors had parabolic relationships with the physiochemical properties of the POPs. Furthermore, a study by Ding et al. (2022) revealed that PCBs and polybrominated diphenyl ethers bioaccumulate in amphibians, with dragonflies being a major contributor to their diet.

8.7 Fate of biochar after organic contaminants removal

COCs undergo degradation processes that lead to the formation of a wide range of by-products. Recent studies, for example, have shown that after volatilization, EOCs can lead to the formation of carcinogenic disinfection byproducts (DBPs) in community drinking water treatment plants after breakdown, posing risks to human health (Chaves et al., 2022; Roque et al., 2023). Additionally, contaminants can degrade into weaker inorganic halide by-products like sodium hypochlorite when ascorbic acid is used as a quencher (Li et al., 2023). AOPs in water treatment systems further convert these contaminants into carbon dioxide and water, showcasing a complete mineralization pathway (Ganiyu et al., 2022).

Moreover, sulfate radical-based AOPs can produce brominated DBPs when contaminants are exposed to sulfate radicals (Xia et al., 2020). Fixed-bed adsorption followed by off-line regeneration using saturated steam and cleanup of steam condensate using fixed-bed photocatalysis results in the breakdown of compounds such as tetrachloroethylene, carbon tetrachloride, p-dichlorobenzene, o-chlorobiphenyl, and methyl ethyl ketone (Suri et al., 1999).

Furthermore, the degradation of COCs can lead to the formation of various intermediate substances and organic radicals, including $CH_3COO^\bullet$ and $CH_3COOO^\bullet$, as well as inorganic anions and other degradation products (Pavel et al., 2023). These diverse by-products highlight the complexity of the degradation processes and the importance of understanding the transformation pathways to manage the potential environmental and health impacts of the by-products effectively.

Sabharwal et al. (2015) concluded that after the breakdown, EOCs exposed to ionizing doses can form various radiolytic byproducts, some of which may be toxic, necessitating toxicity assays for evaluation. Research by Balducci et al. (2012) showed that OCs can volatilize to psychoactive substances in airborne particles in the urban environment after the breakdown. The contaminants can also volatilize to by-products such as polyaromatic hydrocarbons, petroleum compounds, triazine herbicides, chlorinated solvents, and THMs after a breakdown in groundwater (Landmeyer, 2011).

8.8 Biochar and other soil remediation methods

Biochar is often integrated with soil remediation techniques such as phytoremediation. This combination has demonstrated significant potential in enhancing the efficiency of remediation processes for heavy metal-contaminated soils. (Narayanan & Ma, 2022) found that biochar derived from various plant sources can promote the growth and phytoremediation capabilities of native or wild plants grown in metal-polluted soil. Biochar's ability to enhance plant growth is well-documented in the literature, suggesting its potential to boost the yield of phytoremediators (Lu et al., 2014). Additionally, Li et al. (2024) showed that the combination of biochar with phytoremediation, particularly when used with arbuscular-mycorrhizal fungi, significantly increases plant biomass and cadmium remediation efficiency of *Ligustrum lucidum* in contaminated soil, thereby improving overall phytoremediation effectiveness. Moreover, the use of biochar and fulvic acid has been shown to enhance ryegrass phytoremediation of Cd and Zn in sediment by increasing bioavailability while reducing Pb accumulation, highlighting its potential for efficient heavy metal extraction (Zhang et al., 2023).

8.9 Challenges and opportunities of biochar in removal of organic pollutants

8.9.1 Challenges of biochar removal of organic pollutants

Biochar has proven to have great potential in laboratory-scale experiments and pilot plants; however, challenges remain in scaling up the technology for real-world applications. Since in most cases it is made from waste material, its production costs and varying feedstock quality remain uncertain. Additionally, there is a need to look into custom-made treatment systems based on specific contaminants and environmental conditions to guarantee the successful implementation of biochar in removing PPCPs in soils. Biochar is used in the adsorption of a wide variety of toxic pollutants, hence spent biochar is a possible environmental and health hazard which needs to be properly managed. The measures or options that can be used rely on the target pollutant removed by the biochar as in this case, PPCPs. Options include uses as an energy source and landfilling.

8.9.2 Research opportunities

Future research opportunities in this field include:

- Study of biochar's potential for removing other emerging contaminants, such as microplastics and perfluorinated compounds
- Development of novel biochar materials with improved adsorption removal capacities
- Optimization of biochar production parameters to increase adsorption performance
- Assessment of biochar's long-term stability and possibility for reuse or disposal after adsorption

8.10 Conclusion

This chapter discusses the remediation of conventional and EOCs in soil. The degree to which these contaminants are held in the soil depends on the soil properties, as well as their physico-chemical properties. Due to their unique properties such as their hydrophobicity, low degradability, and semivolatility, OCs spread over a large area posing a health hazard to untargeted animals. Bioremediation using biochar has proved to be an eco-friendly option and less expensive compared to traditional remediation methods. It immobilizes the pesticides and other organic pollutants rendering them harmless. Biochar removes organic pollutants in soil via $\pi-\pi$ interaction, hydrophobic interaction, and hydrogen-bond interaction. While the technology is promising, the disposal of spent biochar remains a challenge and future research should focus on the disposal of spent biochar.

References

Abdul, G., Zhu, X., & Chen, B. (2017). Structural characteristics of biochar-graphene nanosheet composites and their adsorption performance for phthalic acid esters. *Chemical Engineering Journal, 319,* 9–20. Available from https://doi.org/10.1016/j.cej.2017.02.074, http://www.elsevier.com/inca/publications/store/6/0/1/2/7/3/index.htt.

Agboola, A. R., Okonkwo, C. O., Agwupuye, E. I., & Mbeh, G. (2022). Biopesticides and conventional pesticides: comparative review of mechanism of action and future perspectives. *AROC in Agriculture, 1*(1), 14–32. Available from https://doi.org/10.53858/arocagr01011432.

Ahmad, M., Rajapaksha, A. U., Lim, J. E., Zhang, M., Bolan, N., Mohan, D., Vithanage, M., Lee, S. S., & Ok, Y. S. (2014). Biochar as a sorbent for contaminant management in soil and water: A review. *Chemosphere, 99,* 19–33. Available from https://doi.org/10.1016/j.chemosphere.2013.10.071, http://www.elsevier.com/locate/chemosphere.

Ali, M., Kazmi, A. A., & Ahmed, N. (2014). Study on effects of temperature, moisture and pH in degradation and degradation kinetics of aldrin, endosulfan, lindane pesticides during full-scale continuous rotary drum composting. *Chemosphere, 102,* 68–75. Available from https://doi.org/10.1016/j.chemosphere.2013.12.022, http://www.elsevier.com/locate/chemosphere.

Aljumaili, M. M. N., & Abdul-Aziz, Y. I. (2023). High surface area peat moss biochar and its potential for Chromium metal adsorption from aqueous solutions. *South African Journal of Chemical Engineering, 46,* 22–34. Available from https://doi.org/10.1016/j.sajce.2023.06.006, https://www.journals.elsevier.com/south-african-journal-of-chemical-engineering/.

Ambaye, T. G., Vaccari, M., van Hullebusch, E. D., Amrane, A., & Rtimi, S. (2021). Mechanisms and adsorption capacities of biochar for the removal of organic and inorganic pollutants from industrial wastewater. *International Journal of Environmental Science and Technology, 18*(10), 3273–3294. Available from https://doi.org/10.1007/s13762-020-03060-w.

Anawar, H. M., Akter, F., Solaiman, Z. M., & Strezov, V. (2015). Biochar: An emerging panacea for remediation of soil contaminants from mining, industry and sewage wastes. *Pedosphere, 25*(5), 654–665. Available from https://doi.org/10.1016/S1002-0160(15)30046-1, http://pedosphere.issas.ac.cn.

Bahlai, C. A., Xue, Y., McCreary, C. M., Schaafsma, A. W., Hallett, R. H., & Johnson, S. J. (2010). Choosing organic pesticides over synthetic pesticides may not effectively mitigate environmental risk in soybeans. *PLoS One, 5*(6)e11250. Available from https://doi.org/10.1371/journal.pone.0011250.

Balducci, C., Perilli, M., Romagnoli, P., & Cecinato, A. (2012). New developments on emerging organic pollutants in the atmosphere. *Environmental Science and Pollution Research, 19*(6), 1875–1884. Available from https://doi.org/10.1007/s11356-012-0815-2.

Bell, K. Y., Wells, M. J. M., Traexler, K. A., Pellegrin, M. L., Morse, A., & Bandy, J. (2011). Emerging pollutants. *Water Environment Research, 83*(10), 1906–1984. Available from https://doi.org/10.2175/106143011X13075599870298, http://docserver.ingentaconnect.com/deliver/connect/wef/10614303/v83n10/s33.pdf?expires = 1320731328&id = 65500755&titleid = 11548&accname = Elsevier&checksum = 3DF3B8B96F651D417E4BCE934003D289.

Bielská, L., Škulcová, L., Neuwirthová, N., Cornelissen, G., & Hale, S. E. (2018). Sorption, bioavailability and ecotoxic effects of hydrophobic organic compounds in biochar amended soils. *Science of the Total Environment, 624,* 78–86. Available from https://doi.org/10.1016/j.scitotenv.2017.12.098, http://www.elsevier.com/locate/scitotenv.

Brausch, J. M., & Rand, G. M. (2011). A review of personal care products in the aquatic environment: Environmental concentrations and toxicity. *Chemosphere, 82*(11), 1518–1532. Available from https://doi.org/10.1016/j.chemosphere.2010.11.018, http://www.elsevier.com/locate/chemosphere.

Buerge, I. J., Keller, M., Buser, H. R., Müller, M. D., & Poiger, T. (2011). Saccharin and other artificial sweeteners in soils: Estimated inputs from agriculture and households, degradation, and leaching to groundwater. *Environmental Science and Technology, 45*(2), 615–621. Available from https://doi.org/10.1021/es1031272.

Byrns, G. (2001). The fate of xenobiotic organic compounds in wastewater treatment plants. *Water Research, 35*(10), 2523–2533. Available from https://doi.org/10.1016/s0043-1354(00)00529-7.

Cabrera, A., Cox, L., Spokas, K., Hermosín, M. C., Cornejo, J., & Koskinen, W. C. (2014). Influence of biochar amendments on the sorption–desorption of aminocyclopyrachlor, bentazone and pyraclostrobin pesticides to an agricultural soil. *Science of The Total Environment, 470-471,* 438–443. Available from https://doi.org/10.1016/j.scitotenv.2013.09.080.

Cao, X., Ma, L., Liang, Y., Gao, B., & Harris, W. (2011). Simultaneous immobilization of lead and atrazine in contaminated soils using dairy-manure biochar. *Environmental Science & Technology, 45*(11), 4884–4889. Available from https://doi.org/10.1021/es103752u.

Cao, Y., Yang, B., Song, Z., Wang, H., He, F., & Han, X. (2016). Wheat straw biochar amendments on the removal of polycyclic aromatic hydrocarbons (PAHs) in contaminated soil. *Ecotoxicology and Environmental Safety*, *130*, 248−255. Available from https://doi.org/10.1016/j.ecoenv.2016.04.033, http://www.elsevier.com/inca/publications/store/6/2/2/8/1/9/index.htt.

Chaturvedi, P., Shukla, P., Giri, B. S., Chowdhary, P., Chandra, R., Gupta, P., & Pandey, A. (2021). Prevalence and hazardous impact of pharmaceutical and personal care products and antibiotics in environment: A review on emerging contaminants. *Environmental Research*, *194*. Available from https://doi.org/10.1016/j.envres.2020.110664, http://www.elsevier.com/inca/publications/store/6/2/2/8/2/1/index.htt.

Chaurasia, P. K., Sharma, N., Nagraj., Rudakiya, D. M., Singh, S., & Bharati, S. L. (2022). Fungal-Assisted Bioremediation of Agricultural Organic Pollutants (Pesticides and Herbicides). *Current Green Chemistry*, *9*(1), 14−25. Available from https://doi.org/10.2174/2213346109666220927121948, benthamscience.com/journals/current-green-chemistry/.

Chaves, R. S., Rodrigues, J. E., Santos, M. M., Benoliel, M. J., & Cardoso, V. V. (2022). Development of multi-residue gas chromatography coupled with mass spectrometry methodologies for the measurement of 15 chemically different disinfection by-products (DBPs) of emerging concern in drinking water from two different Portuguese water treatment plants. *Analytical Methods*, *14*(47), 4967−4976. Available from https://doi.org/10.1039/d2ay01401g, http://pubs.rsc.org/en/journals/journal/ay.

Chen, Z., Wu, Y., Huang, Y., Song, L., Chen, H., Zhu, S., & Tang, C. (2022). Enhanced adsorption of phosphate on orange peel-based biochar activated by Ca/Zn composite: Adsorption efficiency and mechanisms. *Colloids and Surfaces A: Physicochemical and Engineering Aspects*, *651*129728. Available from https://doi.org/10.1016/j.colsurfa.2022.129728.

Cheng, S., Chen, T., Xu, W., Huang, J., Jiang, S., & Yan, B. (2020). Application research of biochar for the remediation of soil heavy metals contamination: A review. *Molecules (Basel, Switzerland)*, *25*(14). Available from https://doi.org/10.3390/molecules25143167, https://www.mdpi.com/1420-3049/25/14/3167/pdf.

Clarke, B. O., & Smith, S. R. (2011). Review of 'emerging' organic contaminants in biosolids and assessment of international research priorities for the agricultural use of biosolids. *Environment International*, *37*(1), 226−247. Available from https://doi.org/10.1016/j.envint.2010.06.004, http://www.elsevier.com/locate/envint.

Das, S., Helmus, R., Dong, Y., Beijer, S., Praetorius, A., Parsons, J. R., & Jansen, B. (2023). Organic contaminants in bio-based fertilizer treated soil: Target and suspect screening approaches. *Chemosphere*, *337*. Available from https://doi.org/10.1016/j.chemosphere.2023.139261, http://www.elsevier.com/locate/chemosphere.

Dechene, A., Rosendahl, I., Laabs, V., & Amelung, W. (2014). Sorption of polar herbicides and herbicide metabolites by biochar-amended soil. *Chemosphere*, *109*, 180−186. Available from https://doi.org/10.1016/j.chemosphere.2014.02.010, http://www.elsevier.com/locate/chemosphere.

Delwiche, K. B., Lehmann, J., & Walter, M. T. (2014). Atrazine leaching from biochar-amended soils. *Chemosphere*, *95*, 346−352. Available from https://doi.org/10.1016/j.chemosphere.2013.09.043, http://www.elsevier.com/locate/chemosphere.

Demisie, W., Liu, Z., & Zhang, M. (2014). Effect of biochar on carbon fractions and enzyme activity of red soil. *Catena*, *121*, 214−221. Available from https://doi.org/10.1016/j.catena.2014.05.020, http://www.elsevier.com/inca/publications/store/5/2/4/6/0/9.

Ding, Y., Zheng, X., Yu, L., Lu, R., Wu, X., Luo, X., & Mai, B. (2022). Occurrence and distribution of persistent organic pollutants (POPs) in amphibian species: Implications from biomagnification factors based on quantitative fatty acid signature analysis. *Environmental Science and Technology*, *56*(5), 3117−3126. Available from https://doi.org/10.1021/acs.est.1c07416, http://pubs.acs.org/journal/esthag.

El-Naggar, A., Awad, Y. M., Tang, X. Y., Liu, C., Niazi, N. K., Jien, S. H., Tsang, D. C. W., Song, H., Ok, Y. S., & Lee, S. S. (2018). Biochar influences soil carbon pools and facilitates interactions with soil: A field investigation. *Land Degradation and Development*, *29*(7), 2162−2171. Available from https://doi.org/10.1002/ldr.2896, http://onlinelibrary.wiley.com/journal/10.1002/(ISSN)1099-145X.

Ersan, G., Apul, O. G., & Karanfil, T. (2016). Linear solvation energy relationships (LSER) for adsorption of organic compounds by carbon nanotubes. *Water Research*, *98*, 28−38. Available from https://doi.org/10.1016/j.watres.2016.03.067, http://www.elsevier.com/locate/watres.

Fang, Y., Nie, Z., Die, Q., Tian, Y., Liu, F., He, J., & Huang, Q. (2017). Organochlorine pesticides in soil, air, and vegetation at and around a contaminated site in southwestern China: Concentration, transmission, and risk evaluation. *Chemosphere*, *178*, 340−349. Available from https://doi.org/10.1016/j.chemosphere.2017.02.151, http://www.elsevier.com/locate/chemosphere.

Ganiyu, S. O., Sable, S., & Gamal El-Din, M. (2022). Advanced oxidation processes for the degradation of dissolved organics in produced water: A review of process performance, degradation kinetics and pathway. *Chemical Engineering Journal*, *429*132492. Available from https://doi.org/10.1016/j.cej.2021.132492.

Gaonkar, O. D., Nambi, I. M., & Govindarajan, S. K. (2019). Soil organic amendments: impacts on sorption of organophosphate pesticides on an alluvial soil. *Journal of Soils and Sediments*, *19*(2), 566−578. Available from https://doi.org/10.1007/s11368-018-2080-6, http://www.springerlink.com/content/1439-0108.

García, J., García-Galán, M. J., Day, J. W., Boopathy, R., White, J. R., Wallace, S., & Hunter, R. G. (2020). A review of emerging organic contaminants (EOCs), antibiotic resistant bacteria (ARB), and antibiotic resistance genes (ARGs) in the environment: Increasing removal with wetlands and reducing environmental impacts. *Bioresource Technology*, *307*. Available from https://doi.org/10.1016/j.biortech.2020.123228, http://www.elsevier.com/locate/biortech.

Garrido, I., Vela, N., Fenoll, J., Navarro, G., Pérez-Lucas, G., & Navarro, S. (2015). Testing of leachability and persistence of sixteen pesticides in three agricultural soils of a semiarid Mediterranean region. *Spanish Journal of Agricultural Research*, *13*(4). Available from https://doi.org/10.5424/sjar/2015134-8339, http://revistas.inia.es/index.php/sjar/article/download/8339/2592.

Ghafoor, A., Jarvis, N. J., Thierfelder, T., & Stenström, J. (2011). Measurements and modeling of pesticide persistence in soil at the catchment scale. *Science of The Total Environment*, *409*(10), 1900−1908. Available from https://doi.org/10.1016/j.scitotenv.2011.01.049.

Ghosh, S., Sharma, I., Nath, S., & Webster, T. J. (2021). *Bioremediation-the natural solution Microbial Ecology of Wastewater Treatment Plants* (pp. 11−40). India: Elsevier. Available from https://www.sciencedirect.com/book/9780128225035, 10.1016/B978-0-12-822503-5.00018-7.

Goswami, L., Kushwaha, A., Kafle, S. R., & Kim, B. S. (2022). Surface Modification of Biochar for Dye Removal from Wastewater. *Catalysts*, *12*(8). Available from https://doi.org/10.3390/catal12080817, https://www.mdpi.com/journal/catalysts.

Gu, J., Zhou, W., Jiang, B., Wang, L., Ma, Y., Guo, H., Schulin, R., Ji, R., & Evangelou, M. W. H. (2016). Effects of biochar on the transformation and earthworm bioaccumulation of organic pollutants in soil. *Chemosphere*, *145*, 431–437. Available from https://doi.org/10.1016/j.chemosphere.2015.11.106, http://www.elsevier.com/locate/chemosphere.

Guarda, P. M., Pontes, A. M. S., Domiciano, Rd. S., Gualberto, Ld. S., Mendes, D. B., Guarda, E. A., & da Silva, J. E. C. (2020). Assessment of Ecological Risk and Environmental Behavior of Pesticides in Environmental Compartments of the Formoso River in Tocantins, Brazil. *Archives of Environmental Contamination and Toxicology*, *79*(4), 524–536. Available from https://doi.org/10.1007/s00244-020-00770-7, link.springer.de/link/service/journals/00244/index.htm.

Guimarães, A. C. D., Mendes, K. F., Campion, T. F., Christoffoleti, P. J., & Tornisielo, V. L. (2019). Leaching of herbicides commonly applied to sugarcane in five agricultural soils. *Planta Daninha*, *37*. Available from https://doi.org/10.1590/s0100-83582019370100029.

Haider, F. U., Wang, X., Farooq, M., Hussain, S., Cheema, S. A., Ain, N. U., Virk, A. L., Ejaz, M., Janyshova, U., & Liqun, C. (2022). Biochar application for the remediation of trace metals in contaminated soils: Implications for stress tolerance and crop production. *Ecotoxicology and Environmental Safety*, *230*. Available from https://doi.org/10.1016/j.ecoenv.2022.113165, http://www.elsevier.com/inca/publications/store/6/2/2/8/1/9/index.htt.

Harju, A. V., Närhi, I., Mattsson, M., Kerminen, K., & Kontro, M. H. (2021). Organic Matter Causes Chemical Pollutant Dissipation Along With Adsorption and Microbial Degradation. *Frontiers in Environmental Science*, *9*. Available from https://doi.org/10.3389/fenvs.2021.666222, journal.frontiersin.org/journal/environmental-science.

Huang, D., Xu, Q., Cheng, J., Lu, X., & Zhang, H. (2012). Electrokinetic remediation and its combined technologies for removal of organic pollutants from contaminated soils. *International Journal of Electrochemical Science*, *7*(5), 4528–4544. Available from http://www.electrochemsci.org/papers/vol7/7054528.pdf, China.

Huang, J., Yu, G., Yang, X., & Zhang, Z. L. (2004). Predicting physico-chemical properties of polychlorinated diphenyl ethers (PCDEs): Potential persistent organic pollutants (POPs). *Journal of Environmental Sciences*, *16*(2), 204–207.

Jablonowski, N. D., Linden, A., Köppchen, S., Thiele, B., Hofmann, D., Mittelstaedt, W., Pütz, T., & Burauel, P. (2012). Long-term persistence of various 14C-labeled pesticides in soils. *Environmental Pollution*, *168*, 29–36. Available from https://doi.org/10.1016/j.envpol.2012.04.022.

Jelic, A., Cruz-Morató, C., Marco-Urrea, E., Sarrà, M., Perez, S., Vicent, T., Petrović, M., & Barcelo, D. (2012). Degradation of carbamazepine by Trametes versicolor in an air pulsed fluidized bed bioreactor and identification of intermediates. *Water Research*, *46*(4), 955–964. Available from https://doi.org/10.1016/j.watres.2011.11.063.

Jeong, C. Y., Wang, J. J., Dodla, S. K., Eberhardt, T. L., & Groom, L. (2012). Effect of biochar amendment on tylosin adsorption-desorption and transport in two different soils. *Journal of Environmental Quality*, *41*(4), 1185–1192. Available from https://doi.org/10.2134/jeq2011.0166, https://www.agronomy.org/publications/jeq/pdfs/41/4/1185, United States.

Jin, J., Kang, M., Sun, K., Pan, Z., Wu, F., & Xing, B. (2016). Properties of biochar-amended soils and their sorption of imidacloprid, isoproturon, and atrazine. *Science of the Total Environment*, *550*, 504–513. Available from https://doi.org/10.1016/j.scitotenv.2016.01.117, http://www.elsevier.com/locate/scitotenv.

Jones, D. L., Edwards-Jones, G., & Murphy, D. V. (2011). Biochar mediated alterations in herbicide breakdown and leaching in soil. *Soil Biology and Biochemistry*, *43*(4), 804–813. Available from https://doi.org/10.1016/j.soilbio.2010.12.015.

Katagi, T. (2013). Soil column leaching of pesticides. *Reviews of Environmental Contamination and Toxicology*, *221*, 1–105. Available from https://doi.org/10.1007/978-1-4614-4448-0_1, http://www.springer.com/series/398.

Keiluweit, M., Nico, P. S., Johnson, M., & Kleber, M. (2010). Dynamic molecular structure of plant biomass-derived black carbon (biochar). *Environmental Science and Technology*, *44*(4), 1247–1253. Available from https://doi.org/10.1021/es9031419.

Khaire, P., Boggala, V., Mamidi, A., & Narute, T. (2022). *The Role of Microbes in Environmental Contaminants' Management* (pp. 117–153). Informa UK Limited. Available from 10.1201/9781003239956-9.

Khalid, S., Shahid, M., Murtaza, B., Bibi, I., Natasha., Asif Naeem, M., & Niazi, N. K. (2020). A critical review of different factors governing the fate of pesticides in soil under biochar application. *Science of the Total Environment*, *711*. Available from https://doi.org/10.1016/j.scitotenv.2019.134645, http://www.elsevier.com/locate/scitotenv.

Khorram, M. S., Zheng, Y., Lin, D., Zhang, Q., Fang, H., & Yu, Y. (2016). Dissipation of fomesafen in biochar-amended soil and its availability to corn (Zea mays L.) and earthworm (Eisenia fetida). *Journal of Soils and Sediments*, *16*(10), 2439–2448. Available from https://doi.org/10.1007/s11368-016-1407-4, http://www.springerlink.com/content/1439-0108.

Kong, L., Gao, Y., Zhou, Q., Zhao, X., & Sun, Z. (2018). Biochar accelerates PAHs biodegradation in petroleum-polluted soil by biostimulation strategy. *Journal of Hazardous Materials*, *343*, 276–284. Available from https://doi.org/10.1016/j.jhazmat.2017.09.040, http://www.elsevier.com/locate/jhazmat.

Kosma, C. I., Lambropoulou, D. A., & Albanis, T. A. (2014). Investigation of PPCPs in wastewater treatment plants in Greece: Occurrence, removal and environmental risk assessment. *Science of the Total Environment*, *466-467*, 421–438. Available from https://doi.org/10.1016/j.scitotenv.2013.07.044.

Kołtowski, M., Hilber, I., Bucheli, T. D., & Oleszczuk, P. (2016). Effect of activated carbon and biochars on the bioavailability of polycyclic aromatic hydrocarbons in different industrially contaminated soils. *Environmental Science and Pollution Research*, *23*(11), 11058–11068. Available from https://doi.org/10.1007/s11356-016-6196-1.

Kumar, P., & Ghosh Sachan, S. (2020). *Exploring microbes as bioremediation tools for the degradation of pesticides Advanced Oxidation Processes for Effluent Treatment Plants* (pp. 51–67). India: Elsevier. Available from https://www.sciencedirect.com/book/9780128210116, 10.1016/B978-0-12-821011-6.00003-7.

Kuśmierz, M., Oleszczuk, P., Kraska, P., Pałys, E., & Andruszczak, S. (2016). Persistence of polycyclic aromatic hydrocarbons (PAHs) in biochar-amended soil. *Chemosphere, 146*, 272–279. Available from https://doi.org/10.1016/j.chemosphere.2015.12.010.

Lan, D., Zhu, H., Zhang, J., Li, S., Chen, Q., Wang, C., Wu, T., & Xu, M. (2022). Adsorptive removal of organic dyes via porous materials for wastewater treatment in recent decades: A review on species, mechanisms and perspectives. *Chemosphere, 293*133464. Available from https://doi.org/10.1016/j.chemosphere.2021.133464.

Landmeyer, J. E. (2011). *Plant Control on the Fate of Common Groundwater Contaminants* (pp. 307–340). Springer Science and Business Media LLC. Available from 10.1007/978-94-007-1957-6_13.

Lapworth, D. J., Baran, N., Stuart, M. E., & Ward, R. S. (2012). Emerging organic contaminants in groundwater: A review of sources, fate and occurrence. *Environmental Pollution, 163*, 287–303. Available from https://doi.org/10.1016/j.envpol.2011.12.034.

Lee, C. K., Chao, H. P., & Lee, J. F. (2004). Effects of organic solutes properties on the volatilization processes from water solutions. *Water Research, 38*(2), 365–374. Available from https://doi.org/10.1016/j.watres.2003.10.009, http://www.elsevier.com/locate/watres.

Li, H., Cao, Y., Zhang, D., & Pan, B. (2018). pH-dependent KOW provides new insights in understanding the adsorption mechanism of ionizable organic chemicals on carbonaceous materials. *Science of the Total Environment, 618*, 269–275. Available from https://doi.org/10.1016/j.scitotenv.2017.11.065, http://www.elsevier.com/locate/scitotenv.

Li, J. X., Wu, L. C., Zhang, J., Wang, C. S., Yu, Q. Q., Peng, Y., & Ma, Y. H. (2015). Research progresses in remediation of heavy metal contaminated soils by biochar. *Ecol. Environ. Sci., 24*, 2075–2081.

Li, L., Zou, D., Xiao, Z., Zeng, X., Zhang, L., Jiang, L., Wang, A., Ge, D., Zhang, G., & Liu, F. (2019). Biochar as a sorbent for emerging contaminants enables improvements in waste management and sustainable resource use. *Journal of Cleaner Production, 210*, 1324–1342. Available from https://doi.org/10.1016/j.jclepro.2018.11.087, https://www.journals.elsevier.com/journal-of-cleaner-production.

Li, T., Yang, H., Zhang, N., Dong, L., Wu, A., Wu, Q., Zhao, M., Liu, H., Li, Y., & Wang, Y. (2024). Synergistic effects of arbuscular mycorrhizal fungi and biochar are highly beneficial to Ligustrum lucidum seedlings in Cd-contaminated soil. *Environmental Science and Pollution Research, 31*(7), 11214–11227. Available from https://doi.org/10.1007/s11356-024-31870-9, https://www.springer.com/journal/11356.

Li, X., Zhao, Z., Qu, Z., Li, X., Zhang, Z., Liang, X., Chen, J., & Li, J. (2023). A review of traditional and emerging residual chlorine quenchers on disinfection by-products: Impact and mechanisms. *Toxics, 11*(5), 410. Available from https://doi.org/10.3390/toxics11050410.

Liang, G., Wang, Z., Yang, X., Qin, T., Xie, X., Zhao, J., & Li, S. (2019). Efficient removal of oxytetracycline from aqueous solution using magnetic montmorillonite-biochar composite prepared by one step pyrolysis. *Science of The Total Environment, 695*133800. Available from https://doi.org/10.1016/j.scitotenv.2019.133800.

Liu, Y., Lonappan, L., Brar, S. K., & Yang, S. (2018). Impact of biochar amendment in agricultural soils on the sorption, desorption, and degradation of pesticides: A review. *Science of the Total Environment, 645*, 60–70. Available from https://doi.org/10.1016/j.scitotenv.2018.07.099, http://www.elsevier.com/locate/scitotenv.

Lu, H., Li, Z., Fu, S., Méndez, A., Gascó, G., Paz-Ferreiro, J., & Haverkamp, R. G. (2014). Can biochar and phytoextractors be jointly used for cadmium remediation. *PLoS One, 9*(4)e95218. Available from https://doi.org/10.1371/journal.pone.0095218.

Mamy, L., & Barriuso, E. (2007). Desorption and time-dependent sorption of herbicides in soils. *European Journal of Soil Science, 58*(1), 174–187. Available from https://doi.org/10.1111/j.1365-2389.2006.00822.x.

Matei, M., Zaharia, R., Petrescu, S. I., Radu-Rusu, C. G., Simeanu, D., Mierliță, D., & Pop, I. M. (2023). Persistent Organic Pollutants (POPs): A review focused on occurrence and incidence in animal feed and cow milk. *Agriculture, 13*(4).

Matter, I. A., Darwesh, O. M., & Matter, H. A. B. (2020). *Nanosensors for herbicides monitoring in soil Nanomaterials for Soil Remediation* (pp. 221–237). Egypt: Elsevier. Available from https://www.sciencedirect.com/book/9780128228913, 10.1016/B978-0-12-822891-3.00011-6.

Mitchell, P. J., Simpson, A. J., Soong, R., & Simpson, M. J. (2015). Shifts in microbial community and water-extractable organic matter composition with biochar amendment in a temperate forest soil. *Soil Biology and Biochemistry, 81*, 244–254. Available from https://doi.org/10.1016/j.soilbio.2014.11.017, http://www.elsevier.com/inca/publications/store/3/3/2.

Mohammadi-Moghadam, F., Khodadadi, R., Sedehi, M., & Arbabi, M. (2022). Bioremediation of polycyclic aromatic hydrocarbons in contaminated soils using vermicompost. *International Journal of Chemical Engineering, 2022*. Available from https://doi.org/10.1155/2022/5294170, http://www.hindawi.com/journals/ijce/.

Mukherjee, S., Tappe, W., Weihermueller, L., Hofmann, D., Köppchen, S., Laabs, V., Schroeder, T., Vereecken, H., & Burauel, P. (2016). Dissipation of bentazone, pyrimethanil and boscalid in biochar and digestate based soil mixtures for biopurification systems. *Science of the Total Environment, 544*, 192–202. Available from https://doi.org/10.1016/j.scitotenv.2015.11.111, http://www.elsevier.com/locate/scitotenv.

Mukome, F. N. D., Buelow, M. C., Shang, J., Peng, J., Rodriguez, M., Mackay, D. M., Pignatello, J. J., Sihota, N., Hoelen, T. P., & Parikh, S. J. (2020). Biochar amendment as a remediation strategy for surface soils impacted by crude oil. *Environmental Pollution, 265*. Available from https://doi.org/10.1016/j.envpol.2020.115006, https://www.journals.elsevier.com/environmental-pollution.

Muskus, A. M., Miltner, A., Hamer, U., & Nowak, K. M. (2022). Microbial community composition and glyphosate degraders of two soils under the influence of temperature, total organic carbon and pH. *Environmental Pollution, 297*. Available from https://doi.org/10.1016/j.envpol.2022.118790, https://www.journals.elsevier.com/environmental-pollution.

Muter, O., Berzins, A., Strikauska, S., Pugajeva, I., Bartkevics, V., Dobele, G., Truu, J., Truu, M., & Steiner, C. (2014). The effects of woodchip- and straw-derived biochars on the persistence of the herbicide 4-chloro-2-methylphenoxyacetic acid (MCPA) in soils. *Ecotoxicology and Environmental Safety, 109*, 93–100. Available from https://doi.org/10.1016/j.ecoenv.2014.08.012, http://www.elsevier.com/inca/publications/store/6/2/2/8/1/9/index.htt.

Narayanan, M., & Ma, Y. (2022). Influences of biochar on bioremediation/phytoremediation potential of metal-contaminated soils. *Frontiers in Microbiology, 13*. Available from https://doi.org/10.3389/fmicb.2022.929730, https://www.frontiersin.org/journals/microbiology#.

Oladipo, A. A., & Ifebajo, A. O. (2018). Highly efficient magnetic chicken bone biochar for removal of tetracycline and fluorescent dye from wastewater: Two-stage adsorber analysis. *Journal of Environmental Management, 209*, 9–16. Available from https://doi.org/10.1016/j.jenvman.2017.12.030, https://www.sciencedirect.com/journal/journal-of-environmental-management.

Oliveira, E. A., Montanher, S. F., Andrade, A. D., Nóbrega, J. A., & Rollemberg, M. C. (2005). Equilibrium studies for the sorption of chromium and nickel from aqueous solutions using raw rice bran. *Process Biochemistry, 40*(11), 3485–3490. Available from https://doi.org/10.1016/j.procbio.2005.02.026.

Ong, S. K., Culver, T. B., Lion, L. W., & Shoemaker, C. A. (1992). Effects of soil moisture and physical-chemical properties of organic pollutants on vapor-phase transport in the vadose zone. *Journal of Contaminant Hydrology, 11*(3-4), 273–290. Available from https://doi.org/10.1016/0169-7722(92)90020-f.

Patel, M., Kumar, R., Kishor, K., Mlsna, T., Pittman, C. U., & Mohan, D. (2019). Pharmaceuticals of emerging concern in aquatic systems: Chemistry, occurrence, effects, and removal methods. *Chemical Reviews, 119*(6), 3510–3673. Available from https://doi.org/10.1021/acs.chemrev.8b00299, http://pubs.acs.org/journal/chreay.

Pavel, M., Anastasescu, C., State, R.-N., Vasile, A., Papa, F., & Balint, I. (2023). Photocatalytic degradation of organic and inorganic pollutants to harmless end products: assessment of practical application potential for water and air cleaning. *Catalysts, 13*(2), 380. Available from https://doi.org/10.3390/catal13020380.

Pazos, M., Rosales, E., Alcántara, T., Gómez, J., & Sanromán, M. A. (2010). Decontamination of soils containing PAHs by electroremediation: A review. *Journal of Hazardous Materials, 177*(1-3), 1–11. Available from https://doi.org/10.1016/j.jhazmat.2009.11.055.

Pereira, L. (2014). Persistent Organic Chemicals of Emerging Environmental Concern. In A. Malik, E. Grohmann, & R. Akhtar (Eds.), *Environmental Deterioration and Human Health: Natural and Anthropogenic Determinants.* (pp. 163–213). Dordrecht: Springer. Available from https://doi.org/10.1007/978-94-007-7890-0.

Pouangam Ngalani, G., Dzemze Kagho, F., Peguy, N. N. C., Prudent, P., Ondo, J. A., & Ngameni, E. (2023). Effects of coffee husk and cocoa pods biochar on the chemical properties of an acid soil from West Cameroon. *Archives of Agronomy and Soil Science, 69*(5), 744–758. Available from https://doi.org/10.1080/03650340.2022.2033733, http://www.tandf.co.uk/journals/titles/03650340.asp.

Qiu, M., Liu, L., Ling, Q., Cai, Y., Yu, S., Wang, S., Fu, D., Hu, B., & Wang, X. (2022). Biochar for the removal of contaminants from soil and water: a review. *Biochar, 4*(1). Available from https://doi.org/10.1007/s42773-022-00146-1, springer.com/journal/42773.

Riedo, J., Wettstein, F. E., Rosch, A., Herzog, C., Banerjee, S., Buchi, L., Charles, R., Wachter, D., Martin-Laurent, F., Bucheli, T. D., Walder, F., & Van Der Heijden, M. G. A. (2021). Widespread occurrence of pesticides in organically managed agricultural soils-The ghost of a conventional agricultural past? *Environmental Science and Technology, 55*(5), 2919–2928. Available from https://doi.org/10.1021/acs.est.0c06405, http://pubs.acs.org/journal/esthag.

Rombolà, A. G., Fabbri, D., Baronti, S., Vaccari, F. P., Genesio, L., & Miglietta, F. (2019). Changes in the pattern of polycyclic aromatic hydrocarbons in soil treated with biochar from a multiyear field experiment. *Chemosphere, 219*, 662–670. Available from https://doi.org/10.1016/j.chemosphere.2018.11.178, http://www.elsevier.com/locate/chemosphere.

Roque, M. I., Gomes, J., Reva, I., Valente, A. J. M., Simões, N. E., Morais, P. V., Durães, L., & Martins, R. C. (2023). An opinion on the removal of disinfection byproducts from drinking water. *Water (Switzerland), 15*(9). Available from https://doi.org/10.3390/w15091724, http://www.mdpi.com/journal/water.

Ruomeng, B., Meihao, O., Siru, Z., Shichen, G., Yixian, Z., Junhong, C., Ruijie, M., Yuan, L., Gezhi, X., Xingyu, C., Shiyi, Z., Aihui, Z., & Baishan, F. (2023). Degradation strategies of pesticide residue: From chemicals to synthetic biology. *Synthetic and Systems Biotechnology, 8*(2), 302–313. Available from https://doi.org/10.1016/j.synbio.2023.03.005, http://www.keaipublishing.com/en/journals/syntheticandsystemsbiotechnology.

Sabharwal, S., Safrany, A., Junior, J. A. O., & Venkatesh, M. (2015). IAEA initiatives in advancement of radiation technologies for environmental remediation. *Journal of Advanced Oxidation Technologies, 18*(2). Available from https://doi.org/10.1515/jaots-2015-0213.

Sadegh-Zadeh, F., Wahid, S. A., & Jalili, B. (2017). Sorption, degradation and leaching of pesticides in soils amended with organic matter: A review. *Advances in Environmental Technology, 3*(2), 119–132. Available from https://doi.org/10.22104/AET.2017.1740.1100, aet.irost.ir.

Sanderson, P., Bahar, M. M., Biswas, B., & Naidu, R. (2023). Remediation of metals and organic contaminants in soil. *Encyclopedia of Soils in the Environment, 1-5*, 3. Available from https://doi.org/10.1016/B978-0-12-822974-3.00247-0, Second Edition: Volume, https://www.sciencedirect.com/referencework/9780323951333/encyclopedia-of-soils-in-the-environment.

Sarangi, N. V., & Rajkumar, R. (2022). *Biodegradation of organic pollutants by microbial process Environmental Microbiology: Emerging Technologies* (pp. 137–160). India: De Gruyter. Available from https://www.degruyter.com/document/doi/10.1515/9783110727227/html?lang=en, 10.1515/9783110727227-006.

Sarkar, B., Mukhopadhyay, R., Mandal, A., Mandal, S., Vithanage, M., & Biswas, J. K. (2020). *Sorption and desorption of agro-pesticides in soils Agrochemicals Detection, Treatment and Remediation: Pesticides and Chemical Fertilizers* (pp. 189–205). United Kingdom: Elsevier. Available from https://www.sciencedirect.com/book/9780081030172, 10.1016/B978-0-08-103017-2.00008-8.

Scow, K. M., & Johnson, C. R. (1996). Effect of sorption on biodegradation of soil pollutants. *Advances in Agronomy, 58*(C), 1–56. Available from https://doi.org/10.1016/S0065-2113(08)60252-7.

Semple, K. T., Morriss, A. W. J., & Paton, G. I. (2003). Bioavailability of hydrophobic organic contaminants in soils: fundamental concepts and techniques for analysis. *European Journal of Soil Science, 54*(4), 809–818. Available from https://doi.org/10.1046/j.1351-0754.2003.0564.x.

Shahzad, B., Tanveer, M., Che, Z., Rehman, A., Cheema, S. A., Sharma, A., Song, H., Rehman, S. U., & Zhaorong, D. (2018). Role of 24-epibrassinolide (EBL) in mediating heavy metal and pesticide induced oxidative stress in plants: A review. *Ecotoxicology and Environmental Safety, 147*, 935–944. Available from https://doi.org/10.1016/j.ecoenv.2017.09.066, http://www.elsevier.com/inca/publications/store/6/2/2/8/1/9/index.htt.

Siddiqa, A., & Faisal, M. (2021). *Microbial degradation of organic pollutants using indigenous bacterial strains* (pp. 625–637). Elsevier BV. Available from 10.1016/b978-0-12-819382-2.00039-9.

Singh, R. P., Pal, S., Rana, V. K., & Ghorai, S. (2013). Amphoteric amylopectin: A novel polymeric flocculant. *Carbohydrate Polymers*, *91*(1), 294–299. Available from https://doi.org/10.1016/j.carbpol.2012.08.024.

Solanki, A., & Boyer, T. H. (2019). Physical-chemical interactions between pharmaceuticals and biochar in synthetic and real urine. *Chemosphere*, *218*, 818–826. Available from https://doi.org/10.1016/j.chemosphere.2018.11.179, http://www.elsevier.com/locate/chemosphere.

Sophia A, C., & Lima, E. C. (2018). Removal of emerging contaminants from the environment by adsorption. *Ecotoxicology and Environmental Safety*, *150*, 1–17. Available from https://doi.org/10.1016/j.ecoenv.2017.12.026, http://www.elsevier.com/inca/publications/store/6/2/2/8/1/9/index.htt.

Sun, K., Ro, K., Guo, M., Novak, J., Mashayekhi, H., & Xing, B. (2011). Sorption of bisphenol A, 17α-ethinyl estradiol and phenanthrene on thermally and hydrothermally produced biochars. *Bioresource Technology*, *102*(10), 5757–5763. Available from https://doi.org/10.1016/j.biortech.2011.03.038.

Suri, R. P. S., Crittenden, J. C., & Hand, D. W. (1999). Removal and destruction of organic compounds in water using adsorption, steam regeneration, and photocatalytic oxidation processes. *Journal of Environmental Engineering*, *125*(10), 897–905. Available from https://doi.org/10.1061/(ASCE)0733-9372(1999)125:10(897).

Tan, G., Sun, W., Xu, Y., Wang, H., & Xu, N. (2016). Sorption of mercury (II) and atrazine by biochar, modified biochars and biochar based activated carbon in aqueous solution. *Bioresource Technology*, *211*, 727–735. Available from https://doi.org/10.1016/j.biortech.2016.03.147.

Tan, H., Wu, Q., Hao, R., Wang, C., Zhai, J., Li, Q., Cui, Y., & Wu, C. (2023). Occurrence, distribution, and driving factors of current-use pesticides in commonly cultivated crops and their potential risks to non-target organisms: A case study in Hainan, China. *Science of the Total Environment*, *854*. Available from https://doi.org/10.1016/j.scitotenv.2022.158640, http://www.elsevier.com/locate/scitotenv.

Tatarková, V., Hiller, E., & Vaculík, M. (2013). Impact of wheat straw biochar addition to soil on the sorption, leaching, dissipation of the herbicide (4-chloro-2-methylphenoxy)acetic acid and the growth of sunflower (Helianthus annuus L.). *Ecotoxicology and Environmental Safety*, *92*, 215–221. Available from https://doi.org/10.1016/j.ecoenv.2013.02.005.

Teixidó, M., Hurtado, C., Pignatello, J. J., Beltrán, J. L., Granados, M., & Peccia, J. (2013). Predicting contaminant adsorption in black carbon (Biochar)-amended soil for the veterinary antimicrobial sulfamethazine. *Environmental Science and Technology*, *47*(12), 6197–6205. Available from https://doi.org/10.1021/es400911c.

Thakur, M., & Pathania, D. (2019). *Environmental fate of organic pollutants and effect on human health Abatement of Environmental Pollutants: Trends and Strategies* (pp. 245–262). India: Elsevier. Available from http://www.sciencedirect.com/science/book/9780128180952, 10.1016/B978-0-12-818095-2.00012-6.

Tran, H. N., Tomul, F., Ha, Nguyen, D. T., Lima, E. C., Le, G. T., Chang, C.-T., Masindi, V., & Woo, S. H. (2020). Innovative spherical BC for pharmaceutical removal from water: insight into adsorption mechanism. *Journal of Hazardous Material*, *394*.

Tseng, R. L., Wu, F. C., & Juang, R. S. (2003). Liquid-phase adsorption of dyes and phenols using pinewood-based activated carbons. *Carbon*, *41*(3), 487–495. Available from https://doi.org/10.1016/S0008-6223(02)00367-6.

Ukalska-Jaruga, A., Debaene, G., & Smreczak, B. (2020). Dissipation and sorption processes of polycyclic aromatic hydrocarbons (PAHs) to organic matter in soils amended by exogenous rich-carbon material. *Journal of Soils and Sediments*, *20*(2), 836–849. Available from https://doi.org/10.1007/s11368-019-02455-8.

Wells, M. J. M., Morse, A., Bell, K. Y., Pellegrin, M. L., & Fono, L. J. (2009). Emerging pollutants. *Water Environment Research*, *81*(10), 2211–2254. Available from https://doi.org/10.2175/106143009X12445568400854, http://docserver.ingentaconnect.com/deliver/connect/wef/10614303/v81n10/s42.pdf?expires = 1261886667&id = 54127184&titleid = 11548&accname = Elsevier + Science&checksum = 60D63897D89F9128CEC366DFABF66114.

Wilkinson, J., Hooda, P. S., Barker, J., Barton, S., & Swinden, J. (2017). Occurrence, fate and transformation of emerging contaminants in water: An overarching review of the field. *Environmental Pollution*, *231*, 954–970. Available from https://doi.org/10.1016/j.envpol.2017.08.032, https://www.journals.elsevier.com/environmental-pollution.

Wu, X., Zheng, X., Yu, L., Lu, R., Zhang, Q., Luo, X. J., & Mai, B. X. (2022). Biomagnification of persistent organic pollutants from terrestrial and aquatic invertebrates to songbirds: associations with physiochemical and ecological indicators. *Environmental Science and Technology*, *56*(17), 12200–12209. Available from https://doi.org/10.1021/acs.est.2c02177, http://pubs.acs.org/journal/esthag.

Xia, X., Zhu, F., Li, J., Yang, H., Wei, L., Li, Q., Jiang, J., Zhang, G., & Zhao, Q. (2020). A review study on sulfate-radical-based advanced oxidation processes for domestic/industrial wastewater treatment: degradation, efficiency, and mechanism. *Frontiers in Chemistry*, *8*. Available from https://doi.org/10.3389/fchem.2020.592056, http://journal.frontiersin.org/journal/chemistry.

Xiong, Y., Wang, C., Dong, M., Li, M., Hu, C., & Xu, X. (2023). Chlorphoxim induces neurotoxicity in zebrafish embryo through activation of oxidative stress. *Environmental Toxicology*, *38*(3), 566–578. Available from https://doi.org/10.1002/tox.23702, http://onlinelibrary.wiley.com/journal/10.1002/(ISSN)1522-7278.

Xu, R.-K., Xiao, S.-C., Yuan, J.-H., & Zhao, A.-Z. (2011). Adsorption of methyl violet from aqueous solutions by the biochars derived from crop residues. *Bioresource Technology*, *102*(22), 10293–10298. Available from https://doi.org/10.1016/j.biortech.2011.08.089.

Xu, T., Lou, L., Luo, L., Cao, R., Duan, D., & Chen, Y. (2012). Effect of bamboo biochar on pentachlorophenol leachability and bioavailability in agricultural soil. *Science of the Total Environment*, *414*, 727–731. Available from https://doi.org/10.1016/j.scitotenv.2011.11.005.

Yang, W., Lampert, D., Zhao, N., Reible, D., & Chen, W. (2012). Link between black carbon and resistant desorption of PAHs on soil and sediment. *Journal of Soils and Sediments*, *12*(5), 713–723. Available from https://doi.org/10.1007/s11368-012-0494-0.

Yi, Y., Huang, Z., Lu, B., Xian, J., Tsang, E. P., Cheng, W., Fang, J., & Fang, Z. (2020). Magnetic biochar for environmental remediation: A review. *Bioresource Technology*, *298*. Available from https://doi.org/10.1016/j.biortech.2019.122468, http://www.elsevier.com/locate/biortech.

Yu, X. Y., Ying, G. G., & Kookana, R. S. (2009). Reduced plant uptake of pesticides with biochar additions to soil. *Chemosphere*, *76*(5), 665–671. Available from https://doi.org/10.1016/j.chemosphere.2009.04.001, http://www.elsevier.com/locate/chemosphere.

Zhang, P., Li, Y., Cao, Y., & Han, L. (2019). Characteristics of tetracycline adsorption by cow manure biochar prepared at different pyrolysis temperatures. *Bioresource Technology*, *285*121348. Available from https://doi.org/10.1016/j.biortech.2019.121348.

Zhang, W., Mao, S., Chen, H., Huang, L., & Qiu, R. (2013). Pb(II) and Cr(VI) sorption by biochars pyrolyzed from the municipal wastewater sludge under different heating conditions. *Bioresource Technology*, *147*, 545–552. Available from https://doi.org/10.1016/j.biortech.2013.08.082, http://www.elsevier.com/locate/biortech.

Zhang, X., Sarmah, A. K., Bolan, N. S., He, L., Lin, X., Che, L., Tang, C., & Wang, H. (2016). Effect of aging process on adsorption of diethyl phthalate in soils amended with bamboo biochar. *Chemosphere*, *142*, 28–34. Available from https://doi.org/10.1016/j.chemosphere.2015.05.037, http://www.elsevier.com/locate/chemosphere.

Zhang, Y., Gong, J., Cao, W., Qin, M., & Song, B. (2023). Influence of biochar and fulvic acid on the ryegrass-based phytoremediation of sediments contaminated with multiple heavy metals. *Journal of Environmental Chemical Engineering*, *11*(2)109446. Available from https://doi.org/10.1016/j.jece.2023.109446.

Zheng, H., Wang, Z., Zhao, J., Herbert, S., & Xing, B. (2013). Sorption of antibiotic sulfamethoxazole varies with biochars produced at different temperatures. *Environmental Pollution*, *181*, 60–67. Available from https://doi.org/10.1016/j.envpol.2013.05.056.

Chapter 9

Biochar for remediation of petroleum hydrocarbons in soil, sediments, and sludge

Musa Manga[1,2], Herbert Cirrus Kaboggoza[1], Swaib Semiyaga[3], Lauren Sprouse[1], Jiahui Guo[1], Anais Gentles[1], Yashraj Banga[1], Sarah Lebu[1] and Chimdi Muoghalu[1]

[1]*Department of Environmental Sciences and Engineering, Gillings School of Global Public Health, University of North Carolina at Chapel Hill, Chapel Hill, NC, United States,* [2]*Department of Construction Economics and Management, College of Engineering, Design, Art and Technology (CEDAT), Makerere University, Kampala, Uganda,* [3]*Department of Civil and Environmental Engineering, School of Engineering, College of Engineering, Design, Art and Technology, Makerere University, Kampala, Uganda*

Chapter outline

9.1 Introduction	149
9.2 Petroleum hydrocarbons in soil, sediments, and sludges	151
9.2.1 Classes and sources of petroleum hydrocarbons	151
9.2.2 Ecological and human health impacts of petroleum hydrocarbons in soils, sediments, and sludges	152
9.3 Biochar for remediation of petroleum hydrocarbons in soils, sediments, and sludges	152
9.3.1 Use and suitability of biochar for PHC remediation	152
9.3.2 Methods of applying biochar to soil, sediments, and sludges	153
9.3.3 Factors that influence the ability of biochar to remediate petroleum hydrocarbon-contaminated solid matrices	153
9.3.4 Mechanisms for remediation of petroleum hydrocarbon contaminated solid matrices using biochar	153
9.3.5 Scale of application for biochar remediation of petroleum hydrocarbon contaminated solid matrices	155
9.3.6 Synergistic interaction of biochar with other remediation technologies	158
9.3.7 Indicators of remediation efficiency of biochar for petroleum hydrocarbons in soil, sediments, and sludge	158
9.4 Ecotoxicity and health risks of biochar-remediated solid matrices	158
9.4.1 Ecotoxicity of biochar	158
9.4.2 Human health effects of biochar	159
9.4.3 Safety measures for biochar application	159
9.5 Future perspectives and directions	160
9.5.1 Key challenges	160
9.5.2 Considerations for future research	160
9.6 Conclusions	160
References	160

9.1 Introduction

The petroleum industry has been the backbone of many economies since the twentieth century as crude oil from it constitutes the primary source of energy in the world (Mukome et al., 2020). However, this industry and its related activities are significant sources of petroleum hydrocarbons (PHCs) in the environment especially on solid matrices like soil, sediments, and sludge (Engelhardt, 1994; Jagaba et al., 2022; Johnson & Affam, 2018). The release of PHCs in the environment stems from activities like crude oil exploration, transportation, storage, and processing, sediment dredging, and improper oil waste management (ATSDR, 2014; Bianco et al., 2021; Engelhardt, 1994; Jagaba et al., 2022; Johnson & Affam, 2018). In soils, PHCs disrupt microbial activities and various biogeochemical processes vital for plant growth (Chikere et al., 2011; Truskewycz et al., 2019). Exposure to PHCs in plants causes enzymatic dysfunction and reduced photosynthesis (Mukome et al., 2020). Also, the mobility of these compounds in soil causes water contamination, highlighting the far-reaching consequences of PHC pollution in terrestrial ecosystems. Beyond its impact on the soil, exposure to PHC is linked to risks of mutations and cancer in humans (Bastami et al., 2013). Therefore, remediation of PHC-contaminated solid matrices is of paramount interest.

Several approaches, such as landfarming/land filling (Harmsen et al., 2007), incineration (Li et al., 1995), bioremediation (Gogoi et al., 2003), phytoremediation (Panchenko et al., 2023), bioaugmentation (Roy et al., 2018a), pyrolysis (Liu et al., 2009), and chemical treatment (Mater et al., 2006) have been used to remediate PHC-contaminated solid matrices. However, these approaches have shortfalls including being cost prohibitive, having low efficacy due to low PHC bioavailability, and the release of noxious byproducts (Lamichhane et al., 2016; Johnson & Affam, 2018). This necessitates the need for other effective and economical alternatives for remediating PHC-contaminated solid matrices. One promising remediation approach for PHC-contaminated solid matrices is adsorption with biochar (Dike et al., 2021). Biochar possesses unique physiochemical properties (such as high specific surface area and high carbon content) as well as rich surface functional groups making it well suited for the remediation of PHC-contaminated solid media by adsorption (Kołtowski et al., 2017; Aziz et al., 2020; Zhang et al., 2020; Kaboggoza et al., 2024). Beyond its adsorption capability, biochar can be used as a soil conditioner for enhancing the physico-chemical and microbial properties of the soil (Gul et al., 2015).

This chapter discusses the use of biochar for remediating PHC-contaminated solid matrices. It specifically discusses the sources and impacts of PHC contamination, efficiency of biochar in removing PHCs, underlying mechanisms involved in the removal as well as factors which influence removal efficiency. In addition, the ecotoxicological impacts of biochar application to the solid matrices are also discussed. The chapter provides valuable insights on how biochar can be strategically applied for effective remediation of PHC-contaminated solid matrices (Fig. 9.1).

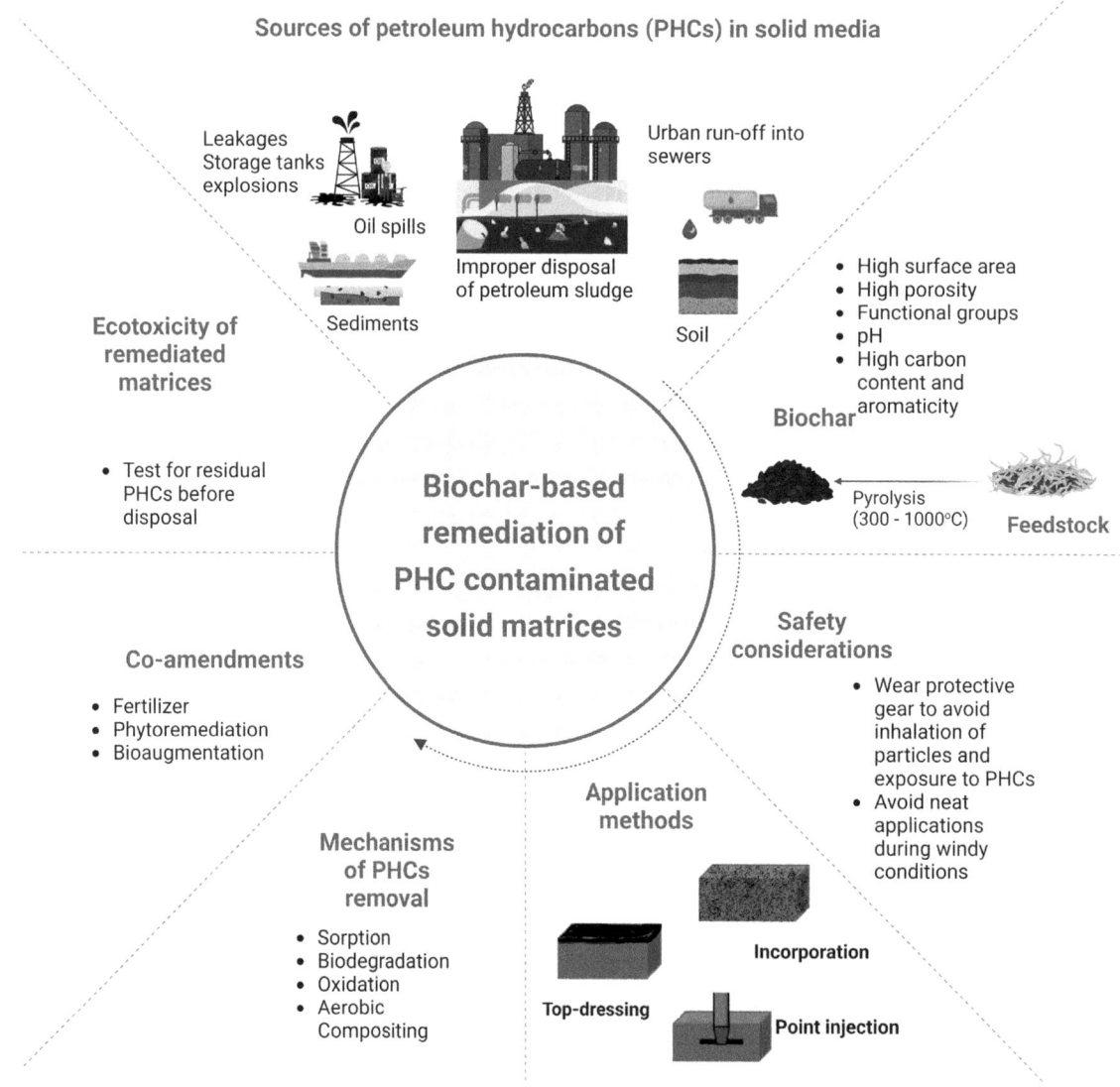

FIGURE 9.1 Summarzes the focus of this chapter.

9.2 Petroleum hydrocarbons in soil, sediments, and sludges

9.2.1 Classes and sources of petroleum hydrocarbons

Petroleum hydrocarbons constitute an extensive class of numerous chemical compounds made up of carbon and hydrogen atoms. PHCs are predominantly found in fuels and petroleum derivatives and are obtained from fractional distillation of crude oil (Weisman, 1998; Irwin et al., 1997). There are two broad classes of PHCs: aliphatic PHCs and aromatic PHCs (Table 9.1) (Kuppusamy et al., 2020; Williams et al., 2006). Aromatic PHCs contain at least one benzene ring while aliphatic PHCs do not contain benzene rings. Aromatic hydrocarbons are more resistant to biodegradation than aliphatic hydrocarbons due to the presence of benzene rings which make them very stable. Common examples of aliphatic hydrocarbons include alkenes, alkanes, and cycloalkanes while examples of aromatic hydrocarbons include monoaromatics like benzene (Table 9.1).

PHC contamination in solid matrices usually emanates from leakages from pipes and wellheads during offshore oil operations (oil spills), leakage/overflow from storage tanks, urban stormwater discharges, and inappropriate management of waste (Srivastava et al., 2019; Koshlaf & Ball, 2017; Perelo, 2010) (Table 9.1). Oil spills are the major sources of PHCs in solid matrices. Oil spills occur due to natural disasters which cause breakage of tanks and pipelines, industrial equipment malfunctions, and human error. The emergence of supertankers that can carry up to 500,000 tons of oil has increased the risk of detrimental oil spills in soils. For sediments, they get contaminated when oil spills into water bodies. In the U.S alone, about 10,000 barrels (158 987.3 L) of oil are spilled into the sea each year (NOAA, 2017). Most oils form a slick and sink down to the seafloor, contaminating sediments (Bagby et al., 2017). Several PHC contamination events in solid matrices due to oil spills have been reported. For instance, the discharge of crude oil (31,800 L) into the environment following explosion of the headspace of an oil tank in Cumberland Plateau, Tennessee resulted in PHC concentration of 500 to 700 mg/kg (Williams et al., 2006; Otton & Zielinski, 2000). PHC concentration of 81,000 mg/kg in soil was reported at a former crude oil and natural gas production

TABLE 9.1 PHC categories and their concentration in solid matrices.

PHC Category	Examples	Description	Concentration (mg/kg)			References
			Soil	Sediment	Sludge	
Aliphatics	Alkanes – paraffins (C_nH_{2n+2}) (Hexane, octane, and decane)	Single bonds between carbon atoms	0.52–35.26	2.50		Emoyan et al. (2020); Yang et al. (2018); Yuan et al. (2014)
	Alkenes – olefins (C_nH_{2n}) (Propene and ethene)	One or more double bonds between carbon atoms				
	Cycloalkanes – naphthalenes (Ethylcyclo-p-hexane and ethylcyclopentane)	Single-bonded carbon atoms arranged in cyclic structures				
Aromatics	Monoaromatics (Benzene, ethylbenzene, toluene, and xylenes (BTEX))	Possess a single benzene ring	0.382–0.733	0.027–2.40	4.20–161.90	Weng et al. (2012); Fu et al. (2018); Yuan et al., (2014); Yuan et al. (2015); Yang et al. (2018)
	Polyaromatics (Polycyclic aromatic hydrocarbons (PAHs); like phenanthrene, anthracene, and acenaphthylene)	Have two or more interconnected benzene rings				

facility (Soukup et al., 2007). These spills are a major environmental concern as hydrocarbons may persist in sediments and soils for long periods of time while affecting human and ecological health.

Urban stormwater discharge constitutes another significant source of PHC contamination in sediments (Raja et al., 2022). The concentration of hydrocarbons within stormwater runoff range from 0.7–6.6 mg/L, which exceeds the maximum recommended levels (0.01–0.1 mg/L) (EPA, 1999). After entering water bodies, due to their immiscibility in water, PHCs get absorbed by sediments which later settle to the bottom of the river. Petroleum sludge, acquired through the extraction, storage, and refining of crude oil (Hui et al., 2020), emerges as an additional contributor to PHC in solid matrices. This occurs when inadequate disposal practices are employed, such as applying petroleum sludge to landfills without appropriate engineering controls, resulting in the leakage of PHCs into soils and their runoff into water sediments (Evans et al., 2017; Truskewycz et al., 2019). Additionally, the improper disposal of gallons of used oil in sewer systems introduces PHCs into sludge.

9.2.2 Ecological and human health impacts of petroleum hydrocarbons in soils, sediments, and sludges

PHC contamination in soil adversely affects soil properties; it decreases soil porosity, enhances resistance to penetration, increases hydrophobicity, and reduces the cation exchange capacity of the soil (Pan et al., 2016; Uzojie & Agunwamba, 2011; Manga, 2017). Additionally, the introduction of PHCs into soil may hinder the growth of many microbial species which are crucial for maintaining the ecological balance of soils (Chikere et al., 2011). PHCs can be taken up by plants via soil, posing risks to grazing livestock, wildlife, and insects. Once in plants, PHCs cause oxidative stress, minimize growth, and cause leaf deformation, leading to enzymatic dysfunction and a reduced ability to photosynthesize (Mukome et al., 2020). When such plants are ingested by humans, PHCs can bioaccumulate and cause adverse health effects due to their mutagenicity and carcinogenicity (Bastami et al., 2013). Human exposure to PHCs has been linked to lung, liver, and skin cancers, and reproductive and neurological defects in humans (Latif et al., 2010; Todd et al., 2002; NIEHS, 2016). In water sediments, PHCs negatively affect aquatic flora and fauna. For example, fish populations and coral reefs significantly declined because of PHC contamination in the Gulf of Mexico after the Deepwater Horizon oil disaster (Incardona et al., 2013). Given the negative impacts of PHC contamination in solid matrices, it is important that release of PHCs in the environment is curtailed.

9.3 Biochar for remediation of petroleum hydrocarbons in soils, sediments, and sludges

9.3.1 Use and suitability of biochar for PHC remediation

Biochar, a carbonaceous material produced from pyrolysis of biomass, is a promising eco-friendly alternative for removing PHCs from soils, sediments, and sludge (Gomez-Eyles & Ghosh, 2018; Dike et al., 2021; Muoghalu et al., 2023). Biochar's physiochemical properties including its high porosity and specific surface area facilitates effective adsorption of PHCs (Aziz et al., 2020). Also, the high specific surface area and porosity enables it to immobilize and harbor microbes thereby facilitating microbial degradation of PHCs (Batista et al., 2018; Zhang & You, 2013; Zhu et al., 2017). In addition, the oxygen-containing functional groups, such as carbonyls and hydroxyls on biochar's surface facilitate different reactions and mechanisms for the removal of PHCs from solid matrices (Tan et al., 2021; Zama et al., 2018)

Biochar has mainly neutral or slightly alkaline pH with biochar produced at high temperatures having very alkaline pH (Wu et al., 2012; Zhao et al., 2013; Ainomugisha et al., 2024a). The neutral to alkaline pH range of biochar improves the carbonate group on its surface as well as the amount of alkali and alkaline earth metals which are suitable for adsorbing PHC. In addition, the neutral to alkaline pH of biochar is optimal for the activity of diverse microorganisms. Further, owing to its high carbon content, biochar is resistant to microbial degradation and chemical recalcitrance indicating that it can be stable for a long period of time when applied to the soil (Domingues et al., 2017; Wiedemeier et al., 2015).

Besides biochar's ability to remove PHC from the soil, biochar improves soil properties by adding nutrients, improving soil water retention capacity and pore structure, thus promoting the growth and activity of microbes (Aziz et al., 2020; Bianco et al., 2021; Hussain et al., 2018; Jagaba et al., 2022; Zhang et al., 2020); and inhibiting the loss of nutrients into the soil by leaching (Atkinson et al., 2010). The use of waste biomass for biochar production contributes to waste management efforts and fosters the adoption of a circular economy framework (Qambrani et al., 2017; Rex et al., 2023; Singh et al., 2022; Ainomugisha et al., 2024b).

9.3.2 Methods of applying biochar to soil, sediments, and sludges

Biochar may be applied to solid matrices via incorporation, injection, and top-dressing methods (Alma & Altikat, 2021; Major et al., 2010). The choice of method depends on factors such as type of soil/sediment, reason for applying biochar, and available equipment and expertise. In incorporation, biochar is mechanically mixed with the media as a slurry or without water (Alma & Altikat, 2021; Husk & Major, 2010; Major et al., 2010). This method is tedious and consumes a lot of energy, however, it ensures effective contact between biochar and the soil particles. Injection involves the introduction of biochar as a slurry at specific points; this eliminates the need for mixing. (Major et al., 2010). Top dressing involves applying biochar on the surface or in-situ of the solid media (Lehmann & Joseph, 2015; Major et al., 2010). It also eliminates mixing and minimizes loss of moisture from the matrix, but biochar can easily be washed away by water or wind if applied on top.

9.3.3 Factors that influence the ability of biochar to remediate petroleum hydrocarbon-contaminated solid matrices

Several factors influence the ability of biochar to remediate PHC-contaminated solid matrices. These factors can be categorized into biochar production factors, biochar application factors, and media-related factors (Dike et al., 2021) (Table 9.2). Biochar production factors include pyrolysis parameters (temperature and time), feedstock type, and particle size (Chi & Liu, 2016; Jia et al., 2020). For the same type of feedstock, biochar produced at a higher temperature usually has more functional groups which makes it more effective for the adsorption of PHC from the soil. This was observed by Kong et al. (2018) who reported that wheat straw-derived biochar produced at a higher temperature (500°C) substantially decreased PAHs compared to that prepared at 300°C. Similarly, Chi and Liu (2016) noted that the biochar they prepared at 400°C was not effective in the adsorption of phenanthrene as it achieved only 11.9% removal efficiency. However, the biochar they produced at 700°C achieved 43.3%. for the same application rate and time.

With regard to feedstock, the source of the feedstock influences the effectiveness of biochar in adsorbing PHCs in the soil. Biochar made from animal manure and sludge usually has higher ash content and is more alkaline than biochar made from wood, agricultural, and food waste. The higher ash content of sludge/manure biochar implies that it is more alkaline and contains alkali metals which help in adsorbing PHCs in the soil. This was observed in the study by Aziz et al. (2020) who used fruit/vegetable waste biochar and sludge biochar for the remediation of PHC-contaminated soil. They observed that the sludge biochar was more effective in adsorbing PHCs than the fruit/vegetable biochar. Further, it has been noted that biochar derived from feedstocks containing heavy metals and organic pollutants can introduce these contaminants into the soil, thereby diminishing the efficacy of PHC adsorption by biochar. For instance, Mukome et al. (2020) reported that walnut shell biochar exhibited low effectiveness in PHC adsorption (less than 5% removal efficiency) after a 60-day incubation period, attributing this ineffectiveness to the presence of dioxins on the biochar surface, which adversely affected microbial activity.

In the context of biochar application factors, it has been noted that the efficiency of PHC adsorption by biochar tends to increase with higher application rates and extended application durations. For instance, in the study by Mukome et al. (2020) where they applied ponderosa pine wood biochar for 30 and 60 days, they observed that biochar remediation after 60 days was higher (69.2%) than after 30 days (58.3%). All the aforementioned examples indicate that in remediating PHC-contaminated soils with biochar, due consideration should be given to the different influencing factors.

9.3.4 Mechanisms for remediation of petroleum hydrocarbon contaminated solid matrices using biochar

9.3.4.1 Hydrophobic adsorption

Hydrophobic adsorption serves as a primary mechanism for the removal of PHCs from soil by biochar (Srivastava et al., 2019) This is primarily attributed to the hydrophobic nature of PHCs, which inherently exhibits low affinity for water. The process of hydrophobic adsorption on biochar entails physical interactions, primarily driven by van der Waals forces and π-π interactions (Chen et al., 2019). The nonpolar nature of PHCs, being hydrophobic compounds, facilitates favorable interactions with the nonpolar regions of the biochar surface. Consequently, these hydrophobic molecules become adsorbed onto the biochar, leading to a notable reduction in their concentration within the solid matrix (Chi & Liu, 2016). However, the adsorption of the PHCs from the contaminated soil onto biochar alone generates an environmental risk. The accumulated PHC on biochar maybe released into the environment causing secondary pollution which may affect the ecosystem (Blenis et al., 2023; Zhang et al., 2019). In the long term, the accumulated

TABLE 9.2 Factors influencing efficiency of biochar in remediating petroleum hydrocarbon-impacted solid matrices.

Factors		Description	Comment	References
Biochar Production	Temperature	500, 700, and 900 °C	Biochar produced at 700°C had superior properties compared to the others and achieved polycyclic hydrocarbon levels fit for soil restoration.	Mukome et al. (2020)
		400 and 600 °C	The removal rate of pyrene was 68.8% for biochar produced at 400 °C and 52.2% at 600°C	Jia et al. (2020)
		300 °C and 500 °C	After 84 days higher reduction (55.6%) in the PAH was achieved by biochar produced at a higher temperature (500 °C) than that at 300°C (47.7%)	Kong et al. (2018)
		400°C and 700°C	Biochar prepared at 400 retarded the degradation of phenanthrene achieving 11.9% after 18 days while that at 700 achieved 43.3%.	Chi & Liu (2016)
	Feedstock type	Ponderosa Pine (PP), Walnut Shell, and Wood Chips	PP had the highest biodegradation rate, while walnut shell derived biochar inhibited TPH biodegradation. Woodchips were found to be less effective than PP for light crude oil-contaminated soil B, but comparably effective for heavy crude oil-contaminated soil.	Mukome et al. (2020)
		Fruit/Vegetable wastes and Sludge	Sludge-derived biochar achieved a higher removal efficiency than biochar from fruit/vegetable wastes	Aziz et al. (2020)
	Particle Size	Biochar of 20, 70 and 480 μm	PHC removal efficiency increased from 480 > 70 > 20 μm particle size	Agarry et al. (2015)
	Modification	Biochar with graphite carbon nitride (g-C3N4) composite	PHC removal rate by the composite was 54.5% was 2.12 and 1.95 times of biochar and g-C3N4, respectively.	Lin et al. (2022)
Application related	Time	30 and 60 days	Biochar remediation after 60 days was higher (69.2%) than after 30 days (58.3%)	Mukome et al. (2020)
	Biochar addition rate	5 and 10%	TPH removal was higher at 5% rate	Mukome et al. (2020)
		0.5 g and 4 g	0.5 g produced the highest CH_4 yield (138.41 mL/gVS), 2.19 times yield without biochar while 4.8 g diminished yield by 32.5% and prolonged lag phase by 5.72 hours.	Shi et al. (2022)
	Co-application	Biochar + Fertilizer and bulking agents	TPH biodegradation rates were higher for amendment with biochar, fertilizer and bulking agents mixed.	Mukome et al. (2020)
		Biochar + Cow dung	Biochar + cow dung amendment was 82.86%, 7.26% more than biochar alone.	Aziz et al. (2020)
		Biochar + Cowpea	Biochar alone achieved 58.3% and 69.2% removal rate of PHC after 30 and 60 days, respectively, while biochar + cowpea achieved 12.8% and 58.2%.	Saliu et al. (2023)
		Biochar (B), biochar + mushroom residue (BM), and biochar + corn straw (BC)	The removal rate of PAHs using BC amendment was higher (72.69%) than B and BM (B - 61.41% and BM - 69.67%)	Bao et al. (2020a)
		Biochar, biosurfactant - rhamnolipid and nitrogen	Biochar and rhamnolipid, biochar and nitrogen, and biochar, rhamnolipid, and nitrogen achieved 32.3%, 73.2%, 80.9% removal of TPHs, respectively.	Liu et al. (2017)
		Biochar with maize, white clover, alfalfa, rye grass, and wheat plants.	Biochar with alfalfa, rye grass, and white clover achieved the highest % removal 65.94, 63.01, 68.1% while maize and wheat achieved 50.44 and 34.3% after 75 days.	Yousaf et al. (2022)
	Application method	Incorporation and injection	Although the comparative was done using different soil samples, incorporation and injections caused a significant decrease after 31 and 334 days, respectively.	Mukome et al. (2020)

toxic substances in biochar maybe adsorbed by the crops which may further pass them to the consumers (Wei et al., 2024). To mitigate this risk, biochar-microbe interaction should be cultivated.

9.3.4.2 Biochar-microbe interactions

In remediating PHC-contaminated solid matrices, biochar acts as a biostimulator, which harbors and promotes the growth of PHC-degrading microbes (Mukome et al., 2020; Bao et al., 2020). The superior adsorption ability of biochar facilitates PHC immobilization from the solid matrices (Iranzi Emile Rushimisha et al., 2022) minimizing the spatial separation between PHCs and the microbes and increasing the availability of PHCs to microbes. (Chen et al., 2019; Xiong et al., 2017). PHC degrading microbes include bacterial strains, such as *Psuedomonas poae, Actinobacter bouvetii* (Hussain et al., 2018), *Proteobacteria* and *Bacteroidetes* (Wei et al., 2020), as well as fungal communities like mycorrhizae (Zhen et al., 2019) These microbes produce hydrolyzing enzymes such as dehydrogenase and urease which can degrade PHCs (Manga et al., 2023; Zhen et al., 2019; Li et al., 2019). These microbes transform complex pollutant compounds into simpler, less harmful compounds during the restoration of contaminated soils (Macaulay & Rees, 2014). It has been suggested that these microbes can completely mineralize PHCs into nontoxic products including, CO_2, H_2O, inorganic compounds, cell proteins, and simple organic compounds found in soil (Das & Chandran, 2011). The action of PHC-degrading microbes occurs via aerobic and anaerobic pathways. Under aerobic conditions, the microbes use oxygen to introduce hydroxyl groups into HC molecules by the action of enzymes, such as monooxygenases and dioxygenases, converting them into intermediates including, alcohols, aldehydes, and ketones (Mekonnen et al., 2024). These intermediates are then via the citric acid cycle (Krebs cycle) and ultimately transformed into carbon dioxide, water, and biomass. In anaerobic conditions, alternative electron acceptors such as nitrate, sulfate, or carbon dioxide are used, initiating the reduction of hydrocarbons to simpler compounds (Cruz Viggi et al., 2023). These compounds are further broken down through fermentation processes into organic acids, alcohols, hydrogen, and carbon dioxide, and can be further converted into methane and carbon dioxide by microbes, such as methanogenic archaea (Wartell et al., 2021).

9.3.5 Scale of application for biochar remediation of petroleum hydrocarbon contaminated solid matrices

The application of biochar for the remediation of PHC-contaminated soils has predominantly been explored through experimental and microcosm studies, with the studies demonstrating its effectiveness in controlled environments. Microcosm studies show that the addition of biochar to the different solid matrices results in high removal rates of PHC with values ranging from 40% to over 90% (Table 9.3). For instance, Mukome et al. (2020) used ponderosa pine wood chips-derived biochar for the remediation of soils contaminated with PHCs (16,000–24,000 mg/kg) and observed over 65% PHC removal rates in 60 days (Table 9.3). Further, the United States EPA total PHC concentration standard (10,000 mg/kg) was achieved after 230 and 30 days in heavily contaminated and light-contaminated soil, respectively. Similarly, in the study by Kong et al. (2018), the addition of biochar derived from sawdust and wheat straw (5% w/w) to PHC-impacted soils resulted in 47.9% to 55.7% PAHs removal rate. Karppinen et al. (2017) investigated the capability of biochar from meat and bonemeal to trigger PHC removal in frozen soil. The addition of biochar to the soil in their study removed over 50% of the PHCs in the soil (Karppinen et al., 2017).

On the other hand, large-scale application of biochar for the remediation of PHC-impacted sites is still in its early stages with only a handful of pilot studies and field trials exploring its scalability. However, the results from these studies indicate that large-scale application is not as effective as laboratory-scale application due to the inability to control environmental conditions (Dike et al., 2021). For example, Karppinen et al. (2017) applied biochar (from wood, fishmeal, and bonemeal) as a remedial amendment for farmlands in Canada using PHC-impacted frozen soil. When they tested the biochar in laboratory settings, biochar amendment achieved between 55%–63% PHC removal rate, while the control (fertilizer amendment) achieved 39.9%. However, there was no significant difference between biochar-based PHC removal rates and those of the control (fertilizer) in field application. To improve its effectiveness in the field, biochar can be co-applied with fertilizer, compost, bioaugmentation, and phytoremediation techniques. Mukome et al. (2020) showed that compared to single amendments, biochar and fertilizer amendments achieved faster PHC reduction and were more effective for heavily than lightly contaminated soil. Higher PHC removal in co-amendments was attributed to more PHC degrading microbial counts and more nutrient availability (Mukome et al., 2020).

TABLE 9.3 Summary of studies on the application of biochar for the remediation of petroleum hydrocarbon-contaminated solid matrices.

Soil type	Contaminant and its Source	Contaminant Concentration (mg/kg)	Biochar preparation conditions				Application Rate(w/w %)	Removal efficiency (%)	Time (days)	Scale of Application	References
			Feedstock	Temp(°C)	Residence Time (hour)	Surface Area (m^2/g)					
Sand/Silt/Clay Soil	Crude oil – Accidental Spills and Waste disposal	24 000	Ponderosa Pine	900		127	5	68.2	60	Lab batch	Mukome et al. (2020)
				900		127	10	61.2	60	Lab batch	
	Diesel – Railway Washing	291 400	Sludge	550	2	46.9	5	75.6	180	Lab batch	Aziz et al. (2020)
			Fruit or Vegetable Waste	550	2	52.5	5	72.2	180	Lab batch	
Sand	PHCs – Accidental Spills	1 200	Titan fishmeal	500		7.1	6^{g2}	57.4	90	Lab batch and field	Karppinen et al. (2017b)
			Zakus Bonemeal	450		110	6^{g3}	61.2	90		
			Zakus Wood	450		78	6^{g4}	55.3	90		
Loamy Sand Soil	PHCs – Lab Spiking	22 867	Wood	400			15^g	69.2	60		Saliu et al. (2023)
	PAH – NA	2.20					5	61.4	77	Lab batch	Bao et al. (2020a)
	PHCs – NA	85 632	Rice Husk	700	3	285.5	2	32.4	60	Lab batch	Zhen et al. (2019)
Light Clay Soil	PAHs – Oil Waste Disposal	3 596.4	Sawdust	300	3	4.8	5	47.8	180	Lab batch	Kong et al. (2018)
		3 596.4	Sawdust	500	3	28.5	5	55.6	180	Lab batch	
Silt/Clay Soil	Crude Oil – Lab spiking	533000	Sugarcane Residue	550		58.9	1	73.2	20	Lab batch	Wei et al. (2020b)
	PHCs – NA	64,228	Wheat Straw	600		53.9	2	42.2	223	Lab batch	Li et al. (2019)
		64,228	Chicken Manure	600		97.2	2	40.7	223	Lab batch	
Silt Loam Soil	PAHs – NA	11.9	Maize Straw	500		36.4	1	47.8	21	Lab batch	Li et al. (2019)
	PHCs – NA	0.04				26.4	5	68.1	75	Lab batch	Yousaf et al. (2022)
River Sediment	PAHs – Accidents Spills	0.1	Macadamia Nut Shells	500	1		1.3	29.0	100	Lab batch	Chen et al. (2019)

Estuary Sediment	PAHs – Urban Run off	15.4	Pine Wood	700–1000	0.2	358	5	98.0	28	Lab batch	Gomez-Eyles & Ghosh (2018)
River Sediment	Phenanthrene – Lab spiking	16	Wheat Straw	700	4	256.04	3	43.2	18	Lab batch	Chi & Liu (2016)
Estuary Sediment	Pyrene – Lab spiking	7.5	Mangrove Plant	400	4	4.4	1	68.8	56	Lab batch	Jia et al. (2020)
Estuary Sediment		11.5	Mangrove Plant	600	4	6.1	1	52.2	56	Lab batch	

9.3.6 Synergistic interaction of biochar with other remediation technologies

Recent studies have employed biochar with other remediation technologies, including composting (Saum et al., 2018), phytoremediation (Han et al., 2016), bioaugmentation (Ren et al., 2020), and phytoremediation-bioaugmentation (Abbaspour et al., 2020) to remediate PHC contaminated solid matrices. Results from these studies indicate that biochar application enhances the effectiveness of other remediation strategies. For example, Ren et al. (2020) used corncob, straw, and sawdust-derived biochar together with immobilized oil-degrading bacteria to remediate PHC-contaminated soil from a shale-gas field. In this study, biochar application increased the PHC removal efficiency - the efficiency of microbial activity alone and combined with corncob biochar were 45% and 70.1%, respectively. Similarly, another study reported that the biochar increased the biodegradation percentages of individual PAHs by 3%–60% compared to the remediation without biochar. In this study, it was suggested that biochar's inorganic minerals were responsible for stimulating microbial abundance and diversity in soil. Further, the use of biochar as a co-remediation agent enhances the effectiveness of phytoremediation by improving the soil's physical, chemical, and biological properties as discussed in Section 8.3.1. By improving soil pH, carbon content, nutrient content, soil aeration, soil fertility, water retention capacity, and microbial diversity, it promotes the growth of plants used for phytoremediation and PHC-degrading microorganisms. This synergistic interaction enhances the overall effectiveness of PHC remediation.

9.3.7 Indicators of remediation efficiency of biochar for petroleum hydrocarbons in soil, sediments, and sludge

Researchers have employed various indicators and endpoints to monitor the remediation efficiency of biochar for PHC in solid matrices, and maybe broadly categorized into chemical, biological, and ecotoxicological assessments. Chemical indicators include the reduction in parent compounds such as total PHCs (TPH) and specific fractions like alkanes, aromatic hydrocarbons, and polycyclic aromatic hydrocarbons (PAHs), along with changes in chemical properties like soil organic matter and nutrient levels. Biological indicators involve assessing microbial activity, including biomass and community structure, as well as enzymatic activities, and evaluating plant growth and health to identify potential phytotoxic effects. Additionally, ecotoxicological bioassays offer a comprehensive understanding of remediation efficiency by examining the overall ecological impact. These bioassays include earthworm toxicity tests to evaluate survival and reproduction, microbial bioassays using luminescent bacteria to detect changes in bioluminescence, aquatic organism tests with species such as Daphnia magna and fish embryos, and plant bioassays for seed germination and root elongation. (Dawson et al., 2007; De Boeck et al., 2001; Deebika, 2021; Ide-Pérez et al., 2020; Ihunwo et al., 2021; Kebede et al., 2021; Mekonnen et al., 2024; Ren et al., 2020; Zawierucha et al., 2022; Zhang et al., 2020).

9.4 Ecotoxicity and health risks of biochar-remediated solid matrices

9.4.1 Ecotoxicity of biochar

Despite the positive impacts of applying biochar to the soil, unintended negative effects might still occur with its application. First, biochar is affected by environmental conditions and ages over time, which can decrease its affinity for certain contaminants and allow their release back into the environment (Kavitha et al., 2018). With aging and adverse environmental conditions, contaminants present on the surface of biochar may leach back into the soil. Additionally, biochar can also impact the chemical, physical, and biological properties in the solid media where it is applied, leading to unfavorable conditions for certain organisms (Godlewska & Oleszczuk, 2021). Application of biochar may cause immobilization of nitrogen and other nutrients due to the large carbon-to-nitrogen ratio, which may decrease bacterial and plant growth (Brewer & Brown, 2012). Treating soils with both biochar and nutrients has been shown to alleviate this problem (Saum et al., 2018; Wang et al., 2017). Concerns have also been raised around the impact of organic contaminants and heavy metals present in biochar for their ecotoxicological effects, suggesting a need to merge chemical analysis with ecotoxicological evaluations (Kuppusamy et al., 2016). Furthermore, studies have highlighted that spent biochar containing accumulated PHC may easily be carried by erosion owing to its lightweight (de Resende et al., 2018). For example, if biochar is applied to PHC-contaminated sites without adequate incorporation into the soil, it could be washed into nearby water bodies during heavy rainfall or irrigation (de Resende et al., 2018; Murtaza et al., 2023). This movement could potentially carry PHC contaminants, posing a risk to aquatic ecosystems. Similarly, biochar can be easily dispersed by wind due to its lightweight nature (de Resende et al., 2018). Therefore, proper handling of biochar is crucial to prevent the dissemination of PHC into the environment as discussed in Section 9.4.3.

9.4.2 Human health effects of biochar

High levels of PHCs present in biochar-amended soils threaten human health, particularly the health of individuals engaged in the production and application of biochar in the soil. Workers are vulnerable not only to contaminated soils but also to biochar itself, as well as to the tiny particles formed during biochar production. Studies have evaluated the increased lifetime cancer risk (ILCR)—a unitless factor representing the increased lifetime incidences of cancer because of exposure to a hazardous substance—of workers exposed to PAHs in soil and found ILCR values ranging from $1.90 \cdot 10^{-7}$ to $1.24 \cdot 10^{-3}$ (Peng et al., 2011; Man et al., 2013), which represent a low to very high increased risk of cancer. This risk is only attributable to PHCs in soils and does not account for workers directly exposed to biochar dust, so the true risk of illness among the biochar workforce may be substantially higher (Kuśmierz & Oleszczuk, 2014).

9.4.3 Safety measures for biochar application

The safety objective of handling and applying biochar is to minimize the amount of biochar in the air, and thus minimize its potential inhalation (Schwab & Hanna, 2012). During field application, it is recommended that biochar be applied as a slurry and should always be mixed with a soil matrix to avoid dust formation (Sigmund et al., 2017). Biochar application should be postponed during windy conditions that may lead to its suspension in the air. Additionally, staying upwind of biochar applications may reduce personal exposure. Workers should always wear individual protective gear, including gloves, googles, long sleeves and pants, and potentially a respirator, according to the conditions (Schwab & Hanna, 2012). To reduce potential harm to ecosystems and human health, additional safety guidelines for storing, handling, applying, and disposing of used biochar are provided below (Table 9.4).

TABLE 9.4 Safety recommendations for storing, handling, applying, and disposing of used biochar and contaminated media.

Safety Category	Recommendation	Goal
Storing biochar	a. Do not store biochar near food or beverages.	a. Avoid ingestion of potential harmful materials.
	b. Store biochar in a dry, well-ventilated area.	b. Prevent moisture adsorption and mold or bacterial growth.
	c. Reseal biochar containers immediately after use.	c. Reduce risk of auto ignition or spontaneous heating.
	d. Keep fire extinguishers readily available and ensure the storage area complies with fire safety regulations. Also, do not store biochar near flammable materials.	d. Minimize the risk of combustion, as biochar is highly flammable.
Handling biochar	a. Dampen the biochar prior to handling or using wetting agents.	a. Minimize dust production and risk of inhalation; prevent loss to wind.
	b. Wear appropriate individual protective equipment (gloves, long sleeves and pants, respirator, eye protection).	b. Protection from irritants or dust particles.
	c. Avoid excessive agitation or unnecessary movement of biochar.	c. Minimize dust production and risk of inhalation.
Applying biochar	a. Consider postponing biochar application during windy conditions.	a. Minimize dust production and risk of inhalation.
	b. Remain upwind of biochar application.	b. Minimize risk of inhalation.
Disposing of used biochar	a. Reuse biochar as an adsorbent for leachate treatment.	a. Reduce contamination from landfills.
	b. Use spent biochar as a landfill cover material.	b. Reduce odors and contamination.

9.5 Future perspectives and directions

9.5.1 Key challenges

Though biochar is a potential adsorbent and biostimulator for remediation of solid media, there are several key engineering, scientific, economic, and regulatory challenges associated with its production and application. One major challenge is reducing the amount of greenhouse gas (methane) released when biochar is applied to the soil. There is a need to find ways of effectively applying biochar to reduce the release of greenhouse gases. Standardization and regulation of biochar production is another challenge because each biochar lot may be unique from other lots, even when the same process conditions and feedstock materials are used (Barquilha & Braga, 2021). Developing guidelines for the production and application of biochar is difficult due to the absence of standardization in terms of both processes and terminology (Barquilha & Braga, 2021). Although the transition to biochar production systems from open-air burns in forestry provides air quality benefits, the regulatory process is much more multifaceted, expensive, and time-consuming compared to the approval process for open burns (Amonette et al., 2021). Additionally, removal by adsorption alone leads to accumulation of PHC in biochar creating a potential source of secondary contamination (Zhang et al., 2019). Moreover, due to the buoyant and lightweight nature of biochar, erosion of spent biochar opens risks associated with the dissemination of PHC compounds (Murtaza et al., 2023). Implementing strategies such as surface covering, vegetation barriers, and using larger biochar particles may reduce the risks associated with erosion of biochar.

9.5.2 Considerations for future research

Several studies have pointed out the need for future research on biochar-microbe interaction to enhance biodegradation of PHCs (Zhang et al., 2019; Zahed et al., 2021; Dike et al., 2022). Combining biochar with PHC-degrading bacteria has shown promise for the restoration of PHC-contaminated soils (Zahed et al., 2021), and has proven to be more effective in breaking down PHCs than the use of either biochar or bacteria alone (Zhang et al., 2019). However, little is known about the environmental toxicity and cost implications of biochar and bacteria co-application for remediation of PHC-contaminated soils (Dike et al., 2022) necessitating the need for research on this. Research is also needed to maximize the synergistic interactions between PHC-degrading bacteria and biochar to capture the full potential of biochar-based remediation of solid media.

Additionally, majority of research conducted thus far on the adsorption of PHCs by biochar has primarily involved laboratory-scale studies, and there is limited data available on field-based processes (Rosales et al., 2017). Recognizing that laboratory conditions may not accurately reflect real-world scenarios, there is a necessity for more comprehensive field-based studies. Further, there is a need for more research on developing modified biochar, such as functionalized biochar, magnetic biochar, and biochar composites (Liu et al., 2023), with better adsorption capacities and long-term sustainability, while also focusing on cost-effectiveness.

Finally, there is also a need for further research to quantitatively predict how biochar amendments will impact plants and other living organisms that rely on soil. Creating predictive models can allow for the development of best management practices for the myriad of prospective settings where biochar may be employed, accelerating uptake of biochar as a mainstream technology (Dai et al., 2020; Jindo et al., 2020).

9.6 Conclusions

PHC contamination of soils, sediments, and sludges (solid matrices) poses significant risks to human and environmental health. Traditional remediation strategies are limited in terms of effectiveness and cost-efficiency. Biochar-based amendment of PHC-contaminated soils, sludges, and sediments is a low-cost and effective alternative. The effectiveness of biochar application in solid media remediation depends on the production characteristics of biochar, the application technique as well as the characteristics of the type of PHCs to be removed. Further research is still required to solve the challenges associated with the application of biochar and there is a need for more studies on large-scale field application.

References

Abbaspour, A., Zohrabi, F., Dorostkar, V., Faz, A., & Acosta, J. A. (2020). Remediation of an oil-contaminated soil by two native plants treated with biochar and mycorrhizae. *Journal of Environmental Management*, 254, 109755. Available from https://doi.org/10.1016/j.jenvman.2019.109755.

Agarry, S., Oghenejoboh, K., & Solomon, B. (2015). Kinetic modelling and half life study of adsorptive bioremediation of soil artificially contaminated with bonny light crude oil. *Journal of Ecological Engineering*, 16, 1–13. Available from https://doi.org/10.12911/22998993/2799.

Ainomugisha, S., Matovu, M., & Manga, M. (2024a). Application of green agro-based nanoparticles in cement-based construction materials: A systematic review. *Journal of Building Engineering, 87*, 108955. Available from https://doi.org/10.1016/j.jobe.2024.108955.

Ainomugisha, S., Matovu, M. J., & Manga, M. (2024b). Influence mechanisms of silica nanoparticles' property enhancement in cementitious materials and their green synthesis: A critical review. *Case Studies in Construction Materials, 20*, e03372. Available from https://doi.org/10.1016/j.cscm.2024.e03372.

Alma, M. H., & Altikat, A. (2021). Methods of Application and Incorporation of the Biochar into Soil. *World Journal of Agriculture and Soil Science, 7*(1), 2021. Available from https://doi.org/10.33552/WJASS.2021.07.000653.

Amonette, J. E., Archuleta, J. G., Fuchs, M. R., Hills, K. M., Yorgey, G. G., Flora, G., Hunt, J., Han, H.-S., Jobson, B. T., Miles, T. R., Page-Dumroese, D. S., Thompson, S., Wilson, K., Baltar, R., Carloni, K., Collins, D. P., Dooley, J., Drinkard, D., Garcia-Perez, M., Hoffman-Krull, K., Kauffman, M., Laird, D. A., Lei, W., Miedema, J., O'Donnell, J., Kiser, A., Pecha, B., Rodriguez-Franco, C., Scheve, G. E., Sprenger, C., Springsteen, B., & Wheeler, E. (2021). Biomass to Biochar: Maximizing the Carbon Value. *Chapter 2 - Key Challenges and Opportunities*. Pullman, WA: Center for Sustaining Agriculture and Natural Resources.

Atkinson, C. J., Fitzgerald, J. D., & Hipps, N. A. (2010). Potential mechanisms for achieving agricultural benefits from biochar application to temperate soils: a review. *Plant Soil, 337*, 1–18. Available from https://doi.org/10.1007/s11104-010-0464-5.

ATSDR, 2014. Total Petroleum Hydrocarbons (TPH). Agency for Toxic Substances and Disease Registry. Retrieved from https://wwwn.cdc.gov/TSP/ToxFAQs/ToxFAQsDetails.aspx?faqid = 423&toxid = 75 (accessed November 13, 2023).

Aziz, S., Ali, M. I., Farooq, U., Jamal, A., Liu, F.-J., He, H., Guo, H., Urynowicz, M., & Huang, Z. (2020). Enhanced bioremediation of diesel range hydrocarbons in soil using biochar made from organic wastes. *Environmental Monitoring and Assessment, 192*, 569. Available from https://doi.org/10.1007/s10661-020-08540-7.

Bagby, S. C., Reddy, C. M., Aeppli, C., Fisher, G. B., & Valentine, D. L. (2017). Persistence and biodegradation of oil at the ocean floor following Deepwater Horizon. *Proceedings of the National Academy of Sciences, 114*, E9–E18. Available from https://doi.org/10.1073/pnas.1610110114.

Bao, H., Wang, J., Zhang, H., Li, J., Li, H., & Wu, F. (2020). Effects of biochar and organic substrates on biodegradation of polycyclic aromatic hydrocarbons and microbial community structure in PAHs-contaminated soils. *Journal of Hazardous Materials, 385*, 121595. Available from https://doi.org/10.1016/j.jhazmat.2019.121595.

Barquilha, C. E. R., & Braga, M. C. B. (2021). Adsorption of organic and inorganic pollutants onto biochars: Challenges, operating conditions, and mechanisms. *Bioresource Technology Reports, 15*, 100728. Available from https://doi.org/10.1016/j.biteb.2021.100728.

Bastami, K. D., Afkhami, M., Ehsanpour, M., Kazaali, A., Mohammadizadeh, M., Haghparast, S., Soltani, F., Zanjani, S. A., Ghorghani, N. F., & Pourzare, R. (2013). Polycyclic aromatic hydrocarbons in the coastal water, surface sediment and mullet Liza klunzingeri from northern part of Hormuz strait (Persian Gulf). *Marine Pollution Bulletin, 76*, 411–416. Available from https://doi.org/10.1016/j.marpolbul.2013.08.018.

Batista, E. M. C. C., Shultz, J., Matos, T. T. S., Fornari, M. R., Ferreira, T. M., Szpoganicz, B., De Freitas, R. A., & Mangrich, A. S. (2018). Effect of surface and porosity of biochar on water holding capacity aiming indirectly at preservation of the Amazon biome. *Science Reports, 8*, 10677. Available from https://doi.org/10.1038/s41598-018-28794-z.

Bianco, F., Race, M., Papirio, S., Oleszczuk, P., & Esposito, G. (2021). The addition of biochar as a sustainable strategy for the remediation of PAH–contaminated sediments. *Chemosphere, 263*, 128274. Available from https://doi.org/10.1016/j.chemosphere.2020.128274.

Blenis, N., Hue, N., Maaz, T. M., & Kantar, M. (2023). Biochar production, modification, and its uses in soil remediation: A Review. *Sustainability, 15*(4), 3442. Available from https://www.mdpi.com/2071-1050/15/4/3442.

De Boeck, G., Vlaeminck, A., Balm, P. H. M., Lock, R. A. C., De Wachter, B., & Blust, R. (2001). Morphological and metabolic changes in common carp, Cyprinus carpio, during short-term copper exposure: Interactions between Cu^{2+} and plasma cortisol elevation. *Environmental Toxicology and Chemistry, 20*(2), 374–381. Available from https://doi.org/10.1002/etc.5620200219.

Brewer, C. E., & Brown, R. C. (2012). 5.18 - Biochar. In A. Sayigh (Ed.), *Comprehensive Renewable Energy* (pp. 357–384). Oxford: Elsevier. Available from https://doi.org/10.1016/B978-0-08-087872-0.00524-2.

Chen, Z., Chen, J., Yang, X., Chen, C., Huang, S., & Luo, H. (2019). Biochar as An Effective Material on Sediment Remediation for Polycyclic Aromatic Hydrocarbons Contamination. *IOP Conference Series: Earth and Environmental Science, 281*, 012016. Available from https://doi.org/10.1088/1755-1315/281/1/012016.

Chi, J., & Liu, H. (2016). Effects of biochars derived from different pyrolysis temperatures on growth of Vallisneria spiralis and dissipation of polycyclic aromatic hydrocarbons in sediments. *Ecological Engineering, 93*, 199–206. Available from https://doi.org/10.1016/j.ecoleng.2016.05.036.

Chikere, C. B., Okpokwasili, G. C., & Chikere, B. O. (2011). Monitoring of microbial hydrocarbon remediation in the soil. *Biotech, 1*, 117–138. Available from https://doi.org/10.1007/s13205-011-0014-8.

Cruz Viggi, C., Tucci, M., Resitano, M., Palushi, V., Crognale, S., Matturro, B., Petrangeli Papini, M., Rossetti, S., & Aulenta, F. (2023). Enhancing the anaerobic biodegradation of petroleum hydrocarbons in soils with electrically conductive materials. *Bioengineering, 10*(4), 441. Available from https://www.mdpi.com/2306-5354/10/4/441.

Dai, Y., Zheng, H., Jiang, Z., & Xing, B. (2020). Combined effects of biochar properties and soil conditions on plant growth: A meta-analysis. *Science of The Total Environment, 713*, 136635. Available from https://doi.org/10.1016/j.scitotenv.2020.136635.

Das, N., & Chandran, P. (2011). Microbial degradation of petroleum hydrocarbon contaminants: an overview. *Biotechnology Research International, 2011*, 941810. Available from https://doi.org/10.4061/2011/941810.

Dawson, J. J. C., Godsiffe, E., Thompson, I. P., Ralebitso-Senior, T. K., Killham, K. S., & Paton, G. I. (2007). Application of biological indicators to assess recovery of hydrocarbon impacted soils. *Soil Biology and Biochemistry, 39*(1), 164–177.

Deebika, P. (2021). Biochar and compost-based phytoremediation of crude oil contaminated soil. *Indian Journal of Science and Technology, 14*, 220–228.

Dike, C. C., Hakeem, I. G., Rani, A., Surapaneni, A., Khudur, L., Shah, K., & Ball, A. S. (2022). The co-application of biochar with bioremediation for the removal of petroleum hydrocarbons from contaminated soil. *Science of The Total Environment, 849*, 157753. Available from https://doi.org/10.1016/j.scitotenv.2022.157753.

Dike, C. C., Shahsavari, E., Surapaneni, A., Shah, K., & Ball, A. S. (2021). Can biochar be an effective and reliable biostimulating agent for the remediation of hydrocarbon-contaminated soils? *Environment International, 154*, 106553. Available from https://doi.org/10.1016/j.envint.2021.106553.

Domingues, R. R., Trugilho, P. F., Silva, C. A., Melo, I. C. N. A. D., Melo, L. C. A., Magriotis, Z. M., & Sánchez-Monedero, M. A. (2017). Properties of biochar derived from wood and high-nutrient biomasses with the aim of agronomic and environmental benefits. *PLoS One, 12*, e0176884. Available from https://doi.org/10.1371/journal.pone.0176884.

Emoyan, O. O., Ikechukwu, C. C., & Tesi, G. O. (2020). Occurrence and sources of aliphatic hydrocarbons in anthropogenic impacted soils from petroleum tank-farms in the Niger Delta, Nigeria. *Ovidius University Annals of Chemistry, 31*, 132–144. Available from https://doi.org/10.2478/auoc-2020-0022.

Engelhardt, F. R. (1994). Limitations and innovations in the control of environmental impacts from petroleum industry activities in the Arctic. *Marine Pollution Bulletin, 29*, 334–341. Available from https://doi.org/10.1016/0025-326X(94)90650-5.

EPA, 1999. Preliminary Data Summary of Urban Storm Water Best Management Practices (No. EPA-821-R-99-012). Retrieved from https://www3.epa.gov/npdes/pubs/usw_a.pdf. (accessed November 6, 2023)

Evans, M., Liu, J., Bacosa, H., Rosenheim, B. E., & Liu, Z. (2017). Petroleum hydrocarbon persistence following the Deepwater Horizon oil spill as a function of shoreline energy. *Marine Pollution Bulletin, 115*, 47–56. Available from https://doi.org/10.1016/j.marpolbul.2016.11.022.

Fu, X.-W., Li, T.-Y., Ji, L., Wang, L.-L., Zheng, L.-W., Wang, J.-N., & Zhang, Q. (2018). Occurrence, sources and health risk of polycyclic aromatic hydrocarbons in soils around oil wells in the border regions between oil fields and suburbs. *Ecotoxicology and Environmental Safety, 157*, 276–284. Available from https://doi.org/10.1016/j.ecoenv.2018.03.054.

Godlewska, P., & Oleszczuk, P. (2021). Effect of biomass addition before sewage sludge pyrolysis on the persistence and bioavailability of polycyclic aromatic hydrocarbons in biochar-amended soil. *Chemical Engineering Journal, 429*, 132143. Available from https://doi.org/10.1016/j.cej.2021.132143.

Gogoi, B. K., Dutta, N. N., Goswami, P., & Krishna Mohan, T. R. (2003). A case study of bioremediation of petroleum-hydrocarbon contaminated soil at a crude oil spill site. *Advances in Environmental Research, 7*, 767–782. Available from https://doi.org/10.1016/S1093-0191(02)00029-1.

Gomez-Eyles, J. L., & Ghosh, U. (2018). Enhanced biochars can match activated carbon performance in sediments with high native bioavailability and low final porewater PCB concentrations. *Chemosphere, 203*, 179–187. Available from https://doi.org/10.1016/j.chemosphere.2018.03.132.

Gul, S., Whalen, J. K., Thomas, B. W., Sachdeva, V., & Deng, H. (2015). Physico-chemical properties and microbial responses in biochar-amended soils: Mechanisms and future directions. *Agriculture, Ecosystems & Environment, 206*, 46–59. Available from https://doi.org/10.1016/j.agee.2015.03.015.

Han, T., Zhao, Z., Bartlam, M., & Wang, Y. (2016). Combination of biochar amendment and phytoremediation for hydrocarbon removal in petroleum-contaminated soil. *Environmental Science and Pollution Research, 23*(21), 21219–21228. Available from https://doi.org/10.1007/s11356-016-7236-6.

Harmsen, J., Rulkens, W. H., Sims, R. C., Rijtema, P. E., & Zweers, A. J. (2007). Theory and Application of Landfarming to Remediate Polycyclic Aromatic Hydrocarbons and Mineral Oil-Contaminated Sediments; Beneficial Reuse. *Journal of Environmental Quality, 36*, 1112–1122. Available from https://doi.org/10.2134/jeq2006.0163.

Hui, K., Tang, J., Lu, H., Xi, B., Qu, C., & Li, J. (2020). Status and prospect of oil recovery from oily sludge: A review. *Arabian Journal of Chemistry, 13*, 6523–6543. Available from https://doi.org/10.1016/j.arabjc.2020.06.009.

Husk, B., & Major, J. (2010). Commercial Scale Agricultural Biochar Field Trial in Quebec, Canada Over two Years: Effects of Biochar on Soil Fertility. *Biology and Crop Productivity and Quality*. Retrieved from https://wiki.opensourceecology.org/images/5/55/BlueLeafBiocharForageFieldTrial-Year3Report.pdf.

Hussain, F., Hussain, I., Khan, A. H. A., Muhammad, Y. S., Iqbal, M., Soja, G., Reichenauer, T. G., Zeshan., & Yousaf, S. (2018). Combined application of biochar, compost, and bacterial consortia with Italian ryegrass enhanced phytoremediation of petroleum hydrocarbon contaminated soil. *Environmental and Experimental Botany, 153*, 80–88. Available from https://doi.org/10.1016/j.envexpbot.2018.05.012.

Ide-Pérez, M. R., Fernández-López, M. G., Sánchez-Reyes, A., Leija, A., Batista-García, R. A., Folch-Mallol, J. L., & Sánchez-Carbente, M. d R. (2020). Aromatic Hydrocarbon Removal by Novel Extremotolerant Exophiala and Rhodotorula Spp. from an Oil Polluted Site in Mexico. *Journal of Fungi, 6*(3), 135. Available from https://www.mdpi.com/2309-608X/6/3/135.

Ihunwo, O. C., Onyema, M. O., Wekpe, V. O., Okocha, C., Shahabinia, A. R., Emmanuel, L., Okwe, V. N., Lawson, C. B., Mmom, P. C., & Dibofori-Orji, A. N. (2021). Ecological and human health risk assessment of total petroleum hydrocarbons in surface water and sediment from Woji Creek in the Niger Delta Estuary of Rivers State, Nigeria. *Heliyon, 7*(8), e07689. Available from https://doi.org/10.1016/j.heliyon.2021.e07689.

Incardona, J. P., Swarts, T. L., Edmunds, R. C., Linbo, T. L., Aquilina-Beck, A., Sloan, C. A., Gardner, L. D., Block, B. A., & Scholz, N. L. (2013). Exxon Valdez to Deepwater Horizon: Comparable toxicity of both crude oils to fish early life stages. *Aquatic Toxicology, 142–143*, 303–316. Available from https://doi.org/10.1016/j.aquatox.2013.08.011.

Irwin, R. J., Mouwerik, M. I., Stevens, L., Seese, M. D., & Basham, W. (1997). Environmental contaminants encyclopedia: Alkanes Entry. Fort Collins. COLORADO, USA: National Park Service.

Jagaba, A. H., Kutty, S. R. M., Lawal, I. M., Aminu, N., Noor, A., Al-dhawi, B. N. S., Usman, A. K., Batari, A., Abubakar, S., Birniwa, A. H., Umaru, I., & Yakubu, A. S. (2022). Diverse sustainable materials for the treatment of petroleum sludge and remediation of contaminated sites: A review. *Cleaner Waste Systems, 2*, 100010. Available from https://doi.org/10.1016/j.clwas.2022.100010.

Jia, H., Jian Li, L. J. G., Jian, L., Li, Y., Lu, H., Liu, J., & Yan, C. (2020). The remediation of PAH contaminated sediment with mangrove plant and its derived biochars. *Journal of Environmental Management*, 268, 110410. Available from https://doi.org/10.1016/j.jenvman.2020.110410.

Jia, H., Li, J., Li, J. G., Jian, L., Li, Y., Lu, H., Liu, J., & Yan, C. (2020). The remediation of PAH-contaminated sediment with mangrove plant and its derived biochars. *Journal of Environmental Management*, 268, 110410. Available from https://doi.org/10.1016/j.jenvman.2020.110410.

Jindo, K., Audette, Y., Higashikawa, F. S., Silva, C. A., Akashi, K., Mastrolonardo, G., Sánchez-Monedero, M. A., & Mondini, C. (2020). Role of biochar in promoting circular economy in the agriculture sector. Part 1: A review of the biochar roles in soil N, P and K cycles. *Chemical and Biological Technologies in Agriculture*, 7, 15. Available from https://doi.org/10.1186/s40538-020-00182-8.

Johnson, O. A., & Affam, A. C. (2018). Petroleum sludge treatment and disposal: A review. *Environmental Engineering Research*, 24, 191–201. Available from https://doi.org/10.4491/eer.2018.134.

Kaboggoza, H. C., Muoghalu, C., Sprouse, L., & Manga, M. (2024). Hydrochar composites for healthcare wastewater treatment: A review of synthesis approaches, mechanisms, and influencing factors. *Journal of Water Process Engineering*, 60, 105222. Available from https://doi.org/10.1016/j.jwpe.2024.105222.

Karppinen, E. M., Siciliano, S. D., & Stewart, K. J. (2017). Application method and biochar type affect petroleum hydrocarbon degradation in northern landfarms. *Journal of Environmental Quality*, 46, 751–759. Available from https://doi.org/10.2134/jeq2017.01.0038.

Kavitha, B., Reddy, P. V. L., Kim, B., Lee, S. S., Pandey, S. K., & Kim, K.-H. (2018). Benefits and limitations of biochar amendment in agricultural soils: A review. *Journal of Environmental Management*, 227, 146–154. Available from https://doi.org/10.1016/j.jenvman.2018.08.082.

Kebede, G., Tafese, T., Abda, E. M., Kamaraj, M., & Assefa, F. (2021). Factors Influencing the Bacterial Bioremediation of Hydrocarbon Contaminants in the Soil: Mechanisms and Impacts. *Journal of Chemistry*, 2021(1), 9823362. Available from https://doi.org/10.1155/2021/9823362.

Kong, L., Gao, Y., Zhou, Q., Zhao, X., & Sun, Z. (2018). Biochar accelerates PAHs biodegradation in petroleum-polluted soil by biostimulation strategy. *Journal of Hazardous Materials*, 343, 276–284. Available from https://doi.org/10.1016/j.jhazmat.2017.09.040.

Koshlaf, E., & Ball, A. S. (2017). Soil bioremediation approaches for petroleum hydrocarbon polluted environments. *AIMS Microbiol*, 3, 25–49. Available from https://doi.org/10.3934/microbiol.2017.1.25.

Kołtowski, M., Hilber, I., Bucheli, T. D., Charmas, B., Skubiszewska-Zięba, J., & Oleszczuk, P. (2017). Activated biochars reduce the exposure of polycyclic aromatic hydrocarbons in industrially contaminated soils. *Chemical Engineering Journal*, 310, 33–40. Available from https://doi.org/10.1016/j.cej.2016.10.065.

Kuppusamy, S., Maddela, N. R., Megharaj, M., & Venkateswarlu, K. (2020). *Total Petroleum Hydrocarbons: Environmental Fate, Toxicity, and Remediation*. Cham: Springer International Publishing. Available from https://doi.org/10.1007/978-3-030-24035-6.

Kuppusamy, S., Thavamani, P., Megharaj, M., Venkateswarlu, K., & Naidu, R. (2016). Agronomic and remedial benefits and risks of applying biochar to soil: Current knowledge and future research directions. *Environment International*, 87, 1–12. Available from https://doi.org/10.1016/j.envint.2015.10.018.

Kuśmierz, M., & Oleszczuk, P. (2014). Biochar production increases the polycyclic aromatic hydrocarbon content in surrounding soils and potential cancer risk. *Environmental science and pollution research international*, 21, 3646–3652. Available from https://doi.org/10.1007/s11356-013-2334-1.

Lamichhane, S., Bal Krishna, K. C., & Sarukkalige, R. (2016). Polycyclic aromatic hydrocarbons (PAHs) removal by sorption: A review. *Chemosphere*, 148, 336–353. Available from https://doi.org/10.1016/j.chemosphere.2016.01.036.

Latif, I. K., Karim, A. J., Zuki, A. B. Z., Zamri-Saad, M., Niu, J. P., & Noordin, M. M. (2010). Pulmonary modulation of benzo[a]pyrene-induced hemato- and hepatotoxicity in broilers. *Poultry Science*, 89, 1379–1388. Available from https://doi.org/10.3382/ps.2009-00622.

Lehmann, J., & Joseph, S. (2015). Biochar for environmental management: an introduction. In J. Lehmann, Joseph, & S. Joseph (Eds.), *Biochar for Environmental Management: Science, Technology and Implementation* (pp. 1–12). Routledge. Available from https://doi.org/10.4324/9780203762264.

Li, C.-T., Lee, W.-J., Mi, H.-H., & Su, C.-C. (1995). PAH emission from the incineration of waste oily sludge and PE plastic mixtures. *Science of The Total Environment*, 170, 171–183. Available from https://doi.org/10.1016/0048-9697(95)04705-X.

Li, X., Song, Y., Wang, F., Bian, Y., & Jiang, X. (2019). Combined effects of maize straw biochar and oxalic acid on the dissipation of polycyclic aromatic hydrocarbons and microbial community structures in soil: A mechanistic study. *Journal of Hazardous Materials*, 364, 325–331. Available from https://doi.org/10.1016/j.jhazmat.2018.10.041.

Lin, H., Yang, Y., Shang, Z., Li, Q., Niu, X., Ma, Y., & Liu, A. (2022). Study on the Enhanced Remediation of Petroleum-Contaminated Soil by Biochar/g-C3N4 Composites. *IJERPH*, 19, 8290. Available from https://doi.org/10.3390/ijerph19148290.

Liu, J., Ding, Y., Ma, L., Gao, G., & Wang, Y. (2017). Combination of biochar and immobilized bacteria in cypermethrin-contaminated soil remediation. *International Biodeterioration & Biodegradation*, 120, 15–20. Available from https://doi.org/10.1016/j.ibiod.2017.01.039.

Liu, J., Jiang, X., Zhou, L., Han, X., & Cui, Z. (2009). Pyrolysis treatment of oil sludge and model-free kinetics analysis. *Journal of Hazardous Materials*, 161, 1208–1215. Available from https://doi.org/10.1016/j.jhazmat.2008.04.072.

Liu, Q., Tang, F., Sun, S., Wang, Y., Su, Y., Zhao, C., Zhang, X., Gu, Y., & Li, L. (2023). Enhanced crude oil degradation and reshaped microbial community structure using straw-sludge biochar-persulfate oxidative system in oil-contaminated soil. *Journal of Environmental Chemical Engineering*, 11, 109690. Available from https://doi.org/10.1016/j.jece.2023.109690.

Macaulay, B. M., & Rees, D. (2014). Bioremediation of oil spills: a review of challenges for research advancement. *Annals of Environmental Science*, 8, 9–37.

Major, J., Rondon, M., Molina, D., Riha, S. J., & Lehmann, J. (2010). Maize yield and nutrition during 4 years after biochar application to a Colombian savanna oxisol. *Plant Soil*, 333, 117–128. Available from https://doi.org/10.1007/s11104-010-0327-0.

Man, Y. B., Kang, Y., Wang, H. S., Lau, W., Li, H., Sun, X. L., Giesy, J. P., Chow, K. L., & Wong, M. H. (2013). Cancer risk assessments of Hong Kong soils contaminated by polycyclic aromatic hydrocarbons. *Journal of Hazardous Materials, 261*, 770–776. Available from https://doi.org/10.1016/j.jhazmat.2012.11.067.

Manga, M. (2017). *The feasibility of co-composting as an upscale treatment method for faecal sludge in urban Africa. PhD*. Leeds, United Kingdom: University of Leeds Retrieved from. Available from http://etheses.whiterose.ac.uk/16997/.

Manga, M., Aragón-Briceño, C., Boutikos, P., Semiyaga, S., Olabinjo, O., & Muoghalu, C. C. (2023). Biochar and its potential application for the improvement of the anaerobic digestion process: A critical review. *Energies, 16*(10), 4051. Available from https://doi.org/10.3390/en16104051.

Mater, L., Sperb, R., Madureira, L., Rosin, A., Correa, A., & Radetski, C. (2006). Proposal of a sequential treatment methodology for the safe reuse of oil sludge-contaminated soil. *Journal of Hazardous Materials, 136*, 967–971. Available from https://doi.org/10.1016/j.jhazmat.2006.01.041.

Mekonnen, B. A., Aragaw, T. A., & Genet, M. B. (2024). Bioremediation of petroleum hydrocarbon contaminated soil: a review on principles, degradation mechanisms, and advancements [Review]. *Frontiers in Environmental Science, 12*. Available from https://doi.org/10.3389/fenvs.2024.1354422.

Mukome, F. N. D., Buelow, M. C., Shang, J., Peng, J., Rodriguez, M., Mackay, D. M., Pignatello, J. J., Sihota, N., Hoelen, T. P., & Parikh, S. J. (2020). Biochar amendment as a remediation strategy for surface soils impacted by crude oil. *Environmental Pollution, 265*, 115006. Available from https://doi.org/10.1016/j.envpol.2020.115006.

Muoghalu, C., Owusu, A., Nakagiri, A., Semiyaga, S., Labu, S., Iorhemen, O. T., & Manga, M. (2023). Biochar as a novel technology for treatment of onsite domestic wastewater: A critical review. *Frontiers in Environmental Science, 11*, 202. Available from https://doi.org/10.3389/fenvs.2023.1095920.

Murtaza, G., Ahmed, Z., Eldin, S. M., Ali, I., Usman, M., Iqbal, R., Rizwan, M., Abdel-Hameed, U. K., Haider, A. A., & Tariq, A. (2023). Biochar as a green sorbent for remediation of polluted soils and associated toxicity risks: A critical review. *Separations, 10*(3), 197. Available from https://www.mdpi.com/2297-8739/10/3/197.

NIEHS, 2016. Polycyclic aromatic hydrocarbons: 15 listings, *Fourteenth Report on Carcinogens*. National Toxicology Program, Department of Health and Human Services.

NOAA, 2017. Largest oil spills affecting U.S. waters since 1969. National Oceanic and Atmospheric Administration - Office of Response and Restoration. Retrieved from https://response.restoration.noaa.gov/oil-and-chemical-spills/oil-spills/largest-oil-spills-affecting-us-waters-1969.html (accessed November 13, 2023).

Otton, J. K., & Zielinski, R. A. (2000). Simple techniques for assessing impacts of oil and gas operations on federal lands - A field evaluation at Big South Fork National River and Recreation Area, Scott County, Tennessee (Report No. 2000–499), *Open-File Report*. https://doi.org/10.3133/ofr2000499.

Pan, S., Zuo, J. Y., Wang, K., Chen, Y., & Mullins, O. C. (2016). A multicomponent diffusion model for gas charges into oil reservoirs. *Fuel, 180*, 384–395. Available from https://doi.org/10.1016/j.fuel.2016.04.055.

Panchenko, L., Muratova, A., Dubrovskaya, E., Golubev, S., & Turkovskaya, O. (2023). Natural and Technical Phytoremediation of Oil-Contaminated Soil. *Life, 13*, 177. Available from https://doi.org/10.3390/life13010177.

Peng, C., Chen, W., Liao, X., Wang, M., Ouyang, Z., Jiao, W., & Bai, Y. (2011). Polycyclic aromatic hydrocarbons in urban soils of Beijing: status, sources, distribution and potential risk. *Environmental Pollution, 159*, 802–808. Available from https://doi.org/10.1016/j.envpol.2010.11.003.

Perelo, L. W. (2010). Review: In situ and bioremediation of organic pollutants in aquatic sediments. *Journal of Hazardous Materials, 177*, 81–89. Available from https://doi.org/10.1016/j.jhazmat.2009.12.090.

Qambrani, N. A., Rahman, M. M., Won, S., Shim, S., & Ra, C. (2017). Biochar properties and eco-friendly applications for climate change mitigation, waste management, and wastewater treatment: A review. *Renewable and Sustainable Energy Reviews, 79*, 255–273. Available from https://doi.org/10.1016/j.rser.2017.05.057.

Raja, P., Karthikeyan, P., Marigoudar, S. R., Venkatarama Sharma, K., & Ramana Murthy, M. V. (2022). Spatial distribution of total petroleum hydrocarbons in surface sediments of Palk Bay, Tamil Nadu, India. *Environmental Chemistry and Ecotoxicology, 4*, 20–28. Available from https://doi.org/10.1016/j.enceco.2021.10.002.

Ren, H.-Y., Wei, Z.-J., Wang, Y., Deng, Y.-P., Li, M.-Y., & Wang, B. (2020). Effects of biochar properties on the bioremediation of the petroleum-contaminated soil from a shale-gas field. *Environmental Science and Pollution Research, 27*, 36427–36438.

de Resende, M. F., Brasil, T. F., Madari, B. E., Pereira Netto, A. D., & Novotny, E. H. (2018). Polycyclic aromatic hydrocarbons in biochar amended soils: Long-term experiments in Brazilian tropical areas. *Chemosphere, 200*, 641–648. Available from https://doi.org/10.1016/j.chemosphere.2018.02.139.

Rex, P., Mohammed Ismail, K., Meenakshisundaram, N., Barmavatu, P., & Sai Bharadwaj, A. (2023). Agricultural Biomass Waste to Biochar: A Review on Biochar Applications Using Machine Learning Approach and Circular Economy. *ChemEngineering, 7*, 50. Available from https://doi.org/10.3390/chemengineering7030050.

Rosales, E., Meijide, J., Pazos, M., & Sanromán, M. A. (2017). Challenges and recent advances in biochar as low-cost biosorbent: From batch assays to continuous-flow systems. *Bioresource Technology, Special Issue on Biochar: Production, Characterization and Applications – Beyond Soil Applications, 246*, 176–192. Available from https://doi.org/10.1016/j.biortech.2017.06.084.

Roy, A., Dutta, A., Pal, S., Gupta, A., Sarkar, J., Chatterjee, A., Saha, A., Sarkar, P., Sar, P., & Kazy, S. K. (2018a). Biostimulation and bioaugmentation of native microbial community accelerated bioremediation of oil refinery sludge. *Bioresource Technology, 253*, 22–32. Available from https://doi.org/10.1016/j.biortech.2018.01.004.

Saliu, A. O., Akinpelumi, B. E., & Najeemdeen, B. A. (2023). Potential of biochar for hydrocarbon degradation of crude oil–contaminated soils. *Journal of Environmental Quality, 52*, 1049–1059. Available from https://doi.org/10.1002/jeq2.20499.

Saum, L., Jiménez, M. B., & Crowley, D. (2018). Influence of biochar and compost on phytoremediation of oil-contaminated soil. *International Journal of Phytoremediation*, 20, 54–60. Available from https://doi.org/10.1080/15226514.2017.1337063.

Schwab, C.V., & Hanna, M. (2012). Master Gardeners' safety precautions for handling, applying, and storing biochar. *Iowa State University*. Retrieved from: https://ag-safety.extension.org/wp-content/uploads/2019/05/MasterGardenerSafetySheet2012Final.pdf (accessed October 10, 2023).

Shi, Y., Liu, M., Li, J., Yao, Y., Tang, J., & Niu, Q. (2022). The dosage-effect of biochar on anaerobic digestion under the suppression of oily sludge: Performance variation, microbial community succession and potential detoxification mechanisms. *Journal of Hazardous Materials*, 421, 126819. Available from https://doi.org/10.1016/j.jhazmat.2021.126819.

Sigmund, G., Huber, D., Bucheli, T. D., Baumann, M., Borth, N., Guebitz, G. M., & Hofmann, T. (2017). Cytotoxicity of Biochar: A Workplace Safety Concern? *Environmental Science and Technology Letters.*, 4, 362–366. Available from https://doi.org/10.1021/acs.estlett.7b00267.

Singh, E., Mishra, R., Kumar, A., Shukla, S. K., Lo, S.-L., & Kumar, S. (2022). Circular economy-based environmental management using biochar: Driving towards sustainability. *Process Safety and Environmental Protection*, 163, 585–600. Available from https://doi.org/10.1016/j.psep.2022.05.056.

Soukup, D. A., Ulery, A. L., & Jones, S. (2007). Distribution of Petroleum and Aromatic Hydrocarbons at a Former Crude Oil and Natural Gas Production Facility. *Soil and Sediment Contamination: An International Journal*, 16, 143–158. Available from https://doi.org/10.1080/15320380601166439.

Srivastava, M., Srivastava, A., Yadav, A., Rawat, V., Srivastava, M., Srivastava, A., Yadav, A., & Rawat, V. (2019). Source and Control of Hydrocarbon Pollution. *Hydrocarbon Pollution and Its Effect on the Environment*. IntechOpen. Available from https://doi.org/10.5772/intechopen.86487.

Tan, X.-F., Zhu, S.-S., Wang, R.-P., Chen, Y.-D., Show, P.-L., Zhang, F.-F., & Ho, S.-H. (2021). Role of biochar surface characteristics in the adsorption of aromatic compounds: Pore structure and functional groups. *Chinese Chemical Letters*, 32, 2939–2946. Available from https://doi.org/10.1016/j.cclet.2021.04.059.

Todd, G. D., Chessin, R. L., & Colman, J. (2002). *ATSDR's Toxicological Profiles: Web Version*. CRC Press. Available from https://doi.org/10.1201/9781420061888.

Truskewycz, A., Gundry, T. D., Khudur, L. S., Kolobaric, A., Taha, M., Aburto-Medina, A., Ball, A. S., & Shahsavari, E. (2019). Petroleum Hydrocarbon Contamination in Terrestrial Ecosystems—Fate and Microbial Responses. *Molecules*, 24, 3400. Available from https://doi.org/10.3390/molecules24183400.

Uzojie, A. P., & Agunwamba, J. C. (2011). Physiochemical Properties of Soil in Relation to Varying Rates of Crude Oil Pollution. *Journal of Environmental Science and Technology*, 4, 313–323. Available from https://doi.org/10.3923/jest.2011.313.323.

Wang, Y., Li, F., Rong, X., Song, H., & Chen, J. (2017). Remediation of Petroleum-contaminated Soil Using Bulrush Straw Powder, Biochar and Nutrients. *Bulletin of Environmental Contamination and Toxicology*, 98, 690–697. Available from https://doi.org/10.1007/s00128-017-2064-z.

Wartell, B., Boufadel, M., & Rodriguez-Freire, L. (2021). An effort to understand and improve the anaerobic biodegradation of petroleum hydrocarbons: A literature review. *International Biodeterioration & Biodegradation*, 157, 105156. Available from https://doi.org/10.1016/j.ibiod.2020.105156.

Wei, Z., Wang, J. J., Gaston, L. A., Li, J., Fultz, L. M., DeLaune, R. D., & Dodla, S. K. (2020). Remediation of crude oil-contaminated coastal marsh soil: Integrated effect of biochar, rhamnolipid biosurfactant and nitrogen application. *Journal of Hazardous Materials*, 396, 122595. Available from https://doi.org/10.1016/j.jhazmat.2020.122595.

Wei, Z., Wei, Y., Liu, Y., Niu, S., Xu, Y., Park, J.-H., & Wang, J. J. (2024). Biochar-based materials as remediation strategy in petroleum hydrocarbon-contaminated soil and water: Performances, mechanisms, and environmental impact. *Journal of Environmental Sciences*, 138, 350–372. Available from https://doi.org/10.1016/j.jes.2023.04.008.

Weisman, W. (Ed.), (1998). *Analysis of petroleum hydrocarbons in environmental media, Total Petroleum Hydrocarbon Criteria Working Group series*. Amherst, Massachusetts: Amherst Scientific Publishers, ISBN 1-884-940-14-5.

Weng, H. X., Ji, Z. Q., Chu, Y., Cheng, C. Q., & Zhang, J. J. (2012). Benzene series in sewage sludge from China and its release characteristics during drying process. *Environmental Earth Sciences*, 65, 561–569. Available from https://doi.org/10.1007/s12665-011-1100-2.

Wiedemeier, D. B., Abiven, S., Hockaday, W. C., Keiluweit, M., Kleber, M., Masiello, C. A., McBeath, A. V., Nico, P. S., Pyle, L. A., Schneider, M. P. W., Smernik, R. J., Wiesenberg, G. L. B., & Schmidt, M. W. I. (2015). Aromaticity and degree of aromatic condensation of char. *Organic Geochemistry*, 78, 135–143. Available from https://doi.org/10.1016/j.orggeochem.2014.10.002.

Williams, S. D., Ladd, D. E., & Farmer, J. J. (2006). Fate and transport of petroleum hydrocarbons in soil and ground water at Big South Fork National River and Recreation Area (U.S. Geological Survey Scientific Investigations Report No. 2005–5104), *Scientific Investigations Report*. Tennessee and Kentucky.

Wu, W., Yang, M., Feng, Q., McGrouther, K., Wang, H., Lu, H., & Chen, Y. (2012). Chemical characterization of rice straw-derived biochar for soil amendment. *Biomass and Bioenergy*, 47, 268–276. Available from https://doi.org/10.1016/j.biombioe.2012.09.034.

Yang, Z., Shah, K., Crevier, C., Laforest, S., Lambert, P., Hollebone, B. P., Yang, C., Brown, C. E., Landriault, M., & Goldthorp, M. (2018). Occurrence, source and ecological assessment of petroleum related hydrocarbons in intertidal marine sediments of the Bay of Fundy, New Brunswick, Canada. *Marine Pollution Bulletin*, 133, 799–807. Available from https://doi.org/10.1016/j.marpolbul.2018.06.047.

Yousaf, U., Ali Khan, A. H., Farooqi, A., Muhammad, Y. S., Barros, R., Tamayo-Ramos, J. A., Iqbal, M., & Yousaf, S. (2022). Interactive effect of biochar and compost with Poaceae and Fabaceae plants on remediation of total petroleum hydrocarbons in crude oil contaminated soil. *Chemosphere*, 286, 131782. Available from https://doi.org/10.1016/j.chemosphere.2021.131782.

Yuan, H., Li, T., Ding, X., Zhao, G., & Ye, S. (2014). Distribution, sources and potential toxicological significance of polycyclic aromatic hydrocarbons (PAHs) in surface soils of the Yellow River Delta, China. *Marine Pollution Bulletin*, 83, 258–264. Available from https://doi.org/10.1016/j.marpolbul.2014.03.043.

Yuan, Z., Liu, G., Da, C., Wang, J., & Liu, H. (2015). Occurrence, Sources, and Potential Toxicity of Polycyclic Aromatic Hydrocarbons in Surface Soils from the Yellow River Delta Natural Reserve, China. *Archives of Environmental Contamination and Toxicology, 68*, 330–341. Available from https://doi.org/10.1007/s00244-014-0085-8.

Zahed, M. A., Salehi, S., Madadi, R., & Hejabi, F. (2021). Biochar as a sustainable product for remediation of petroleum contaminated soil. Current Research in Green and Sustainable Chemistry, 4, Article 100055, 100055. https://doi.org/10.1016/j.crgsc.2021.100055.

Zama, E. F., Reid, B. J., Arp, H. P. H., Sun, G.-X., Yuan, H.-Y., & Zhu, Y.-G. (2018). Advances in research on the use of biochar in soil for remediation: a review. *Journal of Soils and Sediments, 18*, 2433–2450. Available from https://doi.org/10.1007/s11368-018-2000-9.

Zawierucha, I., Malina, G., Herman, B., Rychter, P., Biczak, R., Pawlowska, B., Bandurska, K., & Barczynska, R. (2022). Ecotoxicity and bioremediation potential assessment of soil from oil refinery station area. *Journal of Environmental Health Science and Engineering, 20*(1), 337–346. Available from https://doi.org/10.1007/s40201-021-00780-0.

Zhang, B., Zhang, L., & Zhang, X. (2019). Bioremediation of petroleum hydrocarbon-contaminated soil by petroleum-degrading bacteria immobilized on biochar. *RSC Advances, 9*, 35304–35311. Available from https://doi.org/10.1039/C9RA06726D.

Zhang, G., He, L., Guo, X., Han, Z., Ji, L., He, Q., Han, L., & Sun, K. (2020). Mechanism of biochar as a biostimulation strategy to remove polycyclic aromatic hydrocarbons from heavily contaminated soil in a coking plant. *Geoderma, 375*, 114497. Available from https://doi.org/10.1016/j.geoderma.2020.114497.

Zhang, J., & You, C. (2013). Water holding capacity and absorption properties of wood chars. *Energy Fuels, 27*, 2643–2648. Available from https://doi.org/10.1021/ef4000769.

Zhao, L., Cao, X., Mašek, O., & Zimmerman, A. (2013). Heterogeneity of biochar properties as a function of feedstock sources and production temperatures. *Journal of Hazardous Materials, 256–257*, 1–9. Available from https://doi.org/10.1016/j.jhazmat.2013.04.015.

Zhen, M., Chen, H., Liu, Q., Song, B., Wang, Y., & Tang, J. (2019). Combination of rhamnolipid and biochar in assisting phytoremediation of petroleum hydrocarbon contaminated soil using Spartina anglica. *Journal of Environmental Sciences, 85*, 107–118. Available from https://doi.org/10.1016/j.jes.2019.05.013.

Zhu, X., Chen, B., Zhu, L., & Xing, B. (2017). Effects and mechanisms of biochar-microbe interactions in soil improvement and pollution remediation: A review. *Environmental Pollution, 227*, 98–115. Available from https://doi.org/10.1016/j.envpol.2017.04.032.

Part 3

Biochar for water and wastewater treatment

Chapter 10

Removal of contaminants in drinking water using biochars

Nnanake-Abasi O. Offiong[1], Odunayo T. Ore[2,3], Deborah O. Aderibigbe[4], Ajibola Abiodun Bayode[5,6], Olawale S. Dabo[4] and Olaniran Kolawole Akeremale[7]

[1]*Department of Chemical Sciences, Topfaith University, Mkpatak, Akwa Ibom State, Nigeria*, [2]*Department of Chemical Sciences, Achievers University, Owo, Ondo State, Nigeria*, [3]*Department of Chemistry and Industrial Chemistry, Kogi State University, Kabba, Kogi State, Nigeria*, [4]*Department of Pure and Applied Chemistry, Ladoke Akintola University of Technology, Ogbomoso, Oyo State, Nigeria*, [5]*Department of Chemical Sciences, Redeemer's University, Ede, Osun State, Nigeria*, [6]*Department of Chemical Engineering, Sichuan University of Science and Engineering, Zigong, Sichuan Province, P.R. China*, [7]*Department of Science and Technology Education, Bayero University, Kano, Kano State, Nigeria*

Chapter outline

10.1 Introduction	169
10.2 Removal of inorganic contaminants	170
10.2.1 Removal of radionuclides from water using biochar	170
10.2.2 Removal of nutrients from water using biochar	171
10.2.3 Removal of toxic metals from water	171
10.2.4 Mechanism of removal of inorganic contaminants	173
10.3 Removal of organic contaminants	174
10.4 Removal of microbial contaminants	177
10.5 Conceptual designs of biochar water filters	178
10.5.1 Conceptual designs	178
10.5.2 Overview of performance	179
10.6 Efficient disposal management of spent biochars	180
10.7 Large scale/field applications of biochar in Africa	181
10.8 Future perspectives and conclusion	182
References	182

10.1 Introduction

Access to safe and clean drinking water is a basic human right, yet it continues to be a global concern. The presence of numerous pollutants in drinking water sources, arising from the surge in anthropogenic activities and world population, poses serious health and environmental risks (Palansooriya et al., 2020). Humans, plants, and animals all require clean water to survive and thrive. Water, for example, is essential in human metabolism, assisting in the breakdown of massive molecules and their transit throughout the body. When water becomes contaminated with dangerous substances, it must be treated before it may be consumed (Fu & Xi, 2020). Similarly, tainted water can endanger plants and animals, especially if they consume it and it interferes with their metabolic processes. Water contamination can occur from a variety of causes, including discharge from industries and sewage systems, agricultural activities, illegal waste disposal, and landfill leachate leakage (Akhtar et al., 2021). As a result, addressing these concerns is critical to ensuring the safety of our water resources (Abu Hasan et al., 2016; Abu Hasan et al., 2020).

Inorganic, organic, and microbial pollution pose a serious challenge in surface water. Contact with these dangerous compounds can result in a variety of health problems including the potential for cancer risk, skeletal fluorosis, and arsenical neuropathy among others (Arshad & Imran, 2017; Chakraborti et al., 2016; Lu et al., 2015). Inorganic pollutants in drinking water, such as nutrients and radionuclides, can be caused by either natural geological formations or by human-caused activities (Shannon et al., 2011; Spruill, 1998). An excess of nutrients, for example, can cause eutrophication, a process that reduces oxygen levels in aquatic bodies, resulting in the extinction of marine life. Radionuclides such as uranium-238 and radium-226, on the other hand, are radioactive materials that can cause cancer and other health problems in humans (Ayele & Atlabachew, 2021; Geletu, 2023).

Organic contaminants present a significant difficulty in water treatment. This group of contaminants encompasses pharmaceuticals, pesticides, and volatile organic compounds (Saxena et al., 2020). These toxins are usually by-products of industrial activities and farming practices, and their presence in aquatic habitats can cause a variety of health and environmental issues (Karpińska & Kotowska, 2019). Organic chemicals can accumulate in the human body, resulting in a variety of carcinogenic effects (Dheenadayalan & Thiruvengadathan, 2021). Similarly, microbial contaminants such as parasitic worms, parasites, viruses, and harmful bacteria pose a significant threat to drinking water safety. The most frequently encountered waterborne pathogens include Cryptosporidium, Giardia, and *Escherichia coli* (*E. coli*) (Arslan et al., 2022; Bridle, 2021). These microbes can cause a wide range of ailments, particularly in developing countries where access to clean water and sanitation is generally limited. It is estimated that around 1.7 billion children suffer from diarrhea each year, with roughly 525,000 of these youngsters tragically losing their lives (Kristanti et al., 2022).

Owing to the toxicity of these contaminants, several methods including membrane separations, ion exchange, adsorption, co-precipitation, precipitation, bioremediation, electrochemical methods, advanced oxidation processes, nanotechnology, biologically active carbon filtration, distillation, ultraviolet radiation, and oxidation treatment have been implemented to remove the contaminants from drinking water (Kim et al., 2018; Kristanti et al., 2022). However, in recent years, the diverse applications and unique properties of biochar (BC) have caused it to gain increased attention. BC is a material that is highly rich in carbon usually produced from biomass by thermochemical conversion methods (Iwuozor et al., 2023; Qiu et al., 2022). The efficiency of BC in the removal of contaminants is controlled by its characteristics and synthetic strategies. For example, organic pollutants can be effectively abated using BC produced at relatively high pyrolysis temperature due to the BC's increased microporosity, hydrophobicity, and surface area. On the other hand, inorganic pollutants tend to be favorably abated by BC produced at relatively low pyrolysis temperatures (Qiu et al., 2022). The present chapter is focused on highlighting recent trends in the use of BC for the removal of inorganic, organic, and microbial contaminants in drinking water. In addition, the potential removal mechanisms of each class of pollutants were discussed. Finally, future directions on the use of BC in improving water quality were elucidated.

10.2 Removal of inorganic contaminants

In the last few decades, modern agriculture has relied heavily on the use of synthetic fertilizers (N, P, K), to cope with the rising global population's need for food. In addition, scarce land resources have also encouraged the use of these methods. The use of synthetic fertilizers has resulted in a massive increase in crop yield which has further intensified their use, however, this has also led to a profound negative impact on the environment because a good percentage of these fertilizers are not absorbed by plants and they are released into the environment resulting in ecological effects such as eutrophication of water and elevated concentration of greenhouse gases in the environment (Dai et al., 2020; Marcińczyk et al., 2022; Rombel et al., 2022). Due to concerns over fossil fuels as a source of energy, attention has been shifted to nuclear energy as an alternative source of energy in recent times. However, nuclear power plants are known to generate tons of radioactive wastes during operation which is harmful to the environment if not properly discharged. Some of these radioactive wastes are known to have long half-lives are soluble and very mobile in water bodies can accrue rapidly in soils and sediments, and can end up being concentrated by aquatic plants and animals leading to a potential risk to other animals in the food chain (Guilhen et al., 2021; Pipíška et al., 2020). The use of crude BC in removing these contaminants has led to limited removal efficiency due to challenges such as negative BC surface, and difficulty in separation from solution after the decontamination process. Several researchers have put forward different modifications to enhance BC's affinity towards contaminants and also enhance its adsorption capacity.

10.2.1 Removal of radionuclides from water using biochar

One of the challenges closely associated with the removal of radionuclides from water using BC is the low adsorption capacity resulting from the low content of oxygen-bearing functional groups. To surmount this limitation, several researchers have employed different BC modifications in a bid to positively alter some of its main features such as surface area, and selective adsorption capacities, and also provide additional functional groups on the surface of BC (Ahmed, Mehmood, Núñez-Delgado, Ali, et al., 2021). A review of the current research landscape shows that the current modification methods for radionuclide removal majorly involve chemical modification using chemical reagents (such as HNO_3, KOH, and $KMnO_4$) and preparation of BC-based composite materials. For instance Ahmed, Mehmood, Núñez-Delgado, Qaswar, Ali, et al. (2021) examined the adsorption capacity of both raw rice straws-derived

BC and when also subjected to HNO$_3$ modification in uranium U (VI) removal from aqueous solutions, the results showed that the HNO$_3$-modified BC was enriched with carboxyl and carbonyl functional groups and also showed increased adsorption capacity (145.2 mg/g) towards U (VI) when compared to raw BC (66.3 mg/g). Similar observations were noted by Philipou and his colleagues (Philippou et al., 2023) whose work on the removal of americium; a man-made radioactive metal onto both raw and oxidized sponge gourd BC showed higher efficiency in the oxidized BC due to a higher number of carboxylic components. Some researchers have also recorded success in preparing BC-based composite. Lyu et al. (2021) prepared a phosphate-impregnated BC cross-linked Mg-Al layered composite and studied its adsorption capacity on U (IV) removal. Their results showed a great increase in the surface functional groups which were ascribed to their method of preparing the composite, there was also a significant adsorption capacity (274.15 mg/g) reported by the authors.

Due to shorter activation time and reduced temperature involved in activation, chemical modification has led to greater yield and higher removal efficiency for radionuclides, however, there has been concern about the generation of by-products from chemical modification requiring further treatment and cost of activating agents which might limit the practical application of chemically modified BC. With this in mind, some researchers have favored the use of physical activation. In their earlier study on the use of raw BC in the removal of U (VI) (Guilhen et al., 2021), Guilhen and his colleagues found that although raw BC named (BC-350) had a promising performance in U (VI) removal, the concentration of U (VI) left in solution still exceeded regulatory limits, hence, they carried out physical activation in subsequent studies to ascertain if activation improves the adsorption capacity of macaua BC (BC-350A), a higher U (VI) removal (90%) was recorded over a different range of concentrations (2.5–25 mg/L). The adsorption capacity also increased from 422 mg/g (BC-350) to 488.7 mg/g (BC-350A).

10.2.2 Removal of nutrients from water using biochar

Regular BC has demonstrated considerable adsorption capacity for organic contaminants and positively charged cationic species due to its negatively charged surface. However, adsorption of anions such as ammonium, nitrate, and phosphate which are the common forms of reactive nitrogen and phosphorus in water have low adsorption capacity towards regular BC. Consequently, many researchers have focused on formulating novel modified BCs by incorporating different metal salts such as $AlCl_3$, $MgCl_2$, $LaCl_3$, or $FeCl_3$ to enhance their performance towards anionic contaminants (Vikrant et al., 2018). For example, Jia and his colleagues prepared lanthanum-modified platanus BC taking advantage of lanthanum-specific affinity for phosphate removal of P from actual wastewater (Jia et al., 2020). They recorded a high adsorption capacity of 148.11 mg/g. Min and his colleagues found that impregnation of BC with different percentages of Fe (0%, 1%, 5%, and 10%), enhanced the removal of N and P from actual wastewater (Min et al., 2020). It was noted that the BC modified with 1% Fe (B1) had 50% N removal and B10 recorded over 80% removal for P. To prove the real-world applicability of engineered BC from secondary treated wastewater, a study prepared BC from hickory chips and pretreat with aluminum salt to investigate the removal reclamation, and reuse of phosphorus from wastewater (Zheng et al., 2019). The findings of the study showed that the engineered BC had decent adsorption capacity to P in the secondary treated wastewater. The engineered BC also showed increased seed germination as a value-added fertilizer.

10.2.3 Removal of toxic metals from water

Toxic heavy metal ion removal using BC has demonstrated great promise for Pb, Zn, Cd, Cu, Cr, and other heavy metal ions. The effectiveness of this material in removing heavy metal ions is influenced by several variables, including the water environment and BC properties. The pyrolysis temperature, feedstock makeup, and time all affect the properties of BC. Reaction temperature, pH level, and the binding competition that exists between the chemicals are among the water environmental factors that affect the efficacy of heavy metal ion eradication. A summary of studies carried out on the removal of toxic metals from water using BC is presented in Table 10.1 Ion exchange, complexation, electrostatic contact, and reduction are the main processes of the reaction. When utilizing BC as an adsorbent, the removal of toxic metals from water depends on both the physical and chemical characteristics of the adsorbent. The physical adsorption is accomplished by the development of micropores in the pore structure, high specific surface area (SSA) of the adsorbent, and weak van der Waal force of attraction between the BC (adsorbent) and heavy metals (adsorbate) (Zhang et al., 2020). The type of redox reaction between the adsorbent and metal determines the chemical adsorption. Carboxyl, carbonyl, and hydroxide are examples of functional groups connected to the link which determines how quickly BC adsorbs substances. Since BC

TABLE 10.1 Summary of studies on the removal of toxic metals from water using biochar.

Materials	Feedstock	Pollutants	Adsorption capacity (mg/g)	Interaction mechanism	References
ALB	Alkali lignin	Pb	1003.7	Precipitation, ion exchange, surface complexation	Wu et al. (2021)
FeYBC	Pomelo peel	Cr	39.3	Ion exchange, surface complexation, reduction	Dong et al. (2021)
BC600	Blue algae	Cd	135.7	Ion exchange, surface complexation, precipitation	Li, Xiong, et al. (2021)
PBC$_{KOH}$	Corn straw	Cr	117.0	Electrostatic attraction, complexation, ion exchange, reduction	Qu et al. (2021)
Biochar	Canola straw	Pb	165.1	Precipitation, surface complexation, ion exchange, cation-π interaction	Nzediegwu et al. (2021)
HMB	Wheat straw	Cd	70.9 80.0 61.1	Precipitation, surface complexation, ion exchange, physical adsorption	Fu et al. (2021)
BC	*Tribulus terrestris*	U	49.6	Surface complexation	Ahmed, Mehmood, Núñez-Delgado, Qaswar, Ali, et al. (2021)
PBC@LDH	Bamboo	U	274.2	Complexation, reduction, precipitation	Lyu et al. (2021)

typically has a negative charge on its valency shell due to the dissociation of oxygen-containing functional groups, it can effectively remove harmful heavy metals from water, including Cd, Pb, Al, Cu, Ni, As, Hg, and Mn (Qambrani et al., 2017). Pollutants including Cu(II), Ni(II), Pb(II), Zn(II), Cd(II), and Hg(II) are usually eliminated from the water via the physical adsorption (SSA) mechanism with carboxyl (R-COOH), and hydroxyl group (O-H) groups, according to research on the removal of pollutants from BC on heavy metals (Enaime et al., 2020).

As a key tactic to lessen the negative environmental effects of chromium, the deactivation of aqueous Cr(VI) or transformation from hexavalent to trivalent chromium which is a less toxic species, has been utilized (Li, Xiong, et al., 2021). Cr(VI) usually occurs as oxy-anions, chromate (CrO_4^{2-}) or dichromate ($Cr_2O_7^{2-}$). Using two-step KOH-activated pyrolysis, a porous BC made from maize straw was created, and it demonstrated exceptional elimination ability with a predicted capacity of 116.97 mg/g for Cr(VI) (Qu et al., 2021). Pyrite (FeS_2), a reducing agent that occurs naturally, was combined with BC to create FeS_2@biochar composite (BM-FeS_2@BC), which was then used to eliminate Cr(VI) from wastewater (Tang et al., 2021). The main methods by which Cr(VI) was removed by BM-FeS_2@BC were surface complexation, reduction, and adsorption. Through the use of various BC-based materials and a combination of adsorption, reduction, and co-precipitation methods, Cr(VI) can be effectively removed (Ma et al., 2022; Ri et al., 2022; Wen et al., 2022; Zhou et al., 2022).

The radioactive metal uranium (U) is exceedingly poisonous (Wang, Zhang, et al., 2022). Its presence in water settings above a particular threshold can have long-term health effects that result in severe and irreparable harm (Wang, Shi, et al., 2022). For the immobilization and extraction of U(VI) from water, the sorption-reduction-solidification process works well (Cheng et al., 2021; Li, Hu, et al., 2021; Yang et al., 2021). A variety of BC-based materials were prepared by Ahmed et al. (Ahmed, Mehmood, Núñez-Delgado, Ali, et al., 2021; Ahmed, Mehmood, Núñez-Delgado, Qaswar, et al., 2021; Ahmed, Mehmood, Qaswar, Ali, Khan, et al., 2021; Ahmed, Núñez-Delgado, et al., 2021) and evaluated their efficacy for the removal of U(VI) from wastewater. The pseudo-second-order kinetic model and Langmuir isotherm produced good experimental findings. According to thermodynamics, the adsorption was endothermic, entropy-driven, and characterized by increased unpredictability at the solid-solution interface. Wang et al. synthesized BC modified with polyethyleneimine (PEI) from moso-bamboo and used it to get rid of

U(VI) (Wang et al., 2020). According to the Langmuir model's fitting, the greatest U(VI) elimination capacities were found to be 185.6 mg/g for PEI-acid-BC and 212.7 mg/g for PEI-alkali-BC, respectively. These values were about 9−10 fold higher than those of unmodified BC (20.1 mg/g).

A promising strategy is the adsorption of lead (Pb) from waste and environmental water using BC-based compounds. The type of feedstock, the synthetic procedure, and the reaction temperature are the most important variables to take into account while creating BC to remove Pb(II). Using four feedstocks and three temperatures, Nzediegwu et al. investigated the removal of Pb(II) by BC generated using microwave-assisted pyrolysis (Nzediegwu et al., 2021). The findings showed that the adsorption capability of canola straw BC, which was produced at 500°C, was the highest at 165 mg/g. The precipitation of hydrocerussite and lead oxide phosphate led to an improvement in Pb(II) removal in BC at higher production temperatures. Waste alkali lignin was pyrolyzed to create a novel, inexpensive, and effective BC sorbent to expel Pb(II) (1003.71 mg/g in 5 min) (Wu et al., 2021). Surface complexation made up 8.25% of the reaction process, with mineral precipitation making up the majority (86.72%). According to the study, alkali lignin might be used to make BC adsorbents that would capture and remove Pb(II) from water.

Given its difficulty in degrading and ease of buildup in the human body, the hazardous heavy metal ion cadmium (Cd) has drawn a lot of attention (Zhang et al., 2021). The BC made from blue algae had a Cd(II) adsorption capacity of 135.7 mg/g, which was 66.9% and 85.9% higher than the BC made from rice husk and corn straw, respectively (Liu, Rao, et al., 2021). Su et al. evaluated the approaches used by fresh and aged ramie BC to remove Cd(II) (Su et al., 2021). According to the findings, physisorption and chemisorption both existed, and they were, respectively, the primary mechanisms of fresh and aged BC. Furthermore, fresh BC performed better than aged BC in terms of cation exchange, co-precipitation, and cation-cation interactions. The coordination of new BC was greatly aided by carboxyl and hydroxy in aged BC.

10.2.4 Mechanism of removal of inorganic contaminants

Environmental water pollution from heavy metals, such as As(III), Cr(VI), Ni(II), Zn(II), Cu(II), Cd(II), Hg(II), U(VI), Pu(IV), and metal-like metals, such as As(III), Se(IV), and As(V), has grown to be a significant problem (Cheng et al., 2021; Li, Hu, et al., 2021; Yang et al., 2021). Numerous human disorders, including cancer, can be brought on by heavy metal consumption. An effective and environmentally friendly adsorbent for water purification is BC. The porosity, reactivity, and sorption capacity of BC are significantly improved after surface modification. After the BC is produced, the components that have carbonized and those that have not are separated. While the carbonized portion of the BC helps with surface adsorption, the non-carbonized portion helps with the dispersion of contaminants (Alkurdi et al., 2019). Physical adsorption, surface complexation, surface precipitation, co-precipitation, electrostatic behavior, and ion exchange are the key mechanisms of toxic heavy metal removal from environmental water by BC (Fig. 10.2). Hence, utilizing BC to get rid of heavy metals is quite promising (Liang et al., 2021; Liu et al., 2022; Srivastav et al., 2022). Different modification techniques result in different heavy metal removal mechanisms (Cai et al., 2022). Therefore, if the modification process is carried out properly, the BC will have a sufficient capacity for heavy metal removal as well as significantly increasing the stability of the BC thereby improving the chemisorption and removal mechanism. Through weak van der Wall forces without chemical bonding, heavy metals are adsorbed on the surface of the bioadsorbent during physical adsorption, a diffusion-driven mass transfer. The phenomenon is reversible and temperature-sensitive (Gupta et al., 2020). The two functional groups that are most prevalent on an adsorbent are carboxylic (-COOH) and hydroxyl (-OH). Metal impoundment happens when the anionic or cationic metal ions engage with the functional groups present on the adsorbent. The point of zero charge (PZC), temperature, and pH all have an impact on the mechanism. In the elimination of heavy metals, this mechanism is crucial. It was found that Vetiveria zizanioides (Khus) with H_2SO_4 activation primarily removed toxic metals (Ni, Cu, As, Cr, Cd, and Pb) off the aqueous solution through electrostatic attraction (Tyagi, 2022). Ion exchange involves replacing the exchangeable ions (Na^+, Ca^{+2}, and Mg^{+2}) existing on the adsorbent with metal ions, which results in adsorption. The strength of the ion exchange process depends on the ionic strength and ionic active sites. To remove Cd(II) from the aqueous solution, modified BC made from sewage sludge and calcium sulfate demonstrated an ion exchange mechanism (Liu, Yue, et al., 2021). In the surface complexation mechanism, the central atom is given the ligands, which combine with the metals to create complex ions, and adsorption takes place. Numerous ligand complexes are preferable over mono-ligand complexes because they draw numerous ions from the solution. When As(II) was removed from the aqueous solution using magnetically manipulated wood biomass and $FeCl_3$, it was discovered that surface complexations played a significant role in the biosorption process (Yin et al., 2021). High pH and mineral contents are the main conditions for surface precipitation to take place, where toxic heavy metals encapsulation occurs by solids formation on an adsorbents surface or in solution.

174 PART | 3 Biochar for water and wastewater treatment

It usually happens when the metal content on the surface of the adsorbent exceeds its capacity and is distinguished by a slow adsorption rate (Agarwal et al., 2020).

10.3 Removal of organic contaminants

BC possesses exceptional versatility as a material for sequestering organic contaminants. This is due to its distinctive characteristics, including pH, surface area, and exceptional porosity. The success of the process for adsorption and degradation is contingent upon a variety of factors, including the chemical properties of the BC material and the specific type of contaminant being targeted for removal (Adewuyi, 2020). As shown in Fig. 10.1, some of the major mechanisms involved in the removal of organic contaminants from water using BC include hydrogen bonding, pore filling, hydrophobicity, and electrostatic interactions (Ahmed et al., 2018; Rosales et al., 2017).

Electrostatic interaction is the primary method for eliminating ionizable organic contaminants. The pH level of the organic contaminant affects the extent to which it is adsorbed by BC. The strength of the electrostatic interaction relies on the distance between the atoms and the magnitude of the atomic charges. Bayode and colleagues performed a study on the adsorption of a cationic dye (methylene blue) in water using a modified *Carica papaya* seed BC. This BC had a

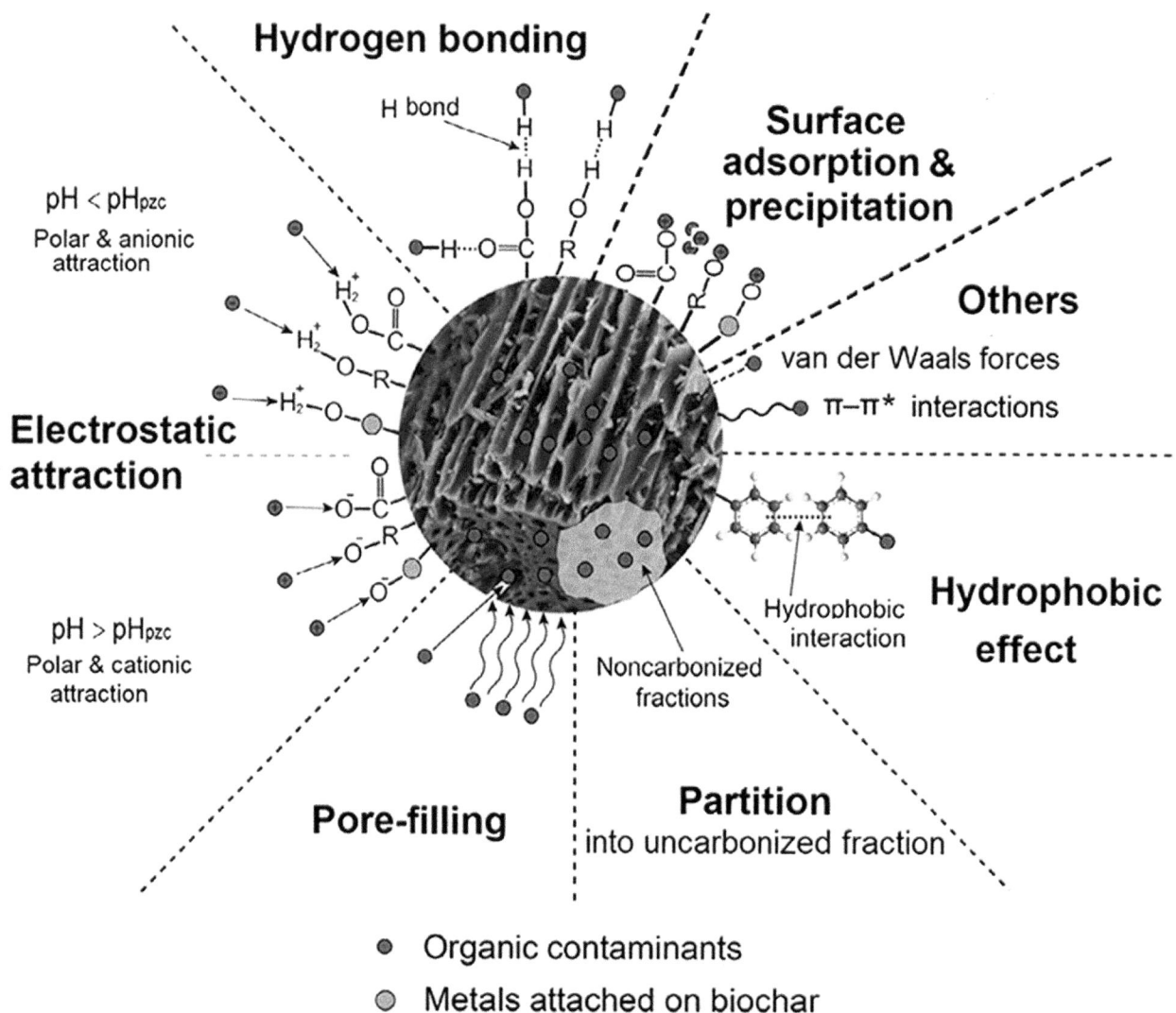

FIGURE 10.1 Mechanisms of removal of organic contaminants using biochar (Guo et al., 2020). *Adapted from Guo, M., Song, W. & Tian, J. (2020). Biochar-facilitated soil remediation: mechanisms and efficacy variations. Frontiers in Environmental Science, 8. https://doi.org/10.3389/fenvs.2020.521512.*

negatively charged surface area that attracted the positively charged MB dye via electrostatic attraction (Bayode et al., 2020). The pore-filling mechanism occurs mainly in BC that has low solute concentration and interestingly it is often accompanied by pore blocking too, this mechanism follows the nonlinear Langmuir adsorption isotherm. For the removal of polar organic contaminants, hydrogen bonding is one of the major mechanisms, BC contains polar functional groups such as carbonyl, ketones, and nitro groups that interact with the hydrophobic organic contaminant that are positively charged leading to the generation of the π-H forces (Bayode, Anthony, et al., 2023; Bayode, Folorunso, et al., 2023; Dai et al., 2019). When an ionizable organic contaminant shares the same pKa value with the BC's pHpzc value, a robust H-bond forms to facilitate the removal of the anionic ionizable organic compound. The efficiency of adsorption is significantly influenced by hydrophobic interaction (Lian & Xing, 2017; Regkouzas & Diamadopoulos, 2019). Extensive research has shown the direct correlation between the adsorption properties of adsorbents and their hydrophobicity. Pristine BC with negligible surface oxidation can effectively eliminate hydrophobic organic contaminants through hydrophobic interaction. The mechanism is closely associated with the aromaticity of the contaminant (Inyang & Dickenson, 2015).

Different BC materials have been employed to remove organic contaminants in water, either in their pristine state or by modification with various metals such as Cu and Zn, which enhances their performance for optimum contaminant removal. BC materials are more cost-effective than synthetic carbon materials or other metal materials due to their natural origin and the fact that they are often made from waste a major source of the conversion of waste to wealth addressing one of the sustainable development goals. Consequently, their affordability and eco-friendliness make them a viable option for industrial production and practical applications. A summary of studies that employed BC in the removal of organic contaminants is presented in Table 10.2

Eggshells (ES), a common household waste are used for the degradation of 2,4-dichlorophenol, a persistent organic pollutant. ES was synthesized by a one-pot pyrolysis method, and treated at different temperatures (400°C, 600°C, and 800°C) in the presence of nitrogen. ES proved to be a suitable catalyst for persulfate activation, it was reported to degrade 90% of 2,4-DCP in 120 mins. The degradation of the contaminant took place via two mechanisms the free radical pathway dominated by the hydroxyl radical and the non-radical pathway. It proved to be sustainable and environmentally friendly as it can be reused and used for real-life water treatments (Liu et al., 2020).

Onion skin-derived sorbent (OSDS) was synthesized by hydrothermal carbonization and used for the removal of methylparaben, a personal care product in water. The XPS, FTIR, and RAMAN confirmed the presence of the SP^2 C = C bond. The sorption process follows a multilayer adsorption mechanism on the heterogeneous surface of the carbon material. Complete removal was observed in under 24 hours, The major interactions involved in this adsorption process are pore filling, hydrogen bonding, π-π stacking and Van der Waals forces (Adeola et al., 2023). In a bid to improve the onion peel BC performance over a shorter period. The onion peel BC was modified with iron oxide, in a simple chemical method, as it has been proven by different researchers that modification enhances and improves the properties of the material. The XRD and the FTIR confirmed the presence of FeO in the synthesized material and it was able to photodegrade 97% methylene blue in 30 min (Abid et al., 2021). The mechanism of adsorption here was ruled by electrostatic interaction.

A novel sewage sludge derived from BC was investigated for the removal of methylene blue in water. The removal efficiency of MB reached almost 100% within 10 hours, with maximum adsorption capacity reported to be 29.85 mg/g at pH 7. The high percentage of removal was attributed to the porous and coarse texture of the adsorbent also the large surface area (Fan et al., 2017). Popular wood chips biochar (WCBC) was prepared by pyrolyzing it at 300°C for 3 h followed by ball milling it and it was used for the degradation of enrofloxacin and antibiotics. WCBC degraded 80% enrofloxacin after 90 mins in the dark and 80.2% upon illumination under 150 min. The presence of more amount of phenol structure and defects on the ball-milled BC adsorbent aided the generation of superoxide. The presence of the O-C = O played a major role in the generation of the reactive oxygen species, The major reactive species aiding the degradation are the holes on the defect which have strong oxidation power and the superoxide will attack the EFA (Xiao et al., 2020). In a further study, BC was modified with graphitic carbon nitride using ball milling, a simple method that is also green and economical. It was found that the material had a dual function of both adsorption and photodegradation. $G-C_3N_4$ is a versatile photocatalyst that can delay the electron pair/hole recombination. The results showed that the $g-C_3N_4$/BC composites had superior adsorption and photocatalytic performance for EFA compared to pure $g-C_3N_4$. 50% $g-C_3N_4$/BC composite performed the best, with optimal efficiency of 45.2% and 81.1% of EFA in the dark and light respectively. When compared to $g-C_3N_4$ alone the removal was 19.0% and 27.3% respectively. Additionally, the 50% $g-C_3N_4$/BC composite showed a high mineralization level of 65.2%, the major reactive species playing an important role in the degradation process were superoxide radical and hole. This proves the material to be very versatile and efficient for water treatment (Xiao et al., 2021).

TABLE 10.2 Summary of studies that employed biochar in the removal of organic contaminants.

Biochar	Pollutant	Removal%	Adsorption capacity (mg·g^{-1})	References
SugBC900	Bisphenol A	100	–	He et al. (2021)
MZBC-600	Ketamine	<90	32.5	Li, Xiong, et al. (2021)
Pig manure solid residue-Biochar	Tetracycline Oxytetracycline Benzene toluene	>95%	51.3 32.9 86 74.1	Liu et al. (2023)
Wheat straw-derived biochar	Bisphenol A		909.1	Shi et al. (2022)
BPB5OO BPB900	Malachite green	–	1790.3 2297.83	Chen et al. (2023)
PiPAC (Pineapple peel activated carbon)	Methylene Blue Methyl Red		165.17 94.87	Rosli et al. (2023)
Activated banana peel carbon	Rhodamine B	81.9%–85.6%	833.1	Singh et al. (2020)
Banana Peel (*Musa sp.*) Biosorbent	Phenol	89.22	–	Mishra et al. (2022)
Activated coconut shell biochar modified with phosphoric acid	Diazinon	98.96	10.33	Baharum et al. (2020)
Orange peel (OP) Banana peel (BP) Modified banana peel (MBP) Modified orange peel (MOP)	Caffeine	95.5 90.5 93.6 89.2	8.88 3.08 5.26 12.17	Almeida-Naranjo et al. (2021)
Banana peel biochar (BPBs)	Reactive Black 5 (RB5)	97	7.58	Kapoor et al. (2022)
CZPP	Ibuprofen Diclofenac	54.2 61.3	136.15 248.82	Bayode, Folorunso, et al. (2023)
CZPPrgo	Ibuprofen Diclofenac	97.3 100	147.77 145.79	Bayode, Folorunso, et al. (2023)
PS-HYCA PP-HYCA PSPP-HYCA	Methylene blue		353.2 206.6 205.6	Bayode et al. (2020)
Adenopus breviflorus seed adsorbent	2,4-dichlorophenol	–	25.94	Adewuyi et al. (2016)
CLY@AM	Phenolphthalein Methyl orange	86 80	43 40	Adewuyi and Oderinde (2022)
CLY (Kaolinite clay)	Phenolphthalein Methyl orange	–	20 22	Adewuyi and Oderinde (2022)
Spent coffee grounds (SCGB)	Norfloxacin	–	69.8	Nguyen et al. (2022)
Fe@PSK@GO	Estrone Estrogen Estriol Ethynyl estradiol	81 89 84 93	– – – –	Bayode, dos Santos, et al. (2021)
K-ZnO/C/GO	Estrone Estrogen Estriol Ethynyl estradiol	89.2 92.9 94.8 98.1	– – – –	Bayode, Vieira, et al. (2021)

10.4 Removal of microbial contaminants

Microbial contaminants otherwise known as pathogens are usually abundant in storm and surface water contaminants and their presence is inferred by elevated concentrations of fecal indicator bacteria (FIB) (Afrooz et al., 2018). FIB concentrations in these two water bodies can be below the detection limit or higher than those observed in sewage. Although there are scanty works of literature on pathogen concentrations in stormwater, the little data suggest their levels are also quite variable and can be elevated meanwhile the presence of pathogens in surface waters can have an adverse effect on public health and affect local economies of communities that rely on clean surface waters. Pathogens are also present in domestic wastewater and can pose a great health risk if not properly treated. These pathogens can be bacterial (e.g., *Escherichia coli*, *Legionella pneumophila*, *Leptospira* spp., *Salmonella* spp., *Shigella* spp., *Vibriocholera*, and *Yersiniaenterocolitis*) and Viral (e.g., adenovirus, enterovirus, hepatitis A, norovirus, reovirus, rotavirus, and echovirus) (Lusk et al., 2017). Biofilters can be used to capture and treat contaminated stormwater before it is discharged to surface waters. However, the conventional biofilter, which typically contains sand and compost, demonstrates inadequate or inconsistent performance in removing pathogens from stormwater, they are now been replaced with BC (Kranner et al., 2019). BC has been reported to have the ability to remove microbial contaminants. Studies have shown that BC used as a filter medium can remove various pathogens from water sources effectively. For example, Perez-Mercado et al. found that BC could remove Saccharomyces cerevisiae (an indicator of fungal contamination) from diluted wastewater (Perez-Mercado et al., 2019).

The mechanisms involved in the removal of microbial contaminants by BC filters include biofilm formation, straining, and adsorption. Biofilm formation is the development of a thin, slimy layer on the surface of the BC, which can supply additional sorption sites for the pathogenic organisms. Straining occurs when the BC physically blocks the movement of the microorganisms because the pore size of the BC is smaller than the size of the bacteria. In addition, adsorption plays a vital role in the removal of pathogens. Pathogen adsorption involves a two-step process. First, reversible adsorption occurs when the pathogen interacts weakly with the porous surface of the BC through electrostatic and van der Waals forces. In this phase, they can detach from the BC surface and return to the bulk water. Second, irreversible adsorption or adhesion occurs when the pathogen's polymers form bridges that connect them to the BC permanently. These polymers, known as biofilms, contribute to the attachment through hydrogen bonding or dipole interactions. Accessibility of sorption sites may also be affected by biofilm formation and they may also act as an additional sorbent (Sasongko et al., 2021). Microorganisms adhere to BC in environmental media due to its rough surface topography, favorable free energy, surface charge, and hydrophobicity. Microorganisms tend to interact with hydrophobic, non-polar surfaces and can penetrate inside through biofilm formation, producing extracellular polymeric substances and forming microbial community structures (Mukherjee et al., 2022). Pathogenic microorganisms are challenging to remove in bioretention systems, which use physical filtration, adsorption, chemical inactivation, predation, and exposure to drought stressors to eliminate them. The removal of pathogens is not the same for all microorganisms, as Escherichia coli, for example, can grow using stormwater nutrients and its desorption will be higher than its adsorption in multiple drying-rewetting conditions, leading to a decrease in pathogen removal efficiency (Chen, Wu, et al., 2022). Some of the characteristics of BC that make it more suitable than sand or gravel include high porosity, many pores of different sizes, organic leaching, and hydrophobicity. BC will not only enhance the adsorption of pathogenic microorganisms but also reduce its mobilization during intermittent infiltration of stormwater (Mohanty & Boehm, 2014).

It is important to note that the effectiveness of BC in removing specific pathogens can vary. Factors such as the size and shape of bacteria, the pore size of the BC, and the presence of biofilms can influence the removal mechanism. Nonetheless, BC-based water treatment systems show promise in providing clean water by effectively reducing the presence of pathogenic organisms (Sasongko et al., 2021). In another study, pathogen removal by a filtration-based onsite wastewater treatment system (OWTS) depends on (1) the efficiency of physical entrapment in the pores of filtration media and (2) the adsorption capacity of the filtration media (Perez-Mercado et al., 2019; Wang et al., 2020; Xiaoqin) meanwhile adsorption was reported to be one of the dominant pathogen removal mechanisms in OWTS studies in another report (Gwenzi et al., 2017).

Because BC has a higher surface area (143 $m^2\ g^{-1}$) than sand particles (<0.004 $m^2\ g^{-1}$), it has been proven to have a superior adsorption capability, making it particularly effective in removing fecal coliform (Kaetzl et al., 2019). Moreover, the growth of *E. coli* was entirely inhibited when a complex of silver (Ag) nanoparticles (NPs) and zero-valent iron (ZVI) was applied to the surface of the BC. In comparison, untreated BC promoted the growth of *E. coli* up to 4×10^8 colony-forming units per milliliter (Zhou et al., 2014). Furthermore, it has been reported that in batch experiments employing 200 mg L-1 of BC, magnetic BC—which contains iron oxide deposited on its surface—possesses antibacterial characteristics and was able to eradicate nearly 100% of E. coli and Staphylococcus aureus. The majority of the scant literature on the effect of BC amendment on pathogen removal from home wastewater concentrates on the elimination of *E. coli* (Boehm et al., 2020). Compared to sand filters, the use of BC in anaerobic biofiltration for treating raw wastewater produced a much greater removal rate of E. coli (99.5%), enterococci (99.6%), and bacteriophages (98.6%) (Kaetzl et al., 2019).

A different study found that the removal of E. coli by the BC-treated bioretention system (BRS) was 96%, and it was not influenced by varying stormwater infiltration rates or concentrations of bacteria (Mohanty & Boehm, 2014). When treated BC was added to the bioretention soil media, the removal percentage of *E. coli* increased significantly from 35% to approximately 92%–98% (Lau et al., 2017). Nevertheless, the E. coli in the pore water of the biofilter stops growing after rainfall, indicating that BC may continue to remove pathogens during inter-event periods even when nutrients are present in stormwater according to another study (Valenca et al., 2021). The data from some studies demonstrated that there was a significant increment in the removal rate of BRS on FIB (e.g., total coliform and *E. coli*) when there was an increase in the amount of BC/pathogen ratio, indicating an enhancement in pathogen removal with the incremental addition of BC (Afrooz et al., 2016; Lau et al., 2017; Mohanty & Boehm, 2014; Suliman et al., 2017). Unlike another study where the BC/pathogen ratio was not increased in the BRS, the maximum removal rate of total coliform and E. coli significantly increased by 11% and 34%, respectively. However, when there was a mixture of BC with compost, compared, there was no significant increase in the removal of FIB (Mohanty & Boehm, 2014). This may be due to the release of large amounts of dissolved organic carbon (DOC) compost in pore water, and DOC may block the pores and hinder adsorption. It may also be due to the competition of organic matter on BC adsorption sites and the increase in electrostatic repulsion between BC and cells after the adsorption of organic matter (Mohanty et al., 2014). In another report, where corn stalk BC was added into a column filter for removal of E. coli and bacteriophage, the percentage removal was 62.59% and 69.82% for E. coli and bacteriophage respectively (Chen, Yang, et al., 2022). BC from corn cobs (CCB) and rice husk (RHB) effectively removed fecal coliform (FC) and fecal streptococci (FS) by a percentage rate of 95.9% and 96.3% for FC on CCB and RHB while FS on CCB and RHB was 96.2% and 99.2% respectively (Visiy et al., 2022). Applying BC as a soil ameliorant was reported in another study and it showed that in stormwater treatment, E coli can be removed by BC at a removal rate of 30%–98% (Chen, Wu, et al., 2022).

10.5 Conceptual designs of biochar water filters

10.5.1 Conceptual designs

Several conceptual designs exist for using BC as an adsorbent medium in drinking water systems (Fig. 10.2). These include (1) single adsorbent BC water filter (Fig. 10.2A), (2) dual adsorbent water filter (Fig. 10.2B), (3) a conventional biosand filter retrofitted with a BC adsorbent (Fig. 10.2C), (4) a BC-ceramic composite filter (Fig. 10.2D), and (5) a multi-barrier BC water filter coupled to a disinfection unit such as solar disinfection (SODIS) (Fig. 10.2E). An overview of each conceptual design is presented below.

1. Single adsorbent BC water filter

    A single adsorbent BC water filter consists of a column packed with BC as an adsorbent. An example of such a filter has been applied to remove micropollutants such as pesticides in drinking water (Kearns et al., 2016).
2. Dual/hybrid adsorbent BC water filter

    A dual adsorbent BC water filter consists of BC as a filter media combined with another adsorbent material such as metallic or zero-valent iron. The two adsorbents can either be in a layered (Fig. 10.2B(i)) or mixed matrix (Fig. 10.2B(ii)). In such filter systems, contaminant removal occurs via adsorption on both filter media. In the case of water filters combining a mixture of metallic iron and BC, the latter may also act as an aggregate that delays the onset of filter clogging and loss of permeability in iron filters.
3. A conventional biosand filter retrofitted with a BC adsorbent

    Biosand filters have limited capacity to remove dissolved contaminants (Gwenzi et al., 2017). Hence, retrofitting a conventional biosand filter could potentially improve the removal performance of dissolved contaminants. An example of a biosand filter retrofitted with a BC layer has been reported in the literature (Eniola & Sizirici, 2023).
4. A BC-ceramic composite filter

    A BC-ceramic filter can also be developed by firing a mixture of clay mixed with biomass aggregates. In such a composite filter, the biomass aggregates act as a porosity or poro-generator (Chaukura et al., 2020). In such a BC ceramic composite filter contaminant removal occurs via adsorption on both ceramic and BC matrices. An example of such a filter was developed and applied to remove contaminants in drinking water (Chaukura et al., 2020).
5. A multi-barrier BC water filter coupled to a disinfection unit

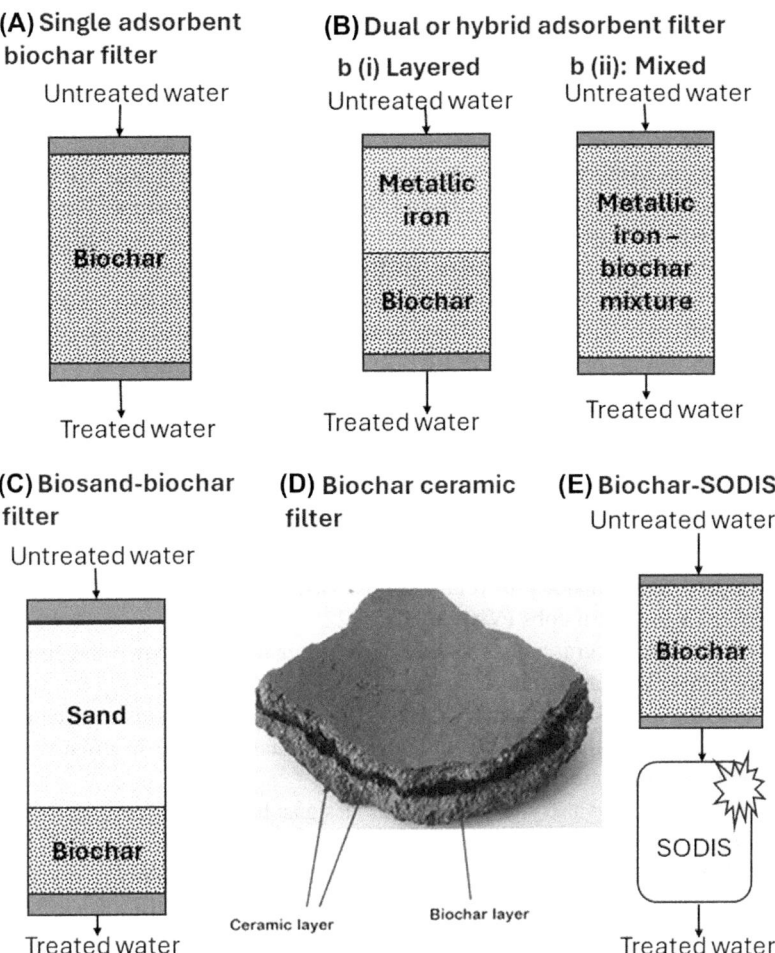

FIGURE 10.2 Conceptual designs of biochar water filters: single adsorent biochar filter (A); dual or hybrid adsorbent filters (B); biosand-biochar filter (C); biochar ceramic filter (D); and biochar-solar disinfectant (SODIS) filter (E). *Adapted with permission from Chaukura, N., Chiworeso, R., Gwenzi, W., Motsa, M. M., Munzeiwa, W., Moyo, W., Chikurunhe, I. & Nkambule, T. T. I. (2020). A new generation low-cost biochar-clay composite "biscuit" ceramic filter for point-of-use water treatment. Applied Clay Science, 185. https://doi.org/10.1016/j.clay.2019.105409.*

A possibility also exists to use BC in a multi-barrier water filter system combining adsorption and disinfection. A conceptual example may entail a BC water filtration unit coupled to a solar disinfection (SODIS) unit. In such a conceptual design, BC will account for the removal of the dissolved contaminants and part of the microbial contaminants, while the disinfection unit will remove the microbial pathogens. However, data on the design and evaluation of such a multi-barrier system is still lacking. Hence, there is a need to design and evaluate such multi-barrier water filtration systems. However, limited data exists on the design principles, evaluation, and field performance of BC filter systems. Therefore, further work is required to (1) develop and document the design principles, and (2) evaluate the performance of such BC systems using raw water with diverse properties indicative of field conditions.

10.5.2 Overview of performance

Several OWTS have been reported to contribute to the introduction of different pollutants into the environment which arises as a result of aging and inadequate maintenance of wastewater treatment facilities (Muoghalu et al., 2023; Perez-Mercado et al., 2018). Onsite wastewater treatment facilities were designed to enhance adequate treatment of wastewater before discharge into the environment, however, challenges arising from design construct and operational limitations have led to poor efficiency consequently making them a source where various pollutants are released into the environment (Perez-Mercado et al., 2018).

In a typical setup for an onsite wastewater treatment facility, a septic tank is immediately followed by a sand filter. Some challenges have however been noted with the use of sand as filters which include the unavailability of well-defined sand particles in certain regions, including the cost associated with transportation among others (García-Ávila et al., 2023; Perez-Mercado et al., 2018). Hence, there is a need for an eco-friendly and efficient substitute which has led researchers to explore the use of BC due to its increased surface area, low density, and porosity (García-Ávila et al., 2023; Perez-Mercado et al., 2018).

Although BC has demonstrated enormous potential for the removal of various contaminants in laboratory-scale experiments, there have been limited studies on the use of BC in wastewater treatment facilities outside the laboratory (Gwenzi et al., 2017). At best, recent works on the use of BC as filters reported recently in literature have focused on the design of such filters and testing their efficiency on synthetic wastewater or wastewater collected from different sources. For instance, Perez-Mercado and co-workers assessed the effectiveness of three different BC (pine-spruce, willow, and activated BC) as filters and assessed the effects of physical, chemical, and hydraulic properties as well as the effect of varying the particle size of the BC filters. According to their results, all BC filters (pine-spruce, willow, and activated BC) effectively facilitated the efficient removal of organic load with no considerable differences between them (94%–99% COD) which was attributed to surface area and porosity. They, however, noted reduced efficiency for the BC filters with different particle sizes, with the small particle-sized BC filters performing better (99% COD removal) and attributed this to the fact that larger particle sized BC filters possess bigger macro-pores which tend to decrease the contact time between the filters and organic load in the wastewater (Perez-Mercado et al., 2018).

Visiy et al. (2022) investigated the efficiency of BC filters synthesized from rice husk and corn cobs in the treatment of domestic wastewater utilizing a vertical flow constructed wetland system with *Echiochloa pyramidalis* spread on it. They noted that the total suspended solid (TSS) removal was of the order corn cobs (90%) > rice husk (89.1%) > sand (71.7%) and also found that nonvegetated filters were more effective in reducing TSS than vegetated ones and attributed this to large surface area of the BC filters aiding their capacity to retain solids. However, for nutrient removal, sand filters had a better performance than both rice husk and corn cobs (Visiy et al., 2022)

Kaetzl and co-workers studied the efficiency of *Miscanthus* grass as an anaerobic biofilter utilizing slow sand filtration technology while using sand as the reference filter. They reported average COD and TOC removal of 74 ± 18% and 61 ± 12% for BC filter and 61 ± 12% and 46 ± 3.8% for sand filters which was credited to the positive influence of BC filter on anaerobic degradation processes. In addition, the BC filter recorded a gradual increase in effectively removing *Escherichia coli* over time in comparison to the sand filter (Kaetzl et al., 2020).

In addition to the use of BC filters in several OWTS, a variety of conceptual designs have also been extended to drinking water which is the main focus of this chapter, examples include BC-clay composite ceramic filters (Fig. 10.2D).

Chan and co-workers employed the use of BC filter in a design similar to biosand with filter for removing organic chemicals from drinking water over a period of 36 days (Chan et al., 2020). It was reported that the BC filters had 37%–97% removal for atrazine, 49%–93% removal for naphthalene, and 100% removal for phenanthrene (Chan et al., 2020).

Chaukura and others, on the other hand, designed "a cost efficient" BC filter incorporated into clay termed 'biscuit ceramic filters (BCF) for the removal of different contaminants from local drinking water treatment plant (Chaukura et al., 2020). Two water samples (before sand filtration and after sand filtration) were collected from the treatment plant and were examined for water quality parameters being passed through the "biscuit ceramic filters" in comparison to ordinary ceramic filter (CF). According to the authors, BCF had a reasonably reduced total hardness than CF with P value greater .05. For turbidity reduction and removal of dissolved organic, BCF had a better performance than CF. A comparison of water samples before sand filtration from the local treatment plant and after being passed through BCF, with the results of the permeate being compared to after sand filtration from the local water treatment plant showed that BCF had a higher performance than the sand filters using parameters such as total hardness, total dissolved solid and reduction in turbidity (Chaukura et al., 2020).

A high-rate filtration system packed with two types of BC made from Eucalyptus and Bamboo (Fig. 10.3) was designed by Garcia-Avila and co-workers to remove suspended solids, reduction of turbidity, and removal of chemical contaminants also. They noted that bamboo BC had the best performance of the two in removing color and turbidity and also reported that BC with reduced particle sizes had the highest removal efficiency (García-Ávila et al., 2023).

10.6 Efficient disposal management of spent biochars

Although BC possesses great benefits in wastewater treatment, discarding spent BC is of huge environmental concern, especially when it has been used for the removal of contaminants such as heavy metals. There is a risk of secondary pollution arising from the disposal of spent BC if proper care is not taken (Gupta et al., 2020). However as noted by Muoghalu et al. (2023), if BC was used for the removal of nutrients (e.g., ammonium, nitrate, and phosphate) devoid of other harmful contaminants such as heavy metals, the spent BC could be utilized as organic fertilizers in order to enhance the fertility of soil (Muoghalu et al., 2023). Additionally, the use of spent BC in agriculture will also enhance

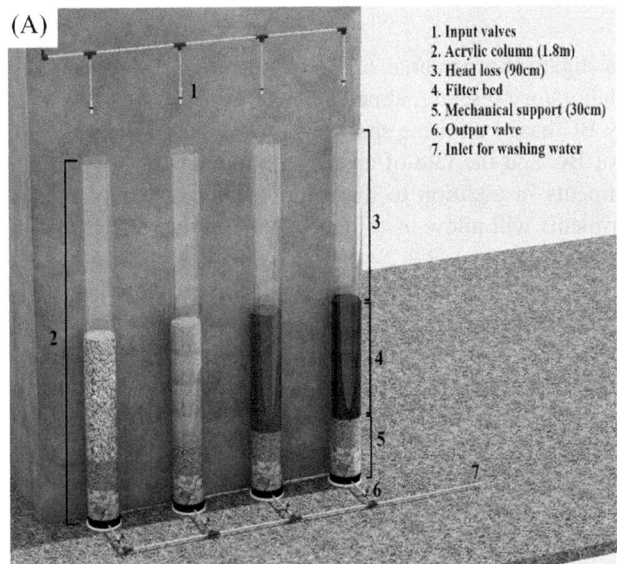 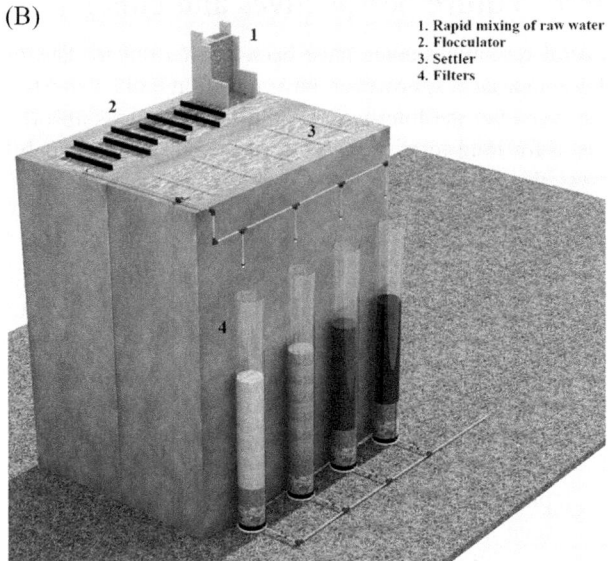

FIGURE 10.3 Image of a high-rate filtration system packed with biochar. *Adapted from García-Ávila, Fernando, Galarza-Guamán, Andrés, Barros-Bermeo, Mónica, Alfaro-Paredes, Emigdio Antonio, Avilés-Añazco, Alex & Iglesias-Abad, Sergio. (2023). Integration of high-rate filtration using waste-derived biochar as a potential sustainable technology for drinking water supply. Biochar, 5(1). https://doi.org/10.1007/s42773-023-00256-4.*

nutrients and organic matter recycling thereby fostering a circular economy, hence, reducing the need for more synthetic fertilizers and mitigating environmental degradation (Samuel Olugbenga et al., 2024).

One other major alternative to the disposal of spent BC comes with its use as carbon sequestrating agent in cement production. It has been reported that the production of cement worldwide leads to carbon dioxide (CO_2) emission into the environment (Senadheera et al., 2023; Suarez-Riera et al., 2020). Biochar is also capable of improving certain crucial features such as strength, and "water tightness" in cement. Additionally, buildings with biochar-incorporated materials are capable of storing carbon for several years, a feature that can't be replicated with the use of steel or concrete (Senadheera et al., 2023; Suarez-Riera et al., 2020).

Gupta et al. (2018) examined the effect of incorporating biochar into mortar while studying its efficiency on mechanical strength, sorptivity of water permeability. It was recorded that the biochar synthesized at 300°C was capable of removing CO_2 and also noted an increase in compressive strength of the mortal while a reduction in sorptivity and water porosity was recorded. Wang et al. (2020) also worked on synergistically combining biochar and CO_2 curing in a bid to boost certain features of cement-based composites. According to the authors, the presence of biochar enhanced the creation of "hydration products" leading to higher compressive strength (16%) in comparison to plain mortal without BC.

10.7 Large scale/field applications of biochar in Africa

The use of BC in recent times in Africa have been centered on improving soil quality. Factors such as erosion, continuous soil usage, dry season irrigation, elevated temperature, and climate change have all led to expedited depletion and disproportionate distribution of carbon in the environment and loss of other essential nutrients in the soil (Obia et al., 2019; Omulo, 2020; Steiner et al., 2018). As a result of this, the attention of the scientific community and farmers on the continent has been drawn to the use of BC in remediating and/or replenishing soil fertility (Offiong et al., 2020, 2022; Offiong et al., 2021; Omulo, 2020).

A participatory study conducted on five farming communities by Steiner and co-workers in Tamale Ghana, evaluated the potency of BC produced from crop wastes from the farm and examined its efficiency when applied to the soil. Lettuce (*Lactuca sativa*) was selected as the crop to be planted on the soil to which BC was applied. They reported an increased diameter of the plants cultivated on lands with BC, fresh matter yields of the lettuce were also greatly enhanced (Steiner et al., 2018). A similar study was also conducted by Kätterer et al. (2022) on nine small-scale farms in Kenya to assess the efficiency of BC synthesized from readily available feedstock and examine its efficiency on the performance of maize. They noted that the maize grain yield was significantly enhanced in plots with BC application (Kätterer et al., 2022).

10.8 Future perspectives and conclusion

Several research studies have been carried out to date to investigate the potential of BC in improving water quality. These approaches, however, have mostly relied on short-term laboratory research aimed at specific pollutants. It is difficult to predict the long-term stability of pollutants absorbed by BC based on these short-term trials. As a result, there is a need for long-term monitoring of the adsorption capability of BC and the fate of adsorbed pollutants. It is critical to assess the co-removal of non-targeted chemicals by BC treatments in addition to the removal of targeted pollutants. Furthermore, conducting large-scale BC water treatment experiments will allow us to uncover issues that arise throughout the process that are not normally visible in laboratory investigations.

Large-scale studies are especially needed to examine BC's purifying capabilities in water contaminated with organic, inorganic, and microbiological contaminants. Furthermore, while the majority of BC research has focused on the removal of organic and inorganic impurities from water, very little is known regarding its efficiency against microbial contaminants such as *E. coli* and other pathogens and pathogenic indicators. Specifically, using advanced spectroscopic and microscopic techniques to analyze the interactions between BC, water, and pollutants is critical for gaining a better understanding of the underlying processes and mechanisms. Some BCs (those generated from waste materials or biosolid/sewage sludge) may introduce secondary pollutants into the water depending on the feedstock. As a result, the development of enhanced technologies for BC treatment may be advantageous in eradicating intrinsic pollutants.

Despite the remarkable results associated with the use of BC in water treatment, there are various possible areas for further investigation in the application of BC for water treatment. To begin, it is critical to investigate the long-term efficacy and sustainability of modified BC in real-world treatment scenarios to assess their practicability and environmental impact. Furthermore, improving BC production and modification techniques to increase cost-effectiveness and reduce environmental impact is critical for its wider application. Furthermore, standardized techniques for characterizing BC materials and their adsorption capacities are required for comparing and selecting the best forms of BC for various water treatment applications. Furthermore, researching the possible synergistic effects of combining BC with other water treatment technologies could improve overall treatment efficiency and lower water purifying costs.

References

Abid, M. A., Abid, D. A., Aziz, W. J., & Rashid, T. M. (2021). Iron oxide nanoparticles synthesized using garlic and onion peel extracts rapidly degrade methylene blue dye. *Physica B: Condensed Matter, 622*. Available from https://doi.org/10.1016/j.physb.2021.413277, http://www.journals.elsevier.com/physica-b-condensed-matter/.

Abu Hasan, H., Sheikh Abdullah, S. R., Tan Kofli, N., & Yeoh, S. J. (2016). Interaction of environmental factors on simultaneous biosorption of lead and manganese ions by locally isolated Bacillus cereus. *Journal of Industrial and Engineering Chemistry, 37*, 295–305. Available from https://doi.org/10.1016/j.jiec.2016.03.038, http://www.sciencedirect.com/science/journal/1226086X.

Abu Hasan, H., Muhammad, M. H., & Ismail, N. 'I. (2020). A review of biological drinking water treatment technologies for contaminants removal from polluted water resources. *Journal of Water Process Engineering, 33*, 101035. Available from https://doi.org/10.1016/j.jwpe.2019.101035.

Adeola, A. O., Oyedotun, K. O., Waleng, N. J., Mamba, B. B., & Nomngongo, P. N. (2023). Onion skin–derived sorbent for the sequestration of methylparaben in contaminated aqueous medium. *Biomass Conversion and Biorefinery*. Available from https://doi.org/10.1007/s13399-023-04332-4, https://www.springer.com/journal/13399.

Adewuyi, A., Göpfert, A., Adewuyi, O. A., & Wolff, T. (2016). Adsorption of 2-chlorophenol onto the surface of underutilized seed of Adenopus breviflorus: A potential means of treating waste water. *Journal of Environmental Chemical Engineering, 4*(1), 664–672. Available from https://doi.org/10.1016/j.jece.2015.12.012, http://www.journals.elsevier.com/journal-of-environmental-chemical-engineering/.

Adewuyi, A. (2020). Chemically modified biosorbents and their role in the removal of emerging pharmaceutical waste in the water system. *Water, 12*(6), 1551. Available from https://doi.org/10.3390/w12061551.

Adewuyi, A., & Oderinde, R. A. (2022). Removal of phenolphthalein and methyl orange from laboratory wastewater using tetraethylammonium modified kaolinite clay. *Current Research in Green and Sustainable Chemistry, 5*, 100320. Available from https://doi.org/10.1016/j.crgsc.2022.100320.

Afrooz, A. R. M. N., Boehm, A. B., & Nerenberg, R. (2016). Escherichia coli Removal in Biochar-Modified Biofilters: Effects of Biofilm. *PLoS One, 11*(12), e0167489. Available from https://doi.org/10.1371/journal.pone.0167489.

Afrooz, A. R. M. N., Pitol, A. K., Kitt, D., & Boehm, A. B. (2018). Role of microbial cell properties on bacterial pathogen and coliphage removal in biochar-modified stormwater biofilters. *Environmental Science: Water Research and Technology, 4*(12), 2160–2169. Available from https://doi.org/10.1039/c8ew00297e, http://pubs.rsc.org/en/journals/journal/ew.

Agarwal, A., Upadhyay, U., Sreedhar, I., Singh, S. A., & Patel, C. M. (2020). A review on valorization of biomass in heavy metal removal from wastewater. *Journal of Water Process Engineering, 38*. Available from https://doi.org/10.1016/j.jwpe.2020.101602, http://www.journals.elsevier.com/journal-of-water-process-engineering/.

Ahmed, M. B., Zhou, J. L., Ngo, H. H., Johir, M. A. H., Sun, L., Asadullah, M., & Belhaj, D. (2018). Sorption of hydrophobic organic contaminants on functionalized biochar: Protagonist role of π-π electron-donor-acceptor interactions and hydrogen bonds. *Journal of Hazardous Materials, 360*, 270–278. Available from https://doi.org/10.1016/j.jhazmat.2018.08.005, http://www.elsevier.com/locate/jhazmat.

Ahmed, W., Mehmood, S., Núñez-Delgado, A., Ali, S., Qaswar, M., Khan, Z. H., Ying, H., & Chen, D. Y. (2021). Utilization of Citrullus lanatus L. seeds to synthesize a novel MnFe2O4-biochar adsorbent for the removal of U(VI) from wastewater: Insights and comparison between modified and raw biochar. *Science of the Total Environment, 771*. Available from https://doi.org/10.1016/j.scitotenv.2021.144955, http://www.elsevier.com/locate/scitotenv.

Ahmed, W., Mehmood, S., Núñez-Delgado, A., Qaswar, M., Ali, S., Ying, H., Liu, Z., Mahmood, M., & Chen, D.-Y. (2021). Fabrication, characterization and U (VI) sorption properties of a novel biochar derived from Tribulus terrestris via two different approaches. *Science of The Total Environment, 780*, 2021.

Ahmed, W., Mehmood, S., Qaswar, M., Ali, S., Khan, Z. H., Ying, H., Chen, D. Y., & Núñez-Delgado, A. (2021). Oxidized biochar obtained from rice straw as adsorbent to remove uranium (VI) from aqueous solutions. *Journal of Environmental Chemical Engineering, 9*(2). Available from https://doi.org/10.1016/j.jece.2021.105104, http://www.journals.elsevier.com/journal-of-environmental-chemical-engineering/.

Ahmed, W., Núñez-Delgado, A., Mehmood, S., Ali, S., Qaswar, M., Shakoor, A., & Chen, D. Y. (2021). Highly efficient uranium (VI) capture from aqueous solution by means of a hydroxyapatite-biochar nanocomposite: Adsorption behavior and mechanism. *Environmental Research, 201*. Available from https://doi.org/10.1016/j.envres.2021.111518, http://www.elsevier.com/inca/publications/store/6/2/2/8/2/1/index.htt.

Akhtar, N., Syakir Ishak, M. I., Bhawani, S. A., & Umar, K. (2021). Various natural and anthropogenic factors responsible for water quality degradation: A review. *Water (Switzerland), 13*(19). Available from https://doi.org/10.3390/w13192660, https://www.mdpi.com/2073-4441/13/19/2660/pdf.

Alkurdi, S. S. A., Herath, I., Bundschuh, J., Al-Juboori, R. A., Vithanage, M., & Mohan, D. (2019). Biochar versus bone char for a sustainable inorganic arsenic mitigation in water: What needs to be done in future research? *Environment International, 127*, 52–69. Available from https://doi.org/10.1016/j.envint.2019.03.012, http://www.elsevier.com/locate/envint.

Almeida-Naranjo, C. E., Aldás, M. B., Cabrera, G., & Guerrero, V. H. (2021). Caffeine removal from synthetic wastewater using magnetic fruit peel composites: Material characterization, isotherm and kinetic studies. *Environmental Challenges, 5*. Available from https://doi.org/10.1016/j.envc.2021.100343, https://www.journals.elsevier.com/environmental-challenges.

Arshad, N., & Imran, S. (2017). Assessment of arsenic, fluoride, bacteria, and other contaminants in drinking water sources for rural communities of Kasur and other districts in Punjab, Pakistan. *Environmental Science and Pollution Research, 24*(3), 2449–2463. Available from https://doi.org/10.1007/s11356-016-7948-7.

Arslan, A. H., Ciloglu, F. U., Yilmaz, U., Simsek, E., & Aydin, O. (2022). Discrimination of waterborne pathogens, Cryptosporidium parvum oocysts and bacteria using surface-enhanced Raman spectroscopy coupled with principal component analysis and hierarchical clustering. *Spectrochimica Acta - Part A: Molecular and Biomolecular Spectroscopy, 267*. Available from https://doi.org/10.1016/j.saa.2021.120475, https://www.journals.elsevier.com/spectrochimica-acta-part-a-molecular-and-biomolecular-spectroscopy.

Ayele, H. S., & Atlabachew, M. (2021). Review of characterization, factors, impacts, and solutions of Lake eutrophication: lesson for lake Tana, Ethiopia. *Environmental Science and Pollution Research, 28*(12), 14233–14252. Available from https://doi.org/10.1007/s11356-020-12081-4, https://link.springer.com/journal/11356.

Baharum, N. A., Nasir, H. M., Ishak, M. Y., Isa, N. M., Hassan, M. A., & Aris, A. Z. (2020). Highly efficient removal of diazinon pesticide from aqueous solutions by using coconut shell-modified biochar. *Arabian Journal of Chemistry, 13*(7), 6106–6121. Available from https://doi.org/10.1016/j.arabjc.2020.05.011, http://colleges.ksu.edu.sa/Arabic%20Colleges/CollegeOfScience/ChemicalDept/AJC/default.aspx, ScienceDirect, http://www.sciencedirect.com/science/journal/18785352.

Bayode, A. A., Agunbiade, F. O., Omorogie, M. O., Moodley, R., Bodede, O., & Unuabonah, E. I. (2020). Clean technology for synchronous sequestration of charged organic micro-pollutant onto microwave-assisted hybrid clay materials. *Environmental Science and Pollution Research, 27*(9), 9957–9969. Available from https://doi.org/10.1007/s11356-019-07563-z, https://link.springer.com/journal/11356.

Bayode, A. A., Anthony, E. T., Ore, O. T., Alfred, M. O., Koko, D. T., Unuabonah, E. I., Helmreich, B., & Omorogie, M. O. (2023). A review on the versatility of Carica papaya seed: an agrogenic waste for the removal of organic, inorganic and microbial contaminants in water. *Journal of Chemical Technology and Biotechnology, 98*(9), 2095–2109. Available from https://doi.org/10.1002/jctb.7415, http://onlinelibrary.wiley.com/journal/10.1002/(ISSN)1097-4660.

Bayode, A. A., Folorunso, M. T., Helmreich, B., & Omorogie, M. O. (2023). Biomass-Tuned Reduced Graphene Oxide@Zn/Cu: Benign Materials for the Cleanup of Selected Nonsteroidal Anti-inflammatory Drugs in Water. *ACS Omega, 8*(8), 7956–7967. Available from https://doi.org/10.1021/acsomega.2c07769, http://pubs.acs.org/journal/acsodf.

Bayode, A. A., dos Santos, D. M., Omorogie, M. O., Olukanni, O. D., Moodley, R., Bodede, O., Agunbiade, F. O., Taubert, A., de Camargo, A. S. S., Eckert, H., Vieira, E. M., & Unuabonah, E. I. (2021). Carbon-mediated visible-light clay-Fe2O3–graphene oxide catalytic nanocomposites for the removal of steroid estrogens from water. *Journal of Water Process Engineering, 40*. Available from https://doi.org/10.1016/j.jwpe.2020.101865, http://www.journals.elsevier.com/journal-of-water-process-engineering/.

Bayode, A. A., Vieira, E. M., Moodley, R., Akpotu, S., de Camargo, A. S. S., Fatta-Kassinos, D., & Unuabonah, E. I. (2021). Tuning ZnO/GO p-n heterostructure with carbon interlayer supported on clay for visible-light catalysis: Removal of steroid estrogens from water. *Chemical Engineering Journal, 420*. Available from https://doi.org/10.1016/j.cej.2020.127668, http://www.elsevier.com/inca/publications/store/6/0/1/2/7/3/index.htt.

Boehm, A. B., Bell, C. D., Fitzgerald, N. J. M., Gallo, E., Higgins, C. P., Hogue, T. S., Luthy, R. G., Portmann, A. C., Ulrich, B. A., & Wolfand, J. M. (2020). Biochar-augmented biofilters to improve pollutant removal from stormwater-can they improve receiving water quality? *Environmental Science: Water Research and Technology, 6*(6), 1520–1537. Available from https://doi.org/10.1039/d0ew00027b, http://pubs.rsc.org/en/journals/journal/ew.

Bridle, H. (2021). *Overview of waterborne pathogens* (pp. 9–40). Elsevier BV. Available from https://doi.org/10.1016/b978-0-444-64319-3.00002-2.

Cai, Y., Zhang, Y., Lv, Z., Zhang, S., Gao, F., Fang, M., Kong, M., Liu, P., Tan, X., Hu, B., & Wang, X. (2022). Highly efficient uranium extraction by a piezo catalytic reduction-oxidation process. *Applied Catalysis B: Environmental, 310*, 121343. Available from https://doi.org/10.1016/j.apcatb.2022.121343.

Chakraborti, D., Rahman, M. M., Ahamed, S., Dutta, R. N., Pati, S., & Mukherjee, S. C. (2016). Arsenic groundwater contamination and its health effects in Patna district (capital of Bihar) in the middle Ganga plain, India. *Chemosphere, 152*, 520–529. Available from https://doi.org/10.1016/j.chemosphere.2016.02.119, http://www.elsevier.com/locate/chemosphere.

Chan., Lari., & Soulsbury. (2020). An intermittently operated biochar filter to remove chemical contaminants from drinking water. *International Journal of Environmental Science and Technology, 17*(6), 3119–3130. Available from https://doi.org/10.1007/s13762-019-02615-w.

Chaukura, N., Chiworeso, R., Gwenzi, W., Motsa, M. M., Munzeiwa, W., Moyo, W., Chikurunhe, I., & Nkambule, T. T. I. (2020). A new generation low-cost biochar-clay composite 'biscuit' ceramic filter for point-of-use water treatment. *Applied Clay Science, 185*. Available from https://doi.org/10.1016/j.clay.2019.105409.

Chen, L., Mi, B., He, J., Li, Y., Zhou, Z., & Wu, F. (2023). Functionalized biochars with highly-efficient malachite green adsorption property produced from banana peels via microwave-assisted pyrolysis. *Bioresource Technology, 376*. Available from https://doi.org/10.1016/j.biortech.2023.128840, http://www.elsevier.com/locate/biortech.

Chen, X., Yang, L., Guo, J., Xu, S., Di, J., & Zhuang, J. (2022). Interactive removal of bacterial and viral particles during transport through low-cost filtering materials. *Frontiers in Microbiology, 13*. Available from https://doi.org/10.3389/fmicb.2022.970338, https://www.frontiersin.org/journals/microbiology#.

Chen, Y., Wu, Q., Tang, Y., Liu, Z., Ye, L., Chen, R., & Yuan, S. (2022). Application of biochar as an innovative soil ameliorant in bioretention system for stormwater treatment: A review of performance and its influencing factors. *Water Science and Technology, 86*(5), 1232–1252. Available from https://doi.org/10.2166/wst.2022.245, https://watermark.silverchair.com/wst086051232.pdf.

Cheng, G., Zhang, A., Zhao, Z., Chai, Z., Hu, B., Han, B., Ai, Y., & Wang, X. (2021). Extremely stable amidoxime functionalized covalent organic frameworks for uranium extraction from seawater with high efficiency and selectivity. *Science Bulletin, 66*(19), 1994–2001. Available from https://doi.org/10.1016/j.scib.2021.05.012.

Dai, Y., Wang, W., Lu, L., Yan, L., & Yu, D. (2020). Utilization of biochar for the removal of nitrogen and phosphorus. *Journal of Cleaner Production, 257*. Available from https://doi.org/10.1016/j.jclepro.2020.120573, https://www.journals.elsevier.com/journal-of-cleaner-production.

Dai, Y., Zhang, N., Xing, C., Cui, Q., & Sun, Q. (2019). The adsorption, regeneration and engineering applications of biochar for removal organic pollutants: A review. *Chemosphere, 223*, 12–27. Available from https://doi.org/10.1016/j.chemosphere.2019.01.161, http://www.elsevier.com/locate/chemosphere.

Dheenadayalan, G., & Thiruvengadathan, R. (2021). *Remediation of Organic Pollutants in Water* (pp. 501–517). Springer Science and Business Media LLC. Available from https://doi.org/10.1007/978-3-030-52395-4_13.

Dong, F.-X., Yan, L., Zhou, X.-H., Huang, S.-T., Liang, J.-Y., Zhang, W.-X., Guo, Z.-W., Guo, P.-R., Qian, W., & Kong, L.-J. (2021). Simultaneous adsorption of Cr (VI) and phenol by biochar-based iron oxide composites in water: Performance, kinetics and mechanism. *Journal of Hazardous Materials, 416*, 2021.

Enaime, G., Baçaoui, A., Yaacoubi, A., & Lübken, M. (2020). Biochar for wastewater treatment-conversion technologies and applications. *Applied Sciences (Switzerland), 10*(10). Available from https://doi.org/10.3390/app10103492, https://res.mdpi.com/d_attachment/applsci/applsci-10-03492/article_deploy/applsci-10-03492.pdf.

Eniola, J. O., & Sizirici, B. (2023). Investigation of biochar- modified biosand filter performance for groundwater treatment for drinking water purposes: A laboratory and pilot scale study. *Journal of Water Process Engineering, 53*. Available from https://doi.org/10.1016/j.jwpe.2023.103914.

Fan, S., Wang, Y., Wang, Z., Tang, J., Tang, J., & Li, X. (2017). Removal of methylene blue from aqueous solution by sewage sludge-derived biochar: Adsorption kinetics, equilibrium, thermodynamics and mechanism. *Journal of Environmental Chemical Engineering, 5*(1), 601–611. Available from https://doi.org/10.1016/j.jece.2016.12.019.

Fu, H., Ma, S., Xu, S., Duan, R., Cheng, G., & Zhao, P. (2021). Hierarchically porous magnetic biochar as an efficient amendment for cadmium in water and soil: Performance and mechanism. *Chemosphere, 281*, 130990. Available from https://doi.org/10.1016/j.chemosphere.2021.130990.

Fu, Z., & Xi, S. (2020). The effects of heavy metals on human metabolism. *Toxicology Mechanisms and Methods, 30*(3), 167–176. Available from https://doi.org/10.1080/15376516.2019.1701594.

García-Ávila, F., Galarza-Guamán, A., Barros-Bermeo, M., Alfaro-Paredes, E. A., Avilés-Añazco, A., & Iglesias-Abad, S. (2023). Integration of high-rate filtration using waste-derived biochar as a potential sustainable technology for drinking water supply. *Biochar, 5*(1). Available from https://doi.org/10.1007/s42773-023-00256-4.

Geletu, T. T. (2023). Lake eutrophication: Control of phytoplankton overgrowth and invasive aquatic weeds. *Lakes and Reservoirs: Science, Policy and Management for Sustainable Use, 28*(1). Available from https://doi.org/10.1111/lre.12425, http://onlinelibrary.wiley.com/journal/10.1111/(ISSN)1440-1770.

Guilhen, S. N., Rovani, S., Araujo, L. Gd, Tenório, J. A. S., & Mašek, O. (2021). Uranium removal from aqueous solution using macauba endocarp-derived biochar: Effect of physical activation. *Environmental Pollution, 272*. Available from https://doi.org/10.1016/j.envpol.2020.116022, https://www.journals.elsevier.com/environmental-pollution.

Guo, M., Song, W., & Tian, J. (2020). Biochar-Facilitated Soil Remediation: Mechanisms and Efficacy Variations. *Frontiers in Environmental Science, 8*. Available from https://doi.org/10.3389/fenvs.2020.521512.

Gupta, S., Kua, H. W., & Low, C. Y. (2018). Use of biochar as carbon sequestering additive in cement mortar. *Cement and Concrete Composites, 87*, 110–129. Available from https://doi.org/10.1016/j.cemconcomp.2017.12.009, http://www.sciencedirect.com/science/journal/09589465.

Gupta, S., Sireesha, S., Sreedhar, I., Patel, C. M., & Anitha, K. L. (2020). Latest trends in heavy metal removal from wastewater by biochar based sorbents. *Journal of Water Process Engineering, 38*. Available from https://doi.org/10.1016/j.jwpe.2020.101561, http://www.journals.elsevier.com/journal-of-water-process-engineering/.

Gwenzi, W., Chaukura, N., Noubactep, C., & Mukome, F. N. D. (2017). Biochar-based water treatment systems as a potential low-cost and sustainable technology for clean water provision. *Journal of Environmental Management, 197*, 732–749. Available from https://doi.org/10.1016/j.jenvman.2017.03.087, https://www.sciencedirect.com/journal/journal-of-environmental-management.

He, L., Liu, Z., Hu, J., Qin, C., Yao, L., Zhang, Y., & Piao, Y. (2021). Sugarcane biochar as novel catalyst for highly efficient oxidative removal of organic compounds in water. *Chemical Engineering Journal, 405*. Available from https://doi.org/10.1016/j.cej.2020.126895, http://www.elsevier.com/inca/publications/store/6/0/1/2/7/3/index.htt.

Inyang, M., & Dickenson, E. (2015). The potential role of biochar in the removal of organic and microbial contaminants from potable and reuse water: A review. *Chemosphere, 134*, 232–240. Available from https://doi.org/10.1016/j.chemosphere.2015.03.072.

Iwuozor, K. O., Emenike, E. C., Ighalo, J. O., Micheal, T. T., Micheal, K. T., Ore, O. T., Saliu, O. D., & Adeniyi, A. G. (2023). Thermomineralization of biomass for metal oxide recovery: A review. *Bioresource Technology Reports, 24*. Available from https://doi.org/10.1016/j.biteb.2023.101664, https://www.journals.elsevier.com/bioresource-technology-reports.

Jia, Z., Zeng, W., Xu, H., Li, S., & Peng, Y. (2020). Adsorption removal and reuse of phosphate from wastewater using a novel adsorbent of lanthanum-modified platanus biochar. *Process Safety and Environmental Protection, 140*, 221–232. Available from https://doi.org/10.1016/j.psep.2020.05.017.

Kaetzl, K., Lübken, M., Nettmann, E., Krimmler, S., & Wichern, M. (2020). Slow sand filtration of raw wastewater using biochar as an alternative filtration media. *Scientific Reports, 10*(1). Available from https://doi.org/10.1038/s41598-020-57981-0.

Kaetzl, K., Lübken, M., Uzun, G., Gehring, T., Nettmann, E., Stenchly, K., & Wichern, M. (2019). On-farm wastewater treatment using biochar from local agroresidues reduces pathogens from irrigation water for safer food production in developing countries. *Science of The Total Environment, 682*, 601–610. Available from https://doi.org/10.1016/j.scitotenv.2019.05.142.

Kapoor, R. T., Rafatullah, M., Siddiqui, M. R., Khan, M. A., & Sillanpää, M. (2022). Removal of Reactive Black 5 Dye by Banana Peel Biochar and Evaluation of Its Phytotoxicity on Tomato. *Sustainability, 14*(7), 4176. Available from https://doi.org/10.3390/su14074176.

Karpińska, J., & Kotowska, U. (2019). Removal of Organic Pollution in the Water Environment. *Water, 11*(10), 2017. Available from https://doi.org/10.3390/w11102017.

Kearns J. Knappe D. Scott Summers R. 2016 Local biochar adsorbent for control herbicide in surface water: laboratory experiments and field experiences http://www.aqsolutions.org/images/2016/10/Kearns-UNC-2016.pdf.

Kätterer, T., Roobroeck, D., Kimutai, G., Karltun, E., Nyberg, G., Sundberg, C., & de Nowina, K. R. (2022). Maize grain yield responses to realistic biochar application rates on smallholder farms in Kenya. *Agronomy for Sustainable Development, 42*(4), 63. Available from https://doi.org/10.1007/s13593-022-00793-5.

Kim, S., Chu, K. H., Al-Hamadani, Y. A. J., Park, C. M., Jang, M., Kim, D. H., Yu, M., Heo, J., & Yoon, Y. (2018). Removal of contaminants of emerging concern by membranes in water and wastewater: A review. *Chemical Engineering Journal, 335*, 896–914. Available from https://doi.org/10.1016/j.cej.2017.11.044, http://www.elsevier.com/inca/publications/store/6/0/1/2/7/3/index.htt.

Kranner, B. P., Afrooz, A. R. M. N., Fitzgerald, N. J. M., Boehm, A. B., & Paz-Ferreiro, J. (2019). Fecal indicator bacteria and virus removal in stormwater biofilters: Effects of biochar, media saturation, and field conditioning. *PLoS One, 14*(9), e0222719. Available from https://doi.org/10.1371/journal.pone.0222719.

Kristanti, R. A., Hadibarata, T., Syafrudin, M., Yılmaz, M., & Abdullah, S. (2022). Microbiological Contaminants in Drinking Water: Current Status and Challenges. *Water, Air, and Soil Pollution, 233*(8). Available from https://doi.org/10.1007/s11270-022-05669-3, http://www.kluweronline.com/issn/0049-6979/.

Kätterer, T., Roobroeck, D., Kimutai, G., Karltun, E., Nyberg, G., Sundberg, C., & de Nowina, K. R. (2022). Maize grain yield responses to realistic biochar application rates on smallholder farms in Kenya. *Agronomy for Sustainable Development, 42*(4). Available from https://doi.org/10.1007/s13593-022-00793-5, http://www.springerlink.com/content/1773-0155.

Lau, A. Y. T., Tsang, D. C. W., Graham, N. J. D., Ok, Y. S., Yang, X., & Li, Xd (2017). Surface-modified biochar in a bioretention system for Escherichia coli removal from stormwater. *Chemosphere, 169*, 89–98. Available from https://doi.org/10.1016/j.chemosphere.2016.11.048, http://www.elsevier.com/locate/chemosphere.

Li, S., Hu, Y., Shen, Z., Cai, Y., Ji, Z., Tan, X., Liu, Z., Zhao, G., Hu, S., & Wang, X. (2021). Rapid and selective uranium extraction from aqueous solution under visible light in the absence of solid photocatalyst. *Science China Chemistry, 64*(8), 1323–1331. Available from https://doi.org/10.1007/s11426-021-9987-1, http://www.springerlink.com/content/1006-9291.

Li, Y., Xiong, W., Wei, X., Qin, J., & Lin, C. (2021). Transformation and immobilization of hexavalent chromium in the co-presence of biochar and organic acids: effects of biochar dose and reaction time. *Biochar, 3*(4), 535–543. Available from https://doi.org/10.1007/s42773-021-00117-y, http://springer.com/journal/42773.

Lian, F., & Xing, B. (2017). Black Carbon (Biochar) In Water/Soil Environments: Molecular Structure, Sorption, Stability, and Potential Risk. *Environmental Science & Technology, 51*(23), 13517–13532. Available from https://doi.org/10.1021/acs.est.7b02528.

Liang, L., Xi, F., Tan, W., Meng, X., Hu, B., & Wang, X. (2021). Review of organic and inorganic pollutants removal by biochar and biochar-based composites. *Biochar, 3*(3), 255–281. Available from https://doi.org/10.1007/s42773-021-00101-6.

Liu, H., Liu, Y., Tang, L., Wang, J., Yu, J., Zhang, H., Yu, M., Zou, J., & Xie, Q. (2020). Egg shell biochar-based green catalysts for the removal of organic pollutants by activating persulfate. *Science of The Total Environment, 745*, 141095. Available from https://doi.org/10.1016/j.scitotenv.2020.141095.

Liu, L., Yue, T., Liu, R., Lin, H., Wang, D., & Li, B. (2021). Efficient absorptive removal of Cd(II) in aqueous solution by biochar derived from sewage sludge and calcium sulfate. *Bioresource Technology*, 336, 125333. Available from https://doi.org/10.1016/j.biortech.2021.125333.

Liu, P., Rao, D., Zou, L., Teng, Y., & Yu, H. (2021). Capacity and potential mechanisms of Cd(II) adsorption from aqueous solution by blue algae-derived biochars. *Science of The Total Environment*, 767, 145447. Available from https://doi.org/10.1016/j.scitotenv.2021.145447.

Liu, Z., Ling, Q., Cai, Y., Xu, L., Su, J., Yu, K., Wu, X., Xu, J., Hu, B., & Wang, X. (2022). Synthesis of carbon-based nanomaterials and their application in pollution management. *Nanoscale Advances*, 4(5), 1246–1262. Available from https://doi.org/10.1039/d1na00843a.

Liu, Z., Xie, S., Zhou, H., Zhao, L., Yao, Z., Fan, H., Si, B., & Yang, G. (2023). Organic contaminants removal and carbon sequestration using pig manure solid residue-derived biochar: A novel closed-loop strategy for anaerobic liquid digestate. *Chemical Engineering Journal*, 471, 144601. Available from https://doi.org/10.1016/j.cej.2023.144601.

Lu, S. Y., Zhang, H. M., Sojinu, S. O., Liu, G. H., Zhang, J. Q., & Ni, H. G. (2015). Trace elements contamination and human health risk assessment in drinking water from Shenzhen, China. *Environmental Monitoring and Assessment*, 187(1). Available from https://doi.org/10.1007/s10661-014-4220-9, https://link.springer.com/journal/10661.

Lusk, M. G., Toor, G. S., Yang, Y. Y., Mechtensimer, S., De, M., & Obreza, T. A. (2017). A review of the fate and transport of nitrogen, phosphorus, pathogens, and trace organic chemicals in septic systems. *Critical Reviews in Environmental Science and Technology*, 47(7), 455–541. Available from https://doi.org/10.1080/10643389.2017.1327787, http://www.tandf.co.uk/journals/titles/10643389.asp.

Lyu, P., Wang, G., Wang, B., Yin, Q., Li, Y., & Deng, N. (2021). Adsorption and interaction mechanism of uranium (VI) from aqueous solutions on phosphate-impregnation biochar cross-linked Mg Al layered double-hydroxide composite. *Applied Clay Science*, 209, 106146. Available from https://doi.org/10.1016/j.clay.2021.106146.

Ma, L., Du, Y., Chen, S., Du, D., Ye, H., & Zhang, T. C. (2022). Highly efficient removal of Cr(VI) from aqueous solution by pinecone biochar supported nanoscale zero-valent iron coupling with Shewanella oneidensis MR-1. *Chemosphere*, 287, 132184. Available from https://doi.org/10.1016/j.chemosphere.2021.132184.

Marcińczyk, M., Ok, Y. S., & Oleszczuk, P. (2022). From waste to fertilizer: Nutrient recovery from wastewater by pristine and engineered biochars. *Chemosphere*, 306. Available from https://doi.org/10.1016/j.chemosphere.2022.135310, http://www.elsevier.com/locate/chemosphere.

Min, L., Zhongsheng, Z., Zhe, L., & Haitao, W. (2020). Removal of nitrogen and phosphorus pollutants from water by FeCl3- impregnated biochar. *Ecological Engineering*, 149, 105792. Available from https://doi.org/10.1016/j.ecoleng.2020.105792.

Mishra, L., Paul, K. K., & Jena, S. (2022). Adsorption Isotherm, Kinetics and Optimization Study by Box Behnken Design on Removal of Phenol from Coke Wastewater Using Banana Peel (Musa sp.) Biosorbent. *Theoretical Foundations of Chemical Engineering*, 56(6), 1189–1203. Available from https://doi.org/10.1134/s0040579522330041.

Mohanty, S. K., & Boehm, A. B. (2014). Escherichia coli removal in biochar-augmented biofilter: Effect of infiltration rate, initial bacterial concentration, biochar particle size, and presence of compost. *Environmental Science and Technology*, 48(19), 11535–11542. Available from https://doi.org/10.1021/es5033162, http://pubs.acs.org/journal/esthag.

Mohanty, S. K., Cantrell, K. B., Nelson, K. L., & Boehm, A. B. (2014). Efficacy of biochar to remove Escherichia coli from stormwater under steady and intermittent flow. *Water Research*, 61, 288–296. Available from https://doi.org/10.1016/j.watres.2014.05.026, http://www.elsevier.com/locate/watres.

Mukherjee, S., Sarkar, B., Aralappanavar, V. K., Mukhopadhyay, R., Basak, B. B., Srivastava, P., Marchut-Mikołajczyk, O., Bhatnagar, A., Semple, K. T., & Bolan, N. (2022). Biochar-microorganism interactions for organic pollutant remediation: Challenges and perspectives. *Environmental Pollution*, 308. Available from https://doi.org/10.1016/j.envpol.2022.119609, https://www.journals.elsevier.com/environmental-pollution.

Muoghalu, C. C., Owusu, P. A., Lebu, S., Nakagiri, A., Semiyaga, S., Iorhemen, O. T., & Manga, M. (2023). Biochar as a novel technology for treatment of onsite domestic wastewater: A critical review. *Frontiers in Environmental Science*, 11. Available from https://doi.org/10.3389/fenvs.2023.1095920, journal.frontiersin.org/journal/environmental-science.

Nguyen, V. T., Vo, T. D. H., Nguyen, T. B., Dat, N. D., Huu, B. T., Nguyen, X. C., Tran, T., Le, T. N. C., Duong, T. G. H., Bui, M. H., Dong, C. D., & Bui, X. T. (2022). Adsorption of norfloxacin from aqueous solution on biochar derived from spent coffee ground: Master variables and response surface method optimized adsorption process. *Chemosphere*, 288. Available from https://doi.org/10.1016/j.chemosphere.2021.132577, http://www.elsevier.com/locate/chemosphere.

Nzediegwu, C., Naeth, M. A., & Chang, S. X. (2021). Lead (II) adsorption on microwave-pyrolyzed biochars and hydrochars depends on feedstock type and production temperature. *Journal of Hazardous Materials*, 412, 2021.

Obia, A., Martinsen, V., Cornelissen, G., Børresen, T., Smebye, A. B., Munera-Echeverri, J. L., & Mulder, J. (2019). *Biochar Application to Soil for Increased Resilience of Agroecosystems to Climate Change in Eastern and Southern Africa*. Climate Change Management (pp. 129–144). Norway: Springer. Available from springer.com/series/8740, https://doi.org/10.1007/978-3-030-12974-3_6.

Offiong, N.-A. O., Inam, E. J., Etuk, H. S., Ebong, G. A., Inyangudoh, A. I., & Addison, F. (2021). Trace Metal Levels and Nutrient Characteristics of Crude Oil-Contaminated Soil Amended with Biochar–Humus Sediment Slurry. *Pollutants*, 1(3), 119–126. Available from https://doi.org/10.3390/pollutants1030010.

Offiong, N. A. O., Inam, E. J., Etuk, H. S., Essien, J. P., Ofon, U. A., & Una, C. C. (2020). Biochar and humus sediment mixture attenuates crude oil-derived PAHs in a simulated tropical ultisol. *SN Applied Sciences*, 2(11). Available from https://doi.org/10.1007/s42452-020-03744-5, springer.com/snas.

Offiong, N. A. O., Inam, E. J., Fatunla, O. K., Ofon, U. A., Abraham, N. A., & Essien, J. P. (2022). Metagenomic signature and total petroleum hydrocarbons distribution in a crude oil contaminated ultisol remediated with biochar–humus sediment slurry. *Remediation*, 32(4), 299–308. Available from https://doi.org/10.1002/rem.21734, http://onlinelibrary.wiley.com/journal/10.1002/(ISSN)1520-6831.

Omulo, G. (2020). *Biochar Potential in Improving Agricultural Production in East Africa*. IntechOpen. Available from https://doi.org/10.5772/intechopen.92195.

Palansooriya, K. N., Yang, Y., Tsang, Y. F., Sarkar, B., Hou, D., Cao, X., Meers, E., Rinklebe, J., Kim, K. H., & Ok, Y. S. (2020). Occurrence of contaminants in drinking water sources and the potential of biochar for water quality improvement: A review. *Critical Reviews in Environmental Science and Technology, 50*(6), 549–611. Available from https://doi.org/10.1080/10643389.2019.1629803, http://www.tandf.co.uk/journals/titles/10643389.asp.

Perez-Mercado, L. F., Lalander, C., Berger, C., & Dalahmeh, S. S. (2018). Potential of Biochar Filters for Onsite Wastewater Treatment: Effects of Biochar Type, Physical Properties and Operating Conditions. *Water, 10*(12), 1835. Available from https://doi.org/10.3390/w10121835.

Perez-Mercado, L. F., Lalander, C., Joel, A., Ottoson, J., Dalahmeh, S., & Vinnerås, B. (2019). Biochar filters as an on-farm treatment to reduce pathogens when irrigating with wastewater-polluted sources. *Journal of Environmental Management, 248*, 109295. Available from https://doi.org/10.1016/j.jenvman.2019.109295.

Philippou, M., Pashalidis, I., & Kalderis, D. (2023). Removal of 241Am from aqueous solutions by adsorption on sponge gourd biochar. *Molecules (Basel, Switzerland), 28*(6), 2552. Available from https://doi.org/10.3390/molecules28062552.

Pipíška, M., Ballová, S., Frišták, V., Ďuriška, L., Horník, M., Demčák, Š., Holub, M., & Soja, G. (2020). Potassium nickel(II) hexacyanoferrate(III)-functionalized biochar for selective separation of radiocesium from liquid wastes. *Journal of Radiation Research and Applied Sciences, 13*(1), 343–355. Available from https://doi.org/10.1080/16878507.2020.1740394.

Qambrani, N. A., Rahman, M. M., Won, S., Shim, S., & Ra, C. (2017). Biochar properties and eco-friendly applications for climate change mitigation, waste management, and wastewater treatment: A review. *Renewable and Sustainable Energy Reviews, 79*, 255–273. Available from https://doi.org/10.1016/j.rser.2017.05.057, https://www.journals.elsevier.com/renewable-and-sustainable-energy-reviews.

Qiu, M., Liu, L., Ling, Q., Cai, Y., Yu, S., Wang, S., Fu, D., Hu, B., & Wang, X. (2022). Biochar for the removal of contaminants from soil and water: a review. *Biochar, 4*(1). Available from https://doi.org/10.1007/s42773-022-00146-1.

Qu, J., Wang, Y., Tian, X., Jiang, Z., Deng, F., Tao, Y., Jiang, Q., Wang, L., & Zhang, Y. (2021). KOH-activated porous biochar with high specific surface area for adsorptive removal of chromium (VI) and naphthalene from water: Affecting factors, mechanisms and reusability exploration. *Journal of Hazardous Materials, 401*, 123292. Available from https://doi.org/10.1016/j.jhazmat.2020.123292.

Regkouzas, P., & Diamadopoulos, E. (2019). Adsorption of selected organic micro-pollutants on sewage sludge biochar. *Chemosphere, 224*, 840–851. Available from https://doi.org/10.1016/j.chemosphere.2019.02.165, http://www.elsevier.com/locate/chemosphere.

Ri, C., Tang, J., Liu, F., Lyu, H., & Li, F. (2022). Enhanced microbial reduction of aqueous hexavalent chromium by Shewanella oneidensis MR-1 with biochar as electron shuttle. *Journal of Environmental Sciences, 113*, 12–25. Available from https://doi.org/10.1016/j.jes.2021.05.023.

Rombel, A., Krasucka, P., & Oleszczuk, P. (2022). Sustainable biochar-based soil fertilizers and amendments as a new trend in biochar research. *Science of The Total Environment, 816*, 151588. Available from https://doi.org/10.1016/j.scitotenv.2021.151588.

Rosales, E., Meijide, J., Pazos, M., & Sanromán, M. A. (2017). Challenges and recent advances in biochar as low-cost biosorbent: From batch assays to continuous-flow systems. *Bioresource Technology, 246*, 176–192. Available from https://doi.org/10.1016/j.biortech.2017.06.084, http://www.elsevier.com/locate/biortech.

Rosli, N. A., Ahmad, M. A., Noh, T. U., & Ahmad, N. A. (2023). Pineapple peel–derived carbon for adsorptive removal of dyes. *Materials Chemistry and Physics, 306*. Available from https://doi.org/10.1016/j.matchemphys.2023.128094, http://www.journals.elsevier.com/materials-chemistry-and-physics.

Samuel Olugbenga, O., Goodness Adeleye, P., Blessing Oladipupo, S., Timothy Adeleye, A., & Igenepo John, K. (2024). Biomass-derived biochar in wastewater treatment- a circular economy approach. *Waste Management Bulletin, 1*(4), 1–14. Available from https://doi.org/10.1016/j.wmb.2023.07.007.

Sasongko, D., Gunawan, D., & Indarto, A. (2021). *Biochar-based Water Treatment Systems for Clean Water Provision. Handbook of Assisted and Amendment-Enhanced Sustainable Remediation Technology* (pp. 77–101). Indonesia: Wiley. Available from https://www.wiley.com/en-us/Handbook+of+Assisted+and+Amendment+Enhanced+Sustainable+Remediation+Technology-p-9781119670360, https://doi.org/10.1002/9781119670391.ch4.

Saxena, R., Saxena, M., & Lochab, A. (2020). Recent Progress in Nanomaterials for Adsorptive Removal of Organic Contaminants from Wastewater. *ChemistrySelect, 5*(1), 335–353. Available from https://doi.org/10.1002/slct.201903542.

Senadheera, S. S., Gupta, S., Kua, H. W., Hou, D., Kim, S., Tsang, D. C. W., & Ok, Y. S. (2023). Application of biochar in concrete – A review. *Cement and Concrete Composites, 143*. Available from https://doi.org/10.1016/j.cemconcomp.2023.105204, http://www.sciencedirect.com/science/journal/09589465.

Shannon, K. L., Lawrence, R. S., & McDonald, D. (2011). *Anthropogenic Sources of Water Pollution: Parts 1 and 2. Water and Sanitation-Related Diseases and the Environment: Challenges, Interventions, and Preventive Measures* (pp. 289–302). United States: John Wiley and Sons. Available from http://onlinelibrary.wiley.com/book/10.1002/9781118148594, https://doi.org/10.1002/9781118148594.ch24.

Shi, W., Wang, H., Yan, J., Shan, L., Quan, G., Pan, X., & Cui, L. (2022). Wheat straw derived biochar with hierarchically porous structure for bisphenol A removal: Preparation, characterization, and adsorption properties. *Separation and Purification Technology, 289*, 120796. Available from https://doi.org/10.1016/j.seppur.2022.120796.

Singh, S., Kumar, A., & Gupta, H. (2020). Activated banana peel carbon: a potential adsorbent for Rhodamine B decontamination from aqueous system. *Applied Water Science, 10*(8). Available from https://doi.org/10.1007/s13201-020-01274-4.

Spruill, T. B. (1998). Water quality in the Albemarle-Pamlico drainage basin, North Carolina and Virginia. *US Geological Survey, 1157*, 1998.

Srivastav, A. L., Pham, T. D., Izah, S. C., Singh, N., & Singh, P. K. (2022). Biochar Adsorbents for Arsenic Removal from Water Environment: A Review. *Bulletin of Environmental Contamination and Toxicology, 108*(4), 616−628. Available from https://doi.org/10.1007/s00128-021-03374-6, https://link.springer.com/journal/128.

Steiner, C., Bellwood-Howard, I., Häring, V., Tonkudor, K., Addai, F., Atiah, K., Abubakari, A. H., Kranjac-Berisavljevic, G., Marschner, B., & Buerkert, A. (2018). Participatory trials of on-farm biochar production and use in Tamale, Ghana. *Agronomy for Sustainable Development, 38*(1). Available from https://doi.org/10.1007/s13593-017-0486-y, http://www.springerlink.com/content/1773-0155.

Su, Y., Wen, Y., Yang, W., Zhang, X., Xia, M., Zhou, N., Xiong, Y., & Zhou, Z. (2021). The mechanism transformation of ramie biochar's cadmium adsorption by aging. *Bioresource Technology, 330*, 124947. Available from https://doi.org/10.1016/j.biortech.2021.124947.

Suarez-Riera, D., Restuccia, L., & Ferro, G. A. (2020). The use of Biochar to reduce the carbon footprint of cement-based materials. *Procedia Structural Integrity, 26*, 199−210. Available from https://doi.org/10.1016/j.prostr.2020.06.023.

Suliman, W., Harsh, J. B., Fortuna, A. M., Garcia-Pérez, M., & Abu-Lail, N. I. (2017). Quantitative Effects of Biochar Oxidation and Pyrolysis Temperature on the Transport of Pathogenic and Nonpathogenic Escherichia coli in Biochar-Amended Sand Columns. *Environmental Science and Technology, 51*(9), 5071−5081. Available from https://doi.org/10.1021/acs.est.6b04535, http://pubs.acs.org/journal/esthag.

Tang, J., Zhao, B., Lyu, H., & Li, D. (2021). Development of a novel pyrite/biochar composite (BM-FeS2@BC) by ball milling for aqueous Cr(VI) removal and its mechanisms. *Journal of Hazardous Materials, 413*, 125415. Available from https://doi.org/10.1016/j.jhazmat.2021.125415.

Tyagi, U. (2022). Enhanced adsorption of metal ions onto Vetiveria zizanioides biochar via batch and fixed bed studies. *Bioresource Technology, 345*, 126475. Available from https://doi.org/10.1016/j.biortech.2021.126475.

Valenca, R., Borthakur, A., Zu, Y., Matthiesen, E. A., Stenstrom, M. K., & Mohanty, S. K. (2021). Biochar Selection for Escherichia coli Removal in Stormwater Biofilters. *Journal of Environmental Engineering, 147*(2). Available from https://doi.org/10.1061/(asce)ee.1943-7870.0001843.

Vikrant, K., Kim, K. H., Ok, Y. S., Tsang, D. C. W., Tsang, Y. F., Giri, B. S., & Singh, R. S. (2018). Engineered/designer biochar for the removal of phosphate in water and wastewater. *Science of the Total Environment, 616-617*, 1242−1260. Available from https://doi.org/10.1016/j.scitotenv.2017.10.193, http://www.elsevier.com/locate/scitotenv.

Visiy, E. B., Djousse, B. M. K., Martin, L., Zangue, C. N., Sangodoyin, A., Gbadegesin, A. S., & Fonkou, T. (2022). Effectiveness of biochar filters vegetated with Echinochloa pyramidalis in domestic wastewater treatment. *Water Science and Technology, 85*(9), 2613−2624. Available from https://doi.org/10.2166/wst.2022.147, https://watermark.silverchair.com/wst085092613.pdf.

Wang, L., Chen, L., Tsang, D. C. W., Guo, B., Yang, J., Shen, Z., Hou, D., Ok, Y. S., & Poon, C. S. (2020). Biochar as green additives in cement-based composites with carbon dioxide curing. *Journal of Cleaner Production, 258*. Available from https://doi.org/10.1016/j.jclepro.2020.120678, https://www.journals.elsevier.com/journal-of-cleaner-production.

Wang, S., Shi, L., Yu, S., Pang, H., Qiu, M., Song, G., Fu, D., Hu, B., & Wang, X. (2022). Effect of Shewanella oneidensis MR-1 on U(VI) sequestration by montmorillonite. *Journal of Environmental Radioactivity, 242*, 106798. Available from https://doi.org/10.1016/j.jenvrad.2021.106798.

Wang, Z., Zhang, L., Zhang, K., Lu, Y., Chen, J., Wang, S., Hu, B., & Wang, X. (2022). Application of carbon dots and their composite materials for the detection and removal of radioactive ions: A review. *Chemosphere, 287*, 132313. Available from https://doi.org/10.1016/j.chemosphere.2021.132313.

Wen, J., Xue, Z., Yin, X., & Wang, X. (2022). Insights into aqueous reduction of Cr(VI) by biochar and its iron-modified counterpart in the presence of organic acids. *Chemosphere, 286*, 131918. Available from https://doi.org/10.1016/j.chemosphere.2021.131918.

Wu, F., Chen, L., Hu, P., Wang, Y., Deng, J., & Mi, B. (2021). Industrial alkali lignin-derived biochar as highly efficient and low-cost adsorption material for Pb(II) from aquatic environment. *Bioresource Technology, 322*, 124539. Available from https://doi.org/10.1016/j.biortech.2020.124539.

Xiao, Y., Lyu, H., Tang, J., Wang, K., & Sun, H. (2020). Effects of ball milling on the photochemistry of biochar: Enrofloxacin degradation and possible mechanisms. *Chemical Engineering Journal, 384*, 123311. Available from https://doi.org/10.1016/j.cej.2019.123311.

Xiao, Y., Lyu, H., Yang, C., Zhao, B., Wang, L., & Tang, J. (2021). Graphitic carbon nitride/biochar composite synthesized by a facile ball-milling method for the adsorption and photocatalytic degradation of enrofloxacin. *Journal of Environmental Sciences, 103*, 93−107. Available from https://doi.org/10.1016/j.jes.2020.10.006.

Yang, H., Liu, X., Hao, M., Xie, Y., Wang, X., Tian, H., Waterhouse, G. I., Kruger, P. E., Telfer, S. G., & Ma, S. (2021). Functionalized iron−nitrogen−carbon electrocatalyst provides a reversible electron transfer platform for efficient uranium extraction from seawater. *Advanced Materials* (51), 2021.

Yin, G., Tao, L., Chen, X., Bolan, N. S., Sarkar, B., Lin, Q., & Wang, H. (2021). Quantitative analysis on the mechanism of Cd2 + removal by MgCl2-modified biochar in aqueous solutions. *Journal of Hazardous Materials, 420*, 126487. Available from https://doi.org/10.1016/j.jhazmat.2021.126487.

Zhang, A., Li, X., Xing, J., & Xu, G. (2020). Adsorption of potentially toxic elements in water by modified biochar: A review. *Journal of Environmental Chemical Engineering, 8*(4), 104196. Available from https://doi.org/10.1016/j.jece.2020.104196.

Zhang, D., Zhang, K., Hu, X., He, Q., Yan, J., & Xue, Y. (2021). Cadmium removal by MgCl2 modified biochar derived from crayfish shell waste: Batch adsorption, response surface analysis and fixed bed filtration. *Journal of Hazardous Materials, 408*, 124860. Available from https://doi.org/10.1016/j.jhazmat.2020.124860.

Zheng, Y., Wang, B., Wester, A. E., Chen, J., He, F., Chen, H., & Gao, B. (2019). Reclaiming phosphorus from secondary treated municipal wastewater with engineered biochar. *Chemical Engineering Journal*, *362*, 460–468. Available from https://doi.org/10.1016/j.cej.2019.01.036, http://www.elsevier.com/inca/publications/store/6/0/1/2/7/3/index.htt.

Zhou, H., Ye, M., Zhao, Y., Baig, S. A., Huang, N., & Ma, M. (2022). Sodium citrate and biochar synergistic improvement of nanoscale zero-valent iron composite for the removal of chromium (VI) in aqueous solutions. *Journal of Environmental Sciences*, *115*, 227–239. Available from https://doi.org/10.1016/j.jes.2021.05.044.

Zhou, Y., Gao, B., Zimmerman, A. R., & Cao, X. (2014). Biochar-supported zerovalent iron reclaims silver from aqueous solution to form antimicrobial nanocomposite. *Chemosphere*, *117*(1), 801–805. Available from https://doi.org/10.1016/j.chemosphere.2014.10.057, http://www.elsevier.com/locate/chemosphere.

Chapter 11

Biowaste-derived biochars for treatment of wastewater contaminated by dyes

Ebuka Chizitere Emenike[1], Hussein K. Okoro[2], Kingsley O. Iwuozor[3], Abel U. Egbemhenghe[4], Kingsley Chidiebere Okwu[5], Adewale George Adeniyi[6], Sujata Paul[7], Akshaya K[7] and Rangabhashiyam Selvasembian[7]

[1]Department of Pure and Industrial Chemistry, Nnamdi Azikiwe University, Awka, Nigeria, [2]Department of Industrial Chemistry, University of Ilorin, Ilorin, Nigeria, [3]Department of Pure and Industrial Chemistry, Nnamdi Azikiwe University, Awka, Nigeria, [4]Department of Chemistry and Biochemistry, College of Art and Science, Texas Tech University, Lubbock, TX, United States, [5]Department of Chemical/Petrochemical Engineering, Rivers State University, Port Harcourt, Nigeria, [6]Department of Chemical Engineering, University of Ilorin, Ilorin, Nigeria, [7]Department of Environmental Science and Engineering, School of Engineering and Sciences, SRM University-AP, Amaravati, Andhra Pradesh, India

Chapter outline

11.1 Introduction	191
11.2 Biowaste-derived biochar	192
11.2.1 Biowaste Feedstock	192
11.2.2 Biochar production from biowaste	193
11.2.3 Activation methods for the production of engineered biochar	195
11.3 Applications and performance	197
11.3.1 Dye removal	197
11.3.2 Adsorption Modeling	200
11.3.3 Mechanism of adsorption	204
11.4 Regeneration and reuse	205
11.5 Future directions and conclusion	207
11.5.1 Future directions	207
11.5.2 Conclusion	208
Disclosure statements	208
Consent to publish	208
Conflict of interest	208
Funding	208
Compliance with ethical standards	208
Data availability	208
References	208

11.1 Introduction

The global demand for food, chemicals, and energy sources, as well as the associated evolution of industrial and agricultural operations, have led to a significant increase in the amount of wastewater produced as a result of these anthropogenic activities (Emenike et al., 2021; Kwon et al., 2020). Over 80% of wastewater was directly discharged to receiving water bodies, with more than 30% occurring in industrialized countries, according to the UN's 2017 assessment on the state of the world's water resources (Water, 2017). The treatment of wastewater contaminated with dyes has continued to attract a great deal of interest as researchers continue to find novel technologies to mitigate the effects (Iwuozor, Emenike et al., 2022; Ogunlalu et al., 2021).

One of the biggest sources of pollution in a number of industries, particularly the textile one, is dyes. Dyes consist of two main components: the chromophone, which adds color, and the auxophone, which serves as a supplement to the chromophone and gives the dye its water-soluble characteristic while also improving the dye's affinity for the fabric (Corda & Kini, 2018). Dyes can be classified based on factors such as charge. Certain dyes, like methylene blue, have cationic properties whereas others, like methyl orange (MO), have anionic properties (Corda & Kini, 2018). The market offers nearly 10,000 diverse commercial dyes and pigments (Iwuozor et al., 2022), with the textile industry alone utilizing around 700,000 tons of dyestuffs each year (Bingul & Adar, 2023). Approximately 30% of these dyes are lost in the dyeing process, and 20% of the lost dyes find their way into wastewater, often being discharged into water bodies without adequate treatment (Boudechiche et al., 2019). Discharging effluent streams containing these dyes into water bodies can have detrimental effects on both human and marine life.

Treatment of dye-containing wastewater largely involves physico-chemical reactions. Some of the technologies employed include membrane filtration, ion exchange, precipitation, and adsorption (Emenike, Adeniyi et al., 2022). Among them, adsorption has been identified as the most effective because of its simplicity, low cost, and environmental friendliness (Iwuozor et al., 2021; Rangabhashiyam et al., 2022). Several adsorbents have been developed by different researchers for the remediation of such wastewater. However, the usage of by-products, such as biochar, is more affordable when compared to other adsorbents due to the elimination of the pricey synthesis process (Emenike et al., 2023).

Biochar is the solid by-product obtained by the pyrolysis of biomass in an oxygen-limited environment. Biochar has drawn more study interest in recent years as an alternative to traditional carbon-based adsorbents such as activated carbon, graphene oxide, and carbon nanotubes because of its unique qualities, which include cheap cost, porosity, and good performance (Ekanayake et al., 2022; Emenike, Ogunniyi et al., 2022; Ighalo et al., 2022b). Biowaste is the term for organic materials derived from a variety of biological sources, such as leftovers from forestry, agriculture, and food production. Crop residues, residual plant material, and organic waste from homes and food industries are all included in this type of waste (Ahmaruzzaman, 2011; Soliman & Moustafa, 2020). Biowaste is an important resource that can be used in a variety of ecologically friendly ways. This chapter provides a brief overview and discussion of the removal of dyes from wastewater using biowaste-derived biochar. This chapter covers isotherm and kinetic studies, thermodynamics, adsorption mechanisms, as well as the regeneration and reuse of the biochars. Additionally, it explores future directions in the study area.

11.2 Biowaste-derived biochar

11.2.1 Biowaste Feedstock

Biowaste feedstock refers to organic materials derived from living organisms that can be used as a source of raw material for various processes. Biowaste feedstock encompasses a diverse range of organic materials derived from biological sources, such as agricultural residues, food waste, livestock manure, and organic municipal solid waste (MSW) (Watson et al., 2018). These materials are rich in carbon and can be harnessed as feedstock for various applications such as biochar production.

11.2.1.1 Agricultural residues

Agricultural residues serve as valuable biowaste feedstock, presenting an abundant and sustainable resource for various applications in bioenergy production and waste management. These residues are by-products of agricultural processes and include crop residues, stalks, stems, leaves, and other organic materials left in fields after harvest. A lot of agricultural waste is generated globally through agricultural activities and is highly underutilized (Biswas et al., 2017). These wastes have a high energy content and are mostly made up cellulose (35%–50%), hemicellulose (15%–40%), and lignin (15%–25%) (Gabhane et al., 2020). Some of the most common agricultural residues used in biochar production include rice husk, rice straw, sugarcane bagasse, wheat straw, corn cob, and coconut shell, among others (Ighalo et al., 2023; Iwuozor, Emenike, Ighalo et al., 2022b; Karam et al., 2022).

11.2.1.2 Municipal solid waste

MSW refers to the varied range of solid materials discarded daily by urban and rural populations as garbage, trash, and refuse. Around 33% of the 2 billion metric tons of MSW that are produced worldwide each year go uncollected by municipalities (Atlas, 2020). Of the entire amount of MSW collected by the municipalities, 11% is used for energy recovery, 19% is recycled, and roughly 70% is dumped in landfills or other disposal sites (Nanda & Berruti, 2021). Food scraps, yard waste, green waste, paper, textiles, and other miscellaneous waste are all included in MSW. Food scraps consist of discarded food items, peels, and kitchen scraps, while green waste includes organic matter from yard trimmings, leaves, and grass. While incineration and landfilling are the usual methods for disposing of MSW, recent initiatives focus on sustainable management practices, such as anaerobic digestion and thermochemical conversion, to create value-added products like biochar. This approach highlights the potential for waste valorization, transforming a significant environmental challenge into a valuable resource for sustainable biochar production.

11.2.1.3 Livestock manure

Globally, the animal production sector raises 30.1 billion livestock and poultry animals annually, resulting in the production of approximately 13 billion metric tons (equivalent to around 5 billion dry tons) of manure waste each year (Guo et al., 2020; Zhang et al., 2017). Livestock manure is a natural and nutrient-rich fertilizer that enhances soil fertility. Produced by various livestock, including cattle, pigs, and poultry, manure is an important resource of organic matters, useful microorganisms, and plant nutrients (Kumar et al., 2013). Its organic content improves soil structure, water retention, and nutrient levels, fostering healthier and more productive crops. The inorganic nutrients present in livestock manure include the primary nutrients (N, P, K, and S), the secondary nutrients (Ca and Mg), and the micronutrients (B, Cl, Cu, Fe, Mn, Mo, and Zn) (Kumar et al., 2013). While the makeup of animal manure may differ based on factors like the species of the animal, its diet, and how it is managed, it generally comprises predominantly of feces, urine, and bedding materials such as straw and sawdust (Guo et al., 2020).

11.2.1.4 Industrial by-products

The growth in the global population and the ensuing rise in industrial output have gradually increased the amount of waste generated throughout various industrial processes. Global production of industrial waste reached up to 7.6 billion tonnes in 2016 (Kwon et al., 2020). Industrial waste contains a high concentration of metallic components, which distinguishes it from municipal waste (Yoon et al., 2019). Their enormous numbers and the presence of this metallic composition make them very challenging to handle (Agrafioti et al., 2013). Although the majority of industrial waste is physically, chemically, and biologically treated before being released, a significant amount of sludge is also produced at the same time as the trash is being treated (Gottumukkala et al., 2016; Iberahim et al., 2019). For example, over 2 billion metric tonnes of red mud, a by-product of the Bayer process in the aluminum industry, are heaped in various locations around the world, wasting valuable land resources and severely polluting the environment (Gao et al., 2021; Zhang et al., 2020), whereas paper mill sludge, an effluent from the pulp and paper industry, generates a significant volume of wastewater after the processes of raw material preparation, mechanical or chemical pulping, washing, and papermaking (Wang et al., 2021; Xu et al., 2021).

Landfill and incineration are the most common methods of disposing of these residual wastes. While incineration releases hazardous gases like SO_2, NO_x, and particulates that are detrimental to both the atmosphere and human health (Monte et al., 2009), landfilling is known to occupy a lot of land and create secondary environmental pollution (Oladejo et al., 2018). Thus, the development of more effective and environmentally acceptable disposal methods is necessary. Recent studies have investigated the thermal pyrolysis of industrial wastes to produce biochar. In order to take advantage of the metallic properties of industrial waste, some studies have also reported the co-pyrolysis of these wastes with lignocellulosic biomass (Cho et al., 2017; Wu et al., 2017; Yoon et al., 2020). Using this method, waste materials and biomass feedstock are combined and co-pyrolyzed to saturate metals into the surface of the biochar (Cho et al., 2017).

11.2.2 Biochar production from biowaste

Biochar can be produced by thermochemical conversion of biomass in an oxygen-limited environment (Adeniyi, Adeyanju, Emenike, Otoikhian et al., 2022). The common biowaste biochar production methods include pyrolysis and hydrothermal carbonization (HTC).

11.2.2.1 Pyrolysis

Pyrolysis, which is the thermal degradation of biomass at temperatures of 200°C–900°C, is the most popular thermochemical process for the production of biochar (Ahmad et al., 2014). A number of additional factors, including residence time, operating temperature, and the type of feedstock, might affect the thermochemical conversion of biomass to biochar (Adeniyi, Adeyanju, Iwuozor et al., 2022; S & P, 2019). Table 11.1 provides a summary of the major parameters that have been reported in the production of biochar from biowaste. Based on the temperature and residence duration, pyrolysis is typically classified as slow, fast, and flash (Mohan et al., 2006). Biochar is the main by-product of slow pyrolysis, which can take a few minutes to several hours or even days to complete (Gabhane et al., 2020). Fast and flash pyrolysis produces bio-oil as the main by-product and has a very short residence time (Ighalo, Iwuchukwu et al., 2022). Table 11.1 shows that the majority of the time, biochar is produced from biowastes at temperatures between 400°C and 900°C, with a heating rate of 5 to 20°C/min and a residence time of between 30 min and 2 h. The yield and characteristics of the biochar are most significantly impacted by temperature. This is due to the thermal

TABLE 11.1 Key parameters in the production of biowaste-derived biochar.

Feedstock	Process	Peak temp. (°C)	Heating rate (°C/min)	Residence time	Surface area (m²/g)	Micropore volume (cm³/g)	Ref.
Paper mill sludge	Pyrolysis	750	10	2 h	17.86	0.0009	Xu et al. (2021)
Cow manure	Pyrolysis	700	20	1 h	10.61	0.620	Zhang, Zhang et al. (2021)
Deinking paper sludge	Pyrolysis	500	10	1 h	22.85	-	Mendez et al. (2014)
Black liquor/red mud	Pyrolysis	850	5	1 h	106.0	0.15	Wang et al. (2020)
$ZnCl_2$-modified pulp sludge	Pyrolysis	700	-	2 h	100.4	-	Zhao et al. (2021)
Maple leave	Pyrolysis	750	10	2 h	191.0	-	Choi et al. (2020)
Fe_2O_3-modified pulp and paper sludge	Pyrolysis	750	-	2 h	15.30	-	Chaukura et al. (2017b)
Pulp and paper sludge	Pyrolysis	750	-	2 h	174.0	-	Chaukura et al. (2017b)
Corn straw/red mud	Pyrolysis	900	10	2 h	24.44	0.064	Gao et al. (2021)
KOH-activated industrial sludge	Pyrolysis	700	5	2 h	157.0	0.119	Jellali et al. (2022)
Modified tannery sludge	Pyrolysis	550	-	-	47.67	0.050	Zhai et al. (2021)
γ-Fe_2O_3/Food waste	Pyrolysis	300	5	7 h	68.56	-	Chu et al. (2020)
$ZnCl_2$-activated cocoa leaves	Pyrolysis	700	10	1 h	957.0	1.35	Jabar et al. (2022)
Coconut shell	HTC	240	-	1–7 h	-	-	Tu et al. (2021)
Sugarcane bagasse	HTC	200	-	18–20 h	1099	0.136	Prasannamedha et al. (2021)
NaOH-impregnated coconut shell	HTC	200	-	2 h	876.1	0.441	Islam et al. (2017)

method used to create biochar, which requires a high temperature to break down the components of biomass (Ighalo, Rangabhashiyam et al., 2022). The heating rate has the lowest impact on the yield of biochar (Ahmad et al., 2014). The amount of time needed for the biomass to elute in the pyrolysis reactor is known as the residence time (Tripathi et al., 2016).

11.2.2.2 Hydrothermal carbonization

HTC is a thermochemical technique that employs elevated temperatures and pressure in the presence of water to transform organic substances into biochar. The process involves both the removal of water and the elimination of carboxyl groups from the fuel, enhancing its carbon content to achieve a greater calorific value (Funke & Ziegler, 2010). Operating within a temperature range of 180°C–350°C, this method submerges biomass in water and subjects it to pressurized conditions (2–6 MPa) for durations ranging from 5 to 240 min (Heidari et al., 2019). The advantages of

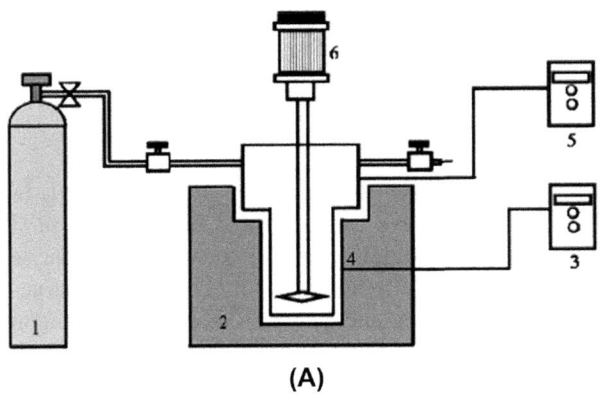

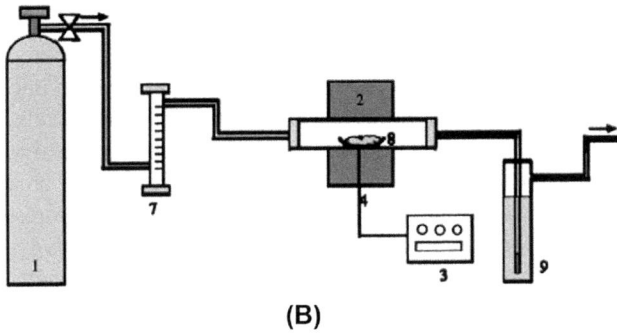

FIGURE 11.1 Diagram (A) illustrates the HTC reactor, while diagram (B) depicts the LTP reactor for biomass enhancement. The components include: (1) N_2 cylinder, (2) heater, (3) temperature controller, (4) thermocouple, (5) pressure meter, (6) stirrer, (7) gas meter, (8) alumina boat, and (9) gas trap. Source: *Liu, Z., & Balasubramanian, R. (2014). Upgrading of waste biomass by hydrothermal carbonization (HTC) and low temperature pyrolysis (LTP): A comparative evaluation. Applied Energy, 114, 857–864. https://doi.org/10.1016/j.apenergy.2013.09.037.*

HTC over other conversion processes include operation at very low temperatures, elimination of the need for energy-intensive drying, and high conversion efficiency (Ighalo et al., 2023). The primary output of HTC is a solid substance known as hydrochar, accompanied by liquid by-products and gaseous by-products, predominantly carbon dioxide (CO_2). Using HTC, Prasannamedha et al. (2021) prepared sugarcane bagasse biochar through a series of steps. Initially, collected bagasse fibers were crushed into a fine powder and sieved. The powder underwent washing, drying, and alkali treatment to eliminate impurities. Subsequently, the treated powder was subjected to HTC at 200°C for 18–20 h. After the reaction, the resulting mixture was separated, and the biochar was obtained through washing, centrifugation, and drying.

In comparing HTC and low-temperature pyrolysis (LTP) for biochar production from raw biomass, Liu and Balasubramanian (2014) found that hydrothermal biochar had higher energy density, whereas pyrolytic biochar exhibited greater energy yield due to higher biochar yield. Pyrolytic biochars retained almost 100% of major ash-forming metals, while hydrothermally prepared biochars retained less than 40%, particularly for Na and K (less than 11% retention rate). Pyrolytic biochars displayed higher reactivities, with main mass loss at lower temperatures, indicating potential for higher thermal efficiency and lower pollutant emissions with hydrothermally prepared biochars. Moreover, hydrothermal biochars displayed a lower activation energy in the temperature range of 150°C–300°C compared to LTP, despite a more extensive biomass decomposition and carbonization process in the HTC process. The schematic diagrams of the HTC and LTP reactors are depicted in Fig. 11.1.

The surface area and micropore volume of biochar (also shown in Table 11.1) are significant characteristics that are crucial in their use as adsorbents for pollutant removal. Both are influenced by the pyrolysis temperature (Liu et al., 2010). The biochar produced by high pyrolysis temperatures often has a large surface area. It is well known that when utilized as an adsorbent, biochar has a high adsorption capacity due to its high surface area and micropore volume. The pore size distribution is likely a key component contributing to the rise in surface area in biochar, as indicated by a positive association between surface area and micropore volume (Ahmad et al., 2014; Downie et al., 2012).

11.2.3 Activation methods for the production of engineered biochar

The activation of biochar involves deliberate processes to enhance its surface properties, making it more effective for specific applications, such as wastewater treatment. Activation techniques produce engineered biochar, which has higher surface area and porosity and thus better adsorption capabilities (Elkhlifi et al., 2022). This section describes various activation methods employed to tailor biowaste-derived biochar for enhanced performance in the removal of dyes from wastewater.

11.2.3.1 Physical activation

Physical activation entails subjecting biochar to elevated temperatures in the presence of an inert gas, usually CO_2 or steam. This process, often referred to as steam activation or gas activation, causes the biochar to undergo structural changes that result in a highly porous substance (S & P, 2019). The specific activation temperature and duration play a crucial role in determining the biochar's final properties. Physical activation results in biochar by distinct pore structures and augmented surface areas, achieved through the elimination of volatiles and internal carbon mass (Sangon et al., 2018). Steam-assisted physical activation of food waste biochar was found to enhance the biochar's porous structure, significantly increasing its surface area and total pore volume (Patra et al., 2021). The steam activation, which was performed at 800°C for 120 min, resulted in impressive BET surface areas of 733 m^2/g for canola hull biochar and 558 m^2/g for oat hull biochar. 0.06 cm^3/g to 0.38 cm^3/g. The high-temperature steam penetrated into the biochar's internal matrix, reacting with unstable carbon and expanding the porous structure (Patra et al., 2021). An innovative single-step pyrolysis method utilizing microwave heating and CO_2 or steam activation to transform orange peel waste into microwave-activated biochar (MAB) was developed by Yek et al. (2020). The process achieves a rapid heating rate, high temperature ($> 800°C$), and short duration (15 min), surpassing conventional methods. MAB exhibits impressive characteristics, including mass yield (31–44 wt %), fixed carbon content (58.6–61.2 wt.%), and BET surface area (158.5–305.1 m^2/g). CO_2 activation results in more micropores, while steam activation generates more mesopores. While physical activation is a straightforward process, it typically requires high temperatures, making it energy-intensive. Consequently, alternative methods such as microwave activation, ultrasound irradiation, plasma treatment, and electrochemical modification have recently gained prominence (Sajjadi et al., 2019). These approaches aim to enhance biochar characteristics, including adsorption capacity, while offering cost-effective alternatives.

Ball milling is another physical activation technique that has gained popularity because of its green approach. Ball milling is a simple solid-state process that operates on the nonequilibrium principle. It makes use of the dynamic energy of moving balls within the machine to grind, break, and generate a new and enhanced surface by consistently blending and grinding the input materials (Amusat et al., 2021). This technique is employed to reduce the size of biochar to the nanometer scale. Studies have shown that ball milling can optimize the physicochemical characteristics of biochar, including specific surface area (SSA), pore volume, and the presence of oxygen-containing functional groups, thereby enhancing its adsorption capabilities (Cheng et al., 2021; Zhang, Yan et al., 2021). Additionally, ball milling represents a mechanical-chemical process that uses mechanical energy to induce chemical and structural alterations in biochar. ZnO/biochar nanocomposites (NCs) were synthesized using ball milling (Yu et al., 2021). The NCs exhibited uniform ZnO dispersion and increased mesopores and macropores and demonstrated efficient methylene blue removal (up to 95.19%) under visible light, showcasing promising adsorption and photocatalytic properties for organic dye wastewater treatment.

Li et al. (2020) developed magnetic ball-milled biochar (MBM-BC) by subjecting hickory chips-derived biochar to ball milling with Fe_3O_4 nanoparticles (Li et al., 2020). This combination of ball milling and magnetite addition resulted in biochar with increased surface area, opened pore structure, functional groups, and aromatic C-C bonds. In another study, a mesoporous Fe-biochar composite with enhanced surface area and positive surface charge was created via Fe-loading and ball-milling techniques (Feng et al., 2021). Ball-milling enhanced the SSA and homogenized pore size distribution, whereas the addition of Fe enhanced the zeta potential and enabled electrostatic adsorption.

11.2.3.2 Chemical activation

Chemical activation involves treating biochar with activating agents, such as acids, bases, or salts, followed by a thermal treatment. This approach initiates a controlled chemical reaction, resulting in the formation of micropores and mesopores within the biochar structure. The most common approach for altering the surface properties of the adsorbent involves acid or alkali modification. This method primarily enhances the SSA and pore structure of biochars, influencing the physical adsorption of pollutants (Cheng et al., 2021). Additionally, the acid-base modification leads to the formation of C-OH and C-H functional groups, crucial in the chemical adsorption process, thereby modifying the adsorption capacity of biochars (Cheng et al., 2021). The modification of 0.5 mol/L $KMnO_4$ to rice husk biochar resulted in an augmentation of its surface area and pore volume, elevating them from 15.77 m^2/g to 20.61 m^2/g and 0.05 cm^3/g to 0.27 cm^3/g, respectively (Li & Li, 2019). Chemically activated cocoa leaves biochar (CLB) was successfully prepared from fallen cocoa leaves (CLs) through $ZnCl_2$ activation and pyrolysis at 700 °C for the removal of toxic crystal violet (CV) dye from aqueous solutions (Jabar et al., 2022). CLB exhibited a high BET surface area (957.02 m^2/g) and mean pore diameter (7.21 nm), indicating enhanced adsorption efficiency.

Increasing pyrolysis temperature significantly enhanced the BET surface area of pulp sludge-derived biochar modified with $ZnCl_2$ under multistep pyrolysis conditions for efficient removal of methylene blue (Zhao et al., 2021). This is as a result of the reaction between $ZnCl_2$ and pulp sludge at high temperatures, leading to the formation of microporous and mesoporous structures. However, beyond 700°C, the SSA decreased, possibly due to pore structure collapse at elevated temperatures. The biochar produced at 700°C exhibited a notably higher surface area (429.175 m^2/g) compared to the raw biochar (78.553 m^2/g). $ZnCl_2$ modification resulted in increased peak strengths for functional groups such as −OH, C=C, and C=O compared to the raw biochar, indicating enhanced surface functional group intensity (Swagathnath et al., 2019). There have also been reports of the biochar produced from biowaste having less surface area following activation. A nano-composite, Fe_2O_3-biochar (Fe_2O_3-BC), was synthesized by Chaukura et al. (2017) through pyrolysis at 750°C using $FeCl_3$-impregnated pulp and paper. The BET surface area of BC, at 174.29 m^2/g, exceeded that of Fe_2O_3-BC (15.32 m^2/g) by an order of magnitude, indicating a reduction in surface area due to Fe_2O_3 activation. This decline was attributed to the obstruction of pores in the biochar matrix by Fe_2O_3 and other inherent metal compounds in the biosorbent.

11.2.3.3 Hybrid activation

While a singular modification of biochar has demonstrated promise in augmenting its adsorption capacity, it has been noted that combining two or more modification methods could yield further improvements in pollutant adsorption capabilities (Foong et al., 2022). This approach involves integrating multiple activation methods synergistically to enhance biochar properties. For instance, a sequence of physical activation followed by a chemical activation process may be employed. Jabar and Odusote (2020) developed a biodegradable adsorbent, pyrolyzed empty fruit bunch fibers, through chemo-physical activation of empty fruit bunch fibres (EFB) biochar for the removal of cibacron blue 3G-A (CB) dye from aqueous solutions. EFB biochar underwent chemical activation through immersion in 72% H_2SO_4 and 0.1 M KOH. Subsequently, it was dried in an oven at 105°C for 1 hour and thermally activated at 600°C under a nitrogen atmosphere for 2 h, resulting in the formation of chemo-physically activated pyrolyzed empty fruit bunch fiber (PEF). PEF exhibited an improved surface area of 362.84 m^2/g, a smoother surface, increased carbon content, and high CB removal efficiency. Cuong and Hou (2022) developed engineered biochar (EBC) from rice husk biochar using three distinct strategies: conventional chemical activation (EBC-1), self-templating coupled with chemical activation (EBC-2), and self-templating coupled with physicochemical activation (EBC-3). EBC-3 activation was achieved by impregnation of biochar with KOH, and then subjected CO_2 atmosphere for 10 min and under N_2 atmosphere for 20 min. Characterization revealed a hierarchical porous structure in EBC-3, with excellent SSA (2344 m^2/g) and mesoporosity (0.70). The large surface area and abundant functional groups provided numerous active sites for CV adsorption.

11.3 Applications and performance

The practical application of biochars produced from biowaste is largely dependent on how well they remove dyes from wastewater. The main factors affecting these biochars' effectiveness are covered in this section, including their adsorption capacities, kinetics, and mechanisms related to dye removal.

11.3.1 Dye removal

Wastewater contaminated with dyes has been treated using biochar made from various biowastes. Tables 11.2 and 11.3 provide an overview of the biochars' maximum adsorption capacities and detail how those capacities were determined. Adsorption capacity (q_{max}) is used as a measure of adsorbent performance because it reflects the inherent potentials of the adsorbent (Ighalo et al., 2020). It is either estimated experimentally or by the use of isotherm models (Emenike, Iwuozor, Agbana et al., 2022). The tables reveal that the Langmuir model is the most frequently employed to remove contaminants using biowaste-derived biochar. Tables also include the pH of the solution. The solubility of the adsorbates, the degree of ionization of the adsorbent during the reaction, and the concentration of the counter ion on the functional groups of the adsorbent are all influenced by the pH of the solution, which is a crucial parameter in the adsorption process (Iwuozor, Oyekunle et al., 2022; Mohammed & Isra'a, 2018; Thanh et al., 2018).

TABLE 11.2 Adsorption performance of different biowaste-derived biochar for cationic dyes.

Biochar	Adsorbate	q_{max} (mg/g)	Optimal pH	Method of q_{max} determination	References
ZnCl$_2$-modified pulp sludge BC	Methylene blue	590.2	8.0	Langmuir	Zhao et al. (2021)
KOH-activated industrial sludge BC	Methylene blue	65.90	6.8	Langmuir	Jellali et al. (2022)
Fe$_2$O$_3$-modified pulp and paper sludge BC	Methylene blue	50.00	12	Langmuir	Chaukura et al. (2017a)
Pulp and paper sludge BC	Methylene blue	33.00	12	Langmuir	Chaukura et al. (2017a)
Modified tannery sludge BC	Cationic blue X-GRL	154.8	-	Langmuir	Zhai et al. (2021)
Magnetic chicken bone BC	Rhodamine B	96.50	10	Langmuir	Oladipo et al. (2017)
Rice straw BC	Basic red 46	22.12	7.0	Langmuir	Sackey et al. (2021)
ZnCl$_2$-activated cocoa leaves BC	Crystal violet	253.3	9.0	Liu	Jabar et al. (2022)
Rice husk BC	Methylene blue	17.97	9.0	Langmuir	Ahmad et al. (2020)
Cow dung BC	Methylene blue	17.50	9.0	Langmuir	Ahmad et al. (2020)
Sludge BC	Methylene blue	19.21	9.0	Langmuir	Ahmad et al. (2020)
Banana waste BC	Methylene blue	526.0	6.0	Langmuir	Amin et al. (2019)
Orange peel BC	Methylene blue	625.0	6.0	Langmuir	Amin et al. (2019)
Rice husk BC	Malachite green	1119	-	Langmuir	Li & Li (2019)
Rice straw BC	Malachite green	295.0	-	Langmuir	Li & Li (2019)
NaOH-activated sugarcane bagasse BC	Methylene blue	114.4	7.0	Langmuir	Moharm et al. (2022)
NaOH-activated sugarcane bagasse BC	Crystal violet	99.50	7.0	Langmuir	Moharm et al. (2022)
Coconut shell BC	Basic blue 41	29.42	8.0	Langmuir	Saravanan et al. (2020)
Coconut shell BC	Basic red 09	65.24	8.0	Langmuir	Saravanan et al. (2020)
Sulfur-doped tapioca peel BC	Malachite green	30.18	8.0	Langmuir	Vigneshwaran et al. (2021)
Sulfur-doped tapioca peel BC	Rhodamine B	32.81	8.0	Langmuir	Vigneshwaran et al. (2021)

TABLE 11.3 Adsorption performance of different biowaste-derived biochar for anionic dyes.

Biochar	Adsorbate	q_{max} (mg/g)	pH	Method of q_{max} determination	Reference
Corn straw/red mud BC	Acid black	69.29	1.0	Experiments	Gao et al. (2021)
Corn straw/red mud BC	Amino black	87.36	1.0	Experiments	Gao et al. (2021)
Modified tannery sludge BC	Active red X-3B	45.13	-	Langmuir	Zhai et al. (2021)
Modified tannery sludge BC	Direct yellow RS	84.10	-	Langmuir	Zhai et al. (2021)
Modified tannery sludge BC	Acid blue 2GL	120.9	-	Langmuir	Zhai et al. (2021)
Sheep manure BC	Methyl orange	45.56	4.0	Langmuir	Lu et al. (2019)
Pulp and paper sludge (PPS) BC	Methyl orange	22.00	8.0	Langmuir	Chaukura et al. (2017b)
$FeCl_3$-impregnated PPS BC	Methyl orange	46.66	3.0	Langmuir	Chaukura et al. (2017b)
Maple leaf BC	Congo red	195.0	7.0	Experimental	Choi et al. (2020)
Fe_3O_4-onion peel BC	Congo red	317.3	2.0	Langmuir	Prajapati & Mondal (2022)
Coconut shell BC	Reactive orange 16	117.5	2.0	Toth	Muralikrishnan & Jodhi (2020)
Coconut shell BC	Reactive black 5	90.54	2.0	Toth	Muralikrishnan & Jodhi (2020)
Coconut shell BC	Reactive blue 19	96.95	2.0	Toth	Muralikrishnan & Jodhi (2020)
Chicken manure BC	Methyl orange	39.47	6.5	Langmuir	Yu et al. (2018)

11.3.1.1 Cationic dyes

Cationic dyes, also known as basic dyes, form a specific category of colorants distinguished by the presence of positively charged functional groups in their molecular configurations (Corda & Kini, 2018). Cationic dyes have a net positive charge, which gives them special qualities and uses above their anionic counterparts. These dyes have an attraction for negatively charged surfaces due to their positive charge, which is usually linked to amine or ammonium groups (Corda & Kini, 2018). This makes them particularly useful in a variety of industrial operations. Biochar derived from biowaste is commonly employed to eliminate various cationic dyes from water, such as methylene blue, malachite green (MG), rhodamine B, and basic yellow, among others. Table 11.2 provides an overview of the adsorption capacities of different biochars derived from biowaste for cationic dyes. It can be observed that high adsorption capacities have been reported in various studies. For instance, biochar produced from pulp sludge and treated with $ZnCl_2$ eliminated more than 99.9% of methylene blue (MB) from water within a 24-h period, exhibiting the highest adsorption capacity of 590.20 mg/g (Zhao et al., 2021). Adsorption of MB and CV from synthetic wastewater using NaOH-activated sugarcane bagasse was investigated under optimized conditions, achieving over 98% removal (Moharm et al., 2022). Langmuir and Freundlich isotherm models represented equilibrium data, showing maximum adsorption capacities for MB and CV of 114.42 and 99.50 mg/g.

In the study carried out by Zhai et al. (2021), tannery sludge biochar was modified to enhance its adsorption capabilities for both cationic and anionic dye pollutants in wastewater. The modification involved introducing polar functional groups (cyano and acylamino groups) through secondary pyrolysis with melamine and KOH. The resulting modified tannery sludge biochar (MBC) exhibited excellent adsorption abilities for various dyes used in textile dyeing processes. The highest removal rates were 98.56% for reactive red X-3B, 99.28% for direct yellow RS, 99.68% for cationic blue X-GRL, and 98.68% for acid blue 2GL. Additionally, iron oxide particles were loaded onto MBC to create magnetic

MBC, allowing for quick separation from the solution under a magnetic field. The modification strategy enhances the practical applicability of tannery sludge biochar for dye pollutant removal, offering a promising solution for wastewater treatment and solid waste pollution in alignment with sustainable development goals.

Similarly, Saravanan et al. (2020) investigated the adsorption capacity of coconut shell-derived biochar for basic dyes, basic blue 41 (BB41) and basic red 09 (BR09), from aqueous solutions. Optimal conditions, including a biochar dosage of 4 g/L, a pH of 8, a temperature of 35°C, and an initial dye concentration of 50 mg/L, achieved 80% dye removal. Isotherm models revealed the Toth model as the best fit for BB41 and the Sips model for BR09, with maximum adsorption capacities of 29.42 mg/g and 65.24 mg/g, respectively. The optimal conditions and successful regeneration of the adsorbent using 0.1 M HCL make it a promising alternative with potential for further environmental impact assessments. Furthermore, a sulfur-doped adsorbent, S@TP biochar, was successfully developed by Vigneshwaran et al. (2021) using tapioca peel as the precursor. S@TP biochar effectively removed cationic dyes, MG, and rhodamine B (RhB), with adsorption efficiencies of 30.18 and 33.10 mg/g, respectively. Adsorption occurred optimally at a pH of around 8, reaching saturation within 120 min. The process followed pseudo-second-order kinetics and adhered to the Freundlich isotherm model. The newly fabricated S@TP biochar demonstrates high removal and reusable efficiency for dye molecules, making it a promising adsorbent for water and wastewater treatment. Table 11.2 provides a more comprehensive summary on biowaste-derived biochar for the removal of cationic dyes.

11.3.1.2 Anionic dyes

Anionic dyes constitute a group of colorants distinguished by the presence of negatively charged functional groups in their molecular compositions. The negative charge found in anionic dyes is frequently linked to carboxylic or sulfonic acid groups, which give them their water solubility and affect how they behave in aqueous solutions. Numerous studies have been conducted on the use of biowaste-based biochar in the treatment of wastewater and water contaminated by anionic dyes; a summary of these studies can be found in Table 11.3. The use of coconut shell-derived biochar for remediating three reactive dyes, reactive orange 16 (RO16), reactive black 5 (RB5), and reactive blue 19 (RB19), in aqueous solutions was investigated by Muralikrishnan and Jodhi (2020). Batch experiments revealed optimal conditions for maximum adsorption at pH 2, 35°C temperature, and 1 g/L biochar dosage. The highest uptake was observed for RO16 (73.03 mg/g), RB5 (56.92 mg/g), and RB19 (57.06 mg/g). Isotherm studies favored the Toth model with a correlation coefficient exceeding 0.9837. Chaukura et al. (2017) synthesized a Fe_2O_3–biochar nano-composite (Fe_2O_3–BC) through the pyrolysis of $FeCl_3$-impregnated pulp and paper sludge (PPS) at 750 °C. Compared to unactivated biochar (BC), Fe_2O_3–BC exhibited nano-sized characteristics confirmed by XRD and SEM. Despite having lower surface area and porosity, Fe_2O_3–BC showed a 52.79% higher MO adsorption capacity than BC. Freundlich model best-represented equilibrium adsorption data with a maximum capacity of 20.53 mg/g at pH 8 and 30 min. The enhanced MO adsorption in Fe_2O_3–BC was attributed to its hybrid composition, utilizing both biochar matrix and Fe_2O_3 nanocrystals. The nano-composite demonstrates efficient MO removal (>97%) from contaminated wastewater.

In their study, Lu et al. (2019) utilized sheep manure to produce biochar through pyrolysis at 600°C, characterized by a large SSA, abundant hole structure, high aromaticity, and polarity. Batch experiments investigated the adsorption of MO in water, considering factors like pH, biochar dosage, adsorption time, and temperature. Optimal conditions (25°C, 20 mg/L MO concentration, initial pH 4.0, and biochar dosage 0.6 g/L) achieved balance at approximately 250 min with a 92.55% MO removal rate. The pseudo-second-order kinetic model and Langmuir model accurately described the adsorption behavior, with a theoretical maximum adsorption capacity of 42.513– 45.563 mg/g. Thermodynamic studies indicated the process as spontaneous, endothermic, and entropy-increasing. Sheep manure biochar demonstrated effectiveness as an adsorption material for MO in water, aligning with the goal of waste control through waste utilization. Prajapati and Mondal (2022) produced Fe_3O_4-onion peel biochar NCs (Fe_3O_4-OPBC NCs) through green pyrolysis in N_2 and CO_2 atmospheres. N_2 favored higher oxygen-containing groups, while CO_2 favored higher surface area. Fe_3O_4-OPBC-1 showed superior adsorption for MB and CR compared to Fe_3O_4-OPBC-2, with a 6.91% and 5.84% higher capacity, respectively. Fe_3O_4-OPBC-1 demonstrated higher adsorption capacity and potential for water pollutant removal, indicating its efficiency in environmental applications.

11.3.2 Adsorption Modeling

11.3.2.1 Isotherm studies

Establishing the most appropriate correlation from the equilibrium data is crucial for improving the design of an adsorption system for the removal of an adsorbate (Chen et al., 2012). The interaction between the adsorbent and the adsorbate is best described using the adsorption isotherm and kinetic models. The degree to which the coefficient of determination (R^2) approaches one indicates which model fits the data the best (Ighalo, Zhou et al., 2022). Nonlinear modeling has been argued to be more accurate than linear modeling due to the linear model's violation of the least squares techniques' normality postulate (Adeniyi et al., 2022).

The adsorption isotherm explains how the adsorbate behaves and is distributed between the liquid and solid phases when an equilibrium condition is established (Igwegbe et al., 2021). The Langmuir and Freundlich models are the most common best-fit isotherm models for the removal of dyes from water by biowaste-derived biochar, as shown in Table 11.4. The Langmuir isotherm is employed to characterize monolayer adsorption on sorbent surfaces (Rangabhashiyam et al., 2014). It posits that adsorption takes place on a uniform surface through monolayer sorption, without any interactions occurring between the adsorbed molecules. Mathematically, the nonlinear and linear forms of the Langmuir isotherm are given as Eqs. (11.1) and (11.2), respectively.

$$q_e = \frac{q_m K_L C_e}{1 + K_L C_e} \tag{11.1}$$

$$\frac{C_e}{q_e} = \frac{1}{K_L q_m} + \frac{1}{q_m} C_e \tag{11.2}$$

where q_e is the maximum adsorption capacity (mg/g), C_e is the equilibrium concentration (mg/L) of the adsorbate, K_L is the Langmuir constant (L/mg), and q_m is the theoretical maximum monolayer coverage (mg/g).

On the other hand, the Freundlich model provides information about the surface heterogeneity of the adsorbent (Wang et al., 2011). The model suggests a scenario of multilayer sorption involving a diverse range of active sites with varying energies, along with interactions among the adsorbed molecules. Eqs (11.3) and (11.4) provide the mathematical expressions for the nonlinear and linear variants of the Freundlich isotherm, respectively.

$$q_e = K_F C_e^{1/n} \tag{11.3}$$

$$In(q_e) = In K_F + \frac{1}{n} In C_e \tag{11.4}$$

where q_e is the maximum adsorption capacity (mg/g), C_e is the equilibrium concentration (mg/L) of the adsorbate, K_f represents the Freundlich constant, $1/n$ represents the heterogeneity factor. $1/n$ and K_f are obtained as slope and intercept from plotting $Log q_e$ versus $Log C_e$

Other isotherm models that have been reported in the literature, as can be seen in Table 11.4, include Liu (Jabar et al., 2022), Toth (Muralikrishnan & Jodhi, 2020; Saravanan et al., 2020), and Sips (Saravanan et al., 2020). However, these models are built from the Langmuir and Freundlich models (Igwegbe et al., 2021). The predominant fitting of the Langmuir and Freundlich isotherms largely arises from the simplification achieved through their linearization and practical applications. This can be linked to the observation that adsorbent materials exhibit homogeneity regardless of their type on a macroscopic scale, and when subjected to agitation, the solution becomes uniformly distributed (Igwegbe et al., 2021).

11.3.2.2 Kinetic studies

Chemical kinetics study entails close observation of experimental setups that influence the rate and path of a chemical reaction as it approaches equilibrium. The kinetic analysis sheds light on the several transition states that arise prior to equilibrium as well as the potential adsorption mechanism. (Es' haghi et al., 2016). From the data presented in Table 11.4, it can be observed that the pseudo-second-order (PSO) model, with a few exceptions, is the best-fit kinetic model for describing the adsorption of dyes onto different biochar derived from biowaste. The PSO model makes the assumption that the chemical adsorption process, which involves shared electrons or electron transfer between adsorbents and adsorbates, influences the reaction rate (Hallaji et al., 2015). The nonlinear and linear forms of PSO kinetic model are given as Eqs. (11.5) and (11.6).

$$\frac{dq_t}{dt} = K_2(q_e - q_t)^2 \tag{11.5}$$

$$\frac{t}{q_t} = \frac{1}{K_2 q_e^2} + \frac{1}{q_e} t \tag{11.6}$$

TABLE 11.4 Best-fit isotherm and kinetic models for dye removal by biowaste-derived biochar.

Biochar	Adsorbate	Isotherm model			Kinetic model			References
		Best-fit	Linear/Nonlinear	R^2	Best-fit	Linear/Nonlinear	R^2	
Fe_2O_3-modified pulp and paper sludge BC	Methyl orange	Freundlich	Linear	0.923	PSO	Linear	0.999	Chaukura et al. (2017b)
$ZnCl_2$-modified pulp sludge BC	Methylene blue	Freundlich	Nonlinear	0.994	PSO	Nonlinear	0.982	Zhao et al. (2021)
KOH-activated industrial sludge BC	Methylene blue	Langmuir	Nonlinear	0.951	PSO	Nonlinear	0.947	Jellali et al. (2022)
Fe_2O_3-modified pulp and paper sludge BC	Methylene blue	Freundlich	Nonlinear	0.801	PSO	Linear	0.958	Chaukura et al. (2017a)
Modified tannery sludge BC	Active red X-3B	Langmuir	Linear	0.990	PSO	Linear	0.990	Zhai et al. (2021)
Modified tannery sludge BC	Direct yellow RS	Langmuir	Linear	0.990	PSO	Linear	0.990	Zhai et al. (2021)
Modified tannery sludge BC	Acid blue 2GL	Langmuir	Linear	0.990	PSO	Linear	0.990	Zhai et al. (2021)
Magnetic chicken bone	Rhodamine B	Langmuir	Nonlinear	0.998	PSO	Nonlinear	0.999	Oladipo et al. (2017)
Rice straw	Basic red 46	Langmuir	Nonlinear	0.996	PSO	Linear	0.998	Sackey et al. (2021)
$ZnCl_2$-activated cocoa leaves	Crystal violet	Liu	Nonlinear	1.000	Avrami fractional	Nonlinear	0.999	Jabar et al. (2022)
Fe_3O_4/onion peel	Congo red	Freundlich	Nonlinear	0.979	PSO	Nonlinear	0.989	Prajapati & Mondal (2022)
Coconut shell	Reactive orange 16	Toth	Linear	0.985	PFO	Nonlinear	0.999	Muralikrishnan & Jodhi (2020)
Coconut shell	Reactive black 5	Toth	Linear	0.984	PFO	Nonlinear	0.995	Muralikrishnan & Jodhi (2020)
	Reactive blue 19	Toth	Linear	0.984	PFO	Nonlinear	0.972	Muralikrishnan & Jodhi (2020)
Rice husk	Malachite green	Langmuir	Linear	0.995	Elovich	Nonlinear	0.998	Li & Li (2019)
NaOH-activated sugarcane bagasse	Crystal violet	Langmuir	Linear	0.999	PSO	Linear	0.999	Moharm et al. (2022)
Coconut shell	Basic red 41	Toth	Nonlinear	0.991	PFO	Nonlinear	Nonlinear	Saravanan et al. (2020)
Coconut shell	Basic red 09	Sips	Linear	0.993	PFO	Nonlinear	Nonlinear	Saravanan et al. (2020)

where K_2 is the rate constant of the PSO model (mg^{-1}min^{-1}). q_e and K_2 can be calculated from the slope and intercept of the plot of t/q versus t.

When using the linear form of the PSO model, however, caution has been urged because the results could lead to false conclusions (Ighalo, Yao et al., 2022). The majority of the studies' best fits the PSO model suggest that the chemical sorption process was essentially the biowaste-derived biochars' rate-limiting step. Notwithstanding, other kinetic models have also been reported in this regard, including the pseudo-first-order (PFO) model (Muralikrishnan & Jodhi, 2020; Saravanan et al., 2020), the Elovich model (Li & Li, 2019), and the Avrami functional order (Jabar et al., 2022).

11.3.2.3 Thermodynamic studies

A thermodynamic study is used to show the relationship between the energy of interaction between the adsorbent and that of the adsorbate (Emenike, Ogunniyi, Ighalo et al., 2022; Kazak & Tor, 2020). The values of the reported thermodynamic parameters, including the change in Gibbs free energy (ΔG), standard enthalpy ($\Delta H°$), and standard entropy ($\Delta S°$) at room or near-room temperature, are presented in Table 11.5. The rate at which adsorption occurs is determined by ΔG. A negative value of ΔG, which is the case for the majority of the studies herein, indicates the process's spontaneity, whereas a positive value (nonspontaneous) indicates an energy barrier in the process (Adeniyi, Abdulkareem et al., 2022). It can be seen that the biochar samples under study adsorb dye contaminants favorably and spontaneously, with values of $\Delta G < 0$. ΔG can be determined using Eq. (11.7):

$$\Delta G = -RT\ln K_c \tag{11.7}$$

where R is the universal gas constant, T is the temperature in Kelvin, and K_c is the equilibrium constant.

TABLE 11.5 Thermodynamic modelling of biowaste-derived biochar for the uptake of dyes.

Biochar	Adsorbate	Temp. (K)	ΔG (KJ/mol)	$\Delta H°$ (KJ/mol)	$\Delta S°$ (KJ/mol)	References
KOH-activated industrial sludge BC	Methylene blue	313	-12.0	78.90	293.0	Jellali et al. (2022)
Banana waste BC	Methylene blue	333	-4.38	8.700	40.00	Amin et al. (2019)
Orange peel BC	Methylene blue	333	-7.51	18.80	80.00	Amin et al. (2019)
Sheep manure BC	Methyl orange	308	-8.795	40.05	0.159	Lu et al. (2019)
Rabbit feces BC	Methylene blue	308	-10.24	14.98	0.083	Huang et al. (2018)
Pig manure BC	Methylene blue	308	-6.694	32.51	0.129	Huang et al. (2018)
Rice straw BC	Basic red 46	308	-12.22	8.74	71.73	Sackey et al. (2021)
ZnCl$_2$-activated cocoa leaves BC	Crystal violet	300	-16.79	-73.79	-0.19	Jabar et al. (2022)
Fe$_3$O$_4$-onion peel BC	Congo red	308	-6.120	24.63	99.94	Prajapati & Mondal (2022)
NaOH-activated sugarcane bagasse BC	Methylene blue	308	-12.55	8.247	66.23	Moharm et al. (2022)
NaOH-activated sugarcane bagasse BC	Crystal violet	308	-9.785	24.80	109.9	Moharm et al. (2022)
Chicken manure BC	Methyl orange	298	-2.240	7.558	32.95	Yu et al. (2018)

The standard enthalpy measures how sensitively the adsorption process responds to a change in temperature (upward or downward) (Ogunlalu et al., 2021). An exothermic process is indicated by a negative value of $\Delta H°$, whereas an endothermic reaction is shown by a positive value (Goh et al., 2019). The findings in Table 11.5 show that adsorption using the biochar samples is dominantly endothermic. It should be noted that the adsorption process is primarily physical and driven by Van der Walls forces when ΔH is less than 40 kJ/mol. Conversely, the adsorption process is chemical when the parameter's value falls between 80 and 400 kJ/mol (Fan et al., 2016). The values of $\Delta H°$ (<40 kJ/mol in some cases and >40 kJ/mol in others) indicate that the adsorption process is dominated by both physical and chemical mechanisms (Fan et al., 2016). The $\Delta S°$ value is used to measure the degree of disorderliness in the system (Sahu et al., 2020). The positive values reported for $\Delta S°$ imply increased randomness and the feasibility of the adsorption process. The values of $\Delta H°$ and $\Delta S°$ can be obtained using the Van't Hoff equation. The relationship is obtained from the free energy equation shown in Eq. (11.8)

$$\Delta G = (\Delta H - T)\Delta S \tag{11.8}$$

Equating the two expressions for ΔG in Eqs. (11.7) and (11.8), we have Eq. (11.9):

$$-RT\ln K_c = (\Delta H - T)\Delta S \tag{11.9}$$

Making $\ln K_c$ the subject of the expression, we have the Van't Hoff equation (Eq. 11.10):

$$\ln K_c = \frac{\Delta S}{R} + \frac{\Delta H}{RT} \tag{11.10}$$

By plotting a graph of $\ln K_c$ against $\frac{1}{T}$, we can obtain the values of ΔS and ΔH from the intercept and slope, respectively.

The adsorption of methylene blue by KOH-activated industrial sludge biochar (IS-KOH-B) rises with temperature, pointing to an endothermic reaction likely caused by an expansion in biochar pore size (Jellali et al., 2022). Thermodynamic parameters show that ΔG was achieved at negative values, indicating a spontaneous and viable MB removal process for the adsorbent, whereas ΔH and ΔS values are positive, indicating an endothermic process with enhanced disorder at the solid particles/solution interface. The specific values obtained signify that IS-KOH-B's MB adsorption encompasses both physical and chemical processes. In contrast, the thermodynamic assessment of CV adsorption onto cocoa leaf biochar (CLB) revealed negativity across all parameters (Jabar et al., 2022). While negative ΔG shows a spontaneous process, negative ΔH signifies that the adsorption process was exothermic, and negative ΔS displays a decrease in randomness at the CV-CLB interface as temperature increases. According to these results, the adsorption of CV dye onto the CLB was not improved by a temperature increase.

11.3.3 Mechanism of adsorption

Different mechanisms are involved when using biochar as an adsorbent to remove various contaminants. This is due to the variety of processes, alterations, and activation techniques that can be used to produce biochar from various carbon-based materials (Barquilha & Braga, 2021). These mechanisms control the attachment of polar substances through the assistance of charged functional groups present on the biochar surface (Abbas et al., 2018). The net charge on the biochar surface is regulated by the solution pH. At low pH, the biochar surface carries a positive charge, whereas at high pH, the surface becomes negatively charged. The pH of the point of zero charge (pH_{pzc}) is formed when the ratio of positive to negative charges on the adsorbent surface is the same. The adsorbent surface is positively charged and protonated when the pH is less than pH_{pzc}, but negatively charged and deprotonated when the pH is more than pH_{pzc} (Emenike, Adeyanju, Iwuozor et al., 2022). Other important factors used in determining the mechanism of adsorption include the isotherm and kinetic studies, thermodynamic studies, and desorption analysis. As mentioned previously, adsorption is primarily governed by physical mechanisms (physisorption) when ΔH is below 40 kJ/mol, and chemical mechanisms (chemisorption) become more prominent at higher ΔH values. Regarding desorption, the effectiveness of eluents like distilled water or organic solvents typically indicates a physical mechanism, while the efficacy of strong acids and/or bases suggests a chemical adsorption mechanism (Igwegbe et al., 2021). In Table 11.6, various mechanisms for the adsorption of dyes by biowaste-based biochar as adsorbents are provided. The results support an earlier observation that adsorption onto biochar made from biowaste involves both chemisorption (such as electrostatic attraction and cation exchange) and physisorption (such as hydrogen bonding and pore fillings).

The adsorption mechanism of CV dye onto CLB involves a three-step process: film diffusion, pore diffusion, and intraparticle diffusion (Jabar et al., 2022). Film diffusion is not the rate-determining step in batch adsorption. Pore

TABLE 11.6 Adsorption mechanisms of biowaste-based biochar for the removal of dyes.

Biochar	Adsorbate	pH$_{pzc}$	Adsorption mechanisms	References
ZnCl$_2$-modified pulp sludge	Methylene blue	2.1	Electrostatic interactions, cation exchange, π-electron interactions	Zhao et al. (2021)
KOH-activated industrial sludge	Methylene blue	-	Pore filling, H-bonds, Van der Waals forces, electrostatic interactions, hydroxyl bonds, π–π stacking	Jellali et al. (2022)
Modified tannery sludge	Active red X-3B	-	Pore diffusion, H-bonds, electrostatic interactions	Zhai et al. (2021)
Sheep manure	Methylene blue	-	Hydrogen bonds, π–π bonds	Huang et al. (2018)
ZnCl$_2$-activated cocoa leaves	Crystal violet	-	Hydrogen bonds, electrostatic attraction	Jabar et al. (2022)
Fe$_3$O$_4$-onion peel	Congo red	7.0	Hydrogen bonds, ion exchange, π–π interaction	Prajapati & Mondal (2022)
KOH-activated industrial sludge	Crystal violet	10.8	Pore filling, hydrogen bonds/van der walls forces, electrostatic interaction, hydroxyl bonds, covalent π–π stacking	Jellali et al. (2022)

diffusion involves the adsorption of dye molecules onto the active pore sites of the adsorbent through mechanical adhesion. Intraparticle diffusion occurs through chemical or electrostatic attraction mechanisms, with chemical adsorption involving covalent bonds and electrostatic attraction depending on the pH. Maximum adsorption occurs at alkaline pH 9, where the anionic surface of CLB exhibits strong electrostatic attraction to positively charged CV dye. Zeta potential analysis of ZnCl$_2$-modified pulp sludge-derived biochar reveals two pH-dependent functional groups affecting carboxyl and hydroxyl group dissociation (Zhao et al., 2021). The biochar, pyrolyzed at 700°C (Zn2PT350–700), exhibits an isoelectric point (pHpzc) of 2.09, influencing surface charge and adsorption preferences. Under strong acidic conditions, ion exchange dominates, while electrostatic interaction becomes prominent with increased alkalinity. Zn2PT350–700 excels in methylene blue adsorption through electrostatic attraction, cation exchange, π-electron interaction, and physical adsorption. Corn stalk biochar prepared at 500°C exhibits increased surface area, pore structure, and improved aromatic structure, contributing to MB removal through electrostatic interaction, hydrogen bonding, and π–π stacking (Liu et al., 2019). Modification with KOH further amplifies the surface area and pore volume, enhancing MB adsorption primarily through π–π stacking. Conversely, H$_3$PO$_4$ modification decreases surface area and pore volume but increases MB adsorption capacity through electrostatic interaction, hydrogen bonding, and π-π stacking, with abundant functional groups contributing to the process. The schematic diagram of this mechanism is depicted in Fig. 11.1.

Additionally, a variety of spectrometric instruments, including XPS, FTIR, and XRD, among others, can be used to elucidate adsorption mechanisms. FTIR and XPS results were used to propose the reaction mechanism of MB and congo red dyes onto Fe$_3$O$_4$-onion peel biochar NCs. The proposed adsorption mechanism was found to involve three steps: (1) Protonation or deprotonation of the surface functional groups of NCs based on the cationic or anionic nature of the dye, leading to electrostatic interaction; (2) Formation of hydrogen bonding and ion exchange between the adsorbent and adsorbate molecules; (3) π–π interaction between the dye and the aromatic and aliphatic groups of NCs. This multistep process incorporates electrostatic interaction, hydrogen bonding, and π–π interaction for effective dye adsorption (Fig. 11.2).

11.4 Regeneration and reuse

Adsorbents that have been used up are often discarded as waste. Though potentially dangerous, these used adsorbents often end up being incinerated. Studies on the regeneration of these adsorbents through desorption have become more common as a result (Adeniyi & Ighalo, 2019). Regeneration and reuse ability of an adsorbent highly determines its industrial applications (Oba et al., 2021). An adsorbent that cannot be regenerated or reused repeatedly is unsuitable for industrial applications as it would increase cost and downtime (Ighalo, Zhou et al., 2022). Table 11.7 shows the results

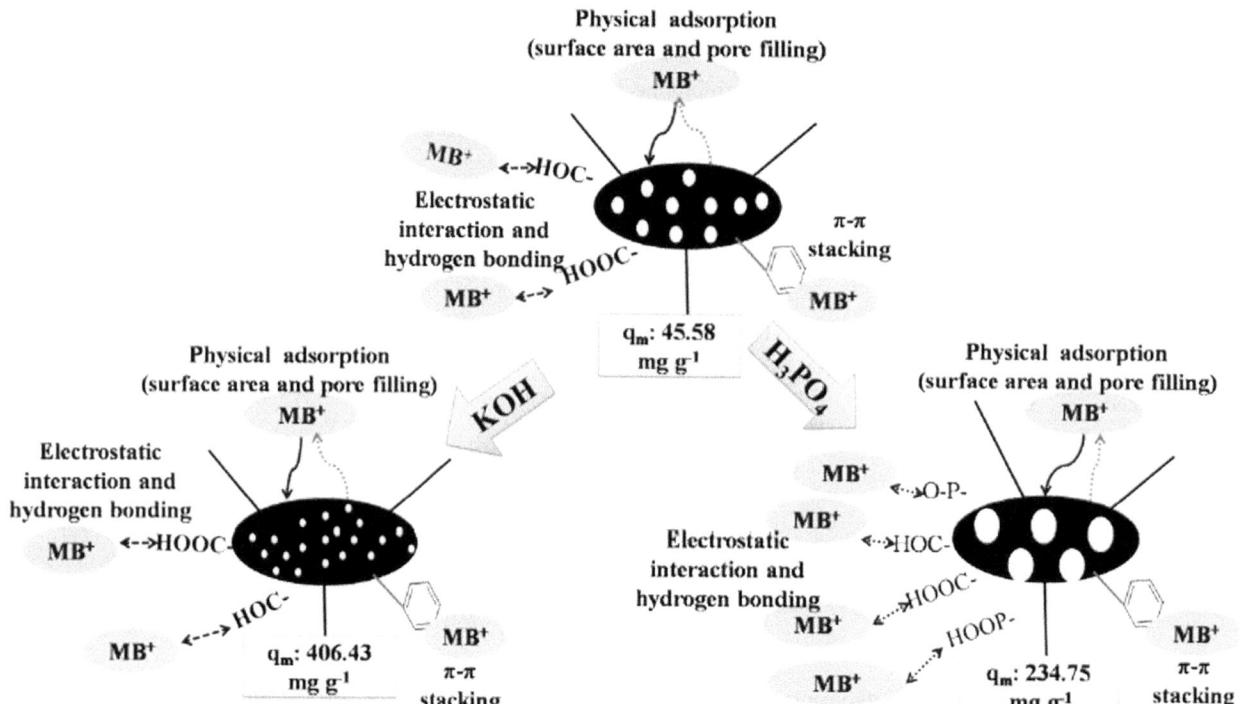

FIGURE 11.2 Diagram showing the removal mechanisms between corn stalk biochar (SCBC), KOH-treated SCBC, and H_3PO_4-treated SCBC. Source: Liu, L., Li, Y., & Fan, S. (2019). Preparation of KOH and H_3PO_4 modified biochar and its application in methylene blue removal from aqueous solution. Processes, 7(12), Article 891. https://doi.org/10.3390/pr7120891. Copyright 2019 MDPI.

of the desorption and reuse of biochar derived from different biowaste. It can be seen that the biochars exhibited consistently high removal efficiencies even after undergoing multiple adsorption-desorption cycles.

The regeneration of coconut shell biochar used for the adsorption of several anionic dyes was investigated using various eluents and studying the sorbent-to-eluent ratio (Muralikrishnan & Jodhi, 2020). The eluents' desorption efficiency was ranked as follows: NaOH > Na_2CO_3 > NH_4OH > HCl > CH_3OH, with NaOH emerging as the most effective eluent, achieving desorption efficiencies of 99.6%, 99.6%, and 99.7% for reactive orange 16, reactive black 5, and reactive blue 19, respectively. Optimal desorption occurred at a sorbent-to-eluent ratio of 5, with a reduction to 95.2%, 95.1%, and 91.4% at a ratio of 6.6. The sorption-elution process demonstrated consistent desorption efficiency (>98.1%) up to the fourth cycle, with a slight decrease to 95.6%, 94.9%, and 95.2% in the fifth cycle. Overall, the biochar proved successful for up to four sorption-elution cycles, highlighting its potential for cost-effective and sustainable treatment processes.

The desorption efficiency of rhodamine B (RB)-loaded magnetic chicken bone biochar (MCBB) was compared using NaOH-spiked H_2O and 0.1 M HCl (Oladipo et al., 2017). Base-spiked H_2O proves more effective, achieving 79.1% desorption after the third cycle. The mechanism involves competition between OH^- and COO^- ions. Reusing MCBB under optimal conditions shows a 90% removal in the first cycle, with a 17% adsorption capacity loss after the third cycle. Despite this loss, MCBB maintains structural stability, making it a promising, recyclable adsorbent for industrial wastewater treatment. Similarly, the desorption and reuse of biochars derived from sheep manure (SMB500), rabbit feces (RFB500), and pig manure (PMB500) indicate that, with increasing cycles, methylene blue adsorption experiences initial fluctuations before gradually decreasing (Huang et al., 2018). The optimal adsorption-desorption cycles for SMB500, RFB500, and PMB500 are 8, 5, and 3, respectively. This pattern implies that repeated use may harm the main functional groups on the adsorbent surface or lead to incomplete degradation of the adsorbed methylene blue, resulting in reduced adsorption capacity. However, after reaching the optimal cycles, the adsorption capacities for methylene blue are 130.80, 38.98, and 23.10 mg/g, respectively, accounting for 82.78%, 79.15%, and 72.77% of the initial equilibrium adsorption amount. Zhao et al. (2021) investigated the desorption of methylene blue from the surface of ZnCl-modified pulp sludge biochar using ethanol as the eluent. After 6 adsorption-desorption cycles, the adsorbent maintained above 90% removal efficiency, while the adsorption capacity was maintained at above 500 mg/g (from 590.2 mg/g), showing the effectiveness of the biochar adsorbent.

TABLE 11.7 Regeneration and reuse of biowaste-derived biochar.

Biochar	Adsorbate	Eluent	Number of cycles (n)	RE (%) after n cycles	References
$ZnCl_2$-modified pulp sludge BC	Methylene blue	Ethanol	6	>90.0	Zhao et al. (2021)
Sheep manure BC	Methylene blue	0.1 mol/L HCl	8	82.8	Huang et al. (2018)
Rabbit feces BC	Methylene blue	0.1 mol/L HCl	5	79.2	Huang et al. (2018)
Pig manure BC	Methylene blue	0.1 mol/L HCl	3	72.8	(Huang et al. (2018)
Magnetic chicken bone BC	Rhodamine B	NaOH-spiked water	3	~74.0	Oladipo et al. (2017)
$ZnCl_2$-activated cocoa leaves BC	Crystal violet	0.1 M NaOH	5	>70.0	Jabar et al. (2022)
Fe_3O_4-onion peel BC	Congo red	NaOH, HCl	5	~60.0	Prajapati & Mondal (2022)
Fe_3O_4-onion peel BC	Methylene blue	NaOH, HCl	5	>65.0	Prajapati & Mondal (2022)
Coconut shell BC	Reactive orange 16	NaOH, Na_2CO_3, NH_4OH	5	95.6	Muralikrishnan & Jodhi (2020)
Coconut shell BC	Reactive black 5	NaOH, Na_2CO_3, NH_4OH	5	94.9	Muralikrishnan & Jodhi (2020)
Coconut shell BC	Reactive black 19	NaOH, Na_2CO_3, NH_4OH	5	95.2	Muralikrishnan & Jodhi (2020)
Coconut shell BC	Basic blue 41	HCl, H_2SO_4, HNO_3	3	~99.0	Saravanan et al. (2020)

11.5 Future directions and conclusion

11.5.1 Future directions

Some knowledge gaps are identified as a result of this study, and intriguing new areas for further investigation are suggested. Findings from this study indicate that augmenting and altering biochar through activation, either before or after pyrolysis, using diverse chemicals, or incorporating other materials to create composites, typically leads to increased surface area and, consequently, higher dye removal efficiencies. Therefore, it is advisable to pursue additional modifications with diverse chemicals to enhance adsorption capacity. Furthermore, the recommendation strongly supports the development of more composite materials utilizing biochar derived from biowaste, including the exploration of magnetic biochars. Industrial scalability of the technique and actual wastewater treatment utilizing biochar based on biowaste are also suggested. The practical implementation of biochars is crucial because, in real-world scenarios, many adsorbents often exhibit significantly lower adsorption capacities than those indicated in laboratory research. This will go a long way toward controlling and recycling the massive amounts of garbage produced by various activities of living organisms, including agricultural and municipal wastes, and remove the negative environmental effects associated with conventional disposal methods like dumping and incineration.

Only a few of the studies in the literature investigated the column adsorption of dyes by biowaste-derived biochar. When such adsorbents are used in industrial processes, understanding the appropriate flow regimes, mechanisms, and best-fit experimental data models is crucial. This is where column adsorption comes in. Therefore, it is advised that more research be done on this. The disposal method for used adsorbents following multiple adsorption-desorption cycles is still a major problem in adsorption systems. How to dispose of used biochar adsorbents and integrate them into other innovative materials should be the subject of further studies. This aspect constitutes a significant component of the study, as improper disposal could undermine the overarching goal of effective adsorption (Emenike, Adeleke, Iwuozor et al., 2022). Furthermore, a cost analysis study would be required to determine the economic benefits of processing biowastes into adsorbent and using the resulting product on a large-scale industrial basis.

11.5.2 Conclusion

This chapter provides a succinct review of the use of biochar made from biowaste to remediate wastewater contaminated with dyes. The biochars were mostly created by pyrolysis at temperatures ranging from 400°C to 900°C with a heating rate of 5 to 20°C/min and a residence time of 30 min to 2 h. When employed as adsorbents, the biochars demonstrated significant removal efficiencies for dyes, with the highest adsorption capacity of 1119 mg/g reported for MG adsorption onto rice husk biochar. The biochars can be further activated and modified to increase their surface area and, thus, their adsorption capability. The adsorption processes involve several mechanisms, including cation exchange, hydrogen bonds, pore fillings, and electrostatic interactions, among others. The pseudo-second-order model fits kinetic data the best, whereas the Langmuir and Freundlich models are the most correlated isotherms. The regeneration and reuse of the adsorbents as well as thermodynamic modeling were also presented and discussed, while potential areas of interest were outlined for future studies.

Disclosure statements

Ethics approval and consent to Participate: Not applicable

Consent to publish

Not applicable.

Conflict of interest

The authors declare that there are no conflicts of interest.

Funding

This work received no external funding

Compliance with ethical standards

This article does not contain any studies involving human or animal subjects.

Data availability

Data sharing is not applicable to this article as no new data were created or analyzed in this study

References

Abbas, Z., Ali, S., Rizwan, M., Zaheer, I. E., Malik, A., Riaz, M. A., Shahid, M. R., Rehman, M. Zu, & Al-Wabel, M. I. (2018). A critical review of mechanisms involved in the adsorption of organic and inorganic contaminants through biochar. *Arabian Journal of Geosciences*, 11, 1–23.

Adeniyi, A. G., Abdulkareem, S. A., Emenike, E. C., Abdulkareem, M. T., Iwuozor, K. O., Amoloye, M. A., Ahmed, I. I., & Awokunle, O. E. (2022). Development and characterization of microstructural and mechanical properties of hybrid polystyrene composites filled with kaolin and expanded polyethylene powder. *Results in Engineering* 100423.

Adeniyi, A. G., Adeyanju, C. A., Emenike, E. C., Otoikhian, S. K., Ogunniyi, S., Iwuozor, K. O., & Raji, A. A. (2022). Thermal energy recovery and valorisation of Delonix regia stem for biochar production. *Environmental Challenges* 100630.

Adeniyi, A. G., Adeyanju, C. A., Iwuozor, K. O., Odeyemi, S. O., Emenike, E. C., Ogunniyi, S., & Te-Erebe, D. K. (2022). Retort carbonization of bamboo (Bambusa vulgaris) waste for thermal energy recovery. *Clean Technologies and Environmental Policy*, 1–11.

Adeniyi, A. G., Emenike, E. C., Iwuozor, K. O., Okoro, H. K., & Ige, O. O. (2022). Acid mine drainage: The footprint of the Nigeria mining industry. Chemistry. *Africa*, 56, 1907–1920.

Adeniyi, A. G., & Ighalo, J. O. (2019). Biosorption of pollutants by plant leaves: an empirical review. *Journal of Environmental Chemical Engineering, 7*3103100.

Agrafioti, E., Bouras, G., Kalderis, D., & Diamadopoulos, E. (2013). Biochar production by sewage sludge pyrolysis. *Journal of Analytical and Applied Pyrolysis, 101*, 72–78.

Ahmad, A., Khan, N., Giri, B. S., Chowdhary, P., & Chaturvedi, P. (2020). Removal of methylene blue dye using rice husk, cow dung and sludge biochar: Characterization, application, and kinetic studies. *Bioresource Technology, 306*123202.

Ahmad, M., Rajapaksha, A. U., Lim, J. E., Zhang, M., Bolan, N., Mohan, D., Vithanage, M., Lee, S. S., & Ok, Y. S. (2014). Biochar as a sorbent for contaminant management in soil and water: a review. *Chemosphere, 99*, 19–33.

Ahmaruzzaman, M. (2011). Industrial wastes as low-cost potential adsorbents for the treatment of wastewater laden with heavy metals. *Advances in Colloid and Interface Science, 166*1-2, 36–59.

Amin, M., Alazba, A., & Shafiq, M. (2019). Comparative study for adsorption of methylene blue dye on biochar derived from orange peel and banana biomass in aqueous solutions. *Environmental Monitoring and Assessment, 191*, 1–14.

Amusat, S. O., Kebede, T. G., Dube, S., & Nindi, M. M. (2021). Ball-milling synthesis of biochar and biochar–based nanocomposites and prospects for removal of emerging contaminants: A review. *Journal of Water Process Engineering, 41*101993.

Atlas, W. 2020. What a waste: an updated look into the future of solid waste management. https://www.worldbank.org/en/news/immersive-story/2018/09/20/what-a-waste-an-updated-look-into-the-future-of-solid-waste-management.

Barquilha, C. E., & Braga, M. C. (2021). Adsorption of organic and inorganic pollutants onto biochars: challenges, operating conditions, and mechanisms. *Bioresource Technology Reports, 15*100728.

Bingul, Z., & Adar, E. (2023). Usability of spent Salvia officinalis as a low-cost adsorbent in the removal of toxic dyes: waste assessment and circular economy. *International Journal of Environmental Analytical Chemistry, 103*18, 6130–6145.

Biswas, B., Pandey, N., Bisht, Y., Singh, R., Kumar, J., & Bhaskar, T. (2017). Pyrolysis of agricultural biomass residues: Comparative study of corn cob, wheat straw, rice straw and rice husk. *Bioresource Technology, 237*, 57–63.

Boudechiche, N., Fares, M., Ouyahia, S., Yazid, H., Trari, M., & Sadaoui, Z. (2019). Comparative study on removal of two basic dyes in aqueous medium by adsorption using activated carbon from Ziziphus lotus stones. *Microchemical Journal, 146*, 1010–1018.

Chaukura, N., Murimba, E. C., & Gwenzi, W. (2017a). Sorptive removal of methylene blue from simulated wastewater using biochars derived from pulp and paper sludge. *Environmental Technology & Innovation, 8*, 132–140.

Chaukura, N., Murimba, E. C., & Gwenzi, W. (2017b). Synthesis, characterisation and methyl orange adsorption capacity of ferric oxide–biochar nano-composites derived from pulp and paper sludge. *Applied Water. Science (New York, N.Y.), 7*5, 2175–2186.

Chen, J.-q, Hu, Z.-j, & Ji, R. (2012). Removal of carbofuran from aqueous solution by orange peel. *Desalination and Water Treatment, 49*1-3, 106–114.

Cheng, N., Wang, B., Wu, P., Lee, X., Xing, Y., Chen, M., & Gao, B. (2021). Adsorption of emerging contaminants from water and wastewater by modified biochar: A review. *Environmental Pollution, 273*116448.

Cho, D.-W., Kwon, E. E., Kwon, G., Zhang, S., Lee, S.-R., & Song, H. (2017). Co-pyrolysis of paper mill sludge and spend coffee ground using CO_2 as reaction medium. *Journal of CO_2 Utilization, 21*, 572–579.

Choi, Y.-K., Gurav, R., Kim, H. J., Yang, Y.-H., & Bhatia, S. K. (2020). Evaluation for simultaneous removal of anionic and cationic dyes onto maple leaf-derived biochar using response surface methodology. *Applied Sciences, 10*9, 2982.

Chu, J.-H., Kang, J.-K., Park, S.-J., & Lee, C.-G. (2020). Application of magnetic biochar derived from food waste in heterogeneous sono-Fenton-like process for removal of organic dyes from aqueous solution. *Journal of Water Process Engineering, 37*101455.

Corda, N. C., & Kini, M. S. (2018). *A review on adsorption of cationic dyes using activated carbon. MATEC Web of Conferences* (p. 02022) EDP Sciences.

Cuong, D. V., & Hou, C.-H. (2022). Engineered biochar prepared using a self-template coupled with physicochemical activation for highly efficient adsorption of crystal violet. *Journal of the Taiwan Institute of Chemical Engineers, 139*104533.

Downie, A., Crosky, A., & Munroe, P. (2012). Physical properties of biochar. *Biochar for Environmental Management, Routledge*, 45–64.

Ekanayake, A., Rajapaksha, A. U., Selvasembian, R., & Vithanage, M. (2022). Amino-functionalized biochars for the detoxification and removal of hexavalent chromium in aqueous media. *Environmental Research, 211*(March)113073. Available from https://doi.org/10.1016/j.envres.2022.113073.

Elkhlifi, Z., Sellaoui, L., Zhao, M., Ifthikar, J., Jawad, A., Shahib, I. I., Sijilmassi, B., Lahori, A. H., Selvasembian, R., Meili, L., Gendy, E. A., & Chen, Z. (2022). Lanthanum hydroxide engineered sewage sludge biochar for efficient phosphate elimination: Mechanism interpretation using physical modelling. *Science of the Total Environment, 803*. Available from https://doi.org/10.1016/j.scitotenv.2021.149888.

Emenike, E. C., Adeleke, J., Iwuozor, K. O., Ogunniyi, S., Adeyanju, C. A., Amusa, V. T., Okoro, H. K., & Adeniyi, A. G. (2022). Adsorption of crude oil from aqueous solution: A review. *Journal of Water Process Engineering, 50*103330.

Emenike, E. C., Adeniyi, A. G., Iwuozor, K. O., Okorie, C. J., Egbemhenghe, A. U., Omuku, P. E., Okwu, K. C., & Saliu, O. D. (2023). A critical review on the removal of mercury (Hg^{2+}) from aqueous solution using nanoadsorbents. *Environmental Nanotechnology, Monitoring & Management*100816.

Emenike, E. C., Adeniyi, A. G., Omuku, P. E., Okwu, K. C., & Iwuozor, K. O. (2022). Recent advances in nano-adsorbents for the sequestration of copper from water. *Journal of Water Process Engineering, 47*102715. Available from https://doi.org/10.1016/j.jwpe.2022.102715.

Emenike, E. C., Iwuozor, K. O., Agbana, S. A., Otoikhian, K. S., & Adeniyi, A. G. (2022). Efficient recycling of disposable face masks via co-carbonization with waste biomass: a pathway to a cleaner environment. *Cleaner Environmental Systems, 6*100094.

Emenike, E. C., Iwuozor, K. O., & Anidiobi, S. U. (2021). Heavy metal pollution in aquaculture: sources, impacts and mitigation techniques. *Biological Trace Element Research*, 1–17.

Emenike, E. C., Ogunniyi, S., Ighalo, J. O., Iwuozor, K. O., Okoro, H. K., & Adeniyi, A. G. (2022). Delonix regia biochar potential in removing phenol from industrial wastewater. *Bioresource Technology Reports*101195. Available from https://doi.org/10.1016/j.biteb.2022.101195.

Es' haghi, Z., Bardajee, G. R., & Azimi, S. (2016). Magnetic dispersive micro solid-phase extraction for trace mercury pre-concentration and determination in water, hemodialysis solution and fish samples. *Microchemical Journal*, 127, 170–177.

Fan, S., Tang, J., Wang, Y., Li, H., Zhang, H., Tang, J., Wang, Z., & Li, X. (2016). Biochar prepared from co-pyrolysis of municipal sewage sludge and tea waste for the adsorption of methylene blue from aqueous solutions: Kinetics, isotherm, thermodynamic and mechanism. *Journal of Molecular Liquids*, 220, 432–441.

Feng, K., Xu, Z., Gao, B., Xu, X., Zhao, L., Qiu, H., & Cao, X. (2021). Mesoporous ball-milling iron-loaded biochar for enhanced sorption of reactive red: Performance and mechanisms. *Environmental Pollution*, 290117992.

Foong, S. Y., Chan, Y. H., Chin, B. L. F., Lock, S. S. M., Yee, C. Y., Yiin, C. L., Peng, W., & Lam, S. S. (2022). Production of biochar from rice straw and its application for wastewater remediation − An overview. *Bioresource Technology*127588.

Funke, A., & Ziegler, F. (2010). Hydrothermal carbonization of biomass: A summary and discussion of chemical mechanisms for process engineering. *Biofuels, Bioproducts and Biorefining*, 42, 160–177.

Gabhane, J. W., Bhange, V. P., Patil, P. D., Bankar, S. T., & Kumar, S. (2020). Recent trends in biochar production methods and its application as a soil health conditioner: a review. *SN Applied Sciences*, 27, 1–21.

Gao, Y., Zhang, J., Chen, C., Du, Y., Teng, G., & Wu, Z. (2021). Functional biochar fabricated from waste red mud and corn straw in China for acidic dye wastewater treatment. *Journal of Cleaner Production*, 320128887.

Goh, C. L., Sethupathi, S., Bashir, M. J., & Ahmed, W. (2019). Adsorptive behaviour of palm oil mill sludge biochar pyrolyzed at low temperature for copper and cadmium removal. *Journal of Environmental Management*, 237, 281–288.

Gottumukkala, L. D., Haigh, K., Collard, F.-X., Van Rensburg, E., & Görgens, J. (2016). Opportunities and prospects of biorefinery-based valorisation of pulp and paper sludge. *Bioresource Technology*, 215, 37–49.

Guo, M., Li, H., Baldwin, B., & Morrison, J. (2020). Thermochemical processing of animal manure for bioenergy and biochar. *Animal manure: Production, characteristics, environmental concerns, and management*, 67, 255–274.

Hallaji, H., Keshtkar, A. R., & Moosavian, M. A. (2015). A novel electrospun PVA/ZnO nanofiber adsorbent for U (VI), Cu (II) and Ni (II) removal from aqueous solution. *Journal of the Taiwan Institute of Chemical Engineers*, 46, 109–118.

Heidari, M., Dutta, A., Acharya, B., & Mahmud, S. (2019). A review of the current knowledge and challenges of hydrothermal carbonization for biomass conversion. *Journal of the Energy Institute*, 926, 1779–1799.

Huang, W., Chen, J., & Zhang, J. (2018). Adsorption characteristics of methylene blue by biochar prepared using sheep, rabbit and pig manure. *Environmental Science and Pollution Research*, 25, 29256–29266.

Iberahim, N., Sethupathi, S., Goh, C. L., Bashir, M. J., & Ahmad, W. (2019). Optimization of activated palm oil sludge biochar preparation for sulphur dioxide adsorption. *Journal of Environmental Management*, 248109302.

Ighalo, J. O., Conradie, J., Ohoro, C. R., Amaku, J. F., Oyedotun, K. O., Maxakato, N. W., Akpomie, K. G., Okeke, E. S., Olisah, C., & Malloum, A. (2023). Biochar from coconut residues: an overview of production, properties, and applications. *Industrial Crops and Products*, 204117300.

Ighalo, J. O., Igwegbe, C. A., Adeniyi, A. G., Adeyanju, C. A., & Ogunniyi, S. (2020). Mitigation of metronidazole (Flagyl) pollution in aqueous media by adsorption: a review. *Environmental Technology Reviews*, 91, 137–148.

Ighalo, J. O., Iwuchukwu, F. U., Eyankware, O. E., Iwuozor, K. O., Olotu, K., Bright, O. C., & Igwegbe, C. A. (2022a). Flash pyrolysis of biomass: a review of recent advances. *Clean Technologies and Environmental Policy*, 1–15.

Ighalo, J.O., Ogunniyi, S., Adeniyi, A.G., Igwegbe, C.A., Sanusi, S.K., Adeyanju, C.A. 2022. Competitive adsorption of heavy metals in a quaternary solution by sugarcane bagasse−LDPE hybrid biochar: equilibrium isotherm and kinetics modelling. Chemical Product and Process Modeling. 18(2)

Ighalo, J. O., Rangabhashiyam, S., Dulta, K., Umeh, C. T., Iwuozor, K. O., Aniagor, C. O., Eshiemogie, S. O., Iwuchukwu, F. U., & Igwegbe, C. A. (2022). Recent advances in hydrochar application for the adsorptive removal of wastewater pollutants. *Chemical Engineering Research and Design*.

Ighalo, J. O., Yao, B., Zhou, Y., Iwuozor, K. O., Anastopoulos, I., Aniagor, C. O., & Rangabhashiyam, S. (2022). Utilization of avocado (Persea americana) adsorbents for the elimination of pollutants from water: a review. *Biomass-Derived Materials for Environmental Applications*, 333–348.

Ighalo, J. O., Zhou, Y., Zhou, Y., Igwegbe, C. A., Anastopoulos, I., Raji, M. A., & Iwuozor, K. O. (2022). A review of pine-based adsorbents for the adsorption of dyes. *Biomass-Derived Materials for Environmental Applications*, 319–332.

Igwegbe, C. A., Oba, S. N., Aniagor, C. O., Adeniyi, A. G., & Ighalo, J. O. (2021a). Adsorption of ciprofloxacin from water: a comprehensive review. *Journal of Industrial and Engineering Chemistry.*, 93, 57–77. Available from https://doi.org/10.1016/j.jiec.2020.09.023.

Igwegbe, C. A., Oba, S. N., Aniagor, C. O., Adeniyi, A. G., & Ighalo, J. O. (2021b). Adsorption of ciprofloxacin from water: a comprehensive review. *Journal of Industrial and Engineering Chemistry*, 93, 57–77.

Islam, M. A., Ahmed, M., Khanday, W., Asif, M., & Hameed, B. (2017). Mesoporous activated coconut shell-derived hydrochar prepared via hydrothermal carbonization-NaOH activation for methylene blue adsorption. *Journal of Environmental Management*, 203, 237–244.

Iwuozor, K. O., Emenike, E. C., Aniagor, C. O., Iwuchukwu, F. U., Ibitogbe, E. M., Temitayo, O. B., Omuku, P. E., & Adeniyi, A. G. (2022). Removal of pollutants from aqueous media using cow dung-based adsorbents. *Current Research in Green and Sustainable Chemistry (Weinheim an der Bergstrasse, Germany)*100300.

Iwuozor, K. O., Emenike, E. C., Ighalo, J. O., Omoarukhe, F. O., Omuku, P. E., & Adeniyi, A. G. (2022). Review on the thermochemical conversion of sugarcane bagasse into biochar. *Cleaner Materials* 100162.

Iwuozor, K. O., Ighalo, J. O., Emenike, E. C., Ogunfowora, L. A., & Igwegbe, C. A. (2021). Adsorption of methyl orange: A review on adsorbent performance. *Current Research in Green and Sustainable Chemistry (Weinheim an der Bergstrasse, Germany)*, 4 100179.

Iwuozor, K. O., Oyekunle, I. P., Emenike, E. C., Okoye-Anigbogu, S. M., Ibitogbe, E. M., Elemile, O., Ighalo, J. O., & Adeniyi, A. G. (2022). An overview of equilibrium, kinetic and thermodynamic studies for the sequestration of Maxilon dyes. *Cleaner Materials* 100148.

Jabar, J. M., Adebayo, M. A., Owokotomo, I. A., Odusote, Y. A., & Yılmaz, M. (2022). Synthesis of high surface area mesoporous ZnCl2–activated cocoa (Theobroma cacao L) leaves biochar derived via pyrolysis for crystal violet dye removal. *Heliyon*, 8 10.

Jabar, J. M., & Odusote, Y. A. (2020). Removal of cibacron blue 3G-A (CB) dye from aqueous solution using chemo-physically activated biochar from oil palm empty fruit bunch fiber. *Arabian Journal of Chemistry*, 13 5, 5417–5429.

Jellali, S., Azzaz, A. A., Al-Harrasi, M., Charabi, Y., Al-Sabahi, J. N., Al-Raeesi, A., Usman, M., Al Nasiri, N., Al-Abri, M., & Jeguirim, M. (2022). Conversion of industrial sludge into activated biochar for effective cationic dye removal: Characterization and adsorption properties assessment. *Water*, 14 14, 2206.

Karam, D. S., Nagabovanalli, P., Rajoo, K. S., Ishak, C. F., Abdu, A., Rosli, Z., Muharam, F. M., & Zulperi, D. (2022). An overview on the preparation of rice husk biochar, factors affecting its properties, and its agriculture application. *Journal of the Saudi Society of Agricultural Sciences*, 21 3, 149–159.

Kazak, O., & Tor, A. (2020). In situ preparation of magnetic hydrochar by co-hydrothermal treatment of waste vinasse with red mud and its adsorption property for Pb (II) in aqueous solution. *Journal of Hazardous Materials*, 393 122391.

Kumar, R. R., Park, B. J., & Cho, J. Y. (2013). Application and environmental risks of livestock manure. *Journal of the Korean Society for Applied Biological Chemistry*, 56, 497–503.

Kwon, G., Bhatnagar, A., Wang, H., Kwon, E. E., & Song, H. (2020). A review of recent advancements in utilization of biomass and industrial wastes into engineered biochar. *Journal of Hazardous Materials*, 400 123242.

Li, X., & Li, Y. (2019). Adsorptive removal of dyes from aqueous solution by KMnO4-modified rice husk and rice straw. *Journal of Chemistry*, 2019.

Li, Y., Zimmerman, A. R., He, F., Chen, J., Han, L., Chen, H., Hu, X., & Gao, B. (2020). Solvent-free synthesis of magnetic biochar and activated carbon through ball-mill extrusion with Fe3O4 nanoparticles for enhancing adsorption of methylene blue. *Science of The Total Environment*, 722 137972.

Liu, L., Li, Y., & Fan, S. (2019). Preparation of KOH and H3PO4 modified biochar and its application in methylene blue removal from aqueous solution. *Processes*, 7 12, 891.

Liu, Z., & Balasubramanian, R. (2014). Upgrading of waste biomass by hydrothermal carbonization (HTC) and low temperature pyrolysis (LTP): A comparative evaluation. *Applied Energy*, 114, 857–864.

Liu, Z., Zhang, F.-S., & Wu, J. (2010). Characterization and application of chars produced from pinewood pyrolysis and hydrothermal treatment. *Fuel*, 89 2, 510–514.

Lu, Y., Chen, J., Bai, Y., Gao, J., & Peng, M. (2019). Adsorption properties of methyl orange in water by sheep manure biochar. *Polish Journal of Environmental Studies*, 28 5, 3791–3797.

Mendez, A., Paz-Ferreiro, J., Araujo, F., & Gasco, G. (2014). Biochar from pyrolysis of deinking paper sludge and its use in the treatment of a nickel polluted soil. *Journal of Analytical and Applied Pyrolysis*, 107, 46–52.

Mohammed, A. A., & Isra'a, S. S. (2018). Bentonite coated with magnetite Fe3O4 nanoparticles as a novel adsorbent for copper (II) ions removal from water/wastewater. *Environmental Technology & Innovation*, 10, 162–174.

Mohan, D., Pittman, C. U., Jr, & Steele, P. H. (2006). Pyrolysis of wood/biomass for bio-oil: a critical review. *Energy & fuels*, 20 3, 848–889.

Moharm, A. E., El Naeem, G. A., Soliman, H., Abd-Elhamid, A. I., El-Bardan, A. A., Kassem, T. S., Nayl, A. A., & Bräse, S. (2022). Fabrication and characterization of effective biochar biosorbent derived from agricultural waste to remove cationic dyes from wastewater. *Polymers*, 14 13, 2587.

Monte, M. C., Fuente, E., Blanco, A., & Negro, C. (2009). Waste management from pulp and paper production in the European Union. *Waste Management*, 29 1, 293–308.

Muralikrishnan, R., & Jodhi, C. (2020). Biodecolorization of Reactive Dyes Using Biochar Derived from Coconut Shell: Batch, Isotherm, Kinetic and Desorption Studies. *ChemistrySelect*, 5 26, 7734–7742. Available from https://doi.org/10.1002/slct.202001454.

Nanda, S., & Berruti, F. (2021). Municipal solid waste management and landfilling technologies: a review. *Environmental Chemistry Letters*, 19, 1433–1456.

Oba, S. N., Ighalo, J. O., Aniagor, C. O., & Igwegbe, C. A. (2021). Removal of ibuprofen from aqueous media by adsorption: A comprehensive review. *Science of the Total Environment*, 780 146608.

Ogunlalu, O., Oyekunle, I. P., Iwuozor, K. O., Aderibigbe, A. D., & Emenike, E. C. (2021). Trends in the mitigation of heavy metal ions from aqueous solutions using unmodified and chemically-modified agricultural waste adsorbents. Current Research in Green and Sustainable. *Chemistry (Weinheim an der Bergstrasse, Germany)*, 4 100188.

Oladejo, J., Shi, K., Luo, X., Yang, G., & Wu, T. (2018). A review of sludge-to-energy recovery methods. *Energies*, 12 1, 60.

Oladipo, A. A., Ifebajo, A. O., Nisar, N., & Ajayi, O. A. (2017). High-performance magnetic chicken bone-based biochar for efficient removal of rhodamine-B dye and tetracycline: competitive sorption analysis. *Water Science and Technology*, 76 2, 373–385.

Patra, B. R., Nanda, S., Dalai, A. K., & Meda, V. (2021). Taguchi-based process optimization for activation of agro-food waste biochar and performance test for dye adsorption. *Chemosphere*, 285 131531.

Prajapati, A. K., & Mondal, M. K. (2022). Green synthesis of Fe3O4-onion peel biochar nanocomposites for adsorption of Cr (VI), methylene blue and congo red dye from aqueous solutions. *Journal of Molecular Liquids, 349*118161.

Prasannamedha, G., Kumar, P. S., Mehala, R., Sharumitha, T., & Surendhar, D. (2021). Enhanced adsorptive removal of sulfamethoxazole from water using biochar derived from hydrothermal carbonization of sugarcane bagasse. *Journal of Hazardous Materials, 407*124825.

Rangabhashiyam, S., Anu, N., Giri Nandagopal, M. S., & Selvaraju, N. (2014). Relevance of isotherm models in biosorption of pollutants by agricultural byproducts. *Journal of Environmental Chemical Engineering, 2*(1), 398–414. Available from https://doi.org/10.1016/j.jece.2014.01.014.

Rangabhashiyam, S., Lins, P. V. do S., Oliveira, L. M. T. d M., Sepulveda, P., Ighalo, J. O., Rajapaksha, A. U., & Meili, L. (2022). Sewage sludge-derived biochar for the adsorptive removal of wastewater pollutants: A critical review. *Environmental Pollution, 293*(May 2021). Available from https://doi.org/10.1016/j.envpol.2021.118581.

S, R., & P, B. (2019). The potential of lignocellulosic biomass precursors for biochar production: Performance, mechanism and wastewater application—A review. *Industrial Crops and Products, 128*(November 2018), 405–423. Available from https://doi.org/10.1016/j.indcrop.2018.11.041.

Sackey, E. A., Song, Y., Yu, Y., & Zhuang, H. (2021). Biochars derived from bamboo and rice straw for sorption of basic red dyes. *PLoS One, 167*e0254637.

Sahu, S., Pahi, S., Tripathy, S., Singh, S. K., Behera, A., Sahu, U. K., & Patel, R. K. (2020). Adsorption of methylene blue on chemically modified lychee seed biochar: Dynamic, equilibrium, and thermodynamic study. *Journal of Molecular Liquids, 315*113743.

Sajjadi, B., Chen, W.-Y., & Egiebor, N. O. (2019). A comprehensive review on physical activation of biochar for energy and environmental applications. *Reviews in Chemical Engineering, 356*, 735–776.

Sangon, S., Hunt, A. J., Attard, T. M., Mengchang, P., Ngernyen, Y., & Supanchaiyamat, N. (2018). Valorisation of waste rice straw for the production of highly effective carbon based adsorbents for dyes removal. *Journal of Cleaner Production, 172*, 1128–1139.

Saravanan, P., Thillainayagam, B. P., Ravindiran, G., & Josephraj, J. (2020). Evaluation of the adsorption capacity of Cocos Nucifera shell derived biochar for basic dyes sequestration from aqueous solution. *Energy Sources, Part A: Recovery, Utilization, and Environmental Effects*, 1–17. Available from https://doi.org/10.1080/15567036.2020.1800142.

Soliman, N., & Moustafa, A. (2020). Industrial solid waste for heavy metals adsorption features and challenges; a review. *Journal of Materials Research and Technology, 95*, 10235–10253.

Swagathnath, G., Rangabhashiyam, S., Murugan, S., & Balasubramanian, P. (2019). Influence of biochar application on growth of Oryza sativa and its associated soil microbial ecology. *Biomass Conversion and Biorefinery, 9*(2), 341–352. Available from https://doi.org/10.1007/s13399-018-0365-z.

Thanh, D. N., Novák, P., Vejpravova, J., Vu, H. N., Lederer, J., & Munshi, T. (2018). Removal of copper and nickel from water using nanocomposite of magnetic hydroxyapatite nanorods. *Journal of Magnetism and Magnetic Materials, 456*, 451–460.

Tripathi, M., Sahu, J. N., & Ganesan, P. (2016). Effect of process parameters on production of biochar from biomass waste through pyrolysis: A review. *Renewable and Sustainable Energy Reviews, 55*, 467–481.

Tu, W., Liu, Y., Xie, Z., Chen, M., Ma, L., Du, G., & Zhu, M. (2021). A novel activation-hydrochar via hydrothermal carbonization and KOH activation of sewage sludge and coconut shell for biomass wastes: Preparation, characterization and adsorption properties. *Journal of Colloid and Interface Science, 593*, 390–407.

Vigneshwaran, S., Sirajudheen, P., Karthikeyan, P., & Meenakshi, S. (2021). Fabrication of sulfur-doped biochar derived from tapioca peel waste with superior adsorption performance for the removal of Malachite green and Rhodamine B dyes. *Surfaces and Interfaces, 23*100920.

Wang, H., Cai, J., Liao, Z., Jawad, A., Ifthikar, J., Chen, Z., & Chen, Z. (2020). Black liquor as biomass feedstock to prepare zero-valent iron embedded biochar with red mud for Cr (VI) removal: Mechanisms insights and engineering practicality. *Bioresource Technology, 311*123553.

Wang, H., Huang, F., Zhao, Z.-L., Wu, R.-R., Xu, W.-X., Wang, P., & Xiao, R.-B. (2021). High-efficiency removal capacities and quantitative adsorption mechanisms of Cd^{2+} by thermally modified biochars derived from different feedstocks. *Chemosphere, 272*129594.

Wang, X. S., Zhu, L., & Lu, H. J. (2011). Surface chemical properties and adsorption of Cu (II) on nanoscale magnetite in aqueous solutions. *Desalination, 2761-3*, 154–160.

Water, U. 2017. The United Nations world water development report 2017: wastewater the untapped resource. Geneva: UN Water.

Watson, J., Zhang, Y., Si, B., Chen, W.-T., & de Souza, R. (2018). Gasification of biowaste: A critical review and outlooks. *Renewable and Sustainable Energy Reviews, 83*, 1–17.

Wu, C., Huang, L., Xue, S.-G., Huang, Y.-Y., Hartley, W., Cui, M.-q, & Wong, M.-H. (2017). Arsenic sorption by red mud-modified biochar produced from rice straw. *Environmental Science and Pollution Research, 2422*, 18168–18178.

Xu, Z., Lin, Y., Lin, Y., Yang, D., & Zheng, H. (2021). Adsorption behaviors of paper mill sludge biochar to remove Cu. *Zn and As in wastewater. Environmental Technology & Innovation, 23*101616.

Yek, P. N. Y., Peng, W., Wong, C. C., Liew, R. K., Ho, Y. L., Mahari, W. A. W., Azwar, E., Yuan, T. Q., Tabatabaei, M., & Aghbashlo, M. (2020). Engineered biochar via microwave CO2 and steam pyrolysis to treat carcinogenic Congo red dye. *Journal of Hazardous Materials, 395*122636.

Yoon, K., Cho, D.-W., Bhatnagar, A., & Song, H. (2020). Adsorption of As (V) and Ni (II) by Fe-Biochar composite fabricated by co-pyrolysis of orange peel and red mud. *Environmental Research, 188*109809.

Yoon, K., Cho, D.-W., Tsang, Y. F., Tsang, D. C., Kwon, E. E., & Song, H. (2019). Synthesis of functionalised biochar using red mud, lignin, and carbon dioxide as raw materials. *Chemical Engineering Journal, 361*, 1597–1604.

Yu, F., Tian, F., Zou, H., Ye, Z., Peng, C., Huang, J., Zheng, Y., Zhang, Y., Yang, Y., & Wei, X. (2021). ZnO/biochar nanocomposites via solvent free ball milling for enhanced adsorption and photocatalytic degradation of methylene blue. *Journal of Hazardous Materials, 415*125511.

Yu, J., Zhang, X., Wang, D., & Li, P. (2018). Adsorption of methyl orange dye onto biochar adsorbent prepared from chicken manure. *Water Science and Technology, 775*, 1303–1312.

Zhai, S., Li, M., Wang, D., Ju, X., & Fu, S. (2021). Cyano and acylamino group modification for tannery sludge bio-char: Enhancement of adsorption universality for dye pollutants. *Journal of Environmental Chemical Engineering, 9*1104939.

Zhang, B., Tian, H., Lu, C., Dangal, S. R., Yang, J., & Pan, S. (2017). Global manure nitrogen production and application in cropland during 1860–2014: a 5 arcmin gridded global dataset for Earth system modeling. *Earth System Science Data, 92*, 667–678.

Zhang, D.-r, Chen, H.-r, Nie, Z.-y, Xia, J.-l, Li, E.-p, Fan, X.-l, & Zheng, L. (2020). Extraction of Al and rare earths (Ce, Gd, Sc, Y) from red mud by aerobic and anaerobic bi-stage bioleaching. *Chemical Engineering Journal, 401*125914.

Zhang, P., Zhang, X., Yuan, X., Xie, R., & Han, L. (2021). Characteristics, adsorption behaviors, Cu (II) adsorption mechanisms by cow manure biochar derived at various pyrolysis temperatures. *Bioresource Technology, 331*125013.

Zhang, W., Yan, L., Wang, Q., Li, X., Guo, Y., Song, W., & Li, Y. (2021). Ball milling boosted the activation of peroxymonosulfate by biochar for tetracycline removal. *Journal of Environmental Chemical Engineering, 9*6106870.

Zhao, F., Shan, R., Li, W., Zhang, Y., Yuan, H., & Chen, Y. (2021). Synthesis, characterization, and dye removal of ZnCl2-modified biochar derived from pulp and paper sludge. *ACS Omega, 6*50, 34712–34723.

Chapter 12

Removal of per- and polyfluoroalkyl substances in environmental matrices by biochars: mechanisms, fate, and research needs

Bashir Adelodun[1,2,3], Oyebankole Agbelusi[4], Qudus Adeyi[2], Abdulhamid Yusuf[5,6,7], Fidelis Odedishemi Ajibade[8], Aminu Abdullahi[9], Golden Odey[2], Pankaj Kumar[10,11], Temitope Fausat Ajibade[8], Tarun Pal[12], Abdulwaheed Mohammed[13] and Timothy Denen Akpenpuun[3]

[1]*Arusha Climate and Environmental Research Center, Aga Khan University, Arusha, Tanzania,* [2]*Department of Agricultural Civil Engineering, Kyungpook National University, Daegu, Korea,* [3]*Department of Agricultural and Biosystems Engineering, University of Ilorin, Ilorin, Nigeria,* [4]*Safeguards and Compliance Department (SNSC), African Development Bank, Abuja, Nigeria (RDNG), Abuja, Nigeria,* [5]*Department of Plant Science and Biotechnology, Federal University Dutsin-Ma, Dutsin-Ma, Nigeria,* [6]*School of Water and Environment, Chang'an University, Xi'an, P.R. China,* [7]*Key Laboratory of Subsurface Hydrology and Ecological Effects in Arid Region, Ministry of Education, Chang'an University, Xi'an, P.R. China,* [8]*Department of Civil and Environmental Engineering, Federal University of Technology, Akure, Nigeria,* [9]*Department of Biotechnology, Modibbo Adama University, Yola, Nigeria,* [10]*Agro ecology and Pollution Research Laboratory, Department of Zoology and Environmental Science, Gurukula Kangri (Deemed to be University), Haridwar, Uttarakhand, India,* [11]*Research and Development Division, Society for AgroEnvironmental Sustainability, Dehradun, India,* [12]*School of Bioengineering and Food Technology, Faculty of Applied Sciences and Biotechnology, Shoolini University, Solan, Himachal Pradesh, India,* [13]*Kwara State Ministry of Agriculture and Natural Resources, Ilorin, Nigeria*

Chapter outline

12.1 Introduction	215
12.2 Occurrence of PFAS and its sources in different environmental matrices	216
12.2.1 Classification and occurrence of PFAS	216
12.2.2 Sources, fate, and transport of PFAS in environmental matrices	217
12.3 Remediation and treatment methods and processes of PFAS in the environment	220
12.3.1 PFAS removal technologies and techniques	220
12.3.2 Biodegradation and transformation of PFAS in the environment	220
12.4 Biochar production processes and its usage for removal of PFAS in the environmental matrices	226
12.4.1 General processes of biochar production	226
12.4.2 Modification of biochar for PFAS removal from environmental matrices	227
12.4.3 Removal mechanism of PFAS from environmental matrices using biochars	229
12.4.4 Performance of biochars on the removal of PFAS	231
12.5 Existing regulations on the PFAS in the environmental matrices and use of biochar	231
12.6 Knowledge gaps, future research needs, and concluding remarks	233
References	234

12.1 Introduction

Per- and polyfluoroalkyl substances (PFAS) are a group of synthetic chemicals that have been widely used in various industrial and consumer applications due to their unique properties, such as high chemical and thermal stability and their resistance to heat, water, and oil (McCarthy et al., 2017; Zhang et al., 2022). According to the Organization for Economic Cooperation and Development (OECD) and the United Nations, PFAS are chemicals that contain at least one aliphatic perfluorocarbon such as a perfluoroalkyl moiety ($-C_nF_{2n}-$) with three or more carbons (n = 3) and a perfluoroalkylether moiety ($-C_nF_{2n}OC_mF_{2m}-$) with two or more carbons (n and m $\geq$ 1) (Organization for Economic Cooperation and Development, 2018; USEPA, 2023). Similarly, PFAS comprises a diverse group of chemicals with the common feature of the fully or per-fluorinated carbon chain (Zheng et al., 2021). The perfluoroalkyl acids, which include perfluoroalkyl sulfonic acids and perfluoroalkyl carboxylic acids, have been the most investigated among the several chemicals that constitute PFAS when it comes to environmental studies (Castiglioni et al., 2015). PFAS are known as "forever chemicals" because of one of the strongest C–F bonds in their molecular structure, which contributes to their resistant nature (Brunn et al., 2023). Due to their persistent property, mobility, bioaccumulation, toxicity, and global distribution, PFAS are categorized among the emerging contaminants of the 21st century that represent potential risks to human and environmental health (Castiglioni et al., 2015; Lee et al., 2020; Zheng et al., 2021).

The safety of the environment from accidental contaminations has gained significant attention for sustainability and regulatory concerns. Much of this attention has recently been focused on PFAS used for varying applications. The widespread nature of PFAS and their presence have been traced to different media at release locations, such as human blood serum, and animal tissue (Domingo & Nadal, 2019; van Larebeke et al., 2022). The Stockholm Convention on Persistent Organic Pollutants (POPs), an international environmental treaty signed in 2001 and taking effect in May 2004, has also designated PFAS and related compounds as target chemicals for regulation. PFAS have been associated with a wide range of harmful health effects, including cancer, liver damage, immunological diseases, and endocrine disruption, in addition to problems with neonatal and child development (Blake & Fenton, 2020; Glüge et al., 2020).

Meanwhile, due to the commercial uses of PFAS, including applications for grease–resistant food contact paper, textiles, carpets, and leather that repel stains and soil (Hepburn et al., 2019; Lyu et al., 2022), the general public may be exposed to PFAS through their nutrition, which includes drinking water, as well as through PFAS–polluted indoor surroundings (Zheng et al., 2020). In fact, Domingo and Nadal opined that diet and drinking water sources are regarded as the main routes of human exposure to PFAS while dust and air have a lesser extent of exposure, especially via indoors (Domingo & Nadal, 2019). The consumption of contaminated food and the migration of PFAS from cookware or food packaging materials are among many other routes through which humans are exposed to PFAS (Shahsavari et al., 2021; Stoiber et al., 2020). Because of their distinct chemical characteristics, such as powerful carbon-fluorine linkages, PFAS are recalcitrant in the environment and have the potential to accumulate in living things. These properties of PFAS compounds pose significant challenges to their management and remediation (Lu et al., 2020).

Moreover, the knowledge of the behaviors of PFAS in the environment is still evolving, especially on how the constituent chemicals behave individually (Lee et al., 2020; McCarthy et al., 2017; Zheng et al., 2020). While the existing knowledge has identified the capability of PFAS to be transported over long distances, the mechanisms and factors influencing these processes still need to be fully understood. Thus, understanding the fate and transport pathways of PFAS compounds in the environment could aid their removal processes. The current chapter provides insight into factors and mechanisms involved in the fate and transport of PFAS in the environment. It also includes an analysis of the physical, chemical, and biological processes that affect how PFAS disperse in various environmental media. Furthermore, it explores the potential pathways through which these substances can impact environmental health. Lastly, this chapter explores the sustainable remediation of PFAS in the environmental media using biochar and existing regulations on PFAS in the environment.

12.2 Occurrence of PFAS and its sources in different environmental matrices

12.2.1 Classification and occurrence of PFAS

PFAS have been more frequently observed in various environmental settings and biota worldwide due to their recalcitrant nature, prompting concerns over their occurrence in the last few decades (Castiglioni et al., 2015). They can be found in drinking water, rivers, groundwater, wastewater, household dust, and soils (Kim & Kannan, 2007; Lu et al., 2020; Murakami & Takada, 2008). Wastewater from several industrial operations was reported as a major source of fluorinated alternatives such as chlorinated polyfluorinated ether sulfonates (Cl–PFESA) and PFAS (Hepburn et al., 2019).

A perfluorinated chain is principally an alkyl chain with all the hydrogen atoms changed to fluorine atoms, and the head can be either a sulfonate or a carboxylate. The PFAS are generically classified into polymers and nonpolymers, and are often characterized based on their functional groups. Fluoropolymers, polymeric perfluoropolyether ethers, and side–chain fluorinated polymers are examples of polymeric PFAS that are regarded to be less hazardous to human and ecological health as compared to some nonpolymers (Hepburn et al., 2019; Lang et al., 2017). In Particular, the

TABLE 12.1 Physiochemical properties of common PFAS (Buck et al., 2011; Shahsavari et al., 2021).

Properties	PFOA	PFOS	PFHxS
Appearance	White to off-white powder	White powder	White crystalline powder
Melting point	54.3°C	>400 °C	–
Boiling point	192.4°C	258°C–260°C	114.7 °C > 400 °C
Density	1.73 g/mL at 20°C	Approximately 0.6	1.84 g/mL at 20 °C
Water solubility	Soluble, 9.5 g/L at 25°C 519 mg/L at 20°C	680 mg/L at 24°C–25°C	Slightly soluble
Organic solvent solubility	Soluble in polar organic solvents	56 mg/L	–
Log Kow	Estimated value of 6.30 in octanol-water mixture	Not measurable	–
pKa	Reported values are 2.8 and 3.8. The estimated value is 0.5.	The estimated value is −3.3	0.14

PFOA, perfluorooctanoic acid; PFOS, perfluorooctane sulfonate.

nonpolymeric PFAS can be grouped into two major subclasses: per− and polyfluoroalkyl substances. In terms of basic chemical structure, PFAS contains the perfluoroalkyl moiety $-CnF_2n + 1$. Perfluoroalkyl substances are distinguished from polyfluoroalkyl substances by being fully fluorinated.

The structure and physicochemical properties of different classes of PFAS and their respective nomenclatures are provided in Buck et al. (2011) and Rahman et al. (2014). PFAS are popularly known to exhibit properties such as heat resistance, non-adhesively (i.e., oil, water, and stain repellency), corrosion resistance, low friction performance, and high surface activity (Lu et al., 2020; Nzeribe et al., 2019). The peculiar physical and chemical characteristics of PFAS (Table 12.1), which are very different from those of their hydrogenated counterparts, are attributed to the perfluorinated tail. As suggested by the word "perfluorinated," fluorine is the most prevalent element in PFAS, and its atomic structure plays a significant role in determining the chemical and physical properties of PFAS.

The physicochemical characteristics of various PFAS substances, such as molecular weight, hydrophobicity, and functional group composition, differ from one another. These characteristics affect their sorption, partitioning, and mobility in the environment. For instance, long-chain PFAS have a more considerable sorption potential than short-chain PFAS, which are most likely more hydrophobic. A perfluorinated chain is essentially an alkyl chain with all the hydrogen atoms changed to fluorine atoms, and the head can be either a sulfonate or a carboxylate. The peculiar physical and chemical characteristics of PFAS, which are very different from those of their hydrogenated counterparts, are attributed to the perfluorinated tail. As suggested by the word "perfluorinated," fluorine is the most prevalent element in PFAS, and its atomic structure plays a significant role in determining the chemical and physical properties of PFAS. They are characterized by high electronegativity due to highly polarizing C−F bonds in the perfluorinated tail (Leung et al., 2023).

12.2.2 Sources, fate, and transport of PFAS in environmental matrices

PFAS chemicals persist in the environment and have raised concerns due to their potential adverse effects on human health and the environment. Because of their toxic and recalcitrant nature, the fate and transport of PFAS in environmental matrices have been a concern and worries (Cui et al., 2020; Nzeribe et al., 2019). The fate and transport of PFAS in environmental matrices are influenced by several sources, including industrial release, firefighting foams, landfill and waste sites, atmospheric deposition, and wastewater treatment plants (Fig. 12.1). These are regarded as the primary sources of PFAS in the environment since they are directly from the manufacturing processes of PFAS, usage of PFAS products, and their ultimate disposal into the environment. However, other indirect sources of PFAS in the environment mainly occur from the transformation of precursor substances, such as chemical reactions of impurities or degradation of chemicals that eventually find their way into the environment or are consumed by human beings (Prevedouros et al., 2006).

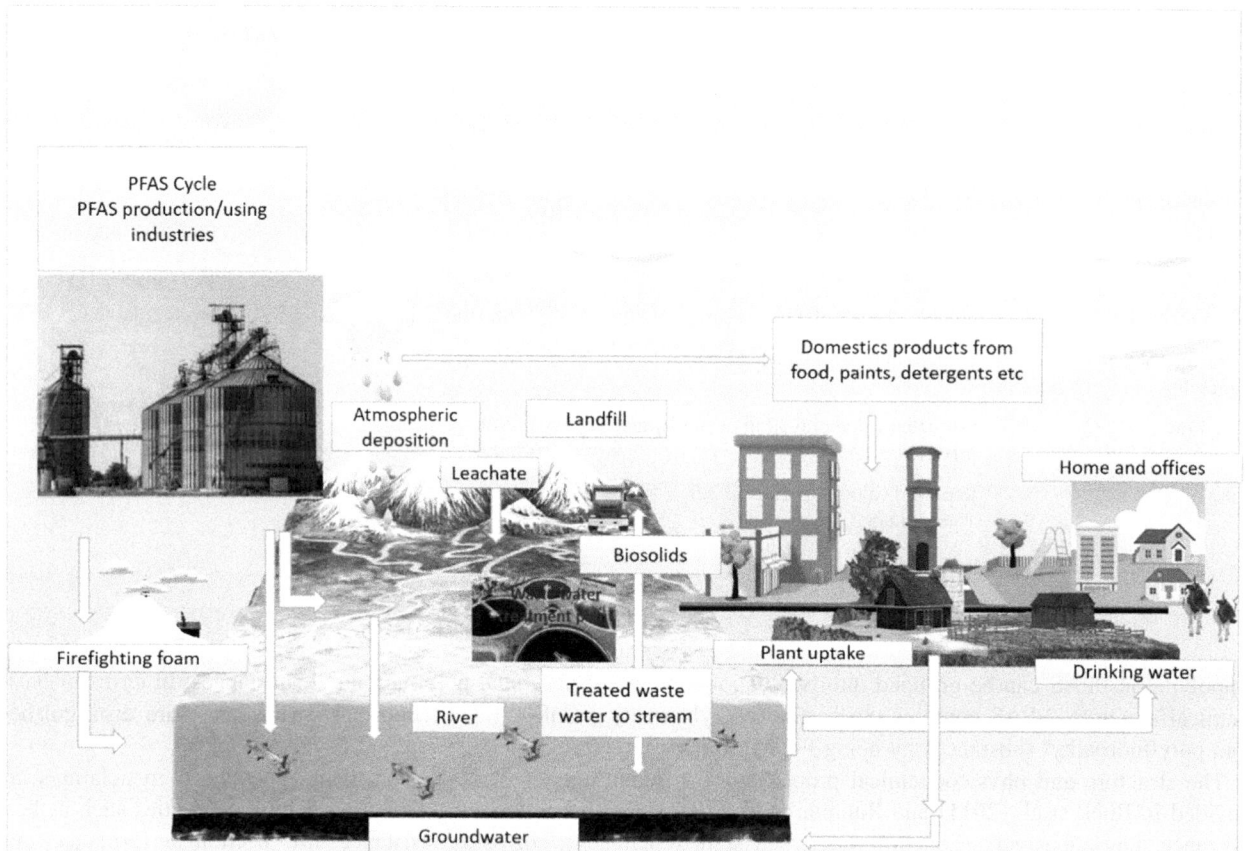

FIGURE 12.1 Source of PFAS and its cycle in the environment.
As shown, PFAS can exhibit persistence, mobility, and bioaccumulation tendencies once released into the environment. They can migrate through soil, dissolve in water, and be transported through groundwater or surface water bodies. PFAS can also bind to organic matter or sediment particles, leading to their accumulation in aquatic sediments.

The environmental fate and transport of PFAS describe the behavior of the chemical compounds of PFAS after their release from various sources into the environment. The properties and composition of the environmental matrix in which PFAS are present play a significant role in their fate and transport. Because PFAS are amphiphilic, they can separate from both polar and non-polar solvents and create their partition (Hepburn et al., 2019). PFAS compounds are generally highly water-soluble, which allows them to be transported in surface waters and groundwater. Highly soluble PFAS can be easily transported over long distances and contaminate water sources far from the original release site. The water solubility is influenced by factors such as molecular structure and the presence of functional groups. It is crucial to understand soil and water partition to forecast the mobility and biological availability of PFASs in the environment and to create more efficient remediation/management techniques.

PFAS can sorb or partition between different phases in the environment, including air, water, soil, sediment, and biota. Sorption processes, such as adsorption onto solid surfaces or partitioning into organic matter, can affect the mobility and bioavailability of PFAS. PFASs Shorter-chain PFASs have lower Kd values and are consequently considerably more mobile in the environment. Environmental factors such as pH, temperature, organic and inorganic matter content, mineral composition, and other chemicals can influence PFAS behavior. The pH plays an observation role in the inconsistent relationship between soil properties and PFAS sorption. An increase in pH value decreases the sorption's property on soil due to the ionization of PFAS. PFAA sorption on soil decreases with an increase in pH. Besides, higher pH makes surface charge on mineral particles less positive, therefore, weakening the electrostatic interaction with anionic PFAA (Wang et al., 2023). Solution pH has the capacity to impact the adsorption of PFAS on soil mainly by influencing the speciation of the PFAS that is pKa-dependent. The pH can also impact the surface charge of soil particles depending on their compositions, which in turn can also influence PFAS sorption. However, observations on the pH effect have not always been consistent for different experiments, even for the same PFAS, indicating that some compounding factors also play an important role in PFAS sorption on soil (Wang et al., 2023).

The distribution of PFAS in various environmental matrices vis-à-vis the migration between and within the environmental compartments such as air–water and surface water–groundwater interactions is influenced by their physical, chemical, and biological processes. The atmospheric and hydrospheric media have been identified as the primary transport pathways for PFAS in the environment (Prevedouros et al., 2006). In the hydrospheric media, for instance, PFAS are transported through precipitation to surface water and groundwater. The fate and transport of PFAS from atmospheric to hydrospheric compartments are important to understand their regional and global distribution (Taniyasu et al., 2013). Due to their characteristic in long-range transportation, PFAS have been commonly found in highly remote environments, which could largely be attributed to atmospheric distribution. The occurrence of PFAS in surface and rainwater samples from isolated islands of Malta, Gozo, and Comino indicated the extent to which PFAS can be transported under the prevailing conditions (Sammut et al., 2017). The primary sources and pathways of PFAS in the environment are further described as follows.

12.2.2.1 Industrial releases and wastewater treatment plants

PFAS can enter the environment through direct and indirect releases from industrial facilities that manufacture or use these compounds. The study by Sha et al. on the concentration profiles and spatial distribution of PFAS around industrial facilities in China showed that all the investigated samples, including tree leaves and bark, were contaminated (Shan et al., 2014). Similarly, a study conducted in Italy reported that PFAS contamination from industrial sources was prevalent over urban sources, with up to 50 times higher than the loads of PFAS in wastewater treatment plants (WWTPs) receiving municipal wastes (Castiglioni et al., 2015). This study and many others have demonstrated the environmental contamination level of PFAS around the industrial areas where the chemicals are produced or used (Table 12.1). Industrial releases of PFAS serve as the primary source through which they enter the environment and can be targeted to address their passage into the environmental matrices and consequently mitigate the potential environmental impact.

The sewer leaks and urban runoff are sources of PFAS in the water environmental compartments, while the WWTPs have been reported to fail to remove PFAS chemicals, leading to their presence in drinking water samples, especially in areas close to industrial activities involving PFAS chemicals (Castiglioni et al., 2015). The effluents of the wastewater treatment plants containing PFAS chemicals that could not be removed are further discharged into the surface water bodies. This can result in the contamination of downstream water sources and the potential for further transport through aquatic systems, considering that PFAS has high stability in the aquifers (Müller et al., 2023). A study conducted in Italy reported mean concentration values ranging from 15 to 1128 ng/L from six WWTPs with higher values of PFAA levels in WWTPs that received most of their wastes from textile-based industries as compared to furniture-based industries (Castiglioni et al., 2015).

12.2.2.2 Firefighting foams

PFAS-containing firefighting foams, known as aqueous film-forming foams (AFFF), have been extensively used for suppressing flammable liquid fires at airports, military bases, and industrial sites. The AFFF is classified into Class A and Class B firefighting foams. The Class A foams are mainly used to suppress Class A fires, such as burning buildings and also vegetation fires. These types of fires have no reported record of perfluorinated chemicals (Dobraca et al., 2015). However, the Class B foams are designed to suppress Class B fires, which involve flammable liquids that routinely contain fluorinated surfactants (Moody & Field, 2000). The presence of large quantities of flammable liquids promotes the use of Class B firefighting foams in the military and industries such as hydrocarbon processing-based industries to extinguish hydrocarbon fuel fires (Moody & Field, 2000). Besides the fluorinated surfactants, the Class B firefighting foams contain water, organic solvents, polymers, hydrocarbon surfactants, and other additives (Sharifan et al., 2021).

The safety measures for using and storing AFFF, referring to Class B, are required by the various institutions and industries involved. However, the use of AFFF under controlled activities, such as during the fire training exercises, and under uncontrolled incidents, such as emergency responses and leaks during transportation, can result in the release of PFAS into the environment, particularly into surface water and groundwater near firefighting training areas or fire incidents (Moody et al., 2002). Reinikainen et al. found that aquatic ecosystems, groundwater usage, and fish consumption have the most significant environmental and health risks from the off-site migration of PFAS from a single fire event in a study conducted in Finland (Reinikainen et al., 2022). Similarly, the major source of PFAS in the soils and subsurface sediments is the AFFF, such that about 97% of the PFAS mass was attributed to the AFFF-impacted soils in the old fire training areas (Nickerson et al., 2020).

A significant distribution of PFAS from a point source of fire–fighting training area via leaching was reported in Australia, which contaminated the nearby groundwater with concentrations <0.17–14 µg/L of Perfluorooctane sulfonate (PFOS) and <0.07–6 µg/L of perfluorohexane sulfonate (Bräunig et al., 2017). The use of PFAS-contaminated groundwater further created an additional human exposure pathway with 38–381 µg/L of PFOS and 39–214 µg/L of perfluorohexane sulfonate found in the human serum of the residents, which is more than 30–fold higher compared to the general Australian population (Bräunig et al., 2017). The extent of contamination of PFAS in the environmental matrices such as soil, groundwater, and sediment in the nearby point source of fire-fighting training ground has been extensively investigated, with the outcomes indicating significant off-site transport such as up to 2 m in the soil cores and widespread contamination of the environment (Baduel et al., 2017; Bräunig et al., 2017).

12.2.2.3 Atmospheric deposition and the hydrospheric compartment

PFAS can be transported over long distances through the atmosphere. They may be released into the air during manufacturing processes, the use of consumer products, and emissions from waste incineration. There is increasing evidence of PFAS and their associated precursors in the atmospheric environment, especially around the industrialized and densely populated areas. The comparison of the presence of PFAS in the samples of street dust collected from residential areas and heavily trafficked areas in Japan showed a significant concentration of the chemicals in fine dust particles containing the PFAS chemicals from the heavily trafficked areas (Murakami & Takada, 2008). Similarly, PFAS chemicals were reportedly found in the air and dust samples collected in the USA (Kim & Kannan, 2007; Wu et al., 2020), from the streets in Japan (Murakami & Takada, 2008), and in indoor and outdoor dust of Fluorochemical Industrial Park in China (Su et al., 2016). Other selected studies where PFAS chemicals were identified in the atmospheric deposition and served as sources of hydrospheric compartments are presented in Table 12.2. Gas deposition of PFAS plays a key role in the deposition of PFAS in marine ecosystems (Wang et al., 2015). These airborne PFAS can be deposited on surface water bodies or soils through precipitation or dry deposition and eventually transported to the groundwater source.

Rainfall or precipitation is the primary source of PFAS in water environments (Sammut et al., 2017). Considering the water-soluble nature of the PFAS and their bioaccumulative property, they tend to persist for a long time in the water environment, thereby promoting their fate and transport via atmospheric medium to hydrospheric compartments. The occurrence of PFAS in water environments, especially in surface water and groundwater can be directly traced to rainwater contaminated with chemicals. Due to the significant role of rainfall in the distribution of PFAS, especially in the water environment, it has been regarded not only as an effective removal of all forms of pollutants from the atmosphere but also as an effective scavenger for the removal of PFAS in the atmosphere and their subsequent deposit into the soil and waterbodies (Sammut et al., 2017; Taniyasu et al., 2013).

12.3 Remediation and treatment methods and processes of PFAS in the environment

12.3.1 PFAS removal technologies and techniques

Several technologies and techniques have been developed to remove PFAS from the environment, such as soil, water, and air, which include adsorption, filtration, chemical oxidation/reduction, thermal treatment, and soil washing, as shown in Fig. 12.2. These remediation methods, especially in soil and water matrices, can be classified into physical, chemical, and biological methods. The biological method has been regarded to have advantages over physical methods due to their applicability to effectively remove POPs, requires less cost, and also leads to less disruption of the soil and water environment (Zhang et al., 2022). The selected existing remediation techniques that were targeted at removing PFAS in different environmental matrices are presented in Table 12.3. Overall, these methods are expensive, impractical for in situ treatment, and use high pressures and temperatures, with most resulting in toxic waste (Shahsavari et al., 2021). However, biomass-based adsorbents using agrowastes are increasingly considered promising alternatives to remove PFAS from the environmental matrices considering their low-cost and green-based environmental friendliness (da Silva Bruckmann et al., 2024). Nevertheless, the removal technologies of PFAS in the environment are still developing and need further study in their application in order to increase their scalability and sustainability.

12.3.2 Biodegradation and transformation of PFAS in the environment

PFAS are known for their persistence in the environment due to their strong carbon-fluorine bonds. However, some PFAS compounds can undergo biodegradation or transformation under specific conditions. Factors such as microbial

TABLE 12.2 Selected references on the sources of PFAS in environmental matrices.

Environmental matrix	Source	Country / Region	Observation	References
Water	Mollusks for the Bohai sea	China	Mollusks contain an average $\sum$PFAS concentration of 41.5 ng/g dw, with PFOA occupying 70.56%.	Meng et al. (2022)
	Pennsylvania Water Quality Network	USA	At least one PFAS was detected in 76% of 161 Pennsylvania streams. Electronic manufacturing facilities in the local catchment and water pollution are the major drivers of $\sum$PFAS contamination.	Breitmeyer et al. (2023)
	Canadian Arctic wastewater	Canada	The potential for wastewater to contribute significantly to local PFAS pollution in the Arctic should not be underestimated.	Stroski et al. (2020)
	Large micro-tidal estuary	Australia	PFOS and PFHxS make the greatest contribution to PFAS concentrations in the Swan Canning Estuary.	Novak et al. (2023)
	WWTPs	United Kingdom	PFHxA and PFPA are potential markers for effluent discharge in rivers, and wastewater effluent is a source of PFAS contamination in surface water.	Müller et al. (2023)
	WWTPs	South Korea	Total (Σ_{13}PFAS) emissions from WWTPs was 4.03 ton/y, with PFOA and PFOS being dominant.	Kwon et al. (2017)
	Municipal WWTPs	Thailand	High concentrations of PFCs, ranging from 662–1383 ngL^{-1}, were found in municipal wastewater and sludge.	Kunacheva et al. (2011)
	Groundwater	Australia	Concentrations of PFOS and PFHxA were <0.17–14 µg/L and <0.07–6 µg/L, respectively.	Bräunig et al. (2017)
	Surface and rainwater	Maltese Islands	Total concentrations of PFAS in rainwater ranged between 0.38 ng/L and 6 ng/L, while PFOS and PFOA were up to 8.6 ng/L and 16 ng/L, respectively, in surface waters.	Sammut et al. (2017)
	WWTPs, river, groundwater, and drinking water	Italy	The presence of PFAS in drinking water samples indicated the inability of WWTPs to remove the chemicals with concentrations of PFAS in WWTPs receiving industrial waste discharge was 50 times more than those from urban sources.	Castiglioni et al. (2015)
	Surface runoff water and lake water	USA	The concentrations of PFAS in surface runoff water, and lake water of urban areas. short–chain were 1.11–81.8 ng/L, and 9.49–35.9 ng/L, respectively.	Kim and Kannan (2007)
Soil	Aqueous film–forming aqueous foam (AFFF)–impacted site	USA	The concentration of Log$_{10}$(ΣPFAS) did not exhibit any notable differences across various sampling events. Nonetheless, there was a discernible rise in Log$_{10}$(ΣPFAS) concentrations as the depth proceeded to become shallower.	Anderson et al. (2022)
	Soil	European countries	A substantial amount of European soil contains PFAS at levels greater than 5000 ng/kg, which suggests a significant level of contamination.	Moghadasi et al. (2023)
	Agricultural soil	Germany	The findings of this study imply that short-chain PFAS are generated and accumulated primarily during the spring and summer seasons and then mobilized, especially during the recharge periods of the fall and winter months.	Röhler et al. (2021)

(Continued)

TABLE 12.2 (Continued)

Environmental matrix	Source	Country / Region	Observation	References
	Topsoil	China	The majority of PFASs in the topsoil are from textile treatment, metal electroplating plants, food packaging, and coating materials.	Ma et al. (2021)
	Soils in industrial areas, airports, landfills, fire stations, and agricultural areas	China	The highest PFAS concentrations were found near fire stations.	Zhu et al. (2022)
Air / Atmosphere	Indoor carpet and dust	USA	The estimated daily intake of PFAAS via dust ingestion for children was found to be 0.023, 0.096, and 1.9 ng/kg body weight in the low–, intermediate–, and high–exposure scenarios, respectively.	Wu et al. (2020)
	Working microenvironments	Greece	Most of the samples, comprising 80%, contained PFHxA, with concentrations ranging between 3.6 and 72.5 ng g^{-1}. Additionally, 90%–95% of the samples showed PFOA, PFDcA, and PFDoDA concentrations ranging from 10–653 ng g−1, 3.2–7.4 ng g^{-1}, and 3.8–13.1 ng g−1, respectively.	Besis et al. (2019)
	Indoor and outdoor dusts in Fluorochemical Industrial Park	China	The median concentration of PFAAs in indoor dust was found to be significantly higher than that in outdoor dust at all locations, ranging from 73 to 13,500 ng/g (median: 979 ng/g) and 5 to 9495 ng/g (median: 62 ng/g), respectively.	Su et al. (2016)
	Gas/particle partitioning	Germany	The concentrations of gas deposition flux, particle deposition flux, and total flux were 1088 ± 611, 189 ± 75, and 1277 ± 627 pg/(m^2 d), respectively.	Wang et al. (2015)
	Air, rain, and snow	USA	The concentrations of PFAS in bulk air, rain, and snow in urban areas were 8.28–16.0 pg/m^3, 0.91–13.2 ng/L, and 0.91–23.9 ng/L, respectively.	Kim and Kannan (2007)
	Sampled street dust from the residential areas and heavily trafficked areas	Japan	The PFOS and PFOA concentrations in sampled street dust ranged from <0.2 to 11 ng g−1 and from 1.2 to 11 ng g−1.	Murakami and Takada (2008)

AFFF, Aqueous film–forming aqueous foam; PFAS, per- and polyfluoroalkyl substances; PFBA, perfluorobutanoic acid; PFOA, perfluorooctanoic acid; PFOS, perfluorooctane sulfonate.

activity, redox conditions, temperature, and the presence of specific enzymes or co-substrates can influence the degradation and transformation potential of PFAS (Zhang et al., 2022). Despite the unsatisfied initial experimental results on biodegradation in PFAS due to their resistance and high stability, researchers, however, have continued to test various microbial species that could be employed or the PFAS degrading capability. During the last decade, however, some studies have reported that various microbial species or mixed cultures were successful. The successful biodegradation cases raise hopes for biological PFAS treatment through the final product transformation. This process is dependent on the "biodegradation triangle," which comprises the structural complexity of PFAS species, the composition of microorganisms present, and the operating environmental conditions (Cui et al., 2020). Biodegradation has the potential to form the basis of a cost-effective, large-scale, in situ remediation strategy for PFAS removal from soils. Both fungal and bacterial strains capable of degrading PFAS have been isolated; however, to date, information regarding the mechanisms of degradation of PFAS is limited. Through the application of new technologies in microbial ecology, such as stable isotope probing, metagenomics, transcriptomics, and metabolomics there is the potential to examine and identify

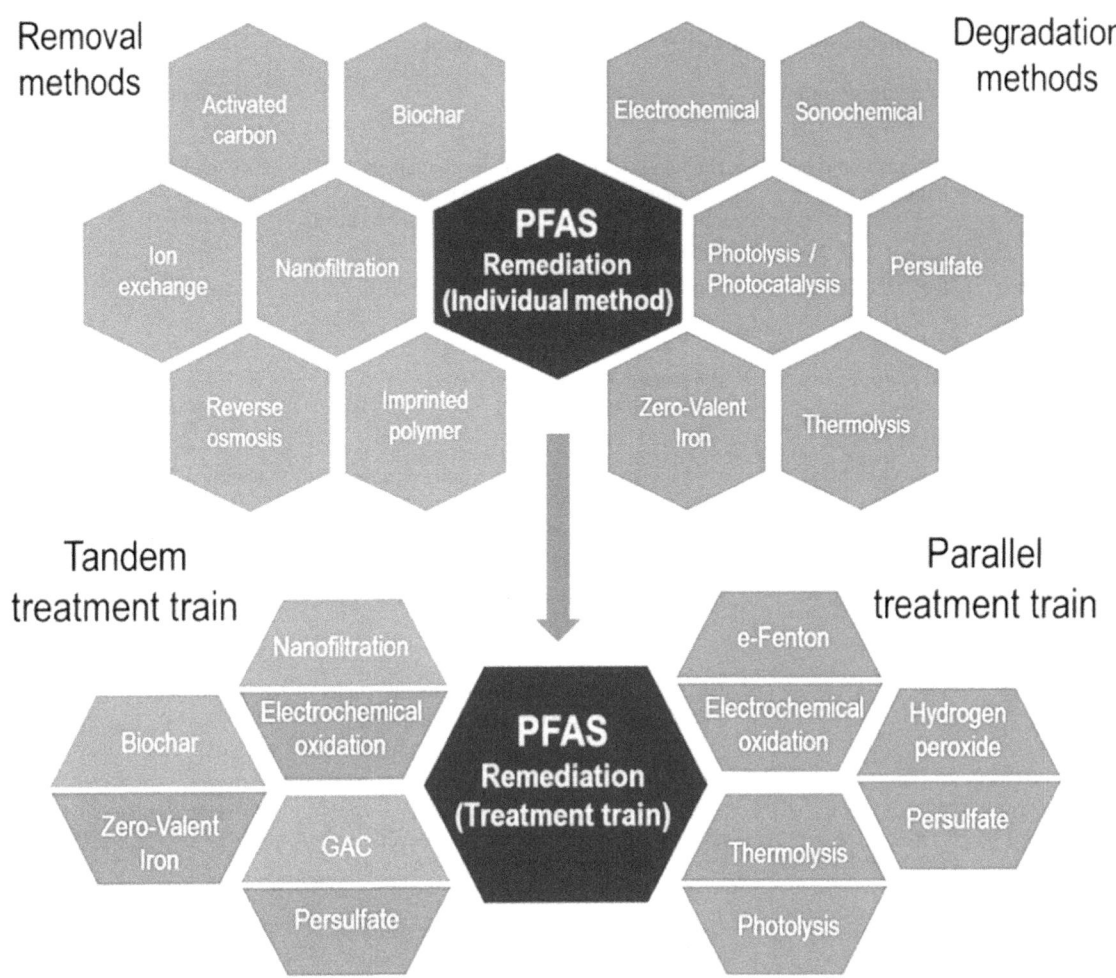

FIGURE 12.2 Treatment strategies for PFAS remediation from the environment. The different types of treatment strategies available in the literature for the removal of PFAS from the environmental matrices. Source: *From Lu, D., Sha, S., Luo, J., Huang, Z. & Zhang Jackie, X. (2020). Treatment train approaches for the remediation of per- and polyfluoroalkyl substances (PFAS): A critical review. Journal of Hazardous Materials, 386. https:// doi.org/10.1016/j.jhazmat.2019.121963.*

the biodegradation of PFAS, a process that will underpin the development of any robust PFAS bioremediation technology (Shahsavari et al., 2021).

12.3.2.1 Microbial remediation

Microbial biodegradation has proven to be an effective remediation method for the removal of certain types of PFAS types. Meanwhile, an increasing number of new PFAS species are being tested. The biodegradation efficiencies vary significantly from slight degradation to complete removal (Zhang et al., 2022). Microorganisms in the soil are essential for preserving the health, function, and quality of the soil. Contamination with PFAS affected the variety, structure, and function of the soil's microbial population, as well as its capacity for nitrification and denitrification, transport of carbohydrates and energy, transcription, signal processing, and cell motility (Sun et al., 2023). Types of PFAS, soil conditions, concentrations, and exposure times all affected the microbial community and diversities in PFAS-contaminated environments. Bacterial species have the tendency to degrade PFAS under certain conditions such as aerobic or anaerobic conditions, i.e., through aerobic biotransformation. The bacterial assemblage and functional diversity in the environmental matrices are essential factors that determine the course of biodegradation and fate of PFAS. These factors result in significant differences in the metabolites types and degradation efficiencies (Zhang et al., 2022). The differences among microbial compositions induced variations in the pathways for biodegradation, thus leading to different degradation efficiencies and terminal metabolites. Among the bacterial species that are capable of degrading PFAS, many of them were identified as *Pseudomonas* sp., a genus that demonstrates metabolic diversities. The biological reduction of

TABLE 12.3 Selected remediation techniques of PFAS in the environment.

Removal techniques	PFAS compounds	Key findings	Environmental media	References
Flocculation, coagulation	PFAS (short and long chain)	The removal efficiency of the coagulants and flocculants varies greatly. The removal efficiency relates to the chain length and head group of coagulants and flocculant	Soil washing water	Hubert et al. (2024)
Adsorption (porous organic polymers)	Five PFAS; PFBS, PFBA, PFHxA, and PFHpA	The PFAS removal efficiency was high (>95%).	Water	Zhang et al. (2024)
Adsorption (ferric hydroxide nanoparticle)	9 PFAS	The nanoparticle achieved the highest removal efficiency of 44%.	Wastewater	Zhang et al. (2021)
Absorption (Foam fractionation)	Six different types of PFAS	The long-chain PFAS were readily removed by all co-foaming surfactants. Cationic surfactants are best suited for the removal of ionic PFAS.	Environment	Buckley et al. (2023)
Photocatalytic ozonation	5 PFAS (PFOA/PFHxS/PFBS/6:2 FTS/GenX)	Photocatalytic ozonation degrades the PFAS.	Water	Lashuk et al. (2022)
Periodically reversing electrocoagulation, PREC	PFBA, PFBS, PFOA, and PFOS	The PREC improves the PFAS removal rate to a level ranging between 10 µg/L and 100 mg/L.	Groundwater	Liu et al. (2022)
Nanofiltration with concentrate foam fractionation	PFPeA, PFHxA, PFHpA, PFOA, PFBS, PFPeS, PFHxS, and PFOS	The novel combination of nanofiltration and concentrate foam fractionation reduced the total detectable PFAS in the drinking water.	Drinking water	McCleaf et al. (2023)
Ion exchange removal and resin regeneration	40 legacy and emerging PFAS	Greater than 90% of the PFAS was removed depending on the resins used and contact time.	Drinking water	Liu and Sun (2021)
Adsorption (Granular activated carbon (GAC))	4 PFCs: PFOS, PFOA, PFBS, and PFBA,	The GAC filter achieved >95% PFOS removal for long-chain PF, and the effectiveness was reduced for shorter chains.	Deionized water (MQ) and landfill groundwater	Zhao et al. (2011)
Foam Separation	37 PFAS	Foam separation increases the mean PFAS removal efficiency by 69% and the median removal efficiency by 92%	Solid waste landfill leachate	Robey et al. (2020)
Gas fractionation enhanced technology	PFAS, PFOS, PFHxS and PFOA	Gas fractionation enhanced the removal efficiency of Total PFAS, PFOS, PFHXS, and PFOA by 528%, 1085%, 15%, and 50%, respectively.	Soil	Pang et al. (2022)
Degradation (High–dose electron beam)	17 PFAS	Ebeam degraded PFAS in 10 out of the 17 soil samples below the detectable limit, and 49%–99.9% removal efficiency was observed in the remaining 7, with more effectiveness in soil with low moisture content.	Groundwater, Soil	Lassalle et al. (2021)
Degradation (photocatalysis)	PFOA, PFOS, and PFNA	The process was able to degrade the PFNA, PFOS, and PFOA at an optimal rate of 90 ± 1%, 88 ± 1%, and 46 ± 2%, respectively.	Wastewater effluents	Xia et al. (2022)
Electrochemical oxidation	PFOA and PFAS	Electrochemical oxidation increased the removal efficiencies of PFOA and PFOS by 80% and 78%, respectively.	Landfill leachates	Pierpaoli et al. (2021)

(Continued)

TABLE 12.3 (Continued)

Removal techniques	PFAS compounds	Key findings	Environmental media	References
Extraction	PFHpA	The extraction efficiency of natural deep eutectic solvents (NADES) was up to 90.7%.	Water	Fortunato et al. (2023)
Adsorption	PFOA, PFBA	Fe–M–BC900 can co–remove long and short-chain PFAS (>96%) and maintain high sorption capacity (>91%).	Natural water and simulated wastewater	Liu et al. (2023)
Adsorption (resin and polymer–based adsorbent)	PFOA and PFOS	High porous anion exchange resins have an effectiveness of up to 99.9% due to their high reaction rate. However, their removal rate was short-chain reduced by about 71%.	Groundwater	Parvin et al. (2023)
Adsorption (Granular or powdered activated carbon)	21 PFAS	Biological activated carbon was found to be ineffective in removing PFAS due to diminished adsorption capacity after long–term use.	Drinking water	Nakazawa et al. (2023)

AFFF, Aqueous film–forming aqueous foam; PFAS, per- and polyfluoroalkyl substances; PFBA, perfluorobutanoic acid; PFOA, perfluorooctanoic acid; PFOS, perfluorooctane sulfonate.

perfluorooctanoic acid (PFOA) and PFOS revealed alternative approaches or methods to the biodegradation of PFAS species.

Furthermore, it is worthy of note that there are vital factors affecting PFAS biodegradation, which include the molecular structure of the PFAS compounds, substrate availability, environmental conditions, microbial community composition, acclimation and adaptation, synergistic effects, and toxicity and inhibitory effects. A brief explanation of these factors is as follows:

Molecular structure: The molecular structure of PFAS compounds plays a crucial role in their biodegradation. PFAS compounds with shorter carbon chains, such as perfluorobutanoic acid (PFBA), are generally more amenable to biodegradation than compounds with longer chains, such as PFOA. Additionally, the presence of functional groups, such as sulfonic acid or carboxylic acid, can also influence the biodegradability of PFAS compounds.

Substrate availability: The availability of suitable carbon and energy sources is essential for microbial growth and PFAS biodegradation. PFAS compounds are not a direct source of carbon and energy for microorganisms, so they require co-substrates, such as organic carbon compounds, to support their growth and metabolic activity. The type and concentration of co-substrates present in the environment can impact biodegradation efficiency.

Environmental conditions: Various environmental factors can influence PFAS biodegradation, among which include temperature, pH, dissolved oxygen, and nutrient availability. Different microorganisms have specific optimal ranges for these factors, and their biodegradation activity may vary accordingly. For example, some microorganisms are more active at higher temperatures, while others can function under colder conditions. pH can also affect the activity of microbial enzymes involved in PFAS biodegradation.

Microbial community composition: The composition and the nature of diverse microbial communities present in the environment can impact the biodegradation of PFAS compounds. Different microorganisms possess varying capabilities to degrade specific PFAS species. Therefore, the presence or absence of specific microbial species can affect the overall biodegradation efficiency. Microbial community structure can be influenced by factors such as environmental conditions, previous exposure to PFAS, and the presence of other contaminants.

Acclimation and adaptation: Microorganisms need to acclimate and adapt to PFAS compounds to enhance their biodegradation capabilities. Prolonged exposure to PFAS can lead to the enrichment of microbial populations with specific enzymes or metabolic pathways for PFAS degradation. Therefore, the history of exposure and previous contact with PFAS can influence the biodegradation efficiency.

Synergistic effects: In some cases, the combination of multiple microbial species or the presence of other environmental factors can enhance PFAS biodegradation. Certain microorganisms may produce enzymes or metabolites that facilitate the degradation of PFAS compounds. Additionally, the presence of other pollutants or co-contaminants in the environment can interact with PFAS and influence their biodegradation (Giesy & Kannan, 2001).

Toxicity and inhibitory effects: Some PFAS compounds may exhibit toxicity toward microorganisms, inhibiting their growth and biodegradation activity. High concentrations of PFAS can be toxic to microbial communities, leading to reduced biodegradation efficiency. It is essential to consider toxicity effects and determine the appropriate PFAS concentration range for effective biodegradation.

12.3.2.2 Phyto-microbial remediation

Phyto-microbial remediation, also known as plant-microbe-assisted remediation, is an emerging approach that combines the benefits of both phytoremediation and microbial remediation techniques. It involves the use of plants and associated microorganisms to degrade, transform, and immobilize contaminants in the environment (Steenland et al., 2010). In the case of PFAS contamination, phyto-microbial remediation can be a sustainable and cost-effective solution. Phyto-microbial remediation of PFAS contamination takes advantage of the natural abilities of plants and their associated microbial communities. Plants can uptake PFAS compounds through their roots via passive and active mechanisms. Passive uptake occurs through processes like transpiration streams or small-molecular diffusion of nonionized PFAS (Steenland et al., 2010). Active uptake involves the use of membrane protein transport or ion channels. Once the PFAS compounds are taken up by the plants, they can be transformed or degraded by the associated microbial communities in the rhizosphere, which is the soil region influenced by the root exudates (Sunderland et al., 2019). These microorganisms, including bacteria and fungi, possess enzymes capable of breaking down PFAS compounds or converting them into less harmful forms. The microbial degradation pathways may involve processes such as enzymatic hydrolysis, oxidation, reduction, or bioaccumulation.

One of the advantages of phyto-microbial remediation is its low capital and maintenance costs. It relies on the natural functions of plants and microbes, reducing the need for expensive infrastructure or energy-intensive processes. Additionally, it is considered an environmentally friendly approach since it does not introduce additional chemicals or disrupt the ecosystem. However, the fate of the plant materials after remediation is an important aspect that requires further investigation. Depending on the extent of PFAS degradation and the specific plant species used, the accumulation of PFAS compounds in plant tissues may vary. It is crucial to assess the potential risks associated with the disposal or reuse of these plant materials to ensure that they do not reintroduce the contaminants into the environment. Ongoing research efforts are focused on optimizing the phyto-microbial remediation process, including identifying plant species with high PFAS uptake and degradation capabilities, improving microbial inoculation techniques, and understanding the factors that influence the efficiency of PFAS removal. Many studies were aimed at enhancing the effectiveness and applicability of phyto-microbial remediation as a sustainable and efficient technology for PFAS-contaminated sites.

In conclusion, phyto-microbial remediation offers a promising approach for the remediation of PFAS contamination. Harnessing the synergistic interactions between plants and microorganisms provides a cost-effective and environmentally friendly method to mitigate the harmful effects of PFAS in soil and water systems. However, further research is needed to fully understand the fate of plant materials and optimize the efficiency of this remediation technique.

12.4 Biochar production processes and its usage for removal of PFAS in the environmental matrices

12.4.1 General processes of biochar production

Biochar has shown great potential as a sustainable alternative for environmental remediation, particularly for removing PFAS contamination in soils, water, and wastewater. It is similarly regarded as a cost-effective, sustainable, and environmentally friendly adsorbent for removing PFAS and other contaminants from environmental matrices (Krahn et al., 2023). Biochar is typically produced using various thermochemical decomposition and carbonization processes, including pyrolysis, torrefaction, gasification, and hydrothermal carbonization (Tripathi et al., 2016). While biochar is commonly used for soil amendment to improve various soil properties, such as physical, chemical, and biological as one of the functions that distinguish it from activated carbon, recent development has allowed biochar to be activated, modified, or engineered using physical, chemical, or biological methods to enhance their applicability in the aspect of pollution remediation including PFAS removal (Kazemi Shariat Panahi et al., 2020; Mohan et al., 2014; Rajapaksha et al., 2016). The effectiveness of biochar for sustainable environmental management stems from its inherent qualities, such as resistance to biodegradation, high cation–exchange capacity, and high porosity (Weber & Quicker, 2018).

Pyrolysis, which is the most common process of biochar production, occurs when the organic feedstock, majorly biomass from agricultural waste and municipal sewage sludge, is thermally decomposed in the absence of oxygen at temperatures ranging from 300°C to 900°C into biochar (carbon-rich solid), bio-oil (liquid), and syngas (gaseous mixture) (Wang & Wang, 2019). The yield and composition of the products of the pyrolysis process are greatly dictated by the characteristics of the feedstock, as well as the reaction operating conditions and parameters (Rajapaksha et al., 2016). Pyrolysis can be carried out in reactors such as paddle kilns, bubbling fluidized beds, fixed beds, wagon reactors, and agitated sand rotating kilns. Various types of feedstocks, such as agricultural residues, wood, sewage sludge, and municipal wastes, have been used in the production of biochar. The characteristics of feedstock, including density, calorific value, moisture content, and proportion of lignocellulosic components, all affect the yield of biochar formation (Suliman et al., 2016). Temperature and atmospheric conditions such as airflow, air limit, and N_2 were identified as critical factors that dictate the biochar properties (Luo et al., 2015). Specifically, carbonization temperature influences physicochemical properties and structures, including pore structure, functional groups, surface area, and elemental composition (Yaashikaa et al., 2020). The yield of biochar decreases with increasing temperature and increases the production of syngas. In fact, temperature is regarded as the main operating condition that determines biochar production efficiency (Wei et al., 2019).

Furthermore, highly lignocellulosic components such as cellulose, hemicellulose, and lignin have also been found to greatly increase the yield of biochar (Ma et al., 2019; Rangabhashiyam & Balasubramanian, 2019). However, a higher moisture content in the biomass inhibits char formation and increases the amount of energy for pyrolysis (Tomczyk et al., 2020; Tripathi et al., 2016). Therefore, a low moisture content in biomass is preferable for biochar formation. Additionally, biochar produced from biosolids is an excellent adsorbent compared to plant-based biochar for the removal of PFAS in water (Kundu et al., 2021). This could be due to the presence of higher metal contents such as Cu, Ca, and Fe and higher content of O (26%–36%), and S (1.03%–3.68%) (Wu et al., 2022). Sewage sludge biochars are also highly effective at adsorbing PFAS from the environmental matrices (Krahn et al., 2023). However, further investigation is still needed to study their effectiveness in the removal of short-chain PFAS.

Pyrolysis can be categorized as fast or slow depending on the residence time, heating rate, temperature, and pressure. Fast pyrolysis is primarily used to produce bio-oil and is majorly used in energy applications (600°C–1000°C) (Wang & Wang, 2019). However, slow pyrolysis is the most efficient method of biochar production; it produces higher yields of biochar (50%–90%) than liquid and gaseous products. Slow pyrolysis typically involves a heating rate of approximately 5°C–7°C/min, residence time ranging from several hours to days, and a temperature range of 300°C–600°C (Liu et al., 2015). Slow pyrolysis is mostly performed in fixed–bed reactors (Chun et al., 2021). Other pyrolysis methods include flash, hydro, vacuum, and microwave pyrolysis.

12.4.2 Modification of biochar for PFAS removal from environmental matrices

Various modification techniques have been applied to tailor the physicochemical properties of biochar for the effective removal of PFAS from environmental matrices. These modification methods or techniques can be categorized into physical, chemical, and biological methods and can be carried out before, during, or after the pyrolytic process. The physicochemical properties such as pore diameter, pore diameter/ volume ratio, contact angle, and nitrogen content are crucial for the removal of PFAS sorption (Krebsbach et al., 2023). Krebsbach reported that biochar pore diameters (7.5–11 nm), pore diameter/pore volume ratios (50–150 nm/cm^3/g), large specific surface area, and high hydrophobicity have high potential to remove PFOS from water. Similarly, the available modification methods and techniques, along with their mechanisms, were reviewed by Rajapaksha et al. and further categorized into four methods, that is, physical, chemical, impregnation with mineral sorbents, and magnetic modifications (Rajapaksha et al., 2016). The application of modified or engineered biochar and the resultant performance of different methods largely depend on the targeted contaminants, environmental conditions, and the goal of remediation of such contaminants (Rajapaksha et al., 2016). An overview of different methods and techniques of biochar modification for PFAS removal is as follows:

12.4.2.1 Physical modification methods

The biochar is sometimes subjected to physical modifications to be able to employ for the removal of PFAS. The common physical modification methods for biochar properties include ball milling, steam/gas activation, microwave activation, and magnetic biochar (Kazemi Shariat Panahi et al., 2020). These methods were specifically carried out to alter the surface area, pore structure, particle size, and surface functional groups to enhance the adsorptive properties of biochar. While physical methods are relatively cheap they require high energy consumption and longer activation times, making chemical modification methods economically preferable.

Ball milling is a physical modification method that combines mechanical energy and chemical reactions to alter the composition and structure of biochar (Amusat et al., 2021). It has been found to adjust the particle size of biochar from millimeter to sub-micron scale, improve the specific surface area and pore structure, and enrich the surface functional groups of the biochar under nonequilibrium conditions (Lyu et al., 2018). The physicochemical and adsorptive properties of the enhanced biochar could be affected by operating parameters, milling methods (wet or dry), milling temperature, reaction temperature, grinding media shape (spherical, ellipsoidal, and cubes), and ball size distribution (Chen et al., 2022).

Microwave activation is an advanced technique that utilizes electromagnetic irradiation between infrared and radio frequencies (300 MHz–300 GHz) corresponding to wavelength range of 1–100 cm to generate heat energy within biomass through induced dipole rotation of molecules at the atomic scale (Falciglia et al., 2018). This allows for the simultaneous heating of both the internal and external surfaces of biochar without direct contact, resulting in increased surface functional groups and a larger surface area compared to the original biochar.

Activating biochar with steam increases the porosity and surface area of pristine biochar by consuming carbon and other impurities in the biomass, leading to the conversion of unstable organic matter into gas (Akhil et al., 2021). Consequently, the formation of tar in the pores is minimized, resulting in larger pores structure within the biochar. This activation method decreases the surface functional groups and results in the formation of less polar biochar. Steam-activated biochar has demonstrated its effectiveness in various environmental remediation applications, such as removing dyes, antibiotics, PFAS, and carbamate pesticides from environmental matrices. Furthermore, biochar can be modified by purging CO_2 into biochar during pyrolysis, which facilitates the formation of micropores and increases the surface area of the biochar (Wang & Wang, 2019).

12.4.2.2 Chemical modification methods

The chemical method of biochar modification involves the activation of biochar in the presence of chemicals and inert gas. These chemical reagents alter the surface chemistry of biochar to enhance the deposition of materials. The chemical modification typically produces new biochar surface functional groups (Kazemi Shariat Panahi et al., 2020). The primary techniques of the chemical method of biochar modification include acid-base modification, organic surfactants, oxidizing treatment, magnetization, carbon nanomaterials, loading of clay minerals, layered double hydroxides, and nonmetallic element doping (Kazemi Shariat Panahi et al., 2020). Modifications are also used to increase the surface area, cation–exchange capability, and a number of functional biochar groups. Acidic oxidizing agents typically lead to biochar with higher concentrations of carboxylic groups, whereas alkaline treatments result in increased surface electrostatic interactions, as well as surface precipitation and complexation (Mulabagal et al., 2021).

12.4.2.3 Biological modification methods

The biological modification of biochar is essentially based on the use of microorganisms. The microbial growth exhibits assimilating characteristics and conversion prospects of organic compounds and various dissolved and sorbed contaminants to produce safe metabolites (Bouwer & Zehnder, 1993). The biochar is required to have a desirable surface area to be able to effectively remove the contaminants, which could be biochemically modified using microorganisms via colonization and biofilm generation (Kazemi Shariat Panahi et al., 2020). The microorganism is able to penetrate the pores of the biochar to generate biofilms with high surface area and other desirable characteristics for contaminants removal. It has been previously reported that biochar is a potential scaffold to produce microbial biofilm and has similar or superior performances on the adsorption and biodegradation of environmental contaminants compared to other conventional sand-active biofilm (Dalahmeh et al., 2018). The high surface area and microporous structure of the biochar produced after the biological-aided modification enhance its potential for contaminant removal. The biochar separation due to small particle size and density has been regarded as a major limitation of biochar produced by the ball milling process, a physical modification method. Thus, biochar produced using the biological modification method with improved surface properties and functional groups, and ease of separation is considered a better option (Kazemi Shariat Panahi et al., 2020).

12.4.3 Removal mechanism of PFAS from environmental matrices using biochars

12.4.3.1 Overview of the removal mechanism of PFAS removal using biochars

Due to the absorptive nature of PFAS, adsorption processes have been identified by several studies as the most effective and cost-effective techniques for the removal of PFAS from environmental matrices (Maimaiti et al., 2018). Different adsorbents have been used to remove PFAS from soil and water; biochar has shown promising results because of its porosity, large surface area, and enriched functional groups. Biochar has removed PFAS from soil and water using different adsorption mechanisms such as pore-filling, π-π interactions, electrostatic interactions, hydrophobic interactions, cation exchange, and hydrogen bonding (Chen et al., 2022; Zhang et al., 2022, 2024). These mechanisms are summarized and depicted in Fig. 12.3. The principal adsorption mechanism depends on the PFAS species (long-chain, short-chain, or aromatic), physicochemical properties of the biochar, and environmental conditions.

PFAS molecules often carry a negative charge due to the ionization of their functional groups, such as sulfonate or carboxylate groups, in an aqueous system (Xiao et al., 2011). This negative charge allows them to interact electrostatically with positively charged surfaces, such as those found on biochar particles. Electrostatic interaction plays a crucial role in the adsorption of PFAS on biochar, particularly for anionic species such as PFOS and PFOA. The electrostatic interaction between PFAS and biochar can be influenced by several factors, including the pH of the solution, the type and concentration of functional groups on the biochar surface, and the molecular structure of the PFAS (Leung et al., 2023). For instance, at acidic pH values, the protonation of functional groups on the biochar surface can enhance the electrostatic attraction with deprotonated PFAS molecules. The absorption capacity for PFOS and PFOA by H-SL (sludge biochar modified by HCl acid) increases as the pH value of the biochar reduces in an experiment conducted by (Zhang et al., 2023).

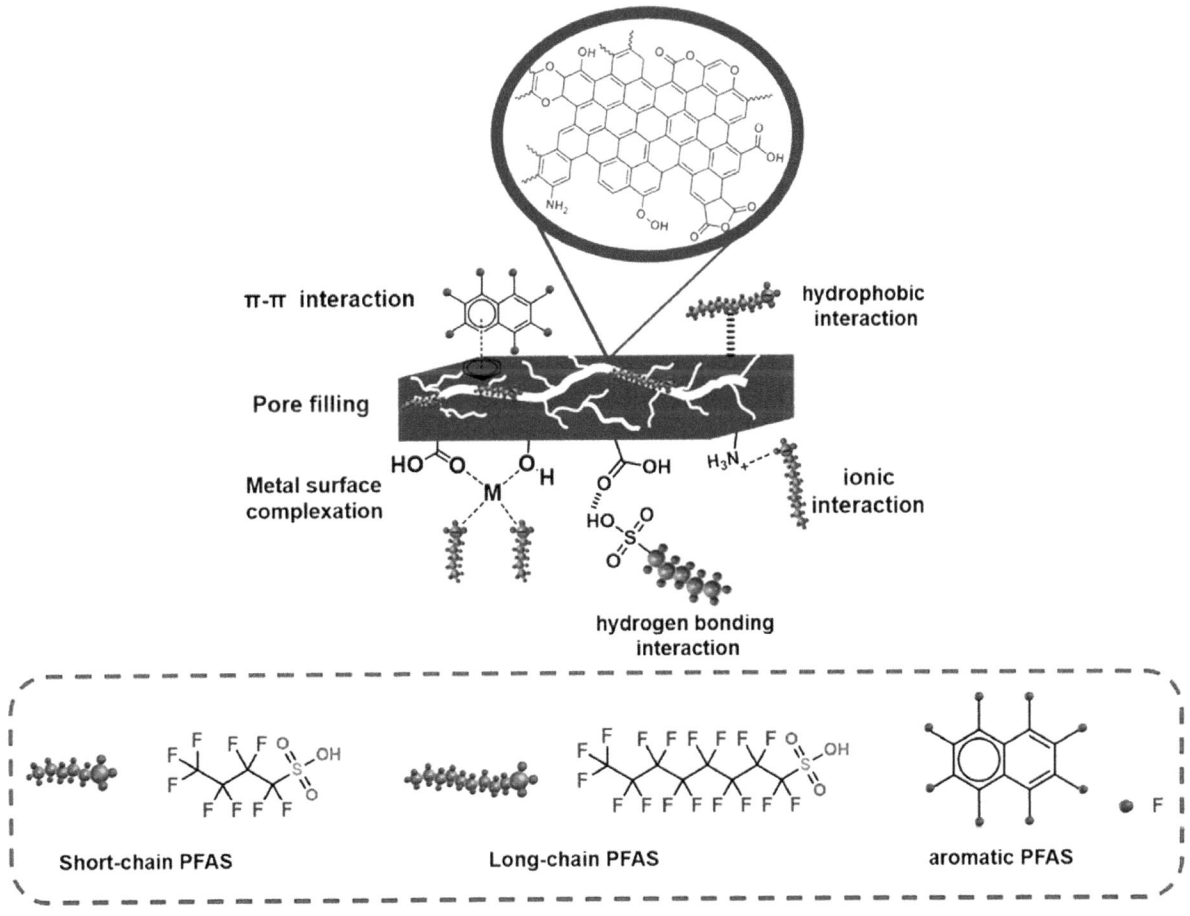

FIGURE 12.3 Summary of the adsorption mechanisms for PFAS removal by Biochar (Behnami et al., 2024). Schematic diagram of different adsorption mechanisms for the removal of PFAS in aqueous form by biochars. Source: *From Behnami, A., Pourakbar, M., Ayyar, A. S. R., Lee, J. W., Gagnon, G. & Zoroufchi Benis, K. (2024). Treatment of aqueous per- and poly-fluoroalkyl substances: A review of biochar adsorbent preparation methods.* Chemosphere, *357. https://doi.org/10.1016/j.chemosphere.2024.142088.*

Hydrophobic interactions contribute to the overall adsorption capacity of biochar for PFAS, especially for longer-chain PFAS molecules such as PFOS and PFOA. Hydrophobic interactions were not affected by the presence of natural organic matter and pH solutions (Militao et al., 2023). Its efficiency as a removal mechanism can be aided by the surface chemistry and hydrophobicity of the biochar. Pore filling also plays a significant role in the adsorption of PFAS on biochar, especially for longer-chain PFAS molecules in aqueous medium. Pore filling occurs when PFAS molecules enter the pores of an adsorbent material, such as biochar, and become trapped within the pore structure. The effectiveness of pore filling as a removal mechanism can be influenced by factors such as the pore size distribution, surface area, and surface chemistry of the biochar. π-π interactions, also known as aromatic stacking, are a significant mechanism for the removal of PFAS molecules, such as PFOA and PFOS, which contain a perfluoroalkyl chain that can interact with aromatic structures, such as those found on biochar particles (Guo et al., 2010).

12.4.3.2 Adsorption kinetics, thermodynamics, and isotherms of PFAS removal by biochars

Understanding the mechanism of the transfer of the PFAS molecules from the aqueous phase to the surface of the biochar is essential to optimize and improve the efficiency of PFAS removal. Adsorption experiments are used to evaluate the effectiveness of adsorbents (biochar in this case) in removing PFAS from environmental matrices and provide valuable insights into their practical application. Generally, the adsorption experiments are in four stages, placing a known mass of adsorbent (biochar) in a series of glass flasks, adding a measured volume of PFAS solution to each flask, and ensuring known initial concentrations. The samples are then agitated using a mixer to ensure uniform mixing and contact between the PFAS molecules and biochar. Samples are then taken at predefined time intervals for further analysis. The kinetic models that are usually used to understand the mechanism and the rate of adsorption of PFAS by biochar include the Pseudo-first-order model, Pseudo-second-order model, Elovich model, and Intra-particle diffusion model. These models have been used to describe the adsorption mechanism for PFAS removal by biochar (Table 12.4).

The Pseudo-first-order model suggests that the adsorption rate is controlled by the movement of adsorbate on the surface of the adsorbent, it can be expressed in Eq. (12.1).

$$q_t = q_e(1 - e^{-k_1 t}) \tag{12.1}$$

TABLE 12.4 Best fit model parameters for sorption kinetics tests of PFAS removal by biochars.

Adsorbents	Adsorbates	Pseudo-first-order		R^2	Pseudo-second-order			References
		K_1 (h^{-1})	q_e (mg/g)		k_2 (g/mg/h)	q_e (mg/g)	R^2	
Sawdust modified biochar (SDN600)	PFOS	0.001	5.20		0.07	185.19		Guo et al. (2010)
Sawdust modified biochar (RMSDN600)		0.001	5.19		0.56	188.67		
Biochar	PFOS	2.28	1.04	0.786	0.017	1.15	0.999	Hassan et al. (2020)
Sediment + 2%BC		15.6	7500	0.956	33.96	8430	0.996	
Sediment + 5%BC		20.4	10600	0.942	24.72	11400	0.991	
HWC	PFBA	2.1	0.58	0.50	3.5	0.6	0.48	Inyang and Dickenson (2017)
PWC		4.5	0.36	0.82	0.009	2.9	-	
HWC	PFOA	0.27	2.2	0.99	0.1	2.4	0.98	
PWC		1.6	2.0	0.57	1.2	2.1	0.80	

HWC, Hardwood; PFAS, per- and polyfluoroalkyl substances; PFBA, perfluorobutanoic acid; PFOA, perfluorooctanoic acid; PFOS, perfluorooctane sulfonate; PWC, pinewood

The Pseudo-second-order model suggests that the adsorption process is controlled by chemisorption (Adane et al., 2015; Marwani et al., 2017). The model can be expressed mathematically in Eq. 12.2. This pseudo-second-order model is frequently more accurate in describing the adsorption data of PFAS (Omo-Okoro et al., 2020).

$$q_t = \frac{k_2 q_e^2 t}{1 + k_2 q_e t} \qquad (12.2)$$

where q_t (mg/g) denotes pseudo-second-order adsorption capacities at time t, q_e (mg/g) denotes adsorption capacities at equilibrium time, k_1 (min^{-1}) denotes first-order rate constant, and k_2 (g/mg/min) denotes pseudo-second-order rate constant.

12.4.4 Performance of biochars on the removal of PFAS

The performance of biochars in removing the PFAS from environmental matrices has been demonstrated in many studies, and the results showed that biochars have excellent adsorption capabilities (Deng et al., 2023). Biochars are effective adsorbents for PFAS with short-chain whose removals have been difficult to achieve with many existing adsorbents (Aboughaly & Fattah, 2023; Behnami et al., 2024). For instance, activated carbon and some other commercially available adsorbents were reported to have a low-performance capability to remove short-chain PFAS compared to long-chain (Zhang et al., 2021). The surface modification of the activated carbon has recently been reported to enhance the adsorption capacity, consequently, the removal of short-chain PFAS (Behnami et al., 2024; Cerlanek et al., 2024), but the high cost of the processes involved has become a limiting factor as compared to the use of biochars (Cerlanek et al., 2024; Ramos et al., 2022).

The choice of biochar materials, that is, biochar properties is an important factor that influences the removal efficiency of PFAS. The results of the removal efficiency of varieties of PFAS using different feedstock materials in making the biochars are presented in Table 12.5. The feedstock materials for the biochars, such as rice straw, pine sawdust, and wood chips in composite form, showed high removal efficiency of PFAS above 90%, including both long-chain and short-chain PFAS with wood chips found to perform better for all the varieties of PFAS. Similarly, biochars made from maize straw with a magnetic porous Fe-doped graphitized modification showed improved removal efficiency (96.5% and 98.7%) of different chain lengths of PFAS from natural waters and wastewater (Liu et al., 2023).

Furthermore, the characteristics of PFAS have also played a significant role in the removal efficiency of the biochars. While biochar made from wood chips showed significant removal efficiency of PFBS, a short-chain type of PFAS synthesized from the laboratory (Deng et al., 2023), the biochars derived from the construction and demolition debris wood were found to be ineffective as an *in-situ* adsorbent for different PFAS from the landfill leachate (Cerlanek et al., 2024). It is worthy of note that the biochars made from the construction and demolition debris wood were not subjected to any treatment unlike the biochars made from wood chips that were composited with polymer, which could have enhanced the adsorption capability of the later and aided the excellent removal efficiency (Yu et al., 2023). Nevertheless, the major identified hindrances of biochars on PFAS sorption from the landfill leachate are the high matrix of organic matter and the low surface area of the biochars (Cerlanek et al., 2024).

The pyrolysis conditions have been identified as one of the major factors influencing the performance of biochars to effectively remove the PFAS from various environmental matrices (Behnami et al., 2024). Aboughaly and Fattah suggested that optimization of pyrolysis reaction conditions of biochars can adsorb PFAS up to the maximum allowable concentration of 70 ng/L in wastewater and biosolids (Aboughaly & Fattah, 2023). Meanwhile, among the various pyrolysis conditions for biochars production targeted at removing PFAS, the temperature has been found to play a critical role (Behnami et al., 2024). For instance, higher performance of biochars made from hardwood and subjected to higher pyrolysis temperature exhibited superior removal performance of various PFAS such as PFOA and PFBA compared to other biochars with lesser pyrolysis temperature (Inyang & Dickenson, 2017).

12.5 Existing regulations on the PFAS in the environmental matrices and use of biochar

Due to the ubiquitous nature of PFAS in the environment and the increasing concerns about their health risk, there have been concerted efforts to regulate PFAS in the environment. For instance, specific legacy PFAS, including perfluorohexane sulfonate, perfluorooctanoate, and PFOS, which are classified as POPs, has been listed for restriction in their use and subsequent elimination under the Stockholm Convention (List perfluorooctanoic acid PFOA, its salts and

TABLE 12.5 Studies of PFAS removal using biochars.

Feedstock	Biochar Type	Biochar Treatment	Initial PFAS Concentration (μg/L)	PFAS Type	Removal efficiency (%)	References
Pinewood	HWC and PWC biochars	None	0.005	PFBA PFOA	87 ± 0.04 (HWC) 18 ± 0.6 (PWC)	Liu et al. (2023)
Rice Straw	Rice straw biochar-alginate beads	Composite	100	PFOS (long-chain) PFBS (short chain)	PFOS (up to 99) PFBS (nearly 40)	Militao et al. (2023)
Pine sawdust	Polypyrrole -coated biochar (PPy/BC) composites	Composite	10000–800000	PFOA, PFOS, PFBA, and PFBS	>95	Yu et al. (2023)
Wood chips	biochar-polymer composite	Composite	2000–100000	PFBA, PFBS, PFOA, PFOS	PFOS (99.7), PFOA (98.3), PFBS (95.2), and PFBA (92.8)	Deng et al. (2023)
Reed Straw	Reed straw-derived biochar (RESCA-900)	None	1 100	PFOA, PFBA, PFBA, PFBS, PFHxA	PFOA 87 ± 6 PFBA (80 ± 2) PFBA (92–96), PFBS (80), PFHxA (89)	Liu and Sun (2021)
Maize Straw	Magnetic porous Fe-doped graphitized biochar (Fe-M−BC$_{900}$)	Modified	100	PFOA, PFBA, (PFHxA), PFOS	96.5~98.7	Liu et al. (2023)
Sawdust	Sawdust derived biochar (SDN)		248480	PFOS	-	Hassan et al. (2020)
Waste wood biochar	Waste wood from timber industry	Modified	1,000	PFOA, PFOS, PFBA, PFBS	PFOA (60), PFOS (83), PFBA (18), PFBS (30)	Zhang et al. (2021)
Spent coffee grounds	Spent coffee grounds (SCG) biochar	Activated	440	PFOS	9.9–99.6	Steigerwald and Ray (2021)
Construction and demolition debris	Reuse of biochar derived from construction and demolition debris (CDD) wood	None	-	Seventeen PFAS	Not significant	Cerlanek et al. (2024)

HWC, Hardwood; PFAS, per- and polyfluoroalkyl substances; PFBA, perfluorobutanoic acid; PFOA, perfluorooctanoic acid; PFOS, perfluorooctane sulfonate; PWC, pinewood.

PFOA-related compounds in Annex A to the Stockholm Convention on Persistent Organic Pollutants with specific exemptions, 2019). The European Commission has included perfluorooctanesulfonic acid to the list of hazardous substances to be prioritized for monitoring in the European Union water bodies with an allowable quantity of 0.65 ng/L as an Environmental Quality Standard for freshwater (Radley-Gardner et al., 2016). The United States Environmental Protection Agency (EPA) has regarded PFAS as harmful substances to human health and the environment, and thus proposed Provisional Health Advisories that range from 200 ng/L and 400 ng/L for PFOA and perfluorooctanesulfonic acid, respectively (Environmental Protection Agency. Provisional Health Advisories for Perfluorooctanoic acid PFOA and Perfluorooctane Sulfonate PFOS. EPA, 2009). There is also an existing regulation that restricts the members of state in the European Union from marketing and using PFOSs, which are considered dangerous substances (Directive 2006/122/EC of the European Parliament and of the Council of 12 December 2006 amending for the 30th time Council Directive 76/769/EEC on the approximation of the laws, regulations and administrative provisions of the Member States relating to,06/122/EC of the European Parliament and of the Council of 12 December 2006 amending for the 30th time Council Directive 76/769/EEC on the approximation of the laws, regulations and administrative provisions of the Member States relating to, 2006). In Australia, the Environmental Health Standing Committee of the Australian Health Protection Principal Committee sets interim national guideline benchmark values of 0.5 µg/L, 5 µg/L, and 0.15 µg/kg/d for the sum of PFOS and PFHxS in drinking water, recreational water, and tolerable daily intake, which are to serve as a guide for the confirmation of water quality supplies from the potentially contaminated areas with PFAS (enHealth Statement: Interim national guidance on human health reference values for per- and poly-fluoroalkyl substances for use in site investigations in Australia, enHealth, 2016).

Despite the tremendous efforts in the regulation, monitoring, and restriction on the use of PFAS, especially in developed countries, there is no proper implementation of a regulatory framework on the PFAS in the environmental matrices in many developing and low-income countries, including Asia and Africa. Even in developed countries, the disposal guidelines and instructions for hazardous materials from industrial and manufacturing activities are not thoroughly followed or are entirely lacking (Zhang et al., 2022). It is crucial to emphasize the urgent need to phase out hazardous PFAS through stringent regulations and implementation in order to avoid major environmental and health risks and fulfill the sustainable development goals (SDGs). It should also be noted that since the regulatory policies on the management of PFAS in the environment are governed by site-specific risk assessment, existing laws and regulations on the benchmarking of PFAS in air, soil, or water cannot be generalized due to the varying drivers of environmental fate and transport of PFAS in the environmental matrices such as off-site migration from the AFFF impacted sites (Reinikainen et al., 2022). Thus, the regulations on the PFAS in the environment should be site–specific to ensure their effective management.

Meanwhile, the use of modified biochar for environmental remediation, though suggested for emerging contaminants such as PFAS, requires adequate care to avoid environmental burdens associated with using biochar for environmental remediation. The modifications of biochar often alter its stability, especially during the physical or chemical activation processes (Rajapaksha et al., 2016). The physical properties such as porosity, density, mechanical strength, particle size distribution, and chemical properties, including organic and chemical characteristics, solid phases and their distribution, and surface chemistry, are to be carefully investigated and monitored, especially during the production process of the biochar, that is, pyrolysis, to ensure they meet the standard required for remediation of PFAS and other contaminants in environmental matrices (Kazemi Shariat Panahi et al., 2020). In general, the techno-economic analysis has been recommended for the biochar produced through modifications, either physical, chemical, or biological methods, to fully grasp the sustainability and potential of environmental and ecological risks of using such engineered biochar, especially in large-scale contaminants remediation like PFAS in the environmental matrices (Kazemi Shariat Panahi et al., 2020; Rajapaksha et al., 2016). Similarly, using biochar for the removal of contaminants, especially in household-scale treatment systems, requires technical details on the loading regimes and implementation modules, which further need to be properly understood (Dalahmeh et al., 2018).

12.6 Knowledge gaps, future research needs, and concluding remarks

While existing studies have continued to focus on the various PFAS in different environmental media, the transformation of these PFAS into new chemicals in the environment has remained less investigated. However, the transformed PFAS could become an alternate potential source in the environment, which could become a risk to the environment and human health. Also, the materials or chemicals that are often used to remediate the PFAS in the environment could cause a further transformation and generate other hazardous chemicals. Few studies have also raised this concern (Baduel et al., 2017; Nickerson et al., 2020), and there is a need to focus more on investigating the composition of various materials and chemicals that are being used to remediate PFAS as well as transformed products. Specifically, the available information on the composition of PFAS breakdown after removal by biochars is limited, which could potentially be transformed into other hazardous products. Besides, the uncertainty associated with the presence of unknown chemicals in the environment could portend a danger. For instance, Baduel et al. detected 42 PFAS in the soil around the fire–fighting training ground, among which eight perfluoroalkane sulphonamides homologues were newly discovered with unknown physico-chemical properties and environmental health effects, but the chemical structure of isomers and homologues indicated the possible transformation due to the use of electrochemical fluorination-based foams (Baduel et al., 2017).

Considering the various regulations on the restrictions on the use of certain PFAS chemicals, there is a likelihood of embarking on the use of other alternative chemicals with similar characteristics to the existing PFAS. Many such chemicals are currently in use, but they are not monitored in the environment to assess their ecological risks. Further studies should be centered on the emerging and novel PFAS chemicals that are often ignored to understand their fate and transport in the environment and, most notably, their environmental and health risks. Similarly, the influence of organic matter on the biochar surface area as it affects the adsorption process of PFAS needs further investigation. Also, the use of modified/engineered biochar, though encouraged, should be thoroughly ensured that the physicochemical properties after modification are appropriate for the targeted contaminants and environmental media.

In conclusion, the future prospects for PFAS involve a range of initiatives aimed at addressing their environmental impact and potential risks. Stricter regulations and phase-outs are being implemented worldwide, while research continues on remediation technologies, especially using a sustainable approach like biochar to clean up PFAS-contaminated sites and remove wastes in various environmental matrices. Advancements in analytical methods are being pursued to improve detection and monitoring capabilities. Understanding the health and environmental effects of PFAS and biochar, as well as promoting safer alternatives, are important areas of focus. International collaboration and knowledge sharing are crucial for developing standardized approaches and fostering a comprehensive understanding of using biochar to remediate PFAS in different environmental matrices.

References

Aboughaly, M., Fattah, I. M. R. (2023). Production of Biochar from Biomass Pyrolysis for Removal of PFAS from Wastewater and Biosolids: A Critical Review. https://doi.org/10.20944/preprints202304.0309.v1.

Adane, B., Siraj, K., & Meka, N. (2015). Kinetic, equilibrium and thermodynamic study of 2-chlorophenol adsorption onto Ricinus communis pericarp activated carbon from aqueous solutions. *Green Chemistry Letters and Reviews*, 8(3-4), 1–12. Available from https://doi.org/10.1080/17518253.2015.1065348, http://www.tandf.co.uk/journals/titles/17518253.asp.

Akhil, D., Lakshmi, D., Kartik, A., Vo, D. V. N., Arun, J., & Gopinath, K. P. (2021). Production, characterization, activation and environmental applications of engineered biochar: a review. *Environmental Chemistry Letters*, 19(3), 2261–2297. Available from https://doi.org/10.1007/s10311-020-01167-7, http://springerlink.metapress.com/app/home/journal.asp?wasp = d86tgdwvtg0yvw9gvkwp&referrer = parent&backto = browsepublicationsresults,140,541.

Amusat, S. O., Kebede, T. G., Dube, S., & Nindi, M. M. (2021). Ball-milling synthesis of biochar and biochar–based nanocomposites and prospects for removal of emerging contaminants: A review. *Journal of Water Process Engineering*, 41. Available from https://doi.org/10.1016/j.jwpe.2021.101993, http://www.journals.elsevier.com/journal-of-water-process-engineering/.

Anderson, R. H., Feild, J. B., Dieffenbach-Carle, H., Elsharnouby, O., & Krebs, R. K. (2022). Assessment of PFAS in collocated soil and porewater samples at an AFFF-impacted source zone: Field-scale validation of suction lysimeters. *Chemosphere*, 308. Available from https://doi.org/10.1016/j.chemosphere.2022.136247, http://www.elsevier.com/locate/chemosphere.

Baduel, C., Mueller, J. F., Rotander, A., Corfield, J., & Gomez-Ramos, M. J. (2017). Discovery of novel per- and polyfluoroalkyl substances (PFASs) at a fire fighting training ground and preliminary investigation of their fate and mobility. *Chemosphere*, 185, 1030–1038. Available from https://doi.org/10.1016/j.chemosphere.2017.06.096, http://www.elsevier.com/locate/chemosphere.

Behnami, A., Pourakbar, M., Ayyar, A. S. R., Lee, J. W., Gagnon, G., & Zoroufchi Benis, K. (2024). Treatment of aqueous per- and poly-fluoroalkyl substances: A review of biochar adsorbent preparation methods. *Chemosphere*, 357. Available from https://doi.org/10.1016/j.chemosphere.2024.142088, https://www.sciencedirect.com/science/journal/00456535.

Besis, A., Botsaropoulou, E., Samara, C., Katsoyiannis, A., Hanssen, L., & Huber, S. (2019). Perfluoroalkyl substances (PFASs) in air-conditioner filter dust of indoor microenvironments in Greece: Implications for exposure. *Ecotoxicology and Environmental Safety*, 183109559. Available from https://doi.org/10.1016/j.ecoenv.2019.109559.

Blake, B. E., & Fenton, S. E. (2020). Early life exposure to per- and polyfluoroalkyl substances (PFAS) and latent health outcomes: A review including the placenta as a target tissue and possible driver of peri- and postnatal effects. *Toxicology*, 443. Available from https://doi.org/10.1016/j.tox.2020.152565, http://www.elsevier.com/locate/toxicol.

Bouwer, E. J., & Zehnder, A. J. B. (1993). Bioremediation of organic compounds - putting microbial metabolism to work. *Trends in Biotechnology*, 11(8), 360–367. Available from https://doi.org/10.1016/0167-7799(93)90159-7.

Bräunig, J., Baduel, C., Heffernan, A., Rotander, A., Donaldson, E., & Mueller, J. F. (2017). Fate and redistribution of perfluoroalkyl acids through AFFF-impacted groundwater. *Science of the Total Environment, 596-597*, 360–368. Available from https://doi.org/10.1016/j.scitotenv.2017.04.095, http://www.elsevier.com/locate/scitotenv.

Breitmeyer, S. E., Williams, A. M., Duris, J. W., Eicholtz, L. W., Shull, D. R., Wertz, T. A., & Woodward, E. E. (2023). Per- and polyfluorinated alkyl substances (PFAS) in Pennsylvania surface waters: A statewide assessment, associated sources, and land-use relations. *Science of the Total Environment, 888*. Available from https://doi.org/10.1016/j.scitotenv.2023.164161, http://www.elsevier.com/locate/scitotenv.

Brunn, H., Arnold, G., Körner, W., Rippen, G., Steinhäuser, K. G., & Valentin, I. (2023). PFAS: forever chemicals—persistent, bioaccumulative and mobile. Reviewing the status and the need for their phase out and remediation of contaminated sites. *Environmental Sciences Europe, 35*(1). Available from https://doi.org/10.1186/s12302-023-00721-8, https://enveurope.springeropen.com.

Buck, R. C., Franklin, J., Berger, U., Conder, J. M., Cousins, I. T., Voogt, P. D., Jensen, A. A., Kannan, K., Mabury, S. A., & van Leeuwen, S. P. J. (2011). Perfluoroalkyl and polyfluoroalkyl substances in the environment: Terminology, classification, and origins. *Integrated Environmental Assessment and Management, 7*(4), 513–541. Available from https://doi.org/10.1002/ieam.258.

Buckley, T., Karanam, K., Han, H., Vo, H. N. P., Shukla, P., Firouzi, M., & Rudolph, V. (2023). Effect of different co-foaming agents on PFAS removal from the environment by foam fractionation. *Water Research, 230*. Available from https://doi.org/10.1016/j.watres.2022.119532, http://www.elsevier.com/locate/watres.

Castiglioni, S., Valsecchi, S., Polesello, S., Rusconi, M., Melis, M., Palmiotto, M., Manenti, A., Davoli, E., & Zuccato, E. (2015). Sources and fate of perfluorinated compounds in the aqueous environment and in drinking water of a highly urbanized and industrialized area in Italy. *Journal of Hazardous Materials, 282*, 51–60. Available from https://doi.org/10.1016/j.jhazmat.2014.06.007.

Cerlanek, A., Liu, Y., Robey, N., Timshina, A. S., Bowden, J. A., & Townsend, T. G. (2024). Assessing construction and demolition wood-derived biochar for in-situ per- and polyfluoroalkyl substance (PFAS) removal from landfill leachate. *Waste Management, 174*, 382–389. Available from https://doi.org/10.1016/j.wasman.2023.12.017, https://www.sciencedirect.com/science/journal/0956053X.

Chen, H., Gao, Y., Li, J., Fang, Z., Bolan, N., Bhatnagar, A., Gao, B., Hou, D., Wang, S., Song, H., Yang, X., Shaheen, S. M., Meng, J., Chen, W., Rinklebe, J., & Wang, H. (2022). Engineered biochar for environmental decontamination in aquatic and soil systems: a review. *Carbon Research, 1*(1). Available from https://doi.org/10.1007/s44246-022-00005-5, http://www.springer.com/journal/44246.

Chun, Y., Lee, S. K., Yoo, H. Y., & Kim, S. W. (2021). Recent Advancements in Biochar Production According to Feedstock Classification, Pyrolysis Conditions, and Applications: A Review. *BioResources, 16*(3), 6512–6547. Available from https://doi.org/10.15376/BIORES.16.3.CHUN, http://www.ncsu.edu/bioresources/Back_Issues.htm.

Cui, J., Gao, P., & Deng, Y. (2020). Destruction of per- and polyfluoroalkyl substances (PFAS) with advanced reduction processes (ARPs): A critical review. *Environmental Science & Technology, 54*(7), 3752–3766. Available from https://doi.org/10.1021/acs.est.9b05565.

Dalahmeh, S., Ahrens, L., Gros, M., Wiberg, K., & Pell, M. (2018). Potential of biochar filters for onsite sewage treatment: Adsorption and biological degradation of pharmaceuticals in laboratory filters with active, inactive and no biofilm. *Science of The Total Environment, 612*, 192–201. Available from https://doi.org/10.1016/j.scitotenv.2017.08.178.

Deng, J., Han, J., Hou, C., Zhang, Y., Fang, Y., Du, W. X., Li, M., Yuan, Y., Tang, C., & Hu, X. (2023). Efficient removal of per- and polyfluoroalkyl substances from biochar composites: Cyclic adsorption and spent regenerant degradation. *Chemosphere, 341*. Available from https://doi.org/10.1016/j.chemosphere.2023.140051, http://www.elsevier.com/locate/chemosphere.

Directive 2006/122/EC of the European Parliament and of the Council of 12 December 2006 amending for the 30th time Council Directive 76/769/EEC on the approximation of the laws, regulations and administrative provisions of the Member States relating to. European Commission. (2006).

Dobraca, D., Israel, L., McNeel, S., Voss, R., Wang, M., Gajek, R., Park, J. S., Harwani, S., Barley, F., She, J., & Das, R. (2015). Biomonitoring in California firefighters: Metals and perfluorinated chemicals. *Journal of Occupational and Environmental Medicine, 57*(1), 88–97. Available from https://doi.org/10.1097/JOM.0000000000000307, http://journals.lww.com/joem.

Domingo, J. L., & Nadal, M. (2019). Human exposure to per- and polyfluoroalkyl substances (PFAS) through drinking water: A review of the recent scientific literature. *Environmental Research, 177*108648. Available from https://doi.org/10.1016/j.envres.2019.108648.

enHealth Statement: Interim national guidance on human health reference values for per- and poly-fluoroalkyl substances for use in site investigations in Australia, enHealth. (2016).

Environmental Protection Agency. Provisional Health Advisories for Perfluorooctanoic acid (PFOA) and Perfluorooctane Sulfonate (PFOS). EPA. (2009).

Falciglia, P. P., Roccaro, P., Bonanno, L., De Guidi, G., Vagliasindi, F. G. A., & Romano, S. (2018). A review on the microwave heating as a sustainable technique for environmental remediation/detoxification applications. *Renewable and Sustainable Energy Reviews, 95*, 147–170. Available from https://doi.org/10.1016/j.rser.2018.07.031, https://www.journals.elsevier.com/renewable-and-sustainable-energy-reviews.

Fortunato, L., Al Fuhaid, L., Murgolo, S., De Ceglie, C., Mascolo, G., Falivene, L., Vrouwenvelder, J. S., Witkamp, G. J., & Farinha, A. (2023). Removal of polyfluoroalkyl substances (PFAS) from water using hydrophobic natural deep eutectic solvents (NADES): A proof of concept study. *Journal of Water Process Engineering, 56*. Available from https://doi.org/10.1016/j.jwpe.2023.104401, http://www.journals.elsevier.com/journal-of-water-process-engineering/.

Giesy, J. P., & Kannan, K. (2001). Global distribution of perfluorooctane sulfonate in wildlife. *Environmental Science and Technology, 35*(7), 1339–1342. Available from https://doi.org/10.1021/es001834k.

Glüge, J., Scheringer, M., Cousins, I. T., Dewitt, J. C., Goldenman, G., Herzke, D., Lohmann, R., Ng, C. A., Trier, X., & Wang, Z. (2020). An overview of the uses of per- And polyfluoroalkyl substances (PFAS). *Environmental Science: Processes and Impacts, 22*(12), 2345–2373. Available from https://doi.org/10.1039/d0em00291g, http://pubs.rsc.org/en/journals/journal/em.

Guo, R., Sim, W. J., Lee, E. S., Lee, J. H., & Oh, J. E. (2010). Evaluation of the fate of perfluoroalkyl compounds in wastewater treatment plants. *Water Research*, *44*(11), 3476–3486. Available from https://doi.org/10.1016/j.watres.2010.03.028, http://www.elsevier.com/locate/watres.

Hassan, M., Liu, Y., Naidu, R., Du, J., & Qi, F. (2020). Adsorption of Perfluorooctane sulfonate (PFOS) onto metal oxides modified biochar. *Environmental Technology & Innovation*, *19*100816. Available from https://doi.org/10.1016/j.eti.2020.100816.

Hepburn, E., Madden, C., Szabo, D., Coggan, T. L., Clarke, B., & Currell, M. (2019). Contamination of groundwater with per- and polyfluoroalkyl substances (PFAS) from legacy landfills in an urban re-development precinct. *Environmental Pollution*, *248*, 101–113. Available from https://doi.org/10.1016/j.envpol.2019.02.018, https://www.journals.elsevier.com/environmental-pollution.

Hubert, M., Meyn, T., Hansen, M. C., Hale, S. E., & Arp, H. P. H. (2024). Per- and polyfluoroalkyl substance (PFAS) removal from soil washing water by coagulation and flocculation. *Water Research*, *249*. Available from https://doi.org/10.1016/j.watres.2023.120888, http://www.elsevier.com/locate/watres.

Inyang, M., & Dickenson, E. R. V. (2017). The use of carbon adsorbents for the removal of perfluoroalkyl acids from potable reuse systems. *Chemosphere*, *184*, 168–175. Available from https://doi.org/10.1016/j.chemosphere.2017.05.161, http://www.elsevier.com/locate/chemosphere.

Kazemi Shariat Panahi, H., Dehhaghi, M., Ok, Y. S., Nizami, A. S., Khoshnevisan, B., Mussatto, S. I., Aghbashlo, M., Tabatabaei, M., & Lam, S. S. (2020). A comprehensive review of engineered biochar: Production, characteristics, and environmental applications. *Journal of Cleaner Production*, *270*. Available from https://doi.org/10.1016/j.jclepro.2020.122462, https://www.journals.elsevier.com/journal-of-cleaner-production.

Kim, S. K., & Kannan, K. (2007). Perfluorinated acids in air, rain, snow, surface runoff, and lakes: Relative importance of pathways to contamination of urban lakes. *Environmental Science and Technology*, *41*(24), 8328–8334. Available from https://doi.org/10.1021/es072107t.

Krahn, K. M., Cornelissen, G., Castro, G., Arp, H. P. H., Asimakopoulos, A. G., Wolf, R., Holmstad, R., Zimmerman, A. R., & Sørmo, E. (2023). Sewage sludge biochars as effective PFAS-sorbents. *Journal of Hazardous Materials*, *445*. Available from https://doi.org/10.1016/j.jhazmat.2022.130449, http://www.elsevier.com/locate/jhazmat.

Krebsbach, S., He, J., Adhikari, S., Olshansky, Y., Feyzbar, F., Davis, L. C., Oh, T. S., & Wang, D. (2023). Mechanistic understanding of perfluorooctane sulfonate (PFOS) sorption by biochars. *Chemosphere*, *330*. Available from https://doi.org/10.1016/j.chemosphere.2023.138661, http://www.elsevier.com/locate/chemosphere.

Kunacheva, C., Tanaka, S., Fujii, S., Boontanon, S. K., Musirat, C., Wongwattana, T., & Shivakoti, B. R. (2011). Mass flows of perfluorinated compounds (PFCs) in central wastewater treatment plants of industrial zones in Thailand. *Chemosphere*, *83*(6), 737–744. Available from https://doi.org/10.1016/j.chemosphere.2011.02.059, http://www.elsevier.com/locate/chemosphere.

Kundu, S., Patel, S., Halder, P., Patel, T., Hedayati Marzbali, M., Pramanik, B. K., Paz-Ferreiro, J., De Figueiredo, C. C., Bergmann, D., Surapaneni, A., Megharaj, M., & Shah, K. (2021). Removal of PFASs from biosolids using a semi-pilot scale pyrolysis reactor and the application of biosolids derived biochar for the removal of PFASs from contaminated water. *Environmental Science: Water Research and Technology*, *7*(3), 638–649. Available from https://doi.org/10.1039/d0ew00763c, http://pubs.rsc.org/en/journals/journal/ew.

Kwon, H. O., Kim, H. Y., Park, Y. M., Seok, K. S., Oh, J. E., & Choi, S. D. (2017). Updated national emission of perfluoroalkyl substances (PFASs) from wastewater treatment plants in South Korea. *Environmental Pollution*, *220*, 298–306. Available from https://doi.org/10.1016/j.envpol.2016.09.063, https://www.journals.elsevier.com/environmental-pollution.

Lang, J. R., Allred, B. M. K., Field, J. A., Levis, J. W., & Barlaz, M. A. (2017). National Estimate of Per- and Polyfluoroalkyl Substance (PFAS) Release to U.S. Municipal Landfill Leachate. *Environmental Science and Technology*, *51*(4), 2197–2205. Available from https://doi.org/10.1021/acs.est.6b05005, http://pubs.acs.org/journal/esthag.

van Larebeke, N., Koppen, G., Decraemer, S., Colles, A., Bruckers, L., Den Hond, E., Govarts, E., Morrens, B., Schettgen, T., Remy, S., Coertjens, D., Nawrot, T., Nelen, V., Baeyens, W., & Schoeters, G. (2022). Per- and polyfluoroalkyl substances (PFAS) and neurobehavioral function and cognition in adolescents (2010–2011) and elderly people (2014): results from the Flanders Environment and Health Studies (FLEHS). *Environmental Sciences Europe*, *34*(1). Available from https://doi.org/10.1186/s12302-022-00675-3.

Lashuk, B., Pineda, M., AbuBakr, S., Boffito, D., & Yargeau, V. (2022). Application of photocatalytic ozonation with a WO3/TiO2 catalyst for PFAS removal under UVA/visible light. *Science of The Total Environment*, *843*157006. Available from https://doi.org/10.1016/j.scitotenv.2022.157006.

Lassalle, J., Gao, R., Rodi, R., Kowald, C., Feng, M., Sharma, V. K., Hoelen, T., Bireta, P., Houtz, E. F., Staack, D., & Pillai, S. D. (2021). Degradation of PFOS and PFOA in soil and groundwater samples by high dose Electron Beam Technology. *Radiation Physics and Chemistry*, *189*. Available from https://doi.org/10.1016/j.radphyschem.2021.109705, http://www.elsevier.com/locate/radphyschem.

Lee, Y. M., Lee, J. Y., Kim, M. K., Yang, H., Lee, J. E., Son, Y., Kho, Y., Choi, K., & Zoh, K. D. (2020). Concentration and distribution of per- and polyfluoroalkyl substances (PFAS) in the Asan Lake area of South Korea. *Journal of Hazardous Materials*, *381*. Available from https://doi.org/10.1016/j.jhazmat.2019.120909, http://www.elsevier.com/locate/jhazmat.

Leung, S. C. E., Wanninayake, D., Chen, D., Nguyen, N. T., & Li, Q. (2023). Physicochemical properties and interactions of perfluoroalkyl substances (PFAS) - Challenges and opportunities in sensing and remediation. *Science of the Total Environment*, *905*. Available from https://doi.org/10.1016/j.scitotenv.2023.166764, http://www.elsevier.com/locate/scitotenv.

List perfluorooctanoic acid (PFOA), its salts and PFOA-related compounds in Annex A to the Stockholm Convention on Persistent Organic Pollutants with specific exemptions. Unep. 9 (2019), 2–4.

Liu, W. J., Jiang, H., & Yu, H. Q. (2015). Development of biochar-based functional materials: Toward a sustainable platform carbon material. *Chemical Reviews*, *115*(22), 12251–12285. Available from https://doi.org/10.1021/acs.chemrev.5b00195, http://pubs.acs.org/journal/chreay.

Liu, Y., Shao, L. X., Yu, W. J., Bao, J., Li, T. Y., Hu, X. M., & Zhao, X. (2022). Simultaneous removal of multiple PFAS from contaminated groundwater around a fluorochemical facility by the periodically reversing electrocoagulation technique. *Chemosphere*, *307*. Available from https://doi.org/10.1016/j.chemosphere.2022.135874, http://www.elsevier.com/locate/chemosphere.

Liu, Y.-L., & Sun, M. (2021). Ion exchange removal and resin regeneration to treat per- and polyfluoroalkyl ether acids and other emerging PFAS in drinking water. *Water Research, 207*117781. Available from https://doi.org/10.1016/j.watres.2021.117781.

Liu, Z., Zhang, P., Wei, Z., Xiao, F., Liu, S., Guo, H., Qu, C., Xiong, J., Sun, H., & Tan, W. (2023). Porous Fe-doped graphitized biochar: An innovative approach for co-removing per-/polyfluoroalkyl substances with different chain lengths from natural waters and wastewater. *Chemical Engineering Journal, 476*146888. Available from https://doi.org/10.1016/j.cej.2023.146888.

Lu, D., Sha, S., Luo, J., Huang, Z., & Zhang Jackie, X. (2020). Treatment train approaches for the remediation of per- and polyfluoroalkyl substances (PFAS): A critical review. *Journal of Hazardous Materials, 386*. Available from https://doi.org/10.1016/j.jhazmat.2019.121963, http://www.elsevier.com/locate/jhazmat.

Luo, L., Xu, C., Chen, Z., & Zhang, S. (2015). Properties of biomass-derived biochars: Combined effects of operating conditions and biomass types. *Bioresource Technology, 192*, 83–89. Available from https://doi.org/10.1016/j.biortech.2015.05.054.

Lyu, H., Gao, B., He, F., Zimmerman, A. R., Ding, C., Huang, H., & Tang, J. (2018). Effects of ball milling on the physicochemical and sorptive properties of biochar: Experimental observations and governing mechanisms. *Environmental Pollution, 233*, 54–63. Available from https://doi.org/10.1016/j.envpol.2017.10.037, http://www.elsevier.com/inca/publications/store/4/0/5/8/5/6.

Lyu, X., Xiao, F., Shen, C., Chen, J., Park, C. M., Sun, Y., Flury, M., & Wang, D. (2022). Per- and Polyfluoroalkyl Substances (PFAS) in Subsurface Environments: Occurrence, fate, transport, and research prospect. *Reviews of Geophysics, 60*(3). Available from https://doi.org/10.1029/2021RG000765, http://agupubs.onlinelibrary.wiley.com/agu/journal/10.1002/(ISSN)1944-9208/.

Ma, D., Zhong, H., Lv, J., Wang, Y., & Jiang, G. (2021). Levels, distributions, and sources of legacy and novel per- and perfluoroalkyl substances (PFAS) in the topsoil of Tianjin, China. *Journal of Environmental Sciences (China), 112*, 71–81. Available from https://doi.org/10.1016/j.jes.2021.04.029, http://www.journals.elsevier.com/journal-of-environmental-sciences/.

Ma, Z., Yang, Y., Wu, Y., Xu, J., Peng, H., Liu, X., Zhang, W., & Wang, S. (2019). In-depth comparison of the physicochemical characteristics of biochar derived from biomass pseudo components: Hemicellulose, cellulose, and lignin. *Journal of Analytical and Applied Pyrolysis, 140*, 195–204. Available from https://doi.org/10.1016/j.jaap.2019.03.015.

Maimaiti, A., Deng, S., Meng, P., Wang, W., Wang, B., Huang, J., Wang, Y., & Yu, G. (2018). Competitive adsorption of perfluoroalkyl substances on anion exchange resins in simulated AFFF-impacted groundwater. *Chemical Engineering Journal, 348*, 494–502. Available from https://doi.org/10.1016/j.cej.2018.05.006, http://www.elsevier.com/inca/publications/store/6/0/1/2/7/3/index.htt.

Marwani, H. M., Albishri, H. M., Jalal, T. A., & Soliman, E. M. (2017). Study of isotherm and kinetic models of lanthanum adsorption on activated carbon loaded with recently synthesized Schiff's base. *Arabian Journal of Chemistry, 10*, S1032. Available from https://doi.org/10.1016/j.arabjc.2013.01.008, http://colleges.ksu.edu.sa/Arabic%20Colleges/CollegeOfScience/ChemicalDept/AJC/default.aspx, http://www.sciencedirect.com/science/journal/18785352.

McCarthy, C. C., Kappleman, W., & DiGuiseppi, W. (2017). Ecological considerations of Per- and Polyfluoroalkyl Substances (PFAS). *Current Pollution Reports, 3*(4), 289–301. Available from https://doi.org/10.1007/s40726-017-0070-8, http://springer.com/environment/pollution+and+remediation/journal/40726.

McCleaf, P. P., Stefansson, W., & Ahrens, L. (2023). Drinking water nanofiltration with concentrate foam fractionation—A novel approach for removal of per- and polyfluoroalkyl substances (PFAS). *Water Research, 232*119688. Available from https://doi.org/10.1016/j.watres.2023.119688.

Meng, L. L., Lu, Y., Wang, Y., Ma, X., Li, J., Lv, J., Wang, Y., & Jiang, G. (2022). Occurrence, temporal variation (2010–2018), distribution, and source appointment of per- and polyfluoroalkyl substances (PFAS) in Mollusks from the Bohai Sea, China. *ACS ES&T Water, 2*(1), 195–205. Available from https://doi.org/10.1021/acsestwater.1c00346.

Militao, I. I. M., Roddick, F., Fan, L., Zepeda, L. C., Parthasarathy, R., & Bergamasco, R. (2023). PFAS removal from water by adsorption with alginate-encapsulated plant albumin and rice straw-derived biochar. *Journal of Water Process Engineering, 53*. Available from https://doi.org/10.1016/j.jwpe.2023.103616, http://www.journals.elsevier.com/journal-of-water-process-engineering/.

Moghadasi, R. R., Mumberg, T., & Wanner, P. (2023). Spatial prediction of concentrations of per- and polyfluoroalkyl substances (PFAS) in European Soils. *Environmental Science & Technology Letters, 10*(11), 1125–1129. Available from https://doi.org/10.1021/acs.estlett.3c00633.

Mohan, D. D., Sarswat, A., Ok, Y. S., & Pittman, C. U. (2014). Organic and inorganic contaminants removal from water with biochar, a renewable, low cost and sustainable adsorbent - A critical review. *Bioresource Technology, 160*, 191–202. Available from https://doi.org/10.1016/j.biortech.2014.01.120, http://www.elsevier.com/locate/biortech.

Moody, C. C. A., & Field, J. A. (2000). Perfluorinated surfactants and the environmental implications of their use in fire-fighting foams. *Environmental Science and Technology, 34*(18), 3864–3870. Available from https://doi.org/10.1021/es991359u.

Moody, C. C. A., Martin, J. W., Kwan, W. C., Muir, D. C. G., & Mabury, S. A. (2002). Monitoring perfluorinated surfactants in biota and surface water samples following an accidental release of fire-fighting foam into Etobicoke Creek. *Environmental Science and Technology, 36*(4), 545–551. Available from https://doi.org/10.1021/es011001+.

Mulabagal, V. V., Baah, D. A., Egiebor, N. O., Sajjadi, B., Chen, W.-Y., Viticoski, R. L., & Hayworth, J. S. (2021). Biochar from Biomass: A Strategy for Carbon Dioxide Sequestration, Soil Amendment, Power Generation, CO2 Utilization, and Removal of Perfluoroalkyl and Polyfluoroalkyl Substances (PFAS) in the Environment. *Springer Science and Business Media LLC, 2*, 1–64. Available from https://doi.org/10.1007/978-1-4614-6431-0_80-2.

Müller, V. V., Kindness, A., & Feldmann, J. (2023). Fluorine mass balance analysis of PFAS in communal waters at a wastewater plant from Austria. *Water Research, 244*120501. Available from https://doi.org/10.1016/j.watres.2023.120501.

Murakami, M. M., & Takada, H. (2008). Perfluorinated surfactants (PFSs) in size-fractionated street dust in Tokyo. *Chemosphere, 73*(8), 1172–1177. Available from https://doi.org/10.1016/j.chemosphere.2008.07.063.

Nakazawa, Y. Y., Kosaka, K., Yoshida, N., Asami, M., & Matsui, Y. (2023). Long-term removal of perfluoroalkyl substances via activated carbon process for general advanced treatment purposes. *Water Research, 245*120559. Available from https://doi.org/10.1016/j.watres.2023.120559.

Nickerson, A. A., Maizel, A. C., Kulkarni, P. R., Adamson, D. T., Kornuc, J. J., & Higgins, C. P. (2020). Enhanced extraction of AFFF-associated PFASs from source zone soils. *Environmental science & technology, 54*(8), 4952−4962. Available from https://doi.org/10.1021/acs.est.0c00792.

Novak, P. P. A., Hoeksema, S. D., Thompson, S. N., & Trayler, K. M. (2023). Per- and polyfluoroalkyl substances (PFAS) contamination in a microtidal urban estuary: Sources and sinks. *Marine Pollution Bulletin, 193*115215. Available from https://doi.org/10.1016/j.marpolbul.2023.115215.

Nzeribe, B. B. N., Crimi, M., Mededovic Thagard, S., & Holsen, T. M. (2019). Physico-chemical processes for the treatment of per- and polyfluoroalkyl substances (PFAS): A review. *Critical Reviews in Environmental Science and Technology, 49*(10), 866−915. Available from https://doi.org/10.1080/10643389.2018.1542916, http://www.tandf.co.uk/journals/titles/10643389.asp.

Omo-Okoro, P. P. N., Curtis, C. J., Karásková, P., Melymuk, L., Oyewo, O. A., & Okonkwo, J. O. (2020). Kinetics, isotherm, and thermodynamic studies of the adsorption mechanism of pfos and pfoa using inactivated and chemically activated maize tassel. *Water, Air, and Soil Pollution, 231*(9). Available from https://doi.org/10.1007/s11270-020-04852-z, http://www.kluweronline.com/issn/0049-6979/.

Organization for Economic Co-operation and Development 2018 2023 6 5 Paris, France Unpublished content Toward a new comprehensive global database of per- and polyfluoroalkyl substances (PFASs): Summary report on updating the OECD 2007 list of per- and polyfluoroalkyl substances (PFASs). https://hero.epa.gov/hero/index.cfm/reference/details/reference_id/5099062

Pang, H. H., Dorian, B., Gao, L., Xie, Z., Cran, M., Muthukumaran, S., Sidiroglou, F., Gray, S., & Zhang, J. (2022). Remediation of poly-and perfluoroalkyl substances (PFAS) contaminated soil using gas fractionation enhanced technology. *Science of The Total Environment, 827*154310. Available from https://doi.org/10.1016/j.scitotenv.2022.154310.

Parvin, S. S., Hara-Yamamura, H., Kanai, Y., Yamasaki, A., Adachi, T., Sorn, S., Honda, R., & Yamamura, H. (2023). Important properties of anion exchange resins for efficient removal of PFOS and PFOA from groundwater. *Chemosphere, 341*139983. Available from https://doi.org/10.1016/j.chemosphere.2023.139983.

Pierpaoli, M. M., Szopińska, M., Wilk, B. K., Sobaszek, M., Łuczkiewicz, A., Bogdanowicz, R., & Fudala-Książek, S. (2021). Electrochemical oxidation of PFOA and PFOS in landfill leachates at low and highly boron-doped diamond electrodes. *Journal of Hazardous Materials, 403*. Available from https://doi.org/10.1016/j.jhazmat.2020.123606, http://www.elsevier.com/locate/jhazmat.

Prevedouros, K. K., Cousins, I. T., Buck, R. C., & Korzeniowski, S. H. (2006). Sources, fate and transport of perfluorocarboxylates. *Environmental Science and Technology, 40*(1), 32−44. Available from https://doi.org/10.1021/es0512475.

Radley-Gardner, O. O., Beale, H., & Zimmermann, R. (2016). *Directive 2013/11/EU of the European Parliament and of the Council Fundamental Texts On European Private Law* (pp. 666−695). Hart Publishing Ltd. Available from 10.5040/9781782258674.0032.

Rahman, M. M. F., Peldszus, S., & Anderson, W. B. (2014). Behaviour and fate of perfluoroalkyl and polyfluoroalkyl substances (PFASs) in drinking water treatment: A review. *Water Research, 50*, 318−340. Available from https://doi.org/10.1016/j.watres.2013.10.045, http://elsevier.com/locate/watres.

Rajapaksha, A. A. U., Chen, S. S., Tsang, D. C. W., Zhang, M., Vithanage, M., Mandal, S., Gao, B., Bolan, N. S., & Ok, Y. S. (2016). Engineered/designer biochar for contaminant removal/immobilization from soil and water: Potential and implication of biochar modification. *Chemosphere, 148*, 276−291. Available from https://doi.org/10.1016/j.chemosphere.2016.01.043, http://www.elsevier.com/locate/chemosphere.

Ramos, P. P., Singh Kalra, S., Johnson, N. W., Khor, C. M., Borthakur, A., Cranmer, B., Dooley, G., Mohanty, S. K., Jassby, D., Blotevogel, J., & Mahendra, S. (2022). Enhanced removal of per- and polyfluoroalkyl substances in complex matrices by polyDADMAC-coated regenerable granular activated carbon. *Environmental Pollution, 294*. Available from https://doi.org/10.1016/j.envpol.2021.118603, https://www.journals.elsevier.com/environmental-pollution.

Rangabhashiyam, S., & Balasubramanian, P. (2019). The potential of lignocellulosic biomass precursors for biochar production: Performance, mechanism and wastewater application—A review. *Industrial Crops and Products, 128*, 405−423. Available from https://doi.org/10.1016/j.indcrop.2018.11.041.

Reinikainen, J. J., Perkola, N., Äystö, L., & Sorvari, J. (2022). The occurrence, distribution, and risks of PFAS at AFFF-impacted sites in Finland. *Science of The Total Environment, 829*154237. Available from https://doi.org/10.1016/j.scitotenv.2022.154237.

Robey, N. N. M., Da Silva, B. F., Annable, M. D., Townsend, T. G., & Bowden, J. A. (2020). Concentrating per- and polyfluoroalkyl substances (PFAS) in municipal solid waste landfill leachate using foam separation. *Environmental Science and Technology, 54*(19), 12550−12559. Available from https://doi.org/10.1021/acs.est.0c01266, http://pubs.acs.org/journal/esthag.

Röhler, K. K., Haluska, A. A., Susset, B., Liu, B., & Grathwohl, P. (2021). Long-term behavior of PFAS in contaminated agricultural soils in Germany. *Journal of Contaminant Hydrology, 241*. Available from https://doi.org/10.1016/j.jconhyd.2021.103812, http://www.elsevier.com/locate/jconhyd.

Sammut, G. G., Sinagra, E., Helmus, R., & de Voogt, P. (2017). Perfluoroalkyl substances in the Maltese environment − (I) surface water and rainwater. *Science of The Total Environment, 589*, 182−190. Available from https://doi.org/10.1016/j.scitotenv.2017.02.128.

Shahsavari, E. E., Rouch, D., Khudur, L. S., Thomas, D., Aburto-Medina, A., & Ball, A. S. (2021). Challenges and current status of the biological treatment of PFAS-contaminated soils. *Frontiers in Bioengineering and Biotechnology, 8*. Available from https://doi.org/10.3389/fbioe.2020.602040, http://journal.frontiersin.org/journal/bioengineering-and-biotechnology#archive.

Shan, G. G., Wei, M., Zhu, L., Liu, Z., & Zhang, Y. (2014). Concentration profiles and spatial distribution of perfluoroalkyl substances in an industrial center with condensed fluorochemical facilities. *Science of The Total Environment, 490*, 351−359. Available from https://doi.org/10.1016/j.scitotenv.2014.05.005.

Sharifan, H. H., Bagheri, M., Wang, D., Burken, J. G., Higgins, C. P., Liang, Y., Liu, J., Schaefer, C. E., & Blotevogel, J. (2021). Fate and transport of per- and polyfluoroalkyl substances (PFASs) in the vadose zone. *Science of the Total Environment, 771*. Available from https://doi.org/10.1016/j.scitotenv.2021.145427, http://www.elsevier.com/locate/scitotenv.

da Silva Bruckmann, F. F., Jemli, S., Ben Amara, F., Adelodun, B., Oliveira Silva, L. F., Bejar, S., Rizwan Khan, M., Ahmad, N., dos Reis, G. S., & Dotto, G. L. (2024). Adsorption of perfluorosulfonic acids (PFSAs) on an ultrafine potato peel waste grafted β-cyclodextrin (UFPPW-β-CD). *Separation and Purification Technology, 350*. Available from https://doi.org/10.1016/j.seppur.2024.127972, https://www.sciencedirect.com/science/journal/13835866.

Steenland, K. K., Fletcher, T., & Savitz, D. A. (2010). Epidemiologic evidence on the health effects of perfluorooctanoic acid (PFOA). *Environmental Health Perspectives, 118*(8), 1100–1108. Available from https://doi.org/10.1289/ehp.0901827, http://ehp03.niehs.nih.gov/article/fetchArticle.action?articleURI = info%3Adoi%2F10.1289%2Fehp.0901827.

Steigerwald, J. J. M., & Ray, J. R. (2021). Adsorption behavior of perfluorooctanesulfonate (PFOS) onto activated spent coffee grounds biochar in synthetic wastewater effluent. *Journal of Hazardous Materials Letters, 2*. Available from https://doi.org/10.1016/j.hazl.2021.100025, https://www.journals.elsevier.com/journal-of-hazardous-materials-letters.

Stoiber, T. T., Evans, S., & Naidenko, O. V. (2020). Disposal of products and materials containing per- and polyfluoroalkyl substances (PFAS): A cyclical problem. *Chemosphere, 260*127659. Available from https://doi.org/10.1016/j.chemosphere.2020.127659.

Stroski, K. K. M., Luong, K. H., Challis, J. K., Chaves-Barquero, L. G., Hanson, M. L., & Wong, C. S. (2020). Wastewater sources of per- and polyfluorinated alkyl substances (PFAS) and pharmaceuticals in four Canadian Arctic communities. *Science of the Total Environment, 708*. Available from https://doi.org/10.1016/j.scitotenv.2019.134494, http://www.elsevier.com/locate/scitotenv.

Su, H. H., Lu, Y., Wang, P., Shi, Y., Li, Q., Zhou, Y., & Johnson, A. C. (2016). Perfluoroalkyl acids (PFAAs) in indoor and outdoor dusts around a mega fluorochemical industrial park in China: Implications for human exposure. *Environment International, 94*, 667–673. Available from https://doi.org/10.1016/j.envint.2016.07.002, http://www.elsevier.com/locate/envint.

Suliman, W. W., Harsh, J. B., Abu-Lail, N. I., Fortuna, A. M., Dallmeyer, I., & Garcia-Perez, M. (2016). Influence of feedstock source and pyrolysis temperature on biochar bulk and surface properties. *Biomass and Bioenergy, 84*, 37–48. Available from https://doi.org/10.1016/j.biombioe.2015.11.010, http://www.journals.elsevier.com/biomass-and-bioenergy/.

Sun, T. T., Wang, F., Xie, Y., Liu, X., Yu, H., Lv, M., Zhang, Y., & Xu, Y. (2023). Biochar remediation of PFOA contaminated soil decreased the microbial network complexity. *Journal of Environmental Chemical Engineering, 11*(1)109239. Available from https://doi.org/10.1016/j.jece.2022.109239.

Sunderland, E. E. M., Hu, X. C., Dassuncao, C., Tokranov, A. K., Wagner, C. C., & Allen, J. G. (2019). A review of the pathways of human exposure to poly- and perfluoroalkyl substances (PFASs) and present understanding of health effects. *Journal of Exposure Science and Environmental Epidemiology, 29*(2), 131–147. Available from https://doi.org/10.1038/s41370-018-0094-1, http://www.nature.com/jes/index.html.

Taniyasu, S. S., Yamashita, N., Moon, H. B., Kwok, K. Y., Lam, P. K. S., Horii, Y., Petrick, G., & Kannan, K. (2013). Does wet precipitation represent local and regional atmospheric transportation by perfluorinated alkyl substances? *Environment International, 55*, 25–32. Available from https://doi.org/10.1016/j.envint.2013.02.005, http://www.elsevier.com/locate/envint.

Tomczyk, A. A., Sokołowska, Z., & Boguta, P. (2020). Biochar physicochemical properties: pyrolysis temperature and feedstock kind effects. *Reviews in Environmental Science and Bio/Technology, 19*(1), 191–215. Available from https://doi.org/10.1007/s11157-020-09523-3.

Tripathi, M. M., Sahu, J. N., & Ganesan, P. (2016). Effect of process parameters on production of biochar from biomass waste through pyrolysis: A review. *Renewable and Sustainable Energy Reviews, 55*, 467–481. Available from https://doi.org/10.1016/j.rser.2015.10.122.

UUSEPA, 2023. PFAS Chemical Lists and Tiered Testing Methods Descriptions.

Wang, J. J., & Wang, S. (2019). Preparation, modification and environmental application of biochar: A review. *Journal of Cleaner Production, 227*, 1002–1022. Available from https://doi.org/10.1016/j.jclepro.2019.04.282.

Wang, Y. Y., Munir, U., & Huang, Q. (2023). Occurrence of per- and polyfluoroalkyl substances (PFAS) in soil: Sources, fate, and remediation. *Soil & Environmental Health, 1*(1)100004. Available from https://doi.org/10.1016/j.seh.2023.100004.

Wang, Z. Z., Xie, Z., Möller, A., Mi, W., Wolschke, H., & Ebinghaus, R. (2015). Estimating dry deposition and gas/particle partition coefficients of neutral poly-/perfluoroalkyl substances in northern German coast. *Environmental Pollution, 202*, 120–125. Available from https://doi.org/10.1016/j.envpol.2015.03.029.

Weber, K. K., & Quicker, P. (2018). Properties of biochar. *Fuel, 217*, 240–261. Available from https://doi.org/10.1016/j.fuel.2017.12.054.

Wei, J. J., Tu, C., Yuan, G., Liu, Y., Bi, D., Xiao, L., Lu, J., Theng, B. K. G., Wang, H., Zhang, L., & Zhang, X. (2019). Assessing the effect of pyrolysis temperature on the molecular properties and copper sorption capacity of a halophyte biochar. *Environmental Pollution, 251*, 56–65. Available from https://doi.org/10.1016/j.envpol.2019.04.128, https://www.journals.elsevier.com/environmental-pollution.

Wu, Y. Y., Romanak, K., Bruton, T., Blum, A., & Venier, M. (2020). Per- and polyfluoroalkyl substances in paired dust and carpets from childcare centers. *Chemosphere, 251*126771. Available from https://doi.org/10.1016/j.chemosphere.2020.126771.

Wu, Y. Y., Qi, L., & Chen, G. (2022). A mechanical investigation of perfluorooctane acid adsorption by engineered biochar. *Journal of Cleaner Production, 340*130742. Available from https://doi.org/10.1016/j.jclepro.2022.130742.

Xia, C. C., Lim, X., Yang, H., Goodson, B. M., & Liu, J. (2022). Degradation of per- and polyfluoroalkyl substances (PFAS) in wastewater effluents by photocatalysis for water reuse. *Journal of Water Process Engineering, 46*102556. Available from https://doi.org/10.1016/j.jwpe.2021.102556.

Xiao, F. F., Zhang, X., Penn, L., Gulliver, J. S., & Simcik, M. F. (2011). Effects of monovalent cations on the competitive adsorption of perfluoroalkyl acids by kaolinite: Experimental studies and modeling. *Environmental Science and Technology, 45*(23), 10028–10035. Available from https://doi.org/10.1021/es202524y.

Yaashikaa, P. P. R., Senthil Kumar, P., Varjani, S., & Saravanan, A. (2020). A critical review on the biochar production techniques, characterization, stability and applications for circular bioeconomy. *Biotechnology Reports, 28*e00570. Available from https://doi.org/10.1016/j.btre.2020.e00570.

Yu, H. H., Chen, H., Zhang, P., Yao, Y., Zhao, L., Zhu, L., & Sun, H. (2023). In situ self-sacrificial synthesis of polypyrrole/biochar composites for efficiently removing short- and long-chain perfluoroalkyl acid from contaminated water. *Journal of Environmental Management, 344*118745. Available from https://doi.org/10.1016/j.jenvman.2023.118745.

Zhang, J. J., Pang, H., Gray, S., Ma, S., Xie, Z., & Gao, L. (2021). PFAS removal from wastewater by in-situ formed ferric nanoparticles: Solid phase loading and removal efficiency. *Journal of Environmental Chemical Engineering, 9*(4)105452. Available from https://doi.org/10.1016/j.jece.2021.105452.

Zhang, Y. Y., Tan, X., Lu, R., Tang, Y., Qie, H., Huang, Z., Zhao, J., Cui, J., Yang, W., & Lin, A. (2023). Enhanced removal of polyfluoroalkyl substances by simple modified biochar: adsorption performance and theoretical calculation. *ACS ES and T Water, 3*(3), 817–826. Available from https://doi.org/10.1021/acsestwater.2c00597, https://pubs.acs.org/page/aewcaa/about.html.

Zhang, Y. Y., Wang, B., Ma, S., & Zhang, Q. (2024). Adsorption of per- and polyfluoroalkyl substances (PFAS) from water with porous organic polymers. *Chemosphere, 346*140600. Available from https://doi.org/10.1016/j.chemosphere.2023.140600.

Zhang, Z. Z., Sarkar, D., Biswas, J. K., & Datta, R. (2022). Biodegradation of per- and polyfluoroalkyl substances (PFAS): A review. *Bioresource Technology, 344*126223. Available from https://doi.org/10.1016/j.biortech.2021.126223.

Zhao, D. D., Cheng, J., Vecitis, C. D., & Hoffmann, M. R. (2011). Sorption of perfluorochemicals to granular activated carbon in the presence of ultrasound. *Journal of Physical Chemistry A, 115*(11), 2250–2257. Available from https://doi.org/10.1021/jp111784k.

Zheng, G. G., Boor, B. E., Schreder, E., & Salamova, A. (2020). Indoor exposure to per- and polyfluoroalkyl substances (PFAS) in the childcare environment. *Environmental Pollution, 258*113714. Available from https://doi.org/10.1016/j.envpol.2019.113714.

Zheng, G. G., Schreder, E., Dempsey, J. C., Uding, N., Chu, V., Andres, G., Sathyanarayana, S., & Salamova, A. (2021). Per- and polyfluoroalkyl substances (PFAS) in breast milk- and trends for current-use PFAS. *Environmental Science and Technology, 55*(11), 7510–7520. Available from https://doi.org/10.1021/acs.est.0c06978, http://pubs.acs.org/journal/esthag.

Zhu, Q. Q., Qian, J., Huang, S., Li, Q., Guo, L., Zeng, J., Zhang, W., Cao, X., & Yang, J. (2022). Occurrence, distribution, and input pathways of per- and polyfluoroalkyl substances in soils near different sources in Shanghai. *Environmental Pollution, 308*119620. Available from https://doi.org/10.1016/j.envpol.2022.119620.

Chapter 13

Application of biochar for the treatment of urban stormwater: processes and future directions

Ahmed Abdelhafez[1,2], Mohamed Abbas[3], Shawky Metwally[4], Ahmed Al-Hossainy[5], Sedky Hassan[6,7], Hassan Abbas[8] and Abdel Aziz Tantawy[8]

[1]*Department of Soils and Water, Faculty of Agriculture, New Valley University, New Valley, Egypt,* [2]*National Committee of Soil Science, Academy of Scientific Research and Technology, Egypt, Giza, Egypt,* [3]*Department of Soils and Water Science, Faculty of Agriculture, Benha University, Benha, Egypt,* [4]*Department of Soils and Water, Faculty of Technology and Development, Zagazig University, Zagazig, Egypt,* [5]*Department of Chemistry, Faculty of Science, New Valley University, New Valley, Egypt,* [6]*Department of Biology, College of Science, Sultan Qaboos University, Muscat, Oman,* [7]*Department of Botany and Microbiology, Faculty of Science, New Valley University, New Valley, Egypt,* [8]*Department of Geology, Faculty of Science, New Valley University, New Valley, Egypt*

Chapter outline

13.1 Introduction	241	
13.2 Urban stormwater challenges	242	
13.3 Role of biochar in stormwater management	243	
13.3.1 Biochar characteristics relevant to stormwater treatment	243	
13.3.2 High surface area and porosity	243	
13.3.3 Adsorption properties	244	
13.3.4 Surface functional groups	244	
13.4 Versatile sorption of contaminants and contaminant-specific adsorption	244	
13.5 Influence of solution chemistry on adsorption	244	
13.5.1 Solution chemistry parameters	245	
13.6 Urban stormwater contaminants	245	
13.6.1 Contaminants in urban stormwater	245	
13.6.2 Inorganic contaminants	246	
13.6.3 Conventional organic contaminants	246	
13.7 Treating emerging contaminants with biochar	247	
13.7.1 Biochar for microbial contaminant removal	248	
13.8 Fate and behavior of contaminants in stormwater	249	
13.9 Biochar catalytic reactions	250	
13.10 Mechanisms of contaminant removal from stormwater using biochars	251	
13.10.1 Ion exchange processes	251	
13.10.2 Precipitation and coprecipitation	251	
13.11 Microbial interactions	252	
13.12 Catalytic degradation	252	
13.13 Adsorption capacity and selectivity	253	
13.14 Pivotal role of biochar in influencing soil structure and water retention	253	
13.15 Challenges and considerations of biochar for treating stormwater	254	
13.15.1 Adsorption capacity and selectivity	254	
13.15.2 Impact on soil structure	254	
13.15.3 Interactions with microbial communities	254	
13.15.4 Long-term stability and persistence	255	
13.15.5 Scale-up and implementation challenges	255	
13.16 Environmental implications of using biochar for treating stormwater	255	
13.17 Future directions	256	
13.17.1 Integrated approaches	256	
13.17.2 Synergies with green infrastructure	256	
13.17.3 Optimization of biochar production methods	256	
13.17.4 Scalability and cost-effectiveness	256	
13.17.5 Collaborative research and policy integration	257	
13.18 Conclusion	257	
References	258	

13.1 Introduction

Urban stormwater comes from rainfall or snowfall that doesn't soak into the ground and instead flows over the surface. Urban areas make this worse because they have impervious surfaces that stop water from seeping into the soil (U.S.

Environmental Protection Agency) (U.S. Environmental Protection Agency EPA., 2021) and also have several sources of contaminants. When natural areas become urban, the natural water cycle changes significantly making stormwater runoff much bigger and faster (National Research Council, 2009). Several factors influence the amount of stormwater. Precipitation patterns such as rain and snow, as well as the intensity and duration of rainfall, have a significant impact (Liptan, 2017). Human activities also play a role, such as the extent of impervious surfaces and land use practices, which can hinder water infiltration into the soil and increase runoff (U.S. Environmental Protection Agency EPA., 2021). Additionally, the soil type, moisture content, and topography contribute to the volume of stormwater (National Research Council, 2009).

Recycling and reusing stormwater present valuable opportunities for conserving and managing water. Methods like collecting rainwater can gather stormwater for nondrinking purposes, which helps lessen the demand for traditional water sources (Liptan, 2017). Green infrastructure strategies, such as green roofs and rain gardens, not only control stormwater at its origin but also offer environmental and social advantages (U.S. Environmental Protection Agency EPA., 2021). Moreover, designed wetlands and ponds can clean stormwater, promoting biodiversity and making treated water available for reuse . Aquifer recharge projects are another way to reuse stormwater, where treated stormwater is used to replenish groundwater basins, ultimately contributing to long-term water sustainability (USDA Forest Service, 2015).

Urbanization has greatly changed the natural environment, resulting in more impermeable surfaces which has led to an increase in the amount and strength of stormwater runoff. Managing stormwater in urban areas is a complex task due to the variety of pollutants that are carried by the runoff, such as heavy metals, nutrients, and organic compounds. Traditional stormwater management methods are often not enough to deal with these complex issues, so there is a need to explore new and sustainable solutions. One promising option for managing urban stormwater is the use of biochar. Biochar is a material rich in carbon that is produced through the pyrolysis of organic matter, and it has gained attention for its unique physical and chemical properties. These properties include a high surface area, porosity, and the ability to absorb a wide range of pollutants. As a result, biochar has the potential to mitigate the negative effects of urban stormwater runoff by effectively capturing and transforming pollutants. Research has shown that biochar can play a crucial role in removing heavy metals, nutrients, organic compounds, and microbial pollutants from stormwater runoff. Despite these promising findings, there are challenges in implementing biochar for stormwater management.

The efficiency of using biochar for stormwater treatment depends on numerous factors, including but not limited to, the age of the biochar, the potential for contaminants to leach from the biochar particles, and the quality variation of the biochar source material. These aspects play a crucial role in the overall effectiveness of biochar-based systems when employed in urban settings. As we continue to make advancements, ongoing studies are dedicating their efforts to addressing these critical issues and enhancing the practicality of using biochar for stormwater treatment. In light of the increasing urbanization, there is an undeniable and growing demand for comprehensive approaches to effectively manage stormwater. Understanding the processes involved in utilizing biochar to treat urban stormwater is of utmost importance. Therefore, this chapter meticulously delves into the depths of this subject matter, scrutinizing the various qualities of biochar that significantly impact stormwater treatment. By understanding the methods through which contaminants are eliminated, we can engineer effective strategies and combat current challenges that may arise. Moreover, this chapter not only explores the current state of affairs but also takes a visionary leap forward. It carves a path toward the future by considering the potential future directions for research and implementing innovative ideas. By adopting a proactive approach, we can maximize the potential of biochar-based systems for stormwater treatment in urban areas. In conclusion, this chapter serves as a comprehensive guide that encompasses a wide range of facets related to the usage of biochar for stormwater treatment. By examining the crucial factors, unraveling the complex processes, tackling existing challenges, and steering towards future directions, we pave the way for the efficient and sustainable management of stormwater in urban environments. This chapter examines biochar's role in urban stormwater issues caused by urban development, highlighting its properties and pollutant absorption capabilities. It analyzes how biochar removes contaminants from stormwater using absorption, ion exchange, and microorganism interactions. The text explores integrating biochar with green infrastructure techniques to improve stormwater management and evaluates its positive effects on soil quality and water retention. Future research paths are outlined to enhance the effectiveness of biochar, emphasizing holistic stormwater management approaches. This chapter also assesses the feasibility and economic viability of using biochar for larger-scale stormwater management.

13.2 Urban stormwater challenges

The process of urbanization has created a range of problems when it comes to managing stormwater, which is causing significant harm to the environment and public health. The rapid increase in nonporous surfaces in cities is disrupting natural water processes and leading to more intense and faster stormwater runoff. This runoff carries a variety of harmful materials, such as heavy metals, nutrients, pathogens, and organic pollutants, which come from human-made sources such as roads, factories, and residential areas. In addition, traditional stormwater management methods are not equipped to handle the complexity and amount of pollutants in urban runoff. The usual infrastructure for quickly getting rid of water often increases the amount of pollutants that end up in bodies of water. As a result, urban stormwater is a major cause of water quality decline in cities, affecting water life and posing risks to human health.

Numerous studies have emphasized the gravity of urban stormwater issues. For instance, research has demonstrated higher concentrations of heavy metals in urban runoff, posing potential harm to ecological systems and human health (Victorian Stormwater Committee, 1999). Furthermore, nutrient runoff from urban areas has been linked to issues such as eutrophication, algal blooms, and the deterioration of aquatic environments (Schueler & Holland, 2000). Due to the complex and diverse nature of urban stormwater challenges, it is crucial to seek innovative and sustainable solutions. Biochar, a carbon-rich substance derived from the pyrolysis of organic matter, has emerged as a promising approach. This chapter will investigate the utilization of biochar for urban stormwater treatment, including the associated processes, the distinctive characteristics of biochar relevant to stormwater treatment, and its potential to alleviate the adverse effects of urban runoff.

13.3 Role of biochar in stormwater management

Urban stormwater management is confronted with the task of lessening the unfavorable effects of runoff, which contains a complicated blend of pollutants. Biochar, a carbon-rich substance created through the pyrolysis of organic materials, has emerged as a promising and innovative tool for enhancing stormwater treatment processes. The critical role of biochar in stormwater management is clarified by its unique physical and chemical properties that make it an effective agent in mitigating the impacts of urban runoff. The use of biochar in stormwater management has been examined in a variety of contexts. Its application in biofilters and constructed wetlands has been found to improve the removal of pollutants such as phosphorus, nitrogen, heavy metals, and organic compounds (Xiang et al., 2020). Biochar's effectiveness in these systems is attributed to its ability to adsorb pollutants and support microbial communities that break down organic pollutants. Despite its potential, the use of biochar in stormwater treatment encounters challenges. The variability in biochar properties due to different raw materials and production conditions can impact its consistency and predictability in treatment performance (Mohanty et al., 2014). Additionally, further research is needed to explore the long-term stability of biochar and its capacity to become saturated with pollutants. Future studies should concentrate on standardizing biochar production for stormwater treatment applications and investigating the regeneration and reusability of biochar.

13.3.1 Biochar characteristics relevant to stormwater treatment

Biochar, which is obtained through the process of pyrolyzing organic matter, possesses unique physicochemical properties that render it an exceptionally promising option for stormwater treatment. This section delves into an in-depth examination of these biochar traits and successfully sheds light on their significance in augmenting the quality of stormwater. Valuable insights from recent research studies are assimilated to provide a comprehensive understanding.

13.3.2 High surface area and porosity

Biochar is a carbon-rich material created by heating organic matter in a process called pyrolysis. Its highly porous and large surface area is essential for effectively treating stormwater. This section delves into the importance of biochar's porosity and surface area, emphasizing how they play a critical role in absorbing pollutants and improving urban stormwater cleanup. Biochar stands out for its extensive surface area and porosity, which come from the complex pore structures formed during pyrolysis (Abdelhafez & Li, 2016; Abdelhafez et al., 2014; Abdelhafez et al., 2020; et al., 2020). These attributes are crucial for treating stormwater as they provide a vast surface for adsorbing contaminants. Several studies have shown that biochar is effective in adsorbing pollutants like heavy metals and organic compounds from water solutions and stormwater runoff (Abdelhafez & Li, 2016; Biswal et al., 2022; Sun et al., 2011). The natural porosity of biochar complements its large surface area, further enhancing its potential for stormwater treatment. The porous structure of biochar includes a network of channels and voids that improve its ability to absorb and retain pollutants (Abdelhafez & Li, 2016). This structural feature is especially beneficial for stormwater management, as it allows for rapid and efficient removal of contaminants, preventing them from entering water bodies.

The expanded surface area and porous characteristics of biochar are instrumental in enhancing its capacity to absorb pollutants. These attributes promote both the physical and chemical interaction between biochar and a range of pollutants present in stormwater runoff, rendering it a highly efficient and sustainable approach to contaminant removal. Moreover, the versatility and adaptability of biochar as a sorbent material further contributes to its efficacy in urban stormwater management (Biswal et al., 2022). The high surface area and porosity of biochar play a pivotal role in stormwater treatment, providing an effective means of capturing and retaining pollutants from urban runoff. Furthermore, the exceptional ability of biochar to adsorb and remove contaminants greatly contributes to its efficacy in enhancing water quality in urban environments. Its unique characteristics make it a sustainable and versatile material, offering numerous potential applications to address various stormwater treatment challenges. As we delve further into this chapter, we will explore additional aspects of biochar's role in stormwater treatment, which will expand our knowledge and offer a comprehensive understanding of its vast potential and benefits in mitigating water pollution and promoting a sustainable urban ecosystem.

13.3.3 Adsorption properties

The effectivity of biochar in treating stormwater is attributed to its exceptional adsorption properties. Its ability to attract and retain a wide variety of contaminants due to its porous structure and functional groups prevents it from entering water bodies (Al Masud et al., 2023). This characteristic makes biochar a highly adaptable and effective material for reducing the impact of urban stormwater runoff. Biochar has the unique capability to adsorb a diverse range of pollutants such as heavy metals, nutrients, and organic and microbial contaminants (Biswal et al., 2022). Various factors, including the type of feedstock, pyrolysis conditions, surface functional groups, pH, temperature, and pollutant concentration, influence biochar's adsorption capacity. Understanding these factors is crucial in optimizing stormwater treatment systems based on biochar (Abdelhafez & Li, 2016; Abdelhafez et al., 2020; Mohanty et al., 2018).

13.3.4 Surface functional groups

Biochar has become known as a versatile adsorbent in stormwater management and water treatment due to its unique surface properties, specifically the presence of a variety of surface functional groups (Ambaye et al., 2020). The surface of biochar is characterized by numerous functional groups, such as hydroxyl (-OH), carboxyl (-COOH), ketone, quinone, and aromatic structures (Ahmad et al., 2014). These functional groups give biochar specific chemical characteristics that impact its ability to adsorb different pollutants. The way biochar interacts with pollutants in stormwater is closely tied to its surface functional groups. The type and abundance of functional groups influence hydrophobic and hydrophilic interactions, ion exchange, and complexation mechanisms, all of which contribute to biochar's effectiveness as an adsorbent (Abdelhafez & Li, 2016; Abdelhafez et al., 2020; Sun et al., 2011). The presence of surface functional groups enhances biochar's capability to remove pollutants and its dynamic sorption properties.

13.4 Versatile sorption of contaminants and contaminant-specific adsorption

The adsorption properties of biochar make it a versatile sorbent for a wide range of contaminants commonly found in urban stormwater runoff. The porous structure and surface functional groups of biochar enable the physical and chemical interactions needed for adsorption, allowing for the capture of pollutants such as heavy metals, nutrients, and organic pollutants (Biswal et al., 2022; Mohanty et al., 2018). Biochar's adsorption capacity is broad and shows specificity toward different types of contaminants (Ulrich et al., 2015). For example, Ashoori et al. (2019) found that woodchip bioreactors amended with biochar effectively removed nitrate (NO_3-), Cd, Cu, and Ni, but were less effective in removing Zn. Koivusalo et al. (2023) also found that the sand-biochar filter effectively controlled both the runoff volume and hydrograph configuration, with long-term retention capturing approximately half of the incoming water flow. However, they also observed increased levels of K, Ca, and Mg in the sand-biochar filter effluent compared to the sand filter alone.

13.5 Influence of solution chemistry on adsorption

The ability of biochar to adsorb substances is influenced by factors related to the chemical composition of the solution, such as pH, ionic strength, the presence of competitive ions, and the specific forms of metal pollutants. It is crucial to comprehend these effects in order to optimize the application of biochar in stormwater treatment, as noted by Biswal et al. (2022). Research has shown that modifications to these factors can impact the adsorption capacity of biochar, underscoring the necessity for a customized approach to effectively eliminate pollutants in varying environmental conditions. With its extensive surface area, porous structure, and functional groups, biochar has emerged as a promising adsorbent for stormwater treatment. The purpose of this review is to examine the intricate connection between solution chemistry and the adsorption capacity of biochar for stormwater pollutants.

13.5.1 Solution chemistry parameters

13.5.1.1 pH

The pH level of the solution is crucial in determining the surface charge of both biochar and pollutants. Research has shown that the pH level affects the adsorption capacity, with the most effective removal occurring under certain pH conditions (Abdelhafez & Li, 2016; Chang et al., 2019; Zhong et al., 2019). Understanding the pH-dependent mechanisms is essential for developing biochar-based stormwater treatment systems that are tailored to specific environmental conditions. Parshetti et al. (2013) found that the pH level has a significant impact on the way dyes are adsorbed onto biochar surfaces. In particular, they discovered that higher alkaline pH levels greatly enhance the adsorption of dyes. This is due to the increased interaction between the negatively charged sites on the biochar's surface and the positively charged dye molecules. On the other hand, under acidic conditions (pH 3), the efficiency of organic dye adsorption decreases. This is because the excess $H+$ ions compete with the positive charges of the dye molecules, leading to reduced adsorption efficiency. The pH-dependent changes in adsorption capacity highlight the importance of considering solution pH as a critical factor in optimizing biochar-based adsorption systems for organic dyes.

13.5.1.2 Ionic strength

The presence of ions in stormwater, originating primarily from salts or other contaminants, impacts the competition for biochar's adsorption sites. Nyamunda et al.'s (2019) research underscores the importance of considering the strength of ions when developing biochar-based water treatment systems, as elevated ion levels can impede the adsorption of pollutants. Their findings demonstrate that as the initial concentration of Zn and Cu ions rises, active sites become saturated, resulting in the majority of metal ions remaining in the solution. Previous studies using other biosorbents have documented similar trends (Abdelhafez & Li, 2016; Moyo et al., 2015).

13.5.1.3 Competing ions

The presence of different ions in stormwater can have a significant impact on how biochar surfaces selectively absorb pollutants. Almanassra, Kochkodan, et al. (2021) and Du et al. (2020) stress the importance of studying how these ions interact with each other to better predict and improve the effectiveness of biochar in removing specific pollutants. When it comes to various types of anions, it is clear that divalent anions with higher charge density, such as SO_4^{2-} and CO_3^{2-}, have a stronger influence on the capacity of phosphate adsorption compared to monovalent anions like $Cl-$ and NO_3-. The higher charge density of divalent anions makes them interact more with the adsorption sites, which has a greater effect on the phosphate adsorption capacity compared to their monovalent counterparts. This provides important insights into the different behavior of anionic species in this context (Almanassra, Mckay, et al., 2021). In the context of various anions, it can be concluded that divalent anions with higher charge density, including species like SO_4^{2-} and CO_3^{2-}, have a stronger impact on phosphate adsorption capacity compared to monovalent anions like $Cl-$ and NO_3-. Several studies have consistently shown that the presence of CO_3^{2-} ions significantly reduces the adsorption capacity of phosphate. This highlights the importance of considering both the valence and charge density of anions when evaluating their impact on phosphate adsorption processes. Therefore, the findings emphasize the importance of thoroughly assessing anion characteristics for a detailed understanding of their effects on phosphate adsorption phenomena (Cai, et al., 2017; Wang, et al., 2016; Yang, et al., 2018).

13.6 Urban stormwater contaminants

13.6.1 Contaminants in urban stormwater

Urban stormwater runoff, resulting from urbanization and industrial activities, carries a wide range of pollutants into water bodies, posing significant environmental challenges. This runoff serves as a critical pathway for various contaminants to enter aquatic systems, leading to water quality degradation and risks to both aquatic life and human health.

Erosion and urban development sites contribute sediments that obstruct water bodies and disrupt aquatic habitats. Nutrients like nitrogen and phosphorus, primarily from fertilizers and sewage, lead to eutrophication and oxygen depletion in water bodies, harming aquatic organisms (Smith & Schindler, 2009). Pathogens from sources such as sewage overflows and pet waste endanger water quality and public health. Additionally, stormwater contains toxic metals like lead and mercury from vehicle emissions and industrial activities, posing a significant threat to marine life (Nagara et al., 2022). Pesticides, herbicides, and hydrocarbons from agricultural and urban usage further deteriorate water quality (Rodrigues et al., 2022). In addition, physical debris, including plastics and litter, not only visually pollute but also directly threaten wildlife. Stormwater can also lead to thermal pollution and increased salinity levels caused by road salt in colder regions, disrupting aquatic ecosystems. Understanding the major contaminants in stormwater is crucial for developing effective management strategies. Addressing these challenges involves implementing measures like green infrastructure and proper waste disposal to mitigate the impact of these contaminants on the environment and health (Ferguson et al., 2003).

13.6.2 Inorganic contaminants

Runoff from urban and agricultural areas often contains high levels of common organic pollutants such as pesticides, pharmaceuticals, personal care products, and industrial chemicals (Phillips et al., 2012). These include herbicides like atrazine and acetochlor, insecticides such as chlorpyrifos and permethrin, antidepressants like fluoxetine and venlafaxine, antibiotics including sulfamethoxazole and ciprofloxacin, disinfectants like triclosan, and plasticizers such as bisphenol A (BPA) and phthalates (Rosi-Marshall & Royer, 2012). When untreated runoff reaches bodies of water, these pollutants can harm aquatic life, build up in sediments, and move up the food chain, posing risks to wildlife and human health (Narwal et al., 2023). Biochar, a solid material rich in carbon, has shown potential for removing these pollutants from runoff. It is made through pyrolysis, the heating of plant biomass in a low-oxygen environment (Ahmad et al., 2014), creating a porous material with a large surface area for trapping organic compounds. Studies have looked at using various materials, such as wood chips, crop residues, and agricultural byproducts, to make biochar (Abdelhafez et al., 2020). Different methods, including physical and chemical processes like steam or CO_2 activation, can enhance biochar's ability to remove pollutants (Demiral et al., 2021). Research has shown promising results, with biochar removing high percentages of pollutants like atrazine, pentachlorophenol, and pharmaceuticals from simulated runoff (Gupta et al., 2011; Nguyen et al., 2014). Some studies have also found that combining biochar with other materials can further improve its pollutant-removal capabilities (Katibi et al., 2021; Yao et al., 2013). Despite these findings, there is still a need for more research to understand how well biochar performs over time and whether it is feasible for widespread use. This includes studying its effectiveness in real-world conditions, refining production methods and activation processes, and assessing its cost-effectiveness compared to other treatment options. Nonetheless, biochar offers a promising eco-friendly solution for reducing harmful organic pollutants in waterways.

13.6.3 Conventional organic contaminants

Stormwater runoff in urban areas may contain a range of organic pollutants. Polycyclic aromatic hydrocarbons (PAHs), which come from vehicle and industrial emissions, are worrisome due to their carcinogenic properties. Pesticides used in urban landscaping and pharmaceutical remnants from household waste are also commonly found in stormwater. Along with various other organic compounds, such as pharmaceuticals and personal care products (PPCPs), BPA, phthalates, alkylphenols, and endocrine disrupting compounds (EDCs). These persistent and hazardous contaminants present significant challenges for traditional water treatment methods (Table 13.1). Biochar has emerged as a promising solution for treating organic pollutants in stormwater. The effectiveness of biochar in stormwater treatment depends on its inherent properties, including its large surface area, porosity, and presence of functional groups. These attributes make biochar a highly efficient adsorbent for a variety of organic contaminants commonly found in stormwater, such as PAHs, pharmaceuticals, and personal care products. Biochar's adsorption capacity is a crucial aspect of its ability to treat stormwater contaminants. The porous structure of biochar provides a large surface area for pollutant adsorption. Studies by Inyang and Dickenson (2015) have highlighted biochar's high adsorption capacity, linking it to its unique surface characteristics and chemical properties. These studies emphasize that biochar's effectiveness in adsorbing contaminants is not solely a result of its physical structure, but also its chemical composition, which can vary based on the biomass source and pyrolysis conditions. Additionally, biochar can catalyze the degradation of contaminants, as demonstrated by the research conducted by Yang et al. (2019). This capability offers a pathway for transforming pollutants into less harmful substances, rather than simply concentrating them. Enhancing biochar's performance in stormwater

TABLE 13.1 Conventional Organic Contaminants in Stormwater.

Organic contaminants	Sources	References
Polycyclic Aromatic Hydrocarbons	Commonly originate from vehicle exhaust, industrial emissions, and the burning of coal and oil	Prabhukumar and Pagilla (2010)
Pesticides	Runoff from urban landscaping, agricultural areas, and residential gardens	Chen et al. (2019)
Pharmaceuticals and Personal Care Products	Source: Includes a wide range of substances such as medications, fragrances, and cosmetics that enter stormwater through sewage overflows and household runoff	Yang et al. (2017)
Bisphenol A	Used in plastics and epoxy resins, BPA can leach into stormwater from various consumer products	Flint et al. (2012)
Phthalates	these are plasticizers found in a variety of consumer products, including PVC piping, cosmetics, and toys	Net et al. (2015)
Alkylphenols	Degradation products of detergents and cleaning agents	Soares et al. (2008)
Endocrine disrupting compounds	A broad group of chemicals, including some pharmaceuticals, personal care products, and industrial chemicals, that can interfere with hormone systems	Khanal et al. (2006)

treatment can be achieved through various modifications. Chemical and physical modifications can increase the adsorption capacity and selectivity of biochar toward specific contaminants. These modifications often involve treating biochar with acids, bases, or specific chemicals to introduce or enhance functional groups that facilitate pollutant interaction, as discussed by Ahmed et al. (2017). Such modifications aim to tailor the adsorption characteristics of biochar to specific types of contaminants, thereby increasing its effectiveness in diverse stormwater treatment scenarios. Another noteworthy aspect of biochar's role in stormwater treatment is its impact on microbial communities. Qiu, Tao, Wang, Li, Ding, Chu (2021) have indicated that biochar can support beneficial microbial communities that contribute to the degradation of organic contaminants. This biotic aspect adds a biological dimension to the chemical and physical processes of pollutant removal, suggesting that biochar can create a more conducive environment for biodegradation processes.

Despite these positive attributes, the practical use of biochar in stormwater treatment faces several challenges. Beesley et al. (2010) stressed the importance of carefully considering factors such as the long-term stability of biochar, potential leaching of contaminants, and its effectiveness in different environmental conditions for it to be effectively implemented. These challenges highlight the necessity of ongoing research and development to improve biochar for use in the field. The environmental and economic benefits of biochar are undeniable. Its integration in stormwater treatment not only addresses pollution concerns but also aligns with sustainable and cost-effective waste management practices. Lehmann and Joseph (2015) have emphasized biochar's role in waste management and its potential for carbon sequestration, highlighting its broader environmental advantages. It is important to recognize that biochar offers a multifaceted and promising approach to addressing organic contaminants in stormwater. Its high adsorption capacity, ability to catalyze degradation, and potential for environmental sustainability make it a compelling option. However, realizing its full potential requires addressing existing challenges and optimizing its application in real-world scenarios.

13.7 Treating emerging contaminants with biochar

New types of pollutants, such as medications, personal care products, and substances that disrupt hormone function, are causing a growing level of concern due to their possible effects on the environment and human health. Research has demonstrated that biochar can effectively capture these new pollutants, which then reduces their ability to move around and their availability in stormwater. The capacity of biochar to capture these pollutants can be affected by factors like the properties of the biochar, the characteristics of the pollutants, and the conditions in the environment. A summary of common emerging pollutants found in stormwater, where they come from, their impact on the environment, and suggested readings is provided in (Table 13.2). These types of pollutants in stormwater make up a varied group of substances that are not typically monitored, but have the potential to enter the environment and cause identified or suspected negative effects on both the environment and human health.

TABLE 13.2 Emerging contaminants in stormwater.

Contaminants	Sources	References
Pharmaceuticals	Medications used in human and veterinary medicine. Can affect aquatic life, antibiotic resistance (Hospitals and pharmaceutical manufacturing). Includes prescription and over-the-counter drugs such as antibiotics, antidepressants, and antiinflammatories. Enter the environment through wastewater effluent and runoff.	LeFevre et al. (2015); Rodak et al. (2019); Verlicchi and Zambello (2014)
Personal care products	Ingredients in consumer products like shampoos, soaps (Household and commercial waste). Can disrupt endocrine system in wildlife. Includes fragrances, cosmetics, sunscreen agents. Enter through wastewater and runoff.	Clark & Pitt (2012); Rodak et al. (2019)
Flame retardants	Used in electronics, furniture, and textiles. Enter through runoff.	Et al. (2021)
Pesticides	Chemicals used to kill pests in agriculture and urban areas, which are toxic to aquatic life. Includes herbicides, insecticides, and fungicides. Enter through runoff from lawns, farms, etc.	Göbel, Dierkes, Coldewey (2007); Rodak et al. (2019)
Plasticizers	Used to make plastics more flexible. Includes phthalates and BPA. Enter through industrial releases and landfill leaching.	Annissa et al. (2021)
PFAS	Used in stain/water repellents, and firefighting foams. Enter via runoff from many sources.	Emily et al. (2022)
Tire wear particles	Generated from tire abrasion on roads. Enter through stormwater runoff.	Annissa et al. (2021). Rosen et al. (2006).
Microplastics	Small plastic particles from textiles, tires, larger plastics, etc. Enter through runoff. Tiny plastic particles (Consumer products and industrial processes), which can be ingested by aquatic organisms and may cause harmful effects for aquatic life	Annissa et al. (2021). Rosen et al. (2006), Li & Davis (2016)
Polyaromatic Hydrocarbons	Organic compounds from fossil fuel combustion (Vehicle and industrial emissions). Carcinogenic and harmful to aquatic life	Beesley et al. (2010)

- Pharmaceuticals: These include a wide range of drugs such as antibiotics, hormones, and painkillers that can enter stormwater through human waste and improper disposal of unused medications (Wicke et al., 2016)
- Personal care products: Substances used in personal care products, such as sunscreens, soaps, and cosmetics, can wash off during bathing or showering and end up in stormwater (Wicke et al., 2016)
- Industrial chemicals: These can include a variety of substances used in manufacturing processes, which can enter stormwater through industrial discharges and runoff from industrial sites (Wicke et al., 2016)
- Endocrine-disrupting compounds: These are chemicals that can interfere with the body's endocrine system and produce adverse developmental, reproductive, neurological, and immune effects. They can come from a variety of sources, including pesticides, plastics, and pharmaceuticals (Wicke et al., 2016). It's important to note that the presence and concentrations of these emerging contaminants can vary depending on a variety of factors, including the local climate, land use, and stormwater management practices (Galella, et al., 2023; He et al., 2022; Michelle, et al., 2021; Wicke et al., 2016).

13.7.1 Biochar for microbial contaminant removal

Biochar has the potential to aid in the reduction of microbial contamination in stormwater. Its porous structure and large surface area create an environment for beneficial microorganisms to thrive, contributing to the breakdown of organic contaminants. Additionally, biochar can absorb microbial pathogens, decreasing their presence in stormwater. However, further research is necessary to comprehend the methods of microbial contaminant removal by biochar and to enhance the efficiency of biochar-based stormwater treatment systems (Galella et al., 2023; Rene et al., 2019). Biochar has

TABLE 13.3 Microbial contaminants in stormwater.

Contaminants	Description	References
Escherichia coli (E. coli)	A bacterium commonly found in the intestines of humans and animals (Human and animal waste, sewage overflows), and can cause gastrointestinal illness	Mohanty et al. (2014)
Enterococcus spp.	Bacteria indicating fecal contamination (Sewage, animal waste) and usually associated with waterborne diseases	Beesley et al. (2010)
Salmonella spp.	Bacteria causing foodborne illness (animal waste, sewage), and can lead to sever gastroenteritis	Clark and Pitt (2012)
Legionella spp.	Bacteria responsible for Legionnaires' disease (Cooling towers, water system and causes respiratory illness, pneumonia)	Göbel, Dierkes, Coldewey (2007)
Cryptosporidium parvum	Protozoan parasite (animal waste, sewage) and causes cryptosporidiosis, a diarrheal disease	Kelly et al. (2021)
Giardia lamblia	Protozoan parasite (sewage, animal waste) and causes giardiasis, a diarrheal disease	Li and Davis (2016)
Norovirus	Highly contagious virus (human sewage and contaminated waters) which causes gastroenteritis	Inyang and Dickenson (2015)

demonstrated effectiveness in treating microbial contaminants in stormwater runoff (El-Sayed et al., 2023; Inyang & Dickenson, 2015). Stormwater can become contaminated with pathogens from animal and human waste, posing risks to human and environmental health if not treated prior to being discharged into water bodies (Asmara et al., 2023). Table 13.3 summarizes some common microbial contaminants in stormwater, their sources, and environmental impacts. Biochar possesses properties suitable for removing microbial contaminants from stormwater, such as high porosity to capture microbes, surface charges for electrostatic interactions, and reactive surface groups for microbial inactivation (Qiu et al., 2022). A recent study demonstrated that biochar filters made from wood effectively removed Escherichia coli from synthetic stormwater under laboratory conditions, achieving bacterial reductions of up to 99% (Inyang & Dickenson, 2015). The high removal efficiency was attributed to microbial adhesion and straining within the biochar's porous structure. Another study tested pine wood biochar columns for removing fecal indicator bacteria from agricultural runoff and found up to 2-log removals of *E. coli* (El-Sayed et al., 2023). They observed greater bacterial removal early on, indicating that sorption sites became occupied over time. Modifying biochar to increase surface charge and oxidation state has been shown to further improve its capability for microbial removal.

While biochar filters have shown the potential to disinfect storm water, it is important to consider that real runoff contains various contaminants that could affect how well the filters work (Qiu et al., 2022). Biochar has the ability to attract and hold onto organic and inorganic particles more than microbes, which could potentially clog up the filter. Additionally, the buildup of organic matter in the filter could provide nutrients for microbes to grow. To determine whether biochar consistently reduces microbes in storm water, more testing under real conditions is necessary. It's possible that multistep treatment methods, involving both filtration and disinfection, may be needed to effectively control contamination. However, because it's affordable and sustainable, biochar has the potential to help reduce microbial pollution in storm water. Using biochar could be a sustainable and cost-effective way to address this issue.

13.8 Fate and behavior of contaminants in stormwater

Stormwater can pick up a wide range of pollutants as it flows through urban areas, including pathogens, nutrients, metals, pesticides, and other harmful chemicals (Masoner et al., 2019; Müller et al., 2020). There is a growing concern about the presence of emerging contaminants such as pharmaceuticals, flame retardants, and microplastics in stormwater (Lambert & Wagner, 2018; Mishra et al., 2023). The movement and reduction of stormwater pollutants is a complex process that depends on the properties of the contaminants, the characteristics of the storm, and the biogeochemical processes that occur in stormwater management systems and the environment.

Microorganisms like *E. coli* and Enterococci, as well as several other bacteria that are commonly present in feces, possess the inherent capability to contaminate stormwater. This contamination can occur due to multiple factors, including the interconnectedness of sewage systems or the natural movement of animal waste. Once these microorganisms find their way into the stormwater, they often establish bonds with particles or various surfaces, and their existence can be influenced by exposure to sunlight and UV radiation within specific stormwater systems, such as wetlands and retention ponds. However, it is crucial to acknowledge that these microorganisms also possess the ability to proliferate within the stormwater environment through a range of processes, including sediment settling and sediment stirring, as well as by utilizing the abundant organic matter available as a source of sustenance.

Certain organic pollutants, such as PAHs and certain pesticides, have a moderate level of being repelled by water and can bind to particles, plants, and organic matter in biofilm (Mukherjee et al., 2022; Ukalska-Jaruga & Smreczak, 2020). The movement of these chemicals can be aided by dissolved organic matter, which can form dissolved complexes. In the removal of organic contaminants from stormwater, photodegradation, and microbial biodegradation play important roles. However, it is worth noting that some compounds may exhibit resistance to these processes or may persist due to continual replenishment in stormwater sources. This persistence of organic pollutants can have significant implications for the overall water quality and the environmental impact of these substances. Therefore, it is crucial to develop comprehensive strategies and technologies that effectively address the persistence and removal of these substances to mitigate their adverse effects on ecosystems and human health. Innovative approaches, including advanced oxidation processes, biofiltration systems, and constructed wetlands, have shown promise in enhancing the efficiency of organic pollutant removal from stormwater. By leveraging the synergistic effects of different treatment methods, it is possible to achieve significant reductions in the concentrations of these pollutants, ultimately leading to healthier and more sustainable water resources. The continued research and implementation of these strategies are essential for ensuring the long-term protection and preservation of our water ecosystems.

Nonorganic pollutants such as metals and nutrients can be reduced by different methods in stormwater best management practices (BMPs), such as sorption, precipitation, plant/microbial uptake, and photochemical transformations (McCabe et al., 2021). However, the effectiveness of these processes differs depending on various factors. For instance, sorption, although not very effective for removing nutrients, happens to be quite efficient in retaining metals on media and soil surfaces due to their strong affinity. Moreover, it is important to consider environmental factors such as pH, temperature, and redox status as they greatly influence the type and distribution of pollutants throughout the system. These factors play a crucial role in determining the overall efficacy of pollutant reduction in stormwater BMPs.

The diverse range of pollutants in stormwater makes it extremely challenging to accurately predict and forecast the behavior of contaminants. With such a vast array of pollutants present, it becomes even more complex to comprehend how these contaminants will behave and react. Furthermore, the variations in water flow that occur during storm events further complicate the accumulation and dispersion of pollutants. These variations directly influence the manner in which pollutants amass and subsequently get washed away. While we rely on BMPs as a means of safeguarding and preserving our environment, it is crucial to acquire a deeper understanding of how overflow discharges and interactions with groundwater impact overall water quality. By doing so, we can gain substantial insights into the efficiency and effectiveness of stormwater treatment methods. This will enable us to assess and evaluate the extent to which stormwater treatment practices successfully protect water quality, particularly under varying conditions, and with diverse types of contaminants and flow rates. Therefore, it is imperative to undertake further research to comprehensively analyze and gauge the efficacy of stormwater treatment in safeguarding water quality in a multitude of scenarios.

13.9 Biochar catalytic reactions

Biochar, a carbon-rich substance created through the pyrolysis of biomass, has demonstrated potential as a sustainable catalyst and adsorbent for eliminating stormwater pollutants through a range of methods (Al Masud, et al., 2023; Hasan et al., 2021; He et al., 2022). Its important characteristics, such as a high surface area, porosity, surface functional groups, and mineral content, allow biochar to efficiently soak up pollutants and offer support for catalytic nanoparticles (Abdelhafez et al., 2020; Kaya, et al., 2022). In stormwater treatment systems like bioretention cells and biofilters, biochar has been discovered to enhance the removal efficiency for nutrients, heavy metals, and organic micropollutants in comparison to standard soil media (Hasan et al., 2021; Kaya et al., 2022; Wijeyawardana et al., 2022). Additional improvements in removal rates for certain contaminants have been achieved by modifying biochar with metal oxides or supporting nanoscale zerovalent iron (Shaheen et al., 2022; Yang et al., 2021; Zhu et al., 2023). The production conditions of biochar, such as pyrolysis temperature and feedstock type, significantly influence its high adsorption capacity and catalytic activity. Gasification-derived biochar typically boasts higher porosity and surface area when compared to slow pyrolysis biochar, leading to improved adsorptive interactions. Hardwoods and nutshells also produce biochars with greater microporosity than softwoods and grasses (Ferraro et al., 2021). Furthermore, chemical and thermal posttreatments can be used to customize biochar properties to target specific stormwater pollutants. While most of the focus has been on biochar's adsorption capacity, recent studies have highlighted its potential as a catalyst for breaking down organic contaminants through advanced oxidation processes in stormwater systems. Biochar-supported iron nanoparticles have been found to generate hydroxyl radicals that degrade pesticides and pharmaceuticals in synthetic runoff. When combined with ozonation, biochar filters have enabled partial oxidation of polycyclic aromatics and chemical dechlorination. There is still much to be learned about the real-world catalytic performance and economic feasibility of biochar, requiring further research.

In general, the diverse capabilities of biochar offer great potential for improving stormwater treatment. The adaptable nature of biochar allows for a wide range of uses in addressing stormwater pollution. Customized production and modification methods can be used to leverage both adsorptive and destructive mechanisms to better meet changing regulatory discharge limits for various stormwater pollutants. The effectiveness of biochar-integrated stormwater control measures has been proven in laboratory-scale studies, but more field-scale demonstrations are needed in order for biochar to be considered a viable solution on a larger scale. Field-scale demonstrations will help confirm the long-term performance and stability of biochar in stormwater control measures. It is important to understand how biochar holds up over time in real-world scenarios and whether it consistently delivers expected results. Based on this information, we can create a comprehensive table summarizing the different studies conducted on stormwater treatment using biochar. This table outlines the type of biochar used in each study, the main mechanism for pollutant removal, and the specific pollutants targeted for treatment. By analyzing and understanding the collective findings from these studies, we can gain valuable insights into the potential uses of biochar in stormwater treatment. This knowledge will help further develop and improve the use of biochar as a sustainable and effective tool for managing stormwater pollution. Table 13.4

13.10 Mechanisms of contaminant removal from stormwater using biochars

Biochar, produced through the pyrolysis of organic matter, has become a promising solution for treating urban stormwater runoff due to its ability to remove a wide range of contaminants. Fig. 13.1 portrays the mechanisms through which biochar eliminates pollutants from stormwater. Adsorption, a key process in contaminant removal, occurs when pollutants attach to the large surface area and intricate pore structures of biochar. The porous nature of biochar allows for extensive physical and chemical interactions with contaminants, while its diverse functional groups, such as carboxyl and hydroxyl groups, further aid in adsorption by promoting chemical bonding. This adsorption process, as demonstrated in multiple studies (Akhtar et al., 2021; Deng, 2020a, 2020b), efficiently captures various contaminants found in stormwater runoff, including heavy metals and organic compounds.

As shown in Fig. 13.1, the main mechanisms of biochar for pollutants removal in stormwater are as follows.

13.10.1 Ion exchange processes

Ion exchange is an important process for removing contaminants using biochar. It works by replacing ions in stormwater with ions on the surface of biochar. The functional groups on biochar's surface, like carboxyl and hydroxyl groups, are crucial for this selective ion exchange process. Many studies have shown how effective biochar is at absorbing positively charged metal ions, especially heavy metals, which helps prevent them from getting into water and improves the overall quality of stormwater (Abdelhafez & Li, 2016; Sun et al., 2011).

13.10.2 Precipitation and coprecipitation

The process of precipitation and coprecipitation in stormwater treatment using biochar involves the creation of insoluble compounds when contaminants interact with components of biochar. During precipitation, contaminants form insoluble compounds with the constituents of biochar, reducing their solubility and ability to move. Coprecipitation further increases the efficiency of contaminant removal as they are precipitated alongside minerals found in biochar. This immobilization process, particularly effective for heavy metals, is well-documented in studies (Abdelhafez et al., 2014; Xu et al., 2018) and plays a significant role in reducing the ecological impact of urban stormwater.

TABLE 13.4 Biochar applications for treating stormwater indicate the percent removal of various contaminants.

Type of biochar	Pollutants targeted	Removal percentage, %	Main mechanism	References
Poultry litter (PL) biochar	Ammonium	90	Adsorption	Tian et al. (2016)
Powdered birch biochar	Phosphorous, organic carbon, K, Ca, Mg	–	Filtration, increased water retention, and storage	Koivusalo et al. (2023)
Not specified	Phosphate and nitrogen	–	Adsorption	Anderson et al. (2023)
Magnetic biochar	Cadmium, arsenic, lead, methylene blue, and phosphate	3.46 to 169.7 mg g^{-1} for methylene blue 1.26 and 474.26 mg g^{-1} for phosphate	Adsorption	Zhao et al. (2021)
Modified forestry wood waste biochar	E. coli	92–98	Surface adsorption	Lau et al. (2017)
Not specified	Indicator bacteria and viruses	–	Adsorption	Torkelson, (2015)
Environmental Ultra, Biochar Supreme	Hydrophilic trace organic contaminants (HiTrOCs), PFASs	–	Sorption-retarded intraparticle pore diffusion transport model	Pritchard et al. (2023)
Biochar-amended woodchip	Nitrate, metals, trace organic contaminants	>80	Not specified	Ashoori et al. (2019)
Not specified	Trace organic contaminants	–	Enhanced trace organic contaminant retention	Mohanty et al. (2014)
Not specified	Trace organic contaminants, metals	–	Not specified	Spahr et al. (2020)
Not specified	Microplastics, organic pollutants, endocrine-disrupting chemicals, pharmaceuticals	75–99	Adsorption	Dong et al., (2023)
Not specified	Various pollutants in water	–	Not specified	Kumar et al. (2023)
Straw derived biochar	Polybrominated diphenyl ethers	–	Sorption, effects on anaerobic biodegradation	Chen et al., (2019)
Lignin-based biochar	Azo anionic dye, methyl orange dye	>85	Photocatalytic and adsorptive remediation	Singh et al. (2023)

13.11 Microbial interactions

The incorporation of biochar into stormwater brings about microbial interactions that can impact how contaminants are processed. The microbial communities linked to biochar play a role in breaking down or changing specific pollutants, ultimately improving the overall remediation process. Recent research by Amoakwah et al. (2022) and Mukherjee et al. (2022) indicates that microbial activity, influenced by biochar, contributes to the removal of contaminants, adding a biological aspect to the process. It is crucial to grasp this aspect in order to optimize the biological processes at play in stormwater treatment.

13.12 Catalytic degradation

The role of biochar in stormwater treatment goes beyond just adsorption and includes the catalytic breakdown of different contaminants. This review takes a close look at how biochar helps to break down pollutants in stormwater, exploring the

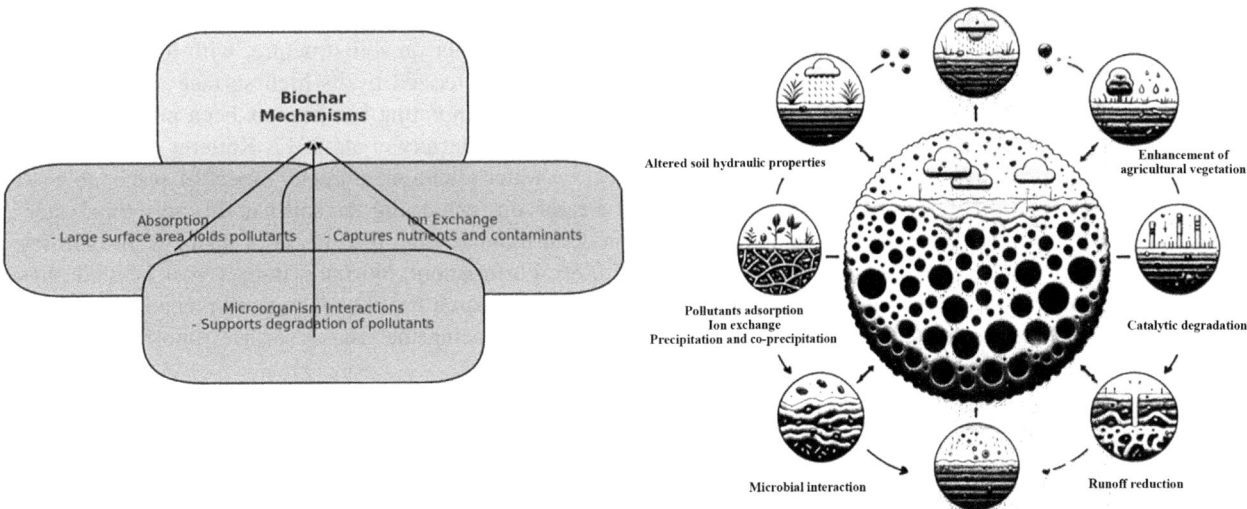

FIGURE 13.1 Main mechanisms of biochar for pollutant removal in stormwater.

mechanisms and effectiveness of this process. Stormwater contamination with organic pollutants is a big environmental problem, and biochar has been getting attention for both its ability to adsorb and its catalytic properties for breaking down pollutants. The review focuses on the process of catalytic degradation by biochar, particularly in stormwater treatment. The catalytic degradation process in biochar is affected by its surface area, porosity, and the presence of functional groups. These properties allow biochar to help with oxidation and reduction reactions that can break down complex organic molecules. Biochar can catalyze vital oxidative reactions that are necessary for breaking down persistent organic pollutants in stormwater through an oxidative degradation process (Qiu, Tao, Wang, Li, Ding, Chu, 2021). Biochar is effective at catalyzing the breakdown of various organic compounds commonly found in stormwater, such as pharmaceuticals and pesticides. Research has shown that the catalytic activity of biochar is successful in breaking down PAHs and endocrine-disrupting compounds, which are especially concerning in urban stormwater runoff (Teixidó et al., 2011).

It should be mentioned that there are multiple factors that have an impact on the catalytic efficiency of biochar. One such factor is surface modification, which involves the impregnation of metals or acids, leading to an enhanced catalytic degradation capacity of biochar (Rajapaksha et al., 2016). This improvement in catalytic efficiency is attributed to the surface modifications that alter the chemical properties of biochar, thereby increasing its ability to facilitate catalytic reactions. The incorporation of metals or acids onto the biochar surface creates active sites that promote the catalytic degradation of various pollutants and contaminants. These modifications not only enhance the overall performance of biochar as a catalyst but also broaden its applicability in diverse environmental remediation processes. Consequently, surface modification has emerged as a promising strategy to optimize the catalytic efficiency of biochar and advance its potential for sustainable waste management and pollution control. By harnessing the capabilities of surface modification, researchers can further enhance the catalytic degradation capacity of biochar, opening up new possibilities for its utilization in various industries and environmental applications.

13.13 Adsorption capacity and selectivity

Although biochar's extensive surface area increases its ability to adsorb substances, there are difficulties associated with its ability to selectively target certain contaminants. This preferential adsorption of some pollutants over others could potentially hinder biochar's overall effectiveness in treating a wide range of stormwater pollutants. By customizing the production and sourcing of biochar, as well as employing surface modification methods, its selectivity can be improved. It is essential to have a thorough understanding of the unique adsorption mechanisms for various contaminants in order to optimize the performance of biochar (Rahman et al., 2021; Sun et al., 2011).

13.14 Pivotal role of biochar in influencing soil structure and water retention

The process of urbanization has resulted in an increase in impervious surfaces, causing changes in natural hydrological processes and exacerbating problems related to stormwater. In the search for sustainable solutions, biochar, which is produced through the pyrolysis of organic matter, has emerged as a versatile tool for treating stormwater in urban areas. The introduction of biochar into soil systems can have a significant impact on soil structure, with long-term implications for stormwater management. The porous nature of biochar, characterized by its high surface area and intricate pore structures, contributes to improved soil aeration and drainage. Incorporating biochar has been proven to alleviate soil compaction, leading to better water infiltration and root penetration (Jeffery et al., 2017; Kätterer et al., 2019). The incorporation of biochar into soil matrices has a positive effect on water retention, a crucial aspect of sustainable stormwater management. The porous structure of biochar acts as a reservoir, enhancing the soil's ability to retain water and reduce surface runoff. This characteristic is particularly beneficial in addressing the effects of changes in hydrological patterns caused by urbanization (Asai et al., 2009; Novak, 2009). Furthermore, biochar's impact goes beyond physical soil characteristics and extends to altering soil hydraulic properties. Research has shown that soil amended with biochar demonstrates improved water retention and hydraulic conductivity, reducing the risk of surface runoff and enhancing groundwater recharge (Abdelhafez et al., 2020; Lehmann & Joseph, 2015).

13.15 Challenges and considerations of biochar for treating stormwater

Urbanization and its associated stormwater runoff pose significant challenges to water quality and ecosystem health. In spite of biochar, a highly effective and innovative solution has gained attention as a promising tool for stormwater treatment. However, the application of biochar in this context is not without challenges and considerations. There are multifaceted aspects of utilizing biochar for treating stormwater, that is, the challenges encountered and important considerations that should shape its implementation. The diverse challenges, ranging from effectively integrating biochar into stormwater management systems to optimizing its performance, demand careful attention and thorough investigation. Additionally, the importance of considering the selection of biochar feedstocks, the specific properties of biochar, and the potential impacts on the environment and human health cannot be overstated. Moreover, understanding the interactions between biochar and other treatment technologies, such as constructed wetlands or permeable pavements, is crucial for maximizing the benefits of biochar in stormwater treatment. Overall, incorporating biochar into stormwater management strategies holds great promise, but comprehensive research and a holistic approach are necessary to overcome the challenges and ensure successful implementation. By dexterously addressing these complexities, the potential for enhancing water quality and restoring ecosystem health in urban areas can be realized, paving the way for a sustainable and resilient future.

13.15.1 Adsorption capacity and selectivity

Although biochar's impressive adsorption capacity is due to its large surface area, there are issues concerning its ability to selectively target specific contaminants. The tendency for biochar to favor certain pollutants over others could hinder its overall effectiveness in treating a wide range of stormwater pollutants. Customizing the production processes and feedstocks of biochar, as well as implementing surface modification techniques, could improve its selectivity. It is essential to comprehend the unique adsorption mechanisms for various contaminants in order to maximize biochar's performance (Tomin et al., 2021).

13.15.2 Impact on soil structure

The addition of biochar to the soil for stormwater control can bring about significant transformations in the soil's composition and properties, which in turn may have profound implications for the rate at which water infiltrates the soil and the intricate cycling of vital nutrients. Consequently, it becomes paramount to meticulously contemplate the long-term repercussions of these alterations to prevent any inadvertent detriment to the environment. By conscientiously evaluating the precise soil conditions at each individual site and duly considering the intricate interplay between biochar and the preexisting soil components, it becomes feasible to mitigate the potential adverse effects on the soil's structural integrity. It is incumbent upon us to adhere to the most efficacious methodologies for managing the soil and conducting pragmatic experiments, as such endeavors hold the key to gaining invaluable insights that can shape our understanding of this domain (Jeffery et al., 2017; Lehmann & Joseph, 2015).

13.15.3 Interactions with microbial communities

Introducing biochar to soil can have an impact on the microbial communities living within it, which in turn can affect the way nutrients are cycled and the overall health of the soil. It is crucial to have a clear understanding of these interactions in order to prevent any disruptions to the delicate balance of the microbial ecosystems. By combining the study of biochar with microbial ecology and using advanced molecular techniques, we can gain insight into how this affects microbial communities. This knowledge is important for creating biochar applications that improve microbial functions, rather than causing any disturbances (Amoakwah et al., 2022; Lehmann & Joseph, 2015).

13.15.4 Long-term stability and persistence

The durability and enduring nature of biochar in soils are crucial and play a pivotal role in determining its effectiveness in various agricultural and environmental applications. Understanding and analyzing the breakdown rates of biochar, the patterns of nutrient release, and the potential risks associated with contaminants leaching from aged biochar are of utmost importance. Thus, it is imperative to conduct extensive and meticulous long-term field studies accompanied by continuous monitoring to accurately evaluate the long-lasting presence of biochar and its profound influence on soil and water quality. Moreover, it is highly beneficial to customize the properties of biochar to align with the specific soil compositions and diverse climatic conditions encountered in different regions. By tailoring biochar characteristics to cater to the unique soil and climate conditions, its sustainable and persistent efficacy can be significantly enhanced, leading to improved agricultural productivity and better environmental outcomes. In recent research conducted by Baronti et al. (2022) and Spokas et al. (2012), they have highlighted the significance of customizing biochar properties to optimize its performance and overall impact on soil health and ecosystem sustainability. Clearly, the durability and enduring nature of biochar, alongside its intricate relationship with soil and water quality, call for in-depth investigations and continuous monitoring through long-term field studies. Additionally, customizing biochar properties based on soil and climatic conditions is a promising approach to maximize its effectiveness and ensure sustainable agricultural practices. The studies conducted by Baronti et al. (2022) and Spokas et al. (2012) shed light on the importance of these considerations and provide valuable insights into the potential of biochar as a transformative tool in sustainable agriculture and environmental conservation efforts.

13.15.5 Scale-up and implementation challenges

Moving from small-scale biochar studies in labs to large-scale applications in urban areas presents logistical and implementation obstacles. It is important to ensure even distribution, find the best application rates, and maintain cost-effectiveness. Pilot programs and cooperation between researchers, practitioners, and policymakers can help with the successful expansion of biochar use. Thorough cost-benefit analyses and flexible management strategies can help deal with challenges related to large-scale implementation (Abdelhafez et al., 2021; Patel & Panwar, 2023)

13.16 Environmental implications of using biochar for treating stormwater

First, the ability of biochar to remove contaminants from stormwater is well-documented. As mentioned above, biochar can effectively adsorb heavy metals, organic pollutants, and nutrients from stormwater, thus reducing their concentration in water bodies. For example, biochar derived from bamboo was found to remove copper and zinc from stormwater effectively. Another notable advantage of using biochar is its potential to improve soil health when used as a soil amendment. The addition of biochar to soil can enhance water retention, reduce runoff, and improve soil fertility, thereby indirectly benefiting stormwater management. This dual functionality of biochar, both as a treatment medium and soil enhancer, presents a sustainable approach to stormwater management. However, the production of biochar itself has environmental implications. The sustainability of biochar production largely depends on the feedstock used and the energy consumed during pyrolysis. Using waste biomass for biochar production can be seen as a form of recycling, but the process must be energy-efficient to ensure overall environmental benefits (Abdelhafez & Li, 2016; Abdelhafez et al., 2020). Concerns have also been raised regarding the potential release of pollutants from biochar into the environment. While biochar is typically stable, under certain conditions it might release adsorbed contaminants or its own constituents, like PAHs, into the stormwater (Chen et al., 2019). This necessitates careful consideration of the biochar characteristics and the conditions under which it is used. Furthermore, the long-term stability and effectiveness of biochar in stormwater treatment systems are subjects of ongoing research. The aging of biochar in natural environments and its interaction with various contaminants over time need to be better understood to ensure its reliability and safety in long-term applications (Beesley et al., 2010). While biochar presents a promising approach for stormwater treatment with its ability to remove various contaminants and improve soil health, the environmental implications of its production and long-term stability warrant careful consideration. Future research should focus on optimizing biochar production from sustainable feedstocks, understanding the long-term behavior of biochar in the environment, and ensuring that its application does not introduce new environmental concerns.

13.17 Future directions

13.17.1 Integrated approaches

In contemplating the future directions of biochar utilization for stormwater treatment, a pivotal focus revolves around the adoption of integrated approaches. The holistic management of stormwater demands a nuanced understanding that transcends individual technologies. The incorporation of biochar into a comprehensive framework, integrating green infrastructure, conventional stormwater management practices, and cutting-edge technologies, presents an opportunity to elevate the overall efficacy of stormwater treatment systems. Integrating biochar into a broader spectrum involving green roofs, permeable pavements, rain gardens, bioinfiltration systems, and constructed wetlands fosters a synergistic effect, not only enhancing pollutant removal but also contributing to the sustainable development of urban areas. The synergies derived from integrated approaches promise to enhance the resilience and multifunctionality of stormwater management, fostering a paradigm shift toward more sustainable urban development. Furthermore, incorporating biochar within stormwater treatment systems aligns with the principles of circular economy and resource recovery, as biochar can be produced from organic waste materials such as agricultural residues, municipal solid waste, and wastewater sludge. By utilizing biochar in stormwater treatment, cities can mitigate the effects of urbanization on water quality, reducing the risks of pollution, flooding, and erosion. The ability of biochar to improve water retention, increase soil fertility, and enhance plant growth further supports its integration into stormwater management. Additionally, the application of biochar in stormwater treatment systems provides an avenue for carbon sequestration, contributing to efforts in mitigating climate change. Moreover, the use of biochar in stormwater design can promote biodiversity and create habitats for various flora and fauna, enhancing the overall ecological value of urban environments. The porous structure of biochar promotes the infiltration of water, facilitating groundwater recharge and reducing the burden on traditional stormwater infrastructure. In conclusion, the integration of biochar into stormwater treatment systems through integrated approaches holds significant potential in revolutionizing urban stormwater management. By harnessing the synergies between biochar, green infrastructure, and advanced technologies, cities can create sustainable and resilient stormwater systems that not only effectively remove pollutants but also contribute to a more balanced and harmonious urban ecosystem. This paradigm shift toward integrated stormwater management will pave the way for a greener, cleaner, and more sustainable future in cities around the world.

13.17.2 Synergies with green infrastructure

The combination of biochar use with green infrastructure shows great promise for improving urban stormwater treatment methods. Green infrastructure, which includes a mix of plants, soil, and other natural elements, is known for its ability to mimic natural water processes. By incorporating biochar into green infrastructure components like bioswales or rain gardens, there is potential to enhance their ability to remove pollutants. Biochar's natural adsorption properties work well with the water retention and nutrient cycling functions of green infrastructure, resulting in a stronger and more versatile stormwater management system. Future research should focus on finding the best ways to combine biochar and green infrastructure to maximize their combined benefits.

13.17.3 Optimization of biochar production methods

The potential success of using biochar for stormwater treatment relies on carefully optimizing production methods to customize biochar properties for specific uses. Different production methods result in biochars with unique physical and chemical traits, such as slow pyrolysis, fast pyrolysis, and gasification. Understanding how variations in production parameters impact biochar properties is crucial for tailoring biochars to effectively address specific stormwater challenges. Future research should focus on refining production methods to create biochars with increased adsorption capacity, porosity, and stability, maximizing their effectiveness for urban stormwater treatment.

13.17.4 Scalability and cost-effectiveness

Propelling biochar-based stormwater treatment into widespread adoption necessitates an unwavering focus on scalability and cost-effectiveness. It is imperative that extensive efforts are directed toward addressing the challenges associated with the scalability of biochar production, encompassing a broad and meticulous exploration of sustainable feedstock sources on a significantly larger scale. Moreover, robust economic assessments need to be conducted in order to thoroughly and rigorously evaluate the cost-effectiveness of seamlessly integrating biochar into the existing stormwater infrastructure, in clear comparison to conventional methodologies. By carefully scrutinizing and analyzing the economic viability and scalability of biochar applications, decision-makers, and urban planners will be equipped with foundational insights that will vastly contribute to the effective and practical incorporation of biochar into overarching stormwater management strategies. Therefore, it is of utmost importance to invest substantial time, resources, and research into these critical aspects, as they hold the key to successful implementation and long-term positive impact.

13.17.5 Collaborative research and policy integration

Steering the future trajectory of biochar application in stormwater treatment necessitates a concerted commitment to collaborative research endeavors and seamless integration into policy frameworks. Collaborative research initiatives, fostering the synergy of scientists, engineers, policymakers, and urban planners, stand as linchpins for fostering a comprehensive understanding of biochar's vast potential and inherent limitations. An interdisciplinary approach becomes imperative for sculpting robust guidelines and best practices governing the seamless integration of biochar into diverse urban settings. Additionally, the strategic integration of biochar-based stormwater treatment into local and regional policies ensures that the accrued benefits of this technology are effectively harnessed, contributing tangibly to broader sustainability objectives. Furthermore, the collaborative efforts of researchers, policy makers, and stakeholders from diverse sectors need to be intensified, promoting a harmonious exchange of knowledge and the exploration of innovative techniques to optimize biochar implementation. This cross-sectoral collaboration not only enhances the efficiency and efficacy of biochar's stormwater treatment potential but also facilitates its widespread adoption. By fostering synergy among various disciplines, including but not limited to environmental science, engineering, and governance, a transformative impact can be achieved, revolutionizing the biochar landscape and propelling it toward a future where stormwater treatment becomes increasingly sustainable and resilient. Consequently, the seamless integration of biochar into urban ecosystems will solidify its position as a pioneering solution, capable of mitigating the adverse effects of urbanization and ensuring the conservation of precious water resources. Moreover, the benefits of biochar-based stormwater treatment must be communicated comprehensively, reaching diverse audiences and stakeholders through targeted educational campaigns and advocacy efforts. This broad dissemination and awareness-raising will catalyze further research, innovation, and investment in the field, consequently expanding the horizons of biochar's application and accelerating its adoption on a global scale. As such, by cultivating a culture of cross-sectoral collaboration, integrating biochar into policy frameworks, and fostering comprehensive understanding through research and education, the future of biochar in stormwater treatment appears promising, with the potential for transformative change that will benefit both urban environments and broader sustainability goals.

13.18 Conclusion

In conclusion, this thorough analysis of biochar's role in treating stormwater highlights its multifaceted potential in addressing the urgent environmental challenges caused by urban runoff. Biochar, with its strong ability to absorb substances, improve soil health, and remove pollutants, emerges as a promising solution. However, using biochar for stormwater treatment is not without difficulties, including its sustainable production, long-term stability, and possible environmental impacts. Integrating biochar into stormwater management systems offers a sustainable approach, especially when combined with green infrastructure practices. This combination can lead to more effective pollutant removal, improved water conservation, and overall environmental resilience. The versatility and adaptability of biochar make it a valuable tool in urban stormwater management, capable of addressing a variety of pollutants, from heavy metals and nutrients to organic and microbial contaminants. Future efforts should focus on optimizing biochar production, ensuring its consistency and safety for environmental applications. Research must continue to explore the long-term behavior of biochar in different environmental settings and its interaction with various pollutants. Additionally, it is important to thoroughly investigate the scalability and cost-effectiveness of incorporating biochar into existing stormwater infrastructure. Collaborative research efforts and policy integration are essential for realizing the full potential of biochar in sustainable urban development. By bridging the gap between scientific research and practical application, biochar can play a crucial role in transforming urban stormwater management, contributing to the health of aquatic ecosystems and the well-being of urban communities.

Biochar is seen as a potential solution for improving stormwater management, although its intricacies pose challenges. Its advantages and drawbacks are both considerable. With urbanization altering environments, the importance of innovative measures such as biochar in lessening environmental effects grows.

References

Abdelhafez, A. A., & Li, J. (2016). Removal of Pb(II) from aqueous solution by using biochars derived from sugar cane bagasse and orange peel. *Journal of the Taiwan Institute of Chemical Engineers, 61*, 367−375. Available from https://doi.org/10.1016/j.jtice.2016.01.005.

Abdelhafez, A. A., Li, J., & Abbas, M. H. H. (2014). Feasibility of biochar manufactured from organic wastes on the stabilization of heavy metals in a metal smelter contaminated soil. *Chemosphere, 117*, 66−71. Available from https://doi.org/10.1016/j.chemosphere.2014.05.086.

Abdelhafez, A. A., Zhang, X., Zhou, L., Cai, M., Cui, N., Chen, G., Zou, G., Abbas, M. H. H., Kenawy, M. H. M., Ahmad, M., Alharthi, S. S., & Hamed, M. H. (2021). Eco-friendly production of biochar via conventional pyrolysis: Application of biochar and liquefied smoke for plant productivity and seed germination. *Environmental Technology and Innovation, 22*, 101540. Available from https://doi.org/10.1016/j.eti.2021.101540.

Abdelhafez, A.A., Zhang, X., Zhou, L., Zou, G., Cui, N., Abbas, M.H.H., Hamed, M.H. (2020). Introductoiry Chapter: Is biochar safe? Ahmed A. Abdelhafez; Mohamed H.H. Abbas, Applications of biochar for Environmental Safety. IntechOpen 10.5772/intechopen.91996 Available from: https://www.intechopen.com/chapters/71729.

Ahmad, M., Rajapaksha, A. U., Lim, J. E., Zhang, M., Bolan, N., Mohan, D., Vithanage, M., Lee, S. S., & Ok, Y. S. (2014). Biochar as a sorbent for contaminant management in soil and water: A review. *Chemosphere, 99*, 19−33. Available from https://doi.org/10.1016/j.chemosphere.2013.10.071.

Ahmed, M. A., Zhou, J. L., Ngo, H. H., Guo, W., Thomaidis, N. S., & Xu, J. (2017). Progress in the biological and chemical treatment technologies for emerging contaminant removal from wastewater: A critical review. *Journal of Hazardous Materials, 323*, 274−298. Available from https://doi.org/10.1016/j.jhazmat.2016.04.045.

Akhtar, L., Ahmad, M., Iqbal, S., Abdelhafez, A. A., & Mehran, M. T. (2021). Biochars' adsorption performance towards moxifloxacin and ofloxacin in aqueous solution: Role of pyrolysis temperature and biomass type. *Environmental Technology and Innovation, 24*, 101912. Available from https://doi.org/10.1016/j.eti.2021.101912.

Al Masud, M. A., Shin, W. S., Sarker, A., Septian, A., Das, K., Deepo, D. M., Iqbal, M. A., Islam, A. R. M. T., & Malafaia, G. (2023). A critical review of sustainable application of biochar for green remediation: Research uncertainty and future directions. *Science of the Total Environment, 904*, 166813. Available from https://doi.org/10.1016/j.scitotenv.2023.166813.

Almanassra, I. W., Kochkodan, V., Mckay, G., Atieh, M. A., & Al-Ansari, T. (2021). Kinetic and thermodynamic investigations of surfactants adsorption from water by carbide-derived carbon. *Journal of Environmental Science and Health, Part A, 56*(11), 1206−1220. Available from https://doi.org/10.1080/10934529.2021.1973822.

Almanassra, I. W., Mckay, G., Kochkodan, V., Atieh, M. A., & Al-Ansari, T. (2021). A state-of-the-art review on phosphate removal from water by biochars. *Chemical Engineering Journal, 409*, 128211. Available from https://doi.org/10.1016/j.cej.2020.128211.

Ambaye, T. G., Vaccari, M., Castro, F. D., Prasad, S., & Rtimi, S. (2020). Emerging technologies for the recovery of rare earth elements (REEs) from the end-of-life electronic wastes: a review on progress, challenges, and perspectives. *Environmental Science and Pollution Research, 27*(29), 36052−36074. Available from https://doi.org/10.1007/s11356-020-09630-2.

Amoakwah, E., Arthur, E., Frimpong, K. A., Lorenz, N., Rahman, M. A., Nziguheba, G., & Islam, K. R. (2022). Biochar amendment impacts on microbial community structures and biological and enzyme activities in a weathered tropical sandy loam. *Applied Soil Ecology, 172*, 104364. Available from https://doi.org/10.1016/j.apsoil.2021.104364.

Anderson, M., Durgesh, V., Baker, M., Yu, P., & Möller, G. (2023). Biomimetic crossflow filtration with wave minimal surface geometry for particulate biochar water treatment. *PLoS Water, 2*(1), e0000055. Available from https://doi.org/10.1371/journal.pwat.0000055.

Annissa, H., Guo, E., Sedlak, M., Sutton, R., Box, C., Lin, D., Gilbreath, A., Holleman, R. C., Fortin, M.-J., & Rochman, C. (2021). Holistic assessment of microplastics and other anthropogenic microdebris in an urban bay sheds light on their sources and fate. *ACS ES&T Water Journal*. Available from https://doi.org/10.1021/acsestwater.0c00292.

Asai, H., Samson, B. K., Stephan, H. M., Songyikhangsuthor, K., Homma, K., Kiyono, Y., Inoue, Y., Shiraiwa, T., & Horie, T. (2009). Biochar Amendment Techniques for Upland Rice Production in Northern Laos 1. Soil Physical Properties, Leaf SPAD and Grain Yield. *Field Crops Research, 111*, 81−84. Available from https://doi.org/10.1016/j.fcr.2008.10.008.

Ashoori, N., Teixido, M., Spahr, S., LeFevre, G. H., Sedlak, D. L., & Luthy, R. G. (2019). Evaluation of pilot-scale biochar-amended woodchip bioreactors to remove nitrate, metals, and trace organic contaminants from urban stormwater runoff. *Water Research, 154*, 1−11. Available from https://doi.org/10.1016/j.watres.2019.01.040.

Asmara, D. H., Allaire, S., van Noordwijk, M., & Khasa, D. P. (2023). The Effect of Biochar Amendment, Microbiome Inoculation, Crop Mixture and Planting Density on Post-Mining Restoration. *Forests, 14*(4), 856. Available from https://doi.org/10.3390/f14040856.

Baronti, S., Magno, R., Maienza, A., Montagnoli, A., Ungaro, F., & Vaccari, F. P. (2022). Long term effect of biochar on soil plant water relation and fine roots: Results after 10 years of vineyard experiment. *Science of the Total Environment, 851*(1)), 158225. Available from https://doi.org/10.1016/j.scitotenv.2022.158225.

Beesley, L., Moreno-Jiménez, E., & Gomez-Eyles, J. L. (2010). Effects of biochar and greenwaste compost amendments on mobility, bioavailability and toxicity of inorganic and organic contaminants in a multi-element polluted soil. *Environmental Pollution, 158*(6), 2282–2287. Available from https://doi.org/10.1016/j.envpol.2010.02.003.

Biswal, B. K., Vijayaraghavana, K., Tsen-tieng, D. L., & Balasubramanian, R. (2022). Biochar-based bioretention systems for removal of chemical and microbial pollutants from stormwater: A critical review. *Journal of Hazardous Materials, 422*, 126886. Available from https://doi.org/10.1016/j.jhazmat.2021.126886.

Cai, R., Wang, X., Ji, X., Peng, B., Ton, C., & Huang, X. (2017). Phosphate reclaim from simulated and real eutrophic water by magnetic biochar derived from water hyacinth. *Journal of Environmental Management, 187*, 212–219. Available from https://doi.org/10.1016/j.jenvman.2016.11.047.

Chang, R., Sohi, Sp, Jing, F., Liu, Y., & Chen, J. (2019). A comparative study on biochar properties and Cd adsorption behavior under effects of ageing processes of leaching, acidification and oxidation. *Environmental Pollution, 254*(B), 113123. Available from https://doi.org/10.1016/j.envpol.2019.113123.

Chen, C., Guo, W., & Ngo, H. H. (2019). Pesticides in stormwater runoff—A mini review. *Frontiers of Environmental Science & Engineering., 13*, 72. Available from https://doi.org/10.1007/s11783-019-1150-3.

Clark, S. E., & Pitt, R. (2012). Targeting treatment technologies to address specific stormwater pollutants and numeric discharge limits. *Water Research, 46*(20), 6715–6730. Available from https://doi.org/10.1016/j.watres.2012.07.009.

Demiral, I., Samdan, C., & Demiral, H. (2021). Enrichment of the surface functional groups of activated carbon by modification method. *Surfaces and Interfaces, 22*, 100873. Available from https://doi.org/10.1016/j.surfin.2020.100873.

Deng, Y. (2020a). Biochar as a novel soil amendment: Properties and mechanisms of action. *Frontiers of Environmental Science & Engineering., 14*(5), 83. Available from https://doi.org/10.1007/s11783-020-1262-9.

Deng, Y. (2020b). Low-cost adsorbents for urban stormwater pollution control. *Frontiers of Environmental Science & Engineering., 14*(5), 83. Available from https://doi.org/10.1007/s11783-020-1262-9.

Dong, M., He, L., Jiang, M., Zhu, Y., Wang, J., Gustave, W., Wang, S., Deng, Y., Zhang, X., & Wang, Z. (2023). Biochar for the Removal of Emerging Pollutants from Aquatic Systems: A Review. *International Journal of Environmental Research and Public Health, 20*(3), 1679. Available from https://doi.org/10.3390/ijerph20031679.

Du, F. J., Bao, J., Hassan, M. A., Isrhod, S., Talib, M. A., & Zheng, H. (2020). Efficient capture of phosphate and cadmium using biochar with multifunctional amino and carboxylic moieties: Kinetics and mechanism. *Water, Air, and Soil Pollution, 231*, 25. Available from https://doi.org/10.1007/s11270-019-4389-1.

El-Sayed, M. M., Mahdy, A. Y., Gebreel, M., & Abdeen, S. (2023). Effectiveness of biochar, organic matter and mycorrhiza to improve soil hydrophysical properties and water relations of soybean under arid soil conditions. *Eurasian Soil Science, 56*, 1055–1066. Available from https://doi.org/10.1134/S1064229323600276.

Emily H.M., U. Ghosh, T.P. Needham, N. Lombard, E. Foss, M. Bokare, S. Joshee, L. Cheung, J. Damond, and M. Lorah. (2022). Refining sources of polychlorinated biphenyls in the Back River Watershed, Baltimore, Maryland, 2018–2020. *Scientific Investigations Report 2022-5012*. Prepared in cooperation with Baltimore City Department of Public Works and Maryland Department of the Environment. https://doi.org/10.3133/sir20225012.

Ferguson, C., Husman, A. M. R., Altavilla, N., Deere, D., & Ashbolt, N. (2003). Fate and Transport of Surface Water Pathogens in Watersheds. *Critical Reviews in Environmental Science and Technology, 33*(3), 299–361. Available from https://doi.org/10.1080/10643380390814497.

Ferraro, G., Pecori, G., Rosi, L., et al. (2021). Biochar from lab-scale pyrolysis: influence of feedstock and operational temperature. *Biomass Conversion and Biorefinery*. Available from https://doi.org/10.1007/s13399-021-01303-5.

Flint, S., Markle, T., Thompson, S., & Wallace, E. (2012). Bisphenol A exposure, effects, and policy: A wildlife perspective. *Journal of Environmental Management, 104*, 19–34. Available from https://doi.org/10.1016/j.jenvman.2012.03.021.

Galella, J. G., Kaushal, S. S., Mayer, P. M., Maas, C. M., Shatkay, R. R., & Stutzke, R. A. (2023). Stormwater best management practices: Experimental evaluation of chemical cocktails mobilized by freshwater salinization syndrome. *Frontiers in Environmental Science., 11*, 1020914. Available from https://doi.org/10.3389/fenvs.2023.1020914.

Göbel, P., Dierkes, C., & Coldewey, W. G. (2007). Storm water runoff concentration matrix for urban areas. *Journal of Contaminant Hydrology, 91*(1-2), 26–42. Available from https://doi.org/10.1016/j.jconhyd.2006.08.008.

Gupta, V. K., Gupta, B., Rastogi, A., Agarwal, S., & Nayak, A. (2011). Pesticides removal from waste water by activated carbon prepared from waste rubber tire. *Water Research, 45*(13), 4047–4055. Available from https://doi.org/10.1016/j.watres.2011.05.016.

Hasan, M. S., Vasquez, R., & Geza, M. (2021). Application of Biochar in Stormwater Treatment: Experimental and Modeling Investigation. *Processes, 9*(5), 860. Available from https://doi.org/10.3390/pr9050860.

He, M., Xu, Z., Hou, D., et al. (2022). Waste-derived biochar for water pollution control and sustainable development. *Nature Reviews Earth & Environment, 3*, 444–460. Available from https://doi.org/10.1038/s43017-022-00306-8.

Inyang, M., & Dickenson, E. (2015). The potential role of biochar in the removal of organic and microbial contaminants from potable and reuse water: A review. *Chemosphere, 134*, 232–240. Available from https://doi.org/10.1016/j.chemosphere.2015.03.072.

Jeffery, S., Abalos, D., Prodana, M., Bastos, A. C., van Groenigen, J. W., Hungate, B. A., & Verheijen, F. (2017). Biochar boosts tropical but not temperate crop yields. *Environmental Research Letters, 12*, 053001. Available from https://doi.org/10.1088/1748-9326/aa67bd.

Katibi, K. K., Yunos, K. F., Man, H. C., Aris, A. Z., Nor, M. Z. M., & Aziz, R. S. (2021). An insight into a sustainable removal of bisphenol a from aqueous solution by novel palm kernel shell magnetically induced biochar: synthesis, characterization, kinetic, and thermodynamic studies. *Polymers, 13*(21), 3781. Available from https://doi.org/10.3390/polym13213781.

Kätterer, T., Roobroeck, D., Andrén, O., Kimutai, G., Karltun, E., Kirchmann, H., Nyberg, G., Vanlauwe, B., & de Nowina, K. R. (2019). Biochar addition persistently increased soil fertility and yields in maize-soybean rotations over 10 years in sub-humid regions of Kenya. *Field Crop Research*, 235, 18–26. Available from https://doi.org/10.1016/j.fcr.2019.02.015.

Kaya, D., Croft, K., Pamuru, S. T., Yuan, C., Davis, A. P., & Kjellerup, B. V. (2022). Considerations for evaluating innovative stormwater treatment media for removal of dissolved contaminants of concern with focus on biochar. *Chemosphere*, 307(4)), 135753. Available from https://doi.org/10.1016/j.chemosphere.2022.135753.

Kelly, E., Cronk, R., Fishetr, M., & Bartram, J. (2021). Sanitary inspection, microbial water quality analysis, and water safety in handpumps in rural sub-Saharan Africa. *npj Clean Water*, 4, 3. Available from https://doi.org/10.1038/s41545-020-00093-z.

Khanal, S. K., Xie, B., Thompson, M. L., Sung, S., Ong, S. K., & Van Leeuwen, J. (2006). Fate, transport, and biodegradation of natural estrogens in the environment and engineered systems. *Environmental Science and Technology*, 40(21), 6537–6546. Available from https://doi.org/10.1021/es0607739.

Koivusalo, H., Dubovik, M., Wendling, L., Assmuth, E., Sillanpää, N., & Kokkonen, T. (2023). Performance of sand and mixed sand–biochar filters for treatment of road runoff quantity and quality. *Water*, 15(8), 1631. Available from https://doi.org/10.3390/w15081631.

Kumar, A., Bhattacharya, T., Shaikh, W. A., Roy, A., Chakraborty, S., Vithanage, M., & Biswas, J. K. (2023). Multifaceted applications of biochar in environmental management: a bibliometric profile. *Biochar*, 5, 1–40.

Lambert, S., & Wagner, M. (2018). Microplastics Are Contaminants of Emerging Concern in Freshwater Environments: An Overview. In M. Wagner, & S. Lambert (Eds.), *Freshwater microplastics. The handbook of environmental chemistry* (vol 58). Cham: Springer. Available from https://doi.org/10.1007/978-3-319-61615-5_1.

Lau, A. Y. T., Tsang, D. C. W., Graham, N. J. D., Ok, Y. S., Yang, X., & Li, X. (2017). Surface-modified biochar in a bioretention system for Escherichia coli removal from stormwater. *Chemosphere*, 169, 89–98. Available from https://doi.org/10.1016/j.chemosphere.2016.11.048.

LeFevre, G. H., Müller, C. E., Li, R. J., Luthy, R. G., & Sattely, E. S. (2015). Rapid Phytotransformation of Benzotriazole Generates Synthetic Tryptophan and Auxin Analogs in Arabidopsis. *Environmental Science & Technology*, 49(18), 10959–10968. Available from https://doi.org/10.1021/acs.est.5b02749, 2015.

Lehmann, J., & Joseph, S. (2015). *Biochar for environmental management: Science, technology and implementation* (2nd ed.). Routledge, ISBN: 978-0415704151.

Li, H., & Davis, A. P. (2016). Heavy metal capture and accumulation in bioretention media. *Environmental Science & Technology*, 50(5), 2247–2255. Available from https://doi.org/10.1021/es702681j.

Liptan, T. (2017). *Sustainable stormwater management: A landscape-driven approach to planning and design*. Portland: Timber Press.

Masoner, J. R., Kolpin, D. W., Cozzarelli, I. M., Barber, L. B., Burden, D. S., Foreman, W. T., Forshay, K. J., Furlong, E. T., Groves, J. F., Hladik, M. L., Hopton, M. E., Jaeschke, J. B., Keefe, S. H., Krabbenhoft, D. P., Lowrance, R., Romanok, K. M., Rus, D. L., Selbig, W. R., Williams, B. H., & Bradley, P. M. (2019). Urban Stormwater: An Overlooked Pathway of Extensive Mixed Contaminants to Surface and Groundwaters in the United States. *Environmental Science & Technology*, 53(17), 10070–10081. Available from https://doi.org/10.1021/acs.est.9b02867.

McCabe, K. M., Smith, E. M., Lang, S. Q., Osburn, C. L., & Benitez-Nelson, C. R. (2021). Particulate and Dissolved Organic Matter in Stormwater Runoff Influences Oxygen Demand in Urbanized Headwater Catchments. *Environmental Science & Technology*, 55(2), 952–961. Available from https://doi.org/10.1021/acs.est.0c04502, 2021.

Mishra, R. K., Mentha, S. S., Misra, Y., & Dwivedi, N. (2023). Emerging pollutants of severe environmental concern in water and wastewater: A comprehensive review on current developments and future research. *Water-Energy Nexus*, 6, 74–95. Available from https://doi.org/10.1016/j.wen.2023.08.002.

Mohanty, S. K., Cantrell, K. B., Nelson, K. L., & Boehm, A. B. (2014). Efficacy of biochar to remove Escherichia coli from stormwater under steady and intermittent flow. *Water Research*, 61, 288–296. Available from https://doi.org/10.1016/j.watres.2014.05.026.

Mohanty, S. K., Valenca, R., Berger, A. W., Yu, I. K. M., Xiong, X., Saunders, T. M., & Tsang, D. C. W. (2018). Plenty of room for carbon on the ground: Potential applications of biochar for stormwater treatment. *Science of the Total Environment*, 256(1), 1644–1658. Available from https://doi.org/10.1016/j.scitotenv.2018.01.037.

Moyo, M., Guyo, U., Mawenyiyo, G., Zinyama, N. P., & Nyamunda, B. C. (2015). Marula seed husk (Sclerocarya birrea) biomass as a low cost biosorbent for removal of Pb(II) and Cu(II) from aqueous solution. *Journal of Industrial and Engineering Chemistry*, 27, 126–132. Available from https://doi.org/10.1016/j.jiec.2014.12.026.

Mukherjee, S., Sarkar, B., Aralappanavar, V. K., Mukhopadhyay, R., Basak, B. B., Srivastava, P., Marchut-Mikołajczyk, O., Bhatnagar, A., Semple, K. T., & Bolan, N. (2022). Biochar-microorganism interactions for organic pollutant remediation: Challenges and perspectives. *Environmental Pollution*, 308, 119609. Available from https://doi.org/10.1016/j.envpol.2022.119609.

Müller, A., Österlund, H., Marsalek, J., & Viklander, M. (2020). The pollution conveyed by urban runoff: A review of sources. *Science of the Total Environment*, 709, 136125. Available from https://doi.org/10.1016/j.scitotenv.2019.136125.

Nagara, N. N., Sarkar, D., Elzinga, E. J., & Datta, R. (2022). Removal of heavy metals from stormwater runoff using granulated drinking water treatment residuals. *Environmental Technology and Innovation*, 28, 102636. Available from https://doi.org/10.1016/j.eti.2022.102636.

Narwal, N., Katyal, D., Kataria, N., Rose, P. K., Warkar, S. G., Pugazhendhi, A., Ghotekar, S., & Khoo, K. S. (2023). Emerging micropollutants in aquatic ecosystems and nanotechnology-based removal alternatives: A review. *Chemosphere*, 341, 139945. Available from https://doi.org/10.1016/j.chemosphere.2023.139945.

National Research Council. (2009). *Urban stormwater management in the United States*. Washington, DC: The National Academies Press. Available from https://doi.org/10.17226/12465.

Net, S., Sempéré, R., Delmont, A., Paluselli, A., & Ouddane, B. (2015). Occurrence, fate, behavior and ecotoxicological state of phthalates in different environmental matrices., *49*(7), 4019−4035. Available from https://doi.org/10.1021/es505233b.

Nguyen, T. A. H., Ngo, H. H., Guo, W. S., Zhang, J., Liang, S., Lee, D. J., Nguyen, P. D., & Bui, X. T. (2014). Modification of agricultural waste/by-products for enhanced phosphate removal and recovery: Potential and drawbacks. *Bioresource Technology*, *169*, 750−762. Available from https://doi.org/10.1016/j.biortech.2014.07.047.

Novak, J. M., et al. (2009). Impact of biochar amendment on fertility of a southeastern coastal plain soil. *Soil Science*, *174*(2), 105−112. Available from https://doi.org/10.1097/SS.0b013e3181981d9a.

Nyamunda, B. C., Chivhanga, T., Guyo, U., & Chingondo, F. (2019). Removal of Zn (II) and Cu (II) ions from industrial wastewaters using magnetic biochar derived from water hyacinth. *Journal of Engineering*, *2019*, 5656983. Available from https://doi.org/10.1155/2019/5656983.

Parshetti, G. K., Hoekman, S. K., & Balasubramanian, R. (2013). Chemical, structural and combustion characteristics of carbonaceous products obtained by hydrothermal carbonization of palm empty fruit bunches. *Bioresource Technology*, *135*, 683−689. Available from https://doi.org/10.1016/j.biortech.2012.09.042.

Patel, M. R., & Panwar, N. L. (2023). Biochar from agricultural crop residues: Environmental, production, and life cycle assessment overview. *Resources, Conservation & Recycling Advances*, *19*, 200173. Available from https://doi.org/10.1016/j.rcradv.2023.200173.

Phillips, P. J., Chalmers, A. T., Gray, J. L., Kolpin, D. W., Foreman, W. T., & Wall, G. R. (2012). Combined Sewer Overflows: An Environmental Source of Hormones and Wastewater Micropollutants. *Environmental Science & Technology*, *46*(10), 5336−5343. Available from https://doi.org/10.1021/es3001294.

Prabhukumar, G., Pagilla, K. (2010). Polycyclic Aromatic Hydrocarbons in Urban Runoff −Sources, Sinks and Treatment: A Review. DuPage River Salt Creek- Workgroup, Naperville, IL. Illinois Institute of Technology. https://drscw.org/wp/wp-content/uploads/2019/01/Review-of-PAHs-Sources-and-Treatment_FINAL_13-Oct2010.pdf.

Pritchard, J. C., Cho, Y.-M., Hawkins, K. M., Spahr, S., Higgins, C. P., & Luthy, R. G. (2023). Predicting PFAS and hydrophilic trace organic contaminant transport in black carbon-amended engineered media filters for improved stormwater runoff treatment. *Environmental Science & Technology*, *57*(38), 14417−14428. Available from https://doi.org/10.1021/acs.est.3c01260.

Qiu, B., Tao, X., Wang, H., Li, W., Ding, X., & Chu, H. (2021). Biochar as a low-cost adsorbent for aqueous heavy metal removal: A review. *Journal of Analytical Applied Pyrolysis*, *155*, 105081. Available from https://doi.org/10.1016/j.jaap.2021.105081.

Qiu, M., Liu, L., Ling, Q., Cai, Y., Yu, S., Wang, S., Fu, D., Hu, B., & Wang, X. (2022). Biochar for the removal of contaminants from soil and water: a review. *Biochar*, *4*, 19. Available from https://doi.org/10.1007/s42773-022-00146-1.

Rahman, M. Y. A., Cooper, R., Truong, N., Ergas, S. J., & Nachabe, M. H. (2021). Water quality and hydraulic performance of biochar amended biofilters for management of agricultural runoff. *Chemosphere*, *283*, 130978. Available from https://doi.org/10.1016/j.chemosphere.2021.130978.

Rajapaksha, A. U., Chen, S. S., Tsang, D. C. W., Zhang, M., & Vithanage, M. (2016). Engineered/designer biochar for contaminant removal/immobilization from soil and water: Potential and implication of biochar modification. *Chemosphere*, *148*, 276−291. Available from https://doi.org/10.1016/j.chemosphere.2016.01.043.

Rene, E. R., Shu, L., & Jegatheesan, V. (2019). Appropriate technologies to combat water pollution. *Environmental Science and Pollution Research*, *26*(33), 33719−33721. Available from https://doi.org/10.1007/s11356-019-06906-0.

Rodak, C. M., Jayakaran, A. D., Moore, T. L., David, R., Rhodes, E. R., & Vogel, J. R. (2019). Urban stormwater characterization, control, and treatment. *Water Environment Research*, *92*(10), 1552−1586. Available from https://doi.org/10.1002/wer.1173.

Rodrigues, P., Oliva-Teles, L., Guimarães, L., & Carvalho, A. P. (2022). Occurrence of Pharmaceutical and Pesticide Transformation Products in Freshwater: Update on Environmental Levels, Toxicological Information and Future Challenges. *Reviews of Environmental Contamination and Toxicology*, *260*(1), 14. Available from https://doi.org/10.1007/s44169-022-00014-w.

Rosen, M. R., Goodbred, S. L., Patiño, R., Leiker, T. A., & Orsak, E. (2006). Investigations of the Effects of Synthetic Chemicals on the Endocrine System of Common Carp in Lake Mead, Nevada and Arizona. *Fact Sheet*, 2006−3131. Available from https://doi.org/10.3133/fs20063131, Prepared in cooperation with the U.S. Fish and Wildlife Service.

Rosi-Marshall, E. J., & Royer, T. V. (2012). Pharmaceutical Compounds and Ecosystem Function: An Emerging Research Challenge for Aquatic Scientists. *Ecosystems*, *15*(6), 867−880. Available from https://www.jstor.org/stable/23253729.

Schueler, T. R., & Holland, H. K. (2000). The Practice of Watershed Protection. *Center for Watershed Protection*, ISBN 10: 097200520X ISBN 13: 9780972005203.

Shaheen, S. M., Mosa, A., Natasha., Abdelrahman, H., Niazi, N. K., Antoniadis, V., Shahid, M., song, H., Kwon, E. E., & Rinklebe, J. (2022). Removal of toxic elements from aqueous environments using nano zero-valent iron- and iron oxide-modified biochar: a review. *Biochar*, *4*, 24. Available from https://doi.org/10.1007/s42773-022-00149-y.

Singh, A. K., Giannakoudakis, D. A., Arkas, M., Triantafyllidis, K. S., & Nair, V. (2023). Composites of Lignin-Based Biochar with BiOCl for Photocatalytic Water Treatment: RSM Studies for Process Optimization. *Nanomaterials (Basel)*, *13*(4), 735. Available from https://doi.org/10.3390/nano13040735, PMID: 36839103; PMCID: PMC9959841.

Smith, V. H., & Schindler, D. W. (2009). Eutrophication science: where do we go from here? *Trends in Ecology & Evolution*, *24*(4), 201−207. Available from https://doi.org/10.1016/j.tree.2008.11.009.

Soares, A., Guieysse, B., Jefferson, B., Cartmell, E., & Lester, J. N. (2008). Nonylphenol in the environment: A critical review on occurrence, fate, toxicity and treatment in wastewaters. *Environment International*, *34*(7), 1033−1049. Available from https://doi.org/10.1016/j.envint.2008.01.004.

Spahr, S., Teixidó, M., Sedlak, D. L., & Luthy, R. G. (2020). Hydrophilic trace organic contaminants in urban stormwater: occurrence, toxicological relevance, and the need to enhance green stormwater infrastructure. *Environmental Science: Water Research & Technology.*, 6, 15−44. Available from https://doi.org/10.1039/c9ew00674e.

Spokas, K. A., Cantrell, K. B., Novak, J. M., Archer, D. W., Ippolito, J. A., Collins, H. P., Boateng, A. A., Lima, I. M., Lamb, M. C., McAloon, A. J., Lentz, R. D., & Nichols, K. A. (2012). Biochar: a synthesis of its agronomic impact beyond carbon sequestration. *Journal of Environmental Quality*, 41(4), 973−989. Available from https://doi.org/10.2134/jeq2011.0069, PMID: 22751040.

Sun, K., Ro, K., Guo, M., Novak, J., Mashayekhi, H., & Xing, B. (2011). Sorption of bisphenol A, 17α-ethinyl estradiol and phenanthrene on thermally and hydrothermally produced biochars. *Bioresource Technology*, 102(10), 5757−5763. Available from https://doi.org/10.1016/j.biortech.2011.03.038.

Teixidó, M., Pignatello, J. J., Beltrán, J. L., Granados, M., & Peccia, J. (2011). Speciation of the ionizable antibiotic sulfamethazine on black carbon (biochar). *Environmental Science & Technology*, 45(23), 10020−10027. Available from https://doi.org/10.1021/es202487h.

Tian, J., Miller, V., Chiu, P. C., Maresca, J. A., Guo, M., & Imhoff, P. T. (2016). Nutrient release and ammonium sorption by poultry litter and wood biochars in stormwater treatment. *Science of the Total Environment*, 553, 596−606. Available from https://doi.org/10.1016/j.scitotenv.2016.02.129.

Tomin, O., Vahala, R., & Yazdani, M. R. (2021). Tailoring metal-impregnated biochars for selective removal of natural organic matter and dissolved phosphorus from the aqueous phase. *Microporous and Mesoporous Materials*, 328, 111499. Available from https://doi.org/10.1016/j.micromeso.2021.111499.

Torkelson, A. (2015). Enhanced granular media filtration of waterborne pathogens: Effect of media amendments for treatment of drinking water and stormwater. UC Berkeley. ProQuest ID: Torkelson_berkeley_0028E_15860. Merritt ID: ark:/13030/m5sb8v0b. Retrieved from https://escholarship.org/uc/item/9fz7g048.

Ukalska-Jaruga, A., & Smreczak, B. (2020). The impact of organic matter on polycyclic aromatic hydrocarbon (PAH) availability and persistence in soils. *Molecules*, 25(11), 2470. Available from https://doi.org/10.3390/molecules25112470.

Ulrich, B. A., Im, E. A., Werner, D., & Higgins, C. P. (2015). Biochar and activated carbon for enhanced trace organic contaminant retention in stormwater infiltration systems. *Environmental Science & Technology*, 49, 6222−6230. Available from https://doi.org/10.1021/acs.est.5b00376, 2015.

USDA Forest Service. (2015). Stormwater to street trees: Engineering urban forests for stormwater management. https://www.fs.fed.us/nrs/pubs/gtr/gtr_nrs121.pdf.

U.S. Environmental Protection Agency (EPA). (2021). Managing stormwater. https://www.epa.gov/green-infrastructure/managing-stormwater.

Verlicchi, P., & Zambello, E. (2014). How efficient are constructed wetlands in removing pharmaceuticals from untreated and treated urban wastewaters? A review. *Science of the Total Environment*, 470, 1281−1306. Available from https://doi.org/10.1016/j.scitotenv.2013.10.085.

Victorian Stormwater Committee. (1999). Urban stormwater: Best practice environmental management guidelines. CSIRO Press. 064310285X, 9780643102859. https://books.google.com.eg/books?id = VS8ImoNhg5QC&printsec = copyright&hl = ar#v = onepage&q&f = false.

Wang, Y., Lu, H., Liu, Y., & Yang, S. M. (2016). Removal of phosphate from aqueous solution by SiO 2−biochar nanocomposites prepared by pyrolysis of vermiculite treated algal biomass. *RSC Advances*, 6, 83534−83546. Available from https://doi.org/10.1039/C6RA15532D.

Wicke, D. Matzinger, A. Caradot, N. Sonnenberg, H. Schubert, R.-L. won Seggern, D., Heinzmann, B., Rouault, P. (2016). Extent and dynamics of classic and emerging contaminants in stormwater of urban catchment types. In *Novatech 2016 - 9th International conference on planning and technologies for sustainable management of Water in the City*, Lyon, France, June 2016. HAL Archives, ffhal-03322070f, 2021.

Wijeyawardana, P., Nanayakkara, N., Gunasekara, C., Karunarathna, A., Law, D., & Pramanik, B. K. (2022). Removal of Cu, Pb and Zn from stormwater using an industrially manufactured sawdust and paddy husk derived biochar. *Environmental Technology and Innovation*, 28, 102640. Available from https://doi.org/10.1016/j.eti.2022.102640.

Xiang, W., Zhang, X., Chen, J., Zou, W., He, F., Hu, X., Tsang, D. C. W., Ok, Y. S., & Gao, B. (2020). Biochar technology in wastewater treatment: A critical review. *Chemosphere*, 252, 126539. Available from https://doi.org/10.1016/j.chemosphere.2020.126539.

Xu, Y., Seshadri, B., Sarkar, B., Wang, H., Rumpel, C., Sparks, D., Farrell, M., Hall, T., Yang, X., & Bolan, N. (2018). Biochar modulates heavy metal toxicity and improves microbial carbon use efficiency in soil. *Science of the Total Environment*, 621, 148−159. Available from https://doi.org/10.1016/j.scitotenv.2017.11.214.

Yang, D., Yang, A. S., Wang, L., Xu, J., & Liu, X. (2021). Performance of biochar-supported nanoscale zero-valent iron for cadmium and arsenic co-contaminated soil remediation: Insights on availability, bioaccumulation and health risk. *Environmental Pollution*, 290, 118054. Available from https://doi.org/10.1016/j.envpol.2021.118054.

Yang, Q., Wang, X., Luo, W., Sun, J., Xu, Q., Chen, F., Zhao, J., Wang, S., Yao, F., & Wang, D. (2018). Effectiveness and mechanisms of phosphate adsorption on iron-modified biochars derived from waste activated sludge. *Bioresource Technology*, 247, 537−544. Available from https://doi.org/10.1016/j.biortech.2017.09.136.

Yang, X., Wan, Y., Zheng, Y., He, F., Yu, Z., Huang, J., Wang, H., Ok, Y. S., Jiang, Y., & Gao, B. (2019). Surface functional groups of carbon-based adsorbents and their roles in the removal of heavy metals from aqueous solutions: A critical review. *Chemical Engineering Journal*, 366, 608−621. Available from https://doi.org/10.1016/j.cej.2019.02.119.

Yang, Y., Ok, Y. S., Kim, K. H., Kwon, E. E., & Tsang, Y. F. (2017). Occurrences and removal of pharmaceuticals and personal care products (PPCPs) in drinking water and water/sewage treatment plants: A review. *Science of The Total Environment*, 596−597, 303−320. Available from https://doi.org/10.1016/j.scitotenv.2017.04.102.

Yao, Y., Gao, B., Chen, J., Zhang, M., Inyang, M., Li, Y., Alva, A., & Yang, L. (2013). Engineered carbon (biochar) prepared by direct pyrolysis of Mg-accumulated tomato tissues: Characterization and phosphate removal potential. *Bioresource Technology*, 138, 8−13. Available from https://doi.org/10.1016/j.biortech.2013.03.057.

Zhao, Q., Xu, T., Song, X., Nie, S., Choi, S. E., & Si, C. (2021). Preparation and Application in Water Treatment of Magnetic Biochar. *Frontiers in Bioengineering and Biotechnology*, *9*, 769667. Available from https://doi.org/10.3389/fbioe.2021.769667, PMID: 34760880; PMCID: PMC8572963.

Zhong, D., Jiang, Y., Zhao, Z., Wang, L., Chen, J., Ren, S., Liu, Z., Zhang, Y., Tsang, D. C. W., & Crittenden, J. C. (2019). pH Dependence of Arsenic Oxidation by Rice-Husk-Derived Biochar: Roles of Redox-Active Moieties. *Environmental Science & Technology*, *53*(15), 9034–9044. Available from https://doi.org/10.1021/acs.est.9b00756.

Zhu, L., Chattopadhyay, S., Elijah Akanbi, O., Lobo, S., Panthi, S., Malayil, L., Craddock, H. A., Allard, S. M., Sharma, M., Kniel, K. E., Mongodin, E. F., Chiu, P. C., Sapkota, A., & Sapkota, A. R. (2023). Biochar and zero-valent iron sand filtration simultaneously removes contaminants of emerging concern and Escherichia coli from wastewater effluent. *Biochar*, *5*(2023), 41. Available from https://doi.org/10.1007/s42773-023-00240-y.

Chapter 14

Treatment of industrial and municipal wastewaters using biochar: performance, mechanisms, and research needs

Abudu Ballu Duwiejuah[1], Abubakari Zarouk Imoro[2], Noel Bakobie[2], Abdul-Aziz Bawa[3], Joseph Payne[1] and Emmanuel Okoampah[4]

[1]*Biotechnology and Molecular Biology, University for Development Studies, Tamale, Ghana,* [2]*University for Development Studies, Environment, Water and Waste Engineering, Tamale, Ghana,* [3]*Spanish Laboratory Complex, University for Development Studies, Tamale, Ghana,* [4]*Biochemistry, University for Development Studies, Tamale, Ghana*

Chapter outline

14.1 Introduction	265
14.2 Wastewater treatment	266
14.3 Biochar technology in wastewater treatment	266
14.4 Type of wastewater treated by biochar	267
14.5 Industrial wastewater treatment	267
14.6 Municipal wastewater treatment	267
14.7 Adsorption of inorganic pollutants onto biochar in wastewater	268
14.8 Adsorption of organic pollutants onto biochar in wastewater	269
14.9 Adsorption mechanisms of biochar for inorganic pollutants in wastewater	269
14.10 Adsorption mechanisms of biochar for organic pollutants in wastewater	270
14.11 Some applications of biochar for wastewater treatment	271
14.12 Current and emerging constraints and gaps	272
14.13 Future research in biochar application for wastewater treatment	273
14.14 Conclusion	274
References	274

14.1 Introduction

Water pollution is a severe global issue because contamination is due to a variety of factors such as population growth, urbanization, industrialization, and so on. Environmental contaminants can cause several illnesses, including cancer and neurological issues. Industries produce waste, which is then released into the atmosphere, water bodies, and soil, resulting in air, soil, and water pollution (Jha et al., 2023). The entire ecosystem is negatively impacted by pollution, especially human life. The majority of environmental threats caused by contaminants are produced by industry.

Rapid population growth, industrialization, the spread of urbanized culture, and excessive material consumption are the leading causes of excess agricultural, municipal, and industrial waste in human civilization. Slaughterhouses, for instance, are a significant source of contamination of water (Quarters et al., 2018). Though abattoirs face significant challenges in enforcement and monitoring, they threaten aquatic life and the quality of natural water (Quarters et al., 2018). Frequent monitoring of pollution point sources, for instance, effluents from industry and treatment facilities, should be done by environmental authorities.

Water quantity and quality difficulties associated with wastewater are the cause of most of humanity's major problems in the 21st century. For proper wastewater treatment plant operation, industrial wastewater must be considered as it makes up a sizable component of municipal wastewater (Ekanayake & Manage, 2022). It is critical to treat wastewater whether it is for reuse or to reduce the recipient's exposure to pollution. In metropolitan regions, traditional wastewater treatment facilities are frequently used to treat wastewater; however, in rural locations, these facilities will be financially unviable (Westholm, 2021).

Technology is constantly being investigated to remediate pollutants in chemical wastewater. Adsorption technology for removing heavy metals from wastewater has recently received much interest because it is convenient and affordable (Xue et al., 2019). Biochar wastewater treatment has become a novel technology owing to its several distinct qualities that set it apart from other alternatives. When used as an adsorbent, biochar, a byproduct of biomass pyrolysis, has several physicochemical properties that make it easier to adsorb and fix heavy metals from water. Biochar is formed when biomass undergoes thermochemical decomposition under limited oxygen conditions. Any organic waste such as algae, wood chips, crop and forest residues, manures, municipal solid waste, and sewage sludge can be applied as feedstock. Water is used widely in many crucial human processes, from daily consumption and cleanliness to its involvement in most industrial manufacturing processes. However, when water is used in industrial and urban settings, it is frequently polluted with various chemicals that pose various threats to human health and the environment.

This chapter was on recent developments in the application of biochar for industrial and municipal wastewater treatment, including a discussion on wastewater treatment and mechanisms of inorganic and organic pollutant adsorption. Moreover, this covers biochar technology in wastewater treatment, including the types of wastewater, industrial and municipal wastewater treatment, adsorption of inorganic and organic pollutants onto biochar in wastewater, and biochar adsorption mechanism. It also discusses some studies on the usage of biochar for wastewater treatment, current and emerging, constraints and gaps, and areas of future research in biochar application for wastewater treatment. The objective of this chapter is to discuss biochar application in treating industrial and municipal wastewater.

14.2 Wastewater treatment

Biochar has high porosity and a large surface area, making it an exciting alternative for treating sewage. Biochar can significantly increase toxin expulsion in wastewater due to its strong adsorption and high porosity qualities, which permit pollutants to adsorb on its surfaces to yield a clean effluent (Yaashikaaa et al., 2020). The usage of raw biowaste and carbonized materials in wastewater treatment is increasing. Many researchers have reported that biochar effectively removes more hazardous pollutants than activated coal (Yaashikaaa et al., 2020). Comparatively, the financial and ecological potential of biochar in terms of removing hazardous pollutants is far better than activated coal. The impact of the production of activated coal is established by the lower greenhouse gas emissions produced during biochar preparation. More energy is needed to produce activated coal (97 MJ/kg) than to produce 6.1 MJ/kg of biochar (Yaashikaaa et al., 2020). Biochar could be preferable to activated coal for removing some hazardous contaminants from wastewater whilst considering energy demand, greenhouse gas emissions, and, consequently, creation costs.

14.3 Biochar technology in wastewater treatment

Biochar is an effective adsorbent for a variety of contaminants removal owing to its characteristics such as its sizeable steady-state estimate, which embodies the reaction rate of a multiple-step reaction, its content in nanomaterials, its abundance in surface functional groups, and its porous structure (Xiang et al., 2020). Biochars have gained prominence in reducing pollution in the agricultural and industrial sectors and enhancing environmental quality due to their unique features as a good adsorbent in removing a variety of pollutants (Wang et al., 2017). Being a result of home, commercial, industrial, or agricultural activity, wastewater has been a problem on a global scale. The ramifications on humans, animals, and plants are worrying, making it an issue of public health interest. The affinity and capability for sorbing organic molecules in biochar are incredibly high. The kind of feedstock, production temperature, and production settings (such as on-site or in a lab) all affect the sorption affinity of biochar. It has been discovered that some chemical classes are predominantly strongly sorbed to biochar surfaces. This has been linked to specific interactions between the rings of polycyclic aromatic hydrocarbons and the molecules of biochar (Lehmann & Joseph, 2009). Toxic metals, nitrate, and phosphorus have all been successfully removed by biochar from wastewater (Xiang et al., 2020). Even at low concentrations, prolonged exposure to toxic metals in the aqueous phase poses significant health risks (Ahmed et al., 2016). Animal and plant wastes biochar can adsorb metals from wastewater and water successfully is evidence that keeps growing (Dai et al., 2017). Biochar can absorb phosphorus and nitrate in the aqueous phase (Xue et al., 2016). With novel biochar modifications and variants constantly developing, biochar is being used more frequently in wastewater treatment every year (Xiang et al., 2020). Biochar can be used for various remediation procedures, resulting in the efficient removal of pollutants, thanks to the material's adaptability and a wide range of technological manufacturing choices (Xiang et al., 2020). Given the abundance of research resources on biochar and its uses, there is a clear need to summarize and analyze this information.

14.4 Type of wastewater treated by biochar

Biochar can efficiently remove pollutants and treat wastewater based on the materials used. Fewer studies (Ok et al., 2018; Zheng et al., 2019) stated municipal or household wastewater as an application ground for biochar for cleanup. In other situations, the effectiveness of biochar in removing certain pollutants frequently connected with unusual industries such as naphthenic acids in the petroleum business and methylene blue dye in the textile sector has been investigated. Another promising application of biochar is wastewater cleanup at cattle farms, as this type of wastewater often contains antibiotics or other potentially hazardous substances that biochar may efficiently remove (Thang et al., 2019).

In the construction of wetlands, biochar can also be incorporated directly aiming at wastewater treatment. In constructed wetlands, the usage of biochar appears to be the material's most common extra application among those mentioned. Because it is a versatile material that can be utilized in a variety of disciplines of environmental management, the biochar used in wastewater is becoming more popular in bioremediation. A constructed wetland is a multifaceted, intricate system mainly used to clean wastewater (Parde et al., 2021). Using currently built wetlands as a wastewater treatment system is not very common. Still, it is gaining popularity as a highly self-sufficient and sustainable on-site treatment that requires little upkeep (Abedi & Mojiri, 2019). Its significant drawbacks, such as expense and reliance on the weather, make it unsuitable for use as a common wastewater treatment technique (Parde et al., 2021). Biochar can be used to improve microalgal growth, leading to better microalgal remediation of wastewater (Sforza et al., 2020).

14.5 Industrial wastewater treatment

Industrial wastewater is produced by several processes such as mining, smelting, battery production, chemical production, leather production, dye production, and others. Additionally, the principal pollutants in industrial wastewater include organic pollutants and heavy metals (Xiang et al., 2020). Industrial effluent has been treated with biochar in some cases. It is effective as an adsorbent for toxic metal adsorption in industrial effluent. According to Hussain et al. (2017), the ratio of chitosan and biochar might influence arsenic, cadmium, lead, copper, and other toxic metals adsorption in industrial effluent. The process of adsorption is influenced by pore volume, pH level, and surface area of biochar (Wathukarage et al., 2017). According to Poonam et al. (2018), the highest adsorption capacity of 13 mg/g, and the adsorptive process is predisposed by the medium's pH value, dosage, and contact time.

Biochar was also utilized to recover nutrients such as nitrate and phosphate, among others, from dairy wastewater containing ammonium and phosphate. According to Ghezzehei et al. (2014), biochar can absorb 19 to 65% of phosphate and 20%–43% of ammonium under 24 hours from flushed dairy manure. Most trials so far have been conducted under laboratory conditions, using biochar to remove pollutants from industrial wastewater. However, more research and application in wastewater treatment are required. To recover labile nitrate and phosphate from urban wastewater, only biochar can be used or in combination with biofilters and other technologies (Cole et al., 2017). Biochar that has been loaded with aluminum oxyhydroxides has been utilized to recover phosphate from wastewater (Zheng et al., 2019). Electrostatic attraction is essentially how phosphorus can be adsorbed.

Ammonium removal was accomplished using digested sludge biochar from municipal wastewater. Biochar produced at 450°C has the maximum removal capacity of ammonium owing to its chemisorption regulates the process, larger surface area, and higher functional groups (Tang et al., 2018). Biochar can stimulate oxidation by forming mineralized petroleum contaminants and hydroxyl radicals, as it contains Si/O structures, metallic oxides, and functional carbon groups (Chen et al., 2019).

14.6 Municipal wastewater treatment

Municipal wastewater treatment can be done with municipal biowaste biochar at the biofiltration stage. Biochar can serve as a biofilter for purifying sewage in a municipal setting because of the huge surface area of its pores. After passing wastewater through a charcoal biofilter, its chemical oxygen demand (COD) decreased by 90%, total suspended solids (TSS) by 89%, total Kjeldahl nitrogen by 64%, and 78% for total phosphorus (TP) (Manyuchi et al., 2018). Adding biochar accelerates the elimination of several polar and hydrophilic chemicals such as methanol, acetone, ethanol, dioxane, isopropanol, propanol, alcohols, acetic acid, urea, and ammonia. Thus, on-site sewage treatment facilities can be improved with affordable adsorbents.

Domestic on-site wastewater includes flush water, urine, human feces, cleaning supplies, and occasionally gray water. It also contains pathogens, pharmaceuticals and personal care products (PPCPs), and heavy metals, as well as organic matter. Pathogens, for instance, *Escherichia coli*, *Leptospira* spp., *Legionella pneumophila*, *Shigella* spp., *Yersinia enterocolitica*, *Salmonella* spp., and *Vibro cholera*, as well as helminthic pathogens such as Ascaris, hookworms, and whipworms, have all been identified in domestic wastewater on-site (Manga et al., 2023).

Detergents and personal care products are the source of heavy metals in domestic wastewater (Akpor et al., 2014). Lead (Pb), cadmium (Cd), copper (Cu), manganese (Mn), mercury (Hg), nickel (Ni), iron (Fe), zinc (Zn), and silver (Ag) are the heavy metals that are most frequently found in domestic wastewater (Manga et al., 2022). Caffeine, ciprofloxacin, carbamazepine, ibuprofen, erythromycin, ketoprofen, and sulfamethoxazole are just a few of the PPCPs that have been identified in domestic wastewater (Miège et al., 2009). The aforementioned toxins can be found in varying concentrations in household wastewater streams depending on the weather, water consumption patterns, and the number of users (Oteng-Peprah et al., 2018). Most contaminants, especially PPCPs, are persistent and difficult to remove using standard on-site treatment methods such as sand filters and soil infiltration treatment systems. It has been demonstrated that these systems are only partially efficient in eliminating the various organic micropollutants, present in household wastewater. In contrast, Blum et al. (2017) reported in soil beds negligible ($>10\%$) acetaminophen removal, whilst Gros et al. (2017) distinguished low to intermediate PPCPs removal in soil beds that are applied for on-site filtration (e.g., 46% ibuprofen removal, 86% diclofenac removal, 44% metoprolol removal, and 73% caffeine removal).

PPCPs are currently removed from wastewater using a variety of techniques, including adsorption, chemical oxidation, and membrane (process) (Acero et al., 2015), electrochemical (Barrios et al., 2016), and microbial (Ferreira et al., 2016). Adsorption is a widespread technique among the aforementioned, because of its low price, high efficiency, and broad processing scope. Activated carbon, alumina, adsorbent resin, silica gel, polyacrylamide, and zeolite are the most often used adsorbents (Jiang et al., 2020). Whilst other adsorbents such as silica gel, alumina, and adsorbent resin have narrow surfaces for adsorption and complicated production processes, activated carbon is expensive and huge specific surface area and hence a large adsorption capacity (Cheng et al., 2021). Also, it is feasible to employ biochar as an effective adsorbent to remove pollutants from aqueous solutions due to its unique qualities, which include enhanced surface functional groups, a large specific surface area, mineral components, and porous structure (Tan et al., 2015). Other practical, affordable, and ecologically acceptable treatment methods such as the utilization of biochar are therefore required. The biochar characteristics such as its rich functional groups, porous structure, mineral components, and large surface area make it likely to be used as a suitable adsorbent for pollutants in domestic wastewater compared to other adsorbents.

14.7 Adsorption of inorganic pollutants onto biochar in wastewater

Adsorption of inorganic salts by biochar in wastewater increases quantities of nutrients, for instance, ammonium, nitrates, and phosphates, which can have major environmental consequences for the ecosystem (Goala et al., 2021). For example, they will enhance photosynthesis, eutrophicate aquatic ecosystems, and degrade water quality (Gizaw et al., 2021). Diverse biomass has been utilized to make biochar, which has been used extensively and successfully as an adsorbent material for organic contaminants in water and wastewater. The quality of the water, aquatic ecosystems, canal transit, and human health can all suffer from small amounts of phosphate (>25 g/L) in surface water (Zhang et al., 2020). In general, electrostatic attraction causes oxyanions from an aqueous solution to bind to the positively charged biochar. Because the biochar surface has an amphoteric character in reaction to solution pH, it is a crucial element in nitrate adsorption (Tareq et al., 2019).

Fluoride is a significant inorganic contaminant that is toxic. Due to more outstanding negative charges on the biochar surface, it was discovered that acidic conditions are more suitable for fluoride uptake (Oh et al., 2012). Low pH positively charged oxidized iron and aluminum species on the sludge biochar increased the electrostatic interaction between fluoride ions. More fluoride ion adsorption by biochar at acidic pH has been reported in drinking water defluoridation (Tareq et al., 2019). Phosphate adsorption was inhibited by competitive adsorption of other anions. This was due to the layered double hydroxide component's acid-base characteristics, electrostatic attraction, and surface complexation, phosphate was also pH-sensitive during adsorption. Mixed metal oxides of Mg and Al (Zhang et al., 2020). Among numerous low-priced adsorbents, biochar has a high promise as an eco-friendly adsorbent. The inorganic pollutants adsorption in wastewater by biochar is determined by the biochar's features, which include active components, mineral content, porous structure, and surface groups (Wang et al., 2024).

14.8 Adsorption of organic pollutants onto biochar in wastewater

Rapid industrialization and agrochemical-based farming produce significant amounts of organic contaminants, which hurt the environment. These pollutants include antibiotics, dyes, fungicides, herbicides, polycyclic aromatic hydrocarbons, pesticides, and polychlorinated biphenyls (PCBs). It is still difficult to get rid of these pollutants from wastewater, drinking water, and soil. According to Tareq et al. (2019), organic pollutants are of special concern. They can linger in the environment for a very long time and harm aquatic ecosystems, human health, and the food chain. Tetracycline in animal farm wastewater is one of the most prevalent antibiotics, and phenols can impact the flavor and odor of fish and drinking water even at deficient concentrations (Zhang et al., 2020).

Water pollution from organic debris has gotten worse in recent years. Antibiotics and dyes are two examples of common organic contaminants (Utami et al., 2023). Significant degrading effects or adsorption of biochar on organic contaminants in water have been demonstrated (Wang et al., 2024). Among the organic contaminants released into wastewater by sectors such as paint, leather, paper, and pharmaceuticals are dyes (Gao et al., 2021). Their intricate biological structures may lead to issues with the skin, gastrointestinal system, and lungs (Rajarathinam et al., 2020). Also, dyes have the potential to improve COD and obstruct light penetration into water bodies when dispersed, seriously harming the aquatic animals' and plants' natural habitats (Georgin et al., 2019). As a result, coloring components must be eliminated or changed into safe materials. Dyes of various chemical structures can be successfully removed by biochar (Wathukarage et al., 2019). To fulfill the role of adsorption, hydrogen bonds are usually established between the functional groups of the dye and the oxygen-containing biochar. Additionally, $\pi-\pi$ interaction and electrostatic attraction are complicated in the dye-removal processes of biochar. Dependent on the kind of dye molecules and how they interact with the particular biochar being utilized, different removal procedures apply. The newly discovered pollutants, such as antibiotics, are frequently used in clinical settings to treat and prevent illnesses in both people and animals. Antibiotics have a poor metabolism and can have hormonal effects, carcinogenic, mutagenic, or teratogenic. As a result, when released into the water bodies can cause ecological disasters (Ma et al., 2019).

Solution via the deprotonation and protonation of oxygen-encompassing groups on the composite surface of biochar, pH primarily impacts the charge properties of biochar, which, in turn, aid the adsorption of pollutants. The pH primarily impacts the charge properties of biochar in solution through deprotonation and protonation of oxygen-encompassing groups on their surface, which, in turn, affect contaminant adsorption. The composite surface is charged positively when the pH of the solution is lower it encourages anion adsorption than the biochar point of zero charge (pzc). On the other hand, cation adsorption is encouraged when pH > pHpzc because of the negative charges on the surface (Wang et al., 2020a, 2020b). The participation of the " + − " electron donor-acceptor interaction is muted, and the adsorption of sulfathiazole by biochar is reduced when the pH rises above a particular point (amino groups are deprotonated). Dyes are a challenge to conventional waste treatment methods because they resist light, aerobic digestion, and oxidation and are thus amenable to adsorption (Zhang et al., 2020). As a result, biochar applications in the remediation of organic have attracted substantial research interest in recent years as a sustainable, eco-friendly, and affordable alternative. According to Khataee et al. (2017), the ZrO_2-biochar composite, for instance, had a pzc of 7.35 and adsorption at pH 6.0 when the ZrO_2-biochar surface was charged positively. The mixture of ZrO_2 positively charged and biochar was also an effective adsorbent for anionic dyes that might be broken down by the hydroxyl formed during the reaction. As the pH of the solution rises, biochars are less effective at adsorbing organic pigments. The phosphorus adsorption by biochars is also significantly influenced by solution pH, with a maximal effect occurring at roughly 3.0 pH (Zhao et al., 2021).

14.9 Adsorption mechanisms of biochar for inorganic pollutants in wastewater

The type of pollutant (adsorbate) and the heterogeneity of the adsorbent's surface both affect the adsorption method used by biochar. Biochar, due to its rich microporous structure and high specific surface area, can offer phosphate ions active sites of adsorption (Wang et al., 2024). The characteristics of biochar influence its mechanism of adsorption of inorganic pollutants and its field application. To remove inorganic pollutants, the adsorption process uses ion exchange, coprecipitation, complexation, and electrostatic attraction. Van der Waals forces and hydrogen bonds through functional groups, filtration, surface complexation, electrostatic contact, and ion exchange all contribute to adsorption to the biochar surface (Cheng et al., 2021; Li et al., 2019). According to Li et al. (2014), coprecipitation and complexation occur when metals in domestic wastewater combine with free hydroxyl and carboxyl functional groups or oxide minerals on the surface of biochar to form complexes, after which they physically adhere to the biochar surface or form layers (precipitation). According to Xiang et al. (2020), biochar typically adsorbed emerging pollutants such as PPCPs, primarily by mechanisms of hydrophobic effects, electrostatic attraction, cationic p-acceptor and aromatic p-donor conjugation, hydrogen bonding, and pore filling. The mechanism of adsorption consists of electrostatic interaction, surface functional group contact, hydrogen bonding, surface complexation, and coprecipitation (Zubair et al., 2022). Using phosphate as an example, the specific surface area, the functional groups, and metal ion composition of the biochar all play a major role in the phosphate adsorption mechanism (Zeng et al., 2013). Phosphate ions will diffuse to the surface of the biochar and cause physical deposition. The biochar functional groups, for instance, C-H, C=O, -COOH, and -OH (Karunanithi et al., 2017), $-NO_2$ and $-NH_2$ (Krishna Veni et al., 2017) react with phosphate ions. Phosphate experiences chemical adsorption on the biochar surface if it has metal elements such as magnesium hydroxides and calcium or oxides on it (Kamyab et al., 2023). For instance, Park et al. (2018) showed that hydroxyapatite (HAP) is formed when phosphate ions mix with $Ca(OH)_2$ or CaO on the surface of crayfish biochar.

14.10 Adsorption mechanisms of biochar for organic pollutants in wastewater

The biochar surface functional groups, the type of contaminants, and the specific surface area all influence the mechanisms of biochar adsorption to antimicrobial contaminants (Zhou et al., 2022). Pollutant elimination occurs when biochar is used as a medium for residential wastewater treatment through different physical, chemical, and biological methods. However, several variables such as the physicochemical features of the biochar, system setup, and operational state affect the dominating removal process (Quispe et al., 2022). Adsorption, ion exchange, biodegradation, and entrapment/pore filling are some of these mechanisms (Fig. 14.1). When organic pollutants need to be eliminated, hydrophobic sorption, hydrogen bonds, electrostatic attraction, electron-donor acceptor interactions, and pore fills all help the adsorption process (Fig. 14.1). Another method of pollutant removal from on-site wastewater via adsorption is pore filling.

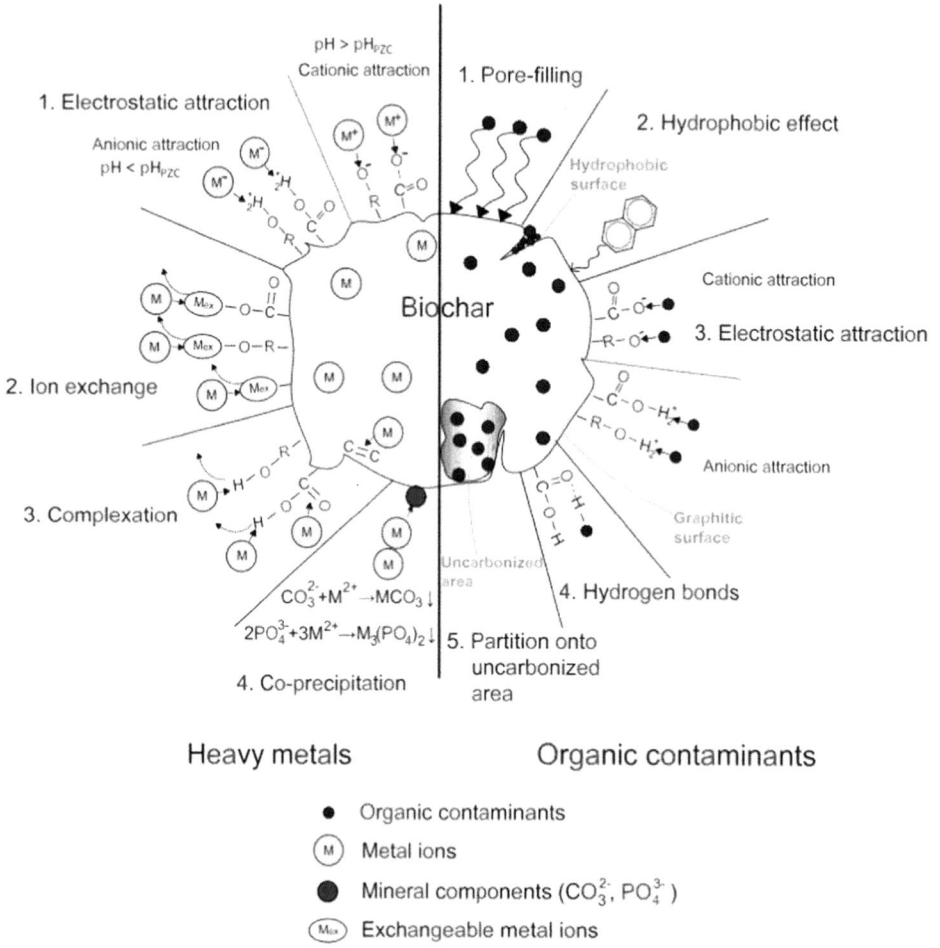

FIGURE 14.1 Adsorption mechanisms of inorganic and organic pollutants on biochar. Source: *Wang, X., et al. (2020b). Recent advances in biochar application for water and wastewater treatment: A review. PeerJ, 8, Article e9164. https://doi.org/10.7717/peerj.9164.*

The small mesopores and micropores (2−20 nm) contribute significantly to the biochar surface area and primarily affect its uptake of organic molecules (Pignatello et al., 2006). The pore network of biochar is typically made of micropores 50 nm. Pore filling is typically used to remove organic contaminants from residential wastewater, which depends on the total volume of micropores and mesopores on the biochar surface (Hao et al., 2013). The organic pollutant's nature, polarity, and biochar type influence the pore filling mechanism. Its primary adsorption processes, as shown in Fig. 14.1, are hydrogen bonding, surface complexation, and electrostatic attraction. Multilayer mechanisms, such as $\pi-\pi$ interaction and pore filling, come next.

14.11 Some applications of biochar for wastewater treatment

Biochar has previously been used successfully in on-site sanitation systems for domestic wastewater treatment. In trials conducted by Mwenge & Seodigeng (2019), activated biochar made from agricultural waste sandwiched between two layers of gravel served as a filter to remove TSS, NH_3-N, NO_3-N, and COD from graywater used for cleaning floors, showers, and laundry. It was discovered that kitchen wastewater had the maximum pollutant content and also recorded the highest adsorption rates for the most contaminants, indicating that the concentration of the starting contaminant used in the biochar impacted the removal effectiveness. According to Mwenge & Seodigeng (2019), the biochar column was effective at removing COD (63% of kitchen graywater), NO_3-N (76% of bathroom graywater), TSS (95% of kitchen graywater), and NH_4-N.

In a laboratory filtering column, Sidibe (2014) evaluated the salix leaves biochar filter efficiency for removing COD, male-specific phages, and *Salmonella* spp. from artificial graywater. The rate of organic loading was 240 g COD m^{-2} day^{-1} (76 mg BOD m^{-2} day^{-1}) and the hydraulic loading rate was 0.03 m^3 m^{-2} day^{-1}. It was discovered that a 60 cm high biochar filter reduced COD by 90% and reduced *Salmonella* spp. by 3 log_{10}. However, the virus only experienced a 2 log_{10} drop for male-specific phages. To prepare home wastewater for irrigation, Perez-Mercado et al. (2019) investigated the efficacy of a hardwood charcoal filter for pathogens removal (*Enterococcus* spp., *E. coli*, *S. cerevisiae*, and bacteriophage (Ms2)). *S. cerevisiae* was successfully removed from graywater using a charcoal filter. Its impact on viral elimination, however, was shown to be minimal. These insufficient virus loads decreased in the laboratory scale when the biochar filters were used, which implies that biochar has a limited ability to eliminate viruses. The pH of the filter medium affects how well graywater filtration systems remove viruses because the charge of viruses varies with pH (Lalander et al., 2013).

Additionally, using biochar media with smaller particle sizes to enable more significant straining might be researched. Most of the investigations were conducted using column filters, as observed from a study of the papers. Instead of column filters, further research must be done utilizing other types of filters such as horizontal flow filters. Using a pilot-scale biochar filter derived from a sand bed to treat residential graywater, Dalahmeh et al. (2016) found that the filter's removal efficiencies for BOD were 93%, TSS was 85%, and 1 log_{10} decrease for *E. coli*.

Niwagaba et al. (2014), who built a biochar filter of 60 cm high gravel for the treatment of graywater, successfully removed BOD, TSS, and fecal coliforms with high removal efficiencies after 36 hours of operation. The respective results were 96%, 85.20%, and 95%. It can be noted that Niwagaba et al. (2014) obtained high clearance of fecal coliform, which contrasts with the study by Dalahmeh et al. (2016). These variations can be directly attributed to the bacteria characteristics, the characteristics of biofilm-supportive materials, environmental factors such as temperature and pH, the other bacteria's presence, and the nutrients in the media (Van Houdt & Michiels, 2010). All of these factors work together to contribute to the removal of fecal coliforms. In a study, biochar of approximately 0.20 m was added to a sand layer of 0.80 m on top of a filtration unit used to recover domestic wastewater with 1.9 mg P/L (Kholoma et al., 2016).

When compared to a sand-only column, the phosphorus removal efficiency of the biochar-amended sand column increased by 25.60%. Gupta et al. (2016) combined horizontal flow-built wetland filter media with oak tree biochar to remediate synthetic wastewater. The use of biochar gives rise to an upsurge in pollutant removal rates of 92% for NO_3-N, 91.30% for COD, 79.50% for TP, 67.70% for PO_4, 58.30% for NH_3, and 58.30% for total nitrogen. The 100% chipped hemp fiber biochar filter addition in a pilot-scale study after a horizontal flow gravel wetland for domestic wastewater treatment improved phosphorus removal over the 5 months (long term); 97% phosphate removal was related to the control of 87% (Bolton et al., 2019). According to the studies described above, biochar can be used as a filter media to improve the ability of manmade wetlands to remove phosphorus, which is usually not very effective. Ok et al. (2018) exposed high nitrogen concentration wastewater to saturated and unsaturated biochar filters for a year. They discovered that the removal rate of NH_4^+ (ammonium ion) in the unsaturated biochar nitrifying column was higher at 0.08−0.10 kg N/m^3.day than in the control gravity column of 0.04−0.09 kg N/m^3.day. Most studies showed adding biochar to treatment units advances the effluent quality produced by these systems, making it a viable choice for on-site domestic wastewater treatment.

Biochar-modified on-site sanitation methods have also been used for improved PPCP adsorption (Cheng et al., 2021). Applying wheat residue and eucalyptus wood biochar to the soil matrix of 0.50 t/ha increased the elimination of specific active pharmaceutical components (carbamazepine and propranol) threefold (Williams et al., 2015). The efficiency of nitrogen removal in a sand column in a different trickling filter system increased from 8% to 42% when biochar was added (Tait et al., 2015). For improved PPCP adsorption, biochar-modified on-site sanitation methods have also been used (Cheng et al., 2021). The elimination of specific active pharmaceutical components (propranol and carbamazepine) increased threefold when wheat residue and eucalyptus wood biochar were added to the soil matrix of 0.5 t/ha (Williams et al., 2015). Another study examined the effectiveness of removing contaminants from wastewater (Table 14.1).

14.12 Current and emerging constraints and gaps

Biochar is still a relatively new wastewater treatment technology that requires extensive testing and research. Some environmental issues and future advancement factors must be researched before further implementation. In addition to the current environmental issues Wang et al. (2020a, 2020b) suggest a few recommended options for future development. The first consideration is cost, as currently producing biochar involves a significant amount of preparation such as preparing the feedstock, pretreating it, pyrolyzing it, and sometimes extra procedures such as chemical or physical

TABLE 14.1 Some studies on wastewater treatment.

Biochar	Type of pollutant	Wastewater	Pollutant	References
Raw corn	Inorganic pollutant	Synthetic wastewater	PO_4-P	Zhang et al. (2013)
Lodgepole pine wood / wood)	Inorganic pollutant	Brewery wastewater (real industrial wastewater)	COD-T, COD-D, TSS, PO_4-P, and NH_4-N	Huggins et al. (2016)
Bamboo	Inorganic pollutant	Synthetic wastewater	NH_4-N and COD	Zhou et al. (2018)
Corn straw	Inorganic pollutant	Domestic wastewater	COD and NH4-N	Sun et al. (2018)
Quercus sp. (Oak tree) mixed with other media	Inorganic pollutant	Synthetic wastewater	N and P	Gupta et al. (2016)
Hardwood mixed with other media	Inorganic pollutant	Secondary clarified wastewater septage	BOD	de Rozari et al. (2015), de Rozari et al. (2016), and de Rozari et al. (2018)
Hardwood mixed with other media	Inorganic pollutant	Secondary clarified wastewater septage	P and N	de Rozari et al. (2015), de Rozari et al. (2016), and de Rozari et al. (2018)
Forestry waste, biosolids	Inorganic pollutant	Wastewater, surface water, and stormwater	Sulfamethoxazole (SMX)	Shimabuku et al. (2016)
Corn straw	Inorganic pollutant	Domestic wastewater	TN	Sun et al. (2018)
Quercus sp. (Oak tree) mixed with other media	Organic pollutant	Synthetic wastewater	Organic matter	Gupta et al. (2016)
Poplar wood (high and low temp. pyrolysis)	Organic pollutant	Wastewater	Organic matter	de Caprariis et al. (2017)
Hardwood mixed with other media	Bacteria	Secondary clarified wastewater septage	SS; *E. coli*	de Rozari et al. (2015), de Rozari et al. (2016), and de Rozari et al. (2018)

modification of the biochar, which raises the cost of the finished product. Therefore, it will be crucial to consider and work with these production prices in future development because they directly affect how widely applicable biochar is as a water treatment technique.

The absence of pilot-scale and industrial adoption of biochar technology in undeveloped nations may lead environmental groups to view it as futuristic and too ambitious. Currently, only laboratory-scale research has been done on using biochar for on-site sanitation, including treating graywater and fecal sludge. It is unclear how using biochar for on-site sanitation would affect the environment.

The primary methods for activating biochar physically and chemically are expensive, energy-intensive, and have environmental consequences. Biochar has a large surface area that can be used as a support material for biofilm colonization and growth where the microbes may stick to its surface and form an extracellular biofilm. Owing to its porous structure, functional groups, and high surface area, biochar adsorbs other pollutants such as heavy metals in this modification system; however, microorganisms will accelerate the decomposition of organic pollutants.

According to a life cycle and cost assessment done by Homagain et al. (2016), wood pyrolysis to produce biochar is an expensive endeavor. This is particularly true when biochar needs to be modified or activated to increase its functionality; pyrolysis, for example, requires the feedstock to be pyrolyzed first before it is activated, and this typically calls for multiple treatment steps, corrosive chemicals, and high energy and electricity inputs, making it a time-consuming, complex method, and expensive (Zhou et al., 2019). Therefore, more methods that can enhance the functioning of biochar, such as simultaneous pyrolysis and activation, should be pursued.

Biochar changed by biological processes is helpful for on-site sanitation systems because of the synergistic elimination action. Through sorption and biodegradation, respectively, organic and inorganic contaminants can be eliminated simultaneously. Finally, although the use of biochar for treating wastewater has been extensively documented in the literature, nothing is known about how it affects antibiotic-resistant gene migration from home wastewater to the environment. Emerging microbial infections that include antibiotic-resistant genes are now a significant public health issue. Once these dangerous genes are released into the environment media, they can change the microbial community makeup, which, in turn, affects the water quality (Feng et al., 2021). They can easily spread through horizontal gene transfer, which causes susceptible bacteria to acquire the essential gene element from the developed resistance and mobile genetic element. Therefore, it is urgently necessary to conduct thorough experiments to determine the potential of biochar in reducing antibiotic-resistant genes found in on-site wastewater.

14.13 Future research in biochar application for wastewater treatment

It is essential to thoroughly investigate the effects of environmental concerns and future development directions of these variables on the performance of biochar and test new potential technologies that could improve it because the performance of biochar is directly related to the production conditions, feedstock, and its modifications. The stability of the biochar, as well as the stability of any alterations, should be taken into account and assessed for practical applications.

Municipal wastes, industrial wastes, and agricultural wastes can all be used as biomass feedstock for biochar production. However, chemicals from the feedstock may hurt the performance of the biochar and even cause additional pollution in some cases. For instance, suppose the biochar was made from sewage sludge and contains toxic metals that can leak out when applied. Therefore, it is essential to carefully assess and explore the safety of using biochar (Ok et al., 2018). The majority of recent studies on biochar focus primarily on particular chemical compounds to gauge how well they removed those substances. However, wastewater influent is frequently polluted with several contaminants. Therefore, it is important to carry out more actual tests and studies about biochar's effectiveness concerning various types of pollutants. Given the number of biochar changes that have been successfully shown, it is possible to develop more potent alterations to boost the material's effectiveness. These modifications typically involve increasing surface area, porosity, or surface sorption sites. Though biochar is considered a reasonably sustainable and renewable process compared to numerous other technologies, more testing and study are still needed to determine its potential uses and cyclical uses.

Additionally, biochar loses mechanical strength over time (Spokas et al., 2014). Therefore, research is necessary to optimize the use of biochar to reduce the emission of hazardous compounds into the atmosphere. Prolonged biochar exposure to stress may result in the production of fragments. To improve the adsorption of contaminants, it is also crucial to determine the ideal biochar dose and the resident times each dose can be applied to managing fecal sludge. Although biochar and modified carbon have certain similarities, it is known that the physicochemical characteristics of biochar can vary greatly depending on the feedstock type and the pyrolysis operating conditions. Introducing biochar technology into on-site sanitation treatment systems still lacks established standardized protocols (Gwenzi et al., 2017).

To use biochar more frequently, advanced pyrolysis procedures with accurate reaction condition control, low environmental effect, and high energy efficiency are necessary. Furthermore, the production of biochar using pyrolysis technology is energy-intensive and inefficient, making it a less appealing option today.

Applying biochar to the environment is fraught with risk because the conditions in the real world are more complex than those fabricated in the lab. Therefore, further *in-situ* research is needed before biochar is used on a large scale to determine the actual impact on the environment.

It is necessary to investigate how the inherent qualities of biochar may affect its capacity to eliminate antibiotic-resistant genes from on-site wastewater. Additionally, adjustments like the employment of transition metals to boost the quantity of persistent free radicals on the biochar surface should be considered.

14.14 Conclusion

The versatile and adaptable biochar technology has a wide range of potential applications. It has been found that adding biochar to on-site wastewater treatment systems increases the systems' capacity for treating wastewater. A developing technique that is becoming more and more common is biochar. A mixture of mechanisms, including adsorption, biodegradation, electrostatic attraction, filtration, ion exchange, and pore filling, is used by biochar to remove various contaminants in wastewater. However, to assess all potential hazards and issues and enable its effective performance in wastewater treatment, much study remains to be done. There is a lack of consensus regarding the use of biochar for pollutant removal. Comparing the qualities of the biochar under investigation is impossible because different varieties of biochar based on feedstock, produced under different circumstances, have been studied. However, studies on biochar and its usage for removing contaminants from wastewater are growing. As a result, the body of knowledge will continue to grow, and it will be helpful to continue this research because biochar appears to have promise. To improve the qualities of biochar and its capability to remove emerging contaminants from home wastewater, more study is also required.

References

Abedi, T., & Mojiri, A. (2019). Constructed wetland modified by biochar/zeolite addition for enhanced wastewater treatment. *Environmental Technology & Innovation*, 16. Available from https://doi.org/10.1016/j.eti.2019.100472, Elsevier BV.

Acero, J. L., et al. (2015). Elimination of selected emerging contaminants by the combination of membrane filtration and chemical oxidation processes. *Water, Air, and Soil Pollution*, 226(5). Available from https://doi.org/10.1007/s11270-015-2404-8, Spain: Kluwer Academic Publishers.

Ahmed, K. M., Baki, M. A., Kundu, G. K., Saiful Islam, M., Monirul Islam, M., & Muzammel Hossain, M. (2016). Human health risks from heavy metals in fish of Buriganga river, Bangladesh. *SpringerPlus*, 5, 1–12.

Akpor, O. B., Ohiobor, G. O., & Olaolu, D. T. (2014). Heavy Metal Pollutants in Wastewater Effluents: Sources, Effects and Remediation. *Advances in Bioscience and Bioengineering*, 2(4), 37–43. Available from https://doi.org/10.11648/j.abb.20140204.11.

Barrios, J. A., et al. (2016). Influence of solids on the removal of emerging pollutants in electrooxidation of municipal sludge with boron-doped diamond electrodes. *Journal of Electroanalytical Chemistry*, 776, 148–151. Available from https://doi.org/10.1016/j.jelechem.2016.07.018, Elsevier BV.

Blum, K. M., et al. (2017). Non-target screening and prioritization of potentially persistent, bioaccumulating and toxic domestic wastewater contaminants and their removal in on-site and large-scale sewage treatment plants. *Science of the Total Environment*, 575, 265–275. Available from https://doi.org/10.1016/j.scitotenv.2016.09.135, Sweden: Elsevier B.V.

Bolton, L., et al. (2019). Phosphorus adsorption onto an enriched biochar substrate in constructed wetlands treating wastewater. *Ecological Engineering*, 142. Available from https://doi.org/10.1016/j.ecoena.2019.100005, Elsevier BV.

de Caprariis, B., et al. (2017). Pyrolysis wastewater treatment by adsorption on biochars produced by poplar biomass. *Journal of Environmental Management*, 197, 231–238. Available from https://doi.org/10.1016/j.jenvman.2017.04.007, Italy: Academic Press.

Chen, C., et al. (2019). Activated petroleum waste sludge biochar for efficient catalytic ozonation of refinery wastewater. *Science of the Total Environment*, 651, 2631–2640. Available from https://doi.org/10.1016/j.scitotenv.2018.10.131, China: Elsevier B.V.

Cheng, N., et al. (2021). Adsorption of emerging contaminants from water and wastewater by modified biochar: A review. *Environmental Pollution*, 273. Available from https://doi.org/10.1016/j.envpol.2021.116448, Elsevier BV.

Cole, A. J., Paul, N. A., De, R. N., & Roberts, D. A. (2017). Good for sewage treatment and good for agriculture: Algal based compost and biochar. *Journal of Environmental Management*, 200, 105.

Dai, L., et al. (2017). Production of bio-oil and biochar from soapstock via microwave-assisted co-catalytic fast pyrolysis. *Bioresource Technology*, 225, 1–8. Available from https://doi.org/10.1016/j.biortech.2016.11.017, China: Elsevier Ltd.

Dalahmeh, S. S., et al. (2016). Quality of greywater treated in biochar filter and risk assessment of gastroenteritis due to household exposure during maintenance and irrigation. *Journal of Applied Microbiology*, 121(5), 1427–1443. Available from https://doi.org/10.1111/jam.13273, Oxford University Press (OUP).

de Rozari, P., Greenway, M., & El Hanandeh, A. (2015). An investigation into the effectiveness of sand media amended with biochar to remove BOD5, suspended solids and coliforms using wetland mesocosms. *Water Science and Technology, 71*(10), 1536–1544. Available from https://doi.org/10.2166/wst.2015.120.

de Rozari, P., Greenway, M., & El Hanandeh, A. (2016). Phosphorus removal from secondary sewage and septage using sand media amended with biochar in constructed wetland mesocosms. *Science of The Total Environment, 569-570*, 123–133. Available from https://doi.org/10.1016/j.scitotenv.2016.06.096.

de Rozari, P., Greenway, M., & El Hanandeh, A. (2018). Nitrogen removal from sewage and septage in constructed wetland mesocosms using sand media amended with biochar. *Ecological Engineering, 111*, 1–10. Available from https://doi.org/10.1016/j.ecoleng.2017.11.002.

Ekanayake, M. S., & Manage, P. (2022). Mycoremediation Potential of Synthetic Textile Dyes by Aspergillus niger via Biosorption and Enzymatic Degradation. *Environment and Natural Resources Journal, 20*(3), 234–245. Available from https://doi.org/10.32526/ennrj/20/202100171, Sri Lanka: Faculty of Environment and Resource Studies, Mahidol University.

Feng, D., Guo, D., Zhang, Y., Sun, S., Zhao, Y., Shang, Q., Sun, H., Wu, J., & Tan, H. (2021). Functionalized construction of biochar with hierarchical pore structures and surface O-/N-containing groups for phenol adsorption. *Chemical Engineering Journal, 410*, 127707. Available from https://doi.org/10.1016/j.cej.2020.127707.

Ferreira, L., et al. (2016). Bacillus thuringiensis a promising bacterium for degrading emerging pollutants. *Process Safety and Environmental Protection, 101*, 19–26. Available from https://doi.org/10.1016/j.psep.2015.05.003, Spain: Institution of Chemical Engineers.

Gao, Y., et al. (2021). Functional biochar fabricated from waste red mud and corn straw in China for acidic dye wastewater treatment. *Journal of Cleaner Production, 320*. Available from https://doi.org/10.1016/j.jclepro.2021.128887, Elsevier BV.

Georgin, J., et al. (2019). Potential of Cedrella fissilis bark as an adsorbent for the removal of red 97 dye from aqueous effluents. *Environmental Science and Pollution Research, 26*(19), 19207–19219. Available from https://doi.org/10.1007/s11356-019-05321-9, Brazil: Springer Verlag.

Ghezzehei, T. A., Sarkhot, D. V., & Berhe, A. A. (2014). Biochar can be used to capture essential nutrients from dairy wastewater and improve soil physico-chemical properties. *Solid Earth, 5*(2), 953–962. Available from https://doi.org/10.5194/se-5-953-2014, Copernicus GmbH.

Gizaw, A., et al. (2021). A comprehensive review on nitrate and phosphate removal and recovery from aqueous solutions by adsorption. *Aqua Water Infrastructure, Ecosystems and Society, 70*(7), 921–947. Available from https://doi.org/10.2166/aqua.2021.146, Ethiopia: IWA Publishing.

Goala, M., et al. (2021). Phytoremediation of dairy wastewater using Azolla pinnata: Application of image processing technique for leaflet growth simulation. *Journal of Water Process Engineering, 42*. Available from https://doi.org/10.1016/j.jwpe.2021.102152, India: Elsevier Ltd.

Gros, M., Blum, K. M., Jernstedt, H., Renman, G., Rodríguez-Mozaz, S., Haglund, P., Andersson, P. L., Wiberg, K., & Ahrens, L. (2017). Screening and prioritization of micropollutants in wastewaters from on-site sewage treatment facilities. *Journal of Hazardous Materials, 328*, 37–45.

Gupta, P., Ann, T. W., & Lee, S. M. (2016). Use of biochar to enhance constructed wetland performance in wastewater reclamation. *Environmental Engineering Research, 21*(1), 36–44. Available from https://doi.org/10.4491/eer.2015.067, South Korea: Korean Society of Environmental Engineers.

Gwenzi, W., et al. (2017). Biochar-based water treatment systems as a potential low-cost and sustainable technology for clean water provision. *Journal of Environmental Management, 197*, 732–749. Available from https://doi.org/10.1016/j.jenvman.2017.03.087, Zimbabwe: Academic Press.

Hao, F., Zhao, X., Ouyang, W., Lin, C., Chen, S., Shan, Y., & Lai, X. (2013). Molecular Structure of Corncob-Derived Biochars and the Mechanism of Atrazine Sorption. *Agronomy Journal, 105*(3), 773–782. Available from https://doi.org/10.2134/agronj2012.0311.

Homagain, K., et al. (2016). Life cycle cost and economic assessment of biochar-based bioenergy production and biochar land application in Northwestern Ontario, Canada. *Forest Ecosystems, 3*(1). Available from https://doi.org/10.1186/s40663-016-0081-8, Canada: SpringerOpen.

Van Houdt, R., & Michiels, C. W. (2010). Biofilm formation and the food industry, a focus on the bacterial outer surface. *Journal of Applied Microbiology, 109*(4), 1117–1131. Available from https://doi.org/10.1111/j.1365-2672.2010.04756.x, Oxford University Press (OUP).

Huggins, T. M., et al. (2016). Granular biochar compared with activated carbon for wastewater treatment and resource recovery. *Water Research, 94*, 225–232. Available from https://doi.org/10.1016/j.watres.2016.02.059, United States: Elsevier Ltd.

Hussain, A., Maitra, J., & Khan, K. A. (2017). Development of biochar and chitosan blend for heavy metals uptake from synthetic and industrial wastewater. *Applied Water Science, 7*(8), 4525–4537. Available from https://doi.org/10.1007/s13201-017-0604-7, India: Springer Verlag.

Jha, S., et al. (2023). Biochar as Sustainable Alternative and Green Adsorbent for the Remediation of Noxious Pollutants: A Comprehensive Review. *Toxics, 11*(2). Available from https://doi.org/10.3390/toxics11020117, MDPI AG.

Jiang, N., et al. (2020). Adsorption of triclosan, trichlorophenol and phenol by high-silica zeolites: Adsorption efficiencies and mechanisms. *Separation and Purification Technology, 235*. Available from https://doi.org/10.1016/j.seppur.2019.116152, Netherlands: Elsevier B.V.

Kamyab, H., et al. (2023). Exploring the potential of metal and metal oxide nanomaterials for sustainable water and wastewater treatment: A review of their antimicrobial properties. *Chemosphere, 335*. Available from https://doi.org/10.1016/j.chemosphere.2023.139103, Ecuador: Elsevier Ltd.

Karunanithi, R., et al. (2017). Sorption, kinetics and thermodynamics of phosphate sorption onto soybean stover derived biochar. *Environmental Technology and Innovation, 8*, 113–125. Available from https://doi.org/10.1016/j.eti.2017.06.002, Australia: Elsevier B.V.

Khataee, A., et al. (2017). Sonocatalytic degradation of Reactive Yellow 39 using synthesized ZrO_2 nanoparticles on biochar. *Ultrasonics Sonochemistry, 39*, 540–549. Available from https://doi.org/10.1016/j.ultsonch.2017.05.023, Iran: Elsevier B.V.

Kholoma, E., Renman, G., & Renman, A. (2016). Phosphorus removal from wastewater by field-scale fortified filter beds during a one-year study. *Environmental Technology (United Kingdom), 37*(23), 2953–2963. Available from https://doi.org/10.1080/09593330.2016.1170888, Sweden: Taylor and Francis Ltd.

Krishna Veni, D., et al. (2017). Biochar from green waste for phosphate removal with subsequent disposal. *Waste Management, 68*, 752–759. Available from https://doi.org/10.1016/j.wasman.2017.06.032, India: Elsevier Ltd.

Lalander, C., et al. (2013). Hygienic quality of artificial greywater subjected to aerobic treatment: A comparison of three filter media at increasing organic loading rates. *Environmental Technology (United Kingdom)*, 34(18), 2657–2662. Available from https://doi.org/10.1080/09593330.2013.783603, Sweden: Taylor and Francis Ltd.

Lehmann, J., & Joseph, S. (2009). Biochar for environmental management. *Science and Technology*, Earthscan.

Li, F., et al. (2014). Effects of mineral additives on biochar formation: Carbon retention, stability, and properties. *Environmental Science and Technology*, 48(19), 11211–11217. Available from https://doi.org/10.1021/es501885n, China: American Chemical Society.

Li, Y., et al. (2019). Phosphorus sorption capacity of biochars from different waste woods and bamboo. *International Journal of Phytoremediation*, 21(2), 145–151. Available from https://doi.org/10.1080/15226514.2018.1488806, China: Taylor and Francis Inc.

Ma, J., et al. (2019). Activated municipal wasted sludge biochar supported by nanoscale Fe/Cu composites for tetracycline removal from water. *Chemical Engineering Research and Design*, 149, 209–219. Available from https://doi.org/10.1016/j.cherd.2019.07.013, China: Institution of Chemical Engineers.

Manga, M., et al. (2022). Recycling of Faecal Sludge: Nitrogen, Carbon and Organic Matter Transformation during Co-Composting of Faecal Sludge with Different Bulking Agents. *International Journal of Environmental Research and Public Health*, 19(17). Available from https://doi.org/10.3390/ijerph191710592, United States: MDPI.

Manga, M., et al. (2023). Effect of turning frequency on the survival of fecal indicator microorganisms during aerobic composting of fecal sludge with sawdust. *International Journal of Environmental Research and Public Health*, 20(3). Available from https://doi.org/10.3390/ijerph20032668, United States: MDPI.

Manyuchi, M. M., Mbohwa, C., & Muzenda, E. (2018). Potential to use municipal waste bio char in wastewater treatment for nutrients recovery. *Physics and Chemistry of the Earth*, 107, 92–95. Available from https://doi.org/10.1016/j.pce.2018.07.002, South Africa: Elsevier Ltd.

Miège, C., et al. (2009). Fate of pharmaceuticals and personal care products in wastewater treatment plants – Conception of a database and first results. *Environmental Pollution*, 157(5), 1721–1726. Available from https://doi.org/10.1016/j.envpol.2008.11.045, Elsevier BV.

Mwenge, P., & Seodigeng, T. (2019). Greywater treatment using activated biochar produced from agricultural waste. *International Journal of Chemical and Molecular Engineering*, 13(3), 140–145.

Niwagaba, C. B., et al. (2014). Experiences on the implementation of a pilot grey water treatment and reuse based system at a household in the slum of Kyebando-Kisalosalo, Kampala. *Journal of Water Reuse and Desalination*, 4(4), 294–307. Available from https://doi.org/10.2166/wrd.2014.016, Uganda: IWA Publishing.

Oh, T. K., et al. (2012). Effect of pH conditions on actual and apparent fluoride adsorption by biochar in aqueous phase. *Water, Air, and Soil Pollution*, 223(7), 3729–3738. Available from https://doi.org/10.1007/s11270-012-1144-2, Japan.

Ok, Y. S., et al. (2018). *Biochar from biomass and waste: Fundamentals and applications, Biochar from Biomass and Waste: Fundamentals and Applications* (pp. 1–462). South Korea: Elsevier. Available from 10.1016/C2016-0-01974-5.

Oteng-Peprah, M., Acheampong, M. A., & deVries, N. K. (2018). Greywater characteristics, treatment systems, reuse strategies and user perception—a Review. *Water, Air, and Soil Pollution*, 229(8). Available from https://doi.org/10.1007/s11270-018-3909-8, Netherlands: Springer International Publishing.

Parde, D., et al. (2021). A review of constructed wetland on type, treatment and technology of wastewater. *Environmental Technology and Innovation*, 21. Available from https://doi.org/10.1016/j.eti.2020.101261, India: Elsevier B.V.

Park, J. H., et al. (2018). Effect of pyrolysis temperature on phosphate adsorption characteristics and mechanisms of crawfish char. *Journal of Colloid and Interface Science*, 525, 143–151. Available from https://doi.org/10.1016/j.jcis.2018.04.078, United States: Academic Press Inc.

Perez-Mercado, L. F., et al. (2019). Biochar filters as an on-farm treatment to reduce pathogens when irrigating with wastewater-polluted sources. *Journal of Environmental Management*, 248. Available from https://doi.org/10.1016/j.jenvman.2019.109295, Sweden: Academic Press.

Pignatello, J. J., Kwon, S., & Lu, Y. (2006). Effect of natural organic substances on the surface and adsorptive properties of environmental black carbon (Char): Attenuation of surface activity by humic and fulvic acids. *Environmental Science and Technology*, 40(24), 7757–7763. Available from https://doi.org/10.1021/es061307m, United States.

Poonam., Bharti, S. K., & Kumar, N. (2018). Kinetic study of lead (Pb2 +) removal from battery manufacturing wastewater using bagasse biochar as biosorbent. *Applied Water Science*, 8(4). Available from https://doi.org/10.1007/s13201-018-0765-z, India: Springer Verlag.

Quarters, T., Plants, M.P. and Violated, W. (2018) Water pollution from slaughterhouses, pp. 2016–2018.

Quispe, J. I. B., et al. (2022). Use of biochar-based column filtration systems for greywater treatment: A systematic literature review. *Journal of Water Process Engineering*, 48. Available from https://doi.org/10.1016/j.jwpe.2022.102908, Elsevier BV.

Rajarathinam, N., et al. (2020). Fenalan Yellow G adsorption using surface-functionalized green nanoceria: An insight into mechanism and statistical modelling. *Environmental Research*, 181. Available from https://doi.org/10.1016/j.envres.2019.108920, Elsevier BV.

Sforza, E., et al. (2020). Bioremediation of industrial effluents: How a biochar pretreatment may increase the microalgal growth in tannery wastewater. *Journal of Water Process Engineering*, 37. Available from https://doi.org/10.1016/j.jwpe.2020.101431, Italy: Elsevier Ltd.

Shimabuku, K. K., et al. (2016). Biochar sorbents for sulfamethoxazole removal from surface water, stormwater, and wastewater effluent. *Water Research*, 96, 236–245. Available from https://doi.org/10.1016/j.watres.2016.03.049, United States: Elsevier Ltd.

Sidibe, M. (2014) Comparative study of bark, bio-char, activated charcoal filters for upgrading greywater from a hygiene aspect.

Spokas, K. A., et al. (2014). Physical Disintegration of Biochar: An Overlooked Process. *Environmental Science & Technology Letters*, 1(8), 326–332. Available from https://doi.org/10.1021/ez500199t, American Chemical Society (ACS).

Sun, Y., et al. (2018). Organics removal, nitrogen removal and N_2O emission in subsurface wastewater infiltration systems amended with/without biochar and sludge. *Bioresource Technology*, 249, 57–61. Available from https://doi.org/10.1016/j.biortech.2017.10.004, China: Elsevier Ltd.

Tait, D. R., et al. (2015). Nutrient and greenhouse gas dynamics through a range of wastewater-loaded carbonate sand treatments. *Ecological Engineering, 82,* 126–137. Available from https://doi.org/10.1016/j.ecoleng.2015.04.082, Australia: Elsevier.

Tan, X., et al. (2015). Application of biochar for the removal of pollutants from aqueous solutions. *Chemosphere, 125,* 70–85. Available from https://doi.org/10.1016/j.chemosphere.2014.12.058, Elsevier BV.

Tang, L., et al. (2018). Sustainable efficient adsorbent: Alkali-acid modified magnetic biochar derived from sewage sludge for aqueous organic contaminant removal. *Chemical Engineering Journal, 336,* 160–169. Available from https://doi.org/10.1016/j.cej.2017.11.048, Elsevier BV.

Tareq, R., Akter, N., & Azam, M. S. (2019). *Biochars and Biochar Composites* (pp. 169–209). Elsevier BV. Available from http://doi.org/10.1016/b978-0-12-811729-3.00010-8.

Thang, P. Q., et al. (2019). Potential application of chicken manure biochar towards toxic phenol and 2,4-dinitrophenol in wastewaters. *Journal of Environmental Management, 251.* Available from https://doi.org/10.1016/j.jenvman.2019.109556, Viet Nam: Academic Press.

Utami, M., et al. (2023). Simultaneous photocatalytic removal of organic dye and heavy metal from textile wastewater over N-doped TiO_2 on reduced graphene oxide. *Chemosphere, 332.* Available from https://doi.org/10.1016/j.chemosphere.2023.138882, Indonesia: Elsevier Ltd.

Wang, B., et al. (2020a). Phosphogypsum as a novel modifier for distillers grains biochar removal of phosphate from water. *Chemosphere, 238.* Available from https://doi.org/10.1016/j.chemosphere.2019.124684, Elsevier BV.

Wang, B., Gao, B., & Fang, J. (2017). Recent advances in engineered biochar productions and applications. *Critical Reviews in Environmental Science and Technology, 47*(22), 2158–2207. Available from https://doi.org/10.1080/10643389.2017.1418580, China: Taylor and Francis Inc.

Wang, X., et al. (2020b). Recent advances in biochar application for water and wastewater treatment: A review. *PeerJ, 8.* Available from https://doi.org/10.7717/peerj.9164, China: PeerJ Inc.

Wang, Y., et al. (2024). Research status, trends, and mechanisms of biochar adsorption for wastewater treatment: a scientometric review. *Environmental Sciences Europe, 36*(1). Available from https://doi.org/10.1186/s12302-024-00859-z, China: Springer.

Wathukarage, A., et al. (2017). Mechanistic understanding of crystal violet dye sorption by woody biochar; implications for wastewater treatment. *Environmental Geochemistry & Health,* 1–15.

Wathukarage, A., et al. (2019). Mechanistic understanding of crystal violet dye sorption by woody biochar; implications for wastewater treatment. *Environmental Geochemistry and Health, 41*(4), 1647–1661. Available from https://doi.org/10.1007/s10653-017-0013-8, Springer Science and Business Media LLC.

Westholm, L.J. (2021) Biochar for wastewater treatment – a minireview. *Proceedings of the NTUU "Igor Sikorsky KPI." Series: Chemical Engineering, Ecology and Resource Saving, 4,* 63–66. Available from https://doi.org/10.20535/2617-9741.4.2021.248945.

Williams, M., Martin, S., & Kookana, R. S. (2015). Sorption and plant uptake of pharmaceuticals from an artificially contaminated soil amended with biochars. *Plant and Soil, 395*(1–2), 75–86. Available from https://doi.org/10.1007/s11104-015-2421-9, Australia: Kluwer Academic Publishers.

Xiang, W., et al. (2020). Biochar technology in wastewater treatment: A critical review. *Chemosphere, 252.* Available from https://doi.org/10.1016/j.chemosphere.2020.126539, China: Elsevier Ltd.

Xue, L., et al. (2016). High efficiency and selectivity of MgFe-LDH modified wheat-straw biochar in the removal of nitrate from aqueous solutions. *Journal of the Taiwan Institute of Chemical Engineers, 63,* 312–317. Available from https://doi.org/10.1016/j.jtice.2016.03.021, China: Taiwan Institute of Chemical Engineers.

Xue, S., et al. (2019). Food waste based biochars for ammonia nitrogen removal from aqueous solutions. *Bioresource Technology, 292.* Available from https://doi.org/10.1016/j.biortech.2019.121927, Elsevier BV.

Yaashikaaa, P. R., et al. (2020). A critical review on the biochar production techniques, characterization, stability and applications for circular bioeconomy. *Biotechnology Reports, 28.*

Zeng, Z., et al. (2013). Sorption of ammonium and phosphate from aqueous solution by biochar derived from phytoremediation plants. *Journal of Zhejiang University: Science B, 14*(12), 1152–1161. Available from https://doi.org/10.1631/jzus.B1300102, China.

Zhang, T., et al. (2013). Application of biochar for phosphate adsorption and recovery from wastewater. *Advanced Materials Research.* Available from https://doi.org/10.4028/http://www.scientific.net/AMR.750-752.1389, China.

Zhang, X., et al. (2020). Bio/hydrochar Sorbents for Environmental Remediation. *Energy and Environmental Materials, 3*(4), 453–468. Available from https://doi.org/10.1002/eem2.12074, China: Blackwell Publishing Inc.

Zhao, C., et al. (2021). Formation and mechanisms of nano-metal oxide-biochar composites for pollutants removal: A review. *Science of the Total Environment, 767.* Available from https://doi.org/10.1016/j.scitotenv.2021.145305, China: Elsevier B.V.

Zheng, Y., et al. (2019). Reclaiming phosphorus from secondary treated municipal wastewater with engineered biochar. *Chemical Engineering Journal, 362,* 460–468. Available from https://doi.org/10.1016/j.cej.2019.01.036, United States: Elsevier B.V.

Zhou, X., et al. (2018). An innovative biochar-amended substrate vertical flow constructed wetland for low C/N wastewater treatment: Impact of influent strengths. *Bioresource Technology, 247,* 844–850. Available from https://doi.org/10.1016/j.biortech.2017.09.044, China: Elsevier Ltd.

Zhou, X., et al. (2019). Nitrogen removal in response to the varying C/N ratios in subsurface flow constructed wetland microcosms with biochar addition. *Environmental Science and Pollution Research, 26*(4), 3382–3391. Available from https://doi.org/10.1007/s11356-018-3871-4, China: Springer Verlag.

Zhou, Y., et al. (2022). Facile preparation of alveolate biochar derived from seaweed biomass with potential removal performance for cationic dye. *Journal of Molecular Liquids, 353.* Available from https://doi.org/10.1016/j.molliq.2022.118623, China: Elsevier B.V.

Zubair, M., et al. (2022). Production of magnetic biochar-steel dust composites for enhanced phosphate adsorption. *Journal of Water Process Engineering, 47.* Available from https://doi.org/10.1016/j.jwpe.2022.102793, Saudi Arabia: Elsevier Ltd.

Part 4

Biochar sytems for clean and renewable energy

Chapter 15

Electrochemical properties of biochar for environmental applications: advances, challenges, and perspectives

Vineet Kumar[1], Shivali Sharma[2], Sunny Sharma[3] and Gaurav Sharma[2]

[1]Department of Microbiology, School of Life Sciences, Central University of Rajasthan, Bandar Sindri, Ajmer, Rajasthan, India, [2]Rani Lakshmi Bai Central Agricultural University, Jhansi, Uttar Pradesh, India, [3]Department of Horticulture, School of Agriculture, Lovely Professional University, Phagwara, Punjab, India

Chapter outline

15.1 Introduction	281
15.2 Electrochemical properties of biochar	283
15.2.1 Conductivity	283
15.2.2 Redox activity	283
15.2.3 Surface chemistry	284
15.3 Environmental applications of biochar	284
15.3.1 Agricultural soil amendment	284
15.3.2 Remediation of polluted wastewater	286
15.3.3 Carbon sequestration	289
15.3.4 Biochar as an electrode material for the development of microbial fuel cells	294
15.4 Direct interspecies electron transfer pathways	295
15.5 Challenges	295
15.6 Future perspectives	298
15.7 Conclusion	298
References	299

15.1 Introduction

In the modern era, the escalating problem of environmental pollution has become a major public concern due to rapid industrialization and urbanization (Kumar & Verma, 2023). Anthropogenic activities release significant amounts of various persistent pollutants, encompassing organometallic, organic, and inorganic substances, including heavy metals. Moreover, the inadequate treatment of emerging pollutants like personal care products, pharmaceuticals, and microplastics poses a significant threat (Ahmed et al., 2018; Chen, Zhang, et al., 2021; Chen, Zhao, et al., 2021; Shen et al., 2022; Wang & Wang, 2019). These contaminants contribute substantially to environmental degradation and present substantial risks to human and environmental health (Jenjaiwit et al., 2021; Zhang et al., 2024). Thus, urgent and effective intervention must be addressed appropriately, safeguarding both the environment and human health (Pham et al., 2023; Roha et al., 2021; Zhang et al., 2024). Currently, the most important real-world issue is finding sustainable ways to clean up polluted environments consistent with the global trend toward sustainability. Biochar is increasingly being recognized as a possible solution to this problem due to its remarkable environmental applications and highly efficient nature. On the other hand, it has promise for mitigating greenhouse gases (GHGs) emissions, soil carbon sequestration, and remediation of various organic and inorganic pollutants. Thus, it could make a difference in terms of reducing environmental impact thus improving overall environmental health (Sadhu et al., 2022; Shukla et al., 2019; Wang & Wang, 2019; Wang, Nie, et al., 2024; Wang, Yang, et al., 2024).

    The word Biochar is a grouping of two words: "bio," mentioning to "biomass," and "char," representing "charcoal." Biochar is a carbonaceous porous by-product with a strong resistance to decomposition. A thermochemical conversion or breakdown of low-cost residual biomass, which can be done through typical hydrothermal carbonization (Ye et al., 2020), slow (Campos et al., 2020) or fast pyrolysis (Shen et al., 2022; Yan et al., 2020), gasification, hydrothermal liquefaction (Parsa et al., 2019), flash carbonization, and torrefaction under a limited oxygen atmosphere are some methods involved in biochar production. According to the European Biochar Foundation's definition biochar is a porous carbonaceous material that comes out as a result of pyrolysis of biomass. International Biochar Initiative (IBI) provided the official definition of biochar as "solid material produced by the thermal decomposition of biomasses under limited supply of oxygen, and at relatively low temperatures." Biomass used to make biochar includes lignocellulosic materials, such as pellets and wood chips (Favre et al., 2022; Singh et al., 2021), agri residues, such as coffee waste (Steigerwald & Ray, 2021), pod waste (Emenike et al., 2022), rice husk (Roha et al., 2021), wheat straw, corncob among others Amen, Bashir, et al. (2020), Amen, Yaseen, et al. (2020) and muskmelon peel (Khan et al., 2017), walnut shell (Xu et al., 2022), corn stover (Zhu, An, et al., 2023; Zhu, Yang, et al., 2023), sugarcane bagasse (Tang et al., 2020) and tea waste (Khalil et al., 2020), as well as substitute biomass sources like petroleum sludge (Bao et al., 2020), wastewater sludge (Chen, Zhang, et al., 2021; Chen, Zhao, et al., 2021; Matheri et al., 2020; Wang & Wang, 2019), plant biomass (Hashem et al., 2020; Yan et al., 2020), poultry litter, animal manures (Zhao & Naeth, 2022; Zheng et al., 2021), bacterial cell (Jenjaiwit et al., 2021), dairy manure, bones, and algal biomass (Parsa et al., 2019; Zhang, Deng, et al., 2023; Zhang, Huang, et al., 2023; Zhang, Zhu, et al., 2023), other waste products, fungi biomass (Zhao et al., 2024). Biochar is predominantly made up of five elements: carbon, sulfur, nitrogen, oxygen, and hydrogen. In addition to its high carbon content, biochar normally contains elevated levels of mineral (ash) components, typically as carbonates, phosphates or oxides usually formed by alkaline earth metals, such as magnesium (Mg) and calcium (Ca) or alkali metals like sodium (Na) and pottasium (K). Carbonaceous materials constitute the major part of biochar with organic carbon often exceeding 70% thereby making it contain more organic than inorganic carbon. This high amount of carbon increases the stability and resistance to degradation of biochar allowing it to work for a longer period in various remediation applications. Moreover, biochar has several other unique physico-chemical and electrochemical properties that render it an ideal material for wider environmental applications. These properties include high surface chemistry; high ion exchange capacity along with excellent electrical conductivity due to the presence of numerous functional groups, such as quaternary-N, pyrrole-N, pyridine-N, phenolic, hydroxyl, ester, carbonyl, and carboxylic. Additionally, biochars have fine-grained structures with large specific surface areas giving rise to well-developed networks consisting of micropores, mesopores, and macropores (Chacón et al., 2017; Schievano et al., 2019). The porous structure offers an extensive surface area, facilitating the adsorption and retention of contaminants (Wang, Nie, et al., 2024; Wang, Yang, et al., 2024). Among them, the electrochemical properties, such as redox activity, electrical conductivity which depends on the superficial electroactivity, electron-donating, and electron-accepting capacity, and electron shuttling offer the promising use of biochar in environmental applications have received extensive attention in recent years (Fig. 15.1).

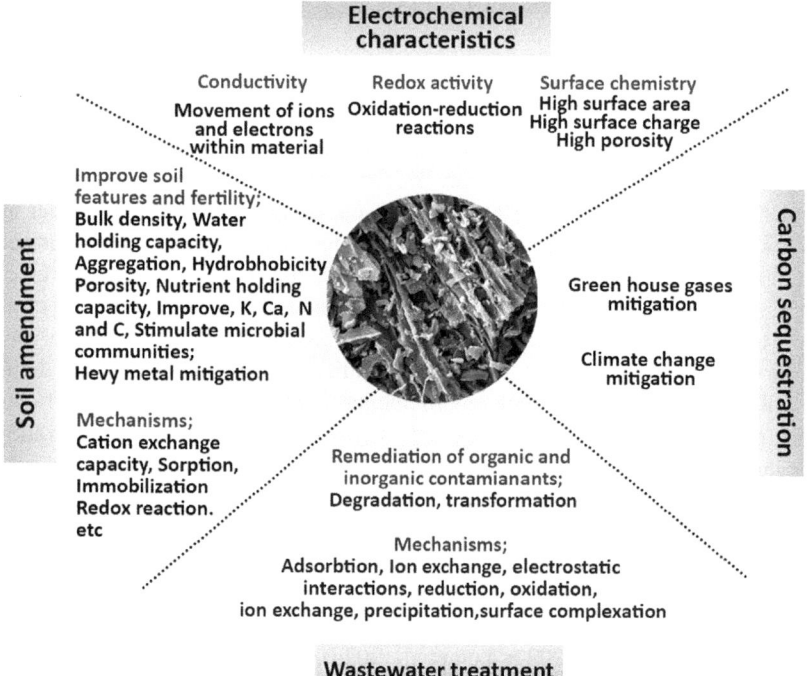

FIGURE 15.1 Schematic representation of the electrochemical properties of biochar and their possible mechanisms for various environmental applications, such as soil amendment, wastewater treatment, and carbon sequestration.

The present chapter aims to provide in-depth knowledge on the role of electrochemical properties of biochar in environmental applications. This review serves as a roadmap for future research endeavors in the effective application of biochar toward environmental sustainability.

15.2 Electrochemical properties of biochar

15.2.1 Conductivity

Conductivity is a measure of how a material allows the flow of electric current. Biochar has developed as a potential electroconductive material because its high carbon content and microporous nature allow the movement of ions and electrons within the material and contribute to its conductivity (Park et al., 2021; Prado et al., 2019). Biochar can facilitate the direct flow of electrons from an electron source to an electron acceptor, eliminating the requirement for storage. In electrochemical uses, such as treating wastewater and soil remediation, biochar's electron conductivity is essential for facilitating electrochemical processes. This enables the disintegration or removal of contaminants, making it a significant factor in the degradation of pollutants (Tian et al., 2021). The conductive features of biochar in soil facilitate the transport of ions, which in turn affects soil structure, nutrient availability, and the dynamics of microbial communities (Barnes et al., 2014; Gelardi et al., 2021). These traits could enable direct interspecies electron transfer (DIET), a process that allows for the flow of electrons between various types of microbes (Chen et al., 2014; Dubé & Guiot, 2015; Zhang, Deng, et al., 2023; Zhang, Huang, et al., 2023; Zhang, Zhu, et al., 2023). Indeed, in this process, certain microbial species can directly transfer electrons to biochar. This transfer allows the electrons to move beyond the individual cells, establishing a connection and facilitating the coupling of metabolic activities between two distinct microorganisms (Zhao et al., 2023). Biochar conductivity is highly influenced by feedstock type and pyrolysis working conditions (Alves et al., 2022; Bartoli et al., 2022; Gabhi et al., 2020). Additionally, the conductivity of biochar is influenced by its porous structure, which generally results in a large surface area (Gabhi et al., 2020; Kane et al., 2021). An increased surface area allows for a greater number of routes through which electrons can flow.

Commercial biochar, as a result of such characteristics, is often observed to have significant variability and somewhat unfavorable structural and mechanical properties compared to other carbon-based materials. Moreover, it usually exhibits low electrical conductivity. This has traditionally limited its technological use as against other high-performance carbon compounds (Schievano et al. 2019). In the microbial electrochemical treatment (MET) field, it is generally accepted that electrical conductivity plays a role in determining the efficiency of electrode materials (Kumar & Verma, 2023). Perhaps, this can be attributed to the fact that most studies on electrochemistry are carried out under these conditions making them more available for research (Li et al., 2021). Traditionally many applications have focused on how electrons could be used outside of the cell through an external circuit similar to nonliving fuel cells like microbial fuel cells (MFCs). The role played by biochar in these applications has been insignificant (Wang et al., 2017). However, it should be noted that MET has various potential applications in the environmental biotechnology field (Bachmann et al., 2016; Qian et al. 2015). The electrical conductivity of biochar is commonly measured by the two-probe method (Chen et al., 2014). In this arrangement, an applied voltage causes a known current to pass through the sample located between two probes.

15.2.2 Redox activity

Redox activity, short for reduction-oxidation activity, refers to the capacity of a material to undergo reduction and oxidation reactions. A redox reaction is a chemical reaction in which electrons reversibly transfer between molecules, showcasing electroconductivity. Redox reactions are fundamental to many biological and chemical processes, including energy production, corrosion, and various metabolic pathways. Carbon is inherently redox-active, and this property is retained in biochar (Xu et al., 2022). Electrochemists have been investigating biochar as an amorphous material with fascinating surface electron transport properties. Under nonliving conditions, biochar has been discovered to undergo irreversible surface biogeochemical redox processes, where it can either take or give electrons. This is made possible by its redox-active components and relatively high conductivity to electricity. The unique attributes of biochar make it a topic of significant interest for a wide range of applications, particularly its electrochemical properties (Wu, Cai, et al., 2022; Wu, Chen, et al., 2022; Wu, Han, et al., 2022).

Biochar surfaces contain functional groups like quinones and oxygen-containing moieties, that can participate in redox reactions, contributing to environmental sustainability (Chang et al., 2022; Min et al., 2023; Pan et al., 2021; Wu, Cai, et al., 2022; Wu, Chen, et al., 2022; Wu, Han, et al., 2022). Redox activity is a crucial factor in the field of biochar and its environmental uses, especially involves in eliminating or altering contaminants (Meng et al., 2023; Seo et al., 2020; Xu et al., 2021; Zheng et al., 2021; Zhu et al., 2018). Due to its adaptable electrochemical properties, such as its ability to provide, absorb, and transport electrons, biochar shows potential to reduce the effects of hazardous compounds in ecosystems (Amen, Bashir, et al., 2020; Amen, Yaseen, et al. 2020; Cuong et al., 2021; Rashid et al., 2022). There has been a significant surge in study attention towards the redox processes and electron transfer effectiveness of biochar in recent years (Wang, Li, et al., 2020; Wang, Ptacek, et al., 2020; Xu et al., 2019). The breakdown of many organic pollutants, including heavy metals, occurs mostly through redox processes, which offer new possibilities for reducing environmental impact. The redox characteristics of biochar can be influenced by various factors, including the conditions under which it is formed, such as temperature and duration. Furthermore, the permeable structure and the surface area accessible at various pore sizes also exert a significant impact on these characteristics (Chacon et al., 2020; Mai et al., 2017; Xu et al., 2021). The higher aromaticity and π- electrons present in biochar contribute to facilitating the electron exchange process. The characteristics of aromaticity and π-electrons can be influenced by various factors, such as the nature of biomass and the pyrolysis temperature during biochar production. Hence, it is vital to acquire a more extensive comprehension of the redox processes of biochar. These insights are essential for improving the effectiveness of in situ bioremediation methods designed to treat polluted environments.

15.2.3 Surface chemistry

Biochar surface chemistry refers to the composition and characteristics of the outer layer of biochar that interrelates with its adjoining environment. The tenability in surface chemistry and porosity are certainly the salient features of biochar (Shaaban et al., 2013; Xiong et al., 2022; Zhang, Deng, et al., 2023; Zhang, Huang, et al., 2023; Zhang, Zhu, et al., 2023). The surface chemistry of biochar plays a crucial role in determining its behavior in many applications, particularly in environmental remediation. Biochar surfaces exhibit a spectrum of functional groups, including as hydroxyl (-OH), carboxyl (-COOH), and phenolic groups. These groups are prevalent and have distinct atomic configurations that confer distinctive chemical capabilities (Fan et al., 2018; Lopez-Tenllado et al., 2021). However, most functional groups are oxygen-containing or alkaline, providing biochar with high absorption, buffering, and ion exchange capacity (Zhai et al., 2021). The presence of functional groups develop adsorption sites on the biochar surface, allowing it to attract and hold onto ions and molecules and play a key role in the sorption of organic and inorganic pollutants (Peng et al., 2017; Wang et al., 2017; Zhai et al., 2021). The surface functional groups of biochar are affected by several factors, including the biomass feedstock used, the composition of biochar, and the pyrolysis temperature during production. These variables collectively contribute to shaping the chemical characteristics of the biochar surface, impacting its interactions with ions and other substances in the environment. Having a comprehensive understanding of these interactions is of utmost importance to customize the properties of biochar for specific purposes, such as enhancing soil quality or addressing environmental issues (Banik et al., 2018; Janu et al., 2021). The physicochemical features of biochar are directly determined by the biomass sources and various process parameters, which in turn affect its usage. The features of redox action, electroactivity, and conductivity are closely linked to the chemical and structural characteristics of a substance. These properties are primarily influenced by the temperature and pressure conditions throughout the manufacturing process.

15.3 Environmental applications of biochar

Due to its unique attributes, including ion exchange capability, microporosity, a particular area of surface, and substantial capacity for adsorption, biochar has attracted considerable attention from researchers. These unique properties make biochar versatile and applicable in a wide range of contexts, including but not limited to environmental remediation, agricultural soil improvement, and as a potential component in various technological solutions. Biochar is mostly used for environmental cleanup. Furthermore, biochar can be employed for the improvement of soil quality and the capture and storage of carbon. The following section focuses on the discussion of the multifaceted electrochemical properties of biochar, which are essential for its wide range of environmental uses for addressing various environmental concerns.

15.3.1 Agricultural soil amendment

The utilization of biochar in agricultural soils has generated substantial interest in recent times, primarily attributed to its high porous structure, high organic carbon content, and its involvement in electrochemical and physicochemical reactions. The properties of biochar make it a highly compelling choice for improving soil fertility, enhancing water retention, and increasing nutrient availability. This, in turn, contributes to the promotion of sustainable and enhanced agricultural practices (Abujabhah et al., 2016; Campos et al., 2020). The biochar's adaptable physico-chemical and electrochemical reactions provide it with a promising soil amendment for agriculture. It has the potential to improve various soil features, including bulk density, water-holding capacity, accumulation, permeability, hydrophobicity, and nutrient-holding capacity. This leads to increased levels of important soil nutrients such as K, Ca, and nitrogen (N), as well as carbon storage. Additionally, biochar stimulates soil microbial communities, which further enhance soil fertility (Abujabhah et al., 2016; Helaoui et al., 2023; Wang et al., 2022; Zhao & Naeth, 2022). The main reason why biochar is used to enhance soil health is its significant cation exchange capacity (CEC). The advantageous aspect of biochar lies in its alkaline composition, which is attributed to the existence of elements, vitamins minerals, and functional groups. Biochar, due to its alkalinity, can be used to increase the CEC of acidic soils and help balance pH levels. This makes biochar a beneficial soil supplement for improving agricultural techniques (Helaoui et al., 2023; Hong et al., 2022).

He et al. (2021) revealed that the soil pH, soil stability of aggregates, and soil organic carbon (SOC) are essential indicators that greatly influence soil functionalities, erosion rate, crop yield, and ecotoxicity. The researchers determined that the utilization of biochar in soils for agriculture can successfully manage these variables. In a study by Joseph et al. (2020), they examined the effects of applying biochar derived from straw on sandy clay loam soil. The results indicated that the use of biochar can enhance the levels of SOC, total organic carbon (TOC), total nitrogen (TN), the carbon-to-nitrogen ratio (C/N ratio), and the distribution of soil aggregates based on size classes. Furthermore, it possesses significant potential to enhance the hydrological qualities of soil and increase water usage efficiency. According to Wu, Chen, et al. (2022), Wu, Cai, et al. (2022), Wu, Han, et al. (2022), the use of biochar improved the total water content (AWC) by 26.8% and the efficiency of water use (WUE) by 4.7 percent.

Heavy metals in soils, which are frequently discharges by anthropogenic activities like factories, mining, and the use of specific agricultural inputs, provide substantial environmental and health hazards. These metals can remain within the soil for long periods, which can result in them seeping into bodies of water and becoming part of the food chain. Utilizing materials, such as biochar is essential for effectively addressing and reducing the consequences of heavy metal pollution in soils. These remediation solutions play a significant role in alleviating the negative effects (Ogunremi et al., 2023). Biochar amendments have been widely employed as a novel method to address soil contamination caused by heavy metals. This technique involves the adsorption, stabilization, and immobilization of contaminants, like toxic metals (Helaoui et al., 2023; Hong et al., 2022; Jia et al., 2017; Lian et al., 2023; Pathy et al., 2023; Wang et al., 2022) and organic contaminants. Multiple studies have consistently shown that the addition of biochar to soil has a beneficial effect on plants. It reduces oxidative stress and counteracts the harmful effects of metals, leading to increased chlorophyll content and promoting plant growth. These positive outcomes are supported by research conducted by Li et al. (2022), Jun et al. (2020), Zhu, Yang, et al. (2023), Zhu, An, et al. (2023), Helaoui et al. (2023). Furthermore, the biomass of plants is significantly increased as demonstrated by Jun et al. (2020), Wang et al. (2022), and Helaoui et al. (2023). As biochars are added to soils, they interact with many components such as soil organic matter (OM), root hairs, plant roots, root exudates, microbes, and water. These interactions result in the creation of complex compounds composed of OM, minerals, and biochar. Redox conditions in the soil-rhizosphere-plant system are essential for the formation and operation of these complexes. They have a significant impact on microbial activity, community composition, soil nutrient availability, as well as nutrient uptake and transformation. Pseudocapacitance capabilities of biochar allow it to serve as an intermediary for electron transfer between bacteria and elements in the soil. The pertinent electrochemical characteristics of biochar for soil remediation encompass its electrical conductivity and redox capacity. The redox capacity of biochars is mostly influenced by the presence of redox-active components, particularly the quinone/phenolic system, metals, and radicals. This redox capacity serves as an electron transfer accelerator in biogeochemical and contaminant redox processes in soil. Biochar can serve as both an electron transporter and electron storage, functioning as an electron receiver or reservoir depending on the soil's redox conditions. The process of limiting the movement of metallic elements is achieved through the electrostatic relationship between metallic ions and charged biochar. The main method by which biochar removes contaminants is through surface sorption, which includes processes such as surface complex formation, hydrogen bonding, complexity, electrostatic attraction, precipitation, acid-base reactions, ion exchange, and π-π interactions. The particular mechanism is contingent upon the characteristics of the biochar generated. The crucial characteristics of biochar for the removal of heavy metals, as well as organic and inorganic pollutants in the soil, are the area of its surface, porosity, and functional groups. The adsorption capacity is influenced by factors, such as permeability, volume of pore size, and surface area. These factors, in turn, are dependent on the thermochemical temperature utilized during the formation of biochar. During surface adsorption or physiological sorption, chemical linkages are established when metal ions permeate into the pores of biochar. In addition, the introduction of biochar can impact the outcome of pesticides in the soil as well, as biochar has a greater ability to absorb pesticides compared to the original soils. The enhanced sorption and diminished dissipation of pesticides can potentially mitigate the hazards of pollution of the environment and human exposure. Biochar has shown its capacity to decrease the availability of organic contaminants in soil and their absorption by plants and microbes. Heavy metals that are absorbed can be transported through vessels with water, eventually reaching leaves and receptacles. These receptacles are transpiration organs that have the highest rate of water dissipation. As the pH values increase, the groups of functional substances and hydrogen ions present on the outside of biochar separate, resulting in a higher rate of heavy metal removal. The ability of biochar to adsorb and form complexes with contaminants is crucial for facilitating electron-mediated activities. Further in-depth study is required to explore the underlying processes of electron transfer and interaction between them within the biochar/microorganism system, as well as to advance research approaches and application domains. When activated correctly, biochar can generate a relatively large specific surface area because of the easily accessible pores in its microstructures. These qualities encompass a permeable structure that offers a sanctuary and ample surface space for bacteria to reside. Biochar possesses notable characteristics, namely its capacity to be adjusted in terms of porosity and surface chemistry. Biochar has been extensively employed to mitigate soil pollution caused by heavy metals, as well as to diminish the movement and accessibility of these metals (see Table 15.1).

15.3.2 Remediation of polluted wastewater

Wastewater is a significant global environmental issue that has harmful impacts on human and ecological well-being (Khalil et al., 2020; Shukla et al., 2019; Ye et al., 2020). Recently, there has been a significant increase in research aimed at exploring the use of biochar to eliminate various organic and inorganic contaminants from polluted settings (Roha et al., 2021; Zhang et al., 2023). Due to its combined environmental and economic advantages, biochar is an excellent choice for wastewater treatment. It is a renewable resource, making it particularly appealing for this purpose (Emenike et al., 2022; Matheri et al., 2020; Wang et al., 2021). Biochar has shown efficacy as a filtering agent due to its potent adsorbent properties and ion exchange capacity. This enables it to efficiently remove various contaminants, such as organic substances, heavy metals, dyes, and emerging pollutants, from water (Chen, Zhang, et al., 2021; Chen, Zhao, et al., 2021; Thuan et al., 2023; Yang et al., 2021). The biochar's extensive surface area and robust interaction with surface functional groups facilitate the efficient elimination of diverse pollutants in the environment. The process of eliminating heavy metals from wastewater using biochar encompasses various mechanisms, such as electrostatic forces, reduction, oxidation, ion exchange, precipitation, and surface complexation (Wang et al., 2022; Yan et al., 2020; Zhu, An, et al., 2023; Zhu, Yang, et al., 2023). The presence of carboxylic acids (-COOH) and groups of hydroxyl (-OH) is essential for the process of heavy metal adsorption. Electrostatic attraction served as an additional mechanism for capturing heavy metal ions. This was due to the presence of numerous functional groups on the biochar surface, which increased the potency of electrostatic adsorption. Particularly, if the surface of biochar became negatively charged during a deprotonation process, it facilitated the movement of heavy metals. The effectiveness of biochar in addressing heavy metal contamination relies on its surface area, carbon composition, pore size, oxygen functional groups, and anion content (Xu et al., 2022). Unlike heavy metals, the elimination of organic pollutants in wastewater mainly takes place through various mechanisms, including pore filling, electrostatic forces, hydrogen bonding, hydrophobic interactions, and π-π bonding involving COOH, OH, and R-OH functional groups. These mechanisms collectively contribute to the removal process. Functional groups on the surface of biochar play a crucial role in pollution remediation by determining the ability to remove specific pollutants and the overall adsorption capacity. Additional mechanisms encompass partitioning in noncarbonized phases, as well as chemical change through reductive reactions and electrical conductivity.

Fig. 15.2 demonstrates the chemical interactions that occur during the elimination of different organic and inorganic contaminants using biochar. The hydrophobic features of biochar, resulting from its aromatic carbon structure and minimal surface oxidation, enable effective adsorption of organic contaminants, leading to their efficient removal. The electrostatic interaction is responsible for managing the presence of ionic organic pollutants. The process of pore-filling is affected by the molecular weight, volume, and surface area of the porous biochar. Additionally, hydrogen bonding plays a crucial role in compounds that contain hydroxyl, amino, and alkoxy functional groups. Due to its adjustable porosity and surface functional groups, biochar is well-suited for regulating interface chemical reactions, making it highly appealing for electrochemical applications. Biochar has been employed in contemporary applications for biological wastewater treatment systems, including anaerobic digesters and treatment wetlands. It promotes microbial activity by enabling the flow of electrons between different species of microorganisms, enhances the ability to resist changes in pH, and provides a material for the creation of biofilms. In addition, biochar has been used as a material for electrodes in microbiological electrochemical processes and as a substrate for enhancing oxygen reduction on cathodes. Biochar

TABLE 15.1 List of some important scientific studies that highlighted the role and mechanisms of biochar derived from various feedstock for the remediation of heavy metals from polluted soil

Biochar	Metals	Adsorption efficiency	Removal efficiency	Mechanisms	References
Lychee biochar	Pb, Cd, As, Zn	—	—	The lychee wood biochar possesses a high degree of microporosity and is enriched with functional groups such as C=C, C–O, and C–H, which chelates Zn, As, Cd, and Pb	Jun et al. (2020)
Waste cotton biocharnovel β-CD/ hydrothermal biochar	Pb^{2+}, Cd^{2+}	Pb^{2+} and Cd^{2+} extended 50.44 and 33.77 mg g^{-1} respectively	92.87% for Pb^{2+}, and 86.19% for Cd^{2+}	The adsorption of Pb^{2+} and Cd^{2+} involves the chemisorption process, where complexation occurs with functional groups containing -OH	Li et al. (2022)
Wheat straw biochar	Cd	—	—	Cd immobilization	Zuo et al. (2023)
Straw (crushed oxalic acid-modified iron biochar composites) (OA-ZVI/BC)	Cr	—	96.7% of Cr^{6+}	Functional groups in the soil absorb Cr^{6+} and then reduce it directly or indirectly. This leads to the formation of co-precipitates of Fe^{3+} and Cr^{3+}, which retain Cr in the soil.	Xie et al. (2023)
Wheat straw, rice husk, pig manure, oyster shell	Cd, Pb	—	—	Biochar exhibited superior potential in the stabilization of Cd and Pb, as well as immproved soil fertility.	Lian et al. (2023)
Rice straw	Cu, Zn, As, Cd, Pb	—	—	Immobilization mechanisms included complexation, oxidation-reduction, coprecipitation, electrostatic attraction, ion exchange, and surface adsorption.	Song et al. (2022)
Cattle slurry	NH_3	—	20% of NH_3 emissions	N sorption of those biochars	Viaene et al. (2023)
alfalfa (Medicago sativa L.)	Cu, Zn	—	27%- 29 of Cu and 18% - 19% of Zn	Adsorption, surface chelation, precipitation, and ion exchange.	Helaoui et al. (2023)
Sulfoaluminate cement-modified straw biochar	Pb, Cd	—	20.45% of Pb and 35.87% Cd	Immobilization	Han et al. (2024)
Manure biochar	Cd	—	Soil Cd decreased by 28% in silt loam and 37% in loamy sand	Absorption, transportation hydrolyzation and decrease in photosynthetic rate caused by heavy metal exposure	Zhao and Naeth (2022)
Brassica chinensis	Pb, Zn, Cr, Cu	Heavy metal immobilization was improved	64.9% of Zn and 82.7%-94.7% of Cd	Adsorption	Wang et al. (2022)
compost, and composted biochar	N	0.29% of N	—	The increased activity of soil microbes	Azman et al. (2023)
Crop straws	Si, O	14.16% of Si,	2%–12%, of O	Hydrolysis, condensation	Chen et al. (2019)

(*Continued*)

TABLE 15.1 (Continued)

Biochar	Metals	Adsorption efficiency	Removal efficiency	Mechanisms	References
Acacia green waste	Carbon	23% of the total organic carbon and 13% of the combined amounts of N, P, K, S, and Ca.	—	—	Abujabhah et al. (2016)
Zea mays L.	Cu^{2+}	—	Mineral Cu^{2+} in the soil of the Cu^{2+} treatment is below the optimal level.	Adsorption, photosynthesis, binding cations	Abideen et al. (2023)
Grazed pastures	—	The total N stocks in biochar-treated soils at the end of the experiment were higher than control	—	Mineralization, solubilization/desorption	Garbuz et al. (2021)
Maize	N, P, C	10% of nitrogen, 45% of phosphorus and 15% of carbon	—	Immobilization, alkalinization, and sorption.	Foster et al. (2016)
Paddy	Nitrous oxide (N_2O) and methane		Biochar significantly decreased the second N_2O flux	Denitrification	Shen et al. (2022)
Compost	N, Cd, lead, and zinc.	37.6%, of total nitrogen and 30.1% of copper	30.1%. Of cadmium, 50.1% lead and 87.3% zinc	—	Zhou et al. (2023)
Rice	N, P	93.4% of P content decreased	12.8% of grain P content	Assimilation, solubilization	Liu et al. (2022)
Enteromorpha prolifera and corn straw	PO_4^{3-}, Na	85.20–200.64 mg kg^{-1} PO_4^{3-} and 4.75–12.67 of NH_4^+ and 39.11%–49.67%	17,478 to 6135 mg kg^{-1} Na	dilution effect	Suo et al. (2021)
Hardwood and sulfurized-hardwood biochars	Hg	10% of Hg in	4.75–12.67 of MeHg	—	Wang, Ptacek, et al. (2020), Wang, Li, et al. (2020)
Theobroma cacao L.	Cd		87% of Cd	Immobilizing, solubilization	Ramtahal et al. (2019)
Wheat (Triticum aestivum)	The elements included are Cr, Cd, Cu, Pb, Ni, Zn, Mg and Fe	—	Cr, Cd, Cu, Pb, Ni, Zn, Mg, and Fe. Absorbed bt 63.08, 78.07, 74.61, 78.11, 75.73, 69.71, 28.78, and 49.26, respectively	Adsorption, reduction, electrostatic adsorption, ion exchange process, mobilization	Pandey et al. (2022)
Selenium–sulfur functionalized biochar	Hg	—	81.12% of Hg	Immobilization and inhibition	Huang et al. (2022)
Contaminated soil	Cd, Zn	Availability of P increased or remained constant	Decreased S, K	Immobilization, reduction, solubilization adsorption	Van Poucke et al. (2020)

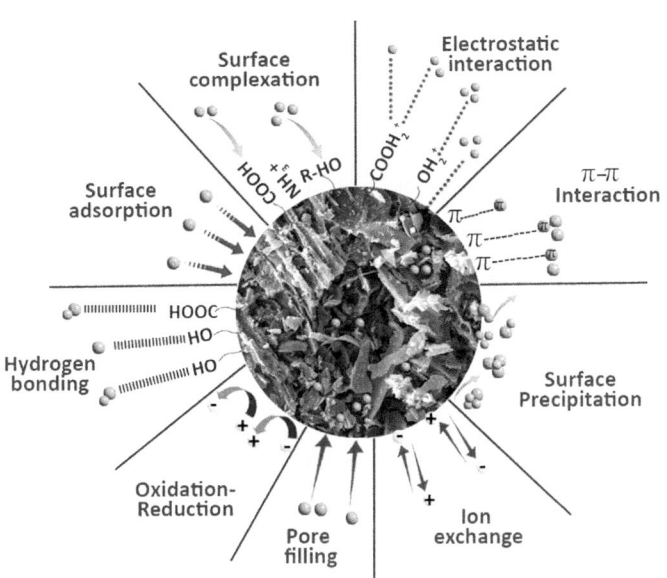

FIGURE 15.2 Schematic diagram of the possible mechanism in mitigation of contaminants from a contaminated environment.

and enhanced biochar materials have demonstrated favorable outcomes in adsorbing dyes and heavy metals, as evidenced by several investigations. The efficacy of different biochar types in eliminating heavy metals and organic contaminants in wastewater systems is outlined in Table 15.2

15.3.3 Carbon sequestration

Carbon dioxide (CO_2), methane (CH_4), and nitrous oxide (N_2O) are potent greenhouse gases (GHGs) that significantly contribute to global environmental issues, such as global warming and climate change (Liu et al., 2022). The use of renewable resources for capturing and storing GHGs, particularly CO_2, has been recognized as a crucial approach to tackling the problems of global warming and climate change (Salem et al., 2021). Biochar has been suggested as a useful method for managing garbage and reducing global warming by sequestering carbon when enters to soil as an environmental amendment. IBI has acknowledged the potential of biochar and it is attributed with the capacity to absorb around 3.67 gigatons of CO_2 annually. The importance of biochar in soil carbon sequestration cannot be overstated. It plays a crucial role in reducing greenhouse gas emissions from fertilized soils, as evidenced by studies conducted by Lai et al. (2013), Yang et al. (2020), and Li et al. (2021). Biochar contributes to enhancing soil health and mitigating the impacts of climate change, particularly in terms of global warming.(Lai et al., 2013; Wang et al., 2023). Carbon sequestration relates to the method of trapping carbon in several reservoirs, including plants, soils, geological formations, and seas. This process elucidates the assimilation of organic carbon into soils, converting it into a durable carbon reservoir that would otherwise be emitted as CO_2. Biochar is believed to aid in carbon sequestration through three primary mechanisms: (1) the introduction of stable carbon into the soil by biochar itself; (2) the potential reduction of the decomposition of existing organic carbon in the soil by biochar; and (3) the promotion of plant growth by biochar, resulting in greater uptake of atmospheric carbon. The scientific community understands the significance of biochar in mitigating the impacts of climate change at a worldwide level. Research has demonstrated that biochar's surface basicity is influenced by nitrogen-containing functional groups. The presence of nitrogen-containing basic surface functional groups is essential for the adsorption of CO_2 by biochar, as it increases its attraction to carbon dioxide. These findings indicate that biochar produced from materials such as sawdust may include a higher concentration of nitrogen-containing functional groups, which can assist in the absorption of CO_2. The existence of biochar in the soil can impact the soil's microbe population by modifying physicochemical parameters. Consequently, this affects the release of CO_2 and the priming impact of soil organic carbon. The priming effect is a phenomenon in which the introduction of an outside carbon source, such as biochar, enhances the microbial breakdown of the existing organic carbon in the soil, hence affecting the total carbon cycle within the soil (Sheng & Zhu, 2018). The effectiveness of biochar in storing carbon

TABLE 15.2 List of some important scientific studies that highlighted the role and mechanisms of biochar resultant from various feedstock for remediation of organic contaminants and toxic metals from the water.

Feedstock for Biochar preparation	Technique	Targeted pollutant	BET Surface area (m²/g)	Adsorption efficiency	Removal efficiency	Mechanism Involved	References
Phenol-formaldehyde resin modified wood	—	Congo red, methylene blue	2301.61	—	Congo red 3472.22 mg per g; methylene blue-112.35 mg per g for	Phenol-formaldehyde resin is used to fill pores. Electrostatic attraction, hydrogen bonding, and π-π interaction work together to achieve this.	Zhang et al. (2024)
Chitosan	Pyrolysis	Phenol	—	55%	Phenol- 95 percent degradation rate and 75 percent mineralization rate of (30 mg/L)	The chitosan-derived biochar adsorbs through the intrinsic pyridinic N and C—C/C==C groups. The adsorption mechanism involves Lewis acid-base interaction, π-π EDA and electrostatic interactions. The C—O groups functioned as the catalytic active sites for PDS activation.	Wang, Yang, et al. (2024), Wang, Nie, et al. (2024)
Corn stover	—	Norfloxacin (NOR), sulfamethoxazole (SMX), tetracycline (TC)			CoPMoV/C/PMS by 99.6, 99.0, and 100 percent for NOR, SMX, and TC, respectively.	The presence of functional groups (—COOH, —COH, and -OH) on the surface of biochar is crucial for the activation of peroxymonosulfate. The production of free radicals in the CoPMoV/C structure is primarily caused by the combined activation of peroxymonosulfate by transition metals and biochar.	Zhu, Yang, et al. (2023), Zhu, An, et al. (2023)
Watermelon rind	—	Fluoride		9.5 mg/g		Adsorption by electrostatic attraction through watermelon rind biochar's protonated basic functions and precipitation at mineral locations	Sadhu et al. (2022)
Delonix regia pods	Carbonization	Phenol		2.59 mg/g		Adsorption	Emenike et al. (2022)
Walnut shell	Pyrolysis	Steroid hormones; estrone, 17β-estradiol; estriol	737.98			Free radicals have been linked to estrogen breakdown Open loop was the primary estrogen degradation route.	Xu et al. (2022)

Feedstock	Method	Pollutant	Value 1	Value 2	Mechanism	Reference
Fermentation residue derived biochar/Spiramycin (SPI) fermentation residue (SFR)	Pyrolysis	Spiramycin	451.68 (BC700)	147.28 mg/g	The adsorption mechanism of biochar includes electrostatic contact, pore filling, π-π interaction, hydrogen bonding, and participation of C-C and O-C—O functional groups. Spiramycin adsorption on biochar was primarily monolayer, involving physical and chemical forces such as electrostatic interaction, pore filling, π-π interaction, hydrogen bonding, and C-C and O-C—O functional groups.	Gao et al. (2022)
Eucalyptus wood	Pyrolysis	Phenolic compounds	2048	95% phenol	π = π interactions interaction, consequential adsorption of phenols	Singh et al. (2021)
Activated sludge (Wastewater treatment plant)	Pyrolysis	Tetracycline	133.4, 114.6	The qmax on HSBC-600 and PSBC-600 achieves 116.9 and 89.3 mg/g, respectively	Pore filling, electrostatic force, hydrogen bond, and π-π interaction Multi-layer adsorption of TC onto HSBC-600 and PSBC-600 involves chemical adsorption.	Chen, Zhang, et al. (2021), Chen, Zhao, et al. (2021)
Cell-immobilized biochar of Pseudomonas fluorescens MC46 (MC46)	Carbonization	Triclocarban (TCC)		potassium hydroxide-modified biochars adsorb 8.43 and 9.17 mg g^{-1} triclocarban.	TCC removals 79.80% (in model TCC solution) 32% (Free MC46 cells), 2% (no MC46 cells)	Jenjaiwit et al. (2021)
Rice husk	Microwave pyrolysis	Phosphate and nitrate		Adsorption capacity for phosphate and nitrate was 71 and 497 mg kg^{-1}, respectively. Adsorption rates were 65% and 75%.	—	Shukla et al. (2019)
Municipal sewage treatment plant secondary sludge	Pyrolysis	Triclosan	157.4	—	Triclosan degradation was caused by hydroxyl, sulfate, and singlet oxygen radicals. Triclosan degraded mostly via dechlorination and hydroxylation. Activation of peroxymonosulfate by biochar to degrade triclosan	Wang and Wang (2019)
Marine macroalgae (C. glomerata and G. gracilis biomass)	Hydrothermal liquefaction	Methylene blue	C. glomeratabiochar (6.45), G. gracilisbiochar (2.11)	226 and C. glomerata and 104 mg/g for G. gracis	MB adsorption onto biochars involves electrostatic interaction, surface complexation, physical function, and π-π stacking interaction.	Parsa et al. (2019)

(Continued)

TABLE 15.2 (Continued)

Feedstock for Biochar preparation	Technique	Targeted pollutant	BET Surface area (m²/g)	Adsorption efficiency	Removal efficiency	Mechanism Involved	References
Functionalized biochar	—	Estrone, 17β-estradiol, estriol, 17α-ethynylestradiol, bisphenol A (BPA), 4-tert-butylphenol (4tBP)	—	—	Removal of ~500 μg/L of each phenolic endocrine disrupting chemicals (EDCs)	Hydrogen bonding and π-π electron-donor-acceptor interactions led to maximum EDC sorption.	Ahmed et al. (2018)
Heavy Metals							
Starch-coated corn cobs		Pb^{2+} and Cd^{2+}		Sorption capacities of SZF@CBC for Pb^{2+} and Cd^{2+} reached 199.98 mg/g and 183.93 mg g^{-1}		Plentiful exposed adsorption sites (–OH, –COOH, Zn-O and Fe-O) on SZF@CBC surface accelerate Pb^{2+} and Cd^{2+} adsorption. Complexation, electrostatic attraction, ion exchange and physical adsorption	Xia et al. (2023)
Auricularia auricula biochar modified by CS2		Cd^{2+}		Adsorption rate and capacity 96.1 percent; and 9.61 mg/g	q_{max} of 450.51 mg/g	CS2 alteration enhances Cd^{2+} adsorption by complexing and precipitating sulfur-containing compounds.	Wang et al. (2022)
Pig manure		Uranium		PMBC-PP, PMBC-HP, PMBC-H$_2$O adsorbed bt 979.3; 661.7; 369.9 mg/g respectively		The abundant active sites on PMBC-PP mostly mediated UVI species-PMBC-PP interaction.	Liao et al. (2022)
Sewage sludge	Pyrolysis	As^{3+}, Cr^{6+}	—	Cr^{6+} uptake by 47.46 mg/g		As^{3+} and Cr^{6+} species adsorption on the charcoal surface was due to physisorption and chemosorption, respectively.	Shen et al. (2022)
Canola straw-based biomass	Pyrolyzed	As		834 and 953 μg/g for OBM and OBC		Hybrid adsorption mechanism governed by chemisorption occurred during the adsorption process.	Zoroufchi Benis et al. (2021)
Halloysite and coconut shell biochar magnetic composites	Co-pyrolysis	Pb^{2+}		833.33 mg/g for Pb^{2+}	90% (Pb^{2+})	Complexation, cation exchange, cation–π interaction, precipitation, pore filling and electrostatic adsorption. Oxygen-containing functional groups introduced on the surface were bound to metallic ions	Wang et al. (2021)

Feedstock	Process	Metals	Efficiency	Mechanism	Reference
Rice husk feedstock, biochar, ethylene-diamine-tetra-acetic acid-modified biochar, mixture	Pyrolysis	Cu^{2+}, Cd^{2+}, Pb^{2+}	98%	EDTA-RHB treatments surpassed RH or RHB in heavy metal adsorption.	Roha et al. (2021)
Luffa rattan	Hydrothermal carbonization	U^{6+}	382 mg g^{-1}	Adsorption mechanism including electrostatic interaction, complexation, and physical adsorption	Ye et al. (2020)
Tea waste and rice husk biochars	Pyrolysis	Cr^{6+}	197.5 mg/g and 195.24 mg/g Cr^{6+} sorbed by tea waste and rice husk biochar, respectively	The –OH, COO–, and –NH2 functional groups that are involved in Cr^{6+} sorption on biochar.	Khalil et al. (2020)
			99.3 and 96.8 percent removal of Cr^{6+} by tea waste biochar and rice husk biochar, respectively		
Lemna minor	Pyrolysis	Ni^{2+}	41.68 mg/g for LM400	The biochar in wastewater interacts with the oxygen-containing functional groups and K^+ ions to form complexation and ion-exchange reactions with Ni^{2+}.	Yan et al. (2020)
			90%		
Eichhornia crassipes	Pyrolysis	Cr^{3+}	—	Adsorption	Hashem et al. (2020)
			99%		
Wastewater treatment plants sludge	Pyrolysis	Cu^{2+}, Co^{2+}, Ni^{2+}	—	Adsorption of metal ions	Matheri et al. (2020)
			77.86%, 75% and 56.25% of Cu^{2+}, Co^{2+}, Ni^{2+}, respectively		
Rice husk, wheat straw, corncob	Pyrolysis	Pb^{2+}, Cd^{2+}	Pb^{2+} adsorption capacity of 96.41%, 95.38%, and 96.92% for rice husk, wheat straw, and corncob, respectively, while for Cd^{2+}, the uptake capacity was found to be 94.73%, 93.68%, and 95.78%.	Redox surface functional groups, when combined, have a crucial function in either donating or absorbing an electron to break down contaminants in wastewater.	Amen, Bashir, et al. (2020), Amen, Yaseen, et al. (2020)
			—	Adsorption occurs when particles adhere to the interior sites of adsorption on the surface of the adsorbent by a specific mechanism called ion diffusion.	
Muskmelon peel	—	Cu^{2+}, Zn^{2+}	78.74; 72.99 mg/g for Cu^{2+} and Zn^{2+}	Metal ions are evenly adsorbed onto the surface of biochar	Khan et al. (2017)

depends on various aspects, such as its ability to remain stable in the soil and its impact on the process of native SOC breakdown. The potential of this is additionally influenced by fluctuations resulting from charcoal manufacturing, becoming older, and the clay composition of the soil. An important factor to examine when evaluating the efficacy of biochar in carbon sequestration is its stability in the soil, how it changes over time, and how it interacts with soil components, particularly clay concentration (Yang et al., 2022). Recent research has emphasized that soil pH has a crucial impact both on the soil's microbe population and the long-term viability of biochar (Sheng & Zhu, 2018). The soil's pH levels are a crucial determinant that can greatly affect the relationships between biochar and the diversity of microbes in the soil, hence altering the overall stability and efficacy of biochar in the soil environment. The addition of biochar to the soil affects CO_2 emission by altering the level of OM and the population of copiotrophic bacteria, with soil pH playing a significant role in this process. The utilization of biochar application offers a threefold advantageous conclusion, which includes improving crop production, augmenting SOC, and reducing greenhouse gas emissions. The effectiveness of this solution is mostly influenced by the characteristics of the biochar and the pH level of the soil. The acidity of soil environments enhances the expulsion of carbonates from biochar, particularly in acidic conditions. The high occurrence of gram-positive fungi and bacteria in acidic soils promotes the breakdown of biochar, which in turn enhances the mineralization of soil organic carbon (SOC). The addition of biochar improves the conditions in acidic soils, allowing for the decomposition of both SOC and biochar. The findings suggest that acidic soils experience a significantly greater release of CO_2 following the application of biochar, in comparison to neutral and alkaline soils. The heightened deterioration of biochar in acidic soils leads to an amplified emission of inorganic CO_2. The presence of a greater percentage of gram-positive microbes within acidic soil (25%–36%) is recognized as a crucial element in the improved breakdown of biochar and the concomitant co-metabolism of SOC. In addition to soil pH, parameters such as clay content and experimental duration also have an impact on the physico-chemical and biotic processes of SOC dynamics (Sheng et al., 2016). The usage of biochar made from rice straw has attracted attention for its capacity to store SOC for a long period. However, the effectiveness of carbon sequestration varies greatly among different agricultural soils, even when the same amount of BC is applied (Bi et al., 2020). The extent of SOC sequestration after biochar application is determined by soil parameters. To optimize long-term SOC sequestration, it is important to carefully analyze the individual soil properties of the site where biochar is applied, as highlighted by Bi et al. (2020).

15.3.4 Biochar as an electrode material for the development of microbial fuel cells

MFCs represent an emerging technology that couples renewable energy with environmental sustainability (Apollon et al., 2024). MFCs certainly hold great promise for sustainable power generation, as they can directly turn the chemical energy found in organic substrates into electrical energy as a result of the metabolic activity of microorganisms (Rashid et al., 2019). Most of the MFC performance materials generally depend largely on the electrodes to which they are attached, facilitating the transfer of electrons from the microbial metabolic processes and leading them through the external circuit (Cao et al. 2022). Recently, biochar was considered one of the upcoming electrode materials due to its properties that offer potential benefits. Biochar, being a carbonaceous material produced at a higher temperature through the process of burning biomass in an environment void of oxygen, has offered a promising potential electrode material for MFCs due to its texture and surface area (Major et al. 2010). The usage of biochar as an electrode material in MFCs because of the High surface area in which a lot of effective space for colonization of microorganisms and electron transfer is available (Van Zwieten et al. 2010). Moreover, enhanced porous structure contributes to increased accessibility of the microbe to the electrode surface, thus increasing the electron transfer (Ding et al. 2020). In addition to that raw biochar has moderate electrical conductivity. However, through some processes and treatments, it can be enhanced to make it biocompatible and desired in the growth and activities of electroactive microbes. It has also good advantages as Biochar in MFCs because of its cost-effectiveness: as compared to traditional electrode materials which include graphite and platinum (Kappler et al. 2014). Usage of biochar in MFCs will lead to waste and carbon sequestration in the environment and, so far, upholds environmental sustainability. Investigations proved that biochar, with its favorable physical and chemical properties, is capable of improving the power output and performance of MFCs (Chu et al. 2022). The potential of biochar as electrode material in microbial fuel cells has gathered a lot of interest in that it is sustainable, low-cost, and of good efficient in comparison with traditional materials. In an MFC, an electroactive bacterium in the anodic chamber oxidizes organic substrates; this oxidation releases electrons and protons. Afterward, these electrons are transferred to the anode, then pass through an external circuit to reach the cathode, where they will reduce with protons and an electron acceptor—usually oxygen—to form water. In the case of the anode material, biochar will have its role in facilitating electron transfer from the microbial cells to the external circuit. It has been shown that BC can greatly improve the performance of MFCs (Kappler et al. 2014). With a large surface area and porous structure, BC offers more sites for microbial adhesion and electron transfer.

In addition, the enhanced electrical conductivity of the activated biochar explains the reduced internal resistance and increased power output. Biochar surface chemistry modifiers have been introduced to enhance the extent of microbial adhesion and kinetics of electron transfer, which further enhances the efficiency of MFCs. In their 2014 study, Kappler et al. noticed that biochar alters the mineral product resulting from the reduction of ferrihydrite, shifting it from magnetite to siderite. Biochar can affect the chemical and structural properties of soil. Additionally, it can directly facilitate the movement of electrons, acting as a mediator in this process. To the present date, research and improvement in biochar-based MFCs are continually being conducted to overcome the prevailing challenges and enhance their performances for various applications. The major challenge of Biochar-based MFCs exhibit impressive performance in the laboratory but are still far from being scalable at the industrial level. The variability of biochar nature due to several influencing factors feedstocks and pyrolysis conditions makes the properties a long list that needs standardization so that repeatability of the performance can be assured. The application of biochar-based MFC has been summarized in Table 15.3

15.4 Direct interspecies electron transfer pathways

Anaerobic digestion (AD) is a cost-effective and well-established technique for digesting organic materials, while also producing fuel and providing useful fertilizer. It additionally serves as an approach to reduce pollution (Pan et al. 2022). The waste valorization technique is the most practical and environmentally friendly option available. AD is facilitated by many groups of bacteria, as described by Valentin et al. (2023). Similarly, AD is a natural process that requires a group of microorganisms capable of efficiently breaking down intricate substances. These bacteria play a crucial role in maintaining certain functions. Neglecting their presence might lead to inhibition or disruption (Fagbohungbe et al., 2017). The biomass that can be processed includes waste from animals, dietary waste, agricultural waste, and the organic fraction of municipal solid trash. The process of breaking down biomass into biomethane consists of acid production, acetate, and methanes, which are performed by specific groups of microorganisms. During the hydrolysis step, hydrolytic microorganisms, such as *Streptococcus* and *Enterobacter* transform organic substrates into basic monomers including lipids, proteins, and carbohydrates (Amin et al., 2021; Zhao et al., 2021). The microbial population within a digester that is anaerobic is distinguished by an intricate network of interactions, in which each bacterium fulfills a distinct role. The microbial population in a digestion facility is extremely dynamic, and alterations in environmental circumstances can impact the structure and functionality of the population. Gaining insight into the microbial population within the AD process is crucial for maximizing the operation and enhancing the effectiveness of organic waste management (Figs. 15.3 and 15.4).

Kutlar et al. (2022) stated that DIET occurs between syntrophic bacteria, specifically acetogens, and archaea, specifically methanogens. The ability of BC to facilitate DIET is influenced by its primary physical and chemical characteristics, such as the dimension of particles, presence of functional groups, conductivity to electricity, and redox-active components. The pyrolysis temperature has a substantial impact on these properties, followed by the residence duration and the kind of biomass. The functional groups in BC have a crucial role in the breakdown of volatile fatty acids (VFAs) and the binding of poisons and heavy metals, as well as their porous structure. Despite having an electrochemical capacity that is 1000 times less than other carbon compounds, the inclusion of redox-active components in BC enhances methanization. The metabolic process of OM in the AD system occurs through the exchange of electrons among syntrophic archaea and bacteria, with each organism providing and absorbing electrons from one another. BC has been identified as a substitute for biological electrical shuttles in facilitating electron transfer. Biochar can support the growth of microorganisms on its surface (in a loose manner), within its small pores (in a tight manner), as well as in the liquid portion above it (supernatants). Specific types of bacteria have exclusive associations with these partitions of biochar. The initial absence of certain bacteria in the AD process can be attributed to their strong adherence to the biochar, which eventually leads to their emergence over time. The dose of biochar is correlated with the AD system's ability to adsorb heavy metals, sulfate, as well as oxidize VFAs. The accumulation of VFAs becomes a significant hindrance when it occurs rapidly and is not efficiently degraded by methanogenic bacteria, which can be measured by the organic loading rate (OLR) and hydraulic retention time (HRT). The biochar functioned as a transient medium for the proliferation of microorganisms (Valentin et al. 2023).

15.5 Challenges

The utilization of biochar for environmental applications faces several challenges. First, the inherent variability in feedstock sources and production methods leads to diverse biochar compositions, impacting its electrochemical performance. Achieving consistent and reproducible electrochemical properties across different biochar types remains a significant hurdle. Furthermore, the understanding of biochar's complex electrochemical behavior is still evolving,

TABLE 15.3 Application of biochar in the development of microbial fuel cell system.

Application of Biochar in MFCs	Materials based Biochar	Power generated	Pollutant removal and effectiveness	References
Anode	Sludge MFC + biochar-modified-anode	108.05 mW/m²	*Longilinea*, *Denitratisoma*, and *Pseudomonas* removed contaminants. Highest COD and TP removal. EC significantly impacted bacterial community organization.	Sun et al. (2024)
—	Straw biochar	Maximum output voltage: 662.64 mV; Open circuit voltage: 798.24 mV; power density: 1.54 W/m²; Columbic efficiency: 50.39%	A COD removal efficiency of 77.45% Exchange current density (4.6142×10^{-4} A/cm²)	Yan et al. (2024a)
—	Corn stalk biochar	Power density: 2.68 W/m²	Coulombic efficiency (70.52%)	Yan et al. (2024b)
—	Peanut shell biochar (PSB) exhibited a	Maximum power production: 165 mW/m²	High surface area, specific capacitance, and inter-facial electron transport increased MFC power output	Maan et al. (2024)
—	Wood biochar	532–457 Mw/m²	—	Qian et al. (2015)
—	Forestry residue	Forestry 457 Mw/m²	—	Huggins et al. 2014
—	Sawdust	3.2 Ma/m³	—	Chaijak et al. (2020)
—	Biochar + NiFe₂O₄	1200 Mw/m²	COD reduction: 28%	Senthilkumar et al. (2019)
Cathode	Banana plant + KOH	528.2 Mw/m²	NA	Yuan et al. (2014)
—	Watermelon peel biochar	0.262 W/m³	—	Mo et al. (2016)
—	Wood-based + manganese oxide	187.8 W/m²	—	Huggins et al. (2015)
—	Alfalfa leaves + Biochar altered with KOH	1328.9 m/Wm²	—	Jiang et al. (2014)
—	Waste-banana-peels	142.2 mW/m² of power density	—	Priyadarshini et al. (2024)

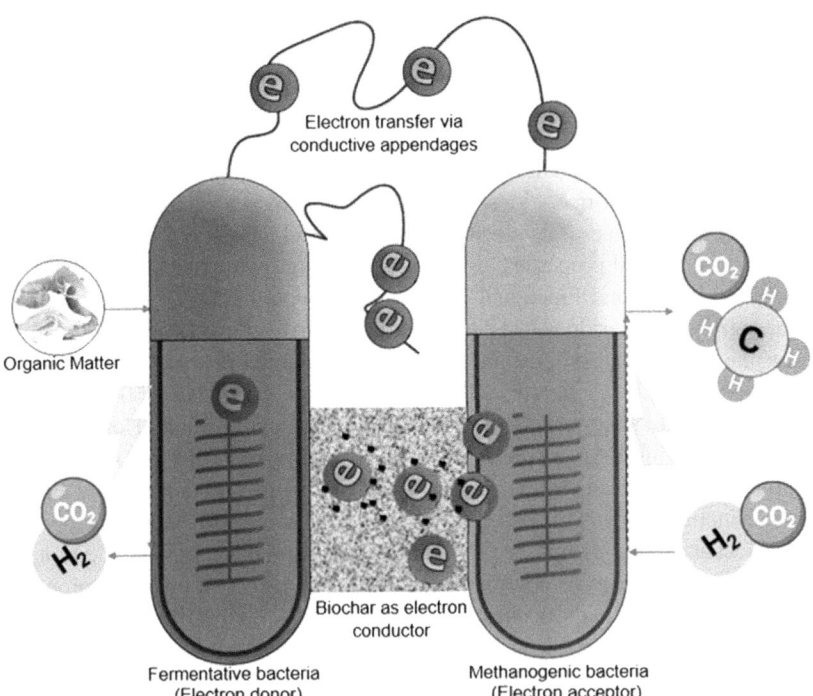

FIGURE 15.3 Direct interspecies electron transfer (DIET) mechanism through the anaerobic bacteria. Source: *Valentin, M. T., Luo, G., Zhang, S., et al. (2023). Direct interspecies electron transfer mechanisms of a biochar-amended anaerobic digestion: A review.* Biotechnology for Biofuels, 16, Article 146. https://doi.org/10.1186/s13068-023-02391-3; *licensed under a Creative Commons Attribution 4.0 International License.*

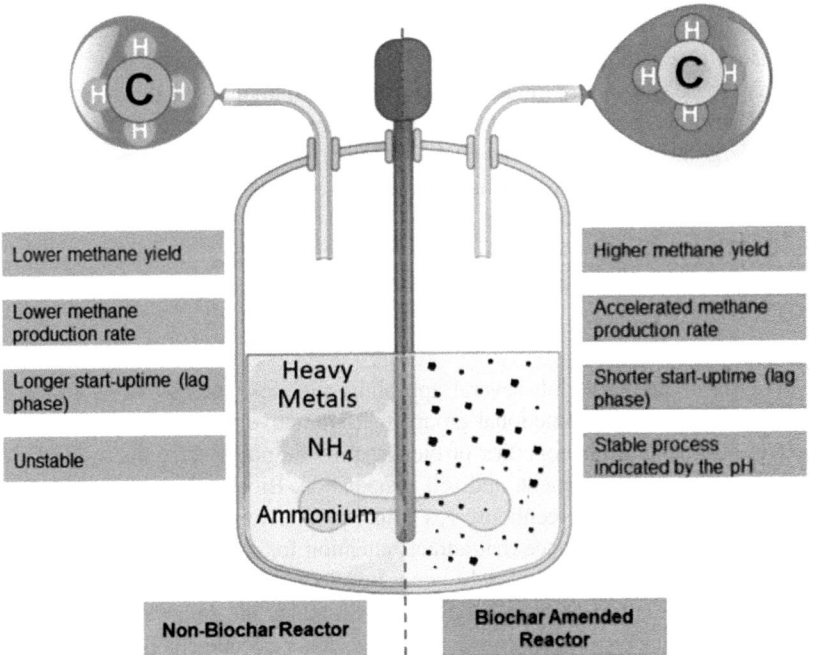

FIGURE 15.4 Direct interspecies electron transfer (DIET) mechanisms: DIET anaerobic digester versus non-DIET reactor. Source: *Valentin, M. T., Luo, G., Zhang, S., et al. (2023). Direct interspecies electron transfer mechanisms of a biochar-amended anaerobic digestion: A review.* Biotechnology for Biofuels, 16, Article 146. https://doi.org/10.1186/s13068-023-02391-3; *licensed under a Creative Commons Attribution 4.0 International License.*

requiring advanced characterization techniques for a comprehensive assessment. The long-term stability and durability of biochar electrodes under varying environmental conditions pose another challenge, demanding research into enhancing their robustness. Overcoming these challenges will be crucial for realizing the full potential of biochar in diverse environmental applications. Additionally, the scale-up of biochar production for widespread environmental use must address economic feasibility, logistical constraints, and the environmental footprint of large-scale production processes. Nevertheless, the limited use of biochar can be attributed to insufficient study and information.

15.6 Future perspectives

Biochar is a unique and renewable resource that has significant potential to tackle a range of environmental concerns that have emerged in recent times. The electrochemical characteristics of biochar have favorable prospects for versatile utilization in various fields, including the removal of contaminants in soil and water, the storage of carbon, and the reduction of greenhouse gas emissions. Additionally, biochar can enhance the quality of soil, water, and air synergistically. The potential of biochar as an adsorbent in wastewater and soil remediation operations, particularly for removing hazardous chemicals released by different industries, has been extensively studied due to its electrochemical capabilities. The proper adoption of feedstock and pyrolysis conditions to tailor the electrochemical properties of biochar is a fast-growing area. In principle, the idea is to develop biochar with high surface area, porosity, and functional groups, which would enhance electrochemical reactions. Lignocellulosic biomass, in particular, agricultural residues—residues from rice husk, corn stover, and sugarcane bagasse—and wood and forestry residues, offer the highest carbon content among the considered raw materials, therefore yielding biochar with desired electrochemical performance. Generally, a higher pyrolysis temperature, 600°C–900°C, enhances the surface area, porosity, and conductivity of biochar required for electrochemical applications. However, very extreme temperatures may lower productivity. Slow heating rates can lead to better development of the porous structure, while fast heating rates can increase surface area but likely decrease the uniformity of the pores. Hybridization of biochar with other materials such as graphene, carbon nanotubes, metal oxides, etc., to enhance its electrochemical properties. However, using blended feedstocks to optimize biochar properties for specific electrochemical applications. In addition, utilizing advanced characterization techniques to better relate pyrolysis conditions and feedstock composition to the properties of biochar. Focus on sustainable and scalable production methods that are low cost and low environmental impact. The feedstock selection and optimization of pyrolysis conditions can result in improved electrochemical performance of biochar materials used in energy storage, environmental remediation, and catalysis. Nevertheless, due to the unfamiliar atomic surface properties of these extremely disordered materials, there is currently a significant lack of accurate theoretical models. The inherent challenges, stemming from the diverse nature of biochar feedstocks and the need for standardized production methods, necessitate concerted efforts from the scientific community. The exploration of the electrochemical properties of biochar represents a dynamic field with both challenges and exciting prospects. Collaborative efforts between researchers, industry, and policymakers are essential to standardize biochar production, ensuring consistent electrochemical performance. In addition, future research should focus on elucidating the fundamental mechanisms governing biochar's electrochemical behavior and developing tailored biochar-based materials to address specific environmental challenges, fostering the evolution of this eco-friendly technology. Therefore, biochar treatment holds the potential to act as a pretreatment method for removing toxic compounds before undergoing subsequent biological treatment. A comprehensive study is necessary to elucidate the removal mechanisms of these toxic compounds. Biochar engineering enables the customization of biochar properties to be optimal for specific applications and conditions. It is anticipated that further research in this domain could unveil new opportunities in science, paving the way for practical applications in diverse fields.

15.7 Conclusion

Biochar is a carbon-rich byproduct of the pyrolysis of biomass with several appealing qualities, such as being low-cost, environmentally friendly, and recyclable. It also possesses highly functional groups that are active and a wide surface area, which are advantageous features. The distinctive electrochemical properties of biochar provide novel solutions in the fields of storing energy, monitoring the environment, wastewater management, and soil remediation. Biochar is receiving growing recognition for its multifaceted uses, such as carbon sequestration, greenhouse gas emission reduction, soil improvement, and environmental cleanup. Biochar is unambiguously a powerful substance that attracts attention for its ability to tackle various environmental issues. The elimination of environmental pollutants is closely connected to their relationship to the active functional groups, such as -COOH, R-OH, and -OH, present in biochar. Organic contaminants are removed by processes such as hydrophobic contacts, electrostatic attraction/repulsion involving π-π electrons donor-acceptor conversations, and partitioning. In contrast, inorganic contaminants are eliminated through mechanisms such as the exchange of ions, surface complexation, precipitating, and interaction between ions. Although biochar demonstrates satisfactory electrochemical performance, there is potential for enhancement. To improve its effectiveness, one must possess a thorough comprehension of surface chemicals and molecular interactions. To attain this comprehension, a comprehensive methodology that incorporates both empirical and theoretical research is required. To make further progress, it will be crucial to develop more advanced analytical tools and get a more profound comprehension of the underlying electrochemical processes that underlie the behavior of biochar. The future shows potential for the further development of biochar-based technologies, making a substantial contribution to environmental sustainability and tackling the intricate challenges faced by our times.

References

Abideen, Z., Koyro, H. W., Zulfiqar, F., Moosa, A., Rasool, S. G., Ahmad, M. Z., Altaf, M. A., Sharif, N., & El-Keblawy, A. (2023). Impact of biochar amendments on copper mobility, phytotoxicity, photosynthesis and mineral fluxes on (*Zea mays* L.) in contaminated soils. *South African Journal of Botany, 158*, 469–478. Available from https://doi.org/10.1016/j.sajb.2023.05.036.

Abujabhah, I. S., Bound, S. A., Doyle, R., & Bowman, J. P. (2016). Effects of biochar and compost amendments on soil physico-chemical properties and the total community within a temperate agricultural soil. *Applied Soil Ecology, 98*, 243–253.

Ahmed, M. B., Zhou, J. L., Ngo, H. H., Johir, M. A. H., & Sornalingam, K. (2018). Sorptive removal of phenolic endocrine disruptors by functionalized biochar: Competitive interaction mechanism, removal efficacy and application in wastewater. *Chemical Engineering Journal, 335*, 801–811. Available from https://doi.org/10.1016/j.cej.2017.11.041.

Alves, Z., Ferreira, N. M., Figueiredo, G., Mendo, S., Nunes, C., & Ferreira, P. (2022). Electrically conductive and antimicrobial agro-food waste biochar functionalized with zinc oxide particles. *International Journal of Molecular Sciences, 23*(14), 8022. Available from https://doi.org/10.3390/ijms23148022.

Amen, R., Bashir, H., Bibi, I., Shaheen, S. M., Niazi, N. K., Shahid, M., & Rinklebe, J. (2020). A critical review on arsenic removal from water using biochar-based sorbents: the significance of modification and redox reactions. *Chemical Engineering Journal, 396*, 125195. Available from https://doi.org/10.1016/j.cej.2020.125195.

Amen, R., Yaseen, M., Mukhtar, A., Klemeš, J. J., Saqib, S., Ullah, S., Al-Sehemi, A. G., Rafiq, S., Babar, M., Fatt, C. L., Ibrahim, M., Asif, S., Qureshi, K. S., Akbar, M. M., & Bokhari, A. (2020). Lead and cadmium removal from wastewater using eco-friendly biochar adsorbent derived from rice husk, wheat straw, and corncob. *Cleaner Engineering and Technology, 1*. Available from https://doi.org/10.1016/j.clet.2020.100006.

Amin, F. R., Khalid, H., El-Mashad, H. M., Chen, C., Liu, G., & Zhang, R. (2021). Functions of bacteria and archaea participating in the bioconversion of organic waste for methane production. *The Science of the Total Environment, 763*, 143007. Available from https://doi.org/10.1016/j.scitotenv.2020.143007.

Apollon, W., Rusyn, I., Kuleshova, T., Luna-Maldonado, A. I., Pierre, J. F., Gwenzi, W., & Kumar, V. (2024). An overview of agro-industrial wastewater treatment using microbial fuel cells: Recent advancements. *Journal of Water Process Engineering, 58*, 104783. Available from https://doi.org/10.1016/j.jwpe.2024.104783.

Azman, N. A. N. M. N., Khalid, P. I. M., Amin, N. A. S., Zakaria, Z. Y., Zainol, M. M., Ilham, Z., Phaiboonsilpa, N., & Asmadi, M. (2023). Effects of biochar, compost, and composted biochar soil amendments on okra plant growth. *Materials Today: Proceedings*, 1–4.

Bachmann, H. J., Bucheli, T. D., Dieguez-Alonso, A., Fabbri, D., Knicker, H., Schmidt, H.-P., Ulbricht, A., Becker, R., Buscaroli, A., Buerge, D., et al. (2016). Toward the standardization of biochar analysis: The COST action TD1107 interlaboratory comparison. *Journal of Agricultural and Food Chemistry, 64*(2), 513–527.

Banik, C., Lawrinenko, M., Bakshi, S., & Laird, D. A. (2018). Impact of pyrolysis temperature and feedstock on surface charge and functional group chemistry of biochars. *Journal of Environmental Quality, 47*(3), 452–461. Available from https://doi.org/10.2134/jeq2017.11.0432.

Bao, D., Li, Z., Liu, X., Wan, C., Zhang, R., & Lee, D. J. (2020). Biochar derived from pyrolysis of oily sludge waste: Structural characteristics and electrochemical properties. *Journal of Environmental Management, 268*, 110734. Available from https://doi.org/10.1016/j.jenvman.2020.110734.

Barnes, R. T., Gallagher, M. E., Masiello, C. A., Liu, Z., & Dugan, B. (2014). Biochar-induced changes in soil hydraulic conductivity and dissolved nutrient fluxes constrained by laboratory experiments. *PLoS One, 9*(9), e108340. Available from https://doi.org/10.1371/journal.pone.0108340.

Bartoli, M., Troiano, M., Giudicianni, P., Amato, D., Giorcelli, M., Solimene, R., & Tagliaferro, A. (2022). Effect of heating rate and feedstock nature on electrical conductivity of biochar and biochar-based composites. *Applications in Energy and Combustion Science, 12*, 100089. Available from https://doi.org/10.1016/j.jaecs.2022.100089.

Bi, Y., Cai, S., Wang, Y., Zhao, X., Wang, S., & Xing, G. (2020). Structural and microbial evidence for different soil carbon sequestration after four-year successive biochar application in two different paddy soils. *Chemosphere, 254*, 126881. Available from https://doi.org/10.1016/j.chemosphere.2020.126881.

Campos, P., Miller, A. Z., Knicker, H., Costa-Pereira, M. F., Merino, A., & De la Rosa, J. M. (2020). Chemical, physical and morphological properties of biochars produced from agricultural residues: Implications for their use as soil amendment. *Waste Management, 105*, 256–267. Available from https://doi.org/10.1016/j.wasman.2020.02.013.

Cao, T. N. D., Mukhtar, H., Yu, C. P., Bui, X. T., & Pan, S. Y. (2022). Agricultural waste-derived biochar in microbial fuel cells towards a carbon-negative circular economy. *Renewable and Sustainable Energy Reviews, 170*, 112965.

Chacón, F. J., Cayuela, M. L., Roig, A., & Sánchez-Monedero, M. A. (2017). Understanding, measuring, and tuning the electrochemical properties of biochar for environmental applications. *Reviews in Environmental Science and Biotechnology, 16*(4), 695–715. Available from https://doi.org/10.1007/s11157-017-9450-1.

Chacon, F. J., Sanchez-Monedero, M. A., Lezama, L., & Cayuela, M. L. (2020). Enhancing biochar redox properties through feedstock selection, metal preloading and post-pyrolysis treatments. *Chemical Engineering Journal, 395*, 125100. Available from https://doi.org/10.1016/j.cej.2020.125100.

Chaijak, P., Sato, C., Lertworapreecha, M., Sukkasem, C., Boonsawang, P., & Paucar, N. (2020). Potential of biochar-anode in a ceramic-separator microbial fuel cell (CMFC) with a Laccase-based air cathode. *Pol J Environ Stud, 29*(1), 499–503. Available from https://doi.org/10.15244/pjoes/99099.

Chang, Y., Liu, W., Mao, Y., Yang, T., & Chen, Y. (2022). Biochar addition alters C: N: P stoichiometry in moss crust-soil continuum in Gurbantünggüt desert. *Plants, 11*(6), 814. Available from https://doi.org/10.3390/plants11060814.

Chen, Q., Zhang, Q., Yang, Y., Wang, Q., He, Y., & Dong, N. (2021). Synergetic effect on methylene blue adsorption to biochar with gentian violet in dyeing and printing wastewater under competitive adsorption mechanism. *Case Studies in Thermal Engineering, 26*. Available from https://doi.org/10.1016/j.csite.2021.101099.

Chen, X., Lin, Q., Rizwan, M., Zhao, X., & Li, G. (2019). Steam explosion of crop straws improves the characteristics of biochar as a soil amendment. *Journal of Integrative Agriculture, 18*(7), 1486–1495. Available from https://doi.org/10.1016/S2095-3119(19)62573-6.

Chen, S., Rotaru, A. E., Shrestha, P., et al. (2014). Promoting Interspecies Electron Transfer with Biochar. *Scientific Reports, 4*, 5019. Available from https://doi.org/10.1038/srep05019.

Chen, W., Zhao, B., Guo, Y., Guo, Y., Zheng, Z., Pak, T., & Li, G. (2021). Effect of hydrothermal pretreatment on pyrolyzed sludge biochars for tetracycline adsorption. *Journal of Environmental Chemical Engineering, 9*(6). Available from https://doi.org/10.1016/j.jece.2021.106557.

Chu, M., Tian, W., Zhao, J., Zou, M., Lu, Z., Zhang, D., & Jiang, J. (2022). A comprehensive review of capacitive deionization technology with biochar-based electrodes: Biochar-based electrode preparation, deionization mechanism and applications. *Chemosphere, 307*, 136024.

Cuong, D. V., Wu, P. C., Chen, L. I., & Hou, C. H. (2021). Active MnO2/biochar composite for efficient As (III) removal: Insight into the mechanisms of redox transformation and adsorption. *Water Research, 188*, 116495. Available from https://doi.org/10.1016/j.watres.2020.116495.

Ding, Y., Wang, T., Dong, D., & Zhang, Y. (2020). Using biochar and coal as the electrode material for supercapacitor applications. *Frontiers in Energy Research, 7*, 159.

Dubé, C. D., & Guiot, S. R. (2015). Direct interspecies electron transfer in anaerobic digestion: a review. In G. Guebitz, A. Bauer, G. Bochmann, A. Gronauer, & S. Weiss (Eds.), *Biogas science and technology. In Advances in biochemical engineering/ biotechnology* (pp. 101–115). Springer.

Emenike, E. C., Ogunniyi, S., Ighalo, J. O., Iwuozor, K. O., Okoro, H. K., & Adeniyi, A. G. (2022). Delonix regia biochar potential in removing phenol from industrial wastewater. *Bioresource Technology Reports, 19*. Available from https://doi.org/10.1016/j.biteb.2022.101195.

Fagbohungbe, M. O., Herbert, B. M. J., Hurst, L., Ibeto, C. N., Li, H., Usmani, S. Q., et al. (2017). The challenges of anaerobic digestion and the role of biochar in optimizing anaerobic digestion. *Waste Manage, 61*, 236–249. Available from https://doi.org/10.1016/j.wasman, 2016.11.028.

Fan, Q., Sun, J., Chu, L., Cui, L., Quan, G., Yan, J., & Iqbal, M. (2018). Effects of chemical oxidation on surface oxygen-containing functional groups and adsorption behavior of biochar. *Chemosphere, 207*, 33–40. Available from https://doi.org/10.1016/j.chemosphere.2018.05.044.

Favre, F., Slijepcevic, A., Piantini, U., Frey, U., Abiven, S., Schmidt, H. P., & Charlet, L. (2022). Real wastewater micropollutant removal by wood waste biomass biochars: A mechanistic interpretation related to various biochar physico-chemical properties. *Bioresource Technology Reports, 17*. Available from https://doi.org/10.1016/j.biteb.2022.100966.

Foster, E. J., Hansen, N., Wallenstein, M., & Cotrufo, M. F. (2016). Biochar and manure amendments impact soil nutrients and microbial enzymatic activities in a semi-arid irrigated maize cropping system. *Agriculture, Ecosystems & Environment, 233*, 404–414.

Gabhi, R., Basile, L., Kirk, D. W., et al. (2020). Electrical conductivity of wood biochar monoliths and its dependence on pyrolysis temperature. *Biochar, 2*, 369–378. Available from https://doi.org/10.1007/s42773-020-00056-0.

Gao, T., Shi, W., Zhao, M., Huang, Z., Liu, X., & Ruan, W. (2022). Preparation of spiramycin fermentation residue-derived biochar for effective adsorption of spiramycin from wastewater. *Chemosphere, 296*, 133902.

Garbuz, S., Mackay, A., Camps-Arbestain, M., DeVantier, B., & Minor, M. (2021). *Biochar amendment improves soil physico-chemical properties and alters root biomass and the soil food web in grazed pastures, Agriculture, Ecosystems & Environment* (319, p. 107517).

Gelardi, D. L., Ainuddin, I. H., Rippner, D. A., Patiño, J. E., Abou Najm, M., & Parikh, S. J. : (2021). Biochar alters hydraulic conductivity and impacts nutrient leaching in two agricultural soils. *Soil, 7*, 811–825. Available from https://doi.org/10.5194/soil-7-811-2021.

Han, T., Liu, K., Huang, J., Khan, M. N., Shen, Z., Li, J., & Zhang, H. (2024). Temporal and spatial characteristics of paddy soil potassium in China and its response to organic amendments: A systematic analysis. *Soil and Tillage Research, 235*, 105894. Available from https://doi.org/10.1016/j.still.2023.105894.

Hashem, M. A., Hasan, M., Momen, M. A., Payel, S., & Nur-A-Tomal, M. S. (2020). Water hyacinth biochar for trivalent chromium adsorption from tannery wastewater. *Environmental and Sustainability Indicators, 5*. Available from https://doi.org/10.1016/j.indic.2020.100022.

He, M., Xiong, X., Wang, L., Hou, D., Bolan, N. S., Ok, Y. S., Rinklebe, J., & Tsang, D. C. (2021). A critical review on performance indicators for evaluating soil biota and soil health of biochar-amended soils. *Journal of Hazardous Materials, 414*, 125378. Available from https://doi.org/10.1016/j.jhazmat.2021.125378.

Helaoui, S., Boughattas, I., Mkhinini, M., Chebbi, L., Elkribi-Boukhris, S., Alphonse, V., Livet, A., Banni, M., & Bousserrhine, N. (2023). Biochar amendment alleviates heavy metal phytotoxicity of *Medicago sativa* grown in polymetallic contaminated soil: Evaluation of metal uptake, plant response, and soil properties. *Plant Stress, 10*, 100–212. Available from https://doi.org/10.1016/j.stress.2023.100212.

Hong, Y., Li, D., Xie, C., Zheng, X., Yin, J., Li, Z., Zhang, K., Jiao, Y., Wang, B., Hu, Y., & Zhu, Z. (2022). Combined apatite, biochar, and organic fertilizer application for heavy metal co-contaminated soil remediation reduces heavy metal transport and alters soil microbial community structure. *Science of the Total Environment, 851*. Available from https://doi.org/10.1016/j.scitotenv.2022.158033.

Huang, P., Yang, W., Johnson, V. E., Si, M., Zhao, F., Liao, Q., Su, C., & Yang, Z. (2022). Selenium−sulfur functionalized biochar as amendment for mercury-contaminated soil: High effective immobilization and inhibition of mercury re-activation. *Chemosphere, 306*, 135552.

Huggins, T., Wang, H., Kearns, J., Jenkins, P., & Ren, Z. J. (2014). Biochar as a sustainable electrode material for electricity production in microbial fuel cells. *Bioresource Technology, 157*, 114–119. Available from https://doi.org/10.1016/j.biortech.2014.01.058.

Huggins, T. M., Pietron, J. J., Wang, H., Ren, Z. J., & Biffinger, J. C. (2015). Graphitic biochar as a cathode electrocatalyst support for microbial fuel cells. *Bioresource Technology, 195*, 147–153. Available from https://doi.org/10.1016/j.biortech.2015.06.012.

Janu, R., Mrlik, V., Ribitsch, D., Hofman, J., Sedláček, P., Bielská, L., & Soja, G. (2021). Biochar surface functional groups as affected by biomass feedstock, biochar composition and pyrolysis temperature. *Carbon Resources Conversion, 4*, 36–46. Available from https://doi.org/10.1016/j.crcon.2021.01.003.

Jenjaiwit, S., Supanchaiyamat, N., Hunt, A. J., Ngernyen, Y., Ratpukdi, T., & Siripattanakul-Ratpukdi, S. (2021). Removal of triclocarban from treated wastewater using cell-immobilized biochar as a sustainable water treatment technology. *Journal of Cleaner Production, 320*. Available from https://doi.org/10.1016/j.jclepro.2021.128919.

Jia, W., Wang, B., Wang, C., & Sun, H. (2017). Tourmaline and biochar for the remediation of acid soil polluted with heavy metals. *Journal of Environmental Chemical Engineering, 5*(3), 2107–2114. Available from https://doi.org/10.1016/j.jece.2017.04.015.

Jiang, H., Zhu, Y., Feng, Q., Su, Y., Yang, X., & Li, C. (2014). Nitrogen and phosphorus dual-doped hierarchical porous carbon foams as efficient metal-free lectrocatalysts for oxygen reduction reactions. *Chem–Eur J, 20*(11), 3106–3112. Available from https://doi.org/10.1002/chem.201304561.

Joseph, U. E., Toluwase, A. O., Kehinde, E. O., Omasan, E. E., Tolulope, A. Y., George, O. O., Zhao, C., & Hongyan, W. (2020). Effect of biochar on soil structure and storage of soil organic carbon and nitrogen in the aggregate fractions of an Albic soil. *Archives of Agronomy and Soil Science., 66*(1), 1–12. Available from https://doi.org/10.1080/03650340.2019.1587412.

Jun, L., Wei, H., Aili, M., Juan, N., Hongyan, X., Jingsong, H., Yunhua, Z., & Cuiying, P. (2020). Effect of lychee biochar on the remediation of heavy metal-contaminated soil using sunflower: A field experiment. *Environmental Research, 188*. Available from https://doi.org/10.1016/j.envres.2020.109886.

Kane, S., Ulrich, R., Harrington, A., Stadie, N. P., & Ryan, C. (2021). Physical and chemical mechanisms that influence the electrical conductivity of lignin-derived biochar. *Carbon Trends, 5*, 100088. Available from https://doi.org/10.1016/j.cartre.2021.100088.

Kappler, A., Wuestner, M. L., Ruecker, A., Harter, J., Halama, M., & Behrens, S. (2014). Biochar as an electron shuttle between bacteria and Fe (III) minerals. *Environmental Science & Technology Letters, 1*(8), 339–344. Available from https://doi.org/10.1016/j.gca.2023.01.027.

Khalil, U., Bilal Shakoor, M., Ali, S., Rizwan, M., Nasser Alyemeni, M., & Wijaya, L. (2020). Adsorption-reduction performance of tea waste and rice husk biochars for Cr(VI) elimination from wastewater. *Journal of Saudi Chemical Society, 24*(11), 799–810. Available from https://doi.org/10.1016/j.jscs.2020.07.001.

Khan, T. A., Mukhlif, A. A., & Khan, E. A. (2017). Uptake of Cu^{2+} and Zn^{2+} from simulated wastewater using muskmelon peel biochar: Isotherm and kinetic studies. *Egyptian Journal of Basic and Applied Sciences, 4*(3), 236–248. Available from https://doi.org/10.1016/j.ejbas.2017.06.006.

Kumar, V., & Verma, P. (2023). A critical review on environmental risk and toxic hazards of refractory pollutants discharged in chlorolignin waste of pulp and paper mills and their remediation approaches for environmental safety. *Environmental Research, 236*, 116728. Available from https://doi.org/10.1016/j.envres.2023.116728.

Kutlar, F. E., Tunca, B., & Yilmazel, Y. D. (2022). Carbon-based conductive materials enhance biomethane recovery from organic wastes: a review of the impacts on anaerobic treatment. *Chemosphere, 290*, 133247. Available from https://doi.org/10.1016/j.chemosphere.2021.133247.

Lai, W. Y., Lai, C. M., Ke, G. R., Chung, R. S., Chen, C. T., Cheng, C. H., Pai, C. W., Chen, S. Y., & Chen, C. C. (2013). The effects of woodchip biochar application on crop yield, carbon sequestration and greenhouse gas emissions from soils planted with rice or leaf beet. *Journal of the Taiwan Institute of Chemical Engineers, 44*(6), 1039–1044. Available from https://doi.org/10.1016/j.jtice.2013.06.028.

Li, S., Ma, Q., Zhou, C., Yu, W., & Shangguan, Z. (2021). Geoderma Applying biochar under topsoil facilitates soil carbon sequestration : A case study in a dryland agricultural system on the Loess Plateau. *Geoderma, 403*, 115186. Available from https://doi.org/10.1016/j.geoderma.2021.115186.

Li, Y., Shao, M., Huang, M., Sang, W., Zheng, S., Jiang, N., & Gao, Y. (2022). Enhanced remediation of heavy metals contaminated soils with EK-PRB using β-CD/hydrothermal biochar by waste cotton as reactive barrier. *Chemosphere, 286*. Available from https://doi.org/10.1016/j.chemosphere.2021.131470.

Lian, W., Shi, W., Tian, S., Gong, X., Yu, Q., Lu, H., Liu, Z., Zheng, J., Wang, Y., Bian, R., Li, L., & Pan, G. (2023). Preparation and application of biochar from co-pyrolysis of different feedstocks for immobilization of heavy metals in contaminated soil. *Waste Management, 163*, 12–21. Available from https://doi.org/10.1016/j.wasman.2023.03.022.

Liao, J., Ding, L., Zhang, Y., & Zhu, W. (2022). Efficient removal of uranium from wastewater using pig manure biochar: Understanding adsorption and binding mechanisms. *Journal of Hazardous Materials, 423*, 127190.

Liu, Y., Li, H., Hu, T., Mahmoud, A., Li, J., Zhu, R., Jiao, X., & Jing, P. (2022). A quantitative review of the effects of biochar application on rice yield and nitrogen use efficiency in paddy fields: A meta-analysis. *Science of the Total Environment, 830*, 154792. Available from https://doi.org/10.1016/j.scitotenv.2022.154792.

Lopez-Tenllado, F. J., Motta, I. L., & Hill, J. M. (2021). Modification of biochar with high-energy ball milling: Development of porosity and surface acid functional groups. *Bioresource Technology Reports, 15*, 100704. Available from https://doi.org/10.1016/j.biteb.2021.100704.

Maan, K. S., Gajbhiye, P., Sharma, A., & Al-Gheethi, A. (2024). Efficient anode material derived from nutshells for bio-energy production in microbial fuel cell. *Journal of Environmental Management, 364*, 121422. Available from https://doi.org/10.1016/j.jenvman.2024.121422.

Mai, D., Wen, R., Cao, W., Yuan, B., Liu, Y., Liu, Q., & Qian, G. (2017). Effect of heavy metal (Zn) on redox property of hydrochar produced from lignin, cellulose, and d-Xylose. *ACS Sustainable Chemistry & Engineering, 5*(4), 3499–3508. Available from https://doi.org/10.1021/acssuschemeng.7b00204.

Major, J., Rondon, M., Molina, D., Riha, S. J., & Lehmann, J. (2010). Maize yield and nutrition during 4 years after biochar application to a Colombian savanna oxisol. *Plant Soil 2010, 333*, 117–128. Available from https://doi.org/10.1007/s11104-010-0327-0.

Matheri, A. N., Eloko, N. S., Ntuli, F., & Ngila, J. C. (2020). Influence of pyrolyzed sludge use as an adsorbent in removal of selected trace metals from wastewater treatment. *Case Studies in Chemical and Environmental Engineering, 2*. Available from https://doi.org/10.1016/j.cscee.2020.100018.

Meng, F., Wang, Y., & Cao, Q. (2023). Synergistic enhancement of redox pairs and functional groups for the removal of phenolic organic pollutants by activated PMS using silica-composited biochar: Mechanism and environmental toxicity assessment. *Chemosphere, 337*, 139441. Available from https://doi.org/10.1016/j.chemosphere.2023.139441.

Min, X. U., Yang, L. I. N., Jing, M. A., Lulu, L. O. N. G., Chao, C. H. E. N., Gang, Y. A. N. G., & Peng, G. A. O. (2023). Biochar regulates biogeochemical cycling of iron and chromium in paddy soil system by stimulating *Geobacter* and *Clostridium*. *Pedosphere, 34*, 929–940. Available from https://doi.org/10.1016/j.pedsph.2023.07.013.

Mo, R.-J., Zhao, Y., Wu, M., Xiao, H.-M., Kuga, S., Huang, Y., et al. (2016). Activated carbon from nitrogen rich watermelon rind for high-performance supercapacitors. *RSC Advances, 6*(64), 59333–59342. Available from https://doi.org/10.1039/C6RA10719B.

Ogunremi, O. O., Ogunkunle, C. O., & Fatoba, P. O. (2023). Characterization and remediation potential of sorghum and rice straw-derived biochars on incubated spent-oil contaminated soil. *Scientific African, 22*. Available from https://doi.org/10.1016/j.sciaf.2023.e01921.

Pan, J., Sun, J., Ao, N., Xie, Y., Zhang, A., Chen, Z., et al. (2022). Factors influencing biochar-strengthened anaerobic digestion of cow manure. *Bioenerg Research, 15*, 10. Available from https://doi.org/10.1007/S12155-022-10396-3.

Pan, S. Y., Dong, C. D., Su, J. F., Wang, P. Y., Chen, C. W., Chang, J. S., & Hung, C. M. (2021). The role of biochar in regulating the carbon, phosphorus, and nitrogen cycles exemplified by soil systems. *Sustainability, 13*(10), 5612. Available from https://doi.org/10.3390/su13105612.

Pandey, B., Suthar, S., & Chand, N. (2022). Effect of biochar amendment on metal mobility, phytotoxicity, soil enzymes, and metal-uptakes by wheat (*Triticum aestivum*) in contaminated soils. *Chemosphere, 307*, 135889. Available from https://doi.org/10.1016/j.chemosphere.2022.135889.

Park, W., Kim, H., Park, H., et al. (2021). Biochar as a low-cost, eco-friendly, and electrically conductive material for terahertz applications. *Science Reports, 11*, 18498. Available from https://doi.org/10.1038/s41598-021-98009-5.

Parsa, M., Nourani, M., Baghdadi, M., Hosseinzadeh, M., & Pejman, M. (2019). Biochars derived from marine macroalgae as a mesoporous by-product of hydrothermal liquefaction process: Characterization and application in wastewater treatment. *Journal of Water Process Engineering, 32*. Available from https://doi.org/10.1016/j.jwpe.2019.100942.

Pathy, A., Pokharel, P., Chen, X., Balasubramanian, P., & Chang, S. X. (2023). Activation methods increase biochar's potential for heavy-metal adsorption and environmental remediation: A global meta-analysis. *Science of the Total Environment, 865*. Available from https://doi.org/10.1016/j.scitotenv.2022.161252.

Peng, H., Gao, P., Chu, G., Pan, B., Peng, J., & Xing, B. (2017). Enhanced adsorption of Cu (II) and Cd (II) by phosphoric acid-modified biochars. *Environmental Pollution, 229*, 846–853. Available from https://doi.org/10.1016/j.envpol.2017.07.004.

Pham, H. T., Đoan, G. L., Hoang, L. B., Hoa, N. T., Cuong, N. C., & Hoang, T. H. T. (2023). Enhancing biochar structure and removal efficiency of ammonium and microalgae in wastewater treatment through combined biological and thermal treatments. *Journal of Water Process Engineering, 56*. Available from https://doi.org/10.1016/j.jwpe.2023.104529.

Prado, A., Berenguer, R., & Esteve-Núñez, A. (2019). Electroactive biochar outperforms highly conductive carbon materials for biodegrading pollutants by enhancing microbial extracellular electron transfer. *Carbon, 146*, 597–609. Available from https://doi.org/10.1016/j.carbon.2019.02.038.

Priyadarshini, M., Ahmad, A., Shinde, A., Das, I., & Madhao Ghangrekar, M. (2024). Anodic degradation of salicylic acid and simultaneous bioelectricity recovery in microbial fuel cell using waste-banana-peels derived biochar-supported MIL-53(Fe)-metal-organic framework as cathode catalyst. *Journal of Electroanalytical Chemistry, 967*, 118451. Available from https://doi.org/10.1016/j.jelechem.2024.118451.

Qian, K., Kumar, A., Zhang, H., Bellmer, D., & Huhnke, R. (2015). Recent advances in utilization of biochar. *Renewable and Sustainable Energy Reviews, 42*, 1055–1064. Available from https://doi.org/10.1016/j.rser.2014.10.074.

Ramtahal, G., Umaharan, P., Hanuman, A., Davis, C., & Ali, L. (2019). The effectiveness of soil amendments, biochar and lime, in mitigating cadmium bioaccumulation in *Theobroma cacao* L. *Science of the Total Environment, 693*, 133563. Available from https://doi.org/10.1016/j.scitotenv.2019.07.369.

Rashid, T. U., Salem, K. S., Islam, M. M., Zaman, A., Khan, M. N., Luna, I. Z., Haque, P., & Rahman, M. M. (2019). Recent developments of microbial fuel cell as sustainable bio-energy sources. *Microbial Catalysts, 2*, 57–115, Nova Science Publishers, Inc.: Hauppauge, NY, USA.

Rashid, M. S., Liu, G., Yousaf, B., Hamid, Y., Rehman, A., Munir, M. A. M., & Song, Y. (2022). Assessing the influence of sewage sludge and derived-biochar in immobilization and transformation of heavy metals in polluted soil: Impact on intracellular free radical formation in maize. *Environmental Pollution, 309*, 119768. Available from https://doi.org/10.1016/j.envpol.2022.119768.

Roha, B., Yao, J., Batool, A., Hameed, R., Ghufran, M. A., Hayat, M. T., & Sunahara, G. (2021). Model sorption of industrial wastewater containing Cu^{2+}, Cd^{2+}, and Pb^{2+} using individual and mixed rice husk biochar. *Environmental Technology and Innovation, 24*. Available from https://doi.org/10.1016/j.eti.2021.101900.

Sadhu, M., Bhattacharya, P., Vithanage, M., & Padmaja Sudhakar, P. (2022). Adsorptive removal of fluoride using biochar – A potential application in drinking water treatment. *Separation and Purification Technology, 278*. Available from https://doi.org/10.1016/j.seppur.2021.119106.

Salem, I. B., Gamal, M. E., Sharma, M., Hameedi, S., & Howari, F. M. (2021). Utilization of the UAE date palm leaf biochar in carbon dioxide capture and sequestration processes. *Journal of Environmental Management, 299*, 113644. Available from https://doi.org/10.1016/j.jenvman.2021.113644.

Schievano, A., Berenguer, R., Goglio, A., Bocchi, S., Marzorati, S., Rago, L., & Esteve-Núñez, A. (2019). Electroactive biochar for large-scale environmental applications of microbial electrochemistry. *ACS Sustainable Chemistry & Engineering, 7*(22), 18198–18212. Available from https://doi.org/10.1021/acssuschemeng.9b04229.

Senthilkumar, N., Pannipara, M., Al-Sehemi, A. G., & Gnana kumar, G. (2019). PEDOT/NiFe2O4 nanocomposites on biochar as a free-standing anode for high-performance and durable microbial fuel cells. *New Journal of Chemistry, 43*(20), 7743–7750. Available from https://doi.org/10.1039/C9NJ00638A.

Seo, Y. D., Oh, S. Y., Rajagopal, R., & Ryu, K. S. (2020). FeS–biochar and Zn (0)–biochar for remediation of redox-reactive contaminants. *RSC Advances, 10*(50), 30203–30213. Available from https://doi.org/10.1039/D0RA05571A.

Shaaban, A., Se, S. M., Mitan, N. M. M., & Dimin, M. F. (2013). Characterization of biochar derived from rubber wood sawdust through slow pyrolysis on surface porosities and functional groups. *Procedia Engineering, 68*, 365–371. Available from https://doi.org/10.1016/j.proeng.2013.12.193.

Shen, C., Gu, L., Chen, S., Jiang, Y., Huang, P., Li, H., Yu, H., & Xia, D. (2022). Sewage sludge derived FeCl3-activated biochars as efficient adsorbents for the treatment of toxic As(III) and Cr(VI) wastewater. *Journal of Environmental Chemical Engineering, 10*(6). Available from https://doi.org/10.1016/j.jece.2022.108575.

Sheng, Y., & Zhu, L. (2018). Biochar alters microbial community and carbon sequestration potential across different soil pH. *Science of the Total Environment, 622–623*, 1391–1399. Available from https://doi.org/10.1016/j.scitotenv.2017.11.337.

Sheng, Y., Zhan, Y., & Zhu, L. (2016). Reduced carbon sequestration potential of biochar in acidic soil. *Science of the Total Environment, 572*, 129–137. Available from https://doi.org/10.1016/j.scitotenv.2016.07.140.

Shukla, N., Sahoo, D., & Remya, N. (2019). Biochar from microwave pyrolysis of rice husk for tertiary wastewater treatment and soil nourishment. *Journal of Cleaner Production, 235*, 1073–1079. Available from https://doi.org/10.1016/j.jclepro.2019.07.042.

Singh, R., Dutta, R. K., Naik, D. V., Ray, A., & Kanaujia, P. K. (2021). High surface area Eucalyptus wood biochar for the removal of phenol from petroleum refinery wastewater. *Environmental Challenges, 5*. Available from https://doi.org/10.1016/j.envc.2021.100353.

Song, P., Ma, W., Gao, X., Ai, S., Wang, J., & Liu, W. (2022). Remediation mechanism of Cu, Zn, As, Cd, and Pb contaminated soil by biochar-supported nanoscale zero-valent iron and its impact on soil enzyme activity. *Journal of Cleaner Production, 378*, 134510.

Steigerwald, J. M., & Ray, J. R. (2021). Adsorption behavior of perfluorooctanesulfonate (PFOS) onto activated spent coffee grounds biochar in synthetic wastewater effluent. *Journal of Hazardous Materials Letters, 2*. Available from https://doi.org/10.1016/j.hazl.2021.100025.

Sun, F., Chen, J., Sun, Z., Zheng, X., Tang, M., & Yang, Y. (2024). Promoting bioremediation of brewery wastewater, production of bioelectricity and microbial community shift by sludge microbial fuel cells using biochar as anode. *Science of The Total Environment, 929*, 172418. Available from https://doi.org/10.1016/j.scitotenv.2024.172418.

Suo, F., You, X., Yin, S., Wu, H., Zhang, C., Yu, X., Sun, R., & Li, Y. (2021). Preparation and characterization of biochar derived from co-pyrolysis of Enteromorpha prolifera and corn straw and its potential as a soil amendment. *Science of the Total Environment, 798*, 149167. Available from https://doi.org/10.1016/j.scitotenv.2021.149167.

Tang, Y. H., Liu, S. H., & Tsang, D. C. W. (2020). Microwave-assisted production of CO_2-activated biochar from sugarcane bagasse for electrochemical desalination. *Journal of Hazardous Materials, 383*, 121192. Available from https://doi.org/10.1016/j.jhazmat.2019.121192.

Thuan, D. V., Thu, T., Chu, H., Do, H., Thanh, T., Vien, M., & Long, H. (2023). Adsorption and photodegradation of micropollutant in wastewater by photocatalyst TiO_2/rice husk biochar. *Environmental Research, 236*, 116789. Available from https://doi.org/10.1016/j.envres.2023.116789.

Tian, R., Dong, H., Chen, J., Li, R., Xie, Q., Li, L., Li, Y., Jin, Z., Xiao, S., & Xiao, J. (2021). Electrochemical behaviors of biochar materials during pollutant removal in wastewater: A review. *Chemical Engineering Journal, 425*, 130585. Available from https://doi.org/10.1016/j.cej.2021.130585.

Valentin, M. T., Luo, G., Zhang, S., et al. (2023). Direct interspecies electron transfer mechanisms of a biochar-amended anaerobic digestion: a review. *Biotechnology for Biofuels, 16*, 146. Available from https://doi.org/10.1186/s13068-023-02391-3.

Van Poucke, R., Meers, E., & Tack, F. M. (2020). Leaching behavior of Cd, Zn, and nutrients (K, P, S) from a contaminated soil as affected by amendment with biochar. *Chemosphere, 245*, 125561.

Van Zwieten, L., Kimber, S., Morris, S., Chan, K. Y., Downie, A., Rust, J., Joseph, S., & Cowie, A. (2010). Effects of biochar from slow pyrolysis of papermill waste on agronomic performance and soil fertility. *Plant and Soil, 327*, 235–246. Available from https://doi.org/10.1007/s11104-009-0050-x.

Viaene, J., Peiren, N., Vandamme, D., Lataf, A., Cuypers, A., Jozefczak, M., & Vandecasteele, B. (2023). Biochar amendment to cattle slurry reduces NH3 emissions during storage without risk of higher NH_3 emissions after soil application of the solid fraction. *Waste Management, 167*, 39–45. Available from https://doi.org/10.1016/j.wasman.2023.05.023.

Wang, A. O., Ptacek, C. J., Blowes, D. W., Finfrock, Y. Z., Paktunc, D., & Mack, E. E. (2020). Use of hardwood and sulfurized-hardwood biochars as amendments to floodplain soil from South River, VA, USA: Impacts of drying-rewetting on Hg removal. *Science of the Total Environment, 712*, 136018. Available from https://doi.org/10.1016/j.scitotenv.2019.136018.

Wang, F., Sun, H., Ren, X., Liu, Y., Zhu, H., Zhang, P., & Ren, C. (2017). Effects of humic acid and heavy metals on the sorption of polar and apolar organic pollutants onto biochars. *Environmental Pollution, 231*, 229–236. Available from https://doi.org/10.1016/j.envpol.2017.08.023.

Wang, G., Li, Q., Li, Y., Xing, Y., Yao, G., Liu, Y., & Wang, X. C. (2020). Redox-active biochar facilitates potential electron transfer between syntrophic partners to enhance anaerobic digestion under high organic loading rate. *Bioresource Technology, 298*, 122524. Available from https://doi.org/10.1016/j.biortech.2019.122524.

Wang, G., Yang, R., Liu, Y., Wang, J., Tan, W., Liu, X., Jin, Y., & Qu, J. (2022). Adsorption of Cd(II) onto Auricularia auricula spent substrate biochar modified by CS2: Characteristics, mechanism and application in wastewater treatment. *Journal of Cleaner Production, 367*. Available from https://doi.org/10.1016/j.jclepro.2022.132882.

Wang, L., Chen, D., & Zhu, L. (2023). Biochar carbon sequestration potential rectification in soils: Synthesis effects of biochar on soil CO_2, CH_4 and N_2O emissions. *Science of the Total Environment, 904*, 167047. Available from https://doi.org/10.1016/j.scitotenv.2023.167047.

Wang, S., & Wang, J. (2019). Activation of peroxymonosulfate by sludge-derived biochar for the degradation of triclosan in water and wastewater. *Chemical Engineering Journal, 356*, 350–358. Available from https://doi.org/10.1016/j.cej.2018.09.062.

Wang, S., Xiao, D., Zheng, X., Zheng, L., Yang, Y., Zhang, H., Ai, B., & Sheng, Z. (2021). Halloysite and coconut shell biochar magnetic composites for the sorption of Pb(II) in wastewater: Synthesis, characterization and mechanism investigation. *Journal of Environmental Chemical Engineering, 9*(6). Available from https://doi.org/10.1016/j.jece.2021.106865.

Wang, X., Yang, Z., Jiang, Y., Zhao, P., & Meng, X. (2024). Adsorption and catalytic degradation of phenol in water by a Mn, N co-doped biochar via a non-radical oxidation process. *Separation and Purification Technology, 330*. Available from https://doi.org/10.1016/j.seppur.2023.125267.

Wang, Z., Nie, Q., Lei, Z., Zhang, Z., Shimizu, K., & Yuan, T. (2024). Enhanced Pb(II) removal from wastewater by co-pyrolysis biochar derived from sewage sludge and calcium sulfate: Performance evaluation and quantitative mechanism analysis. *Separation and Purification Technology, 329*. Available from https://doi.org/10.1016/j.seppur.2023.125124.

Wu, D., Chen, Q., Wu, M., Zhang, P., He, L., Chen, Y., & Pan, B. (2022). Heterogeneous compositions of oxygen-containing functional groups on biochars and their different roles in rhodamine B degradation. *Chemosphere, 292*, 133518. Available from https://doi.org/10.1016/j.chemosphere.2022.133518.

Wu, S., Cai, X., Liao, Z., He, W., Shen, J., Yuan, Y., & Ning, X. (2022). Redox properties of nano-sized biochar derived from wheat straw biochar. *RSC Advances, 12*(18), 11039–11046. Available from https://doi.org/10.1039/D2RA01211A.

Wu, W., Han, J., Gu, Y., Li, T., Xu, X., Jiang, Y., Li, Y., Sun, J., Pan, G., & Cheng, K. (2022). Impact of biochar amendment on soil hydrological properties and crop water use efficiency: A global meta-analysis and structural equation model. *GCB Bioenergy, 14*(6), 657–668. Available from https://doi.org/10.1111/gcbb.12933.

Xie, L., Chen, Q., Liu, Y., Ma, Q., Zhang, J., Tang, C., & Li, S. (2023). Enhanced remediation of Cr (VI)-contaminated soil by modified zero-valent iron with oxalic acid on biochar. *Science of the Total Environment, 905*, 167399.

Xiong, X., Liu, Z., Zhao, L., Huang, M., Dai, L., Tian, D., & Shen, F. (2022). Tailoring biochar by PHP towards the oxygenated functional groups (OFGs)-rich surface to improve adsorption performance. *Chinese Chemical Letters, 33*(6), 3097–3100. Available from https://doi.org/10.1016/j.cclet.2021.09.099.

Xu, H., Han, Y., Wang, G., Deng, P., & Feng, L. (2022). Walnut shell biochar based sorptive remediation of estrogens polluted simulated wastewater: Characterization, adsorption mechanism and degradation by persistent free radicals. *Environmental Technology and Innovation, 28*. Available from https://doi.org/10.1016/j.eti.2022.102870.

Xu, J., Li, C., Zhu, N., Shen, Y., & Yuan, H. (2021). Particle size-dependent behavior of redox-active biochar to promote anaerobic ammonium oxidation (anammox). *Chemical Engineering Journal, 410*, 127925. Available from https://doi.org/10.1016/j.cej.2020.127925.

Xu, X., Huang, H., Zhang, Y., Xu, Z., & Cao, X. (2019). Biochar as both electron donor and electron shuttle for the reduction transformation of Cr (VI) during its sorption. *Environmental Pollution, 244*, 423–430. Available from https://doi.org/10.1016/j.envpol.2018.10.068.

Xu, X., Wu, Y., Wu, X., Sun, Y., Huang, Z., Li, H., Wu, Z., Zhang, X., Qin, X., Zhang, Y., Deng, J., & Huang, J. (2022). Effect of physicochemical properties of biochar from different feedstock on remediation of heavy metal contaminated soil in mining area. *Surfaces and Interfaces, 32*. Available from https://doi.org/10.1016/j.surfin.2022.102058.

Yan, F. L., Wang, Y., Wang, W. H., Zhao, J. X., Feng, L. L., Li, J. J., & Zhao, J. C. (2020). Application of biochars obtained through the pyrolysis of Lemna minor in the treatment of Ni-electroplating wastewater. *Journal of Water Process Engineering, 37*. Available from https://doi.org/10.1016/j.jwpe.2020.101464.

Yan, J., Zhang, M., Chen, X., Chen, C., Xu, X., & Jiang, S. (2024a). Straw-derived macroporous biochar as high-performance anode in microbial fuel cells. *Process Biochemistry*. Available from https://doi.org/10.1016/j.procbio.2024.06.024.

Yan, J., Zhang, M., Chen, X., Chen, C., Xu, X., & Jiang, S. (2024b). Multi-walled carbon nanotubes modified corn straw biochar as high-performance anode in microbial fuel cells. *Journal of Environmental Chemical Engineering, 12*(5), 113316. Available from https://doi.org/10.1016/j.jece.2024.113316.

Yang, H., Kang, J. K., Park, S. J., & Lee, C. G. (2021). Effect of pyrolysis conditions on food waste conversion to biochar as a coagulant aid for wastewater treatment. *Journal of Water Process Engineering, 41*. Available from https://doi.org/10.1016/j.jwpe.2021.102081.

Yang, W., Feng, G., Miles, D., Gao, L., Jia, Y., Li, C., & Qu, Z. (2020). Impact of biochar on greenhouse gas emissions and soil carbon sequestration in corn grown under drip irrigation with mulching. *Science of the Total Environment, 729*(306), 138752. Available from https://doi.org/10.1016/j.scitotenv.2020.138752.

Yang, Y., Sun, K., Han, L., Chen, Y., Liu, J., & Xing, B. (2022). Biochar stability and impact on soil organic carbon mineralization depend on biochar processing, aging and soil clay content. *Soil Biology and Biochemistry, 169*, 108657. Available from https://doi.org/10.1016/j.soilbio.2022.108657.

Ye, T., Huang, B., Wang, Y., Zhou, L., & Liu, Z. (2020). Rapid removal of uranium(VI) using functionalized luffa rattan biochar from aqueous solution. *Colloids and Surfaces A: Physicochemical and Engineering Aspects, 606*. Available from https://doi.org/10.1016/j.colsurfa.2020.125480.

Yuan, H., Deng, L., Qi, Y., Kobayashi, N., & Tang, J. (2014). Nonactivated and activated biochar derived from bananas as alternative cathode catalyst in microbial fuel cells. *The Scientific World Journal, 2014*, 1–8. Available from https://doi.org/10.1155/2014/832850.

Zhai, S., Li, M., Wang, D., Ju, X., & Fu, S. (2021). Cyano and acylamino group modification for tannery sludge bio-char: Enhancement of adsorption universality for dye pollutants. *Journal of Environmental Chemical Engineering, 9*(1), 104939. Available from https://doi.org/10.1016/j.jece.2020.104939.

Zhang, G., Zhu, X., Yu, M., & Yang, F. (2023). Electrochemical activation of peroxymonosulfate using chlorella biochar modified flat ceramic membrane cathode for berberine removal: Role of superoxide radical and mechanism insight. *Separation and Purification Technology, 318*, 124002. Available from https://doi.org/10.1016/j.seppur.2023.124002.

Zhang, K., Deng, Y., Liu, Z., Feng, Y., Hu, C., & Wang, Z. (2023). Biochar Facilitated Direct Interspecies Electron Transfer in Anaerobic Digestion to Alleviate Antibiotics Inhibition and Enhance Methanogenesis: A Review. *International Journal of Environmental Research and Public Health, 20*(3), 2296. Available from https://doi.org/10.3390/ijerph20032296.

Zhang, Z., Huang, G., Zhang, P., Shen, J., Wang, S., & Li, Y. (2023). Development of iron-based biochar for enhancing nitrate adsorption: Effects of specific surface area, electrostatic force, and functional groups. *Science of The Total Environment, 856*, 159037. Available from https://doi.org/10.1016/j.scitotenv.2022.159037.

Zhang, Z., Zhang, M., Zhao, X., & Cao, J. (2024). High-efficient removal and adsorption mechanism of organic dyes in wastewater by KOH-activated biochar from phenol-formaldehyde resin modified wood. *Separation and Purification Technology, 330*. Available from https://doi.org/10.1016/j.seppur.2023.125542.

Zhao, D., Yan, B., Liu, C., Yao, B., Luo, L., Yang, Y., et al. (2021). Mitigation of acidogenic product inhibition and elevated mass transfer by biochar during anaerobic digestion of food waste. *Bioresource Technology, 338*, 125531. Available from https://doi.org/10.1016/j.biortech.2021.125531.

Zhao, S., Wang, X., Wang, Q., Sumpradit, T., Khan, A., Zhou, J., Salama, E. S., Li, X., & Qu, J. (2023). Application of biochar in microbial fuel cells: Characteristic performances, electron-transfer mechanism, and environmental and economic assessments. *Ecotoxicology and Environmental Safety, 267*, 115643. Available from https://doi.org/10.1016/j.ecoenv.2023.115643.

Zhao, Y., & Naeth, M. A. (2022). Lignite derived humic products and cattle manure biochar are effective soil amendments in cadmium contaminated and uncontaminated soils. *Environmental Advances, 8*. Available from https://doi.org/10.1016/j.envadv.2022.100186.

Zhao, Z., Liu, L., Sun, Y., Xie, L., Liu, S., Li, M., & Yu, Q. (2024). Combined microbe-plant remediation of cadmium in saline-alkali soil assisted by fungal mycelium-derived biochar. *Environmental Research, 240*. Available from https://doi.org/10.1016/j.envres.2023.117424.

Zheng, B., Liao, J., Ding, L., Zhang, Y., & Zhu, W. (2021). High efficiency adsorption of uranium in solution with magnesium oxide embedded horse manure-derived biochar. *Journal of Environmental Chemical Engineering, 9*(6). Available from https://doi.org/10.1016/j.jece.2021.106897.

Zhou, S., Jia, P., Xu, W., Alam, S. S., & Zhang, Z. (2023). A novel composting system for mitigating ammonia emissions and producing nitrogen-rich organic fertilizer. *Bioresource Technology, 386*, 129455. Available from https://doi.org/10.1016/j.biortech.2023.129455.

Zhu, L., Yang, F., Lin, X., Zhang, D., Duan, X., Shi, J., & Sun, Z. (2023). Highly efficient catalysts of polyoxometalates supported on biochar for antibiotic wastewater treatment: Performance and mechanism. *Process Safety and Environmental Protection, 172*, 425–436. Available from https://doi.org/10.1016/j.psep.2023.02.037.

Zhu, M., Zhang, L., Zheng, L., Zhuo, Y., Xu, J., & He, Y. (2018). Typical soil redox processes in pentachlorophenol polluted soil following biochar addition. *Frontiers in Microbiology, 9*, 579. Available from https://doi.org/10.1016/j.jhazmat.2019.122002.

Zhu, Y., An, M., Mamut, R., & Wang, H. (2023). Comparative analysis of metabolic mechanisms in the remediation of Cd-polluted alkaline soil in cotton field by biochar and biofertilizer. *Chemosphere, 340*. Available from https://doi.org/10.1016/j.chemosphere.2023.139961.

Zoroufchi Benis, K., Shakouri, M., McPhedran, K., & Soltan, J. (2021). Enhanced arsenate removal by Fe-impregnated canola straw: Assessment of XANES solid-phase speciation, impacts of solution properties, sorption mechanisms, and evolutionary polynomial regression (EPR) models. *Environmental Science and Pollution Research, 28*, 12659–12676.

Zuo, W., Wang, S., Zhou, Y., Ma, S., Yin, W., Shan, Y., & Wang, X. (2023). Conditional remediation performance of wheat straw biochar on three typical Cd-contaminated soils. *Science of the Total Environment, 863*, 160998.

Chapter 16

Biochar in bioelectrochemical systems: applications and future directions

Wilgince Apollon[1], Tatiana Kuleshova[2], Willis Gwenzi[3,4,5], Felipe Caballero-Briones[1] and Sathish Kumar Kamaraj[1]

[1]*Instituto Politécnico Nacional (IPN), Centro de Investigación en Ciencia Aplicada y Tecnología Avanzada (CICATA), Unidad Altamira, Altamira, Tamaulipas, Mexico,* [2]*Agrophysical Research Institute, Department of Plant Lightphysiology and Agroecosystem Bioproductivity, Saint-Petersburg, Grazhdanskiy pr., Russia,* [3]*Formerly Alexander von Humboldt Fellow and Guest Full Professor, Leibniz-Institut für Agrartechnik und Bioökonomie e. V. (ATB), Potsdam, Germany,* [4]*Formerly Alexander von Humboldt Fellow and Guest Full Professor, Grassland Grassland Science and Renewable Plant Resources, Faculty of Organic Agricultural Sciences, Universität Kassel, Witzenhausen, Germany,* [5]*Biosystems and Environmental Engineering Research Group, Marlborough, Harare, Zimbabwe*

Chapter outline

16.1 Introduction	307
16.2 Materials and methods	308
16.2.1 Bioelectrochemical systems: microbial fuel cells	308
16.2.2 Principles of microbial fuel cells	308
16.2.3 Designs and types of microbial fuel cells	308
16.3 Constraints and limitations of bioelectrochemical systems	312
16.3.1 Constraints and limitations	312
16.4 Biochar properties and their potential applications in microbial fuel cells	312
16.4.1 Electrochemical properties of biochar	312
16.4.2 Applications in microbial fuel cells	316
16.5 Performance and Mechanisms of biochar function in microbial fuel cells	318
16.5.1 Overview of performance	318
16.5.2 Mechanisms of biochar enhancement of microbial fuel cells	318
16.6 Future directions and perspectives	319
16.6.1 Knowledge gaps	319
16.6.2 A roadmap for the advancement of biochar-based microbial fuel cells	320
16.7 Conclusions	321
Acknowledgment	321
References	321

16.1 Introduction

Bioelectrochemical systems, or microbial fuel cells (MFCs), are a novel technology for wastewater treatment (WWT) and bioelectricity generation (Mohyudin et al., 2022). Recent interest in clean and renewable bioenergy has led to a rapid increase in research on bioelectrochemical systems and their applications. A substantial body of evidence now exists on the principles, designs, and types of MFCs (Nawaz et al., 2022), as well as their performance in terms of WWT efficiency and bioelectricity generation (Aghababaie et al., 2015; Ahmad et al., 2013).

Despite this progress, the upscaling and adoption of MFCs still face challenges due to several constraints (Apollon et al., 2024). One major issue is the appropriate choice of electrode material to enhance overall performance. To address this, recent attention has turned to biochar, a low-cost, readily available, and renewable biomaterial with unique electrochemical properties. These properties make biochar an ideal electrode material for MFCs.

Several studies have investigated the use of biochar as an anode or cathode material in MFCs (Huggins et al., 2014; Sun et al., 2024). Additionally, there is research on the performance of biochar-based MFCs. However, earlier books on biochar for environmental remediation have paid limited attention to biochar applications in MFCs. Consequently, evidence on the application of biochar in MFCs remains scattered across individual articles.

This chapter aims to bridge this gap by discussing the properties, applications, performance, and mechanisms of biochar in MFCs. The objectives are as follows:

1. To present an overview of the principles, designs, and types of MFCs.
2. To discuss the electrochemical properties of biochar relevant for its use in MFCs.
3. To examine the performance and mechanisms of biochar enhancement in MFCs.
4. To propose future directions and a roadmap to further advance biochar-based MFCs.

16.2 Materials and methods

16.2.1 Bioelectrochemical systems: microbial fuel cells

Bioelectrochemical systems (BESs), that is, MFCs, offer a promising solution to our growing energy needs, providing a sustainable and renewable source of power. These systems showed great potential in reducing/removing contaminants from different wastes, such as municipal wastewater and industrial or agro-industrial wastewater plants (WWPs) (Apollon et al., 2024). However, concerns remained about the long-term viability of the BES's electricity production. Hence, various modifications have been made to BESs to enhance their efficiency (Debnath & Dutta, 2023). One such modification that has garnered considerable attention and has undergone extensive research. The aim of this chapter is to offer a comprehensive analysis of the MFC, including its operational mechanisms, potential advantages, and limitations.

16.2.2 Principles of microbial fuel cells

MFCs are a type of BES that use microorganisms as biocatalysts to convert organic matter into electrical energy. This process involves the transfer of electrons from the organic substrate to the anode of the fuel cell, followed by the transport of these electrons to the cathode through an external circuit. The cathode then reduces oxygen to form water, completing the electrochemical cycle (Logan et al., 2006). It is worth mentioning that this technology is not new. The use of microorganisms to generate bio-electricity was first discovered in the early 20th century ("Electrical effects accompanying the decomposition of organic compounds," 1911). The initial trials involved using *E. coli* bacteria in a reactor, which showed great potential. This makes MFCs particularly attractive because they are environmentally friendly and utilize renewable resources, making them a promising emerging technology for the future (Yaqoob et al., 2022).

The construction of MFCs is based on the fundamental principles of their components. MFC technology is comprised of two main components: an anode electrode and cathode electrodes, which are separated by a membrane. The membrane can be a proton exchange membrane (PEM), an anion exchange membrane (AEM), or a cation exchange membrane (CEM), and plays a critical role in MFCs by physically separating the anode and cathode while enabling protons to traverse and reach the cathode (Akter et al., 2022). Within the anode electrode, a critical process takes place as electroactive bacteria (EAB) engage in the oxidation of organic compounds. This process stimulates the formation of biofilms, which serve as a conduit for the transfer of electrons. These EAB biofilms play a vital role in the production of electricity in BES, and research conducted by Zhu et al., (2022) highlights their importance. Meanwhile, in the cathode, oxygen is reduced to form water, thereby concluding the electrochemical cycle. When constructing an MFC, careful consideration of electrode materials, configurations, and designs is essential for optimal performance.

16.2.3 Designs and types of microbial fuel cells

Recent research indicates that the MFC design is a critical factor in its effectiveness (Jalili et al., 2024). Moreover, other factors, such as membrane selection and electrode materials, are also essential for improving MFC performance. In addition, scientists worldwide have been experimenting with various MFC variations, including SC-MFCs and DC-MFCs, to enhance their productivity. Of these, SC-MFCs have produced remarkable outcomes in both electricity generation and WWT. The principle behind the design of the SC-MFC reactor is the direct contact between the cathode and the membrane (i.e., PEM). In an SC-MFC, the cathode takes the O_2 from the air (i.e., atmosphere), and in the anode chamber, wastewater is treated using the microbial community as a biocatalyst in an anaerobic process (see Fig. 16.1). SC-MFCs have been found to be highly effective in eliminating common pollutants from wastewater. For instance, they can remove 89%−96.3% of COD (organics), as demonstrated by studies conducted by Mirbagheri and Malekmohammadi, (Sato et al., 2023). Kharti et al. (2024) have shown that SC-MFCs can remove 90% of Fe(II), while Omenesa Idris et al. (2023) have found that they can remove 89% of Pb^{2+} and 76.45% of Cd^{2+}. SC-MFCs can also generate electricity, with Wang, Xing,

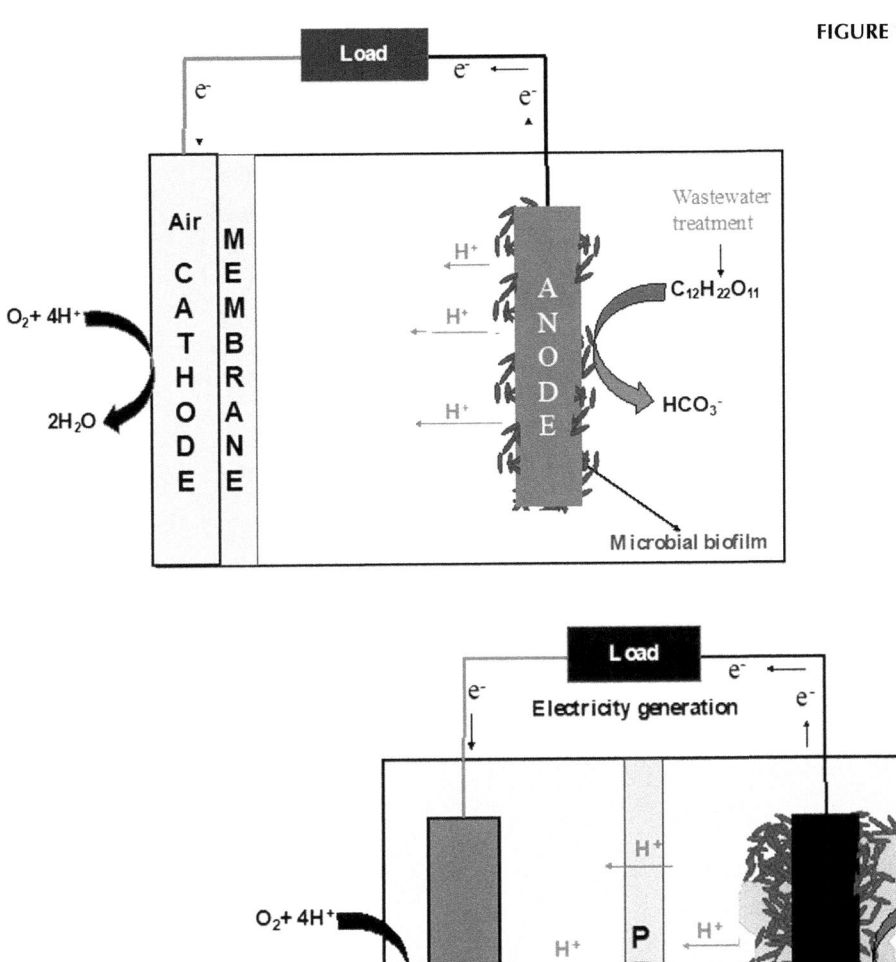

FIGURE 16.1 Typical single chamber MFC.

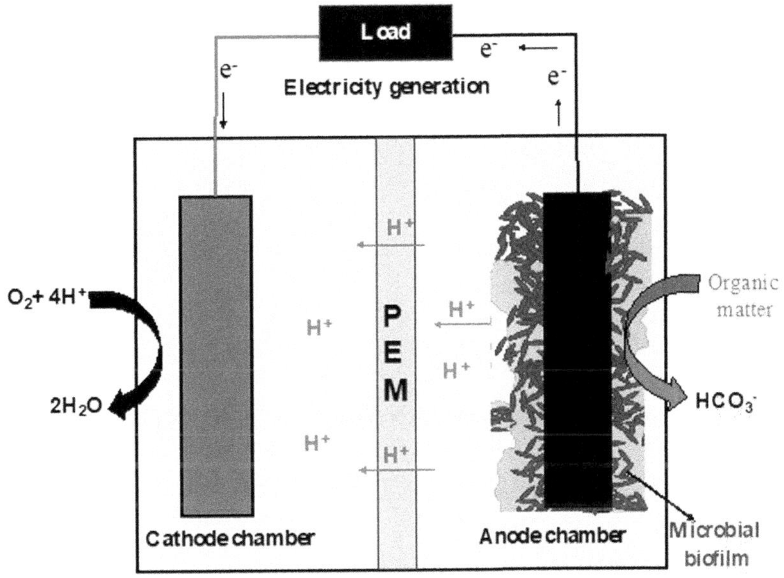

FIGURE 16.2 Typical DC-MFC for power generation.

et al. (2023), Wang, Chai, et al. (2023), Wang, Tai, et al. (2023) reporting a range of 241 to 280 mW/m^2. This breakthrough has significant implications for industries that generate substantial amounts of wastewater, as they can now depend on SC-MFCs for sustainable and effective WWT. SC-MFCs' exceptional performance highlights the potential of innovative technologies to tackle environmental issues while also generating clean energy.

Furthermore, DC-MFCs (see Fig. 16.2), like other BESs, such as MFC technology, have proved useful for a variety of applications such as WWT, soil contaminant elimination, organic nutrient recovery for sustainable agriculture, and electricity generation. Researchers have explored numerous methods to increase these systems' power efficiency, working diligently to create prototypes capable of consistently producing high-power bioelectricity (Choudhury et al., 2017; Varanasi et al., 2016). Table 16.1 presents a comparison of different MFC configurations proposed to enhance power generation performance and value-added product recovery. The table indicates that these systems have recently been utilized to treat various wastes, including wastewater, achieving high removal efficiencies of COD, bioelectrochemical oxygen demand (BOD), sulfate, copper ((Cu^{+2}), uranium [U(VI)], and nickel (Ni), in addition to the highest power density. Moreover, in this comparative Table (i.e., Table 16.1), we consider the working volume and operating time of each MFC. This Table was modified from the reference, Apollon (2023). https://creativecommons.org/licenses/by/4.0/

TABLE 16.1 Application of SC-MFC and DC-MFC in wastewater treatment.

Configuration of BES	Working volume (mL)	Operation (days)	Type of electrolyte	Removal efficiency (%)	Maximum power generation	References
SC-MFC	N/A	N/A	Swine WW	84 (OM)	228 mW/m²	Kim et al. (2008)
SC-MFC	28	N/A	Swine WW	88–92 (sCOD); 88 (ammonia)	261 mW/m²	Min et al. (2005)
SC-MFC	321 (0.321 L)	N/A	WW from chemical process, dyes, pesticides, etc.	47.80–58.98 (COD)	135–186 mA/m²	Raghavulu et al. (2009)
SC-MFC	850	~30	Activated sludge	N/A	105 mW/m²	Papillon et al. (2021)
SC-MFC	150	30	Synthetic WW	89 (COD)	450.36 mW/m²	Kumar et al. (2021)
SC-MFC	250	N/A	Fermented WW		2981 mW/m³	Nam et al. (2010)
SC-MFC	260	70	Dye processing WW	85 (tCOD); 73 (sCOD)	515 mW/m²	Karuppiah et al. (2018)
SC-MFC	90 (90 cm³)	18 (450 h)	Dairy WW	91 (COD)	20200 mW/m³	Mardanpour et al. (2012)
SC-MFC	80	N/A	WW	83 (COD)	548 mW/m²	Kolubah et al. (2023)
SC-MFC	N/A	18	WW	81; 94 (COD)	989 mV	Adnan and Hassoon Ali (2023)
SC-MFC	100	N/A	WW	73.7 HCQ	280 mW/m²	Wang, Xing, et al. (2023)
DC-MFC	120	N/A	Uranium-containing WW	99.0 U(VI)	269.5 mW/m²	Wu et al. (2021)
DC-MFC	1000	N/A	WW	92 (Ni); 87 (Cd)	722 mW/m³	Singh and Kaushik (2021)
DC-MFC	100–200	N/A	WW	90 (COD); 40; 60 (orgN)	1.69 A/m²	Burns and Qin (2023)
DC-MFC	118	30	SS	99.08 (P)	~40 mV	Wang, Tai, et al. (2023)
DC-MFC	250	28	WW	95.7; 94.7; 92.37 (COD)	1696.56 mW/m²	Huang et al. (2023)
DC-MFC	300	30	WW	70–88; 18–44 (COD)	86.9 mW/m²	Verma and Mishra (2023)
DC-MFC	125 (125 cm³)	6 (144 h)	WW	99.16 (Cu^{+2})	24.75 mW/m²	Wang, Chai, et al. (2023)
DC-MFC	850,000 (850 L)	180 (six months)	Real WW	91 (COD); 91 (BOD)	43 mW/m² (0.043 W/m²)	Rossi et al. (2022)

BES, Bioelectrochemical system; *SC-MFC*, single chamber microbial fuel cell; *DC-MFC*, dual chamber microbial fuel cell; *BOD*, bioelectrochemical oxygen demand; *COD*, chemical oxygen demand; *HCQ*, hydroxychloroquine; *MFCs*, microbial fuel cells; *WW*, wastewater.
Source: Wilgince, A. (2023) An Overview of Microbial Fuel Cell Technology for Sustainable Electricity Production, *Membranes*. MDPI AG, 13(11). doi: 10.3390/membranes13110884. Under the terms and conditions of the Creative Commons Attribution (CC BY) license. (https://creativecommons.org/licenses/by/4.0/).

A recent study by Eslami et al. (2023) introduced a novel DC-MFC design that utilized a proton exchange membrane made of polypropylene due to its exceptional properties. The anode and cathode electrodes consisted of graphite rods that were separated by the proposed membrane (polypropylene) in comparison to a Nafion 117 membrane. During an eight-day operation period using a mixed culture and glucose, the DC-MFC generated a power density of 0.7 mW/m^2, which is the highest recorded. However, this performance was significantly lower than the bioelectricity generation reported by Rossi et al. (2022) using a pilot scale MFC with an air-cathode (DC-MFC) for WWT (Table 16.1). To enhance DC-MFC performance with polypropylene membranes, further studies using MFCs with similar characteristics are recommended, such as doubling the cathode surface area and modifying electrode spacing to increase power generation yield, as described (Apollon, 2023). Additionally, the experiment by Eslami et al. (2023) was 97.6% lower than the bioelectricity generation reported in a prior study (Singh and Kaushik, 2021). Research has found that hybrid systems (i.e., combining two interrelated systems) and MFC technologies can produce more power (Paucar & Sato, 2022). Moreover, it can be argued that the effectiveness of an MFC system could be influenced by the operating conditions of such a system. Furthermore, it's possible to enhance MFC technology performance by introducing redox mediators and altering the microorganisms within the system. Such changes have the potential to enhance electron transfer efficiency throughout the MFC (Xie et al., 2022). However, it can be challenging to determine which MFC configuration is better when both have high-efficiency rates. For example, SC-MFCs with air-cathode configurations offer several benefits, including their robustness, low cost, simple structure and use, and low internal resistance, making them more efficient electrochemically (Li et al., 2023). SC-MFCs are also more practical and easier to use and scale for WWT at the laboratory level. In contrast, DC-MFCs are recommended for the recovery of aggregate products and electricity production. However, DC-MFCs, like SC-MFCs, present some limitations despite their advantages (as shown in Table 16.1).

In a recent experiment, researchers such as Choudhury and co-workers constructed two SC-MFCs with identical characteristics. These were designed to hold 300 mL of water and were built with both the anode and cathode made of carbon cloth. The two chambers were separated by a Nafion 117 membrane. It's worth noting that the distance between the anode and cathode was precisely 157 μm, which was considered a critical factor in the system's performance (Choudhury et al., 2022). The system was initialized by introducing a pure strain into the MFCs, with dairy wastewater serving as the primary source of inoculum. This marked the first instance of utilizing the strain *Pseudomonas aeruginosa*-MTCC-7814 in an MFC with these specific characteristics. Throughout the 15-day experimental period, the system achieved its highest power density at 105 mW/m^2, with a current of 313 mA/m^2 when the SC-MFCs were connected in series. The results revealed intriguing findings, leading the authors to hypothesize that the introduction of the pure strain *Pseudomonas aeruginosa* played a pivotal role in enhancing power production within the SC-MFC. However, when alterations were made to the MFC structure, such as changes in the anode electrode material, a maximum power density of approximately 104 mW/m^2 was observed (Chaturvedi et al., 2022), which was statistically similar to the previously reported result by Choudhury et al. (2022). As a result, the variation among these studies can be attributed to differences in materials, system configurations, and operational methodologies as described (Apollon et al., 2024).

DC-MFCs, similar to other types of BESs, are being used for various purposes such as WWT, soil contaminant removal, nutrient recovery, smart agriculture, bioelectricity generation, biosensors, etc. Researchers have employed different methods to enhance these systems' power efficiency, striving to develop prototypes or hybrid systems capable of consistently generating high-power bioelectricity. As can be seen, Table 16.1 presents a comparison of various MFC configurations proposed in the literature in order to enhance power generation performance in these systems. The table also demonstrates the recent use of these systems for treating different types of waste, including wastewater. These systems have achieved high removal efficiencies (%) for uranium [U(VI)], nickel (Ni), COD, sulfide, and copper (Cu^{+2}), as well as the highest power density. In a recent experiment, metal oxides such as MnO_2, SnO_2, and CuO were employed as cathodic catalysts in a DC-MFC to simultaneously recover phenolic compounds and sustainable electricity from industrial wastewater (Yap et al., 2023). Alongside these catalysts, the system utilized carbon felt and carbon plate as anodic and cathodic electrode materials, respectively. Following the completion of the system configuration, it operated for 7 days (168 hours). The outcomes of this study underscored the potential of DC-MFCs to produce sustainable electricity and eliminate phenolics from wastewater. Furthermore, the above study revealed that factors such as pH and temperature profoundly impact the performance of DC-MFCs. For instance, when the catholyte maintained an alkaline pH (8), the system achieved a maximum power density of 29.24 mW/m^2; in contrast, at pH 11, the production of sustainable electricity decreased by 43.50%. The findings from this study indicate that elevated alkalinity has a detrimental effect on the functioning of BESs. In order to gain a comprehensive understanding of the factors influencing BES performance, it is imperative to conduct further research to explore additional contributing factors.

Moreover, an effective approach to enhancing the performance of MFC technology involves the construction of single and/or double-chamber reactors and operating them in a stacked configuration. This method has been extensively researched by scientists worldwide over the past decade for WWT (Dong et al., 2015; Feng et al., 2014; Liang et al., 2018). The results have been very promising, offering the potential for large-scale/real-time implementation of this innovative technology. For instance, a study by Zhang & Liu (2023) reported that stacked MFCs achieved the highest COD removal efficiency of 99.28% in WWT. Furthermore, a domestic WWT system utilizing 18 stacked MFC units demonstrated a maximum COD removal efficiency of 87% and generated 136.55 mW of power (Linares et al., 2019). Although the performance was somewhat lower compared to that reported by Zhang and Liu, the continuous flow operation of the system makes it a practical choice for improving MFC performance.

In summary, the recent advancements in MFC designs, configurations, and operations have significantly improved MFC efficiency in terms of power output and WWT. These advances have focused on enhancing anode surface modification and employing suitable materials for these modifications. It has become evident that the performance of this technology depends not only on the configuration but also on various factors such as pH, temperature, anolyte or catholyte concentration, type of substrate, bacterial community, and environmental conditions during operation. To overcome the limitations associated with low power output, it is recommended that more comprehensive studies on MFC hybrids be conducted to enable their large-scale application. Additionally, comparing the performance of different MFC configurations remains challenging due to variations in electrode materials, wastewater, and operating conditions. Therefore, there is a need for systematic comparative studies of MFC configurations while keeping all other factors or aspects constant.

16.3 Constraints and limitations of bioelectrochemical systems

16.3.1 Constraints and limitations

Following the adoption of MFC technology, researchers have encountered a number of significant constraints and challenges. These issues have prompted active efforts to explore alternative approaches to address these obstacles. The primary limitations associated with MFC technologies include low power density, high cost of electrode materials, the need for continuous electricity generation, and low conductivity of electrode materials (Apollon et al., 2022). Beyond these primary limitations, other challenges related to MFCs include toxicity in anode biofilm (Xing et al., 2021), system instability, membrane biofouling, cost-effectiveness, robustness, longevity, substrate effects, material suitability for design, and issues related to large-scale application (Apollon et al., 2022). Notable limitations also involve electrode contamination and aging over time (Apollon, 2023). Nevertheless, MFC technologies offer significant advantages, as outlined in Table 16.2, various strategies have been implemented to address the limitation issues of MFCs. These alternatives include the modification of electrode materials, the use of appropriate substrates, and the selection of microbial strains, among others. Additionally, the use of potassium dichromate ($K_2Cr_2O_7$) or potassium permanganate ($KMnO_4$) as catholyte or bacterial culture/growth media as anyolite in an MFC could be a promising alternative to enhance its performance, as described in Naha et al. (2023). Throughout the operation of an MFC, the transfer of electrons from the anode chamber to the cathode electrode is facilitated by microbial activity. Consequently, microorganisms present on the anode surface play a pivotal role in the performance of MFCs, as they are solely responsible for the oxidation of organic matter and the enhancement of the electron transfer mechanism (Logan et al., 2006).

NPK – Nitrogen, phosphorus, and potassium.

On the other hand, another strategy to address the limitations of MFCs involves the use of biochar. This feedstock has been extensively researched for its potential to enhance the power output and efficiency of MFC technology. With its cost-effectiveness, porous structure, and high specific surface area, as noted by He et al. (2022), biochar is proving to be a valuable catalyst for improving electricity production in MFCs and for the recovery of pollutants from various wastewater sources and sediments.

16.4 Biochar properties and their potential applications in microbial fuel cells

16.4.1 Electrochemical properties of biochar

Carbon continues to be the most widely used electrochemical agent due to its thermal and chemical stability, high conductivity, and tunable surface characteristics (Stein et al., 2009). Recent advances in the field of nanocarbons make it possible to create materials at the molecular level with a hierarchical pore structure and specified surfaces hence

TABLE 16.2 Pros and cons of SC-MFC and DC-MFC technologies.

Type of design	Advantages	Disadvantages
SC-MFC	Low cost	High price
	Low internal resistance	Formation of precipitate on the cathode surface
	Simultaneous WWT and power generation	Biofouling and substrate diffusion
	Lower maintenance cost	Low cathode potential and scalability
	Compact and space-efficient	Oxygen diffusion
	Others:	Low power efficiency
		Large-scale application and commercialization
	—	Low growth rate of microorganisms
	Production of organic fertilizers, such as NPK.	
	—	
	Greenhouse gas control (reduction)	
	—	
	Biosensor	
	—	
	Smart farming application	
DC-MFC	High power generation and voltage	
	High value-added product recovery	
	Better control over the electroactive bacteria	Oxygen supply
	100% reduction efficiency	High internal resistance
	Improved stability and scalability	Constant aeration of the cathode compartment to provide oxygen
	Biosensor	Decrease of power density over time
	Powered digital clock and LED	Biofouling
	Smart farming application	Large-scale application
	Reduction of greenhouse gas emissions	

MFCs, Microbial fuel cells. *WWT*, wastewater treatment.

tailored for specific functionality (Tang et al., 2014). However, the synthesis of nanocarbon materials still remains quite complex, energy-intensive, and therefore economically unviable.

The surface sensitivity of carbon materials is not always suitable for the harsh operating conditions of fuel cells (Huggins et al., 2015). For example, the use of wastewater in MFCs can reduce the efficiency of carbon structures and generate low power density (Kiely et al., 2011). For applications in bioenergy devices, it is necessary to develop more versatile carbon materials that are simple, less resource-intensive, reliable, environmentally friendly and scalable. Ideal electrodes should not only be energy efficient, but also ensure minimal production costs. One material that meets these criteria is biochar from biomass waste. It is easy to produce as biomass is renewable and low-cost (White et al., 2009). In addition, biochar can be a by-product of biofuel and bio-oil production (Parmar et al., 2014).

Biochar is a solid carbon-containing material with a high content of difficult-to-mineralize aromatic structures produced by pyrolysis — the carbonization of renewable organic biomass at high temperature without access to air (Rawat et al., 2023). Biochar is produced as a result of thermal treatment of biomass in almost the absence of air and its electrical and electrochemical properties depend on the raw materials used, pyrolysis conditions, production temperature and the composition of active minerals present in the organic phase (Kuleshova, 2024).

Carbon yield is defined as the ratio of the mass of activated carbon to the mass of dry raw materials (Eq. 16.1):

$$Y = (W_1/W_0) * 100 \tag{16.1}$$

where W_0 denotes the mass of the precursor, and W_1 denotes the mass of the resulting activated carbon.

For example, the content of element C in porous carbon prepared from cellulose is 90.3%. The surface of carbon materials is prepared by sulfonated lignin and contains rich functional groups such as -COOH, -C-O, etc. At the same time, it is found that the molecular structure of cellulose is unstable, and the chemical bond is easily broken, forming a new small molecular structure and accumulating biochar materials (Zhang et al., 2023). Fig. 16.3 shows some biomass materials reported in previous studies and the corresponding carbon contents.

Biochar is a complex heterogeneous material consisting of mineral phases, amorphous carbon, graphitic carbon, and labile organic molecules, many of which can be either electron donors or acceptors (Joseph et al., 2015). As a result, biochar can act as both a reducing agent and an oxidizing agent (Klüpfel et al., 2014).

Electron transfer processes depend on the surface's structure and the biotic and abiotic redox reactions occurring on it.

C and mineral phases in biochar have different electrochemical potentials (Petter & Madari, 2012). In accordance with the original structure of the biomass, phase C has a number of tubular pores, which can connect different mineral phases and are themselves pores. This porous structure may have properties similar to a semipermeable membrane, in which two parallel tubular C pores with different metal concentrations can act as a galvanic cell (Suda et al., 2007).

The physical and chemical properties of biochar are highly dependent on the raw material and production temperature. Elevated temperatures above 700°C are necessary to order the carbon structure, obtaining high surface area and conductivity (Klüpfel et al., 2014). Charcoal carbonized at <300°C acts as an insulator, at 300°C−800°C as a semiconductor, and at >800°C as a conductor (Ishihara, 1996). Biochar produced at 600 °C and above are conductive, but have a lattice structure containing a significant concentration of micropores (Bourke et al., 2007). This biochar is used in electrochemical devices such as batteries, supercapacitors and MFCs.

High-temperature processing of biomass can be an alternative to carbon black, as the resulting biochar exhibits satisfactory energy performance and high conductivity, which has already found application in lithium-ion batteries, fuel cells, and supercapacitors (Jin et al., 2014).

At high temperatures, the organic components of biomass volatilize and a transition occurs from disordered carbon structures to a crystalline carbon structure (Klüpfel et al., 2014). At a carbonization temperature of about 800°C the breaking of CH bonds and the creation of interconnected turbostratic crystallites leads to an increase in the electrical conductivity of the material due to the formation of free valence electrons (Kumar & Gupta, 1993). Turbostratic sheets also resist volumetric shrinkage which leads to an increase in surface area due to the formation of micropores and microcracks (Kercher & Nagle, 2003).

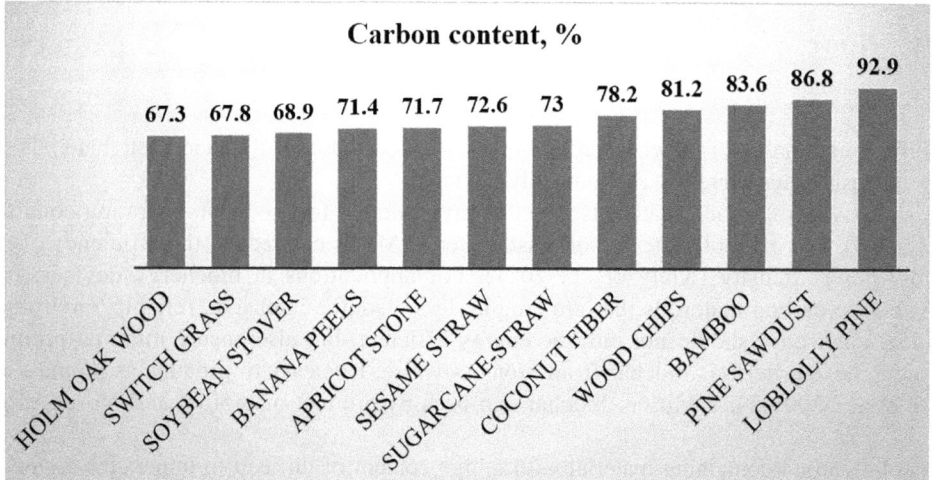

FIGURE 16.3 Carbon content of biochar-based feedstock: holm oak wood (López-Cano et al., 2016), switch grass (Essandoh et al., 2017), soybean stover (Ahmad et al., 2012), banana peels (Zhou et al., 2017), apricot stone (Ismanto et al., 2010), sesame straw (Park et al., 2016), sugarcane-straw (Melo et al., 2016), coconut fiber (Liu et al., 2013), wood chips (Chen et al., 2011), bamboo (Wang et al., 2017), pine sawdust (Reguyal, Sarmah, & Gao, 2017), loblolly pine (Yoo et al., 2018). Elsevier?.

The electrochemical properties of biochar involve at least three different mechanisms (Chacón et al., 2017):

1. Pseudocapacitance is the ability to store electrical energy through Faraday reduction-oxidation reactions associated with the presence of redox-active functional groups, metals, and radicals on the surface of the biochar that can be oxidized or reduced giving the biochar the ability to act as an electron storage material (Klüpfel et al., 2014).
2. Electronic conductivity as a consequence of electron delocalization and the presence of graphite-like sheet structures (Saquing et al., 2016).
3. Double layer capacity is related to the porous structure of biochar (Brousse et al., 2015).

The electron-donating capacity of biochar is generally greater than its electron-accepting capacity due to the loss of oxygen atoms due to the elevated pyrolysis temperature and high hydrogen to carbon (H/C) ratio, which can provide a reducing environment during pyrolysis (Liu et al., 2017).

Conduction in biochar refers to the migration of electrons between energy bands of different electrical potential. If the pyrolysis temperature is high enough, successive growth and compaction of amorphous carbon sheets in biochar will produce graphite-like sheet structures with increased crystallinity and aromaticity (Keiluweit et al., 2010). The accumulation of these conjugated p-electron systems gradually increases the conductivity of the material as the electrons associated with the p-bonds are delocalized and become available as charge carriers. Thus, biochar is able to directly transfer electrons from electron donor to electron acceptor. Unlike mediation by the surface functional group, this electron transfer process does not require a chemical reaction as it occurs very quickly.

The most common cathodic reaction in proton exchange membrane MFCs is the oxygen reduction reaction. The two possible oxygen reduction reaction pathways include a four-electron reduction (Eq. 16.2):

$$O_2 + 4H^+ + 4e^- \rightarrow 2H_2O \tag{16.2}$$

and the two-electron reduction (Eq. 16.3):

$$O_2 + 2H^+ + 2e^- \rightarrow H_2O_2 \tag{16.3}$$

Redox potential derived by combining the standard free energy change of the generic redox reaction (Eq. 16.4):

$$O_x + mH^+ + ne^- \rightarrow Red \tag{16.4}$$

where Ox — the oxidized species, Red — the reduced species.

Redox potential determined in accordance with the Nernst equation (Eq. 16.5):

$$E_h = E^o + R.T.\ln[Red]/n.F[O_x] + 2,303 m.R.T.pH/n.F \tag{16.5}$$

where F — the Faraday constant, R — gas constant, T — temperature (K), n — number of electrons, m — number of protons exchanged.

The zero point E^o is set by the standard hydrogen electrode.

E_h is correlated to pH, as an example, a biochar suspension having an E_h of 500 mV and pH 5 has a higher electron activity than a biochar suspension having an E_h of 400 mV and pH 9 (Huggins et al., 2015).

Typically, platinum nanoparticles are used as catalysts for cathodic reactions, but they significantly deteriorate their properties in the presence of trace impurities (Gasteiger et al., 2005). Therefore, as an alternative for the cathode material, manganese oxides, carbons doped with an iron complex, and various forms of carbon itself without the addition of metal can be used as a catalyst for the oxygen reduction reaction (Watanabe, 2008). For example, nanostructured MnO_2 crystals were successfully immobilized on biomass-based graphite sheets and showed a satisfactory maximum power density of 187,8 W/m^2 (Huggins et al., 2015)

Most biochars contain a redox active based on the minerals iron and manganese, which exist in the form of nanophase particles. For example, iron oxide in biochar can exist as magnetite or hematite. The biosolids content of biochar can be as high as 2.3% by weight (Van Zwieten et al., 2010), but wood-based biochar typically has much lower concentrations: 0.14%–0.34% by weight.

Biochar can be produced from many types of feedstocks, including wood, grass, algae, manure, sewage sludge, etc. Biochar derived from wood consistently has a relatively high carbon content and low ash content compared to biochars of non-wood origin (Mukome et al., 2013). The surface area associated with the capacity of a double layer and pseudocapacity appears to be higher in softwood and derived biochar, followed by hardwood biochar. The less dense structure of softwoods makes them more susceptible to thermal decomposition, producing biochar with more pores. Biochar derived from animal waste has a high nitrogen content and therefore an increased amount of nitrogen-containing functional groups, which can improve its pseudocapacity (Zhao, Baccile, et al., 2010; Zhao, Zhou, et al., 2010). Biochar from sewage sludge is even more effective. the concentration of heavy metals is higher than in the original waste material (Lu et al., 2016). The abundance of lignocellulosic biomass makes biochar of plant origin, the majority of commercial biochar available.

Various studies have also shown that biochar with high mineral content can have high redox activity, likely due to the presence of redox active metals (Joseph et al., 2013). Biochar derived from high-ash biomass (such as grass) typically has greater electron storage capacity than biochar from low-ash feedstock such as wood (Klüpfel et al., 2014). However, sometimes the presence of redox metals may account for only a portion of this increase, since metal ions present in biomass can influence the thermal decomposition of lignocellulose and favor different depolymerization pathways (Harvey et al., 2012).

Biomass of forest plants, agricultural products and waste is used to produce activated carbon. The characteristics of biomass precursors used for the manufacture of electrode materials in electrochemical devices are presented in Fig. 16.4. Specific surface area per unit mass of a material or substance calculated by the Brunauer–Emmett–Teller(BET) method.

Unfortunately, the combination of high electron storage capacity and conductivity is difficult to achieve. If high energy storage capacity is required, biochar with increased amounts of redox-active functional groups, radicals and/or metals must be produced. To improve conductivity, its need to create a graphite structure.

16.4.2 Applications in microbial fuel cells

One of the most promising and effective ways to utilize biomass, including wood and lignocellulosic waste, is thermal processing into biochar. Recently, the scope of biochar application has expanded significantly. Biochar is used to improve soil properties, as sorbents for purifying water, soil and gases from pollutants, as catalysts and carriers for them, for capturing and storing gases, for the production of carbon-containing materials, and as an environmentally friendly and highly efficient fuel for energy. At the same time, biochar is characterized by low cost, its production is considered carbon neutral. The required level of biochar characteristics, such as porosity, sorption capacity, etc., can be achieved by varying the conditions for its production and further modification.

Biochar has a wide range of applications and many positive properties. The physical properties of biochars contribute to their function as a tool for environmental management. One of the important agro-industrial properties of biochar is its ability to influence the structural and aggregate state, water-air properties and soil pH, the availability of nutrients, and organic carbon, which ultimately leads to increased plant productivity. When biochar is present in the soil mixture, its contribution to the physical nature of the system may be significant, influencing depth, structure, texture, porosity, and consistency through changing the bulk surface area, pore-size distribution, particle-size distribution, density and packing (Downie et al., 2012).

Biochar also plays an important role in environmental terms, helping to reduce the amount of soil pollutants, carbon sequestration and reduce greenhouse gas emissions (Hawthorne et al., 2017). The aromatic, high-density crystalline structure of biochar makes it environmentally friendly, stable, and expected to be chemically unchanged over time (Nguyen et al., 2010). Thus, biochar has the following advantages (see Fig. 16.5): (1) "carbon neutral" because carbon

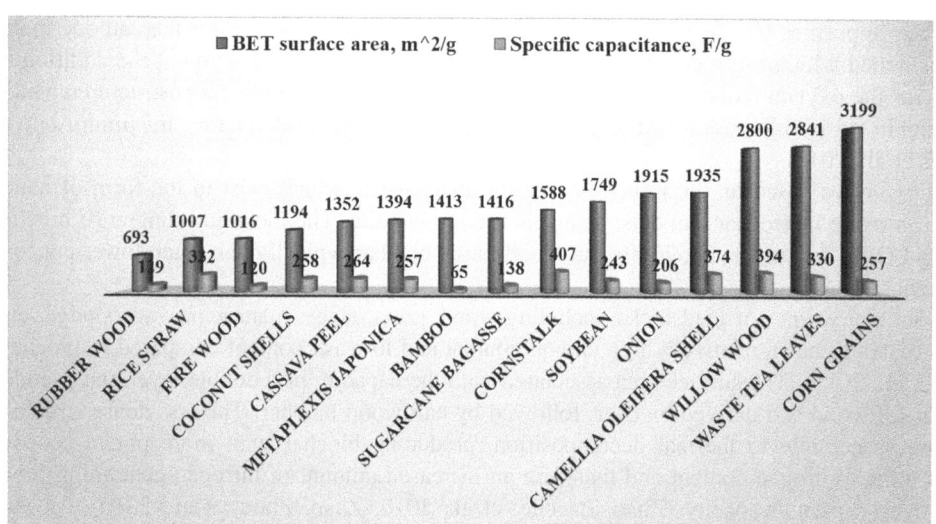

FIGURE 16.4 Specific capacitance and surface area of activated carbons from biomass: rubber wood (Thubsuang et al., 2017), rice straw (Kim et al., 2006), fire wood (Wu et al., 2004), coconut shells (Phiri et al., 2019), cassava peel (Liang et al., 2017), metaplexis japonica (Phiri et al., 2019), bamboo (Si et al., 2011), sugarcane bagasse (Wang et al., 2018), cornstalk (Cossutta et al., 2020), soybean (Sudhan et al., 2017), onion (Zhang et al., 2012), camellia oleifera shell (Balathanigaimani et al., 2008), willow wood (Peng et al., 2013), waste tea leaves (Sun et al., 2017), corn grains (Ismanto et al., 2010).

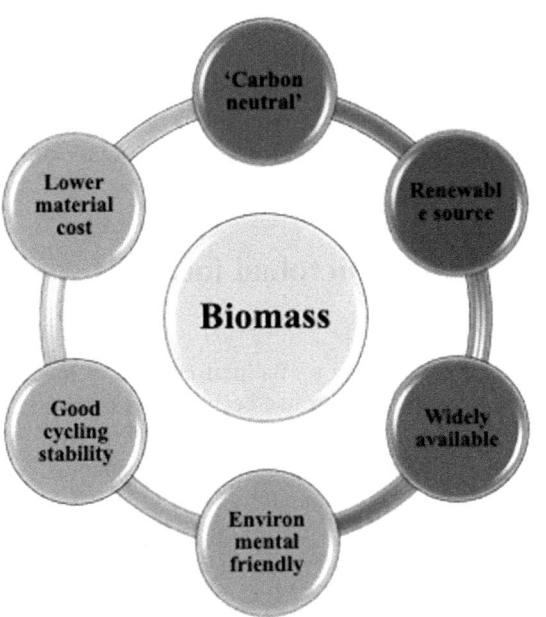

FIGURE 16.5 Advantages of using biomass as electrode material for MFC.

dioxide released from biomass-derived synthesis and application process does not disturb the nature on account of the reuse of released carbon dioxide through photosynthesis (Zhao, Baccile, et al., 2010), (2) renewable source that requires shorter time to be reproduced compared to fossil fuels for use in food, construction, and textile industry (Klass, 2004), (3) widely available (Xu et al., 2015), (4) environmentally friendly, (5) good cycling stability and (6) present lower material cost than metal and synthetic polymer (Peng et al., 2013). However, one disadvantage of biochar is its low energy density in energy storage applications.

Fe^{3+}-forming microorganisms such as *Shewanella* sp. can release electrons, increasing the availability of iron minerals.

Biochar with a high content of iron oxide nanoparticles on the pore surface can significantly increase the reduction rate of NO^{3-} and NO^{2-} (Zwieten et al., 2015). Redox processes involving Fe may include:

1. Electron transfer from organic matter to Fe(III)(hydr) oxides via C oxidation (Davidson et al., 2008).
2. Reduction of NO^{3-} to NO^{2-} with oxidation of Fe^{2+} to Fe^{3+}.
3. Mineralization of organic N to NH_4^+ (Yin et al., 2010) and oxidation of NH^{4+} to NO^{2-} with subsequent reduction of Fe^{3+} to Fe^{2+}.
4. Oxidation of NH_4^+ to NO^{2-} followed by reduction of Fe^{3+}, formation and oxidation of FeS minerals in the sulfur cycle (Li et al., 2012).
5. Cyclic conversion of S from solid to soluble liquid forms caused by the oxidation or reduction of Fe particles (Li et al., 2012).

Biochar can reversibly accept and donate up to 2 mmol of electrons per gram. Moreover, biochar based on mineral ash has a higher electron exchange capacity than biochar made from wood.

Due to their conductive properties, solid biochar particles promote direct electron transfer when used in MFC in cocultures of *Geobacter metallireducens* with *Geobacter sulfurreducens* or *Methanosarcina barkeri* to an extent similar to granular activated carbon (Chen et al., 2014).

The conductive properties of biochar can play an important role in MFC by promoting direct interspecies electron transfer (DIET) and the mechanism of electron exchange between different microorganism species and electrodes. Some microbial species can directly transfer electrons to biochar, transporting them outside individual cells (Chen et al., 2014).

Wood-based biochar has been used as electrodes for MFCs, significantly reducing the cost and carbon footprint. The specific power when using electrodes from biochar obtained from logging waste in MFC was 457 mW/m^2, and when using pressed grinding waste 532 mW/m^2 (Huggins et al., 2014). Biochar produced from forest waste is an effective substitute for activated carbon, which has adsorption capabilities (Zhang et al., 2004).

Biochar's high porosity and surface area, its low cost and low toxicity make it an ideal alternative to traditional carbon materials for the production of high-capacity electrical energy storage devices. In addition to surface area, the conductivity of a material is positively correlated with its double layer capacitance. In some cases, the presence of functional groups on the surface can also increase the double layer capacity of the carbon material by improving its wettability by electrolyte ions. This results in a higher usable surface area and can attract electrolyte ions closer to the surface, reducing the thickness of the double layer and thereby increasing its capacity.

16.5 Performance and Mechanisms of biochar function in microbial fuel cells

16.5.1 Overview of performance

Some studies have investigated the performance of biochar-based MFCs. Detailed data on the performance of biochar-based MFCs are presented elsewhere in this book (i.e., Chapter 15). In summary, biochar has been used as an anode or cathode material in MFCs. For example, a sludge MFC with a biochar-modified anode generated 108.05 mW/m^2 and removed Longilinea, Denitratisoma, and Pseudomonas, showing the highest removal of COD and total phosphorus (Sun et al., 2024). However, electrical conductivity significantly impacted the microbial community (Sun et al., 2024). In another study, an MFC using corn stalk biochar had a power density of 2.68 W/m^2 and a coulombic efficiency of 70.52% (Yan et al., 2024).

Biochar has also been used as a cathodic material. For example, in one study, an MFC with watermelon peel biochar produced 0.262 W/m^3 (Mo et al., 2106). In another study, an MFC with waste-banana-peel biochar had a power density of 142.2 mW/m^2 (Priyadarshini et al., 2024). Further evidence on the performance of biochar-based MFCs is presented in Chapter 15 of this book. However, systematic optimization and comparison of various biochar types in different MFC designs and configurations are still lacking. Hence, further work is required to address this gap.

16.5.2 Mechanisms of biochar enhancement of microbial fuel cells

Biochar enhances MFC performance through several key mechanisms. These mechanisms are reviewed in detail in the literature (Jiang et al., 2019; Li et al., 2018; Wang & Wang, 2020; Zhang & et al., 2019; Zhang, 2021), and include:

1. Enhanced Electron Transfer and Shuttling

    Biochar facilitates electron transfer within MFCs due to its high electrical conductivity and large surface area. Biochar can act as an electron shuttle, facilitating the transfer of electrons from microbial cells to the anode. This reduces the internal resistance of the MFC and increases the rate of electron transfer, leading to improved power generation. These properties improve the efficiency of electron flow from microbes to the anode, thereby increasing the overall power output of the MFC.

2. Improved Microbial Adhesion and Growth

    The porous structure and high surface area of biochar provide an ideal environment for microbial colonization and growth. This increases the density of EAB on the electrode surface, which enhances the bioelectrochemical activity and power generation of the MFC.

3. pH buffering capacity

    Biochar's pH buffering capacity helps maintain a stable pH environment in the MFC, which is crucial for the optimal activity of electroactive microbes. This stability ensures consistent performance and higher efficiency in WWT and electricity generation. This is particularly important when using organic substrates or wastewaters with fluctuating pH conditions.

4. Pseudo-capacitance and double-layer capacitance

    Biochar exhibits pseudo-capacitance and double-layer capacitance, which contribute to the storage and release of charge in the MFC. This improves the anode's ability to accept and release electrons, thus enhancing the overall performance of the cell.

5. Direct Interspecies Electron Transfer.

    Biochar promotes direct interspecies electron transfer between different microbial species. This synergistic interaction enhances the metabolic efficiency of microbial consortia, resulting in higher electricity generation and improved WWT performance.

16.6 Future directions and perspectives

16.6.1 Knowledge gaps

The application of MFCs for WWT and bioelectricity generation, as well as the use of biochar in such systems, is an emerging area of research. Consequently, several knowledge gaps still exist, particularly concerning the use of biochar. To fully realize the potential of biochar in MFCs, future research should focus on several key areas:

1. **Development, functionalization, and optimization of biochars with ideal properties:**
   a. Determine the optimal properties of biochars for use in MFCs by investigating the influence of various biomass feedstocks and pyrolysis conditions on their electrochemical properties.
   b. Enhance biochar's surface area, porosity, and functional groups to improve electron transfer efficiency and microbial adhesion. This may involve activation processes and functionalization with nanomaterials or metal oxides to improve its performance as an electrode material.
2. **Optimum MFC design for various biochar types:**
   a. Limited data exist comparing the effects of different biochar types on MFC performance across various designs. Further work is required to integrate biochar electrodes in different MFC configurations, including single-chamber, double-chamber, and stacked systems.
   b. Studies should monitor the impact of biochar-based anodes and cathodes on overall MFC performance in terms of WWT efficiency, power generation, and longevity.
3. **Elucidation of the contribution of various mechanisms of performance enhancement:**
   a. While mechanisms such as electron shuttling, pH buffering, and direct interspecies electron transfer are known to enhance MFC performance, limited data exist on the dominant mechanisms and their relative contributions.
   b. In-depth studies investigating the interaction between biochar and microbial communities in MFCs to understand synergistic effects are needed.
4. **Scaling up and real-world applications:**
   a. Scaling up and real-world applications of MFCs are still limited, pointing to potential challenges. There is a need to develop protocols for scaling up biochar-MFC systems for real-world applications in WWT and energy generation.
   b. Pilot projects and field trials should be conducted to demonstrate the feasibility and effectiveness of biochar-enhanced MFCs in diverse environmental settings. These studies will also define the biophysical limits of MFCs as a technology for WWT and bioelectricity generation.
5. **Environmental and economic assessments:**
   a. Comprehensive environmental, socio-economic, and techno-economic assessments of biochar-based MFCs are still limited. Life cycle assessments (LCA) and techno-economic analyses are needed to evaluate the sustainability and cost-effectiveness of biochar-MFC systems.
6. **Hybrid systems and Synergies:**
   a. There is potential to integrate biochar-based MFCs with other wastewater and energy generation technologies, though this aspect has received limited research attention.
   b. Research should investigate the integration of biochar-MFCs with other treatment technologies, such as anaerobic digestion, constructed wetlands, and advanced oxidation processes. Additionally, potential synergies between biochar-MFCs and renewable energy systems, such as solar and wind power, should be explored to enhance overall system efficiency and sustainability.
7. **Regeneration and recycling of biochar electrodes:**
   a. The regeneration, reuse, and final disposal of spent biochar electrodes have received limited research attention. Further work is required to develop and evaluate appropriate technologies for the regeneration, recycling, and reuse of spent biochar electrodes to extend their lifespan and reduce waste.
   b. It is also necessary to understand the effects of regeneration methods and multiple regeneration cycles on the performance and properties of biochar electrodes.
8. **Standardization of experimental protocols and best practices:**
   a. The MFC designs, operating conditions, biochar types, and parameters measured vary considerably among studies, making comparison of research results problematic. There is a need to establish standardized experimental protocols for the production, characterization, and testing of biochar for use in MFCs.
   b. Such studies will generate a database of well-characterized biochars with known performance in MFCs. Additionally, best practices and guidelines should be developed for researchers and practitioners to ensure consistent and reliable results.

By addressing these research directions, the potential of biochar to enhance MFCs can be more fully realized, paving the way for more efficient, sustainable, and practical applications in environmental remediation and renewable energy generation.

16.6.2 A roadmap for the advancement of biochar-based microbial fuel cells

To address the gaps highlighted, a 6-step conceptual roadmap is proposed. This roadmap is adapted from a generic one for engineered biochar-based remediation systems (Chapter 25) and an earlier review of biochar-based air filtration systems (Gwenzi et al., 2021). The six steps are as follows (see Fig. 16.6):

1. **Preparation, characterization, and optimization of biochars for MFCs:** This entails the preparation, characterization, and optimization of tailored pristine and engineered biochars for target applications in MFCs. This step is expected to generate a database of well-characterized biochars and their potential applications in MFCs.

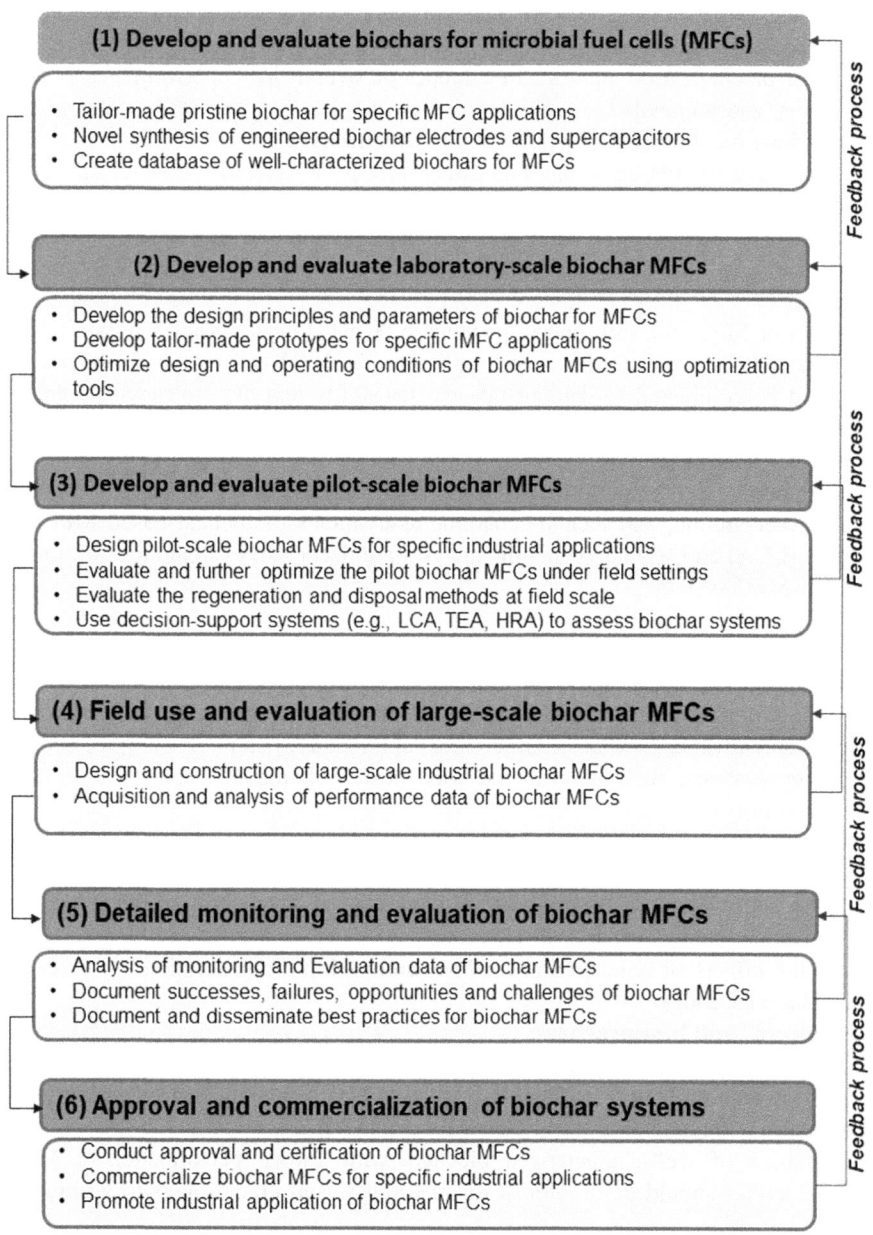

FIGURE 16.6 A conceptual roadmap for the advancement of biochar-based bioelectrochemical or microbial fuel cells. Adapted from a generic framework in Chapter 6 of this book and Gwenzi et al., 2020 (with permission of Elsevier); Gwenzi, W., Chaukura, N., Wenga, T. and Mtisi, M. (2021). Biochars as media for air pollution control systems: Contaminant removal, applications and future research directions. Science of the Total Environment, 753, 142249.

2. **Laboratory-scale Prototyping and Evaluation of biochar MFCs:** This involves designing and evaluating various laboratory-scale prototypes of biochar-based MFCs and determining the optimum design and operating conditions for each MFC design and type.
3. **Pilot-scale prototyping and evaluation of biochar MFCs:** This step focuses on designing and evaluating pilot-scale prototypes of biochar-based MFCs at relevant industrial scales and under environmentally relevant conditions. This will provide realistic data to assess the performance of biochar-based MFCs with respect to WWT and bioelectricity generation. The most promising designs of optimized biochar-based MFCs from this step should be taken to the next step of field-scale application and evaluation.
4. **Field-scale application and evaluation:** This step focuses on the field application and evaluation of large-scale biochar-based MFCs by early adopters or practitioners in real-world conditions. In this step, the operation of such MFCs should be done by the target end-users rather than the researchers.
5. **Monitoring, evaluation, and feedback:** This component is an integral part of all steps and entails systematically monitoring and evaluating the performance of biochar-based MFCs. The results will provide feedback to refine the technology based on performance results.
6. **Approval and commercialization:** The approval and commercialization of biochar-based MFCs are critical for their uptake and adoption. Therefore, this step will focus on the approval, certification, and commercialization of biochar-based MFCs with known design specifications and performance.

This roadmap provides a framework for the further development of MFCs, specifically those using biochar. The roadmap is flexible and enables the user to adapt it to include other relevant aspects, such as decision-support tools for the assessment of biochar-based MFCs. These decision-support systems include: (1) life cycle assessment (LCA), (2) techno-economic assessment, (3) health risk assessment, and (4) regeneration, recycling, and disposal technologies.

16.7 Conclusions

Bioelectrochemical systems, particularly MFCs, represent a promising technology for WWT and bioelectricity generation. Despite their potential, the widespread adoption of MFCs is limited by several constraints. Biochars, with their unique electrochemical properties, present an ideal biomaterial for enhancing the performance of these systems.

In summary, this chapter discussed the following:

1. The fundamental principles and components of bioelectrochemical systems or MFCs.
2. The various designs and types of MFCs, including single and double chamber MFCs.
3. Applications of MFCs in WWT and bioelectricity generation.
4. Biochar is suitable as an electrode biomaterial for MFCs due to its pseudo-capacitance, electronic conductivity, and double-layer capacity.
5. The mechanisms accounting for biochar's enhancement of MFC performance, such as electron shuttling, pH buffering, direct interspecies electron transfer, and functioning as a dispersant.
6. Evidence demonstrates the improved efficiency and effectiveness of MFCs when biochar is utilized.

However, continued research and development focused on optimizing biochar properties and integrating them into bioelectrochemical systems will be crucial for overcoming current limitations and realizing the full potential of this technology. To achieve this, knowledge gaps were highlighted as part of future research directions. Moreover, a conceptual roadmap for the further advancement of MFCs incorporating biochar is proposed.

Acknowledgment

Wilgince Apollon gratefully acknowledges the support from the National Council of Humanities, Sciences, and Technologies (CONAHCYT, for its acronym in Spanish) for the Postdoctoral scholarship granted (Grant No. 878025). The author also acknowledges the National Polytechnic Institute, the Altamira unit, for their support.

References

Adnan, M. Y., & Hassoon Ali, A. (2023). *Investigating the performance of single chamber microbial fuel cell (SCMFC) as sustainable-innovative technique for electricity generation and wastewater treatment, Defect and Diffusion Forum* (Vol. 425, pp. 107–117). Trans Tech Publications Ltd June. Available from https://doi.org/10.4028/p-iy0o22.

Aghababaie, M., Farhadian, M., Jeihanipour, A., & Biria, D. (2015). Effective factors on the performance of microbial fuel cells in wastewater treatment–a review. *Environmental Technology Reviews*, 4(1), 71–89. Available from https://doi.org/10.1080/09593330.2015.1077896, http://www.tandfonline.com/toc/tetr20/current.

Ahmad, F., Atiyeh, M. N., Pereira, B., & Stephanopoulos, G. N. (2013). A review of cellulosic microbial fuel cells: Performance and challenges. *Biomass and Bioenergy*, 56, 179–188. Available from https://doi.org/10.1016/j.biombioe.2013.04.006, http://www.journals.elsevier.com/biomass-and-bioenergy/.

Ahmad, M., Lee, S. S., Dou, X., Mohan, D., Sung, J. K., Yang, J. E., & Ok, Y. S. (2012). Effects of pyrolysis temperature on soybean stover- and peanut shell-derived biochar properties and TCE adsorption in water. *Bioresource Technology*, 118, 536–544. Available from https://doi.org/10.1016/j.biortech.2012.05.042.

Akter, T., Shaha, M., Al Mamun, M., Khan, M. A. S., & Hashem, A. (2022). *The role of the proton exchange membrane (PEM) in microbial fuel cell performance Microbial Fuel Cells: Emerging trends in electrochemical applications* (7). Bangladesh: Institute of Physics Publishing 1. Available from https://iopscience.iop.org/book/edit/978-0-7503-4791-4.

Apollon, W., Rusyn, I., González-Gamboa, N., Kuleshova, T., Luna-Maldonado, A. I., Vidales-Contreras, J. A., & Kamaraj, S.-K. (2022). Improvement of zero waste sustainable recovery using microbial energy generation systems: A comprehensive review. *Science of The Total Environment*, 817, 153055. Available from https://doi.org/10.1016/j.scitotenv.2022.153055.

Apollon, W., Rusyn, I., Kuleshova, T., Luna-Maldonado, A. I., Pierre, J. F., Gwenzi, W., & Kumar, V. (2024). An overview of agro-industrial wastewater treatment using microbial fuel cells: Recent advancements. *Journal of Water Process Engineering*, 58. Available from https://doi.org/10.1016/j.jwpe.2024.104783, https://www.sciencedirect.com/science/journal/22147144.

Apollon, W. (2023). An Overview of Microbial Fuel Cell Technology for Sustainable Electricity Production. *Membranes*, 13(11), 884. Available from https://doi.org/10.3390/membranes13110884.

Balathanigaimani, M. S., Shim, W. G., Lee, M. J., Kim, C., Lee, J. W., & Moon, H. (2008). Highly porous electrodes from novel corn grains-based activated carbons for electrical double layer capacitors. *Electrochemistry Communications*, 10(6), 868–871. Available from https://doi.org/10.1016/j.elecom.2008.04.003.

Bourke, J., Manley-Harris, M., Fushimi, C., Dowaki, K., Nunoura, T., & Antal, M. J. (2007). Do all carbonized charcoals have the same chemical structure? 2. A model of the chemical structure of carbonized charcoal. *Industrial and Engineering Chemistry Research*, 46(18), 5954–5967. Available from https://doi.org/10.1021/ie070415u.

Brousse, T., Belanger, D., & Long, J. W. (2015). To be or not to be pseudocapacitive? *Journal of the Electrochemical Society*, 162, A5185–A5189. Available from https://doi.org/10.1149/2.0201505jes.

Burns, M. K., & Qin, M. (2023). Ammonia recovery from organic nitrogen in synthetic dairy manure with a microbial fuel cell. *Chemosphere*, 325138388. Available from https://doi.org/10.1016/j.chemosphere.2023.138388.

Chacón, F. J., Cayuela, M. L., Roig, A., & Sánchez-Monedero, M. A. (2017). Understanding, measuring and tuning the electrochemical properties of biochar for environmental applications. *Reviews in Environmental Science and Bio/Technology*, 16, 695–715. Available from https://doi.org/10.1007/s11157-017-9450-1.

Chaturvedi, A., Dhillon, S. K., & Kundu, P. P. (2022). 1-D semiconducting TiO2 nanotubes supported efficient bimetallic Co-Ni cathode catalysts for power generation in single-chambered air-breathing microbial fuel cells. *Sustainable Energy Technologies and Assessments*, 53. Available from https://doi.org/10.1016/j.seta.2022.102479, http://www.journals.elsevier.com/sustainable-energy-technologies-and-assessments.

Chen, B., Chen, Z., & Lv, S. (2011). A novel magnetic biochar efficiently sorbs organic pollutants and phosphate. *Bioresource Technology*, 102(2), 716–723. Available from https://doi.org/10.1016/j.biortech.2010.08.067.

Chen, S., Rotaru, A. E., Shrestha, P. M., Malvankar, N. S., Liu, F., Fan, W., Nevin, K. P., & Lovley, D. R. (2014). Promoting interspecies electron transfer with biochar. *Scientific Reports*, 4. Available from https://doi.org/10.1038/srep05019, http://www.nature.com/srep/index.html.

Choudhury, P., Bhunia, B., Mahata, N., & Bandyopadhyay, T. K. (2022). Optimization for the improvement of power in equal volume of single chamber microbial fuel cell using dairy wastewater. *Journal of the Indian Chemical Society*, 99(6)100489. Available from https://doi.org/10.1016/j.jics.2022.100489.

Choudhury, P., Prasad Uday, U. S., Bandyopadhyay, T. K., Ray, R. N., & Bhunia, B. (2017). Performance improvement of microbial fuel cell (MFC) using suitable electrode and Bioengineered organisms: A review. *Bioengineered*, 8(5), 471–487. Available from https://doi.org/10.1080/21655979.2016.1267883, http://www.tandfonline.com/loi/kbie20.

Cossutta, M., Vretenar, V., Centeno, T. A., Kotrusz, P., McKechnie, J., & Pickering, S. J. (2020). A comparative life cycle assessment of graphene and activated carbon in a supercapacitor application. *Journal of Cleaner Production*, 242. Available from https://doi.org/10.1016/j.jclepro.2019.118468, https://www.journals.elsevier.com/journal-of-cleaner-production.

Davidson, E. A., Dail, D. B., & Chorover, J. (2008). Iron interference in the quantification of nitrate in soil extracts and its effect on hypothesized abiotic immobilization of nitrate. *Biogeochemistry*, 90(1), 65–73. Available from https://doi.org/10.1007/s10533-008-9231-6.

Debnath, K., & Dutta, S. (2023). Bio-electrochemical system analysis and improvement: A technical review. *Cleaner and Circular Bioeconomy*, 6100052. Available from https://doi.org/10.1016/j.clcb.2023.100052.

Dong, Y., Qu, Y., He, W., Du, Y., Liu, J., Han, X., & Feng, Y. (2015). A 90-liter stackable baffled microbial fuel cell for brewery wastewater treatment based on energy self-sufficient mode. *Bioresource Technology*, 195, 66–72. Available from https://doi.org/10.1016/j.biortech.2015.06.026.

Downie, A., Crosky, A., & Munroe, P. (2012). Physical properties of biochar. *Biochar for Environmental Management: Science and Technology*, 9781849770552, 13–32. Available from https://doi.org/10.4324/9781849770552, http://www.taylorandfrancis.com/books/details/9781849770552/.

Electrical effects accompanying the decomposition of organic compounds. Proceedings of the Royal Society of London. Series B, Containing Papers of a Biological Character. **84** (571) (1911), 260–276, doi: 10.1098/rspb.1911.0073.

Eslami, S., Bahrami, M., Zandi, M., Fakhar, J., Gavagsaz-Ghoachani, R., Noorollahi, Y., Phattanasak, M., & Nahid-Mobarakeh, B. (2023). Performance investigation and comparison of polypropylene to Nafion117 as the membrane of a dual-chamber microbial fuel cell. *Cleaner Materials, 8*100184. Available from https://doi.org/10.1016/j.clema.2023.100184.

Essandoh, M., Wolgemuth, D., Pittman, C. U., Mohan, D., & Mlsna, T. (2017). Phenoxy herbicide removal from aqueous solutions using fast pyrolysis switchgrass biochar. *Chemosphere, 174*, 49–57. Available from https://doi.org/10.1016/j.chemosphere.2017.01.105, http://www.elsevier.com/locate/chemosphere.

Feng, Y., He, W., Liu, J., Wang, X., Qu, Y., & Ren, N. (2014). A horizontal plug flow and stackable pilot microbial fuel cell for municipal wastewater treatment. *Bioresource Technology, 156*, 132–138. Available from https://doi.org/10.1016/j.biortech.2013.12.104.

Gasteiger, H. A., Kocha, S. S., Sompalli, B., & Wagner, F. T. (2005). Activity benchmarks and requirements for Pt, Pt-alloy, and non-Pt oxygen reduction catalysts for PEMFCs. *Applied Catalysis B: Environmental, 56*(1-2), 9–35. Available from https://doi.org/10.1016/j.apcatb.2004.06.021.

Gwenzi, W., Chaukura, N., Wenga, T., & Mtisi, M. (2021). Biochars as media for air pollution control systems: Contaminant removal, applications and future research directions. *Science of The Total Environment, 753*, 142249. Available from https://doi.org/10.1016/j.scitotenv.2020.142249.

Harvey, O. R., Herbert, B. E., Kuo, L. J., & Louchouarn, P. (2012). Generalized two-dimensional perturbation correlation infrared spectroscopy reveals mechanisms for the development of surface charge and recalcitrance in plant-derived biochars. *Environmental Science and Technology, 46*(19), 10641–10650. Available from https://doi.org/10.1021/es302971d.

Hawthorne, I., Johnson, M. S., Jassal, R. S., Black, T. A., Grant, N. J., & Smukler, S. M. (2017). Application of biochar and nitrogen influences fluxes of CO_2, CH_4 and N_2O in a forest soil. *Journal of Environmental Management, 192*, 203–214. Available from https://doi.org/10.1016/j.jenvman.2016.12.06.

He, M., Xu, Z., Hou, D., Gao, B., Cao, X., Ok, Y. S., et al. (2022). Waste-derived biochar for water pollution control and sustainable development. *Nature Reviews Earth & Environment, 3*(7), 444–460.

Huang, S. J., Dwivedi, K. A., Kumar, S., Wang, C. T., & Yadav, A. K. (2023). Binder-free NiO/MnO_2 coated carbon based anodes for simultaneous norfloxacin removal, wastewater treatment and power generation in dual-chamber microbial fuel cell. *Environmental Pollution, 317*. Available from https://doi.org/10.1016/j.envpol.2022.120578, https://www.journals.elsevier.com/environmental-pollution.

Huggins, T., Wang, H., Kearns, J., Jenkins, P., & Ren, Z. J. (2014). Biochar as a sustainable electrode material for electricity production in microbial fuel cells. *Bioresource Technology, 157*, 114–119. Available from https://doi.org/10.1016/j.biortech.2014.01.058, http://www.elsevier.com/locate/biortech.

Huggins, T. M., Pietron, J. J., Wang, H., Ren, Z. J., & Biffinger, J. C. (2015). Graphitic biochar as a cathode electrocatalyst support for microbial fuel cells. *Bioresource Technology, 195*, 147–153. Available from https://doi.org/10.1016/j.biortech.2015.06.012, http://www.journals.elsevier.com/bioresource-technology/.

Ishihara, S. (1996). Recent trend of advanced carbon materials from wood charcoals. *Mokuzai Gakkai Shi, 42*, 717–723.

Ismanto, A. E., Wang, S., Soetaredjo, F. E., & Ismadji, S. (2010). Preparation of capacitor's electrode from cassava peel waste. *Bioresource Technology, 101*(10), 3534–3540. Available from https://doi.org/10.1016/j.biortech.2009.12.123.

Jalili, P., Ala, A., Nazari, P., Jalili, B., & Ganji, D. D. (2024). A comprehensive review of microbial fuel cells considering materials, methods, structures, and microorganisms. *Heliyon, 10*(3)e25439. Available from https://doi.org/10.1016/j.heliyon.2024.e25439.

Jiang, Y. H., Li, A. Y., Deng, H., Ye, C. H., Wu, Y. Q., Linmu, Y. D., & Hang, H. L. (2019). Characteristics of nitrogen and phosphorus adsorption by Mg-loaded biochar from different feedstocks. *Bioresource Technology, 276*, 183–189. Available from https://doi.org/10.1016/j.biortech.2018.12.079.

Jin, H., Wang, X., Gu, Z., Hoefelmeyer, J. D., Muthukumarappan, K., & Julson, J. (2014). Graphitized activated carbon based on big bluestem as an electrode for supercapacitors. *RSC Advances, 4*(27), 14136–14142. Available from https://doi.org/10.1039/c3ra46037a, http://pubs.rsc.org/en/journals/journalissues.

Joseph, S., Graber, E. R., Chia, C., Munroe, P., Donne, S., Thomas, T., Nielsen, S., Marjo, C., Rutlidge, H., Pan, G. X., Li, L., Taylor, P., Rawal, A., & Hook, J. (2013). Shifting paradigms: Development of high-efficiency biochar fertilizers based on nano-structures and soluble components. *Carbon Management, 4*(3), 323–343. Available from https://doi.org/10.4155/cmt.13.23.

Joseph, S., Husson, O., Graber, E. R., Van Zwieten, L., Taherymoosavi, S., Thomas, T., Nielsen, S., Ye, J., Pan, G., Chia, C., Munroe, P., Allen, J., Lin, Y., Fan, X., & Donne, S. (2015). The electrochemical properties of biochars and how they affect soil redox properties and processes. *Agronomy, 5*(3), 322–340. Available from https://doi.org/10.3390/agronomy5030322, http://www.mdpi.com/2073-4395/5/3/322/pdf.

Karuppiah, T., Pugazhendi, A., Subramanian, S., Jamal, M. T., & Jeyakumar, R. B. (2018). Deriving electricity from dye processing wastewater using single chamber microbial fuel cell with carbon brush anode and platinum nano coated air cathode. *3 Biotech, 8*(10). Available from https://doi.org/10.1007/s13205-018-1462-1, http://www.springerlink.com/content/2190-572x/.

Keiluweit, M., Nico, P. S., Johnson, M. G., & Kleber, M. (2010). Dynamic molecular structure of plant biomass-derived black carbon (biochar). *Environmental Science & Technology, 44*(4), 1247–1253. Available from https://doi.org/10.1021/es9031419.

Kercher, A. K., & Nagle, D. C. (2003). Microstructural evolution during charcoal carbonization by X-ray diffraction analysis. *Carbon, 41*(1), 15–27. Available from https://doi.org/10.1016/S0008-6223(02)00261-0.

Kharti, H., Hitar, M. E. H., Touach, N., Lotfi, E. M., El Mahi, M., Mouhir, L., Fekhaoui, M., & Benzaouak, A. (2024). Generating Sustainable Bioenergy from Wastewater with $Ni_2V_2O_7$ as a Potential Cathode Catalyst in Single-Chamber Microbial Fuel Cells. *Chemistry Africa, 7*(1), 209–218. Available from https://doi.org/10.1007/s42250-023-00761-w, https://www.springer.com/journal/42250.

Kiely, P. D., Rader, G., Regan, J. M., & Logan, B. E. (2011). Long-term cathode performance and the microbial communities that develop in microbial fuel cells fed different fermentation endproducts. *Bioresource Technology, 102*(1), 361–366. Available from https://doi.org/10.1016/j.biortech.2010.05.017.

Kim, J. R., Dec, J., Bruns, M. A., & Logan, B. E. (2008). Removal of odors from Swine wastewater by using microbial fuel cells. *Applied and Environmental Microbiology, 74*(8), 2540–2543. Available from https://doi.org/10.1128/AEM.02268-07.

Kim, Y. J., Lee, B. J., Suezaki, H., Chino, T., Abe, Y., Yanagiura, T., Park, K. C., & Endo, M. (2006). Preparation and characterization of bamboo-based activated carbons as electrode materials for electric double layer capacitors. *Carbon*, *44*(8), 1592–1595. Available from https://doi.org/10.1016/j.carbon.2006.02.011.

Klass, D. L. (2004). *Biomass for Renewable Energy and Fuels* (pp. 193–212). Elsevier BV. Available from 10.1016/b0-12-176480-x/00353-3.

Klüpfel, L., Keiluweit, M., Kleber, M., & Sander, M. (2014). Redox Properties of Plant Biomass-Derived Black Carbon (Biochar). *Environmental Science & Technology*, *48*(10), 5601–5611. Available from https://doi.org/10.1021/es500906d.

Kolubah, P. D., Mohamed, H. O., Ayach, M., Rao Hari, A., Alshareef, H. N., Saikaly, P., Chae, K. J., & Castaño, P. (2023). W2N-MXene composite anode catalyst for efficient microbial fuel cells using domestic wastewater. *Chemical Engineering Journal*, *461*. Available from https://doi.org/10.1016/j.cej.2023.141821, http://www.elsevier.com/inca/publications/store/6/0/1/2/7/3/index.htt.

Kuleshova, T. E. (2024). Production of activated carbon from biomass as electrode material for electrochemical devices (review). *Nauchnoe Priborostroenie*, *34*(1), 43–61.

Kumar, M., & Gupta, R. C. (1993). Electrical resistivity of Acacia and Eucalyptus wood chars. *Journal of Materials Science*, *28*(2), 440–444. Available from https://doi.org/10.1007/bf00357821.

Kumar, V., Rudra, R., & Hait, S. (2021). Sulfonated polyvinylidene fluoride-crosslinked-aniline-2-sulfonic acid as ion exchange membrane in single-chambered microbial fuel cell. *Journal of Environmental Chemical Engineering*, *9*(6)106467. Available from https://doi.org/10.1016/j.jece.2021.106467.

Li, D., Sun, Y., Shi, Y., Wang, Z., Okeke, S., Yang, L., & Xiao, L. (2023). Structure evolution of air cathodes and their application in electrochemical sensor development and wastewater treatment. *Science of The Total Environment*, *869*161689. Available from https://doi.org/10.1016/j.scitotenv.2023.161689.

Li, Y., Yu, S., Strong, J., & Wang, H. (2012). Are the biogeochemical cycles of carbon, nitrogen, sulfur, and phosphorus driven by the "Fe III–Fe II redox wheel" in dynamic redox environments? *Journal of Soils and Sediments*, *12*, 683–693. Available from https://doi.org/10.1007/s11368-012-0507-z.

Li, Y., Zhao, R., Chao, S., Sun, B., Wang, C., & Li, X. (2018). Polydopamine coating assisted synthesis of MnO2 loaded inorganic/organic composite electrospun fiber adsorbent for efficient removal of Pb2 + from water. *Chemical Engineering Journal*, *344*, 277–289. Available from https://doi.org/10.1016/j.cej.2018.03.044.

Liang, C., Bao, J., Li, C., Huang, H., Chen, C., Lou, Y., & Feng, S. (2017). One-dimensional hierarchically porous carbon from biomass with high capacitance as supercapacitor materials. *Microporous and mesoporous materials*, *251*, 77–82. Available from https://doi.org/10.1016/j.micromeso.2017.05.044.

Liang, P., Duan, R., Jiang, Y., Zhang, X., Qiu, Y., & Huang, X. (2018). One-year operation of 1000-L modularized microbial fuel cell for municipal wastewater treatment. *Water Research*, *141*, 1–8. Available from https://doi.org/10.1016/j.watres.2018.04.066.

Linares, R. V., Domínguez-Maldonado, J., Rodríguez-Leal, E., Patrón, G., Castillo-Hernández, A., Miranda, A., Romero, D. D., Moreno-Cervera, R., Camarachale, G., Borroto, C. G., & Alzate-Gaviria, L. (2019). Scale up of microbial fuel cell stack system for residential wastewater treatment in continuous mode operation. *Water (Switzerland)*, *11*(2). Available from https://doi.org/10.3390/w11020217, https://www.mdpi.com/2073-4441/11/2/217/pdf.

Liu, W. J., Li, W. W., Jiang, H., & Yu, H. Q. (2017). Fates of chemical elements in biomass during Its pyrolysis. *Chemical Reviews*, *117*(9), 6367–6398. Available from https://doi.org/10.1021/acs.chemrev.6b00647, http://pubs.acs.org/journal/chreay.

Liu, Z., Quek, A., Kent Hoekman, S., & Balasubramanian, R. (2013). Production of solid biochar fuel from waste biomass by hydrothermal carbonization. *Fuel*, *103*, 943–949. Available from https://doi.org/10.1016/j.fuel.2012.07.069, Singapore.

Logan, B. E., Hamelers, B., Rozendal, R., Schröder, U., Keller, J., Freguia, S., & Rabaey, K. (2006). Microbial fuel cells: methodology and technology. *Environmental Science & Technology*, *40*(17), 5181–5192. Available from https://doi.org/10.1021/es0605016.

Lu, T., Yuan, H., Wang, Y., Huang, H., & Chen, Y. (2016). Characteristic of heavy metals in biochar derived from sewage sludge. *Journal of Material Cycles and Waste Management*, *18*(4), 725–733. Available from https://doi.org/10.1007/s10163-015-0366-y.

López-Cano, I., Roig, A., Cayuela, M. L., Alburquerque, J. A., & Sánchez-Monedero, M. A. (2016). Biochar improves N cycling during composting of olive mill wastes and sheep manure. *Waste Management*, *49*, 553–559. Available from https://doi.org/10.1016/j.wasman.2015.12.031.

Mardanpour, M. M., Esfahany, M. N., Behzad, T., & Sedaqatvand, R. (2012). Single chamber microbial fuel cell with spiral anode for dairy wastewater treatment. *Biosensors and Bioelectronics*, *38*(1), 264–269. Available from https://doi.org/10.1016/j.bios.2012.05.046.

Melo, L. C. A., Puga, A. P., Coscione, A. R., Beesley, L., Abreu, C. A., & Camargo, O. A. (2016). Sorption and desorption of cadmium and zinc in two tropical soils amended with sugarcane-straw-derived biochar. *Journal of Soils and Sediments*, *16*(1), 226–234. Available from https://doi.org/10.1007/s11368-015-1199-y, http://www.springerlink.com/content/1439-0108.

Min, B., Kim, J. R., Oh, S. E., Regan, J. M., & Logan, B. E. (2005). Electricity generation from swine wastewater using microbial fuel cells. *Water Research*, *39*(20), 4961–4968. Available from https://doi.org/10.1016/j.watres.2005.09.039, http://www.elsevier.com/locate/watres.

Mo, R. J., Zhao, Y., Wu, M., Xiao, H. M., Kuga, S., & Huang, Y., et al. (2106). Activated carbon from nitrogen rich watermelon rind for high-performance supercapacitors. *RSC Advances*, *6*(64), 59333–59342. Available from https://doi.org/10.1039/C6RA10719B.

Mohyudin, S., Farooq, R., Jubeen, F., Rasheed, T., Fatima, M., & Sher, F. (2022). Microbial fuel cells a state-of-the-art technology for wastewater treatment and bioelectricity generation. *Environmental Research*, *204*112387. Available from https://doi.org/10.1016/j.envres.2021.112387.

Mukome, F. N. D., Zhang, X., Silva, L. C. R., Six, J., & Parikh, S. J. (2013). Use of chemical and physical characteristics to investigate trends in biochar feedstocks. *Journal of Agricultural and Food Chemistry*, *61*(9), 2196–2204. Available from https://doi.org/10.1021/jf3049142.

Naha, A., Debroy, R., Sharma, D., Shah, M. P., & Nath, S. (2023). Microbial fuel cell: A state-of-the-art and revolutionizing technology for efficient energy recovery. *Cleaner and Circular Bioeconomy*, *5*100050. Available from https://doi.org/10.1016/j.clcb.2023.100050.

Nam, J. Y., Kim, H. W., Lim, K. H., & Shin, H. S. (2010). Effects of organic loading rates on the continuous electricity generation from fermented wastewater using a single-chamber microbial fuel cell. *Bioresource Technology*, *101*(1). Available from https://doi.org/10.1016/j.biortech.2009.03.062, S33-S37S33 Elsevier Ltd South Korea, http://www.journals.elsevier.com/bioresource-technology/.

Nawaz, A., ul Haq, I., Qaisar, K., Gunes, B., Raja, S. I., Mohyuddin, K., & Amin, H. (2022). Microbial fuel cells: Insight into simultaneous wastewater treatment and bioelectricity generation. *Process Safety and Environmental Protection*, 161, 357–373. Available from https://doi.org/10.1016/j.psep.2022.03.039, http://www.elsevier.com/wps/find/journaldescription.cws_home/713889/description#description.

Nguyen, B. T., Lehmann, J., Hockaday, W. C., Joseph, S., & Masiello, C. A. (2010). Temperature sensitivity of black carbon decomposition and oxidation. *Environmental Science and Technology*, 44(9), 3324–3331. Available from https://doi.org/10.1021/es903016y.

Omenesa Idris, M., Al-Zaqri, N., Warad, I., Hossain, A. M. A., Masud, N., & Ali, M. (2023). Impact of commercial sugar as a substrate in single-chamber microbial fuel cells to improve the energy production with bioremediation of metals. *International Journal of Chemical Engineering*, 2023. Available from https://doi.org/10.1155/2023/9741246.

Papillon, J., Ondel, O., & Maire, É. (2021). Scale up of single-chamber microbial fuel cells with stainless steel 3D anode: Effect of electrode surface areas and electrode spacing. *Bioresource Technology Reports*, 13100632. Available from https://doi.org/10.1016/j.biteb.2021.100632.

Park, J. H., Ok, Y. S., Kim, S. H., Cho, J. S., Heo, J. S., Delaune, R. D., & Seo, D. C. (2016). Competitive adsorption of heavy metals onto sesame straw biochar in aqueous solutions. *Chemosphere*, 142, 77–83. Available from https://doi.org/10.1016/j.chemosphere.2015.05.093, http://www.elsevier.com/locate/chemosphere.

Parmar, A., Nema, P. K., & Agarwal, T. (2014). Biochar production from agro-food industry residues: A sustainable approach for soil and environmental management. *Current Science*, 107(10), 1673–1682. Available from http://www.currentscience.ac.in/Volumes/107/10/1673.pdf.

Paucar, N. E., & Sato, C. (2022). Coupling Microbial Fuel Cell and Hydroponic System for Electricity Generation, Organic Removal, and Nutrient Recovery via Plant Production from Wastewater. *Energies*, 15(23), 9211. Available from https://doi.org/10.3390/en15239211.

Peng, C., Yan, X. B., Wang, R. T., Lang, J. W., Ou, Y. J., & Xue, Q. J. (2013). Promising activated carbons derived from waste tea-leaves and their application in high performance supercapacitors electrodes. *Electrochimica Acta*, 87, 401–408. Available from https://doi.org/10.1016/j.electacta.2012.09.082.

Petter, F. A., & Madari, B. E. (2012). Biochar: Agronomic and environmental potential in Brazilian savannah soils. *Revista Brasileira de Engenharia Agrícola e Ambiental*, 16, 761–768.

Phiri, J., Dou, J., Vuorinen, T., Gane, P. A. C., & Maloney, T. C. (2019). Highly Porous Willow Wood-Derived Activated Carbon for High-Performance Supercapacitor Electrodes. *ACS Omega*, 4(19), 18108–18117. Available from https://doi.org/10.1021/acsomega.9b01977, pubs.acs.org/journal/acsodf.

Priyadarshini, M., Ahmad, A., Shinde, A., Das, I., & Ghangrekar, M. M. (2024). Anodic degradation of salicylic acid and simultaneous bio-electricity recovery in microbial fuel cell using waste-banana-peels derived biochar-supported MIL-53(Fe)-metal-organic framework as cathode catalyst. *Journal of Electroanalytical Chemistry*, 967, 118451. Available from https://doi.org/10.1016/j.jelechem.2024.118451.

Raghavulu, S. V., Mohan, S. V., Reddy, M. V., Mohanakrishna, G., & Sarma, P. N. (2009). Behavior of single chambered mediatorless microbial fuel cell (MFC) at acidophilic, neutral and alkaline microenvironments during chemical wastewater treatment. *International Journal of Hydrogen Energy*, 34(17), 7547–7554. Available from https://doi.org/10.1016/j.ijhydene.2009.05.071.

Rawat, S., Wang, C. T., Lay, C. H., Hotha, S., & Bhaskar, T. (2023). Sustainable biochar for advanced electrochemical/energy storage applications. *Journal of Energy Storage*, 63. Available from https://doi.org/10.1016/j.est.2023.107115, http://www.journals.elsevier.com/journal-of-energy-storage/.

Reguyal, F., Sarmah, A. K., & Gao, W. (2017). Synthesis of magnetic biochar from pine sawdust via oxidative hydrolysis of FeCl2 for the removal sulfamethoxazole from aqueous solution. *Journal of Hazardous Materials*, 321, 868–878. Available from https://doi.org/10.1016/j.jhazmat.2016.10.006, http://www.elsevier.com/locate/jhazmat.

Rossi, R., Hur, A. Y., Page, M. A., Thomas, A. O. B., Butkiewicz, J. J., Jones, D. W., Baek, G., Saikaly, P. E., Cropek, D. M., & Logan, B. E. (2022). Pilot scale microbial fuel cells using air cathodes for producing electricity while treating wastewater. *Water Research*, 215. Available from https://doi.org/10.1016/j.watres.2022.118208, http://www.elsevier.com/locate/watres.

Saquing, J. M., Yu, Y. H., & Chiu, P. C. (2016). Wood-Derived Black Carbon (Biochar) as a Microbial Electron Donor and Acceptor. *Environmental Science and Technology Letters*, 3(2), 62–66. Available from https://doi.org/10.1021/acs.estlett.5b00354, http://pubs.acs.org/page/estlcu/about.html.

Sato, C., Apollon, W., Luna-Maldonado, A. I., Paucar, N. E., Hibbert, M., & Dudgeon, J. (2023). Integrating Microbial Fuel Cell and Hydroponic Technologies Using a Ceramic Membrane Separator to Develop an Energy–Water–Food Supply System. *Membranes*, 13(9). Available from https://doi.org/10.3390/membranes13090803, http://www.mdpi.com/journal/membranes.

Si, W. J., Wu, X. Z., Xing, W., Zhou, J., & Zhuo, S. P. (2011). Bagasse-based nanoporous carbon for supercapacitor application. *Journal of Inorganic Materials*, 26(1), 107–112. Available from https://doi.org/10.3724/SP.J.1077.2010.10376.

Singh, A., & Kaushik, A. (2021). Removal of Cd and Ni with enhanced energy generation using biocathode microbial fuel cell: Insights from molecular characterization of biofilm communities. *Journal of Cleaner Production*, 315127940. Available from https://doi.org/10.1016/j.jclepro.2021.127940.

Stein, A., Wang, Z., & Fierke, M. A. (2009). Functionalization of porous carbon materials with designed pore architecture. *Advanced Materials*, 21(3), 265–293. Available from https://doi.org/10.1002/adma.200801492, http://www3.interscience.wiley.com/cgi-bin/fulltext/121519495/PDFSTART, United States.

Suda, F., Matsuo, T., & Ushioda, D. (2007). Transient changes in the power output from the concentration difference cell (dialytic battery) between seawater and river water. *Energy*, 32(3), 165–173. Available from https://doi.org/10.1016/j.energy.2006.04.005.

Sudhan, N., Subramani, K., Karnan, M., Ilayaraja, N., & Sathish, M. (2017). Biomass-Derived Activated Porous Carbon from Rice Straw for a High-Energy Symmetric Supercapacitor in Aqueous and Non-aqueous Electrolytes. *Energy & Fuels*, 31(1), 977–985. Available from https://doi.org/10.1021/acs.energyfuels.6b01829.

Sun, F., Chen, J., Sun, Z., Zheng, X., Tang, M., & Yang, Y. (2024). Promoting bioremediation of brewery wastewater, production of bioelectricity and microbial community shift by sludge microbial fuel cells using biochar as anode. *Science of The Total Environment, 929*172418.

Sun, K., Leng, C. Y., Jiang, J. C., Bu, Q., Lin, G. F., Lu, X. C., & Zhu, G. Z. (2017). Microporous activated carbons from coconut shells produced by self-activation using the pyrolysis gases produced from them, that have an excellent electric double layer performance. *New Carbon Materials, 32*(5), 451–459. Available from https://doi.org/10.1016/S1872-5805(17)60134-3.

Tang, J., Liu, J., Torad, N. L., Kimura, T., & Yamauchi, Y. (2014). Tailored design of functional nanoporous carbon materials toward fuel cell applications. *Nano Today, 9*(3), 305–323. Available from https://doi.org/10.1016/j.nantod.2014.05.003, http://www.elsevier.com/wps/find/journaldescription.cws_home/706735/description#description.

Thubsuang, U., Laebang, S., Manmuanpom, N., Wongkasemjit, S., & Chaisuwan, T. (2017). Tuning pore characteristics of porous carbon monoliths prepared from rubber wood waste treated with H3PO4 or NaOH and their potential as supercapacitor electrode materials. *Journal of Materials Science, 52*(11), 6837–6855. Available from https://doi.org/10.1007/s10853-017-0922-z.

Varanasi, J. L., Nayak, A. K., Sohn, Y., Pradhan, D., & Das, D. (2016). Improvement of power generation of microbial fuel cell by integrating tungsten oxide electrocatalyst with pure or mixed culture biocatalysts. *Electrochimica Acta, 199*, 154–163. Available from https://doi.org/10.1016/j.electacta.2016.03.152, http://www.journals.elsevier.com/electrochimica-acta/.

Verma, M., & Mishra, V. (2023). Bioelectricity generation by microbial degradation of banana peel waste biomass in a dual-chamber S. cerevisiae-based microbial fuel cell. *Biomass and Bioenergy, 168*106677. Available from https://doi.org/10.1016/j.biombioe.2022.106677.

Wang, B., Jiang, Ys, Li, Fy, & Yang, Dy (2017). Preparation of biochar by simultaneous carbonization, magnetization and activation for norfloxacin removal in water. *Bioresource Technology, 233*, 159–165. Available from https://doi.org/10.1016/j.biortech.2017.02.103, http://www.elsevier.com/locate/biortech.

Wang, C., Wu, D., Wang, H., Gao, Z., Xu, F., & Jiang, K. (2018). A green and scalable route to yield porous carbon sheets from biomass for supercapacitors with high capacity. *Journal of Materials Chemistry A, 6*(3), 1244–1254. Available from https://doi.org/10.1039/C7TA07579K.

Wang, C., Xing, Y., Zhang, K., Zheng, H., Zhang, Y. N., Zhu, X., & Qu, J. (2023). Evaluation of photocathode coupling-mediated hydroxychloroquine degradation in a single-chamber microbial fuel cell based on electron transfer mechanism and power generation. *Journal of Power Sources, 559*232625. Available from https://doi.org/10.1016/j.jpowsour.2022.232625.

Wang, H., Chai, G., Zhang, Y., Wang, D., Wang, Z., Meng, H., & Li, H. (2023). Copper removal from wastewater and electricity generation using dual-chamber microbial fuel cells with shrimp shell as the substrate. *Electrochimica Acta, 441*141849. Available from https://doi.org/10.1016/j.electacta.2023.141849.

Wang, H., & Wang, Y. (2020). Biochar-based functional materials as anode in microbial fuel cells. *Bioresource Technology*.

Wang, L., Tai, Y., Zhao, X., He, Q., Hu, Z., & Li, M. (2023). Phosphorus recovery directly from sewage sludge as vivianite and simultaneous electricity production in a dual chamber microbial fuel cell. *Journal of Environmental Chemical Engineering, 11*(3)110152. Available from https://doi.org/10.1016/j.jece.2023.110152.

Watanabe, K. (2008). Recent Developments in Microbial Fuel Cell Technologies for Sustainable Bioenergy. *Journal of Bioscience and Bioengineering, 106*(6), 528–536. Available from https://doi.org/10.1263/jbb.106.528.

White, R. J., Budarin, V., Luque, R., Clark, J. H., & Macquarrie, D. J. (2009). Tuneable porous carbonaceous materials from renewable resources. *Chemical Society Reviews, 38*(12), 3401–3418. Available from https://doi.org/10.1039/b822668g.

Wu, F. C., Tseng, R. L., Hu, C. C., & Wang, C. C. (2004). Physical and electrochemical characterization of activated carbons prepared from firwoods for supercapacitors. *Journal of Power Sources, 138*(1-2), 351–359. Available from https://doi.org/10.1016/j.jpowsour.2004.06.023.

Wu, X., Lv, C., Ye, J., Li, M., Zhang, X., Lv, J., & Xie, W. (2021). Glycine-hydrochloric acid buffer promotes simultaneous U (VI) reduction and bioelectricity generation in dual chamber microbial fuel cell. *Journal of the Taiwan Institute of Chemical Engineers, 127*, 236–247. Available from https://doi.org/10.1016/j.jtice.2021.08.021.

Xie, R., Wang, S., Wang, K., Wang, M., Chen, B., Wang, Z., & Tan, T. (2022). Improved energy efficiency in microbial fuel cells by bioethanol and electricity co-generation. *Biotechnology for Biofuels and Bioproducts, 15*(1). Available from https://doi.org/10.1186/s13068-022-02180-4.

Xing, F., Xi, H., Yu, Y., & Zhou, Y. (2021). Anode biofilm influence on the toxic response of microbial fuel cells under different operating conditions. *Science of The Total Environment, 775*145048. Available from https://doi.org/10.1016/j.scitotenv.2021.145048.

Xu, G., Han, J., Ding, B., Nie, P., Pan, J., Dou, H., Li, H., & Zhang, X. (2015). Biomass-derived porous carbon materials with sulfur and nitrogen dual-doping for energy storage. *Green Chemistry, 17*(3), 1668–1674. Available from https://doi.org/10.1039/C4GC02185A.

Yan, J., Zhang, M., Chen, X., Chen, C., Xu, X., & Jiang, S. (2024). Straw-derived macroporous biochar as high-performance anode in microbial fuel cells. *Process Biochemistry, 145*, 113–121. Available from https://doi.org/10.1016/j.procbio.2024.06.024.

Yap, K. L., Ho, L. N., Guo, K., Liew, Y. M., Lutpi, N. A., Azhari, A. W., Thor, S. H., Teoh, T. P., Oon, Y. S., & Ong, S. A. (2023). Exploring the potential of metal oxides as cathodic catalysts in a double chambered microbial fuel cell for phenol removal and power generation. *Journal of Water Process Engineering, 53*. Available from https://doi.org/10.1016/j.jwpe.2023.103639, http://www.journals.elsevier.com/journal-of-water-process-engineering/.

Yaqoob, A. A., Ibrahim, M. N. M., Ahmad, A., & Alshammari, M. B. (2022). *Basic principles and working mechanisms of microbial fuel cells Microbial Fuel Cells: Emerging trends in electrochemical applications* (2, p. 1). Malaysia: Institute of Physics Publishing. Available from https://iopscience.iop.org/book/edit/978-0-7503-4791-4.

Yin, X. M., Lü, X. G., Jiang, M., & Zou, Y. C. (2010). Research progress of the coupling process of Fe and N in wetland soils. *Huanjing Kexue/Environmental Science, 31*(9), 2254–2259.

Yoo, S., Kelley, S. S., Tilotta, D. C., & Park, S. (2018). Structural Characterization of Loblolly Pine Derived Biochar by X-ray Diffraction and Electron Energy Loss Spectroscopy. *ACS Sustainable Chemistry and Engineering, 6*(2), 2621–2629. Available from https://doi.org/10.1021/acssuschemeng.7b04119, http://pubs.acs.org/journal/ascecg.

Zhang, L. C., Jia, Z., Lyu, F., Liang, S. X., & Lu, J. (2019). A review of catalytic performance of metallic glasses in wastewater treatment: Recent progress and prospects. *Progress in Materials Science*, *105*, 100576. Available from https://doi.org/10.1016/j.pmatsci.2019.100576.

Zhang, J., Gong, L., Sun, K., Jiang, J., & Zhang, X. (2012). Preparation of activated carbon from waste Camellia oleifera shell for supercapacitor application. *Journal of Solid State Electrochemistry*, *16*, 2179–2186. Available from https://doi.org/10.1007/s10008-012-1639-1.

Zhang, Q., & Liu, L. (2023). Pave the way for successful treatment of Nylon wastewater by performance enhancement via unit stacking scale up of dual cathodes up-flow microbial fuel cell and C/N adjustment with urea. *Journal of Environmental Chemical Engineering*, *11*(3), 109930. Available from https://doi.org/10.1016/j.jece.2023.109930.

Zhang, T., Walawender, W. P., Fan, L. T., Fan, M., Daugaard, D., & Brown, R. C. (2004). Preparation of activated carbon from forest and agricultural residues through CO2 activation. *Chemical Engineering Journal*, *105*(1-2), 53–59. Available from https://doi.org/10.1016/j.cej.2004.06.011, http://www.elsevier.com/inca/publications/store/6/0/1/2/7/3/index.htt.

Zhang, X. (2021). Biochar enhances electron transfer and process performance in anaerobic digestion: Mechanisms and applications. *Bioresource Technology*, *332*.

Zhang, Y. R., Wang, R. Q., Chen, W. P., Song, K., Tian, Y., Li, J. X., & Shi, G. F. (2023). Microstructure and electrochemical properties of porous carbon derived from biomass. *International Journal of Electrochemical Science*, *18*(7)100190. Available from https://doi.org/10.1016/j.ijoes.2023.100190.

Zhao, L., Baccile, N., Gross, S., et al. (2010). Sustainable nitrogendoped carbonaceous materials from biomass derivatives. *Carbon*, *48*, 3778–3787. Available from https://doi.org/10.1016/j.carbon.2010.06.0400.

Zhao, L., Zhou, J. H., Sui, Z. J., & Zhou, X. G. (2010). Hydrogenolysis of sorbitol to glycols over carbon nanofiber supported ruthenium catalyst. *Chemical Engineering Science*, *65*(1), 30–35. Available from https://doi.org/10.1016/j.ces.2009.03.026.

Zhou, N., Chen, H., Feng, Q., Yao, D., Chen, H., Wang, H., Zhou, Z., Li, H., Tian, Y., & Lu, X. (2017). Effect of phosphoric acid on the surface properties and Pb(II) adsorption mechanisms of hydrochars prepared from fresh banana peels. *Journal of Cleaner Production*, *165*, 221–230. Available from https://doi.org/10.1016/j.jclepro.2017.07.111.

Zhu, Q., Hu, J., Liu, B., Hu, S., Liang, S., Xiao, K., Yang, J., & Hou, H. (2022). Recent Advances on the Development of Functional Materials in Microbial Fuel Cells: From Fundamentals to Challenges and Outlooks. *ENERGY & ENVIRONMENTAL MATERIALS*, *5*(2), 401–426. Available from https://doi.org/10.1002/eem2.12173.

Zwieten, C., Kammann, M. L., Cayuela, B. P., Singh, S., Joseph, S., & Kimber Spokas, K. A. (2015). Biochar effects on nitrous oxide and methane emissions from soil Biochar for environmental management. In *Biochar for environmental management*, (pp. 489–520). Routledge.

Van Zwieten, L., Kimber, S., Morris, S., Chan, K. Y., Downie, A., Rust, J., Joseph, S., & Cowie, A. (2010). Effects of biochar from slow pyrolysis of papermill waste on agronomic performance and soil fertility. *Plant and Soil*, *327*(1-2), 235–246. Available from https://doi.org/10.1007/s11104-009-0050-x.

Chapter 17

Thermochemical treatment of human excreta to energy and biochar: recent advances, applications, and future directions

Flávio Lopes Francisco Bittencourt[1] and Marcio Ferreira Martins[2]

[1]*Federal Institute of Education, Science, and Technology of Espírito Santo, Piúma, Espírito Santo, Brazil,* [2]*Multiphysics Modeling Laboratories (MM Labs), Department of Mechanical Engineering, Federal University of Espírito Santo (UFES), Vitória, Espírito Santo, Brazil*

Chapter outline

17.1 Introduction	329
17.2 Human excreta as an energy feedstock	330
17.2.1 Overview of thermochemical properties of human excreta	330
17.2.2 Rationale for thermochemical conversion of human excreta	332
17.3 Thermochemical conversion	332
17.3.1 Principles of thermochemical conversion processes	332
17.3.2 Types of thermochemical conversion processes	333
17.3.3 Comparison of thermochemical conversion methods	334
17.4 Current state of applications	336
17.4.1 Case study: a portrait of a Brazilian Amazon community	336
17.4.2 Other thermochemical sanitation technologies	338
17.5 Future research directions and conclusions	339
17.5.1 Future directions	339
17.5.2 Conclusions	340
References	340

17.1 Introduction

There has been an increasing interest in the thermochemical conversion of human excreta in pursuing sustainable and innovative solutions to address the global challenges of waste management and energy production. This multifaceted approach combines chemistry, engineering, and environmental science principles to transform human waste into valuable energy and biochar resources (Bittencourt & Martins, 2022).

Thermochemical treatment involves subjecting human excreta to relatively elevated temperatures in pyrolysis, gasification, and combustion processes. These transformative methods mitigate the environmental impacts of conventional waste disposal while yielding substantial benefits, such as biochar through pyrolysis, thermal energy from combustion, and syngas from gasification. These outputs present a compelling case for integrating such technologies into waste management systems, offering a sustainable means of resource recovery and energy generation. However, as human excreta is considered a waste not readily assessed with a monetary value, the balance between the benefits and drawbacks of conversion technologies still needs clear definitions (Schroeder, 2011; Senecal & Vinnerås, 2017). In fact, even with the incentives for developing decentralized solutions, such as the reinventing the toilet challenge, the knowledge pool converges on sophisticated and expensive solutions (Kennedy, 2013; Saroj, 2023). In contrast, opportunities for developing new low-cost technologies may not find space in the literature or technological investment agencies.

This chapter shows the recent advances, applications, and future directions of thermochemical treatment, aiming to shed light on the potential transformative impact that innovation brings to conversion technologies when the goal is to produce energy and biochar from human excreta.

17.2 Human excreta as an energy feedstock

Sanitation in low-income countries remains a critical and multifaceted challenge, reflecting a complex interplay of economic, social, and environmental factors (Zhou et al., 2018). Access to basic sanitation facilities, such as toilets and proper waste disposal systems, is disproportionately limited in these regions, impacting the health and well-being of millions (Hubbard et al., 2020). The lack of adequate sanitation infrastructure contributes to the spread of diseases, posing a severe threat to public health. In addition, open defecation is still prevalent in many low-income communities, exacerbating contamination of water sources and soil (Kwiringira et al., 2014). The consequences extend beyond health, affecting education, dignity, and overall quality of life. Addressing sanitation in low-income countries requires comprehensive strategies considering socioeconomic disparities, cultural practices, and local contexts (Owusu, 2010). International efforts, collaborations, and innovative solutions are essential to break the cycle of inadequate sanitation and pave the way for improved public health outcomes and sustainable development (Programme & Ryder, 2017).

Traditionally regarded as waste, human excreta contains a significant amount of untapped energy in the form of organic matter (Bittencourt & Martins, 2022). Transforming what is typically considered a societal taboo into a valuable resource for energy production represents a paradigm shift in the renewable energy landscape (Villarín & Merel, 2020). In such a direction, recovering energetic resources from human excreta is promising, but it must be recognizable that the general characteristics of human excreta are strongly dependent on the following (Bittencourt & Martins, 2022):

- where it is being generated, for example, high- or low-income countries
- which sanitation facility is used, for example, sewered toilet, urine-diverting toilet, or pit latrine
- how it is collected, for example, septic tank, pit, or ditches.
- how it is transported, for example, through the sewer, manually, or using trucks.

This section considers these aspects to explore human excreta as an energy feedstock, highlighting the potential benefits and challenges associated with its characteristics.

17.2.1 Overview of thermochemical properties of human excreta

The pit latrines are the most common sanitation option for communities in resource-poor environments of low-income countries, and, therefore, should be responsible for assuring the safety of users (Murungi & van Dijk, 2014). However, paradoxically, the precarious settings in which they exist place the users in unsafe conditions (Gwenzi et al., 2023). For this reason, there are a significant number of people who prefer to practice open defecation whether use improper facilities (Kwiringira et al., 2014); in such cases, feces and urine follow a sanitary pathway somehow completely out of control, breaching the surrounding environment and increasing the hazards from unsafe disposal of these wastes (Capone et al., 2021).

In common pit latrines, the slurry mixture composed of human excrement mixed with other wastes is called fecal sludge (Strande & Brdjanovic, 2014). The "sanitary by-wastes" in fecal sludge may include toilet paper, paper tissues, diapers, and menstrual hygiene materials, among others (Foxon & Buckley, 2008). The presence of other solids in its composition tends to make the properties of fecal sludge different from fresh human feces. For this reason, although very similar, fecal sludge and fresh human feces are not the same waste (Bittencourt & Martins, 2022). Indeed, if the negative aspects of human excreta are evaluated and controlled, we are left with a valuable and extensive resource to recover energy, organic, and inorganic materials (Berendes et al., 2018).

A way to recover and explore the resources from fecal matter can be achieved through thermochemical processes (Torero et al., 2020). These processes use the energy potential of excreta expressed in the energy density, combining heat transfer and chemical reactions to produce energy. Consequently, thermochemical processes eliminate pathogens and mitigate the effects caused by inappropriate disposal of the excreta (Bittencourt et al., 2023). Thermochemical processes include pyrolysis, gasification, and combustion, which will be discussed in more detail in the following sections of this chapter. However, before delineating the specificities of each process, it is essential to explore the properties and characteristics of human excreta. Given the necessity of addressing the outcomes of precarious sanitation in low-income countries, we will characterize two types of waste: fresh human feces and fecal sludge.

Fig. 20.1 represents relevant thermochemical properties of human excreta from several works (Bittencourt & Martins, 2022). In Fig. 17.1A, the moisture content of fecal sludge and fresh feces is grouped to obtain an overall moisture content distribution. Note that the moisture ranges from 70 to 95 wt.%, indicating that addressing the challenges associated with high moisture content is mandatory as a first step in using human excreta as a potential feedstock for energetic processes. High moisture content not only diminishes the calorific value of the raw materials but also demands additional energy inputs for drying, which can substantially increase the operational costs and environmental footprint of thermochemical processes (Somorin et al., 2020).

If we analyze separately the proximate composition of fecal sludge and human feces on a dry basis, we begin to see the real potential of fecal matter as an energetic feedstock. First, compare Fig. 17.1B, expressing the thermochemical composition of fecal sludge, and, Fig. 17.1C, expressing the thermochemical composition of fresh human feces. As aforementioned, note that the properties from fecal sludge are more spread, indicating the effects of other sanitary by-wastes, contributing most to increasing the inorganic content of fecal sludge, reaching up to 70 wt.%. Conversely, the inorganic content of fresh feces is restrained between 10 and 20 wt.%. Following the same trend, the amount of carbon available for thermochemical conversions, i.e., the fixed carbon, is higher in fresh feces than in fecal sludge, 30 wt.% against 10 wt.%, on average. As a consequence, the higher heating value of fecal sludge is lower than fresh feces (Fig. 17.1D). Such

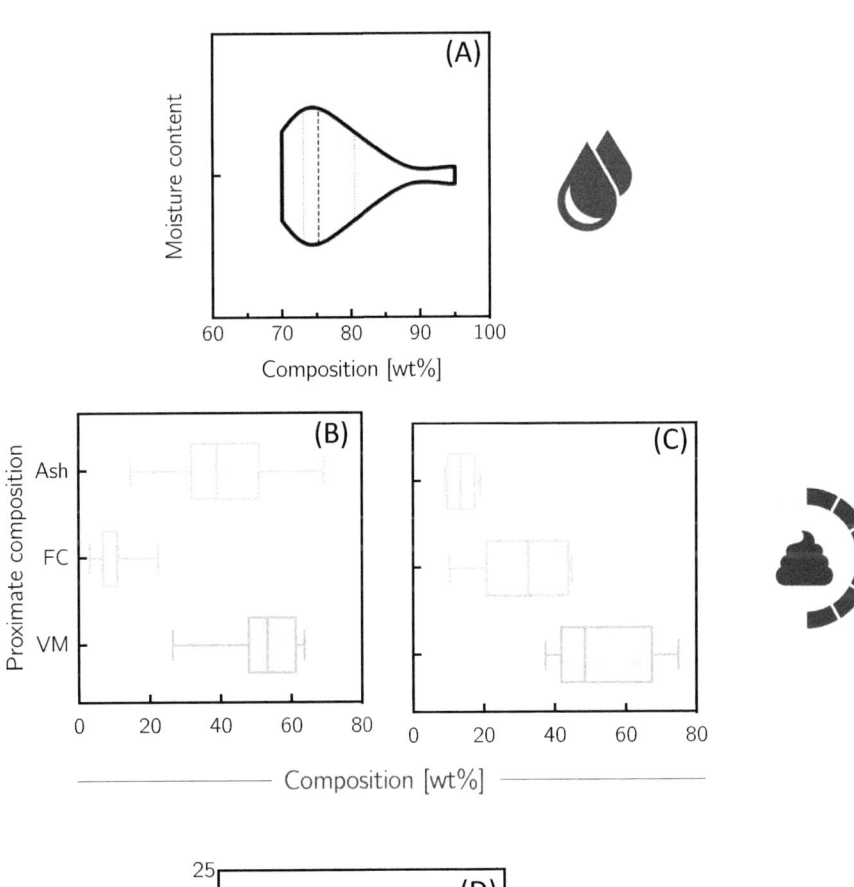

FIGURE 17.1 Overall thermochemical properties. Relevant thermochemical properties from human excreta: (A) moisture content of raw materials, (B) proximate composition in the dry basis of fecal sludge, (C) fresh human feces, and (D) higher heating value of fecal sludge and human feces. The consolidated data used to build this figure can be consulted in Bittencourt & Martins (2022). *From Bittencourt, F. L. F., & Martins, M. F. (2022). Thermochemically-driven treatment units for fecal matter sanitation: A review addressed to the underdeveloped world. Journal of Environmental Chemical Engineering, 10(6), Article 108732. https://doi.org/10.1016/j.jece.2022.108732*

differences underpin that fresh feces are more suitable to be applied in thermochemical systems, and therefore, thermochemical technologies should preserve the desired properties of fecal matter for more efficient conversions.

17.2.2 Rationale for thermochemical conversion of human excreta

Given the properties of feces and fecal sludge explored in the previous section, it is clear that human excreta can be a viable source for energy recovery through thermochemical processes due to several characteristics intrinsic to their composition (Rose et al., 2015). These characteristics make human excreta suitable for conversion into energy and environmental applications.

In summary, the central aspects that support the use of human excreta in thermochemical processes are as follows:

1. Organic matter content: Feces and fecal sludge are rich in organic matter. When the moisture content is controlled, the organic matrix, formed by fixed carbon and volatile materials, provides a feedstock for thermochemical processes, such as combustion, pyrolysis, and gasification (Onabanjo, Kolios, et al., 2016; Onabanjo, Patchigolla, et al., 2016; Koulouri et al., 2023).
2. Nutrient content: Human excreta contains essential nutrients such as nitrogen, phosphorus, and potassium. Thermochemical processes, particularly those that produce biochar, such as pyrolysis, retain these nutrients in a stable form. The resulting biochar can be used as a soil amendment, improving soil fertility (Nuagah et al., 2020).
3. Energy density: Even though the energy content of human excreta is relatively low compared to other biomass sources, it is still substantial. Thermochemical processes can use this energy potential to produce syngas (Bittencourt et al., 2019), biochar (Bittencourt et al., 2022), or heat (Monhol & Martins, 2015), making it suitable for energy recovery and thermal applications.
4. Sanitization: Combustion, pyrolysis, and gasification involve high temperatures that destroy pathogens and harmful microorganisms in human excreta (Mrimi et al., 2020). This is crucial for ensuring public health and safety.
5. Environmental applications: Thermochemical processes and products, especially those producing biochar, offer the opportunity to sequester carbon in stable compounds, either in a concomitant process or by producing materials that can be used as a soil amendment while contributing to carbon capture and storage (Lu et al., 2018).

In the next section, you will find more details about the principles and types of thermochemical conversion processes, as well as the design and operation principles of thermochemical units.

17.3 Thermochemical conversion

17.3.1 Principles of thermochemical conversion processes

The thermochemical conversion processes have been mainly developed to deal with all sorts of biomasses (Tursi, 2019). However, as the world has become alarmingly in themes such as sanitation, waste disposal, and energy transition, and also, some issues were pointed out in using some classes of biomasses for energy (Schueler et al., 2013), all wood-based systems and processes have been readjusted to be employed with all kinds of waste, including human excreta (Bittencourt et al., 2019; Soares et al., 2019, 2020; Duque et al., 2021, 2022, 2023). Still, for the purpose of the present chapter, all apprenticeships from wood-based biomasses persist as unavoidable for those entering this interesting research theme.

Therefore, some specificities related to the principles of thermochemical conversion processes lie in the type of the material that intends to be converted and all the properties of the matter and system that challenge the thermochemical processes, such as morphological (shape, size, structure, and visual aspect) (Ibitoye et al., 2023), thermochemical (moisture, ash, volatile and fixed carbon content, elemental, and heating value) (Martins et al., 2010; Duque et al., 2020; Martins et al., 2021), physical (density) (Klippel & Martins, 2022), and logistics (handling and transportation) (Rentizelas et al., 2009).

Focusing on the properties of matter is essential because, from the knowledge, we can identify it in a given context, as well as know how it behaves at certain times. The inherent properties of the matter determine both the choice of the conversion and any subsequent processing difficulties that may arise (Martins et al., 2022). When selecting a conversion process, the properties of the raw material are one of the more important factors, whether the objective is to convert residues into thermal energy (Bittencourt et al., 2019; Mudasar & Kim, 2017) or a valuable product (Duque et al., 2023).

Another a priori aspect to choosing the thermochemical route is the so-called material potential for feedstock (Somorin, 2020; Maurya, 2012). As bioenergy remains at the top of the political agenda and with general opinion recognizing the important role this form of energy can play in achieving the goals of the global agenda, aspects such as the types of residues currently available and evaluation of the resource in the different world regions have aroused growing interest in scientific circles. For a good example of the human excreta potential as feedstock, see Berendes et al. (2018). The authors reported that in 2014, the total mass of feces was 3.9×10^{12} kg per year, increasing by $>52 \times 10^9$ kg per year since 2003 and anticipated to reach at least 4.6×10^{12} kg in 2030.

Once the matter selection and characterization aspects are overcome, the next important thermochemical conversion principles rely on the compromise of raw material available energy and the energy requirement for converting. Such an assessment is carried out employing thermodynamics principles. The point of departure starts when a given amount of thermal energy is furnished for processing a raw material at a specific temperature or range of temperatures. In doing so, a thermodynamic trigger occurs, a coupling of heat transfer, fluid mechanics, phase change, and chemical reactions occurring in a single control volume (McAllister et al., 2011). Therefore, some concepts are necessary for a proper understanding of the conversion processes, such as enthalpy change, chemical reactions, and calorimetry are the basis for processes of combustion, pyrolysis, gasification, and other variations of these processes. These processes create some amount of char, bio-oil, tar, syngas, and flue gas, using different operating conditions, such as temperature, pressure, heating rate, residence time, and atmosphere, to improve the production and quality of one or more products (Duque et al., 2023; Demirbaş, 1998).

17.3.2 Types of thermochemical conversion processes

17.3.2.1 Combustion

Roughly, we can organize the combustion as a process into four steps, assuming that oxygen is used to promote the combustion front propagation, and aware that an overlapping of the steps occurs to some extent.

Most biomass contains moisture, that is, fecal sludge, to an average of 75 wt.%, so the first step in combustion is evaporating free water (Bittencourt et al., 2022). To ignite a solid such as biosolids, fecal sludge is difficult because water evaporation is an endothermic (energy-requiring) process—energy must be added to start combustion before any energy can be released from it.

Secondly, devolatilization or pyrolysis (no oxygen needed yet) occurs. As heat breaks the chemical bonds within the material, smaller molecules vaporize and escape from the solid particle (or, more widely, from the particle bed). Note that, although pyrolysis is mostly reported as an endothermic process, some materials, such as microalgae (Soares et al., 2020) and human feces (Bittencourt & Martins, 2020), have, on average, an exothermic thermal behavior during pyrolysis.

During the third step, oxidation reactions take place, depending on the reactor type, gas phase oxidation occurs, releasing strong thermal energy and light, and a flame is formed (Torero et al., 2020). If it occurs, that means that the volatile released during pyrolysis was burned (fully or partially oxidized). If the volatile is partially oxidized, it can result in heavy smoke and tar or gas-phase polymerization to soot. After pyrolysis (or devolatilization, in the context of the combustion process), the remaining product left in the solid matrix is char or fixed carbon.

Finally, When all the volatile parts of the material have been oxidized and/or removed, a smoldering front takes place: a slow and flameless form of combustion, sustained by the heat evolved when oxygen directly attacks the surface of a condensed phase fuel (Rashwan et al., 2023)—solid-phase oxidation, or, in terms of chemical reaction nomenclature, heterogeneous combustion. Eventually, all the carbon is oxidized according to $C + \left[\frac{\alpha}{2} + (1 - \alpha)\right]O_2 \rightarrow \alpha CO + (1 - \alpha)CO_2$, and only the noncombustible mineral material, the ash, is left in a solid matrix (Sennoune et al., 2012).

The extent to which each combustion process occurs depends on the amount of energy available, i.e., the temperature; but also on the amount of oxygen, and the residence time and product fractions in the oxidizing atmosphere (Zanoni et al., 2019).

17.3.2.2 Pyrolysis

Biomass pyrolysis has been the subject of intense scientific research for many decades. It is a process that can be defined as the decomposition of matter that occurs due to temperature effects above 250°C, without the partaking of exo-oxygen, such as *Material + Heat → Char + oil + gas*.

Depending on the matter being converted, the pyrolysis, thinking in the whole reaction process, can be predominantly endothermic (Ciuta et al., 2018) or exothermic (Soares et al., 2020; Bittencourt & Martins, 2020; Di Blasi et al., 2017). Also, there are several types of pyrolysis that can be classified according to the heating rate, presence of catalysts, or type of reactor. For example, fast pyrolysis (400°C–700°C), slow pyrolysis (300°C–500°C), catalytic pyrolysis (300°C–650°C), and microwave pyrolysis (450°C–800°C) (Das et al., 2021; Chun et al., 2021). There are three main steps during the pyrolysis process, they are dehydration, devolatilization, and decomposition. Dehydration is carried out at a temperature below 200°C, while devolatilization, which is one of the most important pyrolysis steps, occurs in the temperature range between 200°C–550°C and the final step, the decomposition of solid biomass is carried out above 550°C.

Several factors influence the performance of the process, such as the operating conditions of the pyrolysis, the composition of the raw material, and the types of reactors used. Some other factors that also condition the process are the catalytic charge rate, particle size, carrier gas, and gas flow rate. By properly controlling these factors during the process, product quality and yield can be improved (Duque et al., 2023; Varma et al., 2018).

Some specifics about each type of pyrolysis:

- Coal for heating and other purposes is traditionally produced by slow pyrolysis. The process is characterized by slow heating rates and long residence times. The heat required to initiate and drive the reaction is generally supplied internally by the combustion of a portion of the raw material. The objective of slow pyrolysis is a solid coal product with high carbon content and high energy density. The coproducts are a low-molecular-weight, aqueous acidic liquid called pyroligneous acid or "wood tar" and a low-energy combustible gas (Tan et al., 2021).
- Unlike slow pyrolysis, fast pyrolysis uses very high heating rates (~1000°C/s), short residence times, and the rapid quenching of vapors to maximize the production of the liquid product, bio-oil (Tan et al., 2021).
- Catalytic pyrolysis can be defined as a process that combines pyrolysis and vapor-phase catalytic upgrading (Wrasman et al., 2023), or simply in biomass catalytic pyrolysis the products of the thermal conversion reactions are put into contact with a catalyst to change the composition of the pyrolysis vapors, upgrading the bio-oil fraction (Fermoso et al., 2017).

17.3.2.3 Gasification

The primary product of gasification is the noncondensable gas fraction. The process is performed in the presence of a gasifying agent (air, pure oxygen, steam, or mixtures of these components) at elevated temperatures between 500 and 1400°C and atmospheric or elevated pressures up to 33 bar. The oxygen used is measured in equivalence ratio or the fraction of the amount of oxygen needed for stoichiometric combustion, typically around 0.25 (Ahmad et al., 2016). The product gas, called syngas, consists mainly of carbon monoxide (CO) and hydrogen (H_2), with smaller amounts of carbon dioxide, methane, and other low molecular weight hydrocarbons. Gasification is very similar to combustion. However, due to the limited oxygen, it cannot complete the gas-phase and solid-phase oxidation steps, which would yield CO_2 and H_2O.

The chemical reaction mechanism is thermodynamically controlled in an ideal gasification situation. The gas composition and carbon conversion can be predicted based on temperature and pressure; the only coproduct is char. However, there is insufficient time for the reaction to reach equilibrium in real operation, resulting in sticky, viscous tars that can clog the reactor plumbing and cause significant problems in downstream gas applications. The oxygen needed for the gasification can come from air (air-blown gasification) or a mixture of steam and oxygen (steam/oxygen-blown gasification) (Norman et al., 1995).

The gasification process applies to biomasses with a moisture content lower than 35%. For biomass feedstocks that possess higher amounts of moisture, in the range of 25%–60%, using these feedstocks directly in the gasifier will result in significant energy losses in the overall process. Therefore, it is recommended that the biomasses be preheated or dried to moisture contents between 10% and 20% before they are introduced into the gasifier (Martins et al., 2022; Ahmad et al., 2016).

17.3.3 Comparison of thermochemical conversion methods

As a matter of example, a comparison of technologies driven by the aforementioned thermochemical processes is shown in Table 17.1. Noteworthy, some technologies can combine more than one thermochemical process. For example, the energy from biomass combustion, as a secondary process, can drive the drying, pyrolysis, and oxidation of human excreta. Concerning scalability, it is clear that the technologies promise to work in real-life scenarios, although only a few technologies have worked with a realistic amount of excreta without pretreatment. Indeed, the conditions to treat human excreta in thermochemical units are created by drying, pelletizing feces, and mixing it with other fuels. For instance, the moisture content needs to be under 30 wt.% to gasify human excreta. Regardless of the technology, any of these processes ensure the sanitization of human excreta, given the high peak temperatures achieved during operation.

TABLE 17.1 Comparison of thermochemical conversion methods.

Examples of units/reactors	Main process	Conversion capacity	Peak temperature	Pretreatment	Moisture	Main product	Sanitization	Potential application	References
Smoldering cells	combustion	~100 g	400°C–800°C	Yes	30–75 wt.%	Sand + ashes	Yes	Sand recycle	Yermán et al. (2016)
Downdraft combustor	combustion	~100 g	400°C–600°C	Yes	0–40 wt.%	Ashes	Yes	Soil amendments	Onabanjo, Kolios, et al. (2016)
Microcombustor	combustion	~100 g	400°C–700°C	Yes	0–40 wt.%	Ashes	Yes	Soil amendments	Jurado et al. (2018)
FeD-Latrine	pyrolysis	~1000 g	900°C–1200°C	No	70–80 wt.%	Biochar	Yes	Energy recycle	Bittencourt et al. (2022)
Sol-char toilet	pyrolysis	~2000 g	300°C–750°C	No	70–80 wt.%	Biochar	Yes	Soil amendments	Fisher et al. (2021)
Downdraft gasifier	gasification	~300 g	800°C–900°C	Yes	0–30 wt.%	Syngas	Yes	Electricity	Bittencourt et al. (2019)

Concerning the products from thermochemical units, human feces are transformed into (1) solid products, with the potential to recover valuable chemical elements, such as P, K, and Ca, ranging between 75–95 wt.% (Bittencourt et al., 2022); (2) gaseous emissions mainly composed of CO and CO2, which can be considered controlled emissions when compared to methane typically emitted from pit latrines and feces disposed of in the open environment (Reid et al., 2014).

17.4 Current state of applications

17.4.1 Case study: a portrait of a Brazilian Amazon community

The concept of humanitarian engineering refers to the application of engineering principles and practices to address humanitarian and social problems in developing countries, with a focus on sustainability and improving quality of life (Humanitarian Engineering - Emerging Technologies and Humanitarian Efforts | IEEE Conference Publication | IEEE Xplore, 2023). It involves designing and developing technological solutions that address pressing humanitarian needs such as the design and implementation of sustainable sanitation systems in rural communities.

According to the Brazilian Institute of Geography and Statistics, almost 30 million people in Brazil lack access to improved sanitation facilities, representing around 15% of Brazil's inhabitants. From a local perspective, this share of people can reach more than 50%, for example, in the state of Rio Grande do Norte, northeast of Brazil. In the state of Pará, in the northern region of Brazil, approximately 2.7 million people defecate in rivers, ditches, or precatory facilities without sewer connections (Pesquisa nacional por amostra de domicílios, 2020).

Over the recent years, several actions toward nonsewered sanitation have been made and published scientifically, showing that solving the sanitation crisis in low-income countries is not just as simple as using solutions in high-income countries (Michalak *et al.*, 2023). Considering the humanitarian aspects while developing nonsewered sanitation technologies is unavoidable and implies knowing the peculiarities of each community and its members (Nelson et al., 2022).

From the perspective of low-income settings, thermochemical-driven treatment units for treating human excreta is a promising approach to replace precarious pit latrines and mitigate the hazards from the use of unimproved sanitation facilities, as it has been demonstrated in recently developed concepts and devices, such as the FeD-Latrine (Bittencourt et al., 2022).

The field implementation of thermochemical units must account for several concerns, as there are peculiarities in remote communities, which most need sanitation improvements. For example, in Amazonian communities near igapós (Lobo, Wittmann, & Piedade, 2019), the effect of river basins on the riverine communities is experienced seasonally (Gomes et al., 2015). Thermochemical units in these places must be elevated and sealed (Katukiza et al., 2012). In isolated communities, using natural resources for fuel and energy is necessary to overcome the lack of electricity. In these cases, drying beds and natural conversion systems are mandatory to make the processes feasible (Gold et al., 2014).

The profile of sanitation access in remote communities has been an object of interest in recent research (Gomes et al., 2015; Borges Pedro et al., 2011), which shows that it is common to find riverine communities that lack access to proper sanitation facilities; on the contrary, open defecation and precarious pit latrines are almost consolidated as regular practices among community members (de Andrade et al., 2021), while the attempt to install improved sanitation facilities faced challenges for acceptance (Gomes et al., 2015). For instance, riverine community members may complain about the smell inside the dry toilets' cabins and dislike the idea of containing feces inside boxes.

Recognizing the number of people who can benefit from the development of new thermochemical technologies that treat feces, this section reports a case study conducted in a community living in an igapó region, in Tucuruí, Pará, North Brazil. This case study identifies hygiene conditions, the presence and structure of toilets, and collects relevant information regarding the community dynamics that can guide future improvements of scaled-up thermochemical technologies.

17.4.1.1 The community of Sapolândia

Tucuruí - Pará, North of Brazil, is a city supplied by the Tocantins-Araguaia Basin, 460 km apart from the capital Belém. Tucuruí is currently inhabited by 116,065 residents along approximately 2,000 km^2 of territory. A community called Sapolândia (translated to English as "Frogland") is home to hundreds of families who live in stilt houses above soaked areas that are affected by the flooding of rivers and can remain in such condition for up to months.

The toilet facilities in Sapolândia are, in general, clean and contain a toilet seat, a tank with river water, and a bucket (Fig. 17.2A,B); nonetheless, the toilet seat has no sewer connection; hence, the excreta is dumped directly into the water streams flowing underneath the stilt houses (Fig. 17.2C). With this arrangement, the toilet's dynamic is very straightforward: after defecating and urinating, a bucket filled with water is manually thrown to flush out the excreta.

FIGURE 17.2 **Sanitation profile in Sapolândia.** The typical configuration of a sanitation facility in Sapolândia: (A) a toilet seat without a sewer connection, (B) a water reservoir for bucket flushing, and (C) the soaked soil below the stilt house that receives the flushed excreta.

As a consequence of the nonsewered arrangement, after being flushed out, the excreta can (Zhou et al., 2018) accumulate and degrade under the stilt houses, generating bed smell and insect proliferation, or (Hubbard et al., 2020) in flooding periods, spread throughout the community, affecting vegetable gardens and the soil where children and residents move freely, often with barefoot. In both cases, it represents a significant source of pathogen proliferation and consequential transmission of many infectious diseases.

17.4.1.2 The thermochemical route

After understanding the sanitation dynamics of Sapolândia, the observations lead to searching for evidence to support thermochemical units' potential application and feasibility for sanitation purposes.

Due to the lack of financial resources and the high price of cooking gas, the community gets its cooking energy from locally available solid fuels. Charcoal from nearby local plants is the primary energy resource. They are stocked in bags inside the houses and are smoldered to provide energy to cook in charcoal stoves, which are present in most houses of Sapolândia. The ashes are reused as fertilizer in plantations of vegetables and fruits.

The familiarity of the community members handling charcoal and the solid products underlines the great potential of expanding the existing thermochemical route to encompass sanitation improvements using thermochemical units, such as the FeD-Latrine (see Fig. 17.3). A novel parallel route can explore the benefit of the high thermal energy produced during the oxidation of charcoal to drive transformation processes such as drying, pyrolysis, and complete oxidation of the wastes to eliminate fecal pathogens.

As demonstrated in the previous sections, beyond the clear benefit of inactivating human feces, the final products from thermochemical units can be recovered as (Zhou et al., 2018) dried feces, (Hubbard et al., 2020) biochar with good energy value, or (Kwiringira et al., 2014) metal-rich ashes with fertilizer compounds, such as N, P, Ca, and others. Thus, the CO_2 produced during the process can be captured by low-cost materials, such as calcium oxide, to form calcium carbonate, which can be applied in the community to correct soil acidity and improve the conditions for crops' development (Yadav et al., 2021).

Based on this evidence, Fig. 17.4 shows a conceptual, yet manageable, scenario in which thermochemical units operate in a safe and circular process. Note that four processes can be outlined from the thermochemical route for sanitation:

- A thermochemical device is responsible for inactivating the pathogens from human feces by assuring the thermal energy flux from charcoal combustion to enter the solid waste domain.
- The CO_2 produced is captured by calcium oxide forming calcium carbonate, which is used for soil correction, and improves the quality of crops.
- The remaining inorganic fraction of the converted human feces is rich in soil compounds, and it is used as fertilizer.
- The biochar produced is used as a soil agent or returned to the FeD-Latrine as a complementary energy source.

Given the peculiarities observed in Sapolândia, it is well evident that thermochemical units must meet a few requirements to work properly in such conditions: (Zhou et al., 2018) they should be sealed from water, (Hubbard et al., 2020) they must allow easy maintenance by the users and be accessed from inside the houses, and (Kwiringira et al., 2014) they must separate urine from feces. Hence, when the use of sewer-based sanitation is impracticable, the development

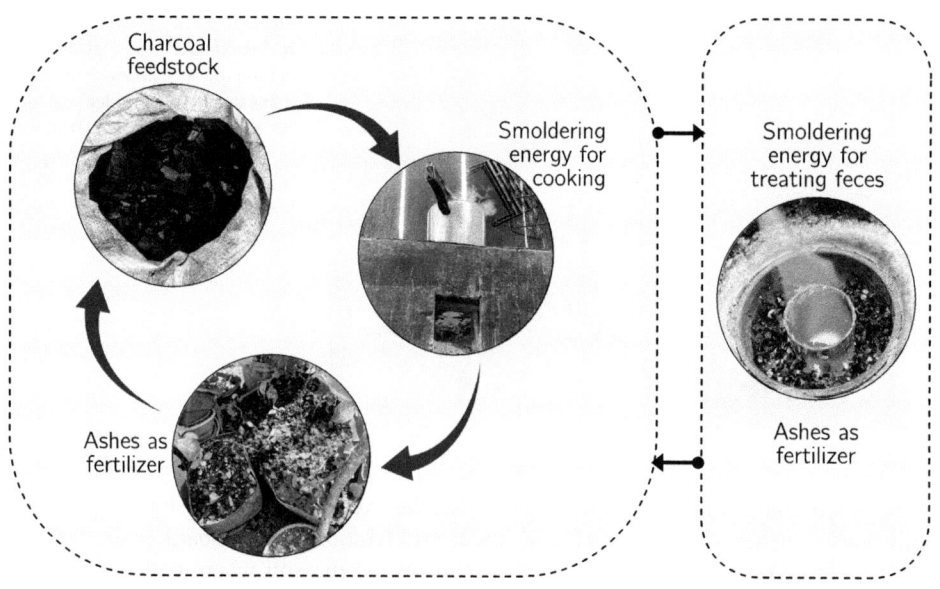

FIGURE 17.3 Evidences of thermochemical processes. Solid fuels and smoldering combustion are present in the daily activities of community members: charcoal is stocked in the houses as a feedstock for generating thermal energy to cook, and charcoal ashes are reused to fertilize crops. *A picture of a charcoal bag and it is used for cooking purposes.*

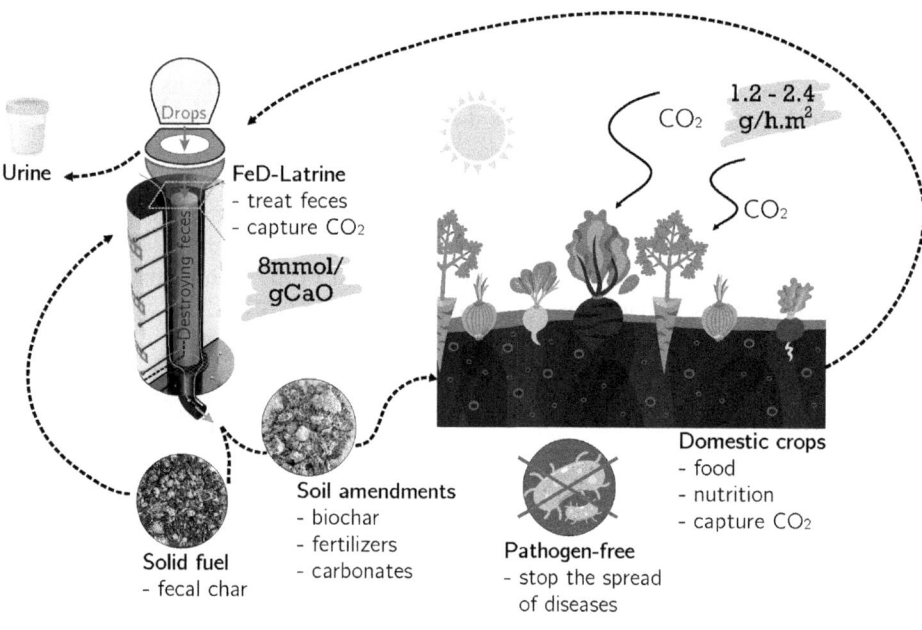

FIGURE 17.4 Safe and circular process of the FeD-Latrine. A conceptual scenario demonstrating the FeD-Latrine inserted in a safe and circular process for pathogen inactivation, carbon capture and utilization, energy recovery, and nutrient recovery. *A schematic showing the FeD-Latrine operating in a circular process.*

of thermochemical solutions must continue and progress toward propositions for the safe reuse of end products developed into the perspective of a circular economy and sustainability. Another determinative aspect is the social acceptance of the technology by community members, which must be assessed to ensure the engagement of all residents (Borges Pedro et al., 2020).

17.4.2 Other thermochemical sanitation technologies

Other thermochemical treatment units based on pyrolysis and combustion have been developed to minimize the harm caused by the unsecured sanitation facilities in rural areas. However, to date, it has not identified any commercial

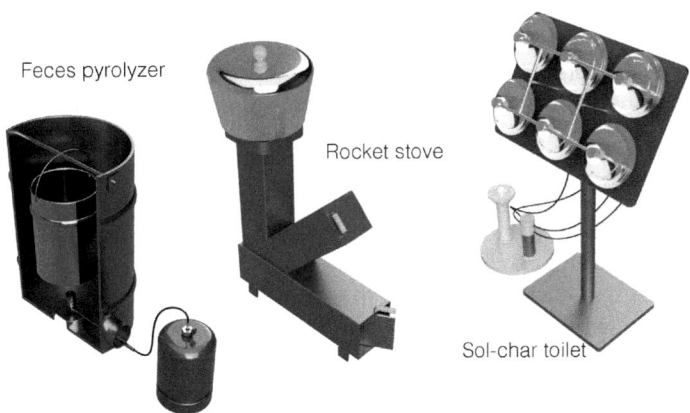

FIGURE 17.5 Thermochemical sanitation technologies. A few designs of thermochemical units for sanitation. *A picture of some designs of thermochemical units.*

technology with a readiness level for the treatment of human feces. Fig. 17.5 illustrates the designs of some pilot-scale thermochemical units, such as the Feces Pyrolyzer (Waste & Pyrolyzer, 2023), the rocket stove (Mrimi et al., 2020), and the Sol-char toilet (Fisher et al., 2021).

The Feces Pyrolyzer and the rocket stove are examples of offsite designs that share easy construction principles fed with available materials. In these units, there is the necessity to transport the excreta, which poses risks if it is manually handled. The Feces Pyrolyzer thermally treats fecal matter in a reproducible domestic device: the heat source is a weed burner placed beneath a metallic bucket containing human excreta. After hours of treatment, the human excreta is transformed into chars suitable for other applications. A more robust device, the rocket stove combines a firing chamber at the bottom, a top-loading feeding system for agricultural biomass, and a metallic pot of accumulated excreta (must be previously dried in solar beds). After a few hours of thermochemical treatment, the excreta is transformed into a pathogen-free product suitable for agricultural applications. The rocket stove was applied in rural Tanzania, Africa, using labor and biomass resources from the local community.

In a more refined and expensive technology, the Sol-char toilet treats fresh feces onsite, using concentrated solar power to pyrolyze the feces into char, which can be used as fuel and soil amendment. The Sol-char toilet demonstrates that raw excreta can be transformed without pretreatment procedures. However, this technology requires specialized labor for construction and a substantial budget.

17.5 Future research directions and conclusions

17.5.1 Future directions

The underdeveloped world's lack of improved sanitation facilities impacts millions of people's lives. Either by spreading diseases, contaminating water and ecosystems, or perpetuating the marginalization of the most vulnerable communities. Treating human feces properly and ensuring adequate sanitation facilities for people represents interrupting the harmful consequences of the uncontrolled fate of human feces in such conditions.

So far, using thermochemical processes to sanitize human feces has been majorly studied for dried feces or mixtures of feces with other dry and bulk materials. The problem is that the transition from fresh to dried feces represents an energy-intensive process and a potential source of contamination. Positively, a few technologies are being built and tested under real-life scenarios. It is crucial to comprehend that fecal matter's thermochemical properties depend on the origin and pathway human excreta will follow, from its generation as a physiological process to the form it will return to the environment, whether safely treated or not. For example, fresh feces and fecal sludge from pit latrines are not identical residues. Such distinction must influence the choice of the treatment route in thermochemical units: choosing centralized units over onsite ones means treating a residue with more water, less fixed carbon, less energy density, and more inorganic content; however, the challenge with onsite units is to dominate a low-cost and applicable thermochemical process capable of destroying fresh feces with almost 80 wt.% of water content.

Special attention must also be placed on the destination and applicability of thermochemical units' end-products. The sustainable use of ashes and biochar combined with controlled emissions will secure the transformation of a pathogenic material into harmless and valuable products that can be completely reused.

17.5.2 Conclusions

The thermochemical treatment of human excreta to produce energy and biochar represents a promising and innovative approach to address multiple challenges in sanitation, energy production, and environmental sustainability. This emerging technology not only tackles the critical issue of waste management but also underpins the potential of human excreta as a valuable resource.

The conversion of human excreta through thermochemical processes, such as pyrolysis, gasification, and combustion, has demonstrated the capacity to generate energy in the form of heat, char, and syngas. This energy can be utilized for various applications, including electricity generation, cooking, or heating. By extracting energy from a traditionally discarded waste stream, this approach contributes to a more circular and sustainable system. Likewise, biochar also offers environmental benefits such as soil amendment, enhancing soil fertility, and water retention. Its application can contribute to carbon sequestration and mitigate the impact of greenhouse gas emissions, thereby promoting environmental conservation.

References

Ahmad, A. A., Zawawi, N. A., Kasim, F. H., Inayat, A., & Khasri, A. (2016). Assessing the gasification performance of biomass: A review on biomass gasification process conditions, optimization and economic evaluation. *Renewable and Sustainable Energy Reviews, 53*, 1333–1347. Available from https://doi.org/10.1016/j.rser.2015.09.030, https://www.journals.elsevier.com/renewable-and-sustainable-energy-reviews.

Berendes, D. M., Yang, P. J., Lai, A., Hu, D., & Brown, J. (2018). Estimation of global recoverable human and animal faecal biomass. Nature Publishing Group, United States. *Nature Sustainability, 1*(11), 679–685. Available from https://doi.org/10.1038/s41893-018-0167-0, http://www.nature.com/natsustain/.

Bittencourt, F. L. F., Lourenço, A. B., Dalvi, E. A., & Martins, M. F. (2019). Thermodynamic assessment of human feces gasification: An experimental-based approach. *SN Applied Sciences, 1*(9). Available from https://doi.org/10.1007/s42452-019-1104-1, springer.com/snas.

Bittencourt, F. L. F., & Martins, M. F. (2020). On the strong exothermicity of fecal matter pyrolysis under an inert atmosphere. *Brazilian Journal of Chemical Engineering, 37*(4), 661–666. Available from https://doi.org/10.1007/s43153-020-00052-8, https://link.springer.com/journal/43153.

Bittencourt, F. L. F., Martins, M. F., Orlando, M. T. D., & Galvão, E. S. (2022). The proof-of-concept of a novel feces destroyer latrine. *Journal of Environmental Chemical Engineering, 10*(1). Available from https://doi.org/10.1016/j.jece.2021.106827, http://www.journals.elsevier.com/journal-of-environmental-chemical-engineering/.

Bittencourt, F. L. F., Martins, M. F., & Orlando, M. T. D. (2023). Integrating in-bed gas looping and CO2 capture in the FeD-Latrine. *Science of the Total Environment, 859*. Available from https://doi.org/10.1016/j.scitotenv.2022.160133, http://www.elsevier.com/locate/scitotenv.

Bittencourt, F. L. F., & Martins, M. F. (2022). Thermochemically-driven treatment units for fecal matter sanitation: A review addressed to the underdeveloped world. *Journal of Environmental Chemical Engineering, 10*(6). Available from https://doi.org/10.1016/j.jece.2022.108732, http://www.journals.elsevier.com/journal-of-environmental-chemical-engineering/.

Borges Pedro, J. P., Gomes, M. C. R. L., & Silva do Nascimento, A. C. (2011). Review of Wastewater Treatment Technologies for Application in Communities in the Amazonian Varzea. *Scientific Magazine UAKARI, 7*(1), 59–69. Available from https://doi.org/10.31420/uakari.v7i1.85.

Borges Pedro, J. P., Oliveira, C. A. da S., Borges de Lima, S. C. R., & von Sperling, M. (2020). A review of sanitation technologies for flood-prone areas. *Journal of Water, Sanitation and Hygiene for Development, 10*(3), 397–412.

Capone, D., Chigwechokha, P., de los Reyes, F. L., Holm, R. H., Risk, B. B., Tilley, E., & Brown, J. (2021). Impact of sampling depth on pathogen detection in pit latrines. *PLoS Neglected Tropical Diseases, 15*(3). Available from https://doi.org/10.1371/journal.pntd.0009176, https://journals.plos.org/plosntds/article?id = 10.1371/journal.pntd.0009176.

Chun, Y., Lee, S. K., Yoo, H. Y., & Kim, S. W. (2021). Recent advancements in biochar production according to feedstock classification, pyrolysis conditions, and applications: A review. *BioResources, 16*(3), 6512–6547. Available from https://doi.org/10.15376/BIORES.16.3.CHUN, http://www.ncsu.edu/bioresources/Back_Issues.htm.

Ciuta, S., Patuzzi, F., Baratieri, M., & Castaldi, M. J. (2018). Enthalpy changes during pyrolysis of biomass: Interpretation of intraparticle gas sampling. *Applied Energy, 228*, 1985–1993. Available from https://doi.org/10.1016/j.apenergy.2018.07.061, http://www.elsevier.com/inca/publications/store/4/0/5/8/9/1/index.htt.

Das, P., Chandramohan, V. P., Mathimani, T., & Pugazhendhi, A. (2021). A comprehensive review on the factors affecting thermochemical conversion efficiency of algal biomass to energy. *Science of The Total Environment, 766*144213. Available from https://doi.org/10.1016/j.scitotenv.2020.144213.

de Andrade, L. C., Borges-Pedro, J. P., Gomes, M. C. R. L., Tregidgo, D. J., do Nascimento, A. C. S., Paim, F. P., Marmontel, M., Benitz, T., Hercos, A. P., & do Amaral, J. V. (2021). The sustainable development goals in two sustainable development reserves in central amazon: achievements and challenges. *Discover Sustainability, 2*(1). Available from https://doi.org/10.1007/s43621-021-00065-4, https://www.springer.com/journal/43621.

Demirbaş, A. (1998). Yields of oil products from thermochemical biomass conversion processes. *Energy Conversion and Management, 39*(7), 685–690. Available from https://doi.org/10.1016/s0196-8904(97)00047-2.

Di Blasi, C., Branca, C., & Galgano, A. (2017). On the experimental evidence of exothermicity in wood and biomass pyrolysis. *Energy Technology, 5*(1), 19–29. Available from https://doi.org/10.1002/ente.201600091, http://onlinelibrary.wiley.com/journal/10.1002/(ISSN)2194-4296.

Duque, J. V. F., Bittencourt, F. L. F., Martins, M. F., & Debenest, G. (2021). Developing a combustion-driven reactor for waste conversion. *Energy, 237*. Available from https://doi.org/10.1016/j.energy.2021.121489, https://www.journals.elsevier.com/energy.

Duque, J. V. F., Martins, M. F., Bittencourt, F. L. F., Debenest, G., Orlando, M. T. D., Profeti, L. P. R., & Profeti, D. (2023). Recovering wax from polyethylene waste using C-DPyR. *Energy, 272*. Available from https://doi.org/10.1016/j.energy.2023.127135, https://www.journals.elsevier.com/energy.

Duque, J. V. F., Martins, M. F., Bittencourt, F. L. F., & Debenest, G. (2022). Relevant aspects of propagating a combustion front in an annular reactor for out-of-bed heat recovery. *Experimental Thermal and Fluid Science, 133*110575. Available from https://doi.org/10.1016/j.expthermflusci.2021.110575.

Duque, J. V. F., Martins, M. F., & Marcos Tadeu, D. A. O. (2020). Biodegradable masterbatch blends: The implications on thermal conversion and recycling stream of polyethylene. *Open Journal of Chemistry, 6*(1), 008–010. Available from https://doi.org/10.17352/ojc.000016.

Fermoso, J., Coronado, J. M., Serrano, D. P., & Pizarro, P. (2017). *Pyrolysis of microalgae for fuel production* (pp. 259–281). Elsevier BV. Available from https://doi.org/10.1016/b978-0-08-101023-5.00011-x.

Fisher, R. P., Lewandowski, A., Yacob, T. W., Ward, B. J., Hafford, L. M., Mahoney, R. B., Oversby, C. J., Mejic, D., Hauschulz, D. H., Summers, R. S., Linden, K. G., & Weimer, A. W. (2021). Solar thermal processing to disinfect human waste. *Sustainability (Switzerland), 13*(9). Available from https://doi.org/10.3390/su13094935, https://www.mdpi.com/2071-1050/13/9/4935/pdf.

Foxon K.M., Buckley C.A., Scientific support for the design and operation of ventilated improved pit latrines (VIPS). Water Research Commission Report. (2008).

Gold M., Niang S., Niwagaba C.B., Eder G., Muspratt A.M., Diop P.S., Strande L., Results from FaME (Faecal Management Enterprises) – can dried faecal sludge fuel the sanitation service chain?. (2014).

Gomes, M. C. R. L., Moura, E. A. F., Borges Pedro, J. P., Bezerra, M. M., & Brito, O. S. (2015). Sustainability of a sanitation program in flooded areas of the Brazilian Amazon. *Journal of Water Sanitation and Hygiene for Development, 5*(2), 261–270. Available from https://doi.org/10.2166/washdev.2015.123, http://www.iwaponline.com/washdev/005/0261/0050261.pdf.

Gwenzi, W., Marumure, J., Makuvara, Z., Simbanegavi, T. T., Njomou-Ngounou, E. L., Nya, E. L., Kaetzl, K., Noubactep, C., & Rzymski, P. (2023). The pit latrine paradox in low-income settings: A sanitation technology of choice or a pollution hotspot? *Science of the Total Environment, 879*. Available from https://doi.org/10.1016/j.scitotenv.2023.163179, http://www.elsevier.com/locate/scitotenv.

Hubbard, S. C., Meltzer, M. I., Kim, S., Malambo, W., Thornton, A. T., Shankar, M. B., Adhikari, B. B., Jeon, S., Bampoe, V. D., Cunningham, L. C., Murphy, J. L., Derado, G., Mintz, E. D., Mwale, F. K., Chizema-Kawesha, E., & Brunkard, J. M. (2020). Household illness and associated water and sanitation factors in peri-urban Lusaka, Zambia, 2016–2017. *npj Clean Water, 3*(1). Available from https://doi.org/10.1038/s41545-020-0076-4, https://www.nature.com/npjcleanwater/.

Humanitarian Engineering - Emerging Technologies and Humanitarian Efforts I IEEE Conference Publication I IEEE Xplore. (2023), 2023.

Ibitoye, S. E., Mahamood, R. M., Jen, T. C., Loha, C., & Akinlabi, E. T. (2023). An overview of biomass solid fuels: Biomass sources, processing methods, and morphological and microstructural properties. *Journal of Bioresources and Bioproducts, 8*(4), 333–360. Available from https://doi.org/10.1016/j.jobab.2023.09.005, https://www.sciencedirect.com/journal/journal-of-bioresources-and-bioproducts.

Jurado, N., Somorin, T., Kolios, A. J., Wagland, S., Patchigolla, K., Fidalgo, B., Parker, A., McAdam, E., Williams, L., & Tyrrel, S. (2018). Design and commissioning of a multi-mode prototype for thermochemical conversion of human faeces. *Energy Conversion and Management, 163*, 507–524. Available from https://doi.org/10.1016/j.enconman.2018.02.065, https://www.sciencedirect.com/science/article/pii/S0196890418301808.

Katukiza, A. Y., Ronteltap, M., Niwagaba, C. B., Foppen, J. W. A., Kansiime, F., & Lens, P. N. L. (2012). Sustainable sanitation technology options for urban slums. *Biotechnology Advances, 30*(5), 964–978. Available from https://doi.org/10.1016/j.biotechadv.2012.02.007.

Kennedy N., Gates' scheme to reinvent the toilet is 'too high-tech'. SciDev. net-Health. (2013).

Klippel, M. S., & Martins, M. F. (2022). Physicochemical assessment of waxy products directly recovered from plastic waste pyrolysis: Review and synthesis of characterization techniques. *Polymer Degradation and Stability, 204*. Available from https://doi.org/10.1016/j.polymdegradstab.2022.110090, https://www.journals.elsevier.com/polymer-degradation-and-stability.

Koulouri, M. E., Templeton, M. R., & Fowler, G. D. (2023). Source separation of human excreta: Effect on resource recovery via pyrolysis. *Journal of Environmental Management, 338*. Available from https://doi.org/10.1016/j.jenvman.2023.117782, https://www.sciencedirect.com/journal/journal-of-environmental-management.

Kwiringira, J., Atekyereza, P., Niwagaba, C., & Günther, I. (2014). Descending the sanitation ladder in urban Uganda: Evidence from Kampala Slums. *BMC Public Health, 14*(1). Available from https://doi.org/10.1186/1471-2458-14-624, http://www.biomedcentral.com/bmcpublichealth.

Lobo, G. D. S., Wittmann, F., & Piedade, M. T. F. (2019). Response of black-water floodplain (igapó) forests to flood pulse regulation in a dammed Amazonian river. *Forest Ecology and Management, 434*, 110–118. Available from https://doi.org/10.1016/j.foreco.2018.12.001, http://www.elsevier.com/inca/publications/store/5/0/3/3/1/0.

Lu, L., Guest, J. S., Peters, C. A., Zhu, X., Rau, G. H., & Ren, Z. J. (2018). Wastewater treatment for carbon capture and utilization. *Nature Sustainability, 1*(12), 750–758. Available from https://doi.org/10.1038/s41893-018-0187-9, http://www.nature.com/natsustain/.

Martins, M. F., Salvador, S., Thovert, J.-F., & Debenest, G. (2010). Co-current combustion of oil shale – Part 1: Characterization of the solid and gaseous products. *Fuel, 89*(1), 144–151. Available from https://doi.org/10.1016/j.fuel.2009.06.036.

Martins, M. F., Soares, R. B., & Gonçalves, R. F. (2022). *Microalgae: The challenges from harvest to the thermal gasification. Algal biotechnology: Integrated algal engineering for bioenergy, bioremediation, and biomedical applications* (pp. 247–258). Brazil: Elsevier. Available from https://www.sciencedirect.com/book/9780323904766, https://doi.org/10.1016/B978-0-323-90476-6.00007-8.

Martins, M. F., Duque, J. V. F., Soares, K., & Orlando, M. T. D. 'A. (2021). *Thermal analysis of recycled plastics* (pp. 213–238). Springer Science and Business Media LLC. Available from https://doi.org/10.1007/978-981-16-3627-1_10.

Maurya, N. S. (2012). Is human excreta a waste? *International Journal of Environmental Technology and Management, 15*(3-6), 325–332. Available from https://doi.org/10.1504/IJETM.2012.049231, http://www.inderscience.com/ijetm.

McAllister, S., Chen, J.-Y., & Fernandez-Pello, A. C. (2011). *Thermodynamics of combustion* (pp. 15–47). Springer Science and Business Media LLC. Available from https://doi.org/10.1007/978-1-4419-7943-8_2.

Michalak, A. M., Xia, J., Brdjanovic, D., Mbiyozo, A.-N., Sedlak, D., Pradeep, T., Lall, U., Rao, N., & Gupta, J. (2023). The frontiers of water and sanitation. *Nature Water*, *1*(1), 10−18. Available from https://doi.org/10.1038/s44221-022-00020-1.

Monhol, F. A. F., & Martins, M. F. (2015). Cocurrent Combustion of Human Feces and Polyethylene Waste. *Waste and Biomass Valorization*, *6*(3), 425−432. Available from https://doi.org/10.1007/s12649-015-9359-2.

Mrimi, E. C., Matwewe, F. J., Kellner, C. C., & Thomas, J. M. (2020). Safe resource recovery from faecal sludge: Evidence from an innovative treatment system in rural Tanzania. *Environmental Science: Water Research and Technology*, *6*(6), 1737−1748. Available from https://doi.org/10.1039/c9ew01097a, http://pubs.rsc.org/en/journals/journal/ew.

Mudasar, R., & Kim, M. H. (2017). Experimental study of power generation utilizing human excreta. *Energy Conversion and Management*, *147*, 86−99. Available from https://doi.org/10.1016/j.enconman.2017.05.052, https://www.journals.elsevier.com/energy-conversion-and-management.

Murungi, C., & van Dijk, M. P. (2014). Emptying, Transportation and Disposal of feacal sludge in informal settlements of Kampala Uganda: The economics of sanitation. *Habitat International*, *42*, 69−75. Available from https://doi.org/10.1016/j.habitatint.2013.10.011, http://www.elsevier.com/inca/publications/store/4/7/9/.

Nelson, S., Thomas, J., Jenkins, A., Naivalu, K., Naivalulevu, T., Naivalulevu, V., Mailautoka, K., Anthony, S., Ravoka, M., Jupiter, S. D., Mangubhai, S., Horwitz, P., Abimbola, S., & Negin, J. (2022). Perceptions of drinking water access and quality in rural indigenous villages in Fiji. *Water Practice and Technology*, *17*(3), 719−730. Available from https://doi.org/10.2166/wpt.2022.022, https://watermark.silverchair.com/wpt0170719.pdf?.

Norman, J. S., Pourkashanian, M., & Williams, A. (1995). *The formation of ammonia in igcc gasifiers and its control* (pp. 109−118). Elsevier BV. Available from https://doi.org/10.1016/b978-0-902597-49-5.50013-5.

Nuagah, M. B., Boakye, P., Oduro-Kwarteng, S., & Sokama-Neuyam, Y. A. (2020). Valorization of faecal and sewage sludge via pyrolysis for application as crop organic fertilizer. *Journal of Analytical and Applied Pyrolysis*, *151*. Available from https://doi.org/10.1016/j.jaap.2020.104903, https://www.journals.elsevier.com/journal-of-analytical-and-applied-pyrolysis.

Onabanjo, T., Kolios, A. J., Patchigolla, K., Wagland, S. T., Fidalgo, B., Jurado, N., Hanak, D. P., Manovic, V., Parker, A., McAdam, E., Williams, L., Tyrrel, S., & Cartmell, E. (2016). An experimental investigation of the combustion performance of human faeces. *Fuel*, *184*, 780−791. Available from https://doi.org/10.1016/j.fuel.2016.07.077, http://www.journals.elsevier.com/fuel/.

Onabanjo, T., Patchigolla, K., Wagland, S. T., Fidalgo, B., Kolios, A., McAdam, E., Parker, A., Williams, L., Tyrrel, S., & Cartmell, E. (2016). Energy recovery from human faeces via gasification: A thermodynamic equilibrium modelling approach. *Energy Conversion and Management*, *118*, 364−376. Available from https://doi.org/10.1016/j.enconman.2016.04.005.

Owusu, G. (2010). Social effects of poor sanitation and waste management on poor urban communities: A neighborhood-specific study of Sabon Zongo, Accra. *Journal of Urbanism*, *3*(2), 145−160. Available from https://doi.org/10.1080/17549175.2010.502001.

Pesquisa nacional por amostra de domicílios 2020 IBGE Unpublished content https://www.ibge.gov.br/estatisticas/sociais/populacao/9127-pesquisa-nacional-por-amostra-de-domicilios.html.

Programme U.W. W.A., Ryder G., The United Nations world water development report. (2017), 2017.

Rashwan, T. L., Zanoni, M. A. B., Wang, J., Torero, J. L., & Gerhard, J. I. (2023). Elucidating the characteristic energy balance evolution in applied smouldering systems. *Energy*, *273*. Available from https://doi.org/10.1016/j.energy.2023.127245, https://www.journals.elsevier.com/energy.

Reid, M. C., Guan, K., Wagner, F., & Mauzerall, D. L. (2014). Global methane emissions from pit latrines. *Environmental Science & Technology*, *48*(15), 8727−8734.

Rentizelas, A. A., Tolis, A. J., & Tatsiopoulos, I. P. (2009). Logistics issues of biomass: The storage problem and the multi-biomass supply chain. *Renewable and Sustainable Energy Reviews*, *13*(4), 887−894. Available from https://doi.org/10.1016/j.rser.2008.01.003.

Rose, C., Parker, A., Jefferson, B., & Cartmell, E. (2015). The characterization of feces and urine: A review of the literature to inform advanced treatment technology. *Critical Reviews in Environmental Science and Technology*, *45*(17), 1827−1879. Available from https://doi.org/10.1080/10643389.2014.1000761.

Saroj, D. P. (2023). 2832523641 *Sustainable sanitation-how can we improve sanitation systems in the global south?* Ladewig Group Membrane Research.

Schroeder E., A study of possible ways to dispose of urine and faeces from slum settlements. (2011).

Schueler, V., Weddige, U., Beringer, T., Gamba, L., & Lamers, P. (2013). Global biomass potentials under sustainability restrictions defined by the European Renewable Energy Directive 2009/28/EC. *GCB Bioenergy*, *5*(6), 652−663. Available from https://doi.org/10.1111/gcbb.12036, http://onlinelibrary.wiley.com/journal/10.1111/(ISSN)1757-1707.

Senecal, J., & Vinnerås, B. (2017). Urea stabilisation and concentration for urine-diverting dry toilets: Urine dehydration in ash. *Science of The Total Environment*, *586*, 650−657. Available from https://doi.org/10.1016/j.scitotenv.2017.02.038, https://www.sciencedirect.com/science/article/pii/S0048969717302796.

Sennoune, M., Salvador, S., & Quintard, M. (2012). Toward the control of the smoldering front in the reaction-trailing mode in oil shale semicoke porous media. *Energy & Fuels*, *26*(6), 3357−3367. Available from https://doi.org/10.1021/ef300479d.

Soares, R. B., Martins, M. F., & Gonçalves, R. F. (2019). A conceptual scenario for the use of microalgae biomass for microgeneration in wastewater treatment plants. *Journal of Environmental Management*, *252*. Available from https://doi.org/10.1016/j.jenvman.2019.109639, https://www.sciencedirect.com/journal/journal-of-environmental-management.

Soares, R. B., Martins, M. F., & Gonçalves, R. F. (2020). Thermochemical conversion of wastewater microalgae: The effects of coagulants used in the harvest process. *Algal Research*, *47*. Available from https://doi.org/10.1016/j.algal.2020.101864, http://www.sciencedirect.com/science/journal/aip/22119264.

Somorin, T., Fidalgo, B., Hassan, S., Sowale, A., Kolios, A., Parker, A., Williams, L., Collins, M., McAdam, E. J., & Tyrrel, S. (2020). Non-isothermal drying kinetics of human feces. *Drying Technology*, *38*(14), 1819–1827. Available from https://doi.org/10.1080/07373937.2019.1670205.

Somorin, T. O. (2020). *Valorisation of human excreta for recovery of energy and high-value products: A mini-review* (pp. 341–370). Springer Science and Business Media LLC. Available from https://doi.org/10.1007/978-3-030-38032-8_17.

Strande, L., & Brdjanovic, D. (2014). *Faecal sludge management: Systems approach for implementation and operation*. IWA Publishing, 2014.

Tan, H., Lee, C. T., Ong, P. Y., Wong, K. Y., Bong, C. P. C., Li, C., & Gao, Y. (2021). A review on the comparison between slow pyrolysis and fast pyrolysis on the quality of lignocellulosic and lignin-based biochar. *IOP Conference Series: Materials Science and Engineering*, *1051*(1), 012075. Available from https://doi.org/10.1088/1757-899x/1051/1/012075.

Torero, J. L., Gerhard, J. I., Martins, M. F., Zanoni, M. A. B., Rashwan, T. L., & Brown, J. K. (2020). Processes defining smouldering combustion: Integrated review and synthesis. *Progress in Energy and Combustion Science*, *81*. Available from https://doi.org/10.1016/j.pecs.2020.100869, https://www.journals.elsevier.com/progress-in-energy-and-combustion-science.

Tursi, A. (2019). A review on biomass: Importance, chemistry, classification, and conversion. *Biofuel Research Journal*, *6*(2), 962–979. Available from https://doi.org/10.18331/BRJ2019.6.2.3, http://www.biofueljournal.com/.

Varma, A. K., Shankar, R., & Mondal, P. (2018). *A review on pyrolysis of biomass and the impacts of operating conditions on product yield, quality, and upgradation. Recent advancements in biofuels and bioenergy utilization* (pp. 227–259). India: Springer Singapore. Available from http://doi.org/10.1007/978-981-13-1307-3, https://doi.org/10.1007/978-981-13-1307-3_10.

Villarín, M. C., & Merel, S. (2020). Paradigm shifts and current challenges in wastewater management. *Journal of Hazardous Materials*, *390*122139. Available from https://doi.org/10.1016/j.jhazmat.2020.122139.

Waste H., Pyrolyzer, (2023), 2023.

Wrasman, C. J., Wilson, A. N., Mante, O. D., Iisa, K., Dutta, A., Talmadge, M. S., Dayton, D. C., Uppili, S., Watson, M. J., Xu, X., Griffin, M. B., Mukarakate, C., Schaidle, J. A., & Nimlos, M. R. (2023). Catalytic pyrolysis as a platform technology for supporting the circular carbon economy. *Nature Catalysis*, *6*(7), 563–573. Available from https://doi.org/10.1038/s41929-023-00985-6, https://www.nature.com/natcatal/.

Yadav, V. K., Yadav, K. K., Cabral-Pinto, M. M. S., Choudhary, N., Gnanamoorthy, G., Tirth, V., Prasad, S., Khan, A. H., Islam, S., & Khan, N. A. (2021). The processing of calcium rich agricultural and industrial waste for recovery of calcium carbonate and calcium oxide and their application for environmental cleanup: A review. *Applied Sciences (Switzerland)*, *11*(9). Available from https://doi.org/10.3390/app11094212, https://www.mdpi.com/2076-3417/11/9/4212/pdf.

Yermán, L., Wall, H., Torero, J., Gerhard, J. I., & Cheng, Y. L. (2016). Smoldering Combustion as a Treatment Technology for Feces: Sensitivity to Key Parameters. *Combustion Science and Technology*, *188*(6), 968–981. Available from https://doi.org/10.1080/00102202.2015.1136299, http://www.tandfonline.com/toc/gcst20/current.

Zanoni, M. A. B., Torero, J. L., & Gerhard, J. I. (2019). Delineating and explaining the limits of self-sustained smouldering combustion. *Combustion and Flame*, *201*, 78–92. Available from https://doi.org/10.1016/j.combustflame.2018.12.004, http://www.elsevier.com/locate/combustflame.

Zhou, X., Li, Z., Zheng, T., Yan, Y., Li, P., Odey, E. A., Mang, H. P., & Uddin, S. M. N. (2018). Review of global sanitation development. *Environment International*, *120*, 246–261. Available from https://doi.org/10.1016/j.envint.2018.07.047, http://www.elsevier.com/locate/envint.

Part 5

Biochar systems for air pollution control

Chapter 18

Clean and efficient biochar cookstoves for mitigation of indoor air pollution

Zia Ur Rahman[1], Abdul Gani Abdul Jameel[1,2] and Li Songlin[3]

[1]*Interdisciplinary Research Center for Refining and Advanced Chemicals, King Fahd University of Petroleum & Minerals, Dhahran, Saudi Arabia,* [2]*Chemical Engineering Department, King Fahd University of Petroleum & Minerals, Dhahran, Saudi Arabia,* [3]*School of Environmental Science and Engineering, Hainan University, Haikou, Hainan, P.R. China*

Chapter outline

18.1 Introduction	347
18.2 Understanding indoor air pollution	348
18.3 Sources of indoor air pollution	348
18.3.1 Traditional cooking methods	348
18.3.2 Biomass combustion and emissions	348
18.4 Biochar cookstoves: an innovative solution	349
18.5 The evolution of cookstove technology	349
18.5.1 From traditional to modern designs	349
18.6 Types and designs of biochar cookstoves	349
18.6.1 Top-lit updraft biochar stoves	349
18.6.2 Rocket stoves	351
18.6.3 Top-lit updraft gasifier stoves	351
18.6.4 Institutional-scale biochar cook stoves	354
18.6.5 Top-lit updraft rice husk gasifier stoves	354
18.7 Biochar cookstoves: efficiency and emission reduction potential	355
18.8 Biochar cookstoves as a multifaceted solution for low-income settings	355
18.9 Environmental benefits of biochar cookstoves	356
18.10 Challenges and solutions	356
18.11 Case studies and success stories	357
18.12 Future directions and innovations	358
18.13 Conclusion	359
18.13.1 Recap of key points	359
18.13.2 Call to action for sustainable cooking practices	359
18.14 AI disclosure	359
References	359

18.1 Introduction

The persistent use of traditional cooking methods, particularly in developing regions, has significantly impacted indoor air quality (IAQ), leading to a global health crisis. Households in these regions heavily rely on solid fuels for cooking, resulting in harmful emissions that jeopardize both individual health and environmental sustainability. To address this issue, there is a growing focus on exploring cleaner and more efficient cooking solutions, with an emphasis on the innovative potential of biochar cookstoves (Chen et al., 1990).

Indoor air pollution (IAP) is a pervasive environmental challenge that significantly impacts the quality of air within enclosed spaces, such as homes, offices, and other indoor environments. Unlike outdoor air pollution, which often receives more attention, IAP results from various sources directly affecting the air quality where people spend the majority of their time. As the global population continues to urbanize, the prevalence of IAP poses a substantial public health concern, underscoring the urgent need for effective mitigation strategies (Edwards et al., 2007).

Fortunately, a promising solution emerges on the horizon: clean and efficient biochar cookstoves. These innovative technologies offer a revolutionary approach to combating IAP. Unlike their traditional counterparts, biochar cookstoves are designed to significantly reduce harmful emissions like smoke, particulate matter (PM), and noxious gases (Edwards et al., 2007).

This chapter delves into the world of biochar cookstoves, exploring their design principles, operational mechanisms, and the critical benefits they offer in the fight against IAP. It will examine, how these innovative stoves can combat

IAP, promoting cleaner cooking practices and empowering healthier lives for millions, particularly women and children who are disproportionately affected by IAP exposure.

However, realizing the full potential of biochar cookstoves requires addressing several challenges. This chapter will also discuss design considerations and variations in biochar stove technologies, factors influencing user adoption, including cultural preferences and affordability, and the need for ongoing research and development to optimize biochar cookstoves and maximize their impact on IAQ.

By exploring the transformative potential of clean and efficient biochar cookstoves, this chapter aims to illuminate a promising pathway towards cleaner air, a healthier planet, and improved livelihoods for millions.

18.2 Understanding indoor air pollution

IAP refers to the degradation of air quality within enclosed spaces, primarily homes and buildings, due to the presence of harmful pollutants. Unlike outdoor air pollution, indoor pollutants stem from various sources related to human activities, with cooking practices being a significant contributor. These pollutants include PM, carbon monoxide (CO), nitrogen dioxide (N_2O), volatile organic compounds (VOC), and others. Prolonged exposure to these contaminants poses serious health risks, particularly respiratory and cardiovascular diseases. Factors such as inadequate ventilation, the use of solid fuels for cooking, and poor housing conditions exacerbate IAP. Understanding the sources, types, and implications of IAP is crucial for developing effective strategies to mitigate its adverse effects on human health and well-being (Dasgupta et al., 2006, 2009).

18.3 Sources of indoor air pollution

18.3.1 Traditional cooking methods

Traditional cooking practices, often deeply embedded in cultural norms, rely on the use of solid fuels like wood, crop residues, or dung. While these methods are cost-effective and readily available, they come at the expense of IAQ. The incomplete combustion of these solid fuels generates a significant amount of PM, carbon monoxide, and other pollutants. Open fires or basic stoves (three stone stoves) as shown in Fig. 18.1, prevalent in many households, particularly in developing regions, contributes to a higher concentration of harmful emissions. The long-standing tradition of using traditional cooking methods underscores the need for targeted interventions to address the associated IAP, considering both cultural nuances and health implications.

18.3.2 Biomass combustion and emissions

Biomass, comprising organic materials such as wood, agricultural residues, and animal dung, is a common source of fuel for cooking. The combustion of biomass releases a complex mixture of pollutants, posing a substantial threat to IAQ. PM, including fine particles and black carbon, is a major byproduct of biomass combustion. Additionally, emissions include CO, VOCs, and other hazardous substances. In households where ventilation is insufficient, these emissions accumulate, creating an environment rich in pollutants. Beyond immediate health concerns for occupants, the emissions from biomass combustion contribute to deforestation, air pollution, and climate change, necessitating comprehensive strategies for cleaner cooking practices.

FIGURE 18.1 Traditional cooking methods. (A) Three stone open fire, (B) Mirt, and (C) Gonze stove respectively (Hailu, 2022). *Reprinted from Hailu, A. (2022). Development and performance analysis of top lit updraft: natural draft gasifier stoves with various feed stocks. Heliyon, 8(8)e10163. Available from https://doi.org/10.1016/j.heliyon.2022.e10163, https://www.sciencedirect.com/science/article/pii/S2405844022014517, with permission from Elsevier.*

18.4 Biochar cookstoves: an innovative solution

Biochar cookstoves represent a groundbreaking solution in the quest to mitigate IAP. These stoves are designed with a dual purpose: not only to provide efficient and clean cooking but also to produce biochar as a byproduct. Biochar is a stable form of carbon that can be used to enhance soil fertility and sequester carbon, contributing to broader environmental sustainability. The innovation lies in the combustion process, which is optimized to reduce harmful emissions associated with traditional cooking methods.

These cookstoves operate on a principle of efficient biomass combustion, ensuring complete burning of fuel and minimizing the release of pollutants. PM and harmful gases like carbon monoxide are significantly reduced, resulting in improved IAQ. The biochar produced during combustion serves as a valuable resource for agricultural practices, promoting soil health and aiding in carbon sequestration.

By adopting biochar cookstoves, households can enjoy the benefits of cleaner air, reduced health risks, and sustainable agricultural practices. The innovation reflects a holistic approach to addressing the intertwined challenges of IAP, health concerns, and environmental degradation, making biochar cookstoves a promising solution for a more sustainable future.

18.5 The evolution of cookstove technology

Cookstove technology has undergone a remarkable evolution, transitioning from traditional to modern designs in response to the imperative of improving efficiency, reducing emissions, and promoting sustainability.

18.5.1 From traditional to modern designs

Traditional Designs: Traditional cookstoves, prevalent for centuries, were characterized by simple structures often relying on open fires or rudimentary enclosures (three stone stoves). While cost-effective, these designs were inefficient in terms of fuel consumption and incomplete combustion, leading to high emissions of PM and other pollutants.

Transition to modern designs: The need for cleaner and more efficient cooking solutions drove the evolution towards modern designs. Advanced combustion techniques, improved insulation, and optimized airflow characterize these stoves. Modern cookstoves are engineered to ensure complete combustion of fuels, significantly reducing emissions of harmful pollutants. Some designs incorporate features such as improved ventilation and insulation, enhancing both safety and efficiency.

18.6 Types and designs of biochar cookstoves

18.6.1 Top-lit updraft biochar stoves

18.6.1.1 Design Features

Primary Combustion Mechanism: The top-lit updraft (TLUD) biochar stove operates on a distinctive combustion principle. Biomass is loaded into the top of the stove, and ignition occurs at the upper surface. As the fuel burns downward, a controlled updraft is created, facilitating efficient combustion, and reducing emissions.

Gasification chamber: TLUD stoves feature a gasification chamber, where the partial combustion of biomass generates combustible gases. These gases are then ignited in the upper part of the stove, creating a clean and efficient flame. The gasification process enhances the overall combustion efficiency, contributing to reduced emissions and increased heat production.

Controlled airflow: The design incorporates a system for regulating airflow, crucial for achieving optimal combustion. By controlling the entry of air into the combustion chamber, TLUD stoves ensure that the combustion process is well-managed, promoting efficiency and minimizing the release of harmful byproducts. Fig. 18.2 shows the basic structure and working principle of the TLUD biochar cook stove.

18.6.1.2 Applications and efficiency

Cooking Applications: TLUD biochar stoves find application in a variety of cooking scenarios, from household cooking in developing regions to outdoor and camping activities. Their portable nature and efficient design make them versatile for different cooking needs.

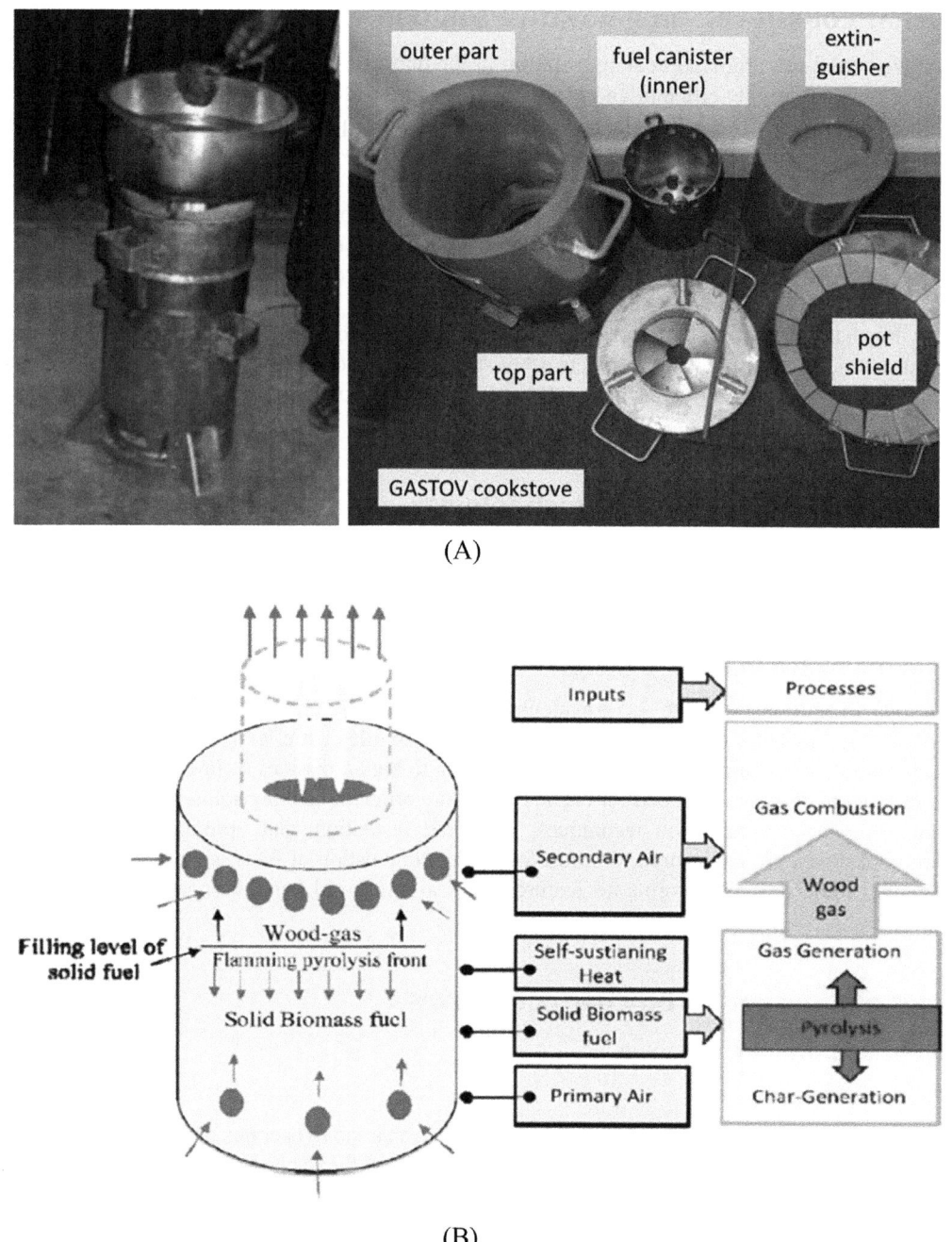

FIGURE 18.2 TLUD stove (Sundberg et al., 2020). (A) Char-producing stove in steel design (left), and the refined version (right) both in ceramic and steel. (B) From the working principle of a typical top-lit updraft (TLUD) stove (Adeniyi et al., 2023).

Biochar Production: One notable feature of TLUD stoves is their ability to produce biochar as a byproduct. The controlled combustion process results in the creation of biochar, a valuable resource for agricultural use. This dual functionality aligns with sustainability goals, offering not only clean cooking but also a means of producing a beneficial soil amendment.

Efficiency and Emissions Reduction: TLUD stoves are renowned for their high combustion efficiency, achieving complete burning of biomass. This efficiency translates to reduced fuel consumption and lower emissions of PM and other pollutants. The controlled combustion process also contributes to a cleaner flame, minimizing environmental impact.

Versatility: Due to their design and efficiency, TLUD biochar stoves are versatile and adaptable. They can burn a variety of biomass materials, including wood chips, crop residues, and agricultural waste. This adaptability makes them suitable for diverse geographic and cultural contexts.

In essence, TLUD biochar stoves combine innovative design features with practical applications. Their efficient combustion process, coupled with biochar production, positions them as sustainable solutions for clean cooking, offering benefits for both households and the environment.

18.6.2 Rocket stoves

18.6.2.1 Compact design

Rocket Principle: Rocket stoves are characterized by a compact and efficient design based on the rocket combustion principle. Biomass fuel is loaded into a vertical combustion chamber, promoting a strong, directed draft. This design allows for intense, controlled combustion in a small space. Different design structures of rocket stoves are shown in Fig. 18.3.

Efficient Airflow: The compact design of rocket stoves optimizes airflow, ensuring that the right amount of oxygen reaches the combustion chamber. This efficient airflow contributes to a more complete and cleaner burning process, reducing emissions and enhancing overall efficiency.

18.6.2.2 Efficiency in resource-constrained settings

Optimized combustion: Rocket stoves excel in resource-constrained settings due to their ability to efficiently burn biomass with minimal fuel consumption. The directed airflow and insulation around the combustion chamber contribute to higher temperatures, ensuring thorough combustion and reducing the need for large quantities of fuel.

Suitability for developing regions: The simplicity and efficiency of rocket stoves make them well-suited for use in developing regions where access to abundant fuel sources may be limited. Their design allows for effective cooking with less fuel, addressing challenges associated with resource scarcity.

In summary, rocket stoves, with their compact design and efficiency, offer a practical solution for clean and resource-efficient cooking, particularly in settings where resources are limited.

18.6.3 Top-lit updraft gasifier stoves

18.6.3.1 Combining top-lit updraft and gasification

Synergy of Technologies: TLUD Gasifier Stoves represent a harmonious fusion of TLUD and gasification technologies. In this innovative design, biomass is ignited from the TLUD, creating a controlled updraft that promotes efficient combustion. Simultaneously, gasification processes occur, utilizing the generated gases for a cleaner and more complete burning process. The basic design and schematic diagram are shown in Fig. 18.4.

Dual combustion benefits: The combination of TLUD and gasification enhances combustion efficiency by leveraging the strengths of both technologies. TLUD initiates the process with controlled updraft, while gasification ensures the utilization of combustible gases, resulting in reduced emissions and increased thermal efficiency.

18.6.3.2 Flexibility for various biomass types

Adaptability: TLUD Gasifier Stoves exhibit flexibility in accommodating various types of biomass for combustion. Whether it's wood chips, crop residues, or other biomass materials, these stoves can efficiently burn a diverse range of fuels.

Versatility in fuel sources: This adaptability makes TLUD Gasifier Stoves suitable for regions with different biomass availability. Users can harness the benefits of clean cooking without being restricted to specific fuel types, enhancing their usability in diverse geographic and cultural contexts.

In summary, TLUD gasifier stoves offer an innovative solution by combining the strengths of TLUD and gasification technologies. This synergy not only enhances combustion efficiency but also provides flexibility in utilizing various biomass sources for clean and sustainable cooking.

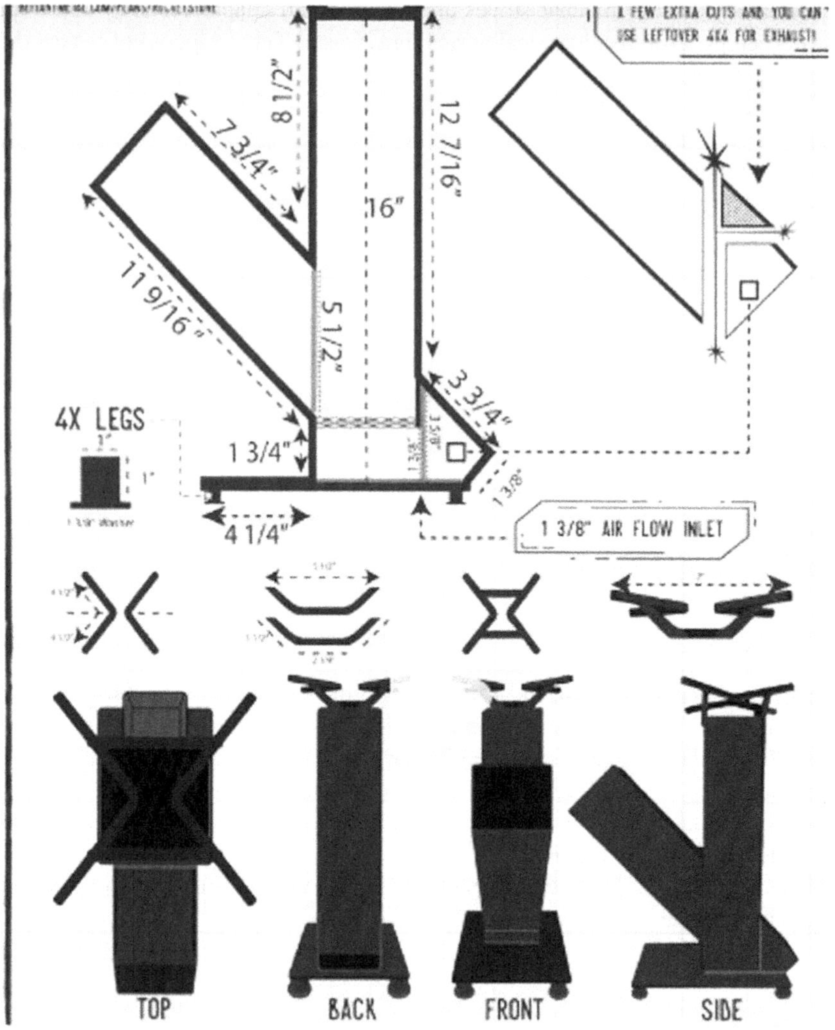

352 PART | 5 Biochar systems for air pollution control

FIGURE 18.3 Rocket stove. (A) Design of different rocket stove. (B) Fabricated rocket stove.

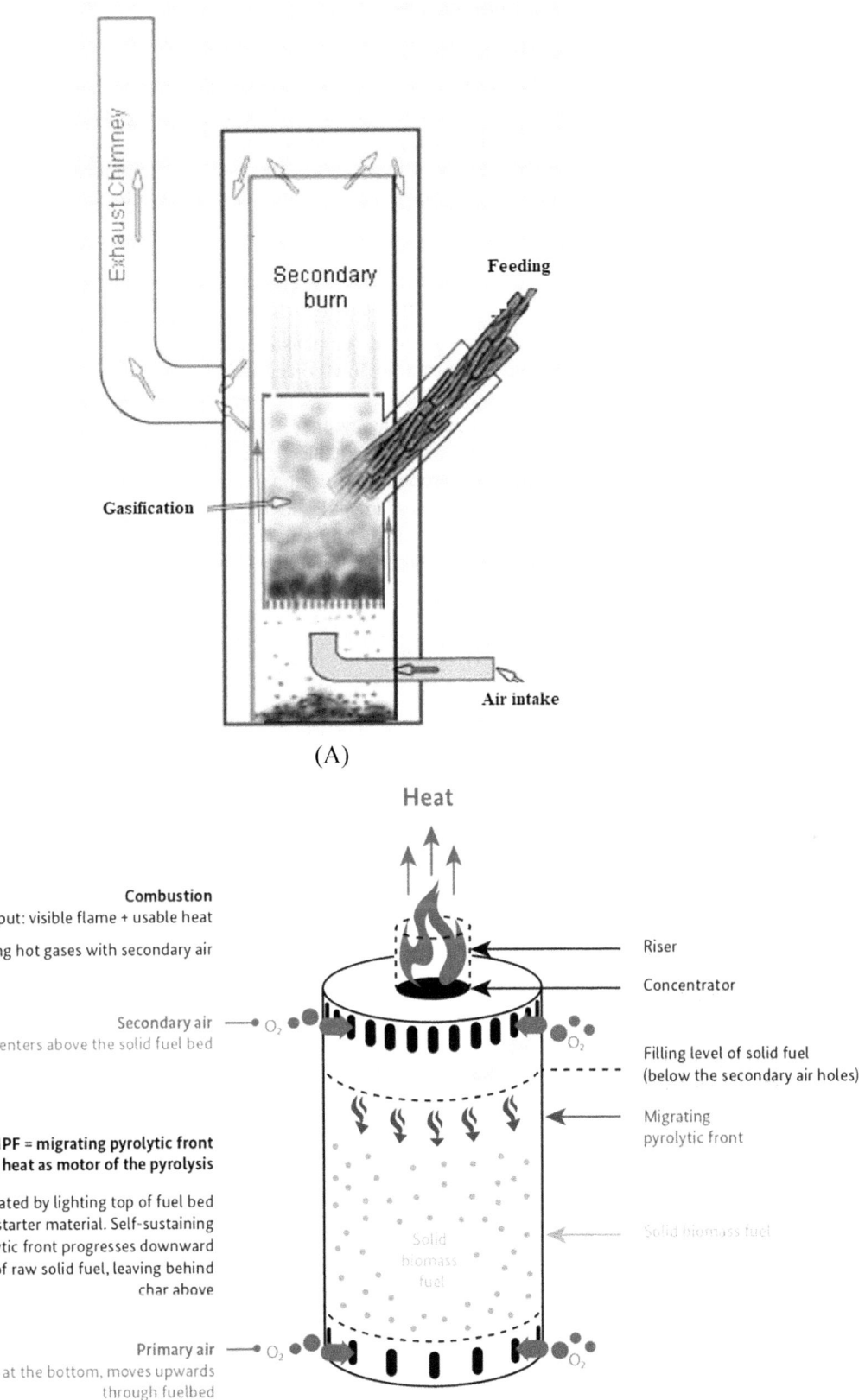

FIGURE 18.4 TLUD gasifier (Adeniyi et al., 2023). (A) Generalized structure of TLUD gasifier. (B) From Detailed schematic diagram of TLUD stove (Roth, 2014).

18.6.4 Institutional-scale biochar cook stoves

18.6.4.1 Larger capacities for community use

Community-centric design: Institutional-scale biochar cook stoves are designed with larger capacities, catering to the cooking needs of communities. These stoves are intended for collective use, making them suitable for schools, community centers, or other institutions where a higher cooking volume is required.

Community empowerment: The larger capacity of these stoves addresses the communal aspect of cooking in institutional settings. They contribute to community empowerment by providing a centralized and efficient cooking solution that serves a larger population.

18.6.4.2 Emphasis on efficiency and environmental impact

Scaled efficiency: Institutional-scale biochar cook stoves emphasize efficiency to meet the demands of larger user groups. Enhanced combustion systems and optimized airflow contribute to increased efficiency, minimizing fuel consumption and reducing emissions.

Environmental considerations: Beyond cooking efficiency, these stoves prioritize environmental impact. The emphasis on biochar production as a byproduct aligns with sustainable practices. The generated biochar can be utilized for agricultural purposes, promoting soil health and carbon sequestration.

In summary, institutional-scale biochar cookstoves are tailored for larger community use, emphasizing both efficiency and positive environmental impact. They provide a communal cooking solution that aligns with sustainable practices and addresses the needs of institutions serving broader populations.

18.6.5 Top-lit updraft rice husk gasifier stoves

18.6.5.1 Specialized design for rice husk utilization

Tailored combustion: TLUD rice husk gasifier stoves are specifically designed to efficiently utilize rice husk as a biomass fuel. The specialized design takes into account the unique characteristics of rice husk, optimizing the combustion process for this particular biomass material. Fig. 18.5 shows the structure of the TLUD rice husk gasifier stove.

Gasification efficiency: The stoves incorporate features that enhance gasification efficiency when using rice husk. This ensures that the gasification process is effective, converting rice husk into combustible gases that contribute to a clean and sustainable flame.

18.6.5.2 Contribution to sustainable cooking practices

Sustainable biomass utilization: TLUD rice husk gasifier stoves play a crucial role in promoting sustainable cooking practices. By efficiently utilizing rice husk, an abundant agricultural residue, these stoves contribute to reducing waste while providing a clean and efficient energy source for cooking.

Environmental impact: The specialized design not only maximizes the utilization of rice husk but also minimizes environmental impact. Through effective gasification and combustion, these stoves help reduce emissions associated with traditional cooking methods, aligning with sustainable and eco-friendly cooking practices.

In summary, TLUD rice husk gasifier stoves showcase a specialized approach to biomass utilization, emphasizing the efficient use of rice husk for clean and sustainable cooking. Their design contributes to both resource efficiency and environmentally friendly cooking practices.

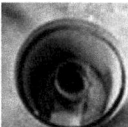  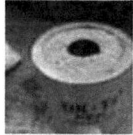  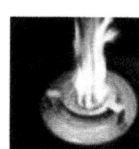

Empty stove | Cover of centre hole during loading | Filled to capacity with rice husk | Fire is started with wood logs | Flame additionally fuelled with gas from rice husk

FIGURE 18.5 Rice husk gasifier stove (Roth, 2014).

18.7 Biochar cookstoves: efficiency and emission reduction potential

A key challenge in addressing IAP from traditional cooking methods is finding stoves that are both efficient and clean-burning. Researchers have conducted studies comparing the performance of various cookstove designs, including biochar stoves. One such study by Jetter et al. (2012) evaluated 22 different stoves, assessing their efficiency and emission reduction capabilities.

The study employed two key metrics:

1. **Modified combustion efficiency (MCE):** This metric indicates how completely the fuel is being combusted. A higher MCE value suggests a cleaner burn with fewer emissions. It's calculated as the ratio of carbon dioxide (CO_2) produced to the sum of CO_2 and carbon monoxide (CO) ($CO_2/CO_2 + CO$).
2. **Overall thermal efficiency (OTE):** This metric combines the MCE with the stove's heat transfer efficiency (HTE). Heat transfer efficiency (HTE) focuses on how effectively the stove transfers heat generated from combustion to the cooking pot. It reflects how effectively the stove utilizes the fuel to generate usable heat for cooking. A higher OTE signifies a more efficient stove that wastes less fuel. The relationship between these efficiencies can be expressed as: $OTE \approx MCE \times THE$

The study's findings, summarized in Table 18.1, revealed that a natural-draft cookstove with a TLUD design achieved the highest MCE and OTE among the tested stoves. However, this specific TLUD stove required pelletized biomass fuel with low moisture content, limiting its applicability in some contexts.

Interestingly, the study also found that four biochar cookstoves (Oorja, Protos, Philips fan, StoveTec TLUD) demonstrated higher combustion efficiency (based on MCE) compared to the traditional three-stone open-fire stove. This suggests that these biochar stoves offer a significant improvement in terms of cleaner burning and reduced emissions.

However, the study also highlights a critical point for future research: comparisons between stove types can be complicated by the different types of biomass fuel used in the tests. This factor needs to be carefully considered when interpreting the results.

Overall, the study by Jetter et al. (2012) provides valuable insights into the potential of biochar cookstoves to improve combustion efficiency and reduce emissions compared to traditional cooking methods. However, further research is needed to explore the performance of these stoves across diverse fuel types and usage scenarios.

18.8 Biochar cookstoves as a multifaceted solution for low-income settings

Biochar cookstoves are emerging as a promising technology for addressing the multifaceted challenges faced by low-income households in developing regions. These challenges include high IAP, excessive fuel consumption due to

TABLE 18.1 Comparison of overall thermal efficiency and indoor emissions of biochar cookstoves and conventional stoves (Jetter et al., 2012).

Fuel	Cookstove	Type	Indoor emissions		Overall thermal efficiency (OTE) %
			CO (g/min)	PM2.5	
Wood	3-stone carefully tended	Open fire	0.885	56.9	14.9
	3-stone minimal tended	Open fire	1.041	93.8	13.7
	Philips HD4012 fan (forced draft)	Gasifier	0.102	6.6	38.4
	Philips HD4008 (natural draft)	Gasifier	0.390	53.8	34.4
Wood pellets	Stove-Tec TLUD	Gasifier	0.042	4.4	53.2
Biomass pellets	Oorja stove	Gasifier	0.087	0.087	34.7
Rice husks	Belonio (forced draft)	Gasifier	0.686	8.2	46.1
	Mayon Turbo	Gasifier	0.845	31.3	28.8
Corn cobs	Jinqilin-CKQ-80I	Gasifier			13.4

inefficient cooking practices, and limited economic opportunities. Here, the technical benefits of biochar cookstoves, focusing on their impact on fuel efficiency, IAQ, and potential economic benefits in low-income settings is describe as:

Fuel efficiency: Traditional cooking methods, such as open fires and three-stone stoves, are notoriously inefficient, with combustion efficiencies (MCE) as low as 10% (Bond et al., 2013). This translates to significant fuel waste and time spent collecting firewood, often a burden disproportionately borne by women and girls (Lusambo & Mbeyale, 2021). Biochar cookstoves, through improved combustion chamber design and airflow management, achieve MCE values ranging from 30% to 70%. This substantial increase in efficiency leads to a dramatic reduction in fuel consumption, with studies showing a decrease of 30%–50% compared to traditional methods (Pratiti et al., 2020).

IAQ: IAP from traditional cooking practices is a major health concern in low-income settings. Smoke and pollutants released during cooking contribute to respiratory illnesses, including pneumonia, chronic obstructive pulmonary disease (COPD), and even lung cancer. Biochar cookstoves, by promoting cleaner combustion and often incorporating chimney designs to vent emissions outdoors, significantly reduce exposure to harmful pollutants. Studies have shown reductions in fine PM (PM2.5) concentrations by as much as 80%–90% within households using biochar cookstoves (Hu & Ran, 2009; Pérez-Padilla et al., 2010; Smith & Pillarisetti, 2017).

Economic Benefits: Traditional cooking methods often require significant time spent collecting firewood. Biochar cookstoves, by significantly reducing fuel needs, free up valuable time for women and girls who are typically burdened with fuel collection tasks (Goldemberg et al., 2018). This freed-up time can be channeled towards income-generating activities, such as small businesses, crafts, or agricultural work, leading to increased household income and economic empowerment. Further, Biochar cookstoves, by reducing harmful emissions, contribute to improved health outcomes. This translates to fewer sick days and increased workforce participation, further boosting household income potential. In addition, some biochar stove designs generate biochar as a by-product. Biochar holds value as a soil amendment, with the potential to improve crop yields and soil health (Okoko et al., 2017). Households can potentially sell surplus biochar to farmers, creating an additional income stream.

18.9 Environmental benefits of biochar cookstoves

Additionally, some biochar cookstoves offer a multifaceted approach to sustainable development. Certain stove designs generate biochar as a by-product during the combustion process, and this charcoal-like material can be utilized in various applications to provide significant benefits. In agroecosystems, biochar's porous structure enhances soil fertility by promoting beneficial microorganisms, retaining nutrients, and improving water-holding capacity, leading to increased crop yields and potentially reduced reliance on chemical fertilizers. Biochar can also be used in household nutrition gardens, where it can be incorporated into raised beds or containers to enhance soil quality and moisture retention, helping to reduce the need for fertilizers and promoting sustainable food production practices. Furthermore, biochar's porous structure shows promise for water treatment applications, as it can act as a filter, adsorbing contaminants like heavy metals and organic pollutants. By effectively utilizing the biochar by-product, biochar cookstoves can create a closed-loop system, offering a multifaceted solution for low-income settings that not only improves fuel efficiency and IAQ but also generates a valuable resource for enhancing soil fertility, improving household nutrition gardens, and potentially treating water sources(Ippolito et al., 2012; Kavitha et al., 2018; Shakoor et al., 2021).

18.10 Challenges and solutions

While biochar cookstoves offer a promising solution to IAP, their widespread adoption faces several challenges. Here are some key hurdles and potential solutions:

Challenges:

- Cost: Biochar cookstoves can be more expensive than traditional options, creating an affordability barrier for low-income households.
- Cultural preferences: Traditional cooking methods are often deeply ingrained in cultural practices, making it difficult to change habits.
- Accessibility: Distribution and availability of biochar cookstoves can be limited in remote areas, hindering access for those most in need.
- Sustainability: Ensuring the long-term sustainability of biochar production and utilization requires careful planning and management.

- Maintenance: Some biochar stoves require specific maintenance procedures that users may not be familiar with or able to perform consistently.
- Solutions:
- Subsidies and microfinance: Offering subsidies or microfinance loans can make biochar cookstoves more affordable for low-income families.
- Community engagement: Culturally sensitive education and awareness campaigns can promote the benefits of biochar cookstoves and address concerns.
- Local production and distribution: Establishing local production and distribution networks can improve accessibility and affordability.
- Integrated approaches: Combining biochar cookstoves with other interventions like improved ventilation can maximize health benefits.
- Training and support: Providing user training and ongoing support can ensure proper maintenance and maximize biochar production.
- Innovation and affordability: Ongoing research and development should focus on reducing production costs and designing even more affordable biochar cookstove models.

Additional considerations:

- Monitoring and evaluation: Regularly monitoring the impact of biochar cookstoves on air quality, health outcomes, and adoption rates is crucial to ensure effectiveness and identify areas for improvement.
- Policy and regulations: Supportive government policies and regulations can incentivize the production and use of cleaner cookstoves.
- Collaboration: Multi-stakeholder partnerships involving governments, NGOs, private companies, and communities are essential for successful implementation and scaling up biochar cookstove initiatives.

By addressing these challenges and implementing effective solutions, biochar cookstoves have the potential to significantly improve IAQ, alleviate respiratory illnesses, empower women, and contribute to a cleaner and healthier future for millions around the world.

18.11 Case studies and success stories

Case Study 1: The Smoke-Free Kitchens of Rwanda

In rural Rwanda, households traditionally cooked over open fires, exposing families to harmful smoke and respiratory illnesses. The introduction of "EcoZoom" biochar cookstoves in 2009 aimed to change that. These efficient stoves reduced PM by 90%, significantly improving IAQ. Within five years, the project reached 55,000 households, leading to a decrease in respiratory illnesses, particularly in children. Additionally, the biochar produced by the stoves improved soil fertility and crop yields, boosting food security (Eltigani et al., 2022).

Success story: Christine, a Rwandan mother, used to suffer from constant coughing and respiratory problems due to cooking smoke. Since adopting the EcoZoom stove, her health has improved, and she spends less time gathering firewood. The biochar has also helped her grow more nutritious food for her family.

Case Study 2: Clean cooking and carbon offsets in ghana

The "BurnStove" project in Ghana distributes improved cookstoves, including biochar models, to replace traditional open fires. These stoves reduce household air pollution by 80%, leading to health benefits similar to the Rwandan case. Additionally, the project generates carbon offsets by capturing carbon dioxide through the use of biochar. These offsets are then sold on the international market, providing additional income for local communities(Gitau et al., 2019; Pope et al., 2017; Sundberg et al., 2020).

Success Story: Kofi, a Ghanaian farmer, received a BurnStove and now enjoys cleaner air while cooking. He uses the biochar to improve his soil fertility, reducing his reliance on chemical fertilizers. The income he earns from selling carbon offsets allows him to invest in his children's education.

Case Study 3: Empowering women and improving lives in cambodia

The "Soot-Free Futures" project in Cambodia focuses on distributing clean cookstoves, including biochar models, to women. These stoves not only improve health but also empower women by reducing the time spent cooking and gathering firewood. Additionally, the project provides training on using biochar in agriculture, boosting women's livelihoods and food security (Sundberg et al., 2020).

Success story: Sreynich, a Cambodian woman, used to spend hours gathering firewood and cooking in a smoky environment. Now, with her biochar stove, she has more time for other activities and enjoys better health. She uses the biochar in her vegetable garden, earning additional income by selling surplus produce.

Beyond the cases:

These are just a few examples of the positive impact biochar cookstoves can have. Ongoing research and development continue to improve their efficiency, affordability, and accessibility. As their adoption expands, so will the benefits for public health, environmental sustainability, and economic development, particularly in vulnerable communities.

Additionally:

- It's important to acknowledge the challenges associated with widespread adoption, such as cultural preferences, affordability, and access to resources.
- Briefly mention other types of clean cookstoves besides biochar for a more comprehensive overview.
- Encourage further research and development to address remaining challenges and optimize the potential of biochar cookstoves.

By highlighting these success stories and the broader potential of biochar cookstoves, we can paint a compelling picture of their significant role in mitigating IAP and creating a healthier and more sustainable future for all.

18.12 Future directions and innovations

(i) Advanced Combustion Technologies:

Smart Combustion Systems: Future innovations may focus on integrating smart combustion systems that dynamically adjust airflow and combustion parameters. This could enhance efficiency, reduce emissions further, and provide real-time monitoring for optimal performance.

(ii) Integration of renewable energy:

Sustainable Energy Solutions: The future of biochar cookstoves may involve the integration of renewable energy sources. This could include hybrid systems that combine biochar combustion with solar or biomass-based power generation, providing a more sustainable and continuous energy supply.

(iii) Internet of Things (IoT) Integration:

Connected Appliances: Innovations may involve IoT integration, allowing users to monitor and control their biochar cookstoves remotely. This could enhance usability, optimize energy consumption, and provide valuable data for ongoing research and development.

(iv) Customized designs for regional needs:

Adaptable designs: Future biochar cookstoves might feature modular and adaptable designs that can be customized based on regional biomass availability, cooking habits, and cultural preferences. This approach ensures that the stoves are well-suited to diverse contexts globally.

(v) Improved biochar utilization:

Enhanced biochar applications: Innovations may focus on optimizing the byproduct of biochar production. Research and development could explore enhanced applications for biochar in agriculture, water filtration, and carbon sequestration, creating a more comprehensive and sustainable solution.

(vi) User-friendly maintenance:

Easy Maintenance Features: Future biochar cookstoves might incorporate user-friendly maintenance features, making it easier for individuals in various settings to clean, repair, and maintain their stoves. This ensures long-term usability and sustainability.

(vii) Accessibility and affordability:

Mass Adoption Solutions: Future innovations will likely address accessibility and affordability, aiming for mass adoption of clean cooking technologies. This could involve cost-effective manufacturing processes, subsidies, or innovative financing models to make these stoves accessible to a larger population.

(viii) Collaborative research and global initiatives:

International Collaborations: The future direction of clean and efficient biochar cookstoves will likely involve collaborative research efforts and global initiatives. International partnerships can accelerate innovations, share best practices, and address challenges associated with diverse cultural and environmental contexts.

In summary, the future of biochar cookstoves lies in advanced combustion technologies, renewable energy integration, IoT connectivity, customized designs, enhanced biochar utilization, user-friendly maintenance, improved accessibility, and collaborative global efforts. These innovations aim to create more efficient, sustainable, and widely adopted solutions for mitigating IAP.

18.13 Conclusion

18.13.1 Recap of key points

In exploring the realm of clean and efficient biochar cookstoves for the mitigation of IAP, several key points emerge. These innovative cooking solutions address the pressing issues associated with traditional cooking methods, offering a sustainable alternative with far-reaching benefits.

From the introduction of various biochar cookstove designs, such as TLUD, gasifier stoves, rocket stoves, to the institutional-scale and rice husk-specific variations, each design serves a unique purpose in promoting clean combustion and minimizing environmental impact. The environmental and health benefits are noteworthy, showcasing a reduction in IAP, lower emissions, and improved air quality. Moreover, the long-term health implications underscore the potential for these stoves to contribute to better respiratory and cardiovascular health, ultimately enhancing the overall quality of life.

18.13.2 Call to action for sustainable cooking practices

As we reflect on the advancements and potential of biochar cookstoves, a clear call to action emerges. Embracing sustainable cooking practices is not just a choice; it is a responsibility. The innovations discussed, from advanced combustion technologies to IoT integration, demonstrate the continuous evolution of clean cooking solutions. However, their impact is contingent on widespread adoption.

The call to action involves promoting awareness and education about the benefits of biochar cookstoves, especially in regions heavily reliant on traditional cooking methods. Governments, organizations, and communities must collaborate to make these technologies accessible and affordable. Investment in research, development, and international partnerships is essential to drive innovation and overcome barriers to adoption.

In our hands lies the power to usher in a future where sustainable cooking is the norm rather than the exception. By choosing and advocating for clean and efficient biochar cookstoves, we contribute not only to environmental conservation but also to the well-being of individuals and communities worldwide. It's time to turn our understanding into action and collectively build a healthier, cleaner, and more sustainable future through responsible cooking practices.

18.14 AI disclosure

During the preparation of this work, the author(s) used Gimini and ChatGPT to improve the readability and paraphrasing. After using this tool/service, the author(s) reviewed and edited the content as needed and take(s) full responsibility for the content of the publication.

References

Adeniyi, A. G., Iwuozor, K. O., Muritala, K. B., Emenike, E. C., & Adeleke, J. A. (2023). Conversion of biomass to biochar using top-lit updraft technology: a review. *Biofuels, Bioproducts and Biorefining, 17*(5), 1411–1424. Available from https://doi.org/10.1002/bbb.2497. Available from, https://doi.org/10.1002/bbb.2497.

Bond, T. C., Doherty, S. J., Fahey, D. W., Forster, P. M., Berntsen, T., DeAngelo, B. J., Flanner, M. G., Ghan, S., Kärcher, B., Koch, D., Kinne, S., Kondo, Y., Quinn, P. K., Sarofim, M. C., Schultz, M. G., Schulz, M., Venkataraman, C., Zhang, H., Zhang, S., ... Zender, C. S. (2013). Bounding the role of black carbon in the climate system: A scientific assessment. *Journal of Geophysical Research: Atmospheres, 118*(11), 5380–5552. Available from https://doi.org/10.1002/jgrd.50171, https://doi.org/10.1002/jgrd.50171.

Chen, B. H., Hong, C. J., Pandey, M. R., & Smith, K. R. (1990). Indoor air pollution in developing countries. *World Health Statistics Quarterly. Rapport Trimestriel de Statistiques Sanitaires Mondiales, 43*(3), 127–138.

Dasgupta, S., Huq, M., Khaliquzzaman, M., Pandey, K., & Wheeler, D. (2006). Indoor air quality for poor families: new evidence from Bangladesh. *Indoor Air, 16*(6), 426–444. Available from https://doi.org/10.1111/j.1600-0668.2006.00436.x.

Dasgupta, S., Wheeler, D., Huq, M., & Khaliquzzaman, M. (2009). Improving indoor air quality for poor families: a controlled experiment in Bangladesh. *Indoor Air, 19*(1), 22–32. Available from https://doi.org/10.1111/j.1600-0668.2008.00558.x.

Edwards, R. D., Liu, Y., He, G., & Yin, Z. (2007). Household CO and PM measured as part of a review of China's National Improved Stove Program. *Indoor Air, 17*(3), 189–203. Available from https://doi.org/10.1111/j.1600-0668.2007.00465.x.

Eltigani, A., Olsson, A., Krause, A., Ernest, B., Fridahl, M., Yanda, P., & Hansson, A. (2022). Exploring lessons from five years of biochar-producing cookstoves in the Kagera region, Tanzania. *Energy for Sustainable Development, 71*, 141–150.

Gitau, J. K., Sundberg, C., Mendum, R., Mutune, J., & Njenga, M. (2019). Use of biochar-producing gasifier cookstove improves energy use efficiency and indoor air quality in rural householdsUse of Biochar-Producing Gasifier Cookstove Improves Energy Use Efficiency and Indoor Air Quality in Rural Households. *Energies, 12*(22), 4285.

Goldemberg, J., Martinez-Gomez, J., Sagar, A., & Smith, K. R. (2018). Household air pollution, health, and climate change: cleaning the airHousehold air pollution, health, and climate change: cleaning the air. *Environmental Research Letters, 13*(3)030201.

Hailu, A. (2022). Development and performance analysis of top lit updraft: natural draft gasifier stoves with various feed stocks. *Heliyon, 8*(8)e10163. Available from https://doi.org/10.1016/j.heliyon.2022.e10163, https://www.sciencedirect.com/science/article/pii/S2405844022014517.

Hu, G., & Ran, P. (2009). Indoor air pollution as a lung health hazard: focus on populous countriesIndoor air pollution as a lung health hazard: focus on populous countries. *Current Opinion in Pulmonary Medicine, 15*(2), 158–164.

Ippolito, J. A., Laird, D. A., & Busscher, W. J. (2012). Environmental benefits of biocharEnvironmental benefits of biochar. *Journal of Environmental Quality, 41*(4), 967–972.

Jetter, J., Zhao, Y., Smith, K. R., Khan, B., Yelverton, T., DeCarlo, P., & Hays, M. D. (2012). Pollutant Emissions and Energy Efficiency under Controlled Conditions for Household Biomass Cookstoves and Implications for Metrics Useful in Setting International Test Standards. *Environmental Science & Technology, 46*(19), 10827–10834. Available from https://doi.org/10.1021/es301693f, https://doi.org/10.1021/es301693f.

Kavitha, B., Reddy, P. V. L., Kim, B., Lee, S. S., Pandey, S. K., & Kim, K.-H. (2018). Benefits and limitations of biochar amendment in agricultural soils: A review. *Journal of Environmental Management, 227*, 146–154.

Lusambo, L. P., & Mbeyale, G. (2021). Development of wood fuel consumption predictive model in Tanzania. *Tanzania Journal of Forestry and Nature Conservation, 90*(2), 115–133.

Okoko, A., Reinhard, J., von Dach, S. W., Zah, R., Kiteme, B., Owuor, S., & Albrecht Ehrensperger, A. (2017). The carbon footprints of alternative value chains for biomass energy for cooking in Kenya and TanzaniaThe carbon footprints of alternative value chains for biomass energy for cooking in Kenya and Tanzania. *Sustainable Energy Technologies and Assessments, 22*, 124–133.

Pope, D., Bruce, N., Dherani, M., Jagoe, K., & Rehfuess, E. (2017). Real-life effectiveness of 'improved'stoves and clean fuels in reducing PM2. 5 and CO: Systematic review and meta-analysisReal-life effectiveness of 'improved' stoves and clean fuels in reducing PM2.5 and CO: Systematic review and meta-analysis. *Environment International, 101*, 7–18.

Pérez-Padilla, R., Schilmann, A., & Riojas-Rodriguez, H. (2010). Respiratory health effects of indoor air pollutionRespiratory health effects of indoor air pollution. *The International Journal of Tuberculosis and Lung Disease, 14*(9), 1079–1086.

Pratiti, R., Vadala, D., Kalynych, Z., & Sud, P. (2020). Health effects of household air pollution related to biomass cook stoves in resource limited countries and its mitigation by improved cookstoves. *Environmental Research, 186*. Available from https://doi.org/10.1016/j.envres.2020.109574.

Shakoor, A., Arif, M. S., Shahzad, S. M., Farooq, T. H., Ashraf, F., Altaf, M. M., Ahmed, W., Tufail, M. A., & Ashraf, M. (2021). Does biochar accelerate the mitigation of greenhouse gaseous emissions from agricultural soil?-A global meta-analysisDoes biochar accelerate the mitigation of greenhouse gaseous emissions from agricultural soil? *Environmental Research, 202*111789.

Smith K.R. Pillarisetti A. (2017). The International Bank for Reconstruction and Development / The World Bank, Washington (DC) Household Air Pollution from Solid Cookfuels and Its Effects on Health. https://www.ncbi.nlm.nih.gov/books/NBK525225.

Sundberg, C., Karltun, E., Gitau, J. K., Kätterer, T., Kimutai, G. M., Mahmoud, Y., Njenga, M., Nyberg, G., Roing de Nowina, K., Roobroeck, D., & Sieber, P. (2020). Biochar from cookstoves reduces greenhouse gas emissions from smallholder farms in Africa. *Mitigation and Adaptation Strategies for Global Change, 25*(6), 953–967. Available from https://doi.org/10.1007/s11027-020-09920-7, https://doi.org/10.1007/s11027-020-09920-7.

Roth, C. Micro-gasification: Cooking with gas from dry biomass. (2014), Available from: https://energypedia.info/images/f/f6/Micro_Gasification_Cooking_with_gas_from_biomass.pdf.

Chapter 19

Biochars for the removal of toxic gaseous contaminants: state-of-the-art and future directions

Robinah Kulabako[1], Swaib Semiyaga[1], Charles Niwagaba[1], Chimdi Muoghalu[2] and Musa Manga[2,3]

[1]Department of Civil and Environmental Engineering, School of Engineering, College of Engineering, Design, Art and Technology, Makerere University, Kampala, Uganda, [2]Department of Environmental Sciences and Engineering, Gillings School of Global Public Health, University of North Carolina at Chapel Hill, Chapel Hill, NC, United States, [3]Department of Construction Economics and Management, College of Engineering, Design, Art and Technology (CEDAT), Makerere University, Kampala, Uganda

Chapter outline

19.1 Introduction	361
19.2 Sources of toxic gaseous contaminants	362
19.3 The need for gaseous contaminant removal	363
19.4 Removal of toxic gaseous contaminants from air/flue gases	363
19.5 Biochar properties and mechanisms for toxic gaseous contaminant removal	365
19.5.1 Biochar properties	365
19.5.2 Mechanisms of toxic gaseous contaminant removal by biochars	367
19.5.3 Adsorption kinetics, isotherm, and thermodynamics	370
19.6 Engineered biochars for enhanced gaseous contaminants removal	371
19.6.1 Physical activation	371
19.6.2 Chemical activation	371
19.6.3 Metal impregnation	371
19.6.4 Heteroatoms doping	371
19.7 Regeneration of spent biochars	372
19.8 Future directions and recommendations	372
19.9 Conclusion	373
References	373

19.1 Introduction

Toxic gaseous contaminants comprised of various harmful gases emitted into the atmosphere have emerged as a key environmental and public health concern. Gaseous contaminants can broadly be categorized into inorganic and organic contaminants (Manga & Muoghalu, 2024; Shahi Khalaf Ansar et al., 2022; Yaashikaa, Kumar, Varjani, Saravanan, 2020). The inorganic contaminants include nitrogen oxides (NO_x), sulfur dioxide (SO_2), toxic heavy metals (Pb, Cr, Hg), carbon monoxide (CO), carbon dioxide (CO_2), and ammonia emissions. Organic contaminants, on the other hand, include volatile organic compounds (VOCs), poly organic pollutants (POPs), polycyclic aromatic hydrocarbons (PAHs), odorous compounds, dioxins, and furans (Gwenzi, Chaukura, et al., 2021; Meij & Winkel, 2007). Thus, polluted air contains an unacceptable concentration of the above-mentioned contaminants.

Exposure to gaseous contaminants at high concentrations is toxic to living organisms, posing a risk to public health and the environment (Shahi Khalaf Ansar et al., 2022). These gaseous contaminants can easily migrate from one place to another; thus, it is necessary that they are removed from the air. Conventionally used techniques for removing gaseous contaminants include incineration, scrubbing, filtration, oxidation, and the use of ultraviolet (UV) radiation (Wang & Ku, 2003). However, the aforementioned techniques have limitations such as high cost, ineffectiveness, and generation of polluted wastewater (Talaiekhozani et al., 2021). This has led to the need for developing more eco-friendly, inexpensive, and innovative techniques for gaseous contaminant removal. One of the most promising approaches is the utilization of biochar. Biochar is derived from the pyrolysis of organic matter and exhibits unique properties for the adsorption and sequestration of a variety of gaseous pollutants, as a safe method for curtailing air pollution (Manga et al., 2023).

The chapter presents information on the sources of gaseous contaminants, as well as their effects on public health, removal efficiency of gaseous contaminants by biochar, mechanisms involved in the removal process, as well as methods of regenerating spent biochar (Fig. 19.1). The insights provided in this chapter are of importance to researchers, policymakers, and environmentalists interested in harnessing biochar technology to improve air quality and reduce industrial emissions, in the quest for cleaner environment and a more sustainable future.

19.2 Sources of toxic gaseous contaminants

Toxic gaseous contaminants are produced by both natural processes and human activities. Human activities mainly include industrial processes, agricultural production, and waste management and/or disposal (Table 19.1). Industrial and waste management processes (such as industrial systems fueled by biomass and coal, and combustion of biomass, solid wastes, and sludges) produce flue gases that contain various toxic contaminants (Chmielewski, 1999; Meij & Winkel, 2007). In addition, the increasing demand for a circular economy has also resulted in the massive adoption of waste-to-energy technologies but these are associated with various toxic gas emissions (Cheng et al., 2020; Manga, Semiyaga, et al., 2024).

A number of common practices are linked to the emission of toxic gas contaminants. For example, processes such as incineration are practiced in the disposal of hazardous wastes such as healthcare wastes and livestock carcasses with infectious diseases, all of which release toxic gas contaminants (Table 19.1). In a number of low- and middle-income countries, industrial boilers and cookers relying on biomass are in common use in institutions such as schools, hospitals, and prisons. Combustion of biomass fuels leads to the release of toxic gases such as VOCs, POPs, SO_2, and NO_x. Fossil fuels, as well as oil, natural gas, and coal account for about 86% of the world's main energy consumption, whereas coal is mainly used by industrial boilers and power plants to generate steam and electricity (Chen et al., 2022). It is crucial to highlight the substantial emissions of particulate matter, sulfur oxides (SO_x), and PAHs stemming from the combustion of coal, contributing to the deterioration of air quality. The interaction between various gaseous precursors such as NH_3, NO_x, and SO_x gives rise to the creation of secondary pollutants, such as photochemical smog (Mochida et al., 2000). There is a need for robust treatment methods for the removal or control of gaseous emissions to abate the negative health effects they pose to the public.

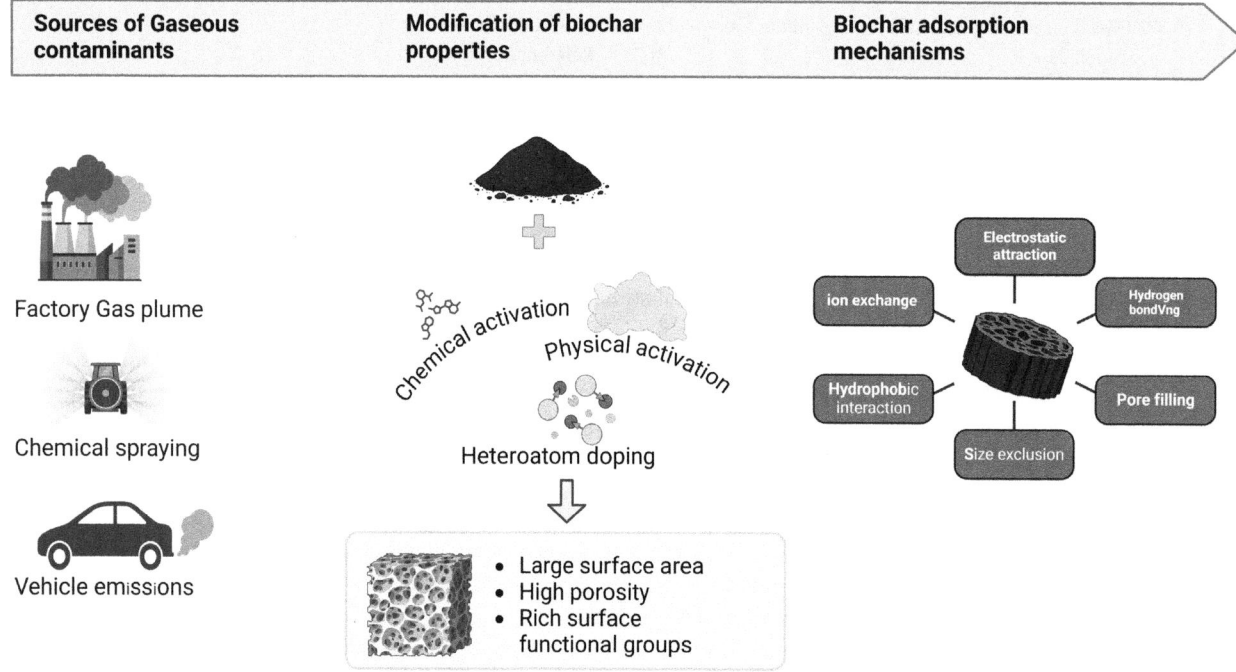

FIGURE 19.1 Summary depiction of sources, modification techniques, and adsorption mechanisms of gaseous contaminants by biochars. *No Permission Required.*

TABLE 19.1 Anthropogenic sources of toxic gaseous contaminants.

Category	Source	Nature of gas contaminant generated	References
Industrial processes	Incinerators	H_2S, SO_2, CO_2, NO_x, and elemental mercury, polychlorinated dibenzodioxins (PCDDs), polychlorinated dibenzofurans (PCDFs), and polychlorinated biphenyls (PCBs),	Gwenzi, Chaukura, et al. (2021); Singh et al. (2008)
	Crude oil and natural gas refinery (combustion of fossil fuels)	H_2S, SO_2, and Hg.	Chmielewski (1999)
	Petrochemical refinery	H_2S, Hg, CO_2, and SO_2	Jumina et al. (2021)
	Smelters	Toxic metals and organic contaminants from burning of coal: PAH, dioxins, and furans.	Meij & Winkel (2007)
	Kilns	Flue gases contain toxic contaminants (PAH and heavy metals).	Yin et al. (2021)
	Cremation	Elemental mercury (from dental amalgamation), POPs, PCDDs, PCBs, and polychlorinated dibenzofurans (PCDFs).	González-cardoso et al. (2020); Gwenzi, (2021)
	Boilers and cookers	H_2S, SO_2, CO_2, NO_x, CO, and fine particles (PM2.5).	Tumwesige et al. (2017)
Agricultural production	Animal husbandry (poultry, cattle, and pig production)	Odors, H_2S, VOCs, and NH_3	Chen et al. (2020)
	Crop drying (such as coffee, tea, and tobacco)	CH_4, NO_X, CO_2, and SO_2	Kaveh et al. (2020)
Waste management and disposal	Liquid/semisolid wastes (Wastewater / FS treatment plants)	Odors, H_2S, VOCs, and methane (CH_4).	Chen et al. (2020)
	Biogas production (landfills and anaerobic digestion facilities)	CH_4 (45–70 vol.%), CO_2 (24–40 vol.%)	Jung et al. (2021); Rivera-Montenegro et al. (2023)
	Combustion of Solid wastes and dried sludges - FS/SS (Waste-to-energy)	VOCs, POPs, SO_2, NO_x, CO_2, CO, and fine particles (PM2.5)	Cheng et al. (2020)

No Permission Required.

19.3 The need for gaseous contaminant removal

To effectively address environmental pollution, it is crucial to identify and comprehend the most prevalent toxic gaseous contaminants, their effects when present, and potential processes for their removal. Some gaseous contaminants such as NO_x and SO_2 contribute to the development of acid rain and ground-level ozone, which can damage ecosystems, aquatic life, and built structures. VOCs play a significant role in smog formation and pose health risks, particularly in urban areas with high population density. Ammonia emissions, originating from agricultural and industrial sources, can lead to air and water quality problems, with consequences, including eutrophication and compromised aquatic ecosystems. Consequently, the imperative to develop effective methods for the removal and mitigation of these toxic gaseous contaminants is more apparent. Common examples of toxic gaseous contaminants and their related effects are presented in Table 19.2.

19.4 Removal of toxic gaseous contaminants from air/flue gases

Toxic gaseous contaminants can either be controlled from the source to reduce their emission or removed from contaminated sources by treatment. The key methods of emissions control include the use of: (i) electrostatic precipitators, (ii)

TABLE 19.2 Relevance of gaseous contaminants and their removal mechanisms.

Contaminant class	Gaseous contaminant(s)	Reason for removal from gas stream?	Reduction or removal mechanisms	References
Acid gases	H_2S	Deleterious to public health corrosive to metal pipelines poisons the gas turbine, and the fuel-processing catalysts and anode catalysts in fuel cell system poison of metal catalysts in chemical plants deactivate metal catalysts in trace concentrations Also, present a safety risk	Adsorption processes and wet processes using basic organic amine solvents	Bamdad et al. (2018); Barbusinski et al. (2021); Wang et al. (2007)
	CO_2	A greenhouse gas which contributes to global warming and climate change	Adsorption, Absorption, and membrane separation	Mochida et al. (2000)
	SO_2	One of the prominent air pollutants globally, it readily undergoes reactions with other substances to produce detrimental components such as sulfuric acid, sulfurous acid, and sulfate particles. Leads to acid rain, respiratory issues, and damage to vegetation	Physical and chemical adsorption	Braghiroli et al. (2019)
Toxic metals	Hg	Gaseous elemental mercury (Hg^0), gaseous oxidized mercury (Hg^{2+}), and particulate mercury (Hg^P). The Hg^0 is usually converted to Hg^{2+} or Hg^P after a sequence of chemical reactions. It is in low concentrations (10 ppm) in natural gas and oil but its accumulation in refinery plants can destroy equipment and catalysts. Also, poses public health and environmental threats.	Amalgamation, physical, chemical, and reactive absorption	Cho, Lee, Kim, Song, Lee, Tsang, Chen, Park, Lee, Jung, Kwon (2023); Zhao et al. (2019)
Odorous compounds	NH_3	Ammonia emissions stem from agricultural and industrial activities. Ammonia is a corrosive, colorless gas characterized by a highly pungent Possesses corrosive and exothermic characteristics, thereby capable of causing harm to the eyes, skin, as well as mucous membranes in the mouth and respiratory tract. In the environment, ammonia can contribute to air and water quality problems, including eutrophication and damage to aquatic ecosystems.	Scrubbing, adsorption, and biofiltration.	Barbusinski et al. (2021)
Volatile organic compounds	Toluene ethyl acetate	originate from storage tanks, venting of process vessels, leaks from piping and equipment, wastewater streams, and heat exchange systems. key contributor to smog formation plays a role in the generation of ground-level ozone. Contributes to change of climate, hinders plant growth	Oxidation, bio-filtration, absorption, adsorption, condensation and membrane separation	Chen et al., (2020); Khan & Ghoshal (2000)
	Siloxanes	Corrosion of pipelines and impairment of combustion engines. Induce significant operational issues in the utilization of biogas.	Adsorption	Rivera-Montenegro et al. (2023)
Nitrogen oxides	NO_x	Contribute to the formation of acid rain and ground-level ozone, impacting air quality and ecosystem health.	Electrostatic precipitation and bag filtration	Mochida et al. (2000)
Poly-halogenated aromatic compounds	PM and PAH	PAHs originate incomplete combustion of fossil fuels and are toxic and carcinogenic.	Bag filtration	Chen et al. (2022)

No Permission Required.

fabric filters, (iv) advanced carbon injection, and (v) adsorption by biochar (Zhao et al., 2019). Biochar, a carbonaceous material produced through pyrolysis of biomass under high temperatures and anoxic conditions, has recently been applied in removal of gaseous contaminants (Cheng et al., 2020; Cho, Lee, Kim, Song, Lee, Tsang, Chen, Park, Lee, Jung, Kwon 2023; Manga et al., 2023). Biochars have been found to removal various gaseous contaminants such as NOx, SO_2, CO_2, H_2S, VOCs, and elemental mercury (Bamdad et al., 2018; Gwenzi, Chaukura, et al., 2021). Biochar has gained prominence as a low-cost, low-tech, and sustainable option for the treatment of gaseous contaminants. In addition, the biochar can be engineered to improve removal efficiency.

Biochar possesses a porous sorptive structure, large surface area, and chemically active sites (Manga, Muoghalu, et al., 2024), making it ideal for the adsorption of diverse toxic gaseous contaminants (Table 19.3). In addition, biochar can be modified or engineered (through physical, chemical, and thermal activation processes) for enhanced removal of selected contaminants air or flue gases (Yang et al., 2010). Therefore, varying properties of biochar from diverse biomass can be engineered to suit biochar for intended use or particular contaminants to be removed. For example, Hg has been noted to be removed through adsorption by the use of sulfur-impregnated activated, which can be achieved through the engineered crop residue biochar (Cho, Lee, Kim, Song, Lee, Tsang, Chen, Park, Lee, Jung, Kwon, 2023).

Biochar can easily be made from biomass such as household organic wastes, animal wastes (cow manure, poultry manure, and pig manure), water hyacinth, algae, agricultural wastes (wheat straw, cotton seed hulls, chili seeds, and bagasse), and sewage sludges (Ainomugisha et al., 2024; Chen et al., 2020; Cho, Lee, Kim, Song, Lee, Tsang, Chen, Park, Lee, Jung, Kwon, 2023; Gwenzi, Chaukura, et al., 2021). It is important to note that given the nature of the feedstock to be used, different pretreatment methods are applied in the production of biochar and these have a bearing on the properties of the resulting biochar to be used in contaminant removal. Therefore, the selection of feedstock significantly impacts the resulting biochar's surface area, pore structure, and chemical composition, and these factors are essential in determining its efficacy in adsorbing specific toxic gaseous contaminants. Table 19.3 presents the various biochars that have been used in adsorbing toxic gas contaminants.

19.5 Biochar properties and mechanisms for toxic gaseous contaminant removal

19.5.1 Biochar properties

Biochar possesses several key properties that are relevant for the efficient removal of toxic gaseous contaminants. These properties collectively render biochar an ideal material for the removal of toxic gaseous contaminants from air or flue gases. Biochar's porous structure, extensive surface area, chemical functionality, thermal stability, and carbon sequestration potential make it suitable for contaminated air remediation.

19.5.1.1 Porosity and surface area

Biochar with increased surface area and high porosity is known for its enhanced sorption properties(Manga, Muoghalu, et al., 2024; Muoghalu et al., 2024). The porosity in biochar is developed during pyrolysis, where water loss occurs during dehydration. Biochar pores can be micro (<2 nm), meso (2–50 nm), or macro (>50 nm), where macropores facilitate the diffusion of gases deep into the biochar, enhancing the removal process (Yaashikaa, Kumar, Varjani, Saravanan, 2020). Commercially available activated carbon generally possesses more surface area than biochar from biomass. In addition, biochar produced without activation has a lower surface area and porosity. To address this, an activation process, involving physical and chemical methods, is employed to increase both porosity and surface area (Li, Dong, da Silva, de Oliveira, Chen, Ma 2017). The porous biochar structure enables physical adsorption, where contaminants are captured within the porous matrix. The more surface area available, the more binding sites are provided for the adsorption of toxic gaseous contaminants. This feature allows biochar to capture and immobilize a substantial quantity of gaseous pollutants.

19.5.1.2 Surface functional groups

Biochar's surface is enriched with vital functional groups, such as carboxylic (-COOH), hydroxyl (-OH), amine, and lactonic groups, all of which contribute significantly to its sorption properties. The key determinants shaping the surface functional groups of biochar are the type of biomass and the temperature used during its production (Li, Dong, da Silva, de Oliveira, Chen, Ma, 2017). Additionally, there is a potential trade-off, where an augmentation of properties such as pH, surface area, and porosity might be accompanied by a reduction in the abundance of functional groups within biochar. These functional groups are actively involved in chemical interactions with toxic gaseous contaminants,

TABLE 19.3 Removal of inorganic and organic gaseous contaminants by use of biochar.

Category	Class	Gas contaminant	Biochar type/Source	Removal efficiency	References
Inorganic	Acid gases	CO_2	Waste biomass (pig manure, sewage sludge, wheat straw, coffee husks, vine shoots, sawdust)	68.0%	Chen et al., 2020; Cho, Lee, Kim, Song, Lee, Tsang, Chen, Park, Lee, Jung, Kwon (2023)
			Sugarcane bagasse and hickory wood biochar	65.6%–72.0%	Creamer et al. (2014)
		H_2S	Compost is followed by biofilter, rice straw, wheat straw, oil palm shell, and coconut shell.	99.0%	Das & Lens (2022)
			Wood chips	98.0%	Kanjanarong et al. (2017)
		SO_2	Wood residue	KOH activation (57.0%)	Braghiroli et al. (2019)
			Corn cob	CO_2 activated (64.0%)	Shao et al. (2018)
	Toxic metals	Mercury (in flue gases)	Rice husk	Modified with hydrogen bromide (HBr) (97%–98.0%); Calcium bromide impregnated (60.0%–78.0%); Ammonium bromide impregnated (80.0%–90.0%)	Zhu et al. (2016)
			Municipal solid waste	Ammonium chloride impregnated biochar (50.0%–100.0%)	Li et al. (2015)
	Nitrous oxide	NO_x	Oakwood chip	CO_2 activation 90.0%–98.0%	Díaz-Maroto et al. (2023)
			oil rape seed	60.0%–99.0%	
Organic	Volatile organic compounds	Toluene	Biomass (grass, Neem)	86.0%	Bhandari et al. (2014)
		Benzene vapor	Cell-immobilized bamboo biochar	50%–100.0%	Liu et al. (2019)
		Dimethyl disulfide (DMDS) and dimethyl trisulfide (DMTS)	poultry litter, swine manure, oak, and coconut shell		Hwang et al. (2018)

No Permission Required.

which can lead to chemical reactions, transformations, or conversions of the contaminants into less harmful compounds. Biochar's chemical composition plays a crucial role in its capacity to neutralize and mitigate the impact of toxic gaseous pollutants.

19.5.1.3 Biochar stability

Biochar contains a high percentage of stable carbon that can be locked away in soils for extended periods. The assessment of biochar's ability to resist both biotic and abiotic soil degradation is crucial in determining its carbon sequestration capacity (Leng, Huang, Li, Li, Zhou, 2019). Biochar contributes to carbon sequestration, which not only addresses toxic gaseous contaminants but also supports broader efforts aimed at lowering greenhouse gas emissions and alleviating the impacts of climate change. The temperature applied during the pyrolysis process has been traditionally used as an indicator of stability. Biochar is known for its exceptional thermal stability, which enables it to withstand a wide range of environmental conditions, including high temperatures, without significant degradation. The elemental composition of biochar is indicative of C-C bonds or aromaticity (Ainomugisha et al., 2024; Yaashikaa, Kumar, Varjani, Saravanan, 2020). Additionally, the stability of biochar is affected by its properties, including porosity, pH, minerals, sorption mechanism, surface area, and particle size (Kaboggoza et al., 2024). Therefore, the ability of biochar to maintain its structural integrity over time is particularly valuable in applications requiring sustained removal of toxic gaseous contaminants.

19.5.2 Mechanisms of toxic gaseous contaminant removal by biochars

Biochar utilizes a variety of mechanisms for the effective removal of toxic gaseous contaminants from air such as electrostatic attractions, hydrogen bonding, hydrophobic interactions, complexation, coprecipitation, ionic attraction, ion exchange, pore filling, and size exclusion (Gwenzi, Chaukura, et al., 2021) (Fig. 19.2). These mechanisms collectively

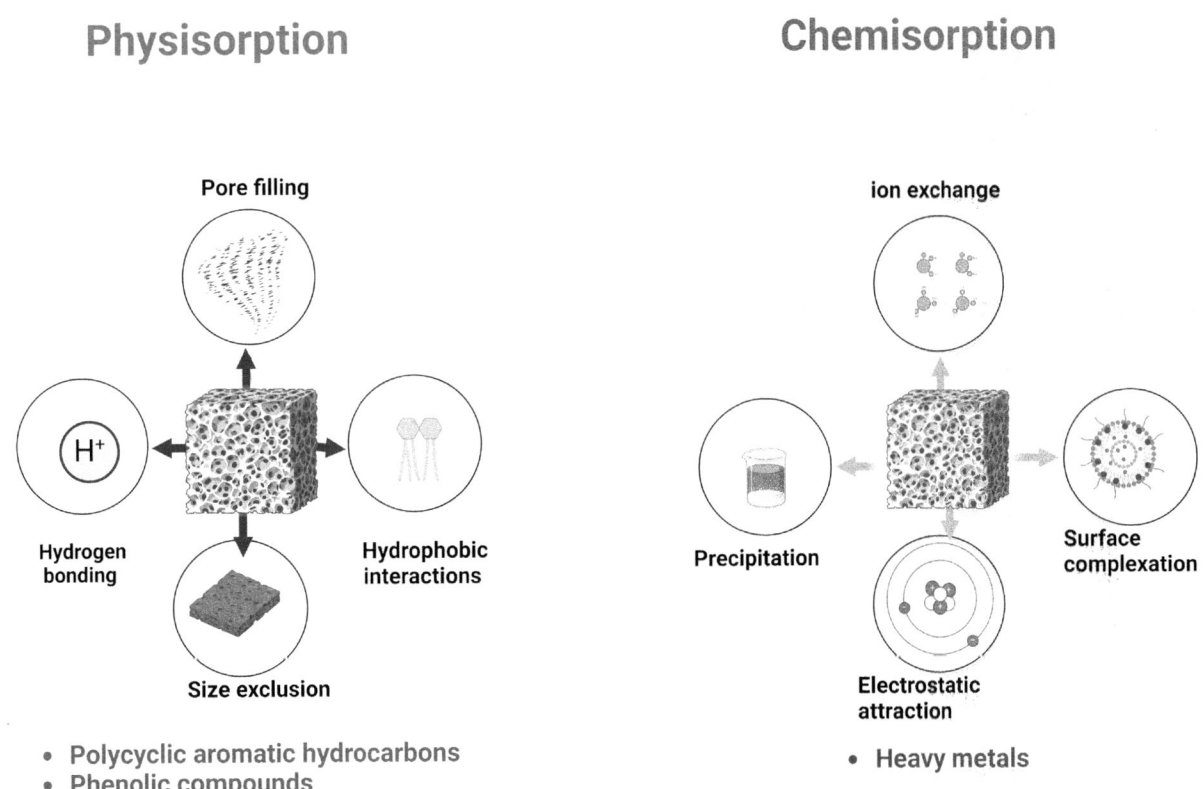

FIGURE 19.2 Mechanisms responsible for contaminant removal by biochars. *No Permission Required.*

support biochar's ability in adsorbing toxic gasses. Additionally, the porous property of biochar enables physical adsorption, the chemical composition allows for chemical reactions, and sorption and catalytic effects further augment its contaminant removal capabilities.

19.5.2.1 Physical adsorption

Physical adsorption is a fundamental mechanism through which biochar captures toxic gaseous contaminants. Biochar's porous structure, characterized by macropores, mesopores, and micropores, provides a matrix that offers an extensive surface area. Gaseous contaminants are drawn to this surface and subsequently trapped within the pores of biochar. Adsorption onto the biochar surface takes place via van der Waals forces and hydrogen bonds involving functional groups, as well as filtration, electrostatic interaction, ion exchange, and surface complexation (Cheng et al., 2021; Li et al., 2019). The porous structure plays a vital role in accommodating and immobilizing gaseous pollutants within the biochar matrix. This physical adsorption mechanism is crucial for adsorbing a gamut of toxic gases, including VOCs, CO_2, elemental mercury, and ammonia (Ahmed et al., 2018).

19.5.2.2 Electrostatic attractions

Electrostatic attractions play a vital role in gaseous contaminant removal by biochar. The surface of biochar contains charged sites, often in the form of functional groups with positive or negative charges. When toxic gaseous contaminants are present in the air, they may carry opposite charges to those on the biochar surface. This creates a strong electrostatic attraction between the charged contaminants and the biochar, leading to their effective removal. The electrostatic interactions result in the contaminants being bound to the surface of the biochar, preventing their release back into the air. This mechanism is particularly effective for contaminants that possess a significant charge, such as ammonia or SO_2.

19.5.2.3 Hydrogen bonding

The complex structure of biochar contains sites capable of forming hydrogen bonds, typically involving oxygen or nitrogen-containing functional groups. When gaseous contaminants possessing hydrogen bond acceptor or donor functionalities are present, such as formaldehyde or VOCs, biochar engages in hydrogen bonding interactions with these contaminants (Ahmed et al., 2018). This interaction facilitates the effective capture and retention of the contaminants on the biochar surface. By forming stable hydrogen bonds, biochar acts as a molecular "trap," immobilizing the contaminants and preventing their diffusion back into the air. The specificity of hydrogen bonding enhances the selectivity of biochar in targeting certain classes of gaseous pollutants that possess a significant charge, such as ammonia or SO_2.

19.5.2.4 Hydrophobic interactions

Within the complex structure of biochar, hydrophobic regions or domains are present, creating an environment that repels water. This hydrophobic nature enables biochar to attract and capture gaseous contaminants with hydrophobic characteristics, such as VOCs and other organic pollutants (Ahmed et al., 2018; Khan & Ghoshal, 2000). When these contaminants encounter the hydrophobic sites on the biochar surface, they undergo hydrophobic interactions, resulting in their adsorption onto the biochar. This physical binding prevents the contaminants from dispersing back into the air.

19.5.2.5 Complexation

The structure of biochar incorporates functional groups with the ability to form stable complexes with specific contaminants, particularly heavy metals and other complex-forming species. When gaseous pollutants carrying these contaminants are present, biochar engages in complexation interactions, forming strong bonds with the contaminant molecules (Li, Dong, da Silva, de Oliveira, Chen, Ma, 2017). This results in the effective sequestration and immobilization of the contaminants, preventing their release back into the air. For example, heavy metal ions, such as lead or mercury, can be bound to biochar through complexation, transforming them into less hazardous forms.

19.5.2.6 Coprecipitation

Biochar facilitates the formation of solid precipitates alongside the contaminants, leading to the effective removal of certain ions. When gaseous contaminants carrying ions susceptible to precipitation, such as SO_2, come into contact with biochar, chemical reactions occur, resulting in the formation of insoluble precipitates (Meij & Winkel, 2007; Yin et al., 2021). These solid compounds encapsulate the contaminants, preventing their dispersion into the air. The coprecipitation mechanism is particularly advantageous for addressing the challenge of removing specific ionic pollutants, as the resulting precipitates offer a stable and easily separable form of the contaminants (Chen et al., 2019).

19.5.2.7 Ionic attraction

The surface of biochar may contain charged sites, allowing it to attract and capture ions present in airborne contaminants. When gaseous pollutants containing ions with opposite charges encounter biochar, ionic interactions come into play. The biochar acts as a magnet, drawing in and retaining the charged ions on its surface (Yuan et al., 2022). This process effectively immobilizes the contaminants, preventing their release back into the air. For example, nitrate ions (NO_3^-) from NO_x can be attracted to positively charged sites on biochar through ionic attraction (La & Hettiaratchi, 2022).

19.5.2.8 Ion exchange

Biochar contains exchangeable cations on its surface, such as calcium (Ca^{2+}), which can undergo a process of exchanging ions with those present in the gaseous contaminants. When biochar comes into contact with gaseous pollutants carrying ions, such as ammonium ions (NH_4^+), ion exchange occurs (Chen et al., 2020). The biochar releases its exchangeable cations and takes up the contaminant ions, effectively capturing and immobilizing them. For instance, biochar's ion exchange capacity can be harnessed for the removal of ammonia, a common gaseous contaminant, through the exchange of ammonium ions with biochar's surface cations (La & Hettiaratchi, 2022).

19.5.2.9 Pore filling

The porous structure of biochar comprises a network of channels and voids, providing ample space for the physical adsorption of contaminants. When gaseous pollutants encounter biochar, they enter these pores and physically fill the available spaces within the material (Pulido-Novicio et al., 2001). This process, known as pore filling, leads to the effective trapping and retention of the contaminants within the biochar matrix. Methane, a greenhouse gas, is an example of a gaseous pollutant that can be captured through pore filling (Zhang et al., 2019). The physical confinement of pollutants within the porous structure prevents their release back into the air.

19.5.2.10 Size exclusion

The porous nature of biochar imparts specific size constraints, allowing the material to selectively exclude contaminants based on their molecular sizes (Cho, Lee, Kim, Song, Lee, Tsang, Chen, Park, Lee, Jung, Kwon, 2023). As gaseous pollutants traverse the biochar matrix, smaller molecules, such as oxygen and nitrogen, may pass through the pores, while larger pollutants, such as benzene or other VOCs, are excluded (Li et al., 2020). This selective sieving effect, known as size exclusion, enables biochar to target and capture contaminants that surpass a certain molecular size threshold.

19.5.2.11 Chemical reactions

The chemical composition of biochar, characterized by diverse functional groups, enables it to undergo specific reactions with certain contaminants. When gaseous contaminants come into contact with biochar, chemical transformations occur, resulting in the conversion of the contaminants into less detrimental compounds. For instance, SO_2 can undergo chemical reactions with biochar, leading to the formation of sulfate compounds, which alteration immobilizes the contaminant (Braghiroli et al., 2019).

19.5.2.12 Sorption and catalytic effects

Sorption and catalytic effects collectively represent a dynamic mechanism by which biochar effectively removes gaseous contaminants from the air (Yuan et al., 2022; Zhang et al., 2019). The porous structure of biochar allows for sorption, encompassing both adsorption and absorption processes, where contaminants physically adhere to the material's surface or are taken up within its matrix (Gwenzi, Chaukura, et al., 2021). Concurrently, biochar may also exhibit catalytic properties, facilitating chemical reactions that transform or degrade contaminants into less harmful forms (Li, Dong, da Silva, de Oliveira, Chen, Ma, 2017). This dual functionality enhances the contaminant removal capabilities of biochar, as it not only physically captures pollutants but also catalytically contributes to their degradation.

19.5.3 Adsorption kinetics, isotherm, and thermodynamics

To better understand the mechanisms through which biochar removes gaseous contaminants, various kinetic and isotherm models are employed to analyze the treatment process. These models are instrumental in illustrating the dynamics of contaminant adsorption and have been extensively used in studies involving wastewater treatment using biochar (de Caprariis et al., 2017; Qiu et al., 2022; Ye et al., 2022). However, their application in the context of gaseous contaminant removal remains underexplored, necessitating further research in this area.

19.5.3.1 Adsorption kinetics

Adsorption kinetics describes the rate and mechanism by which adsorbate molecules are transferred from the gas phase to the surface of adsorbent material, such as biochar (Gwenzi, Chaukura, et al., 2021). Understanding adsorption kinetics is crucial for developing and optimizing adsorption-based treatment technologies. The adsorption kinetics can be studied by fitting experimental data to various models, such as the pseudo-first-order and pseudo-second-order models, or through thermogravimetric analysis (Bamdad et al., 2019; Gwenzi, Chaukura, et al., 2021). The pseudo-first-order and pseudo-second-order models are the most commonly used for studying adsorption kinetics. A better fit with the pseudo-first-order model indicates that the adsorption process is primarily driven by physical adsorption (physisorption), whereas a better fit with the pseudo-second-order model suggests that chemical adsorption (chemisorption) is the dominant mechanism. The extent of physisorption or chemisorption depends largely on the characteristics of the gaseous contaminant and the surface properties of the biochar. For example, CO_2, due to its nonpolar nature and low reactivity is typically removed through physical adsorption, and the experimental data often fits the pseudo-first-order model (Ammendola et al., 2017; Ding & Liu, 2020; Wang et al., 2015).

19.5.3.2 Isotherm modeling

Isotherm models are essential for understanding the interaction between adsorbate molecules and the surface of biochar (Gwenzi, Chaukura, et al., 2021; Li et al., 2021). These models include the Langmuir, Freundlich, Sips, Toth, and Redlich–Peterson equations (Li et al., 2021). Among these, the Langmuir and Freundlich isotherms are the most extensively studied. The Langmuir isotherm describes homogeneous monolayer adsorption, where each adsorption site on the biochar surface has an equal affinity for the adsorbate. In contrast, the Freundlich isotherm accounts for heterogeneous multilayer adsorption, indicating varied affinities across the surface sites.

Isotherm models are influenced by several factors, including adsorption temperature, adsorbate concentration, and the feed gas flow rate. Numerous studies have investigated the isotherm modeling of various gaseous contaminants on biochar surfaces, often finding that the adsorption mechanisms align more closely with the Freundlich model. For example, Li et al. (2021) studied the adsorption of toluene and ethyl acetate (VOCs) on maize straw biochar and found a higher correlation coefficient (R^2) for the Freundlich isotherm compared to the Langmuir model, attributing this to the heterogeneous nature of the maize straw biochar surface. Similarly, Mohamed et al. (2024) examined the adsorption of ammonia from toxic gases with different types of activated carbon. Their isotherm tests at 25°C showed a better fit for the Freundlich model than the Langmuir model. Another study by Bamdad et al. (2019) on the adsorption of CO_2 on woodchip biochar also reported that the Freundlich model ($R^2 = 0.99$) provided a better fit than the Langmuir isotherm ($R^2 = 0.86$), further supporting the prevalence of heterogeneous surface interactions in biochar adsorption processes.

19.5.3.3 Thermodynamics study

The thermodynamic analysis of the adsorption process is typically carried out to determine several critical parameters: Gibbs free energy (ΔG), enthalpy change (ΔH), and entropy change (ΔS) (Gwenzi, Chaukura, et al., 2021). Gibbs free energy (ΔG) is crucial for evaluating the spontaneity of the adsorption process; a negative ΔG indicates that the process occurs spontaneously (Bamdad et al., 2019). Enthalpy change (ΔH) helps to identify the nature of the adsorption process, distinguishing whether it is endothermic (absorbs heat) or exothermic (releases heat). Entropy change (ΔS) provides insight into the degree of disorder or randomness associated with the adsorption process (Bamdad et al., 2019). By comprehensively understanding these thermodynamic parameters, researchers can gain deeper insights into the energetics and feasibility of the adsorption process, ultimately aiding in the optimization of biochar as an effective adsorbent for gaseous contaminants.

19.6 Engineered biochars for enhanced gaseous contaminants removal

Biochar produced without any modification exhibits limited surface functional groups (i.e., only $C-O, C=O$, and OH groups are present) and features a smaller surface area and micropores. These characteristics impede its effectiveness in the removal of a broader spectrum of contaminants (Xiang, Zhang, Chen, Zou, He, Hu, Tsang, Sik, et al., 2020). Biochars can be subjected to surface modification and functionalization to augment the removal of specific toxic gaseous contaminants. This involves changing the biochar's surface properties by adding specific functional groups or coatings, which target particular contaminants. The functionality of biochar can be enhanced or modified through physical and chemical methods, as well as through heteroatom doping and impregnation with nanomaterials (Cho, Lee, Kim, Song, Lee, Tsang, Chen, Park, Lee, Jung, Kwon, 2023).

19.6.1 Physical activation

Physical activation of biochar entails the conversion of surface carbon and hydrogen into carbon dioxide and water vapor (Sajjadi et al., 2018). Physical activation is a two-step process: carbonization of feedstock, followed by activation of biochar with oxidants, mainly CO_2 and water vapor or mixture. The process takes place at extremely high temperatures (800°C–1100°C), which increase the porosity and specific surface area of the biochar (Bansal et al., 1988). For example, subjecting cotton shell biochar to physical activation using steam and CO_2 independently, and subsequently combining them, resulted in an increase in the specific surface area of the activated carbon by over 2000 $m^2 g^{-1}$ (Yang et al., 2010).

19.6.2 Chemical activation

Chemical activation entails incorporating chemical activation agents into biochar. These activation agents are either acids (H_2SO_4, HCl, H_3PO_4, HNO_3, and CO_2), alkalis (KOH, NaOH, and K_2CO_3), or oxidizing agents (H_2O_2 and $KMnO_4$) (Muoghalu et al., 2023). Chemical activation can either be accomplished through single or double-step processes. In a single-step chemical activation process, pyrolysis (biochar production) and activation process take place in one reactor while different reactors are utilized in a two-step process (Cho, Lee, Kim, Song, Lee, Tsang, Chen, Park, Lee, Jung, Kwon, 2023). For example, chemical activation of hickory-wood biochar with $KMnO_4$ increased the surface area from 101 to 205 $m^2 g^{-1}$ (Wang et al., 2015).

19.6.3 Metal impregnation

In impregnation, the surface modification of biochar is achieved by incorporating metal salts, oxides, hydroxides, and nanoparticles. This enhances contaminant adsorption capacity, leveraging their abundance, cost-effectiveness, environmental friendliness, and chemical stability (Li, Dong, da Silva, de Oliveira, Chen, Ma, 2017). Biochar impregnation can be carried out either by soaking the biomass followed by pyrolysis or by pyrolyzing the biochar first and then impregnating it (Ahmed, Zhou, Ngo, Guo, Chen, 2016). Impregnation of biochar improves its properties and hence, enhancement in removal of various toxic gaseous contaminants such as SO_2, H_2S, and elemental mercury from flue gases. For example, the impregnation of crop residue biochars with various metals (Fe, Cu, Ce, Ca, and Mn) was conducted to enhance the adsorption of H_2S, SO_2, and $Hg°$. The impregnated biochars demonstrated improved adsorption capacity, especially for SO_2 (Cho, Lee, Kim, Song, Lee, Tsang, Chen, Park, Lee, Jung, Kwon, 2023).

19.6.4 Heteroatoms doping

Heteroatoms doping involves intentionally introducing atoms of different elements into a biochar's crystal lattice to modify its properties for specific applications. Graphite carbon materials embedded with nitrogen-containing functional groups, such as pyridinic-N, pyrrolic-N, and pyridonic-N, exhibit higher adsorption energy when compared to functional groups lacking nitrogen elements such as SO_2 and CO_2 (Sun et al., 2016). Enhanced adsorption energy for the N-doping is a result of an increase in surface polarity is achieved by redistributing the electronic density on the carbon structure. For example, the presence of N-doping with biochar increased adsorption capacity of SO_2 from 57.8 to 156.2 mg/g, when the binding energy increased from -39.36 to 49.92 kJ/mol (Sun et al., 2016).

19.7 Regeneration of spent biochars

The regeneration of spent biochar used in gaseous contaminant removal is a crucial aspect of ensuring the long-term effectiveness and sustainability of biochar-based air filtration systems. Spent biochar, loaded with adsorbed contaminants, undergoes a process of renewal to restore its adsorption capacity and extend its operational life (Gwenzi, Chaukura, et al., 2021). This helps not only to reduce the raw material demand but also the costs involved and the disposal requirements such as transport, land, and labor costs. Some studies have linked the spent biochar performance to cycles of reuse, where a slight reduction in adsorption on every reuse cycle was realized, but the adsorption capacity was significantly reduced after three regeneration cycles (Dai et al., 2019).

While comprehensive data on the regeneration of biochar used in air filters are still limited, existing literature from biochar applications in aqueous systems demonstrates the potential for regeneration using chemical methods. This involves subjecting the spent biochar to specific chemical treatments, leading to the removal or transformation of the adsorbed contaminants. Understanding the mechanochemical stability of biochars under practical conditions is integral to developing effective regeneration strategies, especially considering the turbulent air flows and potential chemical stresses encountered in industrial settings. Further research is needed to optimize regeneration protocols, ensuring the efficacy and safety of the regenerated biochar for sustained use in gaseous contaminant removal applications. The exploration of regeneration mechanisms contributes to the overall feasibility and efficiency of biochar technology in addressing air pollution challenges.

19.8 Future directions and recommendations

The evolution of biochar in gaseous contaminant removal requires a concerted drive for innovation. Focusing on large-scale applications, educational campaigns, pilot plants, and demonstration projects are critical steps to enhance the acceptance and feasibility of biochar technology.

Efforts to have a competitive advantage over current techniques such as bio-oil production are vital. Highlighting the advantages of biochar, attracting investments, and strategically positioning biochar-based air filters in the market are necessary to drive research progress and ensure competitiveness.

Establishing conducive policy and regulatory frameworks is imperative for supporting the emerging biochar technology. Collaboration between researchers and policymakers is key to addressing knowledge gaps and shaping regulations that promote responsible and effective use.

Research must also tackle concerns related to potential public health risks and carbon footprints associated with biochar facilities. Conducting comprehensive life cycle analyses, focusing on water, carbon, energy, and toxic emissions footprints, is crucial for assessing overall sustainability and safety.

Additionally, studies on the adsorption of toxic gases with biochar have primarily been conducted at the laboratory scale through batch adsorption tests. Notably, there is a significant gap as these studies have not yet been translated into field or large-scale applications of biochar as air filters. Given that laboratory conditions may not accurately reflect real-world scenarios, it is necessary that thorough assessments of biochar's performance at larger scales are conducted.

Future research should explore new biochar-based materials, including catalytic variants and biochar-metal oxide composites. Systematic optimization studies and the use of response surface methodology are needed for tailored applications in air pollution control systems. Striking a balance between adsorption and degradation is crucial for addressing the potential limits of biochar filters. Developing systems capable of both adsorbing and degrading contaminants, especially organics ensures a more sustainable and long-term solution.

Investigating the mechanochemical stability of biochar under turbulent air flows and chemical attacks is essential for ensuring the durability of biochar filters. Concurrently, exploring the regeneration capacity of biochars used in air filters, drawing insights from successful methods in aqueous systems. In addition, developing strategies for disposing of used biochar requires focused research. Understanding sustainable behavior, fate of contaminants, and potential human exposure risks is crucial for responsible waste management.

Performing thorough life cycle analyses that compare the environmental footprints of biochar-based air filters with conventional filters is imperative. This data is crucial for showcasing the cost-effectiveness, sustainability, and renewability of biochar-based technologies.

Guided by a framework for development, research initiatives should progress through stages of synthesizing and characterizing novel biochars, developing and testing laboratory-scale prototypes, designing and evaluating pilot-scale prototypes, and finally, field application with ongoing monitoring and evaluation of large-scale biochar filters. This iterative process ensures continuous improvement through feedback loops.

19.9 Conclusion

Using biochar to eliminate toxic gases presents a promising avenue for air pollution control. The production of biochar from a variety of biomass feedstocks, influenced by factors such as pyrolysis temperature and technology, offers a versatile solution for handling toxic gaseous pollutants. The biochar's properties, shaped by these production parameters, make it an effective sorptive material, with functional groups such ase hydroxyl and carboxyl groups playing a key role in pollutant removal. Biochars stand out for their economic feasibility and easy accessibility. However, the economic viability of biochar production should be considered in future studies, considering activation costs. Moreover, future research should focus on engineered biochars tailored for complex gas mixtures, regeneration, stability, and disposal of biochars containing toxic gases. While this chapter explores the potential of biochar in adsorbing various gases, addressing knowledge gaps and overcoming challenges is essential for realizing the full benefits of biochar technology in air pollution control.

References

Ahmed, M. B., Zhou, J. L., Ngo, H. H., Guo, W., & Chen, M. (2016). Progress in the preparation and application of modified biochar for improved contaminant removal from water and wastewater. *Bioresource Technology, 214*, 836–851. Available from https://doi.org/10.1016/j.biortech.2016.05.057.

Ahmed, M. B., Zhou, J. L., Ngo, H. H., Johir, M. A. H., Sun, L., Asadullah, M., & Belhaj, D. (2018). Sorption of hydrophobic organic contaminants on functionalized biochar: Protagonist role of Π-Π electron-donor-acceptor interactions and hydrogen bonds. *Journal of Hazardous Materials, 360*, 270–278. Available from https://doi.org/10.1016/j.jhazmat.2018.08.005.

Ainomugisha, S., Matovu, M., & Manga, M. (2024a). Application of green agro-based nanoparticles in cement-based construction materials: A systematic review. *Journal of Building Engineering, 87*, 108955. Available from https://doi.org/10.1016/j.jobe.2024.108955.

Ainomugisha, S., Matovu, M. J., & Manga, M. (2024b). Influence mechanisms of silica nanoparticles' property enhancement in cementitious materials and their green synthesis: A critical review. *Case Studies in Construction Materials, 20*, e03372. Available from https://doi.org/10.1016/j.cscm.2024.e03372.

Ammendola, P., Raganati, F., & Chirone, R. (2017). CO2 adsorption on a fine activated carbon in a sound assisted fluidized bed: Thermodynamics and kinetics. *Chemical Engineering Journal, 322*, 302–313. Available from https://doi.org/10.1016/j.cej.2017.04.037.

Bamdad, H., Hawboldt, K., & MacQuarrie, S. (2018). A review on common adsorbents for acid gases removal: Focus on biochar. *Renewable and Sustainable Energy Reviews., 81*, 1705–1720. Available from https://doi.org/10.1016/j.rser.2017.05.261.

Bamdad, H., Hawboldt, K., MacQuarrie, S., & Papari, S. (2019). Application of biochar for acid gas removal: experimental and statistical analysis using CO_2. *Environmental Science and Pollution Research, 26*(11), 10902–10915. Available from https://doi.org/10.1007/s11356-019-04509-3.

Bansal, R.C., Donnet, J.B., Stoeckli, F., 1988. Active Carbon. New York: Marcel Dekker. Retrieved from https://doi.org/10.1080/01932699008943255 (accessed 11.06.23).

Barbusinski, K., Parzentna-Gabor, A., & Kasperczyk, D. (2021). Removal of Odors (Mainly H2S and NH3) Using Biological Treatment Methods. *Clean Technology, 3*, 138–155.

Bhandari, P. N., Kumar, A., Bellmer, D. D., & Huhnke, R. L. (2014). Synthesis and evaluation of biochar-derived catalysts for removal of toluene (model tar) from biomass-generated producer gas. *Renewable Energy, 66*, 346–353. Available from https://doi.org/10.1016/j.renene.2013.12.017.

Braghiroli, F. L., Bouafif, H., & Koubaa, A. (2019). Enhanced SO2 adsorption and desorption on chemically and physically activated biochar made from wood residues. *Industrial Crops and Products., 138*, 111456. Available from https://doi.org/10.1016/j.indcrop.2019.06.019.

de Caprariis, B., De Filippis, P., Hernandez, A. D., Petrucci, E., Petrullo, A., Scarsella, M., & Turchi, M. (2017). Pyrolysis wastewater treatment by adsorption on biochars produced by poplar biomass. *Journal of Environmental Management, 197*, 231–238. Available from https://doi.org/10.1016/j.jenvman.2017.04.007.

Chen, B., Koziel, J. A., Banik, C., Ma, H., Lee, M., Wi, J., Meiirkhanuly, Z., Andersen, D. S., Białowiec, A., & Parker, D. B. (2020). Emissions from swine manure treated with current products for mitigation of odors and reduction of NH3, H2S, VOC, and GHG emissions. *Data, 5*, 1–15. Available from https://doi.org/10.3390/data5020054.

Chen, J., Chen, X., Yan, D., Jiang, M., Xu, W., Yu, H., & Jia, H. (2019). A facile strategy of enhancing interaction between cerium and manganese oxides for catalytic removal of gaseous organic contaminants. *Applied Catalysis B: Environmental, 250*, 396–407. Available from https://doi.org/10.1016/j.apcatb.2019.03.042.

Chen, T., Chen, J., Liu, Z., Chi, K., & Been, M. (2022). Characteristics of PM and PAHs emitted from a coal-fired boiler and the efficiencies of its air pollution control devices. *Journal of the Air & Waste Management Association, 72*, 85–91. Available from https://doi.org/10.1080/10962247.2021.1994483.

Cheng, K., Hao, W., Wang, Y., Yi, P., Zhang, J., & Ji, W. (2020). Understanding the emission pattern and source contribution of hazardous air pollutants from open burning of municipal solid waste in China. *Environmental pollution (Barking, Essex: 1987), 263*, 114417. Available from https://doi.org/10.1016/j.envpol.2020.114417.

Cheng, N., Wang, B., Wu, P., Lee, X., Xing, Y., Chen, M., & Gao, B. (2021). Adsorption of emerging contaminants from water and wastewater by modified biochar: A review. *Environmental Pollution, 273*, 116448. Available from https://doi.org/10.1016/j.envpol.2021.116448.

Chmielewski, A. G. (1999). Environmental effects of fossil fuel combustion. *Interact. Energy/Environment December*, 56–74, Retrieved from. Available from https://inis.iaea.org/search/search.aspx?orig_q = RN:31003493. (accessed 11.09.23).

Cho, S. H., Lee, S., Kim, Y., Song, H., Lee, J., Tsang, Y. F., Chen, W. H., Park, Y. K., Lee, D. J., Jung, S., & Kwon, E. E. (2023). Applications of agricultural residue biochars to removal of toxic gases emitted from chemical plants: A review. *The Science of the Total Environment, 868*. Available from https://doi.org/10.1016/j.scitotenv.2023.161655. Available from, http://www.elsevier.com/locate/scitotenv.

Creamer, A. E., Gao, B., & Zhang, M. (2014). Carbon dioxide capture using biochar produced from sugarcane bagasse and hickory wood. *Chemical Engineering Journal, 249*, 174–179. Available from https://doi.org/10.1016/j.cej.2014.03.105.

Dai, Y., Zhang, N., Xing, C., Cui, Q., & Sun, Q. (2019). The adsorption, regeneration and engineering applications of biochar for removal organic pollutants: A review. *Chemosphere, 223*, 12–27. Available from https://doi.org/10.1016/j.chemosphere.2019.01.161.

Das, J., & Lens, P. N. L. (2022). Resilience of hollow fibre membrane bioreactors for treating H_2S under steady state and transient conditions. *Chemosphere, 307*, 136142. Available from https://doi.org/10.1016/j.chemosphere.2022.136142.

Díaz-Maroto, C., Mašek, O., Pizarro, P., Serrano, D. P., Moreno, I., & Fermoso, J. (2023). Removal of NO at low concentrations from polluted air in semi-closed environments by activated biochars from renewables feedstocks. *Journal of Environmental Management, 341*, 118031. Available from https://doi.org/10.1016/j.jenvman.2023.118031.

Ding, S., & Liu, Y. (2020). Adsorption of CO_2 from flue gas by novel seaweed-based KOH-activated porous biochars. *Fuel, 260*, 116382. Available from https://doi.org/10.1016/j.fuel.2019.116382.

González-cardoso, G., Hernández-contreras, J. M., Valle-hernández, B. L., Hernández-moreno, A., Rosa, N. S., García-martínez, R., & Mugica-álvarez, V. (2020). Toxic atmospheric pollutants from crematoria ovens : characterization, emission factors, and modeling, *7*(35), 43800–43812. Available from https://doi.org/10.1007/s11356-020-10314-0.

Gwenzi, W. (2021). Autopsy, thanatopraxy, cemeteries and crematoria as hotspots of toxic organic contaminants in the funeral industry continuum. *The Science of the Total Environment, 753*, 141819. Available from https://doi.org/10.1016/j.scitotenv.2020.141819.

Gwenzi, W., Chaukura, N., Wenga, T., & Mtisi, M. (2021). Biochars as media for air pollution control systems: Contaminant removal, applications and future research directions. *The Science of the Total Environment, 753*, 142249. Available from https://doi.org/10.1016/j.scitotenv.2020.142249.

Hwang, O., Lee, S.-R., Cho, S., Ro, K. S., Spiehs, M., Woodbury, B., Silva, P. J., Han, D.-W., Choi, H., Kim, K.-Y., & Jung, M.-W. (2018). Efficacy of Different Biochars in Removing Odorous Volatile Organic Compounds (VOCs) Emitted from Swine Manure. *ACS Sustainable Chemistry & Engineering, 6*(11), 14239–14247. Available from https://doi.org/10.1021/acssuschemeng.8b02881.

Jumina., Kurniawan, Y. S., Purwono, B., Siswanta, D., Priastomo, Y., Winarno, A., & Waluyo, J. (2021). Science and Technology Progress on the Desulfurization Process of Crude Oil. *Bull. Korean Chem. Soc., 42*, 1066–1081. Available from https://doi.org/10.1002/bkcs.12342.

Jung, S., Lee, J., Moon, D. H., Kim, K. H., & Kwon, E. E. (2021). Upgrading biogas into syngas through dry reforming. *Renewable and Sustainable Energy Reviews., 143*, 110949. Available from https://doi.org/10.1016/j.rser.2021.110949.

Kaboggoza, H. C., Muoghalu, C., Sprouse, L., & Manga, M. (2024). Hydrochar composites for healthcare wastewater treatment: A review of synthesis approaches, mechanisms, and influencing factors. *Journal of Water Process Engineering, 60*, 105222. Available from https://doi.org/10.1016/j.jwpe.2024.105222.

Kanjanarong, J., Giri, B. S., Jaisi, D. P., Oliveira, F. R., Boonsawang, P., Chaiprapat, S., Singh, R. S., Balakrishna, A., & Khanal, S. K. (2017). Removal of hydrogen sulfide generated during anaerobic treatment of sulfate-laden wastewater using biochar: Evaluation of efficiency and mechanisms. *Bioresource Technology, 234*, 115–121. Available from https://doi.org/10.1016/j.biortech.2017.03.009.

Kaveh, M., Amiri Chayjan, R., Taghinezhad, E., Rasooli Sharabiani, V., & Motevali, A. (2020). Evaluation of specific energy consumption and GHG emissions for different drying methods (Case study: Pistacia Atlantica). *Journal of Cleaner Production, 259*, 120963. Available from https://doi.org/10.1016/j.jclepro.2020.120963.

Khan, F. I., & Ghoshal, A. K. (2000). Removal of Volatile Organic Compounds from polluted air. *Journal of Loss Prevention in the Process Industries, 13*(6), 527–545. Available from https://doi.org/10.1016/S0950-4230(00)00007-3.

La, H., & Hettiaratchi, J. P. A. (2022). Role of biochar in the removal of organic and inorganic contaminants from waste gas streams. In K. Riti Thapar, & P. S. Maulin (Eds.), *BioChar* (pp. 89–116). De Gruyter, https://doi.org/10.1515/9783110734003-005.

Leng, L., Huang, H., Li, H., Li, J., & Zhou, W. (2019). Biochar stability assessment methods: A review. *The Science of the Total Environment, 647*, 210–222. Available from https://doi.org/10.1016/j.scitotenv.2018.07.402.

Li, G., Shen, B., Li, F., Tian, L., Singh, S., & Wang, F. (2015). Elemental mercury removal using biochar pyrolyzed from municipal solid waste. *Fuel Processing Technology, 133*, 43–50. Available from https://doi.org/10.1016/j.fuproc.2014.12.042.

Li, H., Dong, X., da Silva, E. B., de Oliveira, L. M., Chen, Y., & Ma, L. Q. (2017). Mechanisms of metal sorption by biochars: Biochar characteristics and modifications. *Chemosphere, 178*, 466–478. Available from https://doi.org/10.1016/j.chemosphere.2017.03.072. Available from, http://www.elsevier.com/locate/chemosphere.

Li, H., Zhang, J., Cao, Y., Liu, C., Li, F., Song, Y., Hu, J., & Wang, Y. (2020). Role of acid gases in Hg^0 removal from flue gas over a novel cobalt-containing biochar prepared from harvested cobalt-enriched phytoremediation plant. *Fuel Processing Technology, 207*, 106478. Available from https://doi.org/10.1016/j.fuproc.2020.106478.

Li, Z., Li, Y., & Zhu, J. (2021). Straw-Based Activated Carbon: Optimization of the Preparation Procedure and Performance of Volatile Organic Compounds Adsorption. *Materials, 14*(12), 3284. Available from https://www.mdpi.com/1996-1944/14/12/3284.

Li, L., Zou, D., Xiao, Z., Zeng, X., Zhang, L., Jiang, L., Wang, A., Ge, D., Zhang, G., & Liu, F. (2019). Biochar as a sorbent for emerging contaminants enables improvements in waste management and sustainable resource use. *Journal of Cleaner Production, 210*, 1324–1342. Available from https://doi.org/10.1016/j.jclepro.2018.11.087.

Liu, S.-H., Lin, H.-H., Lai, C.-Y., Lin, C.-W., Chang, S.-H., & Yau, J.-T. (2019). Microbial community in a pilot-scale biotrickling filter with cell-immobilized biochar beads and its performance in treating toluene-contaminated waste gases. *International biodeterioration & biodegradation*, *144*, 104743.

Manga, M., Aragón-Briceño, C., Boutikos, P., Semiyaga, S., Olabinjo, O., & Muoghalu, C. C. (2023). Biochar and Its Potential Application for the Improvement of the Anaerobic Digestion Process: A Critical Review. *Energies*, *16*(10), 4051. Available from https://doi.org/10.3390/en16104051.

Manga, M., & Muoghalu, C. C. (2024). Greenhouse gas emissions from on-site sanitation systems: A systematic review and meta-analysis of emission rates, formation pathways and influencing factors. *Journal of Environmental Management*, *357*, 120736. Available from https://doi.org/10.1016/j.jenvman.2024.120736.

Manga, M., Muoghalu, C., Kulabako, R., Kaboggoza, H., Lebu, S., Sprouse, L., Niwagaba, C., & Semiyaga, S. (2024). Biochar Modification for Removal of Inorganic and Organic Contaminants from Industrial Effluent. *Catalytic Applications of Biochar for Environmental Remediation: A Green Approach Towards Environment Restoration* (1, pp. 195–221). American Chemical Society. Available from https://doi:10.1021/bk-2024-1478.ch009.

Manga, M., Semiyaga, S., Lebu, S., Nakagiri, A., Niwagaba, C. B., Salzberg, A., & Muoghalu, C. C. (2024). Bioprocessing of Organic Municipal Solid Waste for Biomethane and Biohydrogen Production. In R. Choudhury A, & G. S. Palani (Eds.), *Material and Energy Recovery from Solid Waste for a Circular Economy* (pp. 212–236). CRC Press. Available from https://doi.org/10.1201/9781003364467-9.

Meij, R., & Winkel, H. (2007). The emissions of heavy metals and persistent organic pollutants from modern coal-fired power stations. *Atmospheric Environment*, *41*, 9262–9272. Available from https://doi.org/10.1016/j.atmosenv.2007.04.042.

Mochida, I., Korai, Y., Shirahama, M., Kawano, S., Hada, T., Seo, Y., & Yasutake, A. (2000). Removal of SOx and NOx over activated carbon fibers. *Carbon*, *38*, 227–239. Available from https://doi.org/10.1016/S0008-6223(99)00179-7.

Mohamed, E. F., El-Mekawy, A., Ahmed, S. A. S., & Fathy, N. A. (2024). High adsorption capacity of ammonia gas pollutant using adsorbents of carbon composites. *Arabian Journal for Science and Engineering*, *49*(1), 261–271. Available from https://doi.org/10.1007/s13369-023-07987-3.

Muoghalu, C. C., Owusu, P. A., Lebu, S., Nakagiri, A., Semiyaga, S., Iorhemen, O. T., & Manga, M. (2023). *Biochar as a novel technology for treatment of onsite domestic wastewater : A critical review* 1–22. Available from https://doi.org/10.3389/fenvs.2023.1095920.

Muoghalu, C., Semiyaga, S., Kaboggoza, H., Yasan, S., Palmer, G., Lui, C, Chandana, N, & Manga, M. (2024). Biochar Application for the Removal of Heavy Metals and Organic Pollutants from Soil. *Catalytic Applications of Biochar for Environmental Remediation: Sustainable Strategies Towards a Circular Economy* (2, pp. 197–223). Washington, DC: American Chemical Society. Available from https://doi:10.1021/bk-2024-1479.ch008.

Pulido-Novicio, L., Hata, T., Kurimoto, Y., Doi, S., Ishihara, S., & Imamura, Y. (2001). Adsorption capacities and related characteristics of wood charcoals carbonized using a one-step or two-step process. *Journal of Wood Science.*, *47*, 48–57. Available from https://doi.org/10.1007/BF00776645.

Qiu, B., Shao, Q., Shi, J., Yang, C., & Chu, H. (2022). Application of biochar for the adsorption of organic pollutants from wastewater: Modification strategies, mechanisms and challenges. *Separation and Purification Technology*, *300*, 121925. Available from https://doi.org/10.1016/j.seppur.2022.121925.

Rivera-Montenegro, L., Valenzuela, E. I., González-Sánchez, A., Muñoz, R., & Quijano, G. (2023). Volatile Methyl siloxanes as key biogas pollutants: Occurrence, impacts and treatment technologies. *BioEnergy Res*, *16*, 801–816. Available from https://doi.org/10.1007/s12155-022-10525-y.

Sajjadi, B., Chen, W., & Egiebor, N. O. (2018). A comprehensive review on physical activation of biochar for energy and environmental applications. *Reviews in Chemical Engineering.*, 1–42. Available from https://doi.org/10.1515/revce-2017-0113.

Shahi Khalaf Ansar, B., Kavusi, E., Dehghanian, Z., Pandey, J., Asgari Lajayer, B., Price, G. W., & Astatkie, T. (2022). Removal of organic and inorganic contaminants from the air, soil, and water by algae. *Environmental Science and Pollution Research*, *30*, 116538–116566. Available from https://doi.org/10.1007/s11356-022-21283-x.

Shao, J., Zhang, J., Zhang, X., Feng, Y., Zhang, H., Zhang, S., & Chen, H. (2018). Enhance SO_2 adsorption performance of biochar modified by CO_2 activation and amine impregnation. *Fuel*, *224*, 138–146. Available from https://doi.org/10.1016/j.fuel.2018.03.064.

Singh, S. P., Tripathy, D. P., & Ranjith, P. G. (2008). Performance evaluation of cement stabilized fly ash-GBFS mixes as a highway construction material. *Waste Management (New York, N.Y.)*, *28*, 1331–1337. Available from https://doi.org/10.1016/j.wasman.2007.09.017.

Sun, F., Cheng, H., Chen, J., Zheng, N., Li, Y., & Shi, J. (2016). Heteroatomic $SenS8-n$ molecules confined in nitrogen-doped mesoporous carbons as reversible cathode materials for high-performance lithium batteries. *ACS Nano*, *10*, 8289–8298. Available from https://doi.org/10.1021/acsnano.6b02315.

Talaiekhozani, A., Rezania, S., Kim, K.-H., Sanaye, R., & Amani, A. M. (2021). Recent advances in photocatalytic removal of organic and inorganic pollutants in air. *Journal of Cleaner Production*, *278*, 123895. Available from https://doi.org/10.1016/j.jclepro.2020.123895.

Tumwesige, V., Okello, G., Semple, S., & Smith, J. (2017). Impact of partial fuel switch on household air pollutants in sub-Sahara Africa. *Environmental pollution (Barking, Essex: 1987)*, *231*, 1021–1029. Available from https://doi.org/10.1016/j.envpol.2017.08.118.

Wang, H., Gao, B., Wang, S., Fang, J., Xue, Y., & Yang, K. (2015). Removal of Pb (II), Cu (II), and Cd (II) from aqueous solutions by biochar derived from $KMnO_4$ treated hickory wood. *Bioresource Technology*, *197*, 356–362. Available from https://doi.org/10.1016/j.biortech.2015.08.132.

Wang, J., Huang, H., Wang, M., Yao, L., Qiao, W., Long, D., & Ling, L. (2015). Direct capture of low-concentration CO_2 on mesoporous carbon-supported solid amine adsorbents at ambient temperature. *Industrial & Engineering Chemistry Research*, *54*(19), 5319–5327. Available from https://doi.org/10.1021/acs.iecr.5b01060.

Wang, W., & Ku, Y. (2003). Photocatalytic degradation of gaseous benzene in air streams by using an optical fiber photoreactor. *Journal of Photochemistry and Photobiology A: Chemistry*, *159*(1), 47–59. Available from https://doi.org/10.1016/S1010-6030(03)00111-4.

Wang, X., Ma, X., Sun, L., & Song, C. (2007). A nanoporous polymeric sorbent for deep removal of H_2S from gas mixtures for hydrogen purification. *Green Chemistry: An International Journal and Green Chemistry Resource: GC, 9*, 695–702. Available from https://doi.org/10.1039/b614621j.

Xiang, W., Zhang, X., Chen, J., Zou, W., He, F., Hu, X., Tsang, D. C. W., Sik, Y., & Gao, B. (2020). Biochar technology in wastewater treatment : A critical review. *Chemosphere, 252*, 126539. Available from https://doi.org/10.1016/j.chemosphere.2020.126539.

Yaashikaa, P. R., Kumar, P. S., Varjani, S., & Saravanan, A. (2020). A critical review on the biochar production techniques, characterization, stability and applications for circular bioeconomy. *Biotechnology Reports, 28*, e00570. Available from https://doi.org/10.1016/j.btre.2020.e00570.

Yang, K., Peng, J., Srinivasakannan, C., Zhang, L., Xia, H., & Duan, X. (2010). Preparation of high surface area activated carbon from coconut shells using microwave heating. *Bioresource Technology, 101*, 6163–6169. Available from https://doi.org/10.1016/j.biortech.2010.03.001.

Ye, Q., Li, Q., & Li, X. (2022). Removal of heavy metals from wastewater using biochars: adsorption and mechanisms. *Environmental Pollutants and Bioavailability, 34*(1), 385–394. Available from https://doi.org/10.1080/26395940.2022.2120542.

Yin, Y., Lv, D., Zhu, T., Li, X., Sun, Y., & Li, S. (2021). Removal and transformation of unconventional air pollutants in flue gas in the cement kiln-end facilities. *Chemosphere, 268*, 128810. Available from https://doi.org/10.1016/j.chemosphere.2020.128810.

Yuan, D., Zhang, L., Wan, S., & Sun, L. (2022). Rational design of microporous biochar based on ion exchange using carboxyl as an anchor for high-efficiency capture of gaseous p-xylene. *Separation and Purification Technology, 286*, 120402. Available from https://doi.org/10.1016/j.seppur.2021.120402.

Zhang, C., Zeng, G., Huang, D., Lai, C., Chen, M., Cheng, M., Tang, W., Tang, L., Dong, H., Huang, B., Tan, X., & Wang, R. (2019). Biochar for environmental management: Mitigating greenhouse gas emissions, contaminant treatment, and potential negative impacts. *Chemical Engineering Journal, 373*, 902–922. Available from https://doi.org/10.1016/j.cej.2019.05.139.

Zhao, S., Pudasainee, D., Duan, Y., Gupta, R., Liu, M., & Lu, J. (2019). A review on mercury in coal combustion process: Content and occurrence forms in coal, transformation, sampling methods, emission and control technologies. *Progress in Energy and Combustion Science., 73*, 26–64. Available from https://doi.org/10.1016/j.pecs.2019.02.001.

Zhu, C., Duan, Y., Wu, C.-Y., Zhou, Q., She, M., Yao, T., & Zhang, J. (2016). Mercury removal and synergistic capture of SO_2/NO by ammonium halides modified rice husk char. *Fuel, 172*, 160–169. Available from https://doi.org/10.1016/j.fuel.2015.12.061.

Part 6

Assessment of biochar systems

Chapter 20

Regeneration, recycling, and disposal of spent biochars

Miranda Mpeta[1], Terrence Wenga[2,3], Kudzanayi Andrew Marondedze[2,4] and Phenias Sadondo[2]

[1]Department of Environmental Engineering, School of Engineering Sciences and Technology, Chinhoyi University of Technology, Chinhoyi, Zimbabwe, [2]Department of Soil Science and Environment, Faculty of Agriculture, Environment and Food Systems, University of Zimbabwe, Mount Pleasant, Harare, Zimbabwe, [3]Key Laboratory of Agro-Forestry Environmental Processes and Ecological Regulation of Hainan Province, School of Environmental Science and Engineering, Hainan University, Haikou, P.R. China, [4]Department of Geography, School of Agricultural, Earth, and Environmental Sciences, University of KwaZulu Natal, Pietermaritzburg, South Africa

Chapter outline

20.1 Introduction	379
20.2 Regeneration of spent biochars	381
20.2.1 Magnetic separation	381
20.2.2 Filtration	382
20.2.3 Thermal desorption and decomposition	382
20.2.4 Chemical desorption	383
20.2.5 Supercritical fluid desorption	383
20.2.6 Advanced oxidation processes	383
20.2.7 Microbial-assisted regeneration	384
20.2.8 Microwave irradiation regeneration	384
20.3 Reuse, recycling, and disposal of spent biochar for a circular economy and environmental sustainability	384
20.3.1 Reuse as a soil amendment	384
20.3.2 Composting	385
20.3.3 Land application	385
20.3.4 Regeneration through pyrolysis	385
20.3.5 Contaminant removal and landfill disposal	385
20.3.6 Incineration	386
20.3.7 Fillers in novel construction materials	386
20.3.8 Catalyst precursors for trace metal-rich spent biochars	386
20.3.9 Incorporation in solid fuels	386
20.3.10 Codisposal of sludge	387
20.4 Behavior and fate of contaminants in spent biochar	387
20.5 Summary and future research directions	388
References	389

20.1 Introduction

Among the technologies that are employed to immobilize and degrade both organic and inorganic contaminants in various polluted environmental compartments, biomass-based biochar has been demonstrated to be an effective and environmentally friendly strategy (Crini et al., 2019; Liao et al., 2022; Nguyen et al., 2023; Rasheed et al., 2020; Wang et al., 2020). In addition, biochar has several other advantages, for instance, high-cost efficiency due to the abundance of low-cost feedstock, as well as simplicity of preparation methods (Barquilha & Braga, 2021).

To further improve the biochar pollutant removal efficiency and capture capacity, increasing efforts are currently being carried out to design and develop biochar with improved properties for the immobilization of both organic and inorganic pollutants (Dutt et al., 2020; Liao et al., 2022). However, the main drawback is the management of the spent biochar loaded and concentrated with various pollutants (Hossain et al., 2020). Several technologies are available for the recovery of spent biochar, including filtration, sedimentation, centrifugation, and magnetic separation (Kar et al., 2022). The spent biochar is then either regenerated for reuse or safe disposal via landfilling or incineration (Kozyatnyk et al., 2020), as illustrated in Fig. 20.1.

Regeneration of spent biochar for several other decontamination cycles further lowers the running and operational cost of remediating the contaminated environmental media. In essence, biochar's high recycling and reuse capabilities are fundamental for the economic management of polluted environments (Herath et al., 2021). Therefore, several

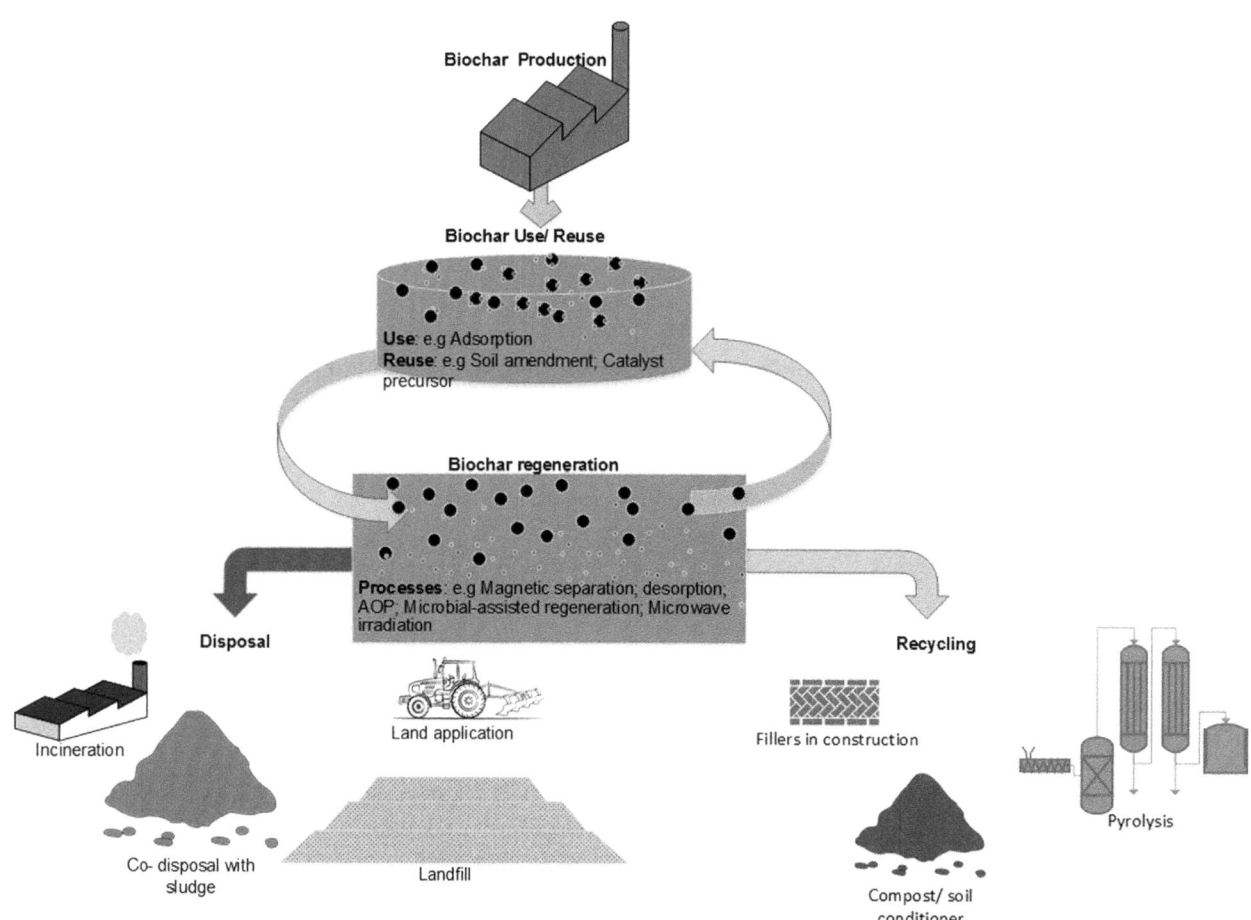

FIGURE 20.1 Reuse, regeneration, recycling, and disposal of spent biochar for a circular economy and environmental sustainability. Biochar used in the process of adsorption can be regenerated and then reused for the same purpose or recycled into other products.

regeneration approaches, such as photodegradation, desorption, and biodegradation of sorbed contaminants, have been proposed, developed, and examined to regenerate and reuse the spent adsorbents (Vakili et al., 2019). The strategies are affected by the type of adsorbent, its properties, and the pollutant type (Alsawy et al., 2022). These approaches can be broadly divided into two main groups (Fig. 20.2), which are the desorption and degradation processes (Omorogie et al., 2016). The degradation/decomposition regeneration process completely depends on mineralizing the adsorbed contaminants and/or converting them into by-products with less toxicity (Liao et al., 2022). This process enables the restoration of the biochar contaminant adsorption capacity.

Ultrasound and oxidation degradation approaches utilize high thermal energy and, as such, are regarded as thermal degradation methods. On the other hand, desorption regeneration techniques employ either thermal or nonthermal approaches to break the bonds between the biochar surface and contaminants. Thermal regeneration methods are destructive techniques deployed only if the adsorbate is a waste, and there is no need for its recovery (Sabio et al., 2004). However, because thermal regeneration consumes a lot of energy (temperature up to 800°C), it elevates the process cost, reaching up to 50% of the total cost of producing new biochar (Alsawy et al., 2022). Nonthermal techniques preserve the biochar and can be reused for several other pollutant decontamination cycles (Iamsaard et al., 2022).

Several good reviews are available and have established the potential benefit of biochar application in various environmental media (Barquilha & Braga, 2021; Kah et al., 2021; Kar et al., 2022; Nguyen et al., 2023). Also, the reactivation of spent biochar concentrated with various pollutants for their reutilization has been demonstrated (Hassan et al., 2020). However, a comprehensive analysis and understanding of the mechanisms for the regeneration, as well as possible safe and sustainable management of spent adsorbents loaded with pollutants, are needed. Therefore, this chapter provides a critical summary of the following:

(1) the detailed mechanism and processes for the regeneration of spent biochar;

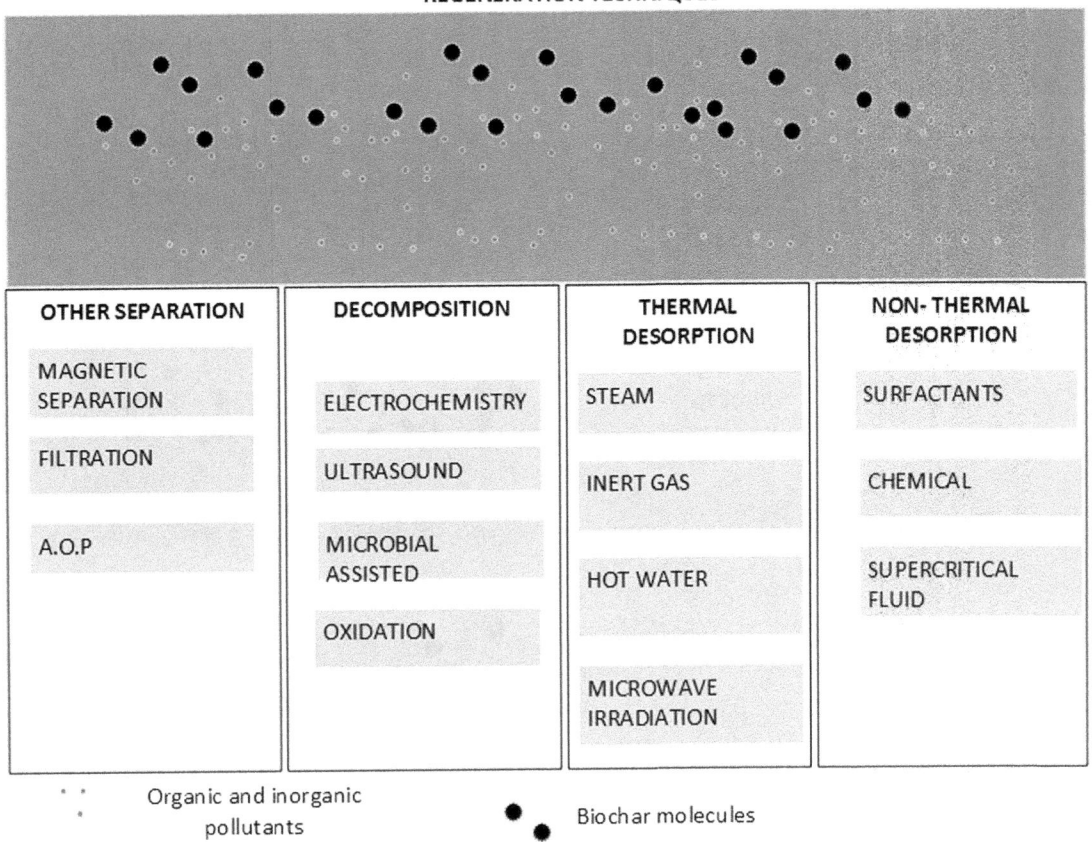

FIGURE 20.2 Different regeneration techniques can be utilized, including separation, decomposition, and degradation (thermal and nonthermal) to recoup biochar.

(2) the recycling and reutilization of spent biochar in various environmental compartments;
(3) the possible safe disposal methods of unrecoverable spent biochar for reutilization; and
(4) challenges and hotspots for future research directions will be discussed.

The findings of this chapter will help readers comprehend the available regeneration techniques and methods for regenerating the adsorbent with minimum capacity loss after several regeneration cycles. Simultaneously, regeneration helps to reduce secondary pollution, while reuse in other processes results in cost-effectiveness and resource recovery.

20.2 Regeneration of spent biochars

Regenerating spent biochar entails reactivating its adsorption and nutrient-retention properties, which may degrade over time or after use. Regeneration is the reverse process of adsorption, which involves adsorbate desorption and adsorbate decomposition (Jia et al., 2016).

From, a good adsorbent should be reusable and recyclable, thereby reducing the cost of purchasing adsorbents. Due to differences in the make-up of the biochar, different regeneration techniques can be used to treat different biochar, as shown in Fig. 20.3. Biochar regeneration efficiencies also vary. For example, in a 2016 study by Shan et al. (2016), the pharmaceuticals that had been adsorbed on spent biochar were degraded by 97% (tetracycline) and 98.4% (carbamazepine). The following section highlights the various regeneration techniques.

20.2.1 Magnetic separation

Depending on the nature of the biochar, magnetic separation can be used as a way of recouping biochar. Adsorbents can be enhanced in terms of specific surface area (SSA), porosity, thermal stability, and other factors by introducing

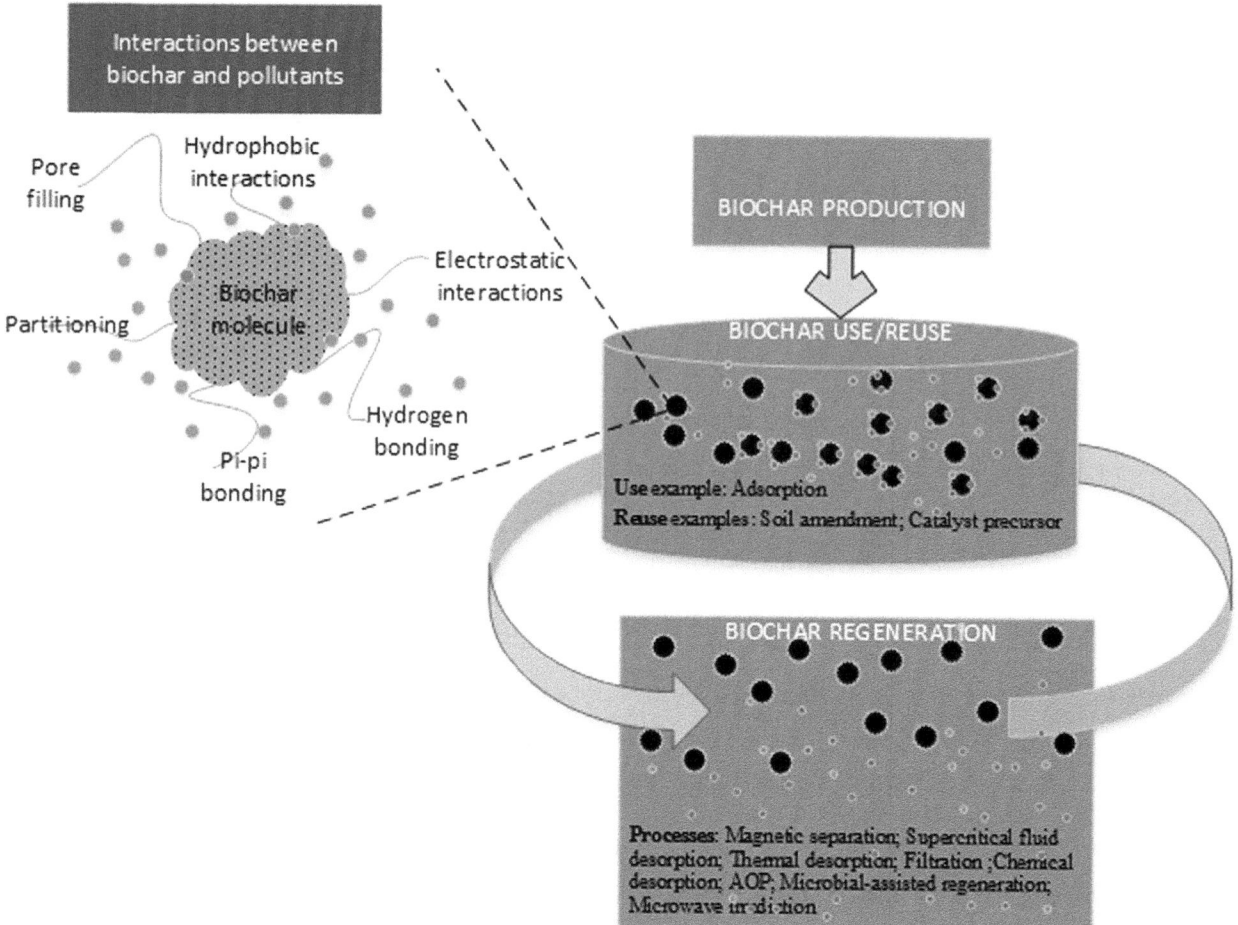

FIGURE 20.3 Schematic illustration showing adsorption and regeneration of adsorbents from wastewater and associated interactions. Adsorption and regeneration of biochar.

metal nanoparticles in the biochar. By so doing, the biochar becomes a magnetic adsorbent. Magnetic biochar has the advantage of improved adsorption efficiency and rate of adsorbent recovery (Gupta et al., 2020). In some cases, biomass substrate is treated with iron (Fe) salts of Fe, such as ferrous chloride, ferrous sulfate, and ferric chloride, are used to make magnetic biochar. Such biochar can be easily regenerated by magnetic separation (Li et al., 2016).

Interestingly, biochar loaded with iron can be pyrolyzed directly to produce magnetic biochar (Ren et al., 2018) and be reused. Such spent magnetic biochar is separated by applying an external magnetic field. These have been mostly used on a laboratory scale (Ren et al., 2018).

20.2.2 Filtration

Filtration as a method of biochar regeneration depends on the particle size of the biochar. The larger the adsorbent particle size, the easier it is to separate. Nowadays, membranes are commonly used for filtration separation, with their recyclability and separation efficiency dependent on the ability of the biochar to disperse. In membrane bioreactors, biochar has the potential to reduce membrane fouling, thereby increasing the design lifespan of the biomembrane (Tan et al., 2016). However, the downside of filtration is that most membranes cannot remove nano-sized biochar. Additionally, filter media often needs to be backwashed.

20.2.3 Thermal desorption and decomposition

Thermal regeneration is another effective method of biochar regeneration. Heat or high temperatures are used on the spent biochar in a controlled environment to remove adsorbed contaminants and restore its surface properties. This process is similar to the initial pyrolysis used to create biochar. Thermal regeneration involves applying heat to biochar to disturb the physical and chemical bonding between biochar and pollutants (Momina et al., 2018). This technique is currently being used at industrial and commercial levels for the regeneration of activated carbon. Heat application to biochar in aerobic conditions at temperatures below 500°C will eliminate the carbon matrix and its volatile compounds (Zhang et al., 2020).

However, conditions for thermal treatment may differ or be subject to the following factors:

- Type of adsorbents
- Contaminants
- Purpose of subsequent use

These conditions require further research and improvement to make the process scalable (Kar et al., 2022).

20.2.4 Chemical desorption

Solvent regeneration mainly applies to high-concentration and low-boiling-point organic matter adsorbents (Zhang et al., 2021). Solvent regeneration uses the equilibrium relationship between biochar, solvent, and adsorbate to break equilibrium by altering the temperature and pH of the solvent so as to necessitate the regeneration process of pollutants from the biochar. Acids and alkali/base compounds that are usable as solvents in the regeneration process are H_2SO_4, HNO_3, HCl, EDTA (ethylene diamine tetra-acetic acid), NaOH, $Ca(NO_3)_2$, and $NaNO_3$(Gupta et al., 2020; Yang et al., 2020). Low pH has been seen to promote desorption, using acids such as H_2SO_4, HNO_3, and HCl as solvents (Gupta et al., 2020). However, very acidic environments have a tendency to deform or destroy the structure of the adsorbents, thereby decreasing the adsorption capacity when reused. As such, treatment using strong acids requires the biochar to be steady in its mechanical structure. Chelators such as EDTA can also be added to improve the rate of desorption. Within their structure, chelators have several functional groups, such as carboxyl (CO-OH) and amine groups, that attract pollutants to form complexes.

20.2.5 Supercritical fluid desorption

Supercritical fluids (SCFs) are used to extract or separate using differences in operating pressure. Momina et al. (2018) defined SCFs as fluids that are formed when a substance is heated to temperatures that are beyond its critical temperature and compressed above its critical pressure. Regeneration using SCFs allows for a short operating cycle, low operating temperature, and reduced biochar loss. Additionally, this technique can recover adsorbate without altering much on the physicochemical properties of the biochar. On the downside, supercritical desorption has high-pressure resistance and high equipment cost. The SCFs are thought to be better substitutes to replace incineration and chemical solvents (Efaq et al., 2015). The most common being CO_2 because CO_2 is not combustible, very economical, and nonhazardous (Noman et al., 2020). CO_2 has reduced surface tension and a high mass transfer rate. This process, however, requires more innovative approaches so that the cost is reduced and is also applicable on an industrial scale level. On the contrary, CO_2 has a lower regeneration capacity for phenol-loaded adsorbents. However, if supercritical water is used in place of supercritical CO_2, there was 100% desorption efficiency achieved on the same phenol-loaded adsorbents (Salvador et al., 2013). Supercritical water can be efficient in some cases, while it needs to be aided with other catalysts such as H_2O_2 and alkali metal in other cases. The use of supercritical water as the SCF entails reduced process duration and lower process cost. However, it requires high pressure. The high-pressure constraint makes its process cost higher, as well as reduces its applicability at the industrial and commercial levels. Therefore, supercritical water regeneration can only be applied on a small scale. In soils, the SCF assists the process of desorption as a typical solvent. The adsorbate is further condensed and collected in a separate small container.

20.2.6 Advanced oxidation processes

In recent years, much acknowledgment has been given to the use of advanced oxidation processes (AOPs) as a technique applicable to spent biochar regeneration (Acevedo-García et al., 2020; Fdez-Sanromán et al., 2020). Studies have shown AOPs to be usable in the regeneration of spent biochar (Acevedo-García et al., 2020), such as coffee biochar. In other studies, viable regeneration by various AOP techniques, including electro-Fenton reaction and Fenton reaction with hydrogen peroxide (H_2O_2), has been successful.

20.2.7 Microbial-assisted regeneration

Microbial-assisted biochar regeneration involves regenerating the adsorbent via microbial action or degrading of biodegradable adsorbates on the biochar surface (Abromaitis et al., 2016). This procedure makes use of either a pure culture or a mixed culture of microbes, such as fungi, bacteria, and even algae. The microbes are allowed to mix in batches, or alternatively, it is done in wastewater treatment during biological treatment of wastewater (Momina et al., 2018). Due to the nature of microbes, the adsorbent surface needs to be nontoxic and conducive to the growth and habitation of the microbes (Baskar, 2017).

Bioregeneration of spent sorbents can be done in two ways:

1. By desorption following a concentration gradient, where the microbes act on or degrade the unconfined organic material, thereby decreasing the amount of the impurities in the aqueous phase. The concentration gradient is set between the biochar surface and the aqueous media (Klimenko et al., 2002).
2. By the action of extracellular enzymes released by microbes, the pollutants are degraded as they diffuse into the porous structure of the biochar, degrade and hydrolyze the trapped pollutants.

The whole procedure is dependent on other aspects, such as microbial diversity, generation time, the microorganism-pollutant concentration ratio, nutrient availability, temperature, and dissolved oxygen level (Klimenko et al., 2003). These can be optimization factors depending on the needs and nature of the microbes.

20.2.8 Microwave irradiation regeneration

This technique utilizes microwave energy to heat the biochar molecules. Desorption takes place due to the rapidness, selectivity, and controlled heating of the adsorbent molecules (Falciglia et al., 2018). Microwave treatment evenly heats the biochar from the outer surface to the inside the core. Microwaves did not change the pores in the biochar molecules; hence, the properties of the adsorbent were preserved during regeneration, unlike with conventional thermal heating. Though the microwave irradiation procedure shows more temperature control, the technique can be costly at an industrial scale. The setup costs, as well as energy requirements, make the technique not very sustainable and uneconomic; hence, it requires more research to make it economically viable (Gagliano et al., 2020).

20.3 Reuse, recycling, and disposal of spent biochar for a circular economy and environmental sustainability

Biochar is a carbonaceous material attained from organic wastes by carbonization under an inert atmosphere (Andrade et al., 2020). After being used for a specific purpose and when no longer needed, spent biochar has to be recycled or disposed of in an environmentally friendly manner (Kar et al., 2022). However, in a bid to promote a circular economy, spent biochar has been utilized in different applications such as soil conditioning. However, recycling and disposal of spent biochar depend on its quality and potential contaminants, which are considerably influenced by the feedstocks, pyrolysis conditions, and pyrolysis technologies (Zahed et al., 2021). In addition, as highlighted by Tan et al. (2016), factors such as stability, potential secondary pollution, effect on carbon sequestration, and economic feasibility have to be considered in the recycling or disposal of spent biochar. Nonetheless, there are several recycling and disposal options for spent biochar, which include using it in heavy metal adsorption, soil amendment, composites, energy storage, tissue engineering, and ammonium adsorption in water, among others. When utilizing the spent biochar, care needs to be taken to ensure that the adverse effects of pollutants are minimized and controlled.

20.3.1 Reuse as a soil amendment

If the spent biochar is free from contaminants and still retains beneficial properties, it can be reused as a soil amendment. Mixing it into garden beds, agricultural fields, or compost piles can improve soil health and fertility. In an experiment, He et al. (2022) utilized spent biochar from the textiles industry as a precursor for compost generation in a process that was conducted using dye-loaded spent biochar, ready compost made of vegetable waste, and sawdust. They concluded that the obtained compost can be used as a soil conditioner for enhancing plant growth. The same conclusions were made by Ramrakhiani et al. (2022), who researched the likelihood of using the spent Zn(II)-laden biochar in the making of plant microfertilizer and soil-fertility improvers. They discovered that the application of the Zn(II)-laden biochar mixed with soil achieved an optimistic germination on the seeds of *Cicer arietinum*, growth parameters of plants, as well as content of protein and chlorophyll a and b. Using spent grain biochar on Hop (*Humulus lupulus L*), Amoriello et al. (2020) noted that biochar significantly improved root growth, and the addition of biochar to the soil further improved yield in terms of the number of cones on the plants. Apart from improving soil nutrient content, spent biochar can also improve soil water retention capacity and drainage (He et al., 2022; Wang et al., 2016). Similarly, Şeker & Manirakiza (2020) concluded that compost and biochar significantly improved aggregation and water retention characteristics of sandy clay loam soil. Therefore, biochar acts as a soil conditioner, improving physical, chemical, and biological properties and enhancing soil fertility and crop yield (Tsolis & Barouchas, 2023), which presents an opportunity for it to be reused as a soil amendment.

20.3.2 Composting

Spent biochar can be added to compost piles as a carbon-rich component that helps balance the carbon-to-nitrogen ratio, improves aeration, and acts as a stable organic matter in the aerobic composting process (Agyarko-Mintah et al., 2017; Liu et al., 2020). Biochar makes organic waste harmless and stable and plays a role in suppressing carbon and nitrogen losses (Liu et al., 2020). This allows the biochar to enhance the quality of the resulting compost. Zhang et al. (2020) noted that straw biochar hastens organic matter degradation and produces nutrient-rich compost, the addition of biochar with the optimal concentration of 10%–15% resulted in more mature and humified, with a decrease in dissolved organic carbon content and NH_4 concentration compared to the control. The concentrations of water-soluble nutrients, including PO_4^{3-}, K, and Ca^{2+}, were also increased with biochar addition. The addition of biochar in composting can reduce the emissions of CO_2, N_2O, and other gases during composting (Li et al., 2016; Liu et al., 2017).

20.3.3 Land application

In some cases, spent biochar can be safely applied to nonagricultural land, such as land reclamation projects or landscaping. Research has shown that spent biochar can be used in land reclamation because of its soil improvement and heavy metals sequestration abilities (Raclavská et al., 2021). The application of biochar in contaminated matrices can also help mitigate the damage caused by agrochemicals and improve crop quality. Additionally, the addition of biochar to composting processes has been shown to reduce volatile organic compound emissions and accelerate the biodegradation process. Biochar can also be used as a stand-alone amendment or in combination with biosolids to enhance tree growth and root development. An assessment of the potential of using an acacia-biochar system for spent mine site rehabilitation by Reverchon et al. (2015) provides insights into the potential benefits of using biochar for mine site rehabilitation. The use of spent biochar in land reclamation can contribute to sustainable agriculture, soil improvement, and environmental remediation. However, it is important to ensure that the biochar is free from contaminants and does not pose any environmental risks.

20.3.4 Regeneration through pyrolysis

Spent biochar can be recycled and regenerated through pyrolysis again. A study by Cui et al. (2022) revealed that pyrolysis of spent biochar can be used to regenerate valuable biochar from exhausted hydrochar sorbent. In addition to increasing the functionality of the biochar, in cases where the biochar was used as an adsorbent, recycling of spent biochar can also immobilize heavy metals such as cadmium (Fan et al., 2023). However, as noted by Tan et al. (2023), regenerating and recycling spent biochar depend on the scale and available technology, with the integration of continuous reactors and catalysis being more efficient and economical for commercialization. As noted in a study by Vercruysse et al. (2021), the regeneration of spent biochar can be done by mixing it with fresh biomass and pyrolyzing it again using slow pyrolysis at different temperatures. Spent biochar can also be used as a feedstock for renewable energy production or converted into bio-oil or syngas (Lee et al., 2020). Therefore, depending on the available technology and type of feedstock, spent biochar can be regenerated through pyrolysis.

20.3.5 Contaminant removal and landfill disposal

If spent biochar contains harmful contaminants, it might need specific treatment to remove or mitigate them. Usually, spent biochar is disposed of after the recovery of heavy metals or immediately after the adsorption process; in both situations, there may be secondary pollution from used biochar and the chemicals used to prepare the biochar (Gupta et al., 2020). Hence, the disposal of spent biomaterials contaminated with harmful organic ions also poses social and environmental problems. It may be necessary to dispose of the biochar in a designated landfill or waste management facility capable of handling contaminated materials. The spent adsorbents can be safely disposed of in landfilling (Kozyatnyk et al., 2020). However, in some cases, pretreatment of the biochar will be necessary before landfilling is done. For example, disposal of peat-based biochar, which is used in the removal of heavy metals, would require pretreatment (Kar et al., 2022). The downside of landfilling spent biochar appears in landfills where it is susceptible to metal-leaching and dispersion (Jambeck et al., 2007). Hence, it may be necessary to dispose of the biochar in a designated landfill or waste management facility capable of handling contaminated materials. Landfilling can be a viable option for final management of biochar. However, it is important for one to determine the concentration of the contaminant before disposing.

20.3.6 Incineration

Spent biochar can be incinerated as a final management technique (Fernández-González et al., 2019). Because biosorbents are rich in biomolecules such as lignin and cellulose, incineration not only greatly reduces the mass and volume of waste biomass but also recovers heavy metals and thermal energy (Ronda et al., 2016). However, this presents problems such as air pollution, which could potentially harm humans and the environment (Hu et al., 2022). Chaukura et al. (2016) suggested other disposal options because incineration will cause more pollution by emitting greenhouse gases (GHGs). Spent biochar can be used to make other products, such as activated carbon, which appears to be a new, potentially cost-effective, and environmentally friendly carbon material with great application prospects in many fields such as water purification (Tan et al., 2016). Activated carbon is produced from any carbon source (fossil, waste, or renewable) and is engineered to be used as a sorbent to remove contaminants from both gases and liquids (Marsh & Rodríguez-Reinoso, 2006; Mukherjee et al., 2022). Spent biochar has been observed to have reduced nitrogen, sulfur, and oxygen while having high carbon levels. These characteristics make spent biochar a good alternative to coal, as it has a high calorific value while at the same time producing less of the toxic emissions and corrosiveness (Kar et al., 2022). Spent biochar can be used to make other products, such as activated carbon, which is used in water purification and other applications.

20.3.7 Fillers in novel construction materials

The cement industry contributes 5%–8% of the GHG emissions (Miller et al., 2021) due to its energy-intensive processes. As such, any alternatives or substitutes to cement in the construction industry are sought-after. Spent biochar has been used in making construction bricks as a geopolymer. Using it as a geopolymer has additional environmental benefits such as decreasing natural resource depletion such as clay, and also the ability to capture GHGs, including CO_2 and CH_4, in its solid phase as carbon. When compared to concrete bricks, biochar bricks (composed of cement, plastic, and biochar) had competitive qualities in terms of compressive strength, insulation, water absorption, and hardness (Nithyalakshmi et al., 2022).

20.3.8 Catalyst precursors for trace metal-rich spent biochars

The physicochemical properties of biochar can be improved through various processes in order to give them properties of catalysts or catalyst supports. This has been attributed to the various surface functional groups present, in addition to their large surface area. Despite all these properties and positive factors, the ability of spent biochar as a catalyst, especially for advanced electrocatalytic processes, hence more study is required in this regard (Ramos et al., 2022).

20.3.9 Incorporation in solid fuels

Biochar can be incorporated into solid fuels for use in different applications (Gwenzi et al., 2021). The biochar has fuel properties with calorific values ranging from 24.12 to 26.24MJ/kg (Vukićević et al., 2024), which was way higher than that of wood (12.5MJ/kg). Ash content of biochar varies depending on the source of the waste charred. Various studies showed different ash values, for example, agricultural waste biochar was found to be in the range of 12%–20% (Vukićević et al., 2024), which was within the range of solid fuels, sawdust was in the range of 0.65%–1.7% and that of sewage sludge was as high as 45.7% (Pulka et al., 2016).

20.3.10 Codisposal of sludge

In various researches, biochar has been mixed with sewage sludge. Its addition even in small quantities proved to yield some benefits such as increased pile temperature and also aid ammonium transformation to nitrite, thereby reducing the total nitrogen losses (Liu et al., 2017). Furthermore, the application of biochar mixed with sludge can help reduce soil contaminants such as polycyclic aromatic hydrocarbons (Stefaniuk et al., 2017) and hence the mixture can be used for different environmental remediation applications, especially involving organic pollutants (Chen et al., 2020).

20.4 Behavior and fate of contaminants in spent biochar

Due to aging and several recycling cycles of the biochar, its chemistry, including the surface functional groups, can be distorted through oxidation or microbial degradation activities (Downie et al., 2009; Khorram et al., 2016). It is also noted that the adsorption capacity of biochar decreases with time as a result of the aging process aided by the clogging of biochar pore space with soil particles and the removal of reactive biochar surfaces due to reduced inactivity by soil mineral elements (Pignatello et al., 2006). This compromises the biochar's capacity as a pollutant removal agent, causing the accumulation of pollutants impacting environmental health and the ecosystem. Additionally, studies have indicated the reversible pesticide adsorption in soils amended with recycled biochar samples through various ways, including sorbent swelling during adsorption process and penetration of the pesticides via active diffusion (George, 2022). This enhances the diameter expansion of the micropores, which could result in the micropore network deformation (Bahia, 2022; George, 2022), and the weak binding between the applied pesticides and spent biochar components (Khorram et al., 2015; Tatarková et al., 2013).

The sorption capacity of biochar particles in amended soils deteriorates on herbicides due to aging (Maestrini et al., 2014; Obia et al., 2017; Mendes et al., 2018). This is attributed to structural changes (Mendes et al., 2018) such as reduced SSA and the porosity of the aging biochar, resulting in the blockage of micropores and the sorption sites, further increasing the bioavailability of contaminants in the environment (Yavari et al., 2015). Other studies reported a reduction in the sorption capacity of diuron by approximately 47%–68% against the control soil, which could be due to the clogging of the pores of the soil particles with spent biocha allowing for the bioaccumulation of the pollutant (Martin et al., 2012). A loss in spent biochar's mass has been revealed over time (Maestrini et al., 2014; Obia et al., 2017), however, Dong et al. (2017) reported an approximated loss of about 40% of biochar in 5 years irrespective of the applied mass during the amendment cycle (30, 60, or 90 t/ha^1). This, however, threatens environmental health as the cumulative effect of contaminants could reach unprecedented levels with the loss of biochar, which significantly plays a pivotal role in immobilizing contaminants in the soil and water ecosystems. Further, studies revealed that structural and chemical changes that transform the biochars' sorptive capacity depend on the type of each material, the pyrolysis temperature, conditions, and the time at which the material will be incubated in the soil, as some mineral elements were observed to have filled pores at the expense of contaminants (Trigo et al., 2014). The spent biochar macropores, which previously served as a route for adsorbate exposure locking up contaminants, may become degraded by selected plants for phytoremediation, whilst providing nutrients to microbes mimicking microbial activity for the degradation of contaminants (Yang et al., 2010).

In order to maintain pesticides and herbicides pollutant concentrations within tolerable limits that do not cause severe environmental health loss following the aging of spent biochars in amended soils, bioremediation processes involving hydrolysis, conjugation, and degradation could be a panacea in providing frugal solutions to the rising levels of contaminants in soil and water (Yang et al., 2010). The enzymatic degradation exhibited by microorganisms, however, requires an understanding of the pollutant's chemical and structural composition, and additionally, the molecular factors that can affect pollutant transformation to a lesser toxic and/or its complete removal from the environment (Mali et al., 2023). The increased regeneration cycles of spent biochar could result in the decline of their adsorption capacity, through the erosion of their mechanical and chemical stability. These functional declines would compromise their capability as contaminants remediation matrices both in soil and water, negatively impacting the environmental health and well-being of humans as contaminants would bioaccumulate in food webs (Gwenzi et al., 2017).

Above all, some disposal techniques of spent biochar and other adsorbents are either desorption by acid and base treatment or landfilling techniques (Gupta & Babu, 2009; Wang et al., 2015). However, the disposal and/or dumping of the spent biochar threaten environmental health as possibilities of leaching, explosions, stinking, and volatilization of unwanted pollutants may occur (Ruaa et al., 2021; Vakili et al., 2019). Due to the affinity for active sites on the adsorbents, the desorption process meant to determine the fate of pollutants usually poses challenges making it not economically viable to use other adsorbents, as they are rendered useless and regarded as secondary waste (Ruaa et al., 2021). Recycled spent biochar also finds its way under the resource utilization strategy on which they can be transformed into valuable electroactive materials (Chen et al., 2021; Sundriyal et al., 2021), and also can be utilized in construction (Selvasembian et al., 2023).

Transitional metals (Cu, Fe, Co, Mn, and Ni) are mostly found as a nuisance in soil and water contamination, however, they can be utilized in designing cost-effective catalysts for the energy-related oxygen evolution reaction (OER), which is essential in the development of green hydrogen technologies (Suen et al., 2017; Zhou et al., 2021). The development and advancement in green hydrogen energy are seen as a panacea and enhanced strategy in the production of efficient, cheap, and sustainable OER electrocatalysts (Wan et al., 2021), while consuming spent biochars, which are likely to contaminate the environment due to their capacity to contain pollutant residues (metal-laden). As such technologies, including the integration of the electrocatalytic transition metal borides with the porous and large SSA spent biochar producing heterostructured electrocatalysts provides both an economic and environmentally based benefit (Chen et al., 2021; Mao et al., 2020). This is also termed the bidding strategy on which the waste is converted to wealth. This is attributed to the increased performance of the electrocatalysts and the removal of pollutant-bound spent biochar from the environment.

Studies have also pointed to the use of recycled Cr-adsorbents as catalysts for the removal of ulfur-containing volatile organic compounds, which is the methyl mercaptan (CH_3SH), as an alternative way of determining the fate of pollutants on spent adsorbents compared to the conventional landfilling and desorption using acid/base treatment (He et al., 2018). Srivatsav et al. (2020) reiterate that biochar's capacity to adsorb pollutants deteriorates when subjected to multiple operational cycles. Due to the high regeneration cost of spent biochar, disposal challenges of metal-laden and other bound pollutants could be hazardous to the environment due to volatilization and pungent smells (Srivatsav et al., 2020). At high temperatures ($>400°C$), heavy metals bound on spent biochar tend to volatilize and contaminant the immediate environment through metal pollution (Lehmann & Joseph, 2009). This is also further supported by the fact that biochars are recalcitrant forms of organic material, therefore, their decomposition can be increased by a further 10°C rise in temperature, compromising their capacity as GHG sequestration media (Davidson & Janssens, 2006; Lehmann & Joseph, 2009). Spent biochar loses weight over multiple regeneration cycles and mineralization could also exacerbate leaching, eluviation, and erosion, resulting in the spread of pollutants due to offsite deposition and lateral export across the drainage pattern (Dong et al., 2017; Lehmann & Joseph, 2009; Rumpel et al., 2006). Through deposition, spent biochar could be buried under both oxygen-free or inert conditions, and this will likely increase underground contamination if metals and other pollutants are desorbed from the biochar. Due to a plethora of impacts on the environment posed by spent biochars at the end of their cycle, they can also be utilized as biocarriers (Srivatsav et al., 2020). Overall, it is worth noting that establishing the relationships between biochar properties and the exposed environmental conditions (e.g., climate and soils) could be fundamental in estimating their stability vis-à-vis decomposition rates, which largely occur over annual to hundreds of years disposing pollutants into the environment.

20.5 Summary and future research directions

Biochar has been demonstrated to be effective in the capture and removal of both organic and inorganic contaminants, as it is widely applied in various environmental media. The spent biochars generally pose a major drawback in the circular economy of the utilization of biochars. However, the spent biochar can be regenerated using techniques such as AOPs, filtration, and chemical and thermal desorption. The current regeneration approaches are dependent on the type of feedstock used for biochar preparation, and contaminant type. Recently, there has been increasing development of advanced biochar with improved contaminant capture capacity and removal efficiency in various environmental compartments. The disposal of spent biochar loaded with contaminants is a major challenge. The spent biochar can be regenerated/activated for reuse as soil amendments, composting, and general land reclamation. If the spent biochar cannot be regenerated, it can be safely disposed of via landfilling or incineration. The reuse of the spent biochar has environmental benefits, as well as reduces the overall cost of the application.

Although the reuse of spent biochar has some environmental benefits, the challenge of secondary pollution is not yet clear, and this aspect is poorly studied relative to biochar applications in remediation. Also, when they are incinerated, the possibility of producing some toxic air pollutants is not clear. Therefore, researchers need to put effort into investigating the effect of incinerating spent biochar on the production of toxic air pollutants. Besides, when landfilled, the possible effects of leaching and polluting underground water need to be further examined.

Owing to the current emphasis on attaining minimum levels of waste generation, circular economy, and enhanced resource efficiency in a bid to achieve environmental sustainability, the following areas could also be pursued for research and improve removal of contaminants using biochar in various environmental media:

- Development of resource-oriented technologies for the reclamation of impurities from spent adsorbents.
- Studies examining the possibilities of groundwater contamination due to long-term leaching of contaminants from spent biochar disposed of in landfills.

- Long-term kinetic studies on the worthiness of recycling spent biochar as a source of nutrients.
- Long-term studies on the stability of reusing and recycling of spent adsorbents.
- Transformation of spent biochar into value-added products for recycling and reuse.
- Studies on the toxicity of adsorbents before their reuse in other applications.
- Regeneration techniques that are energy-efficient for spent adsorbents.
- Identifying new reuse applications for the management of spent adsorbents.
- Life-cycle analysis and risk assessment of recycling and reuse of spent adsorbents.

References

Abromaitis, V., Racys, V., van der Marel, P., & Meulepas, R. J. W. (2016). Biodegradation of persistent organics can overcome adsorption−desorption hysteresis in biological activated carbon systems. *Chemosphere*, 149, 183−189. Available from https://doi.org/10.1016/j.chemosphere.2016.01.085.

Acevedo-García, V., Rosales, E., Puga, A., Pazos, M., & Sanromán, M. A. (2020). Synthesis and use of efficient adsorbents under the principles of circular economy: Waste valorisation and electroadvanced oxidation process regeneration. *Separation and Purification Technology*, 242, 116796. Available from https://doi.org/10.1016/j.seppur.2020.116796.

Agyarko-Mintah, E., Cowie, A., Van Zwieten, L., Singh, B. P., Smillie, R., Harden, S., & Fornasier, F. (2017). Biochar lowers ammonia emission and improves nitrogen retention in poultry litter composting. *Waste Management*, 61, 129−137. Available from https://doi.org/10.1016/j.wasman.2016.12.009, http://www.elsevier.com/locate/wasman.

Alsawy, T., Rashad, E., El-Qelish, M., & Mohammed, R. H. (2022). A comprehensive review on the chemical regeneration of biochar adsorbent for sustainable wastewater treatment. *NPJ Clean Water*, 5(1). Available from https://doi.org/10.1038/s41545-022-00172-3.

Amoriello, T., Fiorentino, S., Vecchiarelli, V., & Pagano, M. (2020). Evaluation of Spent Grain Biochar Impact on Hop (Humulus lupulus L.) Growth by Multivariate Image Analysis. *Applied Sciences*, 10(2), 533. Available from https://doi.org/10.3390/app10020533.

Andrade, T. S., Vakros, J., Mantzavinos, D., & Lianos, P. (2020). Biochar obtained by carbonization of spent coffee grounds and its application in the construction of an energy storage device. *Chemical Engineering Journal Advances*, 4, 100061. Available from https://doi.org/10.1016/j.ceja.2020.100061.

Bahia, W. (2022). Adsorption-Desorption Behavior and Pesticide Bioavailability of Biochar in Soil. *Science Insights*, 41(6), 725−731. Available from https://doi.org/10.15354/si.22.re093.

Barquilha, C. E. R., & Braga, M. C. B. (2021). Adsorption of organic and inorganic pollutants onto biochars: Challenges, operating conditions, and mechanisms. *Bioresource Technology Reports*, 15. Available from https://doi.org/10.1016/j.biteb.2021.100728, https://www.journals.elsevier.com/bioresource-technology-reports.

Baskar, A. V. (2017). Recovery, regeneration and sustainable management of spent adsorbents from wastewater treatment streams: A review. *Science of the Total Environment*, 822(3), 2017.

Chaukura, N., Gwenzi, W., Tavengwa, N., & Manyuchi, M. M. (2016). Biosorbents for the removal of synthetic organics and emerging pollutants: Opportunities and challenges for developing countries. *Environmental Development*, 19, 84−89. Available from https://doi.org/10.1016/j.envdev.2016.05.002, http://www.sciencedirect.com/science/journal/22114645.

Chen, Y.-di, Wang, R., Duan, X., Wang, S., Ren, N.-qi, & Ho, S.-H. (2020). Production, properties, and catalytic applications of sludge derived biochar for environmental remediation. *Water Research*, 187, 116390. Available from https://doi.org/10.1016/j.watres.2020.116390.

Chen, Z., Zheng, R., Graś, M., Wei, W., Lota, G., Chen, H., & Ni, B. J. (2021). Tuning electronic property and surface reconstruction of amorphous iron borides via W-P co-doping for highly efficient oxygen evolution. *Applied Catalysis B: Environmental*, 288. Available from https://doi.org/10.1016/j.apcatb.2021.120037, http://www.elsevier.com/inca/publications/store/5/2/3/0/6/6/index.htt.

Crini, G., Lichtfouse, E., Wilson, L. D., & Morin-Crini, N. (2019). Conventional and non-conventional adsorbents for wastewater treatment. *Environmental Chemistry Letters*, 17(1), 195−213. Available from https://doi.org/10.1007/s10311-018-0786-8.

Cui, X., Wang, J., Wang, X., Du, G., Khan, K. Y., Yan, B., Cheng, Z., & Chen, G. (2022). Pyrolysis of exhausted hydrochar sorbent for cadmium separation and biochar regeneration. *Chemosphere*, 306. Available from https://doi.org/10.1016/j.chemosphere.2022.135546, http://www.elsevier.com/locate/chemosphere.

Davidson, E. A., & Janssens, I. A. (2006). Temperature sensitivity of soil carbon decomposition and feedbacks to climate change. *Nature*, 440(7081), 165−173. Available from https://doi.org/10.1038/nature04514.

Dong, X., Li, G., Lin, Q., & Zhao, X. (2017). Quantity and quality changes of biochar aged for 5 years in soil under field conditions. *CATENA*, 159, 136−143. Available from https://doi.org/10.1016/j.catena.2017.08.008.

Downie, A., Crosky, A., & Munroe, P. (2009). *Physical Properties of Biochar* (2009). Earthscan, Earthscan.

Dutt, M. A., Hanif, M. A., Nadeem, F., & Bhatti, H. N. (2020). A review of advances in engineered composite materials popular for wastewater treatment. *Journal of Environmental Chemical Engineering*, 8(5), 104073. Available from https://doi.org/10.1016/j.jece.2020.104073.

Efaq, A. N., Rahman, N. N. N. A., Nagao, H., Al-Gheethi, A. A., Shahadat, M., & Kadir, M. O. A. (2015). Supercritical Carbon Dioxide as Non-Thermal Alternative Technology for Safe Handling of Clinical Wastes. *Environmental Processes*, 2(4), 797−822. Available from https://doi.org/10.1007/s40710-015-0116-0, http://www.springer.com/earth + sciences + and + geography/environmental + science + /26 + engineering/journal/40710.

Falciglia, P. P., Roccaro, P., Bonanno, L., De Guidi, G., Vagliasindi, F. G. A., & Romano, S. (2018). A review on the microwave heating as a sustainable technique for environmental remediation/detoxification applications. *Renewable and Sustainable Energy Reviews, 95*, 147–170. Available from https://doi.org/10.1016/j.rser.2018.07.031, https://www.journals.elsevier.com/renewable-and-sustainable-energy-reviews.

Fan, Z., Zhou, X., Peng, Z., Wan, S., Fan Gao, Z., Deng, S., Tong, L., Han, W., & Chen, X. (2023). Co-pyrolysis technology for enhancing the functionality of sewage sludge biochar and immobilizing heavy metals. *Chemosphere, 317*, 137929. Available from https://doi.org/10.1016/j.chemosphere.2023.137929.

Fdez-Sanromán, A., Pazos, M., Rosales, E., & Sanromán, M. A. (2020). Unravelling the Environmental Application of Biochar as Low-Cost Biosorbent: A Review. *Applied Sciences, 10*(21), 7810. Available from https://doi.org/10.3390/app10217810.

Fernández-González, R., Martín-Lara, M. A., Moreno, J. A., Blázquez, G., & Calero, M. (2019). Effective removal of zinc from industrial plating wastewater using hydrolyzed olive cake: Scale-up and preparation of zinc-Based biochar. *Journal of Cleaner Production, 227*, 634–644. Available from https://doi.org/10.1016/j.jclepro.2019.04.195.

Gagliano, E., Sgroi, M., Falciglia, P. P., Vagliasindi, F. G. A., & Roccaro, P. (2020). Removal of polyand perfluoroalkyl substances (PFAS) from water by adsorption: role of PFAS chain length, effect of organic matter and challenges in adsorbent regeneration. *Water Research, 171*, 2020.

George, M. (2022). Unravelling the impact of potentially toxic elements and biochar on soil: A review. *Environmental Challenges, 8*, 100540. Available from https://doi.org/10.1016/j.envc.2022.100540.

Gupta, S., & Babu, B. V. (2009). Removal of toxic metal Cr(VI) from aqueous solutions using sawdust as adsorbent: Equilibrium, kinetics and regeneration studies. *Chemical Engineering Journal, 150*(2-3), 352–365. Available from https://doi.org/10.1016/j.cej.2009.01.013.

Gupta, S., Sireesha, S., Sreedhar, I., Patel, C. M., & Anitha, K. L. (2020). Latest trends in heavy metal removal from wastewater by biochar based sorbents. *Journal of Water Process Engineering, 38*, 101561. Available from https://doi.org/10.1016/j.jwpe.2020.101561.

Gwenzi, W., Chaukura, N., Noubactep, C., & Mukome, F. N. D. (2017). Biochar-based water treatment systems as a potential low-cost and sustainable technology for clean water provision. *Journal of Environmental Management, 197*, 732–749. Available from https://doi.org/10.1016/j.jenvman.2017.03.087.

Gwenzi, W., Chaukura, N., Wenga, T., & Mtisi, M. (2021). Biochars as media for air pollution control systems: Contaminant removal, applications and future research directions. *Science of The Total Environment, 753*, 142249. Available from https://doi.org/10.1016/j.scitotenv.2020.142249.

Hassan, M., Naidu, R., Du, J., Liu, Y., & Qi, F. (2020). Critical review of magnetic biosorbents: Their preparation, application, and regeneration for wastewater treatment. *Science of The Total Environment, 702*, 134893. Available from https://doi.org/10.1016/j.scitotenv.2019.134893.

He, D., Zhang, L., Zhao, Y., Mei, Y., Chen, D., He, S., & Luo, Y. (2018). Recycling Spent Cr Adsorbents as Catalyst for Eliminating Methylmercaptan. *Environmental Science & Technology, 52*(6), 3669–3675. Available from https://doi.org/10.1021/acs.est.7b06357.

He, M., Xu, Z., Hou, D., Gao, B., Cao, X., Ok, Y. S., Rinklebe, J., Bolan, N. S., & Tsang, D. C. W. (2022). Waste-derived biochar for water pollution control and sustainable development. *Nature Reviews Earth and Environment, 3*(7), 444–460. Available from https://doi.org/10.1038/s43017-022-00306-8, nature.com/natrevearthenviron/.

Herath, A., Layne, C. A., Perez, F., Hassan, E. I. B., Pittman, C. U., & Mlsna, T. E. (2021). KOH-activated high surface area Douglas Fir biochar for adsorbing aqueous Cr(VI), Pb(II) and Cd(II). *Chemosphere, 269*, 128409. Available from https://doi.org/10.1016/j.chemosphere.2020.128409.

Hossain, N., Bhuiyan, M. A., Pramanik, B. K., Nizamuddin, S., & Griffin, G. (2020). Waste materials for wastewater treatment and waste adsorbents for biofuel and cement supplement applications: A critical review. *Journal of Cleaner Production, 255*, 120261. Available from https://doi.org/10.1016/j.jclepro.2020.120261.

Hu, W., Di, Q., Liang, T., Liu, J., & Zhang, J. (2022). Effects of spent mushroom substrate biochar on growth of oyster mushroom (Pleurotus ostreatus). *Environmental Technology & Innovation, 28*, 102729. Available from https://doi.org/10.1016/j.eti.2022.102729.

Iamsaard, K., Weng, C.-H., Yen, L.-T., Tzeng, J.-H., Poonpakdee, C., & Lin, Y.-T. (2022). Adsorption of metal on pineapple leaf biochar: Key affecting factors, mechanism identification, and regeneration evaluation. *Bioresource Technology, 344*, 126131. Available from https://doi.org/10.1016/j.biortech.2021.126131.

Jambeck, J., Weitz, K., Solo-Gabriele, H., Townsend, T., & Thorneloe, S. (2007). CCA-Treated wood disposed in landfills and life-cycle trade-offs with waste-to-energy and MSW landfill disposal. *Waste Management, 27*(8), S21. Available from https://doi.org/10.1016/j.wasman.2007.02.011.

Jia, Y., Zong, Q., Zhang, M., Wang, Z. M., & Zhang, L. H. (2016). Research progress in regeneration technology of activated carbon from flue gas desulfurization. *Bull. China Ceram. Soc., 35*, 2016.

Kah, M., Oliver, D., & Kookana, R. (2021). Sequestration and potential release of PFAS from spent engineered sorbents. *Science of The Total Environment, 765*, 142770. Available from https://doi.org/10.1016/j.scitotenv.2020.142770.

Kar, S., Santra, B., Kumar, S., Ghosh, S., & Majumdar, S. (2022). Sustainable conversion of textile industry cotton waste into P-dopped biochar for removal of dyes from textile effluent and valorisation of spent biochar into soil conditioner towards circular economy. *Environmental Pollution, 312*, 120056. Available from https://doi.org/10.1016/j.envpol.2022.120056.

Khorram, M. S., Wang, Y., Jin, X., Fang, H., & Yu, Y. (2015). Reduced mobility of fomesafen through enhanced adsorption in biochar-amended soil. *Environmental Toxicology and Chemistry, 34*(6), 1258–1266. Available from https://doi.org/10.1002/etc.2946, http://www.interscience.wiley.com/jpages/0730-7268.

Khorram, M. S., Zheng, Y., Lin, D., Zhang, Q., Fang, H., & Yu, Y. (2016). Dissipation of fomesafen in biochar-amended soil and its availability to corn (Zea mays L.) and earthworm (Eisenia fetida). *Journal of Soils and Sediments, 16*(10), 2439–2448. Available from https://doi.org/10.1007/s11368-016-1407-4, http://www.springerlink.com/content/1439-0108.

Klimenko, N., Smolin, S., Grechanyk, S., Kofanov, V., Nevynna, L., & Samoylenko, L. (2003). Bioregeneration of activated carbons by bacterial degraders after adsorption of surfactants from aqueous solutions. *Colloids and Surfaces A: Physicochemical and Engineering Aspects, 230*(1-3), 141–158. Available from https://doi.org/10.1016/j.colsurfa.2003.09.021.

Klimenko, N., Winthernielsen, M., Smolin, S., Nevynna, L., & Sydorenko, J. (2002). Role of the physico-chemical factors in the purification process of water from surface-active matter by biosorption. *Water Research, 36*(20), 5132–5140. Available from https://doi.org/10.1016/s0043-1354(02)00278-6.

Kozyatnyk, I., Yacout, D. M. M., Van Caneghem, J., & Jansson, S. (2020). Comparative environmental assessment of end-of-life carbonaceous water treatment adsorbents. *Bioresource Technology, 302*, 122866. Available from https://doi.org/10.1016/j.biortech.2020.122866.

Lee, X. J., Ong, H. C., Gan, Y. Y., Chen, W. H., & Mahlia, T. M. I. (2020). State of art review on conventional and advanced pyrolysis of macroalgae and microalgae for biochar, bio-oil and bio-syngas production. *Energy Conversion and Management, 210*. Available from https://doi.org/10.1016/j.enconman.2020.112707, https://www.journals.elsevier.com/energy-conversion-and-management.

Lehmann, J., & Joseph, S. (Eds.), (2009). *Biochar for Environmental Management: Science and Technology* (1st ed.). Routledge. Available from https://doi.org/10.4324/9781849770552.

Li, R., Wang, J. J., Zhou, B., Awasthi, M. K., Ali, A., Zhang, Z., Lahori, A. H., & Mahar, A. (2016). Recovery of phosphate from aqueous solution by magnesium oxide decorated magnetic biochar and its potential as phosphate-based fertilizer substitute. *Bioresource Technology, 215*, 209–214. Available from https://doi.org/10.1016/j.biortech.2016.02.125, http://www.elsevier.com/locate/biortech.

Liao, Y., Jiang, L., Cao, X., Zheng, H., Feng, L., Mao, Y., Zhang, Q., Shen, Q., & Ji, F. (2022). Efficient removal mechanism and microbial characteristics of tidal flow constructed wetland based on in-situ biochar regeneration (BR-TFCW) for rural gray water. *Chemical Engineering Journal, 431*, 134185. Available from https://doi.org/10.1016/j.cej.2021.134185.

Liu, W., Huo, R., Xu, J., Liang, S., Li, J., Zhao, T., & Wang, S. (2017). Effects of biochar on nitrogen transformation and heavy metals in sludge composting. *Bioresource Technology, 235*, 43–49. Available from https://doi.org/10.1016/j.biortech.2017.03.052.

Liu, Y., Ma, R., Li, D., Qi, C., Han, L., Chen, M., Fu, F., Yuan, J., & Li, G. (2020). Effects of calcium magnesium phosphate fertilizer, biochar and spent mushroom substrate on compost maturity and gaseous emissions during pig manure composting. *Journal of Environmental Management, 267*, 110649. Available from https://doi.org/10.1016/j.jenvman.2020.110649.

Maestrini, B., Abiven, S., Singh, N., Bird, J., Torn, M. S., & Schmidt, M. W. I. (2014). Carbon losses from pyrolysed and original wood in a forest soil under natural and increased N deposition. *Biogeosciences, 11*(18), 5199–5213. Available from https://doi.org/10.5194/bg-11-5199-2014.

Mali, H., Shah, C., Raghunandan, B. H., Prajapati, A. S., Patel, D. H., Trivedi, U., & Subramanian, R. B. (2023). Organophosphate pesticides an emerging environmental contaminant: Pollution, toxicity, bioremediation progress, and remaining challenges. *Journal of Environmental Sciences (China), 127*, 234–250. Available from https://doi.org/10.1016/j.jes.2022.04.023, http://www.journals.elsevier.com/journal-of-environmental-sciences/.

Mao, H., Guo, X., Fu, Y., Yang, H., Zhang, Y., Zhang, R., & Song, X. M. (2020). Enhanced electrolytic oxygen evolution by the synergistic effects of trimetallic FeCoNi boride oxides immobilized on polypyrrole/reduced graphene oxide. *Journal of Materials Chemistry A, 8*(4), 1821–1828. Available from https://doi.org/10.1039/c9ta10756h, http://pubs.rsc.org/en/journals/journal/ta.

Marsh, H., & Rodríguez-Reinoso, F. (2006). *Activated Carbon Activated Carbon*. Spain: Elsevier Ltd, Spain Elsevier Ltd. Available from http://www.sciencedirect.com/science/book/9780080444635, 10.1016/B978-0-08-044463-5.X5013-4.

Martin, S. M., Kookana, R. S., Van Zwieten, L., & Krull, E. (2012). Marked changes in herbicide sorption-desorption upon ageing of biochars in soil. *Journal of Hazardous Materials, 231-232*, 70–78. Available from https://doi.org/10.1016/j.jhazmat.2012.06.040.

Mendes, K. F., Júnior, A. F. D., Takeshita, V., Régo, J. A. P., & Tornisielo, V. L. (2018). Effect of Biochar Amendments on the Sorption and Desorption Herbicides in Agricultural Soil. In E. Serpil (Ed.), *Advanced Sorption Process Applications* (p. 5). Rijeka: IntechOpen.

Miller, S. A., Habert, G., Myers, R. J., & Harvey, J. T. (2021). Achieving net zero greenhouse gas emissions in the cement industry via value chain mitigation strategies. *One Earth, 4*(10), 1398–1411. Available from https://doi.org/10.1016/j.oneear.2021.09.011, http://www.cell.com/one-earth.

Momina, M., Shahadat, M., & Isamil, S. (2018). Regeneration performance of clay-based adsorbents for the removal of industrial dyes: a review. *RSC Advances, 8*(43), 24571–24587. Available from https://doi.org/10.1039/C8RA04290J.

Mukherjee, A., Okolie, J. A., Niu, C., & Dalai, A. K. (2022). Techno–Economic analysis of activated carbon production from spent coffee grounds: Comparative evaluation of different production routes. *Energy Conversion and Management: X, 2022*, 14.

Nguyen, T.-B., Sherpa, K., Bui, X.-T., Nguyen, V.-T., Vo, T.-D.-H., Ho, H.-T.-T., Chen, C.-W., & Dong, C.-D. (2023). Biochar for soil remediation: A comprehensive review of current research on pollutant removal. *Environmental Pollution, 337*, 122571. Available from https://doi.org/10.1016/j.envpol.2023.122571.

Nithyalakshmi, B., Soundarya, N., & Praveen, S. (2022). Characterization of Biochar Bricks to be used as a Construction Material. *Journal of Physics: Conference Series, 2332*(1), 012015. Available from https://doi.org/10.1088/1742-6596/2332/1/012015.

Noman, E., Al-Gheethi, A. A. S., Talip, B. A., & Mohamed, R. M. S. R. (2020). *Qualitative Characterization of Healthcare Wastes* (pp. 167–178). Springer Science and Business Media LLC. Available from 10.1007/978-3-030-42641-5_10.

Obia, A., Børresen, T., Martinsen, V., Cornelissen, G., & Mulder, J. (2017). Vertical and lateral transport of biochar in light-textured tropical soils. *Soil and Tillage Research, 165*, 34–40. Available from https://doi.org/10.1016/j.still.2016.07.016.

Omorogie, M. O., Babalola, J. O., & Unuabonah, E. I. (2016). Regeneration strategies for spent solid matrices used in adsorption of organic pollutants from surface water: a critical review. *Desalination and Water Treatment, 57*(2), 518–544. Available from https://doi.org/10.1080/19443994.2014.967726, http://www.tandfonline.com/toc/tdwt20/current.

Pignatello, J. J., Kwon, S., & Lu, Y. (2006). Effect of natural organic substances on the surface and adsorptive properties of environmental black carbon (Char): Attenuation of surface activity by humic and fulvic acids. *Environmental Science and Technology, 40*(24), 7757–7763. Available from https://doi.org/10.1021/es061307m.

Pulka, J., Wiśniewski, D., Gołaszewski, J., & Białowiec, A. (2016). Is the biochar produced from sewage sludge a good quality solid fuel? *Archives of Environmental Protection, 42*(4), 125–134. Available from https://doi.org/10.1515/aep-2016-0043.

Raclavská, H., Růžičková, J., Raclavský, K., Juchelková, D., Kucbel, M., Švédová, B., Slamová, K., & Kacprzak, M. (2021). Effect of biochar addition on the improvement of the quality parameters of compost used for land reclamation. *Environmental Science and Pollution Research*, 1–19, 2021.

Ramos, R., Abdelkader-fernández, V. K., Matos, R., Peixoto, A. F., & Fernandes, D. M. (2022). Metal-Supported Biochar Catalysts for Sustainable Biorefinery, Electrocatalysis and Energy Storage Applications: A Review. *Catalysts*, *12*(2). Available from https://doi.org/10.3390/catal12020207, https://www.mdpi.com/2073-4344/12/2/207/pdf.

Ramrakhiani, L., Ghosh, S., & Majumdar, S. (2022). Heavy metal recovery from electroplating effluent using adsorption by jute waste-derived biochar for soil amendment and plant micro-fertilizer. *Clean Technologies and Environmental Policy*, *24*(4), 1261–1284. Available from https://doi.org/10.1007/s10098-021-02243-4, https://link.springer.com/journal/10098.

Rasheed, T., Hassan, A. A., Bilal, M., Hussain, T., & Rizwan, K. (2020). Metal-organic frameworks based adsorbents: A review from removal perspective of various environmental contaminants from wastewater. *Chemosphere*, *259*, 127369. Available from https://doi.org/10.1016/j.chemosphere.2020.127369.

Ren, X., Zeng, G., Tang, L., Wang, J., Wan, J., Wang, J., Deng, Y., Liu, Y., & Peng, B. (2018). The potential impact on the biodegradation of organic pollutants from composting technology for soil remediation. *Waste Management*, *72*, 138–149. Available from https://doi.org/10.1016/j.wasman.2017.11.032.

Reverchon, F., Yang, H., Ho, T. Y., Yan, G., Wang, J., Xu, Z., Chen, C., & Zhang, D. (2015). A preliminary assessment of the potential of using an acacia—biochar system for spent mine site rehabilitation. *Environmental Science and Pollution Research*, *22*(3), 2138–2144. Available from https://doi.org/10.1007/s11356-014-3451-1, https://link.springer.com/journal/11356.

Ronda, A., Della Zassa, M., Martín-Lara, M. A., Calero, M., & Canu, P. (2016). Combustion of a Pb(II)-loaded olive tree pruning used as biosorbent. *Journal of Hazardous Materials*, *308*, 285–293. Available from https://doi.org/10.1016/j.jhazmat.2016.01.045.

Ruaa, K. M., ALsailawi, H. A., Abdulrasool, M. M., Mudhafar, M., Mays, A. D., Abbas, M., & Bashi. (2021). A Review Study on the Effects of Estrogen Level on Vaginal Candida spp. of Women with Estrogen-Related Receptor Breast Cancer. *Journal of Life Sciences*, *15*(1). Available from https://doi.org/10.17265/1934-7391/2021.01.002.

Rumpel, C., Chaplot, V., Planchon, O., Bernadou, J., Valentin, C., & Mariotti, A. (2006). Preferential erosion of black carbon on steep slopes with slash and burn agriculture. *CATENA*, *65*(1), 30–40. Available from https://doi.org/10.1016/j.catena.2005.09.005.

Sabio, E., González, E., González, J. F., González-García, C. M., Ramiro, A., & Gañan, J. (2004). Thermal regeneration of activated carbon saturated with p-nitrophenol. *Carbon*, *42*(11), 2285–2293. Available from https://doi.org/10.1016/j.carbon.2004.05.007.

Salvador, F., Martin-Sanchez, N., Sanchez-Montero, M. J., Montero, J., & Izquierdo, C. (2013). Regeneration of activated carbons contaminated by phenol using supercritical water. *Journal of Supercritical Fluids*, *74*, 1–7. Available from https://doi.org/10.1016/j.supflu.2012.11.025.

Şeker, C., & Manirakiza, N. (2020). Effectiveness of compost and biochar in improving water retention characteristics and aggregation of a sandy clay loam soil under wind erosion. *Carpathian Journal of Earth and Environmental Sciences*, *15*(1), 5–18. Available from https://doi.org/10.26471/cjees/2020/015/103, http://cjees.ro/actions/actionDownload.php?fileId=1336.

Selvasembian, R., Thokchom, B., Singh, P., Jawad, A. H., & Gwenzi, W. (2023). *Recycling and Disposal of Spent Metal(loid)-Laden Adsorbents. Remediation of Heavy Metals: Sustainable Technologies and Recent Advances* (pp. 275–290). Wiley. Available from 10.1002/9781119853589.ch12.

Shan, D., Deng, S., Zhao, T., Yu, G., Winglee, J., & Wiesner, M. R. (2016). Preparation of regenerable granular carbon nanotubes by a simple heating-filtration method for efficient removal of typical pharmaceuticals. *Chemical Engineering Journal*, *294*, 353–361. Available from https://doi.org/10.1016/j.cej.2016.02.118, http://www.elsevier.com/inca/publications/store/6/0/1/2/7/3/index.htt.

Srivatsav, P., Bhargav, B. S., Shanmugasundaram, V., Arun, J., Gopinath, K. P., & Bhatnagar, A. (2020). Biochar as an Eco-Friendly and Economical Adsorbent for the Removal of Colorants (Dyes) from Aqueous Environment: A Review. *Water*, *12*(12), 3561. Available from https://doi.org/10.3390/w12123561.

Stefaniuk, M., Oleszczuk, P., & Różyło, K. (2017). Co-application of sewage sludge with biochar increases disappearance of polycyclic aromatic hydrocarbons from fertilized soil in long term field experiment. *Science of The Total Environment*, *599-600*, 854–862. Available from https://doi.org/10.1016/j.scitotenv.2017.05.024.

Suen, N. T., Hung, S. F., Quan, Q., Zhang, N., Xu, Y. J., & Chen, H. M. (2017). Electrocatalysis for the oxygen evolution reaction: Recent development and future perspectives. *Chemical Society Reviews*, *46*(2), 337–365. Available from https://doi.org/10.1039/c6cs00328a, http://pubs.rsc.org/en/journals/journal/cs.

Sundriyal, S., Shrivastav, V., Pham, H. D., Mishra, S., Deep, A., & Dubal, D. P. (2021). Advances in bio-waste derived activated carbon for supercapacitors: Trends, challenges and prospective. *Resources, Conservation and Recycling*, *169*. Available from https://doi.org/10.1016/j.resconrec.2021.105548, http://www.elsevier.com/locate/resconrec.

Tan, S., Zhou, G., Yang, Q., Ge, S., Liu, J., Cheng, Y. W., Yek, P. N. Y., Wan Mahari, W. A., Kong, S. H., Chang, J.-S., Sonne, C., Chong, W. W. F., & Lam, S. S. (2023). Utilization of current pyrolysis technology to convert biomass and manure waste into biochar for soil remediation: A review. *Science of The Total Environment*, *864*, 160990. Available from https://doi.org/10.1016/j.scitotenv.2022.160990.

Tan, X. F., Liu, Y. G., Gu, Y. L., Xu, Y., Zeng, G. M., Hu, X. J., Liu, S. B., Wang, X., Liu, S. M., & Li, J. (2016). Biochar-based nano-composites for the decontamination of wastewater: A review. *Bioresource Technology*, *212*, 318–333. Available from https://doi.org/10.1016/j.biortech.2016.04.093, http://www.elsevier.com/locate/biortech.

Tatarková, V., Hiller, E., & Vaculík, M. (2013). Impact of wheat straw biochar addition to soil on the sorption, leaching, dissipation of the herbicide (4-chloro-2-methylphenoxy)acetic acid and the growth of sunflower (Helianthus annuus L.). *Ecotoxicology and Environmental Safety*, *92*, 215–221. Available from https://doi.org/10.1016/j.ecoenv.2013.02.005.

Trigo, C., Spokas, K. A., Cox, L., & Koskinen, W. C. (2014). Influence of soil biochar aging on sorption of the herbicides MCPA, nicosulfuron, terbuthylazine, indaziflam, and fluoroethyldiaminotriazine. *Journal of Agricultural and Food Chemistry, 62*(45), 10855–10860. Available from https://doi.org/10.1021/jf5034398, http://pubs.acs.org/journal/jafcau.

Tsolis, V., & Barouchas, P. (2023). Biochar as Soil Amendment: The Effect of Biochar on Soil Properties Using VIS-NIR Diffuse Reflectance Spectroscopy, Biochar Aging and Soil Microbiology—A Review. *Land, 12*(8), 1580. Available from https://doi.org/10.3390/land12081580.

Vakili, M., Deng, S., Cagnetta, G., Wang, W., Meng, P., Liu, D., & Yu, G. (2019). Regeneration of chitosan-based adsorbents used in heavy metal adsorption: A review. *Separation and Purification Technology, 224*, 373–387. Available from https://doi.org/10.1016/j.seppur.2019.05.040.

Vercruysse, W., Smeets, J., Haeldermans, T., Joos, B., Hardy, A., Samyn, P., Yperman, J., Vanreppelen, K., Carleer, R., Adriaensens, P., Marchal, W., & Vandamme, D. (2021). Biochar from raw and spent common ivy: Impact of preprocessing and pyrolysis temperature on biochar properties. *Journal of Analytical and Applied Pyrolysis, 159*, 105294. Available from https://doi.org/10.1016/j.jaap.2021.105294.

Vukićević, E., Isailović, J., Gajica, G., Antić, V., & Jovančićević, B. (2024). Biochar from agricultural biomass: green material as an ecological alternative to solid fossil fuels. *J. Serb. Chem. Soc., 00*(0). Available from https://doi.org/10.2298/JSC240126048V2024, S.

Wan, G., Freeland, J. W., Kloppenburg, J., Petretto, G., Nelson, J. N., Kuo, D. Y., Sun, C. J., Wen, J., Diulus, J. T., Herman, G. S., Dong, Y., Kou, R., Sun, J., Chen, S., Shen, K. M., Schlom, D. G., Rignanese, G. M., Hautier, G., Fong, D. D., . . . Suntivich, J. (2021). Amorphization mechanism of $SrIrO_3$ electrocatalyst: How oxygen redox initiates ionic diffusion and structural reorganization. *Science Advances, 7*(2). Available from https://doi.org/10.1126/sciadv.abc7323, https://advances.sciencemag.org/content/advances/7/2/eabc7323.full.pdf.

Wang, T., Zhang, L., Li, C., Yang, W., Song, T., Tang, C., Meng, Y., Dai, S., Wang, H., Chai, L., & Luo, J. (2015). Synthesis of core-shell magnetic Fe_3O_4@poly(m-phenylenediamine) particles for chromium reduction and adsorption. *Environmental Science and Technology, 49*(9), 5654–5662. Available from https://doi.org/10.1021/es5061275, http://pubs.acs.org/journal/esthag.

Wang, Y., Zhang, L., Yang, H., Yan, G., Xu, Z., Chen, C., & Zhang, D. (2016). Biochar nutrient availability rather than its water holding capacity governs the growth of both C3 and C4 plants. *Journal of Soils and Sediments, 16*(3), 801–810. Available from https://doi.org/10.1007/s11368-016-1357-x, http://www.springerlink.com/content/1439-0108.

Wang, Y., Liu, Y., Zhan, W., Zheng, K., Wang, J., Zhang, C., & Chen, R. (2020). Stabilization of heavy metal-contaminated soils by biochar: Challenges and recommendations. *Science of The Total Environment, 729*, 139060. Available from https://doi.org/10.1016/j.scitotenv.2020.139060.

Yang, H., Ye, S., Zeng, Z., Zeng, G., Tan, X., Xiao, R., Wang, J., Song, B., Du, L., Qin, M., Yang, Y., & Xu, F. (2020). Utilization of biochar for resource recovery from water: A review. *Chemical Engineering Journal, 397*, 125502. Available from https://doi.org/10.1016/j.cej.2020.125502.

Yang, X. B., Ying, G. G., Peng, P. A., Wang, L., Zhao, J. L., Zhang, L. J., Yuan, P., & He, H. P. (2010). Influence of biochars on plant uptake and dissipation of two pesticides in an agricultural soil. *Journal of Agricultural and Food Chemistry, 58*(13), 7915–7921. Available from https://doi.org/10.1021/jf1011352.

Yavari, S., Malakahmad, A., & Sapari, N. B. (2015). Biochar efficiency in pesticides sorption as a function of production variables—a review. *Environmental Science and Pollution Research, 22*(18), 13824–13841. Available from https://doi.org/10.1007/s11356-015-5114-2, http://www.springerlink.com/content/0944-1344.

Zahed, M. A., Salehi, S., Madadi, R., & Hejabi, F. (2021). Biochar as a sustainable product for remediation of petroleum contaminated soil. *Current Research in Green and Sustainable Chemistry, 4*. Available from https://doi.org/10.1016/j.crgsc.2021.100055, http://www.journals.elsevier.com/current-research-in-green-and-sustainable-chemistry/.

Zhang, X., Li, Y., Wu, M., Pang, Y., Hao, Z., Hu, M., Qiu, R., & Chen, Z. (2021). Enhanced adsorption of tetracycline by an iron and manganese oxides loaded biochar: Kinetics, mechanism and column adsorption. *Bioresource Technology, 320*. Available from https://doi.org/10.1016/j.biortech.2020.124264, http://www.elsevier.com/locate/biortech.

Zhang, Y., Chen, Z., Xu, W., Liao, Q., Zhang, H., Hao, S., & Chen, S. (2020). Pyrolysis of various phytoremediation residues for biochars: Chemical forms and environmental risk of Cd in biochar. *Bioresource Technology, 299*, 122581. Available from https://doi.org/10.1016/j.biortech.2019.122581.

Zhou, Y. N., Yu, W. L., Cao, Y. N., Zhao, J., Dong, B., Ma, Y., Wang, F. L., Fan, R. Y., Zhou, Y. L., & Chai, Y. M. (2021). S-doped nickel-iron hydroxides synthesized by room-temperature electrochemical activation for efficient oxygen evolution. *Applied Catalysis B: Environmental, 292*. Available from https://doi.org/10.1016/j.apcatb.2021.120150, http://www.elsevier.com/inca/publications/store/5/2/3/0/6/6/index.htt.

Chapter 21

Life cycle assessment for biochar systems: a review

Simone Marzeddu[1], Francesca Lazzari[1], Annarita Cepollaro[2], Andrea Cappelli[2] and Maria Rosaria Boni[1]

[1]Department of Civil, Constructional and Environmental Engineering (DICEA), Faculty of Civil and Industrial Engineering, Sapienza University of Rome, Rome, Italy, [2]Department of Chemical Engineering Materials Environment (DICMA), Faculty of Civil and Industrial Engineering, Sapienza University of Rome, Rome, Italy

Chapter outline

21.1 Introduction	395
21.2 Life cycle assessment as a sustainability tool	397
21.3 Comparative life cycle assessment of biochar production systems	400
21.4 Goal and scope definition	402
21.5 Life cycle inventory	406
21.6 Life cycle impact assessment	416
21.7 Conclusion	427
References	428

21.1 Introduction

"Transitioning away from fossil fuels in energy systems, in a just, orderly and equitable manner, accelerating action in this critical decade, so as to achieve net zero by 2050 in keeping with the science" (Wise, 2023).

A lexical sophistry is needed to conclude the chess game played in the 2023 Conference on Parts in Dubai; a terminological quibble that continues to take into account the interests of oil-producing countries, while pleasing a significant number of the participating international delegates.

Although, for the first time in the final document of the Conference, nations are explicitly invited to take the path of an energy transition to avoid a further worsening of the climate crisis, many countries, civil society associations and scientists are strongly disappointed by a conclusion which appears significant in form rather than substance, avoiding making an explicit and definitive commitment, especially from a temporal point of view, toward a phase-out of fossil fuels.

Despite the verbal skirmishes and the uncertain timing of implementation, the road to reducing greenhouse gases (GHG) emissions into the atmosphere and limiting temperature growth to within 1.5°C compared to the pre-industrial era appears to be clear (Dietz et al., 2018; Hoegh-Guldberg et al., 2019; Kuramochi et al., 2018; Roe et al., 2019; Seneviratne et al., 2018): fossil fuels cannot be the development horizon for world economies in the more or less immediate future (Cointe & Guillemot, 2023; Matthews & Wynes, 2022).

A rapid increase in renewable sources for energy production appears unavoidable to meet climate goals, although it opens the door to uncertain scenarios, also in this case related to the timing of the transition. The uncertainty mainly concerns the availability and role of the so-called critical raw materials (CRM), that is, raw materials which, due to their significant technological characteristics, are of fundamental importance in the generation and storage of renewable energy; raw materials whose supply may be associated with such a high risk that *"dependence of Critical Raw Materials may soon replace today's dependence on oil"* (European Commission, 2023) thus creating further doubts on the methods and timing of the energy transition.

It seems clear that the path toward meeting climate goals cannot be based only on the conversion of energy production technologies; great attention must also be paid to technologies that allow the permanent capture of CO_2 (Rodrigues et al., 2023; Yasin et al., 2023): it is in this scenario that biochar takes on a role of great importance.

Biomass, given its abundance, sustainability, and cost-effectiveness, undoubtedly represents an inexhaustible source of renewable energy, playing a fundamental role in the biochar production process (Boni et al., 2020; Lima et al., 2020; Lu et al., 2020; Olabi & Abdelkareem, 2022; Østergaard et al., 2022; Owusu & Asumadu-Sarkodie, 2016; Sokka et al., 2016). Biochar is a carbon-rich solid material, resulting from the thermochemical conversion of biomass in an oxygen-limited environment, through the breakdown of organic matter chemical bonds and the conversion of their intermediates into bio-oil, syngas, and biochar itself (Boni et al., 2021; Boni, Chiavola, Antonucci, et al., 2018; Boni, Chiavola, Marzeddu, et al., 2018; Décima et al., 2021; Viotti et al., 2024; Wang et al., 2022).

This material can be obtained from different types of biomasses, both plant and animal (Boni et al., 2021; p. 202; Chiavola et al., 2020; Marzeddu et al., 2022); for the aim of product quality, its production from wood, agricultural residues and by-products, pruning residues, dry foliage, straw, etc. is of extreme interest (Manikandan et al., 2023; Xia et al., 2024). Furthermore, there are different existing process technologies: slow or fast pyrolysis, also developed with the aid of torrefaction (Potnuri et al., 2023), and gasification (Matuštík et al., 2020). Each of these processes, depending on its operating parameters such as maximum temperature reached and residence time, is characterized by a different balance between the three obtainable products: biochar, bio-oil, and synthesis gas. Bio-oil consists of a liquid product, pyrolytically condensable, which is mainly used to efficiently produce hydrogen and electricity (Ağbulut et al., 2023; Fakayode et al., 2023; Grams et al., 2023). Syngas, composed mainly of H_2, CH_4, and CO, can be used as a fuel in gas turbines or in cogeneration, allowing the production of thermal and/or electric energy (Chavando et al., 2023; Marzeddu et al., 2024; Perreault et al., 2023).

Among the different uses for which biochar is intended, the most interesting is certainly that of an agricultural soil improver; soil degradation (Blenis et al., 2023; Bridges & Oldeman, 1999; Qian et al., 2023), caused among other things by the use of unsustainable agricultural techniques and deforestation processes aimed at expanding exploitable cultivation areas (Wang & Qiu, 2017), appears to be a clear threat to global biodiversity.

The use of biochar in agricultural practices (as well as in other application areas) represents an element of considerable importance for mitigating GHG emissions related to cultivation processes and for addressing the issue of soil degradation and consumption (Jiang et al., 2023; Liu et al., 2023; Mosa et al., 2023). In its application to the soil, in fact, biochar presents fundamental characteristics for the capture and storage of carbon dioxide, allowing the storage of carbon for tens or hundreds of years. Its presence also increases the productivity of crops and the fertility of the soil, allowing a contextual increase in the water retention capacity (Wei et al., 2023) and in essential nutrients for plants (Hussain et al., 2023; Xu et al., 2023), reducing its acidity (Bolan et al., 2023; Huang et al., 2023) and improving the chemistry of the soil through its elemental composition (Singh Yadav et al., 2023). The success of its application, however, depends on the combinations of various factors: its composition, which arises from the type of raw material from which it was obtained and the specific technological process used; the inherent characteristics of the soil to which it is applied; the climate in which it finds itself operating (Agyekum & Nutakor, 2024; Campion et al., 2023; Jayakumar et al., 2023; Li et al., 2023; Safarian, 2023).

The method chosen to characterize the supposed environmental sustainability of the production process and use of biochar, in its various applications, is life cycle assessment (LCA) (Barbhuiya et al., 2024; Ding et al., 2023; Osman et al., 2024). Through this methodology, it is possible to define all the impacts, whether positive or negative, generated during the various phases of the biochar life cycle, as shown in Fig. 21.1, from obtaining the biomass to its transformation, up to the use of the product itself in the various investigated contexts (Viotti et al., 2020).

This study aims to analyze several papers published in the recent past and whose aim was to characterize the life cycle of biochar in various areas of production and use (i.e., soil improver, a constituent of innovative construction materials), to verify, from multiple and diverse perspectives and approaches, the impact indicators according to the different operational parameters of process and use.

The data analyzed are therefore the result of previous studies which were based, in turn, on experimental data, data obtained from databases, and/or data derived from previous literature, which were processed in some cases using open-source software, in others through more complex methodologies. These studies were compared with each other, showing that the application of biochar, but also the use of gasification and/or pyrolysis techniques for its production and for energy production, are strategies that present positive results for the mitigation of process impacts. The present study shows the effectiveness of the LCA methodology, but also the need to standardize the analysis for studies conducted on the production and application of biochar, so as to always make the data comparable.

There are currently a small number of LCA studies aimed at characterizing the environmental sustainability of biochar as a tool for the remediation of contaminated soils (James et al., 2022; Osman et al., 2024), for wastewater treatment (Kumar Mishra et al., 2023) and for improving air quality (Younis & Kim, 2022). However, these studies have not been considered in this review since the impact data are not comparable; in fact, the relevant functional units are different from those used in the applications of the LCA methodology to biochar production processes.

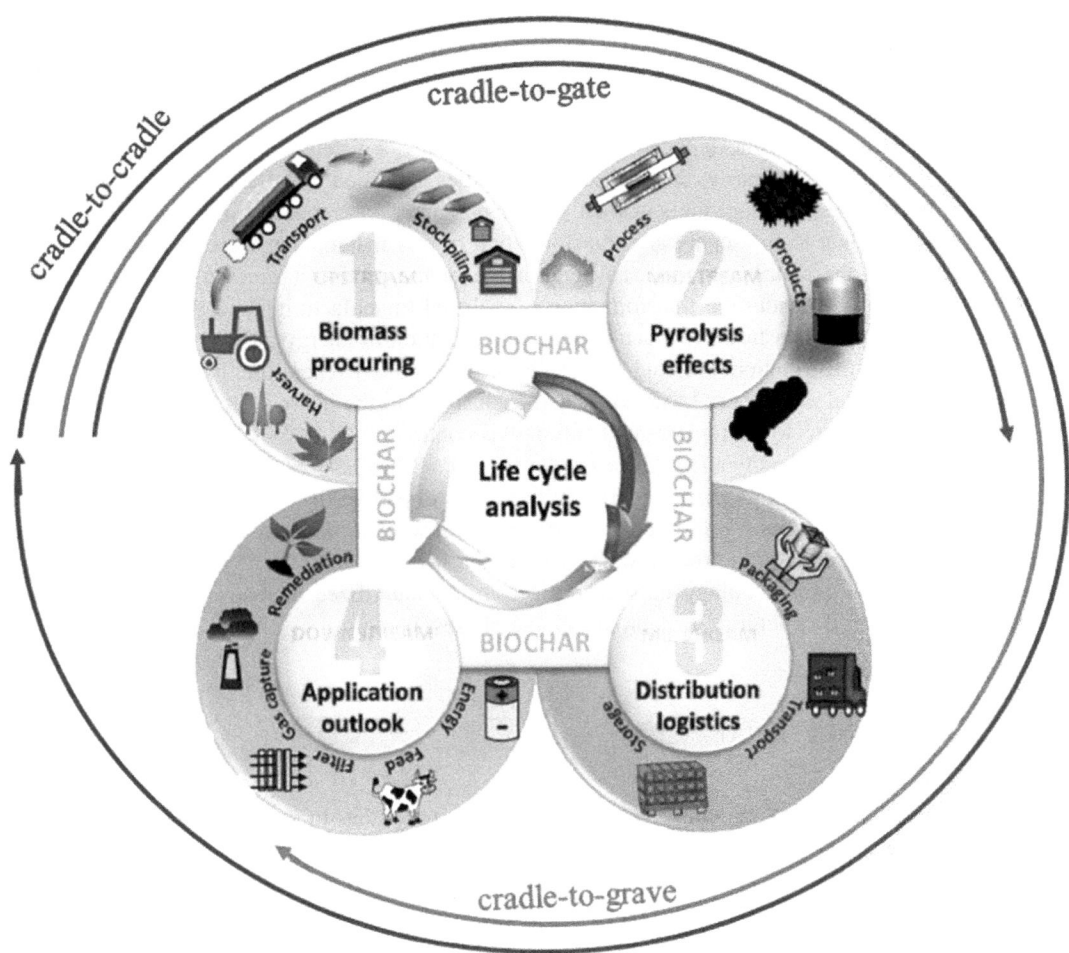

FIGURE 21.1 A life cycle assessment of biochar (Bolan et al., 2022; Patel & Panwar, 2023; Roberts et al., 2010).

21.2 Life cycle assessment as a sustainability tool

LCA is a fundamental methodology on the path toward environmental sustainability of production cycles, activities, and services, and represents the main operational tool of the "*Life Cycle Thinking*" approach, which considers all production phases as interconnected and closely dependent on each other. The aim of LCA is to analyze the production cycle of a good or product by characterizing the incoming and outgoing material flows to quantify the environmental and energy loads associated with these flows during the entire life cycle. Such an approach provides us with the possibility of highlighting the weak points of the process, identifying the phases on which to carry out improvement actions that increase its sustainability, through the reduction of energy consumption, raw materials used, generated waste, or by modifying the production technologies; thus allowing a reduction of environmental impacts and, in economic terms, of production costs and marginal social costs of negative externalities generated in the production process (Pigou, 1933).

LCA is characterized by a cyclical approach: when improvement actions are carried out on the process it is necessary to re-characterize the incoming and outgoing material flows, correlating them with a further evaluation of the environmental impacts generated. The results can be used as decision support, both at a company and political level, to evaluate alternative scenarios and choose more sustainable process solutions (Antwi-Afari et al., 2023; Bruno et al., 2023; Hussin et al., 2023). An LCA applied to a production system is, therefore, a fundamental tool for improving its overall efficiency, allowing the protection of environmental and human health and, at the same time, a rational use of resources within the process.

The LCA approach that provides the greatest level of completeness of the analysis is the one identified with the expression "*from cradle to grave*," which allows the description of the environmental and energy loads of the entire life cycle of the process: from the extraction of the raw materials to the use and end of life of the product (or what remains of it), passing through the transformation of the raw materials, the industrial process, transportation and distribution. However, at times, the development of a complete LCA can be particularly burdensome both in terms of time and money and does not meet the specific objectives required by the study, as it can be conducted at a lower level of detail, without sacrificing completeness, accuracy, and reliability of the obtained results. Possible exemplifying strategies may concern the limitation of the preestablished objectives, the reduction of the quantity of data requested or the reduction of the system boundaries, in the latter case being able to define, for the application of the methodology, a different approach compared to the *"from cradle to grave"* already highlighted previously: this is the case of studies identified with the expression *"from cradle to gate"* and *"from gate to gate."* In the first case, the so-called *"upstream"* and *"core"* phases of a production process are taken into consideration, corresponding to the supply and transformation of raw materials and energy and the subsequent industrial process, thus excluding the distribution, use, and end-of-life of the product (*"downstream"*). In the *"gate to gate"* approach only the *"core"* phase of industrial transformation is analyzed (manufacturing and assembly of the product), thus excluding everything that happens upstream (extraction and transformation of energy and nonenergy raw materials used in the process) and downstream (distribution, use, and end of life of the product). Therefore, its latter approach is suitable for characterizing the environmental and energy loads connected exclusively to the internal processes of the company, seeking possible improvement interventions to be implemented only in this context.

In recent years, a circular approach called *"from cradle to cradle"* has also begun to spread among sector experts, which, compared to the *"from cradle to grave"* approach, has as its objective the maximum valorization of the product (or of what remains) at the end of its life through the recovery of raw materials and energy, with a view to economic circularity and a progressive reduction in process waste sent to landfill.

The application of the LCA methodology must take place in a standardized manner according to the provisions of UNI ISO 14040, through a rigid structuring in four successive phases, namely:

1. Goal and Scope definition;
2. Life Cycle Inventory (LCI);
3. Life Cycle Impact Assessment (LCIA);
4. Life Cycle Interpretation.

Goal and scope definition is the preliminary phase in which it is necessary to outline: the objective of the study, the functional unit of the process, and the boundaries of the production system.

In defining the goals of the study, it must be established a priori what the areas of application of LCA are, the reasons that lead to carrying out the study, and the type of user for whom it is intended; the analysis will have a different approach depending on whether the final aim is, for example, the characterization of the impacts of the industrial process alone (*"gate to gate"*) or the obtaining of an ecological label (*"cradle to gate"* o *"cradle to grave"* depending on the specific procedures required by the certification rules);

The functional unit is the unitary element of the production process that provides a reference to which to link the incoming and outgoing material flows, and therefore the environmental impacts, to allow the comparability of the results of an LCA.

The system boundaries represent the limits of contact and exchange of material between the production system and the external environment. They are defined through the construction of a flow diagram, necessary to delimit the different phases of the process and to plan the collection of data and information necessary for the analysis. The representation of system boundaries is closely related to the goal of the study and therefore to the type of approach used in the LCA application.

In the LCI the production system is modeled by creating an analog model, that is, a simplification of the process itself, identifying the most significant phases from the point of view of the generated impacts. All those phases of the process whose contribution, in terms of impact, is negligible are therefore not taken into consideration. We then proceed to identify the incoming and outgoing flows of material and energy with respect to each production step, making use of both real process data (primary data), when known and monitored, and data coming from databases or from specific literature (secondary data), if the primary data is not available. Once identified, in the next phase, the outputs will be correlated with the impacts generated by the process.

The life cycle impact assessment (LCIA) correlates the outputs deriving from the Inventory phase with the potential environmental impacts. The LCIA therefore calculates the different potential impacts of the process in relation to the specific impact categories considered, chosen according to the objective of the study. The main impact categories used in the analysis are shown below; We will find them again in Chapter 7:

- Acidification potential (AP): It describes the impact of the deposition of acidifying substances such as nitric acid, sulfuric acid, sulfur trioxide, etc. These substances may cause acid rains that are harmful to terrestrial and aquatic species. It is expressed in kg SO_2 eq;

- Depletion of abiotic resources: elements, ultimate reserves (ADPe)—It is related to the extraction of minerals, which are inputs into the system, leading to a reduced availability of these resources for future generations. It is expressed in kg antimony equivalent (kg Sb eq);
- Depletion of abiotic resources: fossil fuels (ADPf)—It is related to the extraction of fossil fuels, which are inputs into the system, leading to a reduced availability of these resources for future generations. It is expressed in MJ;
- Cumulative Energy Demand (CED)—It assesses primary energy usage, as it aims to investigate the energy use throughout the life cycle of a good or a service. This includes the direct uses as well as the indirect or gray consumption of energy due to the use of, e.g., construction materials or raw materials. It is expressed in MJ eq;
- Carbon footprint (CF)—It considers emissions metrics for global warming potential (GWP) and Global Temperature Change Potential (GTP). These numbers are implemented as CFs in the Intergovernmental Panel on Climate Change (IPCC) methods. It expressed in kg CO_2 eq;
- Eutrophication potential (EP)—It includes all impacts due to excessive levels of macro-nutrients in the environment caused by emissions into the air, water, and soil. This leads to a change in the natural equilibrium and to an overgrowth of plants such as algae in rivers and seas, which causes reductions in water quality and animal populations. EP is expressed in kg PO_4^{3-} eq;
- Freshwater Aquatic EcoToxicity Potential (FAETP)—It considers the impact of emissions of toxic substances into the air, water, and soil on freshwater ecosystems, describing the exposure and effects of toxic substances. Results are expressed as kg 1,4-dichlorobenzene eq;
- GWP—It represents the potential contribution of a product to the greenhouse effect for a specific time horizon (i.e., 20/100 years). It is expressed in kg CO_2 eq;
- Human toxicity potential (HTP)—It concerns the effects of toxic substances on the human environment (with the exclusion of work environments). The impact of a compound is affected by the amount emitted, the mobility of the substance, its persistence, exposure patterns and bioavailability, as well as its intrinsic toxicity. Results for human toxicity are expressed as kg 1,4-dichlorobenzene eq;
- Ionizing radiation potential (IRP)—It represents the potential to interact with and change molecules, and damage or even kill cells. It is expressed in kBq Co-60-Eq
- Marine Aquatic EcoToxicity Potential (MAETP)—It refers to the impacts of toxic substances on marine ecosystems. The impact pathway includes exposure, effect, and severity factors. Results are expressed as kg 1,4-dichlorobenzene eq;
- Ozone depletion potential (ODP)—It accounts for the destruction of the stratospheric ozone layer by anthropogenic emissions of ozone-depleting substances. Chlorofluorocarbons (CFCs), halons, and hydrochlorofluorocarbons (HCFC) are the main causes of ozone depletion. Results are expressed as kg CFC-11 eq;
- Photochemical ozone creation potential (POCP)—Ozone is protective in the stratosphere, but on the ground level, it is toxic to humans in high concentrations. Photochemical ozone is formed by the reaction of volatile organic compounds and nitrogen oxides in the presence of both heat and sunlight. Results are expressed as kg ethylene eq;
- Particulate matter formation (PMF)—It refers to a type of air pollution composed of fragments of material on the microscale. Results are expressed as kg PM2.5-Eq;
- Terrestrial Ecotoxicity Potential (TETP)—It indicates the impact of toxic compounds such as heavy metals (zinc, lead, nickel, cadmium, and copper) emitted into terrestrial ecosystems (air, water, and soil), affecting humans, flora, and fauna. It is expressed as kg 1,4-dichlorobenzene eq;
- Water consumption potential (WCP)—It allows you to measure the "quantity of water needed" to produce a certain product, which can be verified and compared by any user and it is a measure of humanity's appropriation of fresh water in volumes of water consumed and/or polluted. It is expressed in m^3;

In the Interpretation phase, the LCI and LCIA results are combined and analyzed to provide a complete, transparent, and unbiased report. The aim is to reach conclusions and recommendations in line with the objective and scope of the study, suggesting possible system improvements that can mitigate the generated impacts.

According to Carvalho et al. (2022), for a production process to be profitable, it is necessary to evaluate the associated costs in addition to the environmental aspects. The LCCA is a methodology that evaluates costs throughout the life cycle of a product/service: costs related to production, use, and maintenance (development, design, consumption of energy and other renewable and nonrenewable resources, costs of the environmental externalities generated—emission of GHG or other pollutants into the atmosphere, soil, and water) and end-of-life costs (collection, landfill, or recycling). LCCA is particularly useful when project alternatives that meet the same performance requirements differ in private and environmental costs and need to be compared to select the one that maximizes net savings.

It must be taken into account that the energy demand has increased due to global population growth. Even waste materials have potential that can be exploited to the fullest. Unmanaged waste collection, poor-quality treatment, and unrestricted landfilling of organic waste, particularly in low- and middle-income countries, have resulted in the loss of potential energy sources that could otherwise be used to supplement and satisfy global energy demand (Gaur et al., 2022).

Today it is estimated that 0.74 kg of waste is generated per person per day; globally, solid waste is expected to increase to 3.40 billion tons by 2050 (Raut et al., 2023).

Biochar, given its countless potential applications, has attracted the interest of several researchers who have investigated its potential also in relation to the possible impacts on the global economy. The technical-economic evaluation of biochar includes an analysis of impact indicators arising from the life cycle, a rigorous economic evaluation, and a characterization of the energy balance of the production process.

Energy efficiency is always a significant factor in determining the economic feasibility of a production process. Energy yield is defined as the energy generated by the product, per unit of energy input through biomass: pyrolysis was found to be the most effective and appropriate method for the production of biochar (Melo et al., 2024; Wang et al., 2024; Zou et al., 2024).

As part of the economic analysis, various factors are taken into consideration, from equipment costs to the cost of wages for labor, transport costs, and those generated by the energy needed to activate the reactor, which can be powered by different sources.

Song et al. (2023) carried out a study regarding the application of biochar within cement materials; it was found that, when evaluating the economic benefits of biochar cement-based materials, it is important to consider costs such as raw materials, processing, production, transportation, sales, and disposal. Furthermore, in the cost analysis, it is crucial to include the environmental benefits of biochar cement-based materials, such as reduction of CO_2 emissions and carbon tax savings, to provide a more comprehensive evaluation. The main cost of biochar-based cementitious materials is attributed to cement. Biochar is usually less expensive because it can be produced from low-cost materials, such as agricultural waste. Additionally, the use of biochar reduces the amount of cement needed, thereby also lowering the cost of cement. The production of cement involves the emission of a significant amount of carbon dioxide and the implementation of a carbon tax. In contrast, the production of biochar-based cementitious materials can significantly reduce CO_2 emissions and save a significant amount of carbon tax.

Chen et al. (2019) demonstrated that a profit of $41.1 per cubic meter of material can be achieved, likely due to significant savings in material costs, carbon tax and waste disposal fees, resulting in greater economic profits. Further research and the creation of related models and systems are obviously needed to accurately evaluate the economic benefits of biochar and cement-based materials.

In this general context, developing a complete LCCA of biochar-based products is an essential step for an economic characterization and to improve the profitability of this technology throughout the lifetime of a project or industrial application period.

21.3 Comparative life cycle assessment of biochar production systems

The present contribution is classified as a review, dealing with LCA analysis carried out on different studies regarding biochar production systems. The main purpose is to compare the different studies, with the intent to emphasize the potential of biochar and its applications and to focus on the use of biomass as a source of biochar.

The articles taken into consideration are all in English and processed between 2020 and 2023 with the exception of one (Sparrevik et al., 2013), prior to this time frame, which is interesting for the issues and the field considered. The keywords to carry out the research were: *"biochar LCA"*; *"biochar Life Cycle Assessment"*; *"biochar from biomass conversion"*; among all the studies obtained from the research were taken into account, those relating to the application of the LCA method to assess the environmental impact of biochar systems, but also taking into account some explaining details on the types of biomass conversion, which allow the product and its use of biochar as soil improver to be obtained.

Out of all the articles found, a total of 15 works were selected that met all the search criteria. While some of the selected articles fully meet the criterion of having the LCA as the main (or sole) objective, others also deal, or primarily, with other issues, such as economic analysis or energy balance. The articles cover studies that have been carried out in various countries of the world, mostly in industrialized countries such as Europe, North America, and China but have also been considered studies that have been carried out in nonindustrialized countries, (Sparrevik et al., 2013) which dealt with the degraded sandy soils of the African continent (present in the countries of Sub-Saharan Africa), where the use of biochar may be particularly important for agricultural purposes.

Biochar systems and the scope of studies generally differ between industrialized countries and rural conditions in developing countries in Asia and Africa.

The authors believe that the limited number of studies on the application of the LCA methodology to biochar production systems in developing countries is mainly attributable to a lack of expertise on the characterization of the life cycle, rather than to the onerousness of the analysis itself. The availability of free, highly performing software in the application of the methodology (i.e., OpenLCA) and freely accessible databases, although often not updated, would allow a robust evaluation of the impacts related to the production processes and use of biochar for its various application purposes. It is also evident that most biochar production processes in developing countries occur in rural areas through the combustion and carbonization of lignocellulosic biomass in large piles These processes, in addition to not being easily monitored for the purposes of collecting primary data and developing the inventory phase, give rise to ash and fumes characterized by high concentrations of harmful substances (CO, CO_2, SO_X, NO_X) with effects that can be dangerous to both humans and the environment.

This review is structured following the steps of the standardized LCA methodology:

1. goal and purpose definition;
2. LCI;
3. LCIA.

It turned out that the most interesting and most proven application of biochar is as a soil improver; The latter can be degraded for various reasons and first of all, this can be included because of aggressive agricultural practices and aimed at continuous exploitation, without taking into account the impoverishment that the soil undergoes. The biochar for its particular characteristics, grain size, ability to interact with minerals and microorganisms of the soil, and its ability to regulate the pH of the latter, is able to condition the state of the soil and the maintenance within it of all the substances necessary for its conservation. Its application in agricultural practices not only avoids an important voice in what are the environmental impacts, that is, the degradation of the soil and subsequent further consumption of the same since, in conventional agriculture to exploitation, Once soil is no longer suitable, we do not try to restore its capacity but we opt for deforestation to obtain more cultivable space, but we improve the yield of crops. The biochar is able to maintain the nutrients in the soil, regulate the quantities of water present in it, and, consequently, avoid the need to use chemical fertilizers, leading to savings in both environmental and purely economic costs.

The analyzed studies reveal all these capabilities in addition to participating in the goal of reducing CO_2 released into the atmosphere. Biochar contributes positively to the balance of emissions released into the atmosphere as, in its production process, it exploits the potential of waste biomass, ensuring that this is not "wasted", for example through incineration practices which contribute to further increasing greenhouse gas emissions. Pyrolysis allows the conversion of biomass into biochar and other by-products (e.g. syngas and bio-oil) that can be used for energy production or other purposes. Biochar, when used as soil improver, can absorb and retain CO_2; this same ability is also exploitable in its applications in other materials such as, for example, construction materials as described in the study of Song et al. (2023). In most of thestudies considered inthis paper, LCA analyses were how all the characteristics of biochar systems, from the production phase to its applications, were analyzed. In the life cycle analysis of biochar it is clear that the savings that are obtained, through the conversion of biomass to obtain the product and through its applications, compensate (generally) the items taking into account emissions due to the transport of both biomass from the collection point to the production plant and product to the site of application.

It should be noted that the studies analyzed in this review do not consider the biomass collection phase. Although the collection, agricultural or industrial extraction phases are conventional entries, there is a common tendency not to consider the collection phase in the environmental balance, since biomass is considered as waste material. In fact, this is a recovery of material that if not used would be lost along with the additional potential that still holds. From the very early stages of its life cycle, therefore, the biochar works positively since it is obtained from a recovery of material. The analyses taken into consideration were based on different data, both primary and secondary, as highlighted in Table 21.1. Primary data are site-specific data, that is, they derive from production plants or experimental activities conducted in the field; the secondary ones are obtained from databases (i.e., Ecoinvent) or from existing scientific literature, and are acquired to make up for any information deficiencies, deriving them from cases similar to those under examination.

The processing of this data took place through the use of different software that were chosen according to the size of the study conducted. The most common were: Open LCA, among the open Source, and SimaPro among the non-open source.

The experimental data before switching to software or as an alternative to these were, in some cases, processed through the use of calculation tools such as Microsoft Excel or by manual calculations. Below are the results extracted from the different papers that have been tabulated and sorted in such a way that they can be compared with each other in a clear and intuitive manner.

21.4 Goal and scope definition

In all the scientific articles analyzed, the guideline is the need to make informed and more useful use of agricultural waste. The common treatment technologies for the disposal of these agricultural wastes are composting or open combustion, which has the disadvantage of producing greenhouse gases (CO_2, CH_4, and N_2O) and pollutants (H_2S, SO_2, and NH_3) thus reducing the added value that agricultural waste represents. In this sense, therefore, the need to identify suitable transformation techniques that lead to a reduction in emissions deriving from the direct combustion of residues, while at the same time maximizing the qualitative characteristics of the products was highlighted. Biochar, as already mentioned, is obtained from the treatment of agricultural waste; applied to soil, it is a powerful soil improver and stands out for a very particular property, namely that of absorbing and retaining carbon dioxide and other contaminants, as well as water and other nutrients that are essential for the soil itself. LCA studies applied to specific production systems have been analyzed to define the actual benefits associated with both thermal conversion technologies and the use of biochar. It was therefore possible to evaluate the actual impacts generated in the different production cycles taken into consideration, also analyzing the possibilities of balancing or mitigating the effects obtained from the use of the biochar itself.

In life cycle assessment, the input parameters required to quantify the environmental impact may be uncertain due to temporal variability or uncertainties in the actual value of emission factors (Ferronato et al., 2023). The change in process impacts caused by the variability of input parameters can be verified through the use of a sensitivity analysis, which allows us to obtain more information on the output variance. A sensitivity analysis can be performed by varying an input parameter and determining its effect on ultimate impacts. In a local sensitivity analysis, we determine the effect of a small change on each of the input parameters taken individually, while in a global sensitivity analysis, we can determine how much each input parameter contributes to the output variance. The latter can be considered as an extension of the uncertainty propagation (Groen et al., 2017).

Of the 15 papers analyzed in this work, only four have developed a systematic sensitivity analysis to address the effect of input uncertainties on outputs. One explanation could be that, although ISO 14044 recommends a sensitivity analysis as part of the LCA framework to identify the importance of input uncertainties, however there is no reference to the application of a specific technique for this purpose.

Some specific study objectives, relating to the LCA of this product, are highlighted in Table 21.1.

In Zhu et al. (2022), the goal was to achieve carbon neutrality by exploiting the biochar characteristics. The starting raw materials, from which biochar was obtained, are heterogeneous residues of agricultural activities, produced through pyrolysis; the study was carried out through the use of data deriving from different countries around the world (among which Australia, EU, North, and South America, including some countries in Asia and Africa), which was processed through different software such as SimaPro and OpenLCA, using different LCA approaches ("cradle to grave," "cradle to gate," and "gate to gate") and defining system boundaries accordingly. The functional unit considered is one ton of raw material or one ton of biochar produced from it.

Through biochar, Patel and Panwar (2023) aimed at optimizing agriculture-related activities, increasing crop productivity, and reducing carbon dioxide emissions. The raw materials used were agricultural waste; data processing was performed using SimaPro, OpenLCA and GaBi; "cradle-to-grave" and "cradle-to-gate" approaches were used.

For Carvalho et al. (2022), the objectives and scope were divided into two different areas of research: assessment of environmental impacts associated with biochar production and use, energy and economic efficiency. The analysis considered data from the processing of both agricultural and forestry waste, transposed from experiments carried out in different countries, and processed using all the different approaches made available by the LCA methodology.

The study by James et al. (2022) aimed at analyzing the positive soil externalities from the use of biochar, obtained through different production methods. In this case, the raw materials used also came from municipal solid waste, energy crops, and sewage sludge; the only approach used was "cradle-to-grave."

Another study set out to highlight the impacts associated with a combined energy-biochar production system and the application of the carbon product to the soil (Marzeddu et al., 2021; Marzeddu et al., 2023). The study concerned a specific Italian plant (Record immobiliare s.r.l.). Data were processed through OpenLCA and Excel, using a cradle-to-grave approach.

Jiang et al. (2021) evaluated studies aimed at using wheat straw as a raw material to obtain biochar for use as a soil conditioner; the functional unit is one ton of biomass, and the approach used was cradle to gate.

Saravanan et al. (2023) in addition to the previously described objectives, investigated both the strategies of attributional LCA and consequential LCA; the former seeks to attribute the environmental loads associated with the

TABLE 21.1 Goal and scope definition for biochar systems.

Goal and scope	Manufacture and feedstock	System boundaries	Functional unit	Sensitivity analysis	References
Reducing carbon emissions; achieving carbon neutrality	Primary and secondary agro-residue pyrolysis	Cradle-to-grave cradle-to-gate cradle-to-cradle gate-to-gate	1t feedstock 1t biochar	soil remediation, biochar yield, change in bio-oil and syngas yield, size of the pyrolysis reactor, transportation distance	Zhu et al. (2022)
Mitigation on agriculture, crop productivity and soil water status; Reducing carbon emissions	Agricultural crop residues	cradle-to-grave; cradle-to-gate.	ND	ND	Patel and Panwar (2023)
Assessment of the impacts of manufacture and use with association of carbon footprint midpoint impacts, air pollutants, acidification and other parameters. The second field of research is focused on energy consumption efficiency and economic aspects.	biomass waste from agriculture and forestry	cradle-to-grave cradle-to-gate Cradle-to-cradle gate-to-gate	1t of dry biomass into biochar	ND	Carvalho et al. (2022)
Potential benefits of applying magnetic biochar derived from PKS catalysts to biodiesel production and its environmental performance.	ND	ND	ND	ND	Anak Erison et al. (2022)
The actual influence of biomass pyrolysis on the environment	ND	cradle to grave	energy-based 1 MJ/1 kWh/1 MW end products and mass-based functional units, (1 kg⁻¹)	ND	Yu et al. (2022)
Determining and comparing the potential environmental impacts associated with the production of Pinus patula wood pellets of raw and Fe-modified biochars.	raw (Pinus patula wood pellets) and Fe-modified	Gate-to-gate	1 kg of biochar	ND	Gallego-Ramírez et al. (2023)
Use of biochar for soil improvement	agricultural residues, municipal solid wastes, energy crops, and wastewater sludge with several methods of biomass convertion: slow pyrolysis, fast pyrolysis, torrefaction, and gasification	Cradle-to-grave	ND	ND	James et al. (2022)
The purpose of this LCA is to show the absolute effects of an energy–biochar production system with subsequent soil application.	Gasification of woody biomass (woodchips from deciduous and coniferous wood)	cradle-to-grave	1000 kg packed biochar	ND	Marzeddu et al. (2021); Marzeddu et al. (2023)

(*Continued*)

TABLE 21.1 (Continued)

Goal and scope	Manufacture and feedstock	System boundaries	Functional unit	Sensitivity analysis	References
Comparatively, assess the environmental impact of wheat production using different fertilizer strategies.	Wheat straw	Gate-to-cradle	1 t wheat grain yield	six parameters, including the field burning rate of wheat straw, wheat grain yield, rate of substitution of the composting product for chemical N fertilizer, CH_4 emission factor of composting, SOC content, and heat recovery rate of biochar production, were analyzed because of their higher uncertainty regarding the research assumptions.	Jiang et al. (2021)
The goal of the LCA is broadened by choosing one of the two primary strategies – consequential LCA or attributional LCA.	Hydrothermal Anaerobic digestion Combustion Thermochemical of biowaste	Cradle to Grave	1 MJ of rice husk pellet	ND	Saravanan et al. (2023)
Quantify the environmental performance of pyrolysis biorefineries fed with PFR, where the pyrolysis process is driven by the production of bio-crude oil	Pyrolysis systems using forestry biomass (Dry wood chips) as feedstock.	use of PFR for pyrolysis and the use of the biorefinery co-products.	1000 kg of dry biomass	ND	Brassard et al. (2021)
environmental impact of CF to that of conventional farming and to estimate whether biochar produced through different methods	organic waste for conservation agriculture techniques instead with conventional agriculture.	Biochar production and soil application.	the functional unit was selected as impact per produced ton of maize per year.	37% yield difference between conventional agriculture and conservation farming substantiated by field data results amendment dose of biochar	Sparrevik et al. (2013)
Energy balance attained for the addition of nano-biochar for the anaerobic digestion.	nano-biochar (lignocellulosic wastes) can be utilized as an electrode in microbial fuel cells (MFCs) and as a catalyst for improving biodiesel and Hydrogen generation.	cradle to grave	ND	ND	Goswami et al. (2022)
Effect of different biochar on carbon sequestration in cementitious materials	biochar cement-based materials obtained from agricultural waste	cradle-to-gate, including raw material acquisition, processing, manufacturing, use, and end-of-life disposal	one kilogram of biochar cement-based material	ND	Song et al. (2023)

Study focus	Scope	Functional Unit		Reference
Seeking a unification of methodology that would be useful in making results on Biochar synthesis and utilization at least partially comparable, as is the reasoning behind most standardizations.	Biochar production from various wastes	cradle to grave cradle to gate		Matuštík et al. (2020)
		1 t of biochar	yes	
		1 kg product þ 1 ha/yr (secondary FU)	yes	
		1 t (Mg) of dry seed (product)	yes	
		1 t oat flakes	No	
		1 t dry matter feedstock	No	
		treatment of 500 m3 liquid sewage sludge per day	yes	
		1 ha agricultural area	yes	
		production of 1 t of biochar	No	
		1 t of feedstock	yes	
		per ha, MWh, odt feedstock, t biochar, per facility (multiple FUs)	yes	
		1 t feedstock (d.w.)	No	
		production of 1 t (Mg) biochar	No	
		9918 kg of Sludge 2777 kg dried sludge 1000 kg of biochar	No	
		1 t dry biomass	yes	
		1 t biosolids (WWTP) used	No	
		1 t biochar	yes	
		treatment of 1 t hardwood logs (green)	yes	
		1 odt straw	yes	
		a unit cultivation area ha; per kg grain produced	No	
		average village household utilizing cocoa over 1 year	yes	
		1 kg of biogenic carbon from biomass	yes	
		1 t biochar	yes	
		production of 1 kg rice	yes	
		management 1 t rice straw	yes	
		one year of project operation þ unit of waste	yes	
		produced t of maize per year	yes	
		preparation and utilization of 1 kg biochar	yes	

ND: not determined.

production and use of a product in a certain time period (present/past) while the latter seeks to identify the environmental consequences of a decision or proposed change in the system under consideration, orienting to the future, taking into consideration the market and economic implications of a decision. Biochar was obtained from different organic wastes and through different methods.

Sparrevik et al. (2014) carried out an LCA, the objective of which was to compare the environmental impacts associated with conventional and conservation agriculture in particular; the product was obtained by means of earthen kilns, retort kilns, and TLUD stoves. This study applies LCA to the use of agricultural biochar in sub-Saharan Africa, perhaps one of the most suitable areas for its use as a means of climate change mitigation and adaptation. Specifically, maize was used as a raw material, a ton of maize was the functional unit.

Matuštík et al. (2020) carried out a synthesis and comparison of several LCA studies aimed at the use of biochar as a soil conditioner. The data, from different countries, were processed through SimaPro and GaBi, and the approaches that defined the system boundaries were cradle-to-grave and cradle-to-gate; biochar was obtained by processing different types of organic waste.

A different study was carried out by Brassard et al. (2021) aimed at analyzing the pyrolysis process related to the production of biochar, in a biorefinery, using forest biomass. The functional unit is one ton of dry biomass and the system boundaries included the effects of the use of PFRs (Primary Forest Residues) for pyrolysis and the use of biorefinery co-products.

Anak Erison et al. (2022) considered the potential benefits of applying magnetic biochar, derived from PKS catalysts, to biodiesel production and its environmental performance.

Gallego-Ramírez et al. (2023) defined and compared potential environmental impacts associated with the production of Pinus patula wood pellets with raw and Fe-modified biochar. The system boundaries were defined by the gate-to-gate approach and the functional unit is one kilogram of biochar.

Goswami et al. (2022) were concerned with studying the energy balance obtained by adding nano-biochar for anaerobic digestion. Nano-biochar was produced from woody waste and the approach used was cradle-to-grave.

Song et al. (2023) defined the effect of different biochars, obtained from agricultural waste, on carbon dioxide sequestration within cementitious materials; the approach used was cradle-to-grave.

In summary, it can be concluded that the majority of LCA analyses conducted were aimed at validating hypotheses regarding the beneficial effects of using biochar as a soil amendment and its potential in improving the utilization of waste materials. The raw materials used to obtain biochar are purely waste materials from agro-forestry processing; the approaches were different, but the boundaries most taken into account were those that have considered the process from cradle to grave; it is important to emphasize that "cradle" does not mean the stage of collection of raw materials, since these are already considered waste materials, but directly the stage of their arrival at processing. The functional unit, generally, is attributed to one ton of raw material and/or biochar. The tools used for the elaboration of the LCA analyses were multiple, depending on the specific study. In most studies, broad and generic areas were considered and only in a couple of cases did they focus on specific realities.

21.5 Life cycle inventory

In the Inventory phase the sources of the data, the typology of raw materials used, and the upstream, core, and downstream processes were taken into consideration (Table 21.2).

In Zhu et al. (2022), data derives mainly from databases, national statistics, reports, and previous studies that have similarities between them. The raw materials are mainly agricultural production waste that was treated, in most cases, inside a pyrolysis reactor. Given the nature of the raw materials, the biochars obtained are all of agricultural origin. When biochar is used for soil remediation or remediation, more factors are involved in the environmental performance of the biochar system, such as its rate of application to soil, the biochar stability, and its effects on soil itself.

The effects of biochar on soil also include increasing its fertility, adsorbing pollutants, improving soil conditions, enhancing soil biota, and increasing crop yield. Yang et al. (2020) highlight that biochar addition increased fertilizer efficiency by 7% and crop yield by 20%. Furthermore, the addition of biochar can reduce methane and N_2O emissions, mitigating the effects on climate change.

In Patel and Panwar (2023), we find as a source of data the heading experiments, in addition to previous databases and literature; the raw materials are always of purely agricultural origin and, again, the most used transformation process is pyrolysis. The analysis of the downstream phase showed that the application of biochar to soil increased nitrogen, potassium, phosphorus, and total soil carbon; moreover, it increased the concentration of K in plant tissues, the

TABLE 21.2 Life cycle inventory for biochar systems.

Data sources	Feedstocks/upstream	Process production/corestream	Biochar characterization	Downstream	References
Literature	Empty fruit bunch	Pyrolysis reactor	Secondary agro-residue biochar	Soil amendment	Zhu et al. (2022)
Experiments; databases	Corn stover	Pilot-scale pyrolyzer	Primary agro-residue biochar		
Databases, simulation, literature, experiments		Small-scale fast pyrolysis	Agro-residue biochar		
Literature, government statistical		Fluidized bed for fast pyrolysis			
Government statistical, literature, reports		Slow pyrolysis reactor	Agro-residue biochar		
Calculations, reports	Oilseed rape straw residues	Pyrolysis plant	Agro-residue biochar		
Literature, government statistics	Crop residues	Pyrolysis plant	Agro-residue biochar		
Literature		Auger-based reactor			
Statistical yearbook		NR			
Literature	Empty fruit bunch	Fluidized bed reactor	Secondary agro-residue biochar		
Literature, databases	Cocoa shells	Retort technology	Secondary agro-residue biochar		
Experiments, literature	Tomato plant residue	Pilot-scale	Primary agro-residue biochar		
Literature, expert opinion	Straw	Pyrolysis	agro-residue biochar		
Literature, government statistical	Rice straw	Fluidized bed for fast pyrolysis	Primary agro-residue biochar		
Experiments, databases	Soybean shells	Electric furnace into a quartz reactor	Secondary agro-residue biochar		
Statistical yearbook, simulation, literature	Agricultural straw	kiln, centralized pyrolysis system	Primary agro-residue biochar		
Literature, databases	Olive solid waste	Mobile pyrolysis units	Primary agro-residue biochar		
Company, simulation	Olive pomace	Pyrolysis plant	agro-residue biochar		
Experiments, literature, databases, company report	Olive-waste cakes	Steel reactor	Secondary agro-residue biochar		
Simulation, databases	Oil palm kernel shell and empty fruit bunches	Modeled pyrolysis reactor	agro-residue biochar		
Literature, databases, estimations, expert opinion	Orange peel waste	Fast pyrolysis reactor	Primary agro-residue biochar		

(Continued)

TABLE 21.2 (Continued)

Data sources	Feedstocks/upstream	Process production/corestream	Biochar characterization	Downstream	References
Databases, report	Rice straw and husk	Pyrolytic cook-stove and drum oven	Primary and Secondary agro-residue biochar		
Questionnaires and interviews, literature, databases, reports	Rice husk	Stove, brick kiln, and BigChar 2200 unit.	Secondary agro-residue biochar		
Literature, company, databases	Coconut shells	Modern facility equipped	Primary agro-residue biochar		
Experiments, databases	Corn pericarp	Tube furnace	agro-residue biochar		
Experiments, company	Cocoa pods	Furnace	agro-residue biochar		
Literature	Wheat, barley, oat straw	Centralized plant, portable systems	agro-residue biochar		
Publicly available data	Sugarcane residues	Industrial slow pyrolyzer	agro-residue biochar		
Field collections, experiments, databases, literature	Oat hulls	Pilot-scale pyrolyzer	agro-residue biochar		
Simulation, Databases	Oil palm empty fruits bunches, kernel shell, and pelletized woody biomass	Co-pyrolysis process (at 600°C)	ND	Soil amendment	Patel & Panwar (2023)
Experiment and opensource data	Woody biomass	Gasification (1000°C)			
54 Literature	Algae, lignocellulosic biomass, and biodegradable organic wastes	Slow pyrolysis (400°C –700°C)			
Experiments and Database	Forest residues	Portable system (BSI, OK and ACBs)			
Experiments	Olive tree pruning	Continue production system (650°C, 15 min)			
Literature	Crop residue	Pyrolysis plant			
Experiment	Rice and wheat straw	ambient from 340°C to 600°C			
Literature	Rice straw and corn stover	Fluidized bed 300°C–550°C 350°C–55°C			
Experiment	Corn stover	Pilot-scale pyrolyzer (400°C, 10 min)			
Experiment	Sewage sludge	Slow (300, 400, and 500°C) fast (400 and 500°C)			
Experiments and Database	Lemon peels	Slow pyrolysis (400°C)			
Experiments and Database		Slow pyrolysis (500°C)			

	Feedstock	Process	Product	Applications	Reference
ND	ND	Slow pyrolysis (300°C–700°C) Fast pyrolysis (350°C–700°C) Gasification (700°C–1200°C) Torrefaction (200°C–300°C)	ND	Catalyst adsorbent Water purification Bio-composites Fuel cells Photovoltaics Carbon sequestration Carbon sink Water retention Plant nutrient Soil Conditioner Biomedical use Pharmaceutical Gas storage	Carvalho et al. (2022)
ND	waste palm kernel shell	Transesterification separation, purification drying stages	impregnated magnetic biochar	biodiesel with impregnated magnetic biochar as a catalyst	Anak Erison et al. (2022)
ND	Corn stover	fast pyrolysis	agro-residue Biochar	Soil amendment	Yu et al. (2022)
	Yard waste (brush, leaves, and grass clippings)		Biochar		
	Various straw		agro-residue Biochar		
	Switchgrass		agro-residue Biochar		
	Miscanthus/short rotation coppice or forestry		Forest waste Biochar		
	Short rotation poplar		Forest waste Biochar		
	Wood waste		Forest waste Biochar		
	T. chui		Biochar		
Ecoinvent database library	Pinus patula wood pellets	gasification process	Fe-modified biochars	ND	Gallego-Ramírez et al., (2023)

(*Continued*)

TABLE 21.2 (Continued)

Data sources	Feedstocks/ upstream	Process production/ corestream	Biochar characterization	Downstream	References
ND	Wastewater sludge	Slow pyrolysis	Biochar	Soil amendment	James et al. (2022)
	Pig manure and willow	Slow pyrolysis			
	Palm shell	Slow pyrolysis			
	Rice straw	Slow pyrolysis			
	Corn stover	Fast pyrolysis			
	Sawdust	Fast pyrolysis			
	Wood	Fast pyrolysis			
	Rice hulls	Gasification			
	Willow wood chips	Torrefaction			
	Peat	Torrefaction			
primary data internet sources free OpenLCA database	Woody biomass produced by forest management activities and was composed of woodchips. Upstream Feedstock collection and pre-treatment processes.	Core process, which consists of the gasification phase of biochar production and energy generation.	RE-CHAR®	Soil amendment	Marzeddu et al. (2021)
ND	compost supply and raw material (i.e., pig manure and wheat straw) collection production of biochar and compost field production, and transportation	Compost Production using the CF, MC, MCB5, and MCB10 strategies.	Biochar	Compost Application	Jiang et al. (2021)

ND	Agricultural waste Corn stover	Thermochemical conversion	Biochar Bioenergy Biocrude	Saravanan et al. (2023)
	Food waste Fruit waste	Composting and Thermochemical		
	Municipal Solid waste Paper waste Wood waste	Combustion Thermochemical		
	Coffee residues and Zilkha black	Hydrothermal liquefaction		
	Xanthium strumarium biomass	Hydrothermal liquefaction		
	Orange peel	Hydrothermal carbonization		
	Food waste digestate	Hydrothermal treatment		
	Anaerobic granular sludge	Hydrothermal carbonization		
	Apple waste	Hydrothermal		
	Food waste and Yard waste	Co-hydrothermal carbonization		
	Barley straw	Hydrothermal liquefaction		
	Olive pulp	Hydrothermal carbonization		
	Pine sawdust	Catalytic liquefaction		
	Pinewood waste	Fixed bed reactor		
	Beechwood	Fluidized bed reactor		
	Bamboo saw dust, rice husk, and Sewage sludge	Batch		
	Maize straw	Tubular reactor		
	Polyalthia longifolia leaves	Batch tubular reactor		

(Continued)

TABLE 21.2 (Continued)

Data sources	Feedstocks/ upstream	Process production/ corestream	Biochar characterization	Downstream	References
Ecoinvent 3.3 database	Dry wood chips	Pyrolysis	Biochar using forestry biomass as feedstock	bio-oil and biochar for the displacement of fossil fuels (heat) in cement manufacturing	Brassard et al. (2021)
ND	ND	pyrolysis	Biochar from organic waste	biochar is used in combination with conservation agriculture techniques instead of combining it with conventional agriculture.	Sparrevik et al. (2013)
ND	lignocellulosic wastes	ND	Nano-biochar produced via the lignocellulosic wastes	electrode in microbial fuel cells (MFCs) and as a catalyst for improving biodiesel and Hydrogen generation.	Goswami et al. (2022)
Ecoinvent 3.0 database; Ets-Esu database	Waste wood, Mixed wood, sawdust, Peanut shells, Maize straw	High-temperature pyrolysis and hydrothermal carbonization	Biochar from agricultural waste	Potential environmental burdens and benefits of biochar cement-based materials	Song et al. (2023)

Scopus database			Biochar from various waste	ND	Matuštík et al. (2020)
	pig manure, willow woodchips	500°C			
	oak residue	slow pyrolysis (650°C)			
	oilseed rape straw	400°C 800°C			
	oat seed side flows, willow woodchips	unspecified			
	miscanthus, sorghum, hemp, black locust	slow pyrolysis			
	sewage sludge	fast pyrolysis (500°C)			
	poplar woodchips	slow pyrolysis			
	tomato plant residue	intermediate pyrolysis (400°C)			
	urban biodegradable wastes	slow pyrolysis fast pyrolysis gasification			
	sewage sludge				
	green waste				
	food waste				
	used cardboard, poultry litter whiskey remains				
	wheat straw				
	barley straw				
	oilseed				
	rape straw				
	sawmill residue				
	forestry residue chips				
	wood chips				
	short rotation coppice and forestry chips				
	miscanthus				
	corn fodder	slow pyrolysis (400°C, 12 min, 20°C/min)			
	forest residue				
	switchgrass	auger reactor (459°C and 591°C)			
	sewage sludge	slow pyrolysis (500°C)			
	pine wood slash	slow pyrolysis (500°C)			
	spent grans from brewery				

(Continued)

TABLE 21.2 (Continued)

Data sources	Feedstocks/ upstream	Process production/ corestream	Biochar characterization	Downstream	References
	yard trimmings	slow pyrolysis, (450°C–500°C, 4–6 h)			
	oat hulls, pine bark	electric pyrolyzer (300, 400, 500°C)			
	hardwood chips (from plantation)	300 and 600°C			
	rice straw	slow pyrolysis			
	wheat straw	vertical kiln 350-500°C			
	cocoa shells	retort kilns 300-400°C			
	biomass waste	flame curtain, retort kilns, earth-mound kilns			
	oil palm waste	pyrolysis reactor "Deorub Liquid Smoke" (280°C)			
	rice straw, rice husk	drum oven (used for cooking) TLUD drum oven			
	rice husk	slow pyrolysis (hypothetical model reactor), no syngas and bio-oil capture			
	maize cobs	earth mound kilns, retort kilns (with gas recirculation), top-lit updraft stoves			
	woody shrub, agricultural residues	earth mound kiln, retort kiln (with gas recirculation), Kon-Tiki flame curtain kiln, micro pyrolytic cook stoves, gasifiers			

nodulation of rhizobia, soil microbial biomass, and crop yield compared to those without biochar application. The application of biochar in soil offers both long-term carbon sequestration and an optimistic low-cost soil amendment.

In Carvalho et al. (2022), the corestream phase is strictly based on the pyrolysis process while in the downstream phase, several fields of application of biochar were considered: as an adsorbent catalyst, in water purification, in Biocomposites, in fuel cells, in photovoltaics, for carbon sequestration, to compensate for the reduction of water and nutrients for plants, as soil conditioner, for biomedical and pharmaceutical use and, finally, for gas storage.

The study by Anak Erison et al. (2022) analyzed corestream phases in which transesterification, separation, purification, and drying are the phases involved in the overall production of biodiesel with magnetic biochar, impregnated as a catalyst.

Yu et al. (2022) have focused exclusively on agro-forestry-derived biochar and on only one transformation process, namely fast pyrolysis.

Gallego-Ramírez et al. (2023), based on data derived mainly from the Ecoinvent database, analyzed Pinus patula wood pellets; biomass was modified with iron leading to the production of crude (Pinus patula wood pellets) and biochar modified with Fe.

James et al. (2022) considered that the raw materials are not only of an agro-forestry nature but also include pig manure and peat. The processes considered are slow pyrolysis, fast pyrolysis, roasting, and gasification. In downstream it turns out that biochar has the potential to improve soil fertility in sandy soils or weathered for agricultural purposes. When mixed with soil, biochar can improve soil quality by increasing soil saturation and water retention. It was reported that biochar can also increase cation exchange capacity in soil and improve mineral absorption in plants by 13%, compared to degraded soils without biochar application. Depending on the production process, biochar can also contribute to acidic and basic soils by neutralizing extreme pH conditions. In addition, biochar is expected to function as a long-term carbon deposit to reduce fertilizer intake, adsorb pollutants from soil, and improve carbon dioxide sequestration by 47% and 55% in paddy fields.

Marzeddu et al. (2021) used technical data from plant operations (primary data), collected through interviews with the company's service manager in the period between May 2020 and December 2020. When data was not available in the published literature, internet sources such as the free OpenLCA database (secondary data) were used. If data from Italian contexts were not available, data from similar contexts were chosen. RE-CHAR® biochar was obtained from woody biomass (hardwood and coniferous wood chips) In this case, harvesting processes are also considered in the upstream, in addition to the upstream pretreatment of raw materials. In downstream, impacts due to transport processes are considered in addition to the effects of using biochar as a soil improver.

In the perimeter of the strategies for compost of Jiang et al. (2021), the upstream stages of compost supply are also taken into account, including the feedstock collection (i.e., pig manure and wheat straw), the production of biochar (only for MCB5 and MCB10 processes) and compost, field production and transport connection steps. In downstream phase, the results obtained from the compost application are considered.

Saravanan et al. (2023) also consider raw materials from food waste and municipal waste, showing that this type of waste can also be optimally used through conversion processes, leading to the production of biochar. Optimization is achieved by two factors: combustion techniques that are less impactful in terms of emissions into the atmosphere and the possibility of obtaining a by-product that is not only usable but that has a great deal of beneficial effects in its use.

Brassard et al. (2021) and Sparrevik et al. (2013) both based their analyses on corestreams focused on the pyrolysis process. In the first study, the downstream phase is focused on the production of bio-oil and biochar for the replacement of fossil fuels, in the production of cement (the objective of this case study is to quantify the environmental performance of pyrolysis biorefineries, fed with PFR, where the pyrolysis process is driven by the production of crude bio-oil); the second took into account the effects of the use of biochar, in combination with conservative farming techniques, instead of combining it with conventional agriculture.

Nano-biochar produced from lignocellulosic waste from Goswami et al. (2022) can be used as an electrode in microbial fuel cells (MFCs) and as a catalyst to improve biodiesel and hydrogen production. From the data obtained from the Ecoinvent database, in Song et al. (2023) only woody raw materials, treated with high-temperature pyrolysis and hydrothermal carbonization from which the biochar was obtained, were considered. In the downstream phase, the effects of the use of biochar inside concrete materials were considered.

The inventory phase showed that most of the processed data are derived from databases rather than directly conducted experiment results; the most widely used database is Ecoinvent. The raw materials, used to obtain biochar, are agricultural and forestry materials for which, being already considered waste materials, and in most cases, the upstream process does not consider the harvesting phase and not even the transport. In the core phase, importance is given to the conversion technologies used; analyses show that the most effective conversion in terms of biochar quality and impact reduction is slow pyrolysis, which is, therefore, the most used. In the downstream phase, the beneficial effects of the application of biochar to soil and the positive effects that this brings when it is used in renewable energy systems (fuel cells) are analyzed.

21.6 Life cycle impact assessment

The analysis carried out by Zhu et al. (2022) shows that system boundaries, functional units, and pyrolysis systems are different throughout the literature analyzed, making it rather complicated to compare the results obtained from different LCA. Climate change is the main impact indicator, while other impact factors (e.g., AP, EP, etc) are considered only in some studies. The results of the LCA report both negative values (CO_2 capture) and positive values (CO_2 emissions), which are related to the boundaries of the LCA system.

Dai et al. (2020) reported a potential reduction in greenhouse gas emissions of 1.41 10^6 t CO_2 eq, based on Chinese crop residue supply data. Clare et al. (2015) stated a GWP of −1.06 t CO_2 eq/t of raw material. Extending the system boundaries to the application of biochar, the LCA results showed that CO_2 capture is achievable. When the boundary of the LCA system excluded the use of biochar, CO_2 emissions have been reported, contrary to what is described in the case of using biochar for soil modification (−386 to −933 kg CO_2 eq/t of raw material) and the energy supply (−240 to −787 kg CO_2 eq/t of raw material).

Pourhashem et al. (2013) detected −217 g CO_2 eq/kWh of electricity for co-combustion of biochar and −84 g CO_2 eq/kWh of electricity for land improvement. The benefits, in terms of reducing carbon emissions are evident when biochar is used for other purposes. Greenhouse gas emissions, caused by biomass pretreatment and biochar production, can be neutralized by the benefits of biochar applications (through carbon sequestration). It has been demonstrated that agro-residues derived from crop collateral flows are suitable raw materials for the production of biochar in terms of environmental benefits. As already said, greenhouse gas emissions from the production of agro-residues are generally not included in the biochar production system. The moisture content of the agro-rice varies during the different phases, such as the harvest and the natural drying phase. A high moisture content in agricultural residues requires higher energy consumption for drying and increases greenhouse gas emissions. For example, when wet orange peel scraps were used as raw material for biochar production, carbon reduction, −5.5 kg CO_2 eq/t of wet raw material, was marginal. Pyrolysis conditions, such as temperature, pyrolysis rate and residence time, also showed a clear impact on the distribution and properties of the product.

Cheng et al. (2020) reported the carbon emissions of biochar produced by plant residues at different temperatures (400°C, 550°C, and 700°C) where the avoided carbon emissions ranged from −200 to −470 kg CO_2 eq/t of raw material.

Thers et al. (2019) found that the carbon reduction of biochar, obtained at 400°C and 800°C, was respectively −392 and −429 kg CO_2 eq/t dry seed. In addition, for rice straw, the CO_2 reductions were respectively 1.14, 1.64, and 1.10 t CO_2 eq/t of raw material through rapid pyrolysis, intermediate pyrolysis, and slow pyrolysis, while for maize stubble they were respectively 1.75, 1.12 and 1.84 × 10^3 kg CO_2 eq/t raw material (Gong et al., 2021).

Pyrolysis conditions affect biochar yield and distribution of co-products (bio-oil and syngas), resulting in variable energy compensation (Matuštík et al., 2020). Furthermore, the percentage of stable carbon in biochar is an important factor for effective carbon sequestration. Different regions have specific requirements for biochar production systems, large-scale centralized pyrolysis plants and small-scale portable pyrolysis reactors need to be balanced. Large-scale pyrolysis plants have higher pyrolysis efficiency and co-product energy compensation, while small-scale reactors may not be able to utilize pyrolysis co-products and even suffer adverse effects such as particulate matter emissions and air pollution. It should be noted that energy requirements are generally not included in the boundaries of the LCA study system. Small-scale pyrolysis reactors require a lot of labor and more effort to produce the same amount of biochar. In addition, transportation distance is an important factor limiting the choice of pyrolysis device. Long transportation distances have a negative effect on net GWP.

Mohammadi et al. (2016) reported that the carbon footprints of the pyrolytic stove and the drum ovens were, respectively, 1.11 and 3.85 kg CO_2 eq/kg of processed rice. Mohammadi et al. (2017) revealed a higher GWP of biochar production equipment, for example −229 kg CO_2 eq/t of dry rice husk for brick kiln, −318 kg CO_2 eq/t of dry rice husk for stove e −360 kg CO_2 eq/t dry rice husk for large-scale pyrolysis plant. In general, advanced pyrolysis equipment provides greater environmental benefits in terms of reducing CO_2 emissions. However, in developing regions, choosing an appropriate pyrolysis technology is crucial. Small-scale portable pyrolysis reactors can provide significant benefits by replacing primitive on-ground furnaces which can present a negative environmental impact.

According to Patel and Panwar (2023), the real benefits of biochar for the reduction of greenhouse gas emissions must quantify the amount of CH_4 and N_2O produced by modified soils, especially in wetlands, which are regularly flooded and drained. In agriculture, CH_4 emissions are mainly caused by land grown on rice, and reducing these emissions would contribute to reducing global warming.

During an experiment conducted by Li et al. (2020), it was observed that the application of biochar to the soil of rice led to a significant reduction in emissions of CH4. The addition of biochar did not inhibit the production of CH_4 bacteria (methanogens) but increased the abundance of CH_4 oxidizing bacteria (methanographs), which led to a decrease in CH_4 emissions.

Based on the LCA, Mohammadi et al. (2016) calculated the effects of outdoor combustion and the conversion of residues into biochar on climate change. An amendment to biochar reduced the carbon footprint of spring rice and summer rice by 26% and 14% respectively compared to open combustion of rice straw. In the global context, nitrogen fertilizers are the main source of anthropogenic N_2O emissions. N_2O is mainly produced during the transformation of nitrogen in soil.

Cayuela et al. (2015) carried out an experimental study, in the laboratory and in the field, involving 56 experimental treatments, finding that the application of biochar reduced N_2O emissions in the field by 28% and in the laboratory by 54%. N_2O emissions are affected by raw materials, pyrolysis parameters, and biochar H:Corg ratio. In total, it has been estimated that the amount of N_2O has decreased by about 31%. The application of biochar has gained popularity as a strategy to reduce the negative effects associated with greenhouse gas emissions in agriculture, such as CH_4, CO_2, and N_2O. The application of soil biochar reduces the amount of CO_2 emitted from the soil and increases the amount of CO_2 captured from the atmosphere. The extent of the reduction varies depending on the specific approach, from about 603 kg CO_2 eq/t for biomass-based building materials to 1173 kg CO_2 eq/t for soil improver incorporating biochar.

In Carvalho et al. (2022), considering all differences in system boundaries, such as functional units and other parameters, LCA results are not comparable. The LCA results presented make a general assessment of greenhouse gases, which vary in value depending on the methodology used.

According to a study by Brassard et al. (2018), biochar was used for soil correction and a GWP impact of -2.561 t CO_2 eq/t biochar was observed; Mohammadi et al. (2020) revealed that the carbon footprint of rice production, with the use of biochar, is 3.85 kg CO_2 eq/kg; the results show similar trends, so a certain periodicity can be defined. It is remarkable that soil biochar correction systems show a clear advantage from a climate change perspective. The application of biochar can significantly contribute to carbon sequestration while increasing yields and simultaneously reducing the use of fertilizers and therefore CH_4 and N_2O emissions.

In Yu et al. (2022), in general, the intermediate categories used to express the environmental impacts of the life cycle include GWP, AP, EP, ETP, WDP, FE, and so on. Through the integration of intermediate categories, the results of the final categories can be obtained. The final categories used in the reviewed studies include human health, ecosystem quality, and resource depletion. Nevertheless, most researchers still use mid-point indicators to show results, believing that mid-point categories can easily describe environmental impacts. GWP is the most commonly used impact category in LCA applications. The GWP expresses the cumulative radiative forcing value caused by an emission of a unit mass of a given greenhouse gas over a defined time horizon, compared to the equivalent value for CO_2. In the LCIA methods selected by the authors, middle point categories are the most commonly used. Since GWP is the most well-known impact value and is included in all reviewed articles, it is discussed in detail.

When biochar is produced by pyrolysis, the net GWP varies from 442 to 1570 CO_2 eq/dry ton. The wide range of net GWP values is due to the diversity of technologies (types of pyrolysis and product yields), the mode of operation (raw materials, temperature, and heating rate), and the limits or decisions regarding the LCA model. Moreover, the treatment of co-products is also a controversial issue. For example, when biochar is used as fertilizer, the value of GWP is different (e.g., 1.570 t CO_2 eq/dry tonne without system expansion, from 1000 to 1200 CO_2 eq/dry tonne and 1250 CO_2 eq/dry tonne with system expansion). In addition, different uses of the same co-product can change the value of the GWP. The biochar used for energy combustion (910 CO_2 eq/dry tonne) had a greater impact than soil modification (1.250 t CO_2 eq/dry tonne). The order of the increasing average GWP is as follows: forest residues < agricultural residues < microalgae. In general, forest residues are the main source of power for electricity generation, while agricultural residues are used for bioenergy products.

As for the production of raw biochar and the impact categories of the methods recipe 2016 and USEtox in Gallego-Ramírez et al. (2023), all environmental impacts are attributed to the energy source used when generating biochar. In fact, wood combustion processes result in emissions of carbon dioxide (CO_2) and carbon monoxide (CO). Furthermore, the gasification process can lead to the formation of polycyclic aromatic hydrocarbons (PAHs), which are widely known for the impacts they can generate on both the environment and living organisms. As regards the category of human toxicity to cancer, from the USEtox method, the emission of formaldehyde is the main hot point of the category, this is because biomass gasification is considered a source of formaldehyde. The toxicity of formaldehyde is attributed to its carcinogenic, mutagenic, and reprotoxic capacity.

For the category of noncarcinogenic human toxicity, the main hot spot is carbon disulfide emissions. This can be attributed to the carbon emissions that are generated in the production of energy through biomass. According to the United States Environmental Protection Agency (EPA), carbon sulfide can cause neurological effects, headaches, dizziness, fatigue and other symptoms when humans are exposed to the compound.

As for ecotoxicity in the production of biochar, the discharge of carbendazim into the soil is the main "hot spot." Carbendazim is a pesticide and a large number of biochar and carbendazim-related research concerns the adsorption of carbendazim in biochar. According to the authors' knowledge, the production of biochar has not been linked to the production of carbendazim or pollution. The ecotoxicity of biochar production may be related to the production of PAHs, which are found to have toxic effects in the environment and can bioaccumulate and biomagnify in the fatty tissues of living organisms.

With regard to the modification of biomass with Fe(II) chloride, it is noted that for the 2016 recipe method, the environmental impacts are mainly attributed to Fe(II) chloride. This can be explained by the environmental impacts of iron extraction, as this industrial activity produces a large amount of wastewater and gas emissions. The use of Fe(II) chloride has the highest influence. The potential environmental impacts do not differ from the production of raw biochar to that of biochar modified with Fe; therefore, the addition and the difference in environmental impacts are obtained in the modification of biomass but not in the production of biochar modified with Fe. The categories of toxicity to humans, freshwater, and ocean ecotoxicity for biomass modification represent the main impacts during this process, which are mostly attributed to the use of Fe. The environmental impacts of Fe-modified biochar production are the same as those of raw biochar production. Therefore, the additional impacts generated in the production of modified biochar are mostly common and appear during the biomass modification phase. This can be attributed to the use of several metals (including Fe) that can cause further environmental pollution.

In James et al. (2022), it emerged that miscanthus biochar has a GWP of about 0.13 tons of CO_2/ton of raw material, which depends on the quality of the raw material and the biochar yield. The LCA study revealed that the growing phase of miscanthus is the focal point of emissions in the biochar life cycle, with 93 kg CO_2 eq/tonne. In addition, the environmental effect (GWP) of biochar has decreased with increased miscanthus yield. The study found that a 10% increase in biochar yield can reduce GWP by 26%.

In Marzeddu et al. (2021), different impact categories were analyzed; as regards the acidification potential, the main positive impact was given by the transport (0,50 kg SO_2 eq), including that of biomass to the processing plant, of bigbags and finally of biochar to the agricultural fields. The electricity produced by the plant, which replaced that generated with the Italian energy mix, was the main avoided product that guaranteed the process a negative acidification potential.(i.e., -27.081 kg SO_2 eq). The electricity mix still contained a significant share of coal electricity (12.34%), associated with high emissions of acidifying substances, especially sulfur dioxide and nitrogen oxides, resulting from coal combustion. The only significant negative impact was due to the reduction of the practice of irrigation of the fields, thanks to the biochar ability to retain water (i.e., 0.020 kg SO_2 eq). In terms of global warming, the energy-biochar system has produced negative greenhouse gas emissions. This result "from cradle to grave" is due to the production of renewable energy through the process of gasification of wood chips, which has replaced, in part, the energy produced by the use of fossil fuels, with a greater environmental impact. A positive value indicates emissions associated with transport operations, direct emissions from the gasification process itself and other processes, such as the production of raw materials. The greatest contribution to climate change was avoided emissions in energy production, with values of -3.844 t CO_2 eq and -3.141 kg CO_2 eq for electricity and heat, respectively. The major contribution was given by biochar carbon sequestration capacity (-1.513 t CO_2 eq). As expected in the downstream process, GWP 100 was found to have a negative net value due to long-term carbon sequestration (-1505,741 kg CO_2 eq). Positive impacts were mainly related to emissions from transport and agricultural machinery. The abiotic resource balance of the cradle-to-grave approach could be considered zero (i.e., -2.8×10^{-4} kg Sb eq), which means that the resources used in the cycle are approximately equal to those avoided. As in the other categories, heat and energy production had a negative impact, while wood production and transport had a positive impact. The balance of abiotic resources in the downstream process could be considered zero ($8,191 \times 10^{-6}$ kg antimony eq). From a "cradle to grave" approach the dominant factor for the depletion of fossil fuels was the negative contribution due to the displacement of fossil fuels from the co-products of biochar (electricity and heat), resulting in a high net reduction in abiotic resource exhaustion. The high net result for this category (−90,758.060 MJ) was the most predictable, as the process replaced the use of natural gas, petroleum products and coal with the gasification of woody biomass. In the downstream process, fossil fuel depletion is obviously the most closely related impact category of emissions from fossil fuel combustion. As a result, the net value is positive and amounts to 177,78 MJ. In the results from cradle to grave, the eutrophication index was −7.140 kg PO_4^{3-} eq per 1 ton of biochar produced and applied to soil. Moreover, in this case, the production of electricity from the gasification process instead of that produced with the Italian energy mix was the process that had the greatest effect (−6.925 t PO_4^{3-} eq), mainly due to ammonia emissions, nitrogen oxides and phosphates.

In the downstream process of the impact of eutrophication, the main positive contribution was the direct combustion of fuel by agricultural machinery (0.017 kg PO_4^{3-} eq) and transport vehicles (0.003 kg PO_4^{3-} eq). A small negative contribution is linked to the impacts avoided thanks to the reduction of water use in irrigation activities (−0.007 kg PO_4^{3-} eq). The category "other" also includes the impacts avoided as a result of the reduction of chemical fertilizers due to the properties of biochar (i.e., reduction of fertilizers based on ammonium nitrate, phosphoric acid and potassium chloride had a negative impact of -7×10^{-4}, -4.7×10^{-4} and -3×10^{-5} kg PO_4^{3-} eq, respectively). In the downstream process, the most impacting element in this category was the production of steel and copper for the manufacture of agricultural machinery, transport-related processes, and the production of diesel.

Regarding the negative impacts, a fundamental role was played by the reduction of irrigation activities (−3.273 kg of 1,4-dichlorobenzene eq). The negative impact of marine aquatic ecotoxicity comes mainly from the decrease in irrigation volumes (−4.340 kg of 1,4-dichlorobenzene eq). In this case, reductions in fertilizers based on ammonium nitrate, phosphoric acid, and potassium chloride had a negative impact of −143.380, −100.070, and −24.310 kg respectively of 1,4-dichlorobenzene eq. In cradle-to-grave results, the photochemical oxidation net balance was negative but very close to zero (−0.510 kg ethylene eq). In this case, the main positive impact is related to the emissions generated by gasification processes: the combustion of syngas, mainly in carbon monoxide (CO), which is then emitted. The negative impacts related to the production of electricity (−0.748 kg eq of ethylene) and heat (−0.181 kg eq of ethylene) through the analyzed process depend strictly on the avoided sulfur dioxide (SO_2) emissions.

As for the results from cradle to grave related to terrestrial ecotoxicity, the net emission balance is still negative (−13.990 kg of 1,4-dichlorobenzene eq). As with the previous ones, the main contribution is due to the avoided emissions related to the production of electricity through the gasification process (−13.179 kg of 1,4-dichlorobenzene eq). Heavy metals were the main class of contaminants that had an impact on terrestrial ecotoxicity, due to their toxicity and persistence in the environment. Electricity production according to the national energy mix has been characterized by high emissions of chromium and mercury in air and soil; the same applies to the production of heat with natural gas boilers. For the latter impact category, in the downstream process, the net emission balance is positive (0.015 kg of 1,4-dichlorobenzene eq). As with previous processes, the main negative contribution is due to the avoided emissions related to the reduced use of water for irrigation. Steel production, characterized by high chromium emissions, was the process with the greatest positive impact (0.052 kg of 1.4-dichlorobenzene eq). Other positive contributions come from transport, diesel, and steel production processes. In this case, reductions of fertilizers based on ammonium nitrate, phosphoric acid and potassium chloride had a negative impact of -1.6×10^{-3}, -1.2×10^{-3} and -2.9×10^{-4} kg of 1,4-dichlorobenzene eq respectively.

In Jiang et al. (2021), the highest GWP value for the life cycle was observed for manure compound (MC) strategy, being estimated at 1.33 t CO_2 eq/t grain, equivalent to 1.563 t CO_2 eq/t MC, which is about six times higher than reported by Zhong et al. (2013) (i.e., 240 kg CO_2 eq/t MC). This great variation could be explained by the different emission factors of CH_4 and N_2O biogenes, assumed during the composting and application of MC to soil. In addition, lower GWP values were observed than those assessed in this study when CF (chemical fertilizer) was partially replaced by other manure-derived fertilizers in a corn-grain rotation system. For example, the estimated GWP values varied between 188 and 567 kg of CO_2 eq/t grain when about half the amount of FC was replaced by fresh solid or liquid manure. Extremely low GWP values (i.e., 45 to 100 kg CO_2 eq/t of cereals) were also found for a case of digested manure application, to which the compensation of greenhouse gases due to the production of bioenergy from the production of digested manure slurry was the main contribution (e.g., bioenergy from CH_4).

In contrast, a much higher GWP of 3270 kg CO_2 eq/t of rice was found for a rice production system with manure fertilizers, compared to that for the MC strategy in this study. Compared to the CF strategy, the value of GWP increased by 9.33% in the MC strategy, while it decreased significantly by 36.9% and 48.2% in the MCB5 and MCB10 strategies. With regard to the application of CF to soil and irrigation, CF products and energy input have been identified as the two main hot spots for GWP, which contributed respectively to 51.4% and 29.5% of greenhouse gas emissions in the life cycle. Overall, the GWP of the field production phase was greatly mitigated by 1.21×10^3 kg CO_2 eq/kg grain in the CF strategy at 254–553 kg CO_2 eq/kg grain in the MC, MCB5, and MCB10 strategies.

These results are supported by those reported in previous studies (Li, Wang, et al., 2020), where GWP was mitigated by 17.8%–48.0% when CF (i.e., 40%–50%) was partially replaced by fresh solid or liquid manure in the corn-grain rotation system. Analysis of the structure of greenhouse gas sources reveals that biogenic greenhouse gases (CH_4 and N_2O) from the compost production phase accounted for 50.4% (i.e., 680 kg/t grain) of the GWP of the life cycle in the MC strategy, while they fell to 26.1% (i.e., 256 kg/t wheat) and 24,2% (i.e., 257 kg/t wheat) in the MCB5 and MCB10 strategies, respectively. The share of greenhouse gas emissions from additional fossil energy accounted for 31.5%–37.4% of total fossil greenhouse gas emissions. This result shows that the additional fossil fuel, consumed in the upstream stages of the compost supply has contributed significantly to the life cycle GWP in the MC, MCB5 and MCB10 strategies. The share of greenhouse gas emissions from additional fossil energy accounted for 31.5%–37.4% of total fossil greenhouse gas emissions. This result shows that the additional fossil fuel, consumed in the upstream stages of the compost supply, contributed significantly to the life cycle GWP in the MC, MCB5 and MCB10 strategies. In addition, the failure to burn wheat straw by harvesting wheat straw and the lack of fossil fuel heat generation replaced by biogenic heat produced by biochar production helped to reduce the GWP life cycle.

In Brassard et al. (2021), the pyrolysis of primary forest residues (PFR) shows a reduction of 906.4 kg CO_2 eq /tonn of feedstock compared to the reference scenario in the category of impact on climate change. The modeled pyrolysis plant treating 4800 tons of biomass (10% of water content) per year would therefore contribute to a reduction of 4264 tons of CO_2 eq per year This corresponds to a reduction of CO_2 of about 10%. This corresponds to about 645 medium-sized cars (a mix of size and use of fossil fuels) traveling 20,000 km, based on the Ecoinvent process "Transport, passenger car, EURO 4 | Conseq, U." The difference between the two scenarios is mainly due to the seizure of ~142 kg of Carbon/ton of dry PFR in the pyrolysis process, whereas 80% of fixed Carbon remains in the soil for more than 100 years. CO_2 emissions from the combustion of biocrude oil (674 kg CO_2 eq) are the major contributor to the category of climate change, as its importance has been defined as crucial.

This is followed by emissions from storage and pretreatment of PFR (295 kg CO_2 eq), which include 206 kg CO_2 eq from storage of wood chips (88.2% CH_4 and 11.6% CO_2) and 87 kg CO_2 eq from drying using the heat of natural gas combustion. The contributions of the PFR supply chain (collection and transport) processes and the pyrolysis and condensation processes are relatively small. Considering the objective of limiting the global average temperature increase, well below 2°C compared to pre-industrial levels, PFRs are therefore clearly better managed as a raw material for pyrolysis for the production of oil and biochar rather than left unharvested to decay in forests. However, Giuntoli et al. (2015) have shown that the efficiency of climate mitigation from the production of bioenergy from forest-cutting residues depends on the rate of decay of the biomass left on the soil. The result obtained for the case study presented here is based on the assumption that 91% of the forest residues left on the soil decompose and emit CO_2 within 100 years.

The reduction of 906,4 kg of CO_2 eq/t of dry PFR is in the range of 0.4–1.2 t CO_2 eq/t dry raw material reported in other studies on the impact of biochar systems on climate change in the life cycle. In this study, only the C sequestration potential of biochar has been considered as a greenhouse gas emission mitigation effect. However, Azzi et al. (2019) reported that considering the cascading effects of biochar could double the benefits provided by the latter in terms of CO_2 emission mitigation. This additional effect can be achieved when biochar is used for feeding cows to reduce enteric CH_4 emissions, mixed with manure, to reduce emissions from manure storage, and applied to soil to reduce N_2O emissions from nitrogen fertilization. A meta-analysis revealed that biochar overall reduces N_2O emissions by 38% (compared to situations without biochar application), but that reductions are greater immediately after application. Biochar can provide additional benefits when added to soil and has positive effects on the physical, chemical and biological properties of soil, which can increase plant growth. However, these effects are specific to the type of biochar, the properties of fertilizer and soil, and environmental conditions.

In Sparrevik et al. (2013), Recipe was used, which operates with a total of 17 different categories of endpoints to describe the outcome of the LCA. In this study, the total normalized values of the final impacts of the different cases were dominated by three different categories: the impacts of climate change, which represent the potential aggregate damage of greenhouse gases, including effects on humans and the ecosystem; particulate emissions, which refer to the effects on humans from the inhalation of fine particulates, and mineral and fossil fuels, which are linked to the cost of extracting minerals and fossil fuels. This category refers to the fact that it is assumed that the extraction of current resources will not only lead to climatic effects but also to an increase in marginal costs of extraction in the future. The other impact categories of the LCA endpoints were grouped into the other categories. Given the objective of this study to assess the impact of biochar as an enhancer of agricultural soil fertility, all reported results are presented as annual figures per ton of corn harvest. The potential negative effect on the climate of atmospheric particulate matter is another topic of debate, currently not considered in the impact model recipe. The effect is regional and living much shorter than other greenhouse gases and therefore difficult to quantify in terms of GWP at 100 years. It is, however, plausible to assume that particulate emissions from biochar production, in some form, adversely affect the climate, underlining the negative impact of PM release.

The spontaneous adoption of the use of biochar in tropical rural areas will most likely occur through traditional production methods, due to their spread and the familiarity of farmers. This impression has been confirmed at sites where spontaneous adopters used only traditional methods (ground ovens or simple holes in the ground) to prepare biochar. The results confirmed that the overall impact of conventional agriculture exceeds that of conservative agriculture. In this study, this was caused by higher yields which reduced the impact per tonne of maize produced, as well as by avoiding the burning of agricultural waste and the more efficient use of fertilizers in conservative agriculture. The impacts of fertilizers arise from the production of ammonia, in which natural gas is catalytically divided into hydrogen and then reacts with nitrogen to produce hydrogen ammonia and then reacts with nitrogen to produce ammonia. This process emits carbon dioxide and consumes nonrenewable resources. Finally, it should be noted that the differences between the three cases were caused by the diversity of harvest results at the three sites. If biochar had been used only to mitigate carbon, this comparison would be less relevant, since the same amount of biochar was applied at all sites. For the use of biochar in tropical rural agriculture it is plausible to assume that an increase in yield, compared to traditional methods, is a prerequisite for its adoption.

In Goswami et al. (2022), an increase in the energy balance should be achieved thanks to the addition of nano-biochar to the Anaerobic Digestion process. This "cradle to grave" approach must also be beneficial from an environmental and economic point of view. LCA of biochar produced from lignocellulosic waste showed that greenhouse gas emissions are between 20–50 g CO_2 eq/MJ while for the shell, bark, etc., they are included between 120–250 g CO_2 eq/MJ. Biochar feedstocks possessing an ash content in the range of 0.1%–2% and a high O/C ratio are mainly related to higher greenhouse gas emissions. Furthermore, compared to petroleum-derived fuels, the use of biofuels leads to an 85% reduction in the release of greenhouse gases, equivalent to 93 g di CO_2 eq/MJ. The biochar-assisted AD process offers encouraging effects compared to anaerobic digestion alone. However, further LCA studies are needed to combine and integrate waste conversion and resource recovery processes. The results of the methodology mainly depend on the biomass type, compositions, reaction conditions, and reactor.

Song et al. (2023) reported that biochar-based cement materials play an important role in environmental protection and carbon sequestration in the construction industry; according to JGJ55−2000 (J64−2000) the construction of a skyscraper requires about 50,000 m^3 of concrete. Also, according to Chen et al. (2019) every cubic meter of biocarbonium can capture 59 to 65 kg of CO_2. If these by-products are used to replace fossil fuels with equivalent thermal content, earlier research has shown that the net carbon dioxide emissions associated with the pyrolysis process can be further reduced.

Considering all the differences in system boundaries, functional units, pyrolysis systems, etc. described in the work of Matuštík et al. (2020), it is obvious that the LCA results analyzed here are not comparable. Summing the GWP results presented in the articles, the overall greenhouse gas balance varies from significantly negative values (CO_2 captured) to positive values (CO_2 emitted). In this way Brassard et al. (2018) report a GWP of 2561 t CO_2 eq/t of biochar, while Mohammadi et al. (2016) report a carbon footprint of rice production with biochar modification of 3.85 kg CO_2 eq/kg of rice. However, when analyzed in more depth, the results show a clear trend. In all cases, soil improvement systems with biochar, as such, show a clear benefit from a climate change perspective. In situations where greenhouse gas emissions are positive, biochar production is seen as a means of neutralizing the impact of plant production which, in fact, significantly reduces. Greenhouse gas emissions caused by the production and manipulation of biochar are always offset by the benefit of carbon capture in biochar and the compensation of energy production, provided by the use of pyrolysis co-products like syngas and bio-oil. Carbon storage in soil biochar is usually the main engine of carbon sequestration, equaled by energy compensation in some cases (Brassard et al., 2018) depending on the assumptions and conditions of pyrolysis. Other factors contributing to some extent to carbon sequestration include increased yield, reduced fertilizer use, and reduced CH_4 or N_2O emissions from biochar (Table 21.3).

TABLE 21.3 Life cycle impact assessment for biochar systems.

Impact method	Impact categories	Impact results	References
ND	GWP	−691 kg CO_2 eq (influence crop yield) −286 kg CO_2 eq (reduce fertilizer requirements)	Zhu et al. (2022)
CML method	GWP	−240 to −787 kg CO_2 eq (for combustion) −386 to −933 kg CO_2 eq (for soil amendment)	
IPCC 2006	GWP	−392 kg CO_2 eq (BC-400) −429 kg CO_2 eq (BC-800)	
CML 2001	GWP	−921.30 kg CO_2 eq	
NR	GWP	−4.46 kg CO_2 eq	
NR	GWP	−200 to −470 kg CO_2 eq	
IPCC 2013	GWP	21–155 kg CO_2 eq	
IPCC 2007	GWP	−1.06 t CO_2 eq	
IPCC 2006	GWP	−217 g CO_2 eq (cofiring biochar) −84 g CO_2 eq (land amendment)	
IPCC 2006	GWP	−1.101 t CO_2 eq (slow pyrolysis) −1.636 t CO_2 eq (fast pyrolysis)	
IPCC 2006	GWP	−1.122 t CO_2 eq under slow pyrolysis −1.839 t CO_2 eq fast pyrolysis	
ReCiPe endpoint	GWP	−5.86 to −47.15 kg CO_2 eq	
MUIO-LCA model	GWP	-1.41×10^6 t CO_2 eq	
CML 2	GWP ADPe AP EP	−0.62 kg CO_2 eq 2.70×10^{-3} kg Sb eq 2.49×10^{-3} kg SO_2 eq 1.53×10^{-3} kg PO_4^{3-} eq	
ReCiPe Midpoint	GWP	−130 kg CO_2 eq	
ReCiPe Mid-point and End-point	GWP HTP FAETP	3.39×10^6 kg CO_2 eq 1.16×10^6 kg 1.4-dB eq 8.65×10^4 kg 1.4-dB eq	
CML 2	GWP ADP AP EP	11.10 kg CO_2 eq 0.079 kg Sb eq 0.108 kg SO2 eq 0.033 kg PO_4^{-3} eq	
NR	GWP HT TETP AP	0.988 kg CO_2 eq 0.003 kg 1,4−DCB eq 0.082 kg 1,4−DCB eq 0.003 kg SO_2 eq	
CML 2002	GWP	−5.5 kg CO_2 eq	
IPCC 2006	GWP	1.11 kg CO_2 eq (pyrolytic cookstove) 3.85 kg CO_2 eq (drum oven)	
IPCC 2006	GWP	−318 kg CO_2 eq for stove −229 kg CO_2 eq for brick kiln −360 kg CO_2 eq for BigChar 2200 unit	
CML-2001	GWP HTTP AP	2.1×10^{-11} person eq 1.2×10^{-10} person eq 4.1×10^{-11} person eq	

(Continued)

TABLE 21.3 (Continued)

Impact method	Impact categories	Impact results	References
ReCiPe Midpoint	GWP	4.63 kg CO_2 eq	
NR	GWP	0.204 t CO_2 eq (centralized system) 0.141 to 0.217 t CO_2 eq (portable systems)	
IPCC 2013	GWP	-6.3 ± 0.5 t CO_2 eq	
IPCC 1996	GWP	-1100 t CO_2 eq/ha for soil -1900 t CO_2 eq/ha for electricity	
ReCiPe midpoint	GWP	-2.59 to -2.70 t CO_2 eq	
ISO 14,000 series (2002)	GWP HTP20 TETP100 AP EP ADP	0.988 kg CO_2 eq 0.003 kg 1,4-DCB eq 0.082 kg 1,4-DCB eq 0.003 kg SO_2 eq 0.001 kg PO_4^{3-} eq 0.088 kg Sb eq	Patel and Panwar (2023)
CML baseline v4.4	GWP100 AP EP	-8267.32 kg CO_2 eq -28.37 kg SO_2 eq -7.14 kg PO_4^{3-} eq	
unspecified	GWP	-470 to -200 kg CO_2 eq (Without Substitution) -1055 to -770 kg CO_2 eq (With substitution)	
ISO14044	GWP	Production 214–1073 kg CO_2 eq (BSI) 553 kg CO_2 eq (OK) 604 kg CO_2 eq (ACBs) Applied to soil -88 to -2017 kg CO_2 eq (BSI) -1623 kg CO_2 eq (OK) -1944 kg CO_2 eq (ACBs)	
CML-IA baseline v3.06	GWP	2.68×10^3 kg CO_2 eq	
CML 2001	GWP	-920 kg CO2 eq/t	
ISO14040 and ISO14044	GWP	258 kg CO_2 eq/ton rice straw -106 kg CO_2 eq/ton wheat straw	
IPCC 2006	GWP	-1101 kg CO_2 eq/ton (slow pyrolysis) 1839 kg CO_2 eq/ton (fast pyrolysis)	
CML	GWP AP EP ADPe	-613 kg CO2 eq 0.598 kg SO2 eq 0.529 kg PO_4^{3-} eq 3.05×10^{-5} kg Sb eq	
ILCD 2011 midpoint	GWP ODP PMF AP HTP EP	3.70 kg CO_2 eq 5.29×10^{-7} kg CFC-11eqss 1.78×10^{-3} kg $PM_{2.5}$ eq 2.89×10^{-2} mol H^+ eq 2.66×10^{-7} CTUh 8.38×10^{-2} mol N eq	
ILCD 2011 midpoint	GWP ODP PMF AP HTP EP	3.66 kg CO_2 eq 5.23×10^{-7} kg CFC-11eq 1.77×10^{-3} kg $PM_{2.5}$ eq 2.88×10^{-2} mol H^+ eq 2.62×10^{-7} CTUh 8.34×10^{-2} mol N eq	

(Continued)

TABLE 21.3 (Continued)

Impact method	Impact categories	Impact results	References
IPCC (2013)	GWP 100 yr, 20 yr	171 kg CO_2 eq/Mg dry seed (400 °C) 111 kg CO_2 eq/Mg dry seed (800 °C) 638 kg CO_2 eq/Mg dry seed (oilseed rape cultivation without biochar)	Carvalho et al. (2022)
IPCC 2007	GWP	Scenario A (459°C): −2110 kg CO_2 eq/Mg biochar, Scenario B (591°C): 2561 kg CO_2 eq/Mg biochar.	
IMPACT 2002 + CML	GWP	−2.063 t CO_2 eq/t biochar (willow) −472 kg CO_2 eq/t biochar (pig manure) −2.089 t CO_2 eq/t biochar (willow) −466 kg CO_2 eq/t biochar (pig manure).	
IPCC (2013)	GWP	350 kg CO_2 eq/t oat flake (biochar potential) 390 kg CO_2 eq/t oat flake (buffer zone biomass)	
IPCC 2007	GWP	−1.35 Mg CO_2 eq/odt straw	
ND	ADPe	0.05432 kg Sb eq	Anak Erison et al. (2022)
	ADPf	31610 MJ	
	GWP100 y	11510 Kg CO_2 eq	
	ODP	0.0299 kg CFC-11 eq	
	HTP	1661 kg 1,4-DB eq	
	FAETP	6629.89 kg 1,4-DB eq	
	MAETP	1245000 kg 1,4-DB eq	
	TETP	2900.38 kg 1,4-DB eq	
	POCP	4.14 kg C_2H_4 eq	
	AP	26.77 kg SO_2 eq	
	EP	123.06 kg PO_4^{3-} eq	
ReCiPe 2016	GWP	ND	Gallego-Ramírez et al. (2023)
USEtox	HTP		
ReCiPe 2016	ODP		
Environmental impact	GWP	0.33 × 10^3 kg CO_2 eq/t of feedstock	James et al. (2022)
Conservation farming		2.44 × 10^3 kg CO_2 eq/t of feedstock	
Replacement of fossil coal in power plants		1.22 × 10^3 kg CO_2 eq/t of feedstock	
Ecoinvent processes CML baseline (v4.4, January 2015)	AP	0.079 kg SO_2 eq	Marzeddu et al. (2021)
	GWP100	−1505.741 kg CO_2 eq	
	ADPe	8.191 × 10^{-6} kg antimony eq	
	ADPf	177.783 MJ	
	EP	0.017 kg PO_4^{3-} eq	
	FAETP	−0.372 kg 1,4-dichlorobenzene eq	
	HTP	2.003 kg 1,4-dichlorobenzene eq	
	MAETP	−611.397 kg 1,4-dichlorobenzene eq	
	ODP	2.599 × 10^{-6} kg CFC-11 eq	
	POCP	0.001 kg ethylene eq	
	TETP	0.015 kg 1,4-dichlorobenzene eq	

(Continued)

TABLE 21.3 (Continued)

Impact method	Impact categories	Impact results	References
ReCiPe 2016% change to the CF of MC strateg % change to the CF of MCB5 strategy % change to the CF of MCB10 strategy	GWP	9.33 kg CO_2 eq −36.9 kg CO_2 eq −48.2 kg CO_2 eq	Jiang et al. (2021)
	SOD	28.7 kg CFC11 eq −10.4 28.7 kg CFC11 eq −11.2 28.7 kg CFC11 eq	
	IR	−64.6 kBq Co-60 eq −66.2 kBq Co-60 eq −67.9 kBq Co-60 eq	
	HTP	−35.1 kg NOx eq −21.4 kg NOx eq −8.08 kg NOx eq	
	FPMF	−34.20 kg $PM_{2.5}$ eq 3.31 kg $PM_{2.5}$ eq 78.5 kg $PM_{2.5}$ eq	
	ODP/	−38.7 kg NOx eq −27.2 kg NOx eq −16.8 kg NOx eq	
	AP/	140 kg SO_2 eq 158 kg SO_2 eq 298 kg SO_2 eq	
	FAETP	9.39 kg Peq 11.2 kg Peq 13.9 kg Peq	
	MAETP	−78.1 kg Neq −78.6 kg Neq −79.1 kg Neq	
	T-ET	−13.2 kg 1,4-DCB −15.6 kg 1,4-DCB −18.2 kg 1,4-DCB	
	F-ET	8.38 kg 1,4-DCB 9.87 kg 1,4-DCB 11.6 kg 1,4-DCB	
	M-ET	10.2 kg 1,4-DC 11.6 kg 1,4-DC 13.3 kg 1,4-DC	
	HCT	−16.8 kg 1,4-DCB −13.0 kg 1,4-DCB −7.55 kg 1,4-DCB	
	HNCT/	23.1 kg 1,4-DCB 29.1 kg 1,4-DCB 37.1 kg 1,4-DCB	
	LU	−53.4 m^2a crop eq −85.4 m^2a crop eq −123 m^2a crop eq	
	MRS	−0.81 kg Cu eq 6.37 kg Cu eq 15.80 kg Cu eq	
	FRS	−10.6 kg oileq −7.01 kg oileq −1.99 kg oileq	
	WC	−16.8 m^3 −16.1 m^3 −16.1 m^3	

(Continued)

TABLE 21.3 (Continued)

Impact method	Impact categories	Impact results	References
ND	GWP	−100 kg CO_2 eq	Saravanan et al. (2023)
ReCipe	GWP	ND	Sparrevik et al. (2013)
ReCiPe andTool for Reduction and Assessment of Chemicals and other environmental Impacts (TRACI)	GWP	20–50 g CO_2 eq MJ^{-1} the utilization of biofuels leads to >85% reduction in GHG release, equivalent to 93 g CO_2 eq MJ^{-1}	Goswami et al. (2022)
ND	GWP	1 mmol CO_2/g 0.75 mmol CO_2/g 2.53 mmol/g (3 bar, 25 °C) 2.6 mmol/g (250 kPa, 40°C) 4.1 mmol CO_2/g (120 kPa, 25°C; Flue gas feed) 3.8 mmol CO_2/g (120 kPa, 0°C) 7.15 mmol CO_2/g (1000 kPa, 30°C) 4.10 mmol CO_2/g (303°K, 1 bar) 18.1 mg CO_2/g (393°K, 1 atm) 1.745 mmol CO_2/g (30°C)	Song et al. (2023)
IPCC (2013)	GWP 100 yr, 20 yr	171 kg CO_2 eq/Mg dry seed (400°C) 111 kg CO_2 eq/Mg dry seed (800°C) 638 kg CO_2 eq/Mg dry seed for oilseed rape cultivation	Matuštík et al. (2020)
IPCC (2013)	GWP	737 kg CO_2 eq/t feedstock dried	
IPCC– unspecified	GWP	11.8 t CO_2 eq/500 m^3 sewage sludge per day	
CML	GWP AP EP	17.73 t CO_2 eq/ha 1.22 t CO_2 eq/t dry biomass feedstock	
IPCC (2013)	GWP	best scenario (45% yield, 80% stability) 156 kg CO_2 eq/t biochar worst scenario (35% yield, 20% stability) 120 kg CO_2 eq/t biochar	
unspecified	GWP	0.07 t CO_2 eq/t feedstock for cardboard (worst performance) 0.79 t CO_2 eq/t feedstock for sewage Sludge 1.07 t CO_2 eq/t feedstock for food waste and green waste (best performance)	
IPCC 2007	GWP	around 1 t CO_2 eq/odt feedstock	
IPCC 1996	GWP	approx. 40 kg CO_2 eq/Mg feedstock for forest residue and corn fodder	
IPCC 2007	GWP	Scenario A (459°C) 2110 kg CO_2 eq/t biochar Scenario B (591°C) 2561 kg CO_2 eq/t biochar	
CML 2015	GWP, FAETP	approx. 250 kg CO_2 eq/t biochar	
IPCC 2007 (CH4 and N2O) þ nonmethane hydrocarbons and PM from	GWP	1.41 Mg CO_2 eq/t dry feedstock	

(Continued)

TABLE 21.3 (Continued)

Impact method	Impact categories	Impact results	References
ReCiPe 2016 (H) midpoint	GWP	agricultural waste −2.59 to 2.70 t CO_2 eq/biochar (300°C–500°C) forestry waste −2.67 to 2.74 t CO_2 eq/biochar (300°C–500°C)	
CML 2001	GWP	−1670 kg CO_2 eq/t feedstock at 300 C, 1100 kg CO_2 eq/t feedstock at 600 °C	
IPCC 2007	GWP	1.35 Mg CO_2 eq/odt straw	
ReCiPe 2016 (H) endpoint	GWP	720 kg CO_2 eq/household per year (mid-point result)	
UNFCC 2015	GWP	average 489 kg CO_2/Mg biochar (many scenarios)	
IPCC (2013)	GWP	1.11 kg CO_2 eq/kg rice (spring) 3.85 kg CO_2 eq/kg rice (summer)	
IPCC (2013)	GWP	0.27 Mg CO_2 eq/1 Mg rice straw (spring) 0.30 Mg CO_2 eq/1 Mg rice straw (summer)	
CML unspecified	GWP	230 t CO_2 eq/year	

ADPe, depletion of abiotic resources: elements, ultimate reserves; *AP*, Acidification potential; *CML*, center of environmental science; *EP*, eutrophication potential; *FAETP*, freshwater aquatic EcoToxicity potential; *FPMS*, fine particulate matter formation; *FRS*, Fossil resource scarcity; *GWP*, global warming potential; *HCT*, human carcinogenic toxicity; *HTP*, human toxicity potential; *IPCC*, intergovernmental panel on climate change; *MAETP*, marine aquatic EcoToxicity potential; *MRS*, mineral resource scarcity; *ODP*, ozone depletion potential; *PMF*, particulate matter formation; *POCP*, photochemical ozone creation potential; *TETP*, terrestrial ecotoxicity potential.

21.7 Conclusion

This chapter is a review of literature that has already been prereviewed and analyzed in detail regarding the LCA of biochar systems. Although soil modification with biochar is, of course, a method of reducing carbon with great potential and prospects, the understanding of the properties of biochar and the consequent effect on soil and crop yield is still an issue that needs to be confirmed and deepened, therefore it is necessary to rely on further studies. An evaluation of several systems in which biochar is applied is still necessary. There is already a substantial body of literature that evaluates the benefits deriving from the production of biochar but, the great methodological and contextual differences prevent us to directly comparing the results and discovering the rules of cause-effect that apply to all biochar systems. Since the contextual differences, including climate, soil type, raw material or pyrolysis unit details, are difficult to eliminate, the methodology could be unified to allow at least a comparison of the results obtained. The greatest opportunity for the consolidation of methodologies lies in the first phase of the LCA, namely the definition of objectives and scope since the selection of the system boundaries and of the functional units is arbitrary; the same applies to the selection of LCA software. In addition, it is necessary to consider the assessment of effects on impact categories other than climate change, which has been the most widely considered.

To accurately assess the impacts related to biochar production processes, it is essential that the application of the LCA methodology takes place both in the design phase of the plant and after its entry into operation, therefore after obtaining the product and its use for the specific purpose; in this way using an iterative approach through which, following careful monitoring of the process parameters and their impacts, it is possible to highlight their interdependencies, thus evaluating any changes to be made to the plant itself to improve its sustainability.

Possible limitations of the application of the LCA methodology may be related to the type of approach used (i.e., cradle-to-gate, cradle-to-grave, cradle-to-cradle, etc.) which, as already underlined, has significant implications on the defined system boundaries and, therefore, on the characterization of the environmental impacts deriving from the production process and possibly from the use of biochar. It is considered appropriate to always include the downstream phase in the assessment process to characterize the impacts of the entire production process as well as the use phase, highlighting the benefits brought to the environment (e.g., carbon sink).

Despite all the differences, a common trend can be defined: the application of biochar in agricultural soil provides significant benefits throughout the life cycle of the system, with respect to the CO_2 storage capacity, the improvement of soil quality, and the capacity of retention of water and other mineral substances. Although the extent of carbon sequestration differs between projects, it is clear that the development of projects involving the application of biochar in soil could be a further step toward a carbon-neutral society. In addition, it is clear that the use of agricultural and forestry waste for the production of biochar is a good way to optimize its use and avoid further negative emissions; the effort in trying to use these residues for the production of biochar has also led to the deepening of what are the conversion techniques to be used to obtain it, leading not only to the conclusion that some types of applications (slow pyrolysis) are better than others in the overall impact balance but are also more easily applicable.

Studies have been analyzed in which biochar was not used as a soil improver but as a component of other materials, such as cement, also bringing good results in terms of carbon dioxide capture in these cases. This conclusion obviously bodes well for what will be the possible future applications of this product.

If the characteristics and capabilities of biochar are consolidated and confirmed by further studies and trials, its applications could extend to much wider horizons than those of today, resulting in a growth of confidence in its use and consequently in demand. The actual benefits that biochar can bring in its various applications will be able to extend, in the near future, the use of this product, also allowing the experimentation of other raw materials to be used in increasingly efficient transformation processes.

Long-term field experiments of biochar are essential to evaluate the real direct and indirect effects of its use in soil and other applications, to confirm its potential and make it more common and widespread in use than it is today.

References

Ağbulut, Ü., Sirohi, R., Lichtfouse, E., et al. (2023). Microalgae bio-oil production by pyrolysis and hydrothermal liquefaction: Mechanism and characteristics. *Bioresource Technology, 376*, 128860. Available from https://doi.org/10.1016/j.biortech.2023.128860.

Agyekum, E. B., & Nutakor, C. (2024). Recent advancement in biochar production and utilization — A combination of traditional and bibliometric review. *International Journal of Hydrogen Energy, 54*, 1137−1153. Available from https://doi.org/10.1016/j.ijhydene.2023.11.335.

Anak Erison, A. E., Tan, Y. H., Mubarak, N. M., et al. (2022). Life cycle assessment of biodiesel production by using impregnated magnetic biochar derived from waste palm kernel shell. *Environmental Research, 214*, 114149. Available from https://doi.org/10.1016/j.envres.2022.114149.

Antwi-Afari, P., Ng, S. T., Chen, J., et al. (2023). Enhancing life cycle assessment for circular economy measurement of different case scenarios of modular steel slab. *Building and Environment, 239*, 110411. Available from https://doi.org/10.1016/j.buildenv.2023.110411.

Azzi, E. S., Karltun, E., & Sundberg, C. (2019). Prospective life cycle assessment of large-scale biochar production and use for negative emissions in Stockholm. *Environmental Science & Technology, 53*, 8466−8476. Available from https://doi.org/10.1021/acs.est.9b01615.

Barbhuiya, S., Bhusan Das, B., & Kanavaris, F. (2024). Biochar-concrete: A comprehensive review of properties, production and sustainability. *Case Studies in Construction Materials, 20*, e02859. Available from https://doi.org/10.1016/j.cscm.2024.e02859.

Blenis, N., Hue, N., Maaz, T. M., & Kantar, M. (2023). Biochar Production, Modification, and Its Uses in Soil Remediation: A Review. *Sustainability, 15*, 3442. Available from https://doi.org/10.3390/su15043442.

Bolan, N., Hoang, S. A., Beiyuan, J., et al. (2022). Multifunctional applications of biochar beyond carbon storage. *International Materials Reviews, 67*, 150−200. Available from https://doi.org/10.1080/09506608.2021.1922047.

Bolan, N., Sarmah, A. K., Bordoloi, S., et al. (2023). Soil acidification and the liming potential of biochar. *Environmental Pollution, 317*, 120632. Available from https://doi.org/10.1016/j.envpol.2022.120632.

Boni, M., Chiavola, A., & Marzeddu, S. (2018a). Application of Biochar to the Remediation of Pb-Contaminated Solutions. *Sustainability, 10*, 4440. Available from https://doi.org/10.3390/su10124440.

Boni, M., Marzeddu, S., Tatti, F., et al. (2021). Experimental and numerical study of biochar fixed bed column for the adsorption of arsenic from aqueous solutions. *Water, 13*, 915. Available from https://doi.org/10.3390/w13070915.

Boni, M. R., Chiavola, A., Antonucci, A., et al. (2018). A novel treatment for Cd-contaminated solution through adsorption on beech charcoal: the effect of bioactivation. *dwt, 127*, 104−110. Available from https://doi.org/10.5004/dwt.2018.22664.

Boni, M. R., Chiavola, A., & Marzeddu, S. (2020). Remediation of lead-contaminated water by virgin coniferous wood biochar adsorbent: batch and column application. *Water, Air, & Soil Pollution, 231*, 171. Available from https://doi.org/10.1007/s11270-020-04496-z.

Brassard, P., Godbout, S., & Hamelin, L. (2021). Framework for consequential life cycle assessment of pyrolysis biorefineries: A case study for the conversion of primary forestry residues. *Renewable and Sustainable Energy Reviews, 138*, 110549. Available from https://doi.org/10.1016/j.rser.2020.110549.

Brassard, P., Godbout, S., Pelletier, F., et al. (2018). Pyrolysis of switchgrass in an auger reactor for biochar production: A greenhouse gas and energy impacts assessment. *Biomass and Bioenergy, 116*, 99−105. Available from https://doi.org/10.1016/j.biombioe.2018.06.007.

Bridges, E. M., & Oldeman, L. R. (1999). Global assessment of human-induced soil degradation. *Arid Soil Research and Rehabilitation, 13*, 319−325. Available from https://doi.org/10.1080/089030699263212.

Bruno, M., Marchi, M., Ermini, N., et al. (2023). Life cycle assessment and cost−benefit analysis as combined economic−environmental assessment tools: Application to an anaerobic digestion plant. *Energies, 16*, 3686. Available from https://doi.org/10.3390/en16093686.

Campion, L., Bekchanova, M., Malina, R., & Kuppens, T. (2023). The costs and benefits of biochar production and use: A systematic review. *Journal of Cleaner Production, 408*, 137138. Available from https://doi.org/10.1016/j.jclepro.2023.137138.

Carvalho, J., Nascimento, L., Soares, M., et al. (2022). Life cycle assessment (LCA) of biochar production from a circular economy perspective. *Processes, 10*, 2684. Available from https://doi.org/10.3390/pr10122684.

Cayuela, M. L., Jeffery, S., & Van Zwieten, L. (2015). The molar H:Corg ratio of biochar is a key factor in mitigating N_2O emissions from soil. *Agriculture, ecosystems & environment, 202*, 135–138. Available from https://doi.org/10.1016/j.agee.2014.12.015.

Chavando, J. A. M., Silva, V., Puig-gamero, M., et al. (2023). Chapter 6 - Simulation of biomass to syngas: Pyrolysis and gasification processes. In M. R. Rahimpour, M. A. Makarem, & M. Meshksar (Eds.), *Advances in Synthesis Gas : Methods, Technologies and Applications* (pp. 159–196). Elsevier.

Chen, W., Zhou, Y., Chen, G., et al. (2019). Alkali Chlorides for the Suppression of the Interfacial Recombination in Inverted Planar Perovskite Solar Cells. *Advanced Energy Materials, 9*, 1803872. Available from https://doi.org/10.1002/aenm.201803872.

Cheng, F., Luo, H., & Colosi, L. M. (2020). Slow pyrolysis as a platform for negative emissions technology: An integration of machine learning models, life cycle assessment, and economic analysis. *Energy Conversion and Management, 223*, 113258. Available from https://doi.org/10.1016/j.enconman.2020.113258.

Chiavola, A., Marzeddu, S., & Boni, M. R. (2020). Remediation of Water Contaminated by Pb(II) Using Virgin Coniferous Wood Biochar as Adsorbent. In V. Naddeo, M. Balakrishnan, & K.-H. Choo (Eds.), *Frontiers in Water-Energy-Nexus—Nature-Based Solutions, Advanced Technologies and Best Practices for Environmental Sustainability* (pp. 363–366). Cham: Springer International Publishing.

Clare, A., Shackley, S., Joseph, S., et al. (2015). Competing uses for China's straw: the economic and carbon abatement potential of biochar. *GCB Bioenergy, 7*, 1272–1282. Available from https://doi.org/10.1111/gcbb.12220.

Cointe, B., & Guillemot, H. (2023). A history of the 1.5°C target. *WIREs Climate Change, 14*, e824. Available from https://doi.org/10.1002/wcc.824.

Dai, Y., Zheng, H., Jiang, Z., & Xing, B. (2020). Combined effects of biochar properties and soil conditions on plant growth: A meta-analysis. *Science of The Total Environment, 713*, 136635. Available from https://doi.org/10.1016/j.scitotenv.2020.136635.

Décima, M. A., Marzeddu, S., Barchiesi, M., et al. (2021). A review on the removal of carbamazepine from aqueous solution by using activated carbon and biochar. *Sustainability, 13*, 11760. Available from https://doi.org/10.3390/su132111760.

Dietz, S., Bowen, A., Doda, B., et al. (2018). The economics of 1.5°C climate change. *Annual Review of Environment and Resources, 43*, 455–480. Available from https://doi.org/10.1146/annurev-environ-102017-025817.

Ding, N., Meng, X., Zhang, Z., et al. (2023). A review of life cycle assessment of soil remediation technology: method applications and technological characteristics. *Reviews EnvContamination (formerly:Residue Reviews), 262*, 4. Available from https://doi.org/10.1007/s44169-023-00051-z.

European Commission. (2023). *Directorate General for Internal Market, Industry, Entrepreneurship and SMEs. Study on the critical raw materials for the EU 2023: final report.* Publications Office, LU.

Fakayode, O. A., Wahia, H., Zhang, L., et al. (2023). State-of-the-art co-pyrolysis of lignocellulosic and macroalgae biomass feedstocks for improved bio-oil production- A review. *Fuel, 332*, 126071. Available from https://doi.org/10.1016/j.fuel.2022.126071.

Ferronato, N., Paoli, R., Romagnoli, F., et al. (2023). Environmental impact scenarios of organic fraction municipal solid waste treatment with Black Soldier Fly larvae based on a life cycle assessment. *Environ Sci Pollut Res, 31*, 17651–17669. Available from https://doi.org/10.1007/s11356-023-27140-9.

Gallego-Ramírez, C., Chica, E., & Rubio-Clemente, A. (2023). Life cycle assessment of raw and fe-modified biochars: contributing to circular economy. *Materials, 16*, 6059. Available from https://doi.org/10.3390/ma16176059.

Gaur, V. K., Gautam, K., Sharma, P., et al. (2022). Carbon-based catalyst for environmental bioremediation and sustainability: Updates and perspectives on techno-economics and life cycle assessment. *Environmental Research, 209*, 112793. Available from https://doi.org/10.1016/j.envres.2022.112793.

Giuntoli, J., Caserini, S., Marelli, L., Baxter, D., & Agostini, A. (2015). Domestic heating from forest logging residues: environmental risks and benefits. *Journal of Cleaner Production, 99*, 206–216. Available from https://doi.org/10.1016/j.jclepro.2015.03.025.

Gong, X., Kung, C.-C., & Zhang, L. (2021). An economic evaluation on welfare distribution and carbon sequestration under competitive pyrolysis technologies. *Energy Exploration & Exploitation, 39*, 553–570. Available from https://doi.org/10.1177/0144598719900279.

Goswami, L., Kushwaha, A., Singh, A., et al. (2022). Nano-biochar as a sustainable catalyst for anaerobic digestion: A synergetic closed-loop approach. *Catalysts, 12*, 186. Available from https://doi.org/10.3390/catal12020186.

Grams, J., Jankowska, A., & Goscianska, J. (2023). Advances in design of heterogeneous catalysts for pyrolysis of lignocellulosic biomass and bio-oil upgrading. *Microporous and Mesoporous Materials, 362*, 112761. Available from https://doi.org/10.1016/j.micromeso.2023.112761.

Groen, E. A., Bokkers, E. A. M., Heijungs, R., & De Boer, I. J. M. (2017). Methods for global sensitivity analysis in life cycle assessment. *Int J Life Cycle Assess, 22*, 1125–1137. Available from https://doi.org/10.1007/s11367-016-1217-3.

Hoegh-Guldberg, O., Jacob, D., Taylor, M., et al. (2019). The human imperative of stabilizing global climate change at 1.5°C. *Science, 365*, eaaw6974. Available from https://doi.org/10.1126/science.aaw6974.

Huang, K., Li, M., Li, R., et al. (2023). Soil acidification and salinity: the importance of biochar application to agricultural soils. *Frontiers in Plant Science, 14*, 1206820. Available from https://doi.org/10.3389/fpls.2023.1206820.

Hussain, A. J., Al-Taey, D. K. A., & Kadhum, H. J. (2023). Biochar application increases the amount of nitrogen, phosphorus and potassium in the soil: A review. *IOP Conf Ser: Earth Environ Sci, 1213*, 012023. Available from https://doi.org/10.1088/1755-1315/1213/1/012023.

Hussin, F., Hazani, N. N., Khalil, M., & Aroua, M. K. (2023). Environmental life cycle assessment of biomass conversion using hydrothermal technology: A review. *Fuel Processing Technology, 246*, 107747. Available from https://doi.org/10.1016/j.fuproc.2023.107747.

James, A., Sánchez, A., Prens, J., & Yuan, W. (2022). Biochar from agricultural residues for soil conditioning: Technological status and life cycle assessment. *Current Opinion in Environmental Science & Health*, 25, 100314. Available from https://doi.org/10.1016/j.coesh.2021.100314.

Jayakumar, M., Hamda, A. S., Abo, L. D., et al. (2023). Comprehensive review on lignocellulosic biomass derived biochar production, characterization, utilization and applications. *Chemosphere*, 345, 140515. Available from https://doi.org/10.1016/j.chemosphere.2023.140515.

Jiang, B.-N., Lu, M.-B., Zhang, Z.-Y., et al. (2023). Quantifying biochar-induced greenhouse gases emission reduction effects in constructed wetlands and its heterogeneity: A multi-level meta-analysis. *Science of The Total Environment*, 855, 158688. Available from https://doi.org/10.1016/j.scitotenv.2022.158688.

Jiang, Z., Zheng, H., & Xing, B. (2021). Environmental life cycle assessment of wheat production using chemical fertilizer, manure compost, and biochar-amended manure compost strategies. *Science of The Total Environment*, 760, 143342. Available from https://doi.org/10.1016/j.scitotenv.2020.143342.

Kumar Mishra, R., Jaya Prasanna Kumar, D., Narula, A., et al. (2023). Production and beneficial impact of biochar for environmental application: A review on types of feedstocks, chemical compositions, operating parameters, techno-economic study, and life cycle assessment. *Fuel*, 343, 127968. Available from https://doi.org/10.1016/j.fuel.2023.127968.

Kuramochi, T., Höhne, N., Schaeffer, M., et al. (2018). Ten key short-term sectoral benchmarks to limit warming to 1.5°C. *Climate Policy*, 18, 287−305. Available from https://doi.org/10.1080/14693062.2017.1397495.

Li, L., Long, A., Fossum, B., & Kaiser, M. (2023). Effects of pyrolysis temperature and feedstock type on biochar characteristics pertinent to soil carbon and soil health: A meta-analysis. *Soil Use and Management*, 39, 43−52. Available from https://doi.org/10.1111/sum.12848.

Li, S., Wu, J., Wang, X., & Ma, L. (2020). Economic and environmental sustainability of maize-wheat rotation production when substituting mineral fertilizers with manure in the North China Plain. *Journal of Cleaner Production*, 271, 122683. Available from https://doi.org/10.1016/j.jclepro.2020.122683.

Li, X., Wang, C., Zhang, J., et al. (2020). Preparation and application of magnetic biochar in water treatment: A critical review. *Science of the Total Environment*, 711, 134847. Available from https://doi.org/10.1016/j.scitotenv.2019.134847.

Lima, M. A., Mendes, L. F. R., Mothé, G. A., et al. (2020). Renewable energy in reducing greenhouse gas emissions: Reaching the goals of the Paris agreement in Brazil. *Environmental Development*, 33, 100504. Available from https://doi.org/10.1016/j.envdev.2020.100504.

Liu, Q., Meki, K., Zheng, H., et al. (2023). Biochar application in remediating salt-affected soil to achieve carbon neutrality and abate climate change. *Biochar*, 5, 45. Available from https://doi.org/10.1007/s42773-023-00244-8.

Lu, Y., Khan, Z. A., Alvarez-Alvarado, M. S., et al. (2020). A Critical Review of Sustainable Energy Policies for the Promotion of Renewable Energy Sources. *Sustainability*, 12, 5078. Available from https://doi.org/10.3390/su12125078.

Manikandan, S., Vickram, S., Subbaiya, R., et al. (2023). Comprehensive review on recent production trends and applications of biochar for greener environment. *Bioresource Technology*, 388, 129725. Available from https://doi.org/10.1016/j.biortech.2023.129725.

Marzeddu, S., Cappelli, A., Ambrosio, A., et al. (2021). A life cycle assessment of an energy-biochar chain involving a gasification plant in Italy. *Land*, 10, 1256. Available from https://doi.org/10.3390/land10111256.

Marzeddu, S., Cappelli, A., Ferraro, A., et al. (2024). The Use of Life Cycle Assessment for the Characterization of the Energy and Environmental Sustainability of a Biochar Production Process: The Case Study of the Nera Biochar™ Plant (Turin, Italy). In M. Ksibi, A. Sousa, O. Hentati, et al. (Eds.), *Recent Advances in Environmental Science from the Euro-Mediterranean and Surrounding Regions* (4th Edition, pp. 587−591). Cham: Springer Nature Switzerland.

Marzeddu, S., Cappelli, A., Paoli, R., et al. (2023). LCA sensitivity analysis of an energy-biochar chain from an italian gasification plant: environmental trade-offs assessment. *CONECT*, 63. Available from https://doi.org/10.7250/CONECT.2023.043.

Marzeddu, S., Décima, M. A., Camilli, L., et al. (2022). Physical-chemical characterization of different carbon-based sorbents for environmental applications. *Materials*, 15, 7162. Available from https://doi.org/10.3390/ma15207162.

Matthews, H. D., & Wynes, S. (2022). Current global efforts are insufficient to limit warming to 1.5°C. *Science*, 376, 1404−1409. Available from https://doi.org/10.1126/science.abo3378.

Matuštík, J., Hnátková, T., & Kočí, V. (2020). Life cycle assessment of biochar-to-soil systems: A review. *Journal of Cleaner Production*, 259, 120998. Available from https://doi.org/10.1016/j.jclepro.2020.120998.

Melo, V. M. e, Ferreira, G. F., & Fregolente, L. V. (2024). Sustainable catalysts for biodiesel production: The potential of CaO supported on sugarcane bagasse biochar. *Renewable and Sustainable Energy Reviews*, 189, 114042. Available from https://doi.org/10.1016/j.rser.2023.114042.

Mohammadi, A., Cowie, A., Anh Mai, T. L., et al. (2016). Biochar use for climate-change mitigation in rice cropping systems. *Journal of Cleaner Production*, 116, 61−70. Available from https://doi.org/10.1016/j.jclepro.2015.12.083.

Mohammadi, A., Cowie, A. L., Anh Mai, T. L., et al. (2017). Climate-change and health effects of using rice husk for biochar-compost: Comparing three pyrolysis systems. *Journal of Cleaner Production*, 162, 260−272. Available from https://doi.org/10.1016/j.jclepro.2017.06.026.

Mohammadi, A., Khoshnevisan, B., Venkatesh, G., & Eskandari, S. (2020). A critical review on advancement and challenges of biochar application in paddy fields: environmental and life cycle cost analysis. *Processes*, 8, 1275. Available from https://doi.org/10.3390/pr8101275.

Mosa, A., Mansour, M. M., Soliman, E., et al. (2023). Biochar as a soil amendment for restraining greenhouse gases emission and improving soil carbon sink: current situation and ways forward. *Sustainability*, 15, 1206. Available from https://doi.org/10.3390/su15021206.

Olabi, A. G., & Abdelkareem, M. A. (2022). Renewable energy and climate change. *Renewable and Sustainable Energy Reviews*, 158, 112111. Available from https://doi.org/10.1016/j.rser.2022.112111.

Osman, A. I., Farghali, M., & Rashwan, A. K. (2024). Life cycle assessment of biochar as a green sorbent for soil remediation. *Current Opinion in Green and Sustainable Chemistry*, 100882. Available from https://doi.org/10.1016/j.cogsc.2024.100882.

Østergaard, P. A., Duic, N., Noorollahi, Y., & Kalogirou, S. (2022). Renewable energy for sustainable development. *Renewable Energy, 199*, 1145–1152. Available from https://doi.org/10.1016/j.renene.2022.09.065.

Owusu, P. A., & Asumadu-Sarkodie, S. (2016). A review of renewable energy sources, sustainability issues and climate change mitigation. *Cogent Engineering, 3*, 1167990. Available from https://doi.org/10.1080/23311916.2016.1167990.

Patel, M. R., & Panwar, N. L. (2023). Biochar from agricultural crop residues: Environmental, production, and life cycle assessment overview. *Resources, Conservation & Recycling Advances, 19*, 200173. Available from https://doi.org/10.1016/j.rcradv.2023.200173.

Perreault, P., Boruntea, C.-R., Dhawan Yadav, H., et al. (2023). Combined methane pyrolysis and solid carbon gasification for electrified CO2-Free hydrogen and syngas production. *Energies, 16*, 7316. Available from https://doi.org/10.3390/en16217316.

Pigou, A. C. (1933). The economics of welfare. *The Economic Journal, 43*, 329. Available from https://doi.org/10.2307/2224491.

Potnuri, R., Rao, C. S., Surya, D. V., et al. (2023). Two-step synthesis of biochar using torrefaction and microwave-assisted pyrolysis: Understanding the effects of torrefaction temperature and catalyst loading. *Journal of Analytical and Applied Pyrolysis, 175*, 106191. Available from https://doi.org/10.1016/j.jaap.2023.106191.

Pourhashem, G., Spatari, S., Boateng, A. A., et al. (2013). Life cycle environmental and economic tradeoffs of using fast pyrolysis products for power generation. *Energy and Fuels, 27*, 2578–2587. Available from https://doi.org/10.1021/ef3016206.

Qian, S., Zhou, X., Fu, Y., et al. (2023). Biochar-compost as a new option for soil improvement: Application in various problem soils. *Science of The Total Environment, 870*, 162024. Available from https://doi.org/10.1016/j.scitotenv.2023.162024.

Raut, N. A., Kokare, D. M., Randive, K. R., et al. (2023). 1 - Introduction: fundamentals of waste removal technologies. In N. A. Raut, D. M. Kokare, B. A. Bhanvase, et al. (Eds.), *360-Degree Waste Management* (Volume 1, pp. 1–16). Elsevier.

Roberts, K. G., Gloy, B. A., Joseph, S., et al. (2010). Life Cycle Assessment of Biochar Systems: Estimating the Energetic, Economic, and Climate Change Potential. *Environmental Science & Technology, 44*, 827–833. Available from https://doi.org/10.1021/es902266r.

Rodrigues, C. I. D., Brito, L. M., & Nunes, L. J. R. (2023). Soil carbon sequestration in the context of climate change mitigation: A review. *Soil Systems, 7*, 64. Available from https://doi.org/10.3390/soilsystems7030064.

Roe, S., Streck, C., Obersteiner, M., et al. (2019). Contribution of the land sector to a 1.5°C world. *Nat Clim Chang, 9*, 817–828. Available from https://doi.org/10.1038/s41558-019-0591-9.

Safarian, S. (2023). Performance analysis of sustainable technologies for biochar production: A comprehensive review. *Energy Reports, 9*, 4574–4593. Available from https://doi.org/10.1016/j.egyr.2023.03.111.

Saravanan, A., Karishma, S., Senthil Kumar, P., & Rangasamy, G. (2023). A review on regeneration of biowaste into bio-products and bioenergy: Life cycle assessment and circular economy. *Fuel, 338*, 127221. Available from https://doi.org/10.1016/j.fuel.2022.127221.

Seneviratne, S. I., Rogelj, J., Séférian, R., et al. (2018). The many possible climates from the Paris Agreement's aim of 1.5 °C warming. *Nature, 558*, 41–49. Available from https://doi.org/10.1038/s41586-018-0181-4.

Singh Yadav, S. P., Bhandari, S., Bhatta, D., et al. (2023). Biochar application: A sustainable approach to improve soil health. *Journal of Agriculture and Food Research, 11*, 100498. Available from https://doi.org/10.1016/j.jafr.2023.100498.

Sokka, L., Sinkko, T., Holma, A., et al. (2016). Environmental impacts of the national renewable energy targets – A case study from Finland. *Renewable and Sustainable Energy Reviews, 59*, 1599–1610. Available from https://doi.org/10.1016/j.rser.2015.12.005.

Song, S., Liu, Z., Liu, G., et al. (2023). Application of biochar cement-based materials for carbon sequestration. *Construction and Building Materials, 405*, 133373. Available from https://doi.org/10.1016/j.conbuildmat.2023.133373.

Sparrevik, M., Field, J. L., Martinsen, V., et al. (2013). Life cycle assessment to evaluate the environmental impact of biochar implementation in conservation agriculture in Zambia. *Environmental Science & Technology, 47*, 1206–1215. Available from https://doi.org/10.1021/es302720k.

Sparrevik, M., Lindhjem, H., Andria, V., et al. (2014). Environmental and Socioeconomic Impacts of Utilizing Waste for Biochar in Rural Areas in Indonesia–A Systems Perspective. *Environmental Science & Technology, 48*, 4664–4671. Available from https://doi.org/10.1021/es405190q.

Thers, H., Djomo, S. N., Elsgaard, L., & Knudsen, M. T. (2019). Biochar potentially mitigates greenhouse gas emissions from cultivation of oilseed rape for biodiesel. *Science of The Total Environment, 671*, 180–188. Available from https://doi.org/10.1016/j.scitotenv.2019.03.257.

Viotti, P., Marzeddu, S., Antonucci, A., et al. (2024). Biochar as alternative material for heavy metal adsorption from groundwaters: lab-scale (Column) experiment review. *Materials, 17*, 809. Available from https://doi.org/10.3390/ma17040809.

Viotti, P., Tatti, F., Rossi, A., et al. (2020). An eco-balanced and integrated approach for a more-sustainable MSW management. *Waste Biomass Valor, 11*, 5139–5150. Available from https://doi.org/10.1007/s12649-020-01091-5.

Wang, B., Hu, J., Chen, W., et al. (2024). Exploring the characteristics of coke formation on biochar-based catalysts during the biomass pyrolysis. *Fuel, 357*, 129859. Available from https://doi.org/10.1016/j.fuel.2023.129859.

Wang, H., & Qiu, F. (2017). Investigating the impact of agricultural land losses on deforestation: evidence from a peri-urban area in Canada. *Ecological Economics, 139*, 9–18. Available from https://doi.org/10.1016/j.ecolecon.2017.04.002.

Wang, J., Ahmad, I., Ghaffar, A., et al. (2022). Activated biochar is an effective technique for arsenic removal from contaminated drinking water in Pakistan. *Sustainability, 14*. Available from https://doi.org/10.3390/SU142114523, 14523 14:14523.

Wei, B., Peng, Y., Lin, L., et al. (2023). Drivers of biochar-mediated improvement of soil water retention capacity based on soil texture: A meta-analysis. *Geoderma, 437*, 116591. Available from https://doi.org/10.1016/j.geoderma.2023.116591.

Wise, J. (2023). COP28 decision to "transition away" from fossil fuels is hailed as milestone but loopholes are decried. *BMJ 383*, p2941. Available from https://doi.org/10.1136/bmj.p2941.

Xia, F., Zhang, Z., Zhang, Q., et al. (2024). Life cycle assessment of greenhouse gas emissions for various feedstocks-based biochars as soil amendment. *Science of The Total Environment, 911*, 168734. Available from https://doi.org/10.1016/j.scitotenv.2023.168734.

Xu, P., Gao, Y., Cui, Z., et al. (2023). Research progress on effects of biochar on soil environment and crop nutrient absorption and utilization. *Sustainability, 15*, 4861. Available from https://doi.org/10.3390/su15064861.

Yang, X., Han, D., Zhao, Y., et al. (2020). Environmental evaluation of a distributed-centralized biomass pyrolysis system: A case study in Shandong, China. *Science of The Total Environment, 716*, 136915. Available from https://doi.org/10.1016/j.scitotenv.2020.136915.

Yasin, G., Nawaz, M. F., Zubair, M., et al. (2023). Role of traditional agroforestry systems in climate change mitigation through carbon sequestration: an investigation from the semi-arid region of Pakistan. *Land, 12*, 513. Available from https://doi.org/10.3390/land12020513.

Younis, S. A., & Kim, K.-H. (2022). Recent advances in biochar-based catalysts: air purification and opportunities for industrial upscaling. *Asian Journal of Atmospheric Environment, 16*, 2022117. Available from https://doi.org/10.5572/ajae.2022.117.

Yu, Z., Ma, H., Liu, X., et al. (2022). Review in life cycle assessment of biomass conversion through pyrolysis-issues and recommendations. *Green Chemical Engineering, 3*, 304–312. Available from https://doi.org/10.1016/j.gce.2022.08.002.

Zhong, J., Wei, Y., Wan, H., et al. (2013). Greenhouse gas emission from the total process of swine manure composting and land application of compost. *Atmospheric Environment, 81*, 348–355. Available from https://doi.org/10.1016/j.atmosenv.2013.08.048.

Zhu, X., Labianca, C., He, M., et al. (2022). Life-cycle assessment of pyrolysis processes for sustainable production of biochar from agro-residues. *Bioresource Technology, 360*, 127601. Available from https://doi.org/10.1016/j.biortech.2022.127601.

Zou, X., Zhai, M., Liu, G., et al. (2024). In-depth understanding of the microscopic mechanism of biochar carbonaceous structures during thermochemical conversion: Pyrolysis, combustion and gasification. *Fuel, 361*, 130732. Available from https://doi.org/10.1016/j.fuel.2023.130732.

Chapter 22

Potential environmental and human health risks of biochar systems: a call for comprehensive health risk assessments

Willis Gwenzi[1,2,3]

[1]Formerly Alexander von Humboldt Fellow and Guest Full Professor, Leibniz-Institut für Agrartechnik und Bioökonomie e.V. (ATB), Potsdam, Germany,
[2]Formerly Alexander von Humboldt Fellow and Guest Full Professor, Grassland Grassland Science and Renewable Plant Resources, Faculty of Organic Agricultural Sciences, Universität Kassel, Witzenhausen, Germany, [3]Biosystems and Environmental Engineering Research Group, Marlborough, Harare, Zimbabwe

Chapter outline

22.1 Introduction	433
22.2 Environmental health risks	434
22.2.1 Soil pollution	435
22.2.2 Water pollution	436
22.2.3 Air pollution	436
22.2.4 Radiative forcing/radiative forcing by black carbon/biochar	436
22.2.5 Ecological health risks	437
22.3 Human exposure and health risks	437
22.4 A call for health risk assessments and mitigation	437
22.4.1 Health risk assessment framework	437
22.4.2 Principles of health risk assessment	438
22.4.3 Qualitative risk assessment	438
22.4.4 Quantitative risk assessment	439
22.4.5 Quantitative human health risk assessment tools	439
22.4.6 Ecotoxicological risk assessments	441
22.4.7 Biochar certification systems	441
22.4.8 Status of research on risk assessment of biochar systems	441
22.5 Future research directions	442
22.6 Conclusions and outlook	442
References	443

22.1 Introduction

Biochar technology, encompassing its production and subsequent industrial applications, is receiving significant public and research attention worldwide. The global interest in biochar is driven by its low cost, ease of production, renewable nature, and abundant availability of biomass feedstock. Furthermore, biochar has diverse industrial applications in agriculture, bioenergy, and environmental remediation, with potential benefits in mitigating greenhouse gas emissions.

Several studies, including original articles, reviews, and books on biochar, have been published. These studies include overview articles on biochar production, activation, and its applications (Gwenzi et al., 2015; Wang & Wang, 2019; Xie et al., 2022). Other reviews have focused on applications of biochars in specific domains, including water and wastewater treatment (Gwenzi et al., 2017; Kamali et al., 2021; Xiang et al., 2020), air pollution control (Zhao et al., 2022; Gwenzi et al., 2021), greenhouse gas emission mitigation (Lyu et al., 2022; Yin et al., 2021), and soil remediation (Ji et al., 2022; Zama et al., 2018). Several studies, including books, have investigated various aspects of biochar and its applications (Ramola et al., 2022; Singh and Singh et al., 2020; Singh & Singh, 2015, 2024). However, the literature on biochar predominantly focuses on its production methods and various industrial applications. Numerous original studies have investigated biomass feedstock and biochar production using various thermochemical and hydrothermal conversion pathways, such as pyrolysis, gasification, and hydrothermal carbonization. Others have investigated biochar applications in agriculture, soil remediation, wastewater and water treatment, soil carbon sequestration, greenhouse gas mitigation, and bioenergy production. Despite these beneficial applications, biochar systems may present potential health hazards.

Compared to its industrial applications, research into the potential health risks and health risk assessments of biochar systems remains limited. Nevertheless, an increasing body of evidence suggests that biochars and their associated contaminants, such as toxic metals and dioxins, may pose health risks. Studies also indicate potential soil, water, and air pollution from biochar particulates and their associated contaminants. Additionally, there are concerns about the potential radiative forcing and radiative impacts of black carbon emitted from biochar systems. These hazards underscore the necessity of conducting health risk assessments of biochar systems.

To date, comprehensive reviews on the health hazards and risk assessment of biochar systems are still scarce. Therefore, this chapter aims to address the following objectives: (1) to discuss the environmental, ecological, and human health hazards associated with biochar systems; (2) to propose risk assessment methodologies based on physicochemical characterization, ecotoxicological bioassays, and qualitative and QRAs; and (3) to highlight future research needs concerning health hazards and the health risk assessment of biochars. Fig. 22.1 presents a summary of this chapter, highlighting the nature of biochar-derived contaminants, potential pollution and health risks, health risk assessments, and future research directions.

22.2 Environmental health risks

Biochar systems, including biomass feedstock, production, logistics, and applications, may pose potential health risks or hazards (Fig. 22.2). Fig. 22.2 presents an overview of the biochar system and its associated health risks. The feedstock system involves the production, harvesting, logistics, and pretreatment. Biochar production entails various thermohydrochemical processes, including pyrolysis systems, gasification, and hydrothermal conversion. The risks associated with each stage of the biochar system are also shown (Fig. 22.2).

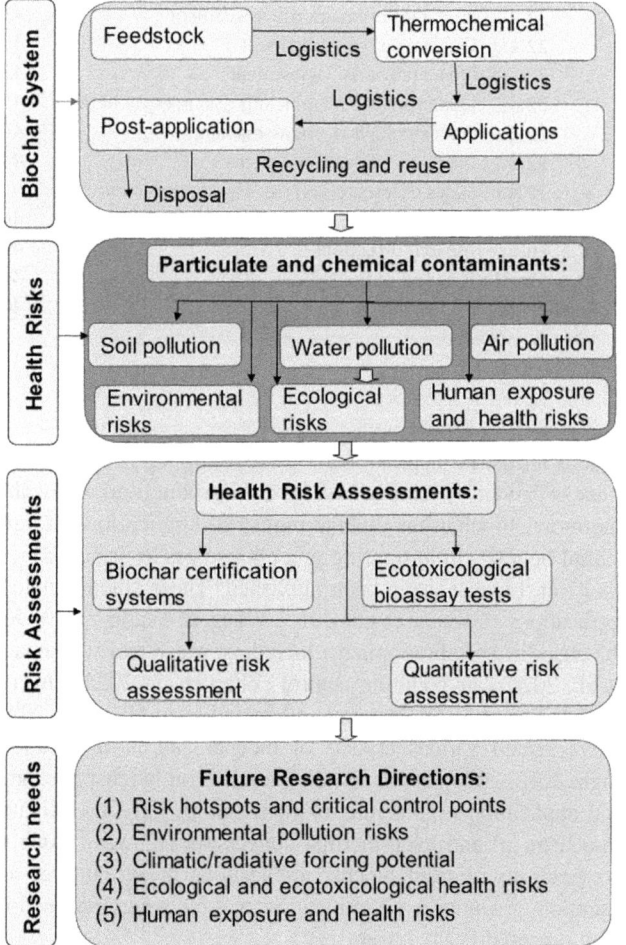

FIGURE 22.1 Overview of potential health risks, risk assessments, and future research directions for biochar systems.

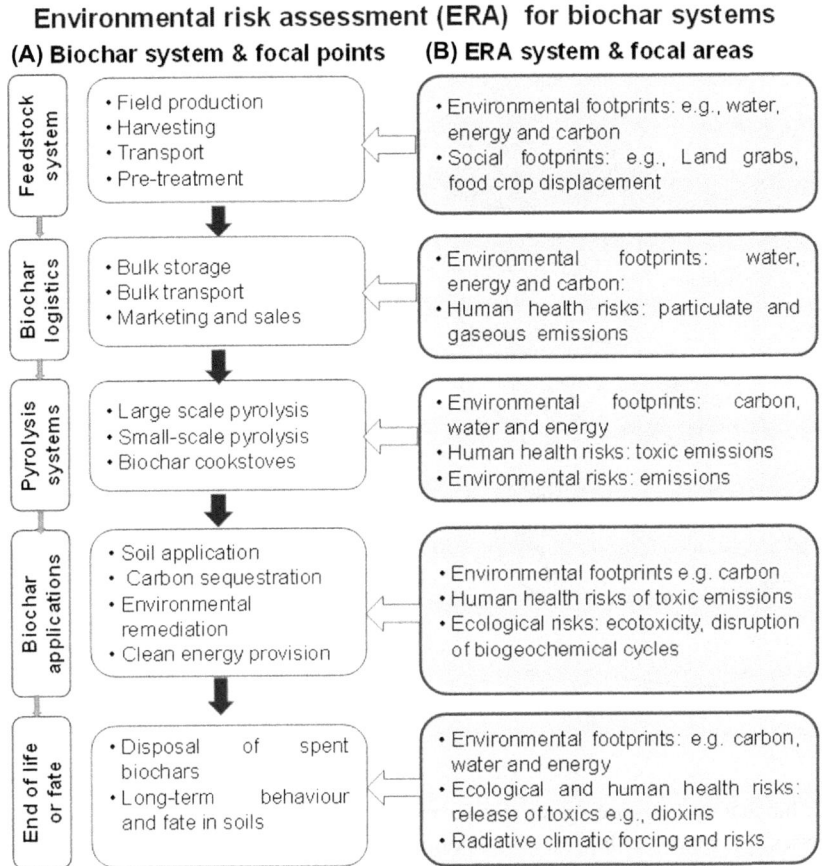

FIGURE 22.2 A conceptual framework for health risks and environment risk assessment in biochar research and applications: (A) Subsystems and focal points in the biochar value chain. (B) Potential environmental, ecological, and human health risks of biochar systems.

The hazards, which may be associated with biomass feedstocks or biochars, include (1) particulates, (2) organic contaminants such as polycyclic aromatic hydrocarbons, and (3) inorganic contaminants such as potentially toxic elements (Fig. 22.3; Xiang et al., 2021). The adverse health effects of biochar systems may occur at various points along the biochar value chain, including feedstock production and pretreatment, biochar production, logistics, biochar application, and postapplication. The potential risks in this chapter include (1) soil pollution, (2) water pollution, (3) air pollution, (4) radiative forcing/radiative forcing, (5) ecological health risks, and (6) human health risks. Here, a summary of these risks, including the status of the research, is presented.

22.2.1 Soil pollution

Biomass feedstocks used for producing biochar, such as sewage sludge, animal manures, or industrial wastes, may contain potentially toxic elements, including trace metals. These metals can be retained and become enriched in biochar after pyrolysis and subsequently introduced into the soil (Zhou et al., 2023). Biochar may also contain organic pollutants such as pesticides, herbicides, or polycyclic aromatic hydrocarbons (PAHs) inherent in the feedstocks or formed during pyrosis (Han et al., 2022). PAHs, such as dioxins, are often formed during the pyrolysis of biomass to produce biochar (Hale et al., 2012; Han et al., 2022). These contaminants may be transferred to soil during the application of biochar to enhance soil fertility and crop productivity. Moreover, biochar may also alter soil pH and affect nutrient availability depending on the pH of the biochar and the receiving soil (Novak et al., 2009). A few studies have also shown that some biochars can be hydrophobic or water-repellent, and their addition to hydrophilic soils may induce water repellency or hydrophobicity (Mao, Zhang and Chen, 2019). In turn, biochar-induced hydrophobicity may alter the hydrologic and hydraulic behaviors of soils. Therefore, the choice of feedstock, control of the pyrolysis process, and physicochemical characterization of the produced biochar are critical to minimize the potential risks of soil pollution.

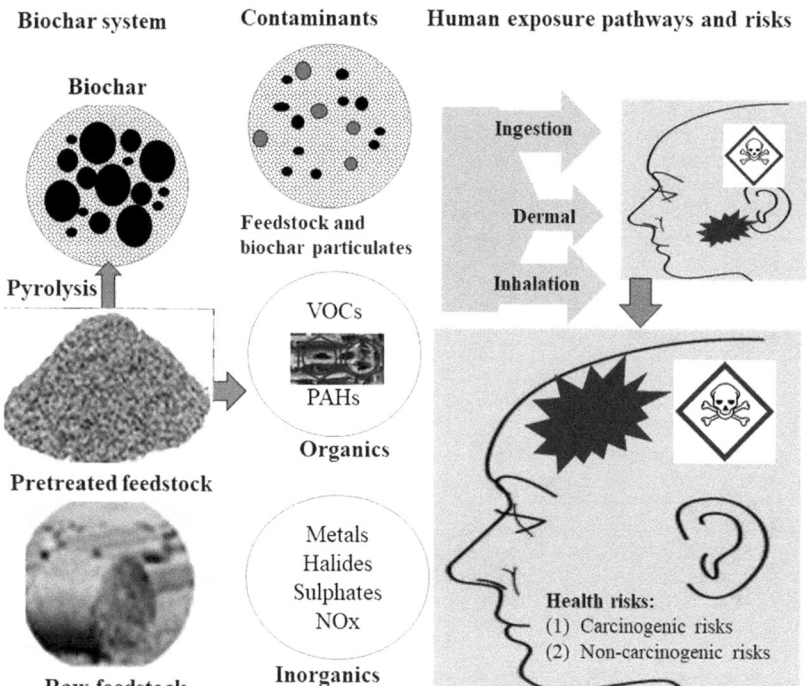

FIGURE 22.3 Conceptual depiction of the contaminants in biochar systems, human exposure pathways, and potential human risks.

22.2.2 Water pollution

Data on the potential water pollution risks of biochar systems are scarce. However, the harvesting and transport of biomass feedstock (e.g., crop residues, litter, and grasses) may expose the soil to erosion and runoff. In turn, this may increase the sediment load and associated contaminants in surface water bodies such as rivers and reservoirs. Moreover, biochars applied as a soil amendment in agroecosystems, degraded lands such as postmining landscapes, and contaminated soils may adsorb various contaminants. Biochar and cargo contaminants may be transported into surface water and groundwater systems, causing pollution. Limited data exist on biochar-mediated contamination of surface water and groundwater systems. However, few studies have reported biochar-mediated transport and cotransport of biochar and contaminants in porous media (Hamed et al., 2021; Ma et al., 2023). Ma et al. (2023) reported that biochar-mediated transport of organic contaminants was dependent on contaminant properties and biochar aging. Therefore, further investigations at the field or pilot scale are needed to understand the potential risk of biochar-mediated pollution and the controlling factors.

22.2.3 Air pollution

Biochar systems may emit particulate and gaseous contaminants into the environment (Dunningan et al., 2018; Ravi et al., 2016). Particulate matter includes feedstock, pristine biochar particles, and particulates laden with contaminants from various biochar applications. Some studies have reported biochar-induced dust and particulate emissions (Gelardi et al., 2019; Ravi et al., 2020), while others have documented particulate emissions from the combustion of biochars (Itoh et al., 2020). Some studies have also investigated gaseous emissions, including greenhouse emissions, from biochar systems (Flatabø et al., 2023; Zhang et al., 2022). Particulate and gaseous emissions are potentially high in poorly designed and operated biochar systems. However, further work comparing particulate and gaseous emissions from biochar systems relative to competing technologies is needed. In the context of the environmental application of biochars, such future studies should focus on typical industrial applications of biochar in energy, water and wastewater treatment, soil remediation, and air quality control. Sustainability assessment tools such as life cycle assessment should be applied for such comparative studies.

22.2.4 Radiative forcing/radiative forcing by black carbon/biochar

Similar to black carbon, biochars applied to the soil may also alter soil thermal properties, including albedo and temperature regimes, and even induce radiative forcing (Kandlikar, Reynolds and Grieshop, 2010; Lal, 2015; Smith, 2016;

Verheijen et al., 2013). However, the potential climatic risks associated with biochar or black carbon-induced radiative climatic forcing in the case of large-scale adoption of biochar technology have received limited attention. This is probably due to methodological challenges associated with understanding such climatic risks and their knock-on effects on ecosystems and human health. Thus, there is a need to investigate the potential impacts of large-scale biochar adoption on climatic forcing and atmospheric feedback at the regional and continental scales. Thus, coupled earth-atmosphere modeling studies accounting for biochar-induced climatic forcing and feedback are required to understand the potential impacts on regional, continental, and global climatic systems.

22.2.5 Ecological health risks

Potential ecological health risks of biochar systems may occur during feedstock collection and subsequent applications of biochar. With respect to feedstock, biomass types such as crop residues, forest/woodland litter, and grasses may play important ecological functions as habitats, refugia, and food sources for fauna/wildlife. For example, forest residues and litter from woodlands and grasslands are critical for biogeochemical cycling and serve as habitats for biodiversity conservation (Behrman et al., 2015; Hättenschwiler et al., 2005). Therefore, the harvesting of these biomass feedstocks for biochar production may result in habitat and biodiversity loss and disrupt ecosystem function and structure. However, the ecological effects of biomass harvesting on biodiversity and habitat loss from natural (woodlands and grasslands) and managed ecosystems (plantations and croplands) have attracted limited research attention. Therefore, field or pilot-scale studies are needed to understand the ecological effects of feedstock collection on various ecological processes and scales. These factors include habitat and biodiversity loss, individuals, population, community, trophic interactions, ecosystem structure and function, and ecosystem goods and services.

Biochar-induced changes in soil physical, chemical, and microbiological properties may include changes in ecosystem properties and function. However, the literature on the ecological effects of biochar is still dominated by studies investigating the beneficial or positive effects of biochar on ecological processes. Potentially toxic elements and organic contaminants (e.g., PAHs) derived from biochars may have adverse ecological effects. Therefore, further work using contaminant-enriched biochars derived from contaminated feedstocks should investigate the ecological health risks of such biochars. Moreover, further work is required to understand the ecological effects of organic contaminants such as PAHs formed during pyrolysis. To provide practically useful information, such studies should be conducted under environmentally relevant conditions concerning biochar application rates and methods, experimental durations, bioassay species, and ecotoxicological endpoints.

22.3 Human exposure and health risks

Human exposure to biochar, as well as other particulates and gaseous contaminants, can occur at various stages along the biochar value chain. This includes feedstock collection and logistics, biochar production via pyrolysis, biochar handling and logistics, industrial applications of biochar, and eventual disposal. The nature of contaminants largely determines human exposure, with inhalation being the primary route for biochar and particulates. High-risk environments for human exposure include occupational settings during feedstock logistics and pretreatment, biochar production and logistics, and applications such as soil amendment and drinking water treatment with biochars. Conceptually, human exposure may occur via inhalation, ingestion, and dermal contact (Fig. 22.3). However, studies on human exposure to biochar and particulates via various exposure pathways during production and application processes are limited, underscoring the need for further research in this area. Moreover, whether biochar systems increase the incidence of certain human health conditions, such as respiratory diseases, under occupational and nonoccupational settings remains unclear. These gaps call for further research using qualitative and QRA tools. QRA tools for human health risks, such as carcinogenic and noncarcinogenic risks, are discussed in subsequent sections.

22.4 A call for health risk assessments and mitigation

22.4.1 Health risk assessment framework

22.4.1.1 The notion of risk

The preceding discussion highlighted the potential environmental, ecological, and human health hazards associated with biochar systems. These hazards underscore the critical need for comprehensive health risk assessments of biochar systems. In this context, the notion of risk comprises two key components:

1. **Likelihood or probability of occurrence**
   This component assesses the chance or probability that a specific adverse event or outcome, such as biochar and associated contaminants, will occur and lead to harm, loss, or damage.
2. **Consequences or impact of the hazard**
   This component evaluates the severity of potential harm or loss resulting from the occurrence of the adverse event. The extent of damage, injury, or negative effects that may arise if a hazard or adverse event occurs is considered.

This understanding of risk forms the basis of most health risk assessments, encompassing both qualitative and quantitative methods. Here, four complementary health risk assessments aimed at understanding and mitigating the risks associated with biochar systems are discussed. The detailed procedures for these assessments are presented elsewhere; hence, only an overview is provided here.

In principle, health risk assessment and mitigation comprises seven main sequential steps (Gwenzi et al., 2022): (1) characterization of hazards, (2) estimation of likelihood, (3) estimation of consequences or impacts, (4) estimation of overall risk, (5) risk evaluation, (6) risk mitigation and management, and (7) documentation and communication. A detailed description of risk assessment and mitigation principles, including those for chemical hazards, is presented in the literature (Gerba, 2019; Gwenzi et al., 2022; Holmberg, 2017; Piper, 2020; Rigaud et al., 2024; Si et al., 2012). Briefly, the key steps are as follows (Gwenzi et al., 2022). An overview of each step is presented below.

22.4.2 Principles of health risk assessment

1. **Characterization of hazards**
   This involves the identification and description of hazards or potential adverse events within a specific context or system.
2. **Estimation of likelihood**
   The likelihood of occurrence of each hazard is estimated based on various techniques, including statistical data, historical records, expert judgment, or modeling techniques.
3. **Estimation of consequences or impacts**
   This involves evaluation of the potential impacts or consequences of each hazard in terms of severity or damage or harm caused if the adverse event occurs.
4. **Risk estimation**
   This integrates the estimated likelihood and consequences to estimate the level of overall risk associated with each hazard.
5. **Risk evaluation**
   This entails comparing estimated risk levels against predefined criteria to determine acceptability and prioritize risks for management actions or mitigation.
6. **Risk mitigation and management**
   This involves developing and implementing strategies to mitigate risks, such as implementing controls, improving safety systems, use of personal protective equipment, or enhancing emergency response plans.
7. **Documentation and communication**
   This involves documenting findings, assumptions, methodologies, and results to communicate risk and mitigation outcomes to stakeholders and decision-makers.

Two main risk assessment methods exist such as (1) qualitative risk assessment and (2) QRA. The two approaches follow the same generic risk assessment framework and principles described earlier. However, as described in detail below, they differ concerning the procedure for estimating likelihood, consequences, risk estimation, and risk evaluation.

22.4.3 Qualitative risk assessment

Qualitative risk assessment utilizes nonquantitative and often subjective methods to assess the likelihood and consequences of hazards (Gwenzi et al., 2022). It combines qualitative assessments of likelihood and consequence to derive a risk-rating scale. Fig. 22.4 illustrates a typical tool for qualitative risk assessment based on likelihood and consequences. The results are then used for risk evaluation by comparing the qualitative risk levels against predefined qualitative criteria to determine acceptability and prioritize risks for mitigation and management. The risks exceeding a set qualitative threshold form the basis for developing and implementing strategies to mitigate risks, such as implementing controls,

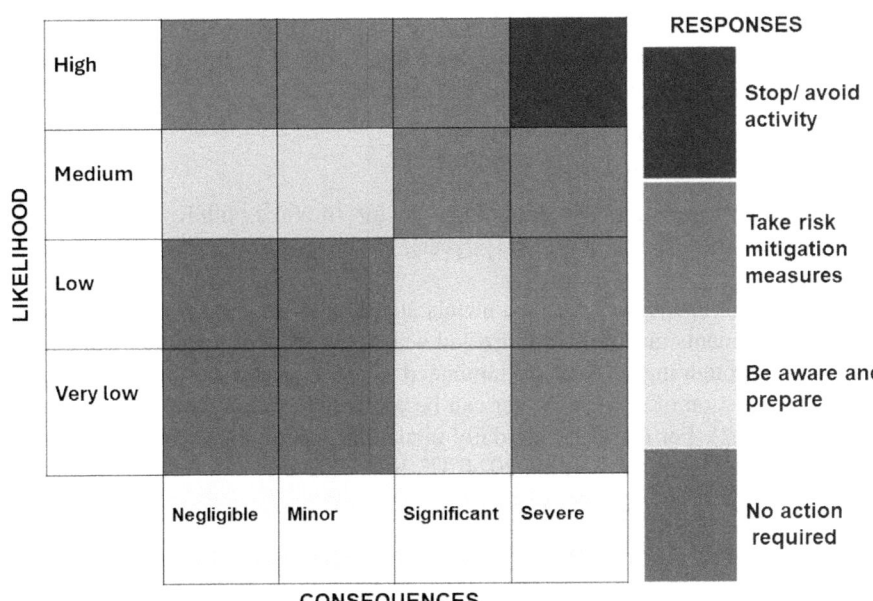

FIGURE 22.4 A qualitative risk assessment tool for evaluating the risks of biochar systems based on the likelihood or probability of occurrence and severity or consequences of an adverse event and corresponding mitigation responses.

improving safety systems, or enhancing emergency response plans. Finally, the findings, assumptions, methodologies, and results are documented to communicate outcomes to stakeholders and decision-makers. Note that the last three steps are also applicable to QRA.

22.4.4 Quantitative risk assessment

QRA is a systematic and numerical approach that assigns numerical values to the likelihood and consequences of identified hazards (Rigaud et al., 2024). A detailed discussion of the QRA of human health risks is presented in subsequent sections. Unlike qualitative methods, it involves quantifying the probability of adverse events occurring and the magnitude of their consequences. The QRA process has been described in detail in the literature and follows the seven generic steps described earlier (Gerba, 2019; Holmberg, 2017; Piper, 2020; Rigaud et al., 2024; Si et al., 2012).

However, unlike qualitative risk assessment, QRA provides quantitative values for the likelihood, consequences, overall risk, and risk rating. For example, statistical data, historical records, expert judgment, or modeling techniques are to quantify the likelihood of each hazard occurring. Moreover, a quantitative estimate of the consequences or impacts is also provided. By combining the estimated likelihood and consequences the overall level of risk associated with each hazard is quantified using models such as risk matrices or curves. The calculated quantitative risk is then compared against predefined criteria to determine acceptability and prioritize risks for management actions. This is then followed by risk mitigation and management, and finally documentation and communication. In subsequent sections, some of the QRA tools for human health risks are discussed.

Both qualitative and QRAs are widely utilized across various industries, including environmental management and occupational safety, to analyze and manage risks effectively. These assessment tools can be applied throughout the biochar value chain, covering production and industrial applications. However, despite their utility, health risk assessments of biochar systems remain limited, emphasizing the need for further research in this area. Such assessments serve as crucial decision-support tools for policy-making and technology selection in biochar applications.

22.4.5 Quantitative human health risk assessment tools

QRA tools exist for the estimation of carcinogenic and noncarcinogenic health risks of airborne contaminants (Chen & Liao, 2006; Chen et al., 2016; Peng et al., 2011; US EPA,1991; Wang et al., 2011). A detailed discussion of the equations and their input data is provided elsewhere in the literature (Kormoker et al., 2021; Rauf et al., 2021; US EPA, 1991). The US EPA provides exposure assessment tools for various exposure routes, including inhalation, ingestion, and dermal intake (US EPA, 2024a, 2024b, 2024c). These tools can be adapted to estimate human exposure to airborne biochar particulates and associated contaminants in occupational and nonoccupational settings. For example, the potential average daily dose (ADD) of a contaminant via inhalation, expressed as the mass of contaminant per unit body weight over time (e.g., mg/kg-day), is the product of contaminant concentration, inhalation rate, exposure time, exposure frequency, and exposure duration divided by the product of averaging time and body weight (US EPA, 2024b, Eq. 22.1):

$$ADD = \frac{C_{air} * Inh.R * ET * EF * ED}{BW * AT} \tag{22.1}$$

ADD: Average daily dose (mg/kg-day), C_{air}: concentration of contaminant in air (mg/m^3), Inh.R: inhalation rate (m^3/hour), ET: exposure time (hours/day), EF: exposure frequency (days/year), ED: exposure duration (years), BW: body weight (kg), AT: averaging time (days).

When biochar contaminated with soluble contaminants such as trace metals and dioxins are used for drinking water treatment, residual concentrations of these contaminants may remain in treated water. Consequently, human exposure to dissolved contaminants from biochar can occur through ingestion of contaminated water. In such cases, human exposure or intake of contaminants such as dioxins via ingestion of drinking water can be estimated as described in the literature (Marumure et al., 2024; US EPA, 2024c) (Eq. 22.2). For example, the daily intake of a contaminant in drinking water via ingestion can be estimated according to Eq. 22.2 (Marumure et al., 2024; US EPA, 2024c):

$$DI = \frac{C_{DW} \times WI_{DW} \times EF \times ED}{BW \times AT} \tag{22.2}$$

where DI is the daily intake of contaminants ($\mu g\ kg^{-1}$ BW day^{-1}), C_{DW} is the concentration of contaminants in drinking water ($\mu g\ L^{-1}$), WI_{DW} is the daily drinking water intake (L kg^{-1} BW), EF is the exposure frequency (days year^{-1}), ED is the exposure duration (years or lifetime), BW is the body weight (kg), and AT is the averaging time (days).

22.4.5.1 Noncarcinogenic human health risks

Some studies have applied the hazard quotient (HQ) and hazard index (HI) to estimate the potential noncarcinogenic risk of airborne contaminants, such as trace metals, via three intake pathways: ingestion, dermal contact, and inhalation (Chen et al., 2016; Eqani et al., 2016; Li et al. 2014; Zhang et al., 2012). The HQ is calculated based on the chronic daily intake (CDI), BAF, and RfD_o as follows (Eq. 22.3):

$$HQ = \frac{CDI * BAF}{RfD_0} \tag{22.3}$$

where CDI is the total chemical daily intake through ingestion, inhalation, and dermal contact; BAF is the bioaccumulation factor, expressed as the ratio between the concentration of each contaminant that is bioavailable and its total concentration in the exposure medium (e.g., biochar in this case); and RfD0 is the oral reference dose.

For noncarcinogenic health risks caused by multiple contaminants, the HI is calculated as the sum of the HQ values for the various contaminants (Eq. 22.4):

$$HI = \sum_{i}^{n} HQ \tag{22.4}$$

where i and n indicate contaminants i to n. The HI values are interpreted as follows: (1) HI values exceeding 1.0 indicate potential adverse noncarcinogenic health risks, and (2) HI values below 1.0 denote no significant noncarcinogenic human health risks.

22.4.5.2 Carcinogenic human health risks

In humans, the carcinogenic risk (CR) is the probability or likelihood of an individual developing cancer in a lifetime following exposure to carcinogenic agents. The CR of a contaminant is estimated to be a product of CDI or ADD and the cancer slope factor (CSF) or potency factor of a specific carcinogenic agent over time (Chen et al., 2016; USEPA, 2013; Li et al. 2014; Zhang et al., 2012). The total cancer risk (TCR) is the sum of the CR values from individual exposure to various contaminants. TCR values are interpreted as follows: (1) TCR values ranging from 1×10^{-6} to 1×10^{-4} indicate low or acceptable risk, and (2) TCR values exceeding 1×10^{-4} indicate unacceptable risk and the possibility of carcinogenic disease in the future. The CR and TCR can be calculated using the following equations (Eqs. 22.5 and 22.6):

$$CR = D * CSF * \frac{ED}{LY} \tag{22.5}$$

Where, CR = cancer risk for an individual contaminant, D = age-specific dose (mg/kg/day), CSF = cancer slope factor $((mg/kg/day)^{-1})$, ED = age-specific exposure duration (years), and LY = lifetime in years (78 years). Note that in some studies, CSF can be calculated as the product of EF-adjusted air concentration ($\mu g/m^3$) (AAC) and inhalation unit risk $((\mu g/m^3)^{-1})$ (IUR) (ATSDR, 2024).

$$TCR = \sum_{i}^{n} CR \qquad (22.6)$$

In the literature, other studies quantifying the cancer risk of contaminants such as PAHs used the incremental lifetime cancer risk using US EPA models (USEPA, 1991, 2011) (Chen & Liao, 2006; Peng et al., 2011; Wang et al., 2011). These QRA tools can be adapted to estimate risk at various points along the biochar value chain.

However, limited data exist on the use of QRA tools to estimate the human health risks of biochar systems. Moreover, the QRA tools (Eqn. 3–22) provide an overall quantitative estimate of noncarcinogenic and CRs of a contaminant or chemical. However, no information is provided on the following aspects of the human health risks: (1) the human receptor organs, (2) the specific nature of the human health risk (e.g., type of cancer or noncarcinogenic health condition), and (3) the timing of onset of the human health conditions.

22.4.6 Ecotoxicological risk assessments

Ecotoxicological risk assessments employ standardized bioassay tests to evaluate the potential ecotoxicity of materials or chemicals present in biochar and associated contaminants to specific organisms. These assessments expose bioassay organisms to biochar and its contaminants through known exposure pathways or routes for defined durations. They measure specific endpoints (e.g., biometric parameters, behavior, reproduction, enzyme, and gene expression) to assess ecotoxicity. Although several ecotoxicological bioassay tests exist, their application in evaluating the ecological health risks of biochars remains limited. Hence, there is a need to expand biochar certification systems to include ecotoxicological assessments to comprehensively evaluate potential ecological risks.

22.4.7 Biochar certification systems

Biochar certification systems are based on the physicochemical characterization of biochar, including the assessment of target contaminants such as toxic metals, against established maximum guidelines. Examples of such systems already exist in the European Community (Schmidt et al., 2016). These systems are often integrated with quality control measures during biochar production to ensure compliance with set standards. Biochar certification systems ensure that only biochar that meets specified standards is used for specific applications, thereby minimizing the risk of environmental pollution by biochar-derived contaminants. While such systems are established in developed regions, comparable frameworks are lacking in low- and middle-income countries, where the risk of biomass feedstock contamination with industrial pollutants may be greater.

22.4.8 Status of research on risk assessment of biochar systems

Overall, the human exposure and health risks of biochar systems are poorly understood. This is because studies on qualitative and QRA of biochar systems are scarce. Moreover, there are insufficient data on the transfer of contaminants from biochar to humans via various media, such as biochar-treated water, and the associated human health risks. Therefore, additional research is necessary to understand human exposure to contaminants such as dioxins from biochar used in drinking water treatment compared to commercial activated carbons. An exposure assessment tool for dermal intake of contaminants also exists (US EPA, 2024a). However, unlike contaminants in bathing and swimming water, the dermal exposure route is likely to be of limited significance in biochar systems. In cases of multiple exposure routes, the total human exposure will be the summation of the various intake routes (e.g., inhalation, ingestion, and skin/dermal contact). Therefore, further research is crucial for human health risk assessment of biochar systems to address the following gaps:

1. Identification and characterization of high-risk settings and points along the biochar value chain.
2. Profiling of high-risk individuals and their susceptibility or risk levels.
3. Understanding human health risks arising from multiple exposure routes for individual contaminants.

4. Understanding the interactive human health effects of biochar-derived contaminants with other contaminants and health stressors.

These future studies should utilize the qualitative and QRA tools discussed in subsequent sections. QRA, for instance, can provide estimates of carcinogenic and noncarcinogenic health risks across different exposure groups.

22.5 Future research directions

The preceding discussion reveals a scarcity of research on health risks and risk assessment within biochar systems. Consequently, several knowledge gaps exist regarding health risks, risk assessment, and mitigation. Therefore, future research should address the following gaps:

(1) **Risk hotspots and critical control points in the biochar value chain**
    There is a need to identify risk hotspots and critical control points throughout the biochar value chain, including feedstock, biochar production, logistics, industrial applications, and subsequent regeneration, recycling, and disposal. Tools such as hazard analysis and critical control points (HACCP) can be adapted and applied to identify and profile hazards and risks in biochar systems. HACCP, a widely used management system in industry, analyses and controls hazards from raw material production through to finished product consumption. Identifying risk hotspots is crucial for developing and implementing risk mitigation measures to ensure the safe and sustainable use of biochars.

(2) **Environmental pollution risks of biochar systems**
    The occurrence of contaminants such as dioxins in biochars points to potential pollution risks associated with the soil application of such biochars. However, limited data exist on the soil, water, and air pollution risks associated with biochar systems following their application in agriculture and environmental remediation. Therefore, empirical studies are needed to investigate the transfer of contaminants via water and wind erosion from application points to environmental receptors such as surface water, groundwater, soils, and air.

(3) **Climatic/radiative forcing potential of biochar systems**
    Concerns have been raised regarding the climatic or radiative forcing potential of black carbon from biochar systems. However, studies on the large-scale effects of such radiative forcing on weather and climatic patterns are lacking. Therefore, further modeling is required to understand how black carbon from biochar systems impacts climate systems at local, regional, and global scales. This information is particularly important if biochar technology is adopted and applied on a large scale.

(4) **Ecological health risks of biochar systems**
    Biochar systems may adversely affect biodiversity and ecological health through the harvesting and collection of biomass feedstock, including forest litter and crop residues, that serve as habitats and refugia for biodiversity. Moreover, biochar may contain toxic contaminants that pose ecotoxicological risks to soil biota. However, there is limited data on the ecological health risks associated with biochar systems, highlighting the need for further research.

(5) **Human exposure and health risks of biochar systems**
    Human exposure and health risks may occur at various points along the biochar value chain. However, there is limited data on human exposure risks to biochar and its associated contaminants in both occupational and nonoccupational settings. Furthermore, studies investigating the human health risks of biochar systems are lacking. Therefore, it remains unclear whether biochar systems increase the risk of human health conditions such as respiratory infections in occupational settings. Further research based on epidemiology and QRA is necessary to understand the human health risks of biochar systems compared to those of other technologies.

22.6 Conclusions and outlook

Biochars have received increasing research and public interest due to their affordability, ease of production, and versatile applications in agriculture, environmental remediation, and bioenergy. However, despite these promising attributes, biochar systems pose potential environmental, ecological, and human health hazards. Currently, the assessment and mitigation of these risks have received limited research attention.

To address this gap, this chapter discusses the health hazards associated with biochar systems. In summary

1. The health hazards of biochar systems are linked to biochar particulates and the presence and emission of toxic contaminants.

2. Potential hazards include soil, water, and air pollution, particularly postapplication.
3. Climatic hazards involve radiative forcing caused by black carbon associated with biochar systems.
4. Ecological health risks arise from biochar-derived contaminants, such as dioxins released during pyrolysis and those originating from contaminated biomass feedstock.
5. Potential human health risks may arise from exposure to biochar and associated contaminants through multiple pathways, such as inhalation and ingestion.

To address these concerns, this chapter proposes comprehensive health risk assessments based on the following:

1. Biochar certification involves physicochemical characterization, quality control, and assurance systems.
2. Ecotoxicological bioassays utilizing standardized protocols.
3. Qualitative risk assessments are based on the likelihood of occurrence and consequences.
4. QRAs based on exposure and probability models.

Further research efforts are essential to deepen our understanding of health hazards and refine risk assessment frameworks associated with biochar. Future research should focus on developing robust methodologies for effectively assessing and mitigating these potential risks. A thorough understanding of the hazards posed by biochar systems is crucial for managing risks and ensuring the safe and sustainable application of biochar in agriculture, environmental remediation, and bioenergy.

References

Behrman, K. D., Juenger, T. E., Kiniry, J. R., & Keitt, T. H. (2015). Spatial land use trade-offs for maintenance of biodiversity, biofuel, and agriculture. Kluwer Academic Publishers, United States. *Landscape Ecology, 30*(10), 1987–1999. Available from https://doi.org/10.1007/s10980-015-0225-1, http://www.springerlink.com/content/103025/.

Chen, S. C., & Liao, C. M. (2006). Health risk assessment on human exposed to environmental polycyclic aromatic hydrocarbons pollution sources. *Science of the Total Environment, 366*(1), 112–123. Available from https://doi.org/10.1016/j.scitotenv.2005.08.047.

Chen, X., Feng, L., Luo, H., & Cheng, H. (2016). Health risk equations and risk assessment of airborne benzene homologues exposure to drivers and passengers in taxi cabins. *Environmental Science and Pollution Research, 23*(5), 4797–4811. Available from https://doi.org/10.1007/s11356-015-5678-x, http://www.springerlink.com/content/0944-1344.

Dunnigan, L., Morton, B. J., Hall, P. A., & Kwong, C. W. (2018). Production of biochar and bioenergy from rice husk: Influence of feedstock drying on particulate matter and the associated polycyclic aromatic hydrocarbon emissions. *Atmospheric Environment, 190*, 218–225.

Eqani, S. A. M. A. S., Kanwal, A., Bhowmik, A. K., Sohail, M., Ullah, R., Ali, S. M., Alamdar, A., Ali, N., Fasola, M., & Shen, H. (2016). Spatial distribution of dust-bound trace elements in Pakistan and their implications for human exposure. *Environmental Pollution, 213*, 213–222. Available from https://doi.org/10.1016/j.envpol.2016.02.017, http://www.elsevier.com/inca/publications/store/4/0/5/8/5/6.

Flatabø, G. Ø., Cornelissen, G., Carlsson, P., Nilsen, P. J., Tapasvi, D., Bergland, W. H., & Sørmo, E. (2023). Industrially relevant pyrolysis of diverse contaminated organic wastes: Gas compositions and emissions to air. *Journal of Cleaner Production, 423*. Available from https://www.journals.elsevier.com/journal-of-cleaner-production, https://doi.org/10.1016/j.jclepro.2023.138777.

Gelardi, D. L., Li, C., & Parikh, S. J. (2019). An emerging environmental concern: Biochar-induced dust emissions and their potentially toxic properties. *Science of the Total Environment, 678*, 813–820. Available from https://doi.org/10.1016/j.scitotenv.2019.05.007, http://www.elsevier.com/locate/scitotenv.

Gerba, C. P. (2019). *Risk Assessment* (pp. 541–563). Elsevier BV. Available from https://doi.org/10.1016/b978-0-12-814719-1.00029-x.

Gwenzi, W., Chaukura, N., Mukome, F. N., Machado, S., & Nyamasoka, B. (2015). Biochar production and applications in sub-Saharan Africa: Opportunities, constraints, risks and uncertainties. *Journal of environmental management, 150*, 250–261.

Gwenzi, W., Chaukura, N., Noubactep, C., & Mukome, F. N. D. (2017). Biochar-based water treatment systems as a potential low-cost and sustainable technology for clean water provision. *Journal of Environmental Management, 197*, 732–749. Available from https://doi.org/10.1016/j.jenvman.2017.03.087, https://www.sciencedirect.com/journal/journal-of-environmental-management.

Gwenzi, W., Chaukura, N., Wenga, T., & Mtisi, M. (2021). Biochars as media for air pollution control systems: Contaminant removal, applications and future research directions. *Science of the Total Environment*, 142249. Available from https://doi.org/10.1016/j.scitotenv.2020.142249.

Gwenzi, W., Muhoyi, E., & Mukura, T. J. (2022). *Health risk assessment and mitigation of emerging contaminants: A call for an integrated approach. Emerging contaminants in the terrestrial-aquatic-atmosphere continuum: Occurrence, health risks and mitigation* (pp. 325–342). Zimbabwe: Elsevier. Available from https://www.sciencedirect.com/book/9780323900515, https://doi.org/10.1016/B978-0-323-90051-5.00021-3.

Hale, S. E., Lehmann, J., Rutherford, D., Zimmerman, A. R., Bachmann, R. T., Shitumbanuma, V., O'Toole, A., Sundqvist, K. L., Arp, H. P. H., & Cornelissen, G. (2012). Quantifying the total and bioavailable polycyclic aromatic hydrocarbons and dioxins in biochars. *Environmental Science and Technology, 46*(5), 2830–2838. Available from https://doi.org/10.1021/es203984k.

Hameed, R., Lei, C., Fang, J., & Lin, D. (2021). Co-transport of biochar colloids with organic contaminants in soil column. *Environmental Science and Pollution Research, 28*, 1574–1586.

Han, H., Buss, W., Zheng, Y., Song, P., Rafiq, M. K., Liu, P., Mašek, O., & Li, X. (2022). Contaminants in biochar and suggested mitigation measures–a review. *Chemical Engineering Journal, 429*, 132287.

Hättenschwiler, S., Tiunov, A. V., & Scheu, S. (2005). Biodiversity and litter decomposition in terrestrial ecosystems. *Annual Review of Ecology, Evolution, and Systematics, 36*, 191−218. Available from https://doi.org/10.1146/annurev.ecolsys.36.112904.151932.

Holmberg, J. E. (2017). Quantitative risk analysis. *Handbook of Safety Principles*, 434−462.

Itoh, T., Fujiwara, N., Iwabuchi, K., Narita, T., Mendbayar, D., Kamide, M., Niwa, S., & Matsumi, Y. (2020). Effects of pyrolysis temperature and feedstock type on particulate matter emission characteristics during biochar combustion. *Fuel Processing Technology, 204*106408. Available from https://doi.org/10.1016/j.fuproc.2020.106408.

Ji, M., Wang, X., Usman, M., Liu, F., Dan, Y., Zhou, L., Campanaro, S., Luo, G., & Sang, W. (2022). Effects of different feedstocks-based biochar on soil remediation: A review. *Environmental Pollution, 294*118655. Available from https://doi.org/10.1016/j.envpol.2021.118655.

Kamali, M., Appels, L., Kwon, E. E., Aminabhavi, T. M., & Dewil, R. (2021). Biochar in water and wastewater treatment - a sustainability assessment. *Chemical Engineering Journal, 420*. Available from https://doi.org/10.1016/j.cej.2021.129946, http://www.elsevier.com/inca/publications/store/6/0/1/2/7/3/index.htt.

Kandlikar, M., Reynolds, C. C. O., & Grieshop, A. P. (2010). A perspective paper on black carbon mitigation as a response to climate Change. *Consensus on Climate*.

Kormoker, T., Kabir, M. H., Khan, R., Islam, M. S., Shammi, R. S., Al, M. A., Proshad, R., Tamim, U., Sarker, M. E., Taj, M. T. I., & Akter, A. (2021). Road dust−driven elemental distribution in megacity Dhaka, Bangladesh: environmental, ecological, and human health risks assessment. *Environmental Science and Pollution Research*, 1−22.

Lal, R. (2015). *Biochar and soil carbon sequestration. Agricultural and environmental applications of biochar: Advances and barriers* (pp. 175−197). United States: wiley. Available from http://onlinelibrary.wiley.com/book/9780891189671, https://doi.org/10.2136/sssaspecpub63.2014.0042.5.

Lehmann, J., & Joseph, S. (2015). *Biochar for environmental management: an introduction* (pp. 33−46). Informa UK Limited. Available from https://doi.org/10.4324/9780203762264-8.

Lehmann, J., & Joseph, S. (Eds.), (2024). *Biochar for environmental management: science, technology and implementation*. Taylor & Francis.

Lyu, H., Zhang, H., Chu, M., Zhang, C., Tang, J., Chang, S. X., Mašek, O., & Ok, Y. S. (2022). Biochar affects greenhouse gas emissions in various environments: A critical review. John Wiley and Sons Ltd, China. *Land Degradation and Development, 33*(17), 3327−3342. Available from https://doi.org/10.1002/ldr.4405, http://onlinelibrary.wiley.com/journal/10.1002/(ISSN)1099-145X.

Ma, P., Qi, Z., Wu, X., Ji, R., & Chen, W. (2023). Biochar nanoparticles-mediated transport of organic contaminants in porous media: dependency on contaminant properties and effects of biochar aging. *Carbon Research, 2*(1). Available from https://doi.org/10.1007/s44246-023-00036-6, http://www.springer.com/journal/44246.

Mao, J., Zhang, K., & Chen, B. (2019). Linking hydrophobicity of biochar to the water repellency and water holding capacity of biochar-amended soil. *Environmental Pollution, 253*, 779−789. Available from https://doi.org/10.1016/j.envpol.2019.07.051, https://www.journals.elsevier.com/environmental-pollution.

Marumure, J., Simbanegavi, T. T., Makuvara, Z., Karidzagundi, R., Alufasi, R., Goredema, M., Gufe, C., Chaukura, N., Halabowski, D., & Gwenzi, W. (2024). Emerging organic contaminants in drinking water systems: Human intake, emerging health risks, and future research directions. *Chemosphere, 356*. Available from https://doi.org/10.1016/j.chemosphere.2024.141699, https://www.sciencedirect.com/science/journal/00456535.

Novak, J. M., Busscher, W. J., Laird, D. L., Ahmedna, M., Watts, D. W., & Niandou, M. A. S. (2009). Impact of biochar amendment on fertility of a southeastern coastal plain soil. *Soil Science, 174*(2), 105−112. Available from https://doi.org/10.1097/SS.0b013e3181981d9a.

Peng, C., Chen, W., Liao, X., Wang, M., Ouyang, Z., Jiao, W., & Bai, Y. (2011). Polycyclic aromatic hydrocarbons in urban soils of Beijing: Status, sources, distribution and potential risk. *Environmental Pollution, 159*(3), 802−808. Available from https://doi.org/10.1016/j.envpol.2010.11.003.

Piper J. W., *Risk Management Framework: Qualitative Risk Assessment through Risk Scenario Analysis*. (2020).

Ramola, S., Mohan, D., Masek, O., Méndez, A., & Tsubota, T. (Eds.), (2022). *Engineered biochar: fundamentals, preparation, characterization and applications* (381). Springer.

Rauf, A. U., Mallongi, A., Lee, K., Daud, A., Hatta, M., Madhoun, W. A., & Astuti, R. D. P. (2021). Potentially toxic element levels in atmospheric particulates and health risk estimation around industrial areas of maros, indonesia. *Toxics, 9*(12). Available from https://doi.org/10.3390/toxics9120328, https://www.mdpi.com/2305-6304/9/12/328/pdf.

Ravi S., Sharratt B. S., Li J., Olshevski S., Meng Z., Zhang J., Particulate matter emissions from biochar-amended soils as a potential tradeoff to the negative emission potential. Scientific Reports. 6 (2016), http://www.nature.com/srep/index.html. https://doi.org/10.1038/srep35984.

Ravi, S., Li, J., Meng, Z., Zhang, J., & Mohanty, S. (2020). Generation, resuspension, and transport of particulate matter from biochar-amended soils: A potential health risk. *GeoHealth, 4*(11)e2020GH000311.

Rigaud, M., Buekers, J., Bessems, J., Basagaña, X., Mathy, S., Nieuwenhuijsen, M., & Slama, R. (2024). The methodology of quantitative risk assessment studies. *Environmental Health, 23*(1). Available from https://doi.org/10.1186/s12940-023-01039-x.

Schmidt, H. P., Bucheli, T., Kammann, C., Glaser, B., Abiven, S., & Leifeld, J. (2016). European Biochar Certificate - Guidelines for a sustainable production of Biochar. *Arbaz (CH): European Biochar Foundation (EBC)*. Available from https://doi.org/10.13140/RG.2.1.4658.7043.

Si, H., Ji, H., & Zeng, X. (2012). Quantitative risk assessment model of hazardous chemicals leakage and application. *Safety Science, 50*(7), 1452−1461. Available from https://doi.org/10.1016/j.ssci.2012.01.011.

Singh, J. S., & Singh, C. (Eds.), (2020). *Biochar applications in agriculture and environment management* (978). Springer International Publishing.

Smith, P. (2016). Soil carbon sequestration and biochar as negative emission technologies. *Global Change Biology, 22*(3), 1315−1324. Available from https://doi.org/10.1111/gcb.13178.

US EPA, 1991. Risk Assessment Guidance for Superfund, Volume 1, Human health evaluation manual (Part B, Development of risk-based preliminary remediation goals). OSWER,[9285.7-01B.EPA/540/R-92/003].

US EPA, 2024a. Exposure Assessment Tools by Routes – Dermal. https://www.epa.gov/expobox/exposure-assessment-tools-routes-dermal. Accessed: 28 June 2024.

US EPA, 2024b. Exposure Assessment Tools by Routes – Inhalation. Available at: https://www.epa.gov/expobox/exposure-assessment-tools-routes-inhalation Accessed: 28 June 2024

US EPA, 2024c. Exposure Assessment Tools by Routes – Ingestion. https://www.epa.gov/expobox/exposure-assessment-tools-routes-ingestion. Accessed: 28 June 2024.

USEPA, 2011. Exposure factors handbook: 2011 edition: USEPA Office of Research and Development Washington. Published online by United States Environmental Protection Agency. Available online at https://cfpub.epa.gov/ncea/risk/recordisplay.cfm?deid = 236252. Accessed 4 July 2024

USEPA, 2013 Integrated risk information system of United States Environmental Protection Agency. Available at http://www.epa.gov/iris Accessed: 7 July 2024.

Verheijen, F. G. A., Jeffery, S., Van Der Velde, M., Penížek, V., Beland, M., Bastos, A. C., & Keizer, J. J. (2013). Reductions in soil surface albedo as a function of biochar application rate: Implications for global radiative forcing. *Environmental Research Letters*, 8(4). Available from https://doi.org/10.1088/1748-9326/8/4/044008, http://iopscience.iop.org/1748-9326/8/4/044008/pdf/1748-9326_8_4_044008.pdf.

Wang, W., Huang, M. J., Kang, Y., Wang, H. S., Leung, A. O. W., Cheung, K. C., & Wong, M. H. (2011). Polycyclic aromatic hydrocarbons (PAHs) in urban surface dust of Guangzhou, China: Status, sources and human health risk assessment. *Science of the Total Environment*, 409(21), 4519–4527. Available from https://doi.org/10.1016/j.scitotenv.2011.07.030.

Wang, J., & Wang, S. (2019). Preparation, modification and environmental application of biochar: A review. *Journal of Cleaner Production*, 227, 1002–1022.

Xiang, L., Liu, S., Ye, S., Yang, H., Song, B., Qin, F., Shen, M., Tan, C., Zeng, G., & Tan, X. (2021). Potential hazards of biochar: The negative environmental impacts of biochar applications. *Journal of Hazardous Materials*, 420126611.

Xiang, W., Zhang, X., Chen, J., Zou, W., He, F., Hu, X., Tsang, D. C., Ok, Y. S., & Gao, B. (2020). Biochar technology in wastewater treatment: A critical review. *Chemosphere*, 252126539.

Xie, Y., Wang, L., Li, H., Westholm, L. J., Carvalho, L., Thorin, E., Yu, Z., Yu, X., & Skreiberg, Ø. (2022). A critical review on production, modification and utilization of biochar. *Journal of Analytical and Applied Pyrolysis*, 161105405. Available from https://doi.org/10.1016/j.jaap.2021.105405.

Yin, Y., Yang, C., Li, M., Zheng, Y., Ge, C., Gu, J., Li, H., Duan, M., Wang, X., & Chen, R. (2021). Research progress and prospects for using biochar to mitigate greenhouse gas emissions during composting: A review. *Science of The Total Environment*, 798149294. Available from https://doi.org/10.1016/j.scitotenv.2021.149294.

Zama, E. F., Reid, B. J., Arp, H. P. H., Sun, G. X., Yuan, H. Y., & Zhu, Y. G. (2018). Advances in research on the use of biochar in soil for remediation: a review. *Journal of Soils and Sediments*, 18(7), 2433–2450. Available from https://doi.org/10.1007/s11368-018-2000-9, http://www.springerlink.com/content/1439-0108.

Zhang, Y., Mu, Y., Liu, J., & Mellouki, A. (2012). Levels, sources and health risks of carbonyls and BTEX in the ambient air of Beijing, China. *J Environ Sci*, 24, 124–130.

Zhang, G., Feng, Q., Hu, J., Sun, G., Evrendilek, F., Liu, H., & Liu, J. (2022). Performance and mechanism of bamboo residues pyrolysis: Gas emissions, by-products, and reaction kinetics. *Science of the Total Environment*, 838156560.

Zhao, Z., Wang, B., Theng, B. K. G., Lee, X., Zhang, X., Chen, M., & Xu, P. (2022). Removal performance, mechanisms, and influencing factors of biochar for air pollutants: a critical review. Springer, China. *Biochar*, 4(1). Available from https://doi.org/10.1007/s42773-022-00156-z. Available from:, springer.com/journal/42773.

Zhou, A., Yu, S., Deng, S., Mikulčić, H., Tan, H., & Wang, X. (2023). Enrichment characteristics and environmental risk assessment of heavy metals in municipal sludge pyrolysis biochar. *Journal of the Energy Institute*, 111101417. Available from https://doi.org/10.1016/j.joei.2023.101417.

Chapter 23

Techno-economic assessment of biochar systems: state-of-the-art and future research directions

Sita Koné[1], Xavier Galiegue[2] and Willis Gwenzi[3,4,5]

[1]International Center for Biosaline Agriculture, Dubai, United Arab Emirates, [2]Faculté de Droit d'Economie et de Gestion, University of Orleans, Orleans, France, [3]Formerly Alexander von Humboldt Fellow and Guest Full Professor, Leibniz-Institut für Agrartechnik und Bioökonomie e.V. (ATB), Potsdam, Germany, [4]Formerly Alexander von Humboldt Fellow and Guest Full Professor, Grassland Grassland Science and Renewable Plant Resources, Faculty of Organic Agricultural Sciences, Universität Kassel, Witzenhausen, Germany, [5]Biosystems and Environmental Engineering Research Group, Marlborough, Harare, Zimbabwe

Chapter outline

23.1 Introduction	447
23.2 Techno-economic assessment of biochar systems: tools and evidence	448
23.2.1 Techno-economic assessment tools	448
23.2.2 Overview of the evidence	449
23.3 Key factors influencing techno-economic feasibility of biochar	450
23.3.1 Biomass feedstock	450
23.3.2 Thermo-hydro-chemical conversion technology	450
23.3.3 Market price of biochar	451
23.3.4 Biochar application	452
23.4 Challenges in techno-economic assessments of biochar systems	452
23.5 Discussion and future perspectives	453
23.5.1 Discussion	453
23.5.2 Future research directions	455
23.6 Conclusion	457
References	457

23.1 Introduction

The pressing challenges of climate change, land degradation, and sustainable resource management are driving innovative solutions. Among these, biochar systems have emerged as a promising technology for sustainable land use development, offering numerous environmental benefits (Li et al., 2017; Qambrani et al., 2017). Recognized by the Intergovernmental Panel on Climate Change as a powerful carbon-negative technology, biochar sequesters carbon in soils for centuries (Rogelj et al., 2018). As interest in sustainable agricultural practices grows, biochar systems are increasingly considered for their roles in agriculture, environmental remediation, and renewable coproduct production.

Biochar is a carbon-rich material produced by pyrolysis of biomass, enhancing soil quality and sequestering carbon (Maroušek et al., 2019; Anderson et al., 2016). Its production technologies have evolved, capturing carbon, improving soil fertility, reducing greenhouse gas emissions, and supporting a circular economy (Palansooriya et al., 2019). Biochar systems also generate valuable coproducts, such as bio-oils and syngas, which contribute to their economic feasibility.

Despite its enormous potential, the widespread adoption of biochar hinges on its technological and economic viability, necessitating comprehensive assessments of its system production costs, economic benefits, and market integration. Techno-economic analyses (TEAs) are essential for evaluating biochar systems, utilizing methodologies such as cost-benefit analysis (CBA), and efficiency analysis (Cervi et al., 2020; Nematian et al., 2021; Sahoo, 2019; Shabangu et al., 2014). However, the TEA of biochar systems is still evolving, with key areas requiring further research.

Critical factors influencing biochar's economic feasibility include feedstock choice, production scale, energy requirements, logistics, applications, and market dynamics. Barriers such as inaccurate data, complex models, regional

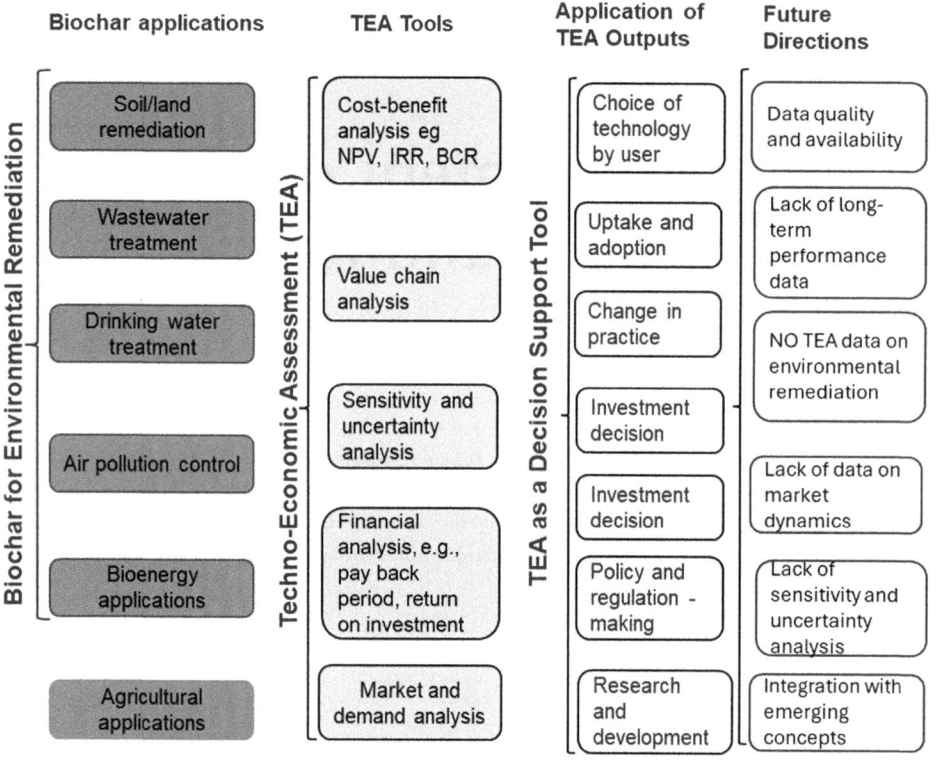

FIGURE 23.1 Biochar application for environmental remediation, TEA as a decision support tool, potential application of TEA outputs, and future research directions. *TEA*, Techno-economic assessment.

variations, policy frameworks, public perception, and market acceptance must be addressed (Koné & Galiegue, 2023; Vochozka et al., 2016; Wu et al., 2017; Meyer et al., 2017).

This chapter reviews the techno-economic evaluation of biochar, surveying current studies and future research directions to optimize its sustainable development. The objectives are to (i) identify studies on biochar TEA, (ii) highlight key factors for economic sustainability, (iii) explore assessment challenges, and (iv) fulfill the propose of future research directions. This framework aims to guide investors, researchers, policymakers, and stakeholders in advancing the biochar industry.

The rest of this chapter is structured as the following sections: (1) an overview of current TEA approaches, (2) key factors affecting economic viability, (3) challenges in TEA, (4) future research directions to promote biochar as a sustainable technology, and (5) conclusion. Fig. 23.1 is a summary depiction of the focus of this chapter highlighting biochar application for environmental remediation, TEA as a decision support tool, potential application of TEA outputs, and future directions.

23.2 Techno-economic assessment of biochar systems: tools and evidence

23.2.1 Techno-economic assessment tools

Techno-economic assessment is a process that uses different approaches to estimate the overall economic returns of a specific technology, enabling decision-makers to objectively compare the benefits against the various costs of applying that technology. It captures the potential returns from an investment allocated to biochar and coproduct production (Petter & Tyner, 2014). Assessing biochar systems from a techno-economic viewpoint is a complex process that necessitates a comprehensive understanding of the interactions between technology, economics, and environmental impact.

Existing approaches to the techno-economic assessment of biochar systems involve various frameworks, models, and methodologies aimed at quantifying the costs, benefits, and overall feasibility of biochar production, from access to inputs to the final use of the product. The economic viability of biochar depends on a range of factors, from feedstock prices to efficient production technologies (Ahmed et al., 2016). In this section, we focus on understanding the different methodologies employed in current and recent advances in the economic evaluation of biochar systems.

Here, the different TEA methodologies used to assess various aspects of the economic profitability of biochar are discussed. The aspects considered include the choice of the right feedstock, pyrolysis method, associated analysis or costs, benefits, energy requirements, coproduct use, environmental impacts of biochar production and application, and the ROI in biochar.

A crucial element in the assessment of biochar systems is the selection of appropriate methodologies that take into account the complex dynamics of the production and use of biochar and its coproducts, as well as its impact on economic and environmental parameters. Most of the existing approaches TEA encompass (CBA), including sensitivity tests, risk analysis, and stochastic techno-economic analysis (e.g., Brown et al., 2011; Shabangu et al., 2014; Zhao et al., 2016; Campbell et al., 2018b).

CBA uses numerous uncertainty assumptions to systematically assess the costs and benefits of biochar synthesis and use. CBA calculates various financial indicators such as cost-benefit ratio (CBR), net present value (NPV), internal rate of return (IRR), payback period, break-even or minimum selling price, and return on investment (ROI) with static inputs, producing deterministic results (Campbell et al., 2018a; Zhao et al., 2016). In addition to CBA variables, Monte Carlo simulation has been used by other researchers (e.g., Campbell et al., 2018a; Petter & Tyner, 2014) to introduce uncertainty in input variables in their economic analyses, allowing the sensitivity of these results to be assessed by quantifying the impact of changes in specific variables.

These methods monetize tangible and intangible benefits, such as increased soil fertility, increased yields and carbon sequestration, reduced greenhouse gas emissions, and reduced prospective revenue streams from the sale and use of biochar. By measuring these costs and benefits, decision-makers can assess the overall attractiveness and economic viability of investing in biochar systems.

23.2.2 Overview of the evidence

Some studies have conducted techno-economic assessments of biochar systems using various TEA tools. Brown et al. (2011) estimated the IRR for feedstock costs and projected revenues for the pyrolysis system in 2015 and 2030. By 2030, projections indicate an increase in the value of biochar as a carbon offset to $60 per metric ton, and gasoline prices are expected to rise from $2.96 per gallon to $3.70, leading to a 26% increase in the rate of return for investors in biochar production. In contrast, Petter & Tyner (2014) computed all the CBA variables, including BCR, IRR, and NPV, combined with Monte Carlo simulation, and found that the uncertainty of the biofuel selling price is the main factor contributing to the risk involved in private investment in a biofuel production facility.

The CBA of biochar field trials on cereals in Northwestern Europe and Sub-Saharan Africa demonstrated a positive NPV for cereal crops in Sub-Saharan Africa in numerous scenarios, with attractive results even over a 30-year timeframe. The Northwestern Europe biochar scenarios, on the other hand, all exhibited a negative NPV in both the short and long term (Dickinson et al., 2015). In their economic analysis of biochar, Wrobel-Tobiszewska et al. (2015) pointed out that the crucial factors responsible for the financial profitability of biochar production from eucalypt plantation residue wood and the product use scenario are biochar price and final product distribution in Tasmania.

Likewise, Zhao et al. (2015, 2016) captured the break-even price and NPV of an investment related to biofuel production and concluded that product market prices had the strongest effect on financial returns, while the break-even price was most sensitive to technical uncertainty associated with feedstock cost and fuel yield. Using fast pyrolysis and hydrotreating, the lowest equilibrium fuel price appears to be $3.11 per gallon gasoline equivalent ($0.82 per liter gasoline equivalent). Given forecast energy prices, investors in the hydrotreating sector can expect a probability of loss of 59%.

Aller et al. (2018) evaluated the NPV of expected private and public net economic benefits of alternative biochar applications over conventional corn and corn-soybean cropping systems and residue removal rate scenarios over a 32-year simulation period in Iowa. The results revealed that the public benefits, quantified as reduced nitrate leaching and increased soil carbon sequestration, significantly offset the private income generated by crop yield benefits. A biochar application rate of 22 Mg ha^{-1} was more profitable per ton than higher biochar rates.

Campbell et al. (2018b) used Monte Carlo simulation to manage uncertainties in critical techno-economic factors and conducted a financial assessment of energy production using regional feedstocks in the western US. The baseline scenarios projected a 20-year average NPV ranging from -$8.3 million for electric power to $76.0 million for liquid biofuel, including a biochar coproduct. Their simulations showed a varied range of NPV results, encompassing both positive and negative values. For electric power, the NPV ranged from -$74.5 million to $51.4 million, while for liquid biofuels it varied from -$21.6 million to $246.3 million, depending on the different scenarios. In a different study by Campbell et al. (2018a) using the same method, sole pyrolysis biochar production had the highest probability of payback with an average NPV of $41.5 million and only 20% of the results showing a net loss. In contrast, 68% of the results for coproduction of biochar and biofuels by auger showed a financial loss, with an average NPV of -$24.2 million.

Furthermore, the economic analysis of date palm waste biochar reveals its superior viability, featuring a payback time of 4 years and 132 days, an IRR of 14.8%, an ROI of 22.9%, and a gross margin of 35.5% (Giwa et al., 2019). The CBA of biochar application at 15t/ha showed an increase in the gross margin by 21% and 53% for specific carbon price scenarios. A comparative TEA and Monte Carlo risk analysis performed by Haeldermans et al. (2020) claim that biochar price is the most important determinant of a biochar production plant's feasibility. Keske et al. (2020) discovered that biochar is profitable in Canada for beetroot production, through a stochastic techno-economic study of the biochar production budget of a slow pyrolysis biochar unit, but not profitable for potato production. They argue that using biochar obtained from forest biomass as a soil improver has the potential to create a local market for biochar while also ensuring food security. Nematian et al. (2021) performed a techno-economic analysis of the conversion of orchard waste to biochar. The results of the sensitivity analysis show that biochar production costs are more sensitive to production rates in California's Central Valley.

These results highlight many uncertain implications regarding the profitability of biochar production and application investments, reflecting the diverse methodologies used in the techno-economic analysis of biochar systems. The use of diverse TEA tools makes it difficult to compare results among studies. Therefore, to enable comparison among studies from various regions, standardized TEA tools may need to be used.

However, compared to studies focusing on biochar applications in environmental remediation and agriculture, data on the TEA of biochar systems are still limited. This is particularly the case in low-income countries such as Africa, Southeast Asia, and Latin America, where biochar could provide potential benefits. Specifically, the TEA of biochar for environmental remediation of soils, drinking water, wastewater, and air pollution is scarce. Therefore, it remains unclear whether biochar applications in these various domains of environmental remediation are economically viable relative to current practices or technologies. This highlights the need for further work on the TEA of biochar systems in various regions. Thus, the proposed roadmap for further advancement of biochar for environmental remediation stresses the need for the application of decision support tools such as TEA for the assessment of biochar systems (Chapter 25, Gwenzi et al., 2015).

23.3 Key factors influencing techno-economic feasibility of biochar

In this section, we will discuss the important factors that significantly impact the feasibility of biochar systems and coproducts from a technological and economic standpoint. These factors include the type of feedstock used, the pyrolysis technique employed, the crop cultivated, and related prices. Other essential factors are the production and usage scale, transportation and logistics, the intended usage of the end product, the dynamics of the carbon market, and the type of soil. We also examined the economic potential of some coproducts (bio-oil, synthesis gas, biogas, and biofuel) from the biochar production process. By considering these factors, stakeholders can make well-informed decisions that optimize the biochar value chain and ensure long-term financial viability in different countries and worldwide.

23.3.1 Biomass feedstock

Firstly, the choice, availability, quality, and purchase prices of the feedstock used in biochar production play a crucial role throughout the life cycle of biochar and affect the outcome's profitability in terms of final product quantity, quality, and prices (Brown et al., 2011; Sessions et al., 2019; Campion et al., 2023; Aller et al., 2018; Campbell et al., 2018a,2018b; Woolf et al., 2021). Feedstock price is an important factor in biochar production (Sessions et al., 2019). However, some scholars argue that the properties of biochar, which affect its performance, depend on variables such as feedstock type and pyrolysis operating temperature (Campion et al., 2023; Woolf et al., 2021). For instance, Ye et al. (2020) found that biochar derived from cereal residues significantly enhanced crop yields compared to biochar from other feedstocks. They also observed that biochar produced at temperatures below 400°C led to the most substantial improvements in crop yield, which is a key factor in biochar utilization. Similarly, Woolf et al. (2021) found that the carbon retained in biochar varies depending on the feedstock used and pyrolysis conditions. Kung et al. (2015) investigated the fast pyrolysis of poplar, maize stover, rice straw, orchard waste, animal waste, and open pasture waste, noting that poplar is preferred due to its carbon sequestration capacity. Biochar derived from hardwood offered enhanced private and external benefits, leading to increased corn yields and elevated carbon sequestration, as well as reduced nitrogen leaching, according to Aller et al. (2018). Case studies show that slow pyrolysis systems using appropriate feedstocks can potentially mitigate up to 1.4 Mg CO2eq per Mg of feedstock consumed, as detailed by Field et al. (2013). However, further work investigating a wide range of feedstocks including biowastes is needed.

23.3.2 Thermo-hydro-chemical conversion technology

Along with the choice of feedstock and related parameters, the second crucial factor affecting the economic profitability of biochar is the thermo-hydro-chemical pathway or technology. Several scholars (e.g., Snyder, 2019; Keske et al.,

2020; Shabangu et al., 2014; Ahmed et al., 2016; Sessions et al., 2019; Campbell et al., 2018b; Salgado et al., 2018) have investigated pyrolysis processes and their economic viability, revealing distinct results that highlight the impact of temperature variations and biochar yield on NPV and overall economic profitability.

Biomass pyrolysis is classified into different categories based on temperature ranges, each producing distinct products including biochar, bio-oils, and synthetic gas: slow pyrolysis at temperatures under 500°C, fast pyrolysis between 500°C and 800°C, and gasification at temperatures above 800°C (Sohi et al., 2009). Biochar production is typically performed using either slow or fast pyrolysis, depending on the temperature, pyrolysis time, and heating rate. Evidence suggests that slow pyrolysis is the most effective for optimizing biochar production, while fast pyrolysis is better suited for maximizing bio-oil production, and gasification for synthetic gas (Duku et al., 2011; Gwenzi et al., 2015).

Studies have shown that pyrolysis techniques affect the profitability of biochar production. Slow pyrolysis generally produces more biochar output compared to fast pyrolysis (Ahmed et al., 2016). Sessions et al. (2019) highlighted that pyrolysis technology is a key parameter for economic profitability, comparing thermal pyrolysis and microwave pyrolysis. Thermal pyrolysis costs USD 485 per Mg of biochar, significantly less than microwave pyrolysis, which costs USD 600 per Mg. Salgado et al. (2018) found that pyrolysis at 450°C was more favorable than at 550°C, attributing this to higher biochar yield and assuming equal energy recovery in both scenarios. They also considered a high biochar price of USD 2,650/Mg. Conversely, Campbell et al. (2018b) found that high-temperature pyrolysis (500°C to 650°C) was more cost-effective compared to lower temperatures (450°C). Similarly, Snyder (2019) found that fast pyrolysis at 500°C yielded an NPV almost three times higher than slow pyrolysis at 400°C, due to significantly higher bio-oil yields and its higher valuation (USD 500/Mg) compared to biochar (USD 35/Mg). Lu and El Hanandeh (2019) identified 550°C as the most desirable temperature for fast pyrolysis, highlighting the importance of temperature in process efficacy. Kung et al. (2015) preferred fast pyrolysis for higher electricity sales, although rising carbon prices might make slow pyrolysis more attractive due to its greater carbon sequestration potential. Aller et al. (2018) compared biochar types from slow pyrolysis of hardwood and fast pyrolysis of corn stover, finding that slow pyrolysis yielded higher private and external benefits due to its greater carbon content and carbon-to-nitrogen ratio. Giwa et al. (2019) noted that solar-based pyrolysis emits only 38% of the CO_2 compared to conventional pyrolysis, demonstrating its environmental benefits.

However, Snyder (2019) and Field et al. (2013) found that biochar is valued lower than bio-oil, making fast pyrolysis more profitable due to lower biochar yields. Field et al. (2013) had earlier found that even fast pyrolysis was economically unviable, consistent with Brown et al. (2011), who did not reach a clear conclusion on profitability based solely on the pyrolysis method. Their study emphasized that profitability was influenced by factors such as feedstock cost, selling prices of by-products (biochar, biofuel, pyrolysis gas, and bio-oil), and potential revenue from carbon offset credits.

Overall, these studies highlight the significant impact of temperature variations on pyrolysis processes, leading to varying conclusions about the most economically advantageous approach. While some studies focus on maximizing yields and bio-oil value, others prioritize increasing biochar production and market value. Despite differing perspectives, slow pyrolysis is generally considered the most efficient process according to most studies.

23.3.3 Market price of biochar

After selecting the feedstock and producing the biochar using a specific pyrolysis process, the next key factor influencing biochar profitability is the selling price of biochar. However, feedstock prices also significantly impact profitability, as noted previously. The selling price of biochar is affected by the type of pyrolysis process used as well as the quantity and type of feedstock employed.

In their economic analysis, Shabangu et al. (2014) identified break-even points for biochar costs at approximately $220 per ton for pyrolysis at 300°C and $280 per ton for pyrolysis at 450°C. Sessions et al. (2019) explored the maximum price farmers would be willing to pay for biochar based on anticipated benefits. They found that wheat producers might be willing to pay up to USD 250 per hectare if biochar application led to a 10% yield increase over 5 years.

Biochar prices have also been compared to carbon pricing, as biochar application sequesters carbon in the soil. Galinato et al. (2011) assessed the economic impact of biochar on agricultural land using three price points: USD 87 per ton, USD 114.05 per ton, and USD 350.74 per ton. They found that wheat growers could only achieve profitability when the biochar price was the lowest (USD 87 per ton) and the carbon dioxide price was the highest (USD 31 per ton CO_2). Similarly, Field et al. (2013) found that a carbon price of USD 50 per ton of CO_2 equivalent would be necessary for biochar to be profitable.

These findings highlight that the market prices of biochar and its associated products have a substantial impact on their financial outcomes. Compared to other critical technical and economic factors, these market prices play a crucial role in shaping future renewable energy policies and influencing upcoming techno-economic analyses. Currently, there is a notable lack of emphasis on the impact of market prices in these analyses, indicating a need for their inclusion in future studies and policy decisions.

23.3.4 Biochar application

Biochar has several applications in environmental remediation, bioenergy, and agriculture. One may posit that the techno-economic feasibility of biochar systems also depends on their application field. For example, in agriculture, the profitability of biochar also depends on soil type and the crop variety being grown. Dickinson et al. (2015) found that biochar was never profitable as a soil amendment for cereal production in northwestern Europe, unlike in Sub-Saharan Africa. Similarly, Meulemans (2016) assessed biochar's impact on two distinct soil types within a similar climate. They observed that biochar positively influenced fertile Mollisol with Napier grass but not with sweetcorn, whereas the opposite effect was noted in the low-fertility Oxisol. Furthermore, when applying biochar in two cropping systems, Aller et al. (2018) found that continuous maize cropping yielded higher results compared to a maize and soybean rotation. These results suggest that the long-term performance and reliability of biochar can be uncertain, depending on soil structure, and investors need to proceed with caution.

The application rate of biochar affects both its public and private profitability (Aller et al., 2018; Pandit et al., 2018). Aller et al. (2018) compared three application rates (22, 45, and 90 Mg/ha) and determined that an application rate of 22 Mg/ha provided the highest benefits per Mg of biochar. This is because higher application rates lead to diminishing private benefits as the increased crop yield does not offset the higher purchase costs. However, external benefits, such as carbon sequestration and reduced nitrate leaching, increase with higher application rates. Conversely, Pandit et al. (2018) evaluated five application rates (5, 10, 15, 25, and 40 Mg/ha) from a farmer's perspective in rural Nepal. They found that, regardless of carbon price scenarios (0, 6, or 42 USD/Mg CO_2), the optimal rate for profitability was 15 Mg/ha. They noted that these results are specific to the particular soil, biochar, and farming system in Nepal.

These findings highlight that the profitability of biochar application rates is not solely determined by the quantity applied but is also influenced by associated costs and carbon market prices. Nonetheless, significant uncertainties remain regarding the profitability and viability of biochar systems.

In summary, the few studies investigating the techno-economic feasibility of biochar applications are limited to agriculture. This is probably because biochar has its origins rooted in agricultural applications as a soil amendment. By comparison, data are scarce on the techno-economic feasibility of biochar for the following applications: (1) soil/land remediation, (2) drinking water treatment, (3) wastewater treatment, (4) air pollution control, and (5) bioenergy applications, such as an amendment in anaerobic digestion or as an electrode in microbial fuel cells. Yet, the applications of biochar for these environmental remediation purposes seem more feasible than in agriculture. This is because smaller quantities are needed for applications in environmental remediation compared to agricultural applications such as a soil amendment. Therefore, further work focusing on the TEA of biochar for environmental remediation is required. Such TEA studies should compare biochar-based remediation systems to the following: (1) existing competing technologies, (2) the "business-as-usual" scenario, and (3) the "do-nothing" scenario.

23.4 Challenges in techno-economic assessments of biochar systems

In this section, we explore factors that may hinder the techno-economic modeling of biochar systems based on the studies analyzed. Biochar systems are recognized for their potential to provide numerous benefits, including soil carbon capture, soil health improvement, yield enhancement, and, ultimately, food security, job creation, and environmental management. Despite these potential benefits, economic assessments of biochar remain uncertain, and the value of the technology has yet to be firmly established. This is due to the disparity between reported short-term agronomic benefits and the projected long-term advantages of biochar as a viable soil improver (Dickinson et al., 2015).

The studies reviewed reveal that assessing the techno-economic feasibility of biochar involves several complex challenges that complicate a comprehensive and accurate evaluation of its feasibility and potential impacts. These challenges include limited and incomplete data, difficulties in modeling, regional disparities, regulatory impediments, market acceptance issues, and underdeveloped value chains. The cost of production—including production technology, ingredient acquisition, and transportation—has also impeded the market development of biochar. Understanding these challenges is crucial for refining methodologies and improving the reliability of assessments in this field. Additionally, recognizing these difficulties can help academics and policymakers develop effective measures to enhance the precision and dependability of TEA for biochar investment.

One of the most significant obstacles in conducting TEA for biochar systems is data availability and quality. The absence of standardized databases—particularly for feedstock parameters, pyrolysis procedures, and biochar use outcomes—poses a serious barrier, leading to economic assessments that rely on numerous assumptions. Variability in data sources and quality among studies reduces the robustness and comparability of TEA outcomes, diminishing evaluation reliability.

The lack of reliable data has also resulted in a lack of consensus on metrics and evaluation frameworks among studies. This absence of adequate evaluation measures and frameworks presents a significant challenge for both biochar technology and economic assessment. The diversity of evaluation methods—including varying approaches such as costing, environmental impact assessment, and social concerns—complicates the comparison and synthesis of findings. This lack of consensus limits stakeholders' ability to make informed judgments based on a structured assessment procedure.

As part of an economic analysis, specifically a CBA, three decisive indicators are generally computed: NPV, IRR, and BCR. These indicators provide a realistic picture of an investment. However, many studies have failed to capture these indicators in biochar economic assessments. Additionally, studies often overlook the production and application costs and profitability of biochar, whether used by the producer, purchased from a supplier, or sold as a product. This shortfall arises from the lack of data availability and the underdeveloped and unstructured supply chain of biochar. Since data on biochar production—from feedstock acquisition to final product use—are not consistently recorded, it is challenging for researchers to capture a comprehensive and effective economic assessment of biochar.

Moreover, some studies fail to include all relevant indicators in the CBA. For instance, in CBA and TEA, NPV is the most commonly used profitability indicator and is generally considered more reliable than IRR. Zhao et al. (2016) suggested that the break-even price is a useful economic indicator compared to NPV or IRR when evaluating emerging technologies using TEA. The CBA often relies on various assumptions. Previous techno-economic analysis studies of biofuel-biochar coproduction have primarily aimed at generating reliable cost estimates with fixed inputs. However, these studies frequently yield deterministic outcomes for NPV and break-even prices, neglecting significant uncertainties inherent in crucial techno-economic factors. While these methods work well for scenarios with known and stable variables, they become inadequate when dealing with volatile input factors, such as market prices for biofuel and biochar, resulting in oversimplified project performance estimates that fail to quantify associated uncertainty or risk. Therefore, Monte Carlo simulation, which accounts for uncertainty in TEA indicators, is not used in many studies, leading to potentially unreliable results.

Other challenges to the TEA of biochar systems include uncertainties arising from the variety of interrelated activities involved, such as feedstock selection, pyrolysis conditions, biochar application methods, and their effects on soil health, crop yield, and carbon sequestration. Even if short-term studies indicate potential benefits such as increased soil fertility and carbon sequestration, questions persist about the long-term stability of biochar, its effects on soil properties over time, and the potential environmental risks of prolonged application. These interactions are complex and variable, making it difficult to develop uniform and standardized techno-economic evaluation frameworks for biochar systems.

The variability and lack of data sources, variations in data quality and computed indicators across research, and the presence of divergent results with multiple assumptions limit the robustness and comparability of TEA outputs, thereby reducing the credibility of TEA assessments for biochar systems.

23.5 Discussion and future perspectives

23.5.1 Discussion

Exploring and evaluating the TEA of biochar systems presents a multifaceted picture characterized by both progress and persistent gaps. This section summarizes the key findings from the reviewed studies, outlines ways to address existing challenges, and offers perspectives for future research and development.

On one hand, some research studies utilize techno-economic modeling to simulate and analyze the economic viability of biochar production and usage. These models assess input and production costs as well as potential earnings associated with biochar by integrating factors such as feedstock costs, pyrolysis techniques, energy inputs, labor expenses, and market pricing. Notably, these models often evaluate multiple scenarios and hypotheses, allowing for sensitivity assessments to understand the impact of various variables on the economic viability of biochar systems. Despite growing research interest, biochar remains underutilized on farms primarily due to its higher cost compared to chemical fertilizers and a lack of awareness or uncertainty about its benefits for agricultural production (Campion et al., 2023; Pourhashem et al., 2019). This situation makes it difficult to obtain the data necessary for a comprehensive economic analysis of biochar systems. For instance, Koné & Galiegue (2023) found that only 12.34% of 384 farmers in western Burkina Faso adopted biochar from traditional pyrolysis. Similarly, Rogers et al. (2021) found that only about 40% of 172 smallholder farmers in Tanzania were aware of biochar. Additionally, biochar prices are often unstable and fluctuate considerably, highlighting the need for economic and commercial studies on biochar pricing to minimize these variations.

On the other hand, among studies on the economic profitability of biochar, few have computed comprehensive CBA indicators, capturing both acquisition and application costs as well as output benefits. Most studies that estimate the monetary profitability of biochar focus primarily on the NPV, which alone cannot fully guide investment decisions. Many studies emphasize environmental, agronomic, and social profitability, often neglecting the monetary economic analysis and investment profitability. For example, Collison et al. (2009) found that biochar application can enhance agricultural profitability by improving crop yields and quality while reducing fertilizer and cropping costs. Similarly, Zhang et al. (2012) reported that biochar application on marginal soils increased crop yields and reduced nutrient leaching and greenhouse gas emissions. In Kenya, biochar introduction led to a significant increase in agricultural production, with an increase of approximately 2.9 tonnes per hectare compared to control plots (Kimetu et al., 2008). In Zimbabwe, biochar improved seed germination and crop establishment in poor soils (Gwenzi et al., 2015). The International Biochar Initiative reports significant increases in coffee yields in countries such as Tanzania, Ghana, Rwanda, Indonesia, Ethiopia, and Brazil, demonstrating biochar's positive impact on agricultural production. Biochar also helps stabilize and sequester carbon in the soil, reducing almost one-eighth of CO_2 emissions (Mukome et al., 2013; Kurniawan et al., 2023).

Furthermore, previous research has shown that the adoption of biochar in developing country agriculture is influenced by socio-economic factors such as age, gender, education, income, occupation, beliefs, farm size, and importantly, availability of raw materials (Fridahl et al., 2020; Rogers et al., 2021). Rogers et al. (2021) found that farmers with lower education and income levels, particularly male farmers aged 40–60, have lower adoption rates of biochar in Tanzania. Barriers include alternative uses for biochar raw materials, lack of government or extension service involvement, and agricultural traditions and customs. Farmers cited improved soil structure, reduced soil acidity and fertilizer needs, increased crop yields, improved food security, increased family income, climate change mitigation, and enhanced resilience to climate change as key reasons for adopting biochar. However, belief in biochar's effectiveness as a soil amendment requires long-term evaluations from investment through to yield and benefits. Conducting such analyses is challenging due to the lack of extensive empirical data over long periods and uncertainties regarding the permanence of biochar.

Additionally, complexities arise during the biochar life cycle concerning input supply, production costs, production methods, application, and benefits, as well as carbon market dynamics (Roberts et al., 2010). These constraints are due to the variability of biochar properties and effectiveness, which are largely determined by feedstock type and pyrolysis conditions. Furthermore, biochar production and application are often not guided by experts or extension service agents, leading to insufficient data collection even in agricultural development projects. For example, a project in the Mbeya and Songwe regions of Tanzania lacked government guidelines or policy support for biochar production, relying instead on knowledge from countries such as Rwanda. Establishing legal frameworks is essential to create a robust and transparent process that highlights the benefits of waste in improving soil health while mitigating potential risks to the environment and human health. It is crucial to differentiate waste types, especially between low-risk waste and mixed waste streams requiring specialized treatment. Defining clear criteria for acceptable feedstock, specifying pyrolysis conditions, and outlining essential biochar characteristics would strengthen these legal instruments. For example, the IFC initiative in partnership with NetZero aims to promote biochar production and distribution in Africa, enhancing greenhouse gas sequestration and agricultural productivity to combat climate change.

To address these challenges, numerous recommendations for improving the efficiency of biochar systems and overcoming TEA obstacles are proposed. These include advancements in pyrolysis technology, integration with circular economy concepts, development of cost-effective raw material sourcing methods, exploration of new biochar uses, and the creation of standardized life cycle monitoring methods. Additionally, understanding farmers' willingness to produce or pay for biochar and addressing risks and uncertainties linked to biochar investments are crucial. Policymakers should focus on collecting and disseminating data across all stages of biochar production and use, including raw material acquisition, production techniques, and results. Collaborative efforts to obtain, verify, and share reliable data are essential for conducting in-depth, country-specific assessments. Furthermore, establishing robust TEA assessment frameworks and conducting long-term studies on biochar's lasting effects are crucial for resolving uncertainties and understanding its long-term impacts. As biochar production also yields by-products such as biofuels, syngas, and heat, analyzing their potential and development opportunities within the circular economy framework is important. Facilitating dialog between stakeholders and researchers to standardize evaluation metrics and frameworks, as advocated by the IBI, can significantly improve the consistency and reliability of the assessment process.

In summary, addressing these challenges is critical to advance the field of TEA in biochar systems. Overcoming data limitations, refining assessment methodologies, and addressing uncertainties will enable more precise evaluations and informed decision-making for integrating biochar systems into sustainable practices. In light of the several knowledge gaps on TEA of biochar systems, future research is warranted. To achieve this, future research directions on TEA of biochar systems are proposed below.

23.5.2 Future research directions

In summary, the existing research on the TEA of biochar technology is limited compared to studies focused on its applications in environmental remediation, bioenergy, and agriculture. Additionally, available evidence has notable limitations, highlighting the need for more data and comprehensive analysis. Future research on the TEA of biochar technology should focus on the following ten thematic aspects:

23.5.2.1 Data availability and quality

Two key limitations pertain to data availability and quality:

- Lack of standardized databases: There is a lack of standardized and comprehensive databases for feedstock parameters, pyrolysis procedures, and biochar use outcomes. This leads to variability in economic assessments and complicates comparisons across studies.
- Incomplete data: Many studies suffer from incomplete or inadequate data regarding production costs, market prices, and application outcomes. This limits the ability to perform robust and reliable TEA, particularly in low-income countries where systematic monitoring, data storage, and quality control are often lacking.

23.5.2.2 Comprehensive cost-benefit analysis

The literature on CBA reveals two limitations:

- Limited economic indicators: Few studies compute all critical economic indicators, such as NPV, IRR, and BCR, which are necessary for a complete assessment of biochar's economic viability.
- Inadequate cost accounting: Many studies fail to capture all costs associated with biochar production and application, including transportation, labor, and feedstock acquisition. These aspects are crucial for determining the techno-economic feasibility of biochar and its applications.

23.5.2.3 Sensitivity and uncertainty analysis

Most TEA studies on biochar technology face two limitations:

- Lack of sensitivity analysis: Not all studies perform sensitivity analysis to assess how variations in key parameters (e.g., feedstock costs, and energy inputs) impact the overall economic viability of biochar systems.
- Uncertainty quantification: There is a lack of systematic approaches to quantify and address uncertainties in techno-economic models. Techniques such as Monte Carlo simulations, which could help in understanding risk and variability, are underutilized.

23.5.2.4 Long-term performance and impact

Biochar research often lacks long-term data, resulting in the following limitations:

- Longevity of Biochar: There is insufficient long-term data on the stability and effectiveness of biochar over extended periods, especially in soils. This impacts the ability to predict its long-term benefits and risks accurately.
- Environmental and Agronomic Effects: More research is needed on the long-term environmental and agronomic impacts of biochar application, including effects on soil health, crop yields, and carbon sequestration.

23.5.2.5 Regional and contextual variability

TEA results should not be generalized across regions with differing biophysical and socio-economic conditions due to the following gaps:

- Regional Disparities: Economic assessments often fail to account for regional variability in feedstock availability, production costs, and market conditions. This limits the generalizability of findings.
- Local Adaptation: There is a need to understand how local factors such as climate, soil type, and agricultural practices influence the economic viability of biochar systems.

23.5.2.6 Policy and regulatory frameworks

Policy and regulatory frameworks are critical for TEA studies, but several countries lack these frameworks for biochar, leading to the following gaps:

- Regulatory Challenges: There is a lack of comprehensive regulatory frameworks and guidelines for biochar production and application, which hinders market development and standardization.
- Incentive Structures: More research is needed to evaluate the effectiveness of policy incentives and subsidies for promoting biochar adoption and market development.

23.5.2.7 Market dynamics and acceptance

The biochar market is still developing, resulting in the following gaps:

- Market Demand: There is limited understanding of market demand and acceptance of biochar products. Studies need to address consumer behavior, market readiness, and potential barriers to adoption.
- Price Fluctuations: The instability and variability in biochar prices need more detailed analysis to understand factors influencing price changes and their impact on economic feasibility.

23.5.2.8 Socio-economic factors

The socio-economic aspects of biochar, especially in low-income settings, require further research:

- Farmer Adoption: Research gaps exist in understanding socio-economic factors affecting farmer adoption of biochar, such as education, income, and access to resources.
- Economic Viability for Smallholders: There is limited information on how biochar systems can be economically viable for smallholder and subsistence farmers, particularly in developing countries.

23.5.2.9 Integration with emerging concepts

TEA studies on the integration of biochar with emerging concepts and technologies are scarce. These include:

- Circular Bioeconomy: More research is needed on the economic potential and development opportunities for biochar coproducts, such as biofuels, syngas, and heat, within a circular economy framework.
- Conservation Agriculture and Agroecology: Further understanding is required on how integrating biochar with conservation agriculture and agroecology affects the techno-economic feasibility of biochar systems.

23.5.2.10 Lack of techno-economic assessment data on biochar for environmental remediation applications

Most TEA studies focus on agricultural applications, with limited research on:

- Environmental Remediation: There is a scarcity of studies investigating the techno-economic feasibility of biochar for environmental remediation of soil, water, wastewater, and air pollution.
- Bioenergy Applications: TEA studies on biochar applications in bioenergy are still lacking. Further work is needed to address these gaps.

Addressing these knowledge gaps requires concerted efforts from researchers, policymakers, and industry stakeholders to develop more comprehensive and standardized methodologies, improve data collection and sharing, and create supportive regulatory environments.

23.6 Conclusion

Biochar systems have the potential to effectively address climate change and soil degradation, as evidenced by various studies. However, the widespread adoption of these systems depends on understanding their feasibility throughout the production and utilization process. This chapter presents a review of advancements in TEA of biochar systems. It examines the studies and methodologies employed, highlights factors influencing the economic viability of biochar products, discusses existing challenges in TEA for biochar systems, provides valuable recommendations, and suggests future research directions.

This review is significant for several reasons. First, it consolidates knowledge on TEA for biochar and identifies research gaps that need to be addressed. Second, our analysis offers policymakers and stakeholders valuable insights into the feasibility of implementing biochar systems and promoting practices within each country's context. Third, considering that biochar systems align with the United Nations Sustainable Development Goals, such as climate action, responsible consumption and production, and sustainable living, our findings can inform actions toward achieving these goals.

Overall, TEA studies on biochar systems have demonstrated progress while also highlighting uncertainties and challenges. The review showcases models, frameworks, and technological advancements, such as CBA indicators and Monte Carlo simulations, used to evaluate the feasibility and potential of biochar production and usage from a techno-economic perspective. Despite notable progress, substantial challenges persist in the field. Data constraints, including issues related to quality assurance, standardization, and accessibility, continue to affect the reliability and comparability of TEA for biochar systems. Additionally, complexities in defining system boundaries, uncertainties surrounding long-term implications, and the absence of established metrics pose considerable obstacles to conduct comprehensive studies.

Numerous factors intricately impact the economic viability of biochar systems, including public perception, feedstock availability, and quality, market prices, pyrolysis methodologies, biochar pricing, and engagement with the carbon market. These factors collectively shape the economic profitability and feasibility of biochar systems.

Looking forward, this chapter examines prospective research avenues to improve the economic feasibility of biochar systems. To this end, ten knowledge gaps in the TEA of biochar technology were proposed. Key issues worthy of attention and investment include advancements in pyrolysis technology, the development of cost-effective and quality raw material sourcing strategies, innovative uses of biochar, and efficient and standardized TEA methods. Advancements in TEA of biochar systems require concentrated effort, collaboration, innovation, and determination to overcome current limitations. Addressing data restrictions, developing robust assessment procedures, and managing uncertainties will enhance the accuracy and informativeness of assessments. This, in turn, will enable stakeholders to make more informed decisions regarding the integration of biochar systems into sustainable agricultural practices, energy production, and environmental management.

Finally, for policymakers and investors, biochar systems often generate valuable coproducts, such as bio-oils and syngas, which hold economic potential and can contribute to the circular economy. Exploring the value of these coproducts and their integration into sustainable practices is crucial to improving the overall techno-economic feasibility of biochar systems.

References

Ahmed, M. B., Zhou, J. L., Ngo, H. H., & Guo, W. (2016). Insight into biochar properties and its cost analysis. *Biomass Bioenergy, 84*, 76–86. Available from https://doi.org/10.1016/j.biombioe.2015.11.002.

Aller, D. M., Archontoulis, S. V., Zhang, W., Sawadgo, W., Laird, D. A., & Moore, K. (2018). Long-term biochar effects on corn yield, soil quality and profitability in the US Midwest. *Field Crops Research, 227*, 30–40.

Anderson, N. M., Bergman, R. D., & Page-Dumroese, D. S. (2016). *A supply chain approach to biochar systems (No. chapter 2* (pp. 25–26). London, UK: Cambridge Press.

Brown, T. R., Wright, M. M., & Brown, R. C. (2011). Estimating profitability of two biochar production scenarios: slow pyrolysis vs fast pyrolysis. *Biofuels, Bioproducts and Biorefining, 5*(1), 54–68.

Campbell, R. M., Anderson, N. M., Daugaard, D. E., & Naughton, H. T. (2018a). Financial viability of biofuel and biochar production from forest biomass in the face of market price volatility and uncertainty. *Applied Energy, 230*, 330–343.

Campbell, R. M., Anderson, N. M., Daugaard, D. E., & Naughton, H. T. (2018b). Techno-economic and policy drivers of project performance for bioenergy alternatives using biomass from beetle-killed trees. *Energies, 11*(2), 293.

Campion, L., Bekchanova, M., Malina, R., & Kuppens, T. (2023). The costs and benefits of biochar production and use: A systematic review. *Journal of Cleaner Production, 408*, 137138. Available from https://doi.org/10.1016/j.jclepro.2023.137138.

Collison, M., Collison, L., Sakrabani, R., Tofield, B., & Wallage, Z. (2009). *Biochar and carbon sequestration: A regional perspective*. Norwich, UK: The Low Carbon Innovation Centre, University of East Anglia. (A report prepared for East of England Development Agency (EEDA), DA1 Carbon Reduction Ref. No: 7049).

Cervi, W. R., Lamparelli, R. A. C., Gallo, B. C., de Oliveira Bordonal, R., Seabra, J. E. A., Junginger, M., & van der Hilst, F. (2020). Mapping the environmental and techno-economic potential of biojet fuel production from biomass residues in Brazil. *Biofuels, Bioproducts and Biorefining, 15*(1), 282–304. Available from https://doi.org/10.1002/bbb.2161.

Dickinson, D., Balduccio, L., Buysse, J., Ronsse, F., & Prins, W. (2015). Cost-benefit analysis of using biochar to improve cereals agriculture. *GCB Bioenergy, 7*(4), 850–864. Available from https://doi.org/10.1111/gcbb.12180.

Duku, M. H., Gu, S., & Hagan, E. B. (2011). Biochar production potential in Ghana—A review. *Renewable and Sustainable Energy Reviews, 15*(8), 3539–3551.

Field, J. L., Keske, C. M., Birch, G. L., DeFoort, M. W., & Cotrufo, M. F. (2013). Distributed biochar and bioenergy coproduction: a regionally specific case study of environmental benefits and economic impacts. *Gcb Bioenergy, 5*(2), 177–191.

Fridahl, M., Haikola, S., Rogers, P. M., & Hansson, A. (2020). Biochar deployment drivers and barriers in least developed countries. *Handbook of Climate Change Management: Research, Leadership, Transformation*, 1–30.

Galinato, S. P., Yoder, J. K., & Granatstein, D. (2011). The economic value of biochar in crop production and carbon sequestration. *Energy policy, 39*(10), 6344–6350.

Giwa, A., Yusuf, A., Ajumobi, O., & Dzidzienyo, P. (2019). Pyrolysis of date palm waste to biochar using concentrated solar thermal energy: Economic and sustainability implications. *Waste Management, 93*, 14–22. Available from https://doi.org/10.1016/j.wasman.2019.05.022.

Gwenzi, W., Chaukura, N., Mukome, F. N., Machado, S., & Nyamasoka, B. (2015). Biochar production and applications in sub-Saharan Africa: Opportunities, constraints, risks and uncertainties. *Journal of Environmental Management, 150*, 250–261.

Haeldermans, T., Campion, L., Kuppens, T., Vanreppelen, K., Cuypers, A., & Schreurs, S. (2020). A comparative techno-economic assessment of biochar production from different residue streams using conventional and microwave pyrolysis. *Bioresource Technology, 318*, 124083. Available from https://doi.org/10.1016/j.biortech.2020.124083.

Keske, C., Godfrey, T., Hoag, L. K., & Abedin, J. (2020). Economic feasibility of biochar and agriculture coproduction from Canadian black spruce forest. *Food and Energy Security, 9*(1), e188. Available from https://doi.org/10.1002/fes3.188.

Kimetu, J. M., Lehmann, J., Ngoze, S. O., Mugendi, D. N., Kinyangi, J. M., Riha, S., & Pell, A. N. (2008). Reversibility of soil productivity decline with organic matter of differing quality along a degradation gradient. *Ecosystems, 11*, 726–739.

Koné, S., & Galiegue, X. (2023). Potential Development of Biochar in Africa as an Adaptation Strategy to Climate Change Impact on Agriculture. *Environmental Management*, 1–15.

Kung, C. C., Kong, F., & Choi, Y. (2015). Pyrolysis and biochar potential using crop residues and agricultural wastes in China. *Ecological Indicators, 51*, 139–145.

Kurniawan, T. A., Othman, M. H. D., Liang, X., Goh, H. H., Gikas, P., Chong, K., & Chew, K. W. (2023). Challenges and opportunities for biochar to promote circular economy and carbon neutrality. *Journal of Environmental Management, 332*, 117429. Available from https://doi.org/10.1016/j.jenvman.2023.117429.

Li, W., et al. (2017). The impacts of biomass properties on pyrolysis yields, economic and environmental performance of the pyrolysis-bioenergy-biochar platform to carbon negative energy. *Bioresource Technology, 241*, 959–968. Available from https://doi.org/10.1016/j.biortech.2017.06.049, United States: Elsevier Ltd.

Lu, H. R., & El Hanandeh, A. (2019). Life cycle perspective of bio-oil and biochar production from hardwood biomass; what is the optimum mix and what to do with it? *Journal of Cleaner Production, 212*, 173–189.

Maroušek, J., Strunecký, O., & Stehel, V. (2019). Biochar farming: defining economically perspective applications. *Clean Technologies and Environmental Policy, 21*(7), 1389–1395. Available from https://doi.org/10.1007/s10098-019-01728-7, Czech Republic: Springer Verlag.

Meulemans, J. (2016). *Linking global warming potential and economics to sustainability of biochar use in Hawaiian agriculture* [Doctoral dissertation, Honolulu, University of Hawaii at Manoa].

Meyer, S., et al. (2017). Biochar standardization and legislation harmonization. *Journal of Environmental Engineering and Landscape Management, 25*(2), 175–191. Available from https://doi.org/10.3846/16486897.2016.1254640, Germany: Taylor and Francis.

Mukome, F. N., Six, J., & Parikh, S. J. (2013). The effects of walnut shell and wood feedstock biochar amendments on greenhouse gas emissions from a fertile soil. *Geoderma, 200*, 90–98.

Nematian, M., Keske, C., & Ng'ombe, J. N. (2021). A techno-economic analysis of biochar production and the bioeconomy for orchard biomass. *Waste Management, 135*, 467–477.

Palansooriya, K. N., Wong, J. T. F., Hashimoto, Y., Huang, L., Rinklebe, J., Chang, S. X., & Ok, Y. S. (2019). Response of microbial communities to biochar-amended soils: A critical review. *Biochar, 1*, 3–22.

Pandit, N. R., Mulder, J., Hale, S. E., Zimmerman, A. R., Pandit, B. H., & Cornelissen, G. (2018). Multi-year double cropping biochar field trials in Nepal: Finding the optimal biochar dose through agronomic trials and cost-benefit analysis. *Science of the Total Environment, 637*, 1333–1341.

Petter, R., & Tyner, W. E. (2014). Technoeconomic and policy analysis for corn stover biofuels. *International Scholarly Research Notices, 4*(1), 515898. Available from https://doi.org/10.1155/2014/515898.

Pourhashem, G., Hung, S. Y., Medlock, K. B., & Masiello, C. A. (2019). Policy support for biochar: Review and recommendations. *GCB Bioenergy, 11*(2), 364–380.

Qambrani, N. A., Rahman, M. M., Won, S., Shim, S., & Ra, C. (2017). Biochar properties and eco-friendly applications for climate change mitigation, waste management, and wastewater treatment: A review. *Renewable and Sustainable Energy Reviews, 79*, 255–273.

Roberts, K. G., Gloy, B. A., Joseph, S., Scott, N. R., & Lehmann, J. (2010). Life cycle assessment of biochar systems: estimating the energetic, economic, and climate change potential. *Environmental Science & Technology, 44*(2), 827–833.

Rogelj, J., Shindell, D., Jiang, K., Fifita, S., Forster, P., Ginzburg, V., & Zickfeld, K. (2018). Mitigation pathways compatible with 1.5 C in the context of sustainable development. In Global warming of 1.5 C (pp. 93-174). Intergovernmental Panel on Climate Change.

Rogers, P. M., Fridahl, M., Yanda, P., Hansson, A., Pauline, N., & Haikola, S. (2021). Socio-economic determinants for biochar deployment in the southern highlands of Tanzania. *Energies, 15*(1), 144.

Sahoo, K., et al. (2019). Techno-economic analysis of producing solid biofuels and biochar from forest residues using portable systems. *Applied Energy, 235*, 578–590. Available from https://doi.org/10.1016/j.apenergy.2018.10.076, United States: Elsevier Ltd.

Salgado, M. A. H., Tarelho, L. A., Matos, A., Robaina, M., Narváez, R., & Peralta, M. E. (2018). Thermoeconomic analysis of integrated production of biochar and process heat from quinoa and lupin residual biomass. *Energy Policy, 114*, 332–341.

Sessions, J., Smith, D., Trippe, K. M., Fried, J. S., Bailey, J. D., Petitmermet, J. H., & Campbell, J. D. (2019). Can biochar link forest restoration with commercial agriculture? *Biomass and Bioenergy, 123*, 175–185.

Shabangu, S., Woolf, D., Fisher, E. M., Angenent, L. T., & Lehmann, J. (2014). Techno-economic assessment of biomass slow pyrolysis into different biochar and methanol concepts. *Fuel, 117*, 742–748. Available from https://doi.org/10.1016/j.fuel.2013.08.053.

Snyder, B. F. (2019). Costs of biomass pyrolysis as a negative emission technology: A case study. *International Journal of Energy Research, 43*(3), 1232–1244.

Sohi, S., Lopez-Capel, E., Krull, E., & Bol, R. (2009). Biochar, climate change and soil: A review to guide future research. *CSIRO Land and Water Science Report, 5*(09), 17–31.

Vochozka, M., Maroušková, A., Váchal, J., & Straková, J. (2016). Biochar pricing hampers biochar farming. *Clean Technologies and Environmental Policy, 18*(4), 1225–1231.

Woolf, D., Lehmann, J., Ogle, S., Kishimoto-Mo, A. W., McConkey, B., & Baldock, J. (2021). Greenhouse gas inventory model for biochar additions to soil. *Environmental Science & Technology, 55*(21), 14795–14805.

Wrobel-Tobiszewska, A., Boersma, M., Sargison, J., Adams, P., & Jarick, S. (2015). An economic analysis of biochar production using residues from Eucalypt plantations. *Biomass and Bioenergy, 81*, 177–182. Available from https://doi.org/10.1016/j.biombioe.2015.06.015.

Wu, H., et al. (2017). The interactions of composting and biochar and their implications for soil amendment and pollution remediation: A review. *Critical Reviews in Biotechnology, 37*(6), 754–764. Available from https://doi.org/10.1080/07388551.2016.1232696, China: Taylor and Francis Ltd.

Ye, L., Camps-Arbestain, M., Shen, Q., Lehmann, J., Singh, B., & Sabir, M. (2020). Biochar effects on crop yields with and without fertilizer: A meta-analysis of field studies using separate controls. *Soil Use and Management, 36*(1), 2–18.

Zhang, A., Liu, Y., Pan, G., Hussain, Q., Li, L., Zheng, J., & Zhang, X. (2012). Effect of biochar amendment on maize yield and greenhouse gas emissions from a soil organic carbon poor calcareous loamy soil from Central China Plain. *Plant and Soil, 351*, 263–275.

Zhao, X., Brown, T. R., & Tyner, W. E. (2015). Stochastic techno-economic evaluation of cellulosic biofuel pathways. *Bioresource Technology, 198*, 755–763.

Zhao, X., Yao, G., & Tyner, W. E. (2016). Quantifying breakeven price distributions in stochastic techno-economic analysis. *Applied Energy, 183*, 318–326.

Part 7

Looking ahead: future perspectives and epilogue

Chapter 24

Biochar for environmental remediation and beyond: a "twin or multiple solutions" heuristic framework for uptake, adoption, and impact

Willis Gwenzi[1,2,3]

[1]*Formerly Alexander von Humboldt Fellow and Guest Full Professor, Leibniz-Institut für Agrartechnik und Bioökonomie e.V. (ATB), Potsdam, Germany,* [2]*Formerly Alexander von Humboldt Fellow and Guest Full Professor, Grassland Grassland Science and Renewable Plant Resources, Faculty of Organic Agricultural Sciences, Universität Kassel, Witzenhausen, Germany,* [3]*Biosystems and Environmental Engineering Research Group, Marlborough, Harare, Zimbabwe*

Chapter outline

24.1 Introduction	**463**
24.2 The biochar "twin or multiple solutions" framework	**465**
24.2.1 Overview of the "twin or multiple solutions" concept	465
24.2.2 A handful of biochar "twin or multiple solutions" targeting the water-energy-food-environment nexus	466
24.3 Discussion and perspectives	**469**
24.3.1 On the role of "twin or multiple solutions" in biochar uptake, adoption, and impact	470
24.3.2 The "twin or multiple solutions" framework as a primer for biochar adoption	470
24.3.3 Moving from biochar as a "silver bullet" to biochar application domains	472
24.3.4 Moving from "one-size-fits-all" and "blanket salesmen" to biochar application domains	472
24.3.5 Confronting biochar critics and skeptics with the "twin or multiple solutions" concept?	472
24.3.6 Coupling biochar systems to address multiple challenges in light of the energy-water-food-environment nexus	473
24.3.7 Linking the twin or multiple solutions framework to emerging concepts	473
24.4 Concluding remarks and outlook	**473**
Acknowledgments	**474**
Declaration of conflict of interest	**474**
References	**475**

24.1 Introduction

Low-income countries, predominantly found in Africa, Southeast Asia, the Caribbean, and Latin America, face a myriad of cooccurring challenges, including lack of clean drinking water, lack of clean energy, food insecurity, and environmental pollution. For instance, according to the World Health Organization, about 2.2 billion people globally lack access to safely managed drinking water services, and the International Energy Agency estimates that 789 million people live without electricity. Moreover, the Food and Agriculture Organization reports that approximately 690 million people suffer from hunger. These issues are intricately linked, forming a water-energy-food nexus. A consensus exists that nexus problems are best solved simultaneously. Yet, limited efforts have been made to address the energy, water, and food problems as a nexus, partly because existing technologies often address only one component of the nexus.

Biochar technology has attracted significant public and research attention as a potential renewable, low-cost, and sustainable technology with the potential to contribute to energy and clean water provision and to improve soil quality and productivity in low-income countries (Gwenzi et al., 2015). Biochar soil application also mitigates climate change through soil carbon sequestration and reduction of greenhouse gas emissions in agroecosystems (Lehmann & Joseph, 2009).

Compared to developed countries, biochar research addressing the energy-water-food nexus in low-income countries is still limited. To date, few studies have investigated the individual aspects of the energy-water-food nexus. For example, several studies have investigated biochar/pyrolytic cookstoves as a potential source of energy for household heating and cooking. Several studies, including reviews, have investigated the application of biochars as a filter media for drinking water and wastewater treatment in developing countries (Gwenzi et al., 2017). Other studies have documented the use of biochar as soil amendments to improve soil quality and crop productivity, thereby enhancing food security. Additionally, some studies have explored soil carbon sequestration, mitigation of greenhouse gas emissions, and remediation of contaminated soils. In summary, biochar research in both low-income and developed countries often tends to focus on one component of the nexus at a time, while overlooking the intricate linkages among the components.

Studies focusing on one component of the energy-water-food nexus are potentially attractive to researchers because they are relatively less complex and hence easier to implement and interpret. However, such studies may underestimate the potential benefits and contributions of biochar to the energy-water-food nexus because they exclude other pertinent aspects. Additionally, estimating the collective contribution of biochar to the energy-water-food nexus from independent studies addressing individual components presents methodological challenges. Furthermore, until now, the field of biochar research has lacked a theoretical framework to guide research and provide a heuristic pathway to uptake, adoption, and impact in low-income countries. Hence, the dominance of studies focusing on individual components of the energy-water-food nexus could reflect the lack of a conceptual framework. Fig. 24.1 shows the biochar ring depicting the circulation and multiple uses of biochar.

Therefore, this chapter proposes a "twin or multiple solutions" framework as a novel heuristic pathway to guide research and enhance the uptake, adoption, and impact of biochar. For brevity, the well-known energy-water-food nexus in low-income countries is used to illustrate the key elements and novelty of the concept. The 'twin or multiple solutions' framework posits that biochar technology should provide tangible and frugal solutions to at least two challenges drawn from the energy-water-food nexus to achieve rapid and sustained uptake, adoption, and impact in low-income countries. However, note that the proposed framework and its core components can be considered generic in nature. Hence, the concept can be adapted and extended to guide biochar research, uptake, and adoption in other regions.

The present chapter first summarizes the "twin or multiple solutions" concept and its core components. Second, a biochar technology "ring" depicting how biochar-based "twin or multiple solutions" are coupled to each other is proposed as part of the framework. Then, a handful of biochar-based "twin or multiple solutions" and "grand questions" addressing the energy-water-food-environment nexus in low-income countries are presented, including evidence corroborating the proposed application. Finally, future perspectives are highlighted, including the need to validate the concept

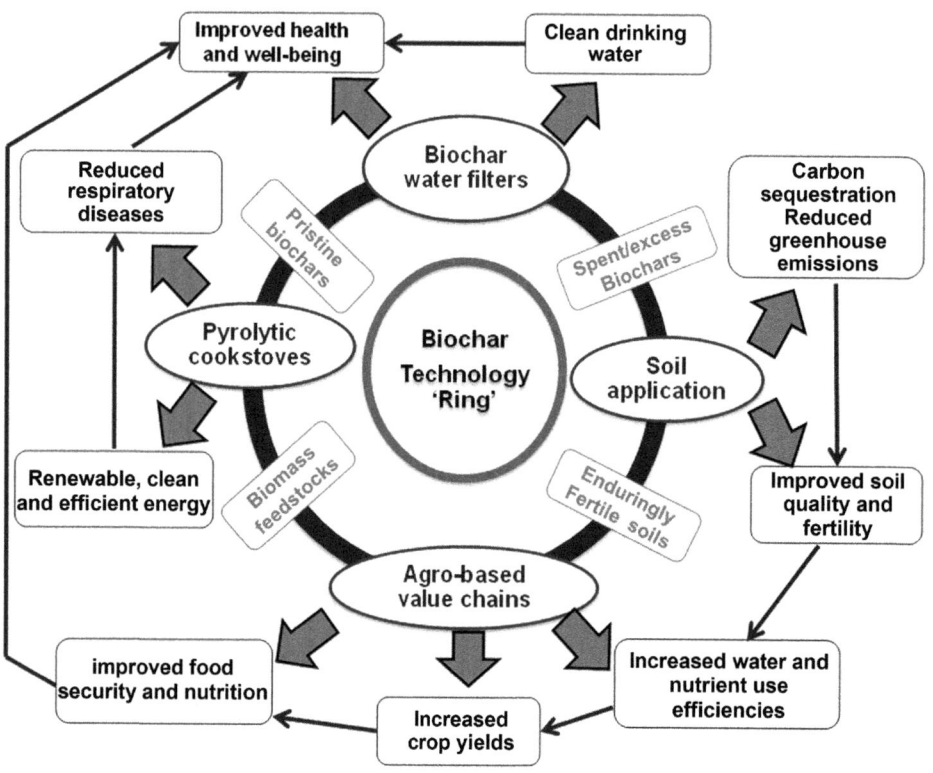

FIGURE 24.1 The biochar technology "ring" depicting twin or multiple solutions to the energy-water-food nexus as a heuristic pathway to uptake, adoption, and impact in low-income countries

and address knowledge gaps to stimulate further debate among the biochar research community. Here, the term "handful" denotes that only a few examples are presented solely for brevity and illustration purposes, but this does not constitute an exhaustive list. Thus, the term should not be misconstrued and interpreted as implying "less significant."

In summary, the proposed "twin-multiple solutions" framework is not designed to be a prescriptive or 'one-size-fits-all' approach. Rather, it seeks to initiate and stimulate further debate among the global biochar research community on the need for a theoretical framework in biochar research in low-income countries and elsewhere. It is envisaged that such a scientific debate, including comments and even rebuttals of the current framework, will ultimately lead to the development of a better and hopefully widely accepted framework.

24.2 The biochar "twin or multiple solutions" framework

24.2.1 Overview of the "twin or multiple solutions" concept

Conceptually, the "twin or multiple solutions" framework seeks to enhance the direct benefits of biochar technology relative to its costs. This concept is akin to the colloquial notion of "killing two or more birds with one stone," which is commonly used in some developing regions such as Africa. The notion entails deriving more than one benefit from a single good or service, in this case, biochar technology. In simple terms, the "twin or multiple solutions" concept proposed here can be considered a formal presentation of this colloquial notion. The existence of an equivalent notion to the "twin or multiple solutions" concept could potentially make it easier for researchers to explain what biochar entails to target end users, practitioners, and decision- and policy-makers. Moreover, several conventional "single solution" technologies exist that can be used to draw a contrast with the biochar "twin or multiple" concept. Overall, it is envisaged that the "twin or multiple solutions" concept will make biochar technology stand out among several competing technologies currently promoted in low-income countries. Fig. 24.2 shows the twin or multiple solutions framework indicating its key components and potential applications as a pathway to biochar uptake, adoption, and impact.

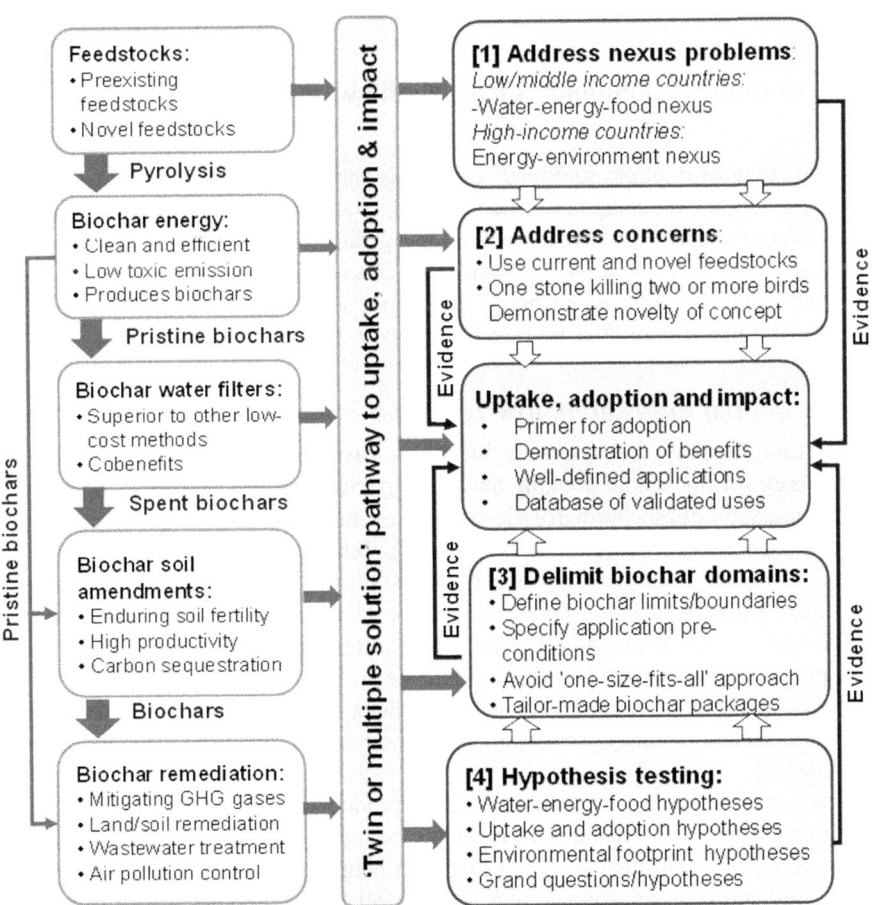

FIGURE 24.2 Conceptual depiction of the "twin or multiple solutions" framework indicating its key components and potential applications as a pathway to biochar uptake, adoption, and impact.

Several examples can be used to contrast biochar technology with existing conventional technologies targeting single components of the energy-water-food nexus. For example, charcoal production and subsequent use as a household fuel source is a well-established technology in Africa and other low-income countries. However, unlike biochar with multiple potential benefits, charcoal is only used for household heating and cooking. Additionally, biochar production can cause deforestation, and typical low-cost charcoal production systems emit toxic particulates and gases, posing human health risks and contributing to greenhouse gas emissions. In contrast, biochar/pyrolytic cookstoves are more energy and fuel-efficient than traditional three-stone biomass cookstoves. These merits provide a strong reason for target end users to consider experimenting with biochar/pyrolytic cookstoves. Positive experiences with such biochar cookstoves could trigger uptake, adoption, upscaling, and ultimately impact.

Similarly, low-cost water treatment methods such as chlorination, solar disinfection (SODIS), boiling, conventional ceramic filters, and biosand filters are only designed to remove pathogenic organisms in drinking water but have no other direct cobenefits. Evidence even shows that, barring the removal of pathogens, such low-cost water treatment methods have several limitations (Gwenzi et al., 2020). For example, chlorination relies on chlorinating agents, which are not readily available for low-income households. High dosages of chlorine may also generate carcinogenic disinfection by-products when they react with natural organic matter, causing cancer. Moreover, unlike biochar, these low-cost water treatment methods have limited or no capacity to remove dissolved geogenic contaminants such as arsenic, fluoride, and radionuclides (e.g., uranium) prevalent in groundwater sources in low-income countries. In fact, boiling evaporates water, thereby increasing the concentration of dissolved toxic geogenic contaminants. This is potentially counterproductive because it could exacerbate the human risks associated with these contaminants.

Regarding food production in agroecosystems, technologies such as the use of organic and inorganic fertilizers only improve crop yields. However, the use of organic and inorganic fertilizers may contaminate aquatic systems such as surface and groundwater sources and also release greenhouse gases contributing to climate change. Similarly, conservation agriculture, touted as a package for Africa's green revolution, targets improving food security. However, the results so far have been variable and inconsistent, and its fit to low-income farming systems has been questioned in the literature. Therefore, in summary, such existing technologies fail to conform to the 'twin or multiple solutions' concept proposed for biochar because they provide little or no direct cobenefits beyond the targeted component of the energy-water-food nexus.

24.2.2 A handful of biochar "twin or multiple solutions" targeting the water-energy-food-environment nexus

In the context of low-income countries, the "twin or multiple solutions" concept is summarized as a biochar technology "ring" (Fig. 24.1). This biochar "ring" is a cyclic loop depicting how biochar-based solutions are interconnected to address the energy-water-food nexus. For brevity, three biochar-based solutions addressing different components of the energy-water-food nexus are discussed: (1) biochar technology as an energy source, (2) biochar-based water filtration systems, and (3) the application of biochar to create enduringly fertile and highly productive anthroposols. For each solution, a summary of current evidence pointing to its potential feasibility is presented to support the proposed application.

24.2.2.1 Biochar technology is a potential renewable, low-cost, and efficient energy

Specifically, biochar/pyrolytic cookstoves using biomass feedstocks, including biowastes from agro-based value chains, provide renewable, clean, and efficient energy for household cooking and heating (Solution 1). An increasing body of literature exists on the development and application of biochar/pyrolytic cookstoves, including (1) cookstove design and operation, and (2) performance with respect to energy and emissions (Getahun, Tessema, & Gabbiye, 2019; Tryner et al., 2016; Panwar, 2009). Chapter 19 of this book discussed gasifier cookstoves as a clean technology for energy provision and mitigation of indoor air pollution. Thus, the existing body of evidence demonstrates the potential feasibility of biochar technology as a clean and renewable source of energy for household heating and cooking.

However, contrary to the framework proposed here, the bulk of these studies (e.g., Kirch et al., 2020) were only limited to energy provision, while the subsequent potential applications of the biochars for drinking water treatment and soil amendment were excluded. Hence, further comprehensive studies of biochar-based energy systems using the 'twin or multiple solutions' framework are required.

For example, Kirch et al. (2020) investigated the energy, emissions, and char production of a reverse-downdraft gasifier cookstove as a function of operating conditions. These conditions included various feedstocks such as wood pellets (WP), wheat straw (WS), sheep manure (SM), and cow manure (CM), fuel bed height, and primary air supply rates. Kirch et al.'s (2020) work showed that high primary air supply rates and a deeper fuel bed significantly reduced producer gas while increasing the combustion efficiency of low-ash-containing biomass fuels such as WP. Specifically, increasing the primary air supply rate by a factor of three almost doubled the PM2.5 emissions for WP and increased the emissions more than tenfold for CM. Conversely, at low air supply rates, the emissions of particulate matter with an aerodynamic diameter of less than 2.5 μm (PM2.5) ranged from 6 to 30 mg/MJ of heat released and significantly varied among the biomass fuels. The particulate emissions of PM2.5 were lowest for WP, followed by WS, SM, and CM. However, the values were lower than those reported in the literature for similar cookstoves. In summary, the best combination of heat energy production, high quantities of char, and low particulate and CO emissions was obtained under low primary air supply rates.

The findings of Kirch et al.'s (2020) work are significant in the context of the 'twin or multiple solutions' framework. First, the study provides insights into how feedstocks and operating conditions can be optimized to achieve an optimum combination of heat energy for household cooking and heating, biochar for drinking water treatment and subsequent soil amendment, and improved indoor air quality to safeguard human health. Moreover, it highlights the opportunity to identify and purposefully select specific biomass feedstocks to maximize the benefits of biochar technology. Further research using similar approaches and other designs of biochar/pyrolytic cookstoves is critical. Ultimately, such research could generate a database of well-characterized and tested biochar-based energy systems ideal for various settings in low-income countries.

Biochar/pyrolytic cookstoves are based on the following design principles: (1) the Anderson and Reed model or top-lit updraft, (2) Anila, and (3) kon-tiki or flame curtain pyrolysis systems. These cookstove designs have the capacity to produce heat energy via the combustion of volatiles or synthetic gas while producing biochar as a residue. Biochar/pyrolytic/gasifier cookstoves have been reported to reduce poverty and improve indoor air quality among smallholders in Kenya. Moreover, biochar/pyrolytic cookstoves are more fuel and energy-efficient than traditional three-stone cookstoves.

However, evidence from Africa and other low-income countries shows that current designs still have limitations (Karanja & Gasparatos, 2019). Depending on the design of the cookstoves, limitations include (1) high initial costs of approximately USD 35 in the case of gasifier cookstoves in Kenya, (2) difficulties in operation, including problems in feeding additional fuel at intervals, (3) low stability of cookstoves that increases the risk of tipping over, and (4) the risk of burns from the hot outer wall during operation (Karanja & Gasparatos, 2019). In some cases, biochar/pyrolytic cookstoves failed to fit socio-cultural and cooking norms in low-income countries (Karanja & Gasparatos, 2019). These challenges call for further research into the design of cookstoves, taking into account the socio-cultural norms, perceptions, and concerns of target users. Failure to ensure compatibility with local socio-cultural norms may force target users to either: (1) abandon biochar/pyrolytic cookstoves altogether and revert to their traditional cooking systems, or (2) use both the biochar/pyrolytic cookstoves and traditional cookstoves for various purposes.

24.2.2.2 Biochar-based water filtration systems for clean drinking water provision

Due to their high surface area and charge, biochars have an excellent capacity to remove a myriad of toxic contaminants in aqueous systems, including pathogens. Hence, biochars produced as a by-product from pyrolytic or gasifier cookstoves can be recycled at a local or household level as filter media for biochar-based water filtration systems, providing decentralized household clean drinking water (Solution 2). Low-income countries in South America, Africa, and Southeast Asia provide ideal settings for the development and application of biochar-based water filtration systems. Biochar-based water filtration systems use biochar as filter media due to their excellent capacity to effectively remove inorganic, organic, and microbiological contaminants in aqueous systems (Gwenzi et al., 2020). This is crucial, as the bulk of the 2 billion people still lacking access to clean drinking water are found in low-income countries. In countries where biosand filtration systems are already in use, the transition to biochar-based water filtration systems could be much easier. This transition involves incorporating a biochar layer in the biosand filtration system or simply replacing the sand medium with biochar.

The potential merits of biochar-based filtration systems relative to existing low-cost water treatment methods have been discussed in an earlier paper (Gwenzi et al., 2020). For example, a large proportion of households in low-income countries rely on biomass fuels for household heating and cooking. Thus, the production of biochar for use as a filter media in biochar-based water filtration systems can be coupled with the provision of clean and renewable energy. This will require a shift from the current traditional three-stone cookstoves that produce ash as a residue to biochar or pyrolytic cookstoves that produce biochar as a residue.

Several low-cost drinking water treatment methods such as chlorination, boiling, SODIS, and ultraviolet (UV) irradiation already exist (WHO, 2020). However, as discussed in an earlier review, current low-cost methods have several limitations compared to biochar-based water filtration systems. For example, current low-cost methods have limited capacity to remove dissolved toxic contaminants such as fluoride, arsenic, and radionuclides prevalent in groundwater sources in low-income countries. In fact, boiling evaporates water, potentially increasing the concentration of toxic dissolved contaminants. Chlorination requires chemical reagents in the form of chlorine dioxide or chlorine, which are often imported, expensive, and not readily available to low-income households. Residual chlorine associated with high dosages may react with natural organic matter in water to release carcinogenic disinfection by-products, causing cancer (Gwenzi et al., 2020). SODIS, boiling, UV irradiation, and chlorination are ideal for nonturbid waters but are not suitable for turbid waters (WHO, 2020). Thus, biochar-based water filtration systems have the potential to overcome these limitations. In addition, biochar technology provides multiple potential cobenefits as a clean energy source and as a soil amendment for improving soil fertility, carbon sequestration, and mitigation of greenhouse gas emissions. By comparison, current low-cost water treatment methods solely targeting clean drinking water provision have no known direct cobenefits. These merits make biochar-based water filtration systems a potential novel low-cost technology for low-income countries.

The focus here has been limited to drinking water treatment given the large number of people lacking access to clean drinking water in low-income countries. However, as evident in the literature summarized here, biochar-based filtration systems can also be extended to wastewater and stormwater (Dalahmeh et al., 2018; Kaetzl et al., 2018; 2020; Mohanty & Boehm, 2014). More recently, biochar filters have been extended to air pollution control systems, including the removal of industrial flue gases (Bamdad et al., 2018; Gwenzi et al., 2020). Other studies have also used biochar for the remediation of contaminated soil (Herath et al., 2015; Ippolito et al., 2017; Kumarathilaka et al., 2018). Thus, biochars can also be used for environmental remediation, including wastewater treatment, air pollution control, and remediation of contaminated soils (Solution 4). This is in addition to the mitigation of climate change via carbon sequestration and the reduction of greenhouse gas emissions from agroecosystems (Gwenzi et al., 2020).

Rather than producing biochars solely for drinking water treatment, the current concept uses residual biochars from biochar/pyrolytic cookstoves as a filter media for biochar-based water filtration systems. The concept can also be extended to cases where a thriving bioenergy industry relying on pyrolysis or gasification already exists. In this regard, the residual biochar from the bioenergy sector can then be used for environmental remediation and soil amendment. A case in point is a study conducted in Sri Lanka, where biochar produced as a by-product of bioenergy was applied to remediate Pb and Cu in shooting range soils (Kumarathilaka et al., 2018). In the same study, biochar reduced the release of lead and copper by 99.5%, demonstrating the potential to couple biochar energy systems to the subsequent biochar application for environmental remediation.

The capacity to enhance contaminant removal in aqueous systems by replacing sand with biochar has been demonstrated in the case of wastewater filtration systems (Kaetzl et al., 2020). Kaetzl et al. (2020) investigated the removal of chemical oxygen demand (COD) and *E. coli* from wastewater effluent from a sedimentation tank under the following experimental conditions: room temperature (22°C), hydraulic loading rate of 0.05 m/h, empty bed contact time of 14.4 h, and a mean organic loading rate of 509 $\pm$ 173 gCOD/m^3/d. In the same study, mean COD removal of biochar filters (74 $\pm$ 18%) was significantly higher than that of sand filters (61 $\pm$ 12%). The E. coli removal for biochar filters (1.35 $\pm$ 0.27 log units) was also significantly higher than that of the sand filters (1.18 $\pm$ 0.31 log units). Unlike in the sand filter, the E. coli removal efficiency of the biochar filter increased with experimental time. Maya et al. (2020) compared metal and fluoride removal in both synthetic and natural groundwater using biochar and sand filters. The authors also reported significantly higher metal and fluoride removal in the biochar filter than in the sand filter. In Ghana, a biochar-based water filtration system using biochar derived from maize cobs effectively removed inorganic contaminants in drinking water. Taken together, the batch studies using synthetic aqueous systems and pilot studies using real water and wastewater demonstrate the potential to use biochar as a filter media for wastewater and drinking water treatment.

The potential to develop biochar-based water filtration systems for drinking water supply, including the mechanisms involved, has been discussed in an earlier review (Gwenzi et al., 2020). Earlier and subsequent laboratory and pilot-scale studies have also demonstrated the capacity to replace sand with biochar as a filter media in water filtration systems (Kaetzl et al., 2020; Maya et al., 2020). Barring studies using synthetic aqueous systems, a few studies have developed and tested biochar-based filtration systems for the treatment of wastewater and drinking water. These include the removal of metals (e.g., Cr), dyes, and fluoride in groundwater (Maya et al., 2020), synthetic pesticides in surface water, and COD and E. coli in wastewater effluent. Kaeltz et al. (2020) and Dalahmeh et al.(2018) applied biochar filters for the removal of pharmaceuticals in wastewater and observed higher removal efficiencies for biochar filters than sand filters. Perez-Mercado et al. (2018) observed higher removal of total nitrogen in biochar filters than in sand filters. Mohanty & Boehm (2014) used an improved sand biofilter augmented with biochar and observed improved removal of *E. coli* (circa 96%) compared to a sand biofilter. Recently, Chaukura et al. (2020) developed a novel "biscuit" biochar filter with a layer of biochar sandwiched between two ceramic layers. The "biscuit" ceramic filter effectively removed total hardness, turbidity, and dissolved organic carbon in real water samples (Chaukura et al., 2020). Detailed reviews on the application of biochar filters for wastewater and water treatment are presented elsewhere (Vikrant et al., 2018; Gwenzi et al., 2020). These studies clearly demonstrate the potential for biochar-based drinking water filtration systems.

24.2.2.3 Biochar soil amendment to create enduringly fertile and productive soils

Biochar-based water filtration systems produce spent biochars laden with contaminants, including essential plant nutrients. Low-income regions, such as Africa, have inherently acidic and infertile tropical soils, resulting in low crop productivity. Hence, the application of spent and pristine biochars could be a novel pathway to develop enduringly fertile soils with high productivity similar to other technosols or anthroposols such as the African Dark Earths in West Africa, the terra preta or terra de índio in Amazonia, and formiguer in South America. As evidenced by the African Dark Earths, terra pretas, and formiguers, such enduringly fertile anthroposols have high productivity and hence could contribute to household food security (Solution 3).

However, the production of biochar for agricultural applications at a large scale using current pyrolysis systems remains problematic and seemingly unachievable, especially in low-income settings. Thus, in such settings, biochar applications in agroecosystems could be limited to household nutrition and herbal gardens, where relatively small quantities are needed given the small landholding (Gwenzi et al., 2020).

24.2.2.4 Biochar-based environmental remediation systems

24.2.2.4.1 Mitigation of climate change in agroecosystems

Moreover, soil application of biochars could also contribute to climate change mitigation via increased soil carbon sequestration and reductions in greenhouse gas emissions from agroecosystems (**Solution 4**). However, compared to **Solutions 1**, **2**, and **3**, one may speculate that in most low-income regions, poor communities grappling with everyday challenges are not likely to prioritize and value Solution 4 as a direct benefit derived from biochar technology.

24.2.2.4.2 Biochar amendments for remediation of contaminated lands and soils

The application of biochar for the remediation of contaminated land and soils, including postmining landscapes is discussed in several papers. In this book, four chapters are dedicated to this application of biochar (Chapters 7, 8, 9, and 10).

24.2.2.4.3 Biochar as media for air pollution control systems

The use of biochar as an air filter media has been discussed in some recent papers, including reviews (Gwenzi et al., 2020). Chapter 20 in this book presents a state-of-the-art on the removal of toxic gaseous contaminants using biochar as a filter medium.

24.3 Discussion and perspectives

The current chapter provides a conceptual framework to guide biochar research and offers a heuristic pathway to promote the uptake, adoption, and impact of biochar technology. Several knowledge gaps still exist regarding biochar technology, as discussed in detail in earlier reviews, including those authored by the current author (Gwenzi et al., 2017, 2020). In summary, future research directions pertaining to the water-energy-food-environment nexus include the following:

1. Lack of socio-economic data.
2. Insufficient design and operation data.
3. Limited technical feasibility data from large-scale studies.
4. Potential ecotoxicological and ecological health risks associated with toxic contaminants in both pristine and spent biochars.

For further details, readers are referred to Chapter 25 in this book. Here, some future perspectives and "grand" questions currently absent in the literature on biochar are presented to guide future research. The term "grand questions" refers to highly relevant and cross-cutting issues that are inadequately addressed in current research. This gap persists due to the exclusion of such aspects in current research approaches or methodological limitations that pose significant challenges in understanding them.

24.3.1 On the role of "twin or multiple solutions" in biochar uptake, adoption, and impact

To this point, the need for demonstrable biochar-based "twin or multiple solutions" has been proposed as a precondition for the uptake, adoption, and impact of biochar technology. Critics and skeptics may wonder, "Why are demonstrable biochar-based 'twin or multiple solutions' a precondition for biochar uptake, adoption, and impact?" The answer to this question lies in how consumers behave concerning a new product or service, in this case, biochar technology. Specifically, a potential consumer or user currently utilizing an existing technology is unlikely to switch to and adopt a new technology unless it has demonstrable additional desirable functionalities. In the context of biochar, a potential user currently producing and using charcoal for household heating and cooking is unlikely to change to biochar solely for household energy. However, if biochar from the cookstoves can act as filter media for drinking water treatment, this additional desirable functionality may encourage adoption.

Regarding biochar-based water filtration systems, individuals are unlikely to shift from current low-cost drinking water treatment methods to adopt biochar technology solely for water treatment unless the biochar technology provides additional desirable functionalities. These additional functionalities act as incentives for potential users to learn and experiment with the new technology. This is particularly true with any new and emerging technology, where users may need to acquire new skills to use the technology effectively. Without a strong incentive to acquire these skills, users may be reluctant to learn and might revert to old technologies. This often happens with technologies driven by funded projects in the NGO or public sectors, where target end users revert to their old technology at the end of the project.

In agriculture, conservation agriculture in Africa is a typical example where "uptake or adoption" is often incentivized or induced using free handouts of food, seed, and fertilizer. Such incentives do not bring long-lasting uptake, adoption, and impact. Instead, demonstrable "twin or multiple solutions" could serve the same purpose more sustainably. This is true for biochar applications aimed at enhancing crop productivity and food security. The author argues that the uptake and adoption of biochar technology for soil application to enhance soil quality, crop productivity, and climate change mitigation are unlikely to appeal to potential users, such as farmers in low-income countries. This is due to the large quantities of feedstock and time required to produce large amounts of biochar using typical low-cost pyrolysis systems such as pit and drum kilns, kon-tiki flame pyrolysis systems, and pyrolytic or gasifier stoves for soil amendment.

Despite some demonstrated biochar benefits reported in experimental studies in Africa and other developing countries, large-scale uptake and adoption are still lacking. The potential role of scale as a key factor determining the uptake and adoption of biochar technology seems to be underestimated, as evidenced by the limited research attention on the topic.

24.3.2 The "twin or multiple solutions" framework as a primer for biochar adoption

The "twin/multiple solutions" framework aligns with the key generic drivers outlined in diffusion theories for technology adoption in low-income settings (Delaney, 2011; Munoz, 2014) These drivers include (1) relative benefits compared to existing technologies or practices, (2) compatibility with local socio-economic settings and cultural norms, (3) the potential of the technology to address various perceived needs, and (4) a history of similar practices (Munoz, 2014). As discussed earlier, the potential benefits and prospects for biochar to address a variety of perceived needs are central to the "twin or multiple solutions" framework. Regarding compatibility, biochar-based water filtration systems could fit well in communities already using biosand filters for water treatment. The transition to biochar-based systems involves incorporating a biochar layer within the preexisting sand layer of the biosand filter or completely replacing the sand with biochar. For soil applications, uptake and adoption could be high where farmers are already applying charcoal and ash to soils. In Africa, the improved soil quality and crop productivity associated with biochar application can be seen as a transition from "slash and burn" shifting cultivation systems to a "slash and char" system.

In addition to potential multiple benefits, there is a need to develop a critical mass of early adopters and change agents using the "twin or multiple solutions" framework to promote the diffusion of biochar technology. Early adopters are crucial in providing demonstrable benefits and identifying constraints requiring further research. Change agents, such as government extensionists and field practitioners from community and nongovernmental organizations, are critical for capacity-building and providing expertise and information on biochar to potential users. Current regulatory and institutional frameworks, and extension programs in agriculture, energy, water, sanitation, and environmental management, including climate change mitigation, are often silent on biochar technology. Thus, there is a need to integrate biochar technology into regulatory, policy, and institutional frameworks in low-income countries. This effort should include (1) a pilot-scale demonstration of biochar systems based on the "twin or multiple solutions" concept, and (2) fair and transparent communication of the biochar technology, its potential benefits, constraints, and aspects still poorly understood.

Public and scientific opinion is still divided on whether biochar application to soils can significantly mitigate greenhouse gas emissions. A significant proportion of the global scientific community considers biochar a potential technology for carbon sequestration and mitigation of greenhouse gas emissions (e.g., Lehmann, 2007). Others have doubted this claim from the onset (e.g., Baveye et al., 2017), while some have highlighted the challenges associated with producing the large quantities of biochar required for soil amendment in agroecosystems (Bach et al., 2016; Dumbrell et al., 2016; Gwenzi et al., 2020). Inconsistent results have also been observed for biochar technology, including variable effects on crop productivity in temperate versus tropical environments and nitrogen immobilization induced by the high C/N ratio of biochar. The benefits of biochar are context-specific, hence, as Baveye et al. (2017) pointed out, caution is needed when presenting biochar as a panacea or "one-size-fits-all" technology. For example, Baveye et al. (2017) and Lorenz et al. (2018) cited a tongue-in-cheek comment attributed to Professor Janice Thies of Cornell University, claiming that in some cases, biochar has been considered a "win-win," "win-win-win," or even "winfinity" solution. The current discussion does not seek to resolve the controversies around the capacity of biochar to offset greenhouse gas emissions. Here, two conjectures are made: first, the currently divided scientific and public opinion, coupled with inconsistent results, points to the existence of biophysical limits or boundaries for biochar technology. Second, the "twin or multiple solutions" framework proposed here still warrants validation as a potential novel pathway to biochar uptake, adoption, and impact in low-income countries. The energy-water-food-environment nexus creates a need at the household and community levels, providing ideal conditions to apply the "twin or multiple solutions" concept to "kill two or more birds with one stone." As pointed out, "killing two or more birds with one stone," and by extension, the "twin or multiple solutions" concept requires that the "two or more birds" are closely juxtaposed. Conceptually, this is the scenario with respect to the provision of clean drinking water, household energy, food production, and even environmental remediation in low-income countries. Note that, biochar aside, approximately 2.4 billion people, most of them in low-income countries, will continue to rely on biomass fuels as a source of energy for household heating, cooking, and, to a lesser extent, lighting (WHO, 2016). Therefore, given the potential advantages of biochar cookstoves relative to traditional cookstoves, scope exists to use biochar cookstoves as an entry point for the uptake and adoption of biochar in low-income communities. In most low-income countries, including those in Africa, household nutrition and herbal gardens are key sources of household income, food security, and nutrition. These gardens can be regarded as intensive cropping systems characterized by high inputs and the production of high-value and nutritional crops, including vegetables and herbs. The current author argues that these household gardens represent ideal niches to demonstrate the potential of biochar to create "enduringly fertile" and highly productive soils. Demonstrating these benefits is critical in most developing regions, including Africa, where soils are highly acidic and inherently infertile. Once the benefits are evident at a small scale, such as in household gardens, this may act as a primer to promote the production and application of biochar on relatively larger scales. In this regard, research aimed at validating the "twin or multiple solutions" concept in low-income countries should focus on pilot demonstrations of the biochar technology at a relatively small scale, but large enough to enable realistic evaluation of the technology by potential end users. Technology transfer approaches based on cognitive and colearning, self-learning, and experimentation, including the concept of farmer field schools, can be used as tools for participatory evaluation of biochar technology.

Overall, the "twin or multiple solutions" framework proposed entails a shift from the often propagated view that biochar is a panacea or "one-size-fits-all" technology. This is achieved by targeting low-income communities in developing countries, where the components of the energy-water-food-environment nexus are closely juxtaposed, and the perceived need is potentially most evident. Moreover, the need to validate the "twin or multiple solutions" concept and define the boundaries or limits of biochar is critical. These aspects are currently excluded from existing literature on biochar technology. Thus, research to validate the "twin or multiple solutions" concept and define the biophysical limits will provide critical evidence to inform the targeting of biochar technology. Ultimately, such evidence will avoid the "one-size-fits-all" or "blanket salesman" approach currently characterizing biochar technology. Moreover, such field evidence will dispel (or even confirm) current misconceptions, myths, and criticisms about biochar. Failure by the global biochar research community to provide irrevocable evidence will create an information void. Such voids can be filled with biochar conspiracy theories and myths, which could be counterproductive to the potential development and application of biochar where it is probably needed most.

Perceived and demonstrable benefits of biochar relative to current competing technologies have been identified as a key driver of the willingness to adopt biochar (Latawiec et al., 2017). In the case of farmers, improved soil quality translating to increased yields, hence household income and food security, could promote uptake and adoption, while high labor requirements and costs may act as constraints (Latawiec et al., 2017). The need for a comprehensive understanding of the socio-economics of biochar, including factors determining its adoption among potential producers and end users, has been highlighted in earlier studies (Latawiec et al., 2017).

24.3.3 Moving from biochar as a "silver bullet" to biochar application domains

To some extent, the inconsistencies and variabilities in research results are expected because biochar has been applied across diverse biophysical environments, including varying climates, soils, cropping systems, and socio-economic and cultural settings. Moreover, biochars are heterogeneous materials whose properties depend on both feedstock and pyrolysis conditions. This necessitates research to define the biophysical limits or boundaries of biochar and the development of a database of well-characterized biochar tailored for specific applications.

This shift away from the current "one-size-fits-all" or "blanket salesman" approach requires identifying biochar application domains or niches and subsequently developing tailor-made biochar solutions for these specific domains.

24.3.4 Moving from "one-size-fits-all" and "blanket salesmen" to biochar application domains

In the literature, biochar is often portrayed as a "silver bullet" or "one-size-fits-all" technology capable of addressing the water-energy-food-environment nexus in both developing and developed countries. Some authors even suggest that biochar technology has 55 potential applications (Schmidt, 2012). However, when this notion is embraced by policymakers, practitioners, biochar enthusiasts, and the media, it can lead to unintended consequences. Moreover, such perceptions can stifle open debate about biochar, labeling anyone questioning its relevance as a heretic.

While acknowledging the potential opportunities biochar presents, it is crucial to recognize that the technology may not deliver the claimed benefits universally. The boundary conditions for biochar applications remain poorly defined, necessitating further research to identify specific niches and define socio-economic and biophysical limits. This effort requires citizen science and effective science communication, fostering open debates that encompass both the opportunities and risks of biochar, including transparent reporting of negative or neutral results in scientific media.

Comprehensive studies investigating contaminant emissions throughout the biochar value chain using life cycle analysis are still limited, largely due to insufficient input data for such analyses. Additionally, few studies have explored the social impacts, ecological implications, and human health risks associated with exposure to contaminants within biochar systems. Key knowledge gaps include a lack of data on (1) environmental footprints concerning raw materials, carbon, energy, and water, (2) the contribution of biochar systems to radiative forcing in climate change, (3) ecotoxicity and ecological health risks linked to the application of spent biochar containing contaminants in soil, and (4) human exposure and health risks from toxic particulate and gaseous emissions during biochar production and application.

Therefore, research aimed at validating the "twin or multiple solution" concept should integrate environmental health risk assessments along the biochar value chain. Chapter 24 discusses potential environmental health risks along this chain. Understanding these risks is essential for developing mitigation strategies that safeguard environmental and human health.

24.3.5 Confronting biochar critics and skeptics with the "twin or multiple solutions" concept?

Global criticism, skepticism, and activism against biochar technology abound, leading to numerous arguments presented against its adoption (ref). Concerns have been raised that the production of biomass feedstocks for biochar could displace food crops and lead to land grabs in vulnerable communities of low-income countries. The proposed framework addresses these concerns in two key ways: (1) by utilizing preexisting biomass fuels and biowastes rather than dedicated biomass feedstocks for biochar production, and (2) through a tightly coupled system where biochar/pyrolytic cookstoves serve as an entry point. Thus, residual biochar from household cooking and heating is repurposed as filter media in biochar-based water filtration systems. Subsequently, spent biochar from these filtration systems and excess biochar are used as soil amendments to enhance soil fertility, improve environmental quality, and increase crop productivity in agroecosystems; Fig. 24.1). In the context of a circular economy, this concept transforms linear and open-ended biomaterial flows into tightly coupled circular loops at the household level. Conceptually, widespread uptake and scaling could create rural or urban symbiosis by integrating agricultural value chains with water treatment and bioenergy industries.

In summary, this approach diverges from the current practice of addressing singular challenges within the water-energy-food-environment nexus in low-income countries. However, a comprehensive understanding of the potential benefits and limitations of this concept at both household and economic levels requires further investigation through pilot-scale studies. Such studies should explore how perceived benefits translate from household to broader spatial scales within frameworks of industrial symbiosis or the circular economy.

Critics may draw parallels between attempting to apply biochar to multiple challenges within the water-energy-food-environment nexus and the proverbial Chinese wisdom of "chasing two rabbits at a time," potentially leading to information overload for end users. Using an investment analogy, this scenario resembles pursuing dual goals, where addressing energy needs with biochar cookstoves precedes water treatment and subsequently soil amendments. This sequential approach mitigates the risk of information overload, supported by biochar technology's reputation for simplicity and low cost. Moreover, prioritization of specific applications can be tailored to biophysical, socio-economic settings, and local needs. However, effective packaging and dissemination of biochar technology for addressing the water-energy-food-environment nexus remain areas warranting further research.

24.3.6 Coupling biochar systems to address multiple challenges in light of the energy-water-food-environment nexus

Coupling household cooking and heating with biochar cookstoves for biochar production in low-income countries is innovative for two reasons. First, it addresses the issue of biomass feedstocks by utilizing conventional biomass fuels currently in use, while also incorporating bioavailable wastes such as crop residues. Second, the quantities of biochar required for a biochar-based water filtration system in a typical household with an average of 8 adults are expected to be relatively small. Although exact figures are not available, the volume of sand used in conventional biosand filters for household water treatment provides a rough estimate. Data from Kenya suggests that a typical biochar cookstove producing approximately 1 kg of biochar per household per day could generate sufficient biochar for drinking water treatment. However, further research on biochar-based water filtration systems is necessary to establish design and operational parameters, including optimal filter media bed depth and replacement intervals (Gwenzi et al., 2020).

The proposal to utilize household nutrition and herbal gardens as pilot-scale field plots to demonstrate the capacity of biochar to enhance soil quality and crop productivity is also innovative. These gardens, typically less than 0.1 ha in urban, peri-urban, and rural areas, are suitable because a household biochar cookstove can produce several tons of biochar annually. This translates into achievable biochar application rates, addressing the challenge often cited in the literature of producing large quantities of biochar (Bach et al., 2016; Baveye et al., 2017; Dumbrell et al., 2016). Focusing on household nutrition and herbal gardens represents a departure from large-scale biochar applications in agroecosystems, which require substantial biochar volumes. The production of such quantities using the available low-cost pyrolysis technologies typical in low-income countries poses significant challenges (Gwenzi et al., 2020).

24.3.7 Linking the twin or multiple solutions framework to emerging concepts

The twin or multiple solutions framework presented here aligns with emerging concepts aimed at promoting the dual or multiple uses of biomass resources. This includes the circular bioeconomy concept, which advocates for the continuous circular use of biomaterials to provide goods and services while minimizing waste generation (European Commission, 2018; Stegmann et al., 2020). A circular bioeconomy aims to achieve several objectives: (1) minimize waste, (2) circulate biomaterials to reduce the extraction of pristine biomass, (3) reduce pollution, and (4) create value-added products and goods from biogenic resources (European Commission, 2018).

In this context, biomass feedstocks undergo various conversion pathways to yield multiple coproducts such as bioenergy, biofertilizer, and biochar. Another emerging concept is the use of dual or integrated process systems such as the integrated generation of solid guel and biogas from biomass (IFBB) (Wachendorf et al., 2009; Bühle et al., 2012). IFBB involves biomass pretreatment by pressing to generate two biomass streams: (1) press fluid and (2) press cake (Wachendorf et al., 2009; Bühle et al., 2012). The press fluid can serve as a feedstock for biogas production through anaerobic digestion or as a source of platform biochemicals such as lactic acid. The digestate produced during anaerobic digestion can be used as a biofertilizer when applied to soil. The press cake can be converted into solid fuel or energy briquettes, or subjected to thermohydrochemical treatment to produce biochar or biogenic activated carbon for water and wastewater treatment (Kaetzl et al., 2018, 2019, 2020).

Despite its novelty, IFBB has received limited research attention outside Germany, where it was originally developed. Therefore, further efforts to validate the conceptual framework should integrate related concepts such as the circular bioeconomy and integrated or dual process systems such as IFBB.

24.4 Concluding remarks and outlook

Biochar technology is currently attracting global attention in research and public discourse due to its potential contributions to climate change mitigation, food security, energy provision, and environmental remediation. It also holds promise for addressing the energy-water-food nexus in low-income countries across Africa, Southeast Asia, the Caribbean, and Latin America. Nexus challenges involving energy, clean drinking water, food security, and environmental degradation are interconnected and can be more effectively tackled simultaneously. Therefore, biochar technology's ability to address multiple challenges makes it particularly suitable for low-income countries.

Existing biochar research has often focused on isolated components of the energy-water-food nexus, neglecting their intricate linkages. Furthermore, until now, a theoretical framework based on the energy-water-food nexus has been lacking. Conceptual frameworks are essential tools in guiding hypothesis formulation, research design, and implementation. In the context of biochar research in low-income countries, such frameworks also offer heuristic pathways for the uptake, adoption, and impact of biochar.

This chapter proposes a novel 'twin or multiple solutions' framework as a conceptual pathway to guide biochar research, uptake, and adoption for impact in low-income countries. At its core, the framework suggests that biochar technology is most likely to be adopted and impactful when it provides tangible and cost-effective solutions to at least two key challenges in these settings. The concept of 'twin or multiple solutions' aims to clearly demonstrate the comparative benefits of biochar over competing technologies in low-income countries.

Illustratively, the well-known energy-water-food nexus in low-income countries is used to outline the core elements of this concept. The concept is visualized through a 'ring' depicting biochar technology systems and their applications to address this nexus. Three potential 'twin or multiple solutions' (Solutions 1, 2, and 3) are highlighted as typical examples in low-income countries:

1. Biochar/pyrolytic cookstoves using readily available and inexpensive biomass feedstock for renewable, clean, and efficient household energy (Solution 1).
2. Use of pristine biochar from pyrolytic or gasifier cookstoves as filter media in biochar-based water filtration systems for household clean drinking water (Solution 2).
3. Application of spent and pristine biochar to inherently acidic and infertile tropical soils to enhance soil fertility and increase crop productivity, thereby contributing to household food security (Solution 3).

The existence of naturally fertile soils such as African Dark Earths in West Africa, terra preta or terra de Indo in Amazonia, and formiguer in South America underscores the potential feasibility of biochar application for soil improvement. Additionally, biochar's potential role in climate change mitigation through increased soil carbon sequestration and reduced greenhouse gas emissions from agroecosystems is noted (Solution 4). However, the direct benefits of climate change mitigation are less immediately tangible to low-income communities compared to other solutions.

Note that the solutions presented here are not exhaustive, and the selection of priorities should be based on local needs and circumstances. This chapter provides a conceptual framework to guide biochar research and offers a heuristic pathway for promoting its uptake, adoption, and impact. However, this framework remains a hypothesis that requires validation through systematic biochar research, including pilot-scale field studies in low-income countries. Regional research collaboration is emphasized to share experiences and pool resources effectively, particularly focusing on the energy-water-food nexus as a primary research area in Africa and other low-income regions.

Acknowledgments

WG thanks the Georg Forster Fellowship for Experienced Researchers funded by the Alexander von Humboldt Stiftung/Foundation, Germany for funding the project entitled, *"Environmental and Climate Footprints of a Biochar and Biogas Technology versus Current Practices in Sub-Saharan Africa: A Life Cycle Analysis."*

I am also grateful to the International Foundation for Science — IFS, Sweden for providing seed research grants for biochar research [grant numbers: C-5266−1, and C/5266−2]. The projects formed the basis and motivated the current paper. However, the views expressed in the current paper are solely those of the author. IFS played no role whatsoever in experimental design, data interpretation, write-up, and decision to submit the manuscript for publication.

Declaration of conflict of interest

The author declares no conflict of interest financially or otherwise that could have influenced his interpretation of the evidence. The author declares that he has published a number of earlier papers on biochar application for soil amendment, drinking water treatment, and environmental remediation, including air pollution control, the bulk of them targeting low-income countries. Therefore, given their relevance to the current manuscript, they are cited accordingly in the current paper.

References

European Commission. (2018), *A sustainable bioeconomy for Europe: Strengthening the connection between economy*. Available from https://doi.org/10.2777/478385.

Bach, M., Wilske, B., & Breuer, L. (2016). Current economic obstacles to biochar use in agriculture and climate change mitigation. *Carbon Management, 7*(3-4), 183−190.

Bamdad, H., Hawboldt, K., & MacQuarrie, S. (2018). A review on common adsorbents for acid gases removal: Focus on biochar. *Renewable and Sustainable Energy Reviews, 81*, 1705−1720. Available from https://doi.org/10.1016/j.rser.2017.05.261.

Baveye, P. C., Berthelin, J., Tessier, D., & Lemaire, G. (2017). The "4 per 1000" initiative: A credibility issue for the soil science community? Letter to the editor. *Geoderma*. Available from https://doi.org/10.1016/j.geoderma.2017.05.005.

Bühle, L., Hensgen, F., Donnison, I., Heinsoo, K., & Wachendorf, M. (2012). Life cycle assessment of the integrated generation of solid fuel and biogas from biomass (IFBB) in comparison to different energy recovery, animal-based and non-refining management systems. *Bioresource Technology, 111*, 230−239. Available from https://doi.org/10.1016/j.biortech.2012.02.072.

Chaukura, N., Chiworeso, R., Gwenzi, W., Motsa, M. M., Munzeiwa, W., Moyo, W., Chikurunhe, I., & Nkambule, T. T. (2020). A new generation low-cost biochar-clay composite 'biscuit' ceramic filter for point-of-use water treatment. *Applied Clay Science, 185*, 105409.

Dalahmeh, S., Ahrens, L., Gros, M., Wiberg, K., & Pell, M. (2018). Potential of biochar filters for onsite sewage treatment: Adsorption and biological degradation of pharmaceuticals in laboratory filters with active, inactive and no biofilm. *Science of the Total Environment, 612*, 192−201. Available from https://doi.org/10.1016/j.scitotenv.2017.08.178.

M.R. Delaney, *An analysis of biochar's appropriateness and strategic action plan for its adoption and diffusion in a high poverty context: The case of central Haiti*. (2011).

Getahun E. Tessema D. Gabbiye N. 2019 Design and development of household gasifier cooking stoves: Natural versus forced draft. In *Lecture Notes of the Institute for Computer Sciences, Social-Informatics and Telecommunications Engineering (LNICST), 298*, 314. Springer Verlag, Ethiopia. doi: 10.1007/978-3-030-15357-1_25.

Dumbrell, N. P., Kragt, M. E., & Gibson, F. L. (2016). What carbon farming activities are farmers likely to adopt? A best−worst scaling survey. *Land Use Policy, 54*, 29−37.

Gwenzi, W., Chaukura, N., Mukome, F. N., Machado, S., & Nyamasoka, B. (2015). Biochar production and applications in sub-Saharan Africa: Opportunities, constraints, risks and uncertainties. *Journal of Environmental Management, 150*, 250−261.

Gwenzi, W., Chaukura, N., Noubactep, C., & Mukome, F. N. (2017). Biochar-based water treatment systems as a potential low-cost and sustainable technology for clean water provision. *Journal of environmental management, 197*, 732−749.

Gwenzi, W., Chaukura, N., Wenga, T., & Mtisi, M. (2020). Biochars as media for air pollution control systems: Contaminant removal, applications and future research directions. *Science of the Total Environment*, 142249. Available from https://doi.org/10.1016/j.scitotenv.2020.142249.

Herath, I., Kumarathilaka, P., Navaratne, A., Rajakaruna, N., & Vithanage, M. (2015). Immobilization and phytotoxicity reduction of heavy metals in serpentine soil using biochar. *Journal of Soils and Sediments, 15*, 126−138.

Ippolito, J. A., Berry, C. M., Strawn, D. G., Novak, J. M., Levine, J., & Harley, A. (2017). Biochars reduce mine land soil bioavailable metals. *Journal of Environmental Quality, 46*(2), 411−419.

Kaetzl, K., Lübken, M., Gehring, T., & Wichern, M. (2018). Efficient low-cost anaerobic treatment of wastewater using biochar and woodchip filters. *Water, 10*, 818.

Kaetzl, K., Lübken, M., Nettmann, E., Krimmler, S., & Wichern, M. (2020). Slow sand filtration of raw wastewater using biochar as an alternative filtration media. *Scientific Reports, 10*, 1229. Available from https://doi.org/10.1038/s41598-020-57981-0.

Kaetzl, K., Lübken, M., Uzun, G., Gehring, T., Nettmann, E., Stenchly, K., & Wichern, M. (2019). On-farm wastewater treatment using biochar from local agroresidues reduces pathogens from irrigation water for safer food production in developing countries. *Science of the Total Environment, 682*, 601−610.

Karanja, A., & Gasparatos, A. (2019). Adoption and impacts of clean bioenergy cookstoves in Kenya. *Renewable and Sustainable Energy Reviews, 102*, 285−306.

Kirch, T., Medwell, P. R., Birzer, C. H., & van Eyk, P. J. (2020). Feedstock dependence of emissions from a reverse-downdraft gasifier cookstove. *Energy for Sustainable Development, 56*, 42−50. Available from https://doi.org/10.1016/j.esd.2020.02.008.

Kumarathilaka, P., Ahmad, M., Herath, I., Mahatantila, K., Athapattu, B. C. L., Rinklebe, J., Ok, Y. S., Usman, A., Al-Wabel, M. I., Abduljabbar, A., & Vithanage, M. (2018). Influence of bioenergy waste biochar on proton- and ligand-promoted release of Pb and Cu in a shooting range soil. *Science of The Total Environment, 625*, 547−554. Available from https://doi.org/10.1016/j.scitotenv.2017.12.294.

Latawiec, A., Królczyk, J., Kuboń, M., Szwedziak, K., Drosik, A., Polańczyk, E., Grotkiewicz, K., & Strassburg, B. (2017). Willingness to adopt biochar in agriculture: The producer's perspective. *Sustainability, 9*(4), 655. Available from https://doi.org/10.3390/su9040655.

Lehmann, J. (2007). A handful of carbon. *Nature, 447*, 143−144.

Lehmann, J., & Joseph, S. (2009). *Biochar for environmental management: Science and technology*. London: Earthscan.

Lorenz, K., Lal, R., Lorenz, K., & Lal, R. (2018). Importance of soils of agroecosystems for climate change policy. *Carbon Sequestration in Agricultural Ecosystems*, 357−386.

Maya, M., Gwenzi, W., & Chaukura, N. (2020). A biochar-based point-of-use water treatment system for the removal of fluoride, chromium and brilliant blue dye in ternary systems. *Environmental Engineering & Management Journal (EEMJ), 19*(1), 143−156.

Mohanty, S. K., & Boehm, A. B. (2014). Escherichia coli removal in biochar-augmented biofilter: Effect of infiltration rate, initial bacterial concentration, biochar particle size, and presence of compost. *Environ. Sci. Technol., 48*, 11535−11542 (2014).

L.C.V. Munoz, *Spreading the char: The importance of local compatibility in the diffusion of biochar systems to the smallholder agriculture community context*. Pomona Senior Theses. (2014).

Panwar, N. L. (2009). Design and performance evaluation of energy efficient biomass gasifier based cookstove on multi fuels. *Mitigation and Adaptation Strategies for Global Change, 14*, 627–633.

Perez-Mercado, L. F., Lalander, C., Berger, C., & Dalahmeh, S. S. (2018). Potential of biochar filters for onsite wastewater treatment: Effects of biochar type, physical properties and operating conditions. *Water (Switzerland), 10*(12). Available from https://doi.org/10.3390/w10121835. Available from, https://www.mdpi.com/2073-4441/10/12/1835/pdf.

Schmidt, H. P. (2012). *55 Uses of Biochar* (pp. 450–454). Arbaz, Switzerland: Ithaka-Journal.

Stegmann, P., Londo, M., & Junginger, M. (2020). The circular bioeconomy: Its elements and role in European bioeconomy clusters. *Resources. Conservation & Recycling: X, 6*, 100029.

Tryner, J., Tillotson, J. W., Baumgardner, M. E., Mohr, J. T., DeFoort, M. W., & Marchese, A. J. (2016). The effects of air flow rates, secondary air inlet geometry, fuel type, and operating mode on the performance of gasifier cookstoves. *Environmental Science and Technology, 50*(17), 9754–9763. Available from https://doi.org/10.1021/acs.est.6b00440. Available from, http://pubs.acs.org/journal/esthag.

Vikrant, K., Kim, K.-H., Ok, Y. S., Tsang, D. C. W., Tsang, Y. F., Giri, B. S., & Singh, R. S. (2018). Engineered/designer biochar for the removal of phosphate in water and wastewater. *Science of The Total Environment, 616–617*, 1242–1260. Available from https://doi.org/10.1016/j.scitotenv.2017.10.193.

Wachendorf, M., Richter, F., Fricke, T., Graß, R., & Neff, R. (2009). Utilization of semi-natural grassland through integrated generation of solid fuel and biogas from biomass. I. Effects of hydrothermal conditioning and mechanical dehydration on mass flows of organic and mineral plant compounds, and nutrient balances. *Grass and Forage Science, 64*(2), 132–143. Available from https://doi.org/10.1111/j.1365-2494.2009.00677.x.

WHO. (2016). *Burning opportunity: clean household energy for health, sustainable development, and wellbeing of women and children*. Geneva: World Health Organization.

WHO. (2020). *Water, sanitation, hygiene and waste management for COVID-19: Technical Brief, 03 March 2020*. World Health Organization.

Chapter 25

Moving ahead with biochar for environmental remediation and beyond: future research directions, roadmap, and epilogue

Willis Gwenzi[1,2,3]

[1]*Formerly Alexander von Humboldt Fellow and Guest Full Professor, Leibniz-Institut für Agrartechnik und Bioökonomie e.V. (ATB), Potsdam, Germany,*
[2]*Formerly Alexander von Humboldt Fellow and Guest Full Professor, Grassland Grassland Science and Renewable Plant Resources, Faculty of Organic Agricultural Sciences, Universität Kassel, Witzenhausen, Germany,* [3]*Biosystems and Environmental Engineering Research Group, Marlborough, Harare, Zimbabwe*

Chapter outline

25.1 Introduction	477
25.2 Biochar for environmental remediation: a synoptic overview of the state-of-the-art	479
25.3 Looking ahead: future research directions and perspectives	480
25.3.1 Twenty knowledge gaps to advance biochar for environmental remediation	481
25.3.2 A handful of "grand questions" on biochar technology	484
25.4 Advancing biochar for environmental remediation: a conceptual roadmap	486
25.4.1 Preparation and characterization of biochars	486
25.4.2 Laboratory-scale prototyping and evaluation	487
25.4.3 Pilot-scale prototyping and evaluation	487
25.4.4 Field-scale application	487
25.4.5 Monitoring, evaluation, and feedback	488
25.4.6 Approval and commercialization	488
25.5 Epilogue and outlook	488
Acknowledgments	488
Declaration of conflict of interest	489
References	489

25.1 Introduction

Biochar technology has emerged as a focal point of policy and research discussions due to its potential as a biomaterial for environmental remediation. Recent years have witnessed a significant shift in biochar research from its traditional focus on agricultural applications to its potential in environmental remediation. This transition has led to a substantial body of evidence comprising both original articles and comprehensive reviews on the subject (Chaukura, Chiworeso, et al., 2020; Chaukura, Masilompane, et al., 2020; Gwenzi et al., 2020, 2021). Despite this wealth of information, the findings remain scattered across numerous publications, creating a challenge for those seeking a comprehensive understanding of biochar's role in remediation. This book aims to address this gap by providing an exhaustive review of biochar applications for the remediation of soils, water, wastewater, and air pollution (Fig. 25.1).

Numerous studies have explored various aspects of biochar, including the biomass feedstocks used for its production, characterization techniques, and the properties of different types of biochars (Duku et al., 2011; Gwenzi et al., 2015, Duku et al., 2016; Wang & Wang, 2019). Recently, emerging tools such as machine learning and in-silico computation methods have been applied to the advanced design, characterization, and evaluation of biochar systems (Khan et al., 2024; Zhu, Li and Wang, 2019).

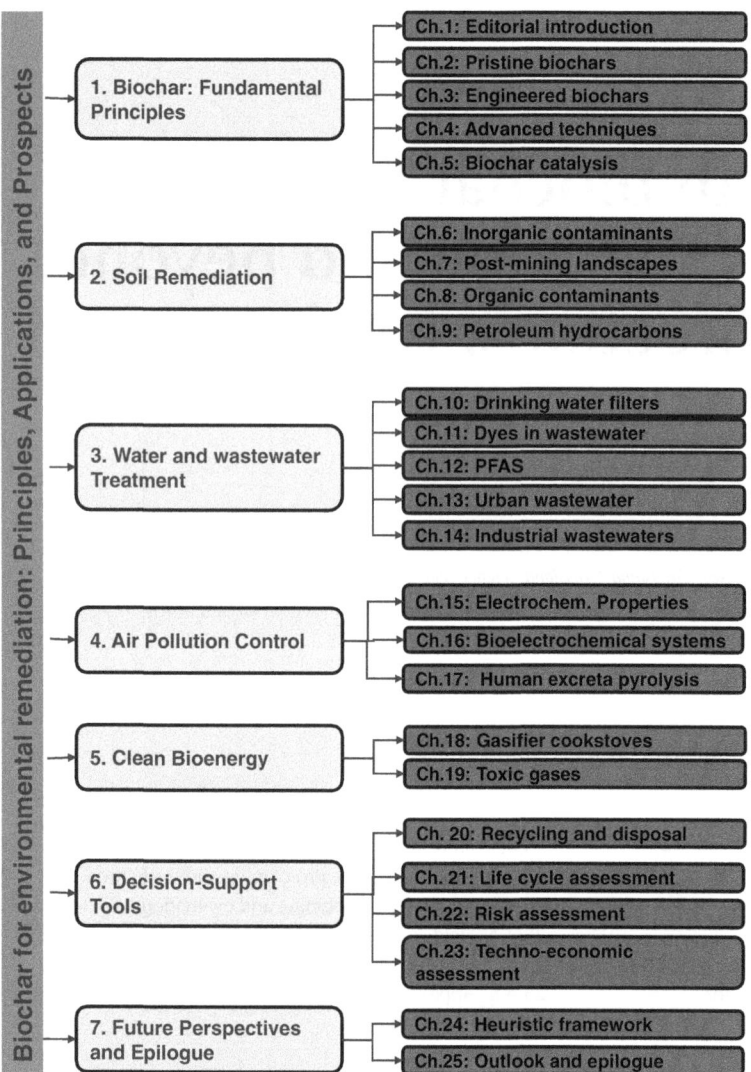

FIGURE 25.1 Overview of the state-of-the-art on biochar for environmental remediation.

Beyond agricultural applications, biochar has been investigated for its efficacy in remediating various contaminants in soils (Sizmur et al., 2016). For instance, studies have examined the use of biochar for rehabilitating post-mining landscapes (Hagner et al., 2021) and remediating soils contaminated with petroleum hydrocarbons (Dike et al., 2022) and various inorganic and organic contaminants such as toxic metals (Forján et al., 2017; Zama et al., 2018).

The use of biochar as an adsorbent for removing contaminants from aqueous systems, including drinking water and wastewater, has garnered considerable research attention. Several studies have investigated biochar-based filtration systems for drinking water and wastewater treatment (Chaukura et al., 2017a, 2017b; Chaukura, Chiworeso, et al., 2020; Chaukura, Masilompane, et al., 2020; Gwenzi et al., 2014, 2017; Maya et al., 2020).

Additionally, biochar has been explored in clean and renewable bioenergy applications, including anaerobic digestion, bioelectrical systems such as microbial fuel cells, and as electrodes and supercapacitors for energy storage (Ahmed & Hussain, 2018; Hu et al., 2024; Ma et al., 2020; Pérez-Rodríguez et al., 2021). The potential of gasifier cookstoves as a clean and renewable energy source for household cooking and heating has also been investigated (Gitau, Mutune, et al., 2019; Gitau, Sundberg, et al., 2019; Sundberg et al., 2020). Recent studies in Brazil have examined the thermochemical conversion of fecal matter to energy and biochar (Bittencourt & Martins, 2022; Bittencourt et al., 2019).

Biochar's potential as an adsorbent for air pollution control has also been explored (Bamdad, 2019; Bamdad et al., 2018; Gwenzi et al., 2021; Mohamed et al., 2022). Studies have investigated the use of biochar from agricultural residues to remove toxic gases emitted from chemical plants (Cho et al., 2023). Additionally, numerous studies have examined biochar's role in mitigating greenhouse gas emissions, particularly in agroecosystems (Gwenzi et al., 2015; Zhang et al., 2019).

The application of decision-support tools for assessing biochar systems has also gained research attention. These tools include techno-economic assessments, environmental and social impact assessments, and life cycle assessments (Azzi et al., 2021; Bergman et al., 2022; Sahoo et al., 2021). Furthermore, some studies have investigated the regeneration, recycling, and disposal technologies for spent biochar (Chen et al., 2022; Jia et al., 2023; Kar et al., 2023).

In summary, while significant research has been conducted on biochar for environmental remediation, the findings have been dispersed across individual articles and reviews. This book consolidates this information, providing thorough and timely coverage of biochar for the remediation of soils, water, wastewater, and air pollution. Despite the progress made, several knowledge gaps remain, and a conceptual roadmap for the development and application of engineered biochar remediation systems is necessary. Therefore, this book presents future research directions, comprising key knowledge gaps and grand questions, and proposes a conceptual roadmap for advancing biochar for environmental remediation.

The specific objectives of this chapter are: (1) to present a synoptic overview of the state-of-the-art on biochar for environmental remediation, (2) to highlight the key knowledge gaps pertaining to various environmental applications of biochar, and (3) to propose higher-order or grand questions concerning biochar technology. Finally, the chapter closes with an epilogue and outlook outlining the achievements, challenges, and future prospects of biochar technology.

25.2 Biochar for environmental remediation: a synoptic overview of the state-of-the-art

Recent years have witnessed a shift in biochar research from agricultural applications to environmental remediation. Consequently, a large body of evidence comprising original articles and reviews exists on biochar applications in environmental remediation (Chaukura, Chiworeso, et al., 2020; Chaukura, Masilompane, et al., 2020; Gwenzi et al., 2020, 2021). However, these findings remain scattered across individual articles and reviews. Until now, readers interested in biochar for remediation have lacked a comprehensive resource on the topic. This book addresses this gap by providing an exhaustive review of biochar for the remediation of soils, water, wastewater, and air pollution. Chapter 1 presents an overview of the principles, applications, and prospects of biochar for environmental remediation. Fig. 25.1 presents an overview of the state-of-the-art biochar for environmental remediation discussed in this book.

Many studies have investigated the biomass feedstocks, production, characterization, and properties of various types of biochars (Duku et al., 2011; Gwenzi et al., 2015; Duku et al., 2016; Wang & Wang, 2019). Some recent studies have also explored the application of emerging tools such as machine learning and in-silico computation methods for the advanced design, characterization, and evaluation of biochar systems (Khan et al., 2024; Zhu, Li and Wang, 2019). Chapters 2–5 provide a comprehensive discussion of the state-of-the-art on the fundamental principles of biochar, covering biomass feedstocks, biochar production, characterization techniques, and properties, including catalysis.

Besides soil applications in agroecosystems, biochars have been used for the remediation of various contaminants in soils (Sizmur et al., 2016). For example, some studies have investigated the use of biochar for the rehabilitation of post-mining landscapes (Hagner et al., 2021). Others have explored the use of biochar for soils contaminated with petroleum hydrocarbons (Dike et al., 2022) and inorganic and organic contaminants such as toxic metals (Forján et al., 2017; Zama et al., 2018). Chapters 6–9 present the state-of-the-art biochar applications for soil remediation, focusing on the rehabilitation of post-mining landscapes, inorganic contaminants, conventional and emerging contaminants, and petroleum hydrocarbons.

The removal of contaminants in aqueous systems, such as drinking water and wastewaters, using biochars as an adsorbent media, has also garnered considerable research attention. Several studies have investigated biochar-based filtration systems for drinking water and wastewater treatment (Chaukura et al., 2017a, 2017b; Chaukura, Chiworeso, et al., 2020; Chaukura, Masilompane, et al., 2020; Gwenzi et al., 2014, 2017; Maya et al., 2020). Chapters 10–14 synthesize and discuss the application of biochars for the remediation of contaminants in aqueous systems, with an emphasis on biochar-based drinking water filters and the treatment of municipal and industrial wastewater, urban stormwater, and dye removal using biochar derived from biowaste.

Other studies have explored the use of biochars in clean and renewable bioenergy applications, including anaerobic digestion, bioelectrical systems such as microbial fuel cells, and as electrodes and supercapacitors for energy storage materials (Hu et al., 2024; Ma et al., 2020; Pérez-Rodríguez et al., 2021; Vivekanandhan & S., 2018). Additionally, gasifier cookstoves have been investigated as a clean and renewable source of energy for household cooking and heating (Gitau, Mutune, et al., 2019; Gitau, Sundberg, et al., 2019; Sundberg et al., 2020). Recently, work in Brazil has examined the thermochemical conversion of fecal matter to energy and biochar (Bittencourt & Martins, 2022; Bittencourt et al., 2019). Chapters 15–18 discuss the electrochemical properties of biochars, their applications in bioenergy, biochar use in bioelectrochemical systems or microbial fuel cells, the thermochemical conversion of human excrement to biochar and energy, and gasifier cookstoves for clean energy provision and indoor air pollution control.

Biochars have also been used as adsorbent media for air pollution control (Bamdad, 2019; Bamdad et al., 2018; Gwenzi et al., 2021; Mohamed et al., 2022). For example, one study investigated the use of biochar from agricultural residues for the removal of toxic gases emitted from chemical plants (Cho et al., 2023). Numerous studies also exist on the use of biochar for greenhouse gas mitigation, with many focusing on agroecosystems (Gwenzi et al., 2015; Zhang et al., 2019).

The use of decision-support tools for the assessment of biochar systems has also received research attention. Some of the decision-support tools applied to assess biochar systems include techno-economic assessment, environmental and social impact assessment, and life cycle assessment (Azzi et al., 2021; Bergman et al., 2022; Sahoo et al., 2021). A few studies have also investigated the regeneration, recycling, and disposal technologies for spent biochars (Chen et al., 2022; Jia et al., 2023; Kar et al., 2023).

In summary, until now, research findings on biochar for environmental remediation have been scattered across individual articles and reviews. Consequently, readers interested in biochar for remediation have lacked a comprehensive resource on the topic. This book addresses this gap by providing thorough and timely coverage of biochar for the remediation of soils, water, wastewater, and air pollution. However, as an emerging technology, several knowledge gaps exist regarding biochar for environmental remediation. Moreover, a conceptual roadmap for the further development and application of engineered biochar remediation systems has been missing. Therefore, future research directions comprising 20 key knowledge gaps and grand questions are presented. Following this, a conceptual roadmap for the further advancement of biochar for environmental remediation is proposed.

25.3 Looking ahead: future research directions and perspectives

In this section, future research directions and perspectives, comprising 20 knowledge gaps, are presented. Fig. 25.2 presents an overview of some of the knowledge gaps. The gaps under each thematic topic are a non-exhaustive list, allowing the reader to identify further research areas.

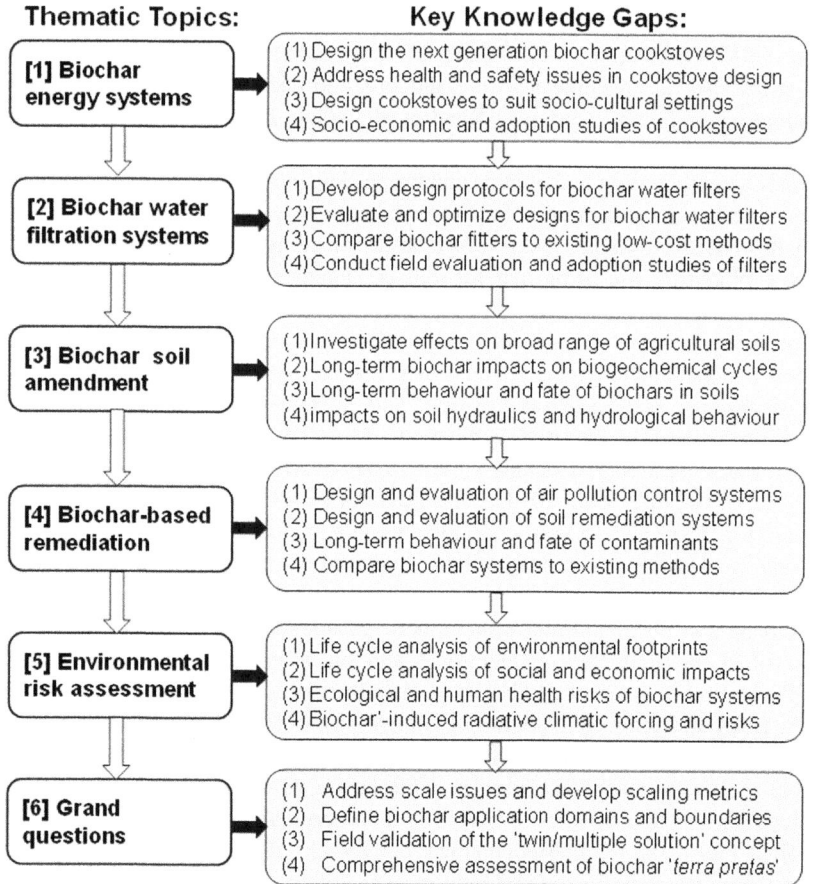

FIGURE 25.2 A summary of some of the key knowledge gaps and "grand questions" on the biochar thematic topics.

25.3.1 Twenty knowledge gaps to advance biochar for environmental remediation

25.3.1.1 Biomass trade-offs: biochar feedstock versus competing current uses

Biomass feedstocks, often considered free and readily available for biochar production, serve multiple ecological, biodiversity, and soil conservation functions. Balancing the use of such biomass for biochar production with existing competing uses creates potential biomass trade-offs. However, biomass trade-offs associated with biochar systems have received limited research attention.

25.3.1.2 Environmental pollution risks

Biochar systems release potentially potent particulate matter and gaseous emissions. Moreover, biochar may act as a medium for transferring contaminants from polluted sources to environmental receptors such as soils, air, and aquatic systems. However, the pollution risks of biochar systems and biochar-mediated transport of contaminants are poorly understood.

25.3.1.3 Ecological and biodiversity risks

Biochar systems may disrupt habitats through biomass collection, and biochars may contain residual toxic chemicals from feedstocks or formed during pyrolysis. To date, limited studies have investigated the ecological and biodiversity risks of biochar systems at various scales of biological organization.

25.3.1.4 Ecotoxicity of free versus biochar-bound contaminants

The application of biochars as adsorbents in soil, drinking water, wastewater, and air pollution involves moving contaminants from a mobile phase to a solid adsorbed phase. However, it is unclear whether biochar-bound contaminants are less ecotoxic than their mobile or free-form counterparts. Comparative studies are needed to understand the ecotoxicity of biochar-bound contaminants versus the free or mobile forms.

25.3.1.5 Long-term fate of biochar-bound contaminants

Research on biochar for environmental remediation of soils, water, wastewater, and air is dominated by short-term studies, providing no information on the long-term fate of biochar-bound contaminants. It is unclear whether such contaminants are permanently adsorbed, released back into the mobile phase, or undergo (bio)degradation over time. Long-term fate studies of biochar-bound contaminants using isotopic tracers are needed.

25.3.1.6 Biochar controls on the environmental resistome

Biochar, like other biosolids, may contain several contaminants of anthropogenic origin, including toxic metals and pharmaceutical residues, which may co-select for antibiotic resistance. Conversely, biomass pyrolysis may inactivate antibiotic-resistant bacteria and their resistance genes. There is a lack of data on whether biochar controls the emergence, proliferation, and persistence of antibiotic resistance in environmental matrices. Further work is required to understand biochar's role in the environmental resistome.

25.3.1.7 Human exposure and health risks in occupational and non-occupational settings

Human exposure to biochar-associated contaminants may occur in occupational and non-occupational settings. However, there is a lack of studies investigating the human exposure and health risks of biochar in these settings. Quantitative epidemiological studies are required to unravel the human exposure and health risks of biochar systems.

25.3.1.8 Health risk assessment

Several qualitative and quantitative tools exist for assessing the health risks of biochar systems. Yet, limited systematic studies have examined the health risks of biochar systems. Moreover, the risk hotspots and critical control points along the biochar value chain are still unknown.

25.3.1.9 Biochar as a novel renewable green biomaterial for mineral recovery

Biochar shares several similarities with activated carbon, including high porosity and surface area. Therefore, like activated carbon, biochar could be a potential alternative green biomaterial for the recovery of minerals and technology-critical elements such as rare earth elements. To date, limited studies have investigated the potential use of biochar as a novel green biomaterial in mineral recovery and processing applications.

25.3.1.10 Design principles and protocols for engineered biochar remediation systems

The use of biochars as a filter medium in water, wastewater, soil, and air pollution remediation lacks design principles and protocols, making the uptake and adoption of such systems problematic. Hence, there is a need to develop design principles and protocols for engineered biochar systems for various environmental remediation applications.

25.3.1.11 Life cycle analysis of the environmental and social footprints of biochar systems

Regardless of the target application, the biochar value chain or life cycle invariably entails: (1) biomass feedstock production, harvesting, transport, and preparation, (2) biochar production via pyrolysis, and logistics including transport and storage, and (3) subsequent industrial applications and post-application disposal/fate. Each of these steps may involve energy, water, and raw materials usage, and emissions of toxic contaminants. To date, only a few studies based on life cycle analysis have investigated the environmental footprints of biochar technology, mostly limited to energy applications. Comprehensive life cycle analyses investigating the environmental footprints of biochars applied in the context of the "twin or multiple solution" framework are still lacking. In addition, the occupational, human exposure, and health risks of toxic emissions in the biochar value chains are still poorly understood.

25.3.1.12 Biochar or black carbon-induced radiative forcing and associated climatic risks

Current literature relating biochar to the climate system is dominated by studies reporting the potential benefits of biochar for climate mitigation via carbon sequestration and reduction of greenhouse gas emissions. However, poorly designed pyrolysis systems, similar to charcoal production pit kilns used in low-income countries, may release highly potent particulate and gaseous emissions into the atmosphere. Similar to black carbon, biochars applied to soil may also alter soil thermal properties, including albedo and temperature regimes, and even induce radiative forcing. However, the potential climatic risks associated with biochar or black carbon-induced radiative forcing in the case of large-scale adoption of biochar technology have received limited attention. This is probably due to methodological challenges in understanding such climatic risks and their knock-on effects on ecosystems and human health. There is a need to investigate the potential impacts of large-scale adoption of biochars on radiative forcing and atmospheric feedbacks at regional and continental scales. Coupled earth-atmosphere modeling studies accounting for biochar-induced climatic forcing and feedbacks are required to understand potential impacts on regional, continental, and global climatic systems.

25.3.1.13 Techno-economic assessment

Techno-economic assessments of biochar systems are critical for decision-making and policy formulation, including technology uptake and adoption. However, existing data are limited to biochar production. Studies investigating the techno-economic assessment of industrial applications of biochar in environmental and bioenergy contexts are scarce. Further work is required to compare the techno-economic performance of biochar for environmental remediation to that of competing technologies.

25.3.1.14 Technology readiness assessment

Technology readiness assessment (TRA) determines the development level of a technology using a standardized scale. TRA is critical for identifying technologies ready for uptake, adoption, and promotion, as well as those still in their infancy stages of development. Biochar applications in environmental remediation and bioenergy are likely at different readiness levels. However, studies on TRA for biochar systems are still lacking. Therefore, TRA studies are required for various environmental remediation domains.

25.3.1.15 Validation of the twin or multiple solution heuristic framework

A novel twin or multiple solution framework was proposed for promoting the uptake, adoption, and upscaling of biochar technology. This framework is premised on the need to promote dual or multiple applications of biochar technology to address related processes such as the water-energy-food-environment nexus. However, further work is required to validate the framework and determine whether using biochar technology to address two or more problems indeed drives its uptake, adoption, and upscaling. This aspect should be linked to adoption studies because large-scale adoption of biochar for environmental remediation and other industrial applications remains low. Yet, systematic adoption studies of biochar applications in environmental remediation are still lacking, highlighting the need for further studies.

25.3.1.16 Integrating biochar into emerging concepts in agroecosystems

Research in agroecosystems, especially in low-income regions, has witnessed the promotion of several emerging technologies, including conservation agriculture, agroecology, and sustainable intensification. However, literature on biochar applications in agroecosystems is unclear on how biochar can be integrated with these emerging technologies. Further work is required to address this gap, given that some of these emerging technologies may potentially conflict. For example, conservation agriculture requires the retention of biomass on soil, while biochar technology requires harvesting and pyrolysis of biomass followed by soil application. Therefore, there is a need to reconcile these potential conflicts between biochar and other emerging technologies.

25.3.1.17 Synergistic interaction of biochar and other remediation technologies

Several conventional and emerging remediation technologies can be integrated with biochar to enhance remediation efficiency. For example, recent studies show that some insect species and earthworms have the capacity to remediate organic and inorganic contaminants and antibiotic resistance. However, the synergistic interactions of biochar and other remediation technologies have received limited attention in soil, water, wastewater, and air pollution remediation. Further work should investigate the synergistic interactive effects of biochar and other conventional and emerging technologies on the remediation efficiency of various contaminants in environmental matrices.

25.3.1.18 Biochar technology as part of a broader circular bioeconomy

Biochar applications in environmental remediation and agriculture should not be considered an isolated technology but should fit within the broader emerging concept of a circular bioeconomy and related principles. The principles of a circular bioeconomy aim to: (1) minimize waste, (2) circulate biomaterials to reduce the extraction of pristine biomass, (3) reduce pollution, and (4) create value-added products and goods from biogenic resources. In a circular bioeconomy, a biomass stream undergoes various conversion pathways to yield multiple co-products such as bioenergy, biofertilizer, and biochar. Related to the circular bioeconomy is the emerging concept of dual or integrated process systems like the Integrated Generation of Solid Fuel and Biogas from Biomass (IFBB). IFBB involves biomass pretreatment by pressing to generate two biomass streams: (1) press fluid and (2) press cake. The press fluid can serve as a feedstock for biogas production through anaerobic digestion or as a source of platform biochemicals such as lactic acid. The digestate produced during anaerobic digestion can be used as biofertilizer when applied to soil. The press cake can be converted into solid fuel or energy briquettes, or subjected to thermohydrochemical treatment to produce biochar or biogenic activated carbon for water and wastewater treatment.

25.3.1.19 Socio-cultural aspects of biochar technology

Biochar technology, such as the use of gasifier cookstoves for household heating and cooking in low-income settings, has socio-cultural implications, including gender roles and changes in cultural norms and practices. For example, in Africa, open fires from traditional three-stone biomass cookstoves serve multiple functions, including as meeting points for family discussions as well as for religious purposes. Therefore, the socio-cultural impacts of transitioning from such traditional cook systems to gasifier cookstoves are unknown. In settings with strong religious and socio-cultural norms, there is a need to understand how biochar fits in.

25.3.1.20 Harnessing biochar to earn carbon credits

Biochar is considered a carbon-neutral or negative technology. Hence, its application in environmental remediation and agriculture should potentially earn users some carbon credits. However, the protocols for claiming such carbon credits for individual users (e.g., households, farmers, enterprises) are not yet well developed. Moreover, even when the protocols are fully developed, several potential biochar users in low-income settings may remain unaware of such carbon credit systems and how they operate. Therefore, as an incentive to promote biochar technology, there is a need to develop and publicize the carbon credit systems applicable to biochar.

25.3.2 A handful of "grand questions" on biochar technology

This section presents some proposed "grand questions" on biochar technology. These "grand questions" address issues that are currently excluded or poorly understood in the literature on biochar but are critical to guide future research. The term "grand questions" refers to highly relevant and cross-cutting issues inadequately addressed in current literature. These questions persist due to (1) the exclusion of such aspects in current research approaches or (2) methodological limitations that pose significant challenges in understanding them. Further work is required to address these most pressing and cross-cutting questions in biochar research to maximize the technology's potential while minimizing its risks.

25.3.2.1 Biochar and the paradox of scale

The bulk of the data on biochar technology and its applications in energy, carbon sequestration, soil quality, crop productivity, water treatment, and mitigation of emissions are based on small-scale and even point-scale measurements often taken over short periods. However, data acquired at relatively large spatial scales over extended periods are ideal for practitioners and policymakers to evaluate biochar technology. Most studies seem to assume that short-term point measurements of biochar's impacts and benefits linearly scale up to large temporal and spatial scales, including ecosystem goods and services such as biogeochemical cycling. This raises several questions pertaining to scale:

- How do short-term impacts and benefits of biochar measured at small and point-scales cascade across spatial and temporal scales?
- What metrics can be used to scale up small-scale and point-scale measurements to larger temporal and spatial scales, considering nonlinearities and complex systems?
- How can spatial and temporal heterogeneities be addressed when scaling small and point-scale data to large scales, and what uncertainties are associated with such scaled-up data?

Currently, no answers exist to these questions, yet the potential impacts and benefits of a technology at scale are critical in decision and policy-making. Interestingly, the problem of scale is not limited to biochar technology but is also pervasive in other disciplines, including earth sciences, hydrology, and ecology.

25.3.2.2 Moving from "one-size-fits-all" to specific biochar application domains

Biochar is often touted as a "silver bullet" or "one-size-fits-all" technology capable of addressing the water-energy-food-environment nexus in both developing and developed countries. Some authors even claim that biochar technology has 55 potential applications (Schmidt and Wilson, 2014). In the hands of policymakers, practitioners, biochar enthusiasts, and the media, this notion may result in undesirable consequences. For example, in the case of climate change mitigation, such notions may divert resources and research efforts from other equally important disciplines to biochar. Moreover, such perceptions stifle open debate about biochar, and anyone questioning its relevance can be labeled a heretic. While there is no doubt about the potential opportunities presented by biochar, the technology is unlikely to achieve the claimed benefits in all cases. Yet, the boundary conditions for biochar applications remain poorly defined.

Biochars are quite heterogeneous materials, with properties depending on both feedstock and pyrolysis conditions. This calls for research to define the biophysical limits or boundaries of biochar and develop a database of well-characterized biochars for tailor-made applications. Overcoming the "blanket salesman" approach requires research to identify biochar application domains or niches and subsequently develop tailor-made biochar solutions for such domains.

Further research is required to identify application niches clearly and define the socio-economic and biophysical boundary limits for biochar. This requires citizen science and effective science communication, as well as open debate about biochar opportunities and risks, including fair and transparent communication of negative or neutral results in scientific media.

25.3.2.3 Creating enduringly fertile soils and the terra preta enigma

The concept of biochar was motivated by observations of the terra preta, terra de indo, and recently the African Dark Earths, which are enduringly fertile and highly productive tropical soils. These soils were created via anthropogenic activities through the addition of carbonaceous materials over a long period. Yet, the exact mechanisms responsible for their formation remain an enigma. For example, current evidence shows that terra preta and Africa dark earths store large stocks of stable carbon. However, the biogeochemical fluxes, including greenhouse gas emissions, that occurred during their formation are poorly understood. Evidence suggests that biochars are highly dynamic; hence, their properties and even the physico-chemical and biological properties of biochar-amended soils are bound to change over time. Yet, data on biogeochemical cycles, including greenhouse gas emissions of biochar-amended soils, are based on short-term studies. This raises several questions:

- How much, how frequent, and in what form were the carbonaceous materials added to create such soils, and over what timescales?
- How do biogeochemical cycles of biochar-amended soils behave in the long term in terms of fluxes and stocks, and partitioning among stable and labile fractions, and what climatic and biogeochemical controls influence such C and N balances?
- How feasible is it to create terra pretas within human timescales relevant for decision and policy-making at typical and feasible biochar application rates and frequencies?
- What global biochar application and adoption levels are required to significantly impact biogeochemical cycles and greenhouse gas emissions and mitigate climate change at regional and global scales?

Finally, similar to carbon sequestration in biomass and even in agricultural soils (Franzluebbers et al., 2012), some may pose the question, "Do current biochars really represent a long-term stable carbon pool or are we simply buying time and delaying the inevitable?" Currently, no clear answers exist to these questions, yet such information is critical in our current and future efforts to create terra pretas through biochar applications.

25.3.2.4 What scale of biochar application is required to significantly influence the climate system?

Current literature relating biochar to the climate system is dominated by studies reporting the potential benefits of biochar to the climate via carbon sequestration and mitigation of greenhouse gas emissions. However, it is unclear what scale of biochar adoption in agriculture and other applications is required to achieve a shift in greenhouse gases and the climate system. Some studies also show that biochar or black carbon induces radiative climatic forcing. However, the extent to which the mitigatory effects of biochar offset the radiative forcing effect is currently unknown. This is probably due to methodological challenges associated with understanding such climatic risks and their knock-on effects on ecosystems and human health. Thus, there is a need to investigate the potential impacts of large-scale adoption of biochars on climatic forcing and atmospheric feedbacks at regional and continental scales. Coupled earth-atmosphere modeling studies accounting for biochar-induced climatic forcing and feedbacks are required to understand potential impacts on regional, continental, and global climatic systems.

25.3.2.5 Long-term impacts of biochar on soil health and fertility

Understanding how biochar affects soil properties, nutrient cycling, and microbial communities over extended periods is crucial, but most available data are based on short-term studies. Research should aim to determine the sustainability and longevity of biochar's benefits in different soil types and climates.

25.3.2.6 Optimization of biochar remediation systems for maximum effectiveness

Investigating the dominant mechanisms by which biochar adsorbs contaminants and identifying the types of biochar that are most effective for various pollutants is critical in developing standardized and highly effective environmental remediation systems. However, optimized systems and design protocols for biochar remediation systems are still lacking for soil, water, wastewater, and air pollution remediation.

25.3.2.7 Sustainable scaling of biochar production systems to meet global demands

Biochar production is still largely limited to small-scale production systems, which have limited capacity to meet large biochar demands, such as those required in agriculture. Developing large-scale sustainable biochar production methods that minimize environmental impact, optimize energy use, and utilize diverse biomass feedstocks without competing with food production or biodiversity conservation is critical for large-scale adoption.

25.3.2.8 Socio-economic implications of biochar technology in low-income countries

The evidence-based on biochar applications in low-income settings is still weak. Yet, such low-income countries often have unique biophysical, socio-economic, and cultural settings. Thus, research findings on biochar from high-income countries cannot be directly extrapolated to low-income settings. Understanding how biochar technology affects local economies, communities, and cultural practices can inform policies and programs that promote equitable and beneficial adoption. This includes evaluating the potential for biochar to contribute to poverty alleviation, job creation, and improved health outcomes.

Moreover, poor people have limited or no access to global commodity markets, and carbon and biochar markets are no exception. The question arises, how can we create viable and inclusive biochar and carbon markets than also work for the poor in low- and middle-income countries?

25.3.2.9 How do we catalyze the global transition to biochar to achieve the perceived benefits?

Biochar technology has several perceived benefits and applications in environmental remediation and agriculture. Yet, practitioners, and decision—and policy-makers tend to "lock in" to existing technologies even in the face of better options. This raises the questions, (1) how to we catalyze a global transition to biochar technology in environmental remediation and agriculture, (2) what models and drivers (processes, triggers, incentives) can be used to achieve the transition?

25.3.2.10 To what extent can large-scale biochar application reduce the environmental footprint of food systems and humans?

Data from small-scale studies show that biochar increases nutrient and water retention, and improves water and nutrient efficiencies, thus reducing water and chemical fertilizer requirements. Moreover, the potential of biochar to sequester carbon and mitigation greenhouse gas emissions in agroecosystems has been documented. However, current unclear is to what extent can large-scale biochar application reduce the environmental footprint of food systems and humans under the "worst" and "best" case scenarios relative to the "business-as-usual" scenario?

25.4 Advancing biochar for environmental remediation: a conceptual roadmap

The literature on biochar for environmental remediation lacks a conceptual roadmap for the further development and industrial application of engineered biochar remediation systems. Here, engineered biochar systems refer to designed and optimized processes and devices for environmental remediation. In environmental remediation, examples of engineered biochar systems include:

1. Soil remediation systems
2. Biochar drinking water filters
3. Wastewater filtration systems using biochar as filter media
4. Biochar air filtration systems

Examples of bioenergy applications include:

1. Biochar electrodes and supercapacitors
2. Biochar-enhanced bioelectrochemical systems or microbial fuel cells
3. Biochar-amended anaerobic digestion and fermentation systems
4. Gasifier or biochar cookstoves

Here, a conceptual roadmap to guide further advancement of biochar for environmental remediation is presented (Fig. 25.3). This framework has been adapted from an earlier paper on the application of biochars as filter media in air quality control systems (Gwenzi et al., 2021). The steps are considered generic, hence ca be adapted and applied to developed engineered biochar systems for soil, water, wastewater and air pollution remediation as well as biochar applications in bioenergy applications.

The conceptual roadmap for the development and application of engineered biochar remediation systems comprises six key steps (Fig. 25.2).

25.4.1 Preparation and characterization of biochars

Prepare and optimize tailored pristine and engineered biochars for the target remediation application. A database of well-characterized biochars and their potential applications should be developed and maintained.

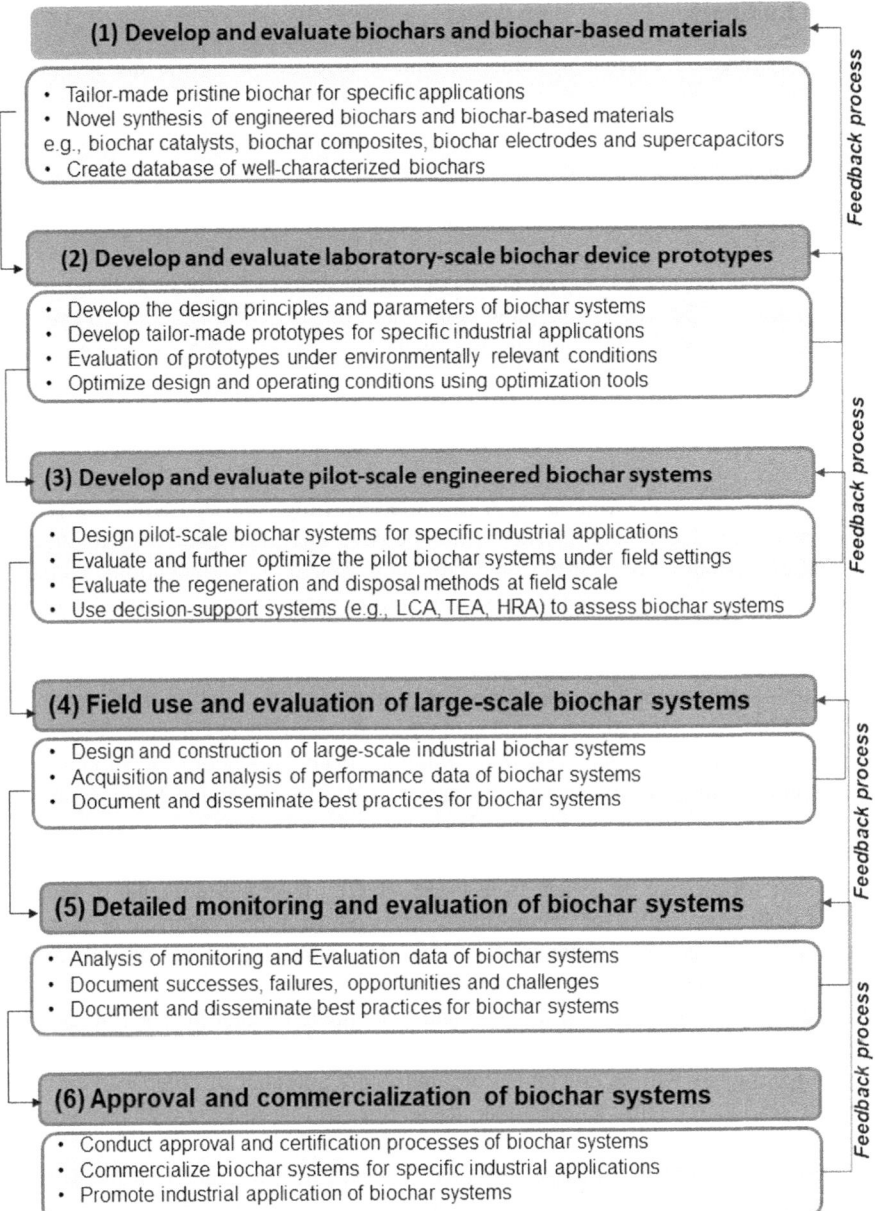

FIGURE 25.3 A conceptual roadmap for the development and application of engineered biochar remediation systems. *LCA*, Life cycle assessment; *TEA*, techno-economic assessment; *HRA*, health risk assessment. *Adapted and drawn based on Gwenzi, W., Chaukura, N., Wenga, T., & Mtisi, M. (2021). Biochars as media for air pollution control systems: Contaminant removal, applications and future research directions.* Science of The Total Environment, 753, *Article 142249.* https://doi.org/10.1016/j.scitotenv.2020.142249

25.4.2 Laboratory-scale prototyping and evaluation

Design and evaluate laboratory-scale prototypes of engineered biochar systems and devices, determining the optimum design and operating conditions.

25.4.3 Pilot-scale prototyping and evaluation

Design and evaluate pilot-scale prototypes of engineered biochar systems at relevant industrial scales and under environmentally relevant conditions.

25.4.4 Field-scale application

Apply and evaluate large-scale engineered biochar systems by early adopters or practitioners in real-world conditions.

25.4.5 Monitoring, evaluation, and feedback

Monitor and evaluate the performance of engineered biochar systems, providing feedback to refine the technology based on performance results.

25.4.6 Approval and commercialization

Achieve approval, certification, and commercialization of engineered biochar systems with known design specifications and performance.

The application of this conceptual roadmap in the further advancement of engineered biochar remediation systems should be integrated into future research to address the knowledge gaps highlighted in the earlier section. To this end, this should include the application of decision-support tools such as life cycle assessment, TRA, techno-economic assessment, and health risk assessment.

25.5 Epilogue and outlook

In recent years, biochar has emerged as a pivotal tool in environmental remediation, shifting from its traditional agricultural applications to address broader environmental challenges. The comprehensive coverage presented in this 26-chapter book underscores the vast potential of biochar across various environmental remediation applications, including soil, water, wastewater, and air pollution. By consolidating disparate research findings into a single, cohesive volume, this book serves as an essential resource for scientists, practitioners, and policymakers aiming to harness biochar's capabilities for environmental health and sustainability.

As I conclude this book through the state-of-the-art applications of biochar, it is evident that the field has made significant strides, yet numerous avenues for future research remain open. The detailed discussions in Chapters 2–18 highlight both the successes and the gaps in current biochar research, ranging from production techniques and characterization to specific remediation applications. The inclusion of advanced tools like machine learning and in-silico computation methods represents a forward-looking approach to biochar system optimization.

The future of biochar research is poised at a critical juncture, with 20 key knowledge gaps identified as priorities for advancing the field. These gaps encompass a wide range of issues, from understanding the long-term fate of biochar-bound contaminants to the development of design principles for engineered biochar remediation systems. The emphasis on integrating biochar into broader concepts such as the circular bioeconomy and addressing socio-cultural impacts underscores the need for holistic and inclusive approaches.

One of the most pressing challenges highlighted is the need for scalable and sustainable biochar production methods. As the demand for biochar grows, particularly in large-scale agricultural and industrial applications, developing efficient, environmentally-friendly production techniques will be crucial. Additionally, the potential of biochar to earn carbon credits presents an exciting opportunity for incentivizing its adoption, particularly in low-income regions where the socio-economic benefits can be profound.

Looking ahead, the conceptual roadmap proposed in the final section of this book offers a strategic framework for guiding future research and development. By addressing the "grand questions" of biochar technology, we can move beyond short-term, small-scale studies to understand the broader, long-term impacts of biochar on environmental and climatic systems. This includes examining the paradox of scale, the feasibility of creating enduringly fertile soils akin to terra preta, and the potential climatic implications of large-scale biochar adoption.

Ultimately, the promise of biochar lies in its versatility and potential to provide multifaceted solutions to some of the most pressing environmental challenges of our time. By continuing to explore and expand upon the foundational research presented in this book, we can unlock new applications and innovations that will drive the field forward. The journey of biochar is far from over, and with continued interdisciplinary collaboration and innovative research, its role in environmental remediation and sustainability will undoubtedly become more prominent in the years to come.

Acknowledgments

I thank Georg Forster Fellowship for Experienced Researchers funded by the Alexander von Humboldt Stiftung/Foundation, Germany for funding the project entitled, *"Environmental and Climate Footprints of a Biochar and Biogas Technology versus Current Practices in Sub-Saharan Africa: A Life Cycle Analysis."* I also thank the University of Kassel and the Leibniz Institute for Agricultural Engineering and Bioeconomy e.V. (ATB) for co-hosting me as a Geord Forster fellow and for providing facilities and a conducive environment to work on the book.

I am also grateful to the International Foundation for Science – IFS, Sweden for providing seed research grants for biochar research [grant numbers: C-5266−1, and C/5266−2]. The projects formed the basis, and motivated the current paper. However, the views expressed in the current paper are solely those of the author. IFS and the Alexander von Humboldt Foundation played no role whatsoever in the research, write-up and decision to publish the work.

Declaration of conflict of interest

The author declares no conflict of interest financially or otherwise that could have influenced his interpretation of the evidence. The author declares that he has published a number of earlier papers on biochar application for soil amendment, drinking water treatment, environmental remediation, including air pollution control, the bulk of them targeting low-income countries. Therefore, given their relevance to the current manuscript, they are cited accordingly in the current paper.

References

Ahmed, S., & Hussain, C. M. (2018). *Biochar supercapacitors: Recent developments in the materials and methods. Green and Sustainable Advanced Materials: Applications* (pp. 223−249). Wiley. Available from 10.1002/9781119528463.ch10.

Azzi, E. S., Karltun, E., & Sundberg, C. (2021). Assessing the diverse environmental effects of biochar systems: An evaluation framework. *Journal of Environmental Management, 286*112154. Available from https://doi.org/10.1016/j.jenvman.2021.112154.

H. Bamdad, *A theoretical and experimental study on biochar as an adsorbent for removal of acid gases (CO_2 and H_2S)*. Doctoral dissertation. (2019).

Bamdad, H., Hawboldt, K., & MacQuarrie, S. (2018). A review on common adsorbents for acid gases removal: Focus on biochar. *Renewable and Sustainable Energy Reviews, 81*, 1705−1720. Available from https://doi.org/10.1016/j.rser.2017.05.261. Available from: https://www.journals.elsevier.com/renewable-and-sustainable-energy-reviews.

Bergman, R., Sahoo, K., Englund, K., & Mousavi-Avval, S. H. (2022). Lifecycle assessment and techno-economic analysis of biochar pellet production from forest residues and field application. *Energies, 15*(4). Available from https://doi.org/10.3390/en15041559. Available from: https://www.mdpi.com/1996-1073/15/4/1559/pdf.

Bittencourt, F. L. F., Lourenço, A. B., Dalvi, E. A., & Martins, M. F. (2019). Thermodynamic assessment of human feces gasification: an experimental-based approach. *SN Applied Sciences, 1*(9). Available from https://doi.org/10.1007/s42452-019-1104-1, springer.com/snas.

Bittencourt, F. L. F., & Martins, M. F. (2022). Thermochemically-driven treatment units for fecal matter sanitation: A review addressed to the underdeveloped world. *Journal of Environmental Chemical Engineering, 10*(6). Available from https://doi.org/10.1016/j.jece.2022.108732, http://www.journals.elsevier.com/journal-of-environmental-chemical-engineering/.

Chaukura, N., Chiworeso, R., Gwenzi, W., Motsa, M. M., Munzeiwa, W., Moyo, W., Chikurunhe, I., & Nkambule, T. T. (2020). A new generation low-cost biochar-clay composite 'biscuit' ceramic filter for point-of-use water treatment. *Applied Clay Science, 185*, 105409.

Chaukura, N., Masilompane, T. M., Gwenzi, W., & Mishra, A. K. (2020). Biochar-based adsorbents for the removal of organic pollutants from aqueous systems. *Emerging Carbon-Based Nanocomposites for Environmental Applications*, 147−174.

Chaukura, N., Murimba, E. C., & Gwenzi, W. (2017a). Sorptive removal of methylene blue from simulated wastewater using biochars derived from pulp and paper sludge. *Environmental Technology & Innovation, 8*, 132−140.

Chaukura, N., Murimba, E. C., & Gwenzi, W. (2017b). Synthesis, characterisation and methyl orange adsorption capacity of ferric oxide−biochar nano-composites derived from pulp and paper sludge. *Applied Water Science, 7*, 2175−2186.

Chen, Z., Zheng, R., Wei, W., Wei, W., Zou, W., Li, J., Ni, B. J., & Chen, H. (2022). Recycling spent water treatment adsorbents for efficient electrocatalytic water oxidation reaction. *Resources, Conservation and Recycling, 178*, 106037.

Cho, S. H., Lee, S., Kim, Y., Song, H., Lee, J., Tsang, Y. F., Chen, W. H., Park, Y. K., Lee, D. J., Jung, S., & Kwon, E. E. (2023). Applications of agricultural residue biochars to removal of toxic gases emitted from chemical plants: A review. *Science of the Total Environment, 868*. Available from https://doi.org/10.1016/j.scitotenv.2023.161655, http://www.elsevier.com/locate/scitotenv.

Dike, C. C., Hakeem, I. G., Rani, A., Surapaneni, A., Khudur, L., Shah, K., & Ball, A. S. (2022). The co-application of biochar with bioremediation for the removal of petroleum hydrocarbons from contaminated soil. *Science of the Total Environment, 849*. Available from https://doi.org/10.1016/j.scitotenv.2022.157753, http://www.elsevier.com/locate/scitotenv.

Duku, M. H., Gu, S., & Hagan, E. B. (2011). Biochar production potential in Ghana - A review. *Renewable and Sustainable Energy Reviews, 15*(8), 3539−3551. Available from https://doi.org/10.1016/j.rser.2011.05.010.

Forján, R., Asensio, V., Guedes, R. S., Rodríguez-Vila, A., Covelo, E. F., & Marcet, P. (2017). Remediation of soils polluted with inorganic contaminants: Role of organic amendments. *Enhancing Cleanup of Environmental Pollutants, 2*, 313−338. Available from https://doi.org/10.1007/978-3-319-55423-5_10. Available from: http://www.springer.com/in/book/9783319554228.

Gitau, J. K., Sundberg, C., Mendum, R., Mutune, J., & Njenga, M. (2019). Use of biochar-producing gasifier cookstove improves energy use efficiency and indoor air quality in rural households. *Energies, 12*(22), 4285.

Gitau, K. J., Mutune, J., Sundberg, C., Mendum, R., & Njenga, M. (2019). Factors influencing the adoption of biochar-producing gasifier cookstoves by households in rural Kenya. *Energy for Sustainable Development, 52*, 63−71. Available from https://doi.org/10.1016/j.esd.2019.07.006.

Gwenzi, W., Chaukura, N., Mukome, F. N. D., Machado, S., & Nyamasoka, B. (2015). Biochar production and applications in sub-Saharan Africa: Opportunities, constraints, risks and uncertainties. *Journal of Environmental Management, 150*, 250–261. Available from https://doi.org/10.1016/j.jenvman.2014.11.027. Available from: https://www.sciencedirect.com/journal/journal-of-environmental-management.

Gwenzi, W., Chaukura, N., Noubactep, C., & Mukome, F. N. (2017). Biochar-based water treatment systems as a potential low-cost and sustainable technology for clean water provision. *Journal of Environmental Management, 197*, 732–749.

Gwenzi, W., Chaukura, N., Wenga, T., & Mtisi, M. (2021). Biochars as media for air pollution control systems: Contaminant removal, applications and future research directions. *Science of the Total Environment, 753*142249.

Gwenzi, W., Musarurwa, T., Nyamugafata, P., Chaukura, N., Chaparadza, A., & Mbera, S. (2014). Adsorption of Zn^{2+} and Ni^{2+} in a binary aqueous solution by biosorbents derived from sawdust and water hyacinth (Eichhornia crassipes). *Water Science and Technology, 70*(8), 1419–1427. Available from https://doi.org/10.2166/wst.2014.391. Available from: http://www.iwaponline.com/wst/07008/1419/070081419.pdf.

Hagner, M., Uusitalo, M., Ruhanen, H., Heiskanen, J., Peltola, R., Tiilikkala, K., Hyvönen, J., Sarala, P., & Mäkitalo, K. (2021). Amending mine tailing cover with compost and biochar: Effects on vegetation establishment and metal bioaccumulation in the Finnish subarctic. *Environmental Science and Pollution Research, 28*(42), 59881–59898. Available from https://doi.org/10.1007/s11356-021-14865-8.

Hu, J., Wachendorf, M., Gwenzi, W., Joseph, B., Stenchly, K., & Kaetzl, K. (2024). Improving acid-stressed anaerobic digestion processes with biochar-towards a combined biomass and carbon management system. *Environmental Research Communications, 6*(3), 035010.

Jia, L., Cheng, P., Yu, Y., Chen, S. H., Wang, C. X., He, L., Nie, H. T., Wang, J. C., Zhang, J. C., Fan, B. G., & Jin, Y. (2023). Regeneration mechanism of a novel high-performance biochar mercury adsorbent directionally modified by multimetal multilayer loading. *Journal of Environmental Management, 326*. Available from https://doi.org/10.1016/j.jenvman.2022.116790, https://www.sciencedirect.com/journal/journal-of-environmental-management.

Kar, S., Dey, S., Chowdhury, K. B., Ghosh, S. K., Mukhopadhyay, J., Kumar, S., Ghosh, S., & Majumdar, S. (2023). Phyto-assisted synthesis of CuO/industrial waste derived biochar composite for adsorptive removal of doxycycline hydrochloride and recycling of spent biochar as green energy storage device. *Environmental Research, 236*. Available from https://doi.org/10.1016/j.envres.2023.116824. Available from: https://www.sciencedirect.com/journal/environmental-research.

Khan, H., Arshad, M., Usama, M., Hussain, N., Khan, M. I., Bibi, A., Hussain, S., Khan, F., & Tariq, R. (2024). In-silico design of metal oxide-biochar composites for enhanced dibenzothiophene adsorption: DFT and statistical physics approach. *Separation and Purification Technology*128642.

Ma, Y., Yao, D., Liang, H., Yin, J., Xia, Y., Zuo, K., & Zeng, Y.-P. (2020). Ultra-thick wood biochar monoliths with hierarchically porous structure from cotton rose for electrochemical capacitor electrodes. *Electrochimica Acta, 352*136452. Available from https://doi.org/10.1016/j.electacta.2020.136452.

Maya, M., Gwenzi, W., & Chaukura, N. (2020). A biochar-based point-of-use water treatment system for the removal of fluoride, chromium and brilliant blue dye in ternary systems. *Gheorghe Asachi Zimbabwe Environmental Engineering and Management Journal, 19*(1), 143–156. Available from: http://www.eemj.eu/index.php/EEMJ/article/view/4040/3978.

Mohamed, G. O., Saleh, M. E., Shalaby, E. A., & Elsafty, A. S. (2022). Using biochar to control nitric oxide air pollution. *Journal of Physics: Conference Series, 2305*(1), 012029. Available from https://doi.org/10.1088/1742-6596/2305/1/012029.

Pérez-Rodríguez, S., Pinto, O., Izquierdo, M. T., Segura, C., Poon, P. S., Celzard, A., Matos, J., & Fierro, V. (2021). Upgrading of pine tannin biochars as electrochemical capacitor electrodes. *Journal of Colloid and Interface Science, 601*, 863–876. Available from https://doi.org/10.1016/j.jcis.2021.05.162. Available from: http://www.elsevier.com/inca/publications/store/6/2/2/8/6/1/index.htt.

Sahoo, K., Upadhyay, A., Runge, T., Bergman, R., Puettmann, M., & Bilek, E. (2021). Life-cycle assessment and techno-economic analysis of biochar produced from forest residues using portable systems. *The International Journal of Life Cycle Assessment, 26*(1), 189–213. Available from https://doi.org/10.1007/s11367-020-01830-9.

Sizmur, T., Quilliam, R., Puga, A.P., Moreno-Jiménez, E., Beesley, L. and Gomez-Eyles, J.L., 2016. *Application of biochar for soil remediation. Agricultural and environmental applications of biochar: Advances and barriers, 63*, pp. 295-324.

Sundberg, C., Karltun, E., Gitau, J. K., Kätterer, T., Kimutai, G. M., Mahmoud, Y., Njenga, M., Nyberg, G., Roing de Nowina, K., Roobroeck, D., & Sieber, P. (2020). Biochar from cookstoves reduces greenhouse gas emissions from smallholder farms in Africa. *Mitigation and Adaptation Strategies for Global Change, 25*(6), 953–967. Available from https://doi.org/10.1007/s11027-020-09920-7, http://www.wkap.nl/journalhome.htm/1381-2386.

Wang, J., & Wang, S. (2019). Preparation, modification and environmental application of biochar: A review. *Journal of Cleaner Production, 227*, 1002–1022.

Zama, E. F., Reid, B. J., Arp, H. P. H., Sun, G. X., Yuan, H. Y., & Zhu, Y. G. (2018). Advances in research on the use of biochar in soil for remediation: A review. *Journal of Soils and Sediments, 18*(7), 2433–2450. Available from https://doi.org/10.1007/s11368-018-2000-9. Available from: http://www.springerlink.com/content/1439-0108.

Zhang, C., Zeng, G., Huang, D., Lai, C., Chen, M., Cheng, M., Tang, W., Tang, L., Dong, H., Huang, B., Tan, X., & Wang, R. (2019). Biochar for environmental management: Mitigating greenhouse gas emissions, contaminant treatment, and potential negative impacts. *Chemical Engineering Journal, 373*, 902–922. Available from https://doi.org/10.1016/j.cej.2019.05.139, http://www.elsevier.com/inca/publications/store/6/0/1/2/7/3/index.htt.

Zhu, X., Li, Y., & Wang, X. (2019). Machine learning prediction of biochar yield and carbon contents in biochar based on biomass characteristics and pyrolysis conditions. *Bioresource Technology, 288*, 121527. Available from https://doi.org/10.1016/j.biortech.2019.121527.

Index

Note: Page numbers followed by "*f*" and "*t*" refer to figures and tables, respectively.

A

AAC. *See* Adjusted air concentration (AAC)
Abiotic resources, 418–419
Abiotic soil degradation, 367
Absorption processes, 369
Abundant materials, 14
Acid-base modification, 196–197
Acidic environments, 383
Acidic soils, 289–294
Acidic treatment, 39
Acidification potential (AP), 398
Actinobacter bouvetii, 155
Activated carbon, 51, 386, 481–482
Activation method, 38–39
AD. *See* Anaerobic digestion (AD)
ADD. *See* Average daily dose (ADD)
Adjusted air concentration (AAC), 440–441
ADPe, 399
ADPf, 399
Adsorbate, 200–201
Adsorbents, 51, 200–201, 205–206, 381–382
Adsorption, 117, 139, 153–155, 192, 268, 270–271
  biochar in wastewater
    inorganic pollutants onto, 268
    organic pollutants onto, 269
  capacity, 285–286
    and selectivity, 253
    and selectivity, 254
  and immobilization mechanisms, 97–99
    complexation, 98
    ion exchange, 98
    physical adsorption, 97–98
    precipitation, 98–99
    reduction-oxidation process, 99
  influence of solution chemistry on, 244–245
  isotherms, 50–51, 201
  kinetics, 370
    of PFAS removal by biochars, 230–231
  mechanism, 50, 134–137, 136*f*, 204–205, 205*t*
    π-π electron-donor acceptor interactions, 136–137
    of biochar for inorganic pollutants in wastewater treatment, 269–270
    of biochar for organic pollutants in wastewater treatment, 270–271
    electrostatic interactions, 137
    hydrogen bonding, 137
    hydrophobic interactions, 136
    of inorganic and organic pollutants on biochar, 270*f*
    partitioning, 136
    pore filling, 137
  methods, 40
  modeling, 200–204
    isotherm studies, 200–201
    kinetic studies, 201–203
    thermodynamic studies, 203–204, 203*t*
  process, 48, 128, 369, 385–386
  properties, 244
  solution chemistry parameters, 245
    competing ions, 245
    ionic strength, 245
    pH, 245
Advanced combustion techniques, 349, 358
Advanced oxidation processes (AOPs), 82–83, 134, 319, 383
AEM. *See* Anion exchange membrane (AEM)
AFFF. *See* Aqueous film-forming foams (AFFF)
Africa, large scale/field applications of biochar in, 181
Agricultural biomass, 14–15
Agricultural processes, 192
Agricultural residues, 16–17, 192, 348
Agricultural waste, 14, 16–17, 85, 192, 273, 386, 400, 402, 421
Agriculture, 26, 452, 470, 483
  soil amendment, 284–286, 290*t*
Agroecosystems, 356, 436, 479
  integrating biochar into emerging concepts in, 483
  mitigation of climate change in, 469
AI. *See* Artificial intelligence (AI)
Air, 8
  air-cathode, 311
  biochar as media for air pollution control systems, 469
  pollution, 386, 434, 436
  quality control, 7–8
  removal of toxic gaseous contaminants from air gases, 363–365
Airborne contaminants, 440
Alcaligenes faecalis, 139–140
Alkaline, 39
Amazonian communities, 336
Amides ($COHN_2$), 62
Amines (NH_2), 62
Ammonia (NH_3), 78–79, 421
  component, 48
  emissions, 361, 363
Ammonium ions, 369
Ammonium removal, 267
Anaerobic digestion (AD), 295, 319, 478–479
Analytical methods, 24–25
Animal dung, 17, 348
Animal manure, 17
Animal production sector, 193
Animal waste, 315–316
Anion exchange membrane (AEM), 308
Anionic dyes, 200
Anions, 171
ANN. *See* Artificial neural networks (ANN)
Anode, 311
  electrode, 308
Anthroposols, 469
Antibiotic resistance genes (ARGs), 131
AOPs. *See* Advanced oxidation processes (AOPs)
Applications
  of biochar, 26–27
  economic feasibility studies of biochar production and, 51–52
  for wastewater treatment, 271–272
Aquatic biomass, 14–15
Aqueous film-forming foams (AFFF), 219
Aqueous systems, 479
ARGs. *See* Antibiotic resistance genes (ARGs)
Aromatic hydrocarbons, 151
Aromatic PHCs, 151
Aromatic structure, 316–317
Arsenic, 466
Artificial intelligence (AI), 25–26, 60, 65
  techniques, 24–25
  tools, 66–67, 67*t*
Artificial neural networks (ANN), 66
AT. *See* Averaging time (AT)
Atmospheric media, 219
Average daily dose (ADD), 439–440
Averaging time (AT), 439–440

B

Bacteria, 289–294
Bacteroidetes, 155
Ball milling, 196, 228
Basic blue 41 (BB41), 200
Basic dyes. *See* Cationic dyes

Index

Basic red 09 (BR09), 200
Bauxite residues, 116
BB41. *See* Basic blue 41 (BB41)
BC. *See* Biochar (BC)
BCF. *See* Biscuit ceramic filters (BCF)
Beetroot production, 449–450
Benthic brown algae, 63–64
BESs. *See* Bioelectrochemical systems (BESs)
Best management practices (BMPs), 250
BET method. *See* Brunauer–Emmett–Teller method (BET method)
Bio-oil, 19, 396
  production, 372
Biochar (BC), 3, 8–9, 13–14, 18, 23, 35, 39, 59, 61, 63–64, 68, 75–76, 92, 97, 110, 114, 134, 141, 150, 152–153, 160, 170–171, 180–181, 192–193, 197, 200, 226, 242–244, 266–267, 281, 286, 289–294, 298, 307, 313, 316, 361–362, 365, 386, 396, 400, 473, 477, 480–482
  activation techniques, 37–40
    chemical activation, 39–40
    physical activation, 38–39
  adsorbents, 40–48
  adsorption kinetics, isotherm, and thermodynamics, 370
    adsorption kinetics, 370
    isotherm modeling, 370
    thermodynamics study, 370
  adsorption kinetics, thermodynamics, and isotherms of PFAS removal by, 230–231, 230t
  adsorption mechanisms of biochar
    for inorganic pollutants in wastewater treatment, 269–270
    for organic pollutants in wastewater treatment, 270–271
  advancing biochar for environmental remediation, 486–488
  amended soils, 132–134
  amendments, 285–286
  applications, 3–4, 26–27
    agriculture, 26
    composite development, 26
    environmental remediation, 26–27, 27f
    for wastewater treatment, 271–272, 272t
  rates, 101
  biochar-amended soils
    behavior and fate of contaminants in, 101–102
    factors influencing bioavailability of contaminants in, 102
    properties influencing behavior of contaminants in, 102
  biochar-associated contaminants, 481
  biochar-based catalysts, 75
  biochar-based cement materials, 421
  biochar-based environmental remediation systems, 469
    amendments for remediation of contaminated lands and soils, 469
    as media for air pollution control systems, 469
    mitigation of climate change in agroecosystems, 469
  biochar-based filtration systems, 467, 478–479
  biochar-based functional materials, 70
  biochar-based land remediation, 120
  biochar-based materials, 372
  biochar-based remediation systems, 48, 110
    depicting feedstocks, 111f
    metal-contaminated soils, 114–116, 115f
    of contaminated lands, 113–119
    revegetation of postmining landscapes, 113–114
    sludge and wastewater-amended soils, 116–117
    toxic geogenic contaminants in serpentines, 116
  biochar-based stormwater treatment, 256–257
  biochar-based systems, 5, 110, 118–119, 242
  biochar-based water filtration systems, 468–470
    for clean drinking water provision, 467–469
  biochar-derived contaminants, 443
  biochar-induced radiative forcing and associated climatic risks, 482
  biochar-microbe interactions, 155
  biochar-modified on-site sanitation methods, 272
  biomass
    fabrication, 36f
    quantification approaches, 16–17
  case for biochar-based land remediation, 110–111
    biochar feedstocks and production systems, 110–111
    rationale and context, 110
  catalysis, 85, 121
  catalysts, 76, 84–85
    nature of, 76
    regeneration, 84–85
  catalytic reaction, 250–251
  cement-based materials, 400
  certification systems, 441
  challenges and considerations of biochar for treating stormwater, 254–255
    adsorption capacity and selectivity, 254
    impact on soil structure, 254
    interactions with microbial communities, 254–255
    long-term stability and persistence, 255
    scale-up and implementation challenges, 255
  challenges and opportunities of biochar in removal of organic pollutants, 141
  characterization of, 22–24, 60–64
    3-D micro-CT analysis, 63
    biochar surface properties/phenomena, 61–62
    future perspectives and research directions, 69–70
    internal microstructure, 62
    synchrotron X-ray microtomography and multifractal analysis, 63–64
  chemical composition, 365–367
  composites, 75
  conceptual designs, 178–179, 179f
    of biochar water filters, 178–180
  confronting biochar critics and skeptics with "twin or multiple solutions" concept, 472–473
  coupling biochar systems to address multiple challenges in light of energy-water-food-environment nexus, 473
  contaminant removal mechanisms, 48–51
    inorganic contaminants, 50–51
    organic contaminants, 48–50
  controls on environmental resistome, 481
  cookstoves, 347–349, 355–356, 358–359
    designs, 359
    emphasis on efficiency and environmental impact, 354
    environmental benefits of, 356
    institutional-scale biochar cook stoves, 354
    larger capacities for community use, 354
    as multifaceted solution for low-income settings, 355–356
    rocket stoves, 351
    top-lit updraft biochar stoves, 349–351
    top-lit updraft gasifier stoves, 351–353
    top-lit updraft rice husk gasifier stoves, 354
    types and designs of, 349–354
  current and emerging, constraints and gaps, 272–273
  depiction of, 4f
  devices, 8
  economic feasibility studies of biochar production and application, 51–52
  ecotoxicity and health risks of biochar-remediated solid matrices, 158–159
    ecotoxicity of, 158
    human health effects of, 159
    safety measures for biochar application, 159, 159t
  effects of pyrolysis type/temperature on changes in functional groups, 19–20, 19f
  electrochemical properties of, 283–284, 312–316
    challenges, 295–297
    conductivity, 283, 287t
    future perspectives, 298
    redox activity, 283–284
    surface chemistry, 284
  electron conductivity, 283
  engineering, 298
  environmental applications, 284–295
    agricultural soil amendment, 284–286
    biochar as electrode material for development of microbial fuel cells, 294–295
    carbon sequestration, 289–294
    of biochar catalysis, 85
    remediation of polluted wastewater, 286–289

environmental implications of using biochar for treating stormwater, 255–256
environmental remediation, 479–480
  applications, 40–48
epilogue and outlook, 488
existing regulations on PFAS in environmental matrices and use of, 231–233
factors affecting capacity of biochar in soil remediation, 100–101
  application rate and particle sizes, 101
  physicochemical characteristics of biochars and removal efficacy, 100–101
  physiochemical attributes of polluted soils, 100
factors that influence ability of biochar to remediate petroleum hydrocarbon-contaminated solid matrices, 153
feedstock vs. competing current uses, 481
filters, 249
functional characterization, 24–26
  analytical methods, 24–25
  artificial intelligence, 25–26
future outlook, 52–53
  regeneration and disposal, 53
  synthesis and fabrication, 53
  testing and evaluation, 53
future perspectives and prospects, 8–9
future research and perspectives, 119–121
  biochar-based extraction and recovery systems for essential elements, 120
  building Africa' biochar research capacity, 121
  increasing Africa's research footprint on biochar-based remediation systems, 119
  large-scale pilot field studies, 120
  long-term behavior and fate of contaminants, 119–120
  metal-enriched biomass from metalliferous substrates as unique biomass feedstock, 121
  need for biochar research funding and collaboration, 121
  remediation of organic contaminants, 120
  repurposing postmining landscapes as biomass sources for circular bioeconomy, 120
  technical and economic feasibility studies, 120
future research directions, 86
future research in biochar application for wastewater treatment, 273–274
harnessing biochar to earn carbon credits, 483
impregnation, 371
industrial wastewater treatment, 267
integrating biochar into emerging concepts in agroecosystems, 483
key factors influencing techno-economic feasibility of, 450–452
  biochar application, 452
  biomass feedstock, 450
  market price of biochar, 451–452

thermo-hydro-chemical conversion technology, 450–451
lack of techno-economic assessment data on biochar for environmental remediation applications, 456
large scale/field applications of biochar in Africa, 181
large-scale remediation of inorganics by, 96–97
life cycle assessment, 400
long-term fate biochar-bound contaminants of, 481
looking ahead, 480–486
  handful of "grand questions" on biochar technology, 484–486
  twenty knowledge gaps to advance biochar for environmental remediation, 481–483
materials, 175, 231
mechanisms for remediation of petroleum hydrocarbon contaminated solid matrices using, 153–155
mechanisms of biochar catalysis, 80–84
  Fenton system, 81–82, 82t
  persulfate activation system, 82–83
  photocatalytic system, 83–84
mechanisms of biochar remediation of mine wastes and metalliferous substrates, 117–118
  design of biochar-based remediation systems, 118–119
  enhancement of microbial activity, 118
  immobilization of trace metals, 117
  nutrient retention, bioavailability, and uptake, 118
  pH modification, 117
  plant growth promotion, 118
  reduction of toxic metal uptake by plants, 118
  soil structure improvement, 117
mechanisms of contaminant removal from stormwater using, 251
mechanisms of toxic gaseous contaminant removal by, 367–369
media, 271
  for microbial contaminant removal, 248–249, 249t
modification methods, 50, 62
moving from "one-size-fits-all" and "blanket salesmen" to biochar application domains, 472
moving from biochar as "silver bullet" to biochar application domains, 472
municipal wastewater treatment, 267–268
nature and extent of contaminated lands, 111–113
  metal-contaminated lands, 112
  postmining landscapes, 111–112
  serpentinitic geological systems, 112–113
  sludge and wastewater-amended soils, 113
novel characteristics of engineered biochars, 48
as novel renewable green biomaterial for mineral recovery, 481–482

after organic contaminants removal, 140–141
performance, 179–180, 181f, 318
  and mechanisms of biochar function in microbial fuel cells, 318
  mechanisms of biochar enhancement of microbial fuel cells, 318
perspectives, 469–473
  role of twin or multiple solutions in biochar uptake, adoption, and impact, 470
pH, 76
physicochemical properties, 22–24
  chemical properties, 24
  physical properties, 24
pivotal role of biochar in influencing soil structure and water retention, 253–254
possesses, 150
preparation and characterization of, 17–22, 76–80, 78t, 486
  biochar pretreatment and catalyst modification, 79f
  biochar-based catalyst's physicochemical, 80f
production, 59, 160, 350, 449, 451, 454–455, 466, 481
  comparative life cycle assessment of, 400–401
  from biowaste, 193–195
  general processes of, 226–227
  hydrothermal carbonization, 194–195, 195f
  life cycle inventory for biochar systems, 407t
  methods, optimization of, 256
  modification of biochar for PFAS removal from environmental matrices, 227–228
  performance of biochars on removal of PFAS, 231
  processes and usage for removal of PFAS in environmental matrices, 226–231
  pyrolysis systems, 17–19, 193–194, 194t
  removal mechanism of PFAS from environmental matrices using, 229–231
  systems, 416
  torrefaction of biomass, 22
properties, 365–367
  biochar stability, 367
  and mechanisms for toxic gaseous contaminant removal, 365–370
  porosity and surface area, 365, 366t
  and potential applications in microbial fuel cells, 312–318
  surface functional groups, 365–367
purpose, motivation, and novelty, 4–6
pyrolytic cookstoves, 466–467
regeneration, 381
remediation, 128
removal
  adsorption and immobilization mechanisms, 97–99
  adsorption mechanisms, 134–137, 136f
  conventional and emerging organic contaminantsfrom contaminated soils, 132–134

Biochar (BC) (*Continued*)
    factors affecting adsorption efficiency, 137–138
    mechanism of PFAS removal using, 229–230
    mechanisms for biochar removal of inorganic contaminants in soils, 97–100
    mechanisms of organic contaminants in soils, 134–138
    of inorganic contaminants, 95–96
    of nutrients from water using, 171
    of radionuclides from water using, 170–171
    soil inorganic pollutants removal by biochar, 96t
    synergistic interactions of biochar with remediation technologies, 99–100
  research
    opportunities, 141
    recent shifts and expansions in, 5
  ring, 466
  roadmap for advancement of biochar-based microbial fuel cells, 320–321
  scale of application for biochar remediation of petroleum hydrocarbon contaminated solid matrices, 155–157
  soils, 68
    amendment to create enduringly fertile and productive soils, 469
  soils, sediments, and sludges
    indicators of remediation efficiency of biochar for petroleum hydrocarbons in, 158
    methods of applying biochar to, 153
    for remediation of petroleum hydrocarbons in, 152–158
    use and suitability of biochar for PHC remediation in, 152
  state-of-the-art on biochar for environmental remediation, 478f
  in stormwater management, 243–244
    adsorption properties, 244
    characteristics relevant to stormwater treatment, 243
    high surface area and porosity, 243–244
    surface functional groups, 244
  surfaces, 61, 68–69
    area, 48
    chemistry modifiers, 294–295
    functional groups, 270–271
    properties, 24
  synergistic interaction, 483
    of biochar with remediation technologies, 158
  synthesis routes, 36–37, 38f
  systems, 9, 400, 434, 436, 447, 481
    artificial intelligence and machine learning tools, 66–67, 67t
    challenges in techno-economic assessments of, 452–453
    current and potential applications, 68–69, 68f
    design and evaluation of, 64–69
    in-silico-computational modeling or computer-aided design approach, 65–66
    life cycle analysis of environmental and social footprints of, 482
    overview of evidence, 449–450
    status of research on risk assessment of, 441–442
    techno-economic assessment tools, 448–449
    techno-economic assessment of, 448–450
  technology, 4–5, 8, 110, 273, 433, 463, 477, 482–483, 486
    as part of broader circular bioeconomy, 483
    biochar and paradox of scale, 484
    biochar application required to significantly influence climate system, 485
    catalyze global transition to biochar to achieve perceived benefits, 486
    creating enduringly fertile soils and terra preta enigma, 484–485
    fundamental principles, 6
    handful of "grand questions" on, 484–486
    in wastewater treatment, 266–267
    large-scale biochar application reduce environmental footprint of food systems and humans, 486
    long-term impacts of biochar on soil health and fertility, 485
    moving from "one-size-fits-all" to specific biochar application domains, 484
    optimization of biochar remediation systems for maximum effectiveness, 485
    origin and evolution of, 4–5
    potential renewable, low-cost, and efficient energy, 466–467
    socio-cultural aspects of, 483
    socio-economic implications of biochar technology in low-income countries, 485–486
    sustainable scaling of biochar production systems to meet global demands, 485
  treating emerging contaminants with, 247–249, 248t
  twin or multiple solutions, 470–471
    concept, 465–466, 465f
    framework, 464f, 465–469
    framework as primer for biochar adoption, 470–471
    targeting water-energy-food-environment nexus, 466–469
  type of wastewater treated by, 267
  types of biomass feedstocks for biochar preparation, 14–16
  in wastewater
    adsorption of inorganic pollutants onto, 268
    adsorption of organic pollutants onto, 269
  wastewater treatment, 266
Biochar and zinc functionalized (ZnBC), 62
Biochar-treated bioretention system (BRS), 178

Biodegradation, 220–223, 379–380
  microbial remediation, 223–226
  phyto-microbial remediation, 226
  and transformation of PFAS in environment, 220–226
Biodiversity risks, 481
Bioelectrical systems, 478–479
Bioelectricity generation, 307
Bioelectrochemical oxygen demand (BOD), 309
Bioelectrochemical systems (BESs), 69, 307–308
  biochar properties and potential applications in microbial fuel cells, 312–318
  constraints and limitations of, 312
    constraints and limitations, 312
  future directions and perspectives, 319–321
    knowledge gaps, 319–320
    roadmap for advancement of biochar-based microbial fuel cells, 320–321
  materials and methods, 308–312
    designs and types of microbial fuel cells, 308–312
    microbial fuel cells, 308
    principles of microbial fuel cells, 308
  performance and mechanisms of biochar function in microbial fuel cells, 318
Bioenergy, 5, 332–333, 456, 473
Biofertilizer, 473
Biofilms, 177
  formation, 177
Biofilters, 177
Biofuels, 456
  production, 449
Biogenic waste, 15
Biogeochemical cycling, 484
Biological modification method, 228
Biomass, 14–16, 18, 35–36, 64–65, 121, 295, 316, 333, 349, 365, 396, 418, 481
  biomass-based adsorbents, 220
  biomass-based biochar, 26
  biomass-to-energy systems, 121
  catalytic pyrolysis, 334
  combustion, 334, 348
  feedstocks, 435, 437, 450, 473, 479
    agricultural, forest, and aquatic biomass, 14–15
    metal-enriched biomass from metalliferous substrates as unique, 121
    plastics, 15–16
    types for biochar preparation, 14–16
  flexibility for, 351–353
  fuel, 351
  gasification of, 21–22
  hydrothermal carbonization of, 21
  pyrolysis, 333, 451, 481
  quantification approaches, 16–17
    agricultural waste, 16–17
    animal manure, 17
    municipal sewage sludge, 17
    municipal solid waste, 17
    repurposing postmining landscapes as biomass sources for circular bioeconomy, 120

trade-offs, 481
torrefaction of, 22
Bioregeneration, 384
Bioretention systems, 177
Biosand filtration systems, 467
Biosolids, 333, 481
  content, 315
Biosorbents, 386
Biotic soil degradation, 367
Biowaste, 36, 110, 192
  activation methods for production of engineered, 195–197
  applications and performance, 197–205
    adsorption modeling, 200–204
    dye removal, 197–200
    mechanism of adsorption, 204–205, 205t
  biochar production from, 193–195
  biowaste-derived biochars, 192–197
  feedstock, 192–193
    agricultural residues, 192
    industrial by-products, 193
    livestock manure, 193
    municipal solid waste, 192
  future directions, 207–208
  materials, 14
  regeneration and reuse, 205–206, 207t
Biscuit ceramic filters (BCF), 180
Bisphenol A (BPA), 246
Black carbon, 8–9, 482
  climatic/radiative forcing by, 436–437
  induced radiative forcing and associated climatic risks, 482
"Blanket salesman" approach, 484
  moving blanket salesmen to biochar application domains from, 472
BM-FeS$_2$@BC, 172
BMPs. See Best management practices (BMPs)
BOD. See Bioelectrochemical oxygen demand (BOD)
Body weight (BW), 439–440
BPA. See Bisphenol A (BPA)
BR09. See Basic red 09 (BR09)
Brazilian Amazon community, portrait of, 336–338
  community of Sapolândia, 336–337
  thermochemical route, 337–338
Brazilian Institute of Geography and Statistics, 336
Brevibacterium iodinum, 139–140
Brunauer–Emmett–Teller method (BET method), 316
  adsorption method, 80
  isotherm model, 23
  porosimetry, 23
"BurnStove" project, 357
BW. See Body weight (BW)

C

C/N ratio. See Carbon-to-nitrogen ratio (C/N ratio)
Cadmium (Cd), 268
Calcium (Ca), 284–285, 369
Calcium carbonate, 337

Calcium oxide, 337
Cancer slope factor (CSF), 440–441
Carbendazim, 418
Carbon, 18, 283, 312–313, 315
  carbon-based materials, 40
  carbon-rich nanomaterials, 39–40
  emissions, 416
  harnessing biochar to earn carbon credits, 483
  materials, 313
  nanotubes, 298
  sequestration, 289–294, 485
  storage, 421
  yield, 314
Carbon content bonds (C-C bonds), 21, 281–282
Carbon dioxide, 396
  activation, 38–39
Carbon dioxide (CO_2), 181, 194–195, 289–294, 355, 361, 417–418
Carbon footprint (CF), 399, 416–417
Carbon monoxide (CO), 334, 348, 355, 361, 417–419
Carbon-to-nitrogen ratio (C/N ratio), 285
Carbonaceous materials, 281–282
Carboxyl functional group, 39
Carboxyl groups (–COOH), 82–83, 171–174, 244
Carcinogenic health risks, 439–440
Carcinogenic human health risks, 440–441
Carcinogenic risk (CR), 440–441
Carica papaya, 174–175
Catalyst precursors for trace metal-rich spent biochars, 386
Catalytic degradation process, 252–253
Catalytic effects, 369
Catalytic pyrolysis, 334
Cathode, 308, 311
  electrodes, 308
  material, 315
Cathodic reaction, 315
Cation exchange capacity (CEC), 23, 60–61, 76, 95, 118, 284–285
Cation exchange membrane (CEM), 308
Cationic dyes, 199–200
CB. See Conduction band (CB)
CBA. See Cost-benefit analysis (CBA)
CBR. See Cost-benefit ratio (CBR)
CCB. See Corn cobs (CCB)
CDI. See Chronic daily intake (CDI)
CEC. See Cation exchange capacity (CEC)
CED. See Cumulative Energy Demand (CED)
Cellulose, 386
CEM. See Cation exchange membrane (CEM)
Cement industry, 386
Ceramic filter (CF), 180
CF. See Carbon footprint (CF); Ceramic filter (CF); Chemical fertilizer (CF)
CFCs. See Chlorofluorocarbons (CFCs)
CFD. See Computational fluid dynamics (CFD)
Characterization of biochar
  and biochar-contaminant systems, 60–64
  preparation and characterization of biochar catalysts, 76–80

Charcoal, 314, 337
  production, 466
Chemical activation, 78–79, 196–197, 371
Chemical desorption, 383
Chemical fertilizer (CF), 419
Chemical indicators, 158
Chemical interactions, 286–289
Chemical kinetics study, 201–203
Chemical modification approaches, 39, 228
Chemical organics, 128–129
Chemical oxygen demand (COD), 267–268, 468
Chemical reactions, 369
  mechanism, 334
Chemisorptions, 50
Chinese herb, common (*Flueggea suffruticosa*), 62
Chlorinated polyfluorinated ether sulfonates (Cl-PFESA), 216
Chlorination, 467–468
Chlorine, 466
Chlorobenzene, 75
2-chlorobiphenyl, 75
Chlorofluorocarbons (CFCs), 399
Chlorophenols, 128
Chromate (CrO_4^{2-}), 172
Chromium (Cr), 113
Chromophore, 191–192
Chronic daily intake (CDI), 440
Chronic obstructive pulmonary disease (COPD), 356
Cicer arietinum, 384–385
Circular bioeconomy, 456, 473, 483
  biochar technology as part of broader, 483
Circular economy, 362, 384
  reuse, recycling, and disposal of spent biochar for, 384–387
CLB. See Cocoa leaves biochar (CLB)
Clean drinking water provision, biochar-based water filtration systems for, 467–469
Climate change, 416, 418–419
  mitigation, 470, 484
  in agroecosystems, 469
Climate system, 482
  scale of biochar application is required to significantly influence, 485
Climatic forcing by black carbon, 436–437
Climatic hazards, 443
Climatic risks, 436–437
CLs. See Cocoa leaves (CLs)
CM. See Cow manure (CM)
Cobalt (Co), 113
Cocoa leaves (CLs), 196–197
Cocoa leaves biochar (CLB), 196–197, 204
Coconut shell biochar, 206
COCs. See Conventional organic contaminants (COCs)
COD. See Chemical oxygen demand (COD)
Codisposal of sludge, 387
Coffee biochar, 383
Combustion, 332
  process, 333, 349
  techniques, 415
  thermochemical conversion, 333

Commercial biochar, 283
Community empowerment, 354
Community engagement, 357
Community-centric design, 354
Complexation, 98, 368
Composting, 385
Computational fluid dynamics (CFD), 65
  simulation models, 65
  theoretical calculations, 65
Computational methods, 65
Computer-aided design approach, 65–66
Conduction band (CB), 83–84
Conductivity, 283
Constructed wetlands, 319
Construction materials, 401
Contaminants, 109, 281, 286, 437
  behavior and fate of contaminants in spent biochar, 387–388
  emissions, 472
  removal, 385–386
Contaminated lands, 109
  biochar amendments for remediation of, 469
Contaminated soils, biochar amendments for remediation of, 469
Contamination pathways, molecular structure of, 132, 132f
Controlled airflow, 349
Controlled combustion process, 350
Conventional "single solution" technologies, 465
Conventional biosand filters, 473
Conventional methods, 13, 65, 109–110
Conventional organic contaminants (COCs), 127–130, 140–141, 246–247, 247t
  biochar removal from contaminated soil, 132–134
  physico-chemical characteristics of, 129, 130t
  sources and pathways of COCs for contaminating soil, 129–130
  types of, 128–129
Conventional organics, 140
Conventional torrefaction, 22
Cooking applications, 349
Cooking efficiency, 354
Cookstoves, 349
  evolution of cookstove technology, 349
  from traditional to modern designs, 349
COPD. See Chronic obstructive pulmonary disease (COPD)
Copper (Cu), 268
Copper ions (Cu^{2+}), 98
Coprecipitation, 368–369
"Core" phases, 397–398
Corestream phase, 415
Corn cobs (CCB), 178
Cost-benefit analysis (CBA), 447
  comprehensive, 455
Cost-benefit ratio (CBR), 449
Coupling biochar systems to address multiple challenges in light of energy-water-food-environment nexus, 473
Covers biochar technology, 266
Cow manure (CM), 466–467
CR. See Carcinogenic risk (CR)

Critical raw materials (CRM), 395
CRM. See Critical raw materials (CRM)
Crop residues, 17, 473
Crystal violet (CV), 196–197, 204–205
CSF. See Cancer slope factor (CSF)
Cultivation processes, 396
Cumulative Energy Demand (CED), 399
CV. See Crystal violet (CV)

D
Daily intake of contaminant, 440
Data, 5
  availability and quality, 455
DBPs. See Disinfection byproducts (DBPs)
DC-SOFCs. See Direct carbon solid oxide fuel cells (DC-SOFCs)
Debye-Scherrer formula, 23
Decision makers, 486
Decision support systems (DSS), 66, 321
  for assessing biochar systems, 7
Decision-support tools, 479–480, 488
Decomposition regeneration process, 379–380
Degradation regeneration process, 379–380
Dehydration, 333–334
Dehydrogenation reactions, 17–18
Density functional theory (DFT), 65
Deposition, 388
Deprotonation process, 286
Desorption, 139, 379–380
  regeneration techniques, 380
DFT. See Density functional theory (DFT)
Dichromate ($Cr_2O_7^{2-}$), 172
DIET. See Direct interspecies electron transfer (DIET)
Dioxins, 434–435, 441–443
Direct carbon solid oxide fuel cells (DC-SOFCs), 69
Direct interspecies electron transfer (DIET), 283, 317
  pathways, 295
Disinfection byproducts (DBPs), 140–141
Disposal of spent biochar for circular economy and environmental sustainability, 384–387
Disposal technologies, 321, 387
Dissolved organic carbon (DOC), 178
DOC. See Dissolved organic carbon (DOC)
Double-layer capacitance, 318
Downstream
  phase, 406–415
  process, 418–419
Drinking water, 479
  treatment, 437
Drying process, 64
DSS. See Decision support systems (DSS)
Dual combustion benefits, 351
Dung, 17
Dye removal, 197–200
  anionic dyes, 200
  biowaste-derived biochar
    for anionic dyes, 199t
    for cationic dyes, 198t
  cationic dyes, 199–200

Dyeing process, 191–192
Dyes, 48–49, 191–192, 269
  dye-containing wastewater, 192

E
EAB. See Electro-active bacteria (EAB)
Earliest production methods, 59
Earth-atmosphere modeling studies, 436–437, 482, 485
EBC. See Engineered biochar (EBC)
Echiochloa pyramidalis, 180
Ecological health risks, 114, 437
Ecological risks, 481
Economic analysis, 449–451
Economic assessments, 456
Economic benefits, 356
Ecosystem, 70
Ecotoxicity of free vs. biochar-bound contaminants, 481
Ecotoxicological risk assessments, 441
"EcoZoom" biochar cookstoves, 357
ED. See Exposure duration (ED)
EDA. See Electron donor acceptor (EDA)
EDAX. See Energy dispersive X-ray analysis (EDAX)
EDCs. See Endocrine disrupting compounds (EDCs)
EDS. See Energy dispersive spectrometer (EDS)
EF. See Exposure frequency (EF)
EFB fibres. See Empty fruit bunch fibres (EFB fibres)
Efficiency of water use (WUE), 285
Efficient airflow, 351
Eggshells (ES), 175
Eichhornia crassipes. See Water hyacinth (*Eichhornia crassipes*)
Electric power, 449
Electrical conductivity, 22–23, 281–282, 318
Electricity, 418–419
  production, 419
Electro-active bacteria (EAB), 308
Electro-Fenton reaction, 383
Electrochemical properties of biochar, 283–284, 312–316
Electrode materials, 69, 308–309
  biochar as electrode material for development of microbial fuel cells, 294–295
Electron donor acceptor (EDA), 48–49, 134
Electron spin resonance (ESR), 60
Electronic conductivity, 315
Electrons (e^-), 81, 294–295, 308
  electron-donating capacity, 315
  transfer, 317
    efficiency, 311
    processes, 314
Electrostatic attraction, 267, 286, 368
Electrostatic forces, 137
Electrostatic interactions, 137, 174–175
Elemental analysis, 80
Elemental analyzer, 23
Elements, 92

Emerging organic contaminants (EOCs), 127
Emissions, 348
Empty fruit bunch fibres (EFB fibres), 197
Endocrine disrupting compounds (EDCs), 246–247
Energy
  density, human excreta, 332
  efficiency, 400
  energy-biochar
    production system, 402
    system, 418–419
  energy-intensive process, 339
  energy-water-food nexus, 464
  energy-water-food-environment nexus, 471
    coupling biochar systems to address multiple challenges in light of, 473
    linking twin or multiple solutions framework to emerging concepts, 473
    human excreta as energy feedstock, 330–332
  production, 330
  technologies, 395
Energy dispersive spectrometer (EDS), 80
Energy dispersive X-ray analysis (EDAX), 80
Engineered biochar (EBC), 70, 197. *See also* Spent biochars
  activation methods for production of, 195–197
    chemical activation, 196–197
    hybrid activation, 197
    physical activation, 196
  design principles and protocols for EBC remediation systems, 482
  for enhanced gaseous contaminants removal, 371
    chemical activation, 371
    heteroatoms doping, 371
    metal impregnation, 371
    physical activation, 371
  novel characteristics of, 48
Enhanced gaseous contaminants removal, engineered biochars for, 371
Enthalpy change (ΔH), 370
Entropy change (ΔS), 370
Environmental benefits, 359
  of biochar cookstoves, 356
Environmental contaminants, 265
Environmental footprint of food systems and humans, large-scale biochar application reduce, 486
Environmental health risks, 434–437
  air pollution, 436
  call for health risk assessments and mitigation, 437–442
  climatic/radiative forcing by black carbon, 436–437
  consequences or impact of hazard, 438
    biochar certification systems, 441
    ecotoxicological risk assessments, 441
    principles of health risk assessment, 438
    qualitative risk assessment, 438–439
    quantitative human health risk assessment tools, 439–441
    quantitative risk assessment, 439
  status of research on risk assessment of biochar systems, 441–442
  ecological health risks, 437
  future research directions, 442
  health risks and environment risk assessment in biochar research and applications, 435f
  human exposure and health risks, 437
  likelihood or probability of occurrence, 438
  potential health risks, risk assessments, and future research directions for biochar systems, 434f
  soil pollution, 435
  water pollution, 436
Environmental impact, 354
  emphasis on, 354
Environmental pollution, 281, 363
  risks, 481
Environmental remediation, 26–27, 27f, 456, 481, 483
  advancing biochar for, 486–488
    approval and commercialization, 488
    field-scale application, 487
    laboratory-scale prototyping and evaluation, 487
    monitoring, evaluation, and feedback, 488
    pilot-scale prototyping and evaluation, 487
    preparation and characterization of biochars, 486
  applications, 40–48
    lack of techno-economic assessment data on biochar for, 456
    removal of inorganic contaminants, 40–48
    removal of organic contaminants, 40
  biochar for, 479–480
  case for biochar in, 5
  twenty knowledge gaps to advance biochar for, 481–483
    biochar as novel renewable green biomaterial for mineral recovery, 481–482
    biochar controls on environmental resistome, 481
    biochar or black carbon-induced radiative forcing and associated climatic risks, 482
    biochar technology as part of broader circular bioeconomy, 483
    biomass trade-offs, 481
    design principles and protocols for engineered biochar remediation systems, 482
    ecological and biodiversity risks, 481
    ecotoxicity of free vs. biochar-bound contaminants, 481
    environmental pollution risks, 481
    harnessing biochar to earn carbon credits, 483
    health risk assessment, 481
    human exposure and health risks in occupational and nonoccupational settings, 481
    integrating biochar into emerging concepts in agroecosystems, 483
    life cycle analysis of environmental and social footprints of biochar systems, 482
    long-term fate of biochar-bound contaminants, 481
    socio-cultural aspects of biochar technology, 483
    synergistic interaction of biochar and remediation technologies, 483
    techno-economic assessment, 482
    technology readiness assessment, 482
    validation of twin or multiple solution heuristic framework, 482–483
Environmental resistome, biochar controls on, 481
Environmental sustainability, reuse, recycling, and disposal of spent biochar for, 384–387
Enzymatic degradation, 387
EOCs. *See* Emerging organic contaminants (EOCs)
EP. *See* Eutrophication potential (EP)
EPA. *See* United States Environmental Protection Agency (EPA)
ES. *See* Eggshells (ES)
Escherichia coli, 170, 180, 268, 468
ESR. *See* Electron spin resonance (ESR)
Essential elements, biochar-based extraction and recovery systems for, 120
Ester, 85
ET. *See* Exposure time (ET)
European Biochar Foundation, 281–282
Eutrophication potential (EP), 399
Exothermic process, 204
Expansion effect, 63
Exposure duration (ED), 439–440
Exposure frequency (EF), 439–440
Exposure time (ET), 439–440
Extracellular enzymes, 384

F

FAETP. *See* Freshwater Aquatic EcoToxicity Potential (FAETP)
Farmers, 454, 471
  adoption, 456
Fast pyrolysis, 18–19, 77
FC. *See* Fecal coliform (FC)
Fecal coliform (FC), 178
Fecal indicator bacteria (FIB), 177
Fecal matter, 330
Fecal sludge, 330–333, 339
Fecal streptococci (FS), 178
Feces pyrolyzer, 339
Feces sludge, 332
Feedstock, 35, 37, 63, 153
  price, 450
  results, 62
  system, 434
Fenton reaction, 383
Fenton system, 81–82, 82t
Ferric chloride, 381–382
Ferrous chloride, 381–382
Ferrous sulfate, 381–382

Fertile soils, creating enduringly, 484–485
Fertilizers, 68
　compounds, 337
FFAs. *See* Free fatty acids (FFAs)
FIB. *See* Fecal indicator bacteria (FIB)
Fillers in novel construction materials, 386
Filtration, 382
Flash pyrolysis, 19
Flexibility for biomass types, 351–353
Flue gases, removal of toxic gaseous contaminants from, 363–365
Flueggea suffruticosa. See Chinese herb, common (*Flueggea suffruticosa*)
Fluoride, 268, 466
Food
　large-scale biochar application reduce environmental footprint of, 486
　scraps, 192
　waste biochar, 196
Food and Agriculture Organization, 463
Forest, 14–15
　residues, 35, 437
Forestry waste, 85
"Forever chemicals", 215–216
Fossil fuels, 362, 395, 418–419
Fourier transform infrared spectrometry (FTIR), 23, 59–60, 80
Fourth Industrial Revolution (4IR), 65
Free fatty acids (FFAs), 85
Free OpenLCA database, 415
Fresh feces, 339
Freshwater Aquatic EcoToxicity Potential (FAETP), 399
Freundlich isotherms, 370
Freundlich models, 201
FS. *See* Fecal streptococci (FS)
FTIR. *See* Fourier transform infrared spectrometry (FTIR)
Fuel efficiency, 356
Functional groups, 36, 76, 99, 251
　effects of pyrolysis type/temperature on changes in, 19–20, 19*f*

G

Gas activation, 78
Gas-enhanced pyrolysis, 62
Gaseous contaminants, 361, 368–369
　removal, 363
Gaseous emissions, 336
Gaseous pollutants, 369
Gasification, 21, 332, 433
　of biomass, 21–22, 22*f*
　chamber, 349
　combining top-lit updraft and, 351
　efficiency, 354
　process, 334, 351
　techniques, 396
　thermochemical conversion, 334
Gasifier cookstoves, 479, 483
GBM. *See* Gradient boosting machines (GBM)
GBR. *See* Gradient boosting regression (GBR)
GCR. *See* Gross crop residues (GCR)

Geobacter
　G. metallireducens, 317
　G. sulfurreducens, 317
Geological systems, 111–112
GHGs. *See* Greenhouse gases (GHGs)
Gibbs free energy (ΔG), 370
Global Temperature Change Potential (GTP), 399
Global warming potential (GWP), 399
Gradient boosting machines (GBM), 66
Gradient boosting regression (GBR), 67
Gram-positive fungi, 289–294
Gram-positive microbes, 289–294
"Grand questions", 469, 488
　handful of grand questions on biochar technology, 484–486
Graphene, 298
　graphene-like polyaromatic crystalline structure, 48–49
Graphite, 294–295
Green catalysts, 76
Green infrastructure, 256
　strategies, 242
　synergies with, 256
Greenhouse gases (GHGs), 65, 289–294, 386, 417
　emissions, 281, 395, 416, 420–421, 486
　mitigation, 4
　sources, 420
Gross crop residues (GCR), 16–17
GTP. *See* Global Temperature Change Potential (GTP)
GWP. *See* Global warming potential (GWP)

H

H/C ratio. *See* Hydrogen to carbon ratio (H/C ratio)
HACCP. *See* Hazard analysis and critical control points (HACCP)
Halogen, 49–50
HAP. *See* Hydroxyapatite (HAP)
Harvesting processes, 415
Hazard analysis and critical control points (HACCP), 442
Hazard index (HI), 440
Hazard quotient (HQ), 440
HCFC. *See* Hydrochlorofluorocarbons (HCFC)
HCl. *See* Hydrochloric acid (HCl)
HDPE. *See* High-density polyethylene (HDPE)
Health benefits, 359
Health hazards, 442
Health risk assessment, 321, 438, 481
　call for health risk assessments and mitigation, 437–442
　framework, 437–438
　　notion of risk, 437–438
　in occupational and nonoccupational settings, 481
　principles of, 438
　　characterization of hazards, 438
　　documentation and communication, 438
　　estimation of consequences or impacts, 438

estimation of likelihood, 438
　risk estimation, 438
　risk evaluation, 438
　risk mitigation and management, 438
Heat, 456
　source, 339
Heat transfer efficiency (HTE), 355
Heavy metal ions, 368
Heavy metals, 93, 95, 173–174, 242–243, 285–286, 419
Herbicides, 127
　pollutant concentrations, 387
Heteroatoms, 59–60
　doping, 371
Heterogenous adsorption process, 63
Heuristic conceptual framework, 8
HI. *See* Hazard index (HI)
High moisture content, 331
High-density crystalline structure, 316–317
High-density polyethylene (HDPE), 15–16
High-rate filtration system, 180
High-risk environments, 437
High-temperature processing, 314
Holes (h^+), 81
HQ. *See* Hazard quotient (HQ)
HRT. *See* Hydraulic retention time (HRT)
HTC. *See* Hydrothermal carbonization (HTC)
HTE. *See* Heat transfer efficiency (HTE)
HTP. *See* Human toxicity potential (HTP)
Human activities, 241–242
Human disorders, 173–174
Human excreta, 7, 329, 331
　comparison of thermochemical conversion methods, 335*t*
　as energy feedstock, 330–332
　rationale for thermochemical conversion of, 332
　thermochemical properties of, 330–332
Human exposure
　and health risks, 437
　in occupational and nonoccupational settings, 481
Human feces, 333
Human health
　call for health risk assessments and mitigation, 437–442
　consequences or impact of hazard, 438
　biochar certification systems, 441
　ecotoxicological risk assessments, 441
　principles of health risk assessment, 438
　qualitative risk assessment, 438–439
　quantitative human health risk assessment tools, 439–441
　quantitative risk assessment, 439
　status of research on risk assessment of biochar systems, 441–442
　effects of biochar, 159
　future research directions, 442
　human exposure and health risks, 437
　likelihood or probability of occurrence, 438
Human toxicity potential (HTP), 399
Humanitarian engineering, 336
Humulus lupulus L., 384–385
Hybrid activation, 197

Hybrid systems, 311
Hybridization, 298
Hydraulic retention time (HRT), 295
Hydrocarbons, 152
Hydrochars, 21, 194–195
Hydrochloric acid (HCl), 84–85
Hydrochlorofluorocarbons (HCFC), 399
Hydrogen (H_2), 65, 334
Hydrogen bonding, 49–50, 137, 269–270, 286–289, 368
Hydrogen chloride (HCl), 15–16
Hydrogen peroxide (H_2O_2), 383
Hydrogen to carbon ratio (H/C ratio), 315
    molecular ratio, 20
Hydrolytic microorganisms, 295
Hydrophobic adsorption, 153–155
Hydrophobic interactions, 48–49, 136, 368
Hydrophobic pollutants, 48–49
Hydrospheric media, 219
Hydrothermal carbonization (HTC), 14, 21, 35–37, 59, 75, 193–195, 195f, 433
    of biomass, 21
Hydrothermal treatment, 77
Hydroxyapatite (HAP), 269–270
Hydroxyl group (O-H), 82–83, 171–174, 244
Hydroxyl radicals (•OH), 82–83

I

IAP. See Indoor air pollution (IAP)
IAQ. See Indoor air quality (IAQ)
IBI. See International Biochar Initiative (IBI)
Ideal electrodes, 313
ILCR. See Increased lifetime cancer risk (ILCR)
In-silico computation methods, 65–66, 477, 479, 488
In-silico methods, 67
Incineration, 386
Incorporation in solid fuels, 386
Increased lifetime cancer risk (ILCR), 159
Indoor air pollution (IAP), 347
    biochar cookstoves, 349, 355
        environmental benefits of, 356
        as multifaceted solution for low-income settings, 355–356
        types and designs of, 349–354
    call to action for sustainable cooking practices, 359
    case studies and success stories, 357–358
    challenges and solutions, 356–357
    evolution of cookstove technology, 349
    future directions and innovations, 358–359
    recap of key points, 359
    sources of, 348
        biomass combustion and emissions, 348
        traditional cooking methods, 348
    understanding IAP, 348
Indoor air quality (IAQ), 347
Industrial by-products, 193
Industrial management processes, 362
Industrial processes, 113
Industrial waste, 193, 273
Industrial wastewater, 265, 267
    treatment, 267

Industrialization, 3
Inhalation unit risk (IUR), 440–441
Inorganic compounds, 92, 101
Inorganic contaminants, 40, 50–51, 92–93, 246, 361, 379, 435. See also Organic contaminants (OCs)
    biochar removal of, 95–96
    removal, 40–48, 170–174
        future perspectives, 182
        mechanism of removal of inorganic contaminants, 173–174, 179f
        nutrients from water using biochar, 171
        radionuclides from water using biochar, 170–171
        toxic metals from water, 171–173, 172t
    in soils
        mechanisms for biochar removal of, 97–100
        metal and nutrient-contaminated soils, 93–94, 94t
        occurrence in munition fields, 95
        occurrence of, 92–95, 93f
        wastewater and sludge-amended soils, 94–95
Inorganic fertilizers, 466
Inorganic fraction, 337
Inorganic pollutants, 91, 169–170
    adsorption mechanisms of biochar for inorganic pollutants in wastewater, 269–270
    adsorption of inorganic pollutants onto biochar in wastewater, 268
Inorganic salts, 268
Institutional-scale biochar cook stoves, 354
Instrumental analytical techniques, 25
Integrated generation of solid fuel and biogas from biomass (IFBB), 473, 483
Intergovernmental Panel on Climate Change (IPCC), 399, 447
Internal rate of return (IRR), 449
International Biochar Initiative (IBI), 281–282, 454
Internet of Things (IoT), 65, 358
Ion exchange, 98, 369
    capability, 284
    processes, 251
Ionic attraction, 369
Ionic strength, 245
Ionizing radiation potential (IRP), 399
IoT. See Internet of Things (IoT)
IPCC. See Intergovernmental Panel on Climate Change (IPCC)
Iron (Fe), 76, 81–82, 113, 268, 381–382
Iron oxide, 315
    nanoparticles, 317
IRP. See Ionizing radiation potential (IRP)
IRR. See Internal rate of return (IRR)
IS-KOH-B. See KOH-activated industrial sludge biochar (IS-KOH-B)
Isotherm, 370
    modeling, 370
    models, 201, 370
    of PFAS removal by biochars, 230–231
    studies, 200–201
IUR. See Inhalation unit risk (IUR)

K

Kinetic models, 50
Kinetic studies, 201–203
Knowledge gaps, 233–234
KOH-activated industrial sludge biochar (IS-KOH-B), 204

L

Laboratory experiments, 40–48
Lactic acid, 473, 483
Lactuca sativa. See Lettuce (*Lactuca sativa*)
Land application, 385
Landfill disposal, 385–386
Langmuir isotherm, 172–173, 370
Langmuir models, 201
Lantana camara, 119
Large-scale pilot field studies, 120
Large-scale pyrolysis
    plants, 416
    systems, 18
Large-scale studies, 182
Large-scale systems, 110
LCA. See Life cycle assessments (LCA)
LCI. See Life cycle inventory (LCI)
LCIA. See Life cycle impact assessment (LCIA)
LDPE. See Low-density polyethylene (LDPE)
Leaching, 139
Lead (Pb), 173, 268, 368
Leptospira spp., 268
    L. pneumophila, 268
Lettuce (*Lactuca sativa*), 181
LEV. See Levofloxacin (LEV)
Levofloxacin (LEV), 62
Lexical sophistry, 395
Life cycle analysis, 372, 401
    of environmental and social footprints of biochar systems, 482
Life cycle assessments (LCA), 86, 319, 321, 396, 402, 436
    comparative LCA analysis of biochar production systems, 400–401
    goal and scope definition, 402–406
    for biochar systems, 403t
    life cycle impact assessment, 416–426
    life cycle inventory, 406–416
    as sustainability tool, 397–400, 397f
Life cycle impact assessment (LCIA), 398–399, 416–426, 422t
Life cycle interpretation, 398
Life cycle inventory (LCI), 398, 406–416
"Life cycle thinking", 397
Lignin, 386
Lignocellulosic biomass, 298, 315–316
Lignocellulosic materials, 281–282
Ligustrum lucidum, 141
Livestock manure, 193
Long-chain PFAS, 217, 230
Low-cost drinking water treatment methods, 467–468
Low-density polyethylene (LDPE), 15–16
Low-temperature pyrolysis (LTP), 195
LTP. See Low-temperature pyrolysis (LTP)

M

MAB. *See* Microwave-activated biochar (MAB)
Machine learning (ML), 60
   methods, 477, 479, 488
   tools, 66–67, 67t
      common AI and ML algorithms, 66f
Macropores, 368
MAE. *See* Mean absolute error (MAE)
MAETP. *See* Marine Aquatic EcoToxicity Potential (MAETP)
Magnetic ball-milled biochar (MBM-BC), 196
Magnetic biochar, 381–382
Magnetic chicken bone biochar (MCBB), 206
Magnetic nitrogen-doped biochar catalysts, 77
Magnetic separation, 381–382, 382f
Malachite green (MG), 199
Manganese (Mn), 113, 268, 315
Manganese oxides, 315
Manure compound (MC), 419
Marine Aquatic EcoToxicity Potential (MAETP), 399, 419
Mass adoption solutions, 358
MB. *See* Methylene blue (MB)
MBM-BC. *See* Magnetic ball-milled biochar (MBM-BC)
MC. *See* Manure compound (MC)
MCBB. *See* Magnetic chicken bone biochar (MCBB)
MCE. *See* Modified combustion efficiency (MCE)
MD Simulations. *See* Molecular Dynamics Simulations (MD Simulations)
Mean absolute error (MAE), 67
Mechanism
   of biochar remediation of mine wastes and metalliferous substrates, 117–118
   of contaminant removal from stormwater using biochar, 251, 253f
   for remediation of petroleum hydrocarbon contaminated solid matrices using biochar, 153–155
   of removal of inorganic contaminants, 173–174, 179f
Mechanochemical stability, 372
Medium-scale pyrolysis systems, 18
Membrane
   bioreactors, 382
   selection, 308–309
Mercury (Hg), 268, 368
Mesopores, 368
Meta-analysis, 420
Metabolic process, 295
Metal chelation, 49–50
Metal impregnation, 371
Metal ions, 49–50
Metal oxides, 24–25, 298, 311
Metal-contaminated lands, 112, 115f
Metal-contaminated soils, 93–94, 94t, 112, 114–116
Metal-enriched biomass from metalliferous substrates as unique biomass feedstock, 121
Metal-loaded biochar, 48
Metal-rich spent biochars, catalyst precursors for trace, 386
Metallic elements, 285–286
Metals, 92–93, 112, 435
Methane (CH_4), 289–294
Methanosarcina barkeri, 317
Methyl mercaptan (CH_3SH), 388
Methyl orange (MO), 191–192
Methylene blue (MB), 199
MFCs. *See* Microbial fuel cells (MFCs)
MG. *See* Malachite green (MG)
Microalgae, 333
Microbial activity, 286–289
   enhancement of, 118
Microbial biodegradation, 223–225
Microbial communities, interactions with, 254–255
Microbial contaminants, 113, 170
   biochar for, 248–249, 249t
   removal of, 177–178
Microbial ecology, 220–223
Microbial fuel cells (MFCs), 69, 294–295, 307–308, 415, 478–479
   applications in microbial fuel cells, 316–318
   biochar
      electrochemical properties of, 312–316
      as electrode material for development of, 294–295
      properties and potential applications in, 312–318
   designs and types of, 308–312, 313t
   mechanisms of biochar enhancement of, 318
   performance and mechanisms of biochar function in, 318
   principles of, 308, 310t
Microbial growth, 228
Microbial interactions, 252
Microbial metabolic processes, 294–295
Microbial remediation, 223–226
Microbial-assisted regeneration, 384
Microcomputed tomography (μ-CT), 63
Microcosm studies, 155
Microorganisms, 85, 139–140, 177, 223–225, 228, 317
Micropores, 368, 387
Microporosity, 284
Microsoft excel, 401
Microwave
   activation, 228
   energy, 384
   heating, 22
   irradiation regeneration, 384
   pyrolysis, 75, 451
   treatment, 384
Microwave-activated biochar (MAB), 196
Military activities, 95
Mineral recovery, biochar as novel renewable green biomaterial for, 481–482
Minerals iron, 315
Miscanthus, 14–15, 180
ML. *See* Machine learning (ML)
MO. *See* Methyl orange (MO)
Modern agriculture, 170
Modified combustion efficiency (MCE), 355
Modified tannery sludge biochar (MBC), 199–200
Moisture content, 332, 334, 416
Molecular Dynamics Simulations (MD Simulations), 65
Monte Carlo simulation, 449, 453, 455
MSS. *See* Municipal sewage sludge (MSS)
MSW. *See* Municipal solid waste (MSW)
Municipal sewage sludge (MSS), 17
Municipal solid waste (MSW), 17, 192
Municipal wastes, 273
Municipal wastewater, 308
   treatment, 267–268

N

Nano zerovalent iron (nZVI), 39–40
Nano-biochar, 406, 415
Nanocarbon materials, 312–313
Nanocomposites (NCs), 196
Nanoparticles (NPS), 177–178
   NPS-modified biochar, 36
Nanoscale zerovalent iron (nZVI), 81–82
Natural areas, 241–242
Natural hydrological processes, 253–254
NCs. *See* Nanocomposites (NCs)
Near edge X-ray absorption fine structure spectroscopy (NEXAFS), 60–61
Nernst equation, 315
Net present value (NPV), 449
Neutral pollutants, 48–49
NEXAFS. *See* Near edge X-ray absorption fine structure spectroscopy (NEXAFS)
Nickel (Ni), 113, 268, 309, 311
Nitrate (NO3-), 244
Nitric acid, 398
Nitrogen, phosphorus, and potassium (NPK), 312
Nitrogen (N), 26, 284–285, 332
Nitrogen dioxide (N_2O), 348
Nitrogenous abundant biomass, 62
Nitrogenoxides (NOx), 361
Nitrous oxide (N_2O), 289–294
Non-polymeric PFAS, 216–217
Nonagricultural land, 385
Nonbiodegradability, 94
Noncarcinogenic health risks, 439–440
Noncarcinogenic human health risks, 440
Noncarcinogenic human toxicity, 418
Nonmetallic impregnation method, 77–78
Nonorganic pollutants, 250
Nonthermal techniques, 380
Novel construction materials, fillers in, 386
Novel renewable green biomaterial for mineral recovery, biochar as, 481–482
Novel sewage sludge, 175
NPK. *See* Nitrogen, phosphorus, and potassium (NPK)
NPS. *See* Nanoparticles (NPS)
NPV. *See* Net present value (NPV)
Nuclear power plants, 170
Nutrients, 242–243
   human excreta, 332
   nutrient-contaminated soils, 93–94, 94t

nutrient-enriched biochar, 118
removal of nutrients from water using biochar, 171
runoff, 243
nZVI. *See* Nano zerovalent iron (nZVI); Nanoscale zerovalent iron (nZVI)

O

OCPs. *See* Organochlorine pesticides (OCPs)
OCs. *See* Organic contaminants (OCs)
ODP. *See* Ozone depletion potential (ODP)
OECD. *See* Organization for Economic Cooperation and Development (OECD)
Oenothera picensis, 95–96
OER. *See* Oxygen evolution reaction (OER)
Official data, 17
OFGs. *See* Oxygen containing functional groups (OFGs)
OFI. *See* Opuntia ficus-indica (OFI)
Oil spills, 151–152
OLR. *See* Organic loading rate (OLR)
OM. *See* Organic matter (OM)
"One-size-fits-all" approach, 471
  moving to biochar application domains, 472
  moving to specific biochar application domains, 484
Onion skin-derived sorbent (OSDS), 175
Onsite wastewater treatment system (OWTS), 177, 179
OpenLCA software, 402
Opuntia ficus-indica (OFI), 68–69
Organic acids, 116
Organic chemicals, 127, 170
Organic components of biomass, 314
Organic contaminants (OCs), 40, 48–50, 127, 138, 170, 361, 379, 435–436. *See also* Inorganic contaminants
  behavior and fate of OCs in soils, 138–140
    accumulation and bioavailability, 140
    adsorption and desorption, 139
    leaching, 139
    persistence, 140
    transformation and degradation, 139–140
    volatilization, 140
    washed away by runoff and erosion, 140
  biochar removal mechanisms of OCs in soils, 134–138
  biochar removal of emerging from contaminated soil, 132–134
  emerging, 131–132
    molecular structure of some pharmaceuticals and personal care products and contamination pathways, 132, 132f
    pharmaceuticals and personal care products, 131
  fate of biochar after, 140–141
  molecular ratio, 20
  remediation of, 120
  removal of, 40, 174–176, 174f, 176t
Organic content, human excreta, 332
Organic fertilizers, 466
Organic loading rate (OLR), 295

Organic material, 37, 348
Organic matter (OM), 94–95, 285–286
Organic molecules, 128, 137
Organic pollutants, 62, 128, 137, 139–140, 170, 242–243, 250, 269, 283–284, 435
  adsorption
    mechanism of biochar for organic pollutants in wastewater treatment, 270–271
    of organic pollutants onto biochar in wastewater, 269
  challenges
    of biochar removal of, 141
    and opportunities of biochar in removal of, 141
Organic waste, 266
Organization for Economic Cooperation and Development (OECD), 215–216
Organochlorine pesticides (OCPs), 140
OSDS. *See* Onion skin-derived sorbent (OSDS)
OTE. *See* Overall thermal efficiency (OTE)
Outdoor air pollution, 347–348
Overall thermal efficiency (OTE), 355
OWTS. *See* Onsite wastewater treatment system (OWTS)
Oxidation degradation approaches, 380
Oxidation reactions, 333
Oxygen, measuring, 334
Oxygen containing functional groups (OFGs), 19, 61, 77, 92, 98, 134
Oxygen evolution reaction (OER), 388
Oxygen-containing groups, 80–81
Ozone depletion potential (ODP), 399

P

P-nitrophenol, 75
PAHs. *See* Polycyclic aromatic hydrocarbons (PAHs)
Particulate matter (PM), 347
Particulate matter formation (PMF), 399
Pathogens, 177, 242–243, 245–246
PCBs. *See* Polychlorinated biphenyls (PCBs)
PCDEs. *See* Polychlorinated diphenyl ethers (PCDEs)
PCM. *See* Phase change materials (PCM)
PCPs. *See* Personal care products (PCPs)
Peat-based biochar, 385–386
Pelagic algae, 63–64
Pellets, 281–282
PEM. *See* Proton exchange membrane (PEM)
Perfluorinated chain, 216–217
Perfluoroalkyl acids, 215–216
Perfluoroalkyl substances, 7, 215–216
  biochar production processes and usage for removal of PFAS in environmental matrices, 226–231
  existing regulations on PFAS in environmental matrices and use of biochar, 231–233
  knowledge gaps, future research needs, 233–234
  occurrence of PFAS and sources in different environmental matrices, 216–220

classification and occurrence of PFAS, 216–217
remediation and treatment methods and processes of PFAS in environment, 220–226
sources, fate, and transport of PFAS in environmental matrices, 217–220
  atmospheric deposition and hydrospheric compartment, 220, 221t
  firefighting foams, 219–220
  industrial releases and wastewater treatment plants, 217t, 219
Perfluorobutanoic acid (PFBA), 225
Perfluorooctane sulfonate (PFOS), 220
Perfluorooctanoic acid (PFOA), 223–225
Persistent free radicals (PFRs), 80–81
Persistent organic pollutants (POPs), 140, 216
Personal care products (PCPs), 40, 131
Persulfate (PS), 82–83
  activation system, 82–83
  relationship between functional structures and mechanisms in, 83t
Pesticides, 127–129, 132–134, 139, 246–247
  pollutant concentrations, 387
  residues, 139
PET. *See* Polyethylene terephthalate (PET)
Petroleum
  industry, 149–150
  petroleum-derived fuels, 421
  sludge, 152
Petroleum hydrocarbons (PHCs), 149–150
  contaminated solid matrices
    biochar-microbe interactions, 155
    factors that influence ability of biochar to remediate, 153
    hydrophobic adsorption, 153–155
    mechanisms for remediation of, 153–155
    scale of application for biochar remediation of, 155–157
  contamination, 152
  ecotoxicity and health risks of biochar-remediated solid matrices, 158–159
  future perspectives and directions, 160
  considerations for future research, 160
  key challenges, 160
  PHC-degrading microbes, 155
  in soil, sediments, and sludges, 151–152
  in soils, sediments, and sludges
    biochar for remediation of, 152–158
    classes and sources of, 151–152, 151t
    indicators of remediation efficiency of biochar for, 158
    use and suitability of biochar for, 152
PFAS. *See* Polyfluoroalkyl substances (PFAS)
PFBA. *See* Perfluorobutanoic acid (PFBA)
PFOA. *See* Perfluorooctanoic acid (PFOA)
PFOS. *See* Perfluorooctane sulfonate (PFOS)
PFR. *See* Primary forest residues (PFR)
PFRs. *See* Persistent free radicals (PFRs)
Pharmaceuticals, 131, 381
Pharmaceuticals and personal care products (PPCPs), 127, 131, 246–247, 268
  molecular structure of, 131f, 132, 132f
Phase change materials (PCM), 69

PHCs. *See* Petroleum hydrocarbons (PHCs)
Phosphoric acid (H_3PO_4), 78–79
Phosphorus (P), 26, 332
Photocatalytic system, 83–84
    relationship between functional structures and mechanisms in, 84t
Photochemical ozone creation potential (POCP), 399
Photodegradation, 379–380
Physical activation, 78, 196, 371
Physical adsorption, 97–98, 367f, 368
Physical modification methods, 227–228
Physico-chemical constraints, 113–114
Phyto-microbial remediation, 226
Phytoextraction, 114
Phytoremediation, 99–100
Pig manure (PMB500), 206
Pit latrines, 330
π-electrons, 283–284
π-π electron-donor acceptor interactions, 136–137
Plant-microbe-assisted remediation. *See* Phyto-microbial remediation
Plants, 226
    growth promotion, 118
    reduction of toxic metal uptake by, 118
    wastes biochar, 266–267
Plastics, 15–16
    pyrolysis of types of plastics to produce biochar, 16t
Platinum, 294–295
    nanoparticles, 315
PM. *See* Particulate matter (PM)
PMF. *See* Particulate matter formation (PMF)
POCP. *See* Photochemical ozone creation potential (POCP)
Point of zero charge (PZC), 62, 173–174, 269
Polar functional groups, 100–101
Policymakers, 454–455, 486
Pollutants, 131, 171–172, 244, 247–248, 250, 269–270, 356, 387
Polluted wastewater, remediation of, 286–289
Pollution, 129–130, 191–192
    sources, 130
Poly organic pollutants (POPs), 361
Polyaromatic hydrocarbons (PAHs), 435
Polychlorinated biphenyls (PCBs), 75, 127, 269
Polychlorinated diphenyl ethers (PCDEs), 140
Polycyclic aromatic hydrocarbons (PAHs), 127, 158, 246–247, 361, 387, 417–418, 435
Polyethylene terephthalate (PET), 15
Polyfluoroalkyl substances (PFAS), 7, 215–216
    biochar production processes and usage for removal of PFAS in environmental matrices, 226–231
    biodegradation and transformation of PFAS in environment, 220–226
    compounds, 218
    existing regulations on PFAS in environmental matrices and use of biochar, 231–233
    knowledge gaps, future research needs, 233–234
    modification of biochar for PFAS removal from environmental matrices, 227–228
        biological modification method, 228
        chemical modification methods, 228
        physical modification methods, 227–228
    molecules, 229
    occurrence of PFAS and sources in different environmental matrices, 216–220
        classification and occurrence of PFAS, 216–217
    performance of biochars on removal of, 231, 232t
    PFAS-containing firefighting foams, 219
    remediation and treatment methods and processes of PFAS in environment, 220–226
    removal mechanism of PFAS
        adsorption kinetics, thermodynamics, and isotherms of, 230–231, 230t
        from environmental matrices using biochars, 229–231
        removal using biochars, 229–230
    removal technologies and techniques, 220
    sources, fate, and transport of PFAS in environmental matrices, 217–220
        atmospheric deposition and hydrospheric compartment, 220, 221t
        firefighting foams, 219–220
        industrial releases and wastewater treatment plants, 217t, 219
Polymeric materials, 15
Polypropylene (PP), 15–16, 311
Polystyrene (PS), 15–16
POPs. *See* Persistent organic pollutants (POPs); Poly organic pollutants (POPs)
Population growth, 3
Pore filling process, 137, 286–289, 369
Porosity, 365
Postmining landscapes, revegetation of, 113–114
Potassium (K), 26, 76, 284–285, 332
Potassium carbonate (K_2CO_3), 78–79
Potassium dichromate ($K_2Cr_2O_7$), 312
Potassium hydroxide (KOH), 78–79
Potassium permanganate ($KMnO_4$), 312
Potentially toxic elements (PTEs), 92
PP. *See* Polypropylene (PP)
PPCPs. *See* Pharmaceuticals and personal care products (PPCPs)
PPS. *See* Pulp and paper sludge (PPS)
Predictive models, 160
Pretreatment method, 22
Primary forest residues (PFR), 406, 420
Production methods, 76–77
Proteobacteria, 155
Proton exchange membrane (PEM), 308
PS. *See* Persulfate (PS); Polystyrene (PS)
Pseudo-capacitance, 318
Pseudo-first-order model, 230–231, 370
Pseudocapacitance, 315
Pseudomonas
    P. aeruginosa, 139–140, 311
    P. fluorescens, 139–140
    P. japonica, 139–140
Pseudo second-order model (PSO model), 172–173, 201–203, 231, 370
PSO model. *See* Pseudo second-order model (PSO model)
Psuedomonas poae, 155
PTEs. *See* Potentially toxic elements (PTEs)
Pulp and paper sludge (PPS), 200
Pyrolysis, 17–18, 61, 64–65, 77, 136, 193–194, 194t, 227, 243, 332–333, 401, 433
    process, 17–18, 367, 435, 450–451
    reactors, 35–36, 52
    regeneration through, 385
    systems, 18, 110
        for biochar production, 17–19
    techniques, 396, 451
    temperature, 19, 102
    thermochemical conversion, 333–334
    types of, 18–19
        fast pyrolysis, 18–19
        flash pyrolysis, 19
        slow pyrolysis, 18
Pyrolytic biochars, 195
Pyrolyzed empty fruit bunch fiber (PEF), 197
PZC. *See* Point of zero charge (PZC)

Q

Qualitative methods, 439
Qualitative risk assessment, 438–439, 439f
Quantitative human health risk assessment tools, 439–441
    carcinogenic human health risks, 440–441
    noncarcinogenic human health risks, 440
Quantitative risk assessment, 439

R

Rabbit feces (RFB500), 206
Radiative forcing by black carbon, 436–437
Radionuclides, 169–171, 466
    from water using biochar, 170–171
Random forest (RF), 66
Raw biochar, 76
Raw biomass, 37
RB. *See* Rhodamine B (RB)
RB19. *See* Reactive blue 19 (RB19)
RB5. *See* Reactive black 5 (RB5)
Reactive black 5 (RB5), 200
Reactive blue 19 (RB19), 200
Reactive orange 16 (RO16), 200
Reactive oxygen species (ROS), 69
Recycled spent biochar, 387
Recycling
    of spent biochar for circular economy and environmental sustainability, 384–387
    technologies, 321
Redox activity, 281–284
Redox capacity, 285–286
Redox reaction, 171–172, 283
Redox-active trace metals, 113
Reduction-oxidation process, 99
Regeneration
    biochar amendments for remediation of contaminated lands and soils, 469

of polluted wastewater, 286–289
through pyrolysis, 385
and reuse, 205–206, 207t
of spent biochars, 372, 381–384
- advanced oxidation processes, 383
- chemical desorption, 383
- filtration, 382
- magnetic separation, 381–382
- microbial-assisted regeneration, 384
- microwave irradiation regeneration, 384
- supercritical fluid desorption, 383
- thermal desorption and decomposition, 382–383
synergistic interaction of remediation technologies, 483
techniques, 109–110, 321
and treatment methods and processes of PFAS in environment, 220–226
- biodegradation and transformation of PFAS in environment, 220–226
- PFAS removal technologies and techniques, 220
Remote communities, 336
Renewable resources, 289–294
Renewable sources, 395
Residual biochar, 472
Return on investment (ROI), 449
Reuse as soil amendment, 384–385
Reuse of spent biochar for circular economy and environmental sustainability, 384–387
RF. *See* Random forest (RF)
RHB. *See* Rice husk (RHB)
Rhodamine B (RB), 200, 206
Rice husk (RHB), 178
- specialized design for, 354, 354f
Risk assessment
- methods, 438
- status of research on risk assessment of biochar systems, 441–442
Risk estimation, 438
Risk evaluation, 438
Risk mitigation and management, 438
RMSE. *See* Root mean squared error (RMSE)
RO16. *See* Reactive orange 16 (RO16)
Rocket stoves, 339, 351
- compact design, 351
- efficiency in resource-constrained settings, 351, 352f
ROI. *See* Return on investment (ROI)
Root mean squared error (RMSE), 67
ROS. *See* Reactive oxygen species (ROS)

S

Saccharomyces cerevisiae, 139–140
Salmonella spp., 268
"Sanitary by-wastes", 330
Sanitization, human excreta, 332
Sapolândia, community of, 336–337, 338f
SBC. *See* Sulfur-modified biochar (SBC)
Scaling biochar technology, 4–5
SCFs. *See* Supercritical fluids (SCFs)
SCR. *See* Surplus crop residue (SCR)

SDGs. *See* Sustainable development goals (SDGs)
Sediments
- biochar for remediation of petroleum hydrocarbons in, 152–158
- indicators of remediation efficiency of biochar for petroleum hydrocarbons in, 158
- methods of applying biochar to, 153
- petroleum hydrocarbons in, 151–152
  - classes and sources of petroleum hydrocarbons in, 151–152
  - ecological and human health impacts of petroleum hydrocarbons in, 152
  - use and suitability of biochar for PHC remediation in, 152
Sensitivity analysis, 402, 455
Serpentines, 112–113
Serpentinitic geological systems, 112–113
Sewage sludge, 17, 94
Sewer-based sanitation, 337–338
SFGs. *See* Surface functional groups (SFGs)
Sheep manure (SM), 206, 466–467
Shewanella sp., 317
Shigella spp., 268
Silver (Ag), 177–178, 268
"Silver bullet" for biochar application domains, moving from biochar, 472
SimaPro software, 402
Size exclusion, 369
Slow pyrolysis, 18, 35–36, 77, 227
Sludge(s), 113, 116–117
- amended soils, 94–95
- biochar for remediation of petroleum hydrocarbons in, 152–158
- codisposal of, 387
- indicators of remediation efficiency of biochar for petroleum hydrocarbons in, 158
- methods of applying biochar to, 153
- petroleum hydrocarbons in, 151–152
  - classes and sources of petroleum hydrocarbons in, 151–152
  - ecological and human health impacts of petroleum hydrocarbons in, 152
  - use and suitability of biochar for PHC remediation in, 152
SM. *See* Sheep manure (SM)
Small-scale biochar studies, 255
Small-scale portable pyrolysis reactors, 416
Small-scale pyrolysis
- reactors, 416
- systems, 18
Smart combustion systems, 358
Smoke, 356
SOC. *See* Soil organic carbon (SOC)
Social footprints of biochar systems, life cycle analysis of, 482
Socio-cultural aspects of biochar technology, 483
Socio-economic implications of biochar technology in low-income countries, 485–486
SODIS. *See* Solar disinfection (SODIS)

Sodium hydroxide (NaOH), 78–79
Soil organic carbon (SOC), 285
Soil(s), 8, 91–92, 96–97, 132, 139–140, 254–255, 434
amendment, 437, 452
- reuse as, 384–385
aquifer treatment, 116–117
behavior and fate of organic contaminants in, 138–140
biochar, 417
- correction systems, 417
- for remediation of petroleum hydrocarbons in, 152–158
- removal mechanisms of organic contaminants in, 134–138
- soil amendment to create enduringly fertile and productive, 469
carbon sequestration, 464
clusters, 68
contaminants, 387
- remedial techniques, 91–92
contamination, 92
degradation, 396
environments, 289–294
improvement systems, 421
indicators of remediation efficiency of biochar for petroleum hydrocarbons in, 158
inorganic contaminants in
- behavior and fate of contaminants in biochar-amended soils, 101–102
- biochar removal of inorganic contaminants, 95–96
- factors affecting capacity of biochar in soil remediation, 100–101
- large-scale remediation of inorganics by biochars, 96–97
- mechanisms for biochar removal of inorganic contaminants in soils, 97–100
- occurrence of inorganic contaminants in soils, 92–95
- sources of inorganic pollutants and contamination, 93f
long-term impacts of biochar on soil health and fertility, 485
methods of applying biochar to, 153
microorganisms, 118
nutrients, 284–285
- content, 384–385
organic matter, 285–286
petroleum hydrocarbons in, 151–152
- classes and sources of petroleum hydrocarbons in, 151–152
- ecological and human health impacts of petroleum hydrocarbons in, 152
pollution, 435, 436f
remediation methods, 141, 283
salinization, 91
soil-rhizosphere-plant system, 285–286
sources and pathways of conventional organic contaminants for contaminating, 129–130
sources-pathway-receptor process, 131f

Soil(s) (Continued)
  structure
    impact on, 254
    improvement, 117
    pivotal role of biochar in influencing, 253–254
    use and suitability of biochar for PHC remediation in, 152
Sol-char toilet, 339
Solar disinfection (SODIS), 178–179, 466
Solar power, 319
Solids, 330, 333
  compounds, 368–369
  fuels, 348
    incorporation in, 386
  residue, 37
Soluble contaminants, 440
Solvent regeneration, 383
"Soot-Free Futures" project, 357
Sorption capacity, 387
Sorption effects, 369
Sorption process, 50, 175, 218
Specific surface area (SSA), 13–14, 23, 67, 97–98, 171–172, 196, 381–382
Spent adsorbents, 385–386
Spent biochars, 385, 388. See also Engineered biochar (EBC)
  behavior and fate of contaminants in spent biochar, 387–388
  efficient disposal management of, 180–181
  future research directions, 388–389
  macropores, 387
  recycling of, 380f
  regeneration, 372, 381–384, 381f
  reuse, recycling, and disposal of spent biochar for circular economy and environmental sustainability, 384–387
    catalyst precursors for trace metal-rich spent biochars, 386
    codisposal of sludge, 387
    composting, 385
    contaminant removal and landfill disposal, 385–386
    fillers in novel construction materials, 386
    incineration, 386
    incorporation in solid fuels, 386
    land application, 385
    regeneration through pyrolysis, 385
    reuse as soil amendment, 384–385
SSA. See Specific surface area (SSA)
Steel production, 419
Sulfate radicals (SO4•-), 82–83, 141
Sulfur dioxide (SO$_2$), 361, 419
Sulfur oxides (SOx), 362
Sulfur trioxide, 398
Sulfur-modified biochar (SBC), 62
Sulfuric acid (H$_2$SO$_4$), 78–79, 398
Supercritical desorption, 383
Supercritical fluids (SCFs), 383
  desorption, 383
Supercritical water, 383
Support vector machines (SVM), 66
Surface area, 365
Surface chemistry, 284

Surface functional groups (SFGs), 25, 37–38, 59–60, 365–367
Surplus crop residue (SCR), 16–17
Sustainability tool, life cycle assessment as, 397–400
Sustainable biomass utilization, 354
Sustainable cooking practices
  call to action for, 359
  top-lit updraft rice husk gasifier contribution to, 354
Sustainable development goals (SDGs), 233
Sustainable energy solutions, 358
Sustainable scaling of biochar production systems to meet global demands, 485
SVM. See Support vector machines (SVM)
Switchgrass, 14–15
Synchrotron X-ray microtomography and multifractal analysis, 63–64
Synergies, 100
Synergistic interaction of biochar and remediation technologies, 483
Syngas, 396, 456
Synthetic fertilizers, 170
Synthetic gases, 37
Synthetic organics, 128–129

T

Tainted water, 169
TCR. See Total cancer risk (TCR)
TEAs. See Techno-economic assessments (TEAs)
Techno-economic assessments (TEAs), 9, 51, 233, 319, 321, 447–449, 453, 457, 482
  of biochar systems, 448–450
  challenges of biochar systems, 452–453
  data on biochar for environmental remediation applications, 456
  future perspectives, 453–456
  future research directions, 455–456
    comprehensive cost-benefit analysis, 455
    data availability and quality, 455
    integration with emerging concepts, 456
    long-term performance and impact, 455
    market dynamics and acceptance, 456
    policy and regulatory frameworks, 456
    regional and contextual variability, 456
    sensitivity and uncertainty analysis, 455
    socio-economic factors, 456
    techno-economic assessment data lack on biochar for environmental remediation applications, 456
  key factors influencing techno-economic feasibility of biochar, 450–452
  studies, 453
  tools, 448–449
Techno-economic feasibility, 452
  key factors influencing, 450–452
Techno-economic modeling, 452–454
Technology readiness assessment (TRA), 9, 482
Technosols, 469
Terra preta, 4–5
  creating enduringly, 484–485

Terrestrial Ecotoxicity Potential (TETP), 399
TETP. See Terrestrial Ecotoxicity Potential (TETP)
TGs. See Triglycerides (TGs)
Thermal desorption and decomposition, 382–383
Thermal energy, 333
Thermal putrefaction, 17–18
Thermal pyrolysis, 451
Thermal regeneration methods, 380, 382–383
Thermal stability, 25
Thermal treatment, 84, 383
Thermo-hydro-chemical conversion technology, 450–451
Thermochemical composition, 331–332
Thermochemical conversion, 329, 332–336
  comparison of, 334–336
  principles of, 332–333
  processes, 3
  rationale for thermochemical conversion of human excreta, 332
  types of, 333–334
    combustion, 333
    gasification, 334
    pyrolysis, 333–334
Thermochemical processes, 330, 332, 339
Thermochemical properties of human excreta, 330–332
Thermochemical route, 332–333, 337–338
Thermochemical sanitation technologies, 338–339
Thermochemical treatment, 329, 338–339
  current state of applications, 336–339
    portrait of Brazilian Amazon community, 336–338
    thermochemical sanitation technologies, 338–339
  future research directions and conclusions, 339–340
  future directions, 339
  human excreta as energy feedstock, 330–332
  overall thermochemical properties, 331f
  thermochemical conversion, 332–336
Thermochemical units, 336
Thermodynamics, 370
  of PFAS removal by biochars, 230–231
  studies, 200, 203–204, 203t, 370
Thermohydrochemical processes, 434
3-D micro-computed tomography (μCT), 60–61
3-D micro-CT analysis, 63
3D visualization model, 63
TLUD. See Top-lit updraft (TLUD)
TN. See Total nitrogen (TN)
TOC. See Total organic carbon (TOC)
Top-lit updraft (TLUD), 349
  biochar stoves, 349–351
    applications and efficiency, 349–351, 350f
    design features, 349
  gasifier stoves, 351–353
    combining top-lit updraft and gasification, 351

flexibility for biomass types, 351–353
rice husk gasifier stoves, 354
contribution to sustainable cooking practices, 354
specialized design for rice husk utilization, 354
Torrefaction, 77
of biomass, 22
process, 22
Total cancer risk (TCR), 440–441
Total nitrogen (TN), 285
Total organic carbon (TOC), 23, 285
Total phosphorus (TP), 267–268
Total surface area (TSA), 61
Total suspended solids (TSS), 180, 267–268
Toxic elements, 92
Toxic gaseous contaminants, 361
biochars for removal of
biochar properties and mechanisms for, 365–370
engineered biochars for enhanced gaseous contaminants removal, 371
future directions and recommendations, 372
need for gaseous contaminant removal, 363
regeneration of spent biochars, 372
removal of toxic gaseous contaminants from air/flue gases, 363–365, 364t
sources of, 362, 362f
mechanisms of toxic gaseous contaminant removal by biochars, 367–369
chemical reactions, 369
complexation, 368
coprecipitation, 368–369
electrostatic attractions, 368
hydrogen bonding, 368
hydrophobic interactions, 368
ion exchange, 369
ionic attraction, 369
physical adsorption, 368
pore filling, 369
size exclusion, 369
sorption and catalytic effects, 369
Toxic heavy metals, 171–172, 361
Toxic metal ions, 50
Toxic metals, 434, 479
reduction of toxic metal uptake by plants, 118
removal of toxic metals from water, 171–173
TP. See Total phosphorus (TP)
TRA. See Technology readiness assessment (TRA)
Trace elements, 93
Trace metals, immobilization of, 117
Traditional carbon-based adsorbents, 192
Traditional catalysts, 76
Traditional cooking methods, 347–348, 348f, 356
Traditional methods, 128
Traditional stormwater management methods, 242
Transitional metals, 388
Triglycerides (TGs), 85

TSA. See Total surface area (TSA)
TSS. See Total suspended solids (TSS)
Turbostratic sheets, 314
Twin or multiple solution
in biochar uptake, adoption, and impact, role of, 470
concept, 465–466
confronting biochar critics and skeptics with, 472–473
framework, 464–465, 471
to emerging concepts, 473
as primer for biochar adoption, 470–471

U

UBC. See Unmodified biochar (UBC)
Ultrasound approaches, 380
Ultraviolet (UV)
irradiation, 467–468
radiation, 361–362
UV-Vis Drs, 80
Uncertainty analysis, 455
United Nations Sustainable Development Goals, 457
United States Environmental Protection Agency (EPA), 231–233, 418
Unmodified biochar (UBC), 62
"Upstream" phases, 397–398
Uranium (U), 172–173, 311
Urban stormwater, 241–242, 248–249
adsorption capacity and selectivity, 253
biochar catalytic reactions, 250–251
catalytic degradation, 252–253
challenges and considerations of biochar for treating stormwater, 254–255
contaminants, 245–247
conventional organic contaminants, 246–247, 247t
environmental implications of using biochar for treating stormwater, 255–256
fate and behavior of contaminants in stormwater, 249–250
future directions, 256–257
collaborative research and policy integration, 257
integrated approaches, 256
optimization of biochar production methods, 256
scalability and cost-effectiveness, 256–257
synergies with green infrastructure, 256
influence of solution chemistry on adsorption, 244–245
inorganic contaminants, 246
management, 243
mechanisms of contaminant removal from stormwater using biochar, 251, 253f
ion exchange processes, 251
precipitation and coprecipitation, 251
microbial interactions, 252
pivotal role of biochar in influencing soil structure and water retention, 253–254
role of biochar in stormwater management, 243–244

runoff, 245–247
treating emerging contaminants with biochar, 247–249, 248t
treatment
biochar characteristics relevant to, 243
systems, 250–251
urban stormwater challenges, 242–243
urban stormwater contaminants, 245–247
versatile sorption of contaminants and contaminant-specific adsorption, 244

V

Valence band (VB), 83–84
Value-chain approach, 6
Van der Waals forces, 269–270
Van der Waals interactions, 48–49
Van't Hoff equation, 204
VB. See Valence band (VB)
Vegetation, 114
Versatile adsorbent, 244
Versatility, 351
VFAs. See Volatile fatty acids (VFAs)
Vibro cholera, 268
VOC. See Volatile organic compounds (VOC)
Volatile fatty acids (VFAs), 295
Volatile organic compounds (VOC), 348, 361

W

Waste
alkali lignin, 173
biomass, 152
CO_2, 52
management
processes, 362
systems, 329
materials, 399–400
valorization technique, 295
waste-banana-peel biochar, 318
Wastewater, 94–95, 191, 197, 313, 479
adsorption
of inorganic pollutants onto biochar in, 268
of organic pollutants onto biochar in, 269
contaminants removal using biochars
conceptual designs of biochar water filters, 178–180
disposal management of spent biochar, 180–181
future perspectives, 182
large scale/field applications of biochar in Africa, 181
mechanism of removal of inorganic contaminants, 173–174
removal of inorganic contaminants, 170–174
removal of microbial contaminants, 177–178
removal of nutrients, 171
removal of organic contaminants, 174–176
removal of radionuclides, 170–171
removal of toxic metals, 171–173

Wastewater (*Continued*)
    type of wastewater treated by biochar, 267
    wastewater-amended soils, 113, 116–117
Wastewater plants (WWPs), 308
Wastewater treatment (WWT), 6, 132, 179, 266, 307, 396
    adsorption mechanisms of biochar for inorganic pollutants in, 269–270
    organic pollutants in, 270–271
    applications of biochar for, 271–272, 272t
    areas of future research in biochar application for, 273–274
    biochar technology in, 266–267
    industrial releases and, 217t, 219
Water, 6, 8, 169, 266, 434
    bodies, 13
    contamination, 169
    pivotal role of biochar in influencing, 253–254
    pollution, 265, 269, 436
    purification, 386
    removal of toxic metals from, 171–173, 172t
    water-energy-food-environment nexus, 3, 472, 482–483
    biochar soil amendment to create enduringly fertile and productive soils, 469
    biochar technology potential renewable, low-cost, and efficient energy, 466–467
    biochar-based environmental remediation systems, 469
    biochar-based water filtration systems for clean drinking water provision, 467–469
    handful of biochar "twin or multiple solutions" targeting, 466–469
Water consumption potential (WCP), 399
Water hyacinth (*Eichhornia crassipes*), 119
WCBC. *See* Wood chips biochar (WCBC)
WCP. *See* Water consumption potential (WCP)
Wet orange peel scraps, 416
Wheat straw (WS), 466–467
Wind power, 319
Wood, 316, 348
    chips, 281–282
    combustion processes, 417–418
    wood-based biochar, 317
Wood chips biochar (WCBC), 175
Wood pellets (WP), 466–467
Woody-based biochar, 19
World Health Organization, 463
WP. *See* Wood pellets (WP)
WS. *See* Wheat straw (WS)
WWPs. *See* Wastewater plants (WWPs)
WWT. *See* Wastewater treatment (WWT)

X

X-ray diffraction (XRD), 23, 59–60, 80
X-ray photoelectron spectroscopy (XPS), 80
XPS. *See* X-ray photoelectron spectroscopy (XPS)
XRD. *See* X-ray diffraction (XRD)

Y

Yersinia enterocoliticia, 268

Z

Zero-valent iron (ZVI), 96–97, 177–178
Zinc (Zn), 113, 268
Zinc chloride ($ZnCl_2$), 78–79
ZVI. *See* Zero-valent iron (ZVI)

9780323998895